NATURAL RESOURCE CONSERVATION

Management for a Sustainable Future

Tenth Edition

DANIEL D. CHIRAS
Colorado College

JOHN P. REGANOLD
Washington State University

Benjamin Cummings
San Francisco   Boston   New York
Cape Town   Hong Kong   London   Madrid   Mexico City
Montreal   Munich   Paris   Singapore   Sydney   Tokyo   Toronto

Library of Congress Cataloging-in-Publication Data

Chiras, Daniel D.
    Natural resource conservation: management for a sustainable future / Daniel D. Chiras,
John P. Reganold.—10th ed.
        p. cm.
    ISBN-13: 978-0-13-225138-9
    ISBN-10: 0-13-225138-8
    1. Conservation of natural resources.   2. Natural resources—Management.   3. Sustainable
development.   4. Environmental protection.   5. Conservation of natural resources—United States.
6. Natural resources—United States—Management.   7. Sustainable development—United States.
8. Environmental protection—United States.   I. Reganold, John P.   II. Title.

S938.O87 2010
333.72—dc22   2008046775

Editor-in-Chief: Beth Wilbur

Editor: Chalon Bridges

Project Manager: Crissy Dudonis

Marketing Manager: Jay Jenkins

Managing Editor, Chemistry and Geosciences: Gina M. Cheselka

Art Production Editor: Connie Long

Art Studio: Precision Graphics

Art Director: Maureen Eide

Interior/Cover Design: Suzanne Behnke

Media Production Manager: Natasha Wolfe

Photo Research Manager: Elaine Soares

Photo Researcher: Yvonne Gerin

Production Service: Elm Street Publishing Services

Composition: Integra Software Services Pvt. Ltd.

Cover Credit: McCarthy, Joanna/Getty Images Inc.

ISBN-10   0-13-225138-8
ISBN-13   978-0-13-225138-9

Benjamin Cummings
is an imprint of

PEARSON

www.pearsonhighered.com

BRIEF CONTENTS

CONTENTS

PREFACE

Natural Resource Conservation is written for the introductory resource conservation course. This book is designed to provide comprehensive coverage of a variety of local, regional, national, and global resource and environmental issues, including population growth, wetlands, wildlife management, sustainable agriculture, and global air pollution.

The first edition of this book was published in 1971, a year after the first Earth Day, by our esteemed colleague, the late Oliver S. Owen. To many observers, Earth Day marked the formal beginning of the environmental movement in the United States. Since that time, impressive gains have been made in air and water pollution control, species protection, forest management, and rangeland management.

Despite this progress, many environmental problems still remain. Many others have grown worse. In 1970, for instance, the world population hovered around 3 billion. Today, it has exceeded the 6.5 billion mark and is growing by more than 80 million people a year. Hunger and starvation have become a way of life in many less-developed nations. An estimated 12 million people die each year of starvation and disease worsened by hunger and malnutrition. Species extinction continues as well. Today, an estimated 100 species become extinct every day. In the United States and abroad, soil erosion and rangeland deterioration continue.

Added to the list of growing problems are a whole host of new ones. Groundwater pollution, ozone depletion, acid deposition, global warming, growing mountains of urban trash, and electronic waste top this list. Yet along with the new problems are new and exciting solutions.

If we work together in solving these problems, there is much hope. However, many experts believe that addressing these problems in meaningful ways will require dramatic changes in the way we live our lives and conduct commerce. We need a way that is sustainable—a way of doing business and living on the planet that does not bankrupt the Earth. Most people call this *sustainable development*. Sustainable development is about creating a new relationship with the Earth. It is about creating a sustainable economy and a sustainable system of commerce. It is about creating sustainable communities and sustainable lifestyles. It requires new ways of managing resources, using the best available scientific knowledge and understandings of complex systems and how they are maintained, even enhanced, over time. It will entail changes in virtually every aspect of our society, including farming, forest management, and energy production.

We believe that establishing a sustainable relationship with the Earth will require us to use resources more frugally—using only what we need and using all resources much more efficiently than we do today. Creating a sustainable way of life will very likely mean a massive expansion of our recycling efforts, not just getting recyclables to markets, but encouraging manufacturers to use secondary materials for production and encouraging citizens to buy products made from recycled materials.

Creating a sustainable society will also very likely mean a shift to clean, economical, renewable energy supplies, such as solar and wind energy. Another vital component of a sustainable society is restoration—replanting forests, grasslands, and wetlands—to ensure an adequate supply of resources for future generations as well as for the many species that share this planet with us.

Essential to the success of our efforts to create a sustainable society are efforts to slow down, even stop, world population growth. But that means stopping population growth in all nations, not just the poorer, less-developed nations. Population growth in the rich nations, combined with our resource-intensive lifestyles, is contributing as much if not more to the current global crisis as population growth in the less-developed nations.

Curtailing population growth also entails efforts to better manage how we spread out on the land—that is, how and where our cities and towns expand. By adhering to judicious growth measures, we can preserve farmland, forests, pastures, wildlands, and fisheries—all essential to our future and often crucial to the well-being of the countless species that share this planet with us.

In this book, we present the case for building a sustainable future based on conservation, recycling, renewable resources, restoration, and population control. We call these the operating principles of a sustainable society. We believe that by putting these principles into practice in all sectors of our society—including agriculture, industry, and transportation—we can build an enduring relationship with the planet.

The operating principles, however, must be accompanied by a change in attitudes. No longer can we afford to regard the Earth as an infinite source of materials meant exclusively for human use. Many of the Earth's resources, upon which human beings depend, are finite. The Earth offers a limited supply of resources. We ignore this imperative at our own risk.

We and many others believe that humans must adopt an attitude that seeks cooperation with, rather than domination of, nature. Our efforts to dominate and control nature are often in vain and sometimes backfire on us. Cooperation may be one of the keys to our long-term success. By cooperation, we mean fitting into nature's cycles—creating production systems, for instance, on farms that more closely correspond with nature's cycles.

Finally, we believe it is time to rethink our position in the ecosystem. Humans are not apart from nature but a part of it. Our lives and our economy are vitally dependent on the environment. The Earth is the source of all goods and services and the sink for all of our wastes. What we do to the environment we do to ourselves. The logical extension of this simple truth is that planet care is the ultimate form of self-care.

Despite the wonderful accomplishments of human society over many centuries, many observers argue that it is time to recognize and respect the rights of other species to exist and thrive alongside humans. They contend that natural resources should be viewed as the Earth's endowment and services to all species—not just to humans. Such a view may mean curbing our demands and finding new ways to live on the planet. In the long run, such changes will benefit all of us.

Focus on Principles, Problems, and Solutions

This book describes many important principles of ecology and resource management, concepts that will prove useful throughout your lifetime. It also outlines many of the local, regional, national, and global environmental problems, and offers a variety of solutions to these problems. Solutions take three basic forms: legislative (new laws and regulations), technological (applying existing or new and improved technologies), and methodological (changing how we do things). Applying these solutions is a responsibility we all have in common. It is not just the domain of government. Citizens, businesspeople, and government officials all have an important role to play in solving the environmental crisis and in building a sustainable society.

On the personal level, what we do or what we fail to do can have a remarkable impact on the future. We encourage you to take active steps to find ways to reduce your impact.

Learning Aids

To help students learn key terms and concepts, we have included a number of learning aids: key words and phrases, chapter summaries of key concepts, and critical thinking and discussion questions. To help students deepen and broaden their knowledge, we have included Ethics in Resource Conservation boxes, a section on critical thinking, Case Studies, A Closer Look boxes, GIS and Remote Sensing case studies, and numerous Suggested Readings. To help promote individual action, we've added green living tips, called Go Green!

Key Terms

At the end of each chapter is a list of key words and phrases. We recommend that students read this list before reading the chapter. After reading the chapter, take a few moments to define the terms.

Summary of Key Concepts

Each chapter in the book also contains a summary of important facts and concepts. These short summaries will help students review material before tests. Before reading the chapter, we

think it is a good idea to read through the summary or study the major headings and subheadings to orient yourself.

Critical Thinking and Discussion Questions

Discussion questions at the end of each chapter also provide a way of focusing on important material and reviewing concepts and crucial facts. We have written many questions that encourage you to tie information together and to draw on personal experience. We have also included questions that ask you to think critically about various issues.

Ethics in Resource Conservation

This book contains seven essays on ethics and resource management. These brief pieces present important ethical issues that confront resource managers and people like yourself on a daily basis. The ethics boxes were designed to encourage you to think about your own values and how they influence your views. They will help you understand others, too.

Critical Thinking

Critical thinking is a vital skill for all of us, but it is especially important in resource conservation and management. In Chapter 2, we present a number of critical thinking rules that will help you analyze the material we present.

Case Studies and Closer Look Boxes

The Case Studies and Closer Look boxes delve into controversial issues or provide detailed information that may be of interest to students pursuing a career in natural resource management.

GIS and Remote Sensing

This edition also includes important information on geographic information systems and remote sensing. Chapter 1, for instance, presents an overview of these resource management tools. GIS and Remote Sensing case studies, researched and written by John Hayes at Salem State College and Dr. Chiras, describe some applications of these tools.

Suggested Readings

The Suggested Readings section in each chapter lists articles and books that are worthwhile reading for students who want to learn more about the environment.

New to the Tenth Edition

Because this field changes rapidly, we have carefully updated the text with recent statistics, recent examples, and new photographs. Thanks to reviewer comments, we have added numerous topics not covered in previous editions and expanded coverage of key topics, including mountaintop mining, peak oil and natural gas, dead zones, e-waste, hydrogen, biodiesel, ethanol, tidal power, and more. We've completely revamped and expanded the discussion of basic economics.

- **Global-warming chapter.** We've also added a new chapter dedicated entirely to global warming and global climate change. This chapter presents the latest scientific evidence on this important topic, showing that the Earth is indeed warming and that humans are the main cause of this change. This chapter also outlines the social, economic, and environmental impacts of climate change and efforts to reduce greenhouse gas emissions and other activities that contribute to climate change.
- **Go Green!** With so many people looking for ways to save energy and live in a more environmentally friendly manner, we've added a new feature to each chapter: simple tips on ways to Go Green. These tips include advice on cutting transportation and home energy use, saving water, recycling

waste, reducing consumption, buying organically grown food, and much more.

- **Closer Look boxes.** In this edition, we've added numerous Closer Look boxes that present exciting case studies highlighting what businesses and governments are now doing to help build a sustainable future. For example, some are installing solar electric systems and producing green products.

We have continued to look for ways to expand the theme of critical thinking and, as we have in previous editions, tried to maintain an objective approach, offering both sides of many issues. The reader will also find useful our new Web page: **http://www.prenhall.com/chiras.**

Finally, we have made a special effort to expand the scope of this book to include more examples of environmental and resource issues and solutions from other countries.

About Our Sustainability Initiatives

This book is carefully crafted to minimize environmental impact. The materials used to manufacture this book originated from sources committed to responsible forestry practices. The paper is FSC certified. The binding, cover, and paper come from facilities that minimize waste, energy consumption, and the use of harmful chemicals.

Pearson closes the loop by recycling every out-of-date text returned to our warehouse. We pulp the books, and the pulp is used to produce items such as paper coffee cups and shopping bags. In addition, Pearson aims to become the first carbon neutral educational publishing company.

Pearson is also supporting student sustainability efforts, through our Sustainable Solutions Awards, our Student Sustainability Summits, and our Student Activity Fund.

The future holds great promise for reducing our impact on Earth's environment, and Pearson is proud to be leading the way. We strive to publish the best books with the most up-to-date and accurate content, and to do so in ways that minimize our impact on Earth.

Mixed Sources

Product group from well-managed forests, controlled sources and recycled wood or fiber
www.fsc.org   Cert no. SCS-COC-00648
© 1996 Forest Stewardship Council

ACKNOWLEDGMENTS

We thank the staff at Prentice Hall, especially our editor, Chalon Bridges. Chalon has been a delight to work with throughout this project, for which we are eternally grateful. We also thank Chalon's extremely capable and amazingly helpful project manager, Crissy Dudonis, for her help throughout the project. Many thanks to Yvonne Gerin for her hard work and persistence in researching photographs. Many thanks to Amanda Zagnoli of Elm Street Publishing Services for seeing this book through production. Amanda handled the production of this title expeditiously and thoughtfully. It was a pleasure to work with her.

Many thanks to Linda Stuart for her diligence and thoroughness in updating the statistics. A special thanks to Professor John Hayes for his contribution of the GIS and Remote Sensing case studies.

Finally, we thank our families for their love and support during the writing and production of this book.

Reviewers

Donald F. Anthrop, *San Jose State University*

Gary J. Atchinson, *Iowa State University*

Thomas B. Begley, *Murray State University*

Ronald E. Beiswenger, *University of Wyoming*

Mikhail Blinnikov, *St. Cloud State University*

Michael Brody, *Montana State University*

Peter T. Bromley, *North Carolina State University*

Conrad S. Brumley, *Texas Tech University*

Neal E. Catt, *Vincennes University*

Thomas Daniels, *SUNY, Albany*

Ray DePalma, *William Rainey Harper College, Illinois*

Karen Eisenhart, *University of Pennsylvania*

Donald Friend, *Minnesota State University*

Eric Fritzell, *University of Missouri*

Ken Fulgham, *Humboldt State University*

Tracy Galarowicz, *Central Michigan University*

Jerry D. Glover, *The Land Institute*

Dale Green, *University of Georgia*

Paul K. Grogger, *University of Colorado*

Jeanne Harrison, *Rockingham Community College*

John Hayes, *Salem State College*

Carol Hazard, *Meredith College*

Bill Kelly, *Bakersfield College*

William E. Kelso, *Louisiana State University*

Linda R. Klein, *Washington State University*

John Lemberger, *University of Wisconsin, Oshkosh*

Amy Lilienfeld, *Central Michigan University*

Michael T. Mengak, *University of Georgia*

Jim Merchant, *University of Kansas*

Frederick A. Montague Jr., *Purdue University*

Steve Namikas, *Louisiana State University*

Gary M. Nelson, *Des Moines Area Community College*

Wanna D. Pitts, *San Jose State University*

Stephen E. Podewell, *Western Michigan University*

Jerry Reynolds, *University of Central Arkansas*

David W. Willis, *South Dakota State University*

Gary W. Witmer, *USDA Animal and Plant Health Inspection Service, Fort Collins, Colorado*

Richard J. Wright, *Valencia Community College, Florida*

BIOGRAPHIES

Dan Chiras earned his Ph.D. in reproductive physiology in 1976 from the University of Kansas Medical School. After graduating, Dr. Chiras pursued interests in environmental science and has become a leading authority on environmental issues and sustainability. He is a Visiting Professor at Colorado College, where he teaches courses on global warming, energy, and the environment. Dr. Chiras is founder and director of The Evergreen Institute's Center for Renewable Energy and Green Building where he teaches workshops on residential renewable energy, including solar energy and wind, and green building. Dr. Chiras has published 25 books and more than 250 articles in journals, magazines, newspapers, and encyclopedias. In the past thirteen years, he has focused much of his attention on environmentally friendly building and residential renewable energy. He has published numerous books on the subject, including *Green Home Improvement, Power From the Wind, The Homeowner's Guide to Renewable Energy, The Natural House, The Solar House: Passive Heating and Cooling, The New Ecological Home, The Natural Plaster Book,* and *Superbia! 31 Ways to Create Sustainable Neighborhoods.* Dr. Chiras also lectures widely on a variety of topics, including ways to build a sustainable society, green building, and residential renewable energy. In addition to his scientific and environmental pursuits, Dr. Chiras is a river runner, cross-country skier, bicyclist, organic gardener, and musician. He and his two sons live in a state-of-the-art environmental passive solar/solar electric home in Evergreen, Colorado, overlooking the snowcapped Rocky Mountains.

John Reganold received his Ph.D. in soil science from the University of California at Davis in 1980. He has worked as a soil scientist with the U.S. Department of Agriculture Natural Resource Conservation Service and as an environmental engineer for Utah International Inc., a worldwide mining company. He joined Washington State University in 1983 and is currently Regents Professor of Soil Science. He teaches courses in introductory soils, land use, and organic farming and conducts research in agroecology and sustainable agriculture. He also advises undergraduate students majoring in soil science and organic agriculture systems. His research with graduate students measures the effects of alternative and conventional farming systems on soil health, crop quality, financial performance, and environmental quality. Dr. Reganold has published more than 130 papers in scientific journals, magazines, and proceedings, including *Science, Nature, Proceedings of the National Academy of Sciences USA, Scientific American,* and *New Scientist.* He is also co-editor of the book *Organic Agriculture: A Global Perspective.* He has given more than 150 invited presentations on his research to international, national, and regional groups of scientists, students, growers, and consumers from around the world. In addition to his teaching and research, he enjoys spending time outdoors, swimming, cycling, and backpacking.

CHAPTER 1

NATURAL RESOURCE CONSERVATION AND MANAGEMENT

PAST, PRESENT, AND FUTURE

The late Aldo Leopold once defined conservation as "a state of harmony between man and the land." Leopold believed strongly that effective conservation depends primarily on a basic human respect for natural resources, which he called a land ethic. Each of us, he said, is individually responsible for maintaining "the health of the land." A healthy land has "the capacity for self-renewal." "Conservation," he concluded, "is our effort to . . . preserve that capacity." It is this concept of conservation that has guided and influenced the writing of this book over the past three decades.

1.1   A Crisis on Planet Earth?

Effective conservation and management of natural resources in the United States and other countries is becoming more and more urgent for many reasons. First and foremost, the human population is growing at an extraordinarily rapid pace. Eighty million new people are added to the planet each year. Second, along with this growth is an unprecedented growth in the human economy. As the world's population and economy expand, human society is causing greater damage to an already-deteriorating environment. The implications for wildlife are enormous, but damage also affects us, for the Earth and its ecosystems provide all of the resources that fuel our economy and support our lives. The natural environment is also a sink for all of our wastes. The damage we create therefore threatens our future and the future of our children and the millions of species that share the planet with us.

Ironically, humankind prides itself on conquering outer space and on its many new technologies that make space exploration possible. Yet after two centuries of technological progress, we still fail to adequately manage the space around us here on planet Earth. This failure has led to an environmental crisis that results from three interrelated problems: (1) a large and rapidly growing human population, (2) excessive resource consumption and depletion, and (3) local, regional, and global pollution.

Before we examine each facet of the crisis, it is important to point out that even if there were no crisis, it would be important to improve how we manage our activities, our natural resources, and the environment. The planet with its living organisms is vital to all life. Our personal health and well-being are intimately tied to the planet's health. So, whether or not you believe there is an environmental crisis, this book will help you understand ways to live on Earth that are essential to creating a healthy planet and a healthy human community.

Population Increase

The human population is growing rapidly. At the current rate of growth, global population will surge from more than 6.67 billion in 2008 to more than 8 billion by 2025. This rapid growth of the human population is the main driving force behind the depletion of resources and the pollution of our planet. Why is **population growth** such an important force?

As a general rule, every increase in population results in an increase in the demand for food, water, clothing, shelter, and other goods and services. In meeting these needs, we draw on Earth's natural resources, many of which are already in short supply or are declining in quality. Meeting our demands for these resources also increases environmental pollution (Figure 1.1). Evidence of this simple but powerful relationship is all around us. For example, the rapid increase in population in many less-developed countries (LDCs), such as those in Africa, has resulted in rampant environmental damage, including deforestation, desertification (the spread of desert), and water pollution (Figure 1.2). Population growth in the wealthier, more-developed nations also has caused serious problems, among them water pollution, air pollution, and loss of prime agricultural land. In fact, there's not an environmental problem that can't be linked to human population growth. Even many social problems, such as drug abuse, mental illness, crime, and suicide, are all thought to increase as a result of overcrowding in urban environments.

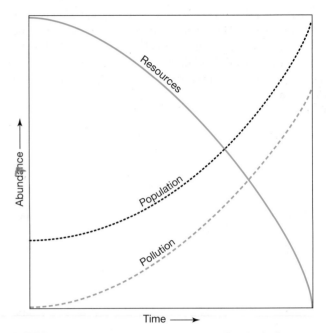

FIGURE 1.1 Population, resources, and pollution. This simple diagram illustrates a very fundamental relationship between people, the resources they require, and environmental destruction and pollution.

The evidence is pretty clear: Rising population is resulting in a decline in the standard of living in virtually all nations of the world. Unless population growth is halted within the very near future, even the most soundly conceived and effectively implemented conservation and environmental practices will be to no avail.

How fast is the human population growing? By this time tomorrow, nearly 220,000 people will join the global family; in a week, over 1.5 million more will be here; and by next year, an additional 80 million will be making demands for food and other necessities. On Memorial Day, the United States honors the memory of those Americans who gave

FIGURE 1.2 Overpopulation and other factors like poverty cause a host of environmental problems from (left) degraded landscapes, as in this desert landscape in the Middle East produced by years of overgrazing and poor land management, to (right) crowded, squalid living conditions that result in pollution of nearby waterways, among other problems.

their lives for their country on the world's battlefields. The fatalities have indeed been numerous—57,000 in the Vietnam War alone. Yet the rate of population growth is so high that all the battlefield deaths of soldiers the world over since the voyages of Christopher Columbus are replaced in about six months.

Resource Consumption and Depletion

All people need resources. The most noticeable demand for those resources comes from the world's industrialized or more-developed nations (MDCs), which are consuming many natural resources (coal, oil, gas, copper, zinc, and cobalt, for example) at an accelerating pace. The United States ranks first in per capita consumption—that is, consumption per person. Although the United States has less than 5% of the global population, it consumes 30% of the world's resources.

Americans consume disproportionately more resources than almost everyone on the planet to feed, house, clothe, transport, and entertain ourselves. Our enormous consumption of cars, high-definition color television sets, automatic dishwashers, air conditioners, cell phones, computers, CD players, and DVD players satisfies a wide range of longings, far beyond basic needs. Through such excessive production and consumption, the United States and other highly industrialized nations such as Canada and Japan are accelerating the depletion of our planet's resources.

Resource demands are extraordinary in the heavily populated less-developed nations, too. But in this instance, demands are generally due to efforts to meet the people's basic needs for food, shelter, and clothing. A large part of the demand is also due to the exportation of raw materials and goods to industrial nations. That said, several less-developed countries, among them China and India, are also experiencing a significant increase in economic growth, and many citizens are now enjoying much higher standards of living. China, which houses the largest population in the world, for example, has posted double-digit economic growth for a decade and shows no sign of slowing (Figure 1.3). Higher standards of living mean high levels of consumption, a much greater demand for resources, and more environmental damage. Lax environmental laws in such countries only worsen the environmental problems created by these rapidly expanding economies.

Whatever the cause, it is clear that the natural environments and resources that less-developed nations need to survive suffer enormously under the strain of large and rapidly growing populations.

Pollution

People produce pollution through the extraction and use of resources, which occurs in rich and poor nations alike. Because of its appetite for resources, the United States, which is the world's most affluent nation, has also become its most effluent (Figures 1.4 and 1.5). Like other industrialized nations, our country has degraded the environment with an enormous variety and volume of contaminants. We have polluted lakes, streams, oceans, and groundwater with sewage,

FIGURE 1.3 China's rapidly growing economy is resulting in a dramatic increase in the consumption of resources.

industrial wastes, radioactive materials, heat, detergents, fertilizers, pesticides, and plastics. Millions of tons of sulfur dioxide and carbon dioxide are spewed into the air each year from the combustion of fossil fuels, especially coal and oil, and are causing serious environmental effects, not only in the United States but also in other nations. Our dependence on nuclear power, as well as on nuclear arms, has led to the accumulation of large amounts of radioactive waste.

Pollution also abounds in LDCs, where sewage, animal waste, and sediment from farms and deforested lands contaminate the air, water, and land. As LDCs industrialize, their industries expand and standards of living improve. The environment is bound to become more degraded.

FIGURE 1.4 The United States and other nations have polluted their lakes and streams with sewage, industrial wastes, radioactive materials, heat, detergents, agricultural fertilizers, and pesticides. Massive fish kills are often the result.

FIGURE 1.5   Like many cities, Los Angeles is often blanketed in a thick layer of pollution from cars, buses, trucks, motorcycles, lawn mowers, factories, power plants, backyard grills, and other sources.

1.2   Differing Viewpoints: Are We on a Sustainable Course?

The Earth's ecosystem is the life support system of the planet. Can the Earth and its ecosystems support through 2050 the reasonably high standard of living many of us now enjoy? Can they support the rising level of affluence in LDCs? Will they be able to support the human population by the year 2100? These important questions are almost impossible to answer with any degree of certainty. Why?

The reason for our inability to answer these questions is that many factors that influence the issue. In the early 1970s, a research group at the Massachusetts Institute of Technology (MIT) led by the late Donella Meadows and Dennis Meadows set out to find an answer to such questions. In 1972, they published their results in a landmark book, *The Limits to Growth.*

The researchers showed through computer analysis that the human population would exceed the planet's carrying capacity (its ability to provide resources and assimilate our wastes over the long haul) within a century if exponential growth continued. Figure 1.6 summarizes the findings. You may want to take a moment to study the graph. As the graph generated by their computer program shows, as the world population expands, resource supplies fall. Growth in the human population is accompanied by a decline in the amount of food available on an individual basis (food per capita). Declining resources will result in a decline in industrial output per capita. In time, the human population begins to decrease, largely as a result of starvation.

What would happen if resource supplies were much larger than the researchers estimated? To address this question, the team doubled their estimated available supply of nonrenewable resources, such as oil and minerals. What they found was that the

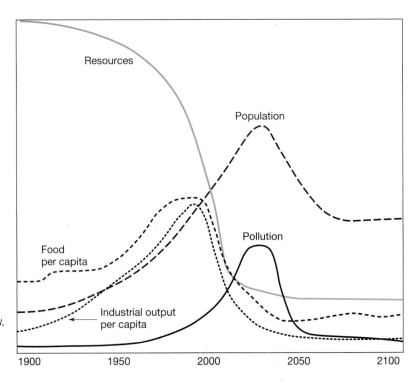

FIGURE 1.6   The Limits to Growth study. Researchers used the computer to predict the fate of human society if current trends continue. This graph shows that if the population continues to grow, resources will decline dramatically. Pollution levels will increase. The combined effect is a decline in human population and considerable environmental damage.

human population would still overshoot the Earth's resource supplies, just a couple of decades later. In another scenario, the authors assumed that world resources were unlimited. Under these conditions, population growth was still halted by rising levels of pollution.

The conclusions of the MIT study were unequivocal. Any way you look at it, if the human population continues to increase, we will exceed the planet's ability to support human life indefinitely—the human carrying capacity of the planet. Although some people disagreed with the conclusions, these startling conclusions increased many people's awareness of the true limits to growth and spurred a great deal of activity in areas such as renewable energy.

In 1992, three members of the original *Limits to Growth* team reexamined their findings and re-analyzed the state of the world, publishing their results in a book entitled *Beyond the Limits.* Their conclusion: Their earlier projections had been wrong. They found convincing evidence that humans have already exceeded critical limits and did so much earlier than they had anticipated. They also found that human society was dangerously close to exceeding other limits. They concluded that their earlier projections had seriously underestimated the hazards of continued population growth with its accompanying rise in resource demand and pollution.

Ultimately, the *Limits to Growth* report suggests that our current path cannot be sustained. That is, our society is on an unsustainable path. Don't get us wrong, however. This does not mean we're doomed and that you should abandon hope. It simply means that our present course cannot be sustained for the long term. It will lead to ruination. But we believe that human society can create a sustainable future. However, steering onto a sustainable course will require efforts by individuals like you, businesses, and governments. That's largely what this book is about—outlining problems and offering sustainable personal, governmental, and corporate solutions proposed by many scientists, government officials, citizens, and activists. But action must occur soon. Although there are areas of improvement, the level of environmental destruction continues at an unsustainable rate.

Moreover, there are signs that we've already overstepped the planet's human carrying capacity in several vital areas.

Viewpoint of the Optimists

Not everyone agrees with these somber projections and the need for swift action. In fact, the *Limits to Growth* study has been severely criticized by many. Especially vocal are those who believe that technology can solve all of our resource and environmental problems. History, they like to point out, is full of examples showing how necessity fosters new inventions and cultural changes. They believe that technology can be brought into play once again. We call such individuals the optimists.

Athelstan Spilhaus of the University of Minnesota is a leading spokesperson for this school of thought. He has suggested that if enough cheap energy is available, all things can be accomplished, pollution can be controlled, food can be made available for everyone, and clothing and shelter can be provided for the needy millions. To increase food production, the optimists suggest a variety of schemes ranging from fish farming to synthesizing food in test tubes, from growing yeast and algal culture to irrigating deserts, from draining swamplands to using genetic engineering to produce miracle wheats and supercorn. Our options are endless, say optimists. They are limited only by our ingenuity. According to the optimists, we can always depend on human ingenuity to pull another rabbit out of our technological hat.

Viewpoint of the Pessimists (or Realists?)

On the other side of the fence is a group we call the pessimists. Some think they're more in touch with the reality of the situation and prefer to call this group the realists.

Whatever they're called, they believe that technology will not—and cannot—solve all of our problems. For one, there is not enough time to find technological fixes to problems that need solutions today.

Why is time so crucial? The key to the answer is in the words *exponential growth.* Figure 1.7 shows a J-curve. It represents exponential growth. **Exponential growth** occurs when something—be it a population, resource demand, pollution, or a bank account—grows by a fixed percentage and the increase is added to the base amount. For example, an interest-bearing bank account is growing exponentially if the interest is added to the principal. Exponential growth is surprisingly deceptive.

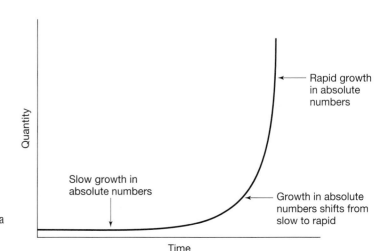

FIGURE 1.7 Exponential growth curve. Exponential growth is a fixed percentage growth. Each increase is added to the base amount. Exponential growth is deceptive. The entity in question grows slowly over a long period, but once it reaches a certain level, even small percentage increases result in remarkable increases in the size.

It begins slowly, but over time, the net increase gets larger and larger as the base amount increases. All of a sudden, the base amount gets so large that each yearly increase becomes enormous. The item being measured is said to have rounded the bend of the growth curve. Human population growth has clearly rounded the bend. Even though the global population is growing at only 1.2% per year, there are 6.7 billion people. Thus, each year adds around 75 million more people.

Global population, resource consumption, and many forms of pollution are all growing in step and are growing exponentially. *Moreover, all three have rounded the bend of the exponential growth curve and are now on the steepest incline.* Even though many of the world's scientists, technologists, ecologists, and economists are struggling to find solutions, the annual increase in human numbers and growth in economic activity are becoming overwhelming. In short, we and the problems we create, such as pollution, habitat destruction, and the loss of forests, are increasing faster than solutions can be found. So not only is technology unable to solve all problems, as the optimists proclaim, it can't keep up with the rapid growth caused by our having rounded the bend and reached the steep ascent on the exponential growth curve.

Viewpoint of the Moderates

Which viewpoint, optimistic or pessimistic, is closer to the truth? Unfortunately, we cannot be sure. Perhaps, however, a moderate viewpoint is more correct. The moderates view our current resource-environmental posture with justifiable concern. Yet they feel that there is still sufficient time, if we start now, to shift from today's spendthrift society to a sustainable society—one that lives within the Earth's limits. A **sustainable society** is defined here as one that meets its needs without preventing future generations and other species from meeting their needs. It lives without robbing Peter (serving future generations) to pay Paul (meeting our needs and desires). It's a simple goal, but given human nature, the ethics of many people, economics, existing laws, and a host of other factors, it could be enormously difficult to achieve.

Fortunately, there's no shortage of writers, teachers, and even some political leaders who are proposing, even experimenting with, strategies for building a sustainable society. In *Lessons from Nature: Learning to Live Sustainably on the Earth* and other writings, Dr. Chiras has outlined six key ecological principles of sustainability based on studies of natural systems. They are (1) conservation, (2) recycling, (3) renewable-resource use, (4) restoration, (5) population control and management, and (6) adaptability. We call them the **biological principles of sustainability,** for they explain in a general way why natural systems sustain themselves. For example, natural biological systems, or **ecosystems,** tend to persist because organisms use what they need and use resources with efficiency. We call this the **conservation principle.** Although there are notable exceptions—for example, at times grizzly bears eat only the eggs of a freshly killed salmon—most species are pretty frugal. Other species don't have the technology we have to exploit resources, either. Another reason why natural systems tend to be sustainable is that life forms recycle wastes for reuse. There is no waste in nature. Virtually everything is used over and over again. Waste from one species is food for another. Natural systems tend to persist because life relies in large part on renewable resources. Natural systems also restore damage. Finally, organisms can adapt to changes.

We believe that these principles can be successfully applied to human societies and that they can help steer us onto a sustainable course—if we act soon and rapidly deploy sustainable solutions. This book attempts to show you how these and other principles can be used to create solutions in a wide range of arenas, including forest management, range management, and waste management.

The task of achieving a truly sustainable society is challenging. It requires the dedicated, highly coordinated, and long-sustained efforts of many different types of people—factory workers and business executives, college students and farmers, scientists and politicians, food specialists and geographers. It requires imaginative and inspirational leadership from government leaders at all levels, from small-town mayors to presidents of the United States and leaders of other industrialized nations, from village tribal chiefs in Africa to benevolent despots in South America. Some of the required changes have already begun. Much of the groundwork for change began many years ago, as you shall soon see.

1.3 A Brief History of the Resource Conservation, Environmental, and Sustainability Movements

Conservation in the 19th Century

The 1700s and 1800s were times of seemingly limitless boundaries in the United States. The new land stretched for miles on end. It appeared to be an endless source of resources. Settlers spread westward, cutting trees to build homes and towns and to make room for farms. Huge forests were leveled, and the prairies were converted to farmland. Swamps were drained, and countless roads cut into the landscape. The prevalent attitude of the times was to use it up and then move on.

In the 1800s, the need for conservation became evident, however, at least for a small number of visionaries. In the early 1800s, for example, George Washington and Thomas Jefferson employed effective methods to control soil erosion on their farms, notably contour farming—plowing and planting across the slope rather than down the fall line of a hill, to reduce soil erosion. In 1864, the diplomat-naturalist George Perkins Marsh wrote *Man and Nature,* a book that did much to draw attention to the fragile nature of our resources and how they can be abused by humans. It served as a catalyst for the fledgling conservation movement.

Another key figure in this period was John Muir. Muir was born in 1838 in Scotland and immigrated to the United States, settling first in Wisconsin at age 11. In 1867, he walked from Indiana to the Gulf of Mexico, recording his observations of plant and animal life in a book. He then moved to California and became enamored with its exquisite wildlands. Devoting his life to conservation, especially the preservation of forest land in the West, Muir lobbied hard for the establishment of

Yosemite and Sequoia National Parks. Muir's love for wilderness, which he expressed in his books and articles, influenced many people, and continues to appeal to many readers.

In large part because of Muir, the United States Congress established three national parks in the 1800s: Yellowstone, the world's first, in 1872, followed by Yosemite and Sequoia in 1890. In 1891, Congress established 28 forest reserves, later to be designated as the United States' first national forests. In 1892, John Muir founded the Sierra Club, an organization that today is one of our country's most politically active conservation groups.

Conservation in the 20th Century

By far the most significant advances in natural resource conservation have been made in the 20th century. They have occurred primarily in four waves. The first (1901–1909) developed under the dynamic and forceful leadership of President Theodore Roosevelt, the second (1930s) occurred during the presidency of Franklin D. Roosevelt, and the third (1970–1980) was given impetus by the Nixon, Ford, and Carter administrations. The newest wave is global in scope.

The First Wave (1901–1909)   In 1908, President Theodore Roosevelt convened the **White House Conference on Natural Resources,** a high-water mark for the cause of conservation (Figure 1.8). Several developments influenced Roosevelt's decision to call the conference. Among them were scientists' deep concern over the severe depletion of timber in the Great Lakes states and a growing apprehension that the United

States' resources were being grossly mismanaged and that severe economic hardship would be the inevitable result.

Roosevelt invited governors, congressional leaders, scientists, anglers, hunters, and resource experts from several foreign nations to the White House Conference. As a result of the meeting, a 50-member **National Conservation Commission** was formed, composed of scientists, legislators, and businessmen. Inspirational leadership was provided by Gifford Pinchot, a professional forester who profoundly influenced the way forests are now managed. Pinchot is credited with introducing scientific principles to forest management. He favored conservation and future growth—the use of forest resources but with measures taken to ensure regrowth so forests can provide a steady supply of resources to people. His personal friendship with President Theodore Roosevelt, himself an avid conservationist, was crucial in forging public policy.

The commission completed the United States' first comprehensive **Natural Resources Inventory.** The White House Conference also resulted indirectly in the formation of 41 state conservation departments, almost all of which are still operating today.

The Second Wave (1933–1941)   Franklin D. Roosevelt is a good example of the right man in the right place at the right time (Figure 1.9). In 1934, Roosevelt established the National Resources Board, which completed the nation's second comprehensive Natural Resources Inventory. In its report, the board identified serious resource problems plaguing the country and described methods for solving them. Roosevelt also created an imaginative nationwide program to create jobs and simultaneously solved many natural resource problems affecting the nation. Much of the impetus for these programs

FIGURE 1.8   President Theodore Roosevelt, outdoorsman, hunter, and ardent conservationist, at Yosemite National Park.

FIGURE 1.9   President Franklin Delano Roosevelt. During his administration (1933–1945), many of the United States' resource problems were dealt with in creative ways.

came from the Dust Bowl era and the Great Depression of the 1930s. The Dust Bowl, described in Chapter 7, resulted from extended drought (1927–1932) that caused crops to fail. In the ensuing years, massive amounts of soil were eroded from farmland laid barren by drought occurring in the Great Plains. Here are some examples of Roosevelt's programs:

1. The **Prairie States Forestry Project** was begun in 1934. Its goal was to establish shelter belts of trees and shrubs (thin bands of trees and shrubs to reduce wind erosion) on farmland along the 100th meridian extending from the Canadian border of North Dakota south to Texas. This project did much to reduce soil erosion from wind.

2. The **Civilian Conservation Corps (CCC),** which was established in 1933 and functioned until 1949, was organized into 2,652 camps of 200 men each. Many were located in the national parks and forests. The forest workers constructed fire lanes, removed fire hazards, fought forest fires, controlled pests, and planted millions of trees. The park workers constructed bridges, improved roads, and built hiking trails. In addition, the CCC made lake and stream improvements and participated in flood-control projects.

3. In 1935, Roosevelt established the **Soil Conservation Service (SCS).** The time was ripe for such a program. The frequent occurrence of severe dust storms over the Dust Bowl of the Great Plains in the late 1920s and the 1930s bore testimony to the vulnerability of the nation's soils. The SCS conducted soil conservation demonstrations to show farmers the techniques and importance of erosion control. (The Soil Conservation Service remains today but is now known as the Natural Resource Conservation Service.)

4. The establishment of the **Tennessee Valley Authority (TVA)** in 1933 was a bold experiment, unique in conservation history, to integrate the use of the resources (water, soil, forests, and wildlife) of an entire river basin. Although highly controversial at the time, it has received international acclaim and has served as a model for similar projects in India and other nations.

5. The **North American Wildlife and Resources Conference** was convened by President Roosevelt in 1936. Attended by wildlife management specialists, hunters, anglers, and government officials, it set out to develop an inventory of the nation's wildlife resources and a statement on wildlife and other conservation problems, including policies by which those problems might be solved. This conference meets annually to this day.

The Third Wave (1960–1980)   During the 1960s, the U.S. conservation and environmental movement really took off. Several highly influential books and essays were written during this time that sensitized the general public to the gravity of the nation's problems. Rachel Carson's *Silent Spring* (1962), a runaway best-seller, alerted the general public to the potentially harmful effects of pesticides such as DDT on both wildlife and humans. Noted ecologist Paul Ehrlich of Stanford University wrote *The Population Bomb,* which warned of the environmental degradation that would result if society did not control the

worldwide surge in population. Garrett Hardin's classic essay, "The Tragedy of the Commons," proposed that any resource shared by many people would eventually be exploited and degraded.

In 1969 Senator Gaylord Nelson (D–WI) called for a nationwide environmental teach-in in an attempt to marshal the energies of the nation's college students "to halt the accelerating pollution and destruction of the environment." Co-organized by Denis Hayes, the event was called Earth Day. It continues to be celebrated each year on April 22 and offers an opportunity to examine new issues and renew one's commitment to environmental protection.

Congress also responded to citizen protests and constituents' letters by enacting many important laws to upgrade our resources and to control pollution. Many of them were passed in the period from 1970 to 1980—so much so that this period has been called the **decade of the environment.** Much of the progress during this period occurred during the Nixon administration. A partial list of these acts is shown in Table 1.1.

One of the major advances in environmental protection during this period was the establishment of the **Environmental Protection Agency (EPA),** which was formed during the early 1970s from environmental branches of key federal agencies. Over the years, the EPA has become a prime mover in environmental protection, with enormous power and oversight over numerous issues. Although the EPA's focus was primarily that of a watchdog and regulatory agency in its early years, in the 1980s and 1990s it began to become more and more proactive, seeking ways to prevent pollution by working cooperatively with businesses in ways that helped them retain, even improve, profits.

Although many great environmental laws were passed during this period, the existence of an environmental law does not ensure environmental protection. A law is only as good as its enforcement, and enforcement requires money for personnel and equipment.

The Fourth Wave (1980–Present): The Beginnings of a Sustainable Revolution?   From an environmental standpoint, the 1980s through the 2000s represent some of the best and worst of times. This has been a period of intense resistance to environmental protection, especially in the United States. After suffering through a crippling period of global inflation in the late 1970s and early 1980s, many politicians and business leaders cast a skeptical eye on environmental protection, perceiving it as counterproductive to economic progress.

Improving on Environmental Protection   In the 1980s and 1990s, a growing number of observers, including key environmental leaders such as Fred Krupp, executive director of the nonprofit organization, Environmental Defense, and Barry Commoner, a professor and author, were beginning to speak out on an important finding: Despite the billions of dollars being spent to protect the environment, many gains were being offset by economic growth and growth in the population. Some solutions, critics noted, only shifted problems from one medium to another. For example, pollution-control devices were installed at many power plants to remove sulfur dioxide, which causes acid rain. Although effective in cleaning up the air, scrubbers simply convert sulfur dioxide into a solid waste that is typically disposed of in

TABLE 1.1	Major Environmental Acts Passed During the Decade of the Environment (1970–1980)
Air Quality	Clean Air Acts of 1970, 1977, 1990
Control of Noise	Noise Control Act of 1972 Quiet Communities Act of 1978
Control of Toxic Substances	Toxic Substances Control Act of 1976 Resource Conservation and Recovery Act of 1976
Control of Solid Wastes	Solid Waste Disposal Act of 1965 Resources Recovery Act of 1970
Energy	National Energy Act of 1978
Land Use	National Coastal Zone Management Act of 1972 Forest Reserves Management Acts of 1974, 1976 Federal Land Policy Management Act of 1976 National Forest Management Act of 1976 Surface Mining Control and Reclamation Act of 1977 Endangered American Wilderness Act of 1978
Water Quality	Federal Water Pollution Control Act of 1972 Ocean Dumping Act of 1972 Safe Drinking Water Act of 1974 Clean Water Act of 1977
Wildlife	Federal Insecticide, Fungicide, and Rodenticide Control Act of 1972 Marine Protection, Research, and Sanctuaries Act of 1972 Endangered Species Act of 1973

landfills, posing a potential threat to groundwater. Some solutions were incomplete. For instance, catalytic converters, pollution-control devices on cars, converted carbon monoxide, a poison to people, into carbon dioxide, a pollutant that traps heat in the atmosphere and is contributing to global warming (Chapter 20). Early catalytic converters burned off unburned hydrocarbons in the exhaust of vehicles, helping to reduce smog, but didn't remove nitrogen oxides, which contribute to acid rain (Chapter 21).

During the 1980s and 1990s, it became increasingly clear that environmental protection was simply not working as well as it could—and it was costing us a fortune. Couldn't we find a way to protect the environment without exacting such a huge price tag? Couldn't we find ways to prevent problems in the first place?

A Shift Toward Sustainable Solutions During this period, many experts began to note that modern society was on a fundamentally unsustainable course. One of the first to call attention to this phenomenon was Lester Brown, an agricultural economist who founded the Worldwatch Institute in Washington, DC (Figure 1.10). Brown's book, *Building a Sustainable Future,* outlined the persistent erosion of the Earth's life support system and proposed a strategy for building a sustainable society. His organization has gone on to produce a series of books and papers on a wide range of environmental topics that are translated into many languages. With uncanny accuracy and vision, they outline problems and sustainable solutions, and have influenced thinking and policy worldwide.

Pioneering work in energy during the 1980s and 1990s was performed by Amory Lovins, a physicist who cofounded the Rocky Mountain Institute. Lovins raised awareness of the environmental and economic importance of energy conservation and served to redirect the thinking of many power companies the world over. In the 1990s, he turned his attention to the development of super-energy-efficient cars, which we'll examine in Chapter 22.

Another highly influential force emerged in 1983, when the United Nations assembled a commission to address the issues and propose actions to promote a sustainable future. Called the **World Commission on Environment and Development,** it produced a book entitled *Our Common Future,* that stirred intense global interest in alternatives to the environmentally unsustainable development taking place worldwide.

This interest, in turn, was largely responsible for the 1992 **United Nations Conference on Environment and Development,** commonly called the **Earth Summit** (see Case Study 1.1). Attended by officials from nearly 180 nations, the Earth Summit was the

FIGURE 1.10   A leader in the sustainability movement, Lester Brown and his staff at the Worldwatch Institute have probably done more to promote an understanding of sustainable development than anyone else in the world.

largest international meeting on the environment in the history of human civilization. Participants produced agreements on a wide range of issues, including global climate change, biodiversity, and deforestation. One of the most impressive outcomes was an 800-page document called *Agenda 21*. It outlined over 4,000 action items for achieving sustainable development.

Since the Earth Summit, numerous cities, towns, states, and national governments have adopted sustainable development as an official policy. **Sustainable development** is defined here as a strategy to meet human needs in ways that do not prevent future generations and other species from meeting their needs. Sustainable development requires a long-term approach to resource management and human development. It requires us to think in terms of systems, too. Systems thinking, which we discuss in this book, requires an understanding of how human and natural systems operate. It requires us to understand our dependence on natural systems and how we affect them, positively and negatively. It is preventive in nature—seeking solutions that prevent problems in the first place. Numerous entities have undertaken sustainable development projects, many of which are outlined in this book.

Important gains have been made in Europe and in the less-developed countries of the world. One of the most promising has been the gradual slowing of the growth in the human population in many parts of the world. Another promising development is the rapid growth of renewable energy, especially large-scale wind power. Numerous wildlife preserves have also been set up in the United States and in less-developed countries, thanks to efforts of the Nature Conservancy and other organizations, protecting the vital habitat of numerous species.

Ecological Justice   As many people grappled with environmental problems in the 1980s and 1990s, it became evident that not only were humans having adverse, unsustainable effects on the environment, not all people bore environmental burdens equally. Certain socioeconomic and ethnic groups, such as poor whites and poor African Americans, often bore much more of the impact of environmental transgressions. For instance, corporations often sited waste dumps, factories, and other environmentally damaging activities near poorer communities, as opposed to more affluent communities. Why? Because poor, economically disadvantaged people can generally not muster either the political or the economic power to fight them. As a

CASE STUDY 1.1   THE EARTH SUMMIT AND BEYOND

All the world's people belong to just one ecological system. This means that whatever environmental damage you create in your hometown may eventually, either directly or indirectly, have a harmful effect on people living elsewhere on this planet, even though they may be half a world away. For example, suppose that you burned up 1,000 gallons of gasoline driving your car last year. As you sped along the city streets, roads, and highways, the exhaust pipe on your car released many billions of molecules of carbon dioxide into the air. These molecules then trapped heat that otherwise would have escaped into outer space. The net result was a warming of the Earth's atmosphere, a phenomenon known as the **greenhouse effect** (Chapter 19). And not only will the atmosphere over your town warm up to a very slight degree, but eventually so will the atmosphere around the entire planet!

It is apparent, therefore, that future resource and environmental management cannot be accomplished effectively without an integrated global effort, rather than a piecemeal and localized approach. To this end, the United Nations convened the first Conference on Environment and Development (UNCED) in Rio de Janeiro, Brazil, in 1992. This conference, popularly referred to as the Earth Summit, was worldwide in scope. It was attended by official environmental delegates from nearly 180 nations. It also attracted many heads of state, including then-president George Bush, more than 8,000 science reporters, thousands of environmentalists, and concerned citizens, who attended a parallel conference put on by the nonprofit community.

Participants in the Earth Summit passed two international agreements, one on global warming and another on protecting nonhuman species (biodiversity). The climate treaty called on the nations of the world to reduce greenhouse gas emissions such as carbon dioxide to 1990 levels by the year 2000. Although some nations took this seriously, most made little progress toward this goal. Since the Earth Summit, a new agreement on climate change has been drawn up. Known as the Kyoto Protocol and signed in 1997, it calls on industrial nations and former Eastern Bloc countries to cut emissions of greenhouse gases by 5.2% below 1990 levels between 2008 and 2012, a goal few nations have taken seriously.

The biodiversity agreement calls on nations to create inventories of their plants and animals and to develop strategies to protect them. Discussion over these documents was intense, and as a result, many critics consider both agreements to be rather weak but a step in the right direction.

Participants of the Earth Summit also created a massive document called *Agenda 21*. This report outlines over 4,000 actions that can be taken to create a sustainable future. In addition, the participants agreed on a set of principles for forest management and protection. It, too, was watered down by certain participating nations. Finally, the Rio conference produced a declaration of legal principles essential to creating global sustainability.

Even though the outcome of the Rio conference was much less than many had hoped for, the fact that the event took place at all is testimony to the heightened environmental awareness and concern of the international community. As Maurice F. Strong, Secretary General of the conference, so eloquently stated in his opening address to the delegates, "The Earth Summit is not an end to itself, but a new beginning. The measures you agree on here will be but first steps on a new pathway to our common future."

International conferences and treaties are only as good as the action they inspire. Fortunately, the Earth Summit has stimulated an incredible amount of action on the part of individuals, communities, cities, states, and nations. In the United States, the interest in sustainable development that emerged in communities has resulted in numerous strategies—for example, community plans that outline steps for sustainable development. Soon after being elected, President Bill Clinton established a national commission, the President's Council on Sustainable Development, which brainstormed policies and actions required to build a sustainable future. In Holland, government agencies and businesses have hammered out a plan to drastically cut pollution without hurting the national economy. Other countries were also spurred to take similar steps. Many stakeholders—individuals, nonprofits, corporations, and governments—have also worked together since the Earth Summit, forming voluntary partnerships to help achieve goals set forth in UNCED.

In 2002, delegates from nearly 110 nations met in Johannesburg, South Africa, for a ten-year reunion to reaffirm their interest in sustainable development. They issued a proclamation calling for additional efforts to create a sustainable society and have once again stirred many nations to take action.

result, many citizens of such communities were subjected to groundwater pollution and air pollution that few, if any, of the more prosperous communities would tolerate.

Activists coined the term **environmental justice** to refer to the inequities suffered by the disadvantaged and the need to ensure that these people do not continue falling victim to such injustices. Over the past two decades, activists such as the Trial Lawyers for Public Justice in Washington, DC, have launched many successful campaigns to stop such activities and compensate victims for health effects and death.

Peak Oil and Global Warming: A New Level of Commitment   Despite the realizations and gains of the 1980s and 1990s, it was not until the 2000s that many citizens, businesses, and governments really understood how serious our problems were and got serious about change. Two key factors were responsible for this shift: rising prices for oil and natural gas and global warming.

The film *The End of Suburbia,* which featured Richard Heinberg, author of *The Party's Over,* helped convince the world that we were running out of cheap oil and natural gas—two vital resources in the modern economy. Rising prices for these fuels also convinced many people that we needed to shift to a sustainable energy supply, notably renewable energy.

During the 2000s, it also became clear that global warming was real and was very likely caused by human activities, primarily the release of carbon dioxide gas from the combustion of fossil fuels. Scientific studies and more violent and costly storms, including deadly tornados and hurricanes, caused by global warming, raised awareness. These events caused billions of dollars' worth of damage and took thousands of lives each year and helped spur action, especially measures to use energy more efficiently and to develop renewable energy resources such as wind (Figure 1.11). A key player in the rising awareness was former vice president Al Gore, whose video *An Inconvenient Truth,* which received an Academy Award for

Documentary Feature, helped raise global awareness of the unsustainability of our current fossil fuel energy path. In response, many large corporations, including Wal-Mart, Microsoft, and Google, have taken significant steps to use energy efficiently and to shift to renewable energy resources. Some U.S. manufacturers, including GE and John Deere, have also become leaders in the production of renewable-energy technologies.

Although progress toward sustainable development is moving slower than many would like, progress *is* occurring. If progress

FIGURE 1.11   Offshore wind farms like this one could supply a steady stream of electricity to power our society while dramatically lowering air pollution, including greenhouse gas emissions.

continues, some believe that the 1980s and 1990s could mark the beginning of a Sustainable Revolution, not unlike the Agricultural and Industrial Revolutions in their impact on human civilization. This change, already under way, seeks strategies that allow people to live well while protecting and enhancing the natural resources and environment so vital to our future and the future of all other life forms.

Continuing Resistance to Environmental Protection   Although many people, government agencies, and businesses have rallied behind sustainable development, powerful antagonism against environmental protection has gained momentum. At the state level, there has been a great deal of backpedaling on environmental protection. Businesses and the powerful lobbying groups that represent them have attempted, often successfully, to weaken environmental laws.

At the national level, anti-environmental efforts have been well funded and extensive. Legislators and environmentalists who have favored renewing many key pieces of environmental legislation such as the Endangered Species Act have found themselves locked in battle with those who see environmental protection as a hindrance to economic progress and personal property rights. Powerful public personalities, including conservative commentator Rush Limbaugh, often with little scientific information to back their arguments, have swayed many people's opinions on key issues. This sentiment so far continues into the 21st century and may be with us for a long time. The George W. Bush administration embarked on a series of highly successful efforts to weaken environmental laws and regulations in virtually every area, including air pollution, species extinction, and wilderness protection. Many environmental supporters think these efforts seriously undermine the nation's previous successes. The administration opened up sensitive wilderness to timber, mining, and fossil fuel interests. It refused to sign agreements with other nations on key issues such as global climate change.

Critics also point out that efforts of the Clinton and Bush administrations and their respective Congressess to globalize the world economy have also had serious detrimental effects on the progress toward a sustainable human future. In the name of free trade, many companies based in the United States and other industrial nations have moved into LDCs. These changes not only cost American jobs, they have significant impact on the environment, because the LDCs frequently have few, if any, environmental regulations on mining, resource extraction, and manufacturing. If environmental regulations do exist, they are poorly enforced or not enforced at all. In such instances, companies often extract resources, refine metals, and produce goods in ways that would be illegal in the United States or other MDCs. This approach, of course, saves companies money, so they can compete better in the global marketplace, but at a huge environmental price.

Efforts to thwart gains continue, despite all the progress that has been made since 1980.

1.4   Classification of Natural Resources

To begin to understand the challenge of building a sustainable society, we must first examine the resource base itself. Conservationists recognize two major kinds of resources: renewable and nonrenewable. Be sure to take a look at Table 1.2 for a more detailed explanation.

TABLE 1.2   Classification of Natural Resources

Renewable

Resources whose continued harvest or use usually depends on proper human planning and management. Improper use and/or management often results in impairment or exhaustion, with resulting harmful social and economic effects. (The only exception is renewable energy, such as solar energy.)

1. Fertile soil. The fertility of soil can be renewed, but the process is expensive and takes time.

2. Products of the land. Resources grown in or dependent on the soil.
   a. Agricultural products. Vegetables, grains, fruits, and fibers.
   b. Forests. Source of timber and paper pulp. Valuable as a source of scenic beauty, as an agent in erosion control, as recreational areas, and as wildlife habitat.
   c. Rangeland. Sustains herds of cattle, sheep, and goats for the production of meat, leather, and wool.
   d. Wild animals. Provide aesthetic value, hunting sport, and food. Examples are deer, wolves, eagles, bluebirds, and fireflies.

3. Products of lakes, streams, and oceans. Examples are black bass, lake trout, salmon, cod, mackerel, lobsters, oysters, and seaweed.

4. Groundwater and surface water.

5. Ecosystems. Natural ecosystems provide many free ecological services. Wetlands, for instance, help reduce flooding, filter surface water, increase groundwater discharge, and provide habitat for animals.

6. Certain energy resources such as wind energy, solar energy, geothermal energy, tidal energy, and hydropower. Many of these are theoretically unlimited and inexhaustible.

Nonrenewable

Amount of resource is finite. When destroyed or consumed, for example, when coal is burned, the resource cannot be replaced.

1. Fossil fuels. Produced by processes that occurred millions of years ago. When consumed (burned), heat, water, and gases (carbon monoxide, carbon dioxide, and sulfur dioxide) are released. The gases may pose serious air pollution problems. These resources cannot be recycled.

2. Nonmetallic minerals. Phosphate rock, glass, sand, and salt. Phosphate rock is of crucial importance as a source of fertilizer.

3. Metals. Gold, platinum, silver, cobalt, lead, iron, zinc, and copper. Without these, modern civilization would be impossible. Zinc is used in galvanized iron to protect it from rusting, tin is used in toothpaste tubes, and iron is used in cans, auto bodies, and bridges. These resources can be recycled.

Renewable resources are those resources that can be renewed by natural processes; they include soils, rangelands, forests, fish, wildlife, air, and water. Although these resources can be regenerated, they can also be depleted as a result of overuse by humans. Overharvesting fish populations, for example, can result in their decline. By the same token, renewal also can be facilitated by humans. That is to say, human actions can enhance fish populations. Some resources may be renewed much more rapidly than others. For example, it may take 1,000 years for just 1 inch of fertile topsoil to form. A pine forest may develop from the seedling stage to maturity in 100 years. By contrast, a herd of deer, under optimal conditions, could build up from 6 to 1,000 in only 10 years!

Several energy resources are also classified as renewable, for example, solar energy, wind energy, and hydropower. Although solar energy comes from a finite source, the sun, we still consider it a renewable resource because the sun's projected life span is huge, perhaps a couple billion years. We won't need to worry about running out of solar energy for a long time.

Wind and hydropower, both discussed in more detail in Chapter 22, are produced largely as a result of solar energy. Winds, for example, result from the differential heating of the Earth's surface. Hydropower, energy captured from flowing water, depends on evaporation and precipitation. Evaporation is powered by sunlight. Both wind and flowing water will continue so long as the sun survives. They are renewed daily by the sun.

Nonrenewable resources occur in a fixed amount. They cannot be renewed by natural processes at all, or not rapidly enough to be usable by current human society. They include fossil fuels (oil, coal, natural gas), nonmetallic minerals (phosphates, silica), and metallic minerals (copper, aluminum). Among nonrenewable resources, there are some major distinctions worth considering. One of the most important is that some nonrenewables are recyclable. Metals, for instance, are finite but can be recycled. Fossil fuels, however, are not. When they are burned to release energy, they are lost forever.

In building a sustainable society, we must find strategies to better manage both renewable and nonrenewable resources. We must also give serious consideration to the exhaustibility of a natural resource. Some, like solar energy and wind energy, are inexhaustible. So long as the sun shines, these sources of energy will be available to us.

Renewable resources are not a panacea, however. Some can be exhausted by overexploitation and poor management practices. Forests, fish, topsoil, groundwater, and many other renewable resources we depend on, while capable of being renewed, can be depleted. If we cut down a rain forest and let the soil wash away, we may have destroyed a valuable renewable resource forever. Other resources, such as coal and oil, are nonrenewable (finite). There's only so much of each in Earth's crust. Coal and oil are therefore clearly exhaustible. When they're burned, they're lost forever. But not all nonrenewable resources are like coal and oil. Minerals, for instance, are nonrenewable or finite, but they can be recycled ad infinitum. Keep these distinctions in mind as you ponder the fate of the planet and strategies to steer our society onto a sustainable path. The next section begins to uncover the strategies that Americans

and others the world over have employed in the past, and then outlines the rudiments of a sustainable management/living strategy, which provides important background for the rest of this book and your study.

1.5 Approaches to Natural Resource Management

As you read this book and talk to others about resource issues, you will find that opinions vary considerably about the value of natural resources and the importance of protecting them as we meet our needs for daily life. During the past two centuries, four resource management ideologies have emerged in the United States and elsewhere: (1) exploitation, (2) preservation, (3) the utilitarian approach, and (4) the ecological or sustainable approach. Understanding them will help you understand where we've been and where we're going as a society. They'll also help you understand different perspectives.

Exploitation: A Human-Centered Approach

The **exploitation approach** is based on a belief that natural resources should be used as intensively as possible to provide the greatest profit. This philosophy prevailed early in the United States' history and continues today in many parts of the world, especially in the less-developed nations that are beginning to industrialize.

Little, if any, concern was given to the adverse effects of exploitation, such as soil erosion, water pollution, or wildlife depletion. For example, the slogan of the early loggers of the 1800s who exploited the United States' primeval forests was "Get in, log off the trees, and get out." When the trees were gone, the loggers moved westward and repeated the process. Why not? The nation's forests seemed inexhaustible (Figure 1.12).

The exploitive ideology seeks to maximize human gain with little, if any, concern for nature and natural resources. It assumes that the supply of resources is unlimited and that nature is here to serve humans. It also assumes that nature is of no importance to us other than as a source of commodities to make our lives better. The montra of the Republican party, "Drill, baby drill" represents the same approach.

Preservation: A Nature-Centered Approach

The **preservation approach** to resource management suggests that resources should be preserved, set aside, and protected. A forest, for example, should not be used as a source of timber. It should be preserved in its natural state as a wilderness. In the 1880s, the naturalist John Muir, who founded the Sierra Club, proposed that federal lands of unique beauty should be withdrawn from exploitation by timber, grazing, and mining interests and converted into national parks. As such, they would be preserved virtually unchanged for the enjoyment of future generations. Partly as a result of Muir's influence, Congress established Yosemite and Sequoia as national parks in 1890 and 1891. Those who promote wilderness protection seek a similar goal for millions of acres in the United States and abroad, often over

FIGURE 1.12 Loggers cut down and removed huge trees in the Pacific Northwest and hauled the massive logs out by train. During the 1800s, many thousands of acres were cleared throughout the United States, then abandoned as timber companies moved on to virgin forests for more timber.

the protests of development interests, especially mining and timber companies.

Preservation ideology is the antithesis of the exploitive approach. In fact, it probably emerged in reaction to the observed effects of exploitation. Preservation remains alive today in various forms. A national environmental movement that calls itself Earth First! proffers such a view. It promotes dramatic actions, including tree spiking, to stop the continued destruction of natural systems. Although most environmental and conservation groups today support the protection of wilderness, their tactics are more mainstream. That is, they prefer to work within the system to protect wild areas. They also recognize the need for resources and wise management of them. This is sometimes called a utilitarian approach.

Utilitarian Approach

The **utilitarian approach** to resource management began during the late 1800s and early 1900s, spearheaded by Gifford Pinchot, who was head of the U.S. Forest Service from 1898 to 1910, and President Theodore Roosevelt. The key organizing principle behind the utilitarian approach is that of **sustained yield.** This concept suggests that renewable resources (soil, rangelands, forests, wildlife, fisheries, and so on) should be managed so that they will never be exhausted. Careful management will ensure their replenishment so they can serve future generations ad infinitum. When a forest has been logged off, the site must be reseeded naturally or artificially so that a new forest can develop over time and provide timber for future generations. Similarly, when fish are harvested from a lake or stream, they must be replaced, either by natural reproduction or by stocking with hatchery-reared fish.

The utilitarian approach is human centered. The idea behind this approach is simple: Protect a nation's resources by harvesting them at rates that can be sustained over the long run, a task that has proved to be much more difficult than once thought. Scientific research has shown that systems are more complex than once realized and are connected to other systems. To manage them sustainably requires many more considerations than ever imagined—and it requires better management of ourselves. A good understanding of the science of ecology is essential to the newest approach, the sustainable approach.

The Sustainable Approach

The sustainable approach relies, in large part, on an understanding of **ecology,** the study of the interrelationships between organisms and their environment. That is to say, modern conservation strategies base their decisions on sound ecological facts and principles. Chapter 3 outlines key principles and describes the ways they're currently applied. The resource management chapters in this book also focus on principles that are essential to sustainable management of the world's natural resources.

The sustainable approach to resource management is evolving, changing as our knowledge expands. One key concept in this approach is that policies promoted to improve management are often designed to protect more than harvestable species. That is to say, efforts are made to protect entire ecosystems, including those that harvestable renewable resources depend on. Soil in a forest must be protected, for example, to ensure a steady supply of lumber from it. If we let it erode away, lumber production will decline.

More than 100 years ago, the American diplomat-naturalist George Perkins Marsh observed how humans had abused agricultural lands in Europe and Asia. He further observed how this abuse resulted in soil erosion, dust storms, water pollution, and a decline in many nations' economic well-being. Upon returning to the United States, Marsh wrote *Man and Nature* (1864), mentioned earlier. In it he argued that humans cannot degrade one part of the environment without harming other parts. Marsh argued that, although our natural environment is highly varied and infinitely complex, it is a dynamic and organic whole. Today's conservationists realize that Marsh was correct. Soil conservation measures taken on forested land not only help to ensure a sustainable yield, but also help to protect neighboring streams, rivers, and lakes. Fish and other organisms that depend on these water bodies are protected by sustainable forest practices.

The **sustainable approach** (also sometimes referred to as the **ecological approach**) therefore requires a view of entire systems. That is to say, it requires managers to think in terms of whole systems, ecosystems, not just isolated parts of them. When working on a problem at any one level (species, population, or ecosystem), resource managers must seek to understand and protect the connections between all levels. For example, efforts to protect a species of fish in a river or lake would encompass more than simply improving in-stream habitat. In fact, such measures may be meaningless if outside influences poison a stream. In wildlife management, this is sometimes referred to as **ecosystems management.** Ecosystems management, therefore, includes steps to protect the surrounding watershed, which contributes to the ecological integrity of the water body. Stream protection begins many miles from stream banks and requires measures to eliminate careless road building, control sources of industrial pollution, or modify ecologically unsound farming practices. It may require measures to eliminate human development, which can have adverse effects on neighboring biological systems. Or it may require ways to better manage our own activities—that is, find alternative ways of meeting our needs that don't alter neighboring ecosystems.

Protecting and managing entire ecosystems to protect species may require broader efforts—protection and management of two or more ecosystems required for a species to survive. For example, grizzly bears in and around Yellowstone National Park depend on a 2-million-hectare (5-million-acre) area that includes the park and surrounding areas. Protecting the grizzlies requires efforts to protect and manage these areas, a portion of which is privately owned.

Another key concept of ecosystem management is carrying capacity. **Carrying capacity** is defined as the ability of an ecosystem to support a population of a given species living in a given manner indefinitely. It is determined by resource availability and the ability of the ecosystem to absorb wastes, known as source and sink functions, respectively. Each ecosystem has a carrying capacity for individual organisms. Altering one component in an ecosystem, we've learned, may help one species to the detriment of another. Single-species management, once a common practice for popular species of fish and huntable wildlife, is falling out of favor. "Protect the ecosystem

and protect the species" is a view growing more and more popular in resource management agencies.

Maintaining carrying capacity and stability within an ecosystem requires steps to maintain its ecological integrity. This often means that efforts must be made to protect the fertility of soils, the purity of water supplies, and the diversity of species, populations, and ecosystems in a given management area. This may require efforts to protect viable populations of native species and permit natural disturbances such as periodic fire. This may also require the removal of nonnative species and the reintroduction of native species.

In sustainable resource management, resources such as soil, water, rangelands, wildlife, fisheries, and forests are used in ways that ensure their long-term health and vitality. In other words, it pursues ways of harvesting the Earth's generous supply of natural resources sustainably without damaging their ability to regenerate. Sustained yield of a desired species is not as important as sustaining the environment and its many interrelationships.

Sustainable resource management requires a long-term view. It embraces the concept of multiple uses. A forest, for example, is not only a source of timber, as the utilitarians suggest, but has many other values, such as wildlife habitat, scenic beauty, and flood and erosion control. Forests must therefore be managed so that these other values can be realized. A forest ecologist would ask the following questions regarding the effects of a proposed timber harvest:

1. Will water runoff be accelerated? If so, will this result in flood damage in the valley towns? Will valley farmers have enough water for their crops and livestock?

2. Will the resultant soil erosion destroy trout spawning beds in the streams below the cut?

3. Will soil fertility at the site be reduced because of erosion, thus jeopardizing the development of seedlings on the logged-off site?

4. What will happen to deer, grouse, and other wildlife when the forest that provided food, cover, and breeding sites is removed?

5. Will the scar left after the stand has been logged off cause visual pollution for the thousands of motorists who travel along the nearby highways?

6. What effect will this timber harvest, and thousands like it throughout the world, have on the buildup of carbon dioxide in the global atmosphere and the progressive warming of the planet?

In the sustainable approach to conservation, resources should be used without adversely affecting the physical and biological environments, or should be used in ways that do not result in irreparable damage. Better yet, they should be used in ways that enhance their potential. Much of this book focuses on this objective.

This approach to resource management embraces the notion of multiple uses, as noted above. However, it also may require actions that restrict or limit human activity, sometimes drastically. In other words, it recognizes the need for preservation. As an example, to protect vital biological resources such as birds and wildlife, it is often necessary to establish core reserves in forested areas. A **core reserve** is a region left exclusively to plants and animals. It is a region that is off-limits to human exploitation. Surrounding these reserves are **buffer zones,** areas of minimal human intrusion and impact. Limited tree harvesting may occur in such areas. Surrounding the buffer zones are regions where human activities are permitted.

Managing ecosystems requires good scientific information and vigilant monitoring of conditions so that if management strategies are not working, they can be adjusted. The use of such information to monitor conditions and make policy changes is called **adaptive management** and is a keystone of good ecosystem management. Unfortunately, it is not yet widely used. In many cases, management agencies establish policies and then manage accordingly. Changes in our understanding of systems may not result in immediate changes in policy because of bureaucratic inertia.

Another exciting development is **collaborative conservation**—resource management decisions made in a collaborative setting with a variety of stakeholders, for example, timber companies, local residents, Native Americans, wildlife enthusiasts, recreationalists, and other conservationists. Good examples of this approach on public lands in the American West are described in the book *Across the Great Divide: Explorations in Collaborative Conservation and the American West.*

As you will see in other chapters, ecosystem management requires the cooperation of various government agencies that control public land, because as noted earlier, the habitat of a species often spans two or more jurisdictions. The involvement of private landholders is also essential to protect ecosystems.

In a sense, a sustainable approach helps us live within limits while sustaining human culture. It is a human-centered and nature-centered approach, giving equal consideration to both. Ecosystems management is part of a larger sustainable-development strategy for human survival and prosperity on a finite planet.

Creating a sustainable society requires better ways of managing resources, but it also requires better management of ourselves—our systems and society. We would be remiss if we were to suggest that all we need to do is find ways to better manage our resources. That's only half of the task. To create an enduring human presence, we must live our lives and conduct our business affairs in ways that do not deplete the Earth's resources or foul its air, water, and soil. Earlier in the chapter, we mentioned this task in conjunction with the *Limits to Growth* study. Many chapters in this book point out the problems created by our own systems, such as transportation and waste management. To create a sustainable society, we must also address these matters. We believe that the biological principles of sustainability, outlined earlier, can be used to put human civilization back on a sustainable course.

GO GREEN!

If you or others dump supplies, desk lamps, clothing, or furniture at the end of each academic year, contact www.dumpandrun.org. They collect discards and sell them to incoming students, donating the proceeds to charity.

These principles include conservation, recycling, renewable-resource use, restoration, and population control.

Because they are so important, we will repeat them here. In this book, the conservation principle refers to two basic notions: using only the resources we need (the frugality principle) and using resources efficiently (the efficiency principle). Recycling, of course, means ensuring that the materials we use are not discarded but rather placed back into manufacturing systems, thus virtually eliminating wastes. Building a sustainable society will also require efforts to use and protect renewable resources. Especially important, say proponents, is the use of renewable energy, which is a relatively clean and safe form of energy. Restoration simply implies restoring damaged ecosystems to enhance their productive capacity and ecological functions, many of which benefit us (for example, oxygen production by plants). It addresses the need to restore millions of acres the world over that have been exploited in the past and abandoned. Finally, population control means limiting human population growth and better managing how we spread out on the land.

These principles apply to the management of natural resources like farmland and forests, but they also apply to the management of human systems such as transportation, energy, waste management, and the like. This book primarily deals with natural resources, but we remind you early on to remember that the challenge of creating a sustainable future entails much more than simply managing farms, forests, rangeland, and fisheries in ways that ensure their ecological integrity and long-term vigor. It requires changes in us. As future chapters will show, these principles hold great promise for redirecting many human systems and bringing our lives and our businesses back in line with the ecological limits of our planet.

1.6 Changing Realities: Environmental Synergies

With the continued growth of the economy and the human population, the need for a new way of conducting human affairs will only increase. This book will present ample evidence to support this claim. But there's another reason for striking out in new directions, and doing it quickly, that's rarely discussed among scientists, conservationists, environmentalists, and others. Professor Chiras has written about it in his books for many years. It is, quite simply, the potential for several problems such as global climate change, acid deposition, and growth in the human population to produce effects that are greater than one might expect from each problem individually. Chris Bright of the Worldwatch Institute in Washington, DC, refers to this as the **nemesis effect.** Put in simple terms, the nemesis effect refers to environmental outcomes that result from the interaction of several changes. For example, acid rain, global warming, and habitat loss all contribute to the loss of species. But they may also combine to drive species more rapidly into extinction.

"If there is comfort to be found in the prospect of environmental decline, it lies in the idea that the process is gradual and predictable," writes Bright. "But this way of thinking is sleepwalking." Many factors will combine forces to produce unanticipated results, effects that occur more rapidly than predicted. "When one problem combines with another problem, the outcome may be not a double problem," writes ecologist Norman Myers, "but a super-problem." Although ecologists, scientists who study ecosystems and the relationships in them, have barely begun to identify potential super-problems, writes Bright, "in the planet's increasingly stressed natural systems, the possibility of rapid, unexpected change is pervasive and growing."

Keep this idea in mind as you observe environmental change in your area, read the newspaper, listen to the television, and study this book. It may become a key idea in the coming years as decades of environmental disturbance begin to interact. This possibility, combined with a host of others, also argues for the need to achieve root-level solutions—solutions that strike at the root of environmental problems.

1.7 New Tools for Resource Management: Geographic Information Systems and Remote Sensing

Solutions to environmental problems are many and varied. There are technological solutions, personal solutions, corporate solutions, and governmental solutions through policies and other measures. You will learn about many viable solutions in upcoming chapters. Two tools that are of increasing value and are used in many areas of resource management are geographic information systems (GISs) and remote sensing. We'll discuss GISs first.

Geographic Information Systems (GISs)

A **geographic information system** is a computer system consisting of hardware and software that is used to assemble and store information, but not just any information. It stores geographically referenced information, that is, data identified according to its location. This includes a vast collection of information about the Earth, including vegetation types, land disturbance, human settlement patterns, and surface water temperature. Geographic information systems do more than assemble and store information, however. They allow operators to display, manipulate, and analyze geographically referenced information. They are a valuable scientific tool and have many practical applications, which we'll be pointing out in the text.

A GIS can use information from many different sources and can help us analyze trends, detect change, and formulate policy. For example, rainfall and stream flow data from your state can be used to determine which wetlands identified by aerial photos (images taken from airplanes) might run dry in times of drought. Federal and state air-quality-control agencies can use

GISs to study the effects of air pollution on childhood asthma by comparing air-quality data and hospital admission records.

GISs identify locations based on coordinates such as latitude and longitude. But that's not all. Even zip codes or postal codes and highway mile markers can be used to identify locations. A GIS can operate using any data so long as it can be located spatially, and the computer software is set up to handle the data. If the data isn't already **georeferenced,** or linked to specific locations, it must be before it can be used. All data used in a GIS program must be georeferenced so it can be easily compared and analyzed.

Data can be entered either from existing maps and aerial photos or in digital form (for example, digital satellite images can be used to generate useful maps). Hydrologic data or air pollution data contained in tables, for instance, can also be used to generate maps so long as it is georeferenced. The GIS can then be used to understand and underscore spatial relationships between mapped variables, such as air pollution and asthma attacks in children.

One of the most time-consuming parts of GIS is called data capture, simply defined as entering georeferenced information into the system. Maps, for instance, can be digitized—that is, converted into a numerical data set or hand-traced with a computer mouse—to collect the coordinates of features. Geographers create an abstract picture of our world, essentially a map in two dimensions that can be viewed on-screen or printed to produce a hard copy. Lines, points, and areas on the map represent features of our world. Points on the map indicate the location of fire hydrants, telephone poles, buildings, and cell phone towers. Lines represent roads, power lines, streams, and trails. Regions (for example, regions of a certain soil type or vegetative type) are indicated by polygons. Three-dimensional representations are also commonly made. Watersheds, for example, may be drawn in three dimensions to show mountains, hills, and stream gradients.

GISs are used by numerous government agencies at the federal level, including the U.S. Forest Service, Bureau of Land Management, U.S. Geological Survey (USGS), Environmental Protection Agency, and National Center for Atmospheric Research. State and local governments also use GISs, as do private companies. A considerable amount of this information is also available to the general public on Web sites maintained by these organizations.

Although a GIS may not seem like much more than fancy map making, "its power of spatial analysis extends far beyond using a computer to do cartography," says John Hayes, a geographer at Salem State College in Salem, MA. According to the USGS, "traditional maps are abstractions of the real world, a sampling of important elements portrayed on a sheet of paper with symbols to represent physical objects." People who use maps must interpret these symbols. Topographic maps, for example, show the shape of land surfaces with contour lines. The actual shape of the land can be seen only in the mind's eye. "Graphic display techniques in GISs make relationships among map elements visible, heightening one's ability to extract and analyze information," according to the USGS. In addition, "a GIS makes it possible to link, or integrate, information that is difficult to associate through any other means. Thus, a GIS can use combinations of mapped variables to build and analyze new variables." For example, "by using GIS technology and water company billing information, it is possible to simulate the discharge of (potentially harmful) materials in the septic systems in a neighborhood upstream from a wetland." Water bills indicate how much water is used at each address. This, in turn, is an approximate measure of the amount of waste that is discharged into each septic tank and leach field. Areas of heavy septic discharge can, therefore, be located using a GIS. These and other spatial combinations are often referred to as **overlay analyses,** or **overlay operations.** Figure 1.13 illustrates this concept.

The computer also makes it easier to integrate information from different sources. For example, property ownership might be indicated on city maps drawn with a different scale than county soil maps prepared by the U.S. Department of Agriculture. The software in a GIS can alter the scale of the property map, manipulating it so that it registers, or fits, with the county soil maps. More conventional maps can also be registered with maps generated from data acquired from satellites. The computer may also perform other manipulations, including projections (described next).

Projection refers to a complex mathematical technique to transfer information from Earth's three-dimensional curved surface to a two-dimensional medium, such as paper or a computer screen. Different projections are used for different types of maps, depending on the needs of the cartographer

(map maker) and the user. One of the most common projections is known as the Mercator projection. It is widely used in textbooks and wall maps of the world because it accurately represents the shapes of the continents. Unfortunately, it distorts their relative sizes (Figure 1.14).[1] In contrast, a planar projection gives an accurate view of size. "Since much of the information in a GIS comes from existing maps," says the USGS, "a GIS uses the processing power of the computer to transform digital information, gathered from sources with different projections to a common projection."

A GIS also allows us to retrieve specific georeferenced information. With a GIS, for example, you can point at a location, object, or area on the screen and retrieve recorded information about it from off-screen files. "Using scanned aerial photographs as a visual guide," notes the USGS, "you can ask a GIS about the geology or hydrology of the area or even about how close a swamp is to the end of a street." A GIS can help you locate activities, past and present, allowing you to make conclusions about potential impacts. Or it can help you determine the suitability of future activities. If the analysis reveals the presence of an abandoned factory with contaminated soils, for instance, the land might be viewed as a potential future factory site with a little cleanup. It might not be a good site for a housing development, however. In summary, a GIS permits users to analyze conditions of adjacency (what is next to what), containment (what is enclosed by what), and proximity (how close something is to something else).

With appropriate software, GISs can simulate the route of movement of hazardous materials through surface water or groundwater. Retrieval functions and modeling of chemical movement allow the system's users to draw conclusions more readily than might otherwise be possible from other data sets, as it puts a great deal of information at their fingertips.

GISs can also be used to map sensitivities of various ecosystems. A map that looks at topography, soil type, and other features in and around wetlands, for instance, could be used to assess the sensitivity of various wetlands to human activities. All of this can be printed on paper, providing a graphic view. This data, in turn, can be used to predict possible outcomes of various development patterns and can profoundly influence land use decisions by state, local, and federal governmental agencies. In one case, wildlife biologists used transmitter collars and satellite receivers to track migration routes of polar bears over a two-year period. This data was then used to determine impacts of projected oil development on the bears.

A GIS can also be used in emergency planning. For example, satellite images of earthquake zones are used to pinpoint the location of emergency services, roads, and a variety of hazards. This information can be used to determine evacuation routes. The GIS may be used to determine areas that are most vulnerable.

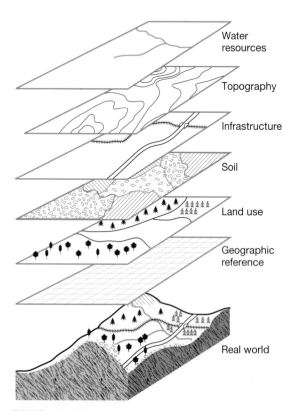

FIGURE 1.13 Overlay analysis allows GIS users to combine a number of georeferenced data sources to produce a more complex and often more useful view of the world we live in. Each data set can be viewed as a thematic layer, each linked to a common georeferencing system.

Water resources

Topography

Infrastructure

Soil

Land use

Geographic reference

Real world

---

[1]Mercator projection is not commonly used in GISs. The more popular projections are the Universal Transverse Mercator and State Plane Coordinate.

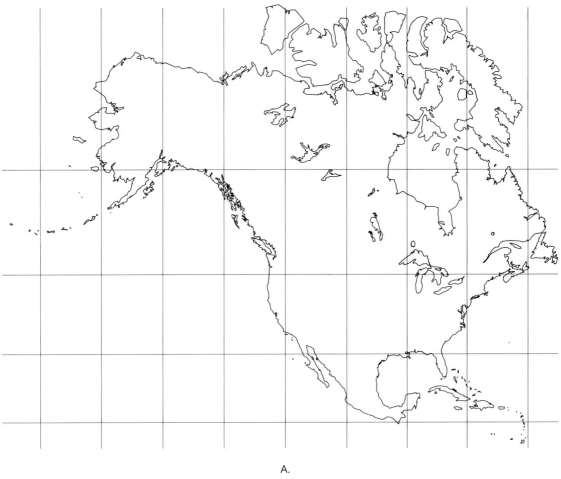

A.

Map of North America, Mercator projection

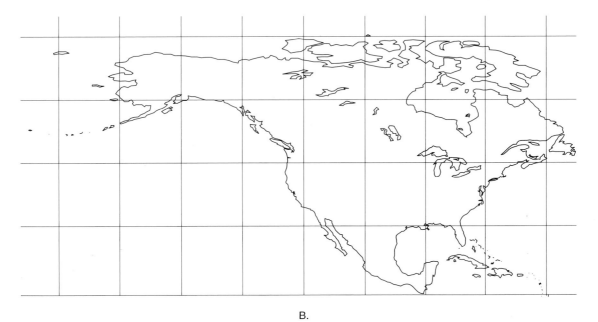

B.

Map of North America, Planar projection

FIGURE 1.14   A World Apart. Maps can be drawn in different ways to project the complex three-dimensional relationships on a two-dimensional surface (paper or computer screen). (A) The Mercator projection is commonly used. However, it distorts the size of land masses at the northern and southern latitudes. (B) A planar projection gives a more accurate view of the size of land masses.

Remote Sensing

A second tool that is used to manage and protect natural resources is known as remote sensing. **Remote sensing** is a means of gathering information on conditions on Earth from remote (distant) locations. Aerial photographs are an example, as are digital images taken from satellites. Satellites, as you well know, are launched into outer space on rockets and then released into orbit. Circling around Earth, satellites are typically powered with electricity generated from solar panels, known as PVs (photovoltaics), described in Chapter 22 of this book.

Aerial photos are usually created using film. To be useful in GISs, they must be digitized, or scanned and converted to digital files. Satellites are equipped with various sensors, such as infrared scanners, that monitor Earth. Information is digitized and then sent to Earth, where it is collected and stored in massive computers. Digital data is processed into maps or other useful images. Today, GISs and remote sensing are used by various government agencies and laboratories.

With GIS, remote sensing, and a host of other tools, we head into the 21st century facing a challenge of enormous magnitude. As Professor Jerry Reynolds of the University of Central Arkansas likes to remind his Conservation and Land Use students, present generations face critical resource problems that have never been faced before. Together, we will have to develop new, perhaps more creative solutions to solve these problems in order to create a sustainable society—one that lives well without undermining the future.

1.8 Risk and Risk Assessment

Another tool commonly used in resource management, especially the management of air and water pollution, is risk assessment. **Risk assessment** is a technique that allows one to analyze existing or potential hazards for the level of danger they pose. With this information, public-policy makers—with public input—can assess the acceptability of a risk.

Hazards are typically categorized as either **anthropogenic** (pronounced an´-throw-poe-gen´-ick)—those created by human beings, such as air pollution—or **natural**—those resulting from natural events such as tornadoes, hurricanes, floods, droughts, volcanoes, and landslides. Natural and anthropogenic hazards may kill people or destroy property. They may cause our health to deteriorate or lead to the extinction of plants and animals.

Although we have just distinguished between natural and unnatural hazards, the lines are not always so clear-cut. In fact, many natural events such as mudslides may be caused by or worsened by human actions. Mudslides, for instance, while natural, may occur with greater frequency in areas where trees have been cut down (so there are few roots to hold together the wet soil) or in hilly or mountainous areas with roads cut through them (because steep road banks are prone to mudslides).

How do we go about assessing hazards and creating policies that minimize or eliminate them in socially and economically acceptable ways?

Three Steps in Risk Assessment

In the mid-1970s, a new science called risk assessment began to take root. Its goal was to help people understand and quantify risks posed by technology, our lifestyles, and our personal habits (smoking, drinking, and diet). Risk assessment involves two steps: hazard identification and estimation of risk. **Hazard identification** involves steps to identify potential and real dangers. **Estimation of risk** involves steps to determine how dangerous a potential hazard is. Estimation of risk first requires a determination of the probability of an event. This process answers the question, "How likely is the event?" Next, one must determine the severity of an event, answering the question, "How much damage could be caused?" To assess the risk of a toxic chemical, for example, scientists must first estimate the number of people (or other organisms) exposed and the levels and duration of exposure. Other complicating factors such as age, sex, health status, personal habits, and chemical interactions must also be taken into account. This information helps us determine how severe the effects might be.

Risk assessment is fraught with uncertainty for many reasons. One of them is that our knowledge is often lacking. Risk assessments of global warming, for instance, are plagued by incomplete understanding of the climate. Risk assessments of toxic chemicals may be uncertain because of our lack of knowledge about the thousands of chemical substances in use today as well as our lack of knowledge of how exposure to two or more chemicals might affect an organism.

Once the potential hazard of a risk is identified, we must determine whether the risk is acceptable. Risk acceptability is one of the most difficult steps in risk assessment. One complicating factor is **perceived harm,** that is, the damage people think will occur. In general, the more harmful a technology and its by-product are *perceived* to be, the less acceptable it is to society. The acceptability of risks is also influenced by **perceived benefit**—how much benefit people think they get from something. In general, the higher the perceived benefit, the greater the risk acceptability. Automobile travel provides the most telling example of the way in which perceived benefits affect our decisions. The risk of dying in an automobile accident in the United States is 1 in 5,000 in any given year. Over your lifetime, however, the risk of dying in a car accident is higher; if you don't wear seat belts, it's about 2 in 100; if you do, it's about 1 in 100. Meanwhile, substances believed to be far less hazardous than driving are banned from public use because their benefit is not so highly valued. Chemical substances that pose a risk of 1 cancer death in 1 million people, for example, are banned.

How Do We Decide if a Risk Is Acceptable?

One of the most commonly used techniques for deciding the acceptability of risk is **cost-benefit analysis,** a process in which one analyzes the many costs and benefits and weighs one group against the other. As simple as it sounds, it is also quite difficult, as it is not always possible to put a price on, for example, the death of a person that results from exposure to urban air pollution. It is much easier to calculate the benefits of not cleaning up air—for example, not putting pollution control devices on power plants. As you will see in this book, one of the most important benefits of a sustainable strategy—including efforts such as energy efficiency, pollution prevention, recycling, and use of renewable resources—is that it offers many benefits at very little cost. It makes the environment and economy complementary, not antagonistic, forces.

For now, however, many decisions must be made about risk acceptability based on an enumeration of costs and benefits. When this is done, one tries to determine if the benefits outweigh the costs. As previously noted, however, as a rule, benefits are generally easier to measure. Financial gain, business opportunities, jobs, and other tangible items can be given dollar values. Many of the costs, however, are less tangible. External costs, costs incurred to society by pollution and other forms of environmental damage, are among the most difficult to quantify. In addition, it is hard to assign a dollar value to human life, environmental damage, and lost species. Moreover, the costs may be long-term and borne by future generations.

Another problem with cost-benefit analysis is that the benefits—most importantly economic gain—often accrue to a select few who have inordinate power to influence the political system and sway people's opinions. Today, a dozen companies control the flow of energy in the world. Many of them profoundly influence the political system by lobbying against actions to combat important issues such as global warming that could affect billions of the world's people. Fortunately, economists have begun to improve techniques so they can assign a dollar value to environmental and health costs. In addition, scientists are also improving efforts to measure the impact of technologies and their by-products on wildlife, the environment, recreation, health, and society in general.

The main purpose of risk assessment is to help policy makers formulate laws and regulations that protect human health, the environment, and other organisms. Ideally, good lawmaking requires that the **actual risk,** or the amount of risk a hazard really poses, be equal to the **perceived risk,** or the risk perceived by the public. When actual risk and perceived risk are equal, public policy yields cost-effective protection.

When the perceived risk is much larger than the actual risk, costly overprotection occurs. In contrast, when the perceived risk is much smaller than the actual risk, underprotection occurs. Underprotection can also be costly to both present and future generations.

GO GREEN!

Check out ongoing efforts to green your campus and take an active role! If your school isn't actively pursuing ways to reduce its environmental impact, consider organizing an effort with key faculty members and administrators. Chances are, there are plenty of environmentally active faculty members and staff who would love to help.

1.9 The Environment and You: The Importance of Citizen Action

As a student with a long future ahead of you, you should clearly recognize as you read this book that many of the problems outlined in the text will affect your life. Some issues, such as urban air pollution, may be affecting you right now! Like many individuals, you may feel that it is prudent to leave the solutions to the scientists, businesses, and government leaders. But government, businesses, environmental experts, and even political activists working for conservation groups are not the only elements of change. Citizens have had and continue to have an impact on the environment.

You can have a profound impact, too. You can make your impact felt by voting for candidates who support sound, sustainable environmental policies—and keep their pledges. Your impact will be felt when you live your life in accordance with your environmental values—for example, by consuming less, recycling, using resources such as water efficiently, and driving an efficient vehicle or walking or riding using forms of transportation that have less impact than even the most energy-efficient cars. Your contribution, added to the efforts of millions of others like you, can have a big impact. You can make changes starting today. The "Go Green!" tips we have provided will help you on your way to a more sustainable lifestyle. As you move into the working world and buy your first home, don't forget your environmental values and responsibility.

You can also affect the political process by writing letters, sending e-mails, or calling your congressional representatives, although you'll usually talk to a receptionist who records your opinion. You can even write to the president or your state governor or foreign leaders. Citizens' voices, when loud and clear, can sway political opinion and result in significant change. Many key pieces of environmental legislation, such as the Clean Air Act and its amendments, are here today because citizens' voices overpowered narrower business concerns and intense lobbying by industry. To make it easier for you, some environmental groups, such as the Natural Resources Defense Council and Environmental Defense, will send you periodic updates on environmental and resource issues that concern you. In their e-mails to you, they'll provide a letter that you can modify and send to the president, key government officials, business leaders, or local government representatives. Be sure your letters are polite and factual and get to the point. And, say activists, never stop writing. Politicians take your input seriously. You can even write to businesses that you want to influence. And don't forget to write letters to thank political leaders or businesses that have adopted environmentally sound policies and voted for important pieces of legislation that advance the cause of sustainability.

In closing, we would like to point out that resource management is improving day by day, thanks to citizen input, scientists who have helped us understand human impacts, and tools such as GISs and risk assessment. In this book, you will learn about many other new developments, such as collaborative planning, that are helping humans better manage our resources and ourselves in an effort to steer society back onto a sustainable course.

Summary of Key Concepts

1. The global environmental crisis is the result of three major factors: (1) a large and rapidly growing human population, (2) depletion of natural resources, and (3) rising levels of pollution that threaten our climate and our planet's ecological health.

2. The global population will surge from more than 6.67 billion in 2008 to 8 billion by 2025. Rising population places increased demands on an already strained resource base.

3. Resource depletion occurs in both the more-developed and less-developed nations. The more-developed nations, including the United States and Canada, whose citizens enjoy high standards of living compared with people in the rest of the world, use a much larger share of the world's resources than less-developed nations. The United States, for instance, has less than 5% of the world's population but consumes 30% of the world's resources.

4. Resource depletion is a threat to economies, people, and the wealth of wild species that share this planet with us.

5. The acquisition and use of resources (for example, the combustion of fossil fuels) results in pollution that threatens ecosystems locally, regionally, and even globally. These ecosystems are vital to the welfare of other species and are also vital to human well-being and the well-being of the global economy. The United States is degrading the environment more rapidly than any other nation on Earth.

6. People's views on solutions to the environmental and resource problems facing the world today vary widely. Some believe that technology can solve virtually any problem and are quite optimistic about the future. Others are less optimistic. They argue that because population, resource use and depletion, and pollution are increasing at an exponential rate, the chance that technological breakthroughs will solve them is rather remote.

7. While technology is clearly a part of the solution, society must also make political and economic changes, and people must take actions and learn to live more sustainably on the Earth.

8. Natural resources can be placed into two broad categories: renewable and nonrenewable. Renewable resources are represented by soil, rangelands, forests, fish, wildlife, air, and water. They also include energy resources such as wind energy, solar energy, hydropower, and geothermal energy. Nonrenewable resources include fossil fuels and metallic and nonmetallic minerals.

9. In a broad sense, conservation can be defined as a state of harmony between humans and the land.

10. The most significant conservation developments in the United States were made in the 20th century in four waves: (1) under the presidency of Theodore Roosevelt (1901–1909), (2) during the presidency of Franklin D. Roosevelt in the 1930s, (3) under the impetus provided by the Nixon, Ford, and Carter administrations during the environmental decade of the 1970s, and (4) in the years following the United Nations Conference on Environment and Development (1992).

11. Although the conservation movement progressed rather slowly during the 19th century, some advances were made. They included (a) the publication of *Man and Nature,* by George Perkins Marsh, (b) the establishment of 28 federal forest reserves, later designated as national forests, and (c) the founding of the Sierra Club by John Muir.

12. The 1960s and 1970s saw many changes, including the establishment of Earth Day and the passage of numerous environmental laws. So many environmental laws were passed by the U.S. Congress from 1970 to 1980 that this period has been called the decade of the environment.

13. Development during the decade of the 1980s led to the realization that, despite substantial progress in environmental protection, many gains were inadequate. Environmental conditions were not improving as was once hoped and in some cases were actually deteriorating. It became clear that most of the world's nations were on an unsustainable course.

14. This realization prompted a flurry of work on sustainable development, including the 1992 United Nations Conference on Environment and Development held in Rio de Janeiro, Brazil. Attended by nearly 180 nations, UNCED resulted in several treaties on key issues and an extensive action agenda for the world community.

15. Attitudes about the environment and the value of resources vary dramatically. Exploitation of resources, a human-centered approach, was commonplace early in U.S. history and is commonplace elsewhere as well. The opposing view, preservation, seeks to set aside nature for its own sake. It is a hands-off view. The utilitarian view is human- and nature-centered, promoting the wise use of resources to ensure lasting supplies. The ecological or sustainable approach seeks to manage resources and ourselves in ways to ensure that both survive well.

16. The ecological/sustainable approach to resource management requires a long-term systems view. Ecosystems are managed to protect multiple components. Protecting the stability and diversity of ecosystems is vital to this approach.

17. Creating a sustainable society will also require changes in human systems and may take many decades to complete, but positive steps are under way to create a more enduring human presence. These involve changes in how we manage natural resources such as forests, farmlands, and fisheries, and also changes in how we design and manage our own systems such as transportation, water supply, energy, and the like.

18. Urgency may be required, given the rapid growth of human population, the economy, and the potential for combined problems to create a greater than anticipated effect, a phenomenon known as the nemesis effect.

19. Two tools that are proving to be helpful in solving local, regional, and even global environmental problems and in improving resource management at these levels are GISs (geographic information systems) and remote sensing. Computer-based GISs allow researchers and others to acquire and depict vast amounts of information about the world we live in. This information is used to analyze spatial relationships and predict future outcomes of current and projected actions. GISs rely on data from many sources, but especially remote sensors aboard satellites.

20. Resource management is also being improved by the science of risk assessment. Risk assessment involves efforts to identify hazards, to estimate the risk they pose, and finally to determine the acceptability of the risk. Risk acceptability is often based on cost-benefit analyses.

Key Words and Phrases

Actual Risk
Adaptive Management
Anthropogenic
Biological Principles of Sustainability
Buffer Zone
Carrying Capacity
Civilian Conservation Corps (CCC)
Collaborative Conservation
Conservation Principle
Core Reserve
Cost-benefit Analysis
Decade of the Environment
Earth Summit
Ecological Approach to Conservation
Ecology
Ecosystem
Ecosystems Management
Environmental Justice
Environmental Protection Agency (EPA)
Estimation of Risk
Exploitation Approach to Conservation
Exponential Growth
Geographic Information System (GIS)
Georeference
Greenhouse Effect
Hazard Identification
Leopold, Aldo
Limits to Growth, The
Man and Nature
Marsh, George Perkins
Muir, John
National Conservation Commission
Natural
Natural Resources Inventory

Nemesis Effect
Nonrenewable Resource
North American Wildlife and Resources Conference
Overlay Analysis
Overlay Operations
Perceived Benefit
Perceived Harm
Perceived Risk
Population Bomb, The
Population Growth
Prairie States Forestry Project
Preservation Approach to Conservation
Projection
Remote Sensing
Renewable Resource
Risk Assessment
Silent Spring
Soil Conservation Service (SCS)
Sustainable Approach to Conservation
Sustainable Development
Sustainable Society
Sustained Yield
Tennessee Valley Authority (TVA)
"Tragedy of the Commons, The"
United Nations Conference on Environment and Development
Utilitarian Approach to Conservation
White House Conference on Natural Resources
World Commission on Environment and Development

Critical Thinking and Discussion Questions

1. Define conservation.
2. Discuss the major theme of the *Limits to Growth* study.
3. Many people think that humankind is in the midst of an environmental crisis. Why do they believe this? What evidence is there to support this assertion? Do you agree with this assertion? Or do you need additional information to make this assessment?
4. In what ways are population growth, resource demand and depletion, and pollution linked? How can they be unlinked, or can they?
5. Debate the following statement: "Even if there were no environmental crisis, there would be good reason to manage our resources more sustainably."
6. Describe four major advances in the conservation movement that occurred during the administration of Franklin D. Roosevelt.
7. Name five major environmental acts that were passed during the decade of the environment (1970–1980).
8. Does the passage of an environmental law guarantee that the goals of the law will be met? Discuss.
9. Discuss the basic difference between renewable and nonrenewable resources. Name four resources of each type.
10. Name and describe the four basic approaches to conservation in the United States.
11. What is sustainable development? Describe the principles of sustainability and how they might be applied to various human systems such as agriculture. How is sustainable development vital to environmental protection?
12. Define the nemesis effect. Can you think of any examples?

Suggested Readings

Assadourian, E. 2007. *Vital Signs 2007–2008.* New York: W. W. Norton. Contains a wealth of information on global social, economic, and environmental trends.

Ausden, M. 2007. *Habitat Management for Conservation: A Handbook of Techniques.* New York: Oxford University Press. An in-depth look at ways habitat can be better managed to protect species.

Black, G. 2005. Patagonia Under Seige. *OnEarth* 28 (3): 14–25. An in-depth look at how economic development and rising demand for energy in Chile could damage one of the wildest places on the planet.

Brick, P., P. D. Snow, and S. Van de Wetering. 2000. *Across the Great Divide: Explorations in Collaborative Conservation and the American West.* Washington, DC: Island Press. Explores a new movement in conservation: collaborative conservation. A must-read for anyone interested in a career in environmental conservation or natural resource conservation.

Bright, C. 2000. Anticipating Environmental "Surprise." In *State of the World 2000,* ed. Linda Starke, pp. 22–38. New York: W. W. Norton, 2000. Extremely interesting and thorough discussion of the way environmental disturbances and alterations combine to produce unanticipated or more rapid deterioration than one might expect.

Carson, R. 1962. *Silent Spring.* Boston: Houghton Mifflin. A classic. Alerted the public to the harmful consequences of the use of pesticides on humans and wildlife.

Chiras, D. D. 1992. *Lessons from Nature: Learning to Live Sustainably on the Earth.* Washington, DC: Island Press. Describes a set of ecological principles of sustainability and their application.

Chiras, D. D. 1993. Toward a Sustainable Public Policy. In *Environmental Carcinogenesis and Ecotoxicology Reviews* C11 (1): 73–114.

Chiras, D. D. 1995. Principles of Sustainable Development: A New Paradigm for the Twenty-First Century. In *Environmental Carcinogenesis and Ecotoxicology Reviews* C13 (2): 143–178.

Chiras, D. D., and D. Wann. 2003. *Superbia! 31 Ways to Create Sustainable Neighborhoods.* Gabriola Island, BC: New Society Publishers. One of the author's most recent books, which describes actions people can take to build a more sustainable world right in their own neighborhood.

Cortner, H. J., and M. A. Moote. 1999. *The Politics of Ecosystem Management.* Washington, DC: Island Press. Examines the political challenges facing ecosystem management as it moves from theory to practice.

Ehrlich, P. R. 1968. *The Population Bomb.* New York: Ballantine Books. A classic. Alerted the public to the urgent need to control the population explosion in order to prevent massive starvation and environmental abuse.

Gardner, G. 2001. Accelerating the Shift to Sustainability. In *State of the World 2001.* New York: Norton. Looks at the role of business, government, and citizens in shaping a sustainable future.

Graham, W. 2007. Dark Side of the New Economy. *Onearth* 29 (1): 14–21. A fascinating look at some of the impacts of rising global trade.

Grumbine, R. E. 1994. What Is Ecosystem Management? In *Conservation Biology* 8 (1): 27–38. Overview of the history of ecosystem management and its underlying principles.

Harris, J. M., T. Wise, K. Gallagher, and N. R. Goodwin. 2001. *A Survey of Sustainable Development: Social and Economic Dimensions.* Washington, DC: Island Press. Broad overview of important topics.

Hertsgaard, M. 1999. *Earth Odyssey: Around the World in Search of Our Environmental Future.* New York: Broadway Books. Fascinating story of one man's journey around the world to observe the deterioration of the environment and its impact on people.

Lee, H., and D. A. Shalmon. 2007. Searching for Oil: China's Initiatives in the Middle East. *Environnment* 49 (5): 8–21. An in-depth look at China's efforts to secure more oil to power rapid economic growth.

Miller, C. 2001. *Gifford Pinchot and the Making of Modern Environmentalism.* Washington, DC: Island Press. A biography of a man who profoundly influenced resource conservation.

Prugh, T., and E. Assadourian. 2003. What Is Sustainability Anyway? In *World-Watch* 16 (5): 10–21. Great overview of the environmental challenges that lie ahead.

Rappaport, A. 2008. Campus Greening: Behind the Headlines. *Environment* 50 (1): 6–17. A detailed look at what colleges and universities are doing to become greener.

Shabecoff, P. 2000. *Earth Rising: American Environmentalism in the 21st Century.* Washington, DC: Island Press. A critique of modern environmentalism.

Shenk, T. M., and A. B. Franklin. 2001. *Modeling in Natural Resource Management.* Washington, DC: Island Press. For students interested in learning about one of the most useful tools in resource management: computer modeling.

Sherbenin, A. de, K. Kline, and K. Taustiala. 2002. Remote Sensing Data: Valuable Support for Environmental Treaties. *Environment* 44 (1): 20–31. Looks at many ways that GISs and remote sensing can be used to help the international community tackle global environmental problems.

Speth, J. G. 2002. A New Green Regime: Attacking the Root Causes of Global Environmental Deterioration. In *Environment* 44 (7): 16–25. Great overview of progress and setbacks in our efforts to create a sustainable future.

Sterner, T., et al. 2006. Quick Fixes for the Environment: Part of the Solution or Part of the Problem? *Environment* 48 (10): 20–30. A look at the need for root-level solutions to address environmental issues.

Taylor, N. H. 2008. *Go Green: How to Build an Earth-Friendly Community.* Salt Lake City, UT: Gibbs Smith. Contains a lot of practical and interesting information on ways individuals can help create a sustainable world.

Trask, C. 2006. *It's Easy Being Green: A Handbook of Earth-Friendly Living.* Salt Lake City, UT: Gibbs Smith. Packed full of useful information you can use to live more sustainably.

Wann, D. 2007. *Simple Prosperity: Finding Real Wealth in a Sustainable Lifestyle.* New York: St. Martins Griffin. A delightful look at things you can do to create a more sustainable lifestyle.

Willers, B. 1999. *Unmanaged Landscapes: Voices for Untamed Nature.* Washington, DC: Island Press. An interesting book well worth reading. It examines the effects and implications of resource management and the need to keep some landscapes free from human interference.

Wondolleck, J. M., and S. L. Yaffee. 2000. *Making Collaboration Work: Lessons from Innovation in Natural Resource Management.* Washington, DC: Island Press. Examines collaborative efforts in natural resource conservation. Well worth reading.

Web Explorations

Online resources for this chapter are on the World Wide Web at: **http://www.prenhall.com/chiras** *(click on the Table of Contents link and then select Chapter 1).*

CHAPTER 2

ECONOMICS, ETHICS, AND CRITICAL THINKING

TOOLS FOR CREATING A SUSTAINABLE FUTURE

In many instances, resource management decisions are based more on economics and politics than science. Therefore, even though scientific findings suggest a particular course of action, political and economic considerations frequently prevail (see Ethics in Resource Conservation, Box 2.1). For example, decisions to cut old-growth forests or to harvest fish above a certain level are often influenced by economic stakeholders, such as by business that exerts considerable influence in the political arena. Such an approach can be dangerous, not only for the ecosystem in question but also, in the long run, for people themselves. Why?

Earth's ecosystems and its web of life are the infra-infrastructure of our society. They provide us with a wealth of free services, and they are the foundation of our economy. History has shown time and again that societies that destroy their topsoil or forests collapse. It may be difficult to believe that a similar fate could befall modern society. Most of us think we're immune to such problems. Part of the reason for this is the globalization of the human economy. Sure, forests are being depleted in many locations, but we can always get the wood we need from another place. The same thinking characterizes the view of many as they ponder resources such as oil and ocean fisheries. Eventually, though, there will be no place to turn to extract resources we need. It would be far better to start using and managing the resources we have more judiciously.

Unfortunately, the debate over resource management is often framed in terms of immediate survival. If loggers cannot cut the remaining old-growth forests, their families will suffer. Local economies will collapse—all to save a few owls or a few trees. Even though such concerns are very real and very important, critics point out that long-term considerations must also be factored into decision making. Sustainable development, which was introduced in the previous chapter, requires a long-term view. Balancing short-term and long-term needs creates enormous frustration and controversy.

This chapter critically examines economics, one of the central issues in debates over resource management. It presents important economic principles that affect both resource management and pollution control. It describes some of the shortcomings of modern economics from an ecological perspective. It also describes ways of creating an economic system that permits us to thrive and prosper without damaging the natural systems we depend on. In other words, we describe an economic system that makes sense from a financial

ETHICS IN RESOURCE CONSERVATION 2.1   ETHICS VERSUS ECONOMICS?

One of the root causes of the current problems facing the environment is economics—more specifically, say many observers, the blind pursuit of economic wealth. That pursuit often results in environmental destruction, pollution, and resource depletion. Our singular dedication to economic growth and our failure to consider environmental concerns are, say many critics, wreaking havoc on the planet.

But it is not fair to blame the environmental crisis solely on economics. Ethics underpin economic decisions.

Economics has a lot to do with the way the world operates, as this chapter points out. Our economic system, however, is based on our ethics—our values. Put another way, economic decisions and our economy are driven, in part, by ethics. In most instances, however, the underlying ethical assumptions are not apparent, and all we see in operation are economic forces. What are some of the ethical forces that influence economic decisions?

One of the predominant ethical positions that drives economic growth is the notion of unlimited abundance, a central tenet of the frontier ethics described in this chapter. When people view the Earth as an unlimited supply of resources, economics is given a kind of moral free rein. This, in turn, often leads to environmentally unsound and environmentally costly actions—overexploitation of a wide assortment of natural resources.

Another ethical underpinning of modern economics is the notion that humans are apart from nature and superior to other life forms. By viewing humans and nature as separate entities, we often act irresponsibly, destroying ecosystems that are vital to our survival.

Another component of frontier ethics is that the key to success is domination and control. Huge levees built along the Mississippi River, for instance, illustrate a persistent desire to tame the river. Over the years, people have tamed other rivers with massive dams and levees. But the domination of nature has led to serious ecological backlashes. One of the most disheartening of these was the channelization of the Kissimmee River in Florida.

Once a meandering river with ample wetlands that supported huge flocks of waterfowl and numerous fish species, the Kissimmee River was channelized by the Army Corps of Engineers in an effort to control flooding. The river was straightened and dammed so that its overall length was cut in half.

As the Corps put the finishing touches on its multimillion-dollar project, the impact was immediately evident. Waterfowl disappeared in droves as wetlands dried up. Fish catches plummeted. Pollution in Lake Ochechobee, into which the river drained, began to increase.

Ecologists realized that the Corps had made a colossal mistake. Human actions had deadened a once-thriving biological system that supported countless species of wildlife, controlled flooding, and absorbed pollutants rather well on its own.

Recognizing this, officials have begun to restore the river, but at an enormous cost. Had humans left the river alone, millions of dollars would have been saved, and a valuable habitat for countless species would have been protected.

Individual ethics also figure in individual economic decision making. For example, the primacy of the individual, a tenet firmly rooted in American culture, drives us to make many ecologically damaging decisions. Viewing our personal pleasure as the paramount concern in our lives, for instance, leads us to acquire wealth and material goods at levels that cannot be sustained. Rare is the individual who says, "Although I can afford a bigger house, I won't buy one, because the Earth cannot sustain such a lavish lifestyle."

In brief, individualism fuels consumption, and consumption is a key driver of our economic system and one of several key underlying factors that have created the environmental crisis. A more inclusive philosophy, in which personal decisions are made on the basis of environmental realities and the welfare of future generations, say critics, might temper consumer desires and could contribute to a more sustainable lifestyle.

Economic competition is yet another cultural norm that, while seemingly good for the economy, spells disaster for the environment. In U.S. society, competition is considered an unquestioned good. But cooperative ventures could save our society and businesses enormous resources. Consider the following example. In Colorado, hundreds of individual water districts have been formed to provide water to cities, towns, and farms. Each district has its own sources, and many have their own dams and reservoirs in the mountains. It is possible that, had the communities joined forces rather than competed for water, they could have provided the same amount of water with far fewer dams and reservoirs, and at a much lower cost. A cooperative ethos might have promoted a more sustainable water supply system.

These are a few of the negative ethical underpinnings of economics. However, positive ethical views could be instilled and could lead to environmentally sustainable economic behavior. For example, if humans viewed themselves as part of nature, environmental protection would be recognized as a means of protecting ourselves and future generations.

We encourage you to take some time to consider your ethics, especially with regard to economics. What are the underlying rules that drive your decisions? Do they promote a sustainable future or detract from it?

standpoint as well as socially and environmentally. This new economic system seeks to satisfy three bottom lines: it's good for people, good for nature, and good for businesses. We refer to this new economics as **sustainable economics.** One point that will become clear is that sustainable economics takes a long-term view, with an eye on what's good for people, the economy, *and* the environment.

This chapter also examines ethics, describing what ethics is and why it is important. It examines different ethical systems and outlines an ecological ethic or, more appropriately, a sustainable ethic. Finally, in this chapter, we present overviews of the scientific method and critical thinking. This discussion provides important background information that will help you analyze complex environmental and resource management issues.

2.1 Understanding Economics

Although economics is a complex science, a few basic principles provide the foundation of economic literacy. First and foremost, **economics** is a science that seeks to understand and explain the production, distribution, and consumption of goods and services. Economics is generally divided into two broad disciplines, microeconomics and macroeconomics.

Microeconomics is concerned with the economic behavior of individuals, households, and businesses. It typically deals with decisions that are made in markets where goods or services are being bought and sold. Microeconomics examines how our decisions and behavior affect the supply and demand for goods and services, which, in turn, determine prices. It also examines how prices affect the supply of and demand for goods and services. Two of the main concerns of microeconomics have been inputs and outputs. **Inputs** include the commodities that companies need to produce goods and services, including raw materials, labor, and energy. **Outputs** include goods and services.

Macroeconomics is a branch of economics that deals with the bigger picture, specifically the performance, structure, and behavior of national or regional economies. It deals with growth, inflation, and unemployment and with national economic policies and governmental actions, such as taxation, that influence these issues.

Command and Market Economics

Economists recognize two basic types of economic systems: command and market. A **command economy,** also known as a planned economy, is one in which the state or government manages the economy. Examples are the economies of Cuba and, until recently, mainland China. In command economies, the state or government controls all major sectors of the economy, from food production to the manufacture of household goods. It makes all decisions about the distribution of income—who gets what—and *what* the economy produces and *how much* is produced.

In **market economies,** the production, distribution, and price of goods and services are made by private individuals—business owners—based upon their own and their customers' interests, rather than on an overarching macroeconomic plan as in command economies. That is to say, in market economies, *what* companies produce and the *quantity* they produce each year are determined by consumer demand and the potential for profit. Although there may be a high demand for a particular product or service, if a company cannot make a profit, it will generally not produce it.

Command and market economies are polar opposites. But in reality, most economies have elements of both. Market economies contain some elements of command economies and vice versa. The market economy of the United States, for instance, is highly influenced by public policy. Some laws provide subsidies for the production of goods. Oil companies, for example, are given special tax breaks to promote further mineral exploration and development. Regardless of whether you agree or disagree with the policy, it clearly represents governmental interference with the free market economy that alters the cost of producing goods. As another example, each year the U.S. military protects oil tankers in the Persian Gulf at a cost of about $50 billion, according to the Rocky Mountain Institute. This oil, which will be refined in the United States, is worth an estimated $10 billion. Military protection such as this represents a means of government intrusion in the free market economy. If oil companies paid the cost of protecting their oil themselves, the cost of fuel oil, diesel, jet fuel, gasoline, and hundreds of other products derived from crude oil would be much higher. Instead, taxpayers pay the bill. To be fair, government policies provide subsidies that help to foster a sustainable future. Companies that install wind turbines to produce electricity in the United States are given a federal production tax credit—a government incentive for each kilowatt-hour of electricity they generate.

Command economies may also contain elements of market economies. China's economy, for instance, is becoming more and more a market economy as the government allows business owners to operate more freely. In 2004, in fact, China's communist leaders passed a constitutional amendment that reinstated private property rights, which have been lacking since 1949, when the Communist Party took over. Today, many businesses in small towns and cities operate by rules of a market economy. Tens of thousands of government-owned businesses were sold to private individuals in the 1990s. (Many of these businesses were performing poorly under government control.) Interestingly, the site where the Communist Party was founded is now home to a shopping and entertainment complex featuring a Starbucks.

Economists also recognize another type of economy: the **subsistence,** or **hunter-gatherer, economy.** For most of human history, the human economy did not deal in dollars and did not require jets, massive ships, and trucks to transport goods and services. People living in small groups acquired or made what they needed to survive from the environment. They generally met their needs without bankrupting the Earth in the process or threatening their own future, although there are some notable exceptions.

Because populations were small and technological development was virtually nonexistent, subsistence economies were rather benign, environmentally speaking. But over time, trade picked up. Money came into use, and the market economies of the world emerged. In remote corners of the world, subsistence economies still exist, although many are now endangered by market economies.

Supply and Demand

One of the key principles of microeconomics is the law of supply and demand. It is a relatively simple idea that explains a great deal of economic activity of businesses and individuals. The **law of supply and demand** describes the market relationship between buyers and sellers. The law of supply and demand says that the price of a good or service determines both the quantity supplied by manufacturers and the quantity demanded by consumers. That is to say, the price determines how much a manufacturer provides and how much is consumed.

On the supply side, the quantity of a good or service is often depicted as directly proportional to price: the higher the price of the product, the more the producer will supply. On the demand side, demand is normally depicted as an inverse relationship: the higher the price of the product, the less the consumer will demand.

Although you may not have studied the law of supply and demand, you are probably familiar with it. You may have seen instances when the demand for a resource exceeded the supply, leading to an increase in price. (This is one of the reasons for rising oil prices in this decade.) If supply exceeds demand, the price falls. Demand is therefore said to be elastic. Examples of the law of supply and demand abound. An expensive car, such as the sporty electric Tesla, shown in Figure 2.1, will fetch a higher price than a mass-produced vehicle simply because, in the case of the former, supply is limited.

Although economics is far more complicated than this discussion may suggest, this is enough information to start with. More basic economics will be introduced as we examine the economy from an ecological perspective.

What's Wrong with the Economic System: An Ecological View

The economy is large, complex, and a major influence in our lives and in the future of the environment. It provides us with a wealth of goods and services as diverse as TV sets, computers, coffee, toothpaste, and bed sheets. It makes our lives better, providing opportunities to work and earn money to pay for our homes, automobiles, food, clothing, and recreation. But economic activity does not come without a cost. In this section, we will take a critical look at the economy with one purpose in mind: To consider ways it can be altered to create an enduring human presence. This is not intended to be a damnation of capitalism or the market economy, but rather simply a way to understand its failings, from an environmental perspective, and fix them to create a sustainable economy—one that meets our needs while ensuring that future generations can meet their needs.

Today, market economies have become the mainstay of the global economy. Command economies have fallen by the wayside for many reasons. The dominance of market economies in the global economy, though, does not mean they are perfect or environmentally benign. We should point out, however, that countries with market economies have done a much better job of protecting the environment than countries with command economies!

Market economies have produced extraordinary wealth and have raised the standard of living of millions of people throughout the world, and they hold great promise for raising the standard of living of millions of people who live in poverty today. One of the problems with market economies and capitalism in general, however, is that they embrace the idea of unlimited growth. Critics point out that unlimited growth is impossible in a finite system. That is to say, our resources are limited, and so

FIGURE 2.1 Demand for a product or service is elastic. It varies with the price. The higher the price, the lower the demand. This sporty electric vehicle, the Tesla, costs around $100,000. Although a few wealthy individuals can buy it, the car is out of the reach of the vast majority of people. The company is currently working on an electric car that the rest of us can afford.

is the planet's capacity to assimilate and detoxify pollution. Economies dedicated to constantly increasing material production and consumption are bound to fail, say critics.

Many others have criticized market economies for other reasons. A brief examination of a few of these criticisms is helpful in forging a sustainable, ecologically sensitive economic system.

Short-Term Planning Horizon   One of the principal complaints of ecologists is that business economists have a rather short time horizon for planning. For a business, a long-term plan might list goals and objectives to be met over a one- to five-year period. Decisions are made within this time frame.

Ecologists, conservationists, and environmentalists, in contrast, take a long-term view in decision making. For any decision to make sense from an ecological standpoint, it must make sense not just 5 years from now but 100, perhaps 200 years from now. They argue that short-term economic decisions often turn out to be ecologically disastrous. For example, it may make monetary sense for a rancher to overgraze his or her land to increase sales, but the ecological costs in the long term, not the least of which is the destruction of once-productive rangeland, are daunting. Resource managers find themselves in this bind over and over again.

The shortsightedness of the economy is also manifested in its reliance on the law of supply and demand to determine prices. Supply and demand economics reflects the immediate picture, that is, the immediate supply of a resource in relation to its demand. Very little consideration is given to long-term supplies. To understand the importance of this point, consider an example.

Oil prices are determined by supply and demand. Although oil supplies appeared to be adequate during the 1990s, they are finite. Many experts predict a peak in global oil production between 2004 and 2010, which would create permanent shortages

as demand exceeds supply. For years, however, the economy has been blind to this possibility. As a result of the interaction of immediate supply and demand, however, oil was priced low throughout the 1990s. These low prices, in turn, encouraged consumption and waste, with many environmental consequences. Since then, however, the price of oil and gasoline, which is made from oil, has skyrocketed, in large part because of high demand and inadequate supplies.

Rangeland, forests, fisheries, and other natural resources have been driven to the brink of ruin the world over in large part because they were once abundant and their services and products were underpriced. Now that many ecosystems are being destroyed, we must spend billions to restore them. In fact, far more money will be needed to restore them than would have been required to manage them properly in the first place. So supply and demand economics has not only resulted in mismanagement of resources, but with pollution and environmental destruction the undesirable outcome, it also ends up costing us more than is necessary in the long run! More careful, sustainable management might have yielded lower returns, but the lower returns would have been offset by a much longer productive period, perhaps even an infinite period of production.

GO GREEN!

When giving gifts, consider giving a little of yourself—for example, a dinner or movie or a little time to help a friend or loved one with a project around the house—rather than material gifts. This will reduce resource consumption and pollution.

Failure to Incorporate Economic Externalities   Another criticism of capitalism as it is currently practiced is its narrow notions of the inputs and outputs of an economy. Critics point out that at least three essential factors are missing from the inputs/outputs analyses of mainstream economics: environmental impact, social and cultural impact, and pollution. In other words, when businesses calculate the costs of manufacturing their products, they typically fail to take into account environmental and social costs (Figure 2.2). Air pollution from factories that

FIGURE 2.2   The price of progress? Waste from an abandoned aluminum factory abutting nearby backyards of homes causes serious health problems in residents, not factored into the cost of manufacturing.

make DVD players, skateboards, CDs, and cars, for instance, is not factored into the cost of doing business or the cost of the product to the consumer. Instead, air pollution costs, including those incurred by damage to human health, buildings, and ecosystems, are ignored and essentially passed on to society at large. Such costs are referred to as **economic externalities.**

Automobiles, for example, create huge economic externalities few of us are aware of. Cars pump out carbon monoxide that affects the health of the aged and infirm, causing an increase in chest pain in those with heart disease. They emit millions of tons of carbon dioxide, a gas that is causing the Earth's temperature to increase, which, in turn, is causing the sea level to rise, which floods islands and coastal regions. Global warming is also upsetting weather patterns and creating much more violent storms, which cause billions of dollars' worth of damage and cause the death of many people each year. Automobiles also produce sulfur dioxide and nitrogen oxides, two pollutants that are converted into acids in the atmosphere. These rain down on the land, causing widespread damage to lakes and the species that depend on them in certain areas of the world. All of these effects have a cost, but the economic externalities are not paid by automobile users at all. Substituting buses and trains for cars, wherever practical, can greatly reduce the economic externalities associated with the transportation of people in automobiles, because they move people more efficiently (with less fuel consumed per passenger mile).

Although efforts are being made to eliminate economic externalities in cars and a host of other pollution sources such as power plants through pollution controls and other measures, many outside costs are still unaccounted for and left unpaid.

GO GREEN!

Walk, ride a bike, carpool, or take a bus or light rail to school or work. These modes of transportation reduce traffic congestion and dramatically reduce energy consumption and pollution.

False Measures of Progress   Yet another shortcoming of traditional capitalism is its single-minded dedication to measuring success by the **gross national product (GNP).** The GNP is the value of all goods and services produced by a nation's economy, including all government expenditure and business activities occurring in other countries. The GNP includes such varied items as washing machines, solar panels, textbooks, and ice-cream cones. It also includes the cost of oil spill cleanups and the billions of dollars spent on cleaning up hazardous waste. It is therefore an aggregate measure of all goods and services. As a side note, the **gross domestic product (GDP)** is a subset of the GNP, as it includes all economic activity within the borders of a nation.

The problem with the GNP is that it fails to distinguish between different economic activities. In other words, as the GNP is structured, a dollar spent on a piano lesson for a future Mozart is just as important as a dollar spent on a pack of cigarettes.

Because the GNP is an indiscriminate measure, it fails to track a nation's true progress. That is, it fails to tell us exactly what we are getting from our economy. To illustrate this point, let us compare two purely hypothetical nations. The first nation has a robust economy with a high output of pollution. Each year 50,000 people die from urban air pollution, and thousands of others are poisoned by pesticides, hazardous wastes that have leaked into the water supply, and workplace chemicals.

Vast sums of money are being spent on producing goods and services. Huge amounts are also being spent on emergency medical help, hospital bills, coffins, and cemetery plots. The net effect is that this economy boasts an exceptional GNP.

The second nation has a robust economy as well, but it is on a sustainable path. Its industries produce products with little pollution. They use clean, renewable energy. As a result, the nation's air is clean, and its water is pure. Pests are controlled on farms without pesticides. Few people die from industrial accidents. Although the per capita GNP, the economic output per person, may be lower than the first nation's, the second nation is far better off. The economy is serving people, not poisoning and killing them.

Are there examples of the failure of economics to distinguish between various economic activities? Actually, yes. Indonesia's economic growth in the 1990s was predicated in part on rapid deforestation—clear-cutting of tropical rain forests. This boosted the economy to the detriment of the nation's future. It has also resulted in massive soil erosion that robbed the land of valuable topsoil and polluted streams. Much of the cleanup and housing of residents displaced by Hurricane Katrina, estimated to exceed $150 billion, adds to the U.S. GNP. How much more productive would the $150 billion be if it were spent on developing a renewable energy supply to reduce greenhouse gas emissions that are causing global climate change that may have been the main reason why this hurricane was so powerful?

Even its inventors argue that the GNP is prone to overvaluing production and consumption of goods and not reflecting improvement in human well-being. Because the GNP fails to make this distinction, critics argue that it is a false measure of success. Distinguishing between the relative values of economic contributions to the GNP, therefore, helps us determine just how sustainable an economy is.

The GNP also fails to take into account natural capital. **Natural capital** is a nation's ecological wealth. It includes topsoil, forests, grasslands, open spaces, farmlands, fisheries, and wild species—resources of great aesthetic and economic importance. In many ways, natural capital is like money in the bank. Nations draw on this capital to fuel their industries and to satisfy human needs for goods and services. But most nations are dipping into the bank account, taking not the interest but the principal itself. Thus, a nation can be rapidly depleting its natural capital while posting a high GNP, and there is no way of knowing it. Economist Herman Daly once wrote that "most nations are treating the Earth as if it were a corporation in liquidation." In other words, most nations are selling off their natural assets such as their forests to make a quick buck. Although their GNPs may be impressive, sooner or later they will have no productive capacity left.

Studies that attempt to sort out contributions to the GNP show that the distinction is more than academic. Herman Daly, for instance, developed a comprehensive **Index of Sustainable Economic Welfare (ISEW)** for the United States for the 1970s and 1980s. The ISEW is now called the **Genuine Progress Indicator (GPI).**

Although the procedure for calculating the GPI is quite complex, the GPI is the outcome of an attempt to measure whether a country's economic growth, resulting from an increase in the production and consumption of goods and services, has improved the health and well-being of the people in the country or caused them to deteriorate. To some, the difference between GNP and GPI is analogous to the difference between the gross profit of a company and the net profit. The net profit is the gross profit (how much money a company brought in) minus the costs incurred. Accordingly, the GPI will be zero if the financial costs of crime and pollution equal the financial gains in production of goods and services, all other factors being constant.

GPI takes into account the enhancement of nature's ability to provide its many services and provide clean air and water, among other things. These factors are therefore part of a more inclusive ideal of progress. The GPI is therefore considered by proponents as a better approximation of a nation's true economic welfare. Growth in GPI, in turn, can be compared with the GNP to determine whether economic progress, measured by growth in GNP, is actually improving our lives or not.

When Daly compared the GPI with the GNP of the United States, he made a startling discovery. In the 1970s, the GNP grew annually by 2% per capita. In the 1980s, the per capita GNP grew by 1.8% per year. The GPI during the 1970s, however, climbed only 0.7% per capita (Figure 2.3A), indicating that only about one-third of the annual growth in GNP improved the lives of Americans. Largely because of environmental degradation, the GPI declined by 0.8% per year in the 1980s. During the 1990s and the 2000s, the growth in the GNP continued, but GPI leveled off. What does this mean?

Although our economic output is increasing, and economists and politicians would have us believe that we are better off, our lot in life is actually declining. In other words, despite a growing economy, which most people think is good news, we are actually worse off.

The United States is not the only country to show a discrepancy between the GNP and GPI, but it is one of the few countries for which the GPI continues to plummet. In Germany, the GPI is actually increasing faster than the GNP (Figure 2.3B).

Myths About Economy and Environment

While we are on the subject of the economy and its faults, it is important to look at the ideas that support environmentally and socially unsustainable economic activities—that is, myths we have about our economy and the environment.

Economics Versus Environment: Rethinking an Old Debate   Several economic myths pervade our society and prevent progress toward a sustainable future. One of those myths is that *environmental protection is bad for the economy.* Many examples show that environmental protection as it has been practiced can be quite costly. But failure to institute environmental safeguards can also result in extremely costly cleanups. An ounce of prevention is worth many pounds of cure.

Experience also shows that environmental protection and economic health need not be antagonistic forces. Careful attention to the design and operation of factories, for example, can minimize pollution. This, in turn, will very likely make costly pollution controls and cleanups unnecessary, which will save companies millions of dollars a year. Using resources more efficiently, yet another sustainable strategy, has many benefits. It can, for instance, make a company more competitive.

All of this is to say, it is important to rethink the old ideas. Critical-thinking rules warn us to avoid oversimplification, to look at the big picture, and to question conclusions. This is a case in point.

The Economic Pursuit of Quality   Another common myth in our society is that *environmental protection is about quality of life, and economics is about survival.* That is to say, environmental protection is a less important endeavor than manufacturing, which provides jobs and income. In even simpler language, environment is a luxury, and economics is a necessity (Figure 2.4). Speaking about wildlife preservation, one local government official in Colorado noted, "Unfortunately, preservation of wildlife falls at the bottom of the list of priorities when it comes to staying ahead of the competition in creating a successful development effort. Wildlife and open space protection efforts are just too costly since they lack revenue-generating components that justify protecting them."

In his book, *The Economic Pursuit of Quality,* Thomas Michael Power, chairman of the Economics Department at the University of Montana, shows that virtually all economics in the industrial nations is about the pursuit of quality. What he means is that most of the money people spend on food, clothing, shelter, and other areas goes to improving the quality of our lives. Power shows that

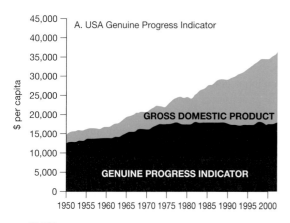

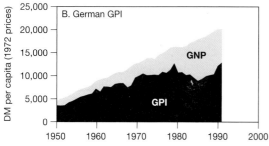

FIGURE 2.3   Economic growth is good for us—or is it? (A) As this graph shows, while the GNP continues to climb, our overall welfare, as measured by the index of sustainable economic welfare, declines. (B) In Germany, the GPI is actually increasing due to many progressive initiatives designed to reduce the impact of the German people on the environment.

FIGURE 2.4 People flock to beautiful areas for vacations but also to live. Nature and beautiful scenery have a very real value.

if we were concerned only about survival, our economic outlay would be much lower, perhaps as much as 70% to 80% lower than it is today. In other words, we can live (survive) on much less. Instead, we lavish on ourselves fashionable clothes, elaborate homes, and foods far beyond our survival needs. The conclusion of this line of reasoning is simple: For most of us, modern economics goes far beyond survival needs. It is about attaining aesthetic qualities.

Power also dispels the notion that environmental quality is far from frivolous from an economic standpoint. He points out that beautiful vistas, open space, clean air, and other environmental values have economic value. Homeowners, for instance, will spend many thousands of dollars to buy beachfront property or a home along a lake, or to live in a low-pollution setting. When they do, they are paying for quality. Many resort towns recognize this connection as well. People travel to visit nature resorts, not just to ski or kayak but to enjoy the unsullied spaces and scenic vistas.

In summary, the environment and its protection are not a frivolous matter beyond the realm of economics. Environmental protection is another quality we pursue—and readily pay for—in our lives.

The Myth of Economic Growth Yet another dominant myth in our society is that *economic growth is good, indeed, essential.* Without a doubt, economic growth has made our lives better than those of our predecessors. It is also essential to raise the standard of living of the world's poor—the more than 2 billion people who barely eke out a living. That said, it is important to note that current economic development comes with a huge cost, and some of its most significant costs are the excessive consumption and depletion of natural resources, the loss of countless species, the pollution of lakes and rivers, and the fouling of the air. From a social perspective, economic growth often fails to deliver on its promises of widespread prosperity. Economic growth, for instance, is touted as a means of eliminating poverty and raising the welfare of the middle class. It

is promoted as a means of solving unemployment. But its success in these areas is questionable in recent years. Statistics show that economic growth often disproportionately benefits the wealthy and fails to deliver on other promises. In *The Economic Pursuit of Quality,* Power cites studies of economic growth (job growth) and per capita income in many western states and finds, almost uniformly, that despite the creation of many new jobs, per capita income increased very little, if at all. From 1950 to 1977, for instance, the number of new jobs in Arizona increased 5.5 times the U.S. average, but the per capita income increased only 2% above the national average. In California, job growth was 2.3 times greater than the national average, but per capita income fell 13% compared with the national average.

If growth cannot deliver the promises its proponents make and singularly undermines our future, we must ask these questions: Can we create an economic system that is not predicated on continual growth? Can we create an economy that exists in harmony with the environment and yet supplies us with the goods and services we need to survive and prosper? Can we develop economically in ways that improve the lives of the world's impoverished? Before we embark on an explanation of an alternative, remember the "we" referred to in this paragraph includes well over 6.6 billion people.

2.2 Creating a Sustainable Economy

The preceding analysis of the economy is not meant to be an indictment of economists or those who participate in the economy, which is all of us. Rather, it is intended to set the stage for a discussion of ways to revamp the economy to create an environmentally sensitive system of commerce, or a sustainable economic system. The goal of sustainable economic development is to allow us to live well but also to raise the standard of living of the world's poor without bankrupting the Earth and its ecosystems.

In general, a **sustainable economy** is one that produces goods and services in a manner that does not foreclose on future generations. A sustainable economy is one that is based in part on the principles introduced in the previous chapter: frugality, efficient resource use, recycling, and use of renewable resources, especially energy. Creating a sustainable economy also hinges on our becoming better able to manage our own population growth and better able to balance our demands for goods and services and living space with our needs for clean air, water, food, and the like. These principles, when seriously applied to all of the basic systems that compose the economy—industrial production, transportation, housing, energy, and waste management—can dramatically reshape how we live on planet Earth.

Much of this book outlines the principles and practices of sustainable natural resource management. Other discussions suggest ways to alter human systems, such as industry and transportation, to make them more sustainable.

To set the stage for this discussion, we will begin with a general discussion of the economics of resource management and pollution control. These sections will help you better understand what a sustainable economy entails, as well as some of the barriers to its creation. We begin with a look at the economics of natural resource management.

Economics of Resource Management

Economics plays a crucial role in the management of natural resources—for example, how farmers manage their farmland or how owners of a logging company manage forests. The future resource manager reading this book will become a practicing economist. Three guiding principles affect our decisions: time preference, opportunity costs, and discount rates.

Time Preference   Time preference is a fairly straightforward concept. What **time preference** refers to is one's preference for economic returns on an investment, expressed in time. An immediate time preference means that one prefers immediate returns.

Time preference is influenced by several factors, with need being one of the most pressing. A less-developed nation that needs money to pay off its debt to a more-developed nation, for instance, is likely to deplete its resources to do so. Little consideration is given to long-term production. Paying off the debt is paramount. Similarly, a farmer with pressing financial needs (perhaps a child in college) may ignore expenditures in soil erosion control and adopt strategies that result in less capital outlay and higher short-term returns.

As the previous examples show, immediate time preference often leads to environmentally, and sometimes economically, shortsighted practices. Creating a sustainable economy may require economic incentives to compensate for lost income resulting from a shift in time preference—that is, a concern for the ecological integrity of forests and other natural resources, which is a key element of ecosystem management and sustainable development. For example, tax credits for farmers willing to shift to sustainable agricultural practices may alter the economic picture sufficiently so that they can afford to invest in environmentally sensible strategies.

The previous discussion, however, is not meant to imply that all ecologically sensible and sustainable practices require a long payback period or exceptionally high initial investments. On the contrary, many practices require only a modest investment and offer rapid and sometimes substantial economic return. As an example, environmentally sound pest-control techniques can offer rapid payback. Thus, a sustainable approach may be economically more beneficial than the traditional approach. So, before one assumes that doing the right thing will cost too much, it is wise to look at the economics.

Opportunity Costs   Another related factor is **opportunity cost**—the cost of lost opportunities resulting from certain policies and actions. Suppose, for instance, you are a farmer and you realized a $10,000 profit from last year's crops. You could invest that $10,000 in a mutual fund and earn 7% per year, perhaps more. You could also invest it in soil conservation measures that may increase your yield by 3%. A strictly economic view would suggest that you would be wiser to invest your money in mutual funds. Therefore, if you invest in your farm, you would be losing an opportunity for slightly higher economic returns.

We like to point out that opportunity costs should be looked at in other ways, too. Consider what happens when a resource is depleted—in other words, consider the cost of lost opportunities by embarking on the exploitive resource management strategy. The large-scale cutting of tropical rain forests, for instance, may make sense from an immediate economic standpoint, but this activity also creates a loss of opportunity for you and future generations. Why? Tropical rain forests have and will continue to provide many life-saving medicines that have been extracted from plants that grow in jungles. New discoveries could be worth billions of dollars and could save thousands of lives. Tourism lost because of clear-cutting (removing all of the trees) could have been worth millions of dollars in revenue each year as well. Tremendous future opportunities are lost when an ocean fishery is depleted or when a farm field is destroyed by erosion caused by farming practices that reflect an immediate time preference and a narrow view of opportunity costs.

In summary, the term *opportunity costs* refers to potential economic losses incurred when one pursues a particular course of action, usually a long-term action rather than an action that yields higher profit in the short term. Short-term exploitation, however, reduces long-term opportunities. Short-term exploitive action may yield higher immediate profits but also rob us of the potential for long-term benefits, both ecological and economic. By opting for short-term profit, we are, in effect, assigning no value to resources that could sustain human life and other species indefinitely.

An understanding of the true meaning of opportunity costs could help business economists shift their time frame. In addition, it could help them venture beyond the narrow realm of economics to consider factors that are not traditionally seen as economic—for example, the aesthetic pleasure we derive from wildlands.

Discounting   Economists use a tool called **discounting** to compare economic returns on various economic strategies. In this technique, they calculate the present economic value of different economic strategies. The **present value** is the dollar value of the stream of income generated by each strategy, taking into account inflation and other factors. This technique can be used to determine whether it makes more sense to clear-cut a tropical rain forest or harvest its wood and other products sustainably over 20 years. Discounting can even be used by individuals, for instance, to compare the economic value of adding a solar electric system to a house with continuing to buy electricity from the utility company.

Through a complicated set of calculations, economists can determine which strategy makes the most sense. The strategy with the highest present value is generally chosen. To understand how it works, consider an example. Suppose you are contemplating installing a wind turbine on your property to generate all of the electricity you and your family consume. Being a smart buyer, you want to determine how this strategy compares with business as usual—buying electricity from the local power company.

To do this, you would first project the cost of all of the electricity you'd buy from the local utility over a period of 30 years, the life of a wind turbine, and compare that to the cost of the wind system. To arrive at an accurate estimate of electrical costs from the utility, however, you'd have to take into account rising electrical costs and the declining value of the dollar due to inflation. (A dollar today is worth more than it will be 10 years from now, due to inflation.)

To determine the value of the electricity from the wind turbine, you'd have to include the cost of the system, including installation, along with maintenance and repair over the 30-year life of the system. You'd also need to calculate the cost of borrowing money to purchase it (interest on loans or interest lost if you took the money out of a savings account). You'd have to calculate inflation, too. When all of this is done, you can determine the net present value of both strategies.

As noted earlier, this technique is used by large business and organizations, such as the World Bank, which funds development in less developed countries. Unfortunately, it is often more rational from an economic standpoint (based strictly on net present value) to liquidate a resource than to harvest at a sustainable rate, which would result in a lower net present value and also a lower rate of return on the investment. The strategy, then, is to make your money and run, investing it in another high-rate-of-return venture. When this happens, economics and ecology clash.

Creating a sustainable system of resource management requires rethinking the way we calculate the value of different opportunities. Education can provide economists and resource managers with a new view of time preference, as well as alternative (sustainable) strategies to achieve respectable economic returns without depleting ecological resources. Bearing in mind that sustainable strategies provide a way of ensuring the long-term prosperity of our species, many critics argue that we must learn to factor in the long-term and hard-to-quantify benefits (like cleaner air) of sustainable strategies that might have lower net present value or yield lower returns. One means of fully valuing such strategies is to calculate the replacement or mitigation cost of various actions. **Replacement cost** is an estimate of the true value of a resource to human society. How much would it cost to replant a tropical rain forest and reestablish the complex ecosystem and the benefits it provides, such as lowering atmospheric carbon dioxide levels and maintaining rainfall patterns that are essential to agriculture?

By factoring these costs into the economic calculus used to determine net present value and opportunity costs, resource managers get a more accurate picture of the economics of various strategies. In this light, many profitable ventures will seem foolish. Sustainable strategies will appear much more attractive under a more ecologically sound means of accounting.

Economics of Pollution Control: Finding New Ways

This book is concerned primarily with natural resource management, but it also delves into other important environmental issues, among them pollution. Because pollution threatens the long-term prospects of humans and other species, we will next examine the economics of traditional pollution control measures and discuss alternative methods of evaluating them.

Pollution control in today's world is heavily driven by economics. The goal of pollution control is to reduce emissions and ambient pollution in the most cost-effective ways. From a purely economic standpoint, then, investment in pollution control devices (capital costs) and their operation should be equal to the benefits received. For example, a factory owner would not want to install a $30 million pollution control device to prevent $1 million worth of damage. In such instances, the cost of control would clearly exceed the benefits.

To begin, take a look at Figure 2.5A. This graph shows the cost of damage caused by different levels of pollution. As illustrated, the higher the pollution level, the higher the cost.

Figure 2.5B shows how the cost of control—that is, reducing pollution—is related to reductions in pollution. It shows that the more money one invests, the more pollution is removed. It also shows that a relatively small initial investment results in a substantial reduction in pollution, but as one tries to reduce pollution emissions further, the cost per unit of pollution removed increases much faster. This is known as the **law of diminishing returns.**

Figure 2.5C overlays the two graphs from parts A and B. The intersection of the two lines is the point at which the cost of control and the economic benefits of pollution reductions are equal. This is the **breakeven point.** If society desires even lower levels of pollution, it must pay more. However, because costs of removing pollution increase disproportionately as one approaches zero emissions, the economic costs can be quite high.

Although balancing costs and benefits may sound simple, it isn't. One of the chief problems with this process, known as **cost-benefit analysis,** lies in determining the full extent of damage—that is, determining *all* of the costs. To do so, we must study and quantify all of the effects an activity has. Many less-obvious effects might be missed. Another problem lies in determining the economic value of the damage, such as the economic cost incurred by polluting a stream or eliminating a species. How can we assign an economic value to a healthy fish population in a stream or a healthy forest? What is the value of your health?

One technique that helps solve the dilemma of underpricing damage is to calculate the costs of mitigation. Thus, while it may be impossible to assign an economic value to the survival of a species of salamander, we can calculate what it would cost to capture the remaining salamanders and raise them in captivity

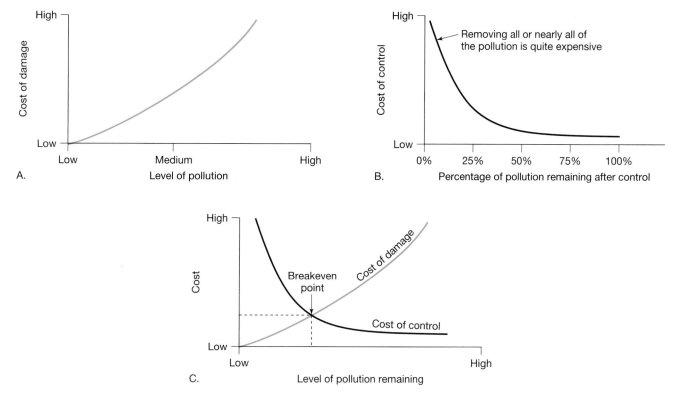

FIGURE 2.5 The Earth in balance. (A) As the amount of pollution increases, the cost of damage increases. (B) Removing pollution also costs money. After a certain point, the cost of removing a unit of pollution climbs dramatically. (C) The goal of pollution control is to find a breakeven point where costs equal benefits. Pollution prevention is an alternative measure. It results in far greater reductions at a lower price.

for eventual release in a cleaner environment. This is the mitigation cost. As another example, a forest in a watershed that supplies drinking water to a city may help keep the water clean. If the forest were cut down, erosion would increase, and that would increase filtration costs. Municipal drinking water facilities might need to expand operations or build new facilities to remove sediment to purify the water for human consumption. These are referred to as mitigation costs. Mitigation costs, however, only indicate part of the value of the forest. What about the loss of wildlife and recreational activities and increased flooding caused by clear-cutting the forest?

Because of these and other problems, officials involved in this economic analysis often fail to include all of the damage in their calculations. They therefore tend to underestimate the real costs to society and the environment of human activities. Investments in pollution control would underprotect us.

If the full costs and extent of human damage were known and could be assigned an economic value, the breakeven point in graph C would very likely shift to the left. It would, therefore, be worth spending much more on pollution control. Until the science of predicting and quantifying costs of human actions such as farming and forestry improve, however, most policy based on cost-benefit analysis will result in an investment that falls short of addressing the true costs.

One fact that many companies are realizing is that there are many economical ways of reducing pollution emissions; they yield substantially greater environmental benefits with minimal economic investment. These strategies fall under the rubric of pollution prevention. **Pollution prevention** is a technique that

differs fundamentally from pollution control, the most common strategy until recently. Pollution prevention seeks to eliminate the production of pollution altogether. **Pollution control** seeks to capture pollutants after they have been produced.

Pollution prevention is achieved in one of several ways. For example, in factories, slight adjustments in manufacturing processes can reduce pollution output enormously. This is an operational change. In some cases, nontoxic chemicals can be substituted in chemical manufacturing or other industrial processes for more toxic substances. This is a design change.

Pollution prevention works and can save enormous sums of money. In Yorktown, VA, for example, Amoco and the Environmental Protection Agency (an agency of the federal government that monitors and regulates many environmental pollutants) dropped their normally adversarial roles to find ways to reduce toxic emissions. They found that a $5 million investment in pollution prevention proposed by the oil company could cut emissions of the toxic chemical benzene five times more than a $35 million traditional pollution control device. In other words, at one-seventh the cost, Amoco could cut pollution emissions five times more than with the traditional strategy. Unfortunately, the law required the company to install the more expensive control technology. This example raised the ire of many people and has resulted in a dramatic shift in policy to avoid such unnecessary burdens in the future.

Similar examples illustrate the economic and environmental benefits of pollution prevention. They also illustrate the need for regulatory flexibility in meeting goals. If, for example, governmental agencies set strict standards for emissions, but companies are allowed to apply their ingenuity in meeting those standards, far greater gains may be made in reducing pollution than is possible with standard pollution control methods. The government of Holland, mentioned in the previous chapter, is embarking on a daring experiment with its many industries. Together, Dutch companies and the government have set strict standards for reducing industrial pollution, but the companies have been left to their own devices to figure out how to do it in the most economical manner.

Measures That Promote Sustainable Economies

Creating a sustainable economy requires the many changes outlined in this book: pollution prevention, sustainable resource management, recycling, energy efficiency, the use of renewable energy, restoration, and the like. Changes in the ways we think about economics will also help promote an enduring human presence.

Alternative Measures of Progress   At the top of the list are new measures of success. The Index of Sustainable Economic Welfare, described earlier, is one example. The GPI provides a more comprehensive picture of the state of a nation than its GNP.

GO GREEN!

In the summer, turn the thermostat up a little at night and while you are away from your home or apartment. In the winter, turn the heat down a little while you are sleeping or away. Doing so can make a substantial dent in your heating and cooling costs.

Several nations, including Germany and France, have developed alternatives to the GNP. The U.N. Statistical Commission has even drawn up guidelines for other nations interested in following suit. Although alternative measures of prosperity are not a cure-all, they are essential to direct corporate and public policy along more sustainable lines. To date, over 200 cities, towns, and states in the United States, among them the state of Oregon; Jacksonville, FL; and Seattle, WA, have developed alternative measures designed to track social, economic, and environmental trends, giving citizens and government officials a more realistic picture of their progress—or lack thereof (Figure 2.6). Similar programs are in place in London, Stockholm, Vienna,

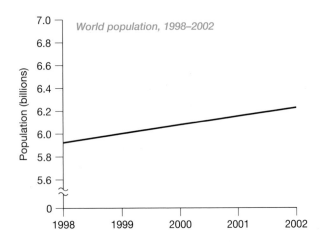

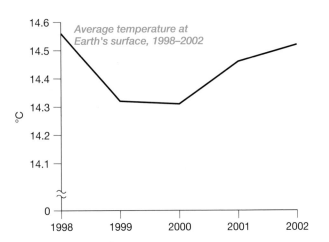

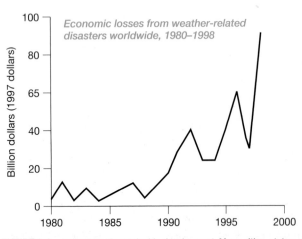

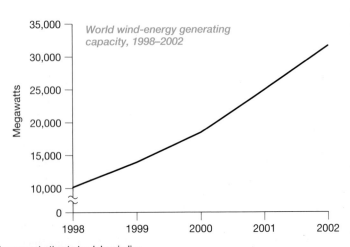

FIGURE 2.6 Indicators of sustainable development. Many cities, states, and nations are starting to track key indicators of their social, economic, and environmental health to determine where they're going and to help them prioritize limited financial and human resources. Such exercises help nations plot their progress toward or away from sustainability. Even information on global trends like these can help governments plot a more sustainable future.

and Zurich. These indicators of sustainable development and quality of life are not aggregate measures like the GPI, however. They consist of dozens of measures of key conditions in various communities that allow people to track their progress toward or away from sustainability. Combined with other measures, they could help steer us back onto a sustainable course.

National Inventories of Natural Resources   Another important movement under way that addresses weaknesses in the GNP is the creation of national inventories of natural resources. Designed in large part to determine if economic growth is depleting natural-capital accounts, natural resource inventories are being conducted in Australia, Canada, France, the Netherlands, and Norway. To create a truly sustainable global economy, such efforts are needed in virtually every nation on Earth.

Green Taxes   Yet another strategy for sustainable economic development invokes user fees or green taxes. **User fees,** or **green taxes,** are taxes levied on raw materials. They are paid by producers, such as mining companies. Green taxes artificially increase the price of a raw material. Although green taxes may seem counterproductive from an economic standpoint, from an environmental viewpoint they make perfect sense. One benefit is that green taxes may help promote more efficient use of resources. Reduced consumption reduces environmental impact and extends the life span of resources. Green taxes can also be levied on undesirable activities—for example, factories that generate pollution or autos that burn excessive amounts of gas and produce inordinate amounts of pollution.

Green taxes offer other benefits in addition to stimulating a more efficient use of resources. Revenues from green taxes, for example, can be used to develop environmentally friendly alternatives or to offset the impact of environmentally damaging activities, such as mining. Several western states, for instance, charge a severance tax (essentially a user tax) on coal. This tax reimburses local communities for the additional cost of infrastructure—roads, schools, and the like—required by the population growth that accompanies resource development. It also provides funds for alternative economic strategies when the coal runs out.

Because the costs imposed on businesses are ultimately passed on to consumers, green taxes provide a useful market signal and help offset one of the principal weaknesses of supply-and-demand economics, namely, the underpricing of goods and services derived from finite resources.

Green taxes are not popular in the United States but are more common in Europe, where an estimated 50 green taxes are currently levied on a variety of products. Nonetheless, there are some U.S. examples. In 1990, the U.S. Congress passed sweeping amendments to the Clean Air Act. One provision of the new law was a substantial tax on industrial use of chemicals that deplete the ozone layer. The **ozone layer** is a region of the upper atmosphere that filters out harmful ultraviolet radiation, which can cause severe burns and skin cancer in humans. It also damages plants. By imposing a tax on chemicals that deplete the ozone layer, the U.S. government hoped to encourage many companies to search for alternative chemicals that perform as well without damaging the ozone layer. These and other efforts have resulted in a dramatic shift to non-ozone-depleting alternatives.

Carefully designed economic disincentives could be implemented to discourage a wide range of activities, among them tropical deforestation, overgrazing, fossil fuel combustion, disposable products, and energy inefficiency.

Full-Cost Pricing   As noted earlier, many goods and services reach consumers at considerable expense to the environment as well as to the communities we live in. That is, they come with unreconciled economic externalities. Thus, what we pay for products is not a reflection of their full cost—especially the environmental costs.

Green taxes can help adjust demand and discourage activities that are environmentally unsustainable, but can also help us pay for external costs. In other words, they help us achieve **full-cost pricing,** defined here as a form of pricing in which the price of a product or service is a reflection of all of the costs associated with its production. For example, a coal tax might be levied to pay for repairing damage to forests and lakes from acid deposition or to offset losses caused by global climate change caused by carbon dioxide emitted from coal-fired power plants.

Far better than taxes such as these applied after the fact are measures that seek to incorporate full-cost pricing at the outset or in the planning stage. Many states, for instance, have regulations that require utility companies to assess the costs of various options needed to produce electricity. When a utility company seeks permission to expand its electrical generating capacity, it must acquire approval from the Public Utilities Commission. But it must select the least costly approach, which frequently does not include environmental costs. To account for economic externalities, several states require that an additional 15% be added to the cost of coal and nuclear energy strategies. Because of this least-cost-planning provision and the adjustment to account for economic externalities, energy conservation and several renewable-energy strategies often appear much more attractive.

Full-cost pricing helps us understand the life cycle costs of various technologies or actitivities. **Life cycle costs** include all costs from the point of origin of a material to its point of disposal—"from cradle to grave." For example, life cycle cost analysis forces business economists to examine long-term environmental damage as well as immediate costs incurred by mining, manufacturing, transportation, and waste disposal, considerations rarely made in most economic decisions.

Full-cost pricing is particularly useful in cost-benefit analysis (described earlier). Cost-benefit analysis is a procedure that companies, government agencies, and legislative bodies use when analyzing a particular strategy or considering various options. As noted earlier, it requires an examination of all costs and all benefits. Historically, most economists have looked only at the cost of inputs and the economic benefits from using these inputs to create goods and services. They have traditionally ignored or undervalued the costs of environmental damage.

Economic Incentives   Economic incentives are another useful market tool used to foster a sustainable economy and sustainable resource management. **Economic incentives** are measures that encourage environmentally compatible goods and services by providing economic benefit. Most governments currently offer a variety of incentives to businesses. In some instances, governments provide outright grants—money to support the research and development of new environmentally friendly products. The Japanese government, for instance, has spent millions of dollars to promote sustainable technologies such as solar energy among businesses. Governments also offer various tax breaks to businesses that are developing or manufacturing environmentally desirable products. These savings on a company's annual tax bill can be used to encourage companies to develop new technologies or to offer environmentally beneficial goods and services. In other instances, governments enter into partnerships with businesses. For example, a local government might donate land it owns to a company to set up a recycling or composting facility. This donation makes it economically feasible for the company to offer a valuable service.

Incentives can also be given to individuals. Federal and state tax credits in the 1980s for the installation of solar-energy systems on homes in the United States, for example, cut the cost of these systems, giving people a huge financial incentive to go solar.

Tradable or Marketable Permits   Another market approach to sustainable development is the **tradable** or **marketable permit** for companies that produce air and water pollution. What is a tradable permit, and how is it different from permits currently issued to businesses?

Many governments currently issue permits to industries to release certain amounts of pollution into the air and water. This is the way government regulates pollution emissions. Marketable permits are similar: they are licenses granted by governments to businesses to release certain amounts of pollution. What makes marketable permits different from other permits already issued by a government is that they can be bought and sold. More importantly, if a company can reduce pollution below the level stipulated on your permit, you can sell the remaining pollution emission allowance to another company. What's the advantage of that? Consider a specific example.

In 1990, the U.S. Congress passed sweeping amendments to the Clean Air Act. One of them established marketable permits for sulfur dioxide, a pollutant that is converted into sulfur acid in the atmosphere, which falls in rain and snow, causing damage to lakes, rivers, crops, forests, buildings, and much more. The marketable permit provides companies an opportunity to buy and sell sulfur dioxide pollution emissions on the Chicago Board of Trade (sales began in 1994).

Under the previous system, state and federal agencies granted companies permits to release certain amounts of sulfur dioxide. Companies had no incentive to release less. The marketable permit, however, provides companies with an economic incentive (profit) to reduce pollution below the permitted level. Suppose you own a factory that releases 2,000 tons of sulfur dioxide into the atmosphere each year. To clean up the air in the city in which your factory is located, the government (usually a state regulatory body such as a department of the environment or natural resources) issues you a permit to release no more than 1,000 tons per year. However, unlike the situation in the past, when the government also required you to install a pollution control device to reach this goal, you are now free to achieve this reduction any way you want. After some deliberation, one of your company's engineers finds a way to reduce the company's annual air emissions to 500 tons per year. Because your annual limit is 1,000 tons, you now have 500 tons of emissions

"credit" that, under the marketable permit system, you can sell (market) at a profit to another company to help it meet its emission allowances. If, for example, another company has an emissions limit of 1,000 tons per year but is unable to lower emissions below 1,500 tons, it can buy your unused tons to meet its legal limit.

The benefit of the marketable permit system is that it allows flexibility in achieving pollution emission goals and at the same time creates an economic incentive to meet or exceed those goals. Not only has your company found a cheap way of reducing pollution, you can make a profit from selling the unused portion of the permit to another company, which would then be able to meet its goals, thanks to the engineers in your company. Your company, in turn, profits by the sale of part of its permit to another company.

In the United States, the first pollutant covered by marketable permits was sulfur dioxide. When marketable permits first appeared on the scene, one enterprising environmental group bought up several hundred thousand tons of pollution emissions. The group did not own a factory, it simply wanted to ensure that those pollutants would be permanently removed from the environment. Another environmental group more recently offered members an opportunity to invest in permits to help rid the air of potential contaminants. Marketable permits are now also being sold for carbon dioxide, as discussed in Chapter 19.

Removing Market Barriers and Subsidies   Another market-based solution that could help us create a sustainable economy is removing market barriers and subsidies. Surprisingly, significant market barriers exist to environmentally responsible businesses. Federal regulations, for instance, make it legal for freight haulers to charge more for scrap metal bound for a recycling plant than for raw ore. Because scrap metal can be fashioned into useful products with much less energy, pollution, and damage to the environment than raw ore, this legal loophole is a major obstacle to creating an economically viable system of recycling in the United States. It gives raw ore an economic advantage over recyclable scrap metal.

Subsidies are ways that governments support private enterprises. They include tax breaks and outright grants, two forms of economic incentive. Subsidies can be used to encourage environmentally useful goods and services, such as wind energy or solar energy. The U.S. government, for instance, currently offers a generous subsidy to the wind industry to help make electricity generated from commercial wind farms more cost competitive. Subsidies to the renewable-energy industry come to about $100 million per year. Unfortunately, most subsidies do not support environmentally sustainable energy, say critics. In fact, by one estimate, the U.S. government provides $200 billion in subsidies to the oil, coal, and nuclear power industries *each year*. Automobiles are subsidized by public taxes that pay for roads, parking lots, traffic lights, police, and emergency protection on the order of $1,500 per year per car. This money comes not from gasoline taxes but rather from general revenues. Whether or not you own a car, you pay the subsidy.

Removing or reducing subsidies that support environmentally unsustainable activities will help create a level economic playing field that will make environmentally compatible industries more competitive with those that are less environmentally beneficial (Figure 2.7).

FIGURE 2.7 In Portland, OR, electric trains whisk passengers to and from the city with a fraction of the air pollution of the private passenger vehicle. Systems such as this are vital to building a sustainable system of transportation. They often can't compete with automobiles, but that may be due to the huge subsidies automobile use receives.

These are just a few of the ideas needed to build a sustainable economy. In many countries, they have already been implemented. Existing efforts, however, are small compared with the task at hand.

2.3 Toward Sustainable Ethics

Decisions about the environment are not all economic. Many are made on the basis of what we think is right or wrong, that is, **ethics.** We learn ethics from our parents, friends, ministers, and teachers. We also learn ethics through experience, reading, television, movies, and thinking. Even the study of science can teach us about Earth ethics—what is right or wrong from an ecological perspective.

Frontier Ethics

Ethics serves a practical purpose in our society: for example, it helps preserve social order by influencing the way we treat one another. In this respect, ethics is utilitarian. That is, it serves us well.

Ethics can also influence how we treat the Earth. In many parts of the world, the modern ethics by which many people live their lives and by which businesses and governments operate threatens the life-support systems of the Earth. In the United States and other industrial nations, for example, many people view the Earth as an unlimited (infinite) supply of resources exclusively for human use. If the Earth's resources are unlimited, so then are our options for tapping into those resources. Another key tenet of modern ethics in many places is that humans are apart from, rather than a part of, the natural world. We are somehow superior to it. Our needs are of paramount importance. To those who hold this view, nature is merely a backdrop for human civilization. It seems therefore of little consequence to their lives, when in fact the future of humankind is intricately linked to the future of the planet. Another key tenet of frontier ethics is that the key to success is the domination and control of nature. To meet our needs, we must conquer nature, leaving hardly a trace of its former self. We reduce complex ecosystems to highly simplified systems that produce food, fiber, and other resources required to meet our needs. We level mountains to access underlying minerals (Figure 2.8). We dam rivers and cut roadways through forests. Our motto has become, "If brute force isn't working, you're not applying enough."

In this book, we refer to this system of ethics as **frontier ethics,** because it reflects the attitudes of the American frontier people who had little, if any, notions of limits. Today, although most of our frontiers have vanished, frontier ethics remains, influencing countless decisions every day. Unspoken ethical tenets drive innumerable environmentally unsustainable decisions of government officials, resource managers, businesspeople, and citizens. In fact, frontier ethics—especially the notion that the world possesses an unlimited supply of resources for human use—is a prime driving force of the national economies of many major economic powerhouses.

Sustainable Ethics

Developing a sustainable, worldwide economy will require a much greater awareness of our problems and a dramatic increase in our understanding of the importance of natural systems to our lives. It will also require a new system of ethics, a sustainable ethic.

Sustainable ethics holds that the Earth has a limited supply of resources. In other words, the Earth's resources are finite and should be managed carefully. Another key tenet of sustainable ethics is that the Earth's resources are not for the exclusive use of humans. Other species require a share of the Earth's wealth to survive and prosper. Some people claim that other species have a right to live and prosper, too. Sustainable ethics also holds that humans are a part of nature—we are dependent on natural systems in many ways. Another key element of this ethical system is that human success requires cooperation with nature, rather than its domination and control. The core value of the new ethical proposition is that humans and nature are inextricably linked: we are dependent on natural systems, and our future depends on a healthy, well-functioning ecosystem. Conversely, nature is also dependent on us. Human society has reached a size and level of technological power so great that we literally hold the future of the planet in our hands.

Sustainable ethics is not a new notion by any means. In fact, it may have been the predominant approach throughout much of the

FIGURE 2.8 This surface mine is a symbol of human domination over the environment. Technological development has allowed us to alter the face of the Earth in dramatic ways.

several-million-year history of the human species. Judging from the practices of Native Americans and other indigenous cultures, sustainable ethics has been central to the survival of people. Only in the past few centuries have we drifted away from this view.

In the 20th century, numerous writers have proposed a return to sustainable ethics. One of the most influential was the late Aldo Leopold, whom we first mentioned in Chapter 1 (Figure 2.9). A wildlife ecologist by trade, Leopold proposed a land ethic in 1933—over 70 years ago. Leopold's **land ethic** held that humans are a part of a larger community that includes the soil, water, plants, and animals—which he referred to as the land. Recognizing that humans are a part of nature, Leopold argued that the role of human beings should shift from "conqueror... to plain members and citizens of it."

FIGURE 2.9 Conservationist Aldo Leopold (shown here) has inspired many individuals over the past 70 years to rethink their role in the environment. His writings, such as *A Sand County Almanac*, published in 1949, continue to sell well today.

Perhaps best known for his book, *A Sand County Almanac*, published in 1949, Aldo Leopold has become an ideological inspiration to many modern conservationists and environmentalists. His ideas go beyond the well-meaning but human-centered ideology of Theodore Roosevelt and Gifford Pinchot, who sought to protect natural resources primarily for their value to humans.

For Leopold, conservation requires equal amounts of reflection and action. In other words, he promoted a sense of responsibility and action. Leopold's suggestions for action reflect the needs of a time when resource management was the primary concern. There was little concern for pollution, and the phrase *global environmental problems* had not been invented. Nonetheless, abiding by his land ethic could have profound implications. That is to say, considering ourselves members of the community of living things that inhabit the biosphere and the physical world that supports this life could have a powerful impact on human activities. Abiding by his ethic, as opposed to acting almost exclusively on behalf of human society and our own personal welfare, might have helped us avoid many of the problems we've created.

Today, the extent of the planet's environmental ills and the threat they pose to our future call for responsibility and action at all levels of society. Sustainable ethics is therefore an extension of the land ethic.

Sustainable ethics rests on four **directive principles,** described earlier: (1) the Earth's resources are limited; (2) humans are a part of nature; (3) the key to success is cooperation with nature; and (4) natural systems are essential to human welfare. But how does one put sustainable ethics into practice?

This book proposes five **operating principles** for putting sustainable ethics into action: (1) conservation, (2) recycling, (3) renewable-resource use, (4) restoration, and (5) population control. These guidelines for action are not all that is needed; they are only some of the most important steps humans can take to create an environmentally sustainable society.

Together, the directive and operating principles could result in a measure of restraint that is necessary to shift away from the resource-intensive society that now threatens its own long-term well-being. Restraint means doing what is right not just for the individual, but for future generations. This concept is known as fairness to future generations, or **intergenerational equity.**

One of the central tenets of sustainable development, intergenerational equity asserts that present generations hold the Earth in common with all generations, past and future. We are, in short, part of a long line of planetary occupants with common rights and obligations. For example, present generations have the right to benefit from the Earth—to profit from its riches and enjoy its beauty. At the same time, we have an obligation to protect the Earth for future generations so that they may benefit as well. Think of the Earth as an heirloom that is handed down from generation to generation. As with an heirloom, the rights and obligations of the Earth's custodians are attached to the gift.

In *Lessons From Nature: Learning to Live Sustainably on the Earth,* the author proposes a notion of **intragenerational equity**— fairness among members of the present generation. The idea behind this concept is that because the Earth's human community shares the global commons—the air, water, and land—and because local actions often have global impact, people have responsibilities to one another in the present. More specifically, we have a responsibility to act in ways that do not adversely affect other people, no matter how distant or unrelated they may be.

As an example, consider the use of fossil fuels by industrialized countries like the United States and Canada. Burning fossil fuels may cause a rise in sea level through the release of carbon dioxide, which in turn causes global warming and the melting of polar ice and glaciers that leads to flooding. Flooding may inundate the coastal farmlands of millions of people in less-developed countries—people who have never burned a drop of oil or an ounce of coal. Do users of fossil fuels have an obligation to these people? Under the doctrine of intragenerational equity, the answer is yes.

One type of intragenerational equity that is of great concern to many is known as **environmental justice** or **environmental equity.** Environmental justice is defined as fair treatment for people of all races, cultures, and socioeconomic levels in the execution of environmental laws, regulations, and policies. It has also been defined as the pursuit of equal justice and equal protection under environmental laws and regulations without discrimination based on race, ethnicity, gender, and/or socioeconomic status. Now spurring an international movement, the notion of environmental justice grew out of concern that minority populations and/or low-income populations (often one and the same) bear a disproportionate amount of adverse health and environmental effects.

Concern for environmental justice began in the early 1980s in Warren County, NC, when a predominantly African-American community was targeted as a proposed site for disposal of soils contaminated by the chemical PCB (an oil once used in electrical insulators) from 14 other sites in the state. Protests emerged over the siting. Although protestors were not able to stop it, the fury sparked further inquiry. The U.S. General Accounting Office studied the issue in eight southern states and found that three out of every four hazardous-waste landfills were located near predominantly minority communities. Since then, numerous studies have shown that hazardous-waste facilities, polluting factories, power plants, incinerators, and other potentially harmful facilities are also frequently sited in or near low-income, minority communities. Why? Because poorer minority communities typically lack the financial resources and political savvy to mount opposition against them.

A study by the *National Law Journal* showed further evidence of environmental discrimination in the government. Their researchers found that it took the EPA 20% longer to nominate an abandoned hazardous-waste site as a priority area in need of cleanup if it was in a minority community rather than a white community. The researchers also found that industrial polluters in minority communities paid fines that were on average 54% lower than fines levied on polluters in white communities.

Since the early 1980s, environmental justice has become a priority in the EPA. Today, in fact, the EPA has an Office of Environmental Equity, which has initiated a number of environmental justice projects in the United States. In 1994, President Clinton issued an executive order requiring all federal agencies to make environmental justice part of their mission. Numerous grassroots organizations have also emerged to address the issue in the United States and in less-developed nations as well, where the practice is also commonplace. The U.S. Congress and state legislators have also begun to grapple with the issue, and two states, Arkansas and Louisiana, have passed environmental justice laws.

A variety of solutions have also been proposed, including pollution prevention and improved stakeholder participation in the public decision-making process—in other words, more participation by minorities in communities that would be affected by proposed projects. Improving access to information could help also (see Case Study 2.1). Yet another solution is improved enforcement and compliance assurance (making sure companies comply with laws and regulations).

Yet another ethical concept is ecological justice. **Ecological justice** holds that people have an obligation to other species, present and future. Put more forcefully, other species have a right to exist, and we must act in ways that ensure their survival.

Intergenerational equity, intragenerational equity, and ecological justice contradict many people's most basic ethical ideas. They fly in the face of the frontierist notions and principles upon which capitalism is based. For example, the idea that we must act in ways that protect the interests of present and future generations contradicts the prevalent ethical notion that the Earth is an infinite source of materials. It also contradicts the economic notion of competition, in which each person, present and future, must compete for his or her own survival and well-being. It calls for cooperation, not competition.

Biocentric and Ecocentric Views

Frontier ethics is anthropocentric—that is, people oriented. It is a system of ethics that serves humans. Unfortunately, it does much damage to the environment and to human society as well.

Sustainable ethics, as we've described it in this book, is anthropocentric and ecocentric. That is, it supports human well-being but not at the exclusion of other living things. It supports actions that are good for people *and* the planet, including the millions of species that share the Earth with us.

| CASE STUDY 2.1 | GEOGRAPHIC INFORMATION SYSTEMS AND ENVIRONMENTAL JUSTICE |

There's a river in Mississippi that pours into the Gulf of Mexico. It is known as the Escatawpa. Along its banks are two communities, one white and rich, the other composed of African Americans who are economically less well off. In the late 1970s, the mostly white affluent community known as Pascagoula decided to build an incinerator to burn its trash. Many residents, however, opposed siting the facility in their town, and the contract eventually went to a chemical company located three miles from Moss Point, the nearby poor, African-American town.

Little controversy was stirred by the decision, until in 1991 the city council of Pascagoula voted to permit the burning of medical waste in the incinerator in addition to the trash. Residents of Moss Point became enraged by the odors from the incinerator. Concern erupted over the potential for contamination from mercury, cadmium, and dioxin released from the facility as well. Local doctors worried about further health problems in an area that had an already high incidence of respiratory disorders. They raised concerns over the potential long-term health effects.

The controversy along the banks of the Escatawpa River is a classic example of environmental injustice, or environmental racism. A wealthy, predominantly white community wants an incinerator, but not in their backyard. The incinerator ends up in an area surrounded by people of color without enough money for health care or legal representation.

Environmental racism is on the decline. Those who are active in the fight for justice are receiving assistance from a high-tech tool: GISs (geographic information systems), specifically a GIS known as LandView™ III. Discussed in Chapter 1, a GIS is a computerized mapping system that permits storage of huge amounts of information in map form to study relationships, make predictions, and better plan human development.

LandView III is described as a community right-to-know software tool by the EPA, which sponsored its development. According to the EPA, LandView III places a "wealth of important environmental information at the fingertips of local decision makers and the public." It provides database extracts from the EPA, Bureau of the Census, the U.S. Geological Survey, the Nuclear Regulatory Commission, the Department of Transportation, and the Federal Emergency Management Agency. These databases are presented on maps that contain jurisdictional boundaries, detailed networks of roads, rivers, and railroads, census information, schools, hospitals, airports, landmark features, and much more.

LandView III is accessible to anyone with a modern computer and contains a tutorial so individuals can learn how to use the system. The system can be used to learn about the demographics of an area—that is, the age, income, and ethnicity of an area—and to learn about potential health effects from a wide variety of sources. The environmental information available to the user includes air pollution emissions, wastewater discharges, toxic-release data, hazardous-waste sites, nuclear sites, watershed assessment, and air-quality monitoring sites. LandView III has the capacity to display multiple layers of information for decision makers, and it is proving to be a useful tool in the effort to develop currently unused industrial sites, known as **brownfields,** that have been contaminated by toxic substances. For example, LandView III can be used to locate potential brownfield sites in a city, county, or state and assess their potential impact on the community and environment. Sites slated for development are typically cleaned up, but such strategies must be communicated to the nearest neighbors. LandView III enables brownfield developers to assess the demographics of the surrounding regions and tailor informational programs to allay fears and initiate public involvement.

LandView III is available online and in CD-ROM from the Bureau of the Census. For information on purchasing a copy through the Bureau of the Census, see their Web page at www.census.gov.

Some people call this stewardship, but we think it is more than this. As we've defined it, other species have a right to exist, too, and not just those that have economic value to humans.

Some people hold a different view, however. Their viewpoint is decidedly biocentric. **Biocentric views** see nonhuman living beings that inhabit the Earth as of primary importance. Human priorities should, according to proponents, take a back seat to ensuring the welfare of the rest of the living world. Thus, maintaining biodiversity is more important than human welfare. In a biocentric viewpoint, other organisms have a right to exist, regardless of their economic value, and moreover, their needs take precedence over ours.

In contrast, others take an **ecocentric view.** To them, ecological integrity and ecological processes are of utmost value. Ecologically centered views see evolution, adaptation, and nutrient cycles—the processes that ensure life's continuation—as most important. These things are valued and protected. In such a system, the whole is more important than its individual

parts. A deer herd could be culled to protect its habitat and the rest of the ecosystem. This book illustrates this viewpoint, and our view of sustainable ethics also incorporates it.

Creating a Global Sustainable Ethic

Creating a global sustainable ethic—which may include elements of anthropocentric, biocentric, and ecocentric views—is among the most pressing tasks of our time. It will require efforts by millions of educators from primary education through college worldwide. It will depend on the efforts of the personnel of museums and nature centers. Religious leaders can play a role. The entertainment industry and media (newspapers, magazines, and television) could participate. It will also require the cooperation of parents, not just in the United States but in all nations, rich and poor.

Given the conflict in many parts of the world (the Middle East, for example) and the decaying conditions under which many people live, convincing people to adopt a long-term view required by sustainable ethics will not be easy. Why?

Meeting immediate needs often takes precedence over urgent environmental concerns. Ironically, meeting immediate needs (an anthropocentric view) may worsen environmental conditions, creating a downward spiral of the Earth's ecosystems, which are the life-support system of the planet.

Nevertheless, fostering sustainable ethics may be one of the most important steps we can take to build a sustainable future. The United Nations Conference on Environment and Development, as described in Chapter 1, was a good start. Many nonprofit organizations, teachers' groups, and business organizations have begun the process of global attitude change.

Critical Thinking and Sustainable Development

This book will help you understand environmental and resource management issues. During your reading, you will encounter many scientific facts and principles that are essential to understanding and solving problems. Understanding issues and solutions requires more than remembering scientific facts and having a grasp of policies and practices that promote sustainable development. To become a resource manager and a responsible citizen, one must learn how to analyze issues and solutions. This section outlines six "rules" for critical thinking.

Critical thinking means many things to many people. We define it as a means of analyzing information to distinguish between beliefs (what people believe is true) and knowledge (facts supported by scientific observations). Critical thinking is used to analyze the results of scientific research, environmental issues, and proposed solutions. It can be used to analyze newspaper articles, speeches, and classroom lectures. Critical thinking is the most ordered kind of thinking. It permits us to look for weaknesses in reasoning. This section presents a few rules of critical thinking (Table 2.1).

Gather Sufficient Information In analyzing issues and assertions or in solving problems, it is important to learn as much as you can about a subject. Hence, the first rule of critical thinking—and the one that is most frequently broken—is to gather as much information as possible. Learn everything you can about a subject *before* you make decisions or criticize.

Don't just gather and analyze information that supports your point of view. Be sure to gather all facts and study them carefully. See what the opposition has to say. Doing so will help you avoid the mistake of cherry picking information that supports your preconceived notions. Many people fall into the trap of seeking to confirm their beliefs and do not question or examine them critically.

While searching for information, be sure to look for statistics, not just stories that support a point of view. Stories or anecdotes can be amazingly convincing but can mislead us. They may not represent reality. Statistics, on the other hand, help us get closer to the truth.

Part of the process of gathering information is to understand all terms and concepts. Many an advocate of a particular viewpoint has a shaky understanding of science and speaks more from emotion than from facts. This is true of environmentalists as well as their critics. The deeper you study many issues, the more you will understand and the better your solutions will be. Whatever you do, do not mistake ignorance on a topic for perspective.

Question the Methods In separating fact from fiction, or belief from knowledge, it is often essential to question the methods by which the information was derived. Many people, for instance, draw sweeping conclusions from casual observations. They may have read an article in the newspaper about a company that closed down because of tighter pollution control regulations. From this observation, they conclude that environmental regulations are too costly to society. Regulations rob us of jobs and income.

A study of business shows that environmental regulations have very little negative impact on most companies. Those that close down are often in danger anyway because they are not operating efficiently or their market is drying up. Environmental regulations are merely the straw that broke the camel's back. So beware of individual case studies and anecdotal information. They may not be representative of the larger truth. This is a good example of how stories can mislead us, a topic just discussed.

Rigorous scientific proof is often necessary to distinguish between knowledge and beliefs regarding many environmental issues—for example, the effects of pesticides on human health and the environment. Such proof comes in the form of carefully conducted experiments. Scrutinize all reports to determine if the experiments were adequately performed. Was the sample size large? If not, the results may be a fluke. Did the experimenters control all variables? For example, did the researchers of a study that demonstrated a particular pollutant caused a negative health effect eliminate the possibility that another factor, such as cigarette smoking, was responsible?

Careful experiments, whether they use laboratory rats or human beings, require two groups. The first is the **control group.** The second is the **experimental group.** In good experiments, the control group and the experimental group are as similar as possible. They differ only in one regard, the **experimental variable.** For example, suppose you want to determine whether a particular pesticide is harmful to mice. You would conduct an experiment with two groups of mice identical in age, sex, weight, diet, and

TABLE 2.1	Critical-Thinking Guidelines
Gather sufficient information and define all terms	
Question the methods by which facts and conclusions were reached	
Question the conclusions of studies	
Question the source of information; look for hidden biases	
Tolerate uncertainty	
Examine the big picture	

so forth. The experimental group would receive daily injections of the chemical under study. The control group would receive injections as well, but the injections would not contain the pesticide under study. Thus, any observed differences could be attributed to the chemical.

To be valid, studies must also use an adequate number of subjects. Generally, ten experimental and ten control animals is enough, but the more, the better. Studies must also be repeated by others to determine if the results are replicable.

Scientists perform experiments to test various ideas—or **hypotheses** (high-poth´-eh-seas). For example, a scientist might hypothesize that a certain chemical in the water of a lake is responsible for birth defects in bird embryos. To test this hypothesis, she might perform an experiment using control and experimental groups. If the results of the experiment do not support the hypothesis, the scientist might change it. For example, she might assume that another chemical is to blame. She would then test this new hypothesis.

Scientific hypotheses that have been supported through experimentation often tell us many facts about the world we live in. Hypotheses related to the same subject may lead to the formulation of a theory. A **theory** is a larger explanation of a phenomenon. For example, the many experiments on the atom have led to a theory that explains what an atom consists of and looks like. This is called atomic theory.

Theories are supported by many observations and, as a rule, cannot be refuted by a single experiment. But theories are not immutable. Some cherished theories have been modified or replaced as new evidence accumulated.

A great deal of information today also comes from computer modeling. The *Limits to Growth* study discussed in Chapter 1 is a good example. Scientists use computer models to predict potential impacts of rising greenhouse gas emissions, which are responsible for global warming and changes in the Earth's climate, too. Creating computer models based on complex mathematics is useful. That is, it helps to explain our world and helps us understand how human activities can alter natural systems and human systems. They are terrific tools for prediction. But they have their shortcomings, too. Computer models are only as good as the assumptions that are made when formulating them. If faulty assumptions are made, then outcomes and predictions will be faulty. Climate models used to predict the effects of increasing concentrations of pollutants in the atmosphere, for example, were primitive at first but have been dramatically improved to include many factors that affect the Earth's climate, although they still are only mathematical models of the atmosphere and climate, an extremely complex system.

Question the Conclusions and the Source Just because an experiment is correctly designed does not mean that the conclusions drawn from its results will necessarily be valid. Scientists are human; they make errors, and they have hidden biases that may taint their interpretation of the results. They may even be funded by special interests such as the pharmaceutical industry or tobacco industry, whose interests can influence the interpretation of results.

Scientists may also fail to take into account contributing factors. This is especially true in studies of the effects of chemicals on people. Unlike commercially available laboratory rats and mice, people are a mixed lot. Our genetic makeup varies. Our life histories vary, as do our exposures to various chemicals. Of course, it is not ethical to expose people to chemicals in controlled experiments. It is not possible to cage a hundred people, control their diet, and intentionally expose them to toxic chemicals to study their effects.

Because of these and other constraints, studies of the effects of chemicals on humans are often difficult to perform. In most cases, scientists locate populations of individuals who have been inadvertently or accidentally exposed—say, at work—and then compare them with a similar population (a control group) that was not exposed. To draw conclusions, scientists must take into account a variety of variables—for example, possible exposure at home or exposure to other factors that might have the same effect as the chemical under study. Such studies on people fall within the realm of **epidemiology** (ep´-eh-deem´-ee-ol´-oh-gee).

Because epidemiological studies are so difficult, it often requires dozens to demonstrate a consistent connection between an environmental pollutant and a health effect. To date, over 40 studies have shown that smoking causes lung cancer, yet some people still dispute the results. This leads to another important point, namely, that people with an axe to grind—for example, representatives of the tobacco industry—are prone to interpreting results in ways that suit their needs.

The lesson here is to question the conclusions. Ask if the facts really support the conclusions, or if a hidden bias or a hidden agenda might be tainting the interpreter's viewpoint.

Tolerate Uncertainty Critical thinkers must tolerate uncertainty as they work through an issue. Formal science is a relatively recent phenomenon in human history, and scientific knowledge is skimpy in many areas. As our culture becomes more complex and our impact grows, scientists struggle to gain a deeper understanding of many issues. In some cases—for instance, global climate—the systems that scientists must understand are so many and so complex that decades may be needed to come to an informed decision. Of course, by that time, action may be too late.

The rule here is to be aware that uncertainty exists. Tidy answers are not always available. When necessary, we must make decisions based on incomplete information. We must, in other words, tolerate some uncertainty and know that, as time passes, our knowledge will improve. That's the nature of human culture. We are constantly learning and refining knowledge. Don't expect all of the answers. Moreover, you should expect scientists to change their mind as new information becomes available.

Understand the Big Picture Too often, political and scientific debates focus on small pieces of the puzzle while ignoring the big picture. For example, in the debate over nuclear energy, many people focus on the issue of reactor safety (Chapter 21). To address this issue, the nuclear industry has proposed smaller, supposedly safer nuclear power plants. Proponents of nuclear power argue that by installing these power plants, we can avoid serious and costly accidents.

Although that logic may be appealing, it misses an important point—one that is evident when you examine the big picture. The nuclear energy issue involves much more than reactor safety. It includes the serious and as yet unresolved issue of waste disposal: How do we safely dispose of waste that will remain radioactive for 10,000 years? And what about radiation exposure to miners and workers in nuclear-fuel-processing plants? Is there enough nuclear fuel to power reactors? These and at least ten other issues are relevant to the debate and can only be appreciated if one examines the issue from the largest possible perspective.

Big-picture analysis also helps one make sense of other issues and come to important conclusions. For instance, many people are currently debating the rate of tropical deforestation. Some say that present estimates of 17 million hectares (about 45 million acres) lost each year are too high. But even if the actual rate of loss is half the estimated rate, it is still too high to be sustained. Stop and look at the big picture before quibbling over details.

Critical thinking requires us to think in larger terms, often in terms of entire systems. By breaking out of narrow perspectives and looking at how humans affect entire systems, we gain a broader perspective that can lead us to more comprehensive and potentially lasting solutions. This is part of a valuable critical-thinking skill called systems thinking.

Systems thinking requires us to think in terms of whole system and the long term. That is, it requires us to look at how human actions affect society and our economy and how they affect the environment, natural systems. It requires us to think about all of the ramifications of human actions, both positive and negative, and to weigh them carefully. It requires that we examine ethics, economics, and environment. It requires actions that result in not just immediate human benefit but long-term benefit of all species.

One area in need of a systems approach is environmental protection. It is clear that we can no longer depend on narrow solutions that strike only at the symptoms of environmental problems. It is time to look at entire systems—for example, forest management, wildlife management, waste management, energy, transportation, housing, and the like—and find ways to revamp them in their entirety, not one piece at a time, which is usually inadequate. Ecosystems management, described in the previous chapter, requires systems thinking.

Economics and ethics are two systems where critical thinking comes in handy and where systems reform is essential. The question is: Do we have the political will to make the changes, or will we continue to muddle along, applying Band-Aids to problems that are often growing worse every day?

Summary of Key Concepts

1. Resource management requires an understanding of economics and ethics and an ability to think critically.
2. Economics is a science that seeks to understand and explain the production, distribution, and consumption of goods and services.
3. Historically, economics has been concerned with inputs and outputs. Inputs include the commodities (raw materials, labor, and energy) companies need to produce goods and services. Outputs include goods and services.
4. One of the guiding principles of market economies is the law of supply and demand, which describes the manner in which the price of a good or service is determined by the interaction of supply and demand.
5. Despite their many successes, market economies have several fundamental weaknesses when viewed from an environmental perspective.
6. One of the principal complaints of ecologists is that business economists have a rather short time horizon for planning. The shortsightedness of the market economy is evident in our almost single-minded dependence on the law of supply and demand. Pricing established by supply and demand typically reflects only the immediate supplies of resources in relation to their demand. Long-term supplies are essential to the future of humankind but are not factored into prices.
7. Oil, rangeland, forests, fisheries, and other resources have been depleted, or driven to the brink of depletion, in large part because they were once abundant and have historically been undervalued.
8. Critics also point out that at least three essential factors are missing from the input/output analyses of mainstream economics: environmental damage, social and cultural impact, and pollution. Thus, when businesses calculate the costs of manufacturing their products, they typically fail to take into account environmental and social costs—that is, economic externalities.
9. Yet another shortcoming of traditional capitalism is its single-minded dedication to measuring success by the gross national product, or GNP. The GNP is the sum total of all goods and services produced by an economy.
10. The problem with the GNP is that it is a crude measure that fails to distinguish between "good" and "bad" economic activities, and thus fails to track a nation's true economic progress. Growth in the GNP may occur at the expense of society and the environment.
11. The GNP also fails to take into account natural capital, or a nation's natural resources and ecological wealth. Nations draw on this capital to fuel their industries and to satisfy human needs for goods and services. But many nations are depleting their natural capital. Although they may boast high GNPs, their long-term prospects for growth are dim.
12. To live sustainably, we need an environmentally sensitive system of commerce, that is a sustainable economic system.
13. In general, a sustainable economy is one that produces goods and services in a manner that does not foreclose on future generations. A sustainable economy uses resources efficiently, recycles materials, and minimizes or eliminates waste. It also depends on sustainable management of natural resources and the restoration of renewable resources.
14. Most advocates of sustainability believe that a sustainable economy will very likely be based on a clean, renewable energy supply. Creating a sustainable economy also hinges on our becoming better able to manage our own population growth and better at balancing our demands for goods and services and living space with our needs for clean air, water, food, and aesthetic pleasure.

15. Three principles affect the management of resources: time preference, opportunity costs, and discount rates.

16. Time preference is the preference for economic returns on an investment, expressed in time. An immediate time preference means that the individual prefers immediate returns. An immediate time preference often leads to practices that are ecologically, and sometimes economically, suicidal.

17. To create a sustainable economy, economic incentives may be needed to compensate for lost income resulting from a shift in time preference caused by a concern for the ecological integrity of forests and other natural resources. However, not all environmentally sensible practices involve a long payback period or exceptionally high initial investments.

18. Opportunity cost is the economic cost of opportunities that are lost when one embarks on a particular path. Many people act in ways to ensure they maximize their economic opportunities, but such actions often lead to resource depletion, which results in lost opportunities in the long term.

19. Economists use a tool called discounting to quantify different strategies. This technique allows them to determine the net present value of different strategies and make choices based on this factor. Unfortunately, many unsustainable activities have a higher immediate value. As a result, it may be more rational for a businessperson to liquidate a resource than to harvest at a rate that brings a lower rate of return.

20. Education can provide economists and resource managers with a new view of time preference, as well as alternative (sustainable) strategies to provide acceptable economic returns without depleting natural resources.

21. One means of taking into account lost opportunity costs of various actions is to calculate replacement costs—that is, how much it would cost to replace all of the topsoil lost on a farm or to replant a tropical rainforest and reestablish the complex ecosystem.

22. The goal of pollution control is to reduce emissions and ambient pollution in the most cost-effective ways. From a purely economic standpoint, investment in pollution control devices (capital costs) and their daily operation should be equal to the benefits received. This is the breakeven point.

23. Although balancing costs and benefits may sound simple, it is not. Two of the chief problems with this process are determining the full extent of damage and the economic value of damage.

24. One alternative to pollution control that brings far greater economic and environmental benefits is pollution prevention. Pollution control seeks to capture pollutants after the fact, when it's too late; pollution prevention seeks to eliminate the production of pollution altogether.

25. Several economic myths pervade our society and prevent progress toward a sustainable future. One of those myths is that environmental protection is bad for the economy. Another myth is that environmental protection is a luxury of lesser value than economic growth. A third myth is that environmental quality is noneconomic. A fourth myth is that economic growth is good,

indeed, essential. This chapter provided evidence to counter these myths.

26. To create a sustainable economy requires the changes outlined in this book: pollution prevention, sustainable resource management, recycling, the use of renewable energy, restoration, and the like. Changes in economics itself will also help promote an enduring human presence. Among such changes are alternative measures of progress, national inventories of natural resources, green taxes, and full-cost pricing.

27. Efforts are also needed to harness market forces through the use of economic disincentives and incentives, tradable and marketable permits, removal of market barriers, and removal of hidden subsidies.

28. We also need a widescale change in ethics. The term *ethics* refers to what people view as right and wrong. Ethics serves a practical purpose in our society—it helps preserve social order by influencing the way we treat one another. Ethics can also influence how we treat the Earth.

29. In the United States and other industrial nations, many people subscribe to frontier ethics, in which the Earth is viewed as an unlimited (infinite) supply of resources for exclusive human use. It also views humans as apart from rather than a part of nature and sees humans as dominators who must control nature to meet their needs.

30. These often-unspoken tenets drive innumerable decisions of government officials, resource managers, businesspeople, and citizens and are at the root of our present crisis of unsustainability as much as the many economic factors described earlier.

31. Steering onto a sustainable path will require a much greater awareness of our problems and will also require a new system of ethics, a sustainable ethic.

32. Sustainable ethics holds that the Earth has a limited supply of resources and that not all of these resources are for humankind. It asserts that other species share the Earth's wealth and have a right to prosper as much as we do. It also holds that humans are a part of nature and that the key to success is through cooperation with nature, not domination. The core value of the new ethical proposition is that humans and nature are inextricably linked: we are dependent on natural systems, and our future depends on a healthy, well-functioning ecosystem.

33. One of the most influential thinkers on ethics was the late Aldo Leopold, a wildlife ecologist who proposed a land ethic in his writings 70 years ago. The land ethic holds that humans are a part of a larger community that includes the soil, water, plants, and animals—in short, the land. Recognition that humans are a part of nature, Leopold argued, would shift the role of human beings from "conqueror...to plain members and citizens of it."

34. This book proposes five operating principles for putting sustainable ethics into action: (1) conservation, (2) recycling, (3) renewable-resource use, (4) restoration, and (5) population control.

35. One of the ethical underpinnings of sustainable ethics is the notion of intergenerational equity. It asserts that the

present generations hold the Earth in common with all generations, past and future, who share certain rights and obligations. Most important, we have the right to benefit from the Earth—to profit from its riches and enjoy its beauty. We also have an obligation to protect the Earth for future generations so that they may benefit as well.

36. Creating a sustainable ethic will require worldwide efforts on the part of citizens, teachers, government officials, businesspeople, and others.

37. Critical thinking is also essential to building a sustainable future. We define critical thinking as a means of analyzing information to distinguish between beliefs (what people believe is true) and knowledge (facts supported by scientific observations).

38. Critical thinking is used to analyze the results of scientific research, environmental issues, and proposed solutions.

39. Table 2.1 summarizes six critical-thinking rules.

40. Systems thinking is vital to critical thinking. It requires us to think in terms of whole systems—human and natural—and to consider the long term.

Key Words and Phrases

Biocentric View	Index of Sustainable
Breakeven Point	Economic Welfare (ISEW)
Brownfields	Inputs
Command Economy	Intergenerational Equity
Control Group	Intragenerational Equity
Cost-Benefit Analysis	Land Ethic
Critical Thinking	Law of Diminishing
Directive Principles	Returns
Discounting	Law of Supply and
Ecocentric View	Demand
Ecological Justice	Life Cycle Cost
Economic Externalities	Macroeconomics
Economic Incentives	Market Economy
Economics	Marketable Permits
Environmental Equity	Microeconomics
Environmental	Natural Capital
Justice	Operating Principles
Epidemiology	Opportunity Cost
Ethics	Outputs
Experimental Group	Ozone Layer
Experimental Variable	Pollution Control
Frontier Ethics	Pollution Prevention
Full-Cost Pricing	Present Value
Genuine Progress	Replacement Cost
Indicator (GPI)	Subsistence Economy
Green Tax	Sustainable Economics
Gross Domestic Product	Sustainable Economy
(GDP)	Sustainable Ethics
Gross National Product	Systems Thinking
(GNP)	Theory
Hunter-Gatherer	Time Preference
Economy	Tradable Permits
Hypothesis	User Fee

Critical Thinking and Discussion Questions

1. What does it mean to say that the Earth's ecosystems are the infra-infrastructure of our society? How different would our world be if every business owner, every citizen, and every government official understood this point?

2. Describe the law of supply and demand. What are its weaknesses from an environmental standpoint? How can these be corrected?

3. What is an economic externality? How can such costs be internalized? Do pollution control devices internalize externalities? How do pollution prevention techniques compare with pollution control techniques?

4. Critically analyze the following statement: "The gross national product is not a measure of our welfare, despite what economists say." Do you agree or disagree with this statement? Why?

5. Outline general measures that could help us build a sustainable economy. What forces or factors serve as obstacles to these ideas?

6. How do time preference, opportunity costs, and discounting influence resource management and environmental decisions?

7. What is discounting, and how is it used? What are its strengths and weaknesses?

8. Not all opportunity costs result in a loss of present income. Explain what this statement means.

9. Draw a graph of pollution costs and the cost of pollution control. What is the breakeven point? Is the breakeven point an accurate representation of the equalization of costs and benefits? Why or why not? If not, how can it be adjusted?

10. Critically analyze the assertion: "Environmental protection is bad for the economy."

11. What is meant when one says that environmental quality is an important economic factor?

12. How could natural resource inventories, green taxes, and full-cost pricing help correct flaws in the economic system?

13. What is meant by harnessing market forces to protect the environment? Give some specific examples.

14. Outline your beliefs regarding humans and the environment. Do they correspond more closely with those of frontier ethics or sustainable ethics?

15. Where have your values come from?

16. In what ways are ethics and economics complementary forces? In other words, how can these two be used together to foster a transition to a sustainable society?

17. Describe the tenets of sustainable ethics. Are these values utilitarian?

18. Define the following terms: intergenerational equity, intragenerational equity, social justice. Do you agree with these principles? How can they be put into action?

19. What is critical thinking? Describe each of the rules outlined in the chapter.

20. Define the following terms: hypothesis, experiment, and theory.

21. What is systems thinking? Give some examples of ways you use this technique when analyzing an issue.

Suggested Readings

Atkisson, A. 1999. *Believing Cassandra: An Optimist Looks at a Pessimist's World*. White River Junction, VT: Chelsea Green. An entertaining and thoughtful book on sustainability.

Ayres, R. U. 2001. How Economists Have Misjudged Global Warming. In *World-Watch* 14 (5): 12–25. Extremely important reading about the myths of economic strength and controls on emissions of carbon dioxide from U.S. industries.

Berry, W. 1987. *Home Economics*. Berkeley: North Point Press. Thoughtful collection of essays on wise stewardship.

Brown, L. R., C. Flavin, and S. Postel. 1999. *Saving the Planet: How to Shape an Environmentally Sustainable Economy*. New York: Norton. Contains considerable information on sustainable economics.

Chapman, A. R., R. L. Petersen, and B. Smith-Moran. 1999. *Consumption, Population, and Sustainability: Perspectives from Science and Religion*. Washington, DC: Island Press. Important reading on ethics.

Chiras, D. D. 1992. *Lessons From Nature: Learning to Live Sustainably on the Earth*. Washington, DC: Island Press. See Chapters 2 and 3 for a detailed discussion of sustainable ethics.

Chiras, D. D. 1992. Teaching Critical Thinking in the Biology and Environmental Science Classrooms. *American Biology Teacher* 54 (8): 464–468.

Chiras, D. D. (Ed.). 1995. *Voices for the Earth: Vital Ideas from America's Best Environmental Books*. Boulder, CO: Johnson Books. A collection of essays on sustainability with several important contributions from ecological economists.

Cobb, C., T. Halstead, and J. Rowe. 1995. If the GDP Is Up, Why Is America Down? *Atlantic Monthly* 276 (4): 59–78. Important reading on economic indicators.

Cole, K. C. 1985. Is There Such a Thing as Scientific Objectivity? *Discover* 6 (9): 98–99. Insightful look at science and the scientific method.

DeVall, B. 1998. *Simple in Means, Rich in Ends: Practicing Deep Ecology*. Salt Lake City: Peregrine Smith. Important discussion of deep ecology and ways to put reverence for nature into action.

Evans, D., and J. A. Kruger. 2007. Where Are the Sky's Limits? Lessons from Chicago's Cap-and-Trade Program. *Environment* 49 (2): 18–35. Looks at an innovative market approach to reducing pollution.

Ford, A. 1999. *Modeling the Environment: An Introduction to System Dynamics Modeling of Environmental Systems*. Washington, DC: Island Press. Introduction to system dynamics designed to help students learn principles of modeling and develop models themselves.

Gardner, G. 2001. Accelerating the Shift to Sustainability. In *State of the World 2001*, ed. L. Starke. New York: Norton. Looks at the role of business, government, and citizens in shaping a sustainable future.

Gardner, G. 2001. The Case for Restraint. *World-Watch* 14 (2): 12–18. Excellent discussion of the differences between development and economic wealth.

Goodstein, E. 1999. *The Trade-Off Myth: Fact and Fiction about Jobs and the Environment*. Washington, DC: Island Press. Examines the deeply held belief that environmental protection threatens jobs.

Gowdy, J. 1998. *Limited Wants, Unlimited Means: A Reader on Hunter-Gatherer Economics and the Environment*. Washington, DC: Island Press. A detailed look at subsistence economies mentioned in the chapter.

Hanna, S., C. Folke, and M. Karl-Goran. 1996. *Rights to Nature: Ecological, Economic, Cultural, and Political Principles of Institutions for the Environment*. Washington, DC: Island Press. Discusses rights, rules, and responsibilities that guide and control human use of the environment.

Karl, H. A., L. E. Susskind, and K. H. Wallace. 2007. A Dialogue, Not a Diatribe: Effective Integration of Science and Policy through Joint Fact Finding. *Environment* 49 (1): 20–34. Examines the ways decisions can be made with imperfect information—ways to reach good decisions based on science and politics.

Krishnan, R., J. M. Harris, and N. R. Goodwin (eds.). 1995. *A Survey of Ecological Economics*. Washington, DC: Island Press. A useful collection of writings on ecological economics.

Leopold, A. 1966. *A Sand County Almanac*. New York: Ballantine. Collection of essays on nature and conservation.

Leopold, A. 1999. *For the Health of the Land: Previously Unpublished Essays and Other Writings*. Edited by J. Baird Callicott and Eric T. Freyfogle. Washington, DC: Island Press. A collection of Leopold's writings that focuses on the notion of land health and ways to protect private land.

Meadows, D. 1991. *The Global Citizen*. Washington, DC: Island Press. Collection of short, insightful essays on a variety of environmental topics, especially values.

Meyers, N., and J. Kent. 2001. *Perverse Subsidies: How Misued Tax Dollars Harm the Environment and the Economy*. Washington, DC: Island Press. Very important reading on the impacts of government subsidies on environmentally unfriendly activities.

Nash, R. F. 1989. *The Rights of Nature: A History of Environmental Ethics*. Madison: University of Wisconsin Press. In-depth analysis of the history of environmental ethics.

Perlman, J. E., and M. O. Sheehan. 2007. Fighting Poverty and Environmental Injustice in Cities. In *State of the World 2007*, ed. L. Starke. New York: Norton. A closer look at environmental justice issues and ways to solve them.

Power, T. M., and R. Barrett. 2001. *Post-Cowboy Economics: Pay and Prosperity in the New American West*. Washington, DC: Island Press. Important reading for those who want to understand some of the myths propagated by anti-environmental factions in the United States.

Renner, M. 2001. Employment in Wind Power. *World-Watch* 14 (1): 22–30. Fascinating look at the employment boom in the wind energy sector, showing that environmentally sound ideas make good economic sense.

Repetto, R., and W. B. Magrath. 1988. *Natural Resources Accounting*. Washington, DC: World Resources Institute. An investigation into the true economic value of our natural resources.

Rock, M. T., and D. P. Angel. 2007. Grow First, Clean Up Later? Industrial Transformation in East Asia. *Environment* 49 (4): 8–19. Looks at a key issue that plagues our world: the misunderstanding that development and environmental protection must go hand in hand.

Rolston, H. 1987. *Environmental Ethics: Duties to and Values in the Natural World*. Philadelphia: Temple University Press. Explains the rights of other creatures.

Sarewitz, D., R. A. Pielke Jr., and R. Byerly Jr. 2000. *Prediction: Science, Decision Making, and the Future of Nature*. Washington, DC: Island Press. Examines predictive science and how it is used and can be better used in making policy.

Schumacher, E. F. 1977. *Small Is Beautiful: Economics as if People Mattered*. New York: Harper and Row. One of the best books ever written on the subject of a sustainable society and new ethical systems.

Sterner, T., et al. 2006. Quick Fixes for the Environment: Part of the Solution or Part of the Problem? *Environment* 48 (10): 20–30. A look at the need for root-level solutions.

 ## Web Explorations

Online resources for this chapter are on the World Wide Web at: **http://www.prenhall.com/chiras** *(click on the Table of Contents link and then select Chapter 2).*

CHAPTER 3

LESSONS FROM ECOLOGY

To understand environmental problems and their solutions, you must have a firm understanding of ecology. **Ecology** is a field of study that attempts to uncover important relationships in the living world—more specifically, relationships between organisms and their environment. Ecologists study how organisms interact with one another and how they interact with their environment. A good understanding of basic ecological concepts will help you appreciate the problems facing conservationists, environmentalists, and resource managers as they grapple with complex resource management issues. It will also help you understand how problems in human systems—for example, transportation, housing, and waste management—might be solved and how society might be nudged onto a sustainable path.

3.1 Levels of Organization

To begin to develop our understanding of ecology, it helps to begin by examining one of the outstanding characteristics of living organisms: their organization. In ascending order of complexity, the organizational levels of an organism are the atom, molecule, cell, tissue, organ, and organ system. Although ecologists are concerned with each of these levels, their attention focuses primarily on levels of organization above that of the organism: the population, community, and ecological system or ecosystem (Figure 3.1). That is to say, they examine interactions occurring in populations, communities, and ecosystems (defined shortly). Let's consider each one.

Population

When laypersons use the term **population,** they are invariably referring to the number of humans in a given locality—for example, the number of people in Michigan or Florida. Ecologists, however, use the term to refer to the number of organisms, human or nonhuman, within an area. They may talk about (and study) the population of white-tailed deer in New York or in an entire country, for instance. Or they may refer to populations of white pine in a forest, pike in a river, or fleas on your pet dog or cat (Figure 3.1A). Some ecologists specialize in studying populations—for example, researching factors that influence their growth and decline. Wildlife ecologists are especially concerned about populations.

Community

Ecologists spend much of their time studying biological communities. A **biological community** is defined as all of the living organisms occupying a given locality. As you can see from Figure 3.1B, a biological community consists of two or more populations that inhabit a certain location. Communities typically contain populations of plants, animals, and microorganisms that interact in many ways. These interactions are often crucial to the survival of the whole. For this reason, more and more resource management plans are

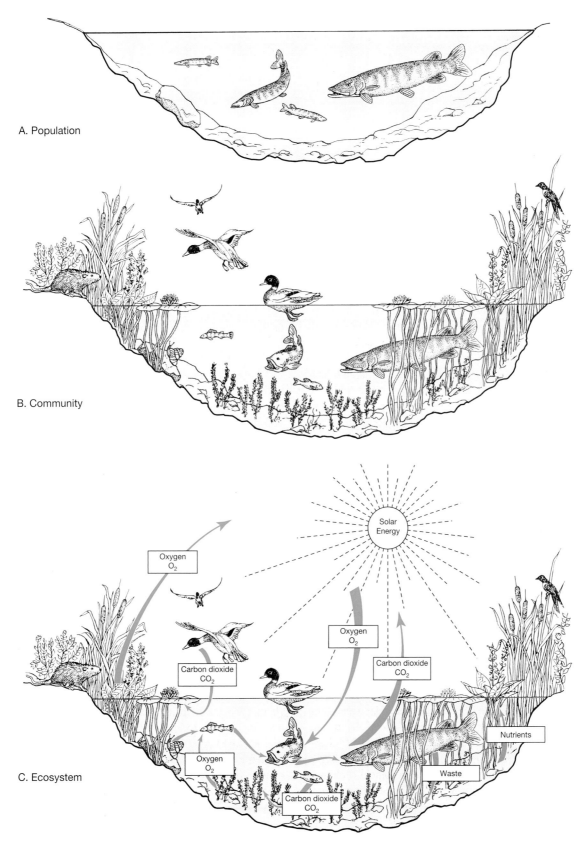

FIGURE 3.1 Levels of organization: (A) population, (B) community, and (C) ecosystem.

GO GREEN!

Work with your parents to create some backyard wildlife habitat. For advice, contact the National Wildlife Federation's Certified Wildlife Habitat program at http://www.nwf.org/backyard/.

being drafted to protect biological communities in a region, rather than focusing narrowly on one or two populations. Ecosystem management, mentioned in Chapter 1, requires community protection.

Ecosystem

You can't protect a community of organisms while ignoring its chemical and physical environment. Together, organisms and their chemical and physical environment constitute an ecological system, or **ecosystem** for short (Figure 3.1C). Therefore all ecosystems are said to consist of two interactive components, the living (biotic) and nonliving (abiotic).

Like so many things, ecosystems are often delineated artificially by humans. For example, an ecologist might study the ecosystem of a lake or the ecosystem of a meadow. As we will see in later chapters, most measures applied by modern science-based resource managers are forms of ecosystem manipulation. A good example is the removal of snow cover from an icebound lake to prevent the winter kill of fish. This permits sunlight to penetrate the ice, making it available to aquatic plants for photosynthesis. The resultant increase in dissolved oxygen helps protect fish populations and prevents wintertime fish mortality.

In reality, all ecosystems are part of one large global ecosystem, the **biosphere** or **ecosphere.** The biosphere extends from the bottom of the oceans to the tops of the tallest mountains, although life is scarce at the extremes. Most life forms exist in a much narrower range where conditions are more conducive to survival.

In the living world, organisms may move from one ecosystem to another, whether immediately adjacent or thousands of miles apart. Many waterfowl, for example, spend their summers in Alaska and northern Canada but winter in the warmer waters of the southern United States. Chemical nutrients and energy, which we'll discuss shortly, also move from one ecosystem to another. This is accomplished by natural biological (animal migration), meteorological (dust storms and hurricanes), hydrological (flowing rivers), and geological (volcanic eruptions) processes. Thus, topsoil containing nutrients from an Oklahoma wheat field may be washed by spring rains into a nearby stream, which flows into larger rivers that eventually flow into the ocean. Phosphorus originating in deep marine sediments may be transferred to terrestrial ecosystems in bird droppings from seabirds that feed on fish, which in turn are nourished by crustaceans that eat algae, which absorb phosphorus from the ocean. Because of this, all ecosystems on Earth are linked, and ecosystem disruption can have far-reaching effects.

3.2   Scientific Principles Relevant to Ecology

With this brief overview of the organization of life and the living world, we turn our attention to scientific principles vital to our understanding of ecology. We begin with a discussion of matter and energy—two key components of the systems you will study in this book.

The Law of Conservation of Matter

Organisms and their physical environment are all composed of matter. **Matter** is defined as anything that occupies space and is perceptible to the senses. Ice cream, for instance, is a form of matter, as are water and lead.

Matter exists in two basic forms: organic and inorganic. Organic matter refers to substances composed primarily of carbon, hydrogen, and oxygen. Proteins in the cells of your body or the starch in rice are two forms of organic matter. Inorganic matter consists of mineral-based substances such as salt or aluminum ore.

Matter can be altered in many ways, but one thing scientists have learned is that it can be neither created nor destroyed. That is, you cannot magically create matter, nor can you destroy it. All you can do is change it from one form to another. This important law is known as the **law of conservation of matter.**

Restated, *the law of conservation of matter states that although matter can be changed from one form to another, it can neither be created nor destroyed by ordinary physical and chemical means.* As an example, the flesh of one organism can be consumed by another, at which time it becomes part of the second organism. The first organism hasn't been destroyed; it has been eaten, and its components have been reassembled to make parts of a new organism. By the same token, the molecules in your body are not new creations; they're made from atoms and molecules that have been around since Earth formed. They originated in outer space. On Earth, these atoms and molecules are recycled over and over again to support new generations of living things. Without this recycling process, life could not exist.

The law of conservation of matter has many important implications. First, it helps us understand the massive pollution problems facing the world. The United States and many other countries such as China and India, for example, are consuming many of their natural resources (matter) at a record-breaking pace. The matter does not disappear, however. Some of it is converted into useful products. When they lose their usefulness or fall out of fashion, these products become waste—for example, the solid waste accumulating in landfills throughout various nations.

The law of conservation of matter reminds us that waste remains, in one form or another, forever. Some ecologists remind us that when we throw something away, there really is no "away." Things we discard do not miraculously vanish. They're always somewhere—and they may come back to haunt us. For example, if a landfill is not properly designed, some of the waste may leach into the groundwater that we drink. Not even burning trash truly gets rid of it. Incineration certainly reduces the volume of waste to a few ashes, but the combustion process generates large amounts of smoke and gases that linger in the atmosphere or rain down upon the land.

The Laws of Energy

The leap of a tiger, the beat of a heart, the scream of an eagle, the turn of a wheel, the dip of a canoe paddle—all of these seemingly diverse events have something in common: they require energy. What is energy?

Physicists define **energy** as the ability to do work or cause change. Lifting a mountain bike to climb a steep path may be your idea of having fun, but to a physicist, it's work.

Energy plays a vital role in the living world. It is used by organisms to make molecules, and it provides motive force—that is, it allows organisms to move about in their environment.

Energy can be found in a highly concentrated state, such as in foods or fuels such as gasoline. It can also be found in more dispersed or disorganized states, such as heat (also known as thermal energy). Energy in a concentrated state can perform a great deal of useful work and is considered to be of high quality. Energy in the dispersed condition, by contrast, cannot perform as much work and therefore is considered to be of relatively low quality. Don't let these labels confuse you. Both forms are essential to life on Earth.

Energy exists in many different forms. The two main types are potential energy and kinetic energy. **Potential energy** is stored energy. Coal, gasoline, jet fuel, and food contain lots of potential energy. The potential energy in fuels is released when it is burned. The potential energy of food is released when it is broken down in cells of the body. **Kinetic energy** is the energy of motion. A moving car has kinetic energy, as does a ball flying through the air or a hammer pounding a nail.

Physicists, engineers, and ecologists apply other labels to different forms of energy. You will read about chemical energy, mechanical energy, thermal energy, nuclear energy, and so on as you work your way through this book.

First Law of Energy Energy, like matter, is governed by several laws, known as the **laws of thermodynamics** or, simply, the **energy laws.** As you shall soon see, these laws profoundly influence the structure of the biosphere—and the living organisms themselves. The **first law of energy** states that *energy can neither be created nor destroyed; it can be converted from one form to another.* Just like matter! Energy locked in the fuel molecules in gasoline, for instance, is released in the car's engine, but it is not destroyed. It is converted into heat and mechanical energy that propels the car along the highway.

Just so you really understand the concept of energy conversions, take a look at Figure 3.2. On the right side of the figure is a lamp. As illustrated, the light given off by the lamp comes from a power plant (generator). The power plant burns coal. When coal is burned at a power plant, the heat (thermal energy) it produces boils water, creating steam that turns a turbine. The turbine is attached to an electrical device known as a generator, so named because it generates electricity by rotating a magnet through a huge mass of copper wires. But where did coal get its energy?

The energy contained in the coal comes from sunlight, more specifically sunlight that shone on Earth several hundred million years ago. This energy was captured by plants in a process known as **photosynthesis** (defined more fully later). During photosynthesis, plants use sunlight, water, and carbon dioxide to produce organic molecules. The chemical bonds in these molecules contain energy that can be used by animals that feed on the plants. In ancient times, the plants growing around swamps that were not eaten were often buried in sediment. Heat and pressure from the accumulating sediment converted the plant matter to coal. Today, when this coal is burned by the electrical power industry, the sunlight energy locked in the organic molecules in the plant matter is released. That is, it is converted into heat energy, which is used to generate steam in power plants.

This example shows the true origin of much of the fuel that powers our society: the sun. It also illustrates the conversion of energy from one form to another. In fact, in this sequence, energy undergoes six conversions.

Take a moment to study Figure 3.2 again to see the changes. As you do, you may notice something else. Note that heat is lost at each conversion. This brings us to the second law.

Second Law of Energy According to the **second law of energy,** whenever energy is converted from one form to another, a certain amount is lost in the form of heat. Heat loss is as inevitable as death and taxes. Put another way, no conversion is 100% efficient. In fact, most conversions are closer to 20% to 40% efficient.

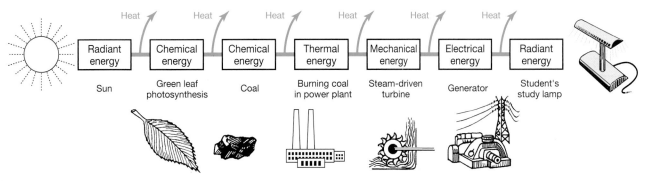

FIGURE 3.2 Different forms of energy. Energy can be changed from one form to another. However, with each change, a certain amount of energy is lost as heat.

When we say that "heat is lost during an energy conversion," what we really mean is that heat is given off. Heat isn't lost per se. It can even be captured to perform additional work. For example, heat given off by a boiler designed to provide hot water to heat an office building can be captured and used to boil water to make electricity. This process is known as cogeneration and is discussed in Chapter 22.

The important point in all of this is that, during the successive energy conversions, heat is given off, and the total amount of high-quality energy tends to decrease, so overall there is less high-quality energy available to do work. Let's test your newfound knowledge. Which would provide more energy for cooking food: natural gas burned directly at the burner of a gas range or electricity made from natural gas burned at an electrical generating plant? If you answered natural gas, you're right, because there's one less conversion.

Before we leave the topic, let's examine another common device, the automobile. You may be surprised to learn that only about 25% of the high-quality chemical energy in the gasoline "consumed" by most car engines is converted into high-quality kinetic energy that propels you and the car along the highway. About 75% of the energy in the gasoline is converted to low-quality heat energy—useless as power but nice to have for warming your car on a trip on a cold winter day. Eventually this heat energy radiates from the car and cannot be used again.

The second law of energy also operates in living systems. For example, during photosynthesis, high-quality solar energy is converted to high-quality chemical energy by green plants. When you eat plants, whether beans or bananas, your body converts the chemical energy of the food into the high-quality chemical energy stored in the cells of your body. Sooner or later, your body converts this energy into kinetic energy (energy of motion) that powers such life-sustaining activities as breathing, the beating of the heart, and the muscular contractions of such organs as the stomach. However, here again, with each conversion, from solar energy to chemical energy and from chemical energy to kinetic energy, a certain amount of energy is lost as heat. Of course, the heat is not wasted—it keeps our body temperature at about 37°C (98.6°F). Eventually, however, all of this heat radiates into the surrounding atmosphere and then into outer space. It isn't lost, but it is no longer available to perform work here on Earth.

Entropy   This brings up another important concept, known as **entropy** (en´-trow-pee). Scientists use the term to refer to matter and energy. More specifically, they use it to refer to the degree of disorder in a system (Table 3.1). All systems move toward maximum disorder or maximum entropy. Think of your room as an example. It grows disordered slowly but surely over time.

TABLE 3.1	Concept of Entropy	
High Organization, Low Entropy		**Low Organization, High Entropy**
Gasoline	→	Movement of car
Electricity	→	Light and heat from study lamp
Volcanic eruption	→	"Rain" of volcanic ash
Candy bar (sugar)	→	Body heat

Biological systems tend to move toward disorder as energy is continuously lost from the bodies of organisms in the form of heat. If the energy in a particular system is largely in a dispersed state—that is, it's in the form of heat—we say the system shows a high degree of disorder.

Biological systems also tend toward disorder with respect to matter. But how is it that living organisms retain their structure, their organization, and their order, all of which are so essential to the continuation of life? They retain their organization because they receive a constant input of energy to keep making molecules needed to maintain the structures of their bodies—to repair and maintain themselves. Animals receive high-quality energy from food they eat. The energy in the food comes from plants that obtain their energy from sunlight.

Matter and Energy Laws: What Can We Learn from Them?   The activities of all organisms—beans and bananas, hummingbirds and humans—are under the control of the basic laws of matter and energy we just described. These laws also control the activities of all ecosystems on Earth. In addition, they affect a whole series of environmental problems. These laws provide us with a key to understanding (1) the urgency of environmental problems facing all of us and (2) ways those problems can be solved or brought under control.

The law of conservation of matter, for example, tells us that there is no "away." What we throw away ends up somewhere. The carbon molecules in the fossil fuels we burn in our cars, jets, factories, and power plants end up in the atmosphere. Here it has the power to alter climate, returning the planet to a much hotter era, similar to that when the dinosaurs roamed the Earth, when the plant matter that was converted to coal and oil and natural gas was growing. Simply put, all that carbon that was in the atmosphere back then and was responsible for the warmer climate is now being released by the combustion of fossil fuels and may well be causing the Earth to warm up again.

The first law of thermodynamics tells us that energy is neither created nor destroyed, but the second law says that energy is degraded at each conversion. For a society that relies primarily on high-quality energy from finite supplies of fossil fuels, the laws of energy tell us that each time we burn fuel, we're depleting our supply. In other words, we're using up valuable supplies that cannot be replenished. Energy cannot be recycled, as it all eventually dissipates into space as heat. Sooner or later, we'll have to find new sources.

3.3 The Flow of Energy Through Ecosystems

The sun is the source of virtually all energy in the biosphere. It heats the Earth and causes winds that turn giant machines that generate electricity. It evaporates water, allowing for rainfall and stream flow. The energy of flowing water can be tapped by dams equipped with turbines and generators to produce electricity. More important for living organisms, sunlight is captured by plants and stored in the molecules they make. This energy is used by plants and a wide assortment of animals and even most microorganisms.

The sunshine that warms you on your way to class reached your skin only eight minutes after leaving the sun's surface, roughly 155 million kilometers (93 million miles) away. The sun releases many forms of energy. As shown in Figure 3.3, the visible light (sunshine) we are familiar with forms only a small portion of the sun's total energy emission, or radiant energy, known as the **electromagnetic spectrum.**

The types of energy represented in the spectrum range from low-energy radio waves to high-energy gamma waves. Each type of electromagnetic radiation has two vital characteristics: wavelength and energy level. Let's begin with wavelength.

Electromagnetic radiation from the sun is transmitted in the form of waves. They are like the waves in the ocean. The distance between two successive wave peaks is known as the **wavelength.** As shown in Figure 3.3, radio waves have a long wavelength. The wavelength of visible light is intermediate. Gamma rays have a relatively short wavelength.

Energy levels also vary. Radio waves are low energy. Light is intermediate. Gamma rays are the highest-energy radiation the sun produces.

Although the sun produces a wide range of electromagnetic radiation, visible light and heat (infrared radiation) are generally the most useful to living organisms. Visible light is detected by animals (although some can perceive ultraviolet light as well). Some colors of visible light are absorbed by plants during photosynthesis. Gamma rays, X-rays, and ultraviolet radiation from the sun, however, are all harmful. We'll learn more about them in later chapters.

Solar Energy Flow

As shown in Figure 3.4, not all of the sun's radiant energy makes it to Earth's surface. In fact, 32% is reflected back into space by dust and clouds in the atmosphere and from Earth's surface—either from land, water, soil, or vegetation. This reflectivity is known as the **albedo** (al-bee´-doe).

As shown in Figure 3.4, 67% of the incoming solar energy is absorbed by Earth's atmosphere, land, water, and vegetation. It is converted to heat. Although this heat will eventually dissipate into space, it performs some vital functions before it escapes. For example, heat from absorbed sunlight warms Earth's surface (both land and water). It also causes evaporation of water from land and water. The water, of course, is later deposited as rain and snow. Heating of Earth's surface also produces winds, waves, and weather. Interestingly, only a tiny fraction of the sunlight striking Earth is used by

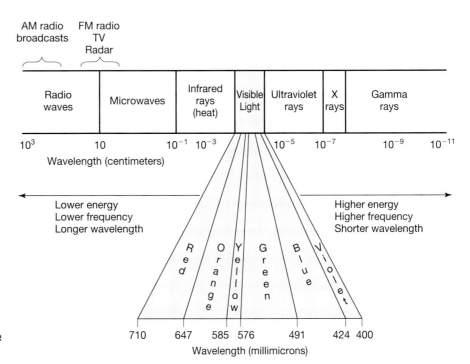

FIGURE 3.3 The electromagnetic spectrum. The sun produces many different forms of radiation, some useful, some potentially dangerous.

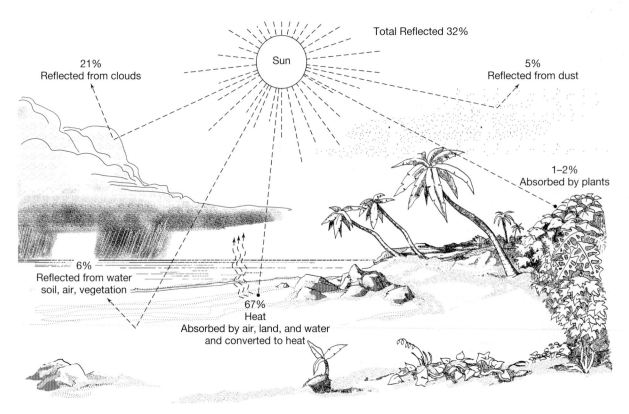

FIGURE 3.4   The global flow of energy. Virtually all energy that strikes Earth escapes as heat. Only a tiny portion of the sun's energy is absorbed by plants for photosynthesis.

photosynthetic organisms—somewhere around 1% to 2%, according to some estimates.

Solar energy flow and heat loss can be altered by human activities. For example, overgrazing of rangeland and pastures can result in a loss of vegetative cover and therefore can decrease albedo, causing Earth's surface to heat up. (Less light is reflected back into space; more sunlight is absorbed by the bare ground.) Tropical deforestation has a similar effect. Pollutants in the atmosphere can also alter these energy flows. Over 100 years ago, scientists found that certain naturally occurring chemicals in the atmosphere, such as carbon dioxide molecules and water vapor, retard the release of heat into outer space. Like a giant blanket surrounding Earth, these chemicals cause heat to be retained in Earth's atmosphere, creating a climate suitable for life. In fact, without them, Earth would be much, much colder and virtually uninhabitable.

In the past 100 years, however, humans have burned massive quantities of fossil fuels. This, in turn, has resulted in a substantial increase in concentration of carbon dioxide in our atmosphere. There's a growing body of scientific evidence that suggests that the release of massive quantities of carbon dioxide (and other factors) principally from human activities is causing the temperature of Earth's atmosphere and oceans to increase, a phenomenon known as **global warming.** This, in turn, could have dramatic effects on ecosystems, economies, and human civilization, as you shall see in Chapter 19. As hard as it is to believe, humans may be profoundly influencing solar energy flow and the climate of the entire planet.

Photosynthesis and Respiration

As noted earlier, all of the energy that powers the living world, our biosphere—from the growth of a cabbage to the beating of the human heart—can be traced back to its original source, the sun. Sunlight energy is captured by plants during photosynthesis. During photosynthesis, plants use solar energy to convert carbon dioxide and water into sugar, mostly glucose. With a few minor exceptions, this process can occur only in the presence of **chlorophyll,** a green pigment found in algae, some bacteria, and plants (Figure 3.5). Chlorophyll acts like a solar collector in the cell, gathering up sunlight energy. The general equation for photosynthesis is

$$\text{solar energy} + 6CO_2 + 6H_2O \rightarrow C_6H_{12}O_6 + 6O_2$$

$$\underset{\substack{\text{(carbon} \\ \text{dioxide)}}}{} \quad \underset{\text{(water)}}{} \quad \underset{\text{(sugar)}}{} \quad \underset{\text{(oxygen)}}{}$$

Photosynthesis produces two important substances, sugar (glucose) and oxygen. Glucose is used as a source of energy by the plant itself and by animals feeding on plants and one another.

Some of the oxygen produced by photosynthesis is used directly by the plant, but the rest passes from the leaf through microscopic pores into the atmosphere. Here the oxygen may be used by other organisms, from bacteria to humans. As you soon shall see, oxygen is used to help us break down sugars to produce energy.

The breakdown of sugars by plants and animals to produce energy is called **cellular respiration.** The general equation for respiration is

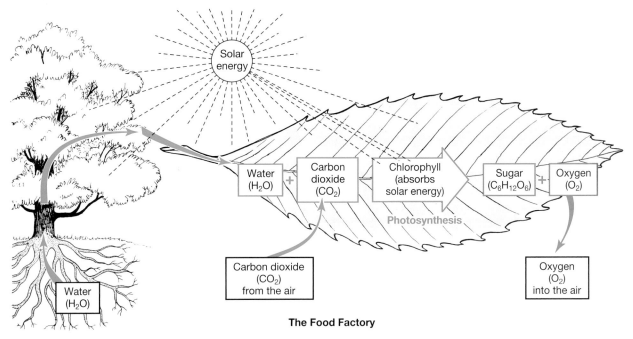

The Food Factory

FIGURE 3.5   The food factory. Leaves absorb carbon dioxide and sunlight. Water is taken up by the roots. Solar energy powers the process of photosynthesis by which water and carbon dioxide are used as raw materials in the production of sugar.

$$C_6H_{12}O_6 + 6O_2 \rightarrow 6CO_2 + 6H_2O + \text{energy}$$
$$\text{(sugar)} \quad \text{(oxygen)} \quad \text{(carbon} \quad \text{(water)}$$
$$\text{dioxide)}$$

If this formula appears familiar, it should. It is exactly the opposite of the equation for photosynthesis. In the living world, then, plants produce glucose and oxygen, which are inputs of cellular respiration. The products of cellular respiration, in turn, become the raw materials of photosynthesis, creating a nifty recycling of materials that is vital to the continuation of life.

An understanding of photosynthesis makes it clear why plants are so important to us and why it is important to protect Earth's terrestrial and aquatic plant life. Unfortunately, many plants are in trouble. Large tracts of forest have been cleared to produce timber or make way for mines, farms, or human settlements, wiping out native vegetation and decreasing the planet's ability to produce oxygen. Some ecologists are concerned that the progressive contamination of the oceans with chemical pesticides and industrial wastes may sharply reduce the photosynthetic activity of marine (saltwater-inhabiting) algae and, over time, diminish Earth's supply of atmospheric oxygen—a harmful consequence for all animals, including humans.

Primary Production and Net Production

Each year, terrestrial and aquatic plants produce an estimated 243 million metric tons of organic matter. This is known as the biosphere's **primary production.** Organic matter serves as a source of organic building blocks and as a source of energy for other organisms. But not all of the organic material that plants produce is available to other organisms. Plants themselves consume some of the material in cellular respiration occurring in their own cells. They use it for growth (for example, root growth) and general cell maintenance (maintenance of cell walls, for instance). For this reason, ecologists distinguish between **gross primary production,** the total amount of biomass produced, and **net primary production,** that which remains after plants use their share to provide food and energy for cellular processes. The equation NPP = GPP − R describes the phenomenon mathematically (R = cellular respiration) Remember, unlike green plants, animals are able to capture energy only by consuming other organisms.

Primary production varies considerably by ecosystem. The most productive are tropical rain forests. They produce on average about 2,200 grams of biomass per square meter per year. Deciduous forests like those of New England produce about half as much—about 1,200 grams per square meter per year. Evergreen forests produce about 800 grams per square meter per year, while farmland averages out at about 650 grams per square meter per year. Deserts produce only about 90 grams per square meter per year.

Food Chains and Food Webs

Now that you understand where food in the biosphere comes from, let's see what happens to it, looking first at food chains. A **food chain** is a feeding sequence in ecosystems. It illustrates who eats who and, in so doing, shows the path through which energy and nutrients move in ecosystems. Ecologists need to know about them to understand the relationships between organisms in the ecosystems they study. Let's consider a food chain that might operate in a wet meadow:

grass → grasshopper → frog → snake → hawk
(producer) (primary   (secondary (tertiary   (quaternary
             consumer)   consumer) consumer)   consumer)

$\underbrace{}$ $\underbrace{}$
herbivore   carnivores

In this food chain, shown in the center of Figure 3.6, grass is classified as a **producer** because it produces organic food molecules that nourish all other organisms. The remaining organisms that either feed directly on grass or feed on other organisms higher in the food chain are known as **consumers.** Consumers are "ranked" by their location in the food chain. The grasshopper in the food chain is known as a **primary consumer.** It is also classified as an **herbivore** because it is a plant eater. The flesh-eating frog, snake, and hawk are all classified as consumers, too, but they're referred to more specifically as **secondary, tertiary,** and **quaternary consumers.** Because they only eat other animals, they're also known as **carnivores.** Food chains like this one that begin with a photosynthetic organism—such as a green plant or algae—are known as **grazer food chains.**

In grazer food chains, plants are the main source of food. The chemical energy in plant material (derived from the sun) and the organic matter it generates during photosynthesis are available to all "members" of the food chain. The plant matter generated in an ecosystem, therefore, contains the energy that powers the activities of all organisms on Earth.

Another type of consumer, not represented in the food chain diagram, is the **detritus feeder** (deh-trite´-us). These organisms obtain their energy and nutrients from waste materials and the dead bodies of plants and animals—that is, **detritus.** Detritus feeders in a forest include microscopic organisms such as bacteria and fungi (mushrooms and molds), as well as macroscopic organisms such as maggots and termites. As a result of the activities of these organisms, large, complex molecules in plant and animal remains and waste products are broken down (decomposed) into smaller molecules such as nitrates. These compounds, in turn, are released into soil and water and are absorbed by plants, ensuring the continuation of life. Detritus feeders, therefore, are said to assist in nutrient recycling. In some shallow lakes, detritus is represented by decaying vegetation. This material is eaten by crayfish and snails, which in turn are consumed by fish.

Any food chain that has its base in the dead remains of plants and animals is known as a **detritus food chain** (Figure 3.7). Together, detritus and grazer food chains account for the movement of all nutrients and energy (except heat) in the biosphere. In an oak forest, roughly 90% of the **biomass** (dry weight of living organisms) eventually dies and enters detritus food chains. The remaining biomass is channeled through the grazer food chains. In the open water of a lake, the primary food chains are grazer food chains. There the floating algae represent the principal producers. Roughly 90 percent of the algal producers are consumed by small crustaceans, which in turn are eaten by fish.

Food chains are easy to understand and help shed light on ecological relationships, but in the real world, food chains virtually never exist as isolated entities. They're usually part of larger **food webs,** or more complex feeding relationships. Consider the food chain corn → pig → human. You might

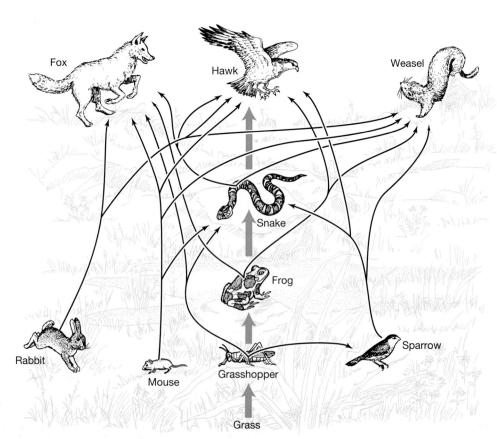

FIGURE 3.6 Food webs and food chains. In the center of the drawing is a food chain. It is really a part of several other food chains, together forming a food web.

FIGURE 3.7   Detritus food web of an oak woods, highly simplified.

assume by the diagram that a pig eats corn and, in turn, is eaten by humans. In reality, pigs eat much more. Pigs roaming the farm will eat rats, mice, insects, grubs, earthworms, baby chicks, grass, weeds, garbage, and even feces. Humans have a varied diet, too. The corn → pig → human food chain is really just part of a much more complex and interconnected food web.

A food web is a collection of interconnected food chains. A food web provides a more accurate picture of the feeding relationships and the flow of energy and nutrients in ecosystems. It illustrates the fact that, in ecosystems, energy and nutrients travel through multiple pathways. The grass → grasshopper → frog → snake → hawk food chain of a wet meadow discussed earlier, for example, is part of a more complex food web in nature (Figure 3.6). Even this food web is oversimplified. The complete food web of a wet meadow includes hundreds of species.

As a general rule, the greater the number of channels through which energy can flow, the greater the stability of the food web and the ecosystem. Why? The reason is simple. The more strands in a food web, the stronger it is. The loss of one strand is not felt. Let us explain by example. In the wet meadow described earlier, foxes prey on rabbits, mice, grasshoppers, sparrows, frogs, and snakes. Suppose, however, that the rabbit population was reduced because of adverse weather during the breeding period or because wild dogs moved into the area. Under those conditions, the fox population would shift to some (or all) of its alternative prey without suffering from nutritional hardship. Reduced predatory pressure on the rabbit population, in turn, might permit it to rebound quickly when breeding conditions become favorable. If the foxes did not have other prey, however, they might exterminate the rabbit population and then might die off themselves or migrate to other areas in search of prey.

As a general rule, the more organisms that are removed from a food web, the more vulnerable it is. Ecologists refer to the loss of species in a food web as **ecosystem simplification.** Consider an example: Meadows contain many different species of plants, among them grasses, herbs, and wildflowers. Plowing the field and planting a single-species crop such as

corn or wheat represents an extreme form of simplification with many ecological ramifications. It not only destroys the plant life, it also eliminates most, if not all, of the animals that depend on those plants. The new single-species ecosystem is also much more vulnerable to insect and disease organisms. The Irish potato famine of the 1840s is an illuminating example of this phenomenon.

The potato was introduced to Ireland in the late 16th century. For many years, it was a staple of the Irish diet and yielded more calories per acre than any other crop. However, in 1845 a fungal disease called the potato blight began attacking Ireland's potato crop. The fungus spread quickly from field to field, causing massive destruction that lasted for the next five years. Because the Irish were highly dependent on only a few varieties of potato, which the fungus attacked, and because few alternative foods were available, more than 1 million people died of starvation or disease. Another 1 million Irish left the country. The overall effect of the blight was a 25% decline in Ireland's population in just five years—all due to ecosystem simplification brought on by the planting of huge crops of a single food that allowed disease organisms to proliferate.

Trophic Levels and the Pyramids of Energy and Biomass

Before we leave the concept of food chains and food webs and move on to nutrient cycles, we must examine a couple of additional concepts and terms. Let's begin with a food web in a lake ecosystem (Figure 3.8). As illustrated, each feeding level in a community is called a **trophic level** (trow´-fic; meaning to nourish). Thus, all the producers (algae, rooted plants, water lilies, cattails, and so on) form the first trophic level. All the plant eaters, known as primary consumers or herbivores, such as crustaceans and insects, form the second trophic level. The secondary consumers (carnivores), such as fish, feed on the primary consumers and form the next trophic level, and so on.

So far, our description of tropic levels has been greatly simplified for clarity. You must not get the idea that a given species of organism can belong to only one trophic level. For example,

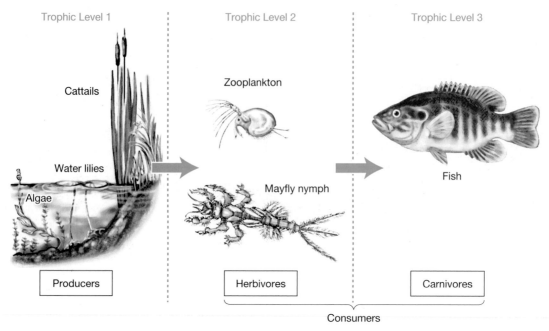

FIGURE 3.8   Trophic levels of a lake ecosystem.

although some fish are either exclusively herbivorous or exclusively carnivorous, other fish, such as the carp, are omnivorous—they feed on both plants and animals. Therefore, carp are both primary and secondary consumers in the food web.

One thing that ecologists have discovered is that not all the energy and biomass (organic matter) in a given trophic level is transferred to the next level. *In fact, as a general rule, only about 5% to 20% of the biomass and energy in one trophic level moves to the next. Ten percent is a typical transfer rate.* Thus, the primary consumers (herbivores) contain only 10% of the biomass and food energy of the producer (green plant)

level. Similarly, the secondary consumers (carnivores) contain only 10% of the biomass and food energy present in the primary consumer level. The energy contained in the various trophic levels of a food web, when graphically represented, forms a pyramid, the **energy pyramid** (Figure 3.9). The biomass forms a **biomass pyramid** (Figure 3.10).

A good example is the corn → pig → human biomass pyramid. For example, 1,000 kilograms (2,203 pounds) of corn are needed to produce 100 kilograms (220 pounds) of pork and ham, which in turn can be converted into 10 kilograms (22 pounds) of human flesh (Figure 3.10).

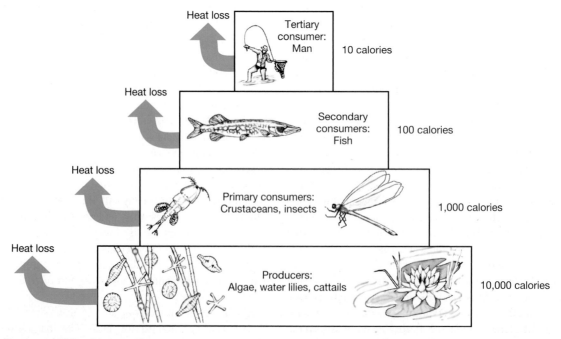

FIGURE 3.9   Energy pyramid for a lake ecosystem.

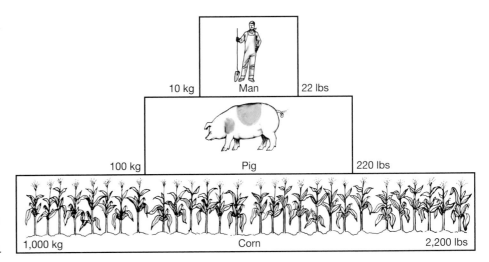

FIGURE 3.10 Biomass pyramid. It takes 1,000 kilograms of corn to produce 100 kilograms of pork needed to produce 10 kilograms of human biomass. Please note that this example assumes that the human eats a 100% meat diet, which is not typical. Most of us also receive calories and other nutrients from starchy foods and vegetables. This diagram is meant to simply illustrate the conversion factors.

Why does each consumer level contain only about 10% of the biomass and food energy of the lower level? At least four reasons account for the low efficiencies.

First, not all producers in a food web are consumed by primary consumers, and not all primary consumers are eaten by secondary consumers. Plants, for example, deter herbivores with spines, thorns, thick protective bark, irritating secretions, or foul odors. Animals such as insects, trout, and antelope fly, swim, or run from predators approaching at high speed. Some animals rely on protective coloration to escape detection by predators. This adaptation is well exemplified in certain species of moths whose colors make them almost invisible against the bark of a tree. Similarly, many ground-nesting birds, like pheasants and grouse, are extremely difficult to detect while they are incubating their eggs, even from a few feet away. If an organism escapes predators, it will eventually die from some other cause, such as disease or starvation. Its carcass will then be decomposed by bacteria and/or fungi and will enter the detritus food web.

The second reason is that not all parts of a food organism's body are digestible. For example, many herbivores such as porcupines, beaver, and deer cannot digest the cellulose in plant cell walls. Therefore, many cellulose fibers are voided from the body as waste.

Third, not all material is accessible. As a rule, the roots of plants are not eaten by herbivores and thus retain much of the energy captured from the sun.

Fourth, the low efficiency in energy transfer can be partially explained by the fact that all organisms respire. You will recall that cellular respiration is the process by which organisms oxidize, or burn, energy-rich organic compounds to release the energy they need to power their life activities. Cellular respiration itself is a relatively inefficient process. For example, only about 40% of the energy in a given sugar molecule may actually be put to work in an organism. The remainder is largely converted to heat that escapes into the environment.

Biomass and energy pyramids may, in some instances, be upside down or inverted (Figure 3.11). This occurs in some aquatic ecosystems where the food webs are based on billions of microscopic algae. The algae reproduce very rapidly, gobbling up nutrients and making new organic matter rapidly. Their populations will double every few days. The consumers (crustaceans), however, feed on the algae almost as rapidly as they are produced. As a result, the algal (producer) trophic level has less biomass than the crustacean (consumer) trophic level (Figure 3.11).

Pyramid of Numbers   In typical food webs based on small green plants or algae and ending in predators such as hawks and fish, the numbers of individuals at each trophic level are frequently greatest at the producer level (base), smaller at the herbivore level, and smallest at the carnivore level (top). When graphically represented, it forms a **pyramid of numbers.** Figure 3.12A illustrates the pyramid of numbers in the food web of 0.4 hectare (1 acre) of bluegrass in Michigan.

In food webs ending in parasites, the pyramid of numbers is upside down. This is evident in the dog → flea → protozoan food chain, where the dog is the food source of the fleas and fleas are the food source of the protozoans (Figure 3.12B).

Pyramid of Energy and Human Nutrition   Over the years, ecologists have consistently found that most terrestrial food chains have only three or four trophic levels. Why? The reason is found in the second law of energy, described earlier in the chapter. As you may recall, the second law says that during each energy conversion, some energy is lost as heat. Therefore, the total amount of high-quality energy decreases with each trophic level.

This simple phenomenon imposes a limit on the length of food chains in nature. For example, in the unusually long food chain clover → grasshopper → frog → snake → hawk, the hawk uses only 0.0001% of the solar energy captured by plants. There's simply not enough energy captured by producers to support many organisms past three or four trophic levels. The reasons cited for the 10% transfer of energy and biomass also come into play.

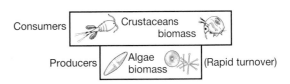

FIGURE 3.11   Inverted biomass pyramid of a lake ecosystem.

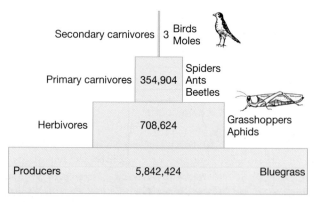

A. Pyramid of numbers

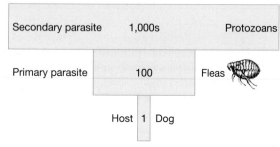

B. Inverted pyramid of numbers

FIGURE 3.12   Two kinds of number pyramids. The upper pyramid (A) is for a weedy field. The lower pyramid (B) is for a dog-parasite system, and is inverted.

Knowledge of these energy relationships enables us to understand the striking difference between Indian and American diets—the former based primarily on grains, the latter based on ample servings of meat. If, as an American, you live to be 70, you probably will have consumed 10,000 pounds of meat, 28,000 pounds of milk and cream, and thousands of pounds of grain, sugar, and specialty foods. However, a 70-year-old Indian probably would have eaten only 1% as much meat as you.

In overpopulated nations such as India and China, it is no accident that humans rely primarily on grains. In such instances, the food chain frequently has only two links: grain → people. Can you guess why?

As shown in Figure 3.13, herbivorous humans have replaced the herbivorous cow, sheep, or pig typically found in the

American diet. The reason is that there is more energy available to the Asian further down on the food chain, closer to the producer base.

Americans still live in a land of milk and honey (as well as pork chops and tuna). However, some futurists believe that the time may well come, as a result of the mushrooming U.S. population (expected to reach 351 million by the year 2025), when we will be faced with the ecological ultimatum: Shorten our food chains or tighten our belts. Eating lower on the food chain uses fewer resources and is also more healthful if done right.

Nutrient Cycles

Earlier in the chapter, you learned that matter is neither created nor destroyed. In the biosphere, scientists have found that matter cycles over and over again in global nutrient cycles, the subject of this section. To understand the importance of nutrient cycles, suppose that Figure 3.14A represents the mass of planet Earth. Now suppose that we draw another figure to represent the total amount of living substance (protoplasm) that has ever existed since the first living organism evolved 3.5 billion years ago (Figure 3.14B). This sphere, of course, would include your own body, as well as those of Stone Age people, ancient tree ferns, and dinosaurs. How big do you think this ball of protoplasm would be? You might be surprised to know that it would be considerably larger than Earth itself.

Of the 103 elements known to science, only about 35 contribute to the formation of protoplasm, or living tissue, of plants, animals, and microorganisms. Of these, carbon, hydrogen, oxygen, and nitrogen form about 96% of the human body. (You can use the letters COHN to remember them.) Elements such as sodium, calcium, potassium, magnesium, sulfur, phosphorus, zinc, iron, iodine, and many others occur in smaller amounts. All the elements forming the bodies of living organisms were

FIGURE 3.13   A theoretical situation in which Americans are exclusively carnivorous and the people of India are exclusively herbivorous. Note that the removal of one link in the food chain permits the Indians to get ten times as many food calories as the Americans receive from equal amounts of grain.

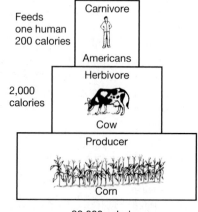

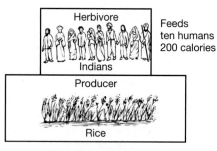

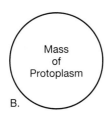

FIGURE 3.14 (A) Mass of Earth. (B) Mass of all organisms that have lived on Earth since the beginning. How can the mass of living protoplasm exceed that of Earth? There's only one possibility: matter is recycled.

derived either from the top few feet of Earth's crust (soil); from the rivers, lakes, oceans, and aquifers (groundwater); or from the atmosphere, the thin blanket of air that envelops Earth. If this is true, then the only way we can explain a cumulative ball of protoplasm being larger than the planet Earth is to assume that the elements forming the protoplasm were used over and over again. It's recycled. An atom of nitrogen that was once part of a dinosaur's jawbone might eventually have formed part of a professor's brain and, at some time in the future, may form part of an apple your great-grandson will have for a snack.

The circular flow of an element from the nonliving (abiotic) environment, such as rocks, air, and water, into the bodies of living organisms and then back into the nonliving environment once again is known as a **nutrient cycle.** The continuation of life depends on the function of these cycles, also known as **elemental** or **biogeochemical cycles.**

For eons, nutrient cycles were in equilibrium. Therefore, the same amount of an element moved into the various elemental **reservoirs** as moved out. However, in the last 200 years or so, humans have caused imbalances of some cycles. The result has been the buildup of large concentrations of certain elements, such as carbon, nitrogen, and phosphorus, in certain parts of the cycle where they can be harmful to humans and other organisms. Imbalances in the carbon cycle, you will soon see, could even be upsetting global climate in a major way. To understand cycles and our impact on them, we will examine three cycles: the nitrogen cycle, the carbon cycle, and the phosphorus cycle.

The Nitrogen Cycle Atomic nitrogen (N) is an essential component of many important compounds (such as chlorophyll) in plants, and hemoglobin, insulin, and deoxyribonucleic acid (DNA, the heredity-determining molecule) in animals. Nitrogen in the bodies of plants and animals comes from atmospheric nitrogen. Nitrogen gas is extremely abundant. It is a colorless, tasteless, and odorless gas and constitutes about 80% of the atmosphere. There are approximately 31,000 metric tons of nitrogen in the atmospheric air column above each 0.4 hectare (1 acre) of the Earth's surface. One would suppose, therefore, that securing adequate supplies of nitrogen would be relatively simple for living organisms. The problem is that nitrogen gas is chemically inactive. It does not combine readily with other elements, and it cannot be used by most organisms in this form.

How, then, can we and the great majority of living organisms make use of gaseous nitrogen to synthesize life-sustaining

proteins? First, the nitrogen has to be converted to a usable form, or **fixed.** There are several mechanisms by which nitrogen is fixed in nature.

One type of nitrogen fixation is **atmospheric fixation.** Atmospheric fixation is a naturally occurring phenomenon caused by lightning or sunlight. Energy from these sources causes the nitrogen to combine with oxygen to form nitrate.

Globally, about 7.6 million metric tons of nitrate are formed annually in this way. This nitrate is then washed to the Earth by rain and snow and is absorbed by the roots of growing plants.

A second type of nitrogen fixation is **biological fixation.** Biological fixation is much more important than atmospheric fixation, for about 54 million metric tons of nitrogen are fixed annually by this process—about seven times more than atmospheric fixation. Biological fixation is accomplished by microscopic organisms, primarily bacteria and cyanobacteria (also known as blue-green algae), which occur abundantly in soil and water.[1] During nitrogen fixation, nitrogen is first combined with hydrogen to form ammonia by certain bacteria. Plants can use this ammonia to produce amino acids.

Nitrogen-fixing bacteria also live inside the root systems of many plants. At least 190 species of plants, including **legumes** (alfalfa, peas, beans, soybeans, and clover), some pines, and alders, contain these nitrogen-fixing bacteria (Figure 3.15). These plants get nitrogen for themselves from the bacteria in their roots. They also produce excesses that are released into the soil and are therefore available for other plants as well. By growing legumes, a farmer may increase the nitrogen content (and hence the fertility) of his or her soil by 90 kilograms per hectare (80 pounds per acre) per year.

The fertility of aquatic ecosystems also may depend on nitrogen fixation. Certain mountain lakes, for instance, may depend on the nitrogen-fixing bacteria living in the roots of the alders along its shores.

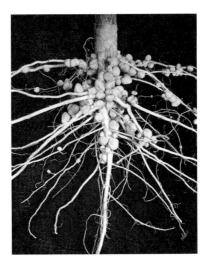

FIGURE 3.15 Root nodules on soybean plants and other legumes incorporate atmospheric nitrogen. Bacteria living inside the nodules convert the nitrogen to forms the plant can use to make amino acids and other biologically important molecules.

---

[1]Cyanobacteria were first called blue-green algae. These tiny bacteria are capable of photosynthesis.

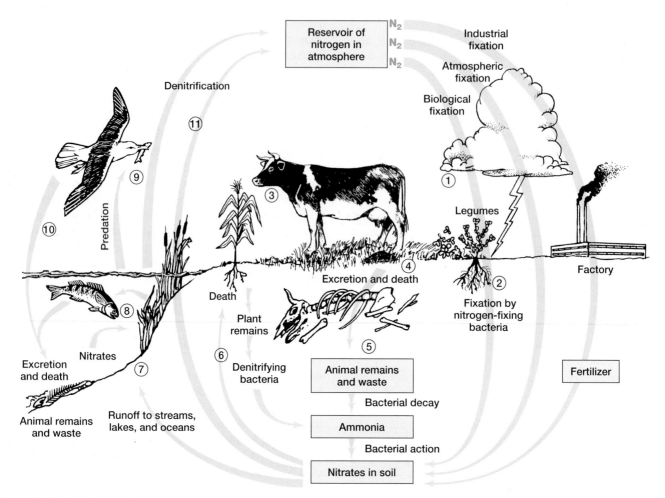

FIGURE 3.16   The nitrogen cycle.

A third type of nitrogen fixation is **industrial fixation,** a process in which nitrogen is combined with hydrogen to form ammonia. Later the ammonia is converted into ammonium salts that can be used as fertilizers. Such commercial production of fertilizer, which requires large amounts of energy (natural gas), has increased enormously since World War II.

Let us now, with the aid of Figure 3.16, follow nitrogen through a hypothetical cycle involving biological fixation by clover. The numbered steps that follow refer to the numbers on the figure and will help you more quickly understand this seemingly complex cycle.

1. Nitrogen (N_2) diffuses from the atmosphere into the air spaces of soil.

2. It enters small swellings on the roots of the clover plants. These are called **root nodules** and are home to the nitrogen-fixing bacteria. The nitrogen-fixing bacteria combine the nitrogen with hydrogen to form ammonia and eventually incorporate the nitrogen into amino acids, the building blocks of proteins. The clover plant builds up its own protein from the surplus ammonia not used by the bacteria.

3. A cow (or other consumer) feeding on the clover digests the plant protein, using the amino acids to make its own proteins. The cow uses protein to build muscle, make

milk, and produce enzymes. Some protein is broken down in the cells of the cow, releasing amino acids.

4. When the amino acids are broken down, nitrogen-containing urea is formed. It is excreted in the urine. The cow's feces are also rich in nitrogen, which comes from undigested protein and bacteria from the cow's digestive tract.

5. The urea and large, complex, nitrogen-containing protein molecules in the wastes (or carcass) are then eventually broken down (decomposed) by successive groups of soil bacteria into nitrates, a process called **nitrification.**

6. A plant, such as corn, wheat, or oak, can now absorb the soluble nitrates through its roots and use them to build up its own essential protein compounds, thus starting the cycle over again.

Nitrogen can flow from one ecosystem to another. For example, it may flow from a terrestrial ecosystem to an aquatic ecosystem and back to a terrestrial ecosystem again, as shown in Figure 3.16. Thus:

7. The soluble nitrate salts formed by the decay of the carcass of a cow may be washed into a stream and eventually carried to the ocean.

8. The nitrates may be absorbed by marine algae. The algae may be consumed by crustaceans that, in turn, are eaten by fish.

9. Fish are consumed by gulls.

10. The gulls then fly back to their nesting colony on the California coast and feed some of the partially digested fish to their young. Or the adult bird may excrete some waste as it flies over the water and the California farmland, thus contributing slightly to the fertility of water or land.

11. The flow of nitrogen is circular. All nitrogen in plants, animals, soil, and water eventually reenters the atmospheric reservoir from which it originally came. This is accomplished by denitrifying bacteria in the soil and water that break down nitrates. The energy released is used to sustain the life processes shown in Figure 3.16. Gaseous nitrogen is given off as a by-product. The nitrogen gas then escapes into the atmosphere from which it originally came. This process is known as **denitrification.** Nitrogen is also released into the atmosphere whenever organic material, such as trees, grasses, and animals, is consumed by fire.

Soil bacteria that fix nitrogen are essential for the continuation of life. Researchers in Florida have shown that some chlorinated hydrocarbon pesticides are detrimental to nitrogen-fixing soil bacteria. This suggests that toxic chemical pesticides should be used cautiously or perhaps even eliminated. Why? If populations of soil bacteria are greatly diminished, the nutrient cycle on which plants, and eventually animals and humans, depend could be severely disrupted.

Humans can disrupt the nitrogen cycle by poisoning the soil. We can also disrupt it by spreading excess fertilizer on the land—a process that overwhelms the cycle. When too much fertilizer is applied, it can run off into streams and lakes, causing serious water pollution. It can also seep into groundwater.

Another form of fixation occurs when fossil fuels are burned. In such instances, atmospheric nitrogen combines with oxygen (a reaction driven by the heat of combustion) to form nitrogen dioxide, which is later converted to nitrates and nitric acid. Nitrates may nourish terrestrial and aquatic plants, causing excess growth and, if severe enough, ecological disruption, a subject discussed in more detail in Chapter 10. Nitric acid can poison many life forms, as you will see in Chapter 19.

For thousands of years before commercial fertilizer production and the modern automobile, the nitrogen cycle was in dynamic equilibrium, a steady state in which the amount of nitrogen leaving the atmosphere by **nitrogen fixation** was balanced by the amount of nitrogen entering the atmosphere from denitrification. Today, however, artificial fixation of increasing amounts of nitrogen has caused a dramatic imbalance in the nitrogen cycle. To create a sustainable future, we must find ways to regain the balance.

The Carbon Cycle   Carbon is a key element in all organisms from bacteria to humans, where it forms the backbone of many biologically important molecules, including DNA, protein, and sugars. In fact, carbon atoms constitute 49% of the dry weight of the human body.

Carbon exists in several different reservoirs in the environment, including the atmosphere, the bodies of organisms, the ocean, ocean sediments, and as calcium carbonate (rocks, shells, and skeletons), as shown in Figure 3.17. Although the actual size of atmospheric and organismic reservoirs is relatively small, the rate of flow of carbon into and out of those reservoirs is relatively high—in other words, carbon is recycled fairly rapidly. Fifteen tons of carbon (in the form of carbon dioxide) are found in the air column above each hectare (2.5 acres) of Earth's surface. One hectare (2.5 acres) of lush vegetation can remove 50 tons of carbon from the atmosphere annually.

To trace the flow of carbon through the carbon cycle, we begin with carbon dioxide in the atmosphere, using the numbers in the text and on Figure 3.17 to follow the process. Carbon dioxide enters tiny pores in the leaves of trees and other plants (1). Inside the leaves, the carbon of the carbon dioxide molecules is combined with hydrogen from water to form sugar and other organic molecules, as illustrated in the equation of photosynthesis on page 56. As noted earlier, energy to drive this reaction comes from sunlight.

When the leaves of clover and grass are consumed by an animal, such as a deer, some of the carbon-containing organic compounds of the clover are digested and converted into deer muscle. The rest is converted to energy in cellular respiration. If humans or coyotes eat the deer meat, the meat is then transformed into human or coyote protoplasm.

In all of these organisms—clover, deer, coyotes, and humans—some carbon is released during cellular respiration (2), as shown in the equation on page 57. In plants, carbon is released as carbon dioxide into the atmosphere via pores in the leaves; in animals, the carbon dioxide is exhaled from the lungs. Carbon also returns to the atmosphere when an organism (clover, deer, humans, and so on) dies and decomposes. In addition, it may reenter via wastes (feces and urine) of animals, which are broken down by the bacteria and fungi that occur abundantly in soil, air, and water. Such decomposition releases carbon into soil and air. It then can reenter the biological world through plant photosynthesis.

About 250 to 300 million years ago, during the Carboniferous Period, giant tree ferns and other plants grew in what is today Pennsylvania, West Virginia, Ohio, Kentucky, Tennessee, Indiana, Illinois, Wyoming, New Mexico, Colorado, and South Dakota (3). Many of those plants were buried by sediment and therefore escaped decomposition. They were eventually converted into coal. Photosynthetic marine organisms that died and settled to the bottom of the ocean were buried in sediment and, over time, converted to crude oil and natural gas. (Some natural gas was also produced on land in conjunction with coal.) The carbon was removed from the carbon cycle by these processes.

Two hundred years ago, however, humankind discovered the remarkable potential of fossil fuels (4). Society today depends almost entirely on these fuels. Humans have also removed enormous amounts of forests, which absorb and store carbon. As discussed in Chapter 19, our accelerated combustion of fossil fuels (5) and the loss of forests have resulted in a dramatic increase in the amount of carbon dioxide in the atmosphere from 1870 to 2008—an increase of more than 30% (6). As noted earlier in this chapter, this increase may result in climate change (changes in average weather conditions) that can have devastating effects on natural systems, human beings, and our economy.

The oceans, which cover 70% of Earth's surface, also serve as a carbon reservoir. Carbon dioxide dissolved in the ocean may move through an algae → crustacean → fish food chain. The fish, in turn, may be eaten by such organisms as sharks,

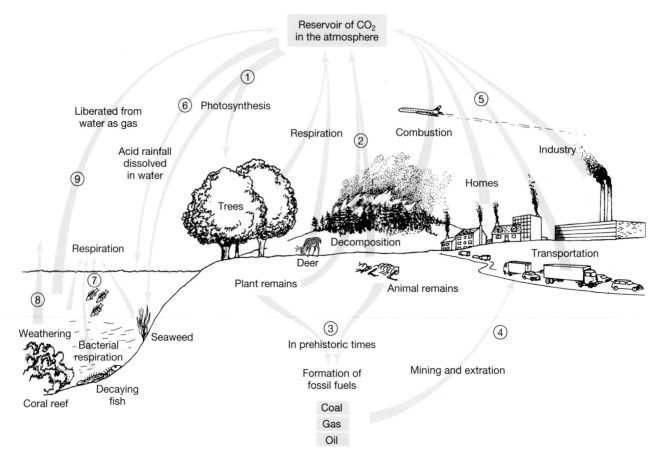

FIGURE 3.17 *The carbon cycle.*

tuna, waterfowl, whales, or humans. Some of this carbon is returned to the ocean by the respiration of marine organisms, and some is used by clams, oysters, scallops, and corals to build limestone shells and skeletons (7). Tremendous quantities of carbon are locked up in coral reefs off the coasts of California and Florida. The Great Barrier Reef off the Australian coast—a mass of limestone 56 meters (180 feet) wide and 2,100 kilometers (1,260 miles) long—is composed of billions of coral skeletons. Eventually, as a result of weathering processes (8) that operate for millennia, small amounts of the carbon from coral reefs and the shells of clams and oysters are returned to the ocean waters (9).

Carbon dioxide is continuously being exchanged between the atmosphere and the ocean. When atmospheric carbon dioxide increases, more is dissolved in the ocean. Conversely, when atmospheric carbon dioxide decreases, carbon dioxide is released from the ocean. By this mechanism, the carbon dioxide in the atmosphere was maintained at a fairly constant level for thousands of years—until the last century, that is.

GO GREEN!

To help reduce carbon dioxide emissions and global climate change, turn off the lights, your computer, and other electronic devices when you leave your dorm room or apartment.

The Phosphorus Cycle Approximately 1% of the human body is composed of phosphorus, an essential component of such compounds as the hereditary molecule DNA and the energy-rich molecule that powers virtually all organisms, adenosine triphosphate (ah-den´-oh-seen´ try-phoss´-fate; also known as ATP).

Like other nutrients, phosphorus is recycled in a giant, global nutrient cycle—the phosphorus cycle (Figure 3.18). To understand how this cycle works, we begin with its reservoir in phosphate rock above the soil (shown as erosion weathering in the diagram, number 1). Raindrops dissolve some of the phosphate in the rock and wash it into the soil (2). There it represents a potential plant nutrient, just like nitrogen. Phosphorus atoms pass into the plants and then into the animals that feed on them (3). When plants and animals die, their bodies are decomposed by bacteria and fungi (4). Phosphorus is released into the soil as a result. Some of this phosphorus may be taken up by other plants, and some of it may be washed into a stream after rainstorms (5). The stream may transport it to a lake or to the ocean. In these aquatic ecosystems, the phosphorus may be absorbed by algae and rooted plants (6). Plants may be eaten by fish (7). When aquatic organisms die and decompose, phosphorus-containing materials are released back into the water. Aquatic animals also excrete phosphorus-containing wastes into the water (8). Phosphorus in freshwater systems may eventually make their way to the oceans of the world. Phosphorus in seawater often settles to the ocean floor, forming part of the sediment. Over a period of millions of years, these sediments

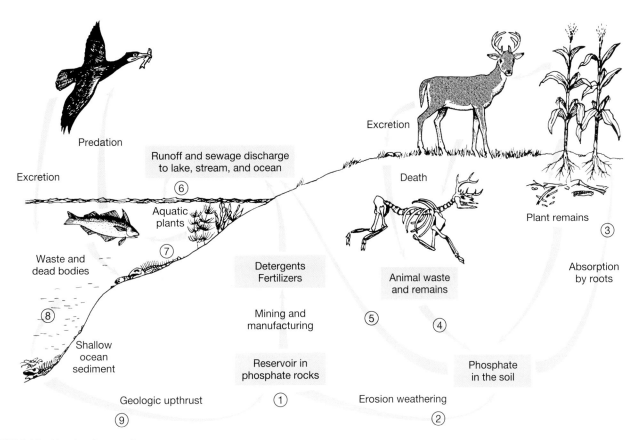

FIGURE 3.18 The phosphorus cycle.

may eventually form phosphate rock (9). Ultimately, geological processes such as upheaval may expose the rock to the atmosphere. Then, by the action of weathering and erosion, some of the phosphorus in the rock may become part of the soil once again.

For eons, the amount of phosphorus moving from its phosphate rock reservoir into the bodies of living organisms was in equilibrium with the amount moving in the reverse direction. The total amount of phosphorus in circulation was relatively small. However, in the past few decades, the intensive fertilization of agricultural lands with phosphorus-containing chemicals to feed a mushrooming human population, as well as the use of phosphorus-containing detergents, has caused an increase in phosphorus in our lakes and streams, discussed more fully in Chapter 10.

3.4 Principles of Ecology

The previous material has provided an overview of scientific principles and the ways energy and matter flow through ecosystems. We now turn our attention to specific ecological principles that will help you deepen your understanding of ecosystems, knowledge essential to creating a more sustainable way of life.

The Law of Tolerance

The survival of any organism depends on many essential factors in its physical environment, such as water, temperature, oxygen, and nutrients. Interestingly, these factors do not remain constant. Temperature, for instance, varies during the day and by season.

Species are adapted to and can tolerate a range of conditions, known as their **range of tolerance.** Figure 3.19 illustrates the concept. Let's consider sunlight and shade tolerance in trees. As you may know, trees vary in the amount of sunlight they can tolerate as seedlings. Some, like pines, are sun-loving species and thrive in open areas such as clear-cuts. Others are shade tolerant: that is, they require shade as seedlings to grow. They grow best in forests.

Consider shade-tolerant trees. As shown in Figure 3.19, the center of the range represents the optimal conditions for survival. In this example, it represents the amount of sunlight in which shade-tolerant trees in a forest thrive. Above and below the optimum range—that is, in places where there is too much sun or too little sun, respectively—the seedlings do not grow as well. Thus, just above and below the optimum range are the zones of physiological stress. Organisms survive, but conditions are not optimal. Heading further outward on the range of tolerance is the zone of intolerance, in which death occurs. Too much sun or too little sun causes death.

The tolerance of a given species for an environmental factor often varies with the age of the organism. Thus, newly hatched salmon are much more vulnerable to such water contaminants as heat, toxic metals, and pesticides than are adults. Tolerance may also vary with the genetic makeup of the organism.

Habitat and Niche

Ecologists also frequently refer to an organism's habitat and its niche. The two are very different. The **habitat** of an organism is its "address"—the place where it lives. For example, the habitat of

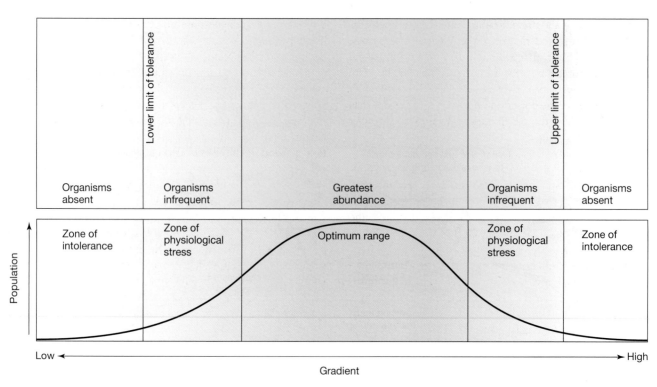

FIGURE 3.19 Range of tolerance.

the red-winged blackbird is a cattail marsh; a cold-water stream is the habitat of a trout. The term **niche** is more difficult to define and has been defined by different people in different ways. Some ecologists consider the niche an organism's functional role—what it does, and its relationship to its food and its enemies. The niche is really much more and can be appreciated by surveying a set of questions that illustrate many aspects of the niche of an organism:

1. What is the animal's range of tolerance to temperature, humidity, solar radiation, and wind velocity?
2. What type of food does it consume?
3. With which species (other than its own) does it compete for food and nesting or breeding sites?
4. Where does it produce its young? In a den? In a nest?
5. If it is a fish, where does it lay its eggs? On the stream bottom? On aquatic vegetation? Or do the eggs simply float on the water's surface?
6. If it is a nest builder, what types of material does it use in nest construction?
7. What parasites plague the animal?
8. To what type of predators is the animal vulnerable?
9. How does the animal protect itself from predators? By "freezing," by its coloration, by hiding under vegetative cover, by fighting, or by swimming, running, or flying away?
10. What type and volume of waste does the animal excrete?
11. How does this waste affect the surrounding plant and animal life?

Field ecologists spend a great deal of time studying organisms in an attempt to find answers to these and other questions.

Understanding an organism's niche is not just a scientific exercise, it is a way of building our knowledge base so that humans can manage other species and the ecosystems they inhabit better and also so we can fashion our systems in ways that do not interfere with other species. With these considerations in mind, let us now examine an important principle based on the concept of the niche.

The Competitive Exclusion Principle

Imagine that you had two species that occupied the same niche and lived in the same habitat. What would happen to them? Studies show that two species of plants or animals cannot occupy the same ecological niche indefinitely. Sooner or later, the population of one of them will decline to zero. This phenomenon is called the **competitive exclusion principle.**

Competitive exclusion, the elimination of one of the species whose niche corresponds to the niche of another species, occurs for two reasons. First, competition between two species with identical niches would be intense. Plants would compete for sunlight, soil moisture, nutrients, and so on. Animals would compete for food, breeding sites, cover, and so on. Second, as a rule, even though the species occupy identical niches, one would expect that the two species would not be equally well adapted to occupy that niche. In other words, we would expect that one of the two species would be better able to meet its needs. For example, one species of plant may be more efficient in photosynthesis. In the case of predators, one species may be able to pursue prey more effectively than another species because of its keener vision. In such instances, the population of the less well-adapted species would eventually decline and be eliminated by the better-adapted organism.

Because of competitive exclusion, organisms that occupy the same habitat often have slightly different niches. Consider an example. The shag and cormorant are both fish-eating birds. They nest on cliffs on the British coast and feed in nearby waters. Studies have shown that although they occupy the same habitat, they do not occupy the same ecological niche. The shag, for example, feeds primarily on eels and herring-like fish in the surface waters. The cormorant preys on shrimp and flatfish that live on or near the ocean bottom. Moreover, the shag nests at lower levels on the cliffs than the cormorant.

In actuality, very few well-documented examples of the competitive exclusion principle have been reported. One such example involved competition between house mice and meadow mice in California. These two species competed for the same space, food (weed seeds), and breeding sites. However, the meadow mouse was much more aggressive in this competition. As a result, the house mouse population declined sharply from about 825 per hectare (330 per acre) to zero in only 14 months.

Competitive exclusion is important when considering introducing a species into a new habitat—for example, introducing a sport fish into waters to which it is not native. The major concern is not just how well the new species will do, but how it will affect native species whose niches may be very similar or identical.

Carrying Capacity

Another important concept in ecology is carrying capacity. Briefly discussed in Chapter 1, *carrying capacity* is defined as the number of organisms an ecosystem can support. An ecologist who speaks of the carrying capacity of an ecosystem for red foxes or white-tailed deer is referring to how many of each species the system can provide for.

Carrying capacity is determined by a variety of factors, such as food supply, nesting sites, water supplies, climatic conditions, and waste assimilation. We think of it as being determined by source and sink functions—the term *source* referring to the resources an organism needs to survive and reproduce, and the term *sink* referring to the ability of an ecosystem to get rid of waste. These factors can be manipulated by resource managers to increase or decrease numbers. But experience shows that ecosystems are complex interactive systems and care should be taken when altering such factors. Altering one factor to benefit a single species can have unforeseen and adverse effects on others.

Bear in mind, too, that carrying capacity is a dynamic number. That is, it changes as conditions change. In dry years, the number of deer a forest and field can support decreases. In wet years, it may increase. In ecosystems, populations change in response to variations in conditions so that organisms don't eat themselves out of house and home. Sustainable management of human ecosystems—for example, rangeland—also requires adjustment. In dry years, for example, ranchers may need to cut back the size of their herd or find more grazing land to accommodate their cattle. If they don't adjust, they risk overgrazing the land in dry years, causing damage that could last for many years, ultimately decreasing the carrying capacity of the land.

Population Growth and Decline

An understanding of ecosystems, communities, and populations requires an understanding of population growth and decline, a topic discussed more fully in the next chapter. As you

will learn, all organisms are governed by the same principles. Their populations grow when conditions are favorable. They decline when conditions are not.

Numerous **biotic factors** (biological) and **abiotic factors** (physical and chemical) stimulate growth, causing a population to reach what ecologists call its **biotic potential**—the maximum reproductive rate. The abilities to find food and hide from predators, for example, are two biotic factors that favor survival and reproduction. They will cause a population to grow, as will favorable light or temperature, which are two abiotic factors (Figure 3.20).

Suppose, for example, that conditions were favorable for American robins. Under such conditions, all American robins could live to be ten years old, and each adult female could annually produce eight young. If the robin population started in 2000, a single breeding pair and its offspring would produce 1.2 billion robins by the year 2030. This population would require 150,000 additional planets the size of Earth to accommodate it.

Fortunately, populations of robins and other species don't generally grow out of control. That's because factors that

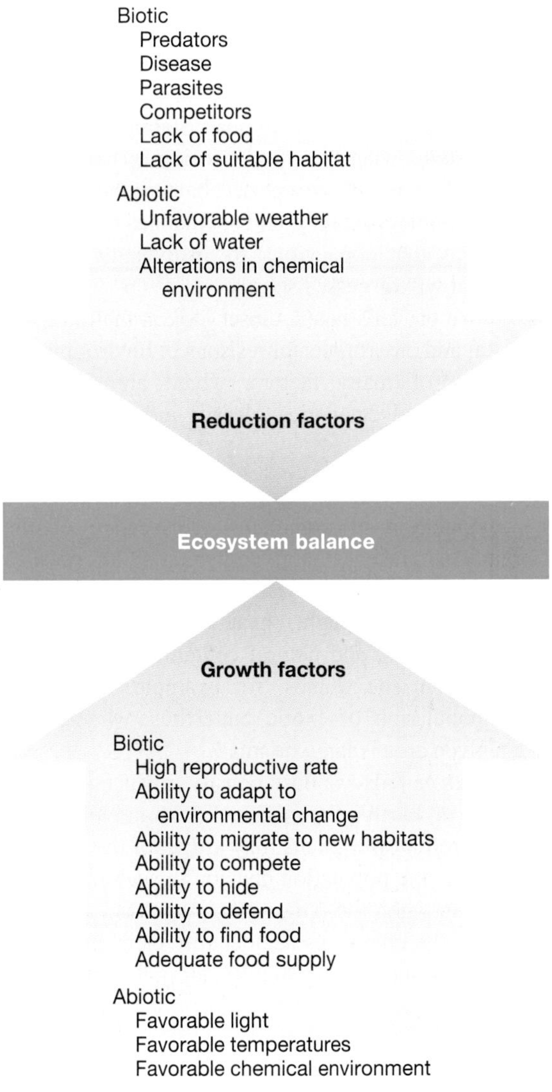

Biotic
  Predators
  Disease
  Parasites
  Competitors
  Lack of food
  Lack of suitable habitat

Abiotic
  Unfavorable weather
  Lack of water
  Alterations in chemical
    environment

Reduction factors

Ecosystem balance

Growth factors

Biotic
  High reproductive rate
  Ability to adapt to
    environmental change
  Ability to migrate to new habitats
  Ability to compete
  Ability to hide
  Ability to defend
  Ability to find food
  Adequate food supply

Abiotic
  Favorable light
  Favorable temperatures
  Favorable chemical environment

FIGURE 3.20 Factors that affect populations. Numerous factors stimulate growth or cause reductions in the size of populations.

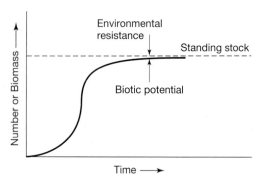

FIGURE 3.21   Environmental resistance and biotic potential. These factors cause populations to decline or check growth or to achieve the full reproductive or biotic potential of an organism.

promote growth tend in populations to be balanced by biotic and abiotic factors that reduce growth in ecosystems (Figure 3.20). The factors that reduce growth are collectively referred to as **environmental resistance.** The size of a population of any organism therefore results from the interaction of factors that contribute to environmental resistance with factors that contribute to its biotic potential (Figure 3.21). In general, when environmental resistance climbs, a population declines. When environmental resistance falls, a population increases. Stability may be achieved in ecosystems when the two opposing sets of factors are in balance. Don't expect to see a perfect balance, however, as conditions in nature shift constantly, and populations rise and fall accordingly. We say that populations are in dynamic equilibrium, a concept that will discussed shortly.

Before we move on, let's take a closer look at the interplay of biotic potential and environmental resistance. Environmental resistance consists of many factors, which are generally grouped into two broad categories: density-independent and density-dependent.

Density-Independent Factors   Any factor that limits the growth of a population of organisms, irrespective of the number of organisms present in a given habitat, is known as a **density-independent factor.** The most common density-independent factors are drought, heat waves, cold spells, tornadoes, storms, floods, and natural contamination such as silt. A rare freeze in the tropics, for example, kills most members of a population of exotic butterflies, whether the population in a given area is large or small.

Human activities can also impose density-independent limits on the growth of Earth's organisms. For example, heated water discharged from power plants may kill all of the fish in a stream, regardless of the population density. In such instances, all organisms perish when the range of tolerance is exceeded. Likewise, many of the chemicals accidentally or intentionally released into the environment—pesticides, air pollutants, water pollutants, and toxic wastes—affect populations regardless of their density.

Habitat destruction also acts independently of population density. Roads, farms, suburbs, factories, cities, and villages all obliterate vital habitat, often destroying all of the species that once lived there, regardless of their density.

It is important to note, however, that some human interventions result in population increases rather than declines. In other words, they unleash biotic potential. For example, the house sparrow population has flourished in the United States because of the abundance of suitable nesting sites and food (waste grain and seeds in manure) near human settlements. Barred owls, once restricted to the eastern United States, have successfully moved westward to the Midwest and the West Coast, largely because of cities and towns. Researchers hypothesize that this successful expansion of habitat has been facilitated in large part by trees planted in and around cities and towns of the previously treeless Midwest.

The predicted increase in global temperature, caused by carbon dioxide and other factors, may shorten winters in North America, making conditions more favorable for insects. As a result, many insect populations could explode, creating even more damage to crops and forests. Disease organisms from the tropics could spread northward and, in fact, already are.

Density-Dependent Factors   Many species have evolved mechanisms that allow them to survive the winter. Birds, for instance, may migrate to warm climates when winter approaches. Some animals grow a thick coat of fur to protect themselves from the cold. Other animals hibernate. Perennial plants, or species that grow year after year from the same roots, such as deciduous trees and grasses, enter a state of dormancy.

For these species, density-dependent factors tend to play a larger role than weather in controlling population growth. **Density-dependent factors** are those whose influence in controlling population size increases or decreases depending on the density of a population. Ecologists recognize at least four major density-dependent factors: predation, competition, parasitism, and disease.

Predation   A host of animals hunt and kill their prey. Known as **predators,** these animals include strictly carnivorous (meat-eating) species, such as mountain lions, coyotes, and cheetahs. They also include herbivores (vegetation eaters), such as cows, deer, elk, and grasshoppers (which "prey" on grass), and **omnivores** ("all-eaters," or consumers of both plants and animals), such as bears and human beings.

Predation is a density-dependent factor. As a general rule, as the population density of a prey species increases, so does the percentage of organisms killed by predators. Why?

It is generally thought that as the density of a prey species increases, individuals become easier to find and attack. Furthermore, increased competition in the dense prey population may result in a larger number of weakened organisms that become easy targets for predators. Prey may also be forced into a less-suitable habitat and may become weakened or diseased, and again are more likely to be captured and killed by predators. Ecologists have also found that when the density of a prey population increases, predators that eat a variety of prey tend to shift their attention to the most concentrated ones.

Competition   Another density-dependent variable is **competition.** Competition occurs at two levels—within a given species and between species. Consider an example of competition within a single species: ponderosa pine trees in the Rocky Mountains. If the number of ponderosa pine trees increases in a given region, creating a denser stand, competition for sunlight, soil nutrients, and water increases proportionately. Trees with shallow, less well-developed root systems receive less water than others. As a result, they are more vulnerable to insects including the pine bark beetle, which carries with it a deadly fungus that clogs the water-transporting vessels in the wood and kills the tree.

An increase in the population density of an animal species intensifies competition for food, water, cover, nesting sites, breeding dens, and space. Competition generally affects the physically unfit—the aged or diseased. In their weakened condition, some may perish from disease or injuries inflicted during territorial battles. Some weakened individuals may be forced to colonize poorer habitats, where death from starvation, disease, or predation awaits them.

Parasitism **Parasites** such as tapeworms are organisms that feed on the bodies of other living organisms, called **hosts,** but generally do so without killing them. Most parasites have limited mobility but are easily spread from organism to organism. The higher the host population density, the more readily they are transmitted. As a result, parasitism is a density-dependent factor. In Canada, for example, a species of parasitic wasp attacks webworm moths. When the moth population density is low, deaths from parasitic wasps are negligible. But when the population density increases, parasitism increases.

In Colorado in the late 1970s, bighorn sheep offspring died in record numbers because of the lungworm parasite. Lungworms weaken sheep and makes them susceptible to pneumonia and other diseases. Passed from mother to fetus, lungworms killed 97% of the newborn sheep in the 1970s. Lungworms are a common parasite that, under normal conditions, live in harmony with the bighorn. But because of human encroachment, the bighorn sheep habitat was reduced, and the population density increased. Animals were forced to calve and feed on the same land. Lungworm eggs are deposited in the sheep's feces, and because the sheep were forced to occupy a smaller habitat due to human encroachment on their natural habitat, the number of lungworm eggs they ingested increased. Record numbers were found in females. The worms passed through the placenta and infected bighorn fetuses. The newborn, too weak to withstand the parasite, perished. Fortunately, wildlife biologists found out about the problem and have captured sheep, treated them, and relocated them to an uninfected habitat.

Disease Infectious diseases also increase when population density increases. The incidence of infectious diseases in human populations, for instance, increases with increasing population density. This was demonstrated during World War I, when American soldiers crowded shoulder to shoulder in the barracks and trenches suffered massive mortality from the influenza virus. Because the organisms that cause infectious diseases (viruses, protozoa, and bacteria) can be transmitted by contact, through food, and by animal vectors (insects), the greater the population density, the more likely it is that the disease will spread. Today's population is extremely vulnerable to infectious disease. In fact, according to one estimate, within

two years of the emergence of a new influenza virus, half the world's population has been exposed. High density is one of the reasons. Of course, modern transportation facilitates the spread of infectious diseases.

Disease organisms that attack crops represent another example of density-dependent factors. When large plots are planted with a single crop species, such as wheat or corn, farmers increase the likelihood that disease organisms will wipe out their crops. The population-reducing effects of wheat rust, corn smut, and other plant diseases are well known.

Although we have grouped the population-regulating mechanisms into two groups—density-independent and density-dependent—in reality these factors operate together on a given population. For example, heavy rains (a density-independent factor) may flood land and reduce the population of a species afflicted with a disabling parasite (a density-dependent factor), causing the population to plummet.

The study of plant and animal **population dynamics**—the changes that occur in populations—is a key tool in resource management, especially fishery and wildlife management. You'll need to understand this information when you read the resource management chapters in this book. Because of this, we'll take a deeper look at the subject now, beginning with growth curves.

The S-Curve Whenever a species is established in a new habitat with adequate resources, its population grows in a characteristic way. The plot of population growth is called an **S-shaped,** or **sigmoidal, curve** (Figure 3.22A). This curve has four distinct phases, listed in chronological sequence: (1) the establishment phase, (2) the explosive (logarithmic) phase, (3) the deceleration phase, and (4) the dynamic equilibrium phase.

In the first phase, the population grows slowly as its new members establish themselves in their new habitat. The population soon begins to expand and fill the unexploited habitat and begins growing at an explosive rate. The rapid growth generally occurs because of the availability of untapped resources (for example, abundant food and shelter) and the lack of environmental resistance such as predation. During the second phase, however, growth of the population begins to reduce the supply of resources. This, in turn, will reduce the rate of growth, resulting in the deceleration phase. Finally, once the population is well established and competition and other factors come into play, the species reaches an equilibrium. At this point, populations shrink and expand but remain more or less constant over the long term. Thus, during years with adverse weather, populations may decrease, but during years with more favorable weather, populations rebound (Figure 3.22B). Such a system is said to be in **dynamic equilibrium.** At this point, no further population surges are possible, for the population has reached the carrying capacity of the habitat (the number of organisms the habitat can support indefinitely).

A number of species introduced into the United States from abroad, known as **exotic** or **alien species,** such as the house (English) sparrow, European starling, and the German carp, have experienced population growth that follows the S-shaped curve. For instance, a few pairs of house sparrows were introduced to the United States in 1899, and within ten years, the sparrow population had increased to many thousands of birds.

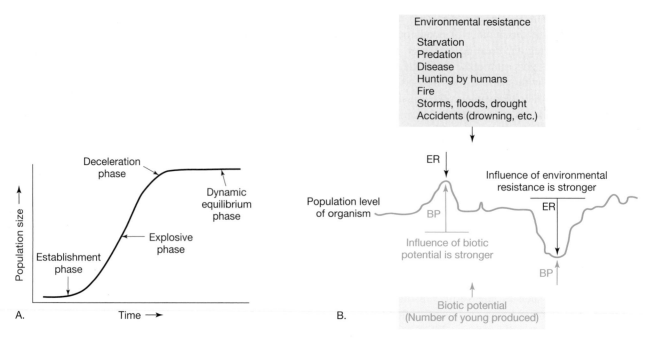

FIGURE 3.22   Sigmoidal curve illustrating the growth of a population of organisms in a new environment. (A) After an initial establishment phase, growth explodes. As environmental resistance rises, growth decelerates and then stops. (B) A dynamic equilibrium occurs in this phase. The population may increase as a result of a decrease in environmental resistance or an increase in the number of offspring (an increase in biotic potential). A decrease in biotic potential or an increase in environmental resistance may cause the population to decline. Despite these shifts, the population remains more or less constant over time.

Today, the sparrow has become a national nuisance, and the U.S. Department of Agriculture is now calling for a nationwide program to exterminate this introduced species.

Although English sparrows and other species were introduced with good intentions (the sparrow was brought over from Europe to control insect pests), many of these exotic or alien species have had an enormous impact on native plants and animals. The sparrow, for instance, outcompetes songbirds for nesting and food in the United States and Canada. Other alien species have destroyed wildlife directly, disseminated diseases, or aggressively outcompeted the more valuable species for food, cover, and breeding sites.

Wildlife managers have used their understanding of the S-curve and population dynamics for years to manage commercially valuable species, and will continue to do so for years to come. Wild populations of deer, cod, and pine trees, for example, may be periodically harvested, much like a commercial crop. Perhaps the most important challenge facing scientists in the fields of forestry, wildlife management, agriculture, ranching, and fisheries is to find ways to secure the **maximum sustainable yield,** or **optimum yield**—the greatest yield possible that does not harm the population's long-term survival. Optimal harvests are generally achieved at a point on the S-curve between the explosive and the deceleration phases. Research has shown that in most species, this point occurs when a population is at roughly 50% of the habitat's carrying capacity. Harvests that drive the population below this level may severely reduce numbers and cause a further population decline, possibly driving the species to extinction.

Unfortunately, managers are discovering that what may appear to be an optimum yield may not in fact be sustainable over a long time frame. Factors not accounted for, such as soil

deterioration caused by periodic harvesting of forests, have encouraged a reexamination of proper harvest levels.

Most plant and animal populations become fairly stable once they reach the plateau of the S-curve—the carrying capacity of the ecosystem. Even so, predators, parasites, competition, climate, and other factors may cause slight upward or downward swings in the population. Thus, the line in the graph in Figure 3.22A that is drawn straight actually has slight upward and downward oscillations.

Irruptive Populations   After fluctuating mildly for many years, some populations may suddenly increase sharply. A sudden increase in population size is called an **irruption.** It may occur after a period of unusually favorable weather that results in an unusually large supply of food or exceptional conditions for the survival of offspring. If a population expands beyond the habitat's carrying capacity, however, it may plummet to a much lower level as its food supply is depleted. Wide oscillations in the population size may continue for years if environmental conditions vacillate. In some instances, the sudden upswing in a population may cause severe damage to the ecosystem—so severe, in fact, that the population crashes.

Cyclic Populations   The brown lemming, a tiny rodent the size of a hamster, lives in the Arctic tundra of North America, a vast, treeless biome above the northern coniferous forest biome (see Section 3.5). Every three or four years, the lemming population peaks and then crashes. A study of this predictable cycle showed that the peak lemming population overgrazes the vegetation that protects it from its prey, making the lemming highly susceptible to its main predators, the Arctic fox and

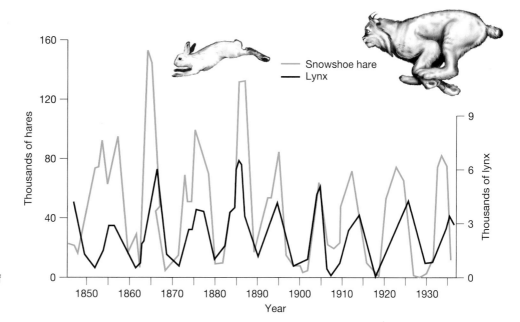

FIGURE 3.23 An example of a wildlife population cycle, based on pelt records of the Hudson Bay Company. Note the approximate ten-year interval.

snowy owl. As you might expect, the populations of these predators vary with the lemming's, increasing in times of abundance and declining in times of scarcity. When lemming populations decline, many snowy owls starve during winters or migrate southward into the United States.

Lemmings, the popular myth goes, migrate en masse to the sea when their populations top out, leaping from cliffs into the perilous waters, where they drown. Although researchers have found that lemmings do indeed migrate down mountainsides from crowded breeding sites, they now know that lemmings do not migrate far. In addition, they disperse individually, not in great numbers. Furthermore, they do not throw themselves into the sea, although they will occasionally cross streams, where some lose their lives. Stream crossings are made, however, only when they are absolutely necessary. In short, there is no suicidal mass march to the sea.

A number of other species experience regular three- to four-year cycles, among them the red-tailed hawk, the meadow mouse, and the sockeye salmon. Protecting these and other cyclic species requires special attention on the part of wildlife managers. Overharvesting commercially important species in crash years could reduce numbers to levels that cannot be sustained.

The populations of snowshoe hares, a species found in the northern coniferous forests, undergo a ten-year boom-and-bust cycle (Figure 3.23). The lynx, which preys largely on the snowshoe hare, also has a ten-year cycle that lags just behind the hare's. The lynx cycle was first discovered by scientists studying the Hudson Bay Company's lynx pelt return records. The cycle is also characteristic of muskrats, grouse, and pheasants.

The original explanation given for the cycles is as follows: When hares are abundant, food for the lynx is abundant. That makes it easier for lynx to raise their young and increases the survival of young as well as the adults. However, the success of the lynx population results in heavier predation on hares, which eventually reduces the hare population. With its food supply reduced, the lynx finds it more difficult to survive, and its numbers dwindle.

However, other factors may play a role in controlling the hare population size. Hares living on islands with no lynx also undergo a ten-year cycle. One explanation is that the rising population of hares reduces its own food supply. Hares begin to die off. This may also occur in populations preyed upon by lynxes. As the hare population declines, the lynx population plummets. A number of other hypotheses have been advanced to explain the ten-year cycle, but as yet, scientists have been unable to determine which, if any, are valid. Among the causative agents in this puzzle are variations in weather, fluctuations in solar radiation, outbreaks of disease, and changes in the nutrient levels of plants.

With these concepts in mind, we now turn our attention to another important concept: biological succession.

Biological Succession

Ecosystems are dynamic entities. Organisms grow and die. Populations increase and decrease. Conditions change over time with the seasons. One form of change is called biological succession. **Biological succession** is the replacement of one community of organisms by another in an orderly and predictable manner. Two types of succession are recognized by ecologists: primary and secondary.

Primary Succession Succession that occurs in areas not previously occupied by organisms—for example, a newly formed volcanic island—is known as **primary succession.** During primary succession, life becomes established on previously lifeless areas such as jagged outcrops of granite, lava-covered slopes, rubble left in the wake of a landslide, or even on the waste heaps of an open-pit mine.

To understand this process, we will trace a primary succession that might occur on a rocky surface of the eastern United States after the retreat of the glaciers (Figure 3.24). The first stage is the **pioneer community.** Organisms of this stage are adapted to withstand great extremes of temperature and moisture. A typical pioneer organism that might become established on a bare, windswept, rocky outcrop is the **lichen,** an odd little

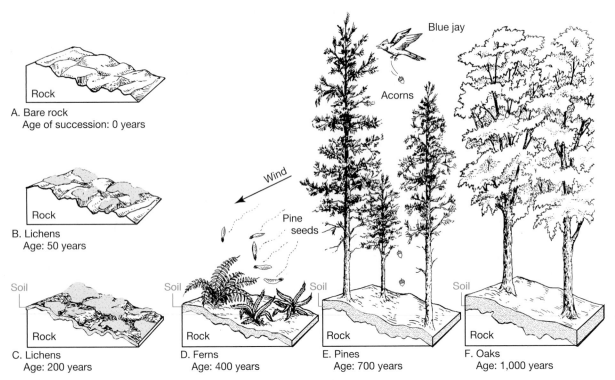

FIGURE 3.24   Stages of primary succession that begins on bare rocks and ends in an oak forest. This process could take 1,000 years to complete.

organism that clings to rocks and other lifeless substrates. A lichen consists of two organisms, one living inside the other. The main body is a fungus. Inside the fungus are many photosynthetic algae. The reproductive bodies of lichens—spores—are blown around in the wind and thus can be blown into the barren, rocky area. The spores settle on the rocks and begin to develop.

The lichens gradually form a crust on the rocks. Many are grayish green, although some lichens are bright orange or even powder blue. Lichens are remarkable organisms. Even if the rock is completely dry, spores can develop so long as adequate atmospheric moisture is available. Once established, lichens begin to modify the immediate environment around them (Figure 3.24). Weak carbonic acid produced by the lichen begins to dissolve the underlying rock. Lichens growing side by side trap particles of windblown sand, dust, and organic debris, which then begin to accumulate. When an occasional lichen dies, bacteria and small fungi cause its decay. The resultant organic material and the excreta of minute lichen-eating insects that have invaded the microhabitat enrich the relatively sterile soil that has accumulated. This soil now acts as a sponge, absorbing water from dew or rain. Once sufficient soil has accumulated, mosses and ferns may become established, also by means of wind-distributed spores. Ferns eventually shade out the lichens and replace them in the succession. The soil becomes further enriched each fall with the decay of the ferns. Eventually, as the decades pass, windblown pine seeds from a hilltop pine forest may fall in the area. They may have been dropped by seed-eating birds, such as the pine siskin, as they flew overhead. The young sun-loving pine seedlings, in turn, compete with the ferns. Eventually the pines shade out the ferns and replace them.

With the passage of time, gray squirrels may enter the area to bury acorns brought in from a neighboring oak woods. Acorns may also be accidentally dropped by a wandering raccoon or a blue jay during a visit to the young pines. The acorns germinate readily in the relatively fertile soil, which by now has been developing for centuries since the succession began. The young oak seedlings grow well in the shade under the pine canopy. The oaks (and other species such as hickory, red maple, and red gum) have long roots and therefore can make use of soil moisture that is unavailable to the shallow-rooted pines. As an occasional pine dies, its position in the forest community is filled by an oak. Over time, an oak forest, with its characteristic complement of plants and animals, will become established as the relatively stable terminal community, or **climax community,** of the succession. Thus, after hundreds of years, the bare, windswept rock outcrop is replaced by a mature oak forest.[2]

Secondary Succession   Succession in an area that was previously occupied by organisms is called **secondary succession.** Much more common than primary succession, secondary succession occurs when a given ecosystem is partially destroyed by natural or human forces. The main causes are fires, volcanoes, hurricanes, deforestation, and agriculture. Chemical contamination caused by smelters or hazardous waste also may cause severe environmental damage that leads to

[2]Most ecologists use the terms *early, intermediate,* and *late stages* of succession to reflect the current thinking on the subject.

secondary succession. Because the topsoil is not generally destroyed, secondary succession usually requires less time to complete than primary succession. Secondary succession is occurring in the forests around Mount St. Helens in Washington. This volcano erupted on May 18, 1980, blowing down forests and spreading ash on much of the surrounding land (see Case Study 3.1).

With the help of Figure 3.25, let us examine the secondary succession that develops in abandoned cotton fields in the Piedmont region of Georgia. During the first year following

---

LIFE RETURNS TO MOUNT ST. HELENS: A DRAMATIC EXAMPLE OF SUCCESSION

Mount St. Helens erupted on May 18, 1980 (Figure 1). It was one of the most spectacular volcanic eruptions ever witnessed and caused extensive damage to plant and animal populations. Some ecosystems were literally knocked back to ground zero and had to start from scratch. Roger del Moral described the dramatic comeback of life in an article entitled "Life Returns to Mount St. Helens." Excerpts from his article follow.

At 8:32 A.M. superheated groundwater close to the magma flashed into steam, resulting in a lateral explosion that pulverized rocks and trees and sent a hurricane-force bolt of ash off the north face and across the Toutle River Valley to the north and west. Temperatures in this inferno were estimated to exceed 9008°F. Comparable to a 400-megaton blast, the explosion blew down trees in a 160° arc up to 14 miles north of the crater. As the summit of the mountain collapsed, two vertical columns of gas and steam were ejected more than 65,000 feet into the air. The ash was eventually deposited in layers up to five inches thick over 49% of Washington State and beyond.

The number of animals killed as a direct result of the explosion was high. Subterranean animals, such as pocket gophers, appear to have survived in many places even within the blast zone, but mammals and birds living above ground had no protection from the blast. An estimated 5,200 elk, 6,000 black-tailed deer, 200 black

bears, 11,000 hares, 15 mountain lions, 300 bobcats, 27,000 grouse, and 1,400 coyotes perished. The eruption also severely damaged 26 lakes and killed some 11 million fish, including trout and young salmon.

Larger vertebrates may form a crucial link in the process of vegetation recolonization. Where heavy ash or mudflows dry to form a hard, uniform crust, there are few cracks to shelter germinating seeds. But large animals wandering in search of food or water make tracks that trap seeds.

Higher terrestrial life may be scarce in the blast zone but dead organic matter is abundant, and such a resource is never unexploited for long. Here, where many humans died, where entire ecosystems ceased to exist in a matter of seconds, Dave Hosford of Central Washington State College has found a mushroom growing from the ash, slowly decomposing organic matter found there and beginning a terrestrial succession.

Of all the regions on the mountain, the most severely affected was the blast zone immediately north of the crater, including Spirit Lake. Every type of volcanic behavior displayed by the mountain has assailed this terrain. All life was obliterated. Trees were pulverized and soil vaporized. A 100-meter wall of volcanic material lies over the surface.

Three decades later, the devastated ecosystems around Mount St. Helens are on the mend. Studies of the recovery (ecological succession) show that the rate of recovery depends on the degree of disturbance. The greater the disturbance, the slower the recovery. In areas protected by snow cover or areas along streams, for example, vegetative recovery has occurred more readily. Another area of notable recovery is the steep terrain where erosion washed ash away and exposed soil and plants. But the region immediately north of the volcano, called the pumice plains, is the slowest to recover.

During the first few years after the blast, recovery was accelerated by surviving plants. These plants spread rapidly and were soon joined by colonizers, plants whose seeds had been blown in by the wind or deposited in bird feces. Since 1990, the number and variety of both survivors and colonizers in the blast zone has increased. Trees have also begun to grow in previously forested areas. Elk and deer have returned, as have a number of bird species. About one-third of the rodent species survived; their populations are expanding as are insect populations.

Despite these successes, the first three decades of vegetative recovery must be viewed as the beginning of a long process of successional change that could take 200 to 500 years, depending on the amount of disturbance. Eventually, the landscape will support a rich old-growth forest as it had prior to the fateful day in 1980.

FIGURE 1 Mount St. Helens after its historic 1980 eruption. Barren land in the foreground will slowly revegetate and over decades return to its previous condition.

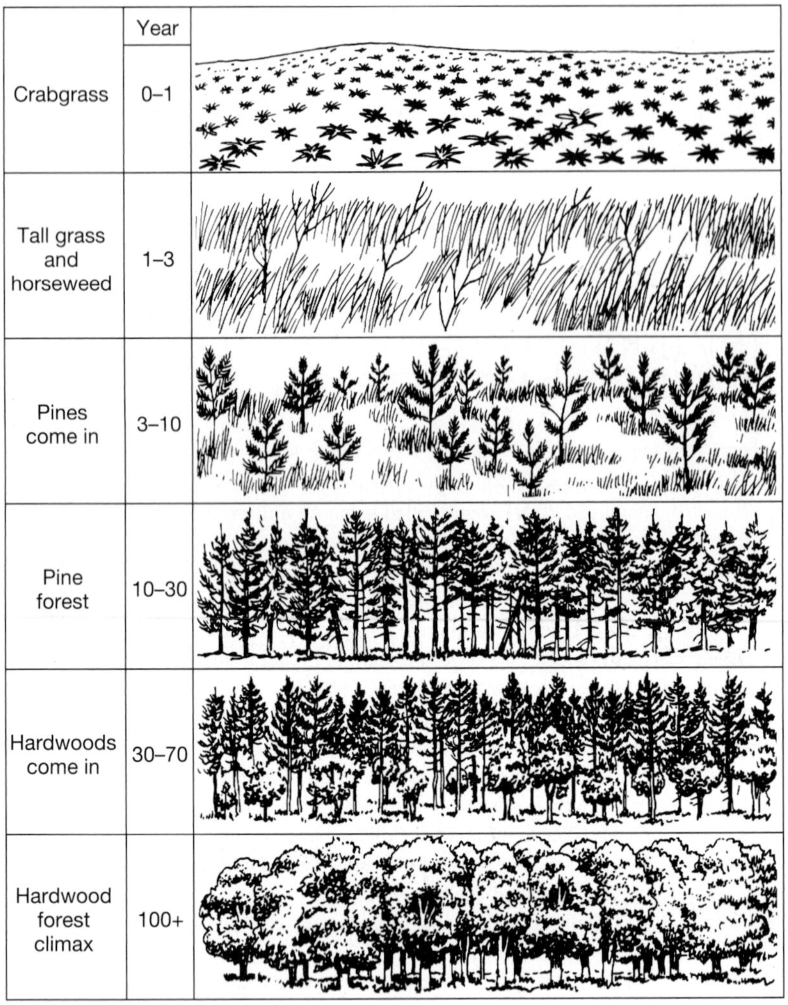

	Year	
Crabgrass	0–1	
Tall grass and horseweed	1–3	
Pines come in	3–10	
Pine forest	10–30	
Hardwoods come in	30–70	
Hardwood forest climax	100+	

FIGURE 3.25   Stages of a secondary succession that begins on abandoned farmland and ends, after only 100 years, in a hardwood forest.

abandonment, the land is populated by two pioneer species: crabgrass and horseweed. During the second year, these species are replaced by aster, a type of wildflower. Tall grass, in turn, replaces the aster by the third year. Shrubs and young pines become established in about the fifth year and gradually shade out the tall grass. The pine forest dominates the succession until it is about 50 years old. At this point, sun-loving pine seedlings can no longer survive in the dense shade on the forest floor. They are replaced by the shade-loving seedlings of oaks and hickories. With the passage of time, mature pines die, one by one, and their places are taken by young oaks and hickories. About 100 years after the succession begins, all remaining pines are shaded out by the oaks and hickories. They eventually become the dominant vegetation in the climax stage of the succession. Note that this secondary succession requires only 100 years to move from the pioneer to the climax stage. By contrast, a primary succession that starts on bare rock or sand may require 500 to 1,000 years before the climax stage is reached. The reason for this is that many of the components of a secondary succession ecosystem, the soil in particular, are already present.

Although we have stressed the succession of plants because vegetational changes are more basic and conspicuous, it should be noted that animal species in the community change as well. This occurs because different animals prefer different types of vegetation—not only as food, but also for protective cover and breeding sites. In the Georgia study, marked changes in breeding bird populations were observed in the various stages. Mammal species also vary during succession. In the example we've been discussing, cottontails and grass snakes of the grass stage were succeeded by such oak–hickory forest inhabitants as the opossum, raccoon, and squirrel. A summary of some of the major trends occurring in a succession is shown in Figure 3.26.

3.5   The Biomes

Anyone who has driven from New England to California is acutely aware of the marked changes in landscape that unfold, from the evergreen forests of Maine to the beech–maple woodlands of Ohio, from the windswept Kansas prairie to the hot Utah desert. Each of those distinctive areas is part of a different life zone, or biome. A **life zone,** or **biome,** is a large terrestrial community characterized by its climate and unique assemblage of plants and animals. The biomes of the world are shown in Figure 3.27 and summarized in Table 3.2. The role of climate in determining biome type is shown in Figure 3.28.

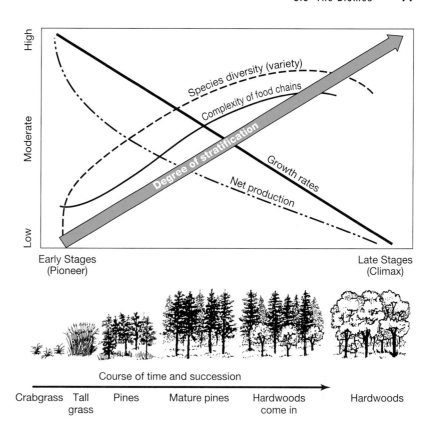

FIGURE 3.26 Summary of some major trends occurring during a secondary ecological succession. The number of species, complexity of food chains, and degree of vegetational stratification (layering) increase as the succession proceeds. However, rates of growth and net production decrease.

Tundra

The Siberian word **tundra** means "north of the timberline." The term is highly appropriate because this biome extends from the timberline in the south to the belt of perpetual ice and snow in the north. The flat to slightly rolling terrain of the tundra, occupying about 10% of Earth's land area, extends around the globe in the northern latitudes.

Because of its relative simplicity, the ecology of the tundra is better understood than that of other biomes. The principal limiting factors are the small amount of solar energy and the bitter winter cold. During June and July, the tundra on the edge of the Arctic Circle is the "land of the midnight sun." In January, however, this same region is the "land of daytime darkness," for the sun never rises. Annual precipitation is less than 10 inches, most of which falls in summer or autumn as rain. Snowfall is scant. Because of the cold weather and the short, six-week growing season, the vegetation in the northern tundra is sparse. As a result, this region is sometimes referred to as an Arctic desert. In fact, plant productivity in the tundra is just slightly higher than in the desert biome. The low temperature slows down the chemical and biological activities that are necessary for the formation of mature soils. Each year during late spring, the upper level of the soil begins to thaw, but the soil beneath it remains permanently frozen and is called **permafrost.** Permafrost occurs at a depth of about 15 to 45 centimeters (6 to 18 inches). In spring and summer the thawed-out ground turns into a quagmire. The meltwaters of late spring form thousands of tiny lakes because of poor drainage and low evaporation. Dwarf willows are characteristic producers. Although only a few meters high, some of these willows may be more than 100 years old. Representative consumer species include herbivores such as the lemming, ptarmigan, cari-

bou, and musk ox (Figure 3.29). The snowy owl and Arctic fox are characteristic carnivores in the area.

The tundra is a fragile ecosystem. One reason is the very slow rate of recovery from human disturbance. Thus, wagon-wheel tracks formed more than 120 years ago by early explorers and inhabitants are still plainly visible.

The tundra is currently threatened by U.S. oil companies' quest for more oil. In 1968, the richest oil deposit in the Western Hemisphere was discovered on Alaska's North Slope. A 1,262-kilometer (789-mile) pipeline was constructed to transport this oil to the port of Valdez on Alaska's southern coast. This caused the destruction of much vegetation and valuable wildlife habitat. With North Slope oil nearly gone, oil companies have set their sights on the Arctic National Wildlife Refuge with hopes of tapping into oil that lies beneath its surface.

In addition to oil, the tundra contains two-thirds of the North American continent's known reserves of coal and substantial amounts of zinc, lead, and copper. These resources could be exploited in the not too distant future. The damage from such development could be severe.

Northern Coniferous Forest

As shown on the global biome map (Figure 3.27), the **northern coniferous forest biome** or **taiga** forms an extensive east–west belt just south of the Arctic tundra in North America, Sweden, Finland, Russia, and Siberia. Characteristic physical features include an annual rainfall of 37 to 100 centimeters (15 to 40 inches), average temperatures of 26.6°C (20°F) in the winter to more than 21°C (70°F) in the summer, and a 150-day growing season. Dominant climax vegetation includes black spruce, white spruce, balsam fir, and tamarack. The evergreen

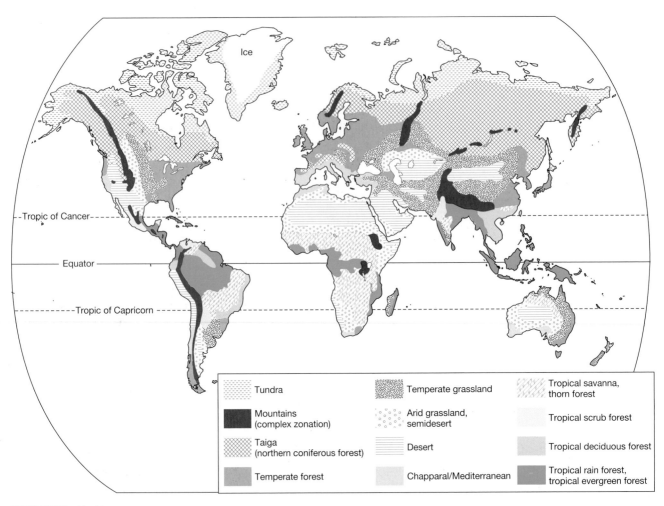

FIGURE 3.27   The biomes.

trees are well adapted to survive; their branches are so flexible that they bend under the burden of a heavy snowfall without snapping. Lightning-triggered crown fires are fairly common because the dead, dry needles, which persist on the branches, are easily ignited. Over 50 species of insects are adapted to feed on the conifers. The populations of some species, such as spruce budworm, tussock moth, pine bark beetle, and pine sawfly, may suddenly erupt, causing considerable mortality of spruce, fir, and pine. White birch and aspen are representative of the earlier successional stages. Typical animals include the moose, snowshoe hare, lynx, pine grosbeak, and red crossbill. The taiga is currently the site of intensive tree cutting and mineral extraction that can have a devastating effect on this biome.

Deciduous Forest

The global biome map shows that the **deciduous** (broad-leafed) **forest biome** covers eastern North America, all of Europe, and parts of China, Japan, and Australia. In the United States, the deciduous forest biome attains its greatest development east of the Mississippi River and south of the northern coniferous forest.

In precolonial times, the deciduous forest of the United States was virtually continuous and unbroken. Today, however, as a result of settlement, agricultural development, logging, mining, and highway construction, the biome has been reduced to a small frac-

tion of its original area. Fingers of deciduous forest extend westward into the prairie country along major river valleys in response to increased levels of soil moisture. Because of abundant precipitation (at least 75 centimeters [30 inches]) and the fairly long growing season, plant productivity is the highest of any biome in North America except the temperate rain forest of the Pacific Northwest. Characteristic trees are oak, hickory, beech, maple, black walnut, black cherry, and yellow poplar. Representative consumers include the gray squirrel, skunk, black bear, and white-tailed deer.

Tropical Rain Forest

The **tropical rain forest biome** is located in the warm, equatorial regions wherever there is more than 200 centimeters (80 inches) of rainfall annually. As shown in Figure 3.27, rain forests are located primarily in Central America, in South America along the Amazon and Orinoco rivers, in the Congo River basin of Africa, in Madagascar, and in Southeast Asia. Heavy rains may fall daily throughout the rainy season. Because so much sunlight is screened out by the canopy, the forest floor is relatively dark and poorly vegetated.

Plant life is highly diverse in the rain forest. Two hundred species of trees may be found in 1 hectare (2.47 acres), compared with only ten species per hectare in the temperate deciduous forests of the United States. The forests are highly layered, or

TABLE 3.2	Biome Summary			
Biome	Temperature	Rainfall	Typical Plants	Typical Animals
Tundra	–57–16°C	10–50 cm	Lichens, mosses, dwarf willows	Ptarmigan, snowy owl, lemming, caribou, musk ox, Arctic fox
Northern coniferous forest	–54–21°C	35–200 cm	Black spruce, white spruce, balsam fir, white birch, aspen	Spruce budworm, tussock moth, moose, snowshoe hare, lynx
Decidous forest	–30–38°C	60–225 cm	Oak, hickory, beech, maple, black walnut, yellow poplar	White-tailed deer, gray squirrel, skunk, opossum, black bear
Grassland	8–17°C	25–75 cm	Little and big bluestem, grama grass, buffalo grass	Meadowlark, burrowing owl, pronghorn antelope, badger, jackrabbit, coyote
Desert	2–57°C	0–25 cm	Prickly pear cactus, saguaro cactus, creosote bush, mesquite, sagebrush	Diamond-backed rattlesnake, Gila monster, roadrunner, kangaroo rat, wild pig
Savannah	13–40°C	25–90 cm	Baobab tree, acacia tree, grasses	Zebra, giraffe, wildebeest, elephant, antelope
Tropical rain forest	18–35°C	125–1250 cm	Great diversity	Great diversity

stratified, with the tree crowns occurring at three or even four different levels.

Animal life is abundant and highly diverse. At least 369 species of birds have been identified in just 2 hectares of Costa Rican rain forest—more than are found throughout Alaska.

Some insects are extremely large. In the Amazon rain forest, the wingspread of one species of moth is nearly 1 foot, and some spiders are large enough to feed on birds caught in their webs!

Tropical rain forests are often cut down or destroyed to make room for human settlements, to acquire wood, to establish farms

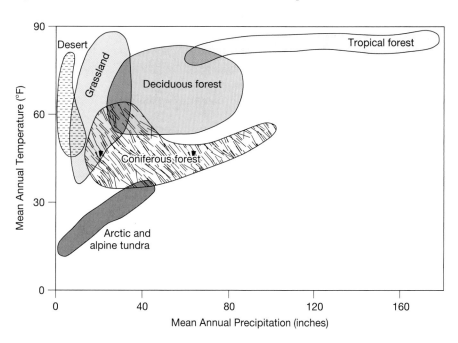

FIGURE 3.28   Influence of temperature and rainfall in determining biome types.

FIGURE 3.29   The ptarmigan, a characteristic tundra herbivore. It is consumed by snowy owls, Arctic foxes, and other predators.

and pastureland, and to clear land for mining operations. Agricultural ventures usually result in failure for two reasons. In many cases, the soil is very infertile and soon stops producing. The soil frequently contains iron compounds that cause it to bake brick-hard under the tropical sun during the dry season. Deforestation has contributed to the atmospheric buildup of carbon dioxide and the warming of the planet (Chapter 19).

Tropical Savannah

A **savannah** (sah-van´-ah) is a warm-climate grassland characterized by scattered trees. As shown in the global biome map (Figure 3.27), savannahs are located primarily in South America, Africa, India, and Australia. Rainfall averages 100 to 150 centimeters (40 to 60 inches) per year. Wet seasons alternate with dry seasons, and fires are common during the prolonged dry spells. Plants and animals that live in the savannah must therefore be drought- and fire-resistant. As a result, the diversity of species is not high.

The African savannah is characterized by the picturesque baobab and thorny acacia trees. This savannah supports the greatest variety and largest number of hoofed herbivores in the world, including the zebra, giraffe, wildebeest, and many species of antelope. In the struggle to produce more food, many people in Africa and India have converted extensive acreages of savannah into farms and livestock pasture. Those disturbances, along with illegal hunting, have caused a rapid decline in the herds of many species.

Grassland

The **grassland biome** is a huge expanse of land characterized, as its name states, by grasses. This land of no trees—except along rivers and where planted by humans around farms, cities, and towns—consists of flat to rolling hills. The main grasslands of the planet include the North American prairie of Canada and the United States, the pampas of South America, the steppes of Eurasia, and the veldts of Africa (Figure 3.27). Major grasslands occur in two regions in the United States: The Great Plains, a vast

area extending from the eastern slopes of the Rockies to the Mississippi River, and the moister portions of the Great Basin—a huge area that nestles between the Sierras to the west and the Rockies to the east. In north temperate latitudes, grasses predominate because average annual precipitation exceeds that typical of the desert biome (over 25 centimeters [10 inches]) but falls below that required for forests (under 75 centimeters [30 inches]). Winter blizzards and summer drought can be severe. Studies suggest that over time, devastating fires have periodically burned the prairie. But the deep root systems of the plants of the grassland, which tap into groundwater supplies during periodic drought, also permit the grasses to regrow after fire.

The dominant vegetation of grasslands includes the big bluestem, little bluestem, buffalo grass, and grama grass. The horned lark, meadowlark, and burrowing owl are characteristic birds. Dominant mammals include the pronghorn antelope, badger, white-tailed jackrabbit, coyote, and prairie dog (Figure 3.30).

Today, only scattered remnants of the climax grassland biome remain. Humans have replaced the original wild grasses with cultivated "grasses" such as corn and wheat. In addition, humans have almost eradicated the bison and replaced it with domesticated herbivores such as cattle and sheep.

Desert

As shown in the global biome map, the world's largest **deserts** are located in the United States, Mexico, Chile, Africa

FIGURE 3.30   A prairie dog, resident of the grassland biome, at the edge of its craterlike burrow entrance. The mound of earth not only serves as a lookout post but also prevents flash-flooding of its burrow.

(Sahara), Asia (Tibet and Gobi), and Australia. American deserts are located in the hotter, drier portions of the Great Basin and in parts of California, New Mexico, Arizona, Texas, Nevada, Idaho, Utah, and Oregon.

Deserts are arid regions that occur primarily leeward (downwind) of prominent mountain ranges, such as the Sierra Nevada and the Rocky Mountains. The prevailing warm, humid air masses from the Pacific Ocean cool as they move up windward slopes and release their moisture as rain or snow. The region leeward of the mountains lies in the rain shadow, where precipitation is minimal. A desert community generally receives less than 25 centimeters (10 inches) of annual precipitation. Rainfall, moreover, may not be uniformly distributed, but may fall periodically in cloudbursts that cause flash floods and severe erosion. An extremely high evaporation rate aggravates the severe moisture problem. For example, the water that theoretically could evaporate from a given hectare in one year may be 30 times the actual amount received as precipitation.

Summer temperatures range from about 10°C (50°F) at night to about 48°C (120°F) during the day. Desert floor temperatures reach 62°C (145°F) in summer. Only organisms that have evolved to survive in extreme heat and dryness can survive in the desert. Characteristic producers are prickly pear cactus, saguaro cactus, creosote bush, and mesquite. Conspicuous among desert consumers in the United States are the rattlesnake, Gila monster, roadrunner, jackrabbit, kangaroo rat, and wild pig.

Altitudinal Biomes

By traveling several thousand miles from Texas northeastward to northern Canada, you pass through a series of different biomes—desert, grassland, coniferous forest, and tundra. These biome changes reflect progressive changes in temperature and precipitation, which in turn result in dramatic differences in primary productivity (Figure 3.31). Such a series of biomes also exists on the slopes of tall mountains, such as the 4,200 meter (14,000-foot) Colorado Rockies, and are know as **altitudinal biomes.** Here again, the gradual changes in biome type are dictated by climatic factors, which in this case change progressively with altitude rather than latitude.

3.6 Ecology and Sustainability

Ecology is a science that examines relationships between organisms and their environment. This is another way of saying that ecology concerns itself with the way natural systems work. Few fields of study could be more important to resource management and sustainable development than ecology. We believe that a solid grounding in basic ecological principles can help resource managers better predict the impacts of various actions—for instance, the elimination of a predator from a given ecosystem. A basic understanding of ecology helps resource managers predict how human actions such as building a dam or bulldozing a road through

a forest will affect species on the site and in neighboring areas. A little knowledge can even help us avoid disastrous consequences. It may suggest alternative approaches or may cause us to rethink plans entirely.

Ultimately, the study of ecology could help human civilization steer onto a sustainable course. For example, an understanding of the laws of energy and the loss of biomass in food chains suggests that we could more easily feed the world's growing population if people ate more fruits, grains, and vegetables. An understanding of the nutrient cycles and of how they can be adversely altered through industrial and agricultural development could help us devise strategies that do not disrupt important cycles of life.

Knowledge of succession and the productivity of different successional communities has proved useful in managing resources for maximum yield. That knowledge, combined with an understanding of factors essential to the long-term health of ecosystems, could help us achieve optimum **sustainable yields**—that is, optimum productive rates that do not threaten the health of farms, forests, and other ecosystems.

The law of tolerance reminds us of the need to act in ways that preserve conditions essential to other life forms. By exceeding the limits, we become agents of destruction. The niche concept helps us better understand a species's full requirements and helps us manage ecosystems more sustainably.

With the understanding that an organism's interactions are complex, we are compelled to examine the big picture before we embark on any course of action. This systems view is essential to sustainable development.

The preceding chapters described a set of biological principles that would ensure the sustainability of natural systems. These principles are conservation, recycling, the use of renewable resources, restoration, and population control. In Chapter 1, we pointed out that natural systems tend to endure because most organisms use what they need and use resources efficiently. This is the conservation principle. Although there are exceptions, such as a grizzly that kills salmon and eats only its eggs, most organisms are fairly frugal. In addition, in natural systems all wastes are recycled. Thus, the waste products of one organism become the food source of another. Consequently, all nutrients cycle continuously through natural systems.

Nature also sustains itself because it relies on renewable resources: the sun, soil, water, and plants. A renewable resource base ensures life's continuation. In addition, natural systems repair damage. They restore themselves. Finally, natural systems control numbers within the carrying capacity of the environment. Collectively, these principles ensure the continuation of life on our finite planet.

But life is not static. Organisms change by evolution. In other words, organisms can adapt to meet changing conditions. This, too, is essential to life's continuation. Humans must also adapt to a changing world, but we do not have the luxury of time needed to adapt biologically. We must make social, economic, political, and even personal changes to ensure our long-term survival. By applying these rules to our own society, we can adapt to the changing conditions of a world threatened by our own actions.

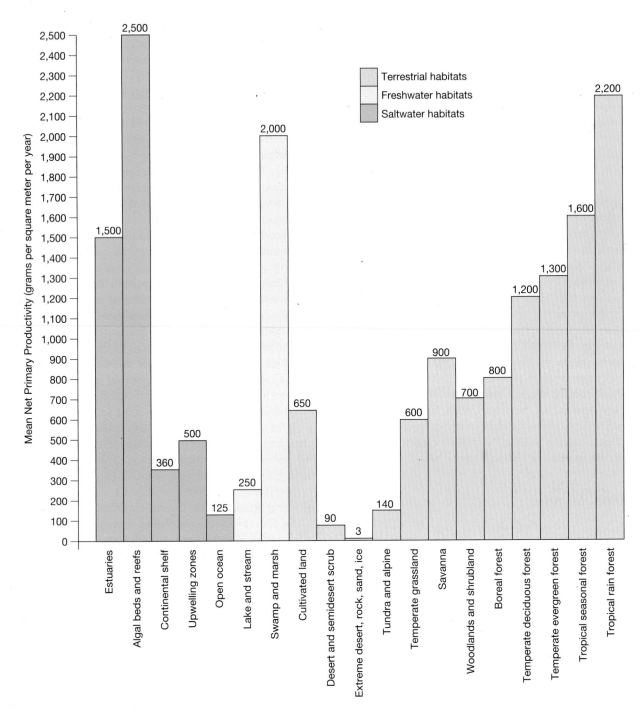

FIGURE 3.31   Biome mean net primary productivity.

Summary of Key Concepts

1. Ecology is the study of interrelationships in the natural world. Ecologists examine relationships between organisms and between organisms and their environment. Ecologists focus their studies primarily on three levels of organization: the population, the community, and the ecosystem.

2. A population is the number of individuals of one species in a given area. A community includes all populations of organisms in a given area—including plants, animals, and microor-

ganisms. An ecosystem is the biological community and the chemical and physical environment with which it interacts.

3. All of the ecosystems of the Earth form the biosphere or ecosphere.

4. All living things consist of matter. Matter is anything that occupies space and is perceptible to the senses. Matter can neither be created nor destroyed, but it can be converted from one form to another. This is known as the law of conservation of matter.

5. Living things depend on energy. Energy exists in many forms. Energy obeys certain laws. The first law of energy states: although energy cannot be created or destroyed, it can be converted from one form to another. The second law of energy states: whenever one form of energy is converted to another, a certain amount is lost as heat.

6. Virtually all of the energy required by living organisms comes from photosynthesis. Photosynthesis is the process by which green plants convert the raw materials carbon dioxide and water into glucose and other organic molecules in the presence of sunlight. Sunlight energy is stored in the chemical bonds of the molecules produced by plants. Some of it is used by the plants; much of the rest will be used by animals and microorganisms in various food chains.

7. A food chain is a sequence of organisms through which nutrients and energy move. In ecosystems, food chains form interconnected networks known as food webs.

8. The stability of a food web increases in proportion to its complexity. Simplifying a food web makes an ecosystem more unstable.

9. During photosynthesis, green plants convert energy from the sun to organic food energy that can be used by other organisms. Because of this process, green plants are called producers. Plant-eating animals, such as the rabbit, are called herbivores. Animals like the wolf, which feed on other animals, are known as carnivores.

10. The total dry weight of living substance (protoplasm) in a given organism, population, or community is known as the biological mass, or biomass. Studies of ecosystems show that biomass decreases at each trophic level in a food chain. This is known as the pyramid of biomass.

11. Because biomass contains potential energy and because living organisms are only about 10% efficient in converting the energy of their food into the energy of their own biomass, there is a progressive reduction in the amount of food energy in successive trophic levels (feeding levels) of a given food web. This is known as the pyramid of energy.

12. The circular flow of an element from the nonliving environment into the bodies of living organisms and then back into the nonliving environment is known as a nutrient cycle. For millions of years, nutrient cycles were in balance. However, in recent years, humans have upset many nutrient cycles, usually by greatly increasing the amount of nutrients in these cycles. The result is the pollution of air, water, and land—often with serious consequences.

13. Each organism possesses a specific tolerance range for any essential environmental factor, known as the range of tolerance.

14. Organisms occupy a specific area, known as their habitat. Their niche is a more complex concept. It consists of many factors, including the foods they eat, their predators, nesting requirements, and so on. The niche describes an organism's unique role in an ecosystem—its requirements and its relationships.

15. Ecologists have found that organisms cannot occupy the same niche. This is known as the competitive exclusion principle.

16. The size of a population results from the interaction of environmental resistance and the factors that create its biotic potential.

17. The biotic potential of a species is the theoretical maximum rate of growth for a population, given unlimited resources. In nature, few organisms reproduce at this rate for any appreciable length of time because of environmental resistance.

18. Environmental resistance refers to all of those factors that reduce population growth—for instance, adverse weather, food shortages, predation, and disease. In general, when environmental resistance climbs, a population declines. When it falls, the population increases.

19. Ecologists group factors that contribute to environmental resistence into two categories: density-dependent and density-independent.

20. Density-independent factors are those that operate irrespective of the population density—for example, drought, cold spells, and storms. Habitat destruction by human populations is another example.

21. Density-dependent factors are those whose influence in controlling population size increases or decreases depending on the density of a population. There are at least four major density-dependent factors: competition, predation, parasitism, and disease.

22. Whenever a species is introduced into a new habitat with adequate resources, its population grows in a characteristic way, following an S-curve. During the second phase, explosive growth occurs because little environmental resistance exists. During the final phase, the population stabilizes as a result of environmental resistance. At this point, the population is said to be in dynamic equilibrium. The population has reached carrying capacity, the population size a habitat can support on a sustainable basis.

23. Wildlife managers use their knowledge of population growth curves to manage wild populations and to achieve the maximum sustainable yield.

24. Not all populations remain in dynamic equilibrium. Outside forces such as highly favorable weather can cause populations to increase abruptly. Such irruptive growth may extend the population past the habitat's carrying capacity and may result in a population crash. Human activities can easily upset the balance, resulting in irruption.

25. A number of species, such as the lemming, naturally experience three- to four-year cycles. Others, like the lynx and snowshoe hare, experience ten-year cycles.

26. The replacement of one community by another under constant climatic conditions is known as biological succession. The initial stage of a succession is known as a pioneer community. The final stage of a succession is known as the climax community. This stage is stable. In theory it will persist indefinitely unless the climate changes or some geological process or human action destroys it.

27. Succession occurring in an area where living organisms were not previously present, such as bare rock, is known as a primary succession. However, a succession that occurs in an area that once supported life, such as a burned-over forest or an abandoned corn field, is known as a secondary succession.

28. A biome is the largest terrestrial community that is easily recognized by a biologist. It is the biological expression of the interaction of climate, soil, water, and organisms. Representative biomes discussed in this chapter are tundra,

northern coniferous forest, deciduous forest, tropical rain forest, tropical savannah, grassland, and desert. Each biome is characterized by a unique assemblage of plants and animals. On the slopes of tall mountains, such as the Rockies, a series of altitudinal biomes can be recognized.

29. An understanding of ecology helps us better understand human impacts on the environment and ways to better manage ecosystems. Ultimately, it can help us steer onto a sustainable course. By practicing ecological principles of sustainability we can create a way of life that respects and operates within the limits imposed by global ecosystems.

Key Words and Phrases

Abiotic Factors
Albedo
Alien Species
Altitudinal Biomes
Atmospheric Fixation
Biogeochemical Cycle
Biological Community
Biological Fixation
Biological Succession
Biomass
Biomass Pyramid
Biome
Biosphere
Biotic Factors
Biotic Potential
Carbon Cycle
Carnivore
Cellular Respiration
Chlorophyll
Climax Community
Community
Competition
Competitive Exclusion
  Principle
Consumer
Cyclic Populations
Deciduous Forest Biome
Denitrification
Desert
Density-Dependent Factors
Density-Independent Factors
Detritus
Detritus Feeder
Detritus Food Chain
Dynamic Equilibrium
Ecology
Ecosphere
Ecosystem
Ecosystem Simplification
Elemental Cycle
Energy
Energy Laws
Energy Pyramid
Entropy
Environmental Resistance
Exotic Species

First Law of Energy
Fixed
Food Chain
Food Web
Global Warming
Grassland Biome
Grazer Food Chain
Gross Primary Production
Habitat
Herbivore
Host
Industrial Fixation
Irruption
Kinetic Energy
Law of Conservation of Matter
Laws of Thermodynamics
Legume
Lichen
Life Zone
Matter
Maximum Sustainable Yield
Net Primary Production
Niche
Nitrification
Nitrogen Cycle
Nitrogen Fixation
Northern Coniferous Forest
  Biome
Nutrient Cycle
Omnivore
Optimum Yield
Parasite
Permafrost
Phosphorus Cycle
Photosynthesis
Pioneer Community
Population
Population Dynamics
Potential Energy
Predator
Primary Consumer
Primary Production
Primary Succession
Producer
Pyramid of Numbers
Quaternary Consumer

Range of Tolerance
Reservoir
Reservation
Root Nodules
S-Shaped (Sigmoidal) Curve
Savannah
Second Law of Energy
Secondary Consumer
Secondary Succession

Succession
Sustainable Yield
Taiga
Tertiary Consumer
Trophic Level
Tropical Rain Forest Biome
Tundra
Wavelength

Critical Thinking and Discussion Questions

1. Define and give an example of each of the following levels of organization: community, population, and ecosystem.

2. Define the biosphere. Why are most life forms restricted to a narrow band within the biosphere?

3. For many years, a lake has been operating as a balanced ecosystem. Suppose now that a disease kills all of the plants. What effect would this have on the animals in this ecosystem? Suppose that instead of killing all of the plants, a disease destroys all the bacteria in the lake. What effect would this have on the lake ecosystem? Discuss your answer.

4. What is energy? What is mass?

5. State the first and second laws of energy. What do they mean to you? How do they affect you? How do they affect ecosystems?

6. Trace the radiant energy from your study lamp backward in time to its ultimate source: the sun.

7. Define photosynthesis with respiration. How are they dependent on each other?

8. What is a food chain? What is a food web?

9. Distinguish between grazing food chains and detritus food chains. Give an example of each type. Which type is more easily observed? Which type of food chain is most important in a forest ecosystem? Which type is most important in a marine ecosystem?

10. How many different species of organisms can you identify that are living either in your dormitory or in your home? Are they producers, herbivores, carnivores, omnivores, or detritus feeders? Compare your results with those of other members of your class.

11. List all the food organisms you consumed, in part or in their entirety, during your last meal. Now construct a food web, using this list and yourself as a starting point. You may include other organisms not actually represented in your meals. Identify the trophic level of each species in the web.

12. The total biomass of the animals living in the English Channel is five times greater than that of the plants in it. Can you offer an explanation for this seeming contradiction of the biomass pyramid concept?

13. Elements are recycled through ecosystems, but energy is not. Suppose, however, that energy were recycled and elements were not. What would be the effects on ecosystems? Discuss your answer.

14. Suppose that the sun stopped shining. What effect would this have on life on Earth? Would life still be possible? Discuss your answer.

15. What are nutrient cycles? Why is it important to know about them?

16. Make labeled diagrams of the shortest nitrogen, carbon, and phosphorus cycles you can devise.

17. Describe three ways in which nitrogen can be fixed.

18. Trace a given carbon atom from the air that existed over Pennsylvania 250 to 300 million years ago to the fried egg you had for breakfast this morning.

19. Pollution can be defined as an imbalance of a nutrient cycle caused by humans. Explain. Can you give one example each of land, air, and water pollution that alters a nutrient cycle?

20. Discuss the ranges of tolerance that you have for various physical conditions in the daily environment to which you are exposed.

21. Why is it important to understand the range of tolerance of species in an ecosystem?

22. What is the competitive exclusion principle?

23. Define the terms *biotic potential* and *environmental resistance*. Give some examples of factors that increase the biotic potential of a species.

24. Describe and give examples of density-dependent and density-independent factors.

25. A new rodent is introduced to an island with no natural predators or diseases. Describe the likely course its population will take. Draw a graph to show the shape of the curve, and explain what happens at each stage. What factors will bring the population into dynamic equilibrium?

26. Define the term *biological succession.*

27. Describe three sites at which a primary succession might start. Describe three sites at which a secondary succession might occur.

28. Construct a chart listing four basic differences between the pioneer and climax stages of an ecological succession.

29. Suppose that a half acre of oak woods in Ohio is removed and replaced with a huge slab of polished marble. One thousand years pass. Will the marble slab still be visible? Why or why not? Give a detailed explanation of the probable events that occurred.

30. What is a biome? In which biome do you live? What are its major characteristics?

31. Describe the physical features of the tundra biome. Why is it sometimes called an Arctic desert? Why is it considered a fragile ecosystem? Would it be a good place to study succession? Why or why not? Explain.

32. Which of the biomes discussed in this chapter has the greatest diversity of animal life? Which features of the biome make this diversity possible?

Suggested Readings

Bebee, A., and A.-M. Brennan. 2007. *First Ecology: Ecological Principles and Environmental Issues.* New York: Oxford University Press. For students interested in learning more about ecology.

Benyus, J. M. 1997. *Biomimicry: Innovation Inspired by Nature.* New York: Perennial. A book that looks at many exciting technological innovations that are inspired by nature's design.

Chiras, D. D. 1992. *Lessons From Nature: Learning to Live Sustainably on the Earth.* Washington, DC: Island Press. Discusses how to build a sustainable society.

Dodson, S. I., et al. 1998. *Ecology.* New York: Oxford University Press. Great reference for students who are eager to learn more about ecology.

Dodson, S. I., et al. (eds.). 1999. *Readings in Ecology.* New York: Oxford University Press. Collection of essays including classical studies and new, interesting thinking about ecology.

Ehrlich, P. R. 1986. The *Machinery of Nature.* New York: Simon and Schuster. Great general discussion of ecology.

Mayhew, P. J. 2006. *Discovering Evolutionary Ecology: Bringing Together Ecology and Evolution.* New York: Oxford University Press. An insightful look at ecology from an evolutionary perspective.

Morgan, S. 1995. *Ecology and Environment: The Cycles of Life.* New York: Oxford University Press. A very general, but very beautiful introduction to basic principles of ecology.

Odum, E. P. 1993. *Ecology and Our Endangered Life-Support Systems.* Sunderland, MA: Sinauer Associates. Good introduction to ecology.

Ricketts, T. H., et al. 1999. *Terrestrial Ecoregions of North America.* Washington, DC: Island Press. A book full of important details on terrestrial ecosystems.

Smith, R. L., and T. M. Smith. 2005. *Elements of Ecology,* 6th ed. Redwood City, CA: Benjamin Cummings. Good coverage of ecology and environmental problems.

Southwick, C. H. 1996. *Global Ecology in Human Perspective.* New York: Oxford University Press. A very readable introduction to ecology.

Wilkinson, D. M. 2007. *Fundamental Processes in Ecology: An Earth Systems Approach.* New York: Oxford University Press. A thought-provoking book for anyone interested in the study of ecology.

 ## Web Explorations

Online resources for this chapter are on the World Wide Web at: **http://www.prenhall.com/chiras** *(click on the Table of Contents link and then select Chapter 3).*

THE HUMAN POPULATION CHALLENGE

Growth of the human population is one of the world's most pressing environmental issues. It is at the root of many other problems, including poverty, unsanitary living conditions, global pollution, and the depletion of resources.

How serious is the problem? Consider a few statistics. In a single day, the world population will expand by over 220,000 people. In a week, the human population grows by 1.5 million. In a year, the Earth and its ecosystems must support another 80 million people. Each of these individuals must be fed, and each one must acquire shelter and clothing. And each one will produce waste.

Fortunately, not all of the news about population is alarming. Although the global human population is increasing, since 1970 the rate of growth has slowly decreased. Most of the slow-down has occurred in the **less-developed countries (LDCs)** of Asia and South America, thanks in part to family-planning programs. The populations in some European countries have stopped growing altogether, a condition called **zero population growth (ZPG),** thanks to family-planning programs and individual actions. The populations in some European countries such as Poland and Lithuania are even on the decline. Russia and the former Eastern European nations have also witnessed a decrease in growth. The decrease in Russia has been attributed in large part to the difficult transition to democracy and capitalism, which left Russia in economic ruin for many years. Growth has slowed in Africa as well, although the decrease has been largely due to increased mortality from starvation, civil war, and infectious disease, notably AIDS.

Important as the decline in population growth is, it does not mean that the Earth's population will stop growing. In fact, the Population Reference Bureau (PRB), a nonprofit organization in Washington, DC, that tracks population growth, predicts that the current population of about 6.7 billion in 2008 could reach 12 billion by 2050 and 24 billion by 2100, *if the current rate of growth continues.* (However, few experts, including researchers at PRB, believe the population will continue to grow. Most predict a leveling off at 8.5 billion to 12 billion.) The reason for continued growth despite a decline in the rate of growth will be made clear soon.

4.1 Understanding Populations and Population Growth

Before we look at the full impact of the exploding human population and suggest some strategies to control it, we will discuss some concepts and key terms that are used to describe populations and population growth. These, in turn, will help you understand the importance of the problem and ways various solutions will help solve it.

Birth and Death Rates

Two of the most fundamental terms you must understand are birth rates and death rates. The **birth rate** of a country is simply the number of persons born per 1,000 individuals in the population in a given year. The birth rate of the United States, for instance, is currently 14. Thus, for every 1,000 people in the population, there are 14 babies born each year.

The **death rate** is the number of persons per 1,000 individuals who die in a particular year. In the United States, the current death rate is 8.

The difference between the birth rate and death rate of a country is known as the **rate of natural increase** (or decrease). **Demographers,** scientists who study populations, use the following equation to determine natural increase:

Natural increase = growth rate − death rate

In the United States, the rate of natural increase is 14 per 1,000 minus 8 per 1,000, which equals 6 per 1,000. That means every year, the population expands by 6 people for every 1,000 people currently living in the country.

This example shows that if the birth rate is higher than the death rate, the population will expand. However, if the birth rate is lower than the death rate, a population will decrease.

Now take a moment to study Figure 4.1. As you can see, the birth rate minus the death rate equals the natural increase. But this figure shows that another factor also influences a country's population growth rate: the **net migration**—that is, the number of people moving into or out of a country. Demographers use the following equation to determine the population growth in a country:

Growth rate = birth rate − death rate + net migration

In the United States, net migration is 4 per 1,000. Net migration is equal to immigration minus emigration. According to several sources 1.5 to 2 million legal and illegal immigrants enter the United States each year. In 2005, an estimated 225,000 people emigrated from the United States to other countries. In the United States, then, the natural increase in the population is about 6 per 1,000, and net migration is 4 per

1,000, bringing the total increase to 10 per 1,000 or slightly under 3 million people per year, according to data from the Population Reference Bureau.

Now that you understand a little about population growth, let's use your knowledge to explore the global human population. According to the Population Reference Bureau, the birth rate of the human population in 2007 was 21 per 1,000. The death rate was 9 per 1,000. This results in a natural increase of about 12 people per 1,000.

The population growth rate of a nation or the world can be calculated *as a percentage,* as follows:

% annual growth rate = birth rate − death rate × 100

Using this formula, the **percent annual growth rate** for the world is determined as follows:

$$\frac{21}{1,000} - \frac{9}{1,000} = \frac{12}{1,000}$$

$$\frac{12}{1,000} \times 100 \quad = 1.2\%$$

Exponential Growth

Figure 4.2 shows the growth of the human population over the past millennium. As illustrated, the human population remained small for many thousands of years. Only in the past 200 years has it begun to climb—reaching 6.7 billion in 2008.

Demographers call this type of growth pattern exponential growth, a term you may have already heard. What exactly is exponential growth, and why all of the fuss over it? **Exponential growth** occurs any time something grows by a fixed percentage if the annual growth (annual increase) is added to the base amount. A bank account growing at 5% a year, for example, is said to be growing exponentially if the interest is retained in the account so you earn interest on your interest, too. As shown in Figure 4.2, exponential growth results in a **J-shaped curve.**

Exponential growth is quite deceiving. As the graph shows, initial growth is quite slow, but there comes a time in the growth process when the population—or whatever else you're

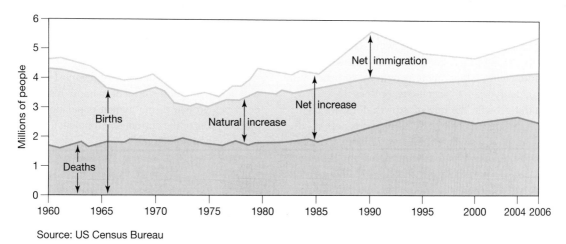

Source: US Census Bureau

FIGURE 4.1 Components of annual population change in the United States, 1960–2006. U.S. population growth is influenced by birth rates, death rates, and net migration.

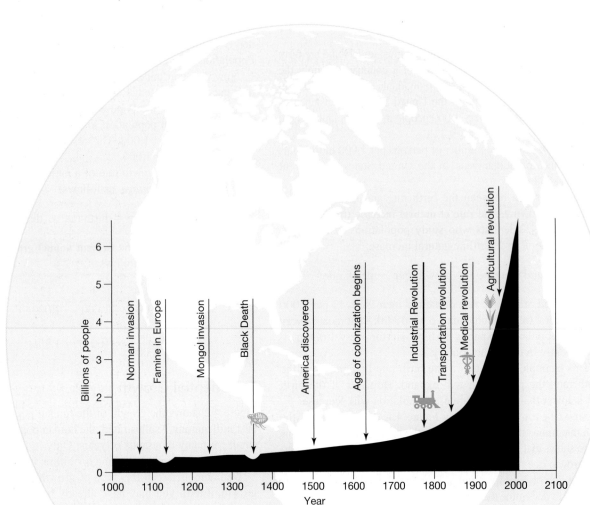

FIGURE 4.2 Exponential growth of the human population. Growth has skyrocketed in the past 200 years as a result of improvements in sanitation, medicine, and agricultural production.

examining—experiences a very rapid upswing. Once the population rounds the bend of the J-curve, the rapid upswing seems to occur from nowhere, taking many people by surprise. Why does growth suddenly seem to accelerate, or does it?

In exponential growth, the rate of growth does not change to produce this mysterious effect. What happens is that, after a long period of slow growth in absolute numbers, the base amount gradually reaches a size at which even small percentage increases result in very large increases in the absolute number. Thus, even though the entity may be growing at a rate that seems insignificant, if the base becomes large, a small percentage increase results in a huge numerical increase. The human population, for example, was 6.7 billion in 2008. Even though it is growing at a rate of 1.2% per year, which seems small, the net increase is huge—80 million

people a year! And as population increases, so does resource demand.

Exponential growth is one of the most important concepts you will learn in your study of natural resource conservation. It describes growth of the human population and growth of many other aspects of the current environmental crisis, such as resource depletion and pollution.

GO GREEN!

If you are contemplating marriage and children, consider delaying motherhood or fatherhood, and consider having only one or two children to help reduce population growth.

Doubling Time

Demographers also use another measure to help understand populations and predict their growth: the doubling time. **Doubling time** is the time it takes a population to double. To determine the doubling time of a population, simply divide 70 (the demographic constant) by the percent annual growth rate. For the global population (in 2008), the doubling time is

$$\frac{70}{1.2} = 58 \text{ years}$$

At its current rate of growth, the world population is projected to double in 58 years.

Doubling times vary considerably from one nation to another. On average, the more-developed nations are growing at a rate of about 0.1% per year, which yields a doubling time of about 700 years. The less-developed nations are growing at a rate of about 1.5% per year, doubling in about 47 years. Table 4.1, on page 96, lists some other important features of more-developed and less-developed nations.

Why Has Global Population Skyrocketed?

As noted earlier in the chapter, the human population has skyrocketed in the last 200 years, increasing from about 1 billion to 6.7 billion. What has caused this phenomenon?

For years, the human population was held in check by disease, starvation, wars, and other factors. Infections caused by viruses, bacteria, and protozoa killed many children early in life and took the lives of many adults, too. In medieval times, for example, plagues killed over 25 million people in Asia and Europe. Smallpox spread like wildfire and killed one out of every four afflicted persons until Jenner developed his vaccine at the close of the 18th century. As a result, a child born in 1550 had a life expectancy of only 8.5 years. Disease continued to cause problems into the early 1900s. In fact, in 1900 tuberculosis was a prime killer, with pneumonia running a close second. As late as 1919, the influenza virus took a toll of 25 million people worldwide.

Although birth rates were high during this time, the human population remained small because of the extremely high death rates. But then a series of advances in agriculture, medicine, and sanitation came along. The invention and perfection of the plow and other farm machinery, for example, made more food available, helping to stave off hunger and starvation. Later on, new drugs and vaccines combated viral and bacterial infections. Today, tuberculosis and pneumonia are largely controlled by antibiotics. Effective vaccines have been developed against smallpox, tetanus, diphtheria, whooping cough, influenza, and many other diseases. Better sanitation methods have had a positive effect, too. In particular, the treatment of human waste and the purification of drinking water have helped to reduce the incidence of infectious disease. Combined, these factors resulted in a dramatic decrease in death rates. Largely because of these changes, the world death rate has fallen from 25 per 1,000 in 1935 to about 9 per 1,000 today.

Chemical pesticides also have aided in the expansion of the human population. Early in the 20th century, for example, the mosquito-borne disease malaria was either directly or indirectly responsible for 50% of all human mortality. After World War II, this disease was held in check by insecticides such as DDT. In Sri Lanka (formerly Ceylon), an intensive malaria-control campaign launched in 1946 resulted in a decline in the mortality rate from 22 per 1,000 to 13 per 1,000 in only six years.

Slowly but surely, humans lowered the death rate, but the decrease in mortality was not accompanied by a decrease in birth rates—at least for 100 to 150 years. And, as noted earlier, the difference between birth rates and death rates determines the growth of a population.

Demographic Transition: Reestablishing the Balance   The nations that benefited from advances in agriculture, medicine, and sanitation eventually did bring birth rates down. Today, many of these nations continue to post low growth rates, and over 90 nations have stable or even shrinking populations. What caused the transition?

Demographers have found that industrialization and rising wealth have had a profound effect on the birth rate of many nations, resulting in a population change called a demographic transition. **Demographic transition** is defined here as a shift in both birth rates and death rates over time, going from high birth and death rates to low birth and death rates. Virtually all of the industrialized European nations, as well as the United States, Canada, and Japan, have undergone the demographic transition.

Demographic transitions occur in four stages, shown in Figure 4.3, which plots data from Finland as an example. In stage I (pre-industrial), large families prevailed because many children were required to work on farms. They also provided financial security for parents in their old age. The birth rates and death rates, however, were both high, so the population remained fairly constant.

In stage II of the demographic transition, death rates began to fall because of improvements in sanitation, food production, and disease control mentioned earlier. The discrepancy between birth rates and death rates resulted in rapid population growth.

In stage III, rising affluence began to drive the birth rates down. The reason for this shift is that as food production techniques improved, fewer farmers were needed to meet demand. Many farmers and their families abandoned their way of life and moved to urban areas to find work. In the city, however, children were no longer needed to help with farm chores. A large family became an economic drain. Couples voluntarily reduced the number of children they had. As a consequence, birth rates and death rates gradually came into balance, and population growth slowed dramatically. This is stage IV of the transition.

The demographic transition began 100 to 150 years ago in Europe, the United States, and the rest of the industrial nations. In the less-developed nations, however, the demographic transition began much later—after World War II—as modern medicine, pesticides, and other advances already enjoyed by the Western world began to be introduced to these nations. But most LDCs never made it through the demographic transition. They remain stuck in stage II (transition), experiencing lower death rates thanks to advances in medicine and sanitation but

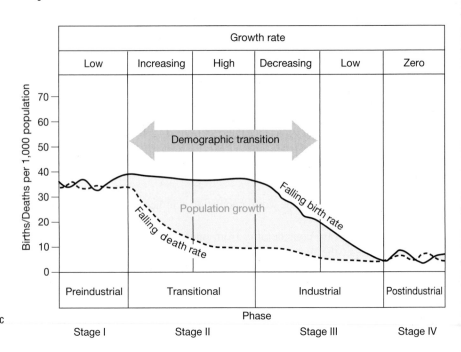

FIGURE 4.3   Characteristic features of the demographic transition.

fairly high birth rates (Figure 4.4). As you would predict, the result is a rapid increase in population among these people. So rapid is their growth that 90% of the global population growth now occurs in the LDCs (Figure 4.5).

The hope among many people was that these countries would industrialize and that this, in turn, would lower the birth rates, as it has in many more-developed nations. However, many experts assert that many LDCs such as Ghana, Chad, and India simply cannot depend on industrial development to carry them through the demographic transition and bring about a reduction in birth rates. Why? There are several reasons:

1. These countries often lack the large number of highly trained people (engineers, for example) on which industrial development depends.

2. They also lack the energy resources such as coal, oil, and natural gas that have been used to fuel industrial development.

3. They do not have enough time. Substantial industrial development cannot be accomplished in less than several decades, even with trained personnel and abundant energy resources. With the populations of many of these nations doubling every 30 to 40 years or less, time is running out!

4. Many countries also lack the financial resources needed to develop economically.

So, what will help these nations reduce their growth rates? Rising death rates, already being seen in some nations, could

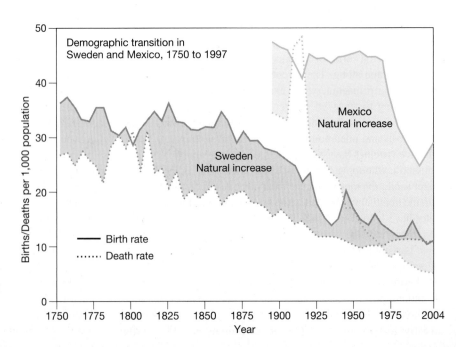

FIGURE 4.4   Demographic transition in Sweden and Mexico. Note that Mexico, like many other LDCs, is stuck in stage II of the transition and is experiencing rapid population growth.

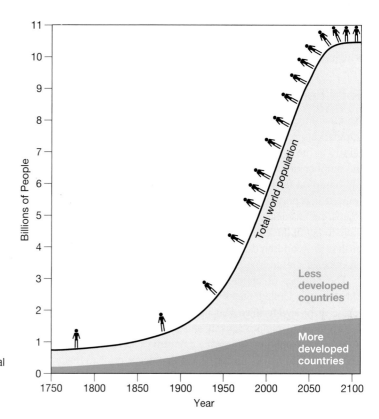

FIGURE 4.5 Growth in the LDCs vs. MDCs. Although the world's annual population growth rate shows a decreasing trend, the number of people in the world is still increasing rapidly. Most growth is occurring in the less-developed nations.

help bring the populations back in balance. As noted earlier, many African nations are already experiencing an increase in death rates due to starvation, infectious disease, civil war, and AIDS. **Family-planning programs,** which allow couples to determine the number and spacing of children they desire, could help stabilize population growth by bringing birth rates down. Sustainable economic development could also assist. Improvements in education and job opportunities for men and women alike also are vital. These options will be described in detail shortly. First, let's take a look at AIDS.

Will AIDS Correct the Imbalance? In the early 1980s, a new human disease known as **acquired immunodeficiency syndrome (AIDS)** began spreading rapidly through the African population and then throughout the rest of the world. AIDS is caused by a virus known as the **human immunodeficiency virus (HIV),** which is transmitted during sexual intercourse and anal intercourse, through the sharing of contaminated needles used for drug injections, through blood transfusions, and from an infected pregnant woman to her fetus. HIV attacks the immune system, greatly diminishing one's protection against infectious diseases such as tuberculosis and pneumonia. Most patients who contract the disease have little hope of surviving it. HIV is nearly 100% fatal, especially in less-developed nations where medical care is limited and access to new drugs to combat it are prohibitively expensive, although aid from the United States and other countries is helping to address this problem.

The spread of this disease is accelerating. Many consider it to be an epidemic. As of December 2006, more than 25 million people worldwide had died from the disease. Fourteen million children have been orphaned as a result of the death of their parents from AIDS. Currently, an estimated 40 million people are infected with the virus. According to the United Nations, 4.3 million people were infected with the AIDS virus in 2006 alone. The hardest hit area is sub-Saharan Africa. Three of every five new infections take place here. Nine of every ten AIDS-related deaths occur in this region as well. In several African nations, 10% to 20% of adults aged 15 to 49 are infected with AIDS. In Botswana and Swaziland, one of every four adults has the virus. It is in these nations that AIDS will have its greatest impact on population growth. In fact, the populations of Botswana and Swaziland are shrinking about 0.1% per year.

Other areas of the world also are witnessing a rise in AIDS. Myanmar, Vietnam, Cambodia, and India are experiencing a rapid increase. Eastern Europe has witnessed a dramatic increase as well. In the United States and Western Europe, AIDS-related deaths have declined, largely due to new antiviral drugs that prolong the onset of the disease after initial infection. However, even here, HIV infections are on the rise.

Not all of the news is bad. Several countries have mounted preventive campaigns to address AIDS and have seen substantial decreases in HIV infections. In about half of the less-developed countries, HIV has not spread into the general public. It is most common among prostitutes, their clients, and addicts.

Unfortunately, a vaccine for AIDS may not be developed for 20 years, and by that time AIDS may have killed 20% of the black population of Africa, and millions of people on other continents as well. Indeed, AIDS ultimately may kill many more people than the Black Death in the 14th century. This devastating disease may eventually be a factor in bringing our global population more in balance with the resource supply on which it depends.

Total Fertility Rate and Population Histograms

Another measure that is extremely helpful in understanding population growth and predicting the future size of the human family is the total fertility rate. The **total fertility rate (TFR)** is the number of children a woman is expected to produce in her life based on fertility rates in various age groups at the present time. It is reported as an average for all women. In the United States, for instance, the TFR in 2007 was 2.1. That is, women alive today in the United States are expected on average to give birth to 2.1 children during their lifetime (every 10 women will produce 21 children).

The number of children required to replace a mother and father in the United States is currently 2.1. This is known as **replacement-level fertility.** This means that each woman must have 2.1 children to replace herself and her husband, or 10 women must have 21 children to replace themselves and their husbands. The extra child makes up for deaths that occur prior to reaching reproductive age. Since 1972, the TFR of American women has been below the replacement-level fertility, but in recent years, it has crept up to replacement level.

Another predictive tool that helps us understand where populations are going is the age structure. The **age structure** is the number of individuals occurring in each age class within the population. Males and females are generally enumerated separately. Plotted as a graph, this forms a **population histogram** (hiss´-toe-gram). By examining the profile of an age structure diagram, it is possible to determine whether a population is growing rapidly, growing slowly, remaining stable, or even declining (Figure 4.6).

When considering age structure, it is useful to divide the population into three major groups: pre-reproductive (ages 1 to 15), reproductive (ages 16 to 45), and post-reproductive (ages 46 to 85 +). Current population growth depends on the number and fertility of the females in the age group 16 to 45. Future population growth depends ultimately on females who are now in the 1- to 15-year-old age group. In other words, the size of this pre-reproductive group tells the future of a population. At the present time, about 28% of the world's population is under 15 years of age. In the less-developed countries, China included, the average is 34%. (In Kenya the figure is 43%!) In the more-developed countries, it is 17%.

An age structure diagram for a typical LDC such as Mexico has a very broad base and a narrow apex. This triangular shape is characteristic of a rapidly growing population. The doubling time for these populations is about 20 to 40 years. The histogram is said to be expansive (Figure 4.6A).

Countries such as the United States are characterized by expansive histograms. Although the U.S. population histogram is shaped like those of other faster-growing nations, its base is not as wide, and the sides of the triangles are steeper (Figure 4.7). This triangular shape indicates that numbers are increasing but more slowly than in fast-growing nations, with doubling times ranging between 40 to 120 years.

Age structure diagrams for slower-growing, nearly stable **more-developed countries (MDCs)** such as Canada have narrow bases and steep sides. The histogram in Figure 4.6B is an example. As you can see, the narrow base and steep sides indicate that the pre-reproductive and reproductive age groups are relatively stable. The histogram is said to be near stationary. The age structure diagram of many European countries has a narrow base, which indicates an extremely slow growth rate, with doubling times of 121 to 3,000 years.

A declining population would show a consistent narrowing at its base. Italy, shown in Figure 4.6C, is an example. Its histogram is said to be constrictive.

Population histograms of countries that are experiencing growth, as are many less-developed nations, indicate the presence

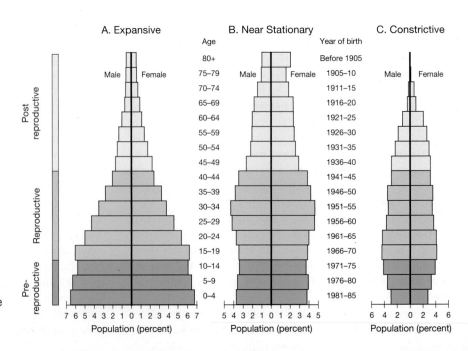

FIGURE 4.6 Characteristic population age structure profiles for (A) expansive, (B) near stationary, and (C) constrictive populations.

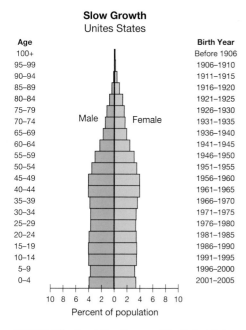

Slow Growth
Unites States

Age		Birth Year
100+		Before 1906
95–99		1906–1910
90–94		1911–1915
85–89		1916–1920
80–84		1921–1925
75–79		1926–1930
70–74	Male Female	1931–1935
65–69		1936–1940
60–64		1941–1945
55–59		1946–1950
50–54		1951–1955
45–49		1956–1960
40–44		1961–1965
35–39		1966–1970
30–34		1971–1975
25–29		1976–1980
20–24		1981–1985
15–19		1986–1990
10–14		1991–1995
5–9		1996–2000
0–4		2001–2005

10 8 6 4 2 0 2 4 6 8 10
Percent of population

FIGURE 4.7 Population diagram of the United States.

of a built-in mechanism for explosive population growth in the near future. From the standpoint of population control, this is an extremely distressing situation. After all, there are already more men and women in the reproductive age group on this planet than there have ever been—about 3 billion.

4.2 The Impact of Overpopulation

With these basic terms and concepts in mind, we now turn our attention to the impacts of population. We will begin by examining a phenomenon commonly referred to as overpopulation.

Overpopulation in the Less-Developed Countries

When discussing environmental issues, many people assert that one of the key problems is overpopulation—too many people. More precisely, **overpopulation** refers to a condition in which the population size exceeds the carrying capacity of the environment. As noted in Chapter 3, carrying capacity refers to the ability of our environment to supply resources and assimilate waste. *Overpopulation, as it pertains to people, means too many people for the available resources. It also means too many people for the planet's waste assimilation/detoxification mechanisms.* What are the symptoms?

In the LDCs, which are generally poor, rural, and agricultural, one of the most obvious symptoms of overpopulation is food shortage and its consequences—hunger, malnutrition, and disease. In 1798, Thomas Malthus, a British economist, stated that populations tend to increase faster than food supplies. He asserted that the only way a population could come into balance with the available food supply would be through a massive die-off resulting from starvation, disease, war, or some other calamity. Although this concept may not be entirely true (population balance has been achieved without

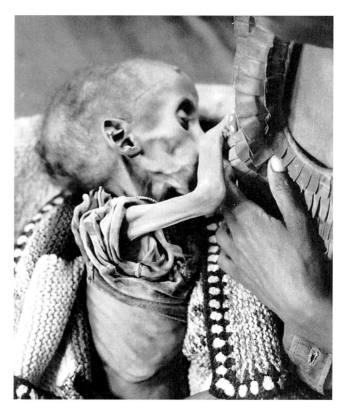

FIGURE 4.8 An African mother with her skeletal child, now close to death from starvation due to Malthusian overpopulation.

these calamities), overpopulation in the less-developed nations is often referred to as **Malthusian overpopulation.** Malthusian overpopulation is described as the result of too many stomachs and not enough food (Figure 4.8). In Kenya, Nigeria, Tanzania, and Uganda, populations will double in 23 years or fewer. Such a population surge has resulted in widespread malnutrition and starvation.

The effects of Malthusian overpopulation are grimly described in the following extract by Lee Ranck, an executive of an international relief organization. The following is merely one of many similar tragedies that he observed while on an extended tour through several of the approximately 100 LDCs, where a majority of people live in appalling poverty and chronic malnutrition (Figure 4.9):

> Hunger is more than cold facts and awesome statistics. Hunger has a face. I know. I have looked into it. Hunger is a Bengali face—a little mother named Jobeda, whom I found in the shade of a tattered lean-to in a refugee camp in Dacca. A small withered form lying close beside her whimpered and stirred. Instinctively she reached down to brush away the flies. Her hand carefully wiped the fevered face of her child. At six years, acute malnutrition had crippled his legs, left him dumb, and robbed him of his hearing. All that was left was the shallow, labored breathing of life itself, and that, too, would soon be gone. But death is no stranger to Jobeda. She has seen starvation take away her husband and five of her seven children.

Today, an estimated 33,000 people will die either directly or indirectly from starvation and malnutrition. Tomorrow

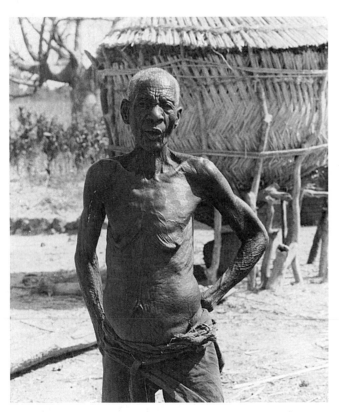

FIGURE 4.9 An aged woman suffering from undernourishment in famine-plagued Africa.

33,000 more will perish. Although the causes of these deaths are multiple, the basic cause is Malthusian overpopulation—too many human stomachs and not enough food, a problem discussed at length in Chapter 5.

Overpopulation in the less-developed nations is much more complex than the preceding discussion suggests. Although hunger and starvation are the most visible symptoms, large populations also take their toll on local rivers and streams, which human waste often makes unfit for drinking. Intense demand for wood for fuel often results in deforestation and erosion in the landscape surrounding population centers. Wildlife species may be exterminated as people struggle to meet the rising demand for food, fiber, and wood. Urban centers in the less-developed nations are often crowded and highly polluted. For this reason, some educators refer to the complex of problems in the LDCs as **population-based resource degradation.** In a phrase, this would be described as the environmental consequences of too many people. Even with meager needs, enormous damage can result.

Overpopulation in the More-Developed Nations

Overpopulation is also a fact of life in the more-developed nations of the world such as Japan, Canada, England, Germany, Russia, and the United States. In these and other industrial nations, starvation and hunger occur, although much less frequently than in the less-developed nations. The symptoms of overpopulation in the more-developed nations are, however, just as serious.

GO GREEN!

Get active. Write letters to the President and your congressional representatives in support of policies that promote family planning and sustainable economic development to help less-developed countries reduce population growth.

To distinguish overpopulation in the industrial nations from that occurring in the less-developed nations, some ecologists use the term **technological overpopulation** because of the technological dependence of such nations and the impacts of the use of many technologies like automobiles on the environment and the resource base. Others prefer the term **consumption-based resource degradation** to reflect the importance of our high rates of consumption. In more-developed nations, then, the most important factors when considering environmental impact are the resources used and the pollution generated per person (per capita). As noted in previous chapters, the United States and other MDCs use more resources per person than citizens of the less-developed nations—often 20 to 40 times more. *A one-mile drive in your automobile to pick up a quart of milk or a six-pack of soda uses more energy than most people in the less-developed nations use in an entire day to perform all of the tasks required to survive.* Because of the higher per capita consumption, a relatively small population may therefore cause considerably more damage. Their environment is threatened by pollution and resource degradation resulting from technology and consumption. Instead of dying from starvation, people are killed by pollutants their technology has generated. Because environmental problems are really based on the level of technology and high rates of consumption, the term **technological and consumption-based resource degradation** might be best.

The effects of overpopulation in the more-developed nations are visible all around us. Highway congestion may be one of the most obvious. Urban air pollution is another visible reminder, as is the loss of open space, farmland, and wildlife at the fringes of major cities and in rural areas as new towns and resort areas expand. Look a little closer, and you will find other, less visible signs of the problem. Species are rapidly going extinct. Rivers turn brown after rains as a result of sediment washed into them from nearby development. The rain has turned acid in many parts of the world, too, as a result of pollutants generated by power plants, factories, cars, trucks, buses, jets, and homes. Even the protective ozone layer above our heads is in decline as a result of the release of chemicals known as chlorofluorocarbons from refrigerators, air conditioners, and other sources. And finally, there's the rapid warming of the planet, known as global warming, resulting in large part from the release of carbon dioxide from a wide range of human sources, including cars, homes, and factories, and from deforestation—the loss of carbon dioxide–assimilating plants.

Clearly, these and other impacts are a result of our resource-intensive way of life, made possible partly by our heavy dependence on certain technologies such as coal-fired power plants and automobiles. Taken together, these impacts constitute an environmental transgression as serious as the starvation and malnutrition problem in the poorer nations of the world.

Because of our resource-intensive way of life, population size and population growth in the MDCs are major contributors to environmental degradation. In fact, *we think that population growth in the MDCs is even more important than in the LDCs.* Consider some facts: As noted earlier, U.S. citizens consume, on average, 20 to 40 times the resources of an average resident of a less-developed nation. They therefore have 20 to 40 times the impact on the environment. Each new child has the environmental impact of 20 to 40 residents in the less-developed world. Each birth averted in this country can make tremendous inroads into environmental problems.

People who don't view population growth as a problem in the more-developed nations often assert that we can solve our problems by reducing levels of consumption—for example, through efficiency, pollution prevention, and recycling. Although that may be true, many people believe that we must also reduce population growth to create a more sustainable society (see Ethics in Resource Conservation, Box 4.1). Such actions have a far greater impact on the overall quality of the environment than our most aggressive energy and resource conservation strategies.

Throughout the remainder of this book, you will see examples of these types of overpopulation in discussions of soil erosion, air and water pollution, wildlife extinction, the energy crisis, the toxic chemical problem, and global starvation. We urge you not to lose track of the fact that overpopulation is at the root of all environmental problems.

We also point out that population size is not the only issue. The rate of growth has profound implications, too. Rapid growth makes matters worse. Problems will escalate and solutions will be more difficult as the number of people increases. This is especially true in the less-developed nations where economic resources are lacking and where corruption in government is prevalent.

GO GREEN!

Get active. Write letters to the President and your congressional representatives in support of policies that support environmentally friendly technologies such as wind and solar energy.

ETHICS IN RESOURCE CONSERVATION 4.1   IS REPRODUCTION A PERSONAL RIGHT?

A pamphlet on wildlife published by the Colorado Division of Wildlife presents engaging descriptions of the state's mammals. The back cover also discusses what has become a central concern among citizens in Colorado: the loss of wildlife habitat. It notes that Coloradans have changed the environment and points out, "What were once elk calving grounds are now shoppettes below ski areas; deer migration routes are cut by six-lane highways." It goes on to say that as many as "30,000 acres of traditional wildlife habitat are converted to human use in Colorado every year." It concludes by saying that, because of this, "we must manage our wildlife resources more carefully than ever."

Leo Tolstoy once wrote something to the effect that everyone dreams of changing the world, but no one dreams of changing himself. The pamphlet displays this fundamental human tendency.

To build a sustainable society, most people agree, we have to do something about ourselves. We have to manage ourselves better. One key area in need of better management is population growth. Most experts agree that, at the very least, we must find ways to slow the rate of human population growth. Eventually, we may have to stop growth altogether. In fact, this was one of the recommendations of the President's Council on Sustainable Development in 1996. After we stabilize, we may even want to find ways to reduce population size humanely—for example, by having smaller families so that the birth rate is lower than the death rate.

Population control brings up ethical issues of incredible magnitude. The overriding question, of course, is this: Is it ethical to control population growth?

As a rule, people tend to line up in one of two camps: those who think it is not ethical to place limits on family size and those who think that it is.

Reasons for opposing limits are many. Some people, for example, object to population control primarily on religious grounds. That is, they object to it because of religious ethics. It is their belief that governments have no right to limit human numbers. The Catholic Church, for instance, opposes population control and all forms of family planning, such as **contraceptives** or birth-control pills. The only form of family planning it deems acceptable is the rhythm method or natural method, which involves sexual abstinence around the time of ovulation.

Some advocates of the viewpoint that population control is unethical base their arguments on personal freedom. The right to reproduce at will is, they say, a basic human freedom. No one should dictate another person's family size. Certain racial groups feel that limits are discriminatory in nature.

Those who argue that it is ethical to limit human population growth say that limitations on population growth will ensure a better quality of life for people who are alive today. In other words, limiting family size is for the greater good. The right to reproduce should be curtailed when it interferes with the welfare of society. Advocates of this viewpoint like to turn the tables and ask, Is it ethical not to limit population growth? In other words, is it ethical to continue as we are, breeding rapidly and quickly overrunning the world and wreaking havoc on the environment?

Unbridled population growth inevitably results in widespread environmental destruction and robs future generations of the opportunities many of us now take for granted. Advocates of such a view note that population growth without adequate resources and services could very well

(continued)

result in disaster. So, not only is it ethical to control numbers, it is prudent.

Advocates of limiting growth also note that measures to reduce population growth prevent unwanted pregnancies and end the cruel fate of those born into families that cannot support and nurture them. Moreover, limitations protect the welfare of other species that share this planet with us.

Advocates of population control and stabilization, such as ourselves, are quick to point out that they need not involve methods that are morally repugnant to people. As noted in the text, improving education, health care, job opportunities, and women's rights can play a huge role in slowing, even stopping, world population growth.

Take a few moments and summarize your views on the subject. What is your position, and why?

4.3 Population Growth in the More-Developed Nations: A Closer Look

With these facts and concepts in mind, let us take a closer look at the populations of more-developed nations. For the most part, the news from the more-developed nations is encouraging (Table 4.1). For example, studies show that the total fertility rate has steadily decreased in the more-developed nations of Western Europe and Canada, although not the United States, over the past three-plus decades. At the current rate of growth, it would take 700 years for the populations of the more-developed nations (now about 1.22 billion) of the world to double.

As noted earlier in the chapter, in many countries, population growth has slowed dramatically, even stopped. Others are experiencing what demographers call "negative growth." In other words, their populations are declining. Today, more than 90 developed nations are stable or declining. The populations of Austria, Greece, and Poland are stable, for example, while Germany, Romania, Bulgaria, and Russia are all on the decline. Overall, Europe's growth rate is −0.1% per year.

As a side note, while many experts view declining population in the more-developed countries as an encouraging trend, the impacts can be quite significant—and should not be ignored. (Better planning can help offset the impacts, too.) Consider an example. For years, birth rates have been falling in Italy. This, in turn, has resulted in an aging of the nation's population. Thus, as time passes, the percentage of the population made up of older adults steadily rises. The aging of Italy's population is having a profound effect on the nation's social security program. Social security (payments paid to retirees from the government) is paid for by deductions from the salaries of people in the workforce. As the percentage of retired individuals increases, deductions from paychecks to cover the pensions of workers' parents and grandparents must rise. Today, Italy's retirement payments consume 15% of Italy's gross domestic product—and will only get bigger.

Small towns are also adversely affected by plummeting birth rates. In fact, scores of small towns in the southern part of Italy are dying because of declining birth rates, but also because more people are leaving to live in larger cities. One town (Laviano) even implemented a baby bonus—a payment of nearly $12,000 made to parents who give birth to a child and remain in the town, paid out over three years' time. Great Britain and France—which once experienced population declines—adopted pronatalist policies, which encourage couples to have children. These programs have enjoyed modest success, and now both nations are posting slight but positive population growth rates.

Although population growth is slowing in many more-developed countries, there are notable exceptions. The U.S. population, for example, is growing naturally at a rate of about 0.6% per year—one of the fastest growth rates in the developed world. When legal and illegal **immigration** are included, the growth rate shoots up to about 1.0% per year. In the sixth edition of this book, we cited U.S. Census Bureau projections that the U.S. population would stabilize at about 300 million by the year 2050. However, because of a shift in trends, the U.S. population topped 300 million in 2007. The Population Reference Bureau now projects a population of 350 million by 2025 and

TABLE 4.1	Comparison of More-Developed and Less-Developed Nations in 2007	
	More-Developed	Less-Developed (Including China)
Population	1.22 billion	5.24 billion
Average Birth Rate	11/1,000	27/1,000
Average Death Rate	10/1,000	9/1,000
Average Population Growth Rate	0.1%	1.8%
Doubling Time	700 years	47 years
Infant Mortality Rate	7/1,000	61/1,000
Total Fertility Rate	1.6	3.3
Life Expectancy at Birth	77 years	64 years
GNP per capita	$29,680/yr	$4,760/yr

420 million by 2050—approximately 50 to 125 million more people than live in the United States today. Zero population growth is no longer in sight.

What caused this change in predictions? Two things: a slight rise in the total fertility rate and an increase in immigration, both legal and illegal. About 1 million people enter the United States legally and illegally each year. Only by maintaining the total fertility rate well below 2.1 and eliminating or drastically curtailing immigration will the United States achieve a stable population size.

4.4 Population Growth in the Less-Developed Nations: A Closer Look

Population growth in the less-developed nations is currently occurring 18 times faster than in the developed nations. Dozens of nations in Central and South America, Africa, and Asia are caught in a kind of demographic trap in which reductions in the death rate have occurred without equal reductions in the birth rate. In most of these nations, natural resources are being depleted, and per capita food supplies and income are declining. Why are birth rates still high?

Large Family Size

In contrast to the MDCs, where the desired number of children per family ranges from zero to three, the LDCs of Asia, Africa, and Latin America, with a few exceptions such as China, prefer to have much larger families. For example, in Saudi Arabia (Middle East) couples are having, on average, 4.1 children and in Niger they're having, on average, 7.1 children.

Large families result from several factors. One of the most important is the desire for them. In agricultural nations, children are still needed to help with household tasks and farm work, such as planting and harvesting, gathering wood and dried cow dung for fuel, and carrying water from distant streams. Children are also needed to provide security to the parents in old age. In contrast, most MDCs, such as the United States, England, and Sweden, have well-developed social security programs to support their older citizens when their wage-earning years end. However, in many of the LDCs, such programs are lacking.

Yet another reason for large family size in the less-developed nations has to do with the macho image of many men. In such cultures, the larger the family, the greater the status of the father. Many children are fathered out of wedlock as well. A man named Denja, from the province of Nyanza in Kenya (Africa), boasts of being the father of 497 children!

Another reason for large family size in the LDCs is that birth control methods such as chemical contraceptives, sterilization, and abortion are strictly forbidden by the religion of the prospective parents. For example, in Mexico, Kenya, and the Philippines, where populations are soaring, the Catholic Church opposes all artificial methods of birth control, although many people ignore the Church's dictates. A similar stand has been taken by the Muslim fundamentalists in Pakistan, Egypt, and Iran.

Lester Brown, former president of the Worldwatch Institute, a highly respected environmental organization, views the continued population surge as a dire threat to the world's resources and the quality of human life. He strongly urges countries to adopt a small-family policy with the number of children averaging one to two per family. To achieve this goal, the governments of the less-developed nations must launch effective mass education programs on birth control in cities and in rural areas. Although such efforts have begun in some nations, they must be greatly expanded and intensified.

Africa: A Continent in Danger

No continent has suffered more from the adverse impact of rapid population increase than Africa. At current growth rates, the African population will increase sharply from over 944 million in 2007 to nearly 2 billion in 2050! Such rapid growth will put enormous strains on wildlife and natural resources. Millions of acres of land will be converted to farmland to feed the growing population, squeezing out wildlife. Already crowded, filthy, and crime-ridden cities will very likely deteriorate even further. Rivers and streams will become even more polluted with human waste.

Demographers believe that Africa's birth rate of 38 per 1,000 will remain dangerously high well into this millennium if current trends continue. Even if strong curbs on population growth could be initiated immediately, the continent's population will continue to soar far into the next century because 41% of Africa's population is under the age of 15 and will soon be entering their reproductive age. The only mitigating factor, discussed earlier, is the death rate, which also is predicted to increase, primarily as a result of AIDS.

4.5 Controlling the Growth of the World's Population

Controlling the growth of the world's population is an enormous task. This section discusses two major approaches: family planning through birth control and abortion, and a host of sustainable-development strategies that have an equally powerful effect.

Birth Control

As we view (and feel) the crush of people on this planet, it may come as a surprise that humans have practiced various methods of birth control for centuries. **Birth control** refers to techniques and devices that prevent births by preventing conception—fertilization of the egg. Birth control was advocated on ancient Egyptian papyri dating back to 5,000 B.C. A great variety of techniques were tried. For example, a combination of wool fibers and alligator dung was used by women as a vaginal barrier to sperm. Men often used condoms fashioned from animal bladders. Foul-tasting concoctions made from bark fibers, weeds, and ground gallbladders were gulped down prior to intercourse in an erroneous belief that they would prevent conception.

According to the Population Reference Bureau, 62% of married women worldwide use some form of birth control. Most are using modern methods. Of course, a much greater variety and more effective methods are available today than in ancient times. Many have been developed in the past 40 years. A major breakthrough in birth control was the **pill**— a chemical contraceptive taken orally that prevents sexually mature females from ovulating (releasing ova or eggs) each month. It is the most widely used method of birth control in the MDCs. The worldwide rate of use ranges from 0.2% in Somalia (eastern Africa) to 59% in Germany. The pill is popular, in part because it is easy to use and highly effective. It does have side effects, including edema (fluid retention) and hair loss, but efforts to reduce other, more serious side effects have proven successful.

In 1988 a French pharmaceutical firm developed a pill known as **RU-486.** Incorrectly referred to as an "abortion pill," RU-486 simply prevents an early embryo from implanting in the wall of the uterus. It works like an IUD, actually. RU-486 has been approved for use in several countries, including France, Sweden, China, Great Britain, and the United States. RU-486 is taken after unprotected sex.

Another option is the "**morning-after pill**," another type of emergency contraception. This pill contains high doses of two female hormones, estrogen and progestin. It can be taken up to 72 hours after unprotected sex or the failure of contraception, but it is most most effective if taken within the first 12 to 24 hours. The morning-after pill primarily functions by stimulating the uterine lining to slough off, either carrying the egg with it or changing the uterine wall so that the egg cannot attach itself. It may also prevent ovulation, so that an egg is not released (in much the same way as lower doses of estrogen and progestin do in birth control pills). Hospitals and doctors can provide these pills.

Another popular birth control method is DepoProvera, a hormone-containing solution that, when injected under the skin, prevents a pregnancy for up to three months. It has been used by women in at least 90 countries, including the United States; the Food and Drug Administration (FDA) approved it for use in 1992.

Norplant is yet another contraceptive that holds considerable promise. It consists of six matchstick-sized capsules that are inserted under the skin of a woman's upper arm (Figure 4.10). These capsules gradually release a hormone that prevents conception for at least five years. The insertion procedure can be completed in less than ten minutes and can be performed by a specially trained nurse. Norplant has been approved for use by at least 57 nations, including Indonesia, Thailand, and China. In 1991 the FDA certified the use of Norplant in the United States.

Another type of contraceptive is a plastic or nylon loop that is inserted into the uterus. Known as **intrauterine devices (IUDs),** these contraceptives prevent the newly formed embryo from embedding in the uterine lining. IUD-manufacturing plants have been built in some of the LDCs so that these contraceptives can be readily available. The late Oliver Owen, who first published this text in 1971, had an acquaintance who wore IUDs as earrings to coffee parties in order to

FIGURE 4.10   Norplant, an effective method of birth control. The 1-inch capsules are inserted into a woman's arm. The chemicals that slowly exude from the capsules prevent the woman from conceiving for at least five years. This birth-control method has been approved by the World Health Organization.

emphasize her position on the importance of birth control! The rate of IUD use among women of childbearing age ranges from 0.1% in several African nations to 0.7% in the United States to 41.5% in Vietnam. IUDs are effective but do cause problems, including uterine perforation. As a result, their use has been on the decline in many countries.

Sterilization is another major method of birth control. In men, sterilization is achieved by cutting and tying off the sperm ducts in the male, a procedure known as a vasectomy. In women, the oviducts or fallopian tubes are cut to prevent sperm from reaching the egg (Figure 4.11). This procedure is known as a tubal ligation.

Sterilization's worldwide use among women of childbearing age ranges from 0.1% in Gambia (Africa) to 24% in the United States to a high of 41% in the Dominican Republic (Caribbean). Sterilization is popular because it is highly effective and requires no effort. Once the surgery has been performed, a person is relatively safe.

Another widely used method of contraception is the barrier method, typified by the condom. A **condom** is a thin rubber sheath worn over the penis during sexual intercourse. It prevents ejaculated sperm from entering the vagina. Condoms also help reduce the spread of sexually transmitted diseases such as genital herpes, syphilis, gonorrhea, chlamydia, and AIDS.

Other barrier methods are the diaphragm, cervical cap, and vaginal sponge. A **diaphragm** is a rubber cap that fits over the cervix, the end of the uterus through which sperm must pass to get from the vagina into the uterus. Diaphragms are generally coated with a thin layer of a sperm-killing (spermicidal) jelly, which greatly increases their effectiveness. A smaller version of the diaphragm, known as a **cervical cap,** fits over the end of the cervix and is held in place by suction. **Vaginal sponges** are small, sponge-like devices that fit over the end of the cervix. They are impregnated with a spermicidal chemical. Barrier methods are easy to use but not as effective as sterilization and the pill.

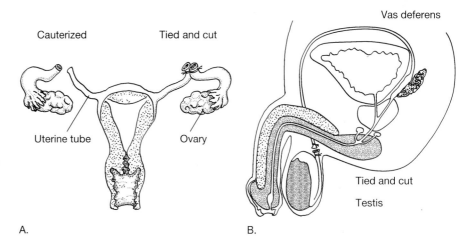

FIGURE 4.11 Female (A) and male (B) sterilization. In females, the oviducts, which transport ova and sperm, are cut and tied. In males, the sperm-transporting ducts, the vas deferens, are cut and tied. Both operations can be reversed through microsurgery, although not with 100% success.

Some couples practice birth control techniques that are behavioral in nature. The **rhythm method** involves timing sexual intercourse so that it occurs well before or after ovulation to prevent sperm and ova from uniting. Body temperature can be used to monitor a woman's time of ovulation. **Abstinence** is a technique often stressed to teenagers and unmarried couples. Unfortunately, both techniques have a high failure rate.

Male Contraception   For years, most contraceptive measures have been aimed at women, primarily because of the ease with which female reproductive processes can be interrupted. Considerable effort is now under way to develop male contraceptives. Some methods are currently being studied in humans, and others are fairly close to being ready to market.

One method that is being used in China is vas occlusion. Unlike vasectomies which seal or remove part of the sperm-carrying ducts, the vas deferens, vas occlusion simply blocks the vas deferens, using an injectable silicone plug. This technique, which is being used in thousands of Chinese men, makes reversal easier. Researchers in Canada and the United States are currently developing a silicone plug called the IVD, which can be implanted in the vas.

Two pharmaceutical companies, working with a dozen universities and nongovernmental organizations, are working on a male contraceptive that will be injected or implanted under the skin. It contains a combination of hormones that suppress sperm production.

A male contraceptive pill is being developed. Researchers are also working with nonhormonal pharmaceuticals that have contraceptive effects. Two researchers in Great Britain, for instance, have found that two common medications used to treat certain diseases (high blood pressure and schizophrenia) disrupt the transport of sperm out of the testes and sperm maturation in the testes.

Abortion

In case of contraceptive failure, a woman may terminate her pregnancy by an abortion rather than give birth to an unwanted child. An **abortion** is a premature expulsion of the fetus from the womb that results in fetal death. Many abortions occur naturally. These so-called spontaneous abortions are believed to be nature's way of eliminating defective fetuses—nearly 40% of the spontaneously aborted fetuses have some struc-

tural impairment or birth defect. Abortions resulting from human intervention result in the death of unborn fetuses.

The U.S. Supreme Court legalized abortions in 1972. They are legal in most other countries of the world as well. Readers know that there are strong ethical, moral, and religious arguments against abortion. Anti-abortion activists ask, "What right does a human who is living today have to take the life of another human, who, except for the violent act of abortion, would be alive tomorrow?" In the eyes of many eminent persons in the law, medicine, and religion, the willful act of abortion is murder, except perhaps in the case of incest or a life threat to the mother. Opponents argue that abortion "gives women a right to control their own bodies." And it prevents unwanted births and suffering. In 1989, the Supreme Court upheld the states' right to restrict abortions funded from state monies, which has touched off what is proving to be a long, drawn-out battle.

On a global basis, abortion is the third most widely used method of birth control after sterilization and the pill. In fact, were it not for the practice of abortion, the resources of this planet would have to sustain another 50 million people every year. Abortion is being used primarily by the women of Asia, Africa, and Latin America. Abortion rates are high in Latin America, even though the act is illegal. Unfortunately, many of the illegal abortions are self-induced with wires, coat hangers, or pointed sticks, and they are 75 times more dangerous to the mother than legal abortions performed by the medical profession.

Sustainable Development May Be the Best Contraceptive

The rapid growth of the human population cannot be solved by just supplying contraceptives. Underlying factors—people's beliefs and their opportunities for education and health care—are also powerful forces that can be enlisted to control population growth worldwide. Economic development is also a key to slowing the growth of the human population. But economic development must be sustainable to counter the adverse effects of economic growth on the environment.

Sustainable development is a strategy that seeks to better people's lives without undermining the environmental life-support systems of the planet. Many argue that it should actually increase the health and vitality of these systems. Among other things, sustainable development seeks to increase the educational level of women and provide men and women with greater economic opportunities. Increasing educational levels has many practical benefits. It will, for instance, help women read and understand instructions on their contraceptive packages. It will also help them seek employment, which may delay childbearing and thus effectively reduce the number of children a woman has in her lifetime. Increasing economic opportunities gives men and women other options as well. Where women can be gainfully employed or operate their own businesses, birth rates decline.

Small-scale economic development, designed to meet the needs of people in ways that do not harm the environment, is vital. It also provides jobs and income. With rising income, say population experts, family size will fall. Several nonprofit organizations are helping in this regard. One effort that seems to be helping provides small business loans to women in rural countries in Asia and Latin America. Women use the loans to start small businesses to support their families and with great success! Women participating in such programs realize that more children make it more difficult to improve their lives. Many opportunities are now available online for people in more-developed countries to lend money to men and women in less-developed countries. So far, the repayment rates have been remarkably high.

Changes in the status of women in a society also have a bearing on the problem. In many cultures, a woman's value derives primarily from her ability to produce offspring for the husband. Cultural shifts, although difficult, may be a powerful force in slowing the growth of the human population.

Improvements in health care, another component of sustainable development, also yield population benefits. Increasing the accessibility and affordability of health care in poor rural villages, for example, reduces infant death and, over time, reduces the number of children a woman must have to produce enough surviving offspring to take care of the mother and father in their old age. Health care clinics are also an avenue for learning about and receiving contraceptives. Because childbirth is a leading cause of maternal death in the less-developed nations, birth control is now also being seen as a means of increasing a woman's long-term prospects.

These changes and several others can help LDCs currently stuck in stage II of the demographic transition move out of it. As with so many environmental problems, the answer is not easy. Deep, systematic changes are needed to reduce the underlying impetus for large families—the social, economic, and cultural factors that create runaway population growth.

The importance of family planning and deeper changes, such as those discussed in this section, were expressed in a document produced by the United Nations' Conference on Population and Development that was convened in

September 1992. More than 20,000 representatives of over 150 nations assembled in Cairo, Egypt, for the conference. The United States was represented by a 35-person delegation. The major conclusions of the conference, contained in a 113-page Voluntary Action Plan, are as follows:

1. To check population growth, family planning services should be advertised and provided to all interested families.
2. Teenagers should be given essential information concerning the use of both chemical and mechanical contraceptives.
3. Women of all nations should have access to safe abortions. (Forty percent of the world's people now have such access.)
4. Sex education should begin within the family, the community, and the school at an early age in the life of the individual.
5. The status of women in all nations should be raised above that of "baby-making machines." Such empowerment of women will reduce their fertility rates.
6. The more developed nations must accept much of the blame for resource exploitation and the pollution of land, water, and air.
7. The less-developed nations of Asia, Africa, and South America must acknowledge their shortcomings in population control.
8. The more-developed nations will assist the less-developed nations to utilize their resources in a sustainable manner so that they can improve the quality of life for their people.
9. There should be a more equitable distribution of the world's wealth, resources, and technology.

4.6 Human Population and the Earth's Carrying Capacity

To create a sustainable society, many think, human society must learn to operate within Earth's carrying capacity. Unfortunately, no one knows Earth's carrying capacity for *Homo sapiens*. Some experts believe this planet can sustain only 500 million; others believe it can sustain 10 billion to 50 billion people.

At the risk of a gross oversimplification, let us assume that our global population can follow two alternative paths, as shown in Figure 4.12. In Scenario A, our population continues to grow exponentially and overshoots Earth's carrying capacity by a wide margin. This would eventually result in a die-off caused by starvation, disease, pollution, and resource depletion, which could reduce our population to the carrying capacity level. In such an instance, the carrying capacity might even be reduced as a result of widespread environmental damage. In other words, the planet might not be able to support anywhere near as many people as it could if we had prevented this calamity.

It is important to note that carrying capacity is about more than food supplies. It is about the planet's ability to provide all resources and process all of our wastes safely. Expanding

GO GREEN!

Consider lending small amounts of money through international nonprofits such as Kiva and MicroPlace that support small businesses in less-developed nations. You'll be paid back with a small amount of interest and could help someone in a developing nation start a business that supports a family.

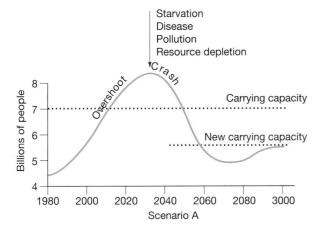

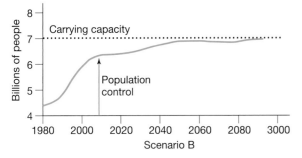

FIGURE 4.12 Population growth and carrying capacity. Scenario A. Population overshoots carrying capacity. Eventually there is a massive die-off due to starvation, disease, pollution, and resource depletion. A new carrying capacity may be established because of massive damage caused by population growth. Scenario B. Before the population reaches carrying capacity, intensive efforts are made at population control (family planning, use of contraceptives, delayed marriage, and so on). As a result, the population gradually reaches carrying capacity, and die-off is prevented.

food production may permit more people to be sustained over the short term. However, the long-term result would be a human-dominated ecosystem that is severely out of balance. Why? In the process of producing more and more food that feeds more and more people, we would very likely deplete natural resources and generate massive amounts of air and water pollution—changes that would cause the extinction of many forms of wildlife.

The preferred alternative, shown in Scenario B, is a gradual reduction in the rate of population growth, long before it reaches the carrying capacity, by means of the strategies discussed in this chapter. Eventually our population would stabilize at the carrying capacity, at which point the human species and the environment with which it interacts would become part of an ecological system that could be sustainable far into the future.

Unfortunately, many scientists believe that human populations have already exceeded the carrying capacity in many regions. Excessive soil erosion, the conversion of vast acreages of grassland to desert, species extinction, loss of fish populations, and other regional problems are blatant signs of a people living beyond the ability of their bioregion to provide for them. Some scientists believe we may have exceeded global limits as well. Acid deposition, ozone depletion, and global warming, discussed in Chapters 19 and 20, are three of the most alarming signs of this transgression.

If these assertions are true, human society is clearly engaged in Scenario A. Our choice may be either to voluntarily control numbers, reduce environmental damage through technological and behavioral change, and restore the Earth to ease our population and its impact to within tolerable limits, or else to face a severe crash. The decision is ours, and time is limited. (See Case Study 4.1.)

CASE STUDY 4.1   CHINA: ONE OF FAMILY PLANNING'S SUCCESS STORIES?

During China's great famine of 1958–1962, 30 million people starved to death. This event was a grim warning to its leaders that the theory advanced by Thomas Malthus is probably correct, although it may have had more to do with the Mao Tse Tung's Cultural Revolution than anything else. However, it was not until the late 1970s that the nation's State Birth Planning Commission initiated the most comprehensive, the most rigidly enforced, and probably the most effective population control program in human history. The results have been impressive: The TFR has been reduced from 5.9 in the late 1960s to only 1.6 in 2007, the crude birth rate has fallen from 36.9 to 12.0, and the annual population growth rate has declined from 2.6 to 0.5. Why was this program so successful?

Several features contributed to the success of China's program. One of the most important is the mass education program launched to foster public understanding of the effects on living

standards if China's population was not curbed. Long-term projections were made public concerning how much food, water, energy, and other resources each person would have if China continued on its course into the demographic trap. The education program emphasized the benefits of postponing marriage to reduce the average family size. Another educational objective was to make the one-child family the norm among recently married couples. Couples who made the commitment to a one-child family received multiple benefits, including (a) cash payments, (b) free family-planning education, (c) larger old-age pensions, so that extra children would not be needed to provide security to the parents in their later years, (d) better housing and employment, and (e) free schooling for children. Family-planning education was made readily available and publicized widely by all the media. Abortions, sterilizations, IUD insertions, and the dispensation of birth-control pills were performed by local nurses and paramedics.

(continued)

China's government took some harsh steps to ensure small family size. For example, penalties were imposed on couples who had more than two children after entering the program. Among the penalties were (a) an increase in taxes, (b) compulsory sterilization for either the father or the mother, (c) the return to the state of all financial benefits awarded to couples who had agreed to curb family size, (d) intense peer pressure for women pregnant with their third child to have an abortion, and (e) reductions in food, employment, and educational benefits for parents and children.

China's family-planning program gained momentum rapidly, especially in the urban areas. By 1982, for example, 70% of the couples in China's three largest cities—Shanghai, Beijing, and Tientsin—with an aggregate population of more than 20 million had agreed to have no more than one child. The goal of the Birth Planning Commission was to achieve zero population growth by 2000, with a population of 1.2 billion, followed by a decline to 800 million by 2100. Currently, however, China's population is over 1.3 billion and is projected to rise to nearly 1.5 billion by 2025. Whether or not China will successfully reduce population size

remains to be seen. At its current rate of growth, its population will double in 117 years. With the one-child-family policy no longer being strictly enforced, it is unlikely that China will see a substantial decline in population.

Although China's population-control program has been eminently successful from a technical standpoint, many democratic nations, including the United States, have been deeply concerned about human rights violations, such as the Chinese government making sterilization mandatory after a woman has had her second child and rumors of forced abortions. The United States had funded population-control programs in the LDCs for many years through such organizations as the World Bank, the Agency for International Development, and the United Nations. However, because of China's coercive sterilization and abortion policy, President Ronald Reagan terminated the United States' support of the United Nations Fund for Population Activity (UNFPA), which provided millions of dollars each year for family planning in less-developed nations. This hurt not only China but also many other countries. Fortunately, President Clinton reinstated support for UNFPA.

Summary of Key Concepts

1. The world population was 6.7 billion in 2008 and is growing at a rate of about 1.2% per year. This results in the addition of 80 million people a year.

2. The birth rate of a country is the number of persons born per 1,000 individuals in a given year. The death rate is the number of deaths per 1,000 individuals in a given year.

3. The rate of natural increase (or decrease) is the difference between the birth and death rates. The world's rate of natural increase at present is about 12 per 1,000.

4. The doubling time for the global population is currently about 58 years.

5. Population increases exponentially. Exponential growth occurs when any entity increases by a fixed percentage, but only if the annual increase is added to the base amount. Growth in absolute numbers occurs very slowly at first, but once the item being measured "rounds the bend," growth is rapid.

6. Human population remained small for many years, as birth rates and death rates were in equilibrium. Decreases in death rates, with accompanying increases in birth rates, resulted in a massive increase in the human population in the past 200 years. Advances in agriculture, medicine (especially the development of vaccines and antibiotics), and sanitation all contributed to the present population explosion on Earth by reducing the death rate.

7. Historically, when nations become industrialized, they undergo a demographic transition. This is characterized by a reduction in both birth and death rates. Many of the LDCs are stuck in stage II of the demographic transition, which is characterized by a falling death rate combined with a high birth rate.

8. One sign of hope is that global population growth has slowed in recent years in many parts of the world. This trend is due to a number of factors, including better family planning, economic prosperity, and rising death rates, especially in Africa.

Rising death rates result from starvation and disease. AIDS is taking its toll on many African nations.

9. Total fertility rate is an estimate of the number of children women will have during their lifetime. It is based on current fertility rates in different age groups. Replacement-level fertility, the number of children required to replace a couple, in developed countries is about 2.1.

10. The age structure of a population is the number of individuals occurring in each age class in a population. It is graphically represented in a population histogram, a useful tool for predicting trends.

11. The age structure diagram of a rapidly growing population has a broad base, whereas the diagram of one that is growing very slowly has a narrow base. Populations that are declining have a narrowing base.

12. At the present time, 28% of the world's population is under 15 years of age.

13. Malthusian overpopulation means too many people for the available food supply. It is characteristic of the LDCs of Asia, Africa, and South America. Although hunger and starvation are the major manifestations of overpopulation in such nations, environmental destruction is often significant. Some scholars prefer to speak of population-based resource degradation to be more inclusive.

14. Technological overpopulation refers to an overabundance of people, usually in industrialized countries, who, because of their use of advanced technology, have a harmful effect on the environment.

15. Technological overpopulation results in depletion of resources; pollution of air, land, and water; defilement of scenic beauty; and wildlife extinction. The terms *consumption-based resource depletion* and *technological and consumption-based resource degradation* also are used.

16. Population growth in the more-developed nations of the world has slowed considerably in recent decades. In many

European countries, population growth has either come close to stabilization, stabilized, or begun to decline.

17. Although growth in the developed nations of the world has slowed, some nations, including the United States and Canada, continue to grow fairly rapidly. Population growth will continue in the United States for many years to come, even though the total fertility rate has been at or below replacement-level fertility for three decades. Two facts are responsible for continued growth: an increase in the number of women in the reproductive age group and immigration.

18. Population growth in the less-developed nations has slowed but still continues at a rapid pace. Growth is especially rapid in Africa.

19. Many women have access to and use modern birth-control measures. Although many birth-control options are available to reduce family size and slow the growth of the human population, changes in the status, educational opportunities, employment opportunities, economic well-being, and health care of women and men also are needed.

20. The major features of China's population-control program, which reduced TFR to replacement level (2.0) in 1994, are the following: (a) mass education on family planning, (b) delayed marriage, (c) multiple educational, health, and economic benefits for one-child families, (d) family-planning services provided by specially trained local people, and (e) mandatory sterilizations or abortions for families having more than two children.

21. The human population may already have exceeded the Earth's carrying capacity. Continued overshoot could result in environmental damage that will ultimately lower the planet's capacity to support life.

Key Words and Phrases

Abortion	J-Shaped Curve
Abstinence	Less-Developed Countries
Acquired Immunodeficiency	(LDCs)
Syndrome (AIDS)	Malthusian Overpopulation
Age Structure	More-Developed Countries
Birth Control	(MDCs)
Birth Rate	Morning-After Pill
Carrying Capacity	Net Migration
Cervical Cap	Norplant
Condom	Overpopulation
Consumption-Based	Percent Annual
Resource Degradation	Growth Rate
Contraceptives	Pill
Death Rate	Population Histogram
Demographer	Population-Based Resource
Demographic Transition	Degradation
Diaphragm	Rate of Natural Increase
Doubling Time	Replacement-Level
Exponential Growth	Fertility
Family-Planning Programs	Replacement Rate
Human Immunodeficiency	Rhythm Method
Virus (HIV)	RU-486
Immigration	Sterilization
Intrauterine Device (IUD)	Sustainable Development

Technological and	Total Fertility Rate (TFR)
Consumption-Based	Vaginal Sponge
Resource Degradation	Zero Population Growth
Technological	(ZPG)
Overpopulation	

Critical Thinking and Discussion Questions

1. Construct a simple graph that shows the buildup of the human population since humans appeared on the Earth. Explain why growth was so slow for so long and then began to increase dramatically in the past 200 years.

2. What is the current estimated size of the global human population? What are its rate of natural increase and its doubling time?

3. What is the current population size of the United States? What are its growth rate and doubling time? What are the global birth, death, and natural-increase rates?

4. How is the percent annual growth rate determined?

5. How do demographers determine the length of time it takes for a population to double?

6. List several reasons why birth rates tend to decrease when the standard of living rises.

7. Does a decrease in the total fertility rate of a nation to replacement-level fertility mean the nation's population is decreasing as well? Why or why not?

8. Using your critical thinking skills, analyze the following statement: "The United States receives an influx of well over 2 million legal and illegal immigrants yearly. This influx is important to us and should not be stopped." Here are some questions to help you grapple with this issue: Do you approve of this rate of immigration? Should it be stopped to stabilize population growth? Why or why not? What benefits might result from continued immigration? What might be the disadvantages to the United States?

9. Describe the shape of age structure diagrams (population histograms) for (a) rapidly growing populations, (b) slowly growing populations, (c) stable populations, and (d) declining populations.

10. Would a study of age structure diagrams of the U.S. population benefit automobile manufacturers? The agricultural industry? School administrators? Why?

11. Give five examples of environmental stress caused by the technological overpopulation experienced by the United States today.

12. Using your critical thinking skills, analyze the following assertion: "Population growth is a problem only in the less-developed nations."

13. Compare the population trends of the MDCs and LDCs.

14. Discuss the major features of China's population-control program. Could any of these be used in the United States?

15. Discuss future world population growth in terms of Earth's carrying capacity. How will it affect the carrying capacity? What are some of the consequences of overshooting Earth's carrying capacity?

16. You are appointed head of a department in the government of a developing country facing rapid growth, poverty, and environmental destruction. Outline features of a population-control policy aimed at reducing growth, stimulating environmental protection, and promoting an improvement in your people's health and welfare.

Suggested Readings

Abramovitz, J. W. 2001. Averting Unnatural Disasters. In *State of the World 2001*. New York: Norton. Looks at disasters brought on by human society, including population growth, and ways to avert them.

Ashford, L. S. 2001. New Population Policies: Advancing Women's Health and Rights. *Population Bulletin* 56 (1): 1–44. Important reading for those who want to see how family planning is evolving—and how it seeks to advance the health and rights of women.

Assadourian, E. 2007. *Vital Signs 2007–2008*. New York: W. W. Norton. A wealth of information on global social, economic, and environmental trends.

Bacci, M. L. 2001. *A Concise History of the World: An Introduction to Population Processes,* 3rd ed. New York: Blackwell. Excellent reading for those wanting to learn more about population.

Brown, L. R., G. Gardner, and B. Halweil. 1998. *Beyond Malthus: Sixteen Dimensions of the Population Problem.* Worldwatch Paper 143. Washington, DC: Worldwatch Institute. A must-read for all students that aptly describes the many effects of overpopulation.

De Souze, R. M., J. S. Williams, and F. A. B. Meyerson. 2003. Critical Links: Population, Health, and the Environment. *Population Bulletin* 58 (3): 1–44. An excellent overview of some of the major problems created by population growth.

Ehrlich, P. R., and A. Ehrlich. 1990. *The Population Explosion.* New York: Simon and Schuster. Updated version of the 1971 classic *The Population Bomb.*

Englemann, R., B. Halweil, and D. Nierenberg. 2002. Rethinking Population, Improving Lives. *State of the World 2002.* New York: W. W. Norton. Very important reading for those interested in understanding the impacts of population growth.

Gelbard, A., C. Haub, and M. M. Kent. 2000. *World Population Beyond Six Billion.* Washington, DC: Population Reference Bureau. A great survey of major population changes and projected changes to 2050.

Hardin, G. 1993. *Living Within Limits: Ecology, Economics and Population Taboos.* New York: Oxford University Press. A comprehensive analysis of the population crisis by a world authority.

Hartman, E. C. 2006. *The Population Fix: Breaking America's Addiction to Population Growth.* An insightful examination of a very important issue.

Kent, M. M., and M. Mather. 2002. What Drives U.S. Population Growth. *Population Bulletin* 57 (4): 1–40. A must-read for anyone interested in understanding U.S. population growth.

Lee, K. N. 2007. An Urbanizing World. *State of the World 2007.* New York: W. W. Norton. An intriguing look at urban growth, another huge population issue.

MacDonald, M., and D. Nierenberg. 2002. Linking Population, Women and Biodiversity. *State of the World 2003.* New York: W. W. Norton. A look at population growth in most ecologically diverse areas and the role women are playing in helping to curb growth.

Malthus, R. 1994. *An Essay on the Principle of Population.* New York: Oxford University Press. A classic; a must-read for all students.

Mastiny, L., and R. P. Cincotta. 2005. Examining the Connections Between Population and Security. In *State of the World 2005.* New York: W. W. Norton. Discusses the implications that population growth and population size of nations have for issues such as violence and armed conflict.

McFalls, J. A. Jr. 2007. Population: A Lively Introduction. *Population Bulletin* 62 (1): 1–31. An excellent overview of demography.

McKee, J. K. 2005. *Sparing Nature: The Conflict Between Human Population Growth and Earth's Biodiversity.* Piscataway, NJ: Rutgers University Press. An interesting look at the effect of population growth on wildlife.

Population Reference Bureau. 2002. *Family Planning Worldwide.* Washington, DC: Population Reference Bureau. A great source of information on family planning and contraception by country.

Population Reference Bureau. 2007. *U.S. Population Data Sheet.* Washington, DC: Population Reference Bureau. A great source for information on the U.S. population, including a great deal of ancillary information for those interested in understanding more about our own population.

Population Reference Bureau. 2007. *World Population Data Sheet.* Washington, DC: Population Reference Bureau. A great source of information on populations of every country in the world.

Population Reference Bureau. 2007. *World Population Highlights: Key Findings from PRB's 2007 World Population Data Sheet.* Washington, DC: Population Reference Bureau. An excellent summary.

United States Bureau of the Census. 2007. *Current Population Reports.* Washington, DC: U.S. Government Printing Office. Published annually. Latest information on births, immigration, and changing age structure of the U.S. population.

United States Bureau of the Census. 2007. *Statistical Abstract of the United States.* Washington, DC: U.S. Government Printing Office. Latest population data for the United States.

Web Explorations

Online resources for this chapter are on the World Wide Web at: **http://www.prenhall.com/chiras** *(click on the Table of Contents link and then select Chapter 4).*

CHAPTER 5

WORLD HUNGER

SOLVING THE PROBLEM SUSTAINABLY

Of all the problems created by the human population, undoubtedly the one with the greatest potential for disaster is a shortage of food. Food shortages result in widespread hunger, inadequate nutrition, starvation, and death.

5.1 World Hunger: Dimensions of the Problem

Although most of us hear very little about world hunger these days, the problem still plagues many countries. Worldwide, an estimated 18 million people, about 10 million of them children, die each year from hunger and starvation or diseases worsened by hunger, according to the United Nations. This death toll is the equivalent of nearly 100 jumbo jets, each carrying 500 passengers, fatally crashing every day of the year.

Undernutrition, Malnutrition, and Overeating

According to recent estimates of the World Health Organization, approximately 850 million people—or nearly one in eight people worldwide—regularly consume less food than they need to stay healthy. Although estimates published by the United Nations Food and Agriculture Organization suggest that the number of undernourished people in the less-developed countries (LDCs) has declined slightly from the 1970s to 2000, current trends in population growth and declining cropland suggest that improvements may be short-lived. Most people experiencing food shortages live in Africa, Asia, and Latin America. The greatest concentration of chronically hungry, however, is found in Africa and South Asia. India also suffers from intense hunger. Hunger is endemic in pockets within the more-developed countries (MDCs), too. In the United States, for instance, an estimated 10% of households go hungry, are on the edge of hunger, or are worried about hunger. Hunger affects one in five American children.

Food shortages result in two basic maladies: undernutrition (insufficient food intake) and malnutrition (a lack of one or more nutrients). In many parts of the world, there's another food-related problem: overnutrition (excessive food intake). Worldwide, the number of overweight people now rivals the number of people who are chronically malnourished (Table 5.1). Let's consider each of these groups.

TABLE 5.1	Overweight Adults and Underweight Children

Country	Percent Adults Overweight
United States	63
Russia	54
United Kingdom	51
Germany	50
Colombia	43
Brazil	31

Country	Percent Children Underweight
Bangladesh	56
India	53
Ethiopia	48
Viet Nam	40
Nigeria	39
Indonesia	34

Source: Worldwatch Institute

GO GREEN!

One of the most important things you can do to help protect the environment is to eat a healthful diet that emphasizes fruits and vegetables. Avoid the tendency to overeat. Food takes a huge amount of resources to grow, process, store, and transport. By cutting down on your caloric intake, you reduce energy and pesticide use.

Undernutrition is a quantitative phenomenon characterized by an inadequate amount of food—or calories—resulting in half-empty stomachs and gnawing hunger pains. Nutritionists assume that the average person requires a minimum of 2,200 to 2,400 calories daily. Western Europeans and North Americans consume 3,200 calories daily on average. They are among the fortunate. The average Ethiopian consumes only 1,600 calories daily—only half of the intake of an average American adult (Figure 5.1). The deficit of 600 calories per day undermines strength, causes severe mental and physical lethargy, and weakens natural resistance to a variety of diseases.

Malnutrition, in contrast, is a qualitative phenomenon characterized by inferior food quality or a lack of certain important foods. This results in a deficient supply of one or more of such key nutrients as proteins, vitamins, and minerals.

Overnutrition results from excess intake of food. In the United States, 65% of all American adults are overweight. The percent of the population classified as severely overweight or obese has climbed from 15% in 1970 to 31% today. Even children are afflicted, with 15% now classified as either overweight or obese. The United States is not alone in this regard. Over 50% of the adults in Russia, Great Britain, and Germany are classified as overweight. Even in less-developed nations, overnutrition is becoming more commonplace among the wealthier classes. While many struggle to get enough to eat, the wealthy eat too much.

Malnutrition and undernutrition often go hand in hand. In many less-developed nations, poor children suffer from a lack of proteins *and* calories. As shown in Figure 5.2, children lacking the calories and protein that their bodies require become thin and emaciated. They gnaw endlessly on their clothing to appease the insatiable hunger that plagues them day and night. This disease, called **marasmus** (pronounced mar-az´-mus), most often afflicts infants who are prematurely separated from their mother's breast milk, which is rich in proteins and calories.

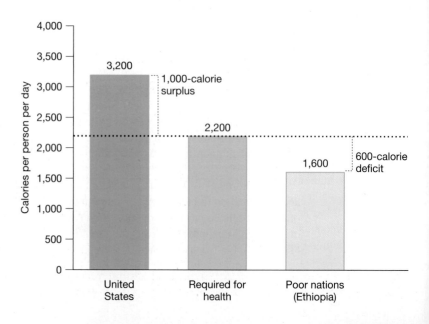

FIGURE 5.1 The average American is overfed; the average person living in a poor nation is underfed. Hunger, diseases worsened by hunger, and starvation kill an estimated 18 million people a year.

FIGURE 5.2   Death from starvation, Parvati Pura, India. This village suffered from a local famine because of an extended drought. The two-year-old boy is almost dead from protein and calorie deficiency (marasmus).

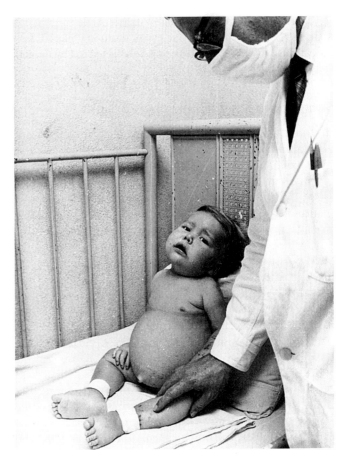

FIGURE 5.3   Guatemalan child suffering from kwashiorkor (protein deficiency). Note the protruding abdomen.

Marasmus may result from maternal death, a cessation of maternal milk production, or the use of milk substitutes, which are often diluted with drinking water by poor families and, therefore, often provide inadequate amounts of nutrients. In some cases, polluted drinking water is used, and children contract infectious diseases, resulting in severe diarrhea and death.

Millions of children under six years of age in the LDCs suffer from the protein deficiency disease **kwashiorkor** (pronounced kwash-ee-or´-core). Kwashiorkor is a West African word that means "the disease the child gets when another baby is born" because the mother can no longer feed the older child with her breast milk. Such children often receive a high-starch, low-protein diet. Although they receive adequate calories, they lack essential protein.

Kwashiorkor usually occurs in children from one to three years of age. Although the disease was first discovered in tropical Africa, it has since been identified among children of Central and South America, the Caribbean, the Middle East, and the Orient.

Children with kwashiorkor are weak and listless. Their limbs are thin and wasted. Their abdomens protrude because of the fluid that has accumulated inside (Figure 5.3). They suffer from skin ulcers and increased susceptibility to the infectious diseases common in many LDCs.

Protein deficiency retards physical as well as mental development, and even if a proper diet is restored several years later, the brain damage cannot be reversed. As a result, protein deficiencies create a real impediment to the success of many LDCs. Countries where protein deficiencies are common are eroding not only their many natural resources, but also their human resources.

Kwashiorkor and marasmus are two clinically recognizable conditions. However, for every child diagnosed with one of these diseases, a hundred suffer from milder forms of malnutrition and undernutrition.

The number of people afflicted by severe hunger could rise catastrophically in the near future in many parts of the world as a result of continued population growth, deterioration of the land, global climate changes, and other factors. The most severe problems will likely occur in Latin America and Africa, where many nations are simultaneously besieged by rapid growth in human population and civil unrest.

Micronutrient Deficiency

A problem that gets less attention than those already discussed is **micronutrient deficiency. Micronutrients** are chemicals required in small amounts. These include vitamins, such as vitamin A, and minerals, such as iron. **Macronutrients** are substances such as protein and carbohydrate that are required in large quantity.

According to one estimate, more than 2 billion of the world's people suffer from micronutrient deficiency. The three most prevalent deficiencies are vitamin A, iodine, and iron. All three can lead to serious medical problems. Iodine deficiency may result in mental retardation. Vitamin A deficiency can result in blindness and, in severe cases, death in children. Iron deficiencies, if severe, lead to anemia, characterized by fatigue, weakness, headaches, and apathy.

Food Trends and Challenges

The world food supply is at an all-time high, yet hunger and malnutrition persist at very high levels. Why? Consider grain production.

World grain production has climbed steadily since the 1960s. Since 1986, however, the production of all three major grain crops—corn, wheat, and rice—has not kept pace with population growth, according to data from the United Nations Food and Agriculture Organization. The result has been a decline in grain production per capita, meaning less food is available per person worldwide.

With increasing amounts of grain going into biofuels (ethanol and biodiesel) and an increase in droughts, fires, violent storms, and flooding, per capita outputs are bound to decrease even more. In 2006, for instance, the United States' ethanol production consumed 20% of the nation's corn crop, up from 6% in 2000. This was equivalent to approximately 55 million tons, about the same amount the country normally exports for food.

The nations of the world face two major food-related challenges. First, in nations where hunger and malnutrition are commonplace, the most immediate challenge is to find ways to end current suffering. The second challenge has to do with meeting food demands resulting from future growth. That is to say, all countries must find ways to provide more food to meet current and future demand. In the face of rapid population growth and rising costs, both tasks may be difficult.

There's a third challenge, too. It is meeting demand for food—whether to solve hunger or meet future demand—in a sustainable manner. That is, we must meet immediate and long-term demands while protecting the soil and water upon which agriculture depends. Rangelands also must be managed in ways that ensure their long-term productivity.

The challenge facing the global community can be best placed in perspective by examining expected global population growth. According to the Population Reference Bureau, a Washington, DC–based nonprofit organization, the world's population of 6.7 billion people in 2008 is expected to increase by roughly 1.2 billion people by 2025. Food production must climb at the same pace just to maintain the status quo, which is woefully inadequate in many parts of the world.

To meet present demands and ensure a long-term supply of food for the world's growing population, the nations must engage in sustainable agriculture and ranching. This chapter and the next three deal primarily with soils and farming. Chapter 12 covers rangeland.

Although there is some difference of opinion regarding the definition of **sustainable agriculture,** it is defined here as a means of producing adequate amounts of high-quality, affordable food while protecting, even enhancing, the soil and other natural resources essential to agricultural production, such as water supplies.

Meeting present and future needs and creating an enduring system of agriculture are essential to building a sustainable future. But how do we increase food supplies sustainably?

5.2 Increasing Food Supplies Sustainably: An Overview

To understand how we can meet our present and future needs while protecting—even enhancing—agricultural land and other resources vital to food production, we must tackle the problem on several fronts. Strategies fall into six major categories: (1) protecting existing soils from destructive forces such as erosion and conversion to nonagricultural uses, (2) increasing productivity (output per hectare) of existing farmland and restoring depleted soils, (3) reducing pest damage, (4) improving food storage and distribution, (5) developing new food sources, and (6) expanding the land under cultivation. The food crisis, however, cannot be solved in the farm fields of the world alone. Strategies that will stabilize the human population are essential. Without them, the technical measures described in this and other chapters will only postpone the day of reckoning predicted by Malthus nearly two centuries ago, a topic discussed in Chapter 4. See Ethics in Resource Conservation, Box 5.1, for a discussion of issues regarding population stabilization.

Protecting Existing Farmland

One of the biggest threats to food production is the deterioration and loss of existing farmland. Farmland is currently being lost at record rates in the United States and most other countries as a result of erosion, nutrient depletion, desertification, salinization, waterlogging, and farmland conversion. How big are these problems? Consider a few examples.

Each year, an estimated 24 billion metric tons of topsoil is washed from the world's farmland. To put this statistic into perspective, 24 billion tons a year equals 240 billion tons of lost topsoil per decade, which is equivalent to about half of the topsoil on all of the farms in the United States. Eroded soil reduces agricultural productivity. Severe erosion, which results in the formation of gullies, may even make farming impossible, permanently sidelining once-productive farmland. You'll learn more about it in Chapter 7. Farming can also deplete the topsoil of nutrients. Without proper fertilization, soil fertility declines over time, further decreasing agricultural productivity.

Desertification is the spread of desert, typically in semiarid regions, largely because of poor land management. It robs us of much arable land and decreases our food supply and our potential for increasing it as the human population increases (Figure 5.4). Desertification results from a number of factors, including global climate change, overgrazing, and deforestation. Together, they have robbed the world of millions of

ETHICS IN RESOURCE CONSERVATION 5.1 FEEDING PEOPLE OR CONTROLLING POPULATION GROWTH?

Rare is the year that a famine in some part of the world does not dominate the news. In 1993, it was the African nation of Somalia. Somalia faced political strife, drought, crop failure, and a massive hungry population, all of which teamed up to produce an incredible famine that killed tens of thousands of people. Many countries and international relief organizations rushed to feed the starving Somalis. The United States and later the United Nations moved soldiers in to quell the civil strife caused by fighting warlords. These measures, combined with an end of the drought, seemed to work. Since that time, famine has struck many other countries. Similar measures have been taken to address these issues.

Important as the measures in Somalia were, many people recognized that they were nothing more than stopgap measures. That is, they afforded a temporary relief to a chronic or recurring problem. Unless something permanent is done,

starvation will very likely return to drought-stricken countries of Africa and elsewhere.

Ending this repetitive cycle, say some observers, will require efforts to stabilize populations. Can the developed nations such as the United States impose this mandate on other nations in exchange for food aid?

Rather than answer the question ourselves, we ask you to make a list of points that support this viewpoint and another list of points that oppose it. Then analyze each argument carefully and decide which side of the debate you agree with. Once you have done this, take a few moments to think about your position and where it has come from. In other words, how has your ethical viewpoint evolved? Were your parents or teachers influential in forging your ethics? Have your friends influenced your thinking? Have books helped to shape your beliefs?

hectares of farmland and rangeland. How do you address these issues to meet demand for food?

After population stabilization, the next most important tactic for meeting present and future demands for food is

FIGURE 5.4 Desertification caused by overgrazing and poor agricultural practices destroys millions of hectares of marginally productive land each year in the United States, South America, Asia, and Africa.

preventive—improving the way existing farms and ranches are managed to reduce soil erosion and other problems such as **salinization** and **waterlogging.** Controls on soil erosion and nutrient depletion are discussed in Chapter 7. Ways to reduce salinization and waterlogging are described in Chapter 9. Desertification of rangelands and range management are outlined in Chapter 12. Measures to reduce global warming are discussed in Chapter 22. This chapter includes ways to reduce farmland conversion.

Reducing Farmland Conversion Each year, millions of hectares of farmland and ranchland are lost to development in LDCs and MDCs (Figure 5.5). The loss of farmland to urban sprawl, shopping centers, new highways, airports, and other human uses is called **farmland conversion.** In the United States, estimates of farmland conversion vary. According to the National Resource Inventory, losses between 1992 and 1997 were estimated to be 6.4 million hectares (16 million acres). Between 1997 and 2003, we lost an additional 3.2 million hectares or 8 million acres, a loss of 460,000 hectares or 1.14 million acres per year. Worldwide, between 1975 and 2000, an estimated 150 million hectares (370 million acres) were lost to nonfarm uses. That figure is nearly equal to all of the land currently farmed in the United States.

Reducing the loss of farmland from development is vital to ensuring a long-term food supply for the growing human population. One way to address the problem is through population stabilization—that is, reducing the growth of the human population.

GO GREEN!

When time comes to buy a house, consider buying within city limits, not in the expanding suburbs, which are typically built on productive farmland.

FIGURE 5.5 Cropland is destroyed by housing developments, highway and airport construction, new shopping malls, and other by-products of urbanization. Prime agricultural land is often taken out of production because it is flat and suitable for building and is often located near expanding cities.

Growth management can also slow down the loss of agricultural land. **Growth management** consists of strategies that reduce the expansion of cities and towns onto arable land. One of the best programs in the world exists in Oregon, which passed growth management legislation in the early 1970s. This legislation directed cities and towns to restrict urban growth to certain areas, thus concentrating growth. This is designed to protect farmland, forests, wetlands, and other ecologically important lands while permitting growth. With pressure from expanding populations, Washington, Florida, New Jersey, and Tennessee passed similar legislation.

Farmland can also be protected by other means. Some states, for example, buy the development rights of farmland in and around cities and towns. Development rights represent the value of a piece of property above and beyond its agricultural use. In other words, it equals the amount of money a farmer would get by selling his or her land to a developer. The purchase of development rights prevents farmers from selling their land, so they remain with the property in perpetuity. This ensures that the land will remain in agricultural production.

Increasing the Productivity of Existing Farmland

So far, we have learned that the first line of defense in the battle to feed the world's people is population stabilization. Second on the priority list are efforts to protect existing farmland from destructive losses such as erosion, desertification, and farmland conversion. The third item on the agenda involves measures to increase the productivity of existing and new farmland through irrigation, fertilization, and the development of new crops and strains of livestock that produce more food (Figure 5.6).

Increasing Irrigation and Irrigation Efficiency Irrigation has helped boost cropland production in the United States and elsewhere in the past four decades. Between 1950 and 2003, for instance, the amount of land under irrigation nearly

FIGURE 5.6 A mechanized irrigation pump at Habu Hat, Egypt, which has replaced the laborious methods of lifting water by a hand-operated or animal-driven waterwheel. Irrigation was made possible in Egypt by the Aswan Dam. An area of 410,000 hectares (1 million acres) formerly dependent on the annual Nile floods for irrigation is now cultivated under a system of perennial irrigation based mainly on lifting water. The new irrigation system makes possible a 40% increase in crop production in Upper Egypt because at least one additional crop can be grown per year. Unfortunately, the increased food production can barely keep up with Egypt's population increase.

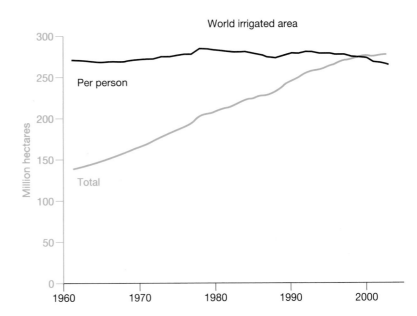

FIGURE 5.7   Irrigated cropland has increased steadily since 1960, but irrigated land per capita has remained more or less constant because of population growth.

doubled (Figure 5.7). Since the late 1970s, however, the amount of irrigated land per capita has steadily declined.

One way to improve the situation is to improve irrigation efficiency—that is, use water more efficiently so that more acres can be irrigated using the available water supply. In the western United States, for instance, open ditches used to transport water from rivers to farm fields can be replaced by cement-lined ditches and pipelines, cutting water losses by 50% to 90%, respectively. Drip irrigation systems that irrigate certain crops, especially fruit trees, use a fraction of the water of sprinkler systems and can be installed to reduce water use. Farmers can install soil moisture sensors to regulate irrigation systems. In recent years, many American farmers have adapted center pivot irrigation systems, shown in Figure 5.8, to reduce water use. These systems once sprayed water upward. By a simple and inexpensive modification, farmers can divert the spray downward, reducing water demand for crops by about 50%. Soil moisture sensors determine the amount of moisture in the soil and send signals to computers that control irrigation systems, thus helping farmers avoid overwatering. Avoiding overwatering not only saves water, it also helps prevent salinization, waterlogging,

and waste. Other, more costly solutions to increase water supplies are discussed in Chapter 9.

Improving Yields Through Selective Breeding: The Green Revolution   Since the dawn of human existence, nearly 4 million years ago, humans have succeeded in domesticating only about 80 species of food plants. Nonetheless, most of our food comes from a handful of species, with three crops—wheat, rice, and corn—accounting for well over 50% of the world's cropland.

After the domestication of wild plants, scientists began to tinker with ways to increase their yield. The most significant advances began in 1943 with the establishment of the **International Maize and Wheat Improvement Center** in Mexico. Sponsored by the Rockefeller and Ford Foundations, the center set out to develop high-yield varieties of wheat and rice under the leadership of Norman Borlaug, an agricultural geneticist (Figure 5.9).

After 25 years of intensive research, the center had amassed an impressive record of new seed varieties. Scientists at the center produced strains of wheat and rice that yielded three to five times as much grain as previous strains. Best publicized, however, was the new high-yield wheat that boosted Mexico's

FIGURE 5.8   Center pivot irrigation systems. (Left) Incredible amounts of water are sprayed into the air to irrigate crops. (Right) Fairly inexpensive modifications of these systems greatly reduce water loss to evaporation, thereby protecting supplies.

FIGURE 5.9   One of the architects of the Green Revolution, Nobel Prize-winning Norman E. Borlaug, recording the vigor and stage of growth of wheat plants on a selective breeding plot in Mexico. Borlaug was successful in developing the so-called miracle wheats that, at least temporarily, greatly boosted wheat production in Mexico and other less-developed nations around the world.

FIGURE 5.10   A new rice strain for hungry India. IR5 (left) is a new high-yielding variety of rice under testing at a research station in Aduthurai. The plant on the right is the traditional variety.

grain production from 780 kilograms per hectare (700 pounds per acre) to 4,700 kilograms per hectare (4,200 pounds per acre). Another important contribution was the development of new, high-yield varieties of rice (Figure 5.10). In addition to wheat and rice, scientists developed blight-resistant potatoes that outperformed previous strains by an amazing 500%. Disease-resistant beans quadrupled yield. Efforts to develop high-yield grains with higher protein content than their predecessors are part of a movement referred to as the **Green Revolution.**

The new wheat and rice developed at the center were primarily developed for use in tropical and subtropical countries. These plants were extremely responsive to fertilizer and irrigation water. They not only grew faster and produced more grain, they were shorter and stouter and thus better able to withstand winds and harvesting than the traditional long-stemmed strains. Their short growing seasons ensured two or three harvests per year.

It took Borlaug and his associates nearly three decades to develop the new "miracle" strains, which won him the Nobel Prize in 1971, a fitting recognition of his service to humankind.

The high-yield varieties spread more quickly and more widely in the more-developed countries than any other agricultural innovation in history. By the mid-1980s, nearly 50% of the wheat cropland and nearly 60% of the rice land in these nations were sown with high-yield seeds. The amount of rice

and wheat grown in the developed world shot up 75% between 1965 and 1980, even though the area in which those crops were planted increased only 20%. The benefits of the Green Revolution also spread to the less-developed countries. India, once crippled by food shortages, became self-sufficient. Mexico and Indonesia boasted similar progress.

Despite the obvious benefits of the new plant crops, critics point out that the Green Revolution has not been a panacea for world hunger. The new high-yield varieties proved to be disappointing for a number of reasons.

Perhaps the most common criticism is that the Green Revolution mostly benefited well-to-do farmers who could afford to irrigate their farmland and buy the fertilizer necessary for the new high-yield varieties. In rural Africa, where food is desperately needed, the Green Revolution had virtually no influence. Worldwide, an estimated 30% to 40% of the world's people live by subsistence farming, according to a report from the National Council for Agricultural Education. They grow barely enough food for themselves and their families, and they cannot afford the expensive seeds or the fertilizer needed to make the new miracle seeds grow.

The high-yield varieties also proved inadequately equipped to ward off pests and disease. The geneticists had inadvertently eliminated the natural resistance of wheat and rice, and had produced a generation of genetic lightweights. To protect their crops, therefore, farmers needed large amounts of pesticides, which further drove up the cost of farming. A farmer who could afford the seed might not have enough money to pay for costly fertilizer and insecticide.

Another problem of the Green Revolution, and of modern agriculture in general, is that it fostered the use of a limited number of genetic strains, thereby reducing **genetic diversity** (the number of genetic strains in use). Where dozens of strains had once been grown, huge fields containing one variety sprang up. The fewer the strains, the more devastating were the emergence of a plant disease or hungry insect pests.

Many critics claimed that the Green Revolution was a failure, but such criticism was premature. Learning from their errors, geneticists have developed new varieties of wheat and rice that grow under less favorable conditions. In Bangladesh, for example, half of the wheat crop is a new high-yield strain that can be grown without irrigation. Research is continuing on strains that can withstand drought and are more resistant to frost, pests, and crop diseases. A frost-resistant strain of winter wheat was developed and is now used in Canada, helping to increase wheat production.

Perhaps one of the most promising developments for the rural poor of Africa and Latin America is the relatively new genetic research aimed at increasing the yield of staples such as yams, potatoes, and various legumes. The Rockefeller Foundation, a key funder of the Green Revolution, is concentrating its agricultural program on research that will lead to genetic improvements in such crops, a great potential benefit to the 1.4 billion subsistence farmers who grow them.

Improving Yield Through Genetic Engineering Plant breeding, while successful in raising the productivity of food crops, is slow, tedious work. Until recently, development of desirable hybrids took 10 to 20 years. Today, a new technology, discovered in 1973, may accelerate genetic enhancement. That technology is genetic engineering. You've probably heard about it in recent years as protesters line up to express their dismay over this new development.

Genetic engineering is a process in which scientists isolate desirable **genes,** segments of the hereditary material of cells (the DNA) that determine various characteristics of organisms such as pest resistance, drought tolerance, fat content, and protein content. Isolated genes can then be reproduced or copied. These genes can be transplanted from one strain of a particular organism—for example, wheat or dairy cow—to another, creating new strains that could potentially outperform their predecessors. It is also used to transplant genes from one species, such as humans, to another, such as pigs, crossing boundaries, which is uncommon in evolution.

Genetic engineering is proving to be a quick but controversial means of altering crops and commercially important species of livestock. In fact, many popular crops today are the result of genetic engineering. They are called **genetically modified, genetically engineered,** or **transgenic crops.** In 1992, China became the first nation to grow transgenic crops commercially. In 2006, 100-million hectares (252 million acres) of transgenic crops were planted in 22 countries worldwide by 10.3 million farmers, according to the Human Genome Project. The majority of these crops were herbicide- and insect-resistant soybeans, corn, cotton, canola, and alfalfa. Researchers are testing a variety of other crops that are being decimated by crop diseases and extreme weather. They are also testing new rice varieties that promise higher levels of iron and vitamins, which could alleviate chronic malnutrition in Asia.

On the horizon, say scientists, are bananas that produce vaccines that could be used to combat infectious diseases such as hepatitis B; fish that grow more quickly, providing more food in aquaculture (fish grown in enclosures); cows that are resistant to mad-cow disease, which may also result in Alzheimer's and related diseases; fast-growing fruit and nut trees that produce food years earlier than conventional trees; and even plants that produce chemicals from which we can make bioplastics with unique properties.

In 2006, countries that grew 97% of the global transgenic crops were the United States (53%), Argentina (17%), Brazil (11%), Canada (6%), India (4%), China (3%), Paraguay (2%), and South Africa (1%), according to the Human Genome Project. Although the project notes that growth is expected to plateau in industrialized countries, the next decade will see exponential progress in genetically modified organisms as researchers gain increasing and unprecedented access to new genes that can be used in crops and animals not grown in the more-developed countries.

Desirable genes for transplantation may come from existing **cultivars,** strains that are currently under cultivation, though often in remote parts of the world. Or they may come from the wild species from which modern crop species were derived many years ago. Many cultivars and wild ancestors of domesticated species are endangered by the expansion of the human population but contain valuable genes that could enhance crop and livestock production. Consequently, many nations have launched aggressive programs to search out the vanishing cultivars and wild species that gave us corn, wheat, beans, and other commercially important plants—species that could benefit greatly from genetic boosts from their untamed cousins. The U.S. Department of Agriculture currently houses over 12,000 species with 483,000 genetically different samples. They're adding approximately 10,000 to 20,000 new samples per year from wild plant species and cultivars. Should either or both vanish from the wild, their genetic potential would theoretically be protected (Figure 5.11). We say "theoretically" because storage facilities are not a perfect answer. Seeds deteriorate in storage even under optimal conditions.

The importance of genetic improvements cannot be overstated. Over a period of 60 years, for instance, corn harvests have increased more than fourfold, from 20 bushels per hectare to 100 to 250 bushels, thanks to genetic infusions.

Geneticists are pursuing two basic approaches when modifying crops through genetic engineering. They're tinkering with "input traits" and "output traits."

Input traits refer to genetically controlled traits that affect tolerance to agricultural inputs such as pesticides or fertilizers. For example, much of the work in genetic engineering has focused on making crops resistant to herbicides. Although herbicides are applied to crops to control weeds, some can also damage crop plants. By tinkering with genes that control herbicide resistance in crops, farmers can use herbicides without affecting the crop.

FIGURE 5.11   Scientist at the National Seed Storage Laboratory at Colorado State University checks seeds stored in liquid nitrogen. This technique allows for better long-term storage than do previous measures.

Much of the future work, some experts predict, will be in output traits, that is, genetic traits that affect a plant's ability to produce food. For example, workers are developing strains of corn that contain more oil. Corn is fed to beef cattle to fatten them. Higher-oil corn could fatten beef cattle faster, reducing costs. Some alterations may affect flavor, nutrient content, or even the color, making food more appealing.

Although transgenic crops have been accepted fairly quickly by North American farmers and many consumers, this is not the case everywhere. Some chefs in the United States, for instance, refuse to serve foods from genetically modified organisms, and some citizens also express their concern. In LDCs, some farmers have protested against their use and have even destroyed fields where they have been planted. Several European nations have even proposed placing a moratorium on genetically modified organisms (transgenic food crops), while more research is done on their effects. What are the possible effects?

Concerns are several. First, there are concerns about human health. The introduction of new genes into food crops could result in allergic reactions among those eating them. Some evidence of this is already coming to light. Some opponents have expressed environmental concerns. Scientists have documented examples in which nontarget species were harmed by pesticide residues in the soil, which were produced by crops that were genetically modified to produce those chemicals. Even more startling, Canadian farmers have discovered that weeds acquire herbicide resistance from nearby transgenic crops two years after the introduction of the transgenic crop. Herbicide-resistant weeds could outcompete other species, spreading uncontrollably.

Clearly, there is much to be learned about transgenic crops. Caution is advised. In fact, in January 2000, an international agreement, known as the Cartagena Protocol, was signed in Montreal, Canada. This treaty regulates the international trade of food from transgenic crops. Its main focus is on sharing of information on existing regulations, risk assessments, and other agreements. It also calls for solicitation of consent from an importing country prior to the introduction of genetically modified organisms—for example, seeds from genetically modified crops. The treaty calls on the MDCs to help other nations develop the skills and the capacity to perform risk assessments.

Increasing Fertilizer Use   Japan has less than 0.07 hectare (0.166 acre) of arable land per capita, 1/13 that of the United States. Only with the most intensive agricultural methods has this tiny island nation been able to feed its nearly 128 million people. Japanese farmers have achieved amazing success, producing 5,300 calories per cultivated hectare (13,200 calories per acre) per year, almost three times the productivity of American farmers. One key to Japan's accomplishments is the large amounts of fertilizer applied to the land: sardine–soybean–cottonseed cakes, animal wastes, green manure, human solid wastes, and artificial fertilizer.

Japan's success with soil enrichment can be repeated in many LDCs. Even without modifying any farming method, 9,500 artificial-fertilizer trials conducted by U.S. agricultural specialists in 14 LDCs have shown an overall average yield increase of 74%. In the next 25 years, however, fertilizer use must increase sevenfold if rapidly growing populations are to be properly fed. How likely is this?

Between 1950 and 1989, world fertilizer use increased from 14 million tons to 146 million tons (Figure 5.12). This was responsible for a dramatic rise in grain production—from 620 million metric tons in 1950 to nearly 1,700 million metric tons in 1989, according to the UN Food and Agriculture Administration. Despite these impressive gains, fertilizer use

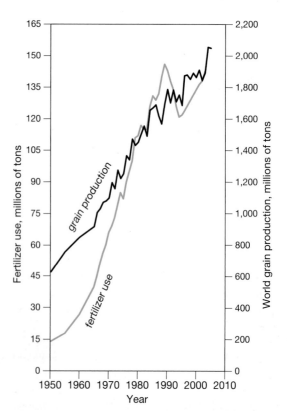

FIGURE 5.12   World fertilizer use and grain harvested per capita, 1950–2005.

has more or less plateaued since then (Figure 5.12). Total grain production has plateaued as well. When population growth is taken into account, it turns out that grain production per capita has actually declined in the past decade.

The long-term prognosis for fertilizer is not very good. Despite its many benefits, many farmers in LDCs simply cannot afford it. Because fertilizer is heavily dependent on fossil fuel energy for its production (it is made from natural gas) and application (oil is used to produce gasoline and diesel fuel used to power farm machinery), the use of fertilizer is becoming much more expensive. As natural gas and oil prices continue to rise, further declines in fertilizer use could occur. To solve this dilemma, some experts argue that what is needed is more efficient application of fertilizer.

Edward Wolf of the Worldwatch Institute, for instance, points out that many countries overfertilize their most productive areas. They would achieve far more benefit from applying that additional fertilizer to marginal land. For instance, Chinese farmers apply most of their fertilizer to the most productive one-third of their cropland. Applying that fertilizer to the marginal cropland—the remaining two-thirds—would yield 3 to 15 times more grain per ton of added fertilizer than it does on the productive land.

In Africa, Latin America, and Asia, where farms could benefit most from fertilizer use, farmers frequently cannot afford it. Therefore, instead of food aid, which is a stopgap solution, some observers believe that the industrial nations should consider donating fertilizer to help hungry nations become self-sufficient. Or they might assist in building fertilizer factories and the transportation networks needed to distribute fertilizer economically and quickly to rural areas. This would be an immensely costly and time-consuming task.

Studies show that low doses of artificial fertilizer applied to nitrogen-fixing plants enhance the plants' ability to fix atmospheric nitrogen. (The plants make their own fertilizer.) Heavier doses, however, impair plants' nitrogen-fixing capabilities. Thus, small donations of nitrogen fertilizer and technical assistance (training) to help farmers achieve a proper balance of artificial fertilizer and nitrogen fixation could help many nations improve production dramatically.

It is important to point out, however, that excess use alters nutrient cycles and pollutes waterways. **Organic fertilizer** such as cow manure, which is discussed in more detail in Chapter 7, may be a better choice. Organic fertilizers have been used in many parts of the world for thousands of years. They represent a more sustainable choice because they do not rely on a finite resource (natural gas) for fertilizer production. However, in many LDCs, one of the chief sources of organic fertilizer, cattle dung, also is a valuable source of fuel. Where it once enriched farmland, it now is dried and packed in cakes that are used for cooking. Population pressures have reduced woodlands in many less-developed nations, forcing the poor to use this alternative fuel. To return to the earlier, more sustainable way, some countries are promoting plans to develop sustainable forests near villages, which could reduce the use of cow dung as fuel, permitting farmers to use this resource to enrich their fields. More efficient cooking stoves are also being widely distributed to reduce the amount of wood and manure burned by villagers. Solar cookers are also being promoted in sunny climates to reduce the demand for wood.

World hunger will not be solved just by pouring more fertilizer on the land. Fertilizer is a costly alternative and, as noted earlier, bound to become more costly in the near future. But improvements in fertilizer application, the use of low doses in combination with nitrogen-fixing plants, better soil management, and the use of organic fertilizers can all help boost cropland production. These approaches are all part of a multifaceted plan needed to increase productivity of existing farmland.

Reducing Pest Damage

Roughly 40 kilograms out of every 100 kilograms of food grown throughout the world are destroyed by pests and disease organisms (mostly fungi), according to most estimates. Rats, insects, and fungi annually destroy enough food each year to feed one-third of India's population. One in every 14 people in the world will starve because of food deprivation caused by agricultural pests. Let's examine the impact of pests more carefully.

Rodents   The rat is one of humankind's greatest competitors. A single rat consumes 40 pounds of grain a year. The estimated 120 million rats in the United States alone destroy several billion dollars' worth of food a year. In India, where rats may outnumber people by ten to one, up to 30% of the crops are lost to rodents.

Disease-Transmitting Insects   In equatorial Africa, 23 varieties of tsetse fly, the transmitter of the protozoan that causes African sleeping sickness in humans and an equally serious disease in livestock, have effectively prevented livestock production in an area larger than the United States. In Southeast Asia, the microscopic malarial parasite *Plasmodium,* which is injected into the human bloodstream by mosquitoes, has been a scourge to the farmers. It incapacitates millions of rice farmers during the critical periods of transplanting and harvesting. To escape the malarial season in northern Thailand, for instance, farmers do not sow a second rice crop.

Locusts and Other Insect Pests   Crop-destroying insects cause billions of dollars of damage worldwide each year. Huge crops inadvertently improve the prospect for harmful pests. Migratory hordes of locusts have plagued farmers since the time of Moses and continue to devastate crops. One locust swarm near the Red Sea, so thick it blocked out the sun, blanketed an area of over 5,000 square kilometers (2,000 square miles). Because of their great mobility, locusts may destroy crops more than 1,600 kilometers (1,000 miles) and several nations away from their hatching sites (Figure 5.13). Swarms of locusts have been known to travel from Saskatchewan, Canada, to Texas. (For more on pests and pest control, see Chapter 8.)

Effective control of agricultural pests in the LDCs would markedly boost food output. But for such measures to be sustainable, they must be inexpensive, reliable, and environmentally safe. Of special interest are the biological techniques that allow farmers to minimize or even eliminate the use of chemical pesticides, long known to have serious environmental effects. Relatively simple measures, such as altering the time of

FIGURE 5.13   Locust swarm threatening crops in Somalia. The operator of the spray plane (left) wished to spray this swarm with insecticides, but the engine would not start because it was clogged with locusts. The locusts are so thick that they blot out the terminal building (right).

planting to avoid the emergence of harmful insects or increasing crop diversity to minimize pest population growth, can dramatically control pest populations without the damaging side effects of chemical pesticides.

Improving Food Storage and Distribution

As noted in the previous section, lots of food fails to reach the end user. It is destroyed by pests or rots while in storage. Improvements in storage, often simple changes in the types of storage bins, can drastically reduce food loss from rats and other pests. The MDCs can assist the LDCs in this simple, cost-effective endeavor by providing financial and technical assistance.

In many LDCs, roads are inadequate, and food distribution is problematic. Trucks transporting food may bog down in flooded roadways and never make it to the end user. Improving the distribution of food through the development of a better transportation system can also help increase the available food supply.

Developing New Food Sources

Much has been said about developing new food sources to feed the world's hungry. Possible candidates include algae and yeast. Despite the promise these new options offer, they are at best only of minor importance. Consider algae.

Algae have been grown for food in the United States, Great Britain, Germany, Venezuela, Japan, Israel, and the Netherlands for many years. One particular species, *Chlorella,* can be grown in small ponds. *Chlorella* is rich in proteins, fats, and vitamins and apparently contains all of the essential amino acids required by humans. Each hectare devoted to algal culture could produce nearly 90 metric tons of dried algae. The protein content of this algae crop may be 40 times greater than the protein content of soybeans and 160 times greater than the yield of beef protein.

So why haven't food-producing nations gone berserk over algae the way they have over hamburgers? For one, algae are extraordinarily expensive to grow. Except perhaps in urban areas, where abundant sewage is available to nourish the microscopic plants, the economic costs are so high that one cannot justify widespread algae culture.

In addition to the cost barrier, new foods like algae face tremendous consumer resistance. Foods not integral to a culture often fail to catch on, making the financial risks of such ventures even greater. In the LDCs, algae culture is an unlikely candidate for raising food production. However, more and more catfish, trout, and other freshwater fish are grown in ponds in the United States. Carp are popular in many LDCs. In Colorado, the Rocky Mountain trout on the menu is likely to have been reared in a nearby fish pond. Today, millions of tons of fish now come from commercial operations.

Livestock Improvements   Man does not live by bread alone. Many people acquire at least some of the protein they need from livestock. Solving world hunger, therefore, requires improvements in livestock production. Chapter 12 outlines methods to improve rangeland management.

In an effort to increase food production, scientists have developed new strains of cattle with meatier carcasses and new strains of pigs that grow faster and produce larger litters. In more-developed nations, scientists have developed strains of dairy cattle with vastly superior milk production compared with the dairy cattle of LDCs. For example, American test breeds of Holsteins produce up to 900 kilograms (2,000 pounds) of milk a year, compared with the 140-kilogram (300-pound) per year production from the yellow cattle of China.

A good part of the success in animal breeding results from artificial insemination, or the injection of sperm from genetically superior bulls into cows. Chickens with greater egg-laying capacity and more efficient feed-to-biomass conversions have been developed as well. Genetic engineering, discussed earlier, may also enhance scientists' ability to improve livestock to increase food production.

Native Grazers   Yet another potential means of increasing food supplies is greater use of **native grazers:** herbivores indigenous to

FIGURE 5.14 Giraffes and zebras in Zimbabwe. Such animals could be raised on large game ranches more profitably and efficiently than traditional livestock such as cattle and goats.

regions, such as Africa, that are adapted to the local climate and the local diseases.

A number of years ago, a team of wildlife biologists compared the meat production of domesticated livestock raised in Africa with native populations of antelope, zebras, giraffes, and even elephants maintained in the wild (Figure 5.14). Their conclusions are illuminating.

First, wild herbivores made exceptional use of the available plant food base. Cattle, in contrast, were very selective, consuming only certain highly palatable grasses and ignoring other apparently less tasty forage. From the viewpoint of a range manager, cattle underutilize the available resources. Fences that protect them also result in overgrazing that could eventually destroy the productive capacity of the land. In sharp contrast, when several native species grazed a given area, they consumed many different plants, making fuller use of the available forage. Antelopes, for instance, fed on grasses and low-level foliage, giraffes consumed foliage higher up on trees, and elephants dined on bark and roots.

Second, wild herbivores were much better adapted to drought than domesticated livestock. When a water hole dried out, they would move to another, often many miles away. Cattle, restricted by fences, cannot wander off in search of water. Native species also require less water per pound than livestock. A zebra, for instance, can get along without drinking water for three days. The gemsbok does not require drinking water at all but survives on the water in the plants it eats and metabolic water. **Metabolic water** is water produced when glucose (a type of sugar found in plants) is broken down by cells to produce energy. Also, native herbivores have evolved to be much better able to avoid native predators. A cow is an easy target for a hungry lion. A healthy antelope has a fighting chance of escape.

Third, wild herbivores are immune to the potentially lethal sleeping sickness transmitted by the tsetse fly. Introduced livestock, however, are not.

For these reasons, ranches that raise wild species for commercial meat production can expect a profit margin six times greater than that of the traditional livestock ranch. This is not to suggest that native grazers are a panacea. Several problems exist. First, because native herbivores migrate in response to available forage and water, ranching operations must include extensive landholdings. This system therefore naturally favors wealthy landowners. But to contain herds, extensive, costly fencing might be required. In addition, ranchers of wild animals need a system to prove and maintain ownership.

Tapping into Farmland Reserves

Many countries have tried to boost agricultural production by farming new land. In an ambitious period from the mid-1950s to the mid-1970s, China, the former Soviet Union, and the United States, for instance, all expanded their farms onto previously untilled land. But much of the expansion was on marginal land, which was often difficult to farm, poor in nutrients, or highly erosive. In the 1980s in the United States, 50% of all the soil eroded from cropland came from a mere 10% of the arable land.

Starting in the late 1970s, the high cost of farming marginal land forced officials in these three countries to withdraw marginal land from production. In the United States, the Food Security Act of 1985 created a program (Conservation Reserve Program) that provides subsidies to American farmers to withdraw about 15 million hectares (35 million acres) of highly erodible farmland from production—over one-tenth of the land then under cultivation. This goal was achieved by 1993. So successful was the program that Congress extended the law to 2008. At then end of 2007, 14.9 million hectares (36.8 million acres) had been set aside.

The prospects for expanding food production in the Third World are mixed. In Southeast Asia, most of the cultivable land is already under cultivation. In Southwest Asia, some researchers believe that more land is currently being farmed than is sustainable. In Africa and South America, however, only a small percent of the land thought to be suitable for agriculture is currently being farmed. Although expansion is feasible on both of these continents, expansion could threaten native wildlife species. The wisest strategy for increasing food production in these areas may be taking better care of existing farmland so that it remains productive indefinitely and enhancing production through sustainable means. It is also important to note that much of the land that agriculturalists view as potential farmland includes tropical rain forests, arid lands, and wetlands. Should they or can they be used?

Tropical Rain Forests: Potential Farmland? Tropical rain forests are blessed with abundant sunshine, rainfall, and perpetually warm days—all conditions conducive to forest growth. To the untrained eye, the dense tropical rain forests might appear to be prime candidates for farmland conversion (Figure 5.15). Just the opposite is true, however: Cleared tropical rain forests are some of the worst farmland in the world.

Unlike deciduous forests of the eastern United States, China, and Europe, where leaves and other plant litter form a thick, spongy layer that decays over time, making the soil rich and productive, the tropical rain forest soils are bare and nutrient poor. Fallen leaves and branches are quickly decomposed by bacteria, insects, fungi, and earthworms. The nutrients released by decomposition into the soil, however, are rapidly absorbed by the roots of the large trees that tower over the forest floor. Thus, the soil that supports the most productive ecosystem on Earth is, paradoxically, among the poorest

FIGURE 5.15 Shifting the type of cultivation in central Sumatra. Agricultural land is opened up by cutting and burning the forests. The cleared area will be intensively cropped for a few years until the soil fertility has been exhausted, and then it will be abandoned. Small plots revegetate and can be used again in 35 years; large plots tend to suffer from extreme erosion, making them permanently unsuitable for agriculture.

known to humankind. Chopping down trees and planting crops is a prescription for disaster. What nutrients are in the soil are quickly taken up by the crops or leached from the soil into the deeper layers, where they are inaccessible to crops. Nutrients are also washed away during the rainy season, impoverishing a land that has little recuperative ability.

Another problem with tropical forest soils is that many are rich in an iron compound that, when exposed to sunlight during dry periods, hardens the soil, creating a bricklike layer impervious to plants and farm equipment alike. Known as **laterites** (*later* is Latin for brick), these reddish-brown soils may have caused the downfall of the Khmer civilization in Cambodia and the Mayas of Mexico. In more recent times, they caused the failure of an agricultural colony started by the Brazilian government in the heart of the Amazon basin.

Today, tropical rain forests fall at an alarming rate. Each year, by several estimates, an area of tropical forest the size of the state of Washington is cleared. This land is cleared for timber and other wood products. Farms and pastures are started on the denuded landscape but, almost without exception, soon fail. Plans to expand agriculture at the expense of forests, most agree, must be stopped. (For more on tropical rain forests, see Chapter 13.)

Arid and Semiarid Lands Deserts and semiarid lands are also viewed as a source of potential farmland. With adequate water and fertilizer, some say, the sandy soils of arid and semiarid lands could be made to produce crops.

However promising, many irrigation projects are riddled with problems. High costs and low returns are key stumbling blocks. Adding to the purely economic barriers are salinization, or the buildup of salts on irrigated farmland, and waterlogging, the saturation of the topsoil in heavily irrigated land, described in Chapter 8. Salinization today threatens every arid region of the world where irrigation is used. In intermediate or final stages of salinization, this land will soon fall into disuse. Thus, plans to expand agriculture into arid and semiarid lands are very likely a short-term answer with high costs and limited utility.

Wetlands Throughout the world, wetlands have been drained to provide living space and valuable farmland. In the United States, agriculture has been a major source of wetland losses. **Wetlands** are lands that are wet for part or most of the year and include swamps, bogs, salt marshes, and mangrove swamps. Lands alongside rivers can be classified as wetlands if they are flooded part of the year. Many agricultural regions in Great Britain, Israel, Italy, and the United States (Mississippi and Florida) grow fruits and vegetables on drained swampland.

Wetlands are productive fish and wildlife habitat. Many species of waterfowl live and breed in wetlands. Many commercially valuable fish and shellfish depend on coastal wetlands. Destroying these areas endangers fish and wildlife populations. But wetlands are also water purifiers that trap sediment and other pollutants. They act as sponges as well, holding back rainwater and reducing flooding while increasing groundwater recharge. (See Chapter 9 for more on wetland functions.) Converting wetlands to farmland robs us of the free ecological services they provide and may require costly engineering solutions such as water-pollution-control facilities to eliminate pollutants and dams to control floods.

In summary, although there is some potential to expand the land under cultivation, it is not as great as the figures suggest. It can only be counted on as part of a mix of solutions designed to expand food output.

5.3   Poverty, Conflict, and Free Trade

Feeding the world's people, like so many challenges facing us today, requires a multifaceted approach—controlling population growth, protecting farmland from erosion and conversion, increasing productivity, boosting crop production in existing land, and so on. The next three chapters outline many solutions, but one aspect they do not touch on is poverty. One of the key stumbling blocks to feeding the world's people is a widespread inability to pay for food. Even in the chronically food-short nations, the well-to-do can buy food while the poor starve to death. Huge numbers of people go hungry because they cannot buy enough food. Some experts believe that not until the standard of living is raised in many LDCs can hunger be eliminated. Even modest improvements in annual income could go a long way toward feeding the world's people. Thus, in addition to the measures described in the previous section, many experts believe that countries must find ways to raise the earning power of their poor. But any economic development strategy must be sustainable. It must provide livable wages and decent work but not at the expense of the environment.

Some proponents of sustainable development argue that one key may be the development of small, sustainable businesses in local villages and cities and towns. Sustainable businesses would use local resources, rather than imported ones, for local consumption rather than for export. Such projects would therefore promote local or regional self-reliance. Bicycle factories or factories that turn out solar cookstoves, for example, could help the poor earn enough money to feed themselves and their children, and the bicycles and stoves could be used locally.

In his 2008 State of the Union address, President Bush said the LDCs should strive to become more self-sufficient in food production, rather than relying on food imports. Ironically, many LDCs were once self-reliant but became indebted to more-developed nations that financed many Western-style development projects. To generate money to pay back loans, many LDCs turned their cropland, which once produced staples, into citrus, coffee, and tea plantations whose products were sold to the West. This reduced the amount of food that LDCs could produce and made them even more reliant on the West, this time for basic food. Free trade, discussed shortly, has also thwarted local food production as farmers find it difficult to compete with large, industrial farmers in the MDCs.

By growing their own food locally at an affordable price, LDCs can achieve self-sufficiency and ease suffering and death.

Self-sufficiency for the less-developed nations may be an essential component of achieving a sustainable future. More-developed nations can lend a hand in this important transition by providing technical assistance and measures to support locally based, sustainable agriculture. Together, these and other measures mentioned in this chapter could bring us closer to the goal of feeding a hungry planet and protecting and enhancing the environment.

Two additional factors also play a key role in hunger and malnutrition: civil conflict and free trade. In numerous countries, warring factions often use food as a weapon. They cut off supplies or destroy crops and force farmers to abandon their farms (people can't farm with bullets flying over their heads) to weaken an opponent and terrorize residents.

War also has negative effects on the economy. For example, it eliminates jobs and makes it difficult for people to afford food. Destruction of roads and other elements of infrastructure such as stores, railroads, storage facilities, and bridges also makes food distribution difficult, if not impossible. Even after a war has ended, many years may pass before agriculture can recover. In Afghanistan, land mines planted in a 1979 conflict remain buried in farm fields. It's estimated that somewhere between one-half and two-thirds of the nation's farmland cannot be worked because of the threat of mines.

Another culprit has emerged in recent years: free trade agreements. Although this may seem like a good thing, it does have a downside. In particular, farmers from industrial regions such as North America and Europe gain access to markets in LDCs, where they sell government subsidized products, often undercutting local farmers. Many local producers subsequently give up farming. They must seek employment elsewhere or go hungry.

Dependence on imported crops for staple foods can be a dangerous proposition for other consumers in LDCs, too. Price fluctuations and currency devaluations, for instance, can cause food prices to rise dramatically. In Mexico, for example, consumers initially benefited from cheap corn imported from the United States. However, in 1995, the peso was devalued, and the price of corn doubled in a year, making it difficult to afford. In 2007, the high price of U.S. corn, resulting from much higher fuel prices and high demand for corn ethanol, resulted in a dramatic increase in food prices of staples such as corn tortillas.

Summary of Key Concepts

1. Each year, an estimated 18 million people, 10 million of them children, die from hunger and diseases worsened by hunger.

2. According to recent estimates, 852 million people are undernourished, 2 billion suffer from micronutrient deficiency, and 1.2 billion are overfed.

3. Undernutrition is a quantitative phenomenon that results from an inadequate intake of food. Malnutrition is a qualitative phenomenon usually resulting in a deficient intake of proteins and vitamins. Overnutrition is an equally dangerous condition caused by overeating.

4. Two clinically identifiable diseases are seen in many of the world's children. Marasmus is a disease that afflicts infants prematurely separated from their mother's milk,

which causes a deficiency of protein and calories. Kwashiorkor occurs in slightly older children and results from an inadequate intake of protein. For every case of kwashiorkor and marasmus, there are a hundred other children who suffer from milder cases of malnutrition and undernutrition.

5. World agriculture faces two major challenges: (a) feeding people who are undernourished and malnourished today and (b) meeting the demands of future citizens. Both challenges must be met in a sustainable way.

6. World population is expected to increase by around 1.2 billion people between 2007 and 2025. Food supplies must increase dramatically to meet the demand of new world citizens and improve the nutrition of those who are undernourished or malnourished today. Fortunately, there are many ways to solve world hunger. The foremost strategy is population stabilization.

7. Another important strategy involves measures to protect existing farmland from such forces as erosion, desertification, nutrient depletion, and farmland conversion.

8. Efforts to increase productivity also can help expand food supplies. These efforts include measures to irrigate more efficiently and to fertilize land with artificial and organic fertilizer. New high-yield plant and animal species developed through selective breeding or genetic engineering also may prove helpful, although there are many concerns about potential ecological and health effects of genetically modified crop organisms.

9. Pest control is also an important means of curtailing pre-harvest and preconsumption losses. Pests destroy an estimated 40% of the world's food supply each year. Through environmentally safe pest control, farmers can raise food production while protecting soil, water, wildlife, and human health.

10. New food sources such as algae and yeast can boost food production but appear to be of limited value. Far more promising are plans to raise native grazers. Native grazers fully utilize the plant food base, tend not to overgraze, are mobile, and are often resistant to diseases.

11. Expanding the amount of land under cultivation is a strategy of limited value, for many countries are farming nearly all of their good land. In the United States and other developed nations, much of the potential farmland is of marginal quality. This land is often too costly and environmentally harmful to farm.

12. In Africa and South America, large amounts of farmland lie in reserve, but much of it, especially the tropical rain forests, is ecologically sensitive land and important wildlife habitat. Before such "reserves" are tapped, more careful land management should be applied to existing farmland to avoid further losses. In addition, tropical rain forests, arid lands, and wetlands should probably be avoided whenever possible. Tropical rain forest soils are poor and highly erodible. Arid lands are easily damaged. Wetlands are an important biological resource better left undisturbed.

13. Reducing poverty through sustainable economic development also is a viable strategy; many people are hungry simply because they cannot afford food.

14. Limiting conflict also is important to ensuring food for the world's people, because food is often used as a weapon in war, and war itself disrupts farming or the distribution of food.

15. Local self-reliance in food production also can help boost food production in an environmentally sustainable fashion.

16. Despite the long list of ways to solve world hunger, little progress will be made without strategies for stabilizing population growth and measures to increase the standard of living of many of the world's poor people.

17. What is needed is an integrated approach that includes the many measures discussed in this chapter and emphasizes self-sufficiency.

Key Words and Phrases

Borlaug, Norman	Kwashiorkor
Brown, Lester	Laterites
Cultivar	Macronutrients
Desertification	Malnutrition
Farmland Conversion	Malthus, Thomas
Farmland Reserves	Marasmus
Free Trade	Metabolic Water
Genes	Micronutrient Deficiency
Genetic Diversity	Micronutrients
Genetic Engineering	Native Grazers
Genetically Engineered	Organic Fertilizer
Crops	Overnutrition
Genetically Modified Crops	Salinization
Green Revolution	Sustainable Agriculture
Growth Management	Transgenic Crops
High-Yield Strains	Tropical Rainforests
International Maize and	Undernutrition
Wheat Improvement Center	Waterlogging
Irrigation	Wetlands

Critical Thinking and Discussion Questions

1. Debate the following statement: "Food surpluses from wealthy nations should be donated to poor, hungry nations now and in the future. Through such donations, world hunger can be solved."

2. Refute this statement with statistics on world hunger: "The world's doing fine. There are very few hungry people."

3. Define *malnutrition* and *undernutrition*. Why are these conditions particularly harmful to infants?

4. Describe the connections between overpopulation and hunger. In what ways does overpopulation worsen hunger? What is the role of poverty in overpopulation and hunger? How can measures that reduce population growth and poverty help solve the hunger crisis?

5. How prevalent is overnutrition? Where is it most common? Why?

6. What are the most immediate challenges facing the world with regard to food production?

7. Define the term *sustainable agriculture.*

8. Do you agree with the following statement? Why or why not? "The answer to world hunger is cropland expansion. We must plow all available land. There's plenty of good farmland left."

9. Give several reasons why the land on which tropical rain forests grow appears to be a prime candidate for new farms. Describe why these impressions are false. What critical thinking rules does this illustrate?

10. Of what value are swamps, salt marshes, lagoons, and other wetlands? Why should or shouldn't they be drained to make more farmland?

11. Describe the importance of protecting farmland from such forces as erosion and conversion as a strategy for meeting the demands of the world's population for food.

12. List ways to increase crop productivity. What are the most sustainable approaches?

13. What is the Green Revolution? In what ways was it initially successful, and in what ways was it unsuccessful? What new developments will help the world's subsistence farmers? How could genetic engineering help increase plant production?

14. Do you agree with the following statement? "Fertilizer has helped farmers the world over increase food production. By applying more fertilizer to cropland, especially in the LDCs, farmers can boost agricultural production." What are the limitations to this strategy? Can you suggest any more-effective strategies?

15. In what ways does war interfere with food production and distribution?

16. Sketch the broad outlines of a master plan to increase food production worldwide.

17. Using your critical thinking skills, debate the following statement: "Poor countries must eventually become self-sufficient in food production. They cannot rely on the wealthy nations of the world to bail them out of their troubles."

Suggested Readings

Assadourian, E. 2007. Vital Signs 2007–2008. New York: W. W. Norton. Lists of key statistics on food and agriculture.

Brown, L. R. 2001. Eradicating Hunger: A Growing Challenge. *State of the World,* ed. L. Starke. New York: W. W. Norton. Discussion of many strategies covered in this chapter.

Conway, G. 2000. Food for All in the 21st Century. *Environment* 42 (1): 8–18. Description of the need for a second Green Revolution and ways it can be achieved.

Durning, A. T. 1993. Supporting Indigenous Peoples. *State of the World,* ed. L. Starke. New York: W. W. Norton. Ways to protect indigenous people and their systems of sustainable agriculture.

Durning, A. T., and H. B. Brough. 1992. Reforming the Livestock Economy. *State of the World,* ed. L. Starke. New York: W. W. Norton. Many ideas for improving livestock production in environmentally sustainable ways.

Gardner, G. 1996. *Shrinking Fields: Cropland Loss in a World of Eight Billion.* Worldwatch Paper 131. Washington, DC: Worldwatch Institute. Excellent update on the current loss of cropland and food trends.

Gardner, G., and B. Halweil. 2000. Nourishing the Underfed and Overfed. *State of the World,* ed. L. Starke. New York: W. W. Norton. A detailed look at two serious global issues.

Gupta, A. 2000. Governing Trade in Genetically Modified Organisms: The Cartagena Protocol on Biosafety. *Environment* 42 (4): 22–33. Description of the provisions of an important international treaty on trade in genetically engineered food.

Halweil, B. 2002. Farming in the Public Interest. *State of the World,* ed. L. Starke. New York: W. W. Norton. Critical analysis of farming with sound ideas for creating a sustainable system of agriculture.

Halweil, B. 2002. *Home Grown: The Case for Local Food in a Global Market.* Worldwatch Paper 163. Washington, DC: Worldwatch Institute. Excellent look at local food production—that is, the production of food by local farmers for local markets—and the benefits it can have for the environment and the economy.

Halweil, B., and D. Nierenberg. 2007. Farming in the Cities. *State of the World,* ed. L. Starke. New York: W. W. Norton. A fascinating look at the rise in small-scale farming in and around cities and the advantages of local agriculture.

Levidow, L. 1999. Regulating Bt Maize in the United States and Europe: A Scientific–Cultural Comparison. *Environment* 41 (10): 10–21. Worthwhile reading for those interested in the controversy over genetically engineered foods.

Moore Lappe, F., J. Collins, P. Rosset, and L. Esparza. 1998. *World Hunger: Twelve Myths.* Union Grove, NY: Grove Press. Key insights into food shortages—for example, how the maldistribution of food results in widespread hunger.

Nierenberg, D., and B. Halweil. 2005. Cultivating Food Security. *State of the World,* ed. L. Starke. New York: W. W. Norton. Coverage of many of the topics discussed in this chapter but in greater depth.

Paarlbert, R. 2000. Genetically Modified Crops in Developing Countries: Promise or Peril? *Environment* 42 (1): 19–27. Candid look at this important subject.

Postel, S. 1993. Facing Water Scarcity. *State of the World,* ed. L. Starke. New York: W. W. Norton. Description of water conservation measures for food production.

Pretty, J. 2003. Agroecology in Developing Countries. *Environment* 45 (9): 8–20. An interesting look at ways agriculture can be made more sustainable in developing countries.

Reganold, J. D., R. I. Papendick, and J. E. Parr. 1990. Sustainable Agriculture. *Scientific American* 262 (6): 112–120. Excellent overview of this important topic.

Renner, M., et al. 2003. *Vital Signs: The Trends That Are Shaping Our Future.* New York: W. W. Norton. Great analysis of trends in agriculture and population.

Rosegrant, M. W., and R. Livernash. 1996. Growing More Food, Doing Less Damage. *Environment* 38 (7): 6–11, 28–32. Good look at policy changes required to promote sustainable agriculture.

Soule, J. D., and J. K. Piper. 1992. *Farming in Nature's Image: An Ecological Approach to Agriculture.* Washington, DC: Island Press.

Southgate, D. D., D. H. Graham, and L. G. Tweeten. 2006. *The World Food Economy.* New York: Wiley-Blackwell. An excellent treatise on world food production and consumption and the forces that influence patterns of consumption.

 Web Explorations

Online resources for this chapter are on the World Wide Web at: **http://www.prenhall.com/chiras** *(click on the Table of Contents link and then select Chapter 5).*

CHAPTER 6

THE NATURE OF SOILS

In this chapter we examine part of the "skin" of planet Earth—the soil. This thin layer of organic and inorganic matter is vital to our survival and that of all the species that share this planet with us. It supports plant life, which, in turn, supports animal life.

6.1 Value of Soil

Unfortunately, much of society's attention on environmental problems focuses on air and water while ignoring the soil. This is a dangerous oversight, however, not just because the condition of the Earth's soils affects our future food supply, but because soils serve many other vital functions. High-quality soils not only promote the growth of plants but also prevent water and air pollution by resisting erosion and by degrading and immobilizing agricultural chemicals, organic wastes, and other potential pollutants.

Most city dwellers refer to soil as "dirt," but to the farmer, soil is the essence of survival. The farmer's economic well-being is inextricably linked with the quality of the soil. Nations, like individuals, also are dependent on their soils. Nations blessed with good soils tend to be economically prosperous. When these resources are exhausted—because of mismanagement or the mounting demands of a swelling population—the future of most nations becomes tenuous. In fact, numerous historians assert that the demise of many great civilizations over the course of history resulted from poor land management and thus the destruction of their soil resource base that originally made possible their rise to prominence.

6.2 Characteristics of Soil

Soil is a natural system consisting of four components: mineral matter, organic matter, water, and air. Soil inherits **mineral matter** from its parent rock and **organic matter** from its living and decaying organisms. As shown in Figure 6.1, the composition by volume of a healthy loam surface soil is usually about 50% solids (minerals and organic matter) and 50% pore space (occupied by water and air). Under optimal conditions for crop yields, the pore space should be occupied by equal amounts of water and air.

In this chapter, we'll examine the major characteristics and components of the world's soils—their texture, structure, organic matter, living organisms, aeration, moisture content, pH, and fertility. An understanding of these characteristics is an essential prerequisite to the study of soil profiles, soil types, soil productivity, and soil management.

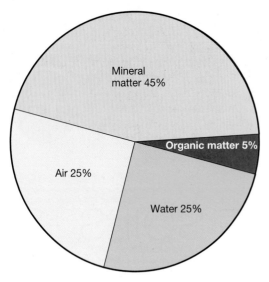

FIGURE 6.1 The composition (by volume) of a healthy loam surface soil. Mineral and organic matter constitute 50% of its volume, and pore space, which contains air and water, makes up the other 50%.

Particle Name	Diameter Range
TABLE 6.1	**USDA Division of Soil Particles Into Fine-Earth and Coarse Fractions**

Particle Name	Diameter Range
Fine-Earth Fraction	
Clay	<0.002 mm
Silt	0.002–0.05 mm
Sand	0.05–2.0 mm
Coarse Fraction	
Gravel	2 mm–7.6 cm (0.08–3 in.)
Cobble	7.6–25 cm (3–10 in.)
Stone	25–60 cm (10–24 in.)
Boulder	>60 cm (>24 in.)

Texture

Mineral particles in soils vary considerably in size, ranging from submicroscopic clay particles to large rock fragments such as stones or gravels. For convenience and efficiency, the U.S. Department of Agriculture (USDA) divides soil particles into two major groups based on size: the **coarse fraction,** consisting of particles greater than 2 millimeters in diameter, and the **fine-earth fraction,** consisting of particles equal to or less than 2 millimeters in diameter (Table 6.1). As you will soon see, the fine-earth fraction has a greater effect on soil behavior than the coarse fraction because it is the most chemically and biologically active. As shown in Table 6.1, soil scientists further break down the fine-earth fraction into three main size groups or "separates." From largest to smallest, the separates are **sand** (2.00 to 0.05 millimeter), **silt** (0.05 to 0.002 millimeter), and **clay** (less than

0.002 millimeter). The relative proportions of sand, silt, and clay in a particular soil determine its **soil texture.**

Since soils are composed of different percentages of sand, silt, and clay, specific terms are used to convey some idea of their textural makeup. For this, **textural classes** are used, such as sandy loam, loam, or silty clay. There are 12 major soil textural classes, which are defined by the percentages of sand, silt, and clay, as shown in the textural triangle (Figure 6.2). These percentages are on a weight, not volume, basis. The textural triangle is used to determine the soil textural name after the percentages of sand, silt, and clay are determined from a laboratory analysis. Organic matter and coarse fragments are not included in determining soil texture.

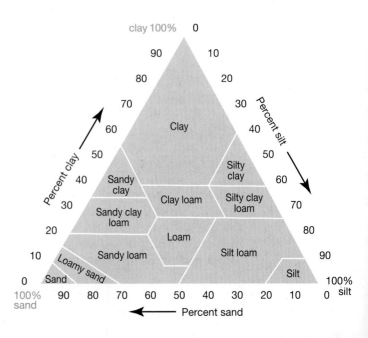

FIGURE 6.2 Standard USDA textural triangle for classifying soil textures, based on percentages of sand, silt, and clay.

Soils that are predominantly clay are called *clay,* while those with a high sand percentage are *sand.* A soil that does not exhibit the dominant physical properties of sand or silt or clay may fall into the textural class *loam.* A **loam** soil is the most desirable soil texture for many crops. As you can see from Figure 6.2, loam soils do not contain equal percentages of sand, silt, and clay, but they exhibit equal properties of sand, silt, and clay. That's because it usually takes a large amount of sand or silt particles to exert as much influence on soil properties as a comparatively small quantity of clay particles.

Soil texture is one of the most important characteristics of soil, because it helps to determine soil **permeability** (the ease with which air and water pass through a layer of soil) and water storage. It also helps determine the ease with which the soil can be tilled and its level of aeration, soil fertility, and root penetration. A coarse sandy loam or loamy sand, for example, generally has rapid permeability. It is also easy to till, has plenty of aeration for good root growth, and is easily wetted. However, it dries quickly and can easily lose plant nutrients, which are drained away in the rapidly lost water. High-clay soils (more than 30% clay) have very small particles that fit tightly together, leaving little open pore space. Therefore, there is little room for water to flow into the soil. Because of this, high-clay soils can be difficult to wet, drain, and till.

In general, the larger the exposed or **surface area** of the soil particles, the greater the potential for chemical reactions to occur between their surfaces and the air and soil solution (water plus solutes). Clay particles are the most chemically reactive, followed by silt and then sand particles. Table 6.2 summarizes the influences that the soil separates—clay, sand, and silt—have on water and air in soil.

It should be emphasized that soil particles are relatively stable. Despite the physical, chemical, and biological activities that are continuously transforming it and despite the soil management activities of the farmer, sand will not change to silt, nor silt to clay, within the average human life span.

Clay is an important reservoir of plant food, a function that largely depends on two characteristics of clay particles:

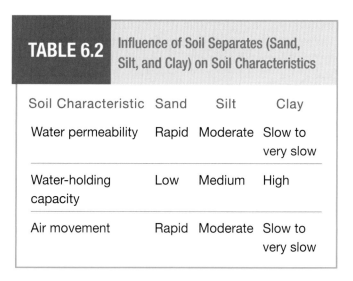

TABLE 6.2	Influence of Soil Separates (Sand, Silt, and Clay) on Soil Characteristics		
Soil Characteristic	Sand	Silt	Clay
Water permeability	Rapid	Moderate	Slow to very slow
Water-holding capacity	Low	Medium	High
Air movement	Rapid	Moderate	Slow to very slow

(1) their very large surface area and (2) their negative electrical charge. Soil scientists estimate that the aggregate surface area of the clay particles in the topsoil of 2 hectares (5 acres) of an Iowa cornfield down to a depth of 30 centimeters (1 foot) is roughly equal to the entire surface area of the North American continent. The negatively charged surface of the clay particle attracts positively charged ions (cations) of nutrient elements such as calcium, potassium, magnesium, zinc, and iron (Figure 6.3). These nutrients form a loose chemical bond with the clay particles. This process is called **adsorption.**

Adsorption, in turn, prevents the leaching (removal) of nutrients from the soil by water. The nutrient ions on clay particles are, therefore, available for plants. Just 10 grams (or 0.35 ounces) of dry Iowa topsoil may have 1.2 quintillion ion exchange sites that can hold nutrients for use by crops! Clay is, therefore, important in maintaining soil fertility.

At the same time, clay particles do not retain negatively charged nitrate ions very well. As a result, if too much nitrogen fertilizer is applied to a crop, nitrates may leach through the

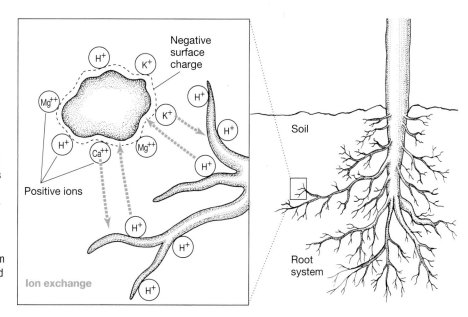

FIGURE 6.3 Ion exchange between clay and plant roots. The positive ions of nutrient elements, such as potassium (K^+), calcium (Ca^{++}), and magnesium (Mg^{++}), are attracted to the negatively charged surface of the clay particle. These nutrient ions on clay particles are available to be taken in by the plant. Note that as these nutrient cations are absorbed by plant roots, H^+ ions are excreted into the soil solution from the root systems of plants (or more organic acid anions are produced inside the cell) to balance the absorbed cations.

soil and into the groundwater, possibly causing contamination. The judicious use of nitrogen fertilizers by farmers and gardeners should ease this problem.

Structure

Soil structure is the arrangement or grouping of soil particles into clusters, or **aggregates.** These aggregates may have a variety of shapes, such as granular, blocky, or prismatic, as shown in Figure 6.4. Aeration, water movement, heat transfer, and root growth in a soil are all dependent to some degree on its structure. Physical changes that occur when farmers plow, cultivate, lime, and apply manure to the land influence soil structure. Other factors affecting soil structure include alternate freezing and thawing, wetting and drying, plant root penetration, and burrowing by animals such as worms and pocket gophers. The addition of slimy secretions from animals, bacterial decay of plant and animal remains, and compaction by farm equipment and off-road vehicles also affect soil structure.

When farmland has good soil structure, such as granular structure, crop production is enhanced. Soil with granular structure has an abundance of pores through which life-sustaining water and oxygen can move to the roots of plants. Such soil allows water to infiltrate after a rainfall or snowmelt (Figure 6.4). A soil with poor structure, such as platy structure, has a minimum number of pore spaces for air and water because of the closely packed soil aggregates. Water permeability is greatly reduced in poorly structured soils.

Some soils lack structure. That is, they exhibit no observable aggregation. Such structureless soils are either single-grained, as in very sandy soils, or massive, as in soils high in silt or clay, where the natural structure has been destroyed or become puddled. Soil structure can be improved by adding organic matter, such as crop residues, compost, and animal manures. The growth and decay of grasses and legumes stimulate aggregation and enhance soil structure, too.

Organic Matter and Soil Organisms

Soil organic matter is the living or dead plant and animal materials in the soil. It includes plant and animal residues at various stages of decomposition, small animals, microorganisms, and a material known as humus. **Humus** is a semistable, dark-colored organic material that consists of the decomposed products of plant and animal wastes and residues (remains of dead plants and animals), as well as materials synthesized by soil microorganisms. Humus is vital to healthy soil. Humus serves many functions. It (1) improves soil structure, (2) increases pore space so that air and water can penetrate more readily, (3) buffers the soil against a rapid change in pH, (4) reduces erodibility of the soil, (5) minimizes leaching of nutrients, (6) increases the nutrient storage ability and water-holding capacity of the soil, and (7) provides a suitable habitat for valuable soil organisms, such as bacteria and earthworms.

Humus formation is favored in moist and cool climates, not in dry and hot climates. Soil humus is richer under native grassland than under forest vegetation. However, humus content in both native grassland and forest environments is higher than under cultivated conditions. This is due to the harvesting of crops, to soil erosion, and to the increased oxidation of organic matter as a result of tillage.

GO GREEN!

Make your own humus by building a compost pile in your backyard. The finished product of composting your organic materials is humus. Adding compost to your garden or flowerbeds enriches the soil and simultaneously saves energy by diverting these organic materials from landfills.

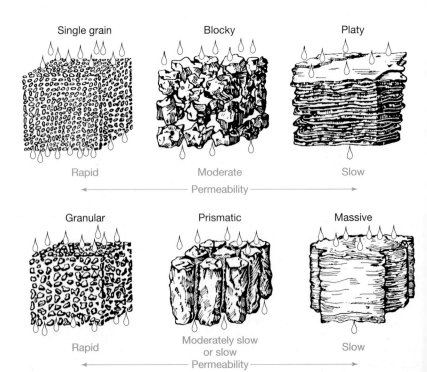

FIGURE 6.4 Effect of structure on water permeability. Note the variety of shapes occurring in soil aggregates and the relative rates at which water can move through them.

Three percent of organic matter in the soil consists of **biomass,** or living organisms. Soil organisms can be large or small. Large organisms, such as insects and earthworms, are known as soil **macroorganisms.** Small organisms, such as fungi, bacteria, and protozoa, are known as soil **microorganisms** (Figure 6.5).

Some soil organisms have little effect on plant growth, some are absolutely essential to plant growth in natural systems, and others cause damage by reducing or destroying plant yields and by spreading animal infections and disease. Some, for instance, aerate and aggregate soils—that is, give the soil structure. Soil microbes convert toxic carbon monoxide to carbon dioxide, fix atmospheric nitrogen (convert it to chemical compounds that plants can use), and are a major source of antibiotics used for disease control. Soil microbes are also instrumental in decomposing organic matter and releasing available plant nutrients.

Although they are tiny, microbes are found in great number. In fact, billions of soil microbes are found in a single gram of soil (one-fifth teaspoon). A hectare (2.5 acres) of healthy topsoil can contain 680 kilograms (1,500 pounds) of microbes. This means that a steer grazing one hectare of rangeland could be outweighed by the microbes grazing within it.

Aeration and Moisture Content

Healthy soil "inhales" and "exhales" continuously. That is, it absorbs oxygen and gives off carbon dioxide. Oxygen, which is present in greater concentration in the atmosphere than in the soil, diffuses into the soil's pores. Here, it is absorbed by plant roots and microorganisms that use the oxygen to break down glucose to produce energy. Carbon dioxide produced by this process moves in the opposite direction: from the soil into the atmosphere. This constant "breathing" action of the soil depends on numerous soil pores, which serve as air reservoirs and passageways.

The **pore space** of a soil is the space between particles of soil. It is is filled with air and water. The amount of pore space in a soil is determined largely by its texture and structure. With smaller surface area, as in sands, or with closeness of particles,

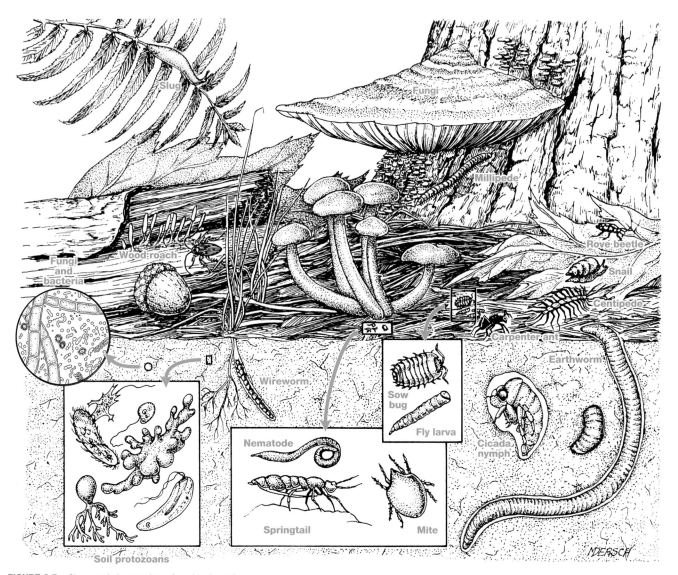

FIGURE 6.5 Characteristic organisms found in the soil.

as in compact subsoils, the total pore space is low. The pore spaces of soils are large if the surface area of individual soil particles is large or the particles are arranged in porous aggregates, as is often the case in medium-textured soils high in organic matter.

Two types of pores occur in soils: macropores and micropores. **Macropores** (larger than 0.08 millimeter in diameter) readily drain after a heavy rain and then fill with air. In contrast, **micropores** (less than or equal to 0.08 millimeter) drain slowly. In micropores, attractive forces bind the water within the fine soil pores. This, in turn, reduces air movement into the pore. Sandy soils contain many macropores, so the movement of air and water is surprisingly rapid, whereas soils with many fine clay particles have mostly micropores and thus allow relatively slow gas and water movement. Aeration in micropores is frequently inadequate for satisfactory root development and desirable microbial activity. Soil conditions are generally ideal where macro- and micropores are in equal proportions.

Water serves several important plant functions. It is, for example, an essential raw material for photosynthesis. It is also a solvent in which minerals are transported upward to the leaves and sugar is transported downward to the roots. Moreover, water is an essential component of protoplasm, forming 90% of the weight of actively growing plant organs, such as buds, roots, and flowers.

The amount of water in the soil is critically important to crop health and survival. When the pore spaces are completely filled with water to the exclusion of air, the soil is said to be at **saturation** (Figure 6.6). This condition frequently occurs after a heavy rain or an intensive irrigation. Two to three days later, after water has drained from the macropores and is held in the micropores, the soil is at **field capacity.** This water is readily absorbed by plants. When plants have removed all the water they possibly can from the soil and begin to wilt, the soil is at the **permanent wilting point.** The only moisture left in the soil is in the form of thin films that line the pores. These films are held so tenaciously by the soil that roots cannot absorb them—a condition experienced by many farmers and backyard gardeners who have seen their beans and lettuce shrivel during a midsummer drought.

Soil pH

An important property of the soil solution is its **reaction;** that is, whether it is acid, neutral, or alkaline (basic). The acidity, neutrality, or alkalinity of a soil depends on the relative amounts of hydrogen (H^+) and hydroxide (OH^-) ions. Soil scientists use the **pH scale** to determine a soil's reaction. The pH scale ranges from 0 to 14, as shown in Figure 6.7. A pH of 7 is neutral, a condition in which the number of hydrogen ions equals the number of hydroxide ions. At a pH below 7, hydrogen ions outnumber hydroxide ions, and the soil is **acidic;** at a pH above 7, hydroxide ions outnumber hydrogen ions, and the soil is **alkaline.**

Soils vary in pH from 4.5 to a little above 7 in humid regions, and from a little below 7 to 9 in arid regions. Most vegetables, grains, trees, and grasses grow best in soils that are slightly acidic to neutral, having a pH between 6.0 and 7. Soils under hardwood (oak, maple, and beech) forests are usually more alkaline than soils under coniferous (spruce, fir, and pine) forests.

Soil pH is the soil chemical property most commonly measured by farmers and urban homeowners alike. It can be determined with color indicators purchased at a nursery or by a pH meter in the laboratory. Lime ($CaCO_3$) may be applied if the soil is too acidic, as indicated in the reaction in Figure 6.8. As illustrated, if lime is mixed with acid clay, each calcium ion (Ca^{++}) from lime replaces two hydrogen ions (H^+) attached to the clay particle. Water and carbon dioxide are formed during the reaction. The overall reaction reduces the acidity of the soil.

Soil Fertility

Soil fertility is a measure of the nutrients found in soil for plant growth. Fertile soils contain an adequate number of nutrients in the right form for maximum plant growth. The fundamental components of soil fertility are the essential nutrients absorbed by plants and utilized for various processes. Most plants need at least 16 essential nutrient elements to grow, although they are capable of absorbing more than 90 elements. An **essential element** is a chemical element required for the normal growth of plants. Each essential nutrient plays one or more special roles in plant growth and metabolism.

Essential nutrients required in large amounts by plants are called **macronutrients.** These include carbon, hydrogen, oxygen, nitrogen, phosphorus, potassium, sulfur, calcium, and

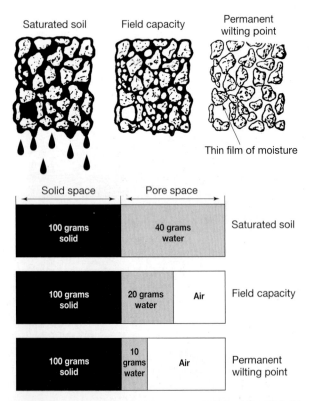

FIGURE 6.6 Variation in water content of a soil. The water content can vary considerably. Note that as the water content of a soil decreases, the air content increases.

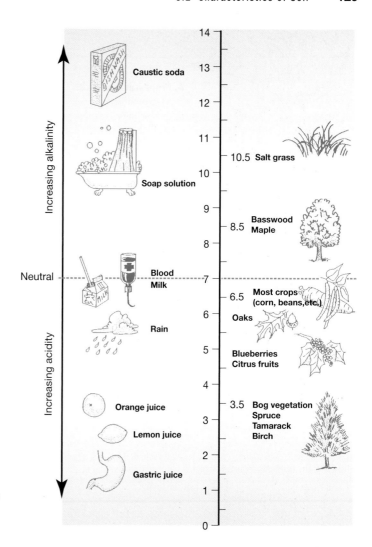

FIGURE 6.7 The pH scale. The pH values of certain well-known substances are shown on the left. The preferred soil pH values of some plants are shown on the right.

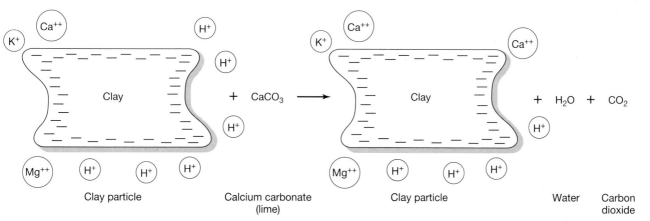

FIGURE 6.8 How lime neutralizes soil acidity.

magnesium (Figure 6.9). The other essential nutrients (manganese, copper, chlorine, molybdenum, zinc, iron, and boron) are used in only small quantities by plants and are known as **micronutrients.** As an example of the trace amount of a micronutrient needed by a plant, only 70 grams of the micronutrient molybdenum is sufficient to satisfy the needs of 1 hectare (2.5 acres) of clover.

Research has shown that each essential element must be present in a specific concentration range for optimum plant

growth (Figure 6.10). If the concentration of a given element in the plant root zone is too low, a deficiency of that element occurs, and plant growth is restricted. Likewise, if the root zone concentration of that element is too high, toxicity occurs, and plant growth is similarly limited. Only in a specific middle range of concentration (the sufficiency range) is optimum plant growth attained. One of the principal objectives of good farmers is to manage the fertility of their fields so that the concentration of each essential element remains in the proper range.

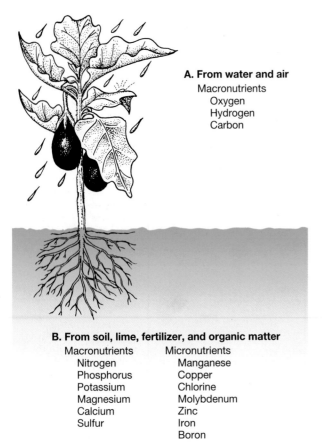

A. From water and air
Macronutrients
    Oxygen
    Hydrogen
    Carbon

B. From soil, lime, fertilizer, and organic matter

Macronutrients	Micronutrients
Nitrogen	Manganese
Phosphorus	Copper
Potassium	Chlorine
Magnesium	Molybdenum
Calcium	Zinc
Sulfur	Iron
	Boron

FIGURE 6.9   Nutrients required by plants.

Soils gain nutrients from a variety of sources, including (1) nitrogen fixation, (2) decomposition of plant and animal remains, (3) animal wastes, (4) weathering of parent materials, and (5) fertilizers (Figure 6.11). Soils lose nutrients mostly by (1) root absorption, (2) leaching due to the downward movement of water, (3) soil erosion, and (4) volatilization (conversion to a gas).

6.3 Soil Formation

The development of a mature soil is a complex process that may require centuries to a million years to complete. If we are to care for our soils so that they will last for future generations, we should understand how they formed and how they relate to their environment. Soil formation depends on five major factors called **soil-forming factors.** These five factors are responsible for the kind of soil that develops in an area, the rate of soil formation, and the extent of soil development. They are (1) climate, (2) parent material, (3) organisms, (4) topography, and (5) time.

Climate

Climate (the average weather conditions of an area) is a dominant factor in soil formation. The most important climatic influences on soil formation are temperature and precipitation. They exert profound influences on the rates of biological, chemical, and physical processes within parent material (defined in the next section) and soil. For example, for every 10°C rise in temperature, the rate of chemical reactions in soils doubles. In addition, many soil processes are controlled by the activity of soil organisms, which in turn is affected by temperature and moisture. Temperature and precipitation also influence soil development indirectly by determining the extent of plant growth.

Temperature and precipitation also influence mineral **weathering**—the breakdown of rock or other inorganic minerals that helps to form soil. Mineral weathering occurs through physical, chemical, and biological processes influenced by temperature and precipitation. When precipitation and the other four factors remain constant, an increase in temperature causes an increased rate of weathering and clay formation. An increase in the presence of water also increases the rate of mineral weathering (up to a point, of course). Thus, high average temperatures and precipitation, as in tropical and subtropical areas, tend to encourage rapid weathering and clay formation. Leaching and water erosion are also more intense and of longer duration in warm and humid regions. In cold and dry climates, weathering, leaching, and water erosion tend to be minimal.

FIGURE 6.10   Relationship between plant growth or yield and concentration of an essential element in the plant tissue. For most nutrients, there is a relatively wide range of values associated with optimum plant growth, called the "sufficiency range." At the upper half of this range is "luxury consumption," which is a plant's intake of an essential nutrient in amounts exceeding what it needs. Outside the sufficiency range, plant growth is reduced from either too little (a deficiency) or too much (a toxicity) of the nutrient.

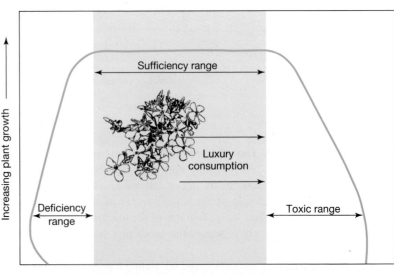

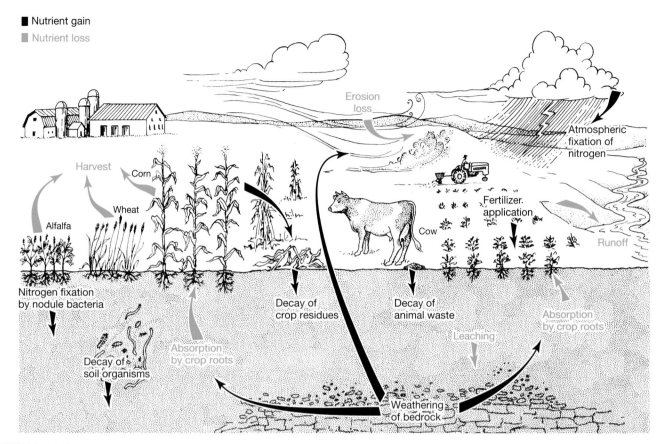

- ■ Nutrient gain
- ■ Nutrient loss

FIGURE 6.11 Nutrient sources for crops.

Parent Material

Parent material is the unconsolidated (soft and loose) material from which soils form. It may be mineral or organic.

Parent materials fall into three major groups: residual, transported, and organic. **Residual parent materials** are formed from the weathering of rock at a specific location. For example, a residual parent material and thus its soil can form from granite bedrock on a hillside. Scientists have determined that soil production rates on residual parent materials in temperate regions range from 0.02 to 0.08 mm a year (0.0007 to 0.0030 inch a year). This rate is equivalent to saying it would take 300 to 1,500 years to form 2.54 cm (1 inch) of soil. Because soil forms so slowly, it's clear why protecting it from erosion is important (Chapter 7).

Transported parent materials are, as their name implies, transported from their place of origin and deposited in a new location. They are the most extensive type of parent material on the Earth. Geologists have identified four transporting agents: ice, water, wind, and gravity. As illustrated in Figure 6.12, glaciers, or huge masses of ice, transport and deposit materials, known as glacial till. Meltwater from glaciers also carries

glacial material away from the glacier, forming deposits called "outwash plains." Water flowing in streams and rivers can also transport parent materials washed from the land, depositing them in floodplains (along the banks of rivers) or deltas (the mouths of rivers where they meet oceans). These deposits are referred to as alluvial deposits. Parent materials that are deposited in lakes are called lake or lacustrine deposits. Deposits may also form along coastal plains or beaches and are known as marine deposits. Wind can transport and deposit materials such as sand, forming large sand dunes that can develop into soils. Wind deposits of mostly silt-sized particles are known as loess, while volcanic ash blown in the wind forms tephra deposits. Rock fragments and soil material accumulated at the base of steep slopes as a result of gravity are known as colluvium.

Most **organic deposits** accumulate in stagnant water (lakes or swamps), as shown in Figure 6.13. Stagnant waters in temperate climates allow for good plant growth. These plants, in turn, shed leaves into the water. With time, the organic material accumulates (due to the lack of oxygen in the waters, which slows the decay process). Over centuries, debris and dead plants form thick organic accumulations that may reach depths of many meters.

Organisms

The activity of plants, animals, and soil microorganisms and the decomposition of their organic wastes and residues have marked influences on soil development. The development of a

GO GREEN!

Protect your garden soil and landscaped areas from erosion with mulch, a protective layer put on top of the soil. There are many types of garden mulch, including wood chips, grass cuttings, gravel, and fabric.

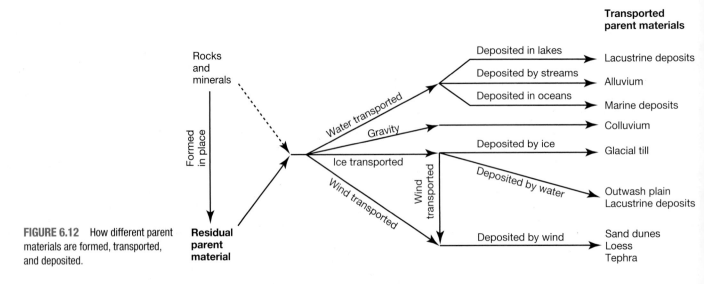

FIGURE 6.12 How different parent materials are formed, transported, and deposited.

mature soil depends on the activity of a great number and diversity of **organisms.** Perhaps the most important soil organisms are the microorganisms, especially the fungi and bacteria. They decompose organic matter and beneficially influence soil structure, aeration, and fertility. Organisms like **lichen** (a symbiotic association between a fungus and one or more algae and/or

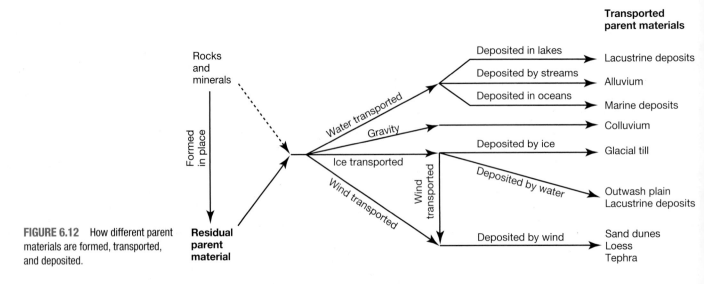

FIGURE 6.13 Stages of development of a bog (an organic deposit): (A) Aquatic plants slowly grow around the edges of the pond. (B) Water-loving plants continue to invade more of the pond. (C) With time, different layers of organic debris fill the bottom of the pond. (D) Trees and shrubs eventually cover the area, forming a woody peat bog.

cyanobacteria) secrete very dilute carbonic acid (H_2CO_3) that slowly dissolves rock, thus adding inorganic material to the developing soil. Upon their death, the lichens decompose, releasing nutrients that enrich the soil.

Rocks may be splintered by the roots of trees and other plants. Rooted vegetation absorbs mineral nutrients from lower levels. These nutrients become incorporated into plant matter, such as the leaves. When leaves, fruits, and nuts fall, and when the plant dies, these nutrients are released, contributing to soil formation. Earthworms, beetle larvae, bull snakes, pocket gophers, moles, and ground squirrels burrow through the soil, aiding the movement of air and water through the ground. Different types of vegetation have a marked effect on soil formation. For example, forests tend to develop acidic soils of moderate to low fertility, whereas grassland soils tend to produce thick topsoils rich in organic matter and fertility (Figure 6.14).

Topography

Topography, or the shape or contour of the land surface, largely determines how water moves in a landscape and how susceptible soils are to water erosion. For example, soils on steeply sloping surfaces tend to be washed away by rainfall and runoff. Soils are thin in such areas, but on the relatively flat valley floor, soils are much thicker because there is less erosion and the valley receives soil particles and organic material that erode from the hillsides above. However, flat valley floors can have waterlogged soils because water moves too slowly through the soil for proper drainage.

Time

When the Earth first formed about 4.5 billion years ago, it was completely devoid of soil. With the passage of **time,** however, physical, chemical, and—when organisms appeared on Earth—biological processes resulted in the formation of the first thin envelope of soil on this planet. Soils technically start forming at time zero, a point in time after new parent material is exposed at the land surface, initiating a new cycle of soil formation. Such an event may be a water deposit of alluvium, a wind deposit of loess, or lava.

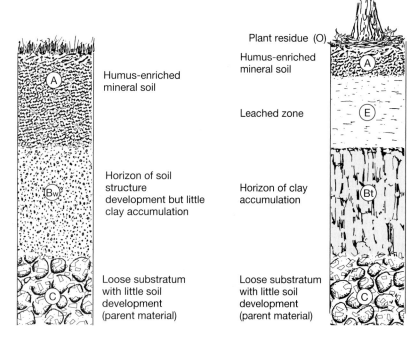

Humus-enriched mineral soil

Horizon of soil structure development but little clay accumulation

Loose substratum with little soil development (parent material)

Plant residue (O)

Humus-enriched mineral soil

Leached zone

Horizon of clay accumulation

Loose substratum with little soil development (parent material)

FIGURE 6.14 Two soil profiles. The one on the left shows a soil from a subhumid grassland, and the one on the right illustrates a soil from a humid forested region.

The length of time required for a soil to form depends on the combined effect of climate and organisms, as modified by topography, acting on parent material. A given period of time may produce much change in one soil and little in another (due to the other four soil-forming factors). For soil development from consolidated hard rock, the time may be very great. In contrast, fresh alluvium may develop into crop-supporting soils in just a few decades.

6.4   The Soil Profile

When one looks at the exposed face of a road cut, it is apparent that many soils are organized into layers, or **horizons.** Each horizon is characterized by a specific thickness, color, texture, structure, and chemical composition. A cross-sectional view of the various horizons is known as the **soil profile** (Figure 6.15). When people travel, they can make their trip more interesting if they pay attention to the changing soil profiles in the road cuts along the way.

The major layers from the ground surface downward to bedrock are called **master horizons** and are designated as horizons O (organic layer), A (topsoil), E (subsurface), B (subsoil), C (parent material), and R (bedrock). These horizons may not exist in all soil types. For instance, in immature soils, where weathering has not fully progressed, some horizons will be missing.

The soil profile is the product of the actions of vegetation, temperature, rainfall, and soil organisms on parent material located on a specific land surface operating for many thousands of years. The soil profile, therefore, tells us a great deal about soil history or genesis. It represents a kind of soil autobiography. From a practical standpoint, the soil profile is of great importance, for it can tell the soil scientist immediately whether the soil is suited for agricultural crops, rangeland, timber, or wildlife habitat and recreation. The profile also reveals the suitability of the soil for various urban uses, such as home

sites, highways, sewage disposal plants, sanitary landfills, and septic tank fields.

We shall examine the basic characteristics of a soil profile, beginning with the uppermost horizon and moving downward to bedrock (Figure 6.15).

O Horizon

The **O horizon** is an organic layer that forms above the mineral soil. It results from organic litter that came from dead plants and animals. O horizons are common in forests but generally absent from grasslands.

A Horizon

Human survival depends on the thin layer of topsoil, or **A horizon,** that covers much of the Earth. In the United States, its thickness ranges from 2.5 centimeters (1 inch) on the slopes of the Rockies to more than 1 meter (more than 3 feet) in the Palouse region of Washington State. This layer is rich in humus. It is from the topsoil that crop roots absorb much of their water and nutrients. Most soil organisms live within this layer.

E Horizon

The **E horizon** is called the **zone of leaching,** because much of the material in this layer is dissolved and carried downward to the B horizon by water. The E horizon comes from the word *eluvial,* meaning washed out.

B Horizon

The **B horizon,** commonly called the **subsoil,** is a **zone of accumulation** that receives and stores clays, soluble salts, and humus that are leached downward from the A and E horizons. After many years of abusive farming, the topsoil may be

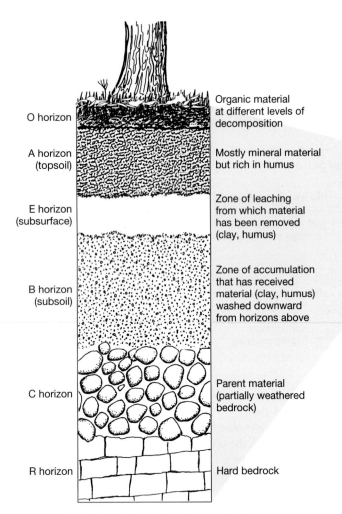

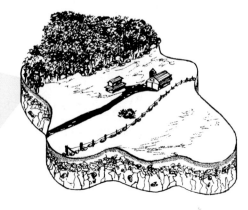

O horizon — Organic material at different levels of decomposition

A horizon (topsoil) — Mostly mineral material but rich in humus

E horizon (subsurface) — Zone of leaching from which material has been removed (clay, humus)

B horizon (subsoil) — Zone of accumulation that has received material (clay, humus) washed downward from horizons above

C horizon — Parent material (partially weathered bedrock)

R horizon — Hard bedrock

FIGURE 6.15 A soil profile showing the major or master horizons.

completely eroded away. The B horizon, which now lies at the surface, usually reduces crop yields because of its poorer physical, chemical, and biological properties.

C Horizon

The **C horizon** is composed of unconsolidated parent material. Usually this material has been transported to a site by glaciers, wind, or water, as noted earlier. However, the parent material in this horizon can also be derived from underlying bedrock.

The parent material in the C horizon determines many soil characteristics—its texture, water-holding capacity, nutrient levels, and pH level. For example, soils formed from parent material derived from granite tend to mature slowly and be acidic. The quartz mineral grains in granite are resistant to weathering and become sand particles in the soil. The less resistant feldspars and ferromagnesium minerals in granite can form clay particles. Thus, depending on the degree of weathering, soils formed from granite can have a wide range of textures. In contrast, soils formed from parent material derived from limestone develop rapidly and tend to be alkaline. Soils that develop from limestone tend to be more productive than those derived from granite.

R Horizon

The **R horizon** consists of bedrock with little evidence of weathering. As bedrock weathers, it forms the parent material in the C horizon above. The R horizon is absent in loess and alluvial soils, where the parent material was transported by wind or water from another location, instead of being derived from bedrock.

6.5 Soil Classification

Soil scientists recognize many thousands of kinds of soils in the United States and around the world. To simplify matters, soil scientists in the United States place soils into 12 major groups. Classifying soils into groups with similar properties minimizes the problem of locating information about any one soil and allows scientists to map soils—that is, to identify where soils are located.

The **soil classification** system used in the United States (and some other countries), called **soil taxonomy,** was developed by the USDA but includes all soils of the world. Other countries and international organizations have developed their own classification systems.

TABLE 6.3	Common Diagnostic Horizons

Diagnostic Horizon*	Main Characteristics
Mollic epipedon (A horizon)	Thick, well-structured, dark-colored surface horizon with base saturation >50%
Ochric epipedon (A horizon)	Light-colored surface horizon or thin, dark-colored surface horizon
Argillic horizon (Bt horizon)	Subsoil layer with an accumulation of silicate clays
Cambic horizon (Bw horizon)	Weakly developed B horizon
Spodic horizon (Bhs horizon)	Subsoil layer with an accumulation of humus and Al and Fe oxides
Oxic horizon (Bo horizon)	Subsoil layer with an accumulation of Fe and Al oxides and kaolinite

*Lowercase letters after master horizon symbols designate subordinate distinctions within the master horizons. For example, a Bt horizon is a subsoil layer with an accumulation of silicate clays.

Diagnostic Horizons

The USDA soil taxonomy recognizes 12 major soil groups in the world, known as **soil orders.** Most soil orders are defined on the basis of having certain **diagnostic horizons,** or layers with distinct physical and chemical properties. These, in turn, are the result of distinct soil-forming processes. Diagnostic horizons can be divided into surface horizons and subsurface or subsoil horizons. The most common diagnostic horizons are shown in Table 6.3.

Specific limits have been placed on the properties of these horizons to help the observer define and recognize them. Some of the properties, such as color and structure, can be determined by observation, while others, such as organic matter content, require laboratory analysis.

The Soil Orders

Each of the 12 major soil orders ends in *sol,* which is derived from the Latin word *solum,* meaning soil. Seven of the 12 soil orders occur in broad zones that depend largely on two soil-forming factors: climate and organisms (mainly vegetation) (Figure 6.16). For example, **Aridisols** form in hot, dry climates under desert vegetation like cactus, mesquite, and sagebrush, whereas **Oxisols** form in warm, wet climates under tropical rain forests. Two more

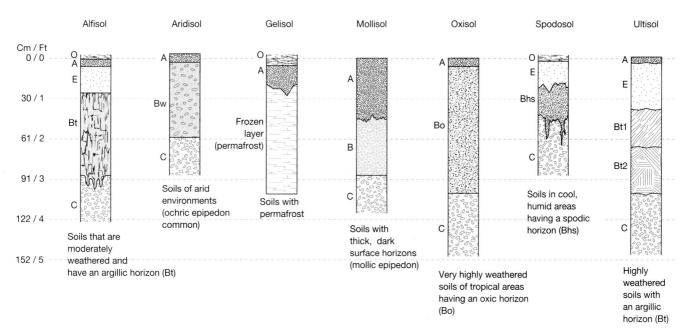

FIGURE 6.16 Typical soil profiles of the seven soil orders that depend largely on two soil-forming factors: climate and organisms (mainly vegetation).

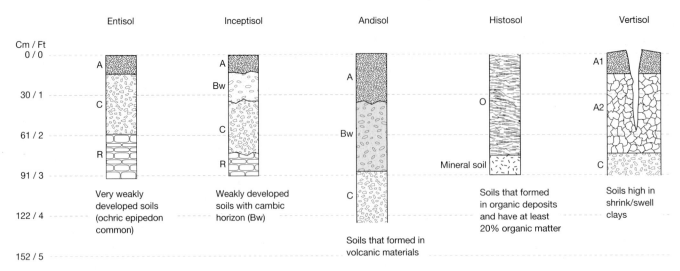

FIGURE 6.17   Typical soil profiles of Entisols, Inceptisols, Andisols, Histosols, and Vertisols.

TABLE 6.4   Percent Land Area and Major Uses of Soil Orders

Order	Percent Land Area of the United States	Percent Land Area of the World	Major Land Use	Fertility
Alfisols	14.5	9.6	Cropland, forest	High
Andisols	1.7	0.7	Cropland, forest	Moderate
Aridisols	8.8	12.1	Rangeland	Low
Entisols	12.2	16.3	Rangeland, forest, cropland	Moderate to low
Gelisols	7.5	8.6	Bogs, tundra	Moderate
Histosols	1.3	1.2	Wetland, cropland	Moderate
Inceptisols	9.1	9.9	Cropland, forest	Moderate to low
Mollisols	22.4	6.9	Cropland, rangeland	High
Oxisols	<0.01	7.5	Cropland, forest	Low
Spodosols	3.3	2.6	Forest	Low
Ultisols	9.6	8.5	Forest, cropland	Low
Vertisols	1.7	2.4	Cropland, forest	High
Miscellaneous lands*	7.8	14.0	Wildlife, recreation	
Total	100%	100%		

*Miscellaneous lands are made up mostly of rocky lands and shifting sands.

orders, **Entisols** (young soils) and **Inceptisols** (weakly developed soils), can be found under any moisture and temperature conditions (Figure 6.17). The remaining three orders, known as **Andisols, Histosols,** and **Vertisols,** are more related to distinctive parent materials (Figure 6.17). For example, Histosols form in organic materials that build up in stagnant waters.

Mollisols are some of the world's most productive soils. As shown in Table 6.4, 7% of the world's land area is covered by them, while 22% of the land mass of the United States is overlain by Mollisols, making the United States one of the world's richest agricultural nations. In comparison to the world average, the United States has less than the average share of Aridisols and Entisols but about the normal share of Inceptisols and Histosols. The United States has almost no Oxisols (tropical soils), compared with 7.5% in the world's land area. Oxisols in the United States can be found in Hawaii.

The distribution of the major soil orders in the United States is shown in Figure 6.18. The extensive soil orders in the eastern part of the United States are Spodosols, Inceptisols, Alfisols, and Ultisols. **Spodosols** are found mainly on the sandy deposits in the northern Great Lakes states, New England, and Florida. They formed in a cool, relatively humid climate under forest vegetation, mostly conifers. They are characterized by a subsoil **spodic horizon** in which iron, aluminum, and humus have accumulated.

Inceptisols are weakly developed soils that occur mainly on sloping mountainsides extending from southern New York through central and western Pennsylvania, West Virginia, and eastern Ohio (Figure 6.18). They also can be found on the floodplains of the larger U.S. rivers, especially the Mississippi. Inceptisols have more significant profile development than Entisols but less development than the other mineral soil orders.

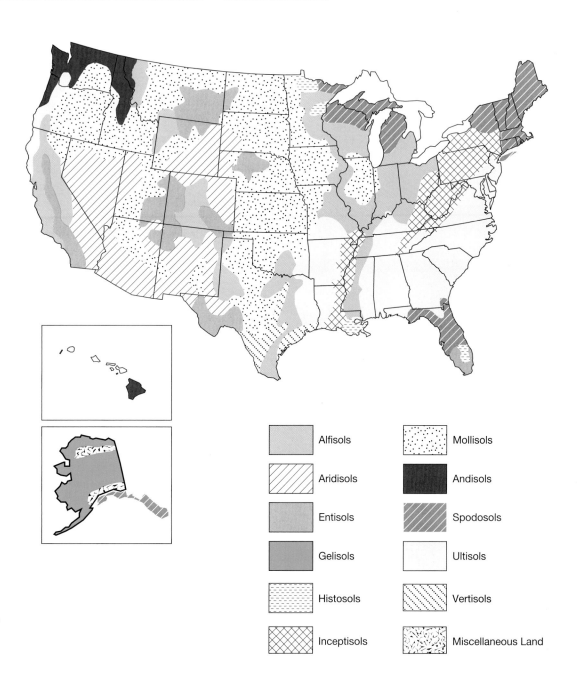

FIGURE 6.18 Distribution of major soil orders in the United States.

Alfisols and **Ultisols** formed in humid climates and are characterized by a subsoil **argillic horizon** in which silicate clays have accumulated from the layer(s) above. Alfisols formed mostly under deciduous forests, although some formed under grasslands. A large area of Alfisols can be found in the Midwest in Ohio, Indiana, Wisconsin, Minnesota, and Michigan They also occur in a narrow belt east of the Mississippi River, where they formed in loess. Alfisols seem to be more strongly weathered than Inceptisols, but less weathered than Spodosols. Ultisols typically formed in forested subtropical and tropical areas. They are more leached, weathered, and acidic and thus less fertile than Alfisols. Most of the soils in the southeastern part of the United States fit in the Ultisol order.

The extensive soil orders of the western United States are Mollisols, Aridisols, Entisols, Alfisols, and Gelisols. Most of the Mollisols developed under grass vegetation, mainly in the Great Plains states in the central part of the United States and in the Northwest in Oregon, Washington, and Idaho. They also extend eastward through Iowa, Illinois, and eastern Indiana. Mollisols are characterized primarily by the presence of a **mollic epipedon,** a thick, humus- and nutrient-rich surface horizon (Figure 6.19). The mollic epipedon is intrinsically more fertile than any other surface soil in the United States. This is partly because of the dense mesh of roots extending down through the A horizon. When a grass plant dies, its root system decomposes in place and releases nutrients that are available to future plant generations.

Aridisols are dry soils found in desert areas where the natural vegetation is desert shrubs and short grasses. They commonly have an **ochric epipedon** that is generally light in color and low in organic matter. Large areas of Aridisols can be found in California, Nevada, Arizona, and New Mexico. Aridisols can be made highly productive for growing cultivated crops when irrigation water and fertilizers are made available. For example, the Aridisols of California's Imperial Valley produce a great variety of high-value crops ranging from citrus to celery, from walnuts to dates. However, they generally require large amounts of water and must be carefully managed to prevent the buildup of soluble salts.

Entisols are young soils with very little soil profile development. They can be found under various environmental conditions. They formed where the soil is very steep, as in the Rocky Mountains, very sandy, as in the Sand Hills of Nebraska, or very young, as in the California trough. In the western United States, Alfisols occur mostly in forested areas in California.

Gelisols, like Entisols, are young soils with little profile development. However, they differ from Entisols in their principal defining feature: the presence of a **permafrost** layer, a layer of material that remains at temperatures below 0°C (32°F) for more than two consecutive years. Gelisols cover most of Alaska, mostly supporting tundra vegetation, such as lichens, grasses, and low shrubs.

Histosols, Vertisols, and Andisols are less extensive in area because they formed in unique parent materials. Histosols are organic soils with at least 20% organic matter. They form in organic deposits that accumulate in shallow lakes, marshes, bogs, and swamps. They can be found in the northern Great Lakes states, Florida, and the Mississippi Delta. Vertisols are found where clayey materials are in abundance and are characterized by a high content (more than 30% percent) of shrink/swell clays. They are most extensive in Texas. Andisols formed on volcanic deposits in the Pacific Northwest. Recent eruptions, like that of Mount St. Helens in 1980, give rise to the formation of new Andisols.

Summary of Key Concepts

1. Soil is one of the most important components of terrestrial communities. High-quality soils not only promote the growth of plants, but also prevent water and air pollution by resisting erosion and by degrading and immobilizing agricultural chemicals, organic wastes, and other potential pollutants.

2. Soil is a natural system consisting of four components: mineral matter, organic matter, water, and air.

3. Soil texture is determined by the relative proportions of sand, silt, and clay. Soil texture helps to determine soil water permeability and storage, the ease of tilling the soil, the amount of aeration, soil fertility, and root penetration.

4. Clay particles have negatively charged surfaces, attracting positively charged ions (cations) of nutrient elements such as calcium, potassium, magnesium, zinc, and iron.

FIGURE 6.19   Profile of Mollisol soil formed from glacial till.

5. Soil structure is the arrangement or grouping of soil particles into clusters or aggregates.

6. Soil organic matter is the living or dead plant and animal materials in the soil.

7. Humus is the semistable, dark-colored material formed from the decomposition of organic residues of plants and animals as well as materials synthesized by microorganisms. Humus improves soil structure, buffers the soil against a rapid change in pH, reduces soil erodibility, and increases the nutrient storage ability and water-holding capacity of the soil.

8. Soil microorganisms (such as fungi and bacteria) are instrumental in decomposing organic matter and releasing available elemental plant nutrients.

9. The pore space of a soil is that portion occupied by both air and water.

10. Macropores (larger than 0.08 millimeter in diameter) are readily drained of water and filled with air after a heavy rain, whereas micropores (less than or equal to 0.08 millimeter) retain water within the fine pores and impede air movement.

11. Soil scientists use the pH scale to determine whether a soil is acidic, neutral, or alkaline. In general, in humid regions, soils vary in pH from 4.5 to a little above 7; in arid regions, soils vary from a little below 7 to 9.

12. Soil fertility refers to a soil's ability to supply nutrients in amounts and forms required for maximum plant growth.

13. An *essential element* is a chemical element required for the normal growth of plants. Most plants are known to need at least 16 essential nutrient elements to grow.

14. Soil formation depends on five soil-forming factors: (1) climate, (2) parent material, (3) organisms, (4) topography, and (5) time. They are responsible for the kind, rate, and extent of soil development.

15. When precipitation and the other four factors remain constant, an increase in temperature causes an increased rate of weathering and clay formation. The rate of weathering increases as precipitation increases (up to a point).

16. Parent material is the unconsolidated (soft and loose), weathered mineral or organic material from which soils form. Parent materials can be classified into three major groups: residual, transported, and organic.

17. Residual parent materials form from the weathering of rock in place.

18. Transported parent materials are transported from their place of origin and redeposited in a new location. The main transporting agents are ice, water, and wind.

19. Most organic deposits accumulate in stagnant water (lakes or swamps), where decomposition of plant residues is retarded by the limited supply of oxygen.

20. The development of a mature soil depends on the activity of a great number and diversity of organisms.

21. Topography, or the shape or contour of the land surface, largely determines how water moves in a landscape and how susceptible the soils are to water erosion.

22. The length of time required for a soil to form depends on the combined effects of climate and organisms, as modified by topography, acting on parent material. The major horizons from the ground surface downward to bedrock are designated as horizons O (organic layer), A (topsoil), E (subsurface), B (subsoil), C (parent material), and R (bedrock).

23. A soil profile is a vertical section of the soil extending through all its horizons and into the parent material.

24. The soil classification system used in the United States (and some other countries) is called soil taxonomy and was developed by the U.S. Department of Agriculture.

25. Soil taxonomy recognizes 12 major soil groups in the world, known as orders. Most soil orders are defined on the basis of diagnostic horizons that result from distinctive soil-forming processes and have specific physical and chemical properties.

26. Spodosols form in a cool, relatively humid climate under forest vegetation, mostly conifers. They are characterized by a subsoil spodic horizon in which humus along with aluminum and iron oxides have accumulated.

27. Inceptisols are weakly developed soils that have more significant profile development than Entisols, but less development than the other mineral soil orders.

28. Alfisols and Ultisols form in humid climates and are characterized by a subsoil argillic horizon in which silicate clays have accumulated from the layer(s) above. Ultisols are more leached, weathered, and acidic and thus less fertile than Alfisols.

29. Mollisols include some of the world's most productive soils. They are characterized primarily by the presence of a mollic epipedon, a thick, humus- and nutrient-rich surface horizon.

30. Aridisols are dry soils found in desert areas where the natural vegetation consists of desert shrubs and short grasses, whereas Oxisols form in warm, wet climates under tropical rain forests.

31. Gelisols are young soils with little profile development because they form in cold temperatures and frozen conditions for much of the year.

32. Histosols, Vertisols, and Andisols form in unique parent materials. Histosols are organic soils with at least 20% organic matter. Vertisols are soils characterized by a high content (more than 30%) of shrink/swell clays. Andisols form in volcanic deposits.

Key Words and Phrases

A Horizon	Diagnostic Horizons
Acidic	E Horizon
Adsorption	Entisols
Aggregates	Essential Element
Alfisols	Field Capacity
Alkaline	Fine-Earth Fraction
Andisols	Gelisols
Argillic Horizon	Histosols
Aridisols	Horizons
B Horizon	Humus
Biomass	Inceptisols
C Horizon	Lichen
Clay	Loam
Climate	Macronutrients
Coarse Fraction	Macroorganisms

Macropores
Master Horizons
Micronutrients
Microorganisms
Micropores
Mineral Matter
Mollic Epipedon
Mollisols
O Horizon
Ochric Epipedon
Organic Deposits
Organic Matter
Organisms
Oxisols
Parent Material
Permafrost
Permanent Wilting Point
Permeability
pH Scale
Pore Space
R Horizon
Reaction
Residual Parent Materials
Sand

Saturation
Silt
Soil Classification
Soil Fertility
Soil-Forming Factors
Soil Orders
Soil Profile
Soil Structure
Soil Taxonomy
Soil Texture
Spodic Horizon
Spodosols
Subsoil
Surface Area
Textural Classes
Time
Topography
Transported Parent Materials
Ultisols
Vertisols
Weathering
Zone of Accumulation
Zone of Leaching

Critical Thinking and Discussion Questions

1. Is there any correlation between soil fertility and the power and influence of nations like the United States? Discuss your answer.

2. Discuss five major characteristics or properties of soils.

3. A farmer has two level fields, each 2 hectares in size. One field is a clay loam soil, and the other field is a sandy loam. If the farmer grows the same crop on both pieces of land, should she fertilize and irrigate the same amount on both fields, or should she manage them differently? Discuss your answer.

4. What characteristics of clay are undesirable for crop production? What characteristics are desirable?

5. Is there a difference between humus and organic matter?

6. Describe three beneficial effects of microorganisms on soil.

7. Why is water permeability faster in a sandy-textured soil than in a clayey-textured soil?

8. Give three reasons why a plant cannot survive without water.

9. What does pH mean? What pH is desirable for most trees and food crops?

10. Suppose that a farmer has soil with a pH of 5.0 but wishes to raise a crop that requires a pH of 6.0. What can he do?

11. What is an essential element? Is a macronutrient any more essential for plant growth than a micronutrient?

12. Describe the five soil-forming factors involved in the development of soils. What is the importance of each?

13. How could a climatic change eventually cause a change in the type of soil occurring in a given region?

14. List three forces that play an important role in the transport of parent materials for soil development.

15. Discuss the following statement: "A soil profile represents a kind of autobiography." Is this statement valid? Why or why not?

16. What is the difference between an E horizon and a B horizon?

17. What is the basis for naming a soil order? Give an example.

18. Compare the general features of an Entisol and an Oxisol. What are the major differences?

19. In your own backyard or at a nearby road cut, write a short summary of the soil structures and colors you see, and what plants are growing in these soils.

Suggested Readings

Brady, N. C., and R. R. Weil. 2008. *The Nature and Property of Soils,* 14th ed. Upper Saddle River, NJ: Prentice Hall. A classic work on soils, which provides a comprehensive and in-depth treatment of the subject.

Buol, Stan. 2008. *Soils, Land, and Life.* Upper Saddle River, NJ: Prentice Hall. For use by non-agricultural majors, an examination of how and why various types of soil and land are used for human food production.

McLaren, R. G., and K. C. Cameron. 1996. *Soil Science: Sustainable Production and Environmental Protection,* 2nd ed. Auckland, New Zealand: Oxford University Press. Excellent introductory soils text with emphasis on sustainability and environmental issues in New Zealand.

Miller, R. W., and D. T. Gardiner. 2008. *Soils in Our Environment,* 11th ed. Englewood Cliffs, NJ: Prentice Hall. Very good, comprehensive introductory soils text.

Nardi, J. B. 2007. *Life in the Soil: A Guide for Naturalists and Gardeners.* Chicago: University of Chicago Press. A worthy and meaningful introduction to the soil biota and their unique ecosystem, with fine graphics showing the diversity of the major organisms that inhabit soils.

Plaster, E. J. 2009. *Soil Science and Management,* 5th ed. Albany, NY: Delmar. Easily readable and practical soils text with emphasis on management.

Singer, M. J., and D. N. Munns. 2006. *Soils: An Introduction,* 6th ed. Upper Saddle River, NJ: Prentice Hall. Excellent, simplified beginning soils text.

Soil Survey Staff. 2006. *Keys to Soil Taxonomy.* Washington, DC: USDA Natural Resource Conservation Service. Description of the latest U.S. soil classification system, available for free downloading at **http://soils.usda.gov/technical/classification/tax_keys/**.

Web Explorations

Online resources for this chapter are on the World Wide Web at: **http://www.prenhall.com/chiras** *(click on the Table of Contents link and then select Chapter 6).*

CHAPTER 7

SOIL CONSERVATION AND SUSTAINABLE AGRICULTURE

Soil conservation, the subject of this chapter, is vital to the long-term future of humankind. Soil conservation helps us preserve a resource crucial to this and all future generations. Why is soil so important? Simply put, it's because we and future generations depend on soil for a variety of essential products, such as food, feed for animals we eat, fuel, and fiber (for, example, clothing), which are derived from plants that grow in it. As pointed out in previous chapters, the demand for these products is growing rapidly because of population growth and the higher standard of living now enjoyed by many of the world's residents.

The easiest way to protect soil from erosion is to leave it undisturbed. In other words, do not expose it directly to rain or wind. But this is often not practical. Growing crops typically involves significant exposure of soils to the elements. Over the years, researchers and farmers have devised many methods to protect cropland soils from erosion.

Soil erosion control is only one indicator of a farm or agricultural system being sustainable. As you soon will see, sustainable agricultural systems are designed to produce adequate yields of high-quality food in a way that is economically viable, environmentally safe, resource conserving, and socially responsible.

7.1 The Nature of Soil Erosion

Over the past three centuries, U.S. farmland soils have undergone considerable deterioration. Much of the deterioration has been caused by erosion. The word *erosion* derives from the Latin word *erodere,* meaning "to gnaw out." **Soil erosion** is, therefore, defined as the process by which soil particles are detached from their original site, transported, and eventually deposited at new locations (Figure 7.1). The principal agents of erosion are wind and water. In humid areas, water is the main cause of soil erosion, whereas in semiarid areas, wind and water are responsible for most soil erosion.

Geological, or Natural, Erosion

Geological, or natural, **erosion** is a process that has occurred at an extremely slow rate ever since the Earth formed 4.5 billion years ago. It wears down high places and fills up low places. It has worn down mighty mountain systems and filled sea basins with sediment. Scientists estimate that geological erosion rates, although they vary depending on the terrain, roughly average about 0.02 millimeter (0.0008 inch) per year.

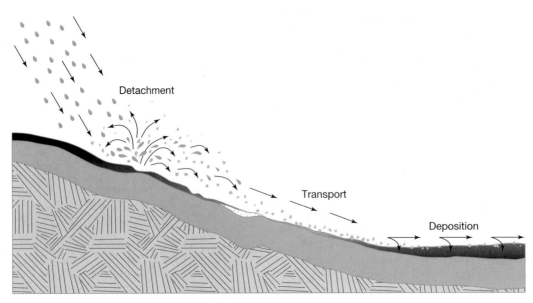

FIGURE 7.1 The three-step process of soil erosion by water. Raindrop impact destroys soil structure, detaching soil particles. The detached soil particles are then transported and eventually deposited downhill.

The mountains, valleys, canyons, coastlines, and deltas on the Earth's surface have been sculpted by water and wind erosion working over the eons. The Appalachian Mountains were once as tall and rugged as the Rocky Mountains, but since their formation 200 million years ago, they have been gradually worn down by erosive forces. Were it not for geological erosion, New Orleans would be resting on the bottom of the Gulf of Mexico, for the delta on which it was built was formed by a deposit of soil transported by the Mississippi River from sites as far as 1,600 kilometers (1,000 miles) away. The Grand Canyon originated as a shallow channel 100 million years ago. It was ultimately scoured to its awesome 1.6 kilometer (1-mile) depth by rain and the churning waters of the Colorado River (Figure 7.2). As a final example of geological erosion, wind-blown silt and fine sand, known as **loess,** have been deposited in thick sheets on North America, China, and Europe, forming some of the Earth's most fertile soils.

FIGURE 7.2 Geological erosion caused by water: Grand Canyon of the Colorado River.

FIGURE 7.3 Huge water-carved gullies on a North Carolina farm, resulting from accelerated erosion caused by humans.

Accelerated Erosion

Geological erosion, then, has continued to operate at a slow, deliberate pace for millions of years. However, with the appearance of humans, a different type of erosion, known as **accelerated erosion,** began. Studies show that this form of erosion is often 10 to 100 times faster than geologic erosion. Accelerated erosion occurs when people disturb the soil or natural vegetation by plowing hillsides, cutting forests, grazing livestock, or constructing buildings or roads. An example of gully erosion, which is one type of accelerated erosion, is shown in Figure 7.3.

Worldwide, most soil removal by wind and water erosion occurs on agricultural land. It is this accelerated erosion with which the conservationist is primarily concerned, and for good reason. The removal of soil degrades arable land and can eventually render it unproductive.

Soil erosion rates are highest in Asia, Africa, and South America, and the lowest rates are in Europe and North America. However, the relatively lower rates in Europe and North America exceed the average rate of soil formation or replacement. Some of the most destructive erosion the United States has ever experienced was caused by heavy winds that occurred on the Great Plains during the 1930s. The part of the Great Plains most affected by this cataclysm was in the Southern Plains, encompassing large areas of Texas, Oklahoma, Kansas, Colorado, and New Mexico. Because of the frequency and severity of the dust storms in the Southern Plains, this region was called the **Dust Bowl.**

7.2 The Dust Bowl

Throughout history, the Great Plains have experienced alternating periods of drought and adequate rainfall. Drought visited the Great Plains in 1890 and again in 1910. During each dry spell, crops withered and died. Farms and ranches were abandoned, only to be reoccupied during the ensuing years of adequate rainfall.

In 1931, there was no better place to be a farmer than in the Great Plains. Farmers grew wheat and turned the plains of rich prairie grass into one of the most productive areas in the world. Then drought and high summer temperatures began to settle in the Northern Plains. The drought then spread to the Southern Plains.

Droughts and windstorms had visited the plains before. But never before in the history of the North American Prairie had the land been more vulnerable to their combined assault. Gone were the profusely branching root systems of the native grasses—buffalo grass, grama grass, big bluestem, and little bluestem—which had originally kept the rich brown soil in place. Gone were the high concentrations of decomposing organic material that aided in building stable soil aggregates. On the ranches, soil structure deteriorated under the pounding of millions of cattle. On the wheat and cotton farms, soil structure broke down under the wheels of heavy machinery.

Many agricultural scientists believe that the Dust Bowl was the result of an inevitable clash of a drought with the agricultural practices used in the region—practices that had been successful in the wetter East. The drought prevented the wheat from growing, and because of extensive tillage, no ground cover blanketed the soil. The stage was set for the "**black blizzards,**" as the blinding, dusty winds of the Dust Bowl came to be called.

In 1932, 14 dust storms occurred in the Great Plains. In 1933, the number of dust storms increased to 38. Farmers, confident that rain would return, continued to plow. But the storms in 1934 and 1935 kept coming with gale-force velocities and alarming frequency (Figure 7.4). In western Kansas and Oklahoma, as well as in the neighboring parts of Texas, Colorado, and New Mexico, the wind whirled soil particles upward into the prairie sky as high as 3.3 kilometers (2 miles) (Figure 7.5). A storm that hit on May 11, 1934, lifted about 275 million metric tons (300 million tons) of fertile soil into the air. (This roughly equals the total soil tonnage scooped from Central America to form the Panama Canal.) In many areas, the wilted wheat was uprooted by the strong winds. In the Amarillo, TX, area during March and April 1935, 15 windstorms raged for 24 hours; 4 lasted for over 55 hours.

Dust from Oklahoma prairies came to rest on the deck of a steamer 330 kilometers (200 miles) out in the Atlantic. Dust sifted into the plush offices of Wall Street and smudged the luxury apartments of Park Avenue. When it rained in the blow area, the drops would sometimes come down as dilute mud. In Washington, DC, mud-splattered buildings of the U.S. Department of Agriculture (USDA) were a rude reminder of the problem facing it and the nation. About 1,600 kilometers (1,000 miles) westward, people stuffed water-soaked newspapers into window cracks, to no avail (Figure 7.6). The dust sifted into kitchens, forming a thin film on pots, pans, and fresh-baked bread. Blinded by swirling dust clouds, ranchers got lost in their own backyards. Motorists pulled off to the side of the highways.

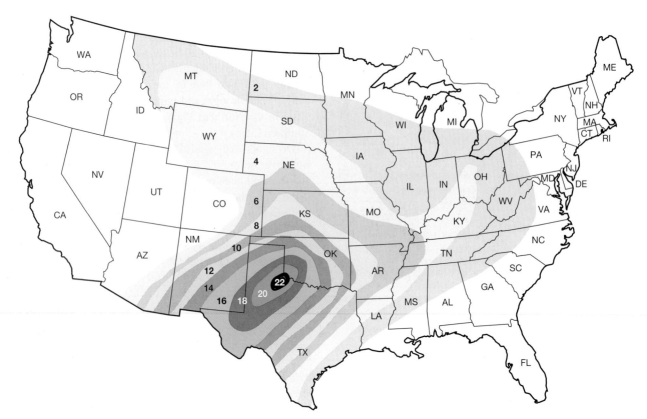

FIGURE 7.4   Number of dust storms in March 1936 in the Great Plains. The greatest number, 22, was recorded in northern Texas. Because the frequency and severity of the dust storms occurred mostly in the Southern Plains states of Texas, Oklahoma, Kansas, Colorado, and New Mexico, this region was called the Dust Bowl.

FIGURE 7.5   Dust storm approaching Springfield, CO, on May 21, 1937. This storm reached the city limits at 4:47 P.M. Total darkness lasted about one-half hour.

Hundreds of airplanes were grounded. Trains were stalled by huge drifts. Hospital nurses placed wet cloths on patients' faces to ease their breathing.

When the rains finally returned and the winds finally subsided in the fall of 1939, ranchers and farmers wearily emerged to survey the desolation. Five to 30 centimeters (2 to 12 inches) of fertile topsoil, mostly composed of silt and clay particles, had been carried to the Atlantic seaboard. Sand particles, too coarse and heavy to be airborne, bounced across the land, sheared off young wheat, and finally accumulated as dunes to the leeward side of homes and barns. Machinery became shrouded in sand.

The dust storms of the 1930s (1931–1939) inflicted both social and economic suffering. Yet a few ranchers and farmers were philosophical about their misfortunes and even cracked jokes about birds flying backward "to keep the sand out of their eyes" and about prairie dogs "digging burrows 100 feet in the air." However, for most Dust Bowl victims, the dusters were not funny. Many victims were virtually penniless. The 275 million metric tons of topsoil removed in a single storm on May 22, 1934, represented the equivalent of taking 3,000 farms of 40 hectares (100 acres) each out of crop production. Up to 1940, Dust Bowl relief alone cost American taxpayers over $1 billion; $7 million (more than was paid for Alaska) was pumped into a single Colorado county. The only recourse for many of these ill-fated farmers was to find a new way of life. They piled their belongings into rickety cars and trucks and moved out—some to the Pacific coast, some to the big industrial cities of the Midwest and East. However, our nation was still in the throes of a depression, and many an emigrating family found nothing but frustration, bitterness, and suffering at the end of the road.

FIGURE 7.6 Abandoned Oklahoma farmstead, showing the disastrous results of wind erosion.

7.3 The Shelterbelt Program

In an attempt to prevent future dust bowls, the federal government launched a massive **shelterbelt** system in 1935. More than 218 million trees were planted on 30,000 farms across the Great Plains from North Dakota to Texas. The green checkerboard patterns formed by the 32,000 kilometers (20,000 miles) of **windbreaks** added color and variety to the prairie landscape. On the Central Plains, a typical shelterbelt or windbreak consisted of one to five rows of trees planted on the western margin of a farm in a north–south line, to intercept winter's prevailing westerly winds (Figure 7.7). Conifers, such as red cedar, spruce, and pine, provide the best year-round protection. By planting a few rows of grain between the rows of trees, farmers further reduced wind erosion. A properly designed shelterbelt of adequate height and thickness can reduce a wind velocity of 50 kilometers (30 miles) per hour to only 13 kilometers (8 miles) per hour leeward (downwind) of the trees.

Although windbreaks occupy valuable land that could be used for crop production, are relatively slow to grow, and must be fenced from livestock until the stands are well established, the accrued benefits far outweigh these disadvantages. In addition to controlling wind erosion, properly designed windbreaks provide aesthetic benefits, increase soil moisture by reducing evaporation and trapping snow,

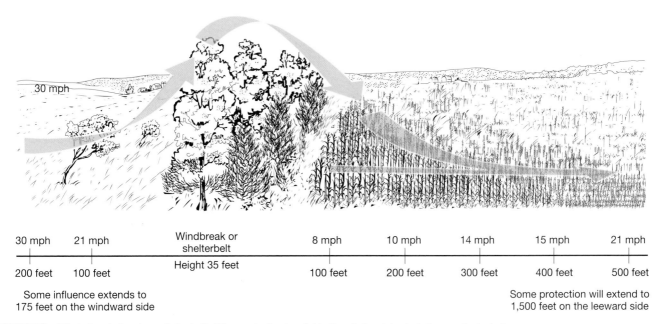

30 mph	21 mph	Windbreak or shelterbelt		8 mph	10 mph	14 mph	15 mph	21 mph
200 feet	100 feet	Height 35 feet		100 feet	200 feet	300 feet	400 feet	500 feet

Some influence extends to 175 feet on the windward side

Some protection will extend to 1,500 feet on the leeward side

FIGURE 7.7 Effect of a windbreak on wind velocity. When protecting farmfields, the windbreak is planted perpendicular to the prevailing winds.

FIGURE 7.8   A North Dakota farm that is well protected from wind and snow by a 17-year-old windbreak of conifers, fruit trees, and shrubs.

GO GREEN!

Plant a windbreak of trees and/or shrubs if you live in an area with lots of wind, whether in the city or out in the country. For assistance, start with your local nursery or USDA Natural Resource Conservation Service office.

and provide a habitat for wildlife. Moreover, the fuel requirement for heating and cooling nearby homes is significantly reduced (Figure 7.8). Unfortunately, many of the shelterbelts planted in the 1930s have been removed—in some cases so that the farmer could use the wood as fuel, and in other cases to make room for crops and to facilitate the use of heavy farm machinery and sprinkler irrigation systems.

7.4   Soil Erosion Today

The Dust Bowl period was devastating to the United States' soil resources, economy, and the emotional well-being of millions of Americans. In the years that have passed since that critical period in American agriculture, the federal government has spent tens of billions of dollars to control erosion. Scientists at many major universities have conducted erosion-control research with taxpayer support. Hundreds of scientific publications have been written on the subject. The USDA has established more than 3,000 conservation districts whose prime function is to assist the farmer with erosion problems. So, after devoting all of this time, energy, and money to soil erosion control, we, as taxpayers, are justified in asking, "Is the American farmer doing any better in controlling erosion today than during the Dust Bowl years?" Generally speaking, the answer is yes.

Since the 1930s, soil conservation practices have significantly reduced the rate of wind and water erosion on U.S. croplands. These practices, which include terraces, residue management, conservaton tillage, cover crops, crop rotations, contour farming, strip farming, windbreaks, and putting highly erodible farmlands into grass or trees, have made significant inroads into soil erosion. However, although many farmers are doing a good job in controlling erosion, there's more work to be done.

In the early 1930s, soil erosion in the United States was estimated at 3.6 billion metric tons (4.0 billion tons) of soil removed each year. According to the USDA's National Resources Inventory data, in 1982 the figure for soil erosion on U.S. cropland was 2.8 billion metric tons per year (3.1 billion tons per year), and by 2003, the most recent year for which data is available, the figure dropped to 1.6 billion metric tons per year (1.7 billion tons per year) (Figure 7.9). That's a decrease in water and wind erosion by 43 percent in 21 years. These estimates are more significant when you take into account that more land was being cultivated in 1982 or 2003 than in the 1930s. About 56% of annual soil erosion is due to water and 44% due to wind.

The good news is that not only the total amount of soil erosion, but also annual rates of soil erosion (measured in metric tons per hectare), have declined in the United States. The bad news is that about 28% of U.S. cropland is losing soil through water and wind erosion above tolerance rates ranging from 2 to 11 metric tons per hectare per year, as determined by the USDA. In some regions, soil loss is or has been extremely high. In some farm fields of Washington, Oregon, and Idaho, 120 to 250 metric tons per hectare per year have been lost because of unsustainable farming practices on steep slopes and highly erodible soil types (Figure 7.10). Such rates are detrimental to the long-term future of agriculture in these areas.

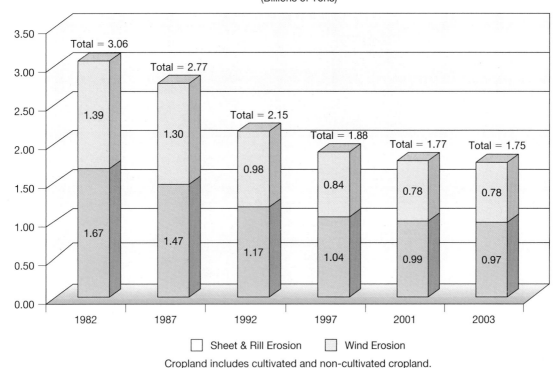

Erosion on Cropland by Year
(Billions of Tons)

Total = 3.06
1.39
1.67

Total = 2.77
1.30
1.47

Total = 2.15
0.98
1.17

Total = 1.88
0.84
1.04

Total = 1.77
0.78
0.99

Total = 1.75
0.78
0.97

1982   1987   1992   1997   2001   2003

☐ Sheet & Rill Erosion   ☐ Wind Erosion

Cropland includes cultivated and non-cultivated cropland.

FIGURE 7.9   Annual rate of soil erosion from water (sheet and rill) and wind on cropland in the U.S. between 1982 and 2003. In this 21-year period, although soil erosion on cropland decreased by 43%, it is still unsustainable in many places in the U.S. (Source: USDA Natural Resource Conservation Service, 2007)

FIGURE 7.10   Rill erosion (causing shallow, narrow channels) in the Palouse region of southeastern Washington State, one of the most productive dry-farmed wheat areas in the world. Notice the sediment accumulating at the bottom of the slope.

Topsoil depth in the United States ranges from a few centimeters in some locations to more than 1 meter in the richest farmlands. The USDA has determined that soils with a thick layer of topsoil can tolerate erosion losses of 11 metric tons per hectare (5 tons per acre) per year without losing their ability to support crops economically and indefinitely. This is known as the **tolerance value** or **T value.** Although there is no sound basis for this figure, this amount of soil, say USDA scientists, is normally generated by the natural processes of soil formation. However, other researchers estimate that soil formation from the weathering of rock (residual parent materials) takes place at an average global rate of about 1.1 metric tons per hectare (0.5 ton per acre). Soil formation rates are higher on most transported parent materials, such as alluvium and glacial moraines. These researchers argue that the T values of 2 to 11 metric tons per hectare accepted by the USDA exceed the regenerative capacity of agricultural soils and that the so-called tolerable levels of soil erosion are not sustainable, because soil is not being produced as fast as these rates require (even under some of the best conditions). You can see why they are concerned when, as shown in Figure 7.11, the average erosion rate of 10.5 metric tons per hectare (4.7 tons per acre) in the United States in 2003 is more than 9 times the average residual soil formation rate of 1.1 metric tons per hectare (0.5 ton per acre) and is also likely greater than most transported soil formation rates.

Soil erosion removes valuable topsoil and reduces the water held in soil, thereby reducing the productivity of farmers' fields. These are the on-site effects of soil erosion. However, scientists have also identified numerous off-site effects, such as the pollution of air and water with soil particles. Dust in the air can be

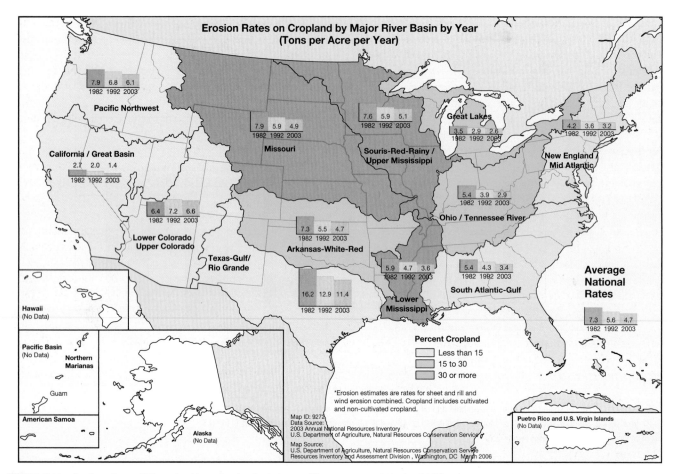

FIGURE 7.11 Erosion rates on U.S. cropland by major river basin by year. The average erosion rate of 4.7 tons per acre (10.5 metric tons per hectare) per year in 2003 is for water (sheet and rill) and wind erosion combined.

unhealthy. Sediment clogs rivers and lakes, reducing their potential for recreation and navigation. Can we put a value on the on-site and off-site damages caused by water and wind erosion?

In one 1995 study in the United States, researchers found these costs to be exceptionally high. Using an average value for wind and water erosion in the United States of 17 metric tons per hectare per year (a 1992 USDA estimate of 14 metric tons per hectare per year could have been used instead), they estimated total annual on-site costs at $27 billion. This estimate is based on the fertility or nutrient value of a 0.9 metric ton (1 ton) of eroded soil being worth about $5, which does not account for other soil components like organic matter and soil organisms. Since the researchers estimated that 4 billion metric tons of soil were lost each year in the United States (a 1992 USDA estimate of 2.2 billion metric tons of soil could have been used instead), they put the replacement value of nutrients in the eroded soil at $20 billion (of the $27 billion). The remaining $7 billion was for lost water (130 billion metric tons of water are lost annually) and lost soil depth. Using similar techniques, these researchers estimated the off-site costs for items such as dredging harbors and waterways, reduced water storage in reservoirs, loss of wildlife habitat, flooding, health costs, and cleaning up domestic water supplies. Off-site costs totaled $17 billion, which brings the grand total to $44 billion per year.

This erosion cost study has been criticized for using erosion rates higher than those generally accepted as average for the United States. However, even if we cut the costs of erosion in half, the estimated annual cost of erosion to our nation is still a staggering $22 billion. Since erosion rates are the lowest in the United States and Europe, one can only imagine what the costs of erosion are in Asia, Africa, and South America, where erosion rates—the highest on the planet—annually average 30 to 40 metric tons per hectare. The United Nations–affiliated International Food Policy Research Institute reported in 2000 that nearly 40% of global farmland is seriously degraded. This human-induced soil degradation has been caused by erosion, loss of organic matter, hardening of soil, chemical penetration, nutrient depletion, and excess salinity. The problem is the worst in less-developed countries, where populations are the highest.

7.5 Factors Affecting the Rate of Soil Erosion by Water

Rainfall

Annual precipitation in the continental United States ranges from almost nothing in some areas of Death Valley, CA, to 360 centimeters (140 inches) in parts of Washington State.

The amount of precipitation a region receives greatly affects erosion rates. However, even more important factors are rainfall intensity and the seasonal distribution of the rainfall. Hard-beating, intensive rains can cause serious erosion, even in areas of low annual rainfall. In contrast, gentle rains of low intensity may cause little erosion, even if the total annual precipitation is high.

A town in Florida once experienced a deluge of 60 centimeters (24 inches) of rain in only 24 hours. The soil loss resulting from **runoff** waters (the flow of water from rain, snowmelt, or other sources over land) must have been severe. Had this 60-centimeter rainfall resulted from daily 1-centimeter (0.4-inch) drizzles occurring over 48 consecutive days, the erosion threat would have been negligible, because the soil would have had sufficient time to absorb the water. Surprisingly, even in the arid deserts of Nevada and Arizona, where annual rainfall averages 13 centimeters (5 inches), excessive erosion occurs because the entire annual precipitation occurs in a few torrential cloudbursts. As a result, the desert floor is dissected by gullies gouged out by runoff waters.

GO GREEN!

Mulch the soil in your yard with bark chips or grass clippings to help protect it from the impact of raindrops. In addition, as organic mulch decomposes, it adds more organic matter to the soil.

Soil Erodibility and Surface Cover

Soil structure, discussed in Chapter 6, greatly influences a soil's erodibility. To resist erosion, soil should be composed of **water-stable aggregates** so that the soil particles are grouped into clusters that cling together even when they are submerged in water. Soil structure can be improved and the erosion hazard decreased by plowing under a crop of clover or alfalfa (**green manure**), by using conservation tillage methods where more crop residue is left on the soil, or simply by adding organic material such as compost or cow manure. A green manure crop is typically a grass or legume that is plowed into the soil or surface-mulched at the end of a growing season to enhance soil productivity and tilth. Increasing the level of organic matter in soils makes them more resistant to erosion. In Iowa, for example, soil loss from nonmanured land was over five times that from heavily manured land.

The addition of organic material also improves the soil's water-absorbing ability, or infiltration capacity. When more water enters the soil, less runoff is available to transport soil particles. This trait, in turn, results in a denser, more vigorous growth of vegetation, which further protects the soil. A developing corn root system, for example, penetrates the soil particles and binds them in place, thereby reducing erosion.

The influence of agriculture on soil erosion depends mostly on the amount of surface cover and the intensity of tillage operations. A dense cover of vegetation with plant residue on the ground, as found under bluegrass pasture or undisturbed natural forests, provides excellent protection from erosion. Clean tillage systems, where crop residues are turned under, leave the soil unprotected for considerable periods and can result in accelerated erosion. Thus, a cultivated field without crop cover or residues has a much greater potential for erosion than a dense pasture or woodland or a field with crop residues left on the surface. Between these two extremes, there are numerous intermediate conditions of soil cover, some of which are shown in Figure 7.12.

Topography

The **slope** of the terrain greatly affects the intensity of surface runoff and soil erosion. Increasing the steepness of a slope greatly increases the velocity of runoff and the rate of erosion. The steepness of a slope is indicated as a percentage. Thus, a 10% slope is one that drops 10 meters over a horizontal distance of 100 meters. In many parts of potato-growing Aroostook County, ME, where slopes may be as steep as 25%, more than 60 centimeters (24 inches) of topsoil has already been removed by erosion since farming began. On farms planted to row crops such as corn and cotton, a doubling of the slope results in a doubling of soil erosion by water. Also, as the length of slope increases, so does the amount of runoff water accumulating downslope. The more water that runs off, the greater the erosion hazard.

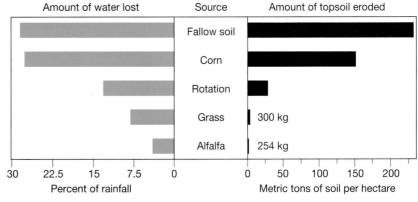

FIGURE 7.12 Influence of plant cover on soil erosion and water runoff. Data is from research conducted by the USDA Soil Conservation Service (now the Natural Resource Conservation Service) at Bethany, MO, on Shelby loam on land with an 8% slope. Average annual rainfall in the area is 102 centimeters (40 inches).

7.6   Controlling Soil Erosion by Water

Erosion-Control Practices

If the factors that contribute to soil erosion are altered, soil erosion can be controlled. Some important erosion control practices include (1) contour farming, (2) strip cropping, (3) terracing, (4) gully reclamation, (5) conservation tillage, and (6) removal of cropland from production.

Contour Farming   **Contour farming** may be defined as plowing, seeding, cultivating, and harvesting at right angles to the direction of the slope, rather than up and down it (Figure 7.13). It was practiced by Thomas Jefferson, who wrote in 1813, "We now plow horizontally, following the curvature of the hills...scarcely an ounce of soil is now carried off." Jefferson, however, was an exception. In the early days of American agriculture, the farmer who could plow the straightest furrows (usually up and down slopes) was considered a master plowman and was praised by his neighbors.

An experiment conducted on a Texas cotton field with a 3% to 5% slope (3- to 5-meter drop per 100 meters) revealed that the average annual water runoff from a noncontoured plot was 12 centimeters (4.6 inches), whereas that for a contoured plot was 65% less, or 4 centimeters (1.6 inches). The lower the water runoff, the lower the erosion rate.

Strip Cropping   **Strip cropping** is the practice of growing crops that require different types of tillage, such as a row crop and a cover crop, in alternate strips. Crops are sown along the contours or across the prevailing direction of the wind to decrease erosion by water and wind (Figure 7.14). When viewed from a distance, a strip-cropped contoured farm looks like a series of slender, curving belts of color. As an example of strip cropping, the row crop could include corn, cotton, tobacco, or potatoes; the cover crop could include hay or various legumes such as soybeans. Strip cropping is frequently combined with crop rotation, so that a strip planted to a soil-depleting corn crop one year will be sown to a soil-enriching legume crop the next. For a study of erosion, see Case Study 7.1.

Terracing   **Terracing** has been practiced by humans for centuries. It was used by the Incas of Peru and by the ancient Chinese. Plagued with relatively dense populations and a scarcity of arable land, those civilizations were forced to till extremely steep slopes, even mountainsides, to prevent widespread hunger. The flat, step-like bench terraces that these ancient agriculturists constructed, however, are not amenable to today's farming methods. A variety of terraces are in use around the world today (Figure 7.15). Broad-based terraces permit the entire surface to be planted. The steep back-slope terrace allows cropping on most of the surface area, except for the backslope, which is planted to permanent grass. Flat-channel terraces permit large volumes of water to move off the soil without causing erosion. Bench terraces can keep water on the cropped area (no runoff) and are mostly used for rice production.

A channel terrace is formed by digging a channel across the slope. It is frequently used in the Tennessee and Ohio River valleys, as well as the Southeast and Mid-Atlantic states, where rainfall is high but the water-absorbing capacity of the soil is poor.

Terraces must be carefully maintained from year to year, and not destroyed during tilling. To be effective, terraces must check water flow before it attains a velocity of 1 meter per second (3 feet per second), which is sufficient to loosen and transport soil.

Gully Reclamation   **Gullies** are erosion channels too large to be erased by ordinary farm operations. They are active when their vertical walls are free of vegetation, and inactive when they are stabilized by vegetation. They are especially common in the southeastern United States, due to a long history of soil abuse and intense rains. Gullies are danger signals indicating

FIGURE 7.13   Aerial view of contour farmed terraces with grassed waterways for erosion control in Kansas.

FIGURE 7.14   Strip cropping in northwestern Illinois. Alternating contour strips of corn and alfalfa reduces erosion and pesticide demands (because of increased diversity) and helps to enhance soil fertility.

that land is eroding rapidly and that the area may become a wasteland unless erosion is promptly controlled. Some gullies work their way up a slope at the rate of 5 meters (15 feet) a year (Figure 7.16A). In North Carolina, a gully 45 meters (150 feet) deep was gouged out in only 60 years, swallowing up fence posts, farm implements, and buildings in the process.

If relatively small, a gully can be plowed and then seeded to a quickly growing "nurse" crop of barley, oats, or wheat. This way, erosion will be checked until sod can be established.

In cases of severe gullying, small check dams of manure and straw constructed at 6-meter (20-foot) intervals may be effective. Silt collects behind the dams and gradually fills the

channel, allowing plants to take root. Dams may be constructed of brush or stakes held securely with a woven wire netting. Earth, stone, and even concrete dams may be built at intervals along the gully. Once dams have been constructed and water runoff has been restrained, soil may be stabilized by planting rapidly growing shrubs, vines, and trees. Willows are effective. Not only does such pioneer vegetation discourage future erosion, it also obliterates the ugly scars and provides food, cover, and breeding sites for wildlife (Figure 7.16B).

Conservation Tillage   Before we discuss the agricultural development known as conservation tillage, we must describe the traditional procedures involved in cultivating the soil. Refer to Figure 7.17 as you read the following steps:

1. One pass is made over the field with a moldboard plow. During this process, the crop residues remaining after the previous harvest are plowed under, and the upper 15 to 25 centimeters (6 to 10 inches) of the soil is inverted and broken up. In this process, up to 90% of surface residues are buried.

2. Three or four passes over the field are then made to break up any clods and prepare a seedbed for the next crop; implements used for this task are disks, field cultivators, and harrows.

3. After row crops such as corn or cotton have sprouted, a row cultivator is used to remove weeds that compete with the crop for moisture and nutrients. Herbicides may also be applied at strategic times after planting.

Traditional cultivation results in considerable disturbance of the soil, disrupting the soil structure and exposing soil to wind

CASE STUDY 7.1   A 100-YEAR STUDY OF THE EFFECTS OF CROPPING ON SOIL EROSION

Just imagine a scientific study that ran continuously for 100 years! It seems hard to believe. But that's exactly what has been done by soil scientists at the University of Missouri. Three generations of workers have collected research data on the rates of soil erosion from plots at Sanborn Field, the oldest agricultural experiment station west of the Mississippi River. The initial data were obtained in 1888, when Grover Cleveland was in the White House, the world's first skyscraper was being built in Chicago, and the nation was agog with the invention of the automobile. It was also a year when many thousands of primitive mule-drawn plows were slicing into the virgin prairie so that it could be planted to grain.

Of course, the University of Missouri research team, headed by Clark Gantzer, knew that the intrinsic nature of the soil and the degree of slope were factors in determining the degree of erosion. But they were primarily interested in finding out the role played by a row crop such as corn, which leaves much land exposed, and by a dense cover crop such as timothy grass. The soil on the study plots was a sandy loam, characteristic of more than 4 million hectares (10 million acres) of cropland in Missouri, Kansas, and Illinois.

The slope on the gently rolling plots varied from 0.5% to 3.0%. Gantzer and his coworkers compared the amount of erosion that had occurred on three different plots over the 100-year period. Plot A had been planted continuously to corn. Plot B was planted to a six-year rotation of corn, oats, wheat, clover, and two years of timothy grass. Plot C was planted to timothy grass. The amount of vegetative cover was least in plot A, greater in B, and greatest in C.

The results of this important research were reported in the *Journal of Soil and Water Conservation.* The most severe erosion, a rate of about 46 metric tons per hectare per year (20 tons per acre per year), occurred on the corn plot. Indeed, this plot had 56% less topsoil than the plot planted to timothy grass. This degree of erosion cut corn yields by 60%. The rotation plot lost 21.0 metric tons of topsoil per hectare per year (9.4 tons per acre per year) and retained 30% less topsoil than the timothy grass plot. The scientists found that during the 100-year study period, timothy grass was 54 times more effective in controlling soil erosion than was corn. They further concluded that cover management was about 35 times more relevant than soil erodibility and 28 times more relevant than slope in explaining the differences in erosion between the plots.

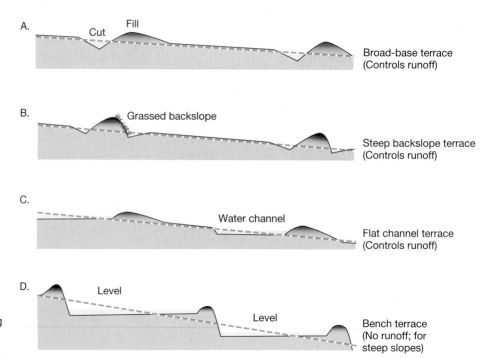

FIGURE 7.15 Four common types of terraces in use around the world. Terraces are effective at controlling water runoff and erosion. To save moisture, terraces are designed to cause ponding of water on the terrace, allowing water time to infiltrate the soil.

and rain. Heavy machinery tends to compact the soil. The process requires a great deal of energy—usually diesel fuel—which produces air pollution.

To address these and other concerns, many farmers practice conservation tillage. **Conservation tillage** restricts plowing of the soil to reduce erosion. Such restriction may vary from slight to complete. For example, in a conservation tillage system, like mulch-till, a chisel plow may be used instead of a moldboard plow, burying up to 50% of the crop residues between 10 and 15 centimeters (4 to 6 inches). If tillage is omitted completely, the term **no-till** is applied. By definition, conservation tillage systems leave enough of the previous crop residues so that at least 30% of the soil surface is covered when the next crop is planted. No-till can leave 70 percent or more of the soil surface covered from harvest to planting, depending on the crop residue. Soil

can be conserved, time and fuel consumption reduced, and comparable yields obtained with various forms of conservation tillage.

In 2004, roughly 22% of all U.S. planted cropland was in no-till, more than tripling from 6% in 1990. If we include all conservation tillage systems, such as mulch-till, ridge-till, and no-till, then the figure jumps to 41%, compared with 26% in 1990. Most of the growth in conservation tillage since 1990 has come from expanded adoption of no-till. Farmers in the United States are encouraged to meet the definition of conservation tillage in order to participate in government subsidy and other programs.

In the no-till form of conservation tillage, farmers use special machinery that cuts narrow slits into the ground, in which the seeds of the next crop are sown. This method is also called direct drilling or seeding. All operations are done in a single

FIGURE 7.16 (A) Gully erosion on a Minnesota farm. (B) The same area planted with protective vegetation, primarily locust trees, to prevent further erosion. Five growing seasons later, the locust trees averaged 4.5 meters (15 feet) in height and not only served to control erosion, but also provided wildlife cover and beautified the landscape.

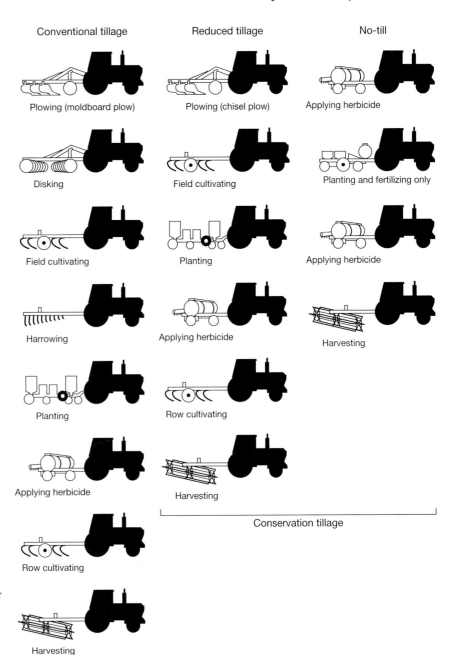

FIGURE 7.17 Differences in equipment use and number of passes over the field among conventional tillage, reduced tillage, and no-till systems. Conventional tillage often requires about eight passes, reduced tillage six passes, and no-till four passes over the field.

pass over the field. No seedbed preparation is necessary. Corn and soybeans, for example, can be sown directly in wheat stubble. Almost all of the residue from the previously harvested crop remains on the surface, compared with only 1% to 10% with conventional tillage. There is a trade-off, however. Weeds usually increase with no-till methods, resulting in greater use of herbicides.

Research conducted on farms in Illinois has shown that no-till farming controls erosion on slopes as steep as 9%. On many farms, the layer of crop residues on the surface can reduce soil loss from erosion by 90%. In fact, a recent experiment conducted by the Natural Resource Conservation Service in Georgia showed that no-till farming reduced erosion losses from 58 metric tons per hectare (26 tons per acre) per year to only 0.2 metric ton per hectare (0.1 ton per acre)—more than a 99% reduction.

No-till farming provides a number of other benefits. There also are some disadvantages. The pros and cons of this farming technique are summarized in Table 7.1.

Removing Cropland From Production In 1985, Congress passed the **Food Security Act** (or Farm Bill). It established funding for the **Conservation Reserve Program (CRP).** It called for the removal of 18 million hectares (45 million acres) of marginal cropland from the cropland base—land that was most vulnerable to erosion when cultivated. This land was to be planted with grasses, trees, and other long-term cover to stabilize the soil.

The objective of the CRP is to control erosion on highly erodible cropland. Under the terms of this program, the farmer makes a contract with the USDA to withdraw erodible farmland from crop production for ten years and to

TABLE 7.1	Pros and Cons of No-Till Farming

Pros	Cons
1. Labor may be reduced by 30%–50%.	1. Farmers require greater management skills, so the transition to no-till is difficult.
2. Use of diesel fuel is reduced by 30%–50%.	2. Special expensive equipment is needed to plant seeds directly in crop residues.
3. Soil erosion is generally reduced by 90%.	3. Seeds may not make contact with soil if the seed planter is not perfectly level, resulting in less seed germination.
4. Soil retains more moisture because of reduced rates of evaporation and surface runoff.	4. Plant diseases may be more abundant or shift in unexpected ways.
5. Sediment and fertilizer pollution of lakes and streams is reduced.	5. Populations of crop-destroying insects and rodents are often greater.
6. Soil health is improved.	6. Weed populations are often greater, resulting in heavier reliance on herbicides.
7. Soil carbon sequestration is increased.	7. Initially more nitrogen fertilizer may be required.
8. Crop residues provide food and cover for wildlife.	8. Soil temperatures are cooler in spring, often resulting in slower germination and more stand problems.
9. Air pollution is diminished because of the decrease in fuel consumption and potential dust from wind erosion.	

establish vegetative cover (such as grass or trees) on this land to stabilize the soil. The USDA (with money from the taxpayers), in turn, makes payments to the farmer during this period. Not only are the plantings valuable in checking erosion, they also are useful in providing food and cover for wildlife.

Although the Food Security Act called for the removal of 18 million hectares of highly erodible land, by the early 1990s only 14.7 million hectares (36.4 million acres) had been removed, cutting annual erosion on U.S. farmland by 410 million metric tons per year. (This is enough topsoil to fill a convoy of dump trucks, 58 side by side, stretching from Los Angeles to New York.) As of 2006, 14.9 million hectares of highly erodible land were in CRP, making it the nation's largest private-lands conservation program.

The Food Security Act also established national incentives for erosion control on actively farmed lands. For example, conservation compliance requires growers to develop and implement erosion-control plans for certain highly erodible lands to remain eligible for federal price support programs, such as subsidies. These erosion-control plans are developed in conjunction with the USDA Natural Resource Conservation Service (formerly the Soil Conservation Service). In addition to these provisions is the "Swampbuster" provision, which protects wetlands by denying eligibility for other USDA programs to growers who drain and farm certain wetlands.

Enforcement of the erosion-control provisions of the Food Security Act poses a formidable challenge for the USDA. However, this problem may be solved by the work of USDA scientists, who have developed regionally adapt-

able computer models of erosion rates (called the Revised Universal Soil Loss Equation, or RUSLE). These models, based on the **Universal Soil Loss Equation** (see A Closer Look 7.1), will enable the USDA to know instantly how much soil erosion a particular farming practice on a certain soil type will cause in a specific type of climate anywhere in the United States. Suppose, for example, that the USDA learns, with the aid of computer models, that a particular farmer's uphill-and-downhill plowing method on a silt loam soil in northern Illinois is causing the erosion of 74 metric tons of soil per hectare (30 tons per acre) annually—63 metric tons per hectare (25 tons per acre) more than is considered acceptable. If the farmer refuses to shift to contour plowing, the USDA could reduce or even eliminate his or her federal subsidies, as required by the Food Security Act.

The Natural Resource Conservation Service and Its Program

The black blizzards of the 1930s alerted a hitherto apathetic nation to the plight of its soil resources more forcefully than thousands of urgent speeches. In 1934, the newly organized Soil Erosion Service set up 41 soil and water conservation demonstration projects. The labor force for these projects was supplied by Civilian Conservation Corps workers drawn from about 50 camps. The projects impressed Congress so much that in 1935 it established the Soil Conservation Service, which was renamed the **Natural Resource Conservation Service (NRCS)** in 1994. The NRCS provides farmers and ranchers with technical assistance that allows them to better utilize land in ways that are as

A CLOSER LOOK 7.1 — Universal Soil Loss Equation

The Universal Soil Loss Equation (USLE) was developed by soil scientists after many decades of research to be able to predict soil losses from water erosion. The equation is:

$$A = RKLSCP,$$

where

A = number of metric tons (tons) of soil lost per hectare (acre) per year

R = rainfall erosivity

K = erodibility of soil

L = length of slope

S = steepness of slope

C = cover type (grass, wheat, forest, etc.)

P = practice used in erosion control (strip cropping, contour farming, etc.)

The USLE has been widely used since the 1970s. In the early to mid-1990s, the equation was revised into a modern, computerized tool called the Revised Universal Soil Loss Equation (RUSLE). RUSLE uses the same factors of USLE, although now some of the factors are better defined, which improves the accuracy of predicting soil loss from water erosion. RUSLE is a readily available computer software package. Farmers or soil scientists can use it to estimate soil loss for any farm in the United States.

Suppose that a farmer in southern Ohio, whose soil is a silt loam, would like to know her erosion losses. From information in the RUSLE software package or even from the original USLE tables available from the Natural Resource Conservation Service office, the farmer determines the following data about her farm:

$$R = 150$$

$$K = 0.33$$

$$LS = 0.40$$

Suppose now that the farmer's land is almost devoid of cover from the time of the harvest in autumn to the planting of the next crop the following spring. In this case, $C = 0.9$. Suppose further that she does not use any soil erosion-control practices, such as terracing or strip cropping. In this case, $P = 1.0$. The expected soil loss on her farm can then be calculated by using the equation:

$$A = (150)(0.33)(0.40)(0.90)(1.00) = 17.8 \text{ tons per acre,}$$
$$\text{or } 40.2 \text{ metric tons per hectare, annually}$$

Obviously, erosion is excessive on the farmer's land, the rate being roughly 3.5 times the tolerable limit of 11.2 metric tons per hectare (5 tons per acre) per year. Since factors $RKLS$ remain fairly constant, the only practical way for the farmer to reduce erosion losses on her farm would be to reduce the values of C (cover) and P (practices). She decides to substitute conservation tillage for traditional tillage on her farm. This way, some crop residues are always on her land, even between the harvest and the next planting. As a result, the value for C drops to 0.1 in the RUSLE. She further decides to till and plant on the contour. Adoption of this practice, in turn, reduces the value of P to 0.4 in the RUSLE.

Now let us recalculate the soil loss on this Ohio farm after substituting the new values for C and P in the RUSLE:

$$A = (150)(0.33)(0.40)(0.10)(0.40) = 0.8 \text{ ton per acre, or}$$
$$1.8 \text{ metric tons per hectare, per year}$$

It is clear that conservation tillage and contour farming have been extremely effective, reducing erosion losses on the farm from 40.2 to 1.8 metric tons per hectare (17.8 to 0.8 tons per acre) annually.

---

consistent with the needs of the soil as with those of the landowner.

The NRCS has regional offices throughout the United States. Each one oversees a particular region, known as a **conservation district.** Each conservation district is organized and run by farmers and ranchers. Each district is staffed by a professional conservationist and several aides who work directly with farmers on their land. The highly diversified types of assistance provided by the NRCS to the farmer are indicated by the kinds of specialists on its staff: agricultural engineers, botanists, chemists, ecologists, foresters, irrigation engineers, land appraisers, land use specialists, soil scientists, and wildlife biologists. Any farmer in a particular district can request assistance to set up and maintain a sound conservation program on his or her farm. Participation in the NRCS program is voluntary.

Today nearly 3,000 conservation districts have been organized, embracing roughly 2 million hectares (5 million acres)

and 96% of the nation's farms and ranch lands. In a typical year the NRCS assists more than 900,000 farmers and ranchers. This assistance includes:

1. Conducting and publishing soil surveys
2. Promoting conservation tillage
3. Controlling salinization
4. Identifying important farmlands, such as prime and unique farmlands
5. Constructing thousands of terraces and farm ponds

The NRCS has even helped to establish plant cover on the lava-covered slopes of Mount St. Helens in Washington State.

The NRCS prepares **soil surveys** used by farmers and ranchers, as well as by homeowners, highway engineers, and land use planners. A soil survey is a systematic examination, description, classification, and mapping of the soils in a given

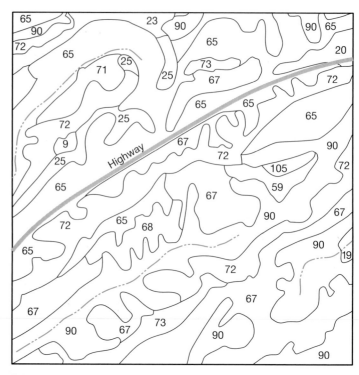

FIGURE 7.18 Part of a soil survey map from the USDA Soil Survey of Whitman County, Washington. The different mapping units (designated by numbers) indicate specific soil types with unique physical, chemical, and biological characteristics, the details of which can be found in the survey report.

9	Athena silt loam, 7–25% slopes	67	Palouse silt loam, 25–40% slopes
19	Caldwell silt loam	68	Palouse silt loam, 25–40% slopes, eroded
20	Caldwell silt loam, drained	71	Palouse-Thatuna silt loams, 7–25% slopes
23	Caldwell silt loam, 5–25% slopes	72	Palouse-Thatuna silt loams, 25–40% slopes
25	Caldwell silt loam, 25–40% slopes, eroded	73	Palouse-Thatuna silt loams, 40–55% slopes
59	Naff silt loam, 7–25% slopes	90	Snow silt loam, 7–15% slopes
65	Palouse silt loam, 7–25% slopes	105	Thatuna silt loam, 25–40% slopes

area (usually a county). It can help users understand the kinds of soil uses possible on a given piece of property. The NRCS uses aerial photographs to form a base map. As shown in Figure 7.18, the base map shows the boundaries of the different soil types and slopes.

Completed soil maps become part of a soil survey report. The report contains four components: (1) a set of soil maps, (2) map legends that explain the map symbols, (3) descriptions of the soils, and (4) use and management evaluations for each soil. Soil reports are available free to the public from the NRCS offices or land grant universities. Many survey reports are now online.

One of the most significant accomplishments of the NRCS early in its history was the development of the **Land Capability Classification System.** This system evaluates land according to its limitations for agricultural use. It classifies land into eight categories, with Class I the least susceptible to erosion and Class VIII the most susceptible (Figure 7.19). Classes I through IV are suitable for cultivation, whereas Classes V through VIII generally are not. Class I land is most suitable for crop production, being flat and fertile. Classes II and III lands may be used for growing crops, but the necessity of conservation measures and careful management increases. Class IV land is suitable only for limited cultivation. Classes V through VII are most suited for pasture, range, forest, wildlife, or recreation. Class VIII land, being stony, infertile,

or very steep, is suitable only as wildlife habitat, wilderness, or recreational areas. The NRCS uses the Land Capability Classification System in preparing conservation (farm) plans for farmers.

Development of an NRCS Conservation Plan

A **conservation plan** is a document that describes what the farmer has agreed to do at the time the plan is developed to address conservation issues on the farm. The plan can be a whole farm plan or a progressive plan to address some (but not all) of the concerns. It documents existing practices and decisions, and includes a schedule to deal with natural resource concerns. As part of the planning process, NRCS can help the farmer identify federal and state programs that may provide financial assistance for needed practices.

When a farmer requests technical assistance from his or her NRCS district, four steps are followed to create and execute the conservation plan for the farm. First, the technician and the farmer do an intensive survey of the farm. On the basis of such criteria as slope, fertility, stoniness, drainage, topsoil thickness, and susceptibility to erosion, the technician maps each parcel of land on the basis of its capability. Each plot is given a land capability class symbol in the form of a Roman numeral or color. This map is then superimposed on an aerial photograph.

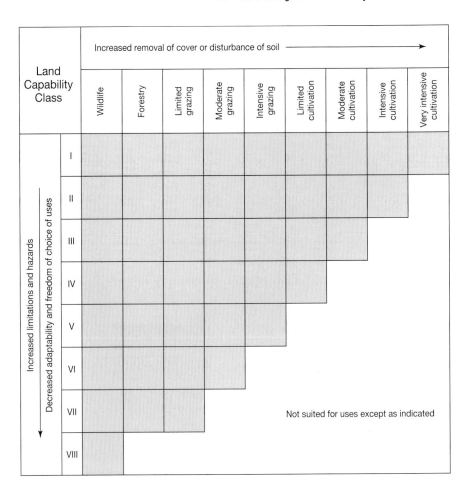

Land Capability Class		Increased removal of cover or disturbance of soil →								
		Wildlife	Forestry	Limited grazing	Moderate grazing	Intensive grazing	Limited cultivation	Moderate cultivation	Intensive cultivation	Very intensive cultivation
I										
II										
III										
IV										
V										
VI										
VII							Not suited for uses except as indicated			
VIII										

(Left axis: Increased limitations and hazards ↓ / Decreased adaptability and freedom of choice of uses)

FIGURE 7.19 Intensity of land use for each land capability class. Class I lands are the most versatile. The limitations and needed conservation practices increase from Class I lands to Class VIII lands.

Second, the farmer draws up a farm plan with assistance from the technician. This plan involves decisions about how each acre will be used and how it will be improved and protected. For example, should a given acre be used for crops, pasture, forests, or wildlife? Usually alternative uses and treatments are considered.

Third, the treatment and uses called for in the plan are applied. Although much of this application can be completed by the farmer alone, he or she may find it helpful to enlist the aid of NRCS technicians for more complex conservation measures such as terracing and strip cropping. Recommended management practices for a Georgia cotton farm (Figure 7.20) are shown in Figure 7.21.

FIGURE 7.20 Harvesting a cotton crop on a Georgia farm.

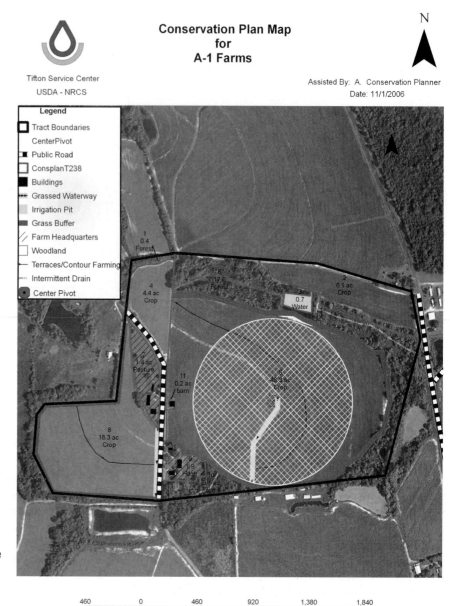

FIGURE 7.21 Conservation plan for a Georgia cotton farm. (Harvesting of the cotton crop is shown in Figure 7.20). This map is tied to a conservation plan document that describes the conservation practices to be installed in each designated field on the map. Conservation practices on this cotton farm include residue management, pest management, nutrient management, contour farming, grassed waterways, terraces, field borders, irrigation water management, and access roads.

The final and most important phase of the program is its maintenance from year to year with the assistance of conservation technicians. As time passes, agricultural geneticists might develop a new strain of rust-resistant wheat or a tick-resistant breed of cattle. A new variety of bluegill may be discovered that thrives in farm ponds. These new developments can gradually be incorporated into the conservation program.

7.7 Alternative Agriculture

For nearly four decades after World War II, U.S. agriculture was the envy of the world, almost annually setting new records in crop production and labor efficiency. During this period, U.S. farms became highly mechanized and specialized. They also became heavily dependent on fossil fuels, borrowed money (loans to farmers), and chemical fertilizers and pesticides. Today, many of these farms are now troubled by declining soil productivity,

deteriorating environmental quality, reduced profits, and threats to human and farm animal health. A growing number of people are concerned about the environmental, economic, and social impact of conventional agriculture and are seeking alternative practices to create a more sustainable agriculture (Table 7.2).

Long before alternative agriculture became popular, Franklin King published a book titled *Farmers of Forty Centuries: Permanent Agriculture in China, Korea and Japan*. This report, published in 1911, documented how farmers in parts of East Asia worked fields for 4,000 years without depleting the fertility of their soil. This text and others of the early 20th century focused on the complex interactions within farming systems and took a systems approach, considering many factors when determining how to grow crops. Yet, around this same time, U.S. agriculture was in the early stages of industrialization. New technologies and scientific methods were developed to help farmers meet the growing demands of expanding urban populations. By substituting

TABLE 7.2	Concerns About Conventional Agriculture

Increased costs and uncertain availability of energy and farm chemicals

Increased resistance of weeds and insects to herbicides and insecticides

Decline in soil productivity from erosion

Decrease in number of farms, particularly family farms

Pollution of surface water and groundwater with sediment and agrochemicals

Destruction of wildlife and beneficial insects

Hazards to human and animal health from pesticides and food additives

Depletion of finite reserves of plant nutrients

mechanical power for horses, for example, farmers could increase their grain acreage by 20% to 30%, because they could plow more ground in less time and did not need to devote some of their land to growing food to feed their horses.

Many groups and individuals continued to believe that biology and ecology rather than chemistry and technology should govern agriculture. Their efforts helped to give birth to the soil conservation movement of the 1930s, the ongoing organic farming movement, and considerable related research. Nevertheless, by the 1950s, technological advances had caused a shift in mainstream agriculture, creating a system that relied on agrochemicals, new varieties of crops, and labor-saving but energy-intensive farm machinery. This system has come to be known as **conventional farming.**

As pesticides, inexpensive fertilizers, and high-yielding varieties of crops were introduced, it became possible to grow a crop on the same field year after year—a practice called **monocropping**—without depleting nitrogen reserves in the soil or causing serious pest problems. Farmers began to concentrate their efforts on fewer crops. Government programs promoted monoculture by subsidizing only the production of wheat, corn, and a few other major crops. Unfortunately, these practices set the stage for extensive soil erosion and for pollution of water by agrochemicals.

In the United States between 1950 and 1985, the cost of farming—including the cost of interest, equipment, and farm inputs such as chemical fertilizers, pesticides, and equipment—almost doubled, jumping from 22% to 42%, when considered as a share of total production cost. During this period, labor and on-farm input expenses declined from 52% to 34%. During most of this period, relatively little research on alternative ways of farming was conducted because of lack of funding and public interest.

By the late 1970s, however, concern was mounting that rapidly rising costs were endangering farmers nationwide. In response, the USDA in 1979 commissioned a study to assess the extent of organic farming in the United States, as well as the technology behind this farming and its economic and ecological impact. The study's report, *Report and Recommendations on Organic Farming,* published in 1980, was based heavily on case studies of 69 organic farms in 23 states. The USDA report concluded that organic farming is energy-efficient, environmentally sound, productive, and stable, and helps promote long-term sustainability. In addition, the report stimulated interest, nationally and internationally, in alternative agriculture. Its recommendations provided the basis for the alternative-agriculture initiative passed by Congress in the Food Security Act of 1985, which called for research and education on sustainable farming systems.

The alternative-agriculture movement received a further boost in 1989, when the Board on Agriculture of the National Research Council released another study report, *Alternative Agriculture.* Although controversial, the report found that well-managed farms growing diverse crops with little or no synthetic chemicals are as productive as and often more profitable than conventional farms. Although the report was controversial at the time, a number of studies since have reported similar results. *Alternative Agriculture* also asserted, "Wider adoption of proven alternative systems would result in even greater economic benefits to farmers and environmental gains for the nation."

Today, the term **alternative agriculture** is used to describe several types of agriculture meant to promote more-sustainable food production systems. These include organic, biological, biodynamic, integrated, low-input, natural systems, and no-till farming. **Organic,** or **biological, farming** excludes the use of agrochemicals, notably synthetic pesticides and fertilizers. In their place, organic farmers use natural methods, such as crop rotations, and naturally derived products, such as natural pesticides, to control pests. They use natural fertilizers, such as manure, to fertilize their fields, and modern technologies. Before a product can be labeled "organic" in the United States, a government-approved certifier inspects the farm where the food is grown to make sure the farmer is following all the rules necessary to meet USDA organic standards.

Biodynamic farming is similar to organic farming, but mainly differs in that biodynamic farmers use a series of eight soil and plant amendments, called preparations, made from cow manure, silica, and various plant substances to fertilize fields. Biodynamic farming also places greater emphasis on (1) the integration of animals to create a closed nutrient cycle, (2) the effect of crop-planting dates in relation to the calendar, and (3) the awareness of spiritual forces in nature. In the United States, the certification program for food and feed produced by strictly biodynamic farming methods is run by a nonprofit organization called Demeter USA.

Integrated farming systems, successfully adopted on a wide scale in Europe, combine methods of conventional and organic production systems in an attempt to optimize both environmental quality and economic profit. For example, integrated farmers build their soils with composts and green manure crops but also use some synthetic fertilizers. In addition, they combine biological, cultural, and mechanical pest-control practices with the use of some synthetic and natural pesticides.

Low-input farming is based on a reduction of materials from the outside, such as commercially purchased chemicals and fuels. Whenever possible, external resources are replaced by resources found on or near the farm. Such internal resources include biological pest controls, solar or wind energy, biologically fixed nitrogen, and other nutrients released from green manures, organic matter, or soil reserves. In many places around the world, farmers are discovering that low-input agriculture can be more profitable (lower yields but lower costs and greater net returns) and involve lower risk than conventional high-input systems.

Natural systems agriculture is an ecology-based approach to agricultural production in which the vegetative patterns of natural ecosystems are used as guides to the composition and management of agricultural crops. For example, at the Land Institute in Kansas, located in the North American tallgrass prairie region,

researchers are working to develop systems of **perennial grains** grown in mixtures to mimic the prairie ecosystem. They are transforming the major annual (single-season) grain crops, such as wheat, into perennials, which can live for many years. The idea, actually decades old, may take decades more to realize, but significant advances in plant-breeding science are bringing this goal within sight at last. Replacing annual grain crops with perennials would create large root systems capable of preserving the soil and would allow cultivation in areas currently considered marginal (Figure 7.22).

No-till farming, another method of alternative agriculture, has already been discussed in this chapter.

Alternative farming systems do not represent a return to pre–Industrial Revolution methods. Rather, alternative farming combines traditional conservation-minded farming techniques with modern technologies. Alternative production systems use modern equipment, certified seed, soil and water conservation practices, genetically improved crop varieties, and the latest innovations in feeding and handling livestock. Emphasis is placed on rotating crops, building up soil, diversifying crops and livestock, and controlling pests naturally. Whenever possible, external resources—such as commercially purchased chemicals and fuels—are reduced or replaced by resources found on or near the farm. These internal resources include solar or wind energy, biological pest controls, and biologically fixed nitrogen and other nutrients released from green manures, organic matter, or soil reserves. In some cases, external resources (such as herbicides for no-till systems) may be essential for reaching sustainability. As a result, alternative farming systems can differ considerably from one another, because each tailors its practices to meet specific environmental and economic needs.

Alternative farmers who practice soil conservation and reduce their dependence on fertilizers and pesticides generally

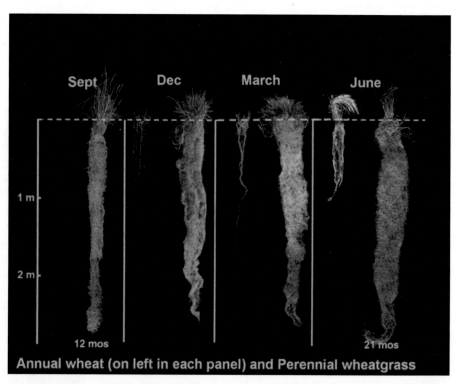

FIGURE 7.22 Root and top growth of annual wheat (at left in panels) and its perennial relative, wheatgrass (at right in panels), at four different times of year. Perennial crops have deeper root systems than annuals, providing access to more water and nutrients. Perennials also have a longer growing season, allowing more sunlight to be captured by the crop. To develop high-yield perennial crop plants such as wheat, scientists and breeders can either domesticate a wild perennial plant to improve its traits or hybridize an annual crop plant with a wild perennial relative to blend their best qualities. Each method requires time- and labor-intensive plant crossbreeding and analysis. High-yielding perennial wheat is likely feasible within the next 25 years. (Courtesy of The Land Institute.)

report that their production costs are lower than those of nearby conventional farmers. Sometimes the yields from alternative farms are lower than those from conventional farms, but they are frequently offset by lower production costs, leading to equal or greater net returns. Organic farmers are often more profitable than their conventional counterparts largely due to price premiums for organically certified products. In fact, research comparing the agronomic, economic, and ecological performance of organic and conventional farming systems confirms this observation. Furthermore, studies generally have found that organic systems have somewhat lower yields but less variability in production from year to year, are equally if not more profitable, cause less erosion and pollution, have better soil quality, are more energy efficient, and rely less on government subsidies than their conventional counterparts (Figure 7.23). In other words, the organic systems seem more sustainable.

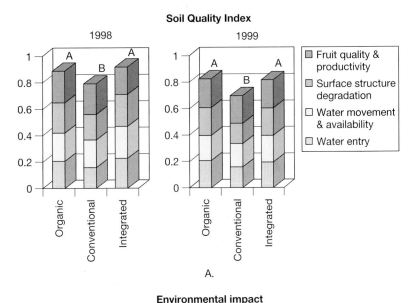

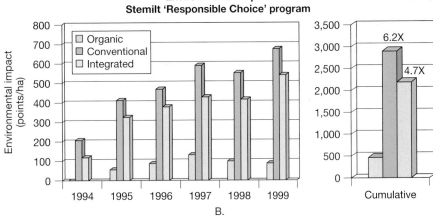

FIGURE 7.23 Soil quality, potential pesticide impact, and energy efficiency of organic, integrated, and conventional apple production systems, as measured in a six-year sustainability study in Washington State. As shown, the organic system had the highest soil quality, lowest potential negative environmental impact, and greatest energy efficiency, followed by the integrated and then the conventional system. Although not shown, scientists also found that all three systems had comparable apple yields, tree growth, and leaf and fruit nutrient contents. When compared with the conventional and integrated systems, the organic system had sweeter and less tart apples and higher profits. Based on all of these properties, the organic system ranked first in overall sustainability, the integrated system second, and the conventional system last.

7.8 Sustainable Agriculture

Principles and Practices

Alternative farming may help promote a more sustainable system of agriculture, often termed **sustainable agriculture,** but it does not mean that an alternative farm is necessarily sustainable. For a farm to be sustainable, it must produce adequate amounts of high-quality food, conserve resources, and be environmentally safe. It must also be profitable and socially responsible (Table 7.3). If any single element of this equation is missing, the farm cannot be considered sustainable. Thus, for example, if an environmentally friendly organic farm that produces high yields of nutritious food is not profitable, it cannot be considered sustainable. Likewise, a conventional farm that meets all the sustainability criteria but pollutes a nearby river with sediment because of soil erosion is not sustainable. To reiterate, for any farm to be considered sustainable, whether it be conventional or alternative, it should meet all the sustainability criteria as listed in Table 7.3.

Sustainable agriculture addresses many serious problems afflicting U.S. and world food production: high energy costs, groundwater contamination, and soil erosion. Sustainable agriculture addresses the continual deterioration of soil fertility, declining farm productivity, and depletion of fossil fuel resources. It addresses low farm incomes and the many risks to human health and wildlife habitats posed by conventional agriculture. It is not so much a specific farming strategy as it is a systems-level approach to understanding the complex interactions within agricultural ecologies.

As noted in Chapter 5, which discussed world hunger, protecting farmland from erosion and other factors is vital to creating a sustainable system of agriculture. Unfortunately, much farmland is being lost, particularly to development (urbanization) (Figure 7.24A). For example, the rate of prime farmland

development increased from an average of 162,000 hectares (400,000 acres) per year between 1982 and 1992 to 243,000 hectares (600,000 acres) per year between 1992 and 2001. Both rates are unacceptable if we want to preserve our best soils for future generations. Most agricultural experts agree that we need to protect the best or prime agricultural lands for use as cropland, pasture, or forestland, and to build houses and so forth on our lower-quality farmland (Figure 7.24B).

A central component of almost all sustainable farming systems is **crop rotation**—a planned succession of various crops grown on one field. When crops are rotated, the yields are usually about 10% higher than when they grow in monoculture. In most cases, monocultures can be perpetuated only by adding large amounts of **fertilizer** (organic or inorganic compounds given to plants to promote growth) and pesticide. Rotating crops provides better weed and insect control, less disease buildup, more efficient nutrient cycling, and other benefits.

Crop rotation can take many forms. Alternating two crops, such as corn and soybeans, for example, is considered a simple rotation. More complex rotations require three or more crops on a four- to ten-year rotation cycle. In growing a more diversified group of crops in rotation, a farmer is less affected by price fluctuations of one or two crops. This may result in more year-to-year financial stability. There are, however, disadvantages. They include the need for more equipment to grow a number of different crops, the reduction of acreage planted with government-supported crops, and more time and information needed to manage more crops.

TABLE 7.3	Examples of Evaluation Criteria for Farm Sustainability	
Economic	Environmental	Social
Farm profitability	Energy efficiency	Adequate yields
Operating costs	Soil, water, and air quality	Food and fiber quality
Income variability	Soil and water conservation	Farmland protection from urbanization
Financial risks	Wildlife protection	Farmworker salaries and benefits
Food costs	Food and feed safety	Quality of life for farmers
Return on investment	Farm safety	Ethics of farming practices

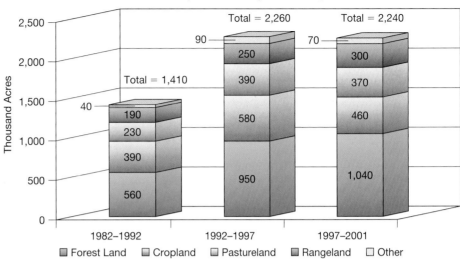

Source of Developed Land
(Annual Acres by Time Period)

FIGURE 7.24A Rate of development of forest land, cropland, pastureland, and rangeland. The rate of development between 1997 and 2001 averaged 2.2 million acres (0.9 million hectares) per year, according to USDA National Resources Inventory data. This was the same average rate experienced between 1992 and 1997, but up from 1.4 million acres (0.6 million hectares) per year in the previous decade (1982–1992). Between 1982 and 2001, more than 34 million acres (almost 14 million hectares)—an area the size of Illinois—were converted to developed uses (urban and built-up areas and rural transportation land) (Source: USDA Natural Resource Conservation Service, 2003).

Besides a diverse assortment of crops, **diversity** also results from mixing species and varieties of crops and from systematically integrating crops, trees, and livestock. When most of North Dakota experienced a severe drought during the 1988 growing season, for example, many monocropping wheat farmers had no grain to harvest. Farmers with more diversified systems, however, had sales of their livestock to fall back on or were able to harvest their late-seeded crops or drought-tolerant varieties.

Regularly adding crop residues, manures, and other organic materials to the soil is another central feature of sustainable farming. Organic matter improves soil structure, increases its water storage capacity, enhances fertility, and promotes the **tilth,** or physical condition, of the soil. The better the tilth, the more easily the soil can be tilled or direct seeded (as with no-till), and the easier it is for seedlings to emerge and for roots to extend downward. Water readily infiltrates soils with good tilth, thereby minimizing surface runoff and soil erosion. Organic materials also feed earthworms and soil microbes.

The main sources of plant nutrients in some sustainable farming systems are animal and green manures (Figure 7.25). Green manures help control weeds, insect pests, and soil erosion, while also providing forage for livestock and cover for wildlife. Some sustainable farming systems rely mostly on synthetic chemical fertilizers but use them in the most appropriate and efficient amounts. Other sustainable farming systems may take an integrated farming approach, in which synthetic fertilizers are considered a supplement to the nutrients made available from biological nitrogen fixation, organic fertilizers, crop residues, and decomposition of soil organic matter.

Controlling insects, diseases, and weeds without chemicals also is a goal of sustainable farming, and the evidence in support of this approach is encouraging. One broad approach to limiting use of pesticides is called **integrated pest management (IPM).** IPM involves many different, nonintrusive methods for controlling pests, such as crop rotation or timing of plantings to avoid the emergence of pests. It also relies on natural pest controls (**biological control**),

FIGURE 7.24B Amounts and types of developed land that was prime farmland (the best soils). The rate of prime farmland development increased from an average of 400,000 acres (162,000 hectares) per year between 1982 and 1992 to 619,000 acres (251,000 hectares) per year between 1992 and 2001. In the period between 1992 and 2001, about 5.6 million acres (2.3 million hectares) or 28% of the new land developed was prime farmland. Between 1982 and 1992, about 4 million acres (1.6 million hectares) or 29% of the new land developed was prime farmland (Source: USDA Natural Resource Conservation Service, 2003).

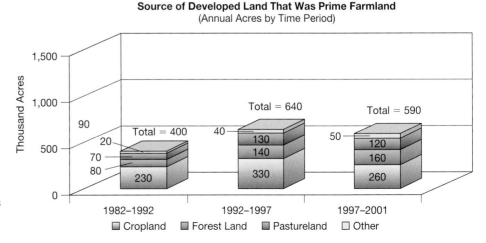

Source of Developed Land That Was Prime Farmland
(Annual Acres by Time Period)

FIGURE 7.25   Green manure crops (cercal rye and crimson clover) being flattened into a surface mulch by a red crimper roller on a Mississippi farm. A cash crop, such as soybeans, will then be no-tilled into the mulch. Green manure crops, when rolled over or plowed under, enrich the soil, improve future crop productivity, and reduce the need for fertilizers.

such as predatory insects that eat crop-destroying pests (see Figure 7.26). For more information on this approach, see Chapter 8.

Barriers to Farmers Adopting Sustainable Methods

What forces inhibit farmers from adopting sustainable methods? One obstacle is the federal farm programs, which generally support prices for only a handful of crops. Corn and other feed grains, wheat, cotton, and soybeans receive roughly three-fourths of all U.S. crop subsidies and account for approximately two-thirds of cropland use. The lack of price support for other crops discourages farmers from

diversifying and rotating their crops and from planting green manures. Suppose, for example, that a farmer was growing wheat on a steep, severely eroding hillside and replaced the wheat with erosion-controlling grass. The benefits from government subsidies for that farmer from the grassed area would be nil.

Government subsidies have given farmers powerful incentive to practice monoculture or short rotations to achieve maximum yields and profits. However, farm legislation called the Freedom to Farm Act, passed in 1996, was supposed to snap the decades-old link between crop prices and government subsidies. The law was to end government-guaranteed prices for corn and other feed grains, soybeans, cotton, rice, and wheat. Instead, farmers were to receive

FIGURE 7.26   Ladybug beetle, a natural predator of cherry aphids (shown here) and other insect pests. Integrated pest management (IPM) programs now in use on many farms take advantage of natural predator–prey relationships or other biological-control mechanisms to reduce the need for chemical pesticides. Farmers who practice biological control encourage the proliferation of beneficial microbes and insects, such as ladybugs.

guaranteed payments that would decline over seven years and would put an immediate end to most planting controls. The payments were to total $36 billion over seven years and terminate at the end of 2002. The idea behind "Freedom to Farm" was that farmers would flourish in an unfettered market. However, the subsidies have not been drying up. Since the act was passed, Congress has repeatedly passed increases in spending for agriculture, causing the subsidies to flow faster than ever. In fact, in the first three years after the act passed, government payments to American agriculture tripled. Some critics complain because the subsidies have primarily benefited the biggest producers (corporate-owned farms), packers, and processors, but not the small family farmers of America.

The long-term economic benefits of sustainable agriculture may not be evident to a farmer faced with having to meet loan payments. Many conventional farmers are greatly in debt, partly because of heavy investments in specialized machinery and other equipment, and their debt constrains the shift to more sustainable methods. For example, contour farming requires more tractor time (as well as costly diesel fuel) than straight-row farming. It takes considerably more time (and skill) for the farmer to follow the curving topographical pattern than to plow straight up and down the hills. It is estimated that the application of soil conservation measures may add 10% to 20% to the farmer's operating expenses—at a time when profit margins on many farms are thin.

A significant portion of today's farmers lease farmland and practice what is called **tenant farming.** An owner-operator may be willing to invest time, energy, and money in effective soil conservation measures, even if the payoff in enhanced crop yields may not materialize for several years. However, a tenant farmer, who often rents a given farm, does not have the same motivation.

More than half the farmland in the United States is farmed by people over age 55, with a significant portion of them being older than 65. Simply put, there is a shortage of young farmers. Furthermore, a recent study showed that most farmers who practice sustainable agriculture adopted the practices when they were under age 40. In other words, young farmers are more apt to make changes.

Also, in some areas of the United States, little information on sustainable practices is available to farmers. Government-sponsored research has inadequately explored alternative farming and has focused instead on agrochemically based production methods. Agribusinesses also greatly influence research by providing grants to universities to develop chemical-intensive technologies for perpetuating grain monocultures.

Legislative support for change in the U.S. agricultural system is growing, but financial support for sustainable agricultural projects is still only a small part of the total outlay for agriculture. The U.S. Congress created the **Sustainable Agriculture Research and Education Program (SARE)** in 1988 to implement research and education programs in sustainable farming. SARE has a number of objectives: to reduce reliance on fertilizer, pesticide, and other purchased resources; to increase farm profits and agricultural productivity; to conserve energy and natural resources; to reduce soil erosion and the loss of nutrients; and to develop sustainable farming systems. Since 1988, the amount of funds that Congress has allocated to SARE is less than 1% of the total USDA research and education budget during that same time. Shifting mainstream agriculture toward more sustainable methods will require a much larger research and education effort. Fortunately, universities and the USDA are slowly beginning to place more emphasis on research into sustainable agriculture.

Future Research and Education

Cropland in the United States and other industrial countries has been receiving more nitrogen than plants can effectively use. For example, only a third of the nitrogen in fertilizer applied to cereal crops globally is harvested in the grain. Excess nitrogen needs to be reduced if nitrogen pollution is to be controlled. The development of specific cropping systems that produce and consume nitrogen more efficiently should be a high priority. The use of slow-release fertilizers provides another means of reducing nitrogen losses from fertilized soils. More, however, must be learned about alternatives to fertilizers and the cycling of nutrients through the agricultural ecosystem.

Producers on farms and ranches face higher energy prices for their direct energy supplies and energy-intensive inputs. Energy conservation practices on the farm can reduce energy demands. Conservation practices such as crop residue management, irrigation water management, pesticide and nutrient management, windbreaks, contour farming, and rotation grazing, among others, can contribute to protecting soil and water resources and help reduce the nation's dependence on fossil fuels (Table 7.4). However, research on new technologies, such as solar power, wind turbines, and environmentally safe biomass energy production, also is needed to replace our petroleum-based economy.

Integrating modern technologies such as global positioning systems with several other computer-based technologies (such as geographic information systems) shows promise in improving the efficiency of fertilizer and pesticide applications on large agricultural fields. The resulting **precision farming** system attempts to treat each small soil unit (say, 1 hectare or less) of a larger farm field (say, 20 hectares) according to its particular properties, such as contents of clay, organic matter, and nutrients. Where relevant soil properties vary considerably from each 1-hectare cell to the next, precision farming should be more efficient than the traditional method of averaging soil properties of all 20 1-hectare cells of a large field and treating the entire field with the same fertilizer and pesticide rates according to the average properties. It remains to be seen whether the expense of all the computerized equipment and extra soil sampling and analysis required for precision farming systems can be offset by the increase in crop yields and reduction in inputs. For more on precision farming, see the GIS and Remote Sensing box.

TABLE 7.4	Energy Savings and Production Potential From Conservation Practices and Measures			
Conservation Practice	Conservation Measure	Resource Savings		Energy Cost Reduction
		On Farm	Total	
Crop residue management	25.3 million hectares (62.4 million acres) of no-till	$4.70/hectare ($11.70/acre)	921 million liters (243 million gal.)	$730 million
	Conversion of additional 20 million hectares (50 million acres) to no-till	$4.70/hectare ($11.70/acre)	739 million liters (195 million gal.)	$585 million
Irrigation water management	Improvement of pumping system efficiency by 10% on 6.5 million hectares (16 million acres)	$6/hectare ($15/acre)	303 million liters (80 million gal.)	$240 million
	Conversion of medium-pressure sprinkler systems to low	$16/hectare ($40/acre)	816 million Joules/hectare (560 kWh/acre)	$390 million
	Conversion of high-pressure sprinkler systems to low	$22/hectare ($55/acre)	1122 million Joules/hectare (770 kWh/acre)	$120 million
Pesticide and nutrient management	Reduction in spray overlap by 5%	Variable	101 million hectares (250 million acres) of cropland	$1 billion

(Source: USDA Natural Resource Conservation Service, 2006)

GIS AND REMOTE SENSING

GIS, Remote Sensing, and Precision Farming

The integrated use of field observations, computers, geographic information systems (GISs), global positioning systems (GPSs), remote sensing (aerial photography), and farm implements is termed *precision farming*. It allows a growing number of farmers to practice site-specific management of their production operations, which can improve efficiency, lower costs, increase yields, and reduce environmental hazards. Site-specific management allows the farmer to selectively apply pesticides, fertilizers, and other agricultural chemicals within a field, as opposed to a uniform application of these substances across a field. This means, for example, that a grower can prevent some parts of a field from being over-fertilized or underfertilized. This alone translates into cost savings, because fertilizers are not wasted. In addition, there is less potential of polluting surface or ground waters with nitrates from fertilizers.

The elements that allow all of this to happen are as follows: (1) the installation of modern GPS receivers on farm equipment, which enables the grower to establish positions within a field to about 1 meter (or less); (2) data collection devices that can be "location stamped" with geographic coordinates and that can even include on-the-fly digitizing capability so that a grower can sketch conditions on a computer map or digital aerial photo regarding pest infestations or yield variability; (3) point sampling in the field, where technical measurements are made of soil nutrient content, soil moisture status, and other physical or chemical properties, all of which are precisely located in the spatial database; (4) the use of GIS in a mobile setting to display information, to derive and evaluate relationships among the data, and to conduct spatial modeling (including the use of remote sensing in the form of large-scale aerial photography) that incorporates data layers of spatial information about the environmental and agricultural properties of the field; (5) the technological development of variable-rate farm implements, which allow farmers to incorporate and continuously locate a tractor's position in the field and on GIS maps, and to vary the application rate of field inputs, such as fertilizers, seed spacing, and seeding rates according to precise instructions for each location.

This combination of the technologies of GIS, GPS, aerial photography, ground information, and farm implements allows the modern farmer, for the first time, to truly attempt to incorporate field variability into the day-to-day management of the farm. Not all problems have been solved, however. Issues such as how to incorporate the point sampling of technical data into continuous surface information and how to characterize and visualize error and uncertainty in all facets of the operation are the focus of much theoretical and applied research today by geographers, soil scientists, and other specialists. Additionally, this technical system of data collection, data integration and management, spatial analysis, and output for decision making is complicated and evolving rapidly. Since precision farming is a dynamic enterprise, farmers and land managers need to be well informed to keep abreast of scientific and technological changes.

Effective strategies must also be developed for controlling pests, weeds, and diseases biologically. The strategies may rely on beneficial insects and microorganisms, allelopathic crop combinations (which discourage weed growth), diverse crop mixtures and rotations, and genetically resistant crops. More research on soil erosion should be dedicated to understanding the relative benefits of various cover crops and tillage practices and the integration of livestock into the cropping system.

U.S. farmers now use only a fraction of the thousands of crop species in existence. They may benefit from increasing cultivation of alternative crops such as triticale, amaranth, ginseng, and lupine, which are grown in other countries. In addition, plant breeders must continually collect and preserve germplasm (seeds, root stocks, and pollen) from traditional crops and their wild relatives. Well-managed collections of germplasm will give plant breeders a broader genetic base for producing new crops with greater resistance to pests, diseases, and drought. Today much of the germplasm that U.S. plant breeders use to improve crops comes from undeveloped countries.

New breeds of crops being developed by biotechnology, such as grains that fix nitrogen in the soil, thus eliminating the need for nitrogen fertilizer, may eventually become popular on sustainable farms. No-till systems that do not require increased use of herbicides also may find their way to modern sustainable farms. But neither biotechnology nor any other single technology can fix all the problems addressed by a balanced ecological approach. The success of sustainable agriculture hinges on a blending of conservation practices and modern technologies.

Better education is as important as further research. Farmers need to know clearly what sustainable agriculture means, and they must see proof of its profitability. The USDA and the Cooperative Extension Service should be continually funded to provide farmers with information that is up-to-date, accurate, practical, and applicable to local farming conditions. Farmers and the public also need to be better educated about the potentially adverse environmental and health consequences of soil erosion and the pollution created by certain agrochemical practices. Some universities already offer undergraduate (B.S.) and graduate (M.S. and Ph.D.) programs in sustainable agriculture. In 2006, Washington State University opened the first undergraduate major in organic agriculture at a U.S. university, and more schools are likely to follow.

One of the most effective methods for communicating practical information about sustainable agriculture is through farmer-to-farmer networks, such as the Practical Farmers of Iowa. Farmers in this association have agreed to research and demonstrate sustainable techniques on their lands. They meet regularly to share information and compare results. Because such networks have aroused growing interest and have proved effective, agricultural universities should promote their development. Farmer-to-consumer networks, called **community supported agriculture (CSA),** also are becoming more widespread in the United States. With CSAs, individuals pay farmers a flat fee or a monthly fee in exchange for a share of their food production. During the growing season, subscribers receive one or two boxes of fruits and vegetables from a nearby farm each week. In the process, farmers bypass intermediaries such as wholesalers and stores, so they ultimately get paid a higher price for their products. Subscribers help keep small farmers, many of whom use sustainable practices, in business.

Some scientists and environmentalists have recommended levying taxes on fertilizers and pesticides to offset the environmental costs of agrochemical use, to fund sustainable agricultural research, and to encourage farmers to reduce excessive use of agrochemicals. This approach is precisely how funding for the Leopold Center for Sustainable Agriculture at Iowa State University was established by the Iowa legislature in 1987.

GO GREEN!

Subscribe to a CSA to support a local farmer and receive healthy fruits and vegetables each week. Supporting farmers' markets is another way of supporting local farmers and getting to know who is growing your food.

The Leopold Center is a research and education center that has a three-fold mission for Iowa State: to conduct research into the negative impacts of agricultural practices; to assist in developing alternative practices; and to work with Iowa State University Extension to inform the public of Leopold Center findings.

Agriculture is a fundamental component of the natural resources on which rests not only the quality of human life but also its very existence. If efforts to create a sustainable agriculture succeed, farmers will profit, and society in general will benefit in many ways. More importantly, the United States will protect its natural resources and move closer to attaining a sustainable society.

Summary of Key Concepts

1. Soil erosion is a three-step process by which soil particles are detached from their original site, transported, and eventually deposited at other locations. The principal agents of erosion are wind and water.

2. Geological erosion is the wearing away of the Earth's surface by water, wind, ice, or other natural agents under natural environmental conditions.

3. Accelerated erosion is erosion that is much more rapid than normal geological erosion and results primarily from human activities and sometimes other animals' activities.

4. Soil erosion rates are highest in Asia, Africa, and South America and lowest in Europe and the United States. However, the relatively lower rates in Europe and the United States exceed the average rate of soil formation, or replacement.

5. The severe dust storms that ravaged the Plains (called the Dust Bowl) in the 1930s resulted from a combination of factors, including severe drought, high-velocity winds, overgrazing, and poor soil management. When rain returned and the winds finally subsided, 5 to 30 centimeters (2 to 12 inches) of topsoil had been removed. Millions of farmers and ranchers in Oklahoma, Colorado, Kansas, Texas, New Mexico, and other states suffered severe economic hardship and were forced to seek employment in urban centers.

6. A shelterbelt or windbreak is a row of living trees and shrubs along the edge of farm fields to protect farmland from the drying and erosive effects of the wind.

7. The USDA considers soil loss tolerance to be the maximum combined water and wind erosion that can take place on a given soil without degrading that soil's long-term productivity. However, soil loss tolerance values are greater than measured soil renewal rates.

8. The rate of water erosion is influenced by such factors as (1) rainfall intensity and seasonal distribution, (2) soil erodibility, (3) surface cover, (4) tillage practices, and (5) topography.

9. Effective methods of erosion control include (1) contour farming, (2) strip cropping, (3) conservation tillage, (4) terracing, (5) shelterbelts, (6) gully reclamation, (7) removal of land from production, and (8) conservation tillage.

10. Conservation tillage systems restrict plowing of the soil to reduce erosion and leave enough of the previous crop residues so that at least 30% of the soil surface is covered when the next crop is planted. If tillage is omitted completely, the term *no-till* is applied.

11. The Conservation Reserve Program authorizes farmers to remove, in exchange for government payments, marginal croplands that are most vulnerable to erosion when cultivated. This land is to be planted with grasses, trees, and other long-term cover to stabilize the soil.

12. The Revised Universal Soil Loss Equation (RUSLE) is $A = RKLSCP$, where A = number of tons of soil lost per hectare (or acre) per year, R = rainfall and runoff, K = erodibility of soil, L = length of slope, S = steepness of slope, C = cover type, and P = practice used in erosion control. Using RUSLE gives fairly good estimates of soil loss on any farm in the United States.

13. More than 3,000 soil conservation districts, the administrative and operative units of the NRCS, have been organized to provide technical and financial assistance (by developing conservation plans) to farmers so that they can effectively manage each acre of land according to its capability.

14. The USDA Land Capability Classification System evaluates land according to its limitations for agricultural use. It classifies land into eight categories, with Class I having the least limitations and Class VIII the most.

15. Alternative agriculture embraces several variants of nonconventional agriculture, including organic, biological, biodynamic, integrated, low-input, natural systems, and no-till farming. It combines traditional conservation-minded farming techniques with modern technologies.

16. Researchers have found that most organic systems have somewhat lower yields but less variability in production from year to year. Moreover, they are equally, if not more, profitable and cause less erosion and pollution. They have better soil quality, are more energy efficient, and rely less on government subsidies than conventional farming systems.

17. Just because a farm is alternative does not mean that it is sustainable. For a farm to be sustainable, it must produce adequate amounts of high-quality food, conserve resources, and be environmentally safe, profitable, and socially responsible.

18. Sustainable agriculture addresses many serious problems afflicting U.S. and world food production: high energy costs, groundwater contamination, soil erosion, loss of productivity, depletion of fossil fuel resources, low farm incomes, and risks to human health and wildlife habitats.

19. Central components of most sustainable-farming systems are crop rotation; the addition of organic materials such as crop residues, composts, and animal and green manures; diversity of crops and/or livestock; and integrated pest management, with an emphasis on biological control techniques. When commercial fertilizers are used, they

are to be added in the most appropriate and efficient amounts.

20. Barriers to farmers adopting more sustainable methods include government subsidies; an unwillingness of older farmers to change; the additional time, skill, and initial investment required to apply sustainable farming methods; tenant farming; and a lack of information available to farmers on sustainable practices.

21. The U.S. Congress created the Sustainable Agriculture Research and Education (SARE) program in 1988 to implement research and education programs in sustainable farming. Since 1988, the U.S. Congress has allocated less than 1% of the total USDA research and education budget to SARE.

22. Shifting mainstream agriculture toward more sustainable methods will require a much larger research effort. Some future research priorities include developing (1) specific cropping systems that produce and consume nitrogen more efficiently, (2) cost-effective precision-farming systems, (3) new biological-control strategies, (4) organic or reduced-herbicide, no-till systems, (5) more land in alternate crops, and (6) genetically improved crop varieties.

23. Better education of farmers and consumers through, for example, extension courses, farmer-to-farmer networks, and farmer-to-consumer networks also is needed to move us toward sustainable agriculture.

Key Words and Phrases

Accelerated Erosion	Low-Input Farming
Alternative Agriculture	Monocropping
Biodynamic Farming	Natural Resource Conservation
Biological Control	Service (NRCS)
Biological Farming	Natural Systems
Black Blizzards	Agriculture
Community Supported	No-Till Farming
Agriculture (CSA)	Organic Farming
Conservation District	Perennial Grain
Conservation Plan	Precision Farming
Conservation Reserve	Runoff
Program (CRP)	Shelterbelt
Conservation Tillage	Slope
Contour Farming	Soil Erosion
Conventional Farming	Soil Survey
Crop Rotation	Strip Cropping
Diversity	Sustainable Agriculture
Dust Bowl	Sustainable Agriculture
Fertilizer	Research and Education
Food Security Act	Program (SARE)
Geological Erosion	Tenant Farming
Green Manure	Terracing
Gullies	Tilth
Integrated Farming Systems	Tolerance Value (T Value)
Integrated Pest Management	Universal Soil Loss Equation
(IPM)	(USLE)
Land Capability	Water-Stable Aggregates
Classification System	Windbreaks

Critical Thinking and Discussion Questions

1. What is the difference between geological and accelerated erosion? Give one example of each.

2. Were the black blizzards of the Dust Bowl era caused directly by drought? Could they have been prevented despite the drought? Discuss your answer.

3. Make a list of the on-site and off-site effects of soil erosion.

4. Is the level of soil erosion in the United States higher or lower today compared with the Dust Bowl era? Discuss your answer.

5. What does it mean to have erosion on a soil at less than its soil loss tolerance value?

6. Discuss the advantages and disadvantages of conservation tillage.

7. What is the difference between contour farming and strip cropping? Can both be done at the same time?

8. What are the benefits of windbreaks? Why are they sometimes taken out?

9. Discuss the advantages and disadvantages of no-till farming.

10. What is the Revised Universal Soil Loss Equation? How can the farmer use this equation to control soil erosion on the land?

11. What is the Natural Resource Conservation Service? What are some of its responsibilities?

12. Why is there such an interest in alternative agriculture?

13. What is monocropping? What is crop rotation? What are green manures, and can they be incorporated into a crop rotation?

14. What are the benefits of conventional farming?

15. Is there a difference between alternative agriculture and sustainable agriculture? Discuss your answer.

16. What are the barriers that inhibit farmers from adopting sustainable practices?

17. Do you think U.S. agriculture is on a sustainable path for the future? If so, discuss your answer. If not, discuss the types of changes that are needed to make it more sustainable.

18. Suppose that all commercial synthetic fertilizers were banned for agricultural use. Discuss the advantages and disadvantages of this development.

19. Suppose that all synthetic pesticides (insecticides, herbicides, fungicides, etc.) were banned for agricultural use. Discuss the advantages and disadvantages of this development.

Suggested Readings

Blanco-Canqui, H., and R. Lal. 2008. *Principles of Soil Conservation and Management.* New York: Springer. Comprehensive review of the state of knowledge on soil erosion and management, addressing the implications of soil erosion with emphasis on global hot spots.

Chiras, D., and D. Wann. 2003. *Superbia! 31 Ways to Create Sustainable Neighborhoods.* Gabriola Island, BC: New Society Publishers. Description of actions people can take to build a more sustainable world in their own neighborhoods, including ideas for creating community gardens and a more edible landscape.

Gantzer, C. J., S. H. Anderson, A. L. Thompson, and J. R. Brown. 1990. Estimating Soil Erosion After 100 Years of Cropping on Sanborn Field. *Journal of Soil and Water Conservation* 45: 641–644. The reference for the 100-year case study of the effects of cropping on soil erosion.

Gillman, J. 2008. *The Truth About Organic Gardening: Benefits, Drawbacks, and the Bottom Line.* Portland, OR: Timber Press. Enjoyable and useful book on the importance of growing more of our own food, with an examination of the safety of garden inputs such as natural and synthetic fertilizers and pesticides.

Glover, J. D., C. M. Cox, and J. P. Reganold. 2007. Future Farming: A Return to Roots? *Scientific American* 297(2): 82–89. Very interesting article on breeding perennial grains and their ecological importance to the planet's future.

Halweil, B. 2002. Farming in the Public Interest. In *State of the World 2002.* New York: W. W. Norton. An important read for students interested in learning more about problems facing agriculture and methods to create more-sustainable farming systems.

Halweil, B. 2002. *Home Grown: The Case for Local Food in a Global Market.* Worldwatch Paper 163. Washington, DC: Worldwatch Institute. Excellent look at local food production—that is, the production of food by local farmers for local markets—and the benefits it can have for the environment and the economy.

Huggins, D. R., and J. P. Reganold. 2008. No-Till: The Quiet Revolution. *Scientific American* 299(1): 70–77. Balanced overview of no-till agriculture, including its importance in protecting the soil.

Klinkenborg, V. 1995. Farming Revolution. *National Geographic* 188(6): 60–89. Wonderfully readable article on sustainable agriculture in the United States.

Mann, C. C. 2008. Our Good Earth. *National Geographic* 214(3): 80–107. Good article with beautiful pictures on the importance of soil and its global protection.

Montgomery, D. R. 2008. *Dirt: The Erosion of Civilizations.* Berkeley: University of California Press. Persuasive argument that soil is humanity's most essential natural resource and directly linked to the survival of past and present civilizations.

Nowak, P., and F. J. Pierce. 2007. The Disproportionality Conundrum. In *Managing Agricultural Landscapes for Environmental Quality: Strengthening the Science Base.* Schnepf, M., and C. Cox (eds.). Ankeny, IA: Soil and Water Conservation Society, pp. 104–111. Great discussion of how agricultural policies and conservation practices need to address the leakier agricultural systems so as to enhance the effectiveness of any conservation efforts in the U.S.

Pimentel, D., et al. 1995. Environmental and Economic Costs of Soil Erosion and Conservation Benefits. *Science* 267: 1,117–1,123. The reference for the comprehensive study of the costs of on-site and off-site erosion and how much it would cost to remedy the effects of erosion.

Reganold, J. P., J. D. Glover, P. K. Andrews, and H. R. Hinman. 2001. Sustainability of Three Apple Production Systems. *Nature* 410: 926–930. Results from a six-year study, as discussed in this chapter, examining the sustainability of organic, integrated, and conventional apple production systems.

Smolik, J. D., T. L. Dobbs, and D. H. Rickerl. 1995. The Relative Sustainability of Alternative, Conventional, and Reduced-Till. *American Journal of Alternative Agriculture* 10 (1): 25–35. One of the best scientific papers to analyze the major sustainability indicators of three different farming systems.

Trimble, S. W., and P. Crosson. 2000. U.S. Soil Erosion Rates—Myth and Reality. *Science* 289: 248–250. Solid discussion of what the scientific data can tell us about soil erosion rates in the United States.

Troeh, F. R., J. A. Hobbs, and R. L. Donahue. 2003. *Soils and Water Conservation for Productivity and Environmental Protection,* 4th ed. Englewood Cliffs, NJ: Prentice Hall. Full coverage of soil erosion and conservation methods.

Web Explorations

Online resources for this chapter are on the World Wide Web at: **http://www.prenhall.com/chiras** *(click on the Table of Contents link and then select Chapter 7).*

INTEGRATED PEST MANAGEMENT

Worldwide, approximately 2.7 million metric tons of pesticides are applied each year to crops, golf courses, lawns, gardens, and other lands, according to the U.S. Environmental Protection Agency (EPA) Office of Pesticides. As its name implies, a **pesticide** is a chemical substance that kills pests. Pests include weeds, insect pests, animal pests such as rodents, and various microorganisms. Pesticides are used primarily to protect fruit and vegetables growing in gardens and in farm fields. They are also used to treat livestock. But pesticides are also sprayed on swamps, pastures, forests, golf courses, lawns, trees, and even entire neighborhoods the world over in an effort to control potentially harmful pests—for example, mosquitos that spread disease.

Despite their popularity early on, pesticides have proved to be a mixed blessing. Concern for their use in the United States began in 1962 after the publication of biologist Rachel Carson's *Silent Spring.* Despite increased awareness of the dangers of chemical-pesticide use and efforts to reduce our dependency on them, total pesticide use has nearly tripled from the early 1960s to the present date in the United States. This increase was largely due to a rise in the use of **herbicides,** chemicals designed to control weeds on crop-land to feed the country's and the world's ever-increasing demand for food. In fact, herbicide use has increased sixfold in that period, while the use of insecticides (chemicals used to combat insect pests) has remained more or less constant. In recent years, pesticide use in the United States has begun to decline slightly, according to the EPA.

In this chapter, we'll examine the use of pesticides. We'll explore the effectiveness of pesticides and the potential risks they pose to people and the enviornment. We'll also examine pesticide regulations and discuss several potentially safer alternatives that are part of a strategy known as *integrated pest management.*

8.1 Where Do Pests Come From?

In undisturbed ecosystems, naturally occurring regulatory mechanisms tend to keep populations of all organisms, including pests, in a dynamic equilibrium (Chapter 15). That is to say, they are generally kept low and in balance. They don't cause problems. Imbalances and troubles begin, however, when humans alter natural systems by plowing up prairies and planting large fields of wheat, or leveling forests and planting a single commercial species (Figure 8.1). Such actions destroy the complex web of life in ecosystems, replacing it with highly simplified systems meant to maximize the production of "useful" species (foods) to meet our needs. Ecosystems that once housed several dozen species, perhaps hundreds, are replaced by artificial systems in which a single species dominates. This is called a **monoculture.**

At least two changes of extraordinary importance occur in such instances. First is the loss of natural insect predators—for example, birds and predatory insects that keep

FIGURE 8.1 Monoculture—a pest's paradise. Modern agriculture is based on the planting of monotypes or monocultures, a vegetation plot made up of a single species of plant, such as the wheat fields shown in this photograph of eastern Washington. Monocultures often cover many hundreds of square kilometers.

potential pests in check. Second, monocultures often present vast expanses of genetically identical plants that provide an enormous supply of food for pests. In the simplified ecosystems, species that were once held in check by predators and by a limited food supply can and often do proliferate rapidly, causing extensive damage. By reducing biological diversity, human civilization has inadvertently unleashed forces that it now struggles to control. In the United States alone, 19,000 agricultural pests exist. Of these, 1,000 are considered major pests (Figure 8.2).

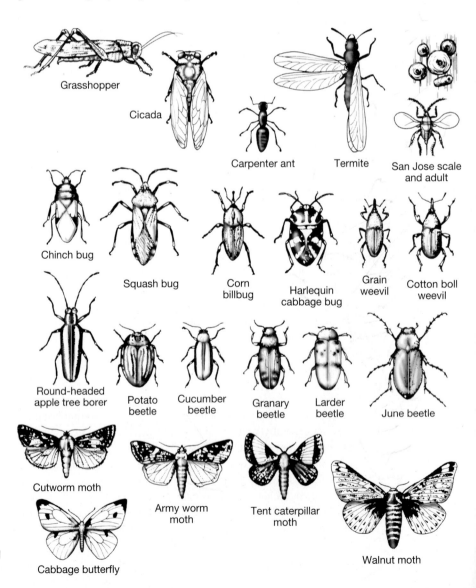

FIGURE 8.2 Insecticides are employed against the harmful species of insects shown here. Only 0.1% of the 800,000 species of insects in the world are considered pests.

Pests may also arise from accidental or intentional introduction of insects or other harmful organisms. In their new habitat, these alien species often face little environmental resistance. Populations explode. Consider some examples.

In 1869, the pupae of the gypsy moth, a native of Europe, were shipped from France to Medford, MA, at the request of French astronomer Leopold Trouvelot of Harvard University. Trouvelot was interested in developing a disease-resistant silkworm moth. Unfortunately, a few of the gypsy moths escaped from captivity and made off into the woods, where they lived in relative obscurity for many years. Twenty years after their escape, however, the town of Medford was crawling with gypsy moth caterpillars—and now so are many other areas (Figure 8.3).

Released from density-dependent control agents such as predators and parasites, which keep this species in check in their native habitat, the gypsy moth population exploded in its new environs. One observer wrote, "The street was black with them...they were so thick on the trees that they stuck together like cold macaroni...the foliage was completely stripped from all the trees...presenting an awful picture of devastation. On a quiet summer night one could actually hear the sound of thousands of tiny mandibles shredding foliage in the trees. Pellets of waste excreted by the larvae rained down from the trees in a steady drizzle."

A single caterpillar can devour a square meter of foliage in a day. Although the larvae prefer oak leaves, they will also consume birch and ash foliage, and when fully grown, they even eat pine needles. Deciduous trees can withstand a single defoliation. Several repeated seasons of defoliation, however, almost always kill them. Why? The loss of leaves eliminates a tree's ability to photosynthesize—to produce the food it needs to grow and survive. Without leaves, plants can't make and store the food molecules they need to survive. Defoliation also makes trees more susceptible to fungi, winds, and drought.

Since its escape from captivity, the gypsy moth has spread throughout the northeastern states and westward into Michigan, Wisconsin, Colorado, and California. In recent years, the spread of the moth to distant forests has been facilitated by recreational vehicles (RVs), on which the moth frequently deposits its eggs.

Fearing further spread, California officials now inspect private and commercial vehicles, such as trucks and RVs, especially those coming from states with gypsy moth problems, at 16 border stations on major highways to check the vehicles for eggs (and other insect pests). California and other states that strictly enforce controls on gypsy moths require people moving to the state from gypsy moth–infested states to be certified free of gypsy moths *before* they can enter the state. Such controls were initiated in the wake of the severe outbreaks that occurred in 1980 in California and the northeastern United States. In the Northeast, trees were defoliated over a 2.5-million-hectare (5-million-acre) area stretching from Maine to Maryland. According to the U.S. Department of Agriculture, since 1980, the gypsy moth has defoliated close to a million acres or more of forest each year. In 1981, it defoliated nearly 13 million acres (5.3 million hectares).

Another exotic or alien species that has had a huge impact on trees is the fungus that causes Dutch elm disease. Accidentally introduced from Europe around 1933, the fungus infects the American elm, a tree that, unlike its Dutch counterpart, is not resistant to this organism.

The American elm is a stately tree that once graced parks, boulevards, college campuses, cities, and suburbs throughout much of the eastern United States. It was considered an ideal species because it was not only beautiful but also long-lived, fast-growing, and tolerant of compacted soils (common in urban environments) and air pollution. Today, most elms are either dead or dying of Dutch elm disease. In one of nature's cruelest ironies, the fungus itself does not kill the tree; the tree kills itself in trying to fight off the organism. Specifically, elm trees produce chemicals to ward off the fungus, but these substances clog the vessels carrying water from the roots to the limbs and leaves. As a result, photosynthesis stops, and the trees die.

The spores of the fungus are spread by bark beetles and also from the root of one tree to the roots of adjacent trees, killing off one tree after another along a city street. Because it was once customary to plant elms in rows along residential streets, the fungus spreads rapidly from tree to tree. Attempts to stop it by killing the beetles with the insecticide DDT only succeeded in killing many robins and other songbirds. By 1976, despite vigorous efforts to stop the disease, it had spread from Massachusetts south to Virginia and west to California. Virtually gone are the magnificent elms in St. Paul

FIGURE 8.3 Leaf-eating caterpillars of the gypsy moth damage hundreds of thousands of dollars' worth of forest and shade trees in the northeastern states annually. They hatch in April from eggs laid the previous year.

and Minneapolis. The lovely elms of Santa Rosa, CA, once a tourist attraction, are now succumbing rapidly, although efforts are being made to save the trees whenever possible, there and in other cities.

Alien species are a major problem in the United States, Canada, and many other countries. In fact, more than half of the weeds in the United States and many of the most destructive insect pests, such as the cotton boll weevil and the Mediterranean fruit fly, are foreigners—species that were sometimes intentionally, sometimes accidentally imported into the United States.

8.2 Types of Chemical Pesticides: A Historical Perspective

In early agricultural societies during the years up to World War II, farmers used a variety of chemical pesticides such as arsenic, ashes, and hydrogen cyanide. Although some of these substances were effective in combating pests, many were ineffective and also highly toxic to people. In 1939, Swiss scientist Paul Müller discovered that a synthetic chemical known as DDT was a powerful insecticide and started a revolution in agriculture with far-reaching impacts on agriculture, people, and the environment.

Chlorinated Hydrocarbons

DDT is a member of a class of chemicals called **chlorinated hydrocarbons**—organic compounds that contain chlorine atoms. DDT's insecticidal activity spurred research that led to the discovery of a string of chlorinated hydrocarbons, such as chlordane, aldrin, lindane, endrin, dieldrin, mirex, heptachlor, and Kepone—all of which have since been banned or severely restricted in the United States. They're all nerve toxins that kill pests by altering the function of their nervous system.

Before being banned in the United States, DDT was widely used. It proved to be extraordinarily effective in killing lice that infested many soldiers in Europe during World War II. It also proved to be effective in controlling the malaria-carrying mosquito in the tropics, so much so that the incidence of malaria worldwide dropped from as many as 50 million cases a year to almost zero. In India, for instance, the number of cases of malaria fell from 1 million per year in the 1950s to just 50,000 in 1961.

DDT also proved to be an effective means of controlling a variety of insect pests. Few people questioned the wisdom of the Nobel Prize selection committee when it announced that Müller would receive the award for physiology and medicine in 1948.

While DDT gained popularity and acclaim, studies showing the biological impacts of DDT and other chlorinated hydrocarbons led some scientists and public-policy makers to question whether the damage caused by these insecticides was worth their benefits.

Scientists found that, first and foremost, DDT and all other chlorinated hydrocarbons persist in the environment for many years, because bacteria lack enzymes needed to

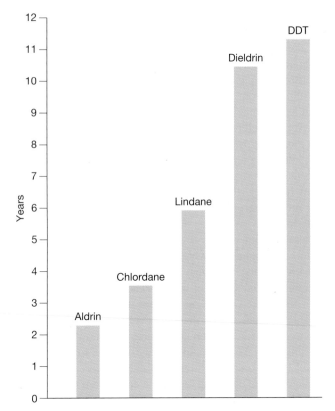

FIGURE 8.4 Average persistence of pesticides in the soil.

break them down. Studies suggest that DDT and its harmful breakdown products remain for 15 to 25 years (Figure 8.4). Today, despite its having been banned in 1972, DDT and its breakdown product, DDE—which is equally harmful—can still be found in the mud on the bottoms of American lakes and rivers.

DDT and similar pesticides are fat soluble and therefore tend to **bioaccumulate**—that is, they concentrate in body tissues, especially fat. Because of DDT's fat solubility, it can remain in fat tissues for decades. Making matters worse, DDT and other chlorinated hydrocarbons build up in food chains, so the highest-level consumers have levels many hundreds of thousands, sometimes millions, of times higher than the environment (Figures 8.5 and 8.6). This process is known as **biomagnification.**

These problems and others, which are described in the section on the hazards of pesticides later in the chapter, caused a furor the world over. As the evidence grew, it became clear that the chlorinated hydrocarbons were indeed too risky to use. Consequently, chemists introduced a new variety of pesticides, the organic phosphates.

Organic Phosphates

Malathion and parathion are the two best known **organic phosphates.** Like the chlorinated hydrocarbons, organic phosphates are neurotoxins. That is, they kill by damaging or interrupting the nervous system of insects. Unlike DDT, however, they are much more quickly degraded in the environment than their predecessors. Unlike DDT and chlorinated hydrocarbons, they are water soluble and thus less

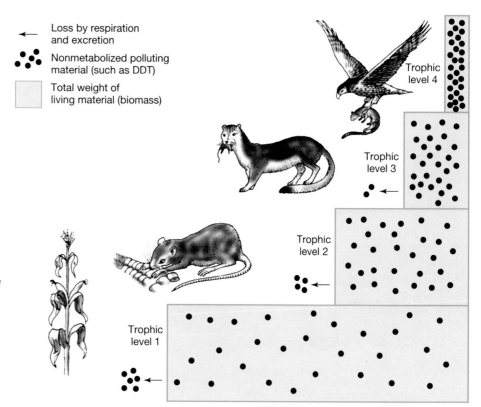

Loss by respiration and excretion

Nonmetabolized polluting material (such as DDT)

Total weight of living material (biomass)

Trophic level 4

Trophic level 3

Trophic level 2

Trophic level 1

FIGURE 8.5   Biomagnification aka biological magnification: the increasing concentration of certain toxic chemicals, such as DDT, in the food chain. A given organism takes in large amounts of contaminated food. Much of the food may not be converted into protoplasm but may be burned up as fuel during respiration or excreted as waste. However, pollutants such as DDT that are taken into the body along with the food may remain inside the cells of the organism. As a result, the concentration of the pollutant increases progressively in the food chain.

likely to biomagnify. Because of these properties, organic phosphates were originally thought to be safe substitutes for the chlorinated hydrocarbons. However, experience soon showed that even at low levels, organic phosphates are hazardous to humans. Low-level exposure, for example, often leads to dizziness, vomiting, cramps, headaches, and difficulty breathing. Higher levels lead to convulsions and death. Symptoms were most common in farm workers and people living around farms.

Like the chlorinated hydrocarbons, many of the organic phosphates have been banned or have been severely restricted in the United States.

Carbamates

To create a safer class of chemicals that biodegrade even more rapidly, pesticide manufacturers developed an entirely new line of chemicals, the **carbamates.** They are nerve poisons, like the chlorinated hydrocarbons and organic phosphates. Perhaps the best known is the commercial preparation called Sevin (carbaryl). Carbamates persist in the environment but for only a few days to two weeks at most. As a result, they are referred to as **nonpersistent pesticides.**

GO GREEN!

When time comes to buy a house, look into natural pest controls for your lawn, trees, shrubs, and gardens. If your parents use pesticides around their home, help them find alternatives or locate companies that rely on natural pest-control techniques.

8.3 How Effective Are Pesticides?

Each year pests, including weeds, insects, fungi, bacteria, and birds, consume or destroy approximately 42% of the annual food production in the United States. By various estimates, pests annually destroy or consume 33% of crops in the field and 9% of food at various stages after harvest. Damage is worse in the tropics, where two or three crops are grown on a field in a single year and where conditions are ripe for insect growth. If this damage could be prevented, it could greatly increase the available food supply. Hence the popularity of chemical pesticides.

But how effective are they?

Studies show that chemical pest controls do indeed work, but not as well as one might suspect. One's perspective on the issue depends on whom one talks to. Conventional farmers and pesticide manufacturers, for example, convincingly argue that chemical pesticides have helped farmers produce much more food than would have been possible without their arsenal of chemicals. In reality, only part of that increase is due to pesticide use. Other factors, such as irrigation, fertilizers, and genetic improvements, have made much larger contributions to production.

According to agricultural economists, each dollar invested in pesticides results in about $2 to $4 in improved yields. The U.S. Office of Technology Assessment estimates that without pesticides, American farmers would lose an additional 25% to 30% of the annual crop, livestock, and timber production (Figure 8.7). David Pimentel, a Cornell University expert on insect pest control, however, believes these figures are exaggerated. A complete ban on pesticides, he estimates, would increase preharvest losses in the United States from 33% lost with pesticides to 45% without them.

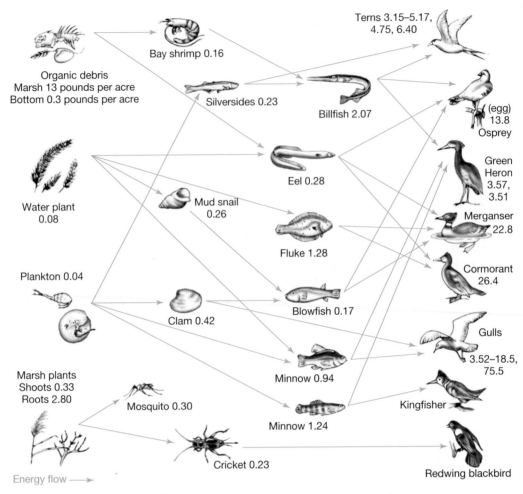

FIGURE 8.6 Food web of the marsh ecosystem off Long Island, NY, which had been sprayed with DDT for mosquito control. Note the biomagnification of DDT in parts per million as it moved up the food web. The greatest concentrations were found in the fatty tissues of fish-eating birds, such as gulls, cormorants, and mergansers.

FIGURE 8.7 Benefits derived from insecticides. Untreated cotton on the left yielded only 0.25 bale per hectare. Cotton on the right, treated with insecticide, yielded 2.5 bales per hectare.

Although pesticides have probably helped to increase crop yields in the United States by less than some assert, they have helped save tens of millions of lives worldwide by killing insects such as mosquitoes, lice, fleas, and tsetse flies. They can carry fatal diseases, including malaria, dengue fever, and the West Nile disease, a type of encephalitis or brain infection. West Nile disease, which is caused by a virus, was unintentionally introduced into the eastern United States in the early 1990s. Today, cases of West Nile infection are reported in Canada and every state in the United States, with the largest numbers occurring in Colorado, North Dakota, and California. In 2007, there were nearly 580 cases in Colorado, nearly 370 in North Dakota, and 380 in California. The virus is spread by mosquitoes to birds to mosquitoes to people and livestock, such as horses. It has killed thousands of birds and hundreds of people, although the disease is usually not fatal.

Unfortunately, in the past decade, scientists have found that the effectiveness of chemical pesticides has steadily dwindled. In fact, despite continued pesticide use, insect damage to U.S. crops has doubled in the past 30 years. Why?

The reason is twofold. First, many insect pests have become resistant to insecticides. Each time a field is sprayed, a small percentage of the insect population survives because it is genetically resistant to the insecticide. This subset of the original population flourishes in the absence of competition. To kill them and their offspring, farmers must spray again at higher doses. The next application kills off much of the once genetically resistant population but leaves behind a small number of insects that are even more resistant. Over time, they may flourish, creating a harder-to-control pest. To combat this genetic strain, higher doses or more frequent applications may be necessary. The net effect is that new, resistant strains develop, and pesticide use skyrockets. This phenomenon is often called the **pesticide treadmill.**

Once on the pesticide treadmill, farmers find it difficult to get off. They may switch to different pesticides to control resistant species, but genetic resistance to the new formulation invariably starts them on the treadmill again. In the 1960s, Nicaraguan farmers sprayed cotton fields 5 to 10 times a year. Today, these same fields are sprayed 30 times a year to control insect pests genetically resistant to Malathion.

Today, over 550 species of insects and mites are resistant to at least one insecticide. More than 20 species are resistant to most types of insecticide. Farmers are also finding that weeds and a variety of organisms that cause crop diseases are becoming resistant. By one estimate, 230 plant diseases and 220 weed species have developed resistance to at least one chemical designed to control them. What is most alarming is the fact that the number of pesticide-resistant insect species is increasing and increasing fairly rapidly.

The rise in insect damage previously noted also results from the destruction of **beneficial insects**—that is, natural predators that help control pest populations. Numerous species, such as ladybugs, praying mantises, spiders, and wasps, rid farm fields and forests of insect pests free of charge. By accidentally poisoning these beneficial species with pesticides when attempting to control insect pests, farmers destroy helpful natural allies that are a part of environmental resistance. Populations of pest species, now unleashed from their natural controls, may surge. In California the spider mite was once an innocuous insect held in check by its natural predators. Today it is the state's leading insect pest. How did it become such a pest? The answer is that pesticide use destroyed its natural predators.

One of the consequences of genetic resistance is that farmers the world over spend tens, if not hundreds of millions of dollars more each year on pesticides to battle resistant insect pests. They also spend tens, if not hundreds of millions more each year to destroy pests whose natural predators have been destroyed by pesticide. Pesticides also destroy native bees and bee colonies in the United States. Bees pollinate many commercial crops, such as fruit trees, worth an estimated $14 billion per year. The destruction of these pollinators reduces crop yields and costs farmers many millions of dollars in lost production.

8.4 How Hazardous Are Pesticides?

Pests cause an enormous amount of economic damage each year. By one estimate, rodents, weeds, and insects in the United States annually cause $2 billion, $5 billion, and $7 billion worth of damage, respectively. As noted earlier, in the United States, pests of various sorts consume or destroy about 42% of the annual food production. Ten percent of the average annual cotton crop in the United States is destroyed by a single insect species, the cotton boll weevil (Figure 8.8). According to the U.S. Forest Service, pests destroy 5 billion board-feet of timber annually.

Controlling such pests is imperative, but so is an understanding of some of the problems posed by chemical pesticides. Such an understanding will help us create a sustainable approach to pest management.

Each year Americans apply 560 million kilograms (1,230 million pounds) of pesticides to their farms, forests, golf courses, roadways, rivers, lawns, and gardens (Figure 8.9).

FIGURE 8.8 Cotton boll weevil attacking cotton boll. This weevil destroys 10% of the average cotton crop in the United States. Pesticides have been used to control its populations.

FIGURE 8.9 Many acres of crop and forest are sprayed with pesticides each year to protect against insects. Unfortunately, much of this potentially toxic spray drifts into neighboring areas.

That is about 3.3 pounds per acre (600 grams per hectare), if it were evenly applied to all land in the United States. In reality, however, not all land is treated. Sprayed fields, for instance, may receive 3 to 18 times that amount. In addition, according to the EPA, homeowners often apply pesticides at a rate 5 to 10 times greater than farmers.

Pesticides end up on American soils but also in our water, our food, our wildlife, and people. It is this widespread contamination that worries biologists and health officials. Are these concerns justified?

Human Health Effects

Western Colorado fruit grower Dorsey Chism's health took a turn for the worse in 1984. Once a happy, optimistic man, he became irrational and depressed. His face and body bloated, and he was constantly short of breath. Local doctors thought he had emphysema.

In August 1984, after years of spraying his fruit trees with pesticides, sometimes without wearing a protective mask or clothing, Chism went into convulsions and was rushed to Aspen Valley Hospital, where he spent 14 days in intensive care. A doctor familiar with the health effects of chemical toxins diagnosed his illness as chronic pesticide poisoning. Chism died in 1985 after months of lingering near death, hooked up to an oxygen tank.

Chism is not alone. According to Lewis Regenstein, author of *America the Poisoned,* at least 100,000 Americans, mostly farmers and farmworkers, are poisoned each year. The National Coalition Against the Misuse of Pesticides puts the number higher, at 300,000; however the National Agricultural Chemicals Association, an industry group, claims that the figure is no higher than 20,000 a year. Most people affected are farm and factory workers who have been exposed to high levels of the agents.

Unfortunately, no precise figures on farmworker poisonings are available—the government does not keep them. Nonetheless, there is good reason to believe that the higher estimates are accurate. For instance, in California, the only state that tracks pesticide poisonings, 1,400 people are poisoned seriously enough to be reported each year. These official figures may grossly understate the real rate. Many cases may go unreported. By some estimates, less than 1% of California's poisonings are reported.

All told, 200 to 1,000 Americans like Chism die each year from pesticide poisoning. Worldwide, half a million people are poisoned by pesticides. These poisonings result in an estimated 5,000 to 14,000 deaths and numerous chronic and fatal illnesses.

One of the most vulnerable groups is farmworkers in less-developed nations. They are often exposed to high levels of pesticides as a result of lax or nonexistent laws and regulations and a lack of training and safety precautions on the part of employers. Even illiteracy in the LDCs contributes to the problem, as illiterate workers are unable to read safety instructions on packages.

Another problem of concern is the ingestion of pesticides in foods. Some doctors believe that general maladies, such as dizziness, insomnia, indigestion, and frequent headaches, may be caused by ingesting pesticides in food. This prompted a *U.S. News and World Report* journalist to write, "When an American sits down to a typical breakfast, chances are the menu includes bug spray, weed killer, an embalming agent, and arsenic." Dr. Marshall Madell summed up this situation aptly: "Future historians will say we were foolhardy. We sprayed our food with poison, then ate it."

In 2004, the Pesticide Action Network North America (PANNA), a nonprofit organization, published a report (*Toxic Trespass*) that analyzed data collected by the Centers for Disease Control and Prevention (CDC) on pesticide levels in Americans. According to the report's authors, "The results showed that many U.S. residents carry toxic pesticides in their bodies at levels above the government's 'acceptable' thresholds." The authors warn, "Many of the pesticides found in the test subjects have been linked to serious short- and long-term health effects, including infertility, birth defects, and childhood and adult cancers." The authors go on to say, "The effects of pesticides may not be immediate, and when they do occur, they may be subtle (or blamed on other causes). How much pesticide exposure a child can take before it begins to affect him or her varies—factors include the level of exposure and the type

GO GREEN!

Eat organic fruits and vegetables. Several studies show that organic fruits and vegetables—those grown without pesticides and synthetic fertilizers—have higher levels of vitamins, minerals, essential fatty acids, and antioxidants. Studies also show that eating organic fruits and vegetables lowers blood levels of pesticides, often quite dramatically. Scientists don't know if this will lower your risk of disease, but it will lower the amount of pesticide circulating in your body.

of chemicals involved, as well as the child's constitution and nutritional status."

A study by the National Academy of Sciences estimates that pesticides contaminating the most common American foods could cause as many as 20,000 cases of cancer per year in the United States, costing $1.1 billion in health care costs in 1980 dollars. Tomatoes, beef, potatoes, oranges, lettuce, apples, and peaches topped the list.

Pesticide residues on food may be minute, but these chemicals can accumulate in body tissues and may cause cancer, birth defects, and other problems. In 1970, a sampling of 1,400 Americans showed they had nearly 8 parts per million (ppm) of DDT in their body fat (Figure 8.10). In 1983, 11 years after DDT was banned, the levels were 1.67 ppm. What the long-term effects are, no one knows. Many pesticides long since banned or restricted persist in our bodies for years. Continued exposure to chlorinated pesticides may lead to cancer. According to one study, exposure to certain chlorinated pesticides may have led to a doubling in the incidence of breast cancer in Danish women since the late 1960s. Women with the highest concentrations of the estrogen-like pesticide dieldrin, which is banned in the United States, were more than twice as likely to develop breast cancer as those whose blood levels showed little or no pesticide.

Pesticides may also damage children. Mexican children belonging to the Yaqui Indian community who are exposed to heavy doses of insecticides used on nearby farm fields and in their homes appear to suffer from a dramatic impairment in development, according to one study. In this area, individual crops may be sprayed 45 times from planting to harvest, and two crops may be grown each year. The researchers found children who lived in the region where pesticide use was common, while showing no signs of pesticide poisoning, performed poorer on tests of recall, stamina, gross and fine eye–hand coordination, and drawing ability than children from areas that were relatively pesticide-free. Some researchers worry that effects could be irreversible and that children in other tropical areas where pesticide-dependent crops are grown year round may suffer enormously.

Herbicides, applied to fields to destroy weeds, also have been linked to a growing number of problems. For instance, the herbicides 2,4-D and 2,4,5-T have been linked to one form of cancer in farmworkers. Shelia Hoar, a University of Kansas researcher, found that farmers and farmworkers exposed to the herbicide 2,4-D 20 days or more per year were six times more likely to develop non-Hodgkin's lymphoma (a type of cancer) than nonfarm workers. She also discovered those who were most in contact with the chemical—for example, those who mixed the chemicals—were eight times more likely to develop this disease than nonfarm workers. In 1978, Bonnie Hixl, a young mother who lived in Alsea, OR, complained to the EPA that she and seven other women had miscarried ten times in five years. Researchers sent to study the problem found that spontaneous abortions occurred most frequently in these women shortly after the forests were sprayed with the herbicide 2,4,5-T to control brush growth. Although it could not be sure that a cause-and-effect relationship existed, the government banned the use of this chemical for brush control.

Far better known are the multitude of health problems that have resulted from the use of **Agent Orange,** a 50:50 mixture of 2,4-D and 2,4,5-T, during the Vietnam War. These chemicals were used to defoliate trees along rivers and roadways and around camps to reduce the chances of ambush of U.S. soldiers and their allies. In some cases, the chemicals were sprayed on crops that could have been used to feed enemy soldiers.

FIGURE 8.10 Avenues for the dispersal of pesticides, such as DDT, once they are released. Average values for DDT concentrations are indicated in parts per million (ppm) and parts per billion (ppb).

Millions of kilograms of Agent Orange were sprayed during the war, and it was not too long after aerial sprayings commenced that problems began to arise. U.S. soldiers heavily exposed to Agent Orange complained of headaches, nausea, dizziness, and diarrhea. Many men were afflicted with an irritating skin rash, called chloracne, over large parts of their bodies. Others became uncontrollable and depressed.

Studies showed that Agent Orange was contaminated with a toxic substance known as **dioxin,** which is now believed to be largely responsible for the many health problems reported by soldiers and Vietnamese citizens. Dioxin is a potent toxic substance that causes birth defects and cancer in mice and rats.

In 1969, a Saigon newspaper reported that Vietnamese soldiers and villagers had developed serious health effects, and linked them to the chemical defoliants liberally sprayed on their land. The newspaper report claimed that miscarriages and birth defects had increased in villages. Initially dismissed as propaganda, the report soon caused a furor in the United States, prompting steps to ban the use of Agent Orange in the war and to ban some domestic uses of 2,4,5-T.

Soon, doctors found that rates of certain cancers (for example, testicular cancer) were higher in Vietnam veterans than in the general public and that many men who had been exposed to Agent Orange fathered children with birth defects. For many years, though, health officials denied the validity of veterans' complaints of dizziness, attributing many of them to stress caused by the war. Growing evidence, however, strongly suggests that Agent Orange was indeed the culprit. One study of 40,000 Vietnamese couples, performed by Vietnamese doctors, showed that women whose husbands had fought in areas sprayed with the defoliant were 3.5 times more likely to miscarry or give birth to babies with birth defects than women whose husbands had been lucky enough to avoid the regions.

In 1984, the Veterans Administration (VA) reached an out-of-court settlement with U.S. veterans. The VA established a $180 million fund to compensate victims.

Recent studies suggest that dioxin may also suppress immune functions. One form of dioxin, TCDD, suppresses the immune system in mice at least 100 times more effectively than corticosterone, one of the body's glucocorticoids. In addition, studies indicate that TCDD may bind to the cell membrane and cytoplasmic receptors that normally bind to hormones, upsetting body functions. Some researchers are calling dioxin an *environmental hormone.*

TCDD also appears to have a variety of direct biological effects—not necessarily involving hormones. In some cells, it causes rapid cell growth. Nonetheless, its effect on the immune system may be far more important than its impact on cancer.

Studies suggest that a number of common herbicides (the thiocarbamates) may upset the thyroid's function and result in the formation of goiter, enlargement of the thyroid, and at higher doses, thyroid cancer. (The thyroid gland is located in the neck and produces hormones that stimulate metabolism.)

Pesticides are now becoming a major health concern in cities and suburbs, where they are used on trees, gardens, parks, and golf courses to control insect pests. For years, pest-control workers have sprayed suburban areas to control mosquitoes, but usually at night, so only an occasional protest was raised. The lawn care industry, now a lucrative endeavor worth several billion dollars a year, also is responsible for poisonings. One woman, for instance, was sprayed by a careless applicator as she rode by on her bicycle. Her tongue went numb, and her head ached for three days after the incident. Children are generally more sensitive and sometimes experience extreme reactions to pesticides, including difficulty breathing. Playing on a recently treated lawn can cause severe chemical burns as well. Adults suffer from headaches, dizziness, and nausea after their lawns are sprayed with pesticides.

In 1982, Navy Lt. George Prior, an avid golfer, died after a severe reaction to a chemical pesticide (chlorothalonil) that had been applied to a golf course. In response to this incident and to reports of bird kills and groundwater contamination, the EPA launched a study of pesticides on golf courses. The EPA's report noted that the United States' 13,000 golf courses annually receive 5,500 metric tons (12 million pounds) of 126 different pesticides applied to control weeds and insects.

As these and other examples show, pesticides pose a risk to the health of the world's people, especially chemical workers and farmworkers. People living near farms also may be exposed to toxic levels, because 50% to 75% of the spray applied by planes and other methods drifts off target, contaminating surrounding fields and homes. It can travel even further.

Although these and other examples suggest that pesticides are highly toxic, many toxicologists note that the overall impact on human health of toxic chemicals released into the environment is probably small, especially compared with the impact of other risk factors, most notably tobacco smoke. Natural carcinogens may be far more harmful than chemicals applied intentionally to our food crops. Nevertheless, an impact on humans is only one part of the complex ecological equation. We must also consider impacts on other species.

Effects on Fish and Wildlife

Pesticides affect fish and wildlife in many ways, often profoundly. Consider **tributyl tin (TBT),** for example. TBT is a potent biotoxin added to paint that is applied to the hulls of ships to prevent the buildup of marine algae and barnacles. These organisms increase a ship's drag and therefore decrease its speed and fuel efficiency. At one time, before high fuel costs, the U.S. Navy estimated that if the entire fleet were painted with TBT-containing paints, its fuel bill would fall by 15%, saving well over $150 million a year. (You can only imagine the savings now!) Underwater cleanings also would be reduced, saving additional money.

Unfortunately, TBT is released from the paint and pollutes bays and harbors. In France, TBT from recreational watercraft caused massive reproductive failure in commercially important oyster beds in 1980 and 1981. Banning the use of this chemical on pleasure craft less than 25 meters long resulted in an abrupt turnabout. Oyster reproduction resumed in 1982 and has continued ever since. In sections of San Francisco Bay, TBT levels were once as high as 500 parts per trillion. Mussels, barnacles, and other marine organisms vanished from the waters as a result. These and other findings led the United States, France, and Australia to ban the use of TBT paint, although it is still used on large oceangoing vessels and trace amounts of it are found in virtually all major harbors.

Pesticides and other chemical pollutants are also believed responsible for an epidemic of fish cancers in waters of the United States. In Puget Sound, for instance, 70% of the English sole have liver cancer. Similar findings have been made in a number of U.S. rivers.

Pesticides—specifically insecticides—may also be partly responsible for the worldwide disappearance of amphibians. Studies in the laboratory show that exposure of eggs and tadpoles of two species of frog and one species of toad to varying concentrations of a common insecticide (endosulfan) at concentrations similar to those found in ditches and other breeding sites near farm fields had profound effects on behavior. In particular, exposure caused a depressed avoidance response, making hatchlings potentially more susceptible to predation.

As noted earlier, pesticides kill beneficial insects, such as honeybees and praying mantises, and birds. In the 1960s and 1970s, DDT was responsible for drastic reductions in the populations of a number of predatory birds. Hardest hit were peregrine falcons, ospreys, brown pelicans, and bald eagles (Figures 8.11 and 8.12). Biomagnified in the food chain, DDT reached high levels in fish and insect-eating birds. Being highest on the food chain, predatory birds that prey on fish and other birds ended up with extremely high concentrations of DDT in their tissues. DDT levels in the predatory birds, however, were not high enough to kill adults, but they were high enough to decrease calcium deposition during eggshell

FIGURE 8.12 America's national symbol, the bald eagle, has been on the official list of endangered species. Its decline was correlated with the amount of chlorinated hydrocarbon pesticides, such as DDT, occurring in its environment. Now that DDT has been banned, the eagle is recovering nicely.

formation. As a result, the eggshells of DDT-contaminated peregrine falcons and other species became thin and were easily broken by the parents (Figure 8.13). The fragile eggs cracked during incubation and killed the embryos. As a result, hatching decreased. By the time scientists had discovered the problem, none of the 200 breeding pairs east of the Mississippi River was able to produce young. Peregrine populations in Europe and the western United States had fallen by 60% to 90%. By 1970, the outlook for this regal bird was bleak. The bald eagle and osprey were similarly threatened.

Today, thanks to captive-breeding programs, hundreds of peregrine falcons have been released into the wild. Free of DDT, the birds may well be able to reestablish their population. Ospreys, bald eagles, and brown pelicans also have made remarkable recoveries.

DDT, which was used in a vain attempt to kill elm bark beetles thought to be responsible for the spread of Dutch elm disease, has also been linked to the deaths of thousands of robins and other insect-eating songbirds. Many birds died in convulsions shortly after eating insects from freshly sprayed trees. Others died from eating earthworms that had been contaminated by DDT. How were earthworms contaminated?

Researchers found that DDT remains on the lower sides of leaves throughout the summer, despite rains. The leaves shed in the fall, decomposed, and released their DDT to the soil. It was taken up by earthworms, which feed on organic matter including leaves in the soil. A hungry bird returning in the spring would succumb from eating only 11 worms—an hour's snack for a hungry robin. Up to 744 ppm of DDE has been found in the tissues of dead robins.

Birds have also been poisoned by carbofuran, a pesticide that eradicates insects and other pests from corn, rice, and other

FIGURE 8.11 Down and out? Pesticides are responsible for the drastic decline of the peregrine falcon population in the United States. Captive-breeding programs and a ban on DDT have helped this magnificent bird recover.

FIGURE 8.13 Newly hatched bald eagle in nest. One egg has not hatched. DDT contamination of the eagle's food chain has caused an eggshell thinning. Sometimes the embryos inside the abnormal eggs are crushed under the weight of the incubating female.

crops. Although this pesticide apparently poses no threat to humans when applied to crops, it is lethal to songbirds, even in minute quantities. In the late 1980s, approximately 2 million birds died from carbofuran poisoning each year, according to EPA records. Poisonings of wildlife and people continue in areas where this pesticide is still used.

In the United States, more than 36.4 million kilograms (80 million pounds) of chemical pesticides are used each year by homeowners to control pests on our lawns and gardens and in our homes. Commercial and industrial use of pesticides to control pests on lawns, golf courses, and other areas comes to 97 million kilograms (213 million pounds) a year, according to the EPA. This includes use by commercial pesticide applicators who apply chemicals to lawns in suburban and urban settings.

In summary, although chemical pesticides may reduce pests and save crops, they do pose a threat to humans and wildlife, even domestic animals.

8.5 Are Pesticides Adequately Regulated?

Pesticides are commonly used in our chemical- and energy-intensive system of agriculture and will be here for many years to come. In the meantime, society must be sure that pesticide manufacture and use are properly regulated. Unfortunately, critics argue that the current system of regulation is inadequate in the United States and abroad.

Federal Regulation

In the United States, formal regulation of pesticides at the federal level began in 1947, when the Congress passed the **Federal Insecticide, Fungicide, and Rodenticide Act (FIFRA).** This law requires manufacturers to register pesticides being shipped across state borders with the USDA, and requires them to be labeled as well. No attempt was made to control the use of pesticides or to limit potentially harmful chemicals. In actuality, the law provided little protection.

Because of an outpouring of public concern regarding the harmful effects of pesticides in the 1960s, Congress amended FIFRA in 1972, 1975, and 1978. Broadening the scope of the act, the amendments required chemical companies to submit to the EPA information on all new pesticides intended to be used in the United States. Each application specifies the crops and insects on which the pesticide is to be used, supported by research data. The EPA then analyzes the costs and benefits of each pesticide and approves pesticides that are deemed effective and safe. These pesticides are technically referred to as registered. Those that are not considered safe or worth the risk are denied registration and are effectively banned from use.

To gain some measure of control over the use of pesticides, FIFRA also requires the EPA to classify pesticides as either general or restricted. **General pesticides** can be used by anyone. **Restricted pesticides** can be applied only by state-certified applicators—special applicators or farmers who have been certified. Farmers are required to obtain a permit to apply a restricted pesticide and must keep records of the date, the type of pesticide used, the crop, and the amount applied. In addition, the EPA or local agricultural department can inspect applicators' facilities to be certain that pesticides are being used according to directions.

Although this may sound good, critics argue that certification is lax and inspections are rare. To become certified, applicators must take a course or read a book on pesticide use and then take a test, often an open-book test. Certification, say critics, does not ensure that applicators will use the product as instructed.

FIFRA permits the EPA to cancel its registration of pesticides when new information indicates they are highly likely to threaten human health and the environment. Cancellations may require months or years of red tape. In the meantime, production and sale can continue. If, however, the EPA believes that a chemical being considered for cancellation is imminently hazardous to human health and the environment, it can suspend its use while the cancellation procedure lumbers on. In this process, the EPA weighs the economic, social, and environmental benefits against all costs.

The United States has taken the lead in pesticide management. On December 31, 1972, for instance, the EPA officially banned all uses of DDT except for emergencies. In the years since the ban, the concentration of DDT in the soil, water, and wildlife has declined substantially.

In August 1974, after two years of hearings, the EPA also banned the general use of two other chlorinated hydrocarbons, aldrin and dieldrin, considered by some experts to be even more toxic than DDT. The EPA also suspended the use of heptachlor, endrin, lindane, Kepone, toxaphene, and other pesticides in the same family as DDT.

Is the Public Adequately Protected?

Despite laws and regulations on U.S. pesticide manufacturing and production, these substances continue to cause problems. As noted in the previous discussion, one of the biggest problems is the enforcement of safe use. The EPA dictates who can apply what chemicals, but enforcement is generally left to the states, and the level of enforcement varies widely from state to state. Poorly educated farmworkers are especially at risk. Unable to read and understand labels, they often misuse products or are not provided with protective clothing and headgear. Farmers and farmworkers often apply chemicals excessively without protective gear, and often suffer the consequences. In 1996, two Louisiana men were arrested for applying Malathion to dozens of houses for pest control. They did not have a permit and sprayed this highly toxic substance liberally in the homes, causing considerable sickness.

Another problem comes with unanticipated adverse effects. Invariably, products once deemed safe on the basis of toxicity data provided for EPA registration can turn out to have adverse effects on human health and wildlife that were not anticipated by EPA officials. In such instances, public outcry may be needed to convince the EPA to reverse a pesticide's registration.

Since 1945, approximately 1,500 chemical pesticides and more than 35,000 different formulations have been introduced to the U.S. pesticide market. A U.S. congressional investigation found that 60% of those in use lacked adequate information on their potential to cause birth defects, 80% lacked adequate cancer data, and 90% had not been sufficiently tested for possible mutations. Insufficient testing remains a major stumbling block in the U.S. government's efforts to protect human health. Another major problem is accurately monitoring pesticide levels in the food Americans consume every year, 15% of which is imported. Studies have shown that small percentages of domestically produced and imported food are contaminated by unacceptable levels of pesticides.

Closing the Circle of Poisons   Nine years after the EPA's ban on DDT, USDA officials turned back a shipload of beef headed for U.S. consumers from Central America because the meat contained unacceptable levels of DDT. This action pointed up a major problem in U.S. pesticide policy: Substances banned here were being sold abroad and reimported in food products, something called the boomerang effect or the circle of poisons.

Today, a dozen chemical companies supply about 90% of the world's pesticides. These European and American companies often manufacture and sell pesticides that are banned or restricted in their country of origin to less-developed countries, which use 30% of the world's pesticides and are the fastest-growing market. When the substances started showing up in food imported from these countries and began to be detected in migratory birds that winter in the warm equatorial countries, many people grew alarmed.

Because of pubic interest, Congress amended FIFRA. Under the new amendment, the EPA is required to notify all governments and international organizations worldwide each time it cancels or suspends a pesticide's registration. The EPA also requires manufacturers and exporters of pesticides that have been banned or have failed registration in the United States to notify a foreign purchaser of the status of the pesticide. Other countries now send out similar notifications.

Notification is only a small step, however, in protecting the farmers and farmworkers of the less-developed nations. Countries must have an intact system of internal regulation as well. However, a recent survey by the International Pesticide Industry Association found that only 51 countries place strict controls on pesticides: 43 have less stringent controls, and 41 have no controls whatsoever. Mistakes, carelessness, and ignorance resulting from lack of information or from poor training are responsible for the injuries to farmworkers and deaths. Even though container labels describe ways to apply the chemicals safely in the language of the importing country, farmworkers often ignore the warnings or may not be able to read them.

Controlling the Lawn Care Industry   Although many view agricultural pesticide use as a serious problem, homeowners and lawn care companies apply far more pesticide per acre than farmers. Throughout the nation, citizens have been pressing their representatives to regulate the lawn care industry—and with growing success. Numerous states and a host of cities, towns, and counties in the United States have passed right-to-know laws and ordinances that require applicators to notify residents in advance of sprayings and to post warning signs after application. Maryland, Rhode Island, Massachusetts, Minnesota, Iowa, Oregon, and numerous other states have right-to-know laws. Some critics say that this is not enough. Signs are too small and placed at ground level, making them difficult to see. The information they contain is vague and not really a warning.

To avoid possible problems, some companies in the lawn care industry have made the switch away from pesticides to safer and more people- and environment-friendly pest-control strategies, known as integrated pest management. Chem-Free Lawns, Inc. of Lancaster, NY (outside Buffalo), for instance, has a thriving business that relies on naturally occurring pesticides and biological controls. "It all works a little slower," says their president, Jay Kolby, "but in the long run it's actually better, you get a better lawn."

Modern society has come a long way in regulating pesticides, but much more needs to be done. Stronger laws, much better enforcement, and worker education are badly needed, especially in the LDCs.

Many experts also believe that we must reduce our use of pesticides, creating a sustainable and safe pest-control strategy, discussed next.

8.6   Sustainable Pest Control

Pesticides pose many problems for people and the environment. Fortunately, there are ways to reduce—even eliminate—their use to create a more sustainable system of agriculture, livestock production, silviculture, lawn care, and home gardening.

Reducing and Eliminating Pesticide Use by Careful Monitoring

Farmers the world over are finding many ways to reduce pesticide use, saving enormous sums of money in the process. One discovery they have made is that they have been applying pesticides far in excess of what is needed to do the job.

One of the main reasons for such widespread overuse is that many farmers fail to perform careful analyses prior to pesticide application. In other words, they do not determine pest population levels or the distribution of pests in their fields. A cursory examination of their crops may turn up a pest or some pest damage, which triggers an automatic response: spraying a whole field.

In many cases, farmers apply chemical pesticides on the basis of predetermined schedules, often established by chemical-pesticide salespeople. They do this without first checking to see if there is an insect problem.

With costly crop loss at stake, farmers can hardly be blamed for such cautious steps. However, with the cost of farming escalating and the environmental impacts of pesticide use becoming widely known, some farmers are abandoning set schedules, applying pesticides only when needed. Farmers are also assessing whether the presence of an insect pest represents a real threat. For example, they may find that a pest infestation is limited to a small portion of the field, which is then sprayed. This reduces overall pesticide use. Or, they may find that the pest populations are at a level that will not create problems, so they do not spray at all. Farmers are also applying pesticides much more sparingly, reducing use by 50% with no increase in crop loss.

For a discussion of ways that GIS and remote sensing are being used to monitor pest damage in crops and forests, see the GIS and Remote Sensing boxes in this chapter and Chapter 7.

Integrated Pest Management

Farmers are also adopting new and more sustainable measures to control insect damage. So important are these that the USDA's pest control program now spends nearly 70% of its money for pest control on nonpesticide approaches. Even some large chemical companies, such as Monsanto, have begun to look for potentially less harmful ways to control pests.

Pest control today often takes an integrated approach, known as **integrated pest management (IPM).** IPM capitalizes on four major strategies: environmental, genetic, chemical, and cultural controls.

Environmental Controls   Environmental controls are measures that alter the biotic and abiotic environment of the pest. The most important means of environmental control are (1) crop rotation, (2) heteroculture, (3) trap crops, and (4) the use of natural predators, parasites, and disease-causing organisms. These measures are often simple, highly effective, inexpensive, and environmentally benign, and they all are preventive in nature.

Crop Rotation   In Chapter 7, we noted that farmers use crop rotation to reduce soil erosion and increase soil fertility. It also helps control pests. How?

Many pest species are highly specialized; they feed on one or a few species of crop plants. For example, the alfalfa weevil feeds mainly on alfalfa, whereas the corn rootworm feeds primarily on corn, and so on. If a farmer plants corn on the same plot year after year, the corn rootworm population grows. Damage can be costly. If, however, a farmer plants corn and oats in alternating years, rootworm populations are kept low. During the years in which the plot is oats, the corn rootworm's food supply is greatly diminished. Food therefore becomes a limiting factor. As a result, crop loss tends to remain low without the use of chemical pesticides.

Heteroculture   Monoculture (single-crop) farming simplifies the ecosystem, creating something of a "Garden of Eden" for pests, as noted earlier in the chapter. **Heteroculture,** or planting several crops on a farm, reduces food supplies for pests and consequently reduces pest outbreaks. Heteroculture is often combined with crop rotation. Pests stand little chance against this combination. Not only are their food supplies limited, but the supplies change from year to year, so that pest populations cannot get firmly established.

A novel approach to heteroculture is called **intercropping,** planting two different crops in alternating strips. For instance, intercropping corn and peanuts has been shown to reduce corn borers by 80%. Intercropping reduces the amount of food available for specialized insect pests. However, researchers also believe that this combination works because the peanuts provide habitat for predatory insects that feed on corn borers.

Intercropping can also dramatically increase crop production. In Nebraska, for example, farmers intermix strips of soybeans and corn. This opens up the corn patch, greatly increasing the amount of sun that strikes each plant and raising production by a remarkable 150%. The corn protects the soybeans from the drying effects of the wind and increases output by about 11%. So, not only do farmers reduce their pesticide use, which saves them money, they also increase their production, which makes them more money per hectare!

Trap Crops   Farmers can lure insects from commercially valuable crops by planting low-value **trap crops** nearby. For example, alfalfa can be used to lure the lygus bug from cotton, where it can cause substantial damage. In Hawaii, fields of melons and squash are bordered by rows of corn, which attract melon flies. Insecticides can then be sprayed sparingly on the trap crop to kill pests. Or, to avoid pesticide use, farmers may plow under the insect-infested trap crop or burn it. In Nicaragua, a major exporter of cotton, farmers are required by law to plow under cotton stalks after harvest to reduce boll-weevil damage. Many also plant postharvest trap crops of cotton to attract the harmful boll weevil, which is then destroyed by insecticides. Insecticide use is greatly reduced, and environmental contamination is minimized.

GIS AND REMOTE SENSING

USING SATELLITE REMOTE SENSING TO DETECT PEST DAMAGE IN FORESTS

Each year, insects cause enormous damage to the forests of the world. Measuring this damage is vital to forest management but is time-consuming and expensive. Recent developments in GIS, however, could make the task much simpler and much more precise. To understand the new techniques, let us look first at the old.

For years, forest managers relied on manual mapping of forests to indicate the extent and severity of insect damage. Information from field inventories was simply transferred onto paper maps. Forest managers then turned to a more sophisticated technique, known as aerial sketch mapping. In this technique, trained observers flew over forests and estimated the extent and severity of tree damage visually from the cockpit of planes or helicopters. This data was then transferred to paper maps.

Today, researchers and forest managers have begun to explore a more high-tech and precise solution: remote sensing and GIS. The process begins with infrared satellite images of forests. They are used to assess insect damage, which, in turn, is used to create forest management plans.

One way satellite images can be used to assess damage is by comparing photos taken from different times side by side. With side-by-side infrared photographs to work from, forest managers can examine changes in forest cover over large areas. A decline in green on the more recent satellite image, for example, indicates a loss of tree cover from insect damage. An increase in magenta (bright pink) indicates the possibility of a decrease in vegetative cover. From this qualitative examination, researchers can estimate the amount of change that has taken place over time.

Another way of assessing damage over time is by combining multitemporal imagery (images from different times)

to create a single image, which scientists call a change detection image. A change detection image shows areas where vegetation is lost or gained or where it has not changed at all. Although more useful than the previous method, it still provides a qualitative look at change. In other words, no measurements of actual change are represented.

For a more precise, quantitative assessment of change, researchers have developed software that generates a change detection map—an actual map of an area that shows the percent of change in the vegetative cover of the landscape. Although this technique may sound complicated, it is basically fairly simple. Scientists use an automated image-processing technique that subtracts pixel values of one satellite image from the pixels of a second image taken at a later time. Land cover changes are represented by differences in the pixel values and are shown on a second image, the change detection map. This map shows the approximate range of forest cover loss.

Using remote sensing (satellite images) and GIS equipped with computerized digital imaging processing software, forest managers can obtain fairly accurate depictions of the amount of forest cover lost to insect damage over time in large regions. This technique is cost-effective and easy to update.

From this map, forest managers can devise forest management plans that stipulate possible salvage logging operations as well as forest fire prevention and suppression. This new approach also helps managers assess the impacts of insect damage on future harvests, wildlife habitat, and recreational uses of forests. It can also be used to assess other important aspects of the forest, such as habitat change, fire hazards, and forest health.

Introducing Predators, Parasites, and Disease-Causing Organisms By alternating crops, intercropping, and planting trap crops, farmers can greatly reduce and in some cases even eliminate pesticide use. Reducing pesticide use helps restore bird populations, which may feed on insect pests.

Farmers can also intentionally introduce biological control agents to help reduce pest problems. The deliberate introduction of natural control agents, such as predatory insects, is one form of environmental control called **biological control.** This approach artificially alters the biotic environment of a pest, creating a more natural condition in which the environmental resistance on the pest population keeps it under control. Moreover, once a natural predator or parasite is established in the more complex farm ecosystem, it may survive and prosper from year to year, holding pest species in check without the use of costly pesticides (Figure 8.14).

Several hundred biological control species—insects, viruses, and bacteria—are currently used to control pests on a

wide variety of crops. In California, for instance, entomologists from the University of California, Davis, introduced several parasitic wasps from the Middle East to control a pesty insect called olive scale, which once caused severe and costly crop damage. Today, these natural enemies have managed to keep the olive scale completely under control. In Colorado, peach growers have long enjoyed the protection of a natural insect predator that is released in their orchards by state officials. The predator is released each year and feeds on the Oriental fruit moth, an insect that ruins peaches by boring into the fruit's interior, where it is also safe from pesticides. Colorado's insectary also releases eight species of parasites to control the alfalfa weevil. Collectively, they control about 40% of the pest population, which by one estimate, increases alfalfa hay production by as much as $12 million annually.

Bacteria and other microorganisms can also be used to control harmful pests. Farmers, gardeners, and foresters, for instance, often use a bacterium known as Bt (short for *Bacillus thuringiensis*)

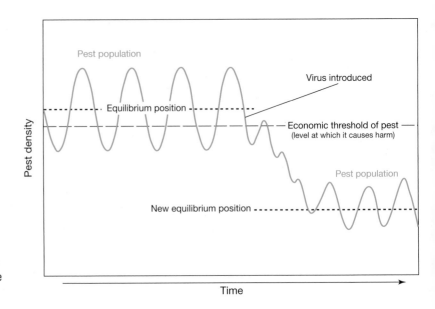

FIGURE 8.14 Biological control agents, such as viruses, can reduce pest populations below the economic threshold, i.e., the level at which it causes harm.

to control leaf-eating caterpillars. Bt spores are applied in a powder to fields and forests, where they are consumed by caterpillars as they dine on the leaves of trees and crops. Inside the caterpillars' stomachs, the spores hatch into bacteria that release a toxic protein that kills the pests.

Bt has been used successfully to control pine caterpillars and cabbage army worms in China. In the northeastern United States, it is being used to help control gypsy moths, described earlier. California has used Bt for three decades to control a number of caterpillars and mosquitoes, helping the state cut its pesticide use.

In Australia, government officials and scientists used a virus to control its burgeoning rabbit population. Rabbits were introduced in the early 1900s, apparently by European immigrants who longed to indulge once again in their favorite sport of hunting the elusive brushland "bounders." Because there were no natural predators to control rabbit populations, their numbers increased sharply. As their population swelled, the rangelands rapidly deteriorated. Grasses were clipped to the ground by sheep and hordes of ravenous rabbits. Faced with economic ruin, Australian ranchers tried numerous conventional control methods: poisons, traps, hunts, and fencing. Their efforts were to no avail.

In 1950, government biologists introduced the myxoma (mix-oh´-mah) virus, which is lethal to rabbits exclusively, into the target area. It is transmitted to healthy rabbits by mosquitoes that only bite rabbits. The immediate results were spectacular. One year later, 99.5% of Australia's rabbits had succumbed to the virus. Unfortunately, the rabbits gradually developed resistance to the virus, and by 1958, the mortality rate from the virus had dropped to only 54%. Rabbit populations are once again on the rise.

Biological control, while effective in many instances, has at least some important limitations. First, it tends to be slower than conventional pesticides. When an insect population explodes, farmers may not have enough time to wait for natural predators to be introduced into their fields and to gain control. By the time

GO GREEN!

When buying fruits and vegetables, purchase the organic varieties, especially apples, bell peppers, celery, cherries, imported grapes, nectarines, peaches, pears, potatoes, red raspberries, spinach, and strawberries. Studies show that these nonorganic fruits and vegetables (the "dirty dozen") contain higher levels of pesticides than others, even after careful washing.

they do, the field may be devastated. To compensate for this, farmers must carefully monitor pests and time the release of biological control agents to avoid outbreaks. In addition, farmers can create conditions conducive to beneficial insects to ensure their continued presence.

Second, alien species introduced as biological control agents can become pests in their own right, as described in Chapter 15. Years of research are needed to avoid creating additional pests. Regardless, biological control agents, combined with crop rotation, heteroculture, and other techniques, can provide an important measure of protection.

Genetic Controls Yet another method of controlling pests involves genetic manipulations. Called **genetic control,** it consists of at least two methods: genetic resistance and the sterile-male technique.

Genetic Resistance Chapter 5, on world agriculture, described ways in which scientists are developing high-yield crops through plant-breeding programs and genetic engineering. Scientists are also working on ways to make plants resistant to insect pests. Today, thanks to this research, much of the wheat planted in the United States is resistant to the Hessian fly, once a notoriously destructive insect that caused several hundred million dollars' worth of damage a year. Scientists have also developed strains of cotton, soybeans, alfalfa, and potatoes that are resistant to leafhoppers.

Combined with other control measures, **genetic resistance** provides an environmentally safe method of reducing pest damage while protecting the environment. Through genetic engineering, described in Chapter 5, new strains can be produced to help facilitate the transition to a global sustainable system of agriculture.

Sterile-Male Technique In their quest to safely control pests, scientists have also been able to capitalize on a fluke in insect biology, notably that many female insects breed only once in their lifetime. If these matings are infertile, perhaps because the male is sterile, the female produces no young. Using this bit of information, scientists have devised an ingenious method of control known as the **sterile-male technique,** which they have used to control several species of harmful insects. The most notable is the screwworm fly, an insect approximately three times the size of a housefly.

The screwworm fly is widely distributed in South America, Central America, and Mexico and ranges northward into the southern United States, including Georgia, Florida, Alabama, Texas, Arizona, New Mexico, and California. Shortly after mating, the adult female deposits about 100 eggs in open wounds of warm-blooded animals, such as cattle or deer. The eggs soon hatch into parasitic maggots that feed ravenously on the flesh and blood of the host (Figure 8.15). As the feeding continues, the wound discharges a fluid that attracts more adult flies. Eventually, in severe infestations, several thousand may feed in a single wound, killing a full-grown, half-ton steer within ten days. After five days of intense feeding, the larvae drop to the ground and pupate. Soon after, they emerge as adults. After mating with a male fly, the female lays its eggs in another open wound, thus completing the life cycle. In a single year, ten generations of flies may be born.

Screwworm flies are a significant pest. In fact, experts estimate that livestock damage in the United States caused by the flies ranges between $40 million and $120 million per year. In the 1930s, Edward Knipling, chief of the USDA's Entomology Research Branch, had an idea. He reasoned that screwworm flies could be controlled by sterilizing and releasing captive-raised male flies. After highly successful preliminary tests on a Caribbean island, he decided to try this method in the United

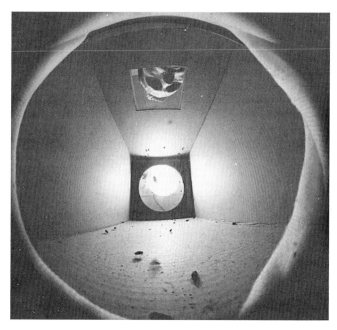

FIGURE 8.16   Sterilization of male screwworm flies. During the sterilization process, canisters containing 30,000 flies are exposed to a cobalt-60 radiation source at the sterilization factory.

States. Starting in 1958, Knipling set up a sterilization factory in an old airplane hanger. There, he and other workers sterilized 50 million flies grown in captivity each week by exposing them to radioactive cobalt (Figure 8.16). Over 2 billion sterilized males were packed in boxes and dropped over target areas in the southeastern United States. Boxes dropped from planes popped open when they struck the ground, releasing the flies (Figure 8.17).

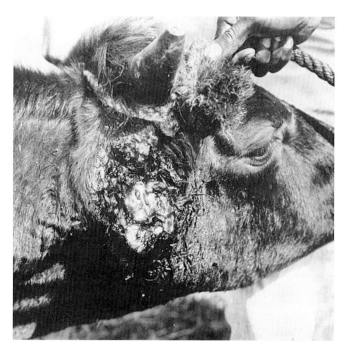

FIGURE 8.15   Screwworm infestation in the ear of a steer. An untreated, fully grown animal weighing about 1,000 pounds may be killed by several thousand maggots feeding in a single wound.

FIGURE 8.17   Nearly ready for release. Mexican technicians unload containers containing sterile screwworm flies in preparation for their release from an airplane over an infested area.

TABLE 8.1	Population Reduction of a Pest Population When a Constant Number of Sterilized Males Are Released in a Pest Population of 1 Million Males and 1 Million Females			
Generation	Number of Virgin Females	Number of Sterile Males Released	Ratio of Sterile to Fertile Males	Number of Fertile Females in the Next Generation
1	1,000,000	2,000,000	2:1	333,333
2	333,333	2,000,000	6:1	47,619
3	47,619	2,000,000	42:1	1,107
4	1,107	2,000,000	1,807:1	Less than 1

The sterilized males competed with fertile wild males for mates. When the ratio of sterile to fertile males is 9:1, over 80% of the matings are infertile (Table 8.1). As a result, the screwworm fly population gradually declined. Eighteen months after the project was begun, the screwworm fly had been eradicated from the Southeast. The screwworm fly still persists in the Southwest, because it continues to migrate in from Mexico. However, the continued release of sterile males in Mexico and the United States will likely keep the population of screwworm flies at manageable levels.

The sterile-male technique has been used in California to control Mediterranean fruit flies, or medflies, which were introduced from Hawaii. The medfly lays its eggs in over 230 fruits, nuts, and vegetables. The larvae that hatch from these eggs then consume the fruit, with potentially devastating effects. The program has worked well for a number of years when combined with other techniques.

This technique has also been used in Africa on the island of Zanzibar off the coast of Tanzania to control disease-carrying tsetse flies. These flies carry a parasitic disease known as trypanosomiasis, which affects people and livestock in much of the African continent.

The sterile-male technique offers many advantages over traditional pest-control strategies. It is species-specific, eliminates or greatly reduces the need for pesticides, and can work when the density of the pest population is low. Researchers note, however, that the technique has its problems. Most important, sterilized males may be less sexually active than fertile males, hence reducing the effectiveness of this approach.

Natural Chemical Controls   Over millions of years, many plants have evolved a number of chemical substances to ward off potential enemies. Biodegradable and nonpersistent, many of these chemicals are now being considered for widespread use. For example, rotenone (ro´-teh-known), derived from the roots of certain Asiatic legumes, is used today by gardeners against an army of insect pests. Pyrethrum, extracted from chrysanthemums and daisylike flowers by pesticide manufacturers, also is used in home gardens.

Aside from these chemicals, researchers are experimenting with two additional chemical groups: pheromones and insect hormones.

Pheromones   Insects and other animal species release a number of chemicals into the environment that influence other members of the same species. These chemical substances are called **pheromones** (fair´uh-moans). One class of pheromones, and perhaps the most important, is the **sex attractants,** chemicals that attract males to females for mating.

Sex attractants help ensure the survival of various species and are an important biological adaptation. Consider how they help the gypsy moth. Although the male gypsy moth has strong, functional wings, the female is far too heavy-bodied for effective flight. After emerging from her pupa case, the virgin female flutters about near the ground or creeps up tree trunks. Soon after emerging, she is ready to breed. To attract males, she secretes minute quantities of a sex attractant pheromone, called gyptol, from glands in her abdomen (Figure 8.18). The sensitive antennal receptors of the male moth detect this scent, and after he locates the female, he mates with her.

Scientists at the USDA synthesized a chemically similar compound, called **gyplure,** which has proved as effective as gyptol as a sex attractant. This and two dozen other synthetic pheromones are now used in a variety of ways to control insect pests. One of the most common uses is in **pheromone traps** (Figures 8.18 and 8.19). A minute amount of the pheromone is placed in a trap laced with insecticide or some sticky substance that entraps unwary males. Deluded into thinking that a willing female waits in the trap, the male enters and is immobilized or killed by the insecticide. Each year, farmers install thousands of traps to control a variety of insects.

Traps can also be used to determine when pest species emerge in the spring, so that insecticide use can be carefully synchronized to do the most good. In addition, traps can be used to monitor the population levels, so that farmers know when to release predatory insects or whether an outbreak is occurring.

Unsuspecting male insects can also be duped by another ingenious technique (Figure 8.18). Farmers impregnate wood chips or cardboard shards with sex attractants. Then, flying

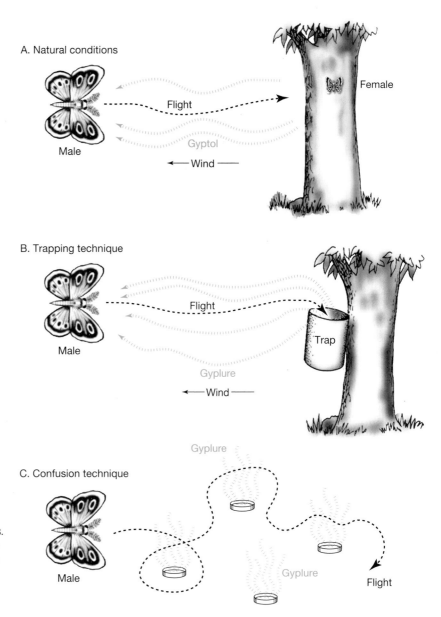

A. Natural conditions

Female

Flight

Gyptol

Male

← Wind

B. Trapping technique

Flight

Male

Trap

Gyplure

← Wind

Gyplure

C. Confusion technique

Male

Gyplure

Flight

FIGURE 8.18 Sex attractants as natural chemical controls. (A) Male gypsy moth attracted to a female by the sex attractant gyptol, which she emits. (B) Male gypsy moth lured into a trap baited with synthetic gyplure. (C) Male gypsy moth, confused by multiple sources of gyplure, is unable to find a female gypsy moth with which to mate.

over infested fields and orchards, they drop their payload. Alerted by the sudden presence of female sex attractant, eager males fly off in search of females but find only wood chips. Undaunted by the lack of physical similarity, many males mount the wood chips and try to mate with them. Alternatively, farmers may simply spray the sex attractant over the infested area. The males spend much of their time and energy tracking down nonexistent females.

In either case, pheromones decrease the likelihood that a male will find a female, and this technique is appropriately called a **confusion technique.** In one experiment, researchers found that gyplure applied at only 5 grams per hectare (12 grams per acre) reduced mating in gypsy moths by 94%.

Sex attractants are nontoxic, species-specific, and biodegradable. Moreover, it is impossible for an insect to develop resistance to them without at the same time developing resistance to the very act of reproduction. Since its first use on the gypsy moth, the pheromone technique has been used successfully on many other insect pests, including the cabbage borer, European

corn borer, cotton boll weevil, Japanese beetle, tomato hornworm, and tobacco budworm.

Insect Hormones Many insects hatch from eggs and pass through larval and pupal stages before maturing (Figure 8.20). Caterpillars, for instance, are larvae that develop into moths and butterflies. During the larval stage, the caterpillar produces a hormone, called **juvenile hormone,** that keeps it in its immature state. When levels fall, the insect pupates.

By spraying insects with juvenile hormone, farmers can prevent them from maturing. However, the larval forms of most insects do the greatest damage. The advantage of this method is that it prevents insects from maturing and reproducing, which reduces pest populations over the long haul.

Juvenile hormone has been successfully used to control mosquitoes in Central and South America and could be used on a variety of other pests. Environmentally safe because it is biodegradable, nonpersistent, and nontoxic, juvenile hormone is unfortunately not as species-specific as pheromones. As a

FIGURE 8.19   A typical gypsy moth trap. It contains captured gypsy moths lured into the trap by gyplure, a synthetic attractant that confuses male moths into thinking a female is inside the trap. Once inside, the moth becomes entangled in a sticky substance and is unable to extricate itself.

result, it may kill predatory insects and nonpest species, seriously upsetting the ecological balance. In addition, juvenile hormone acts slower than traditional chemical pesticides. A week or two may be required before hungry crop-eating larvae die, and by that time, they may have devastated a crop or forest. Furthermore, juvenile hormone is relatively unstable in the environment—so much so that it often breaks down before it can act. Chemists hope to find ways to make it last longer in the environment. Finally, juvenile hormone must be applied at the precise time that levels of the hormone in the caterpillar fall. Only if it is applied at that time will the larva remain immature.

To avoid these problems, scientists are now developing **hormone inhibitors,** or chemicals that block the secretion of juvenile hormone by larvae. If they are successful, this new line of nontoxic chemicals may cause larvae to mature early, disrupting their life cycle and killing the insects.

Cultural Control   **Cultural control** is a catchall term covering all of the techniques that do not fit into the other categories of pest control. Examples of cultural control methods include scarecrows and noisemakers that frighten birds away from crops, electrocution devices that zap unsuspecting bugs, and agricultural inspection stations that monitor fruits and vegetables transported across state and national borders for pests. All of these are essential elements of integrated pest management.

Integrated Pest Management: Combining Measures

Environmental, genetic, chemical, and cultural controls offer abundant opportunities to get off the pesticide treadmill and to clean up our environment. Which of these solutions is best?

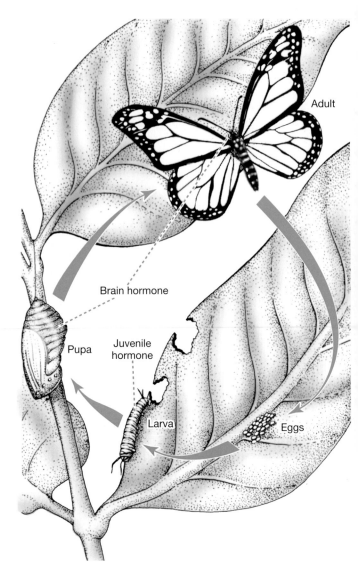

FIGURE 8.20   Life cycle of insects and the role of hormones in insect metamorphosis. Juvenile hormone, secreted mainly during the larval stage, keeps the caterpillar in this immature stage until it is ready to metamorphose into a pupa and adult. The hormones must be secreted in the right amounts at the right time.

In practice, the best strategy often involves a combination of methods, an integrated pest management approach. Farmers may combine measures; for instance, they may plant pest-resistant crops, rotate crops, use trap crops, and apply **insect hormones** or pheromones to reduce pests to economically acceptable levels. Integrated pest management emphasizes control, not the complete eradication of pests, which is expensive and often environmentally harmful. Natural pest-control strategies may be used alone or, as just noted, in combination, either simultaneously or in sequence, depending on the pest. This approach requires ingenuity and knowledge of the life cycles of pests. No doubt insecticides will continue to be used to control insects, but integrated pest management could greatly reduce their use.

Numerous studies show that integrated pest management works. In Texas, for example, researchers applied a variety of techniques to cotton boll weevils once controlled by heavy doses of pesticides. By planting an early-fruiting variety of cotton that was unappealing to the weevil and by reducing the spacing between rows, reducing fertilizer use, and reducing irrigation water, researchers were able to reduce pest populations and cut back on costly pesticide applications—from 12 per growing season to virtually none. Banks that once balked at lending farmers money for cotton because of average crop losses of $4.60 per hectare ($1.88 per acre) became willing to lend money to farmers, who enjoyed a $900 per hectare ($364 per acre) profit.

Summary of Key Concepts

1. Worldwide, 2.7 billion metric tons of pesticides are used per year. Despite increased awareness of the dangers of pesticide use and efforts to reduce our dependency on these substances, total pesticide use has nearly tripled since the early 1960s in the United States. Insecticide use remained more or less constant during that period and has recently begun to fall, but herbicide use has climbed sixfold.

2. Insects often become pests when humans simplify ecosystems, destroying the complex ecological control mechanisms that hold populations in balance. Planting large expanses of a single or a few crop species also provides insects with an enormous supply of food. Pests may also arise from the accidental or intentional introduction of insects (e.g., gypsy moths) or other organisms (e.g., rabbits) into environments in which there are few natural controls.

3. Each year, pests destroy or consume about 42% of the food produced by the United States. Damage is much worse in the tropics, where two or three crops are grown on a field in a single year and where conditions are ideal for insects.

4. Pesticides help control the damage caused by pests. Pesticides also help control disease-carrying insects such as mosquitoes, which transmit malaria.

5. Scientists are finding that the effectiveness of pesticides is rapidly dwindling because hundreds of insects have become genetically resistant to insecticides. The widespread use of insecticides also destroys beneficial insects and other organisms, such as birds, that help hold pest populations in check.

6. Today, more than 550 insects are resistant to insecticides. Twenty species are resistant to all insecticides, and that number is rapidly rising.

7. Before World War II, farmers used a variety of chemical pesticides, including ashes, arsenic, and hydrogen cyanide. In 1939, however, Swiss scientist Paul Müller discovered that DDT, a chlorinated hydrocarbon, was a powerful insecticide. He started a revolution in agricultural pest control.

8. DDT was followed by numerous chlorinated hydrocarbons. Unfortunately, chlorinated hydrocarbons persist in the environment for many years, bioaccumulate, and biomagnify. All of them have since been banned or severely restricted in the United States.

9. As a result of growing discontent with chlorinated hydrocarbons, chemical manufacturers created a new line of pesticides, the organic phosphates. They decompose more rapidly in the environment and are less likely to bioaccumulate and biomagnify. These substances, however, are potent nerve toxins.

10. Pesticide manufacturers now produce potentially safer insecticides called carbamates, which are more quickly degraded than the organic phosphates and chlorinated hydrocarbons.

11. Pesticides end up on American soils and in our water, food, wildlife, and people. Each year approximately 100,000 to 300,000 Americans, mostly chemical or farm workers, are poisoned by pesticides. All told, 200 to 1,000 Americans die yearly from pesticide poisoning. Worldwide, at least 500,000 people are poisoned by pesticides each year, and an estimated 5,000 to 14,000 die.

12. Pesticides are commonly found in the bloodstream and tissues of humans in developed countries and pose a threat to children. Some doctors believe that general maladies such as dizziness, insomnia, indigestion, and headaches may be caused by ingesting insecticides in food. Some pesticides may also cause cancer and birth defects.

13. Herbicides, applied to fields to destroy weeds, also have been linked to a growing number of problems, such as cancer, birth defects, and nervous disorders. Agent Orange, an herbicide used in the Vietnam War, was found to be contaminated with dioxin, which is believed largely responsible for many health problems reported by veterans and Vietnamese citizens.

14. Pesticides affect wildlife as well. Tributyl tin, a potent biotoxin added to paint to retard the buildup of algae and barnacles, is released from paint on ships' hulls and pollutes bays and harbors, killing algae and shellfish.

15. Pesticides have been implicated in the rash of fish cancers reported in American lakes and rivers. And each year, insecticides destroy an estimated 400,000 bee colonies.

16. DDT has been linked to the decline in bald eagle, osprey, brown pelican, and peregrine falcon populations in the United States. DDT, scientists found, disrupted calcium deposition in eggshells, resulting in eggshell thinning and low embryonic survival.

17. Formal regulation of pesticides at the federal level began in 1947, when Congress passed the Federal Insecticide, Fungicide, and Rodenticide Act. Since that time, the act has been substantially strengthened to improve control. The amendments require all manufacturers to submit applications to the EPA to produce all new pesticides to be sold and used in the United States. The EPA approves only those substances for which it believes the benefits outweigh the potential risks. The EPA also classifies registered pesticides as either general or restricted, and can cancel or suspend the registration of pesticides that are later found to be unsafe.

18. To protect overseas users, the EPA must now notify all governments when it cancels or suspends a pesticide registration. It also requires exporters to notify customers as well if they are importing a substance whose use has been banned or restricted in the United States.

19. Despite strict laws and regulations, pesticides continue to cause problems. Enforcement of proper use is virtually nonexistent. Additionally, many products deemed safe on the basis of toxicity data provided for EPA registration have caused unexpected damage when used in

the field. Many chemicals in use today have not been adequately tested.

20. Scientific research has yielded numerous alternatives to pesticides. Pest control today often combines many techniques, sometimes including the judicious use of pesticides. This approach is called integrated pest management. It capitalizes on four major control strategies: environmental, genetic, chemical, and cultural.

21. Environmental controls are measures that alter the biotic and abiotic environment of the pest. The most important ones are (1) crop rotation, (2) heteroculture, (3) use of trap crops, and (4) the introduction of natural biological control agents, notably insect predators, parasites, and disease-causing organisms. These measures are often simple, effective, inexpensive, and environmentally benign.

22. Genetic controls include (1) the sterile-male technique, in which sterilized males of the pest species are released into the wild in infested areas to breed with fertile females, which results in an infertile mating, greatly reducing pest populations; and (2) genetic resistance, or efforts to improve the genetic resistance of plants and animals to pests through genetic engineering and conventional animal and plant breeding programs.

23. Chemical controls include (1) natural chemical substances produced by plants to ward off insects; (2) pheromones, the sex attractants produced by females to attract males for mating, which can be synthesized and applied to infested fields to confuse males; (3) insect hormones, which can be applied to crops to alter the life cycle of pests and, ultimately, reduce their number; and (4) the judicious use of chemical pesticides.

24. Cultural controls include all other techniques, such as scarecrows, noisemakers, electrocution devices, and agricultural inspection stations.

Key Words and Phrases

Agent Orange	Herbicide
Beneficial Insects	Heteroculture
Bioaccumulation	Hormone Inhibitors
Biological Control	Insect Hormones
Biological Diversity	Integrated Pest
Biomagnification	Management
Carbamates	Intercropping
Chlorinated Hydrocarbons	Juvenile Hormone
Confusion Technique	Monoculture
Crop Rotation	Nonpersistent Pesticides
Cultural Control	Organic Phosphates
DDT	Pesticide
Dioxin	Pesticide Treadmill
Environmental Controls	Pheromone
Federal Insecticide,	Pheromone Trap
Fungicide, and Rodenticide	Registration of Pesticides
Act (FIFRA)	Restricted Pesticide
General Pesticide	Sex Attractants
Genetic Control	Sterile-Male Technique
Genetic Resistance	Trap Crops
Gyplure	Tributyl Tin

Critical Thinking and Discussion Questions

1. In what ways does planting monocultures create problems with diseases and pests? How does this activity affect environmental resistance of the environment and the biotic potential of a pest species?

2. Why do alien species often become pests? Give some examples.

3. Describe the pros and cons of chemical pesticide use.

4. What is the pesticide treadmill?

5. Discuss why insect damage has increased dramatically over the past three decades despite the use of insecticides.

6. Name the three types of chemical pesticides, and give an example of each one. In what ways are they similar, and in what ways are the three classes of pesticides different? Which group is the safest to humans and the environment, and why?

7. Given what you know about the harmful effects of pesticides, describe a perfect chemical pesticide. What characteristics would it have? Can you think of any potential candidates?

8. List and describe some of the health effects of pesticide use. Which sector of our population is most likely to be affected?

9. What is Agent Orange? Where was it used? Why? What are some of the consequences of its use?

10. Describe the effects of pesticides on wildlife. Give some examples.

11. Describe the major provisions of the Federal Insecticide, Fungicide, and Rodenticide Act.

12. Define the term *integrated pest management*. What are the major strategies of this technique? Explain each one, and give an example.

13. "Insecticides are the only way to control insects," says one midwestern wheat farmer. Do you agree or disagree? What suggestions might you make to this farmer?

14. Describe how crop rotation and heteroculture keep pest populations in check.

15. What are trap crops? Give some examples. Do trap crops eliminate or reduce insecticide use?

16. Describe the pros and cons of biological control.

17. How can scientists alter the genetic resistance of crops and livestock to reduce pest damage?

18. Describe the principle behind the sterile-male technique. In other words, why is it effective?

19. Define the terms *pheromone* and *insect hormone*. How are they different, and how are they similar? List some pros and cons of pheromones and insect hormones as pest-control tools.

20. You are a farmer. You want to grow corn, alfalfa, and potatoes without using pesticides. How would you go about it?

Suggested Readings

Brady, C. 1992. The Perfect Poison for the Perfect Banana. *Global Pesticide Campaigner* 2(3): 8–9. Alarming interview with a Honduran farm worker who applies herbicides.

Carson, R. 1962. *Silent Spring.* Boston: Houghton Mifflin. The book that raised worldwide alarm over the use of pesticides.

Di Silvestro, R. 1996. What's Killing the Swainson's Hawk. *International Wildlife* 26(3): 38–43. Description of the effects of pesticide use on a migratory species.

Dreistadt, S. H., and D. L. Dahlsten. 1986. California's Medfly Campaign: Lessons from the Field. *Environment* 28(6): 18–20, 40–44. A sobering look at pest control.

Gardner, G. 1996. Preserving Agricultural Resources. In *State of the World 1996,* ed. L. Starke. New York: W. W. Norton. Covers several important topics, including pest management.

Goodstein, C. 1996. Stood Up by the Birds and the Bees. *Amicus Journal* 18 (1): 26–30. A look at the ecological consequences of disappearing birds and insects, both pollinators of flowering plants.

Halweil, B. 1999. The Emperor's New Crops. *World-Watch* 12(4): 21–29. Examination of the controversy over genetically modified crops.

Halweil, B. 2003. This Old Barn, This New Money. *World-Watch* 16(4): 24–29. A fascinating look at how prosperous tobacco farmers are making a shift to lucrative organic farming.

Ignacimuthu, S., and S. Jayaraj. 2004. *Green Pesticides for Insect Pest Management.* Oxford, England: Alpha Science International. An in-depth look at safe alternatives to common chemical pesticides used today.

Kinley, D. H. 1998. Aerial Assault on the Tsetse Fly. *Environment* 40(7): 14–18, 40–41. Tells of a successful effort to control disease-carrying tsetse flies using the sterile male technique.

Mattoon, A. 2000. Amphibia Fading. *World-Watch* 13(4): 12–23. Disturbing article about the worldwide disappearance of amphibians.

McGinn, A. P. 2000. POPs Culture. *World-Watch* 13(2): 26–36. Examination of the problems created by persistent organic pollutants, including pesticides.

McGinn, A. P. 2003. Combatting Malaria. In *State of the World 2003,* ed. L. Starke. New York: W. W. Norton. Important insight into pesticide use to combat insects and disease.

Sachs, J. S. 2004. Poisoning the Imperiled. *National Wildlife* 42 (1): 22–29. A look at endangered species whose future is threatened by pesticides.

Siedenburg, K. 1992. Philippine Pesticide Bans Show NGO Strength. *Global Pesticide Campaigner* 2(3): 1, 6–7. Excellent case study that shows how nongovernmental organizations can have an impact on pesticide use.

Tuxill, J. 2000. The Biodiversity that People Made. *World-Watch* 13(3): 25–35. Superb article on the loss of genetic diversity in crops.

Weber, P. 1992. A Place for Pesticides? *World-Watch* 5(3): 18–25. Superb overview of the pesticide issue, with a frank discussion of options.

Wilcox, F. A. 1983. *Waiting for an Army to Die: The Tragedy of Agent Orange.* New York: Vintage Books. A superb book on the troubles facing Vietnam veterans.

Youth, H. 1994. Flying into Trouble. *World-Watch* 7(1): 10–19. Documentation of pesticides' role in the startling and tragic decline in bird species.

Web Explorations

Online resources for this chapter are on the World Wide Web at: **http://www.prenhall.com/chiras** *(click on the Table of Contents link and then select Chapter 8).*

AQUATIC ENVIRONMENTS

Humans rely on aquatic ecosystems for many reasons. First and foremost, aquatic ecosystems provide food such as fish and shellfish. They also provide water for domestic, agricultural, and industrial use. In addition, they provide recreational opportunities such as canoeing, waterskiing, and fishing. Aquatic ecosystems are also home to many wild species.

Over the years, scientists have found that human use of aquatic ecosystems has numerous, far-reaching effects—some minor, some quite serious. To ensure that these valuable ecosystems continue to serve humans and other species, most experts agree that we must manage them sustainably. To manage them sustainably, however, we must understand them. More specifically, we must understand their physical, chemical, and biological components. We must also understand how these systems function. And, finally, we must understand the many relationships that exist within and between these ecosystems.

In this chapter, we will introduce you to aquatic ecosystems. We will explore the characteristics and key processes in five important aquatic ecosystems: wetlands, lakes, streams, coastlines, and oceans. We also discuss the many effects of human use, bearing in mind that the health and productivity of aquatic habitats are essential for our survival and the well-being of many other species as well. The information you learn here will be particularly important in your study of fisheries management.

9.1 Wetlands

Definition

Wetlands are highly diverse ecosystems. They include swamps, salt marshes, and bogs, among others. Wetlands usually occupy a transitional zone within the landscape. That is, they lie between well-drained upland areas and a permanently flooded deepwater habitat, as shown in the center of Figure 9.1. Mangrove swamps are a good example. Wetlands also exist in depressions, also shown in Figure 9.1. In many wetlands, the land is permanently or intermittently flooded with shallow water. However, some wetlands are less obvious—characterized instead by water tables at or near the soil surface.

In legal contexts, however, the term *wetland* is not easily defined, so it continues to cause controversies. In the United States, the difficulties began in the late 1970s, when laws designed to preserve wetlands were first passed by the Congress. All of a sudden, it became necessary to delineate the boundaries of wetlands. Defining the exact boundary of a wetland on a piece of land suddenly became an economic issue for owners of private lands and for managers of public lands. Under recent law, if land is defined as a wetland, it cannot be drained, farmed, or built on.

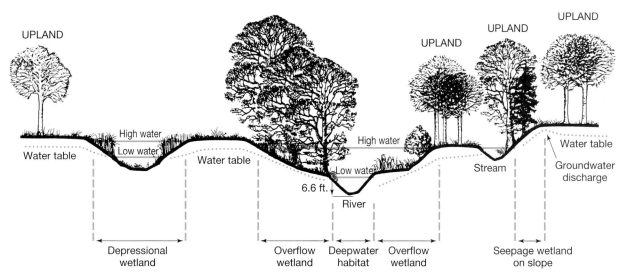

FIGURE 9.1 Diversity of wetlands. Most wetlands occupy a transitional zone between well-drained uplands and deepwater habitats, as illustrated by the overflow wetlands in this diagram. A permanent water depth of at least 2 meters (6.6 feet) is usually considered the boundary between a wetland and a deepwater habitat. Depressional wetlands can form between two upland areas. Seepage wetlands can form on hill slopes exposed to groundwater discharge.

Currently, the legal definition used by most people involved in the regulation and management of wetlands is the one adopted by the Environmental Protection Agency [taken from the EPA Regulations listed at 40 CFR 230.3(t)]:

> The term "wetlands" means those areas that are inundated or saturated by surface or ground water at a frequency and duration sufficient to support, and that under normal circumstances do support, a prevalence of vegetation typically adapted for life in saturated soil conditions. Wetlands generally include swamps, marshes, bogs, and similar areas.

Professionals involved in wetland protection nowadays define the legal boundaries of a wetland by using this definition along with thorough field investigations. Their investigations allow them to identify the presence of certain "diagnostic" environmental characteristics of wetland soils, hydrology, and vegetation. Nonetheless, **wetland delineations** are difficult for several reasons. For one, boundaries are inherently diffuse, almost never abrupt. Water levels vary from season to season. In addition, human land uses, such as farming and logging, alter vegetation, soils, and water regimes, changing the original land characteristics.

Classification

To aid in its efforts to inventory, evaluate, and manage wetlands, the U.S. Fish and Wildlife Service (USFWS) developed a wetlands classification system in 1979. This system, known as the **Cowardin Classification System,** has become the standard for identifying and classifying wetlands throughout the world. Wetlands and their adjacent deepwater habitats are divided into five major categories, shown in Figure 9.2. These five wetland systems are marine, estuarine, riverine, lacustrine, and palustrine. (We'll define each one shortly.)

Deepwater habitats are defined as aquatic ecosystems that are more than 2 meters (6.6 feet) deep. Wetlands are further classified into subsystems, classes, and subclasses (not shown), using a variety of other criteria, such as freshwater depths and velocities, dominant vegetative form, and composition of substrate.

Marine Systems   **Marine wetland systems** include the open ocean overlying the continental shelf and associated high-energy coastlines that are subject to waves and currents of the open ocean. Water salinity exceeds 30 parts per thousand (ppt). Within marine systems are wetlands such as coral reefs, rocky shoreline cliffs, and sandy shorelines that lack appreciable freshwater influence.

Estuarine Systems   Estuaries are typically the tidal mouths of rivers, where rivers flow into oceans. **Estuarine wetland systems,** such as tidal salt marshes and mangrove swamps, are adjacent to the open ocean but semi-enclosed by land. These semi-enclosed coastal bodies of water areas are protected from high-energy waves and currents of the open ocean. However, they are strongly influenced by tidal fluctuations. Saltwater here is frequently diluted by freshwater runoff via coastal rivers. The lower limit for water salinity in an estuarine system is 0.5 ppt.

Tidal salt marshes form near river mouths, in bays, on coastal plains, around lagoons, or behind barrier islands (described shortly). In tidal marshes, the shorelines are gently sloped and are protected from high-energy waves and storms. Sediment accumulates in these areas and serves as as a substrate for the growth of rooted plants.

Organisms that thrive in salt marshes are those tolerant of salinity, fluctuations in water levels (inundation at high tide and dry conditions during low tide), and extreme changes in daily and seasonal temperatures. Plants common to the salt marshes of the United States are salt-tolerant grasses and rushes and mud algae. A network of **tidal creeks** and shallow ponds throughout

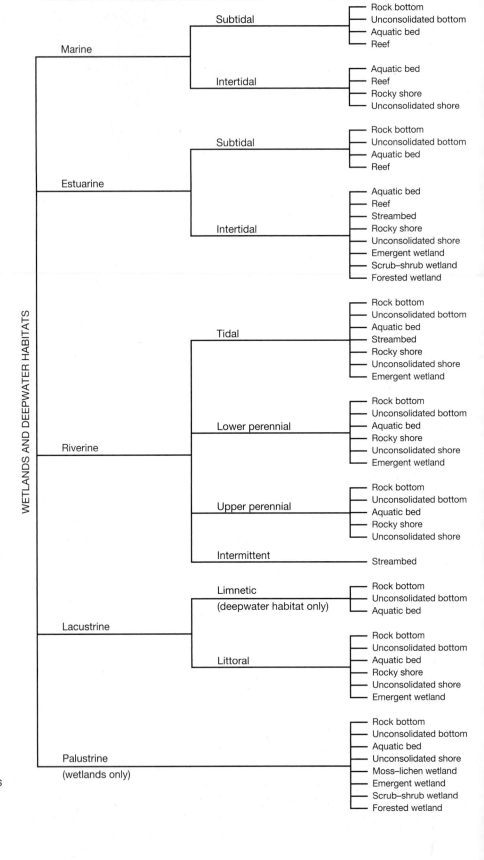

FIGURE 9.2 Cowardin Classification System hierarchy of wetlands and deepwater habitats. This diagram shows the first three categories: system, subsystem, and class.

FIGURE 9.3 A tidal salt marsh off the coast of Florida. Note the branchlike patterns created by the tidal creeks.

the marsh provides passage, spawning grounds, food, and shelter for fish and numerous other marine organisms (Figure 9.3).

Tidal marshes are found in mid- to high-latitude coastal regions. The majority of the salt marshes in the United States are found in states bordering the Atlantic Coast and the Gulf of Mexico and along Alaska's coastline. Strips of salt marshes on the country's Pacific Coast are fewer and usually narrower due to the steepness of the shoreline.

Mangrove swamps are named after their dominant vegetation type, the mangrove tree, and are actually forested wetlands. They consist of dense, impenetrable stands of these salt-tolerant trees and shrubs, as well as ferns. Vegetation in mangrove swamps has adapted to thrive in a relatively wide range of salinity and water inundation frequencies. Tidal creeks are also a common feature of these wetlands.

Mangrove swamps are commonly found along the coasts of tropical and subtropical countries. Mangrove swamps have a limited distribution in the United States, the majority located in southern Florida and in Puerto Rico. Like salt marshes, mangrove wetlands require protection from high-energy waves and need sediment deposits for root growth, although a few mangroves are found on rocky substrate.

Riverine Systems As the name implies, **riverine wetland systems** are associated with rivers and streams. A riverine system begins upstream where streams originate or where the channel departs from a freshwater lake or pond. It ends downstream where freshwater enters a freshwater lake or mixes with saltwater, as in the case of rivers that empty into oceans or associated bodies of water.

Riparian zone is a common term used to define the entire strip of land on either side of a river or stream. The riparian zone includes **riparian wetlands** as well as land that may not meet the criteria for a legally defined wetland (Figure 9.4). The

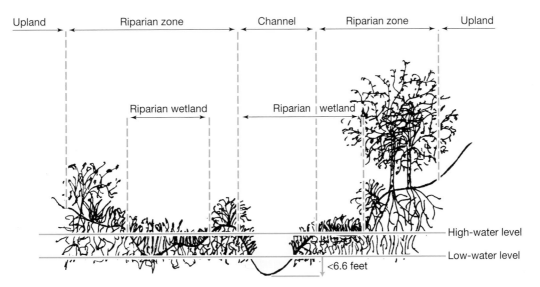

FIGURE 9.4   Relationship between riparian zone and riparian wetland. The boundary of a riparian zone is not easily delineated but often includes drier sites that do not conform to the legal definition of a wetland.

riparian zone exhibits distinct soil, vegetative, and hydrologic characteristics and is considered an **ecotone,** or transitional area, between the aquatic (river or stream) and upland habitats. Riparian zones are characterized by high species diversity, high species densities, and high productivity. Many animals use these areas for refuge, food, and travel.

All riparian zones exhibit a linear form but vary considerably in width. They include broad alluvial valleys many miles wide in the southeastern United States and narrow strips of vegetation along stream banks in the arid regions of the West.

Geographic location also influences the vegetative community composition in riparian zones due to variations in climate, soils, topography, and geology. For example, large expanses of bottomland hardwood forests dominate the broad, flat floodplain of the Mississippi River and the floodplains of the Southeast's smaller rivers. Many of these riparian zones are connected hydrologically to deepwater swamps. But unlike the permanently flooded deepwater swamps, bottomland hardwood forests consist of species that tolerate shallower water depths and shorter periods of inundation, including willow, cottonwood, oak, water hickory, green ash, maple, and birch. In contrast, riparian zones of the arid and semiarid western states are narrow and steep. Trees and shrubs of alder, elm, box elder, hawthorn, sycamore, walnut, and mesquite dominate these areas, as well as willows and cottonwoods, which are found throughout the country.

Lacustrine Systems The term *lacustrine* means of or related to lakes. Lacustrine systems are situated in a topographical depression, as in the case of a lake or a pond, or may be formed by damming a river, which then fills a valley to form a reservoir. **Lacustrine wetland systems** are wetlands associated with lakes, reservoirs, and large ponds (Figure 9.5).

Palustrine Systems The majority of all wetlands fall under the category of **palustrine wetland systems.** These systems include bogs, fens, freshwater marshes, wet meadows, swamps, pocosins, and small shallow ponds, described shortly. Palustrine wetlands are dominated by vegetation and are bound by upland habitat or any of the other four aquatic systems. Water depth in the deepest part of the basin is less than 2 meters (6.6 feet) at low flow, and water salinity is less than 0.5 ppt. Some riparian wetlands that persist due to periodic flooding of a river channel may also be considered palustrine wetlands.

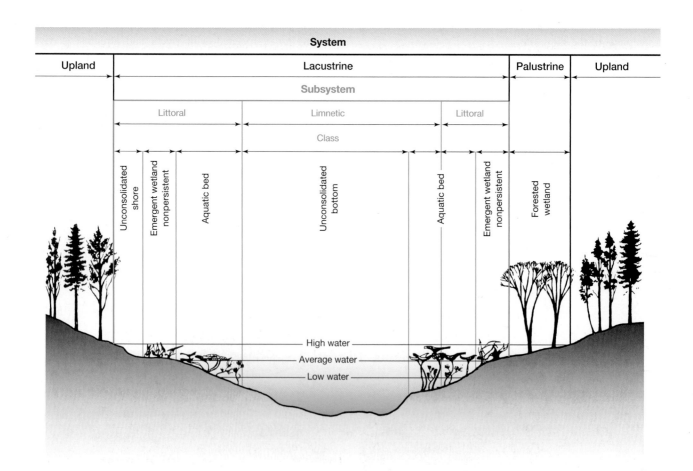

FIGURE 9.5 A typical cross section of a lacustrine wetland system delineated into subsystems, limnetic and littoral, and classes within those subsytems. This particular lacustrine system is bordered on one side by upland habitat and on the other by a palustrine wetland.

Bogs and fens are the major types of **peatlands** and are found primarily in the colder northern, humid regions of the Northern Hemisphere. In general, bogs are associated with relatively stagnant, water-filled depressions and support an abundance of acid-loving plants (such as mosses of the genus *Sphagnum*). Fens are considered a successional stage in the development of bogs. That is, they are a transition between shallow, open lake systems and bogs. In the United States, most peatlands are located in Minnesota, Michigan, Wisconsin, Alaska, and Maine. Canada has the largest peat resources in the world. Extensive areas of peatlands also occur in Scandinavia, eastern Europe, and western Siberia. **Peat** is a generic term used to describe organic soils. Organic soils form in cool, wet climates with an abundance of plant material and anaerobic soil conditions due to standing water or lack of drainage. This type of environment is conducive to extremely slow plant decomposition and can result in peat accumulations several feet thick. Peat is a valuable source of fuel in some parts of the world and is commonly added to garden and potting soils to improve soil structure and water-holding capacity.

In North America, **freshwater marshes** are scattered throughout inland Canada and the United States. In the north-central United States and south-central Canada are numerous prairie pothole marshes. They originated as millions of depressions formed by the scouring action of glaciers. These depressions filled with meltwaters as the glaciers retreated. Along the Great Lakes are extensive marshes. In southern Florida are the **Everglades,** a massive freshwater marsh formed by water flowing out of Lake Okeechobee onto the flat, limestone deposit that forms most of southern Florida.

The vegetative community within freshwater marshes such as these typically consists of nonwoody, water-loving plants known as **hydrophytes.** These include grasses, reeds, cattails, sedges, rushes, broad-leaved plants, and various floating and submergent aquatic species (Figure 9.6). Generally, the freshwater marshes are characterized by shallow water depths and shallow peat formation.

Freshwater swamps occur along the Mississippi floodplain and Atlantic coastal plain of the southeastern United States and support woody vegetation, primarily various species of cypress, gum, and tupelo trees, the roots of which have adapted to thrive under anaerobic conditions. Swamps remain flooded all or most of the year and can occur in a number of hydrologic settings, including: (1) isolated, water-filled depressions fed primarily by rainwater and surface inflow, (2) edges of lakes fed by lake overflow and upland runoff, and (3) broad floodplains permanently flooded by rivers or streams. Although the canopy vegetation in swamps is dominated by cypress, gum, and tupelo trees, the understory vegetation consists of many other tree, shrub, herbaceous, and aquatic plant species. In the Deep South, Spanish moss is frequently associated with these swamps and grows in abundance, hanging from tree branches and stems.

Pocosins are shrub–scrub wetlands usually located on relatively flat terrain between stream systems. These wetlands are found primarily along the southern Atlantic coastal plain. Coastal precipitation and seasonal flooding regimes of nearby stream systems influence the frequency and duration of inundation periods.

Functions and Values

As you can see, there are many types of wetlands. Wetlands cover only about 6% of the Earth's land surface. They are found in every climate regime from the tropics to the tundra and on every continent except Antarctica. Since the detailed study of wetland ecosystems began in the 1960s, scientists

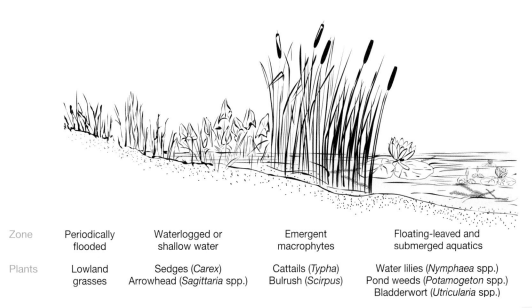

Zone	Periodically flooded	Waterlogged or shallow water	Emergent macrophytes	Floating-leaved and submerged aquatics
Plants	Lowland grasses	Sedges (*Carex*) Arrowhead (*Sagittaria* spp.)	Cattails (*Typha*) Bulrush (*Scirpus*)	Water lilies (*Nymphaea* spp.) Pond weeds (*Potamogeton* spp.) Bladderwort (*Utricularia* spp.)

FIGURE 9.6   A cross section through the edge of a freshwater marsh. Plants are shown as distributed according to their water depth tolerances. Typical plant species are listed below each water inundation zone.

TABLE 9.1 Wetland Functions and Their Ecosystem Services	
Wetland Functions	**Corresponding Ecosystem Services**
Hydrologic Processes	
Flood/Stormwater Control Vegetation binds and protects soils. Intercepts, stores, and slowly releases floodwaters, reducing the amount and velocity of surface runoff.	• Buffers human coastal settlements from storm-driven waves. • Reduces flood damage to human property. • Minimizes soil erosion caused by wave action and stream flow.
Base Flow/Groundwater Support Stores water and then releases it slowly to groundwater deposits.	• Maintains aquifers. • Supports stream base flows during the dry season, necessary for the survival of many aquatic species.
Water Quality Improvement	
Pollutant Removal, Detainment, and Transformation Filters out or detains sediment, inorganic and organic nutrients, and toxic chemicals. Detoxifies some pollutants.	• Reduces the amount of pollutants in surface and groundwaters. • Purifies drinking-water supplies. • Maintains or enhances water quality for domestic, commercial, and recreational uses.
Wildlife Habitat Produces an abundance of organic material that supplies food for a diverse group of organisms. Provides nesting and breeding sites. Provides resting sites for migrating species. Can function as pathways between various ecosystems.	• Provides renewable sources of food, fiber, and energy. • Maintains plant and animal populations. • Preserves biological diversity. • Supports threatened and endangered species. • Supplies game for hunting and fishing and subjects for bird-watching and photography.
	Human (Ecosystem) Services Unrelated to a Specific Wetland Function Scenic beauty, aesthetic enjoyment, hiking/picnic areas, educational/research opportunities, archeological/historical sites, open space, and wilderness preserves.

have discovered that wetlands perform many important functions essential to life on Earth and to human society as well (Table 9.1). Keep in mind that not all functions are performed by every kind of wetland and that some wetlands perform the same function better than other wetlands.

Wetland functions can be grouped into three broad categories: (1) hydrologic processes, (2) water quality improvement, and (3) wildlife habitat. Let's examine each one separately, beginning with hydrologic processes.

Wetlands influence water flows primarily by intercepting surface runoff from neighboring land and, acting like a sponge, storing the water for slow release, reducing flooding. Wetlands also stabilize stream banks, lakeshores, and coastlines from the flowing water in streams and rivers or from waves. This reduces erosion of shorelines and streambanks. Additionally, wetlands contribute to the recharge and discharge of groundwater. That is, they tend to absorb surface water and let it sink into groundwater. These hydrologic functions are of value to humans not only in reducing flood

damage, minimizing erosion, and maintaining aquifers, but also in helping to ensure base stream flows during the dry season, which are vital for many aquatic species.

Wetlands also help purify water. For example, wetlands remove sediment, inorganic and organic nutrients, and toxic materials from surface waters. They physically trap these materials and then may detoxify them chemically or biologically, thus helping with the purification process.

Wetlands are home to a wide variety of flora and fauna. Wetland ecosystems are characterized by high primary productivity and thus produce an abundance of plant material that fuels an extensive food chain. They also provide shelter and nesting sites vital to many species, such as fish, waterfowl, muskrats, songbirds, and others (Figure 9.7). Wildlife abundance and diversity are of value to humans for hunting and fishing. In many countries, wetland plants and wildlife are harvested for food, fiber, and energy. Currently, a disproportionately high percentage of endangered and threatened species depend on wetlands for their survival.

FIGURE 9.7 King rail feeding in the wetlands of the Clarence Cannon National Wildlife Refuge, Annada, MO. Wetlands provide valuable wildlife habitat.

Not related to a specific wetland function, but important to humans, are the cultural values wetlands provide. These benefits include aesthetics, open space, recreational opportunities, outdoor classrooms, and research and historical sites.

Wetland Protection

Once thought to be sinister wastelands and of no economic or ecological importance to humans or the environment, wetlands were drained, filled, dredged, and/or flooded for agriculture, navigation, or land development. At the time of European settlement in the United States (during the 1780s), the lower 48 states contained approximately 87 million hectares (215 million acres) of wetlands—an area approximately twice the size of California. According to the U.S. Fish and Wildlife Service, about 43.6 million hectares (107.7 million acres), or half of the original wetland acreage, remained in 2004 in the conterminous United States. What caused this dramatic decline?

U.S. wetlands declined for a number of reasons. Inland wetlands, for example, were primarily drained and filled to expand farmland. Coastal wetlands were drained and filled primarily for urban and industrial development.

Fortunately, widespread recognition of the valuable benefits of wetlands has slowed the loss of wetlands. In fact, all wetlands are protected by state and federal law. In the 1980s, the U.S. government adopted a **"no net loss" wetland policy** that utilizes a stringent permitting system under Section 404 of the Clean Water Act. Today, government officials attempt to regulate activities in and around wetlands to protect them from degradation. They also offer protection from development.

Today, five federal agencies share responsibility for wetland protection. The U.S. Army Corps of Engineers protects wetland resources in relation to navigation and water supply. The EPA's authority is related to protecting wetlands for their contribution to water quality. The U.S. Fish and Wildlife Service is concerned with protecting wetlands for fish and wildlife, including species that are hunted and fished, as well as threatened and endangered species. The National Oceanic and Atmospheric Administration's (NOAA) authority over wetlands relates to its mission to manage the nation's coastal resources. And the Natural Resources Conservation Service (NRCS) is charged with protecting wetland resources on agricultural lands.

Together, these agencies administer a myriad of programs that provide a variety of mechanisms to protect wetlands. These mechanisms include the acquisition of wetlands, that is, the purchase of wetlands that are set aside in perpetuity. They also include better land-use planning and the restoration and creation of new wetlands. They may include mitigation efforts to lessen our impact on wetlands, as well as incentives to prevent degradation and destruction and disincentives to stop conversion to other land uses. These agencies provide technical assistance, education, and research. According to the Association of State Wetland Managers, a nonprofit membership organization that promotes and enhances protection and management of wetland resources, 35 states today have their own wetland protection programs, some of which exceed the federal programs.

Between the mid-1950s and the mid-1970s, prior to wetland protection efforts, wetland acreage was converted at a rate of 185,400 hectares (458,000 acres) per year, according to the first wetland trends report by the U.S. Fish and Wildlife Service. Since then, however, wetland losses have slowed considerably. From the mid-1970s to the mid-1980s, for example, the estimated rate of wetland loss declined to 117,400 hectares (290,000 acres) per year, according to the second wetland trends report. Between 1986 and 1997, the annual conversion rate declined even more to 23,700 hectares (58,500 acres). According to the most recent report, the area in wetlands actually *increased* by 12,900 hectares (32,000 acres) per year between 1998 and 2004. The net gain in wetland area was attributed to wetlands created, enhanced, or restored through regulatory and nonregulatory restoration programs. Improvements in public education and outreach also helped, as did improvements in monitoring and protecting coastal environments.

Environmental groups and individuals have also waged successful campaigns to prevent the loss of important wetlands. In addition, the federal government requires any necessary losses to be mitigated. That is, if wetlands are lost due to development, steps must be taken to create new wetlands or prevent losses elsewhere.

Efforts to protect wetlands have increased in many nations. Despite these changes, wetlands continue to be lost to development. Losses are particularly high in the less-developed nations, but the United States and Canada still experience what some view as unacceptably high losses.

Further wetland drainage here or abroad must be viewed with caution. The effects on wildlife and fish, stream flow, and water quality often far outweigh the benefits realized by converting wetlands to farmland or other uses. In fact, limiting farmland expansion onto wetlands and other ecologically sensitive regions in many African nations may prove to be more economically profitable than farming, for the wealth of wild species in these nations attracts tourists, who pour millions of dollars into national economies. Revenues from such activities could help support programs of improved soil management and population control, reducing the present pressures on undeveloped land.

Wetland benefits have been recognized, and wetland management policy has shifted from wetland drainage and conversion to protection, restoration, and creation. Although the rate of wetland loss is slowing, these highly sensitive ecosystems will continue to be subject to alteration by human land uses and pollution. Controversy arises because conflicts of interest still exist between landowners and the general public and between developers and conservationists. Nonetheless, through the continued enforcement of wetland protection regulations and the cooperative efforts of informed citizens, we can keep moving toward our "no net loss" wetland goal.

GO GREEN!

Find out where wetlands exist near your home, and try to learn more about them. Encourage your neighbors, developers, and state and local governments to protect the function and value of wetlands in your watershed.

9.2   Lake Ecosystem

Lakes can be formed as a result of many natural forces. Tectonic activity (movements in the Earth's tectonic plates), for example, has created a number of inland seas (saltwater lakes), such as the Caspian Sea, which is located between southeastern Europe and southwestern Asia. Lakes can also be formed by volcanic activity; for example, they may form in a collapsed crater following a volcanic explosion, as in the case of Crater Lake, Oregon. Lakes are also produced by glaciers. Ice scouring by glaciers, ice dams, and moraine depositions (rock and earthen deposits of glaciers) from repeated glaciation have created many of the water-filled depressions and valleys in our northernmost states. In addition, the power of flowing water may cause river channels to shift direction, creating oxbow lakes. Even meteorites striking the Earth can hollow out huge depressions that later fill with water. And, of course, lakes can be produced by human activity, particularly dam construction. Lake Mead in Arizona and Nevada—at 637 square kilometers (246 square miles), the largest artificial lake in the world—was formed when Hoover Dam was built on the Colorado River.

The largest and deepest lakes are those formed by tectonic activities or major glacial processes. However, smaller, shallower lakes are most common. All lakes, regardless of size or origin, have several features in common. For example, they comprise the same zones, our next topic.

Lake Zonation

Figure 9.8 shows the zones of a typical lake: the littoral, limnetic, and profundal zones.

Littoral Zone   As illustrated, the shallow margin of the lake is known as the **littoral zone.** It is characterized by rooted vegetation. These plants are usually arranged in a well-ordered sequence similar to that of freshwater marshes (Figure 9.6). In the littoral zone, moving from shore toward open water, one first finds emergent plants, followed by floating and submergent. **Emergent plants** are those that extend above the water, such as cattails and bulrushes. As their name implies, **floating plants** float on the surface, although they may send roots to the bottom. Examples are water lilies and duckweed. Plants that are completely underwater, such as pondweed and milfoil, are called **submergents.**

Because sunlight penetrates to the lake bottom in the littoral zone, it sustains a high level of photosynthetic activity. In this zone, floating microorganisms, known as **plankton,** frequently give the water a faint greenish brown color. Plankton consist of two groups: plants (chiefly algae), known as **phytoplankton,** and animals (primarily crustaceans and protozoa), known as **zooplankton.** Plankton are largely incapable of independent movement, so they are passively transported by water currents and wave action.

Limnetic Zone   The **limnetic zone** is the region of open water beyond the littoral zone. This zone extends from the surface to the maximum depth at which there is sufficient

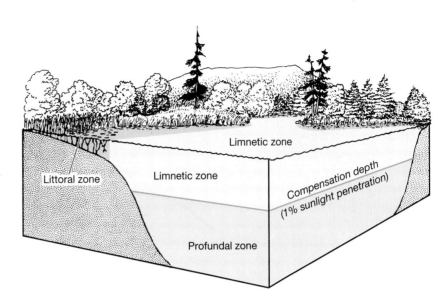

FIGURE 9.8 Principal zones of the lake ecosystem.

sunlight for photosynthesis (Figure 9.8). At this depth, known as the **compensation depth,** photosynthesis balances respiration; the light intensity here is only 1% of that of full sunlight. Although rooted plants are absent, the limnetic zone frequently contains numerous phytoplankton, mostly algae. In large lakes, the phytoplankton may actually produce more total biomass than the much larger rooted plants of the littoral zone. In spring, when nutrients and light are optimal, phytoplankton populations can "explode" to form **blooms.**

The limnetic zone derives its oxygen from the photosynthetic activity of phytoplankton and from the atmosphere immediately over the lake's surface, primarily when the water's surface is disturbed by wind and waves. Suspended among the phytoplankton are the zooplankton (animal plankton), primarily tiny crustaceans that form a trophic link between the phytoplankton food base and larger aquatic animals such as fish (Figure 9.9).

Profundal Zone The **profundal zone** lies beneath the limnetic zone and extends to the bottom of the lake (Figure 9.8). Because of the limited penetration of sunlight, green plants are absent. In north temperate latitudes, where winters are severe, this zone has the warmest water in the winter and the coldest water in the summer. Large numbers of bacteria and fungi occur in the bottom ooze, sometimes up to 1 billion per gram. These organisms constantly decompose the **organic matter** (plant and animal remains and wastes) that settles on the bottom. Nitrogen and phosphorus are released during decomposition and are put back into circulation as soluble salts. In winter, the metabolism of aquatic life is reduced, and the colder water has greater oxygen-dissolving capacity. At this time, oxygen is usually not a limiting factor for fish if the ice cover remains clear of snow. In midsummer, however, the metabolic rates of aquatic organisms are high, the oxygen-dissolving capability of the warm water is relatively low, and the oxygen-demanding processes of bacterial decay proceed at high

levels. Under these conditions, oxygen depletion, or stagnation, of the profundal waters may cause extensive fish mortality.

Thermal Stratification In temperate latitudes, lakes show marked seasonal temperature changes. These, in turn, dramatically affect dissolved-oxygen and nutrient levels in different parts of the lake—information that is vital to resource managers. Let's take a look at the seasonal changes in a typical temperate lake.

Winter In the winter, as temperatures drop below freezing, ice begins to form on lakes in temperate regions. Because ice is less dense than liquid water, it floats on the surface of the lake, sealing off the surface and preserving an area for fish to spend their winters. As shown in Figure 9.10, during the winter, water temperature increases slightly at increasing depth. Throughout the winter, however, the water temperature remains relatively stable from top to bottom (Figure 9.10).

Spring In the spring, the ice on lakes in temperate areas begins to melt in the warm sun. Following the ice melt, the surface water begins to warm. When it reaches 4°C (39°F), all of the water achieves a uniform temperature and density from the surface to the bottom. Strong spring winds typically agitate the water, causing a complete mixing of water, dissolved oxygen, and nutrients from the lake surface to the lake bottom. This phenomenon is known as the **spring overturn** (Figure 9.11). As spring progresses, however, the surface water becomes warmer and lighter than the water at lower levels. As a result, the lake becomes thermally stratified again, as shown in Figure 9.12.

Summer In summer, the warm upper layer of the lake, known as the **epilimnion** ("upper lake"), usually has the highest oxygen concentration. Temperature within the epilimnion falls less than 1°C per meter (3.3 feet). Beneath the epilimnion is a layer called the **thermocline.** The thermocline is characterized by a more rapid fall in temperature—more than 1°C per meter. You can feel it if you

FIGURE 9.9 Characteristic plants and animals of the lake ecosystem: (1) reed; (2) cattail; (3) yellow flag; (4) adult caddis fly; (5) moth; (6) warbler; (7) damselfly; (8) coot; (9) dragonfly; (10) mallard; (11) mayfly; (12) heron; (13) frog; (14) phytoplankton; (15) water lily; (16) minnow; (17) duckweed; (18) snail; (19) snail; (20) hydra; (21) mayfly (immature); (22) water boatman; (23) diving beetle; (24) leech; (25) stickleback; (26) black bass; (27) tadpoles; (28) water strider; (29) diving beetle larvae; (30) gnat larvae; (31) crayfish; (32) amphipod; and (33) caddis fly larva.

dive toward the bottom. The bottom layer of water is the **hypolimnion** ("bottom lake"). Temperature falls within this layer, but more slowly than in the thermocline, that is, less than 1°C per meter (Figure 9.12).

During the summer, the hypolimnion of many lakes becomes depleted of oxygen, a phenomenon caused by a combination of factors: (1) the biological oxygen demand (BOD) of bacterial decomposers, (2) a lack of photosynthetic activity, due to the absence of sunlight, and (3) minimal mixing

with upper waters as a result of the density differences of water in these areas (Figure 9.12). Fish tend to be distributed in the upper levels of the lake, where oxygen is more abundant. (If these upper layers become depleted of oxygen—for example, as a result of certain types of water pollution—fish may perish.)

Autumn In the fall, the surface waters begin to cool again. Eventually, the temperature of the lake becomes uniform from

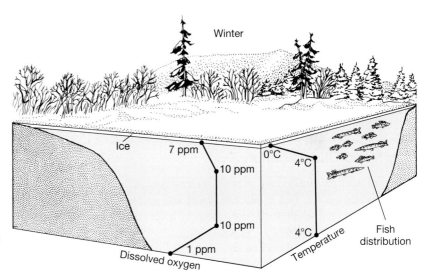

FIGURE 9.10 Lake ecosystem in the northern states in winter. Because water becomes less dense as its temperature drops below 4°C (39°F), the ice that forms at 0°C (32°F) is relatively light and floats at the surface. Note that the concentration of dissolved oxygen drops off sharply in the deeper part of the lake from about 10 ppm to only 1 ppm. Fish remain where the level of dissolved oxygen is high.

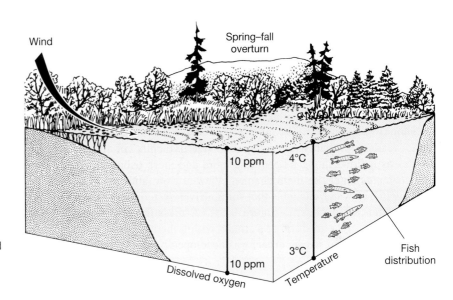

FIGURE 9.11 Lake ecosystem in spring and autumn. Note the uniform distribution of temperature and dissolved oxygen from top to bottom due to overturn, or thorough mixing of the water. Fish are also uniformly distributed in the vertical dimension.

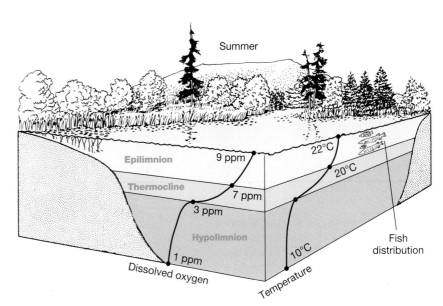

FIGURE 9.12 Lake ecosystem in summer. Cross-sectional view shows thermal stratification and the vertical distribution of dissolved oxygen.

top to bottom. As in the spring, the even temperature results in uniform density. Wind and waves begin to stir the waters, resulting in the **fall overturn** (Figure 9.11). As in the spring, the fall overturn distributes nutrients, dissolved oxygen, and plankton uniformly through the water column.

9.3 Stream Ecosystem

Origin and Classification

The continents are drained by streams, waterways that empty into lakes and oceans. Fed by surface runoff and groundwater that seeps into them along their banks (usually below the water surface), streams fall into two general categories, ephemeral and perennial. **Ephemeral streams** are those that flow only during one part of the year, typically the wet season when rain or snowmelt is abundant. Streams that flow year-round are called **perennial streams.** They are fed by rain (and snowmelt in colder areas) but are typically sustained by groundwater during dry periods.

A stream that flows into a larger stream is called a **tributary.** The area of land, confined by topographic divides, in which all tributaries drain into a single river system is termed a drainage basin, or **watershed** (Figure 9.13). Large watersheds are composed of several smaller **subwatersheds,** which contribute runoff to different locations but ultimately converge into the common river system.

As readers probably know, stream size typically increases as it flows along its path and more and more tributary segments enter the main channel. Viewed from above, the network of water channels that form a river system displays a branchlike pattern (Figure 9.14). A classification system based on the position of a stream within the network of tributaries, called **stream order,** was developed by Robert

E. Horton and later modified by Arthur Strahler. In general, the smaller and fewer the tributaries emptying into it, the lower the stream order number. For example, first-order streams (Order 1) are the smallest streams within the network, having no tributaries. Second-order streams (Order 2) have only first-order streams as tributaries, and so on (Figure 9.14).

Another stream classification system, the Rosgen method, describes in detail short lengths of a stream channel based on (1) gradient, (2) bed material, (3) bankfull width-to-depth ratio, (4) extent of meandering, and (5) composition of bank material. The proper use of this classification system requires extensive training and experience.

Physical Features

Channel Shape The cross-sectional shape of a particular section of stream channel depends primarily on water discharge (erosional force), sediment load, and the composition of bed and bank materials (degree of resistance to erosional forces). In general, sections of stream with densely vegetated banks and cohesive silt or clay bed and bank materials are narrower than sections with sparsely vegetated banks and noncohesive, sandy bed and bank materials.

During high-water events, when the force of the water flowing in a stream is great enough to erode banks and transport bed materials, stream channels are carved out of the bank and underlying beds. Most river cross sections become somewhat trapezoidal in straight sections. However, they tend to be asymmetric in the curves and bends. As the river gets larger downstream, width increases faster than depth, resulting in a larger width-to-depth ratio than an upstream tributary; that is, they tend to be wide and shallow.

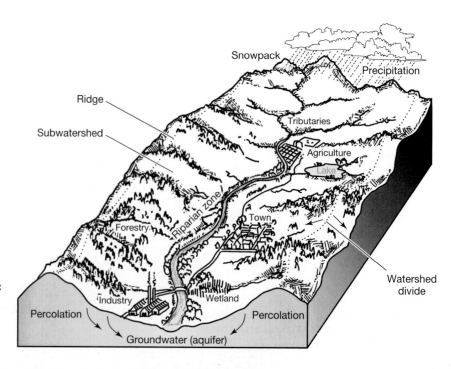

FIGURE 9.13 Water sources, tributaries, and human land uses within a single watershed. Note the topographic divides that define the boundaries of the watershed. Numerous subwatersheds of tributaries drain the area between the secondary ridges situated at nearly right angles to the divides.

Channel Pattern Viewed from above, a river channel can take on three different patterns: meandering, straight, or braided. The position of a stream in the watershed influences its channel pattern. For example, **headwater streams** (the beginning segments of a river system) are usually straighter, being confined by narrow valleys, shallow bedrock, and coarse sediments. In contrast, **lowland streams** (higher order) tend to meander. They have more freedom to carve a meandering path through the finer sediments of the floodplain.

The **meandering pattern** is by far the most common. In the meandering pattern, river channels are straight only for relatively short distances between curves, resembling a continuous string of S shapes. Meandering results when a channel naturally migrates laterally, eroding one bank and depositing material on the opposite bank. The depositional material forms what is called a **point bar.** The sinuosity of a channel— the lateral extent to which a channel meanders—varies (Figure 9.15).

Pools and Riffles Along the length of a channel bed are regularly spaced deep and shallow areas that create an undulating topography. Storm flows scour out deep areas that become **pools** in nonstorm periods; sediments are deposited as mounds on the streambed, forming shallow-water areas called **riffles.** Water flowing over these areas forms tiny waves, also known as riffles. At high flow, the visible "white water" caused by a riffle is obscured. In meandering channels, pools are also associated with the outside of a bend, where erosional forces are greatest. Steeper mountainous streams, having bed materials of boulders and large rocks, exhibit step pools—short (deep or shallow) pools spaced between the boulders. Kayakers call them pool drops.

Floodplains and Terraces **Floodplains** are the flat areas of fine sediment extending outward from the banks of lowland streams. Deposited by the streams over many years, these sediments support a rich variety of plant life, including many crops. As their name implies, floodplains receive river overflow whenever water in a stream or river spills over the top of the banks, a

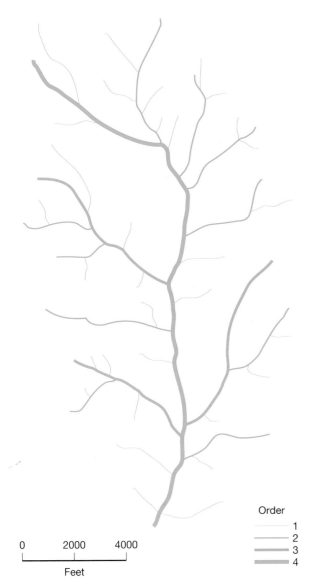

FIGURE 9.14 A network of water channels forming the branchlike pattern of a typical river system. Network channels have been assigned a stream order number based on Horton's classification system.

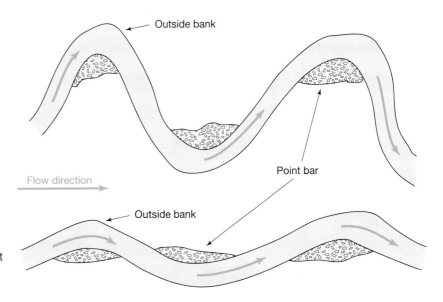

FIGURE 9.15 Lateral extent of channel meanders and deposition of sediment. The lateral extent of channel meanders is variable, but the erosional and depositional processes are similar. The outside bank is eroded, and point bars are created by the deposition of sediment on the opposite, inside bank.

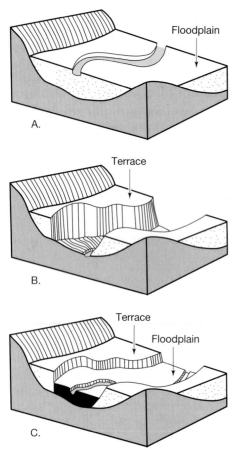

FIGURE 9.16   Formation of a terrace, or abandoned floodplain. (A) Numerous flood events cause a buildup of sediment, or floodplains, extending laterally from the tops of the riverbanks. (B) Changes in climate over time cause the river to downcut into the original floodplain, creating a terrace. (C) The number of flood events increases, causing new floodplain deposits.

condition technically called **bankfull discharge.** As climate changes occur over relatively long periods of time, rivers cut deeper into the floodplains. Called downcutting, this process can continue until the rivers are so far below the floodplains that they no longer overflow into them, as shown in Figure 9.16. These "abandoned" floodplains are then called **terraces.** Terraces lie adjacent to but at a higher elevation than a current floodplain.

Biological Community and Energy Flow

Organisms that live in the stream environment can be divided into three categories based on the function they perform: producers, consumers, and decomposers. Aquatic plants are the **producers.** As pointed out in Chapter 3, producers are photosynthetic organisms that provide energy to the community through photosynthesis. In streams, producers include diatoms, algae, and macrophytes (cattails, bulrushes, water lilies, river weeds, and so forth). **Consumers** are organisms that feed on plants and other organisms. These include invertebrates (aquatic insects, snails) and fish. Those that feed directly on plants are called **herbivores.** Another important group of consumers, the **predators** (fish, birds, mammals), feed on other consumer groups to acquire food and energy they need to survive and reproduce. Yet another important group of consumers

is the **decomposers,** that is, organisms that feed on wastes, recycling the wastes back into nutrient cycles.

Streams are rather open ecosystems. Thus, in addition to in-stream sources of energy, stream food chains receive a considerable portion of their basic energy supply from materials originating on land—outside of the stream ecosystem. In many streams, nutrients are constantly being received from the terrestrial ecosystems that border them (riparian zones). Dissolved organic matter, for example, may enter the stream from groundwater. Organic matter may enter through surface erosion of stream banks. Organic matter in the form of leaves falls into the water, especially in autumn. Organic debris (stems, nuts, twigs, needles, seeds, dead weeds, animal waste, and the bodies of insects, worms, and mice) is washed into streams during the spring runoff. Many of the primary consumers in a stream consume **detritus** (decomposing organic material) rather than living aquatic vegetation.

While streams are generally fed from the riparian zone, lowland streams of wider widths depend less on energy input from adjacent riparian vegetation than headwater streams. Why? Because lowland streams tend to be supplied by food and nutrients from upstream sources.

Riparian vegetation is the source of in-stream large woody debris. Logs, branches, and root wads fall into the stream because of wind or erosional forces. Such debris can shape channels by deflecting and slowing water flow. Large woody debris also creates prime fish habitat by providing cover, creating pools, and trapping organic matter. Some debris can have both deleterious and beneficial effects on fish. Logjams, for instance, can disrupt the migration of spawning fish while forming backwaters, sloughs, and marshes that increase the amount of rearing habitat for juvenile fish.

Roots from riparian vegetation stabilize stream banks against erosion and maintain **undercut banks,** another excellent feature of fish habitat (Figure 9.17). Undercuts provide shelter, while the overhead canopy consisting of trees or grasses shades the stream from the sun, reducing stream temperatures in the summer.

The composition of the biological community in any given section of a stream is determined by channel characteristics and energy sources. Slow-flowing, open streams, for example, support a different community of organisms than fast-flowing, shaded streams.

Clearly, riparian vegetation plays a huge role in the life of a stream and all the organisms that live there. As you may learn in future studies in resource management, the type, amount, and function of riparian vegetation vary from one geographic location to another. They are determined, in part, by climate and geology. But land uses also can affect streamside vegetation for better or for worse.

GO GREEN!

Join or start a local stream restoration group or a local Adopt-a-Stream program. These volunteer groups monitor their local streams and help with stream cleanup, riparian vegetation planting, and other stream restoration activities.

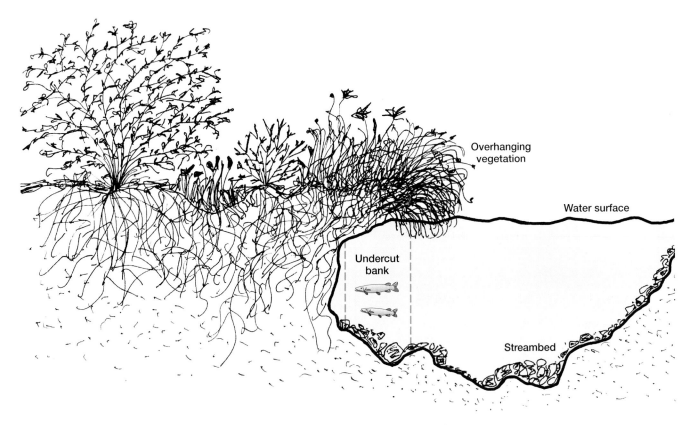

FIGURE 9.17 Impact of riparian vegetation. Rooted riparian vegetation stabilizes stream banks and maintains undercut banks. The undercut banks and overhanging vegetation provide cover, shade, and resting places for fish.

9.4 Coastal Environment

Structure of Coastlines

Coastal areas are dynamic environments. They are continually subjected to wind, rain, waves, tides, currents, and sea spray. These forces shape the world's coastlines primarily through erosion and deposition. If you have had a chance to visit the coastlines of North America, you know that they vary considerably. In this section, we'll examine various types of coastline and some of the forces that influence them.

Dunes and Beaches   **Sand dunes** and **beaches** can be found on any coastline in the world where there is a source of eroding mineral or a place that sediment can deposit. Sand dunes, in particular, develop in areas with an active supply of sand, strong, regular winds, and large tidal ranges. For example, windy conditions and extensive glacial deposits of sand have resulted in the formation of extensive areas of sand dunes along the Pacific Coast of North America from Alaska to southern California. The largest dunes occur near Coos Bay, OR. Here dunes are up to 50 meters (162 feet) high with ridge lengths up to 1,200 meters (3,900 feet) long. The dunes cover 72 kilometers (45 miles) of shoreline. In the United States, the longest and widest stretches of sandy beaches are found on the Atlantic and Gulf Coasts. They're formed by sediments eroded from inland sources that empty into streams that drain into the ocean.

Rocky Cliffs   Stretches of **rocky coastlines** consisting of various cliff formations, platforms, and intertidal rocks are common in many parts of the world. In the United States, rocky coasts dominate the Pacific Coast. The profiles of rocky coastlines are highly diverse, depending on resistance of the rock materials to erosive forces, past and present climatic conditions, and geologic history.

Hard rocks, such as granite or basalt, often form steep cliffs that plunge directly into the sea or rise up sharply behind beaches. Because these rocks are highly resistant to weathering, they have experienced very little erosion over time.

In contrast, cliffs composed of unconsolidated or weakly consolidated material, such as clays or glacial deposits, are receding at relatively high rates. These cliffs are prone to failure and highly susceptible to erosion by wind and water. Cliffs prone to failure frequently experience mass movements such as landslides, rockfalls, and the like. As a result, they are more likely to slope gradually or form steplike slopes.

Vegetation thriving on hard rock cliffs consists of those species adapted to high salinity and rocky substrate. Inaccessible to humans and many predators, the cracks, ledges, and crevices of rocky cliff faces provide protected nesting habitat for many seabirds. In contrast, cliffs composed of weakly consolidated materials support vegetation for only short durations before it is torn away by erosional processes.

In this discussion, we have placed rocky cliffs into two categories: stable (hard rock) and unstable (weak, unconsolidated rock materials). Of course, the real world is more complicated. Cliffs are rarely made up of uniform rock materials. They more often consist of both strong and weak structure materials, so the variety of rocky coast profiles is enormous.

Barrier Islands   Skirting much of the eastern coastline of the United States and the Gulf of Mexico is a string of long, narrow islands known as **barrier islands** (Figure 9.18). Formed by waves, wind, and currents, barrier islands are accumulations of coastal sediments, principally sand. These islands run parallel to and near the shoreline. Structurally, barrier islands usually consist of sandy beaches and sand dunes and a few inlets. They are typically covered with grasses and sparsely distributed pine and oak trees and are usually separated from the mainland by salt marshes or lagoons. Barrier islands are a popular place for resorts, vacation homes, and year-round residences. Barrier islands are popular sites for recreation, and some have been designated as national seashores by the U.S. Congress. Most, however, are privately owned.

Unfortunately, barrier islands are highly dynamic environments, constantly changing shape and size as a result of erosion caused by wind, waves, and offshore currents. Sand on the upcurrent side of the islands is constantly being eroded away, so homes built in these areas may collapse into the ocean.

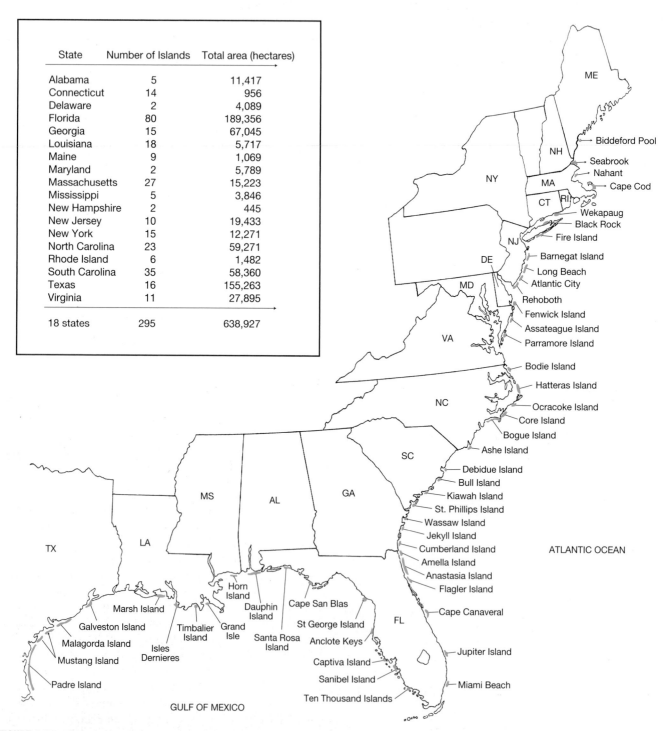

State	Number of Islands	Total area (hectares)
Alabama	5	11,417
Connecticut	14	956
Delaware	2	4,089
Florida	80	189,356
Georgia	15	67,045
Louisiana	18	5,717
Maine	9	1,069
Maryland	2	5,789
Massachusetts	27	15,223
Mississippi	5	3,846
New Hampshire	2	445
New Jersey	10	19,433
New York	15	12,271
North Carolina	23	59,271
Rhode Island	6	1,482
South Carolina	35	58,360
Texas	16	155,263
Virginia	11	27,895
18 states	295	638,927

FIGURE 9.18   Major barrier islands along the Atlantic and Gulf Coasts of the United States. These islands were built up from sediments by the action of waves, wind, and sea currents.

Barrier islands are also first in line for hurricanes striking the coasts. Because the islands are so dynamic and so vulnerable to hurricanes and other violent storms, the buildings and roads—and residents—frequently fall victim to storm surges, flooding, and high winds. Millions of dollars' worth of damage can be caused by a single storm.

Much damage caused by storms is repaired by insurance, although insurance companies are increasingly reluctant to cover homes and other buildings built on barrier islands. Some companies even redline barrier islands—that is, refuse to insure them. For years, the federal government has also stepped in, providing assistance to rebuild after storms. Recognizing how costly this was, in 1982 Congress passed the Coastal Barrier Resources Act, which explicitly prohibits the expenditure of federal money to rebuild storm-struck barrier islands. Nonetheless, development continues.

Coral Reefs   **Coral reefs** are formed from animals having calcareous skeletons (hard structures made from calcium materials) and certain species of algae that provide the sediment or "cement" to seal the coral framework. Coral animal offspring bud out and attach to the side of the parent. As animals die, new generations build onto the dead skeletons, often creating massive structures. Coral reefs are considered the largest biological constructions on Earth. The largest and best-known chain is the **Great Barrier Reef,** located off the eastern coast of Queensland, Australia, which reaches a length of over 2,000 kilometers (1,260 miles) and contains more than 2,500 individual reefs. Coral reefs are primarily found in tropical and subtropical waters where temperatures never go below 20°C (68°F) and depths are less than 100 meters (325 feet).

In the United States, coral reefs are found off the coasts of Hawaii, Florida, and the Virgin Islands. Coral reefs are highly productive environments and home to an abundant and diverse group of plants and animals that live in, upon, and around them. These environments are also highly sensitive to natural and human-induced disturbances. Erosion from nearby land surfaces can smother the life out of coral reefs. Ships running aground can damage delicate coral reefs, as can the fins of careless divers. Perhaps the biggest threat comes from warming seas, which some scientists link to global warming. Worldwide, many coral reefs have become bleached due to rising ocean temperatures that kill the algae required by the coral to survive, leaving behind white skeletons.

Estuarine Ecosystem

Estuaries and their associated coastal marshes are prominent features of the U.S. coastline. As mentioned earlier in the chapter, **estuaries** are semi-enclosed inlets (bays) that form transitional zones between rivers and the sea (Figure 9.19). In one sense, they represent a river–ocean hybrid, possessing some of the characteristics of each ecosystem. Nevertheless, the estuary has some distinctive properties and therefore must be considered a unique ecosystem.

Characteristics   The water level in an estuary rises and falls with the tides. The water is a mixture of freshwater from the stream and saltwater from the ocean. The salinity of estuarine water is highly variable, changing by a factor of ten within a 24-hour period; salinity is highest when the tide comes in and lowest when the tide recedes.

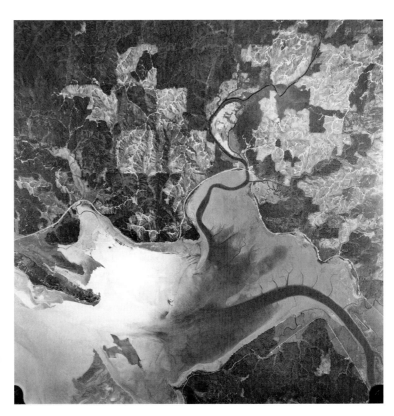

FIGURE 9.19   Aerial map of Willapa Bay, off the coast of Washington State. This bay is a transitional zone between the saltwater of the Pacific Ocean and the freshwater rivers that drain into the bay. The two major coastal rivers forming this estuary are the Willapa River (lower right) and the North Lower Salmon River (upper center). Notice the large clear-cuts (light gray areas in upper half of photo).

The concentration of dissolved oxygen is relatively high because estuaries are shallow and the water is generally quite turbulent. Because of the stirring action of the tides, **turbidity** (the concentration of fine sediment suspended in the water column) is characteristically high. The reduced penetration of sunlight due to this turbidity limits phytoplankton populations. However, nutrient levels are high. Nutrients carried down to the estuary by stream flow and those carried up by the incoming tides are concentrated in the estuaries, making them highly productive ecosystems.

Two food chains operate in the estuary. The first is the grazer food chain, where dissolved nutrients may be absorbed directly by phytoplankton and rooted plants, then passed into consumers. The second is the decomposer food chain, where inert organic material (decayed bodies of marsh grasses, crustaceans, worms, fishes, bacteria, and algae), known as detritus, is consumed directly by detritus feeders, such as clams, oysters, lobsters, and crabs (Figure 9.20).

Because of the abundant supply of nutrients and the high oxygen levels, the estuarine habitat produces more organisms than any other ecosystem except the coral reef.

Values   Estuaries and their associated coastal marshes are indispensable to commercial and recreational fishing industries worth $56 billion annually. Sixty percent of the marine fish harvested by American commercial fishing interests spend part of their life cycle in estuaries. Of every 100 fish taken in the Gulf of Mexico, 98 are dependent on estuaries and salt marshes. Many marine species use the estuary as a "nursery" in which they spend their larval period immediately after hatching from the egg. Other species, such as the Pacific salmon, pass through estuaries twice during their stream–ocean–stream migration.

Estuarine ecosystems provide food, shelter, and breeding sites for millions of waterfowl and fur-bearing animals, such as muskrats. Almost 45% of all U.S. threatened and endangered species are dependent on coastal habitats.

Coastal wetlands in the estuarine zone play an important role in flood and erosion control. They absorb the shock of storm-driven waves, reducing destruction of property and human life in heavily populated areas (Figure 9.21). Estuaries and coastal wetlands are also natural pollution-filtering systems. They cleanse the water of industrial and domestic sewage delivered to the estuaries by rivers.

Detritus food chains of an estuary

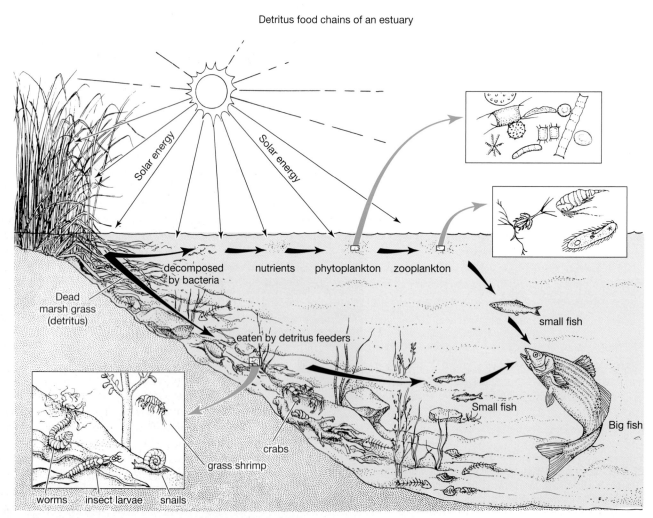

FIGURE 9.20   Detritus food chains of an estuary. The major food chains (and webs) of the estuary are based on detritus—the decomposed bodies of plants and animals.

power plants, and factories have sprung up along the nation's coastlines. Harbors have been dredged to make room for ships. Wetlands have been filled with soil to increase land for development. Sediment has been mined for sand and gravel. Fishing operations have sprung up alongside sewage treatment plants and offshore oil- and gas-drilling rigs.

During the past decade, 17 of the 20 fastest-growing counties and 19 of the 20 most densely populated counties were located along the coast. In addition, coastal areas include 16 of the 20 counties with the most new housing units under construction. In the next decade, the largest increases in population are expected in the coastal areas of Southern California, Florida, Texas, and Washington. Economic development, relocation by retirees, and the proliferation of vacation homes are among the factors fueling this rapid population growth.

Coastal Problems

Coastal environments are fragile and highly sensitive to human activities and natural changes. Human development, for example, can result in habitat destruction and pollution, upsetting the ecological relationships in the coastal environment. Natural events such as hurricanes and tsunamis can also have devastating effects that are often worsened as human development increases in coastal areas. Let's examine a few common problems.

Destruction of Estuarine Habitat and Coastal Wetlands One of the most obvious—and irreversible—effects of human activity is habitat destruction. According to the U.S. Fish and Wildlife Service, estuarine and marine wetlands (all intertidal wetlands) in the conterminous United States decreased from about 2.4 million hectares (6.0 million acres) in 1950 to 2.1 million hectares (5.3 million acres) in 2004 (Figure 9.22). Contributing to this decline are several interrelated factors: deficiencies in sediment deposition, canals and artificially created waterways, wave erosion, land subsidence, and saltwater intrusion causing marsh disintegration. Yet the destruction of estuarine and marine wetlands has fallen dramatically over the years, thanks to greater awareness of the value of wetlands, as well as state and federal laws. However, as encouraging as this trend is, another important part of the reason for the decline in losses is that there is very little marine wetland left.

Another, less obvious problem afflicting estuarines is shoreline erosion caused by waves (called wakes) produced by ships and boats. Although shorelines are pounded by waves naturally, ships and boats augment the impact, destroying precious fish and wildlife habitat.

In addition to causing physical habitat destruction, human activities also affect the chemical balance of estuaries. Dams, for instance, reduce the influx of freshwater into estuaries. Saltwater moves higher up the stream, affecting freshwater fish species that live and reproduce in the mouths of freshwater rivers.

Human activites along the nation's watersheds—for example, farming—may also increase the influx of plant nutrients into estuarine ecosystems, upsetting the ecological balance. Pollutants from factories, sewage treatment plants, and a host of other sources result in a different problem: They can poison fish and other aquatic species.

FIGURE 9.21 Salt marsh grass growing along the Cape Cod shoreline near North Chatham, MA. This vegetation protects the coastline from erosion.

According to one study, an area of only 5.6 hectares (14 acres) of estuary has the same pollution-reducing effect as a $1 million waste treatment plant!

Human Development of Coastal Environments

Coastal areas have many attributes that make them attractive for human development. Not only are shorelines rich in life, they are extremely beautiful. Many have mild climates and offer a wealth of recreational opportunities. Coastlines are prime locations for shipping, receiving, and transporting raw materials and goods by sea. Not surprisingly, the human population near our nation's coastlines has grown faster than in any other area of the country. Today, more than half of all Americans live in this delightful region where the land meets the sea—a narrow fringe of land that represents only 17% of the nation's contiguous land area. Worldwide, two-thirds of the largest cities (those with population greater than 2.5 million) are located near estuaries.

Coastline development began shortly after World War II as enterprising developers launched a multi-billion-dollar building boom. Over the years, they've built many expensive beachfront homes, condominiums, hotels, and resorts. Ample federal subsidies given for water supply systems, sewers, roads, bridges, and shoreline stabilization helped to spur the construction. But coastal development has included much more than homes and resorts. Over the years, ports,

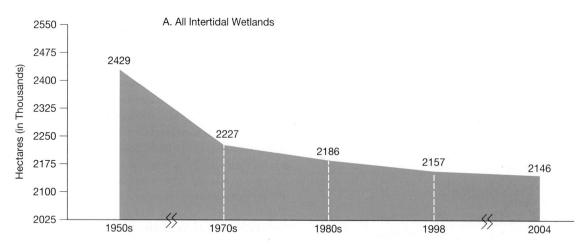

FIGURE 9.22 Long-term trend in all (estuarine and marine) wetlands. (Source: U.S. Fish and Wildlife Service.)

Damage and Loss of Life from Storms As noted earlier in the chapter, salt marshes and mangrove swamps in coastal areas act as buffers against storms, absorbing the high-energy wind, rains, and waves before the storms move inland. As you might suspect, mushrooming development on the coastlines and barrier islands and the associated decrease in coastal wetlands increase our vulnerability to—and very likely increase property damage and loss of life caused by—hurricanes and other ocean storms (Figure 9.23).

Erosion Coastal erosion is a natural process, resulting from the forces of waves, winds, and storms. Erosion, as noted earlier, is especially damaging along coastlines composed of sand and other weakly consolidated materials. These processes, although natural, continually threaten developed coastal areas. In the United States, all 30 states

bordering the ocean and the Great Lakes report problems from erosion. Twenty-six of the 30 report a net loss of shoreline. The coastline on some of the barrier islands in the southeastern United States recedes an average of 7.5 meters (25 feet) per year. Erosion rates as high as 15 meters (50 feet) per year have occurred along the shores of the Great Lakes.

Examples The following are just a few examples of erosion problems in some of our coastal states:

1. *New York.* Aerial photographs show that Long Island's shoreline retreated 30 meters (100 feet) in the latter half of the 20th century.

2. *North Carolina.* The Cape Hatteras lighthouse on the North Carolina coast was once at least 460 meters (1,500 feet)

FIGURE 9.23 Flooded neighborhoods near the coastline of New Orleans caused by Hurricane Katrina in 2005.

inland from the high-tide line. However, so much beach erosion has taken place that the wind-whipped waves in the 1990s began to surge within 54 meters (180 feet) of this picturesque landmark, forcing officials to move the massive structure a mile inland.

3. *California.* More than 86% of California's 1,771 kilometers (1,100 miles) of exposed Pacific shoreline is receding at an average rate of 15 to 60 centimeters (6 inches to 2 feet) per year. Monterey Bay, south of San Francisco, loses 152 to 300 centimeters (5 to 10 feet) annually.

4. *Washington.* Cape Shoalwater, on Washington's Olympic Peninsula, has eroded 30 meters (100 feet) per year since the early 1900s. The cape's sparsely settled sand dunes have also retreated more than 3.2 kilometers (2 miles) since that time.

Accelerated Erosion   Although erosion of coastlines is a natural phenomenon, human activities are causing the world's shorelines to erode at a much faster rate than that due to natural forces. How? According to scientists, humans have "altered the dynamics of coastal sediment deposition." That is, we have altered natural processes responsible for maintaining our shorelines. Let's look at this issue in more detail.

Sand along the nation's shorelines is replenished each year by new sand flowing in from rivers. However, over the years, thousands of dams have been built along the nation's waterways to generate hydroelectric power or to store water for irrigation and municipalities. Unfortunately, dams block the flow of water into the ocean—water that formerly carried millions of tons of sediment to river mouths to form deltas. Many deltas and beaches, therefore, are no longer being replenished.

Certain human activities may also increase shoreline erosion. **Seawalls,** for example, are built to protect the land behind them from erosion. Over time, however, waves and currents often erode the beaches in front of seawalls. In fact, erosion of the "front beach" can be so severe that the seawall may be undercut and may collapse.

Another structure that increases erosion is the groin, shown in Figure 9.24. **Groins** are piers of stone spaced about 30 meters (100 feet) apart along the coastline. They extend into the sea at right angles to the shoreline. As illustrated in Figure 9.24, these structures trap sand washed against them by currents that move parallel to the shore, protecting the beaches. However, sand moves naturally along coastlines, carried by prevailing longshore currents. Groins therefore deprive shorelines immediately downcurrent of them from this beach-replenishing sand. The only way to overcome this problem is to build additional groins downcurrent.

Other human activities can affect the coastal environment. For example, the extraction of oil, natural gas, and groundwater along the nation's coastlines often results in a problem known as **subsidence.** Subsidence is a sinking of the Earth's surface, caused when underground resources such as oil and groundwater are removed. Subsidence has been particularly noticeable—and troublesome—in Louisiana, Texas, and California, in areas that were already at or near sea level. In Louisiana, the land has subsided 1 meter (3.3 feet) in the past 100 years. When coastal lands subside, ocean waves surge inland and wash soils away.

Rise of the Oceans   All of the problems discussed so far are worsened by rising sea levels. Studies have shown that the oceans have been rising since the end of the last Ice Age, about 11,000 years ago. However, in the past century or so, the rate has quickened. In New York City, for example, sea level has

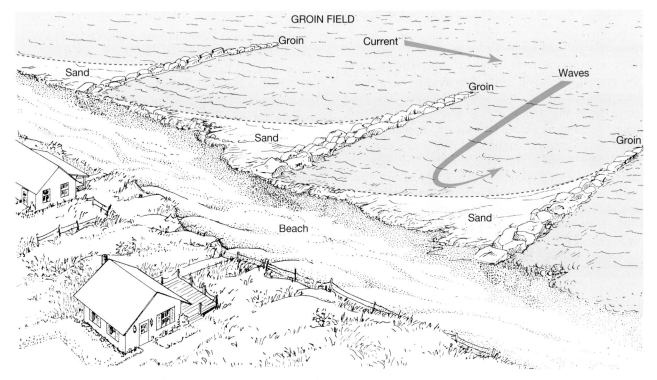

FIGURE 9.24   Use of groins in an attempt to structurally control beach erosion.

risen about 40.0 cm (15.7 inches) since 1850 at a rate of about 2.7 cm (a little more than 1 inch) each decade, with small interannual fluctuations. Why? The primary cause, many scientists believe, is *climate change* resulting from the increasing levels of carbon dioxide in the atmosphere. (See Chapter 19 for more details.) Increases in carbon dioxide have resulted primarily from the combustion of fossil fuels such as coal, oil, gasoline, and natural gas. For more than 100 years, scientists have known that carbon dioxide traps heat in the Earth's atmosphere, warming the Earth's surface. The increase in temperature caused by rising levels of carbon dioxide, in turn, has resulted in the melting of many glaciers and the polar ice caps. This, in turn, has increased sea levels. Scientists also believe that the increase in sea levels is a result of the expansion of the world's oceans—as water warms, it expands.

As you will learn in Chapter 19, carbon dioxide levels are expected to increase in the coming years, raising sea levels even further. As a result, ocean waves—especially during storms—might surge inland hundreds of meters in low-lying coastal regions, washing away beaches, causing serious property damage, and inflicting considerable mortality of both wildlife and humans.

Pollution In addition to all of the factors just described, pollution is affecting the nation's coastlines. Pollution—and the damage it creates to coastal environments—increases with the increasing use of coastal areas, not just from human development but also from recreational activities. Pollution comes from a variety of sources, including sewage treatment plants, stormwater drain outflows, and industries. It also comes from farmland, ships at sea that dump wastes into the ocean, and offshore oil drilling. Even litter left by beach visitors affects our coastlines. In some areas of the country, beaches have been closed for public use because levels of bacteria from human sewage have exceeded recognized health safety standards.

"Dead zones" are major forms of coastline pollution caused by an excess of farm fertilizers (especially nitrogen) and the burning of fossil fuels. The number of oxygen-deprived dead zones in the world's oceans has been increasing since the 1960s and is now more than 400, threatening fisheries as well as humans who depend on fish. A **dead zone** is an area of hypoxic (less than 2 ppm dissolved oxygen) waters found at the mouth rivers. For example, fertilizers from farming operations, which run off into rivers and then flow into oceans, trigger the proliferation of plankton, which in turn depletes oxygen in the water. While some fish might flee this suffocation, slow moving, bottom-dwelling creatures like clams, lobsters, and oysters are less able to escape and die. Dead zones range in size from as small as a square kilometer (0.4 square mile) to more than 70,000 square kilometers (27,000 square miles). One of the largest and most notorious dead zones is a 22,126 square kilometer (8,543 square mile) region in the Gulf of Mexico, where the Mississippi River dumps high-nutrient runoff from its vast drainage basin, which includes the heart of U.S. agriculture, the Midwest, affecting important shrimp fishing grounds in the Gulf waters.

Sustainable Coastline Management

With all of the forces affecting coastlines, there's clearly a need to manage this fragile and valuable environment. Sustainable coastline management seeks to guide human activities that affect coastal environments in ways that protect the function of these vital resources for present and future generations—and the species that depend on them—while allowing human communities and economies to thrive.

In 1972, the U.S. Congress passed the landmark **Coastal Zone Management Act (CZMA).** This law helps support sustainable coastal management by establishing voluntary partnerships between the federal government and state, territory, and commonwealth governments. With federal and state matching funds, states and local governments develop and implement coastal-zone management programs to achieve the following national objectives identified in the CZMA:

1. Protect natural resources.
2. Manage coastal development to reduce the impact of natural hazards.
3. Protect and restore coastal water quality.
4. Provide public access to the coast.
5. Give priority consideration to coastal-dependent uses and orderly siting of major facilities.
6. Encourage urban waterfront and port redevelopment and historic and cultural preservation and restoration.
7. Support comprehensive planning and management for living marine resources.
8. Plan for the effects of land subsidence and sea level rise.
9. Coordinate and simplify government decision making.
10. Encourage public participation in coastal-management decisions.

Amendments in 1990 and 1996 broadened the scope of the CZMA, including additional funding to address nonpoint-source–polluted runoff into coastal wetlands and waters and to establish a National Estuarine Research Reserve System. The act was further amended in 1998 and 2004 to establish a program for the prevention and control of harmful algal blooms and hypoxia in coastal areas.

Estuarine and Coastal Wetland Protection Today, states employ a variety of techniques to protect, enhance, or restore their wetlands. In fact, most states have their own "no net loss" policy to prevent wetland acreage from declining. These policies are backed up by local zoning ordinances, land-use plans, and permit systems that seek to protect remaining wetlands. Twenty-six states also acquire or buy conservation easements to protect wetlands, and 25 states have programs that actively restore, enhance, and/or create wetlands. One of the largest restoration projects is in the Florida Everglades.

Touted by some scientists as the largest environmental restoration project in human history, restoration of the Everglades is aimed at reversing damage to this massive wetland created by well-meaning efforts to control flooding and make room for human activities. A century of human modifications in the southern part of Florida resulted in the conversion of more than 1 million hectares (2.5 million acres) of this "river of grass" into farmland, housing developments, and recreational areas. While the farmland is some of the most productive in the United States, the hydrologic modifications had severe negative effects on water flow through the Everglades, with devastating effects on fish and wildlife, including waterfowl, wading birds, and the Florida panther. Today, engineers have modified the flood-control structures to direct much of the water back into the swamplands, breathing life into the wetlands once again. Cost estimates for the project range from $3 billion to $5 billion over a 10- to 15-year period.

Louisiana contains a whopping 35% to 40% of the nation's coastal wetlands, including ubiquitous old-growth cypress swamps and vast stretches of salt marshes formed in the sediment deposits of the great Mississippi River delta. However, flood control, navigation, oil and gas exploration, and urban and agricultural land development projects have had devastating effects on the wetlands that border Louisiana's Gulf of Mexico coastline. Deprivation of river sediment supplies, accelerated coastal erosion rates, saltwater intrusions, conversion to deepwater habitats caused by dredging, and drainage for other uses have decreased Louisiana's coastal wetlands by 25% since the early 1900s.

Over the years, the loss of Louisiana's wetlands has resulted in a parallel decline in shellfish and finfish. With this decline in ecological diversity, there have been decreases in associated economic activity and in recreational opportunities. Interestingly, oil

and gas wells in coastal wetlands are threatened by the conversion of the wetlands to open water.

In response to these changes, the state and federal governments have dedicated as much as $30 million annually to large-scale wetland creation and enhancement projects (Figure 9.25). In 1988, the U.S. Army Corps of Engineers launched a $25 million project to divert freshwater from the Mississippi River into Louisiana wetlands and estuaries to check saltwater intrusion. Sediment dredged from river channels has been used to create marshes in the coastal region of the Gulf of Mexico. Since 1993, funding has been allocated to restore nearly 20,235 hectares (50,000 acres) of Louisiana's coastal wetlands.

Erosion-Control Techniques   We noted earlier that certain shoreline structures designed to protect beaches from erosion actually increase erosion. In another vain attempt to protect beaches, some communities have trucked sand onto eroded beaches to replenish material lost during storms or as a result of normal beach erosion. Although numerous states have used this approach, they have found that the benefits of **beach nourishment,** as it is called, are temporary, lasting only two to seven years, while the costs are extremely high. In some cases, the benefits are even more short-lived. A $2 million nourishment project at Ocean City, MD, for example, lost 60% of its sand to violent storms only two weeks after it was completed!

FIGURE 9.25   Point au Fer Wetland Restoration Project off the Gulf Coast of Louisiana. Boulders were placed along a 1,220-meter (4,000-foot) stretch of this narrow strip of beach to prevent storm overwash and potential breaching into the wetland situated behind the beach. The salt marsh is now protected from accelerated erosion, saltwater intrusion, and hydrologic alteration.

Many states have begun to take a more natural—and potentially more sustainable—approach to controlling erosion along shorelines: planting native vegetation. Consider an example. In Texas, patches of salt-tolerant cordgrass have been planted by conservation workers in the shallow waters near the coastline of Galveston Bay. This grass serves as a living buffer against the waves. If done correctly, this technique seems to work well and establishes more or less permanent protection. However, experience shows that planting schemes that fail to mimic the natural patterns of succession and spacing may have limited success. As one example, overplanting can lead to overstabilization of certain areas, causing excessive erosion and deposition elsewhere.

Hazard Reduction Techniques Besides protecting coastlines, many states have enacted programs to protect people, notably coastal residents and vacationers, who are vulnerable to the hazards of long-term erosion and damaging forces of natural, catastrophic events such as hurricanes, tsunamis, and earthquakes. Coastal states are utilizing various approaches to prevent or lessen losses of life and property and to protect shoreline resources.

Restriction of Coastal Development States have restricted development and other human activities on or near high-energy beaches, dunes, and cliffs. They do so by requiring setbacks (requiring houses and other buildings to be a certain distance from the shoreline) and eliminating government subsidies for private development. They have also passed regulations to prohibit the construction of shoreline stabilization structures, such as groins. In some cases, they have even restricted vehicle access to beaches. In North Carolina, for example, new buildings must be located at least 40 meters (120 feet) inland from the first line of dunes. In Delaware, state law forbids the construction of any industrial plants within 3 kilometers (2 miles) of the seashore. New Jersey's Waterfront Development Act (1988) empowers the Department of Environmental Protection to fine waterfront owners $1,000 if they illegally place structures too close to the shoreline and to impose a $100-a-day fine for each day the violation continues. Besides helping to ensure human safety and reduce property damage, such measures help to reduce erosion, preserve scenic beauty, and protect vulnerable coastal wetlands and estuaries.

Planning and Public Education Planning and education also help lessen the loss of life and damage to existing coastal development from various natural and human-induced hazards. State and local governments are identifying areas susceptible to storm damage, developing improved hurricane evacuation plans and emergency communications systems, and ensuring public awareness of appropriate evacuation routes and shelter locations. In addition, education and public outreach programs aim to increase public awareness of the dangers associated with building in dynamic shoreline areas (Figure 9.26).

Protecting Coral Reefs Coral reefs have historically exhibited the capacity to recover rapidly from natural stresses including catastrophic tropical storms, predation, diseases, and freshwater runoff that causes sedimentation and upsets the natural

FIGURE 9.26 Beach house located on the North Carolina coast. This building is much too close to the sea. The damage resulted from the house's vulnerability to flooding and beach erosion.

nutrient and salinity balances. However, increases in human populations in coastal subtropical and tropical areas during the past 60 years have resulted in new stresses that have accelerated damage to coral reefs and valuable reef fisheries resources. These stresses include polluted stormwater runoff, agricultural and industrial point and nonpoint source pollution, sewage effluent, and overfishing. Direct contact by humans via snorkeling and diving in popular coral reef destinations has led to physical damage to the reefs and removal of coral and other marine life for souvenirs or for sale. Recent coral-eating starfish plagues in the Pacific and coral diseases in the Caribbean have been particularly devastating in some areas, escalating speculation that both of these stresses may be exacerbated by human activities.

Many coral reefs are located in remote locations away from human development or are—as in the case of the Great Barrier Reef in Australia—under good management. However, in its 2007 annual report, the Coral Reef Alliance, an international nonprofit organization working exclusively to save coral reefs, stated that more than 27% of the world's coral reefs have been lost or severely damaged and another 32% may be destroyed by human activities in the next 30 years. Several management techniques are in place in the United States and other countries to help lessen or reverse human impact on these fragile and valuable resources. Protection techniques include prohibition of coral removal, restrictions on touching coral and anchoring boats in coral reef areas, limitations on recreational activities and boating traffic, and regulations that control polluted runoff. Public outreach and education are important components of many coral reef protection plans and include partnerships with dive shops and clubs to increase the awareness of activities that harm coral reefs. Proactive efforts such as installation of mooring buoys along the outskirts of some coral reefs allow visiting boaters to tie up without anchoring on the reef itself. Several coral reefs have been set aside for research or restoration or have been established as marine protected areas, including the Florida Keys National Marine Sanctuary and the Gray's Reef National Marine Sanctuary off the coast of Georgia.

9.5   The Ocean

General Features

The ocean covers 70% of the Earth's surface and is divided into three major basins: the Pacific, Atlantic, and Indian Oceans. The Pacific Ocean is the deepest and largest ocean basin. The deepest point is the 11-kilometer (6.5-mile) Mariana Trench southeast of Japan. The Atlantic Ocean is the warmest ocean and is relatively shallow. The land area draining into the Atlantic Ocean is four times that draining into the Pacific Ocean. The Indian Ocean is the smallest of the major basins and lies primarily in the Southern Hemisphere.

Ocean water comprises approximately 96.5% water, 3.5% salt, and tiny amounts of living things, dust, and dissolved organic matter. The oceans are about 70 times as salty as a lake or stream. The salts come from volcanic eruptions and the chemical weathering of rocks on the ocean bottom and on exposed land surfaces. The four ions that account for 98% by weight of the ocean's salt are Na^+ (64%), Cl^- (29%), Mg^{2+} (3%), and SO_4^{2-} (2%).

The ocean is continuously circulating (Figure 9.27). The Alaskan Current brings cold water down the Pacific Coast; the Gulf Stream brings warm water upward along the Atlantic Coast. These currents modify the temperatures of nearby coastal regions. For example, New York City has a relatively moderate climate, even in winter, whereas summer evenings in San Francisco may be quite chilly. Vertically moving currents, or **upwellings,** bring nutrient-rich cold water from the ocean bottom to the surface (Figure 9.28).

The sea is relatively infertile compared with freshwater. Nitrates and phosphates are scarce. Two exceptions are the areas of upwelling and estuaries, where streams discharge massive loads of sediment.

Ocean Zonation

Like lakes, the ocean can be divided into zones based on physical and chemical characteristics.

Neritic Zone   The **neritic zone** is the marine counterpart of the littoral zone of the lake. This zone is a relatively warm, nutrient-rich, shallow region that overlies the continental shelf, a submerged extension of the continent. In North America, the neritic zone borders the Atlantic, Pacific, and Gulf Coasts, and is 16 to 320 kilometers (10 to 200 miles) wide and up to 60 to 180 meters (200 to 600 feet) deep. The neritic zone ends where the continental shelf abruptly terminates and the ocean bottom plunges to great depths (Figure 9.29).

The nutrients of the neritic zone are supplied primarily by upwelling and the sedimentary discharge of streams. Sunlight normally penetrates to the ocean bottom, permitting considerable photosynthetic activity in a large population of floating and rooted plants. Animal populations are rich and varied (Figure 9.30). Oxygen depletion is not a problem because of photosynthesis and wave action. The total amount of biomass in the neritic zone is greater per unit volume of water than in any other part of the ocean.

Euphotic Zone   The **euphotic zone** is the open-water zone of the ocean that corresponds to the limnetic zone of a lake

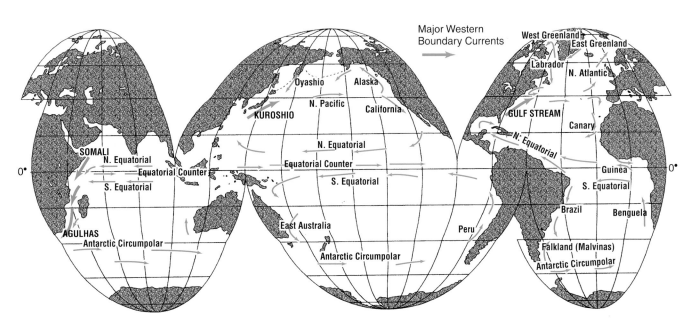

FIGURE 9.27   Major ocean surface currents. The thicker the arrow, the stronger the current.

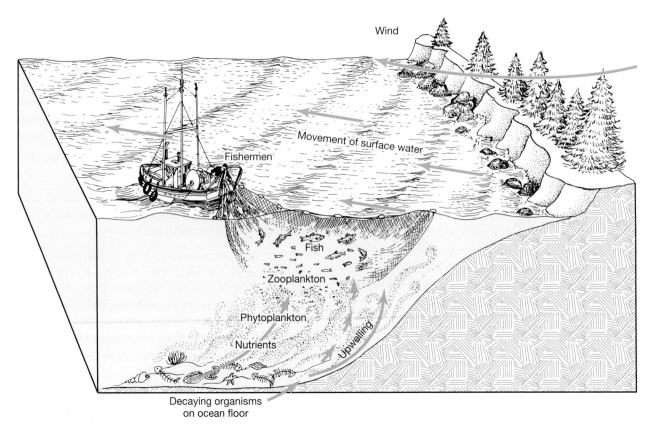

FIGURE 9.28   The upwelling phenomenon. Wind moves the surface water outward. As a result, cold water from the ocean bottom moves upward and shoreward, carrying dissolved nutrients with it. Consequently, the area of upwelling is biologically productive.

(Figure 9.29). The term *euphotic,* which means "abundance of light," is appropriate to this zone, for it has sufficient sunlight to support photosynthesis and a considerable population of phytoplankton. In turn, the phytoplankton support a host of tiny grazing herbivores such as the small crustaceans, part of the zooplankton. The total energy available to animal food chains from euphotic phytoplankton is much greater than that produced by plants of the neritic zone; this is largely because of the vast area of the euphotic zone, which extends for thousands of miles across the open sea. The degree to which light penetrates, of course, depends on the transparency of the surface waters. Because sunlight cannot penetrate deeper than 200 meters (650 feet) in most marine habitats, this depth is frequently considered the lower limit of the euphotic zone.

Bathyal Zone   Beneath the euphotic zone is the **bathyal zone** (Figure 9.29). This zone is a region of semidarkness. Photosynthesis cannot occur here, so no producer organisms can survive.

Abyssal Zone   Beneath the bathyal zone is the **abyssal zone.** The abyssal zone is the cold, dark-water zone of the ocean depths that roughly corresponds to the profundal zone of the lake habitat. This zone lies immediately above the ocean floor (Figure 9.29). Animal life is rather sparse. Any animal living in the abyssal zone must be adapted to low

light and intense cold, because the abyssal water frequently approaches the freezing point. Organisms must also be adapted to extremely low levels of dissolved oxygen (because no photosynthesis can occur here), intense water pressure, and a scarcity of food.

In many areas, the sediment of the abyssal zone is rich in nutrients. These nutrients come from the decaying bodies of organisms from the sunlit waters above and from excretions of living animals. For example, in certain regions of the western Atlantic, the phosphate concentration at a depth of 900 meters (3,000 feet) may be ten times greater than the concentration at 90 meters (300 feet). Because neither phytoplankton nor herbivorous animals can exist in the abyssal zone, most consumers are either predators or detritus feeders. A number of the deep-sea fish of the abyss have evolved luminescent organs that may aid them in attracting food and mates.

Marine Food Chains

Because the ocean covers 70% of the Earth's surface, it receives 70% of the Earth's solar energy. Except for the anchored green plants of the neritic zone, solar energy is trapped primarily by the phytoplankton (producers) in the open water of the oceans. Scientists estimate that 18 billion metric tons of living plant matter (mostly phytoplankton)

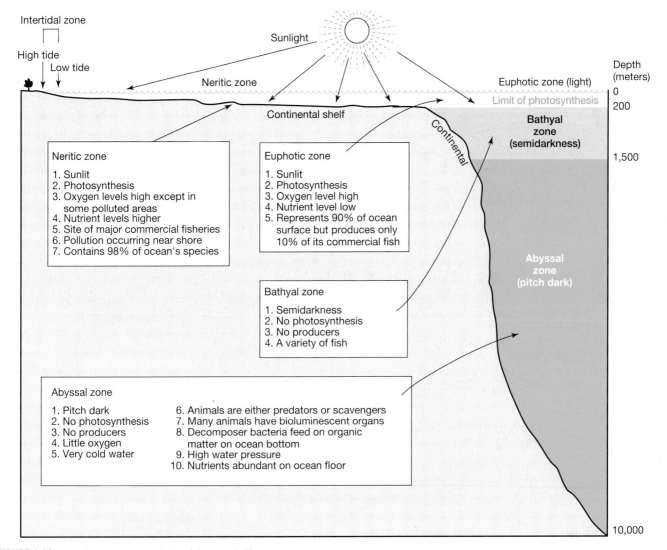

FIGURE 9.29 Locations and characteristics of the ocean's life zones.

are produced annually. This plant matter, in turn, supports 4.5 billion metric tons of zooplankton. Marine zooplankton may be consumed by a variety of filter feeders, including medusae, herring, anchovies, and whales. Fish-eating predators such as the shark, barracuda, cod, salmon, cormorants, pelicans, and humans represent the terminal link of the marine food chain (Figure 9.30). As in the terrestrial food chains, the shorter the chain, the more biomass is available to the top-level organisms.

Ocean Resources

The ocean serves human needs in a great variety of ways. The ocean's 133-million-cubic-kilometer (32-million-cubic-mile) volume is a virtually limitless water supply for all organisms on this planet, including humans. Trillions of tiny algae in sunlit ocean waters have aided in replenishing the oxygen supply of the Earth's atmosphere, upon which the survival of all life depends. The ocean absorbs 20 times more carbon

dioxide (responsible for global warming) than all the Earth's vegetation.

Ever since early humans scooped fish from its tidal pools with their bare hands, the ocean has provided people with abundant supplies of essential protein. The ocean serves as a highway for national and international transport and communication. And the ocean is valued for recreational uses, including boating, sailing, recreational fishing, and cruising.

The ocean bottom is a treasure house of valuable minerals such as manganese, nickel, copper, and cobalt. Only a very small area of the ocean has been surveyed for these deposits. At this time, extraction technology is too expensive to make exploitation of these mineral reserves economically feasible.

Over 26% of all oil production is from offshore wells. Natural gas production from offshore wells accounts for about 17% of the world's production. Additional ocean oil and natural gas deposits have yet to be exploited. The ocean also provides the potential for renewable-energy production if we can

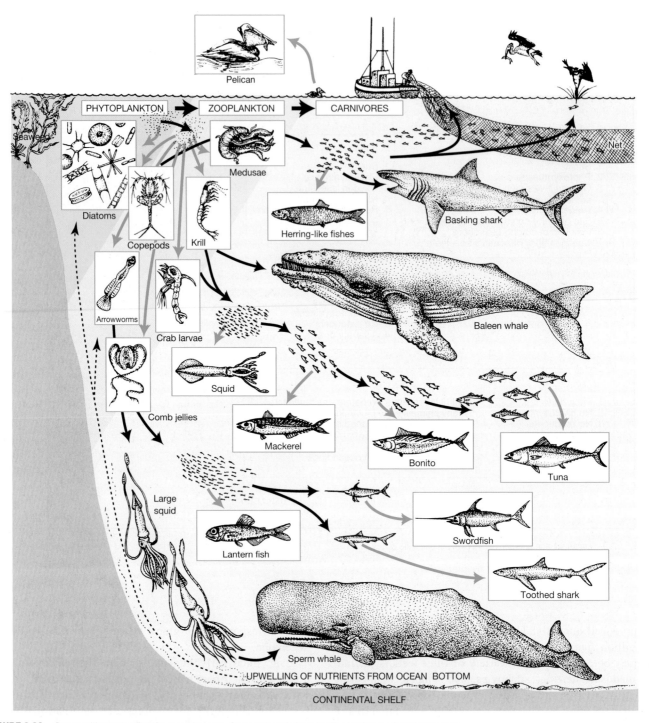

FIGURE 9.30 Communities occupying the oceanic zones. Seaweeds and algae are the producers, forming the base of the food chains and webs. Tiny zooplankton feed on the producers. In turn, zooplankton are consumed by fish, medusae, and whales. The top consumers are the seabirds, sharks, tuna, swordfish, whales, and humans. The bottom-dwellers, which crawl along or cling to rocks or are buried in the sand or mud, include clams, snails, lobsters, shrimp, crabs, and bacteria. They feed on detritus. The dashed arrows on the left indicate upwelling of nutrients from the ocean bottom.

find an effective means to harness the energy from waves, tides, and thermal columns. Research sites in several parts of the world are experimenting with these types of renewable-power generation.

These abundant ocean resources have the potential to provide for the long-term needs of a growing human population, but only if they are regulated to prevent excessive exploitation and disruption of productive ecosystems.

Summary of Key Concepts

1. Wetlands are highly diverse and usually occupy a transitional zone between a well-drained upland area and a permanently flooded deepwater habitat.

2. Wetland and deepwater habitats are classified into five major categories by the Cowardin Classification System: marine, estuarine, riverine, lacustrine, and palustrine.

3. Marine and estuarine wetland systems comprise saltwater environments such as coral reefs, rocky shorelines, tidal salt marshes, and mangrove swamps.

4. Riverine wetland systems include riparian wetlands situated along rivers and streams. Lacustrine wetlands are those bordering the edges of lakes, large ponds, and reservoirs.

5. Palustrine wetland systems contain the majority of all wetland habitats, including bogs, fens, freshwater marshes, freshwater swamps, and small, shallow ponds.

6. Wetland functions can be grouped into three broad categories: (1) hydrologic processes, (2) water quality improvement, and (3) wildlife habitat. Many ecosystem services are derived from the functions performed by wetlands.

7. The United States has a "no net loss" wetland policy. Wetland management policy has shifted from wetland drainage to wetland preservation, restoration, and creation schemes. Activities in and around wetlands are regulated by Section 404 of the Clean Water Act.

8. Ecologists recognize three major lake zones: littoral, limnetic, and profundal.

9. The littoral zone is the shallow marginal region of a lake and is characterized by rooted vegetation.

10. Plant species colonize a lake margin according to specific hydrologic requirements, with those less tolerant of water inundation situated closest to the shore.

11. The suspended, floating microorganisms in a lake are known as plankton.

12. The limnetic zone is the region of open water beyond the littoral zone down to the maximum depth at which there is sufficient sunlight for photosynthesis. The profundal zone lies beneath the limnetic zone.

13. In spring and autumn, the waters of temperate zone lakes undergo a thorough mixing known as the spring and fall overturn.

14. Ephemeral streams dry up during the dry season, whereas perennial streams flow year-round.

15. A smaller channel that flows into a larger channel is called a tributary of that channel.

16. A watershed is the entire land area, confined by topographic divides, in which all tributaries drain into a single river system.

17. A first-order (Order 1) stream is the smallest stream within a river system, having no tributaries; a second-order (Order 2) stream has only first-order streams as tributaries; and so on. In general, the smaller and fewer the tributaries emptying into a stream, the lower that stream's order number.

18. A stream channel takes shape during high-water events, when the force of water flow is enough to erode banks and transport bed materials.

19. The pattern of a river channel can be meandering, straight, or braided. The meandering pattern is the most common.

20. Pools and riffles are regularly spaced along the length of a channel bed.

21. Overflow water and suspended sediment are received by a river's floodplain. Downcutting by a river results in an abandoned floodplain, called a terrace.

22. The stream ecosystem receives a considerable portion of its energy supply from materials like leaves and twigs that originate in the riparian zone.

23. Riparian vegetation is also important to streams and wildlife as a source of large woody debris, erosion control, shade, cover, food, and nesting grounds.

24. The structure of coastlines is highly variable and may include one or more of the following features: beaches, sand dunes, rocky cliffs, barrier islands, coral reefs, coastal salt marshes, and estuaries.

25. A typical estuary is characterized by (1) brackish water, (2) water levels that rise and fall with the tides, (3) high levels of dissolved oxygen, (4) high turbidity, and (5) high nutrient levels.

26. Because of the rapid development of the coastlines and barrier islands, hurricanes and other oceanic storms threaten human life and valuable property.

27. Accelerated coastal development has (1) degraded scenic beauty, (2) destroyed coastal wetlands, (3) accelerated beach erosion, (4) increased pollution, and (5) damaged fish and wildlife habitats.

28. Among the factors contributing to coastal erosion problems are (1) storm events, (2) the reduction of sediment discharge into the ocean due to dam construction, (3) subsidence, and (4) engineered erosion-control structures.

29. Among the methods that have been used to control coastal erosion are (1) beach nourishment projects, (2) vegetative plantings, and (3) restriction of coastal development.

30. The National Coastal Zone Management Act (1972, 1990, and 1996) has provided financial assistance to coastal states, territories, and commonwealths to develop sound strategies for protection of the sensitive environments along coasts and barrier islands.

31. The ocean covers 70% of the Earth's surface, contains vertical and horizontal currents, is about 70 times as salty as a lake or stream, and is relatively infertile.

32. The four major zones of the ocean are (1) neritic, (2) euphotic, (3) bathyal, and (4) abyssal.

33. The neritic zone is a relatively warm, shallow, nutrient-rich region adjoining coasts.

34. The euphotic zone is a region in the open ocean that extends from the surface to the deepest area where there is sufficient sunlight for photosynthesis.

35. Below the euphotic zone is the bathyal zone, a region of semidarkness.

36. The abyssal zone lies immediately above the ocean floor. It is characterized by darkness, intense cold, low levels of dissolved oxygen, and a scarcity of food.

37. In areas of upwelling, fish production is highly efficient because many of the food chains have only two links: the producer phytoplankton and the consumer fish. In the deep waters of the open sea, however, fish production is much less efficient because the food chains may have as many as six links.

Key Words and Phrases

Abyssal Zone	Marine Wetland System
Bankfull Discharge	Meandering Pattern
Barrier Islands	Neritic Zone
Bathyal Zone	"No Net Loss" Wetland
Beach Nourishment	Policy
Beaches	Organic Matter
Blooms	Palustrine Wetland System
Bog	Peat
Coastal Zone Management	Peatlands
Act (CZMA)	Perennial Streams
Compensation Depth	Phytoplankton
Consumer	Plankton
Coral Reefs	Pocosins
Cowardin Classification	Point Bar
System	Pool
Dead Zone	Predator
Decomposer	Producer
Detritus	Profundal Zone
Ecotone	Riffle
Emergent Plants	Riparian Wetlands
Ephemeral Streams	Riparian Zone
Epilimnion	Riverine Wetland System
Estuarine Wetland System	Rocky Coastline
Estuary	Sand Dunes
Euphotic Zone	Seawall
Everglades	Spring Overturn
Fall Overturn	Stream Order
Fen	Submergents
Floating Plants	Subsidence
Floodplain	Subwatershed
Freshwater Marshes	Terraces
Freshwater Swamps	Thermocline
Great Barrier Reef	Tidal Creek
Groin	Tidal Salt Marsh
Headwater Streams	Tributary
Herbivores	Turbidity
Hydrophytes	Undercut Banks
Hypolimnion	Upwelling
Lacustrine Wetland System	Watershed
Limnetic Zone	Wetland Delineation
Littoral Zone	Wetland Functions
Lowland Streams	Wetlands
Mangrove Swamps	Zooplankton

Critical Thinking and Discussion Questions

1. Define the words and phrases in the list of key words and phrases.

2. Explain the difficulties in determining wetland boundaries.

3. Describe the general characteristics of each of the five wetland systems, and give an example of each.

4. All riparian wetlands are included in the riparian zone, but not all land within the riparian zone is wetland. Explain.

5. Distinguish between wetland functions and their value to humans, and give two examples of each.

6. The annual rate of wetland conversion in the United States has been decreasing in the past 50 years. Does this mean we should "relax" our wetland regulation and protection policies? Explain your answer.

7. Describe the characteristic distribution of rooted plants in the littoral zone.

8. Identify three insects, three fish, and three birds that are characteristic of the littoral zone.

9. Suppose that the light intensity on the bottom of a lake is 10% of that at the compensation depth. What percentage of full sunlight would that be?

10. Identify two sources of the dissolved oxygen present in a lake.

11. Give three reasons why oxygen might be a limiting factor for fish in northern states during the winter.

12. What causes the fall overturn? What causes the spring overturn?

13. Do the fall and spring overturns have any significance for the survival of aquatic organisms? Explain your answer.

14. Describe the stream characteristics that control a stream's shape and pattern. Compare the shape and pattern of a mountainous headwater stream with the shape and pattern of a lowland, valley stream.

15. What defines the boundaries of a watershed?

16. How does a healthy riparian zone contribute to the health of the stream ecosystem?

17. In what ways does an estuary differ from an ocean? From a river? Why are estuaries so productive?

18. How do coastal wetlands protect the shore from erosion?

19. Loss of life and property from coastal storms and flooding is likely to increase in the future. Why?

20. What effect(s) will the predicted rise in sea level have on coastal and barrier island habitats and communities?

21. Engineered structures designed to control beach erosion have actually caused additional erosion problems. Explain.

22. Discuss the various strategies states have implemented to restrict coastal development in high-energy or erosion-prone areas. Which strategy do you think has the greatest chance for success?

23. Characterize the neritic, euphotic, and abyssal zones of the ocean.

24. Describe at least four ways in which the ocean provides value to humans or satisfies human needs.

Suggested Readings

Abell, R. A., D. M. Olson, E. Dinerstein, P. T. Hurley, et al. 1999. *Freshwater Ecoregions of North America: A Conservation Assessment.* Washington, DC: Island Press. Important information on America's freshwater habitats.

Alan, J. D., and M. M. Castillo. 2007. *Stream Ecology: Structure and Function of Running Waters*, 2nd ed. New York: Springer. A superb compilation of the ecology of fluvial systems at scales ranging from small mountain brooks to large, continental-sized river basins.

Beatley, T., D. J. Brower, and A. K. Schwab. 2002. *An Introduction to Coastal Zone Management*, 2nd ed. Washington, DC: Island Press. Comprehensive overview of coastal planning and management issues.

Bergen, L. K., and M. H. Carr. 2003. Establishing Marine Reserves: How Can Science Best Inform Policy? *Environment* 45 (2): 8–19. A must-read for anyone interested in a resource management career.

Cowardin, L. M., et al. 1979. *Classification of Wetlands and Deepwater Habitats of the United States.* Washington, DC: U.S. Department of the Interior, Fish and Wildlife Service, Publication FWS/OBS-79/31, U.S. Government Printing Office. The national and international standard for identifying and classifying wetland and deepwater habitats throughout the world; includes detailed descriptions of hierarchical systems and numerous photographs as examples.

Dahl, T. E. 1990. *Wetlands: Losses in the United States, 1780s to 1980s.* Washington, DC: U.S. Department of the Interior, Fish and Wildlife Service, U.S. Government Printing Office. Report to Congress documenting historical wetland losses over a 200-year time span for each state in America.

Dahl, T. E. 2006. *Status and Trends of Wetlands in the Conterminous United States, 1998 to 2004.* Washington, DC: U.S. Department of the Interior, Fish and Wildlife Service, U.S. Government Printing Office. The most recent and comprehensive report on the status and trends of wetlands on a national scale.

Diaz, R. J., and R. Rosenberg. 2008. Spreading Dead Zones and Consequences for Marine Ecosystem. *Science.* 321: 926–929. Excellent summary of the significance of dead zones and their detrimental effect on marine ecosystems.

Fujita, R. 2003. *Heal the Ocean.* Gabriola Island, BC: New Society. A popular book for those interested in learning more about the plight of the world's oceans and ways to protect them.

Hobbie, J., ed. 2000. *Estuarine Science: A Synthetic Approach.* Washington, DC: Island Press. Essays by leading scientists summarizing the current state of knowledge about estuaries that is important for proper management.

Kemper, S. 1992. The Beach Boy Sings a Song Developers Don't Want to Hear. *Smithsonian* 23 (7): 72–85. Nontechnical discussion of the beach erosion problem.

Leopold, L. B. 2006. *A View of the River.* Cambridge, MA: Harvard University Press. Transformation of lengthy, mathematical, and inherently complex descriptions of the physical forms and processes of rivers into a compact, easy-to-read, interesting resource book, written by one of the most creative scholars in river research.

Mitsch, W. J., and J. G. Gosselink. 2007. *Wetlands*, 4th ed. New York: John Wiley & Sons. Wonderful, comprehensive text on all aspects of wetland science, restoration, and management.

Murphy, M. L., and W. R. Meehan. 1991. Stream Ecosystems. In *Influences of Forest and Rangeland Management on Salmonid Fishes and Their Habitats*, American Fisheries Society Publication 19, ed. W. R. Meehan. Bethesda, MD: American Fisheries Society. Extremely informative description of stream ecology and the implications for successful fisheries management.

Sobel, J., and C. Dahlgren. 2004. *Marine Reserves: A Guide to Science, Design, and Use.* Washington, DC: Island Press. An in-depth look at marine reserves and their importance to fisheries management.

Trujillo, A. P., and H. V. Thurman. 2007. *Essentials of Oceanography*, 9th ed. Upper Saddle River, NJ: Prentice Hall. In-depth and rigorous discussions of oceanographic concepts, demystifying the science for the layperson.

Viles, H., and T. Spencer. 1995. *Coastal Problems: Geomorphology, Ecology and Society at the Coast.* London: Edward Arnold. Excellent descriptions of coastal formations and processes and current discussion of human activities and their effects on coastal environments.

Weber, M. L. 2002. *From Abundance to Scarcity: A History of U.S. Marine Fisheries Policy.* Washington, DC: Island Press. An in-depth look at U.S. policy related to marine fisheries.

Web Explorations

Online resources for this chapter are on the World Wide Web at: **http://www.prenhall.com/chiras** *(click on the Table of Contents link and then select Chapter 9).*

CHAPTER 10

MANAGING WATER RESOURCES SUSTAINABLY

Water is a vital resource to humankind. It is used in agriculture to boost crop production and by industry to make products, including energy. Personally, humans rely on water for nourishment. Water is vital to cell function and to our health. Without it, a human being can survive only a few days.

For our society, water is one of our biggest headaches: water shortages plague societies in some areas, while in others, the challenge is to manage surpluses in the form of rain, snow, and subsequent flooding. In this chapter, we'll explore both of these issues, as well as agricultural irrigation issues, from social, economic, and environmental perspectives. Before we explore the issues and solutions, let's take a look at the water cycle.

10.1 The Water Cycle

Water is an intensively used, life-sustaining resource. It moves in a circular path through the ecosphere, a phenomenon called the **water cycle** or **hydrological cycle.** The water cycle is a nutrient cycle, as are the carbon and nitrogen cycles discussed in Chapter 3. Understanding the water cycle is essential to understanding both problems and solutions to the water challenges we face. Before you read this material, however, you may want to take a few minutes to study the water cycle shown in Figure 10.1.

The important thing to remember about water is that it is recycled over and over. Because of the cyclical nature of water movement, a given water molecule may be reused thousands of times throughout the centuries. The bathwater used by Cleopatra more than 2,000 years ago, for example, has flowed to the sea and been mixed with ocean water. Some has already evaporated and fallen on the continents as rain. A few molecules from her bath may be present in your next bath.

When thinking about the water cycle, it is important to differentiate between the components (oceans, rivers, lakes, and groundwater) and the processes (evaporation, precipitation, and so on). The components are the physical locations of water in the cycle. The processes are the driving forces that determine how water moves through the cycle.

To begin, we should point out that the water cycle is powered by two forces: solar energy and gravity. Solar energy causes water to evaporate from the soil, plants, oceans, lakes, and streams. Warm air rises and lifts the water into the atmosphere, where it forms clouds. Clouds release their moisture, and gravity pulls it down to Earth as rain or snow. Rainwater and melting snow run into streams or seep into the ground.

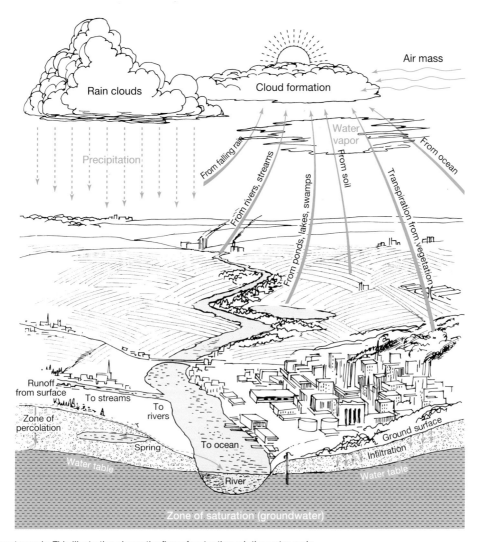

FIGURE 10.1   The water cycle. This illustration shows the flow of water through the water cycle.

All along the way, water can evaporate, once again becoming atmospheric moisture.

Although water moves rapidly and constantly, some water may be stored for varying lengths of time. Water may be stored in the Earth's crust as groundwater, defined later in this chapter. Or it may be stored as ice on the Earth's surface as polar ice caps or glaciers. It can also be stored in oceans and other water bodies. Huge amounts are also stored in the atmosphere as clouds.

Table 10.1 shows how much water resides in various locations such as in ice caps, the ocean, or the atmosphere at any one time. As shown, the vast majority of the world's water resides in the oceans. This table also shows the time required for the complete (100%) replacement of the water in various parts of the cycle, known as the **replacement period** or **renewal time.** As illustrated, the average replacement period ranges from 9 days for water in the atmosphere to 37,000 years for water in the deep oceans.

To understand the water cycle more fully, let's examine its components and processes one at a time, beginning with the oceans.

Oceans

When the astronauts peer down at the Earth from outer space, the planet appears blue with just a few large patches of brown and green, the continents. The blueness of our planet can be explained by the fact that oceans cover 70% of the Earth's surface. The oceans therefore form a major reservoir of liquid water (Figure 10.2). It is from this reservoir that vast amounts of water evaporate.

Precipitation

How does the ocean give up its water to the atmosphere? As the sun shines down on the ocean, water molecules at the surface are warmed by solar radiation. These molecules break loose from the liquid water and rise into the atmosphere as a gas. This process is known as **evaporation.** As water suspended in air—also known as water vapor—rises, it gradually cools and condenses. This forms clouds that can be transported many miles by wind currents. Clouds filled with liquid and frozen

TABLE 10.1   Water Cycle Facts

Location of Reservoir	Total Water Supply (%)	Renewal Time
On the Land		
Ice caps	2.225	16,000 years
Glaciers	0.015	16,000 years
Freshwater lakes	0.009	10–100 years (varies with depth)
Saltwater lakes	0.007	10–100 years (varies with depth)
Rivers	0.0001	12–20 days
Subsurface		
Soil moisture	0.003	280 days
Groundwater		
To 1/2-mile depth (.8 km)	0.303	300 years
Beyond 1/2-mile depth (.8 km)	0.303	4,600 years
Other		
Atmosphere	0.001	9–12 days
Oceans	97.134	37,000 years
Total	100.000	

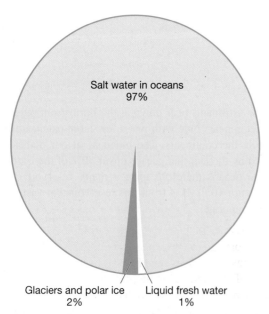

FIGURE 10.2   The oceans represent an enormous reservoir of water, containing 97% of all the water passing through the water cycle.

water eventually give up their moisture in the form of rain or snow, which falls on land and sea.

For humans, it is the precipitation that falls on the Earth's continents that is of greatest importance. This water nourishes our crops and fills rivers and lakes that are used for recreation, supplying water to cities and towns. Rain and melting snow also recharge groundwater. As pointed out earlier, huge amounts of rain fall on the United States every day. In fact, were it distributed evenly throughout the country, annual daily rainfall would be sufficient to cover the nation (if it were perfectly level) to a depth of 1 meter (3.3 feet). As noted earlier, however, rainfall is not uniformly distributed.

As shown in Figure 10.3, the average annual precipitation varies considerably. Death Valley in the arid Southwest, for example, receives only 4.3 centimeters (1.7 inches) annually, while the southeastern United States receives 40 to 60 inches of precipitation per year. The western slope of the Cascades receives 350 to 400 centimeters (140 to 150 inches).

Evaporation and Transpiration

What happens to water that falls on the Earth? Of the 0.75 meter (30 inches) of average annual rainfall, about two-thirds evaporates. As noted earlier, evaporation may take place directly from streams, lakes, puddles, ponds, oceans,

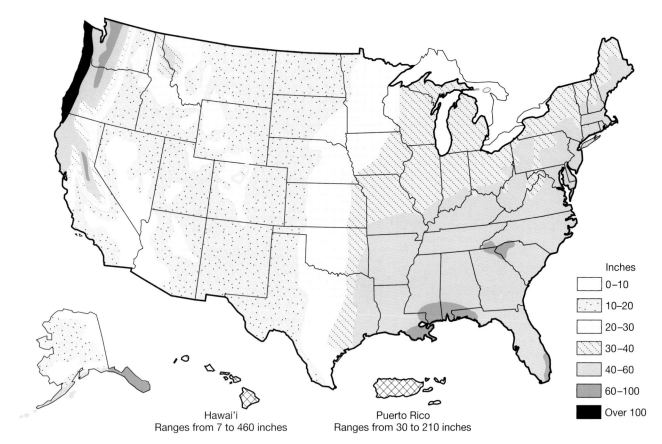

	Inches
	0–10
	10–20
	20–30
	30–40
	40–60
	60–100
	Over 100

Hawai'i
Ranges from 7 to 460 inches

Puerto Rico
Ranges from 30 to 210 inches

FIGURE 10.3   Average annual precipitation in the United States.

soils, vegetation, and the bodies of animals and their wastes (Figure 10.4). Plants lose water through **transpiration,** the escape of water from a plant through pores in its leaves. This process is essential to the plant's survival, for it draws dissolved nutrients from the soil up through the stem (or trunk) to the leaves. One mature oak tree may transpire 380 liters (100 gallons) per day—more than 150,000 liters (40,000 gallons) in a year.

Surface Water

A little over 30% of annual rainfall in the United States flows into **surface waters**—that is, into ponds, lakes, and streams. Surface water and groundwater are of great concern to conservationists, for these are the sources that are usable for domestic, industrial, and recreational purposes. Surface waters are also habitat for many species. Surface water and groundwater are often polluted with toxic chemicals, pesticides, human waste, and many other contaminants—a problem we discuss in the next chapter.

Stream flow in the United States averages 4,560 billion liters (1,200 billion gallons) a day. It may be in the form of a tiny mountain stream or a mighty river such as the Mississippi, which drains 40% of the land area in the United States and flows 3,800 kilometers (2,300 miles) across mid-America to the Gulf of Mexico. Surface water, in the form of streams, ponds, and lakes, satisfies about 75% of our water requirements.

Groundwater

Some of the rainfall and snowmelt (about 3%) seeps down through the soil in a process called **infiltration** (Figure 10.5). This water first moves through the **zone of aeration.** This zone, which includes both the topsoil and subsoil, is characterized by pore spaces that contain water and air. The soil moisture in this zone is known as **capillary water.** Some of this water is absorbed by plant root systems. It then passes up plant stems and tree trunks to the leaves. Most of this water then evaporates (transpires) from the leaves, entering the atmosphere as water vapor. A small amount of the water in the leaves is used as a raw material in photosynthesis.

As the water moves down through the soil, many pollutants adhere to the surfaces of the soil particles. As a result, the water quality is often improved as it flows. Beneath the zone of aeration lies the **zone of saturation**—so named because the pores in the soil are filled (saturated) with water. This water is known as **groundwater.** Groundwater can be contaminated by pollutants seeping down through superficial layers through the zone of aeration.

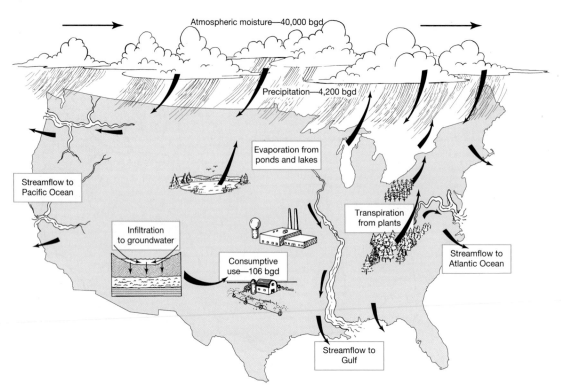

FIGURE 10.4   What happens to the rain and snow? Water continuously evaporates into the atmosphere, most of it from the oceans. About 154 trillion liters (40,000 billion gallons) per day (bgd) passes over the United States as water vapor, even in times of drought. Roughly 4 liters in 40—16 billion liters per day—falls to the surface of the conterminous United States. That works out to an average of 76 centimeters (30 inches) a year, of which 66 centimeters (26 inches) arrive as rainfall and the rest as snow, sleet, and hail. But few places receive the average precipitation. In the conterminous United States, precipitation ranges from less than 10 cm (4 inches) a year in the Great Basin to more than 510 cm (200 inches) a year along the Pacific Northwest coast. More than two-thirds of the precipitation returns to the atmosphere, but 206 cm (9 inches) either soaks down to the water table or runs into lakes or streams, from which it eventually moves to the ocean. Only a small fraction of the precipitation, 410 billion liters per day (106 bgd), is consumed.

Eventually, the downward movement of the water through the zone of saturation is stopped by a layer of impermeable rock (Figure 10.5). As a result, the water accumulates in the porous materials such as sand, sandstone, and limestone above the impermeable layer, filling the spaces, pores, and cracks and creating an underground body of water known as an **aquifer.** The upper level of the zone of saturation is known as the **water table.** Groundwater may be replaced by water percolating downward from large areas overlying aquifers or from restricted areas, known as **aquifer recharge zones.**

Although it is not shown in Figure 10.5, the water table may intersect the surface and form marshes, ponds, or springs. In other cases, the water table may be more than 1.2 km (1 mile) deep. Since time immemorial, humans have tapped groundwater by drilling wells into aquifers or siphoning off water from springs.

The water table may rise and fall depending on the amount of water leaving and entering the zone of saturation. After heavy rainfall, for example, the water table rises. (An exception would be flash floods in deserts or sudden downpours in highly urbanized areas, where most of the water runs off instead of infiltrating.) However, during periods of drought, the water table drops. In some cases, drought may cause wells to run dry (Figure 10.6).

Humans also affect the water table. Heavy withdrawals can cause the water table to drop. If the water withdrawn from an aquifer is replaced at the same rate, an aquifer can provide a continual supply of water. If water is withdrawn from the aquifer faster than it is replenished, the water table and amount of water in an aquifer both drop, leaving shallower wells high and dry. This problem is particularly acute in deep aquifers that contain water that took hundreds of thousands of years to fill, sometimes referred to as fossil groundwater. In cases where recharge is very slow and rates of withdrawal are high, groundwater can be thought of as a nonrenewable resource.

It may be hard to believe, but most (97%) of the world's supply of liquid freshwater (8.34 million cubic kilometers or 2 million cubic miles) is in aquifers. In the United States, the groundwater in aquifers in the upper 0.84 km (0.5 mile) of Earth's crust is equal to all the water that will run off into the oceans during the next 100 years! However, water in these great underground reservoirs often comes from

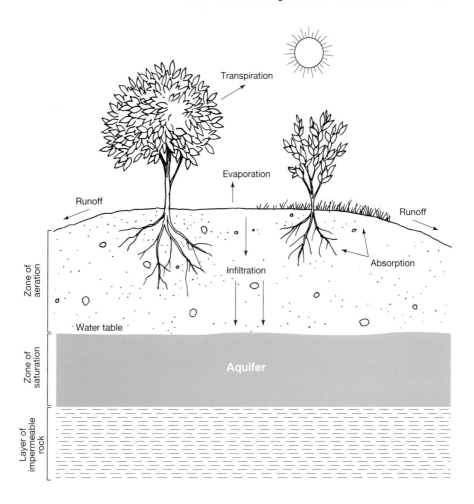

FIGURE 10.5 Fate of water that has fallen on the ground as precipitation. This drawing also shows how an aquifer is formed.

precipitation that fell hundreds of years earlier. For example, the aquifer that supplies a large amount of Chicago's water came from rain and snow that fell on the Great Plains far to the west a million years ago and then slowly trickled through rocks at a rate of a few centimeters or meters a year. The Ogallala Aquifer, described in more detail shortly, lies under 572 square kilometers (225,000 square miles) of

land that extends through eight states from Nebraska south to Texas (Figure 10.7). This aquifer contains 2.5 million billion liters (650 trillion gallons). It provides drinking water for several million people and is a major source of irrigation water, supporting a multi-billion-dollar economy. Unfortunately, this once-massive aquifer is being drained much faster than it can recharge.

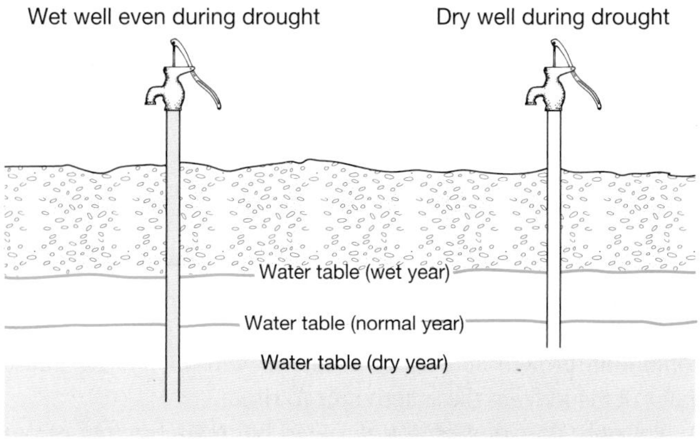

10.2 Water Shortages: Issues and Solutions

The United States receives, on average, 16 trillion liters (4.2 trillion gallons) of precipitation per day. That's about 57,000 liters (15,800 gallons) a day for every man, woman, and child. Yet severe water shortage is one of the most serious problems facing many parts of the United States today. Moreover, many other countries are facing similar, often crippling water shortages. How can this be?

What Causes Water Shortages?

The problem is that, despite this enormous amount of precipitation, rain and snow are not equally distributed throughout the world. Some regions receive an abundance of water, but many others receive less than adequate supplies. Chronic water shortages in the southeastern United

Wet well even during drought

Dry well during drought

Water table (wet year)

Water table (normal year)

Water table (dry year)

FIGURE 10.6 An aquifer serves as a source of well water. Continued withdrawal of water would eventually deplete the supply if the rate of withdrawal exceeded the rate of recharge.

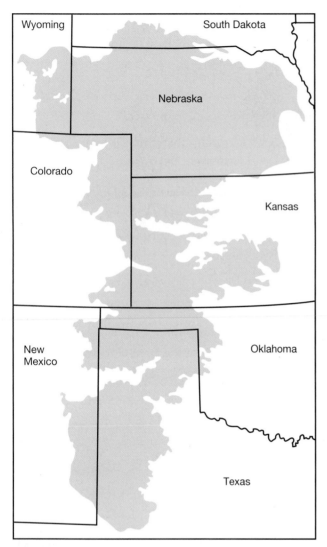

FIGURE 10.7 The Ogallala Aquifer. This huge body of water is quickly being depleted, primarily by farmers. It took hundreds of thousands of years to fill and therefore is essentially being mined.

States and the West result in part from the intensive concentration of people and industries in areas with marginal water supplies.

Expanding demand also contributes to water shortages. The ever-expanding human population, especially in arid or semi-arid areas, and rising demand for water by agriculture, industry, and cities contribute to severe shortages like the one in the southeastern United States in 2007 and 2008. But that's not all. The continual inefficient use of water by many sectors also plays a role in creating water shortages. To understand these factors and how they contribute to water shortages, let's review a few facts about the United States' water situation, starting with population growth.

In the United States, population growth is occurring at a rate of about 1.1% a year as a result of natural increase and legal and illegal immigration. Although that may not sound like much to worry about, continuing growth results in a continual increase in demand for water. Regional population growth often exceeds the national growth rate. In the South, Southwest, and West, for instance, some areas have posted growth rates of

4% per year for many years! Florida, Colorado, Utah, and Arizona are four of the most rapidly growing states that are facing severe water shortages. Such regional booms in population growth put tremendous strains on already overtaxed water supplies. The accelerated migration of people to the Southwest, for example, has reduced water supplies in this region to the vanishing point.

Growth doesn't just increase demand, it also can decrease supplies. In a startling report, *Paving Our Way to Water Shortages,* several prestigious environmental groups presented compelling evidence suggesting that severe water shortages may be partly due to the paving over of America resulting from continued urban sprawl. In the greater Atlanta region, for instance, nearly 85,000 hectares (213,000 acres) were paved and "roofed" over between 1982 and 1997. This, in turn, results in 495 billion liters (132 billion gallons) of rainwater being diverted into storm drains each year. The water no longer naturally seeps into the ground to water trees, replenish aquifers, and filter back into rivers, streams, and lakes. Boston loses 382 billion liters (102 billion gallons) a year.

To meet needs in some areas, water is often transported long distances at great cost from water-rich regions. Making matters worse, in recent years, many of these parts of the country have experienced severe drought that is very likely a result of climate change, which results in shifting rainfall patterns. Increasingly strict conservation measures are being put in place to address water shortages.

The demand for water is also on the rise as a result of an increase in agricultural water use. Irrigated crop acreage in the United States has more than tripled from the 1960s to the 1990s, reaching about 27 million hectares (67.5 million acres). It has subsequently begun to decline. As mentioned earlier, one of the largest supplies of water used for agriculture is a huge deposit of groundwater known as the **Ogallala Aquifer.** This massive aquifer extends from Nebraska to Texas and is currently being depleted by withdrawals that exceed its replenishment—primarily by farmers. In some areas, the water table is dropping 1.5 meters (5 feet) per year. **Groundwater overdraft,** removal of groundwater faster than it can be replenished, is not just a U.S. phenomenon. It is occurring worldwide.

Industries also use water—lots of water—for a variety of purposes. Industries use water to cool equipment and to transport minerals for processing. Water is used to make paper. Producing one copy of the Sunday newspaper, for example, requires 570 liters (150 gallons), and producing an automobile takes 247,000 liters (65,000 gallons).

Water is also used to generate steam to heat buildings and to generate electrical power. In the United States, electrical power plants use half of all water drawn from surface and ground waters, around 735 billion liters (190 billion gallons) a day. Population growth and rising affluence will very likely cause water demand from these activities to rise.

Not only do we use a lot of water, but many sectors of our society use water inefficiently. The average American, for example, uses 230 liters (60 gallons) per day. European families use only half as much. In places where summer lawn irrigation is common, such as Denver, CO, a family can use 733 liters (190 gallons) per day per person.

Water pollution also can decrease water availability. Groundwater, for instance, is increasingly becoming contaminated. This problem is especially severe in Florida, Wisconsin, and California. Because groundwater is so costly and difficult to clean up, contamination can reduce future supplies while demand increases.

Drought and Climate Change

The problems just outlined are already causing regional shortages that experts think will very likely worsen as the U.S. population and economy expand. But as noted, these problems are not unique to the United States. Most countries experience similar problems, so much so that supplying clean water for all end users could be one of the most difficult challenges of the coming decades. Drought caused by climate change may make it even more difficult.

According to the U.S. Weather Bureau, a **drought** exists whenever rainfall for a period of 21 days or longer falls substantially below the average. The Great Plains, from Texas to Montana, averages about 35 consecutive drought days each year and 75 to 100 successive days of drought once in ten years. Up to 120 consecutive rainless days have been recorded for the southern Great Plains, or Dust Bowl, region.

J. Murray Mitchell, a research climatologist with the National Oceanic and Atmospheric Administration, studied drought patterns as revealed by tree rings. (Tree rings represent new growth; they're thinner in dry years.) Going back to 1600, he found that the western states have experienced a prolonged, extended drought about every 22 years. Much to his surprise, Mitchell found that the drought cycle correlated with solar activity.

Drought is a far more common event than most people realize. It causes severe water shortages in cities and agricultural regions. This can cause dramatic loss in crops and severe financial hardship. Drought is accompanied by record heat, which takes a toll on people and livestock. In recent years, the deaths of around 300 to 600 people a year in the United States are attributed to record temperatures. In the summer of 2003, nearly 40,000 people perished in Europe as a result of unseasonably hot weather. Most of the victims in the United States and abroad are elderly persons unable to cope with or escape from the heat. Drought also affects wildlife populations.

While drought—and shortages of water—are natural phenomena, strong evidence suggests that current dry spells and water shortages are at least partly caused by human-induced global warming. Global warming is an increase in the temperature of the Earth's atmosphere and surface waters, caused by the release of greenhouse gases and other activities. Warming leads to changes in global climate, including rainfall patterns. Drought and water shortages are global phenomena that plague every continent. If predictions of global warming are correct, drought and water shortages may occur more often, resulting in widespread damage and suffering.

Increasing Water Supplies

Even if per capita use of water were to remain the same, total water use in the world will increase substantially. In the United States, population will increase by nearly 120 million people between 2007 and 2050 (Chapter 4). In the less-developed nations, population is projected to increase by about 30% or 2.6 billion. Where will additional water come from to supply their needs?

To find enough water, several strategies can be pursued. Among the most important is population control, discussed in Chapter 4. Numerous additional measures, some of dubious value, have been proposed. They include (1) water conservation (using water much more efficiently), (2) reclamation of sewage water, (3) development of groundwater resources, (4) desalination of seawater, (5) rainmaking, (6) diversion of surface water to water-short regions, and (7) changes in crops—for example, efforts to develop **salt-** and **drought-resistant crops** that permit farmers to continue producing crops needed to feed the expanding human population. The following sections outline the pros and cons of each of these solutions and provide a good opportunity for you to practice your critical thinking skills and select the most sustainable approaches—that is, the solutions that make sense from social, economic, and environmental perspectives.

Conservation: Using Water Much More Efficiently

For years, water supplies have been expanded by building new dams to impound river water. This water was then stored and used during the year. A far more cost-effective and environmentally sound approach, say many experts, is water conservation.

On the Farm   Nearly 70% of all water used in the United States is used by farmers, mainly for irrigation. Crop growth requires enormous amounts of water. Producing 0.45 kilograms (1 pound) of cotton, for instance, requires 600 liters (150 gallons) of water; 60,000 liters (15,000 gallons) are required to produce a single bushel of wheat. Conservation on the farm, discussed next, can greatly expand water supplies throughout the world.

Reducing Seepage Losses   Each year large amounts of irrigation water are lost by **seepage**—seeping out of dirt-lined irrigation ditches. Seepage loss can be minimized by lining irrigation canals with concrete or plastic or by replacing ditches with pipes. For instance, the city of Casper, WY, now has larger supplies of water available for domestic and industrial purposes because irrigation canals in the surrounding farmland were lined with impervious materials to reduce losses. These measures can cut water loss by half. Pipes can cut water loss by 90%.

Increasing the Use of Drip Irrigation   At least 1.1 million acre-feet of water[1] could be saved annually nationwide if farmers made maximal use of **drip irrigation.** In this method, water is slowly released at the base of plants by means of perforated plastic pipes. It then seeps into the ground, supplying

---

[1]An acre-foot of water is the amount of water that would cover 1 acre 1 foot deep.

the roots with ample water without much loss. Drip irrigation can cut water loss during irrigation by half. Unfortunately, however, only crops that don't require planting each year—that is, permanent crops, such as fruit trees, grapevines, and nut trees—can be irrigated by this method. It won't work for corn or wheat or the vast majority of commercially grown crops.

Heat Sensors and Soil Moisture Sensors Infrared (heat) sensors have been developed by the U.S. Water Conservation Laboratory in Phoenix, AZ. These devices detect the amount of heat being given off by crops, which indirectly determines soil moisture content, because plants become progressively warmer as they dry out. Farmers can use this instrument to find out whether a given crop requires irrigation, so they can avoid excess irrigation.

Farmers can also use soil moisture sensors, devices that monitor the amount of water in soils in their fields (Figure 10.8). Measuring soil moisture helps farmers determine the timing and the quantity of water they need to apply, potentially saving many millions, perhaps many billions, of liters of irrigation water each year in the United States and other countries.

In Industry For many years, industry has been flagrantly wasteful of water, especially in the water-rich eastern states, where it has been assumed that water is virtually inexhaustible. Contributing to the waste of water in this region is water's extremely low cost. For example, one study showed that when water costs only 1 cent per 4,000 liters (1,000 gallons), a coal-fueled electrical power plant uses 200 liters (50 gallons) of water for each kilowatt-hour of electricity produced. (An average home consumes 800 to 900 kilowatt-hours per month, and power plants produce billions of kilowatt-hours per year.) In contrast, when the price of water rises to 1.25 cents per liter (5 cents per gallon), the same power plant reduces water use to 3.2 liters (0.8 gallon) per kilowatt-hour.

With a blend of newly developed technology and creativity, a number of industries are making great progress in conserving

water. As water-saving technology improves, the savings will also very likely increase.

On the College Campus and at Home As a student in a resource conservation class, you have by now developed sensitivity to, and an understanding of, the overriding importance of protecting Earth's precious resources, among them water. Practical methods by which you, your fellow students, and your family can save water are listed in Table 10.2. One of the most effective measures is the installation and use of a water-efficient showerhead. These devices can cut water consumption by half or more, depending on the existing showerhead.

Water conservation in homes can also be encouraged through new laws and pricing policies. Water meters, for instance, monitor water use by homes and businesses, and customers are charged per gallon. However, in some cities, customers are not metered and are charged a flat fee. They pay one price regardless of how much water they use. When water meters are installed in such households, consumption often drops dramatically—by 15% or more.

In metered homes, water often gets cheaper the more a customer uses, providing no incentive for conservation. Recognizing this, some cities and towns have implemented new pricing policies in which rates increase as water consumption rises beyond a certain point. This cuts down on the wasteful use of water (especially for lawn watering) and can make huge differences in a city's water consumption.

Homeowners can reduce domestic water use by capturing rainwater and using it to water trees, bushes, gardens, and lawns. Rainwater can be captured by rain barrels attached to downspouts attached to gutters that receive water from roofs (Figure 10.9). In some cases, homeowners divert rainwater to large tanks or cisterns. Cisterns in many climates can collect tens of thousands of liters that can be used for a variety of purposes. They are often buried for aesthetic reasons and to protect them against freezing in colder climates.

Overcoming Legal Barriers to Conservation It may come as a surprise to you that certain laws impede water conservation efforts in the United States. In the western United States, for example, water in rivers and streams is allocated by a permit system. Two principles guide the allocation and use of this water.

The first, known as the doctrine of prior appropriation, is colloquially known as *first in time, first in line*. What this means is that the oldest permit holders hold senior rights to water in a stream. Later permittees hold junior rights. Each year, the water in all streams is allocated on the basis of how much is available and who has rights. But holders of senior water rights get their water first; they were the first in time and so get their water before anyone else. If any water is left over, holders of junior water rights get their water. In drought years, then, farmers with junior water rights may not get any water at all.

The second and more important doctrine is colloquially known as the *use it or lose it* principle. What this means is that a holder of water rights must use his or her water each year or risk losing rights to that water. For example, if a

FIGURE 10.8 Specialist taking soil moisture readings from a delicate instrument operated by the Bureau of Reclamation. With such information sustainable, farmers irrigate only when and where needed. Such irrigation efficiency reduces not only the amount of water used but also the severity of salt buildup.

TABLE 10.2	Water Conservation Strategies

In the Bathroom

1. Take shorter showers.
2. Don't use the toilet as a wastebasket.
3. Don't let the water run while brushing your teeth.
4. Don't run the water while shaving. Plug and partly fill the basin to rinse your razor.
5. Repair leaks promptly.
6. If you replace your toilet, install a low-flow (1.6 gal.) unit.
7. A few drops of food coloring or dye tablets in your toilet tank can help you spot a leak. If color appears in the bowl, you have a leak. Fix it promptly.
8. Place bricks or plastic bottles in your old high-flow toilet tank and save 1–2 liters every time you flush.
9. Install a low-flow showerhead.

In the Kitchen

1. Wash only full loads in your washing machine and dishwasher.
2. If you wash dishes by hand, don't let the water run.
3. Cool your drinking water in the refrigerator, not by letting the water run.
4. Don't use water-wasting garbage disposals.
5. Stop those leaks (New York City alone wastes 800 million liters each day because of them). Leaking faucets and wasteful people are robbing Americans of scarce water supplies. A small leak (80 drips per minute) wastes 26.5 liters (7 gal.) of water daily.

Outside Your Home

1. Use a broom, not a hose, to clean driveways, sidewalks, and steps.
2. Use an on–off spray nozzle on your hose.
3. Wash your car with a bucket of water; use a hose only to rinse.
4. Water your lawn and garden only during the cool of the day or during the evening. Older trees and shrubs often do not require irrigation. Plants are frequently overwatered.
5. Remove water-stealing weeds from the lawn and garden.
6. Use less fertilizer; it increases a plant's need for water.
7. Apply mulch between rows in your garden to hold the soil moisture.

FIGURE 10.9 Rain barrel. A simple rain barrel such as this can collect water from the roof of a house. This water can be used to irrigate shrubs, trees, and gardens.

farmer with 100 acre-feet of water[2] finds a way to conserve this water—say, by cutting use by half—he or she could lose the 50 acre-feet allocation. It will be assigned to another user. This archaic system obviously dissuades ranchers and farmers from conserving. Each year, they are essentially forced to use their full allocation, whether they need it or not.

Reclamation of Sewage Water

Sewage produced by homes contains human waste—solid and liquid waste. Huge amounts are generated each day. This waste is delivered to sewage treatment plants that remove many of the pollutants and then release the effluent into nearby water bodies—lakes, streams, or the ocean. Sewage treatment plant

---

[2]An acre-foot of water is the amount of water that would cover one acre, one foot deep.

FIGURE 10.10   A golf course irrigated with treated sewage water. This measure conserves water and supplies valuable nutrients.

effluent is 99% water. If the 1% pollution is removed, the effluent may be purer than the stream water itself.

Processed sewage water is already being used for a variety of purposes. Factories such as steel mills use water to cool steel. Golf courses in San Francisco, Las Vegas, and Santa Fe use it to water grass (Figure 10.10). Colorado College, where Dr. Chiras teaches part-time, uses treated wastewater to irrigate trees and grass. Treated wastewater is also used to irrigate crops in numerous locations, including San Antonio, TX, and Fort Collins, CO. Ornamental shrubs along highways in San Bernardino, CA, and the grounds at the Denver International Airport are watered with it.

Los Angeles daily discharges 68 million liters (17 million gallons) of processed sewage water over sewage-spreading beds at the edge of town. Eventually, this water seeps into aquifers that supply the town's wells. This water is of higher quality than the water from the Colorado River.

For wastewater to be used successfully–for example, to be applied to crops–it must be free of toxic substances such as heavy metals and organic pollutants. These come from factories and other businesses and have, for years, been commingled with sewage from households. Fortunately, water pollution regulations in the United States have helped divert factory wastes from the municipal wastewater systems.

Wastewater can also be recycled by homeowners. For example, some homeowners recycle **graywater** from sinks, showers, baths, and washing machines. This water, which actually appears gray, contains soaps and dirt and is used to supply outdoor vegetation. Because the continued use of detergents for such purposes may damage soils, some homeowners use biocompatible soaps, that is, soaps containing substances that are actually good for plants.

Many creative ways are being tried as well to treat **blackwater,** that is, water from toilets. Artificial wetlands, known as constructed wetlands, for example, are proving to be highly successful. In these systems, blackwater is diverted to small ponds, often lined by impermeable liners, filled partially with rocks and gravel, and then planted with aquatic vegetation. There it is broken down by bacteria and other microorganisms. Soil may be placed over the top of the rock bed, filling it entirely, so wastewater is not visible. Plants grown in the soil send their roots into the rock bed, absorbing water and other nutrients from the waste. When the water exits from such a system, it is often as pure as drinking water.

Composting toilets are also used in some homes. Composting toilets, like the one shown in Figure 10.11, receive solid and liquid waste. The water evaporates from the waste and

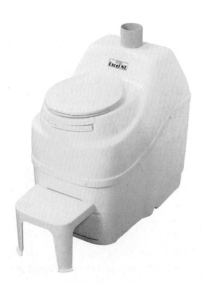

FIGURE 10.11   Composting toilet. Composting toilets capture and compost waste without odors and produce a rich soil supplement.

escapes by a vent pipe through the roof along with any odors. The solid waste then slowly decomposes, producing a rich organic matter that can be buried in flower gardens or on one's property to enrich the soil.

Developing Groundwater Resources

Conservation and water reclamation are two highly sustainable solutions to water supply problems. They promote key principles of sustainability, efficiency, and recycling, respectively. Combined, they could provide a substantial amount of water for human society.

The more traditional and perhaps less sustainable approach to meeting rising demand involves the development of new water sources. In the United States alone, estimates show that there are approximately 53,000 cubic miles of freshwater in the aquifers located in the upper 0.8 kilometer (0.5 mile) of Earth's crust. Wells drilled into these aquifers 150 to 600 meters (500 to 2,000 feet) deep could help alleviate impending water shortages, provided withdrawals do not exceed replenishment.

Some cities are looking at replenishing groundwater by tapping into rivers during years of surplus. The aquifer then becomes an elaborate underground reservoir. Although this option sounds good, it does require pipelines to transport water to recharge zones and a fair amount of energy to pump water to them. Not all of the water would be recoverable, either. Some experts assert that only a small portion of the water intentionally injected into groundwater from surface water supplies would actually be recoverable.

Desalination

As noted earlier, 70% of the Earth's surface is covered by oceans, in some places up to 10 kilometers (6 miles) deep, yet a water shortage plagues civilization from New York to New Delhi. The problem, of course, is that seawater is salty and thus not drinkable without treatment.

A number of U.S. communities are now operating **desalination** plants, facilities that remove salt from seawater to produce potable water. The nation's first desalination plant was built in California and produces 100,000 liters (28,000 gallons) of freshwater daily. Numerous desalination plants have been built on the west coast of Florida alone. According to recent statistics, there are currently nearly 1,600 desalination facilities in the United States, producing 7.6 billion liters (nearly 2 million gallons) of water a day. Although desalination plants still produce only a small fraction of the United States' total freshwater, several countries in the Middle East rely heavily on desalination for drinking water.

Although desalination may seem like a viable option for meeting future demands, it does have some problems. For one, it is a rather inefficient and expensive process. In general, it is much cheaper to pump fresh water—if it's available. However, if the freshwater source must be pumped more than 150 kilometers (90 miles) to the site of consumption, desalination becomes economically feasible. Older desalination plants also require large quantities of energy. Fortunately, new desalination technologies are much more energy efficient and produce freshwater at a much lower cost than older facilities. Even so, desalination plants produce brine—salt waste that requires proper disposal.

Developing Salt-Resistant Crops

After six years of research, two scientists from the University of California, Davis announced the development of a new strain of barley that will grow well even though it was irrigated with seawater. Barley plants have been grown on a tiny windswept beach at Bodega Bay in northern California, just north of San Francisco. They have achieved yields of 1,480 kilograms per hectare (1,320 pounds per acre), equal to the average global per acre yield of barley provided with freshwater. One of the researchers, Emanuel Epstein, noted, "We have shown that sea water is not pure poison to crops."

Their success is highly significant. Millions of acres of once prime agricultural land the world over—1.8 million hectares (4.5 million acres) in California alone—have been rendered worthless because of salinization, the accumulation of salts on irrigated cropland. As discussed later in the chapter, the salts are deposited not by saltwater but by irrigating with freshwater containing lesser amounts of salts. They build up in soils over time. Until now, farmers have been advised to cease cropping salinized soils or to flush out the salt, at considerable expense, with huge volumes of freshwater. However, the new saltwater barley and other crops now under development might do well in these salty soils, rather than cropland irrigated with saltwater, which would result in massive buildup of salt in the soils.

Developing Drought-Resistant Crops

Food production in a water-short world could also be increased by developing new varieties of crops that are resistant to drought. George G. Still, a scientist for the USDA, is optimistic about the water-saving potential of such plant-breeding projects: "Sorghum is a very important grain crop in the arid portion of the Third World. There is something inherent in the plant, the germ plasm or the genes, that causes sorghum to put itself on idle during a dry spell and then go on and yield a crop when the rains come," he says. With the aid of recently developed techniques in genetic engineering, it may be possible to breed plants that require considerably less water.

Rainmaking

Rainmaking is another potential approach to increasing our water supply. One technique is cloud seeding, dispersing tiny crystals of silver iodide or sodium chloride (salt) in moist air. These crystals serve as **condensation nuclei,** particles around which moisture collects until raindrops form. The U.S. Bureau of Reclamation is confident that the weather modification techniques now available can increase the water supply of the San Joaquin River Basin by 25%, that of the Upper Colorado River Basin by 44%, and that of the Gila River Basin (Arizona) by 55%. In the United States, there are about 40 cloud-seeding programs, over half of them in California and Nevada.

Despite the popularity of cloud-seeding programs, there is conflicting evidence as to whether they really work. They also pose some potential problems.

Cloud seeding, say critics, could create some legal, political, economic, and environmental problems. One drawback is the high cost. Another problem is the possibility that while cloud seeding may increase rainfall in one area, it could reduce rainfall in downwind areas. Thus, artificially enhanced rain in one region could result in drought in others. Another problem is the lack of control over the amount and precise location of the precipitation. For example, a late-July rainfall might benefit corn but might damage the alfalfa that is awaiting the baler in a nearby field. Even more serious, lack of control over precipitation might lead to flooding, soil erosion, property damage, and even loss of life.

Long-Distance Transport: The California Water Project

Water diversion, or the transport of water from one region to another through pipelines and canals, is viewed by some developers and government officials as a way to provide water to water-short regions. Because of costs and environmental impact, however, the idea is frequently controversial. The **California Water Project (CWP)** is an example of what can be done.

California has long been victimized by the curious fact that 70% of its potentially usable water falls on the northern third of the state (in the form of relatively abundant rainfall and the snowmelt of the High Sierras), while 77% of the demand is located in the semiarid southern two-thirds, where only 12 centimeters (5 inches) of rain falls per year. The CWP was built to rectify this problem. The most complex and expensive water diversion project in the history of the world, the CWP includes 21 dams and reservoirs, 22 pumping plants, and 1,140 kilometers (685 miles) of canals, tunnels, and pipelines (Figure 10.12). Looking down on Earth, a bluish sphere far below, America's moon-bound astronauts could identify only two artificial structures: the Great Wall of China and the main aqueduct of the CWP. An expensive "faucet," the project cost well over $2 billion—at the time, enough money to build six Panama Canals.

Despite the obvious benefits derived from this colossal project, certain aspects have drawn criticism from environmentalists. They claim that the CWP was built at an excessive cost, not only in tax dollars but in energy costs, losses in scenic beauty, and the destruction of fish and wildlife habitat. Environmentalists also criticize the proposal to dam up other free-flowing wild rivers in northwestern California, such as the Eel, Klamath, and Trinity. In their view, too many of such unharnessed streams have already been lost to irrigation and power production.

Regardless of these criticisms and the enormous cost to the people of California, the CWP is an accomplished fact and is helping to alleviate southern California's recurring water

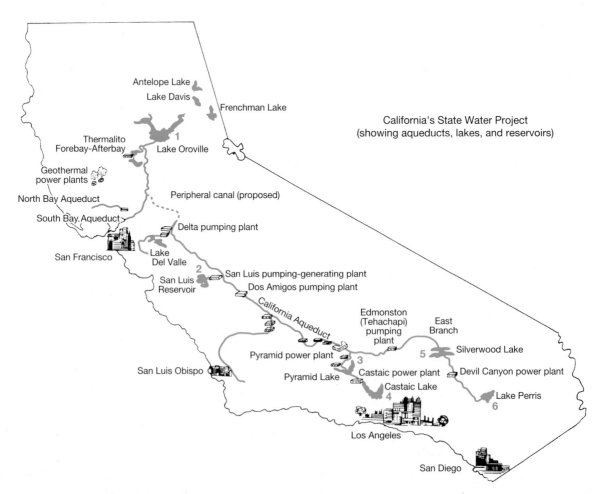

FIGURE 10.12 California Water Project. Major lakes and reservoirs are indicated by numbers.

shortages. Impressive as this project may sound, it is likely to be one of the last of its kind. Far more cost-effective—and more sustainable—means of meeting future demands must be employed, such as water efficiency measures in homes, offices, and businesses, which have been discussed in this chapter. Also essential to creating an environmentally sustainable water supply system are measures to recycle water and recharge groundwater.

10.3 Flooding: Problems and Solutions

When most people think about water issues, they think about water shortages. But many nations are plagued with the equally serious problem of too much water from time to time—flooding. Throughout history humans have suffered from destructive floods (Table 10.3 and Figure 10.13).

TABLE 10.3	Examples of Flood Disasters		
Date	Location	Deaths and Injuries	Property Damage
1900	Galveston, TX (hurricane-spawned floodwaters)	6,000 dead	3,000 buildings destroyed
1936–1937	Mississippi River	500 drowned 800,000 injured	$200 million
1965	Upper Mississippi River	16 drowned, 330 injured	$140 million
1979	Zambezi River (Mozambique)	45 drowned	250,000 homeless
1980	Southeastern Brazil	700 drowned	350,000 homeless
1981	Northern India	1,500 dead	Extensive crop losses
1993	Mississippi River	40 dead	$10 billion; 42,000 homes destroyed or isolated; 70,000 temporarily homeless; Farming on 70 million acres severely disrupted
1997	Red River	Not reported	45,000 evacuated in Grand Rapids, ND
1998	Bangladesh	1,300 dead	More than $3.4 billion; 31 million temporarily homeless
1999	North Carolina and other states	57 dead	6,000 homes destroyed $4.5 billion
2003	Europe, especially Germany, Czech Republic, Austria, Slovakia, Russia, and Romania	100 dead	Hundreds of thousands forced to leave their homes
2007	Bangladesh	3,000–4,000 dead	At least 1.5 million displaced
2008	China	57 dead	1.27 million forced to evacuate their homes; No estimate of economic cost available
2008	Myanmar	Over 100,000 dead	Over 1 million lost their homes; No estimate of economic cost available

FIGURE 10.13 Flooding is becoming a more common occurrence. It costs billions of dollars' worth of damage and takes many lives each year. Much of the world's flooding is caused by humans.

Flooding is a natural event. It happened long before humans appeared on the planet. But in recent years, flooding has grown worse, in large part because of the ways people alter the landscape. Changes in rainfall patterns resulting from a shift in climate caused by certain air pollutants also may be contributing to this problem. This may be the reason for the increase in flooding in many parts of the world, for example, in Texas in 2007 and earlier years, in the upper Midwest and China in 2008, and in Europe throughout the 2000s. We'll consider the ways people affect landscapes here. Climate change is discussed in Chapter 20.

Humans alter the landscape in many ways. We cut trees, graze our cattle on the land, and plow meadows under to grow crops. Deforestation, overgrazing, and farming all decrease the vegetative cover. Vegetated land acts as a sponge, sopping up water that falls on it. The water then percolates into the ground, where it feeds plants and replenishes groundwater. The loss of vegetation increases the amount of water flowing over the surface of the land—known as **surface runoff.** The increase in surface runoff causes streams and rivers to fill with water and flow over their banks, causing flooding (Figure 10.13).

Destruction of wetlands also increases flooding. Scientists believe that devastating flooding along the Mississippi River in 1993 could have been largely prevented had wetlands along this massive waterway not been destroyed. (For a discussion of this flood, one of most costly floods in human history, see Case Study 10.1.) Scientists also believe that the loss of coastal wetlands south of New Orleans could have played a huge role in the devastation to the city after Hurricane Katrina in 2005. These wetlands, had they been left intact, could have buffeted the storm surge that devastated the city and surrounding areas. The UN Food and Agriculture Organization contends that the destruction of

coastal mangrove swamps in Myanmar to increase wood production and to make room for people, farms, and aquaculture (fish farms) made flooding, the destruction of property, and the loss of lives much worse after the nation was hit by a powerful cyclone (hurricane) in early 2008. Mangrove swamps help hold rainfall and reduce the force of waves. At least 78,000 people died as a result of this storm, many of them drowning from massive flooding.

Impervious surfaces such as roads, parking lots, and rooftops in urban and suburban areas also increase surface runoff and flooding in watersheds. That's why cities are so prone to flooding. When heavy rains fall, water runs down streets and parking lots in a torrent, filling streams with water. In an undisturbed ecosystem, much of the water would have been absorbed by the ground.

Flooding is an important resource management issue because it damages homes, cities, towns, and farm crops, causing billions of dollars' worth of damage (including erosion) each year. It can also be lethal to humans, domestic animals, and wildlife. And it can be dramatically reduced by better management of farms, forestry, grazing, and construction and by preservation of natural areas, including forests and wetlands. For a map of the most flood-prone areas in the United States, see Figure 10.14.

Controlling and Preventing Floods

Although humans cannot prevent all floods, we can prevent or reduce the magnitude of many of them. Even some of the monstrous floods of the 1990s and 2000s, experts say, could have been greatly reduced had certain measures been taken during development.

Flood-control measures vary in complexity, cost, and effectiveness. Measuring the **snowpack** each winter helps in fighting floods, as it allows officials to predict future flooding and to take evasive action. Among the most popular approaches are levees, dredging, and dams, all of which are described shortly. As is often the case, one of the simplest and most effective approaches is preventive: restricting development and settlement in floodplains. Floodplains are the low-lying regions surrounding rivers and river deltas, and they are, by nature, prone to flooding. Efforts such as this minimize human damage. Flooding can also be minimized by protecting and restoring wetlands and vegetation within all watersheds—for example, by maintaining a normal vegetative cover to reduce surface runoff. Let's take a look at each of these measures, considering the environmental repercussions of each one.

Measuring the Snowpack to Predict Floods The U.S. Geological Survey (USGS), an agency of the federal government, measures the depth of the snowpack at more than 1,000 locations in the western mountains. Surveys can help the agency predict impending floods, so they can warn government officials and citizens to prepare.

Although measurements of snowpack can be helpful, they do not always work. Other factors may complicate matters. Snow may melt faster than anticipated, or a heavy rainstorm may occur when snows are melting. Further measures are therefore generally needed. For a discussion of ways that geographic

CASE STUDY 10.1   THE GREAT MISSISSIPPI FLOOD OF 1993

Some people have called it the flood of the century. Others give it even more status, claiming that it was the biggest flood ever witnessed in America. No matter where the Great Mississippi Flood of 1993 ranks among the deluges of history, it certainly is one that millions of Midwesterners would like to forget. The almost incessant thunderstorms that released torrential rains on Wisconsin, Minnesota, Iowa, Illinois, and Missouri swelled the Mississippi River to record heights. Indeed, the first seven months of 1993 were the wettest in Iowa for the 121-year period that records had been kept. Some regions received up to ten times their normal amount of rainfall.

The rampaging waters of Ol' Miss swept downstream at a speed of 8 miles per hour. At Hannibal, Missouri, the river crested at 32 feet or nearly 10 meters—2 feet or 0.6 meters higher than the 500-year flood mark. At St. Louis, the river was above flood stage for a record-breaking 80 consecutive days, and in early August, it reached an all-time high of 49.5 feet (15 meters). Eventually the boisterous river spilled over into low-lying farmlands and swamped an area twice the size of New Jersey.

Much of the river from Minneapolis south to St. Louis was bordered by levees—walls of rock and concrete that were effective in containing the annual spring rise of the river from rain and snowmelt. Unfortunately, however, many levees were not designed to handle the flood of 1993. In fact, more than 800 of the levees either crumbled under the force of the swollen waters or were overrun. Paul Schloesser, Chief Petty Officer with the Coast Guard, was standing within a few yards of a levee at West Quincy, IL, when it broke. He recalled water pouring through the opening "with a raging roar like that of a freight train." The waves shot 5 feet (1.5 meters) into the air. Nearby workers ran for safety.

In many a river town where levees were either ruptured or nonexistent, the residents erected sandbag barricades to hold back the water. The dedication and energy expended on sandbagging were impressive. For example, in Cape Girardeau, MO, thousands of residents, in addition to hundreds of volunteers who came from a 400-mile radius, built a barricade of more than 500,000 bags. The sandbaggers ranged from schoolchildren to old-timers, from carpenters to lawyers, from farmers to college professors. In the process of defending the town from the river, these workers exhibited a magnificent degree of cooperation, self-sacrifice, and community bonding.

In countless spots along the upper Mississippi Valley, the floodwaters swirled over banks and then surged over and through levees and the frantically constructed sandbag barriers. The water kept getting higher and higher. It swept into fields of soybeans and corn. It raged into downtown shopping areas. Finally, it made its way into the living rooms, kitchens, bedrooms, and bathrooms of hundreds of riverfront homes, leading television commentator Tom Brokaw to declare that "Mid-America was under siege."

In many of the inundated towns, grocery shopping could only be done with the aid of a rowboat or canoe. One elderly native of Grafton, IL, who was proud of his tomato garden did his best to save his prized red beauties. He dug them up, put them in baskets, and strung them along his wife's clothesline, where they stayed "high and dry." One distraught woman, while wading in hip-deep water, came across a 3-foot water snake swimming in her bedroom. She tried to calm herself by imagining it was only a gigantic earthworm.

The flood raised havoc with transportation. Water swamped low-lying airport runways and severely disrupted flight schedules. Rail service on the transcontinental railways in areas gripped by flooding came to a halt. In Quincy, IL, the Bayview Bridge—the only Mississippi crossing within a 200-mile span of the river—was closed for a month. During the height of the flood, the U.S. Army Corps of Engineers closed the river to all commercial traffic from Dubuque to St. Louis. The shutdown forced at least 200 barges out of action, for an aggregate financial loss of $1 million per day.

At the peak of the crisis, Secretary of Agriculture Mike Espy made a helicopter survey of crop damage. Ironically, farm flooding was so extensive that in some areas there was very little land to see. The U.S. Department of Agriculture (USDA) estimated that at least 8 million farm acres (3.2 hectares) either never got planted or were under water, and on another 12 million acres (nearly 5 hectares), the corn and soybean crops were severely stunted. Eventually, several hundred counties in Wisconsin, Minnesota, Iowa, Illinois, and Missouri were declared official disaster areas and eligible for emergency federal aid.

The flood of 1993 left in its wake an estimated 40 dead, at least 42,000 homes destroyed or isolated by water, about 70,000 people temporarily homeless, the productivity of 20 million acres of cropland severely disrupted, and overall financial damage of at least $10 billion. No wonder many authorities consider the flood of 1993 to be the greatest natural disaster in the history of the United States!

The flood, while devastating and costly, had some benefits, too. It has begun a movement to return floodplains to rivers. Rather than facing devastating floods again, some communities have opted to relocate to higher ground. Not only does this save taxpayers money, as they're the ones who typically foot the bill to rebuild after devastating floods, but land once occupied by people is being returned to wildlife. Natural habitat, nearly gone from the banks of many larger rivers, is being slowly restored. Wildlife biologists hope eventually to create a "string of pearls," that is, a series of natural habitats along large rivers, helping provide valuable habitat for ducks and other species. This will also reduce damage from floods, because it puts people out of harm's way and provides safe places for water to go in case of floods. Restored wetlands could help reduce the intensity of floods as well.

information systems (GISs) and remote sensing are being used to monitor snowpack and predict water supplies, see the GIS and Remote Sensing box.

Levees   Levees have been a popular flood-control measure. **Levees** are dikes constructed of earth, stone, or mortar at varying distances from the riverbank to protect residential, industrial, and agricultural property from floodwaters (Figure 10.15). In 1936 Congress passed the landmark Flood Control Act to protect homes, farms, and businesses along the Mississippi River and its tributaries against frequent floods. The U.S. Army Corps of Engineers was given the responsibility for doing the work. Since then, similar flood-control measures have been applied to many other rivers.

Flooding causes major damage to agricultural, urban, and other developments

Problem not considered major

FIGURE 10.14   Areas of the United States with flooding problems. Flood damage is expected to increase in the future on 175 million acres of land that is flood prone. (A flood-prone area is land adjoining rivers, streams, or lakes where there is a 1% chance of flooding during any given year.) Forty-eight million acres of flood-prone land are cropland, 102 million acres are pasture, range, and forest, and 21 million acres represent other land, including built-up areas. Twenty-one thousand communities are subject to floods, including 6,000 towns or cities with populations exceeding 2,500. The potential damage caused by floods in any given year in the United States is about $4 billion. Because the number and value of buildings are increasing, flood damage will also increase.

In recent years, however, structural control methods such as levees have increasingly come under attack. Indeed, a debate is currently being waged over the merits of expanding the levee system, especially along the upper Mississippi River. Let's briefly consider some of the arguments pro and con.

The Pros

1. The levees and dams built by the U.S. Army Corps of Engineers since 1938 have effectively controlled flooding along the Mississippi River system and have prevented at least $250 billion in damage—a ten-to-one return on the cumulative cost of $25 billion.

2. In 1973 and again in 1993, the Mississippi River reached record once-in-a-century flood levels. Nevertheless, many thousands of Americans remained safe behind the levees. Without this protection, the loss of life would have been considerable.

3. The levees help maintain a river depth that permits extensive barge traffic, which is invaluable to the nation's economy and to the U.S. position in world trade.

4. The protection afforded by the levees has made possible extensive agricultural, industrial, commercial, and residential development very close to the riverbanks.

The Cons

1. Since 1938, the per capita flood damage has actually increased by a factor of 2.5.

2. Despite the ambitious levee-building program of the Corps, the annual number of flood-related deaths has remained constant on a per capita basis from 1916 to the present.

3. The dam and levee system is not structurally perfect. During the 1993 flood, more than 800 of the 1,400 levees that had been built by the Corps or by state and local agencies either crumbled or could not withstand the surging waters.

4. The "straightjacketing" of a river between levees may actually increase flooding severity downstream. This is because restricting flow causes the water level to rise

GIS AND REMOTE SENSING

GIS Aids Snow Monitoring and Modeling at the National Weather Service

In the mountainous states of the western United States and in the western provinces of Canada, winter snows are a vital source of liquid water for many urban and rural areas. So massive is the snowmelt in western states, in fact, that the spring runoff provides a very large percentage of the annual stream flow of many rivers. In the spring when the snows begin to melt, huge reservoirs depleted by farmers, industry, and residences in the previous year begin to fill once again, creating a supply of water that often receives little additional input until the next spring snowmelt begins.

Unfortunately, things don't always go as we'd like them to. Sometimes the snowpack, the amount of snow accumulating in the mountains over the winter, is excessive. Other times, the snow melts much more rapidly than anticipated. The result of both occurrences is flooding—accompanied by loss of life and property damage. Sometimes the snowpack is below average, and water shortages emerge.

In the United States, the National Weather Service (NWS) seeks to help those who are dependent on spring snowmelt and those who live in the floodplains of rivers. The NWS's National Operational Hydrologic Remote Sensing Center (NOHRSC), headquartered in Chanhassen, MN, uses remote sensing and GIS, combined with other techniques (such as mathematical snow modeling and field-gathered data) to provide vital information needed to estimate snowpack and to help predict water supplies and potential flooding.

NOHRSC monitors the extent of snow cover in the United States, its water content, and air temperature and then uses GIS to create detailed maps for display and analysis. NOHRSC obtains data from numerous sources. For example, it receives snow tube measurement data collected at fixed sites along hundreds of snow courses throughout the West from the Natural Resources Conservation Service, another government agency. Data is also supplied by a network of automated snow telemetry stations. Located at remote sites, these stations relay snowpack and weather data via satellites to ground receiving stations, which, in turn, transmit the data to the NOHRSC. The center also obtains data on the water content and the depth of snow cover across the United States by conducting surveys from airplanes whose onboard instruments measure gamma radiation emitted by soils and snowpacks. (Researchers have developed mathematical equations that allow them to estimate the liquid water content of snow based on the amount of gamma radiation detected by the airborne sensors.)

The center processes the data using computer programs and GIS. Much of the data is provided to an extensive network of river forecast centers operated by another branch of the NWS.

Timely information about the quantity and location of winter snowpack is vital to many end users. Farmers, irrigation district managers, hydroelectric power generators, wildlife managers, and flood managers, however, are the main benefactors of these new and exciting technologies. If floods are predicted, evacuation plans can be examined and fine-tuned. If water supplies will fall short, water departments can implement water conservation programs. Given the lead time to take actions, GIS becomes a valuable tool in managing ourselves and our resources.

FIGURE 10.15 Flooding along rivers often results when protective levees break.

downstream more than if a river is allowed to spill over its banks in upstream areas. The Mississippi River near St. Louis is actually 3.3 meters (10 feet) higher today than it was in 1881, largely due to the levees that have been built along its banks. In addition to causing rivers to rise, the levees increase their velocity and water pressure. Therefore, if a levee does rupture, the property damage and loss of life will be much greater at that point than if the levee had never been built in the first place.

5. The levee system causes the drying out and death of any wetlands that once bordered the river. Of course, all the important functions of the wetland, such as a purifier of groundwater, a habitat for wildlife, and a living sponge that soaks up floodwaters, are lost forever.

6. The levee system prevents surrounding croplands from receiving nutrient-rich sediment from the river during annual low-level spring floods.

7. Levees are unsightly barricades of earth or cement. Because many are 12 to 15 meters (40 to 50 feet) tall, they effectively block the view of the scenic river valley.

8. Once the levees are constructed, agricultural, industrial, commercial, or residential development often takes place nearby. In other words, development accelerates in areas once prone to flooding. As a result, if the levee is breached by a massive flood, the property damage and human misery can be extensive and severe.

9. Levee maintenance is very expensive. For example, in just one 89-square-kilometer (35-square-mile) drainage district along the Rock River in northern Illinois, the 1993 disaster caused at least ten ruptures and the flooding of 4,050 hectares of (10,000 acres) of farmland.

The repair of these breaks cost American taxpayers at least $500,000.

10. Finally, levees also tend to trap water behind them. Water falling on the watershed behind a levee may not be able to reach the river, and water pools behind the levee, causing extensive damage.

As in many issues, both sides present convincing arguments. Today, there is a movement underfoot to restrict farming, construction, and other activities in floodplains. At least one town inundated in the 1993 flood moved to higher ground. In recent years, several federal agencies involved in flood protection, such as the Federal Emergency Management Agency and the Army Corps of Engineers, have begun to spend huge sums helping communities along rivers relocate to higher ground (Figure 10.16). This not only protects them from flooding, but also reverts floodplains to their rightful owners, the rivers. Land once occupied by houses is now being preserved for wildlife and natural flood control.

Dredging   Huge amounts of soil are washed into streams from watersheds disturbed by human activities, causing sediment to accumulate in channels. Called **streambed aggradation,** this buildup decreases the depth of waterways and thus increases the probability and severity of flooding. The size of this problem can be appreciated if we note that the Mississippi River transports roughly 2 million metric tons of sediment daily. To cope with this problem (as well as to deepen the channels for navigation), the Mississippi River and a host of others are periodically dredged by the U.S. Army Corps of Engineers.

The importance of **dredging** (removing sediment from river bottoms) is emphasized by the 1852 Yellow River catastrophe in China. As the channel of this river became choked with silt,

FIGURE 10.16   Headed for higher ground. This church is being moved 500 feet away to locate it out of the flood zone.

FIGURE 10.17 Hoover Dam, one of the world's largest dams. It is located on the Arizona–Nevada border, about 30 kilometers (25 miles) southeast of Las Vegas. The electrical power generators at the dam supply most of the power needs of southern California, Arizona, and Nevada. The dam restrains the turbulent waters of the Colorado River and is useful in controlling downstream floods. Water from Lake Mead, the 150-kilometer-long (125-mile-long) impoundment behind the dam, irrigates 0.4 million hectares (1 million acres) and has increased crop production 120% in this region.

levees were built higher and higher, until the Yellow River was flowing above the rooftops. Eventually, a massive surge of floodwaters crumbled the retaining walls and drowned 2 million people.

Dams For many years, the United States was engaged in a vast program of dam construction. The largest was the Hoover Dam on the Colorado River (Figure 10.17). Over the years, dams and other flood control measures have prevented enormous property damage and have saved hundreds of lives. Despite the value of dams in flood control and in generating hydroelectric power, big dams result in a number of problems (Table 10.4).

One major drawback associated with the construction of big dams is the speed with which reservoirs fill with sediment. The rate at which a reservoir fills with sediment depends on several factors. The topography, vegetative cover, and the degree to which soil erosion is controlled within the watershed are important determinants of sediment flow in streams that feed reservoirs. Because the Columbia River is relatively sediment free, dams such as the Grand Coulee and Bonneville might have a storage life of 1,000 years. Others have intermediate lifespans. For example, the huge Lake Mead Reservoir behind Hoover Dam on the Colorado River in Nevada is filling with silt at a rate sufficient to destroy the operation of this multi-million-dollar structure in less than 250 years. The life span of dams constructed across muddier streams may be quite short (Figure 10.18). California's Mono Reservoir, for example, was designed to provide a permanent water source for the people of Santa Barbara. However, the reservoir filled up with sediment in 20 short years. Biological succession has proceeded so rapidly in the sediment that a thicket of shrubs and saplings became firmly established. For all practical purposes, Mono Reservoir has been reclaimed by nature.

Another problem with dams, especially in hot, arid regions, is extensive evaporation. For example, the top 2 meters (7 feet)

TABLE 10.4	Disadvantages of Big Dams

1. Construction is extremely expensive, costing hundreds of millions to billions of dollars. (Example: The Hoover Dam on the Arizona–Nevada border cost $120 million.)

2. Prime agricultural land is flooded.

3. Scenic beauty is destroyed. (Example: The Rainbow Bridge National Monument in Arizona is threatened by the Glen Canyon Dam on the Colorado River.)

4. The resulting saltwater intrusion in coastal areas destroys cropland and pollutes freshwater aquifers. (Example: This has happened in both Florida and California.)

5. Drawdowns periodically eliminate the shallow-water areas where fish frequently spawn.

6. The natural habitat of endangered species is destroyed. (Example: The snail darter at the Tellico Dam.)

7. The upstream migration of adult salmon is blocked, interfering with reproduction.

8. There are excessive water losses from reservoirs because of evaporation.

9. The life of a dam is shortened because of siltation of the reservoir.

10. The collapse of the dam is possible as a result of faulty construction.

of water in Lake Mead in Nevada evaporate every year. About 6 million acre-feet are lost each year from 1,250 large western reservoirs, an amount sufficient to supply all the domestic water needs of 50 million people.

On rare occasions, a dam will collapse because of faulty design. Such was the case with the mighty Teton Dam in Idaho (Figure 10.19). This dam was built by the U.S. Bureau of Reclamation, an agency that has constructed more than 300 dams, including the world-famous Grand Coulee and Hoover Dams. On June 5, 1976, hours after the first fissure appeared in the Teton Dam, the monstrous earthen structure gave way. A mammoth wall of water surged down the valley. At least 14 people died, and estimates of property damage approached $1 billion.

Dams cause a host of other problems, too. They inundate fertile valleys. They destroy streams that are vital habitat for fish and nonaquatic species and are valuable for recreational benefits such as fishing, camping, canoeing, and rafting.

Stream Channelization The Natural Resource Conservation Service (NRCS) has built up a good reputation during many years of valuable service in soil erosion and flood control. However, a storm of criticism has swirled around the service's channelization efforts. **Channelization** is the intentional deepening and straightening of streams to control flooding.

FIGURE 10.18 The death of the Lake Ballenger, TX, reservoir. Although the original depth of this lake was more than 10 meters (35 feet), it eventually had to be abandoned because of siltation.

FIGURE 10.19 The death of a dam: An aerial view of the Teton Dam in Idaho shortly after its rupture, looking upstream. The site of the break is roughly in the center of the photograph. The torrential waters released from the reservoir above the dam caused the deaths of at least 14 people.

There are two main steps in channelization. First, all vegetation is bulldozed away on either side of the stream. Second, bulldozers and draglines deepen and straighten the channel. In essence, the steam is converted into a water-filled ditch that moves water quickly out of a watershed. The denuded banks are then planted with a cover crop to stabilize the soil and prevent erosion.

The benefits of channelization are several. First, as previously noted, deepened channels help remove water from a stream, which reduces flooding. Second, because of this, cropland, towns, and buildings along the channelized portion of the stream are protected. Another benefit is that small lakes for recreation and wildlife were often constructed by the NRCS as an adjunct to the main channelization process.

The disadvantages of channelization are many. In many instances, a picturesque meandering stream is converted to an unsightly ditch prone to erosion. During channelization, much valuable hardwood timber, already in short supply, is destroyed. This, in turn, reduces wildlife habitat. Channelization also causes water temperatures to increase as a result of the removal of overarching trees that formerly intercepted the sunlight. This alters the conditions in streams and rivers, making them unsuitable for many species of fish, a topic discussed in more detail in the next chapter. Channelization also decreases stream enrichment—that is, the provision of nutrients typically from leaves that fall into the water. Finally, channelization often increases flooding downstream along nonchannelized sections.

In its long-range plans, the NRCS proposed to channelize several thousand small watersheds in the United States by 2000. That would have meant degrading nearly half of the nation's small watersheds. Fortunately, the controversy over channelizations has greatly slowed this process.

Protecting Watersheds One of the most economical and sustainable of all approaches to flood control is watershed protection. As noted in Chapter 1, a **watershed** is an area of land drained by streams and rivers. It is also sometimes called a **drainage basin.** A watershed delivers water, as well as sediment and nutrients, via small streams to a larger stream or a river. Watersheds vary in size from extremely small to extremely large.

All watersheds, large or small, convert precipitation (rain and snow) into stream flow and groundwater. Even during a light shower of only 0.25 centimeter (0.1 inch) of rain, a 2.5-square-kilometer (1-square-mile) watershed would produce 6.6 million liters (1.7 million gallons) of stream flow.

Watersheds vary considerably in vegetative type and topography. These characteristics, combined with precipitation, determine how much water seeps into the ground and how much flows over the surface. For example, in the West, where vegetation is not very dense, rainstorms often result in increased surface flow and increased soil erosion. The lack of vegetation in such areas, even in undisturbed watersheds, often results in floods, even sudden flash floods. In the more heavily vegetated East, in undisturbed ecosystems, heavy

vegetative cover reduces surface flow. This, in turn, reduces stream flow and flooding. When watersheds are disturbed, either by natural factors (fires, for example) or human factors (farming, for instance), two things typically happen: Surface water flows increase, and erosion increases. This results in flooding and the buildup of sediment in streams and lakes. The effect of sediment on rivers and reservoirs was discussed earlier.

Watershed protection is to natural resource conservation what preventive medicine is to health care. In the United States, some degree of protection is provided by federal law, notably the **Watershed Protection and Flood Prevention Act.** Under this law, protection of small watersheds is assigned to the Natural Resource Conservation Service of the USDA. According to the USDA, nearly 62% of the 13,000 small watersheds (smaller than 100,000 hectares or 250,000 acres) in the United States suffer from flooding and erosion. While protection of small watersheds is given to the NRCS, protection of large watersheds is assigned primarily to three agencies: the Bureau of Reclamation, the U.S. Army Corps of Engineers, and the Tennessee Valley Authority (TVA).

Although the primary objective of the act is to prevent flooding by better watershed management, it also seeks to reduce erosion, increase and protect water supplies, promote better wildlife management, and protect recreational uses of waterways. Within watersheds that are used for agriculture, for example, soil erosion measures are promoted by the various agencies (discussed in Chapter 7), which help reduce flooding. Erosion-control measures around mining operations and new construction also help. Controls on grazing that permit the emergence of healthy vegetation protect the land and reduce surface runoff. Over time, studies have shown, not only are preventive measures often highly effective, they are extremely cost-effective.

Zoning of Floodplains The U.S. Army Corps of Engineers, the Bureau of Reclamation, and the Natural Resource Conservation Service have spent more than $15 billion on structural flood-control projects (dams, levees, seawalls, etc.) since 1925. However, despite this enormous expenditure of tax money, property damage from flooding continues to rise. Annual costs have increased considerably, from $3 billion in the early 1980s to more than $10 billion by 2000. More and more experts now believe that nonstructural flood controls like watershed protection (discussed in Chapter 1) and **floodplain zoning** are the most effective and economical strategies.

The **floodplain** is the low-lying area along a stream. These areas are called floodplains because they are subject to periodic flooding. To many ecologists, floodplains belong to rivers. Unfortunately, floodplains have been invaded by humans. The flatness and fertility of the land, its natural beauty, and the availability of the river for cheap transportation and water have attracted people since the beginning of time. Today, more than 2,000 cities in the United States, including Harrisburg, PA, Des Moines, IA, and Phoenix, AZ, are located at least partially on floodplains. Most of these cities experience flooding every two to three years.

Interestingly, of the 42,000 homes destroyed or damaged by the great Mississippi flood of 1993, most were located on a floodplain.

In 1973, Congress passed the **Federal Flood Disaster Protection Act** to regulate floodplain development and control flood damage. This law encourages the zoning of floodplains for use as parks, golf courses, bicycle trails, nature preserves, and parking lots. The construction of buildings on the floodplains is discouraged. Under the terms of the act, any home, store, or factory built in a hazardous area on the floodplain is denied federal flood insurance. Unfortunately, the act has had little effect. People have continued to build on floodplains. Cities and towns along the nation's rivers have been repeatedly flooded. In the early 1990s, though, federal agencies began to encourage towns to relocate out of floodplains, converting flood-prone land into natural vegetation.

10.4   Irrigation: Issues and Solutions

In many parts of the world, especially arid and semi-arid regions, farmers irrigate their crops with water to ensure sufficient production. Irrigation demands a high degree of planning, field preparation, and technical skill. It also requires a lot of water. Up to 2.5 million liters (750,000 gallons) of water may be needed to irrigate a single acre in one growing season. It is also costly. Nevertheless, the high market value of many crops makes it worthwhile. In warm, arid regions such as southern California, irrigation allows farmers to plant several crops per year.

Today, more than 10% of the United States' cropland is now under irrigation. As you might suspect, though, four of every five irrigated acres are in the West (Figure 10.20). Until recently, the amount of irrigated farmland increased steadily.

Methods of Irrigation

Irrigation water is either pumped or gravity fed through the main irrigation canal or piped to laterals, which convey the water to individual farms. The laterals frequently follow field borders and fence lines. The major methods are sheet, furrow, sprinkler, and drip irrigation.

Sheet and Flood Irrigation   The **sheet irrigation** method, usually used for hay, grains, and pastures, is suitable on land with a slight grade (slope). For proper sheet irrigation, the topsoil must be carefully prepared in advance so that water infiltrates properly. Water is released into a field and then gradually flows downslope in a sheet (Figure 10.21). If not carefully executed, sheet irrigation can result in soil erosion and the leaching of nutrients from topsoil.

Flood irrigation is also used on some farms, especially in the West, for growing rice. It is also used in cranberry-growing regions. As its name implies, farmers flood fields with irrigation water. Water may stand in the fields for several months as rice plants grow.

Furrow Irrigation   In the **furrow irrigation** method, water may be drawn from laterals by siphon tubes that empty into furrows between crop rows (Figure 10.22). This method is used primarily on row crops such as corn, cabbage, and sugar beets.

Sprinkler Irrigation   The **sprinkler irrigation** method may be used in places where the previous methods are undesirable, for example, on fields with steep slopes that are prone to erosion. It is also used in flatlands where water is acquired from deep wells. It involves costly equipment and is restricted primarily to crops of high cash value. Considerable amounts of energy are required to power the pumps.

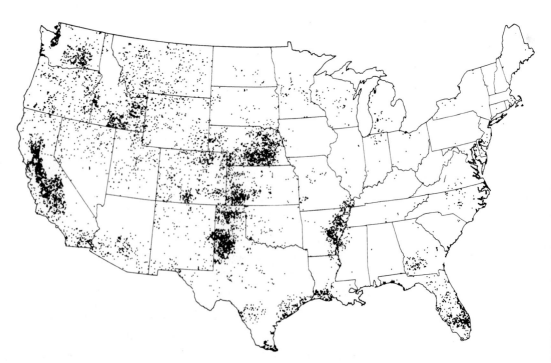

FIGURE 10.20   Irrigated acreage in the United States. One dot equals 8,000 acres of irrigated land.

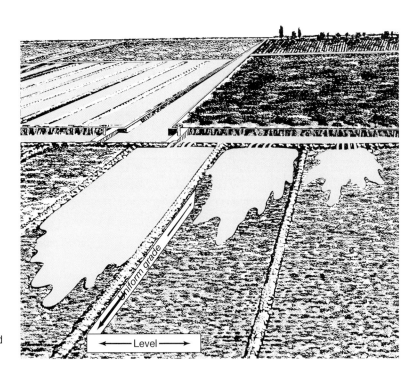

FIGURE 10.21 Sheet or flood irrigation. This field has been prepared for irrigation by building small levees around each leveled area. The areas are then flooded in rotation to irrigate them.

FIGURE 10.22 Furrow irrigation. Water is transferred from the lateral in the foreground to a series of parallel furrows to irrigate this lettuce crop in the Palo Verde Valley in California.

Sprinklers may be either stationary or rotary (Figure 10.23). Center pivot **rotary sprinklers** are very popular. In this technique, water is sprinkled from a raised lateral pipe that is in continuous slow rotary motion around a central pivot point. One model, known as the big gun, can propel a stream of water half the length of a football field. Most models spray water from numerous sprinkler heads spaced evenly along the length of the irrigation pipe. A circular area of 53 hectares (130 acres) can be irrigated in one revolution in 33 hours with a 400-meter (1,300-foot) pipe. Evaporation losses with all sprinkler systems may be considerable. Because of this, many center pivot irrigation models have been retrofitted with downward-facing spray heads that are suspended a relatively short distance above the crop. This greatly reduces water loss through evaporation.

Drip Irrigation Drip irrigation is used in orchards and vineyards where crops are permanent. With the drip method, water is delivered through perforated or highly porous plastic pipes (Figure 10.24). The pipes may either be placed on the surface or buried underground. Water drips from the pipes and seeps into the ground, watering the roots while minimizing evaporation. Since drip irrigation requires 20% to 50% less water than sprinkler systems, it can save considerable amounts of water. Bear in mind, however, that drip irrigation is not suitable for all crops. It cannot be used for wheat or corn, because of the need to lay considerable amount of pipe year after year. It is quite suitable for permanent crops such as fruit and nut trees.

Irrigation Issues

Irrigation clearly helps increase productivity and is essential to meeting future demands (Figure 10.25). But irrigation is not without its problems.

Water Loss Many irrigated fields in the United States receive their water from reservoirs or streams located hundreds of

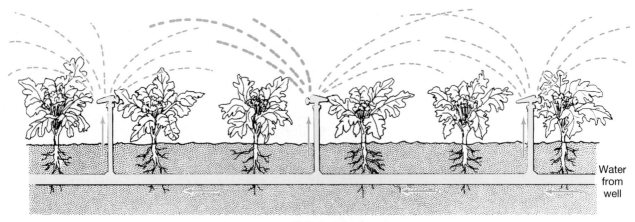

Buried pipe

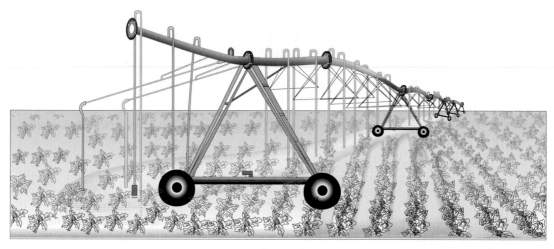

Center pivot

FIGURE 10.23 Two types of sprinkler irrigation. In center pivot irrigation, the perforated pipe moves slowly in a circle and may irrigate an area of about 150 acres.

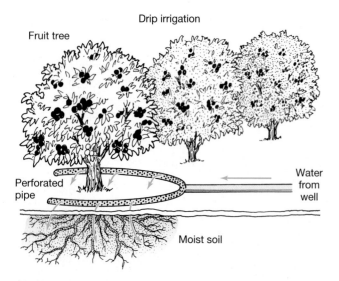

FIGURE 10.24 Drip irrigation. This method is used primarily on fruit trees and high-cash crops. Water is applied directly to the roots of the fruit tree. The perforated plastic pipes may either lie on the ground's surface or be buried.

kilometers away. The fruit-raising Central Valley of California, for example, gets its water from the Colorado River, 500 kilometers (300 miles) to the east. During transit, considerable amounts of water are lost. According to the USDA, only 1 of every 4 gallons drawn for irrigation is actually absorbed by crop root systems. The remaining 3 gallons are lost to evaporation, to water-absorbing weeds, or to ground seepage. In some areas, seepage may be sufficient to raise the water table and form a marsh. Water efficiency measures discussed earlier in the chapter can help solve this problem.

Salinization and Waterlogging Would you believe that bringing freshwater to a desert might be destructive to crops? Sounds odd, doesn't it? However, even freshwater is slightly salty. River water used for irrigation acquires dissolved sodium, calcium, and magnesium salts as it flows down mountain slopes and through valley bottoms. In some cases, water flows through regions that greatly increase its salt concentrations. Salts in agricultural soils may also come from water that has drained from farm fields. Water that

FIGURE 10.25 Irrigation transforms the desert! Water from the Colorado River conveyed by the All American Canal system makes it possible for crops to be produced in Imperial Valley, CA. Note the stark contrast between the nonirrigated desert on the right and the irrigated fruit orchards on the left.

flows off fields into streams may be used further downstream as irrigation water, and then again even farther along the course of the river. Each time, it gets more salty. The Colorado River, for example, receives large amounts of salty water that has been flushed from thousands of irrigated farms in western Colorado, Utah, and Arizona. As a result, the river's water becomes progressively saltier as it flows southwestward across the Mexican border toward the Gulf of California. As a result of irrigation practices, the salinity of the lower Colorado has increased by 30% in the past few decades.

When salt-laden water is applied to farm fields in hot, arid regions, where percolation into the soil is poor and evaporation rates are very high, much of the water spends little time in the soil. It quickly evaporates, leaving behind the salts. They form a white crust on the ground. Additional salt may be deposited because of the evaporation of groundwater that has been drawn to the surface by capillary action.

The buildup of salt in the soil is called **salinization.** As time passes, the salt deposits increase, becoming toxic to crops. Agricultural experts estimate that 30% of the West's irrigated land has salinity problems. After 20 years of irrigation, some land may have a salt load of more than 30 metric tons per hectare (80 tons per acre). Even in California's **Imperial Valley,** where crop harvests have been bountiful, salinization has caused many farms to be abandoned. Salinization is also a problem in other agricultural nations where semiarid and desert lands are irrigated. The Colorado's saltiness jeopardizes cotton production in the Mexicali district of Mexico, where farmers have used the river as a source of irrigation water for decades. The economy of the region was threatened to such an extent that Mexican presidents frequently conveyed their concern to the U.S. government. The United States was eventually forced to construct a desalination plant near the Mexican border so that the water would be usable by Mexican farmers.

Salinization can be reduced by installing underground drainage pipes that collect excess salty groundwater. These porous pipes connect to a master drainpipe from other fields. This fairly effective measure, however, creates a waste disposal problem. The leftover water is rich in salts and potentially toxic heavy metals leached from the soil. If dumped in natural water courses, it could cause serious effects. Contaminated water from farms in the San Joaquin Valley of California caused extensive birth defects and mortality to waterfowl living in and around specially built evaporation ponds at the Kesterson National Wildlife Refuge and other locations in the 1990s.

Efficient irrigation, which reduces water use, can also reduce salinization. Salt-resistant crops that replace traditional strains can also be used. Finally, it may be necessary to convert areas plagued with recurring salt problems from cropland to grazing land—a solution to which many farmers are resistant.

Another related problem is waterlogging. **Waterlogging** occurs when soil receives too much moisture. This may occur naturally or as a result of applying too much irrigation water. When soil becomes saturated with water, that is, when the air spaces are filled with water, oxygen flow to the roots is impaired. Plants suffocate and die. In addition, excess water moves up through the soil via capillary action. This water evaporates, leaving behind once-dissolved salts. In other words, excess irrigation, especially by water with excess salts, not only suffocates plants, it can result in salinization.

Depletion of Groundwater and Surface Water Yet another problem caused by our heavy use of irrigation water is groundwater depletion. Over the years, the number of hectares irrigated by farmers in the United States, Canada, and many other nations has increased sharply. On the high plains, a vast region extending from Nebraska south to Texas, most of the irrigation water is pumped out of the Ogallala Aquifer (Figure 10.7). In 1946, there were only 2,000 irrigation wells in western Texas; today there are more than 70,000. The rate of withdrawal around Lubbock, TX, is 50 times the rate at which the aquifer is naturally recharged by rain and stream flow.

As a result of this enormous overdraft for irrigation, groundwater levels have dropped significantly in some western states—more than 30 meters (100 feet) in portions of Kansas, Oklahoma, Texas, and New Mexico, and more than 120 meters (400 feet) in parts of Arizona. As a result, new wells must be drilled more deeply at considerable expense. And, of course, it takes more fossil fuel or electricity to pump water from the deeper wells. If this trend continues, irrigation farming could become prohibitively expensive. West Texas could experience a 95% decline in irrigation and a 70% decrease in crop harvests by 2015. Serious declines in irrigation farming are expected over an 11-state area of the Great Plains and the Southwest—during an era of ever-increasing food demand.

Groundwater depletion or groundwater mining, as it is sometimes called, may worsen in years to come as agricultural production increases. Projections of climate change caused by greenhouse gases and deforestation indicate that the mid-continental areas of North America, which are heavily dependent on groundwater for food production already, will become hotter and drier.

Many of these regions rely on the Ogallala Aquifer, which is being rapidly depleted. Natural recharge zones for this aquifer, shallow ephemeral lakes, known as **playa lakes,** are also being plowed over and destroyed. Playa lakes fill with water in times of abundance and provide habitat for many species (Figure 10.26). These lakes also release water into the Ogallala Aquifer, replenishing it. If the water from the Ogallala Aquifer continues to be depleted, the world-renowned breadbasket of

the world could face devastating losses. The effects on U.S. and global food supplies, famine and hunger, and economics could be substantial.

Groundwater withdrawals have other, more immediate effects, too. When large volumes of water are removed from fine-grained, porous aquifers, the weight of the overlying soil and rock occasionally causes compression or collapse of the aquifer. As a result, the earth above the aquifer sinks, or subsides (Figure 10.27).

Groundwater overdraft causes significant land sinking or **subsidence** (sub-sidé-ence). Subsidence has occurred in at least 11 states in the South and West due to groundwater overdrafts. **Water mining** caused the dramatic appearance of a sinkhole at Winter Park, FL, in 1981. The huge pit was 37.5 meters (124 feet) deep and 120 meters (396 feet wide). It swallowed up a house, a swimming pool, six sports cars, and a camper. In the San Joaquin Valley of California, an 11,000-square-kilometer (4,200-square-mile) area sank more than 0.3 meter (1 foot). Some regions subsided more than 9 meters (30 feet).

Subsidence damages irrigation facilities such as canals and underground pipes. The Department of the Interior spent $3.7 million in a single year to repair the damage that subsidence caused to federal irrigation projects.

In the Galveston Bay region near Houston, TX, subsidence over a 10,000-square-kilometer (4,000-square-mile) area resulted in widespread flooding of shoreline properties. When subsidence occurs under urban areas, damage may be very extensive: Sewer and water pipes burst, wells are destroyed, building foundations crack, and concrete highways buckle.

Groundwater overdraft can be reduced by efficient water distribution and use, methods described earlier in the chapter. Protection of groundwater recharge zones, areas where aquifers are replenished by water percolating from the surface, by restricting growth and other activities can also help (Figure 10.28). Wastewater applied to the ground in such areas also can help renew groundwater supplies.

Streams and rivers in many parts of the world are greatly depleted each year by water withdrawals for irrigation and other

FIGURE 10.26 Aerial view of playa lakes. Some 50,000 shallow depressions in six states replenish the Ogallala Aquifer and provide valuable habitat for birds and other wildlife. Unfortunately, the lakes are often plowed under or filled with sediment eroded from nearby farmland.

A. Before land sinking

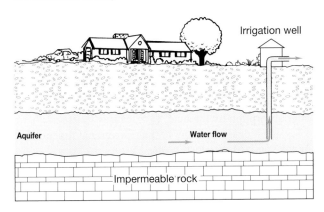

B. After land sinking

FIGURE 10.27   Subsidence. The land-sinking phenomenon, called subsidence, is caused by groundwater overdraft or water mining.

uses. Some streams run completely dry because of the massive withdrawal of water. For an aquatic organism or an organism dependent on the water body, the dewatering of its habitat can be lethal.

In the United States, some states have attempted to rectify the problem by maintaining minimum stream flows—that is, modest flow levels below which a stream or river cannot drop in order to protect aquatic species, especially fish. Scientists have developed numerous ways of determining minimum flows. However, minimum stream flows are often set very low—so low that they greatly reduce a stream's full potential to support life. Fish and other aquatic organisms often require more water to thrive and reproduce than is allowed, but powerful interest groups including farmers and water departments also need the water. Their interests often prevail over the interests of wildlife. Protecting wildlife is socially and politically difficult, and further troubles lie ahead as populations expand and the demand for water by people increases. Many states in the West are continuously involved in litigation over minimum flow, being sued by environmentalists and wildlife agencies.

Sale of Water Rights   Some farmers and ranchers in the arid states are discovering that their irrigation water is more valuable than the wheat, fruits, vegetables, and beef cattle they produce. As a result, they are now beginning to sell their land and water rights to nearby communities that need the water. Surface water is also growing more costly. For example, in Reno, NV, the cost of water from the Truckee River, which flows right past the town, surged from a mere $50 per acre-foot in 1976 to $3,000 per acre-foot in 1991—a 60-fold increase in 15 years.

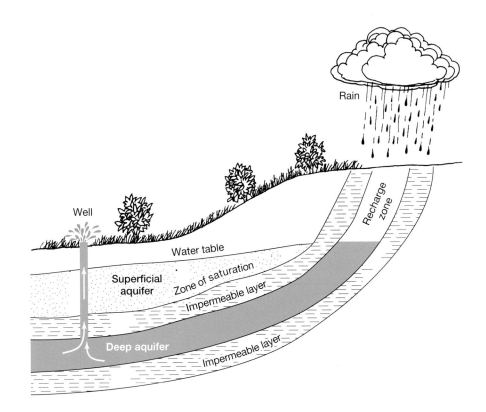

FIGURE 10.28   An aquifer serves as a source of well water. Continued withdrawal of water will eventually deplete the supply if the rate of withdrawal exceeds the rate of recharge.

Summary of Key Concepts

1. Water moves continuously from oceans to air, to land, to rivers, and back to oceans, in what is known as the water cycle. This cycle is powered by solar energy and gravity. Water evaporates from land and water bodies, forms clouds, and then is redeposited as rain or snow.

2. Water that falls on the ground as rain runs into streams or percolates down into the soil and rocks to become groundwater. The uppermost level of the zone that is saturated with groundwater is called the water table.

3. Ninety-seven percent of the world's supply of freshwater is held in porous layers of sand, gravel, and rock known as aquifers. These are important sources of freshwater for drinking, irrigation, and other uses.

4. Water shortages occur with great frequency throughout the world. As the world population increases, many experts predict more severe water shortages arising in many nations.

5. Drought is a natural occurrence but may also be worsening as a result of changes in global climate brought on by air pollution from human sources and deforestation.

6. Many factors contribute to the occurrence of water shortages in drought and normal times: (1) a rapidly increasing population, (2) increasing demands by agriculture, cities, and industry, (3) inefficient use of water, (4) unequal distribution of water, and (5) pollution.

7. Satisfying present and future water demands requires a multifaceted approach with an eye toward sustainability. Possible methods include (1) using water much more efficiently, (2) purifying and reusing the aqueous effluents of sewage treatment plants, (3) developing new sources of groundwater, (4) desalinizing seawater, (5) developing drought-resistant and salt-resistant crops, (6) rainmaking, and (7) water diversion projects, that is, transferring surplus water to water-short areas.

8. Floods are a major problem the world over, costing billions of dollars a year and taking the lives of many people and livestock.

9. Floods are natural events but occur with greater frequency and in greater magnitude in disturbed environments—areas in which the natural vegetative cover has been stripped or reduced. Impervious surfaces such as roadways, parking lots, and roofs also increase the incidence and severity of floods. Damage from floods is more likely as people settle in floodplains.

10. Floods can be controlled by measuring the snowpack to predict flood conditions, protecting watersheds, restricting development and other human activities in floodplains, building levees, dredging rivers, and building dams.

11. Big dams can control flooding but have several major drawbacks, among them: (1) high economic costs, (2) endangerment of life because of possible collapse, (3) reservoir evaporation losses, (4) flooding of prime agricultural land, (5) siltation of reservoirs, which reduces their life span, (6) saltwater intrusion into coastal areas, (7) destruction of scenic beauty and recreational uses such as rafting and kayaking, and (8) destruction of the habitats of species.

12. Stream channelization increases the speed with which water is removed from a watershed, thus reducing the potential for floods in the channelized section. Like other measures, it has disadvantages: (1) it destroys wildlife habitats, (2) it may increase flooding downstream, (3) it may lower the water table of groundwater as it reduces water absorption by the land, (4) it causes severe aesthetic losses, (5) it destroys many of the recreational functions of streams, and (6) it is costly.

13. Many experts believe that watershed protection and floodplain zoning are the most effective, economical, and sustainable methods of flood control.

14. Irrigation permits a flourishing agricultural economy in arid portions of the world and accounts for a very large percentage of the world's food output. Obviously, then, it is a very important way of producing food.

15. Four methods of irrigation are (1) sheet and flood, (2) furrow, (3) sprinkler, and (4) drip or trickle. These vary greatly in their efficiency, with the last two being the most efficient.

16. Serious problems associated with irrigation include water loss during transit along canals, inefficient practices during the application of water to the crops, salinization of the soil, and depletion of groundwater.

Key Words and Phrases

Aquifer	Ogallala Aquifer
Aquifer Recharge Zone	Playa Lakes
Blackwater	Rainmaking
California Water Project (CWP)	Renewal Time
Capillary Water	Replacement Period
Channelization	Rotary Sprinkler
Condensation Nuclei	Salinization
Desalination	Salt-Resistant Crops
Drainage Basin	Saltwater Intrusion
Dredging	Seepage
Drip Irrigation	Sheet Irrigation
Drought	Snowpack
Drought-Resistant Crops	Sprinkler Irrigation
Evaporation	Streambed Aggradation
Federal Flood Disaster Protection Act	Subsidence
Flood Irrigation	Surface Runoff
Floodplain	Surface Water
Floodplain Zoning	Transpiration
Furrow Irrigation	Water Cycle
Graywater	Water Diversion
Groundwater	Water Mining
Groundwater Overdraft	Water Table
Hydrological Cycle	Waterlogging
Imperial Valley	Watershed
Infiltration	Watershed Protection
Irrigation	Watershed Protection and Flood Prevention Act
Levees	Zone of Aeration
	Zone of Saturation

Critical Thinking and Discussion Questions

1. What are the major water-related issues facing the United States and other countries?

2. What is the water cycle? What powers the water cycle? What are the main water reservoirs? What are the renewal times for the atmosphere, rivers, lakes, glaciers, and oceans?

3. Discuss four major factors that contribute to water shortages occurring throughout the world.

4. Describe the many ways we can meet our needs for clean, fresh water. What are the most sustainable approaches? Why? What are the least sustainable approaches? Why?

5. Using your critical thinking skills, discuss this statement: "The United States does not really have a water shortage; it is plagued with a water distribution problem. We can easily solve our problems by diverting water from areas blessed with abundant supplies to areas less fortunate."

6. Discuss the following statement: "Big dams, like the Hoover, are engineering masterpieces that have been indisputable successes in boosting human welfare."

7. Suppose the dams that now exist in the United States had never been built. What would have been the disadvantages? Would there have been any advantages? Discuss your answer in terms of the American economy, human safety, agricultural production, wildlife preservation, scenic beauty, and endangered species.

8. Using your critical thinking skills, discuss the following statement: "Flooding is a natural occurrence. The best way to control it is by building more dams."

9. Using your critical thinking skills, analyze the following assertion: "To protect against flooding, we should build more levees along riverbanks and channelize more rivers."

10. Describe the most sustainable means of reducing flooding. What features of these approaches make them sustainable? What are the least sustainable measures? Why?

11. Describe some of the positive and negative effects of dredging operations.

12. List the disadvantages of stream channelization to agriculture and to the stream ecosystem.

13. What causes salinization? Briefly describe five ways in which salinization can be controlled.

Suggested Readings

Alley, W. M. 2006. Tracking U.S. Groundwater: Reserves for the Future? *Environment* 48(3): 10–25. This article explains how better monitoring and management of groundwater could lead to a more sustainable water supply in the future.

Annin, P. 2006. Great Lakes Water Wars. Washington, DC: Island Press. A detailed look at controversy over plans to divert water from the Great Lakes to supply demand elsewhere.

Abramovitz, J. N. 1996. Sustaining Freshwater Ecosystems. In *State of the World 1996,* ed. L. Starke. New York: W. W. Norton. Important reading.

Brown, L. R. 1991. The Aral Sea: Going, Going…*World-Watch* 4(1): 20–27. Powerful story about the consequences of abusing one's water resources.

Cutter, S. L., L. A. Johnson, C. Finch, and M. Berry. 2007. The U.S. Hurricane Coasts: Increasingly Vulnerable. *Environment* 49(7): 8–21. An eye-opening piece on the vulnerability of U.S. coastlines and populations to flooding and other problems caused by hurricanes.

Dzurik, A. A. 1990. *Water Resources Planning.* Totowa, NJ: Rowman and Littlefield. Comprehensive treatise on water resource planning and management.

Folger, T. 2008. Requiem for a River. *Onearth* 30(1): 24–35. Examination of water use and misuse along the mighty Colorado River.

Funk, N., K. Nortje, K. Findlater, M. Burns, A. Turton, A. Weaver, and H. Hattingh. 2007. Redressing Inequality: South Africa's New Water Policy. Discussion of how early water policies under colonial rule and Apartheid were used as a means of oppression and control and how new policies could provide justice and sustainable management of water resources.

Gardner, G. 1996. Preserving Agricultural Resources. In *State of the World 1996,* ed. L. Starke. New York: W. W. Norton. Excellent in-depth view of agricultural problems, including water supply.

Gleick, P. H. 2000. *The World's Water 2000–2001: The Biennial Report on Freshwater Resources.* Washington, DC: Island Press. Very valuable resource with tons of useful information.

Glennon, R. 2002. *Water Follies: Groundwater Pumping and the Fate of America's Fresh Waters.* Washington, DC: Island Press. An insightful and startling look at groundwater overdraft and problems associated with surface waters in the United States.

Khagram, S. 2003. Neither Temples nor Tombs: A Global Analysis of Large Dams. *Environment* 45(4): 28–37. An excellent look at the effects and performance of large dams.

Levin, T. 2004. Saline Solutions: Turning Oceans into Tap Water. *Onearth* 26(2): 28–33. A look at the pros and cons of desalination as a source of drinking water.

Olmstead, S. M. 2003. Water Supply and Poor Communities. *Environment* 45(10): 22–35. A look at problems of supplying water to poor communities throughout the world—and potential solutions.

Platt, R. H. 2006. Urban Watershed Management: Sustainability, One Stream at a Time. *Environment* 48(4): 26–42. An in-depth view of what cities can do and have done to reduce flooding and protect water quality and fish and other aquatic species.

Platt, R. H., P. K. Barten, and M. J. Pfeffer. 2000. A Full, Clean Glass? Managing New York City's Watersheds. *Environment* 42(5): 8–20. Useful case study for those interested in learning more about watershed management.

Postel, S. 1996. Forging a Sustainable Water Strategy. In *State of the World 1996,* ed. L. Starke. New York: W. W. Norton.

Very important reading for students who want to learn more about this topic.

Postel, S. 2000. Redesigning Irrigated Agriculture. In *State of the World 2000,* ed. L. Starke. New York: W. W. Norton. Excellent overview of what can be done to improve the efficiency of irrigation.

Reisner, M. 1993. *Cadillac Desert: The American West and Its Disappearing Water.* New York: Penguin. Fascinating book for anyone interested in learning about water and the political struggles that surround it in the arid American West.

Rosengrant, M. W., X. Cai, and S. A. Cline. 2003. Will the World Run Dry? Global Water and Food Security. *Environment* 45(7): 24–36. A look at global water demand and ways to meet rising demands sustainably.

Sampat, P. 2000. Poisoning the Well. *World-Watch* 13(1): 10–22. A must-read for those interested in learning more about groundwater contamination.

Satterthwaite, D., and G. McGranahan. 2007. Providing Clean Water and Sanitation. In *State of the World 2007,* ed. L. Starke. New York: W. W. Norton. Discusses problems in supplying potable water to cities throughout the world and what can be done.

Schaffer, D., E. Masood, Y. Nedge, P. Bagla, and K. Mantell. 2007. Dry: Three Stories of Adaptation to Life Without Water. *Environment* 49(1): 8–18. Excellent case studies on ways people in dry lands are meeting their need for water.

Sheaffer, J. R., J. D. Mullan, and N. B. Hinch. 2002. Encouraging Wise Use of Floodplains with Market-Based Incentives. *Environment* 44(1): 32–43. Insightful examination of ways that floodplains can be better managed.

Stauffer, J. 2003. *The Water You Drink.* Gabriola Island, BC: New Society Publishers. A popular book for those interested in learning more about drinking water.

White, G. 2000. Water Science and Technology: Some Lessons from the 20th Century. *Environment* 42(1): 30–38. Good overview of changes in water policy, many of which were discussed in this chapter.

Wolf, A. T., A. Karmer, A. Carius, and G. D. Dabelko. 2005. Managing Water Conflict and Cooperation. In *State of the World 2005,* ed. L. Starke. New York: W. W. Norton. Excellent overview of global water supply issues and conflicts and ways to solve them.

Web Explorations

Online resources for this chapter are on the World Wide Web at: **http://www.prenhall.com/chiras** (*click on the Table of Contents link and then select Chapter 10).*

CHAPTER 11

WATER POLLUTION

Widespread illness in Jackson Township, NJ, ranging from skin rash and kidney malfunction to death; warnings posted by Minnesota and New York state wildlife officials not to eat fish because of mercury contamination; the skeletons of perch loaded with radioactive chemicals; maggot-infested fish rotting on a Lake Erie beach; eight youngsters sick with typhoid fever after eating a watermelon they found floating in the Hudson River; 140 million fish deaths in U.S. waters in a single year; the Cuyahoga River, OH, bursting into flames; 18,000 people in Riverside, CA, stricken with fever and vomiting; an outbreak of hepatitis in New York and New Jersey; cholera spreading through South America—all these events have one thing in common: They were caused by water pollution (Figure 11.1).

11.1   Types of Water Pollution

Water pollution can be defined as any contamination of water that lessens its value to humans and other species, aquatic and nonaquatic. For humans, water pollution may impair swimming, fishing, and other forms of recreation. It may also render water unsuitable for irrigation or use in factories. It can also make water unfit for consumption. To regain its utility, polluted water must often be filtered or purified—a step that can be quite costly.

Water pollution affects both groundwater and surface waters. Water pollution can be classified in two ways: by the source and by the chemical type. We begin with a discussion of the source classification by examining two major types of water pollution, point and nonpoint (Figure 11.2).

Point Source Water Pollution

Point source water pollution has its source in a well-defined location (a point) such as the pipe through which a **sewage** treatment plant or a factory discharges waste into either surface water or groundwater. As described in Chapter 10, the term *surface water* refers to bodies of water in direct contact with the atmosphere. It's the type of water with which most of us are familiar: lakes, rivers, streams, springs, and oceans. *Groundwater* refers to water found in the ground, either in saturated soil or in deeper deposits known as aquifers.

Nonpoint Source Water Pollution

As the name implies, **nonpoint source water pollution** does not arise from distinct point sources such as factories. Instead, it comes from areas such as farmland where chemicals like pesticides or artificial fertilizer are used. These chemicals may drain into groundwater or into surface waters. Other nonpoint sources include pastures, barnyards, and construction sites.

The urban/suburban landscape—streets and lawns of our homes, for instance—also is a major nonpoint source of water pollution. It releases huge amounts and enormous varieties

FIGURE 11.1 A sign of the times. Although considerable progress has been made in cleaning up our nation's waters, signs like this still appear.

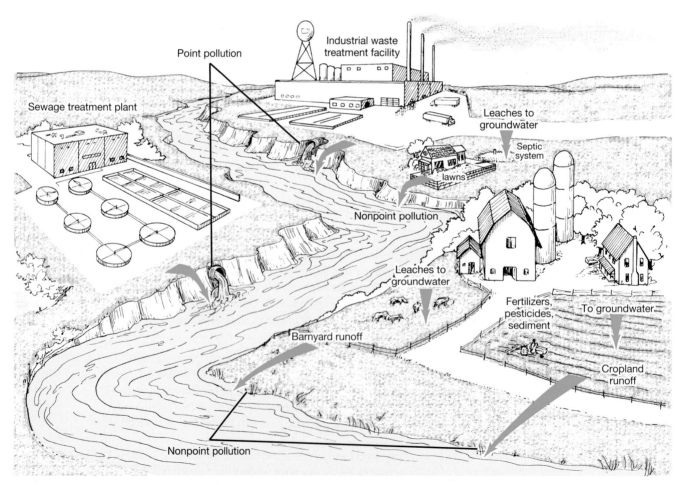

FIGURE 11.2 Examples of point and nonpoint source pollution.

of chemical water pollutants. These pollutants are washed from rooftops, roadways, parking lots, and so on after heavy rains or snowmelt. Even lawns are an important nonpoint source of water pollution, releasing potentially toxic herbicides, insecticides, and other chemicals into storm sewers. Storm drains empty into streams, rivers, and other surface waters, introducing a wide variety of pollutants to the water we swim in and drink. In many cities, storm runoff and sewage from homes and small businesses flow into the same system of pipes. Under normal conditions, this combined waste is delivered to sewage treatment plants. During heavy rainstorms, however, sewage treatment plants may be overwhelmed. Excess human waste and runoff is diverted from the sewage treatment plants and flows directly into surface waters.

11.2 Major Pollutants, Prevention, and Control

As noted, water pollutants can also be classified according to their chemical type. The major categories in this classification system are (1) sediment, (2) inorganic nutrients, (3) thermal pollution, (4) disease-producing microorganisms, (5) toxic organic chemicals, (6) heavy metals, and (7) oxygen-demanding organic wastes. This section describes each of these pollutants, their effects, and ways to control them or prevent their production altogether.

Before we examine the major types of water pollutants, however, let's take a look at measures for ensuring clean water supplies. We'll examine two basic approaches: water pollution control and pollution prevention. This information will help you understand options for controlling or eliminating the various pollutants you're about to study.

For years, government agencies and businesses have reduced or eliminated water pollutants via **pollution-control devices**. These technologies remove pollutants from the effluents of factories and sewage treatment plants and are usually applied to point sources. The pollutants are then either treated so they become less harmful or else concentrated and disposed of in another matter (for example, in a landfill). Water pollution experts consider pollution control such as sewage treatment plants a type of **output control**. That is, they deal with pollutants *after* they have been produced.

For many years, pollution control has been the primary focus of regulators. Most of the money spent on water pollution and most of the regulations focused on this method. While important, output controls are often really nothing more than an expensive way of diverting pollution from one medium to another—that is, capturing wastes bound for one medium (such as water) and diverting them to another (landfill). For example, much of the organic wastes removed from sewage treatment plants is converted to **sludge**, which is disposed of in landfills. There, wastes may leach into the groundwater.

In recent years, many forward-thinking companies and government agencies have been developing and using ways to reduce pollution production or eliminate it entirely. This approach, called **pollution prevention**, involves a

variety of measures applied in factories, water treatment plants, and nonpoint sources. Reducing or eliminating pollutants is often highly cost-effective, so it helps individuals and companies meet environmental goals at a fraction of the cost of many pollution-control devices. Pollution prevention is a type of **input control**. That is, it eliminates pollutants by changing inputs at a factory. For example, chemists may find nontoxic substitutes for chemicals used to produce various products, or engineers might find ways to alter manufacturing processes so that they no longer produce toxic waste. Either way, the input of toxic chemicals can be reduced or even eliminated, thus reducing pollution production.

Some pollution prevention measures are **throughput controls**. They alter the production of waste by adjustments of substances flowing through a system. For example, chemical engineers may find ways to recycle wastes, using them again in other processes.

Input and throughput measures—or combinations of them—help to prevent pollution, which is generally a more desirable approach than pollution-control measures at point sources. Nonpoint pollutants can also be reduced or eliminated by input controls—preventive actions. Terracing a farm field, for example, can reduce surface runoff, eliminating the pollution of nearby streams with farm chemicals. Obviously, input controls and throughput measures are far better and more sustainable than output controls, for they eliminate problems in the first place. They're often a lot less expensive, too. In many cases, these measures save companies enormous amounts of money, which leads to greater profits.

Now that you understand some of the options for controlling pollution, let's take a look at the major types of water pollution.

Sediment Pollution

Sediment includes sand, silt, and clay—inorganic soil particles—eroded from soils and washed from roadways and building sites. It is one of the most destructive and costly water pollutants. Each year, approximately 0.9 billion metric tons of sediment flow into aquatic ecosystems in the United States. The Mississippi River alone carries 210 million metric tons of sediment into the Gulf of Mexico each year. To transport just one year's load would require a train of boxcars 37,000 miles long—sufficient to circle the Earth one and one-half times at the equator! It is a curious paradox that the very soil that makes the production of life-sustaining food possible suddenly becomes one of our nation's most destructive pollutants when washed into our lakes and streams.

Where Does Sediment Come From? Sediment comes from natural sources, for example, natural stream bank erosion. It also comes from human sources—activities that lead to water and wind erosion, such as farmland and construction sites that aren't managed to eliminate soil erosion.

As a rule, in a well-vegetated, undisturbed watershed, sediment pollution in surface waters is minimal. But in disturbed watersheds, erosion and sediment pollution can reach elevated levels. Erosion and sediment pollution are particularly troublesome in agricultural

GO GREEN!
Use natural and nontoxic cleaning agents when cleaning your residence, especially when cleaning sinks and toilets, to reduce water pollution.

regions in which farmers fail to practice good soil conservation measures (Figure 11.3). In extreme cases, for example, in certain Iowa streams, sediment levels in nearby waterways can be as high as 270,000 parts per million (ppm)!

As noted, sediment also washes into streams and other surface waters from construction sites for homes, shopping centers, roadways, and the like. Because the land is laid bare in construction and not revegetated until the building is complete, erosion rates can be as much as ten times higher than from cropland. Although many construction companies now practice better erosion control, they're not always adequately installed.

Timber harvesting, if not carried out carefully, also can produce excessive amounts of soil erosion, leading to sediment pollution in nearby streams and lakes. This is especially common on steeply sloping land that is clear-cut to remove trees, a common practice in the northwestern United States and southwestern Canada. Strip-mining operations, in which the soil mantle is scooped away by giant machines in order to expose the coal, also may result in severe sediment pollution of streams, particularly in Appalachia.

Harmful Effects   According to estimates by the U.S. Army Corps of Engineers, sediment pollution in the nation's rivers causes about $1 million of a damage in the United States. How? Sediment damages turbines in hydroelectric plants, clogs irrigation canals, and fills in navigable rivers. Turbine blades must be repaired or replaced. Drainage ditches must be cleaned. Harbors and river channels must be routinely dredged to remove sediment so ships can pass. Soil particles also carry nutrient pollution (described shortly) and toxic chemicals, such as pesticides. They can cause additional damage in aquatic ecosystems. Suspended sediment in waterways blocks sunlight, which kills microscopic organisms known as algae. These photosynthetic organisms form the base of many aquatic food chains. They also produce the oxygen required by many other organisms. Their demise, therefore, not only reduces food supplies in aquatic ecosystems, it also reduces the levels of dissolved oxygen in the water. The loss of oxygen, in turn, harms other aquatic organisms such as fish and bacteria. Bacteria that live in water decompose organic material that enters from natural sources (leaves) and human sources such as canneries, slaughterhouses, and pulp mills. When oxygen levels fall, wastes from these plants tend to accumulate instead of being decomposed by oxygen-requiring bacteria.

Sediment also smothers the breeding grounds of fish and buries shellfish habitat. Sediment has, for example, destroyed many valuable clam beds in the Mississippi River. This reduces food for humans and other species dependent on aquatic ecosystems. Sediment also kills fish, as you will learn in Chapter 12.

Sediment poses problems to municipalities, too. To supply clean drinking water for their citizens, cities and towns must filter over 7,000 billion liters (1,800 billion gallons) of silt-polluted water each year.

Sediment also threatens reservoirs that store drinking water, help control floods, and provide recreational opportunities. The life span of thousands of reservoirs throughout the United States and other countries has been shortened as a result of sedimentation (Figure 9.13). Finally, as noted in Chapter 9, sediment fills streams and makes them more prone to flooding.

Control of Sedimentation

Input Control   Sediment can be controlled by a variety of measures. Preventive measures are typically the most cost-effective. On croplands, sedimentation can be effectively reduced by the proper application of the erosion-control

FIGURE 11.3 Poor soil conservation measures on farms result in massive soil erosion, destroying farmland and polluting rivers.

strategies described in Chapter 7. These include conservation tillage, strip cropping, and contour farming as well as terracing, gully reclamation, and the removal of marginal land from production. Highly erodible sites should be avoided altogether.

On construction sites, erosion can be reduced or eliminated by a host of measures. Careful site selection is one of the most effective measures. Conscientious builders avoid steeply sloped sites. Controls on land disturbance also help prevent soil erosion. For example, workers can be kept from bulldozing any more soil than is absolutely necessary. This reduces the amount of land that must be disturbed and also protects vegetation. Prompt revegetation also is vital. Denuded areas can be seeded or sodded with grass as soon as possible or covered with straw mulch to prevent erosion. Seeding on steep road cuts can be done with a **hydroseeder,** a machine that blows a slurry of seeds, fertilizer, straw mulch, and water onto the slope (Figure 11.4). New trees can be planted to replace those lost during construction. Water erosion controls such as small dams made of straw bales can be placed across drainage ditches to slow the flow of water. In many sites, workers are now applying "sediment fences" made of a fabric erected around the perimeter of the site. The material is partially buried and extends about 0.3 to 0.6 meters (1 to 2 feet) above the surface, helping stop the flow of water and sediment from the site. Steep slopes can be regraded to reduce their slope and thus reduce runoff and erosion. Additional preventive measures are discussed at the end of the chapter in the section on watershed management.

Output Control   Many output-control methods also help reduce sediment pollution. Muddy water coming from construction sites or farms, for example, can be directed to swamps and marshes. Output-control methods filter out the sediment, including toxic chemicals. However effective sedimentation ponds are at keeping sediment from entering

GO GREEN!

When hiking, stay on the trails, especially in grassy areas. Walking on the side of a trail or taking shortcuts destroys natural vegetation that holds soil in place, reducing sediment erosion.

streams, they do not prevent the loss of valuable soils from the site through erosion. Sediment can also overwhelm wetlands and destroy them. Dredging rivers and harbors, mentioned earlier, is another output control. However, dredging requires a lot of energy and is costly and environmentally harmful, for the dredged sediment must be placed elsewhere. Sediment is removed from drinking water by chemical coagulation, followed by a rapid sand filtration at municipal water treatment plants.

Although output methods work, preventive measures are typically far more cost-effective and desirable in the long run. They not only reduce costs but also help to protect valuable topsoil and the environment.

Inorganic Nutrient Pollution

Human society releases a variety of inorganic chemicals into surface waters and groundwaters waterways. Two inorganic substances, nitrates and phosphates, are nutrients to plants, especially aquatic plants, and are therefore known as **inorganic nutrient pollutants.**

Sources   Nitrates and phosphates in groundwater and surface waters come from three major sources: (1) agricultural fertilizers, (2) domestic sewage, and (3) livestock wastes (Figure 11.5). Let's take a look at each one.

Farmers throughout the world fertilize their fields with commercial fertilizers (also known as artificial or synthetic

FIGURE 11.4   Worker using a hydroseeder. This device propels a stream of water, grass seeds, and fertilizer to establish grass cover on steeply sloping land that has been denuded by construction.

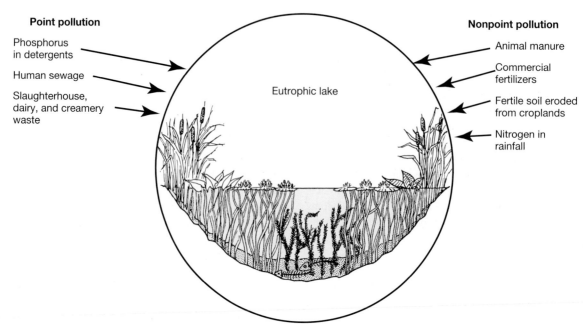

Point pollution

Phosphorus
in detergents

Human sewage

Slaughterhouse,
dairy, and creamery
waste

Eutrophic lake

Nonpoint pollution

Animal manure

Commercial
fertilizers

Fertile soil eroded
from croplands

Nitrogen in
rainfall

FIGURE 11.5   Sources of nutrients causing eutrophication.

fertilizers) and animal wastes or manure. Fertilizers increase crop growth because they are rich in nitrates and phosphates (Chapter 5). The amount of agricultural fertilizer used in the United States has increased from less than 5 million metric tons in 1950 to 25 million metric tons in the 1990s, a five-fold increase. However, much of the fertilizer that is not absorbed by crop roots is washed by runoff waters into lakes and streams. Scientists estimate that more than 0.45 billion kilograms (1 billion pounds) of agriculture-generated phosphates enter America's aquatic ecosystems yearly. In aquatic systems, these chemicals stimulate a population explosion of aquatic plants, the impact of which is discussed shortly.

Domestic sewage containing human wastes and household detergents also contributes millions of tons of nitrates and phosphates to aquatic ecosystems yearly.

Nitrates also come from nonpoint sources such as livestock operations. Each of the more than 1.5 billion cattle in the world, according to the UN's Food and Agriculture Organization, produce ten times as much waste per day as a human being. In other words, the world's cattle produce the waste equivalent of more than 15 billion people—well over twice the world population. That does not include the waste produced by other types of livestock, such as hogs, horses, sheep, pigs, goats, chickens, ducks, and turkeys. Especially noteworthy are concentrated livestock facilities such as pig farms, and **feedlots** where cattle are held for fattening before slaughtering. Unbeknownst to many, a large portion of the nation's cattle are maintained in feedlots. Wastes from these types of operations may wash into rivers during storms. The wastes can also seep into groundwater. One large lot, containing as many as 10,000 cattle at any one time, yields 180 metric tons (200 tons) of manure *daily*. The total amount of waste generated by U.S. feedlots is estimated to be about 720 million metric tons (800 million tons) annually.

Without proper control measures, much of this manure ends up in waterways.

Hogs have become especially troublesome in the southeastern United States, where the rapid increase in hog farming, frequent hurricanes, and improper siting of facilities (they're often built on floodplains) result in flooding of facilities and waste detention ponds (huge ponds into which waste is pumped and stored). Hurricane Denis in 1999, for instance, caused hundreds of detention ponds to break, spilling millions of gallons of hog waste into flooded waterways.

Nitrates and phosphates from manure also end up in surface waters as a result of standard farming practices. During the winter, for example, farmers in the northern states clean barns and other enclosures and often spread animal manure on the frozen ground to provide nutrients for the soil when the snow melts. (Nitrogen and phosphorus are absorbed by crop root systems in the spring.) Unfortunately, however, almost half of this manure may be washed by spring runoff into aquatic ecosystems. Because of this, the state of Vermont has banned this practice.

Interestingly, domestic pets are also a significant source of nitrates and phosphates. The waste deposited on city lots, sidewalks, and streets by the United States' more than 163 million pet dogs and cats releases nitrates that are washed into waterways during rainstorms. In New York City alone, 500,000 dogs produce 18,000 metric tons (20,000 tons) of feces and 4 million liters (1 million gallons) of urine annually. (Feces are supposed to be scooped up and bagged by owners, but not all are.) It is not surprising, therefore, that urban runoff may carry up to 5 ppm of both nitrogen and phosphorus.

Effects of Inorganic Nutrients on Aquatic Ecosystems   All aquatic organisms require a number of key elements to survive, notably carbon, hydrogen, oxygen, nitrogen, phosphorus, sulfur, and a host of others. While all of these are important, nitrogen and phosphorus play a particularly important role in determining growth of aquatic plants. An essential element that plays a key role in growth of individuals and populations is referred to as a **limiting factor.** It exerts its effect because it is typically found in low concentration. In aquatic ecosystems, nitrogen and phosphorus are the most important limiting factors. In small quantity, they're essential; however, in large quantity they can become pollutants, disrupting aquatic ecosystems, as is described in more detail shortly.

Nitrogen is usually available in the form of nitrate (NO_3^-) ions or ammonia (NH_3); phosphorus usually is available as phosphate (PO_4^{-3}) ions. Because phosphorus is less abundant and is required in such minute quantities, it is more likely to be a limiting nutrient than nitrogen in freshwater lakes and rivers.

When nitrogen and phosphorus become abundantly available, aquatic plants proliferate.

Eutrophication The nutrient enrichment of aquatic ecosystems is known as **eutrophication** (you-trow-feh-kay'-shun). This process occurs naturally as a lake or river ages over a period of hundreds or even thousands of years. Eutrophication that occurs as a result of natural nutrient enrichment is called **natural eutrophication.** The release of excessive amounts of nutrients into aquatic ecosystems as a result of human activities speeds up the process and therefore is called **cultural** or **accelerated eutrophication.** It can lead to serious problems.

Classifying Lakes Based on Their Productivity To understand the problems, let's look briefly at freshwater lakes. Ecologists recognize three major types of lakes: oligotrophic (nutrient-poor), mesotrophic (middle-nutrient), and eutrophic (nutrient-rich). The characteristics of each lake type are summarized in Table 11.1.

The **oligotrophic lake** (awl-i-go-troh'-fic) type is represented by Lake Superior, Lake Huron, the Finger Lakes of central New York, and many glacial lakes in northern Minnesota, Wisconsin,

and Michigan. Oligotrophic lakes are frequently beautiful, clear-water lakes, surrounded by pine and spruce forest. Their waters are clear because of a scarcity of suspended algae (phytoplankton), resulting from the low levels of dissolved nutrients. Food chains are largely based on bottom-dwelling producers (green plants) growing in shallow water near the lake margins. The total biomass per unit volume of water is very low.

A **mesotrophic lake** (mez-oh-troh'-fic) is characterized by moderate levels of nutrients. (*Meso* means middle.) Fertility, clarity, levels of dissolved oxygen, and total biomass are intermediate between oligotrophic and eutrophic lakes. These lakes are good for swimming, boating, and fishing. Fish such as northern pike and walleyes may be plentiful in northern lakes; bass may abound in southern lakes. Most of the lakes in the United States were probably mesotrophic at the turn of the 19th century.

As time passes, the nutrient enrichment process (described earlier) continues naturally. As a result, a mesotrophic lake is gradually converted into an extremely fertile **eutrophic lake** (you-troh'-fic). When the average concentration of soluble inorganic nitrogen exceeds 0.3 ppm and the soluble inorganic phosphorus content exceeds 0.01 ppm, algal populations may "explode." Such proliferations of algae, which turn lakes and ponds green, are called **algal blooms.** Typically occurring during the summer, algal blooms convert once-clear water into "pea soup" and restrict visibility to 0.3 meter (1 foot).

Algal blooms have many effects: (1) They may destroy the aesthetics of lakes and ponds, rendering them repulsive to swimmers and other sports enthusiasts. Canoe paddles, motorboat propellers, water skis, and fishing lines (as well as human arms doing the crawl stroke) become tangled in green filamentous algae. (2) Algal blooms also give the water a bad taste and odor. If the lake is a source of drinking water, considerable expense may be involved in improving its quality (Figure 11.6). (3) As a result of wind and wave action, huge masses of algae (and even rooted plants that have been torn loose from the lake bottom) can pile up along the shore and decompose, giving off hydrogen sulfide (H_2S) gas, which smells like rotten eggs. At high concentrations, it is toxic.

TABLE 11.1	Characteristics of Oligotropic, Mesotrophic, and Eutrophic Lakes	
Oligotrophic Lake	Mesotrophic Lake	Eutrophic Lake
Poor in nutrients	Intermediate	Rich in nutrients
Deep basin	Intermediate	Shallow basin
Gravel or sandy bottom	Intermediate	Muddy bottom
Clear water	Intermediate	Turbid water
Plankton scarce	Intermediate	Plankton abundant
Rooted vegetation scarce	Intermediate	Rooted vegetation abundant

FIGURE 11.6 Biologist at the Amarillo, TX, water treatment plant obtaining a water sample to determine algal population. If algal populations are high, water will have a bad taste. Much of Amarillo's drinking water comes from Lake Meredith, 72 kilometers (45 miles) northeast of town.

(4) Algal blooms may result in a proliferation of **blue-green algae.** Some blue-green algae release chemicals that are poisonous to fish and humans.

Dense blooms of algae and other photosynthetic microorganisms can shade out bottom-dwelling plants that normally produce large amounts of oxygen via photosynthesis. Although phytoplankton are also photosynthetic, they usually occupy the upper levels of the lake. Much of the oxygen they release escapes into the atmosphere. As a result, levels of dissolved oxygen in eutrophic lakes are lower than in mesotrophic or oligotrophic lakes, despite the much greater abundance of phytoplankton.

Eutrophic lakes have many other undesirable features, such as muddy bottoms that are poor spawning surfaces for most desirable fish. Eutrophic lakes tend to support populations of less desirable fish such as carp. Carp are often abundant in eutrophic lakes because they are adapted to live in warm, shallow, turbid (cloudy), oxygen-poor waters. When conditions are bad, these fish often rise to the surface to gulp air.

GO GREEN!

Use no- or low-phosphate detergents when washing clothes, and use only as much as the instructions call for to reduce water pollution.

Cultural Eutrophication Human activities are greatly accelerating eutrophication of lakes the world over. Scientists estimate, for example, that roughly 80% of the nitrogen and 75% of the phosphorus entering lakes and streams in the United States come from human activities. As a result, the eutrophication of many lakes is proceeding 100 to 1,000 times faster than it would under natural conditions. An estimated 33,000 medium-sized to large lakes and about 85% of the large lakes located in urban areas in the United States are undergoing cultural eutrophication. The effects are summarized in Table 11.2.

Stream Eutrophication Although eutrophication can occur in streams, the effects are usually not as severe as in lakes. Many rivers are naturally oligotrophic at their headwaters, where the waters flow clear and cold. Such is the case of the small northern streams feeding into the Mississippi River or the mountain streams of the western portion of the United States. However, by the time waters near the stream's mouth, they have received so many nutrients that they become turbid, warm, and muddy, harboring bullheads and carp rather than trout. Stream eutrophication is more easily reversed than lake eutrophication; because the nutrient input is often stopped, the current purges the stream.

Groundwater Pollution As noted briefly, groundwater can also be contaminated by phosphates and nitrates. One source of groundwater pollution is the septic tank used in many rural homes. Septic tanks, discussed shortly, are household waste disposal systems that are common in areas not served by municipal waste treatment plants. Leach fields that drain the tanks contribute nitrate and phosphate to underlying groundwater, especially in areas where porous layers lie over aquifers and in regions with high water tables. Agricultural fertilizer also can contaminate groundwater, as can animal waste.

Control of Eutrophication Eutrophication and its many effects may be controlled by both input and output methods. Consider some output methods first.

1. Output Measures and Controls

   A. Upgrading many wastewater treatment plants to the tertiary (advanced) level, discussed shortly. These facilities would remove a higher percentage of the phosphorus and nitrogen. Wastes from facilities can be used to fertilize rangeland, pastures, and farm fields, recycling nutrients back to the soil they came from.
   B. Using well-built detention basins on feedlots to collect animal wastes that are currently washed by runoff water into lakes and streams (Figure 11.7). Waste in detention basins could be used to fertilize farmland.
   C. Employing weed-cutting machines to remove excess vegetation from aquatic ecosystems. Unfortunately, this would have to be done several times during the year and is rather costly.
   D. Destroying plant growth in aquatic ecosystems with herbicides. Great care must be taken to prevent fish kills and the destruction of spawning beds.

TABLE 11.2	Adverse Effects of Nutrient Pollution on a Lake

Lake aesthetics are destroyed.

The recreational values of a lake are destroyed.

Water quality is impaired by foul tastes and odors.

Gases that emanate from rotting algae have foul odors, tarnish silverware, and discolor painted houses.

Toxins given off by algae result in gastric disturbances if ingested.

Dense algal blooms at the surface reduce penetration of sunlight to the lake bottom.

Decomposing algae at the lake bottom reduce dissolved oxygen levels.

Pollution contributes to the winterkill of fish in northern lakes.

Rooted weeds interfere with navigation and recreation.

Game fish are replaced with less desirable fish.

The lake basin is gradually filled in, and the lake becomes extinct.

E. Dredging the bottom sediment from lakes to remove the nutrients. This is extremely costly and is impractical in deep lakes.

F. Avoiding dumping of treated sewage in lakes susceptible to eutrophication. At Lake Tahoe, NV, and in Cambridge, WI, officials successfully bypassed the sensitive lakes by pumping treated wastewater to less vulnerable waters downstream.

G. Applying treated wastewater (still rich in plant nutrients) to golf courses and other grassy areas along highways.

H. Treating wastewater biologically in artificially constructed wetlands, where aquatic vegetation and microorganisms utilize the nutrients and purify the water.

2. Input Controls in Urban and Suburban Areas

A. Banning the use of phosphate detergents or reducing phosphate levels in them. Bans have been imposed by several cities (Akron, Chicago, Miami, and Syracuse) and states (Indiana, Michigan, Minnesota, Wisconsin,

FIGURE 11.7 Control of feedlot runoff at Boys Town, NE. This basin "catches" the manure and urine that would ordinarily be washed into a stream or lake after a rainfall. Catch basins, however, can release pollutants into groundwater.

New York, Maryland, and Vermont). Several states have limited the phosphorus content in detergents to less than 9%. Detergent manufacturers have subsequently lowered the phosphate content of their products, and many low-phosphate detergents are now available in grocery stores in the United States and abroad.

B. Using newly developed phosphate-free natural detergents. Perhaps the ultimate solution to the phosphate detergent problem is the use of phosphate-free natural soaps and detergents developed by the USDA. These products perform as well as or better than phosphorus detergents in hot or cold, soft or hard water. Moreover, they are relatively inexpensive, are biodegradable, and do not cause eutrophication.

C. Imposing an excise tax on lawn and garden fertilizers to reduce sharply the volume of use.

D. Educating citizens to use less detergent and fertilizer.

3. Input Controls in Rural Areas

A. Minimizing the use of fertilizers on cropland. Farmers, like homeowners, typically apply more fertilizer than is needed.

B. Injecting liquid fertilizer directly into the soil rather than spreading it on the surface, where it can be carried into surface waters by surface runoff.

C. Postponing application of manure in cold regions until after the spring melt of ice and snow.

D. Planting vegetative barriers between fields and waterways to trap pollutants.

E. Implementing soil erosion controls on farms (terracing, contour farming, etc.) to minimize surface runoff.

F. Building hog farms and other areas of concentrated livestock production out of floodplains.

Take a few moments to compare the strategies and think about their social, economic, and environmental costs and benefits.

Thermal Pollution

Biscayne Bay, FL, is an unusually productive ecosystem that supports many species of aquatic organisms, including lobsters, crabs, fish, and wading birds. More than 270,000 kilograms (600,000 pounds) of seafood is harvested annually from the bay. In the early 1970s, however, scientists discovered a 30-hectare (75-acre) region that was virtually lifeless—a biological desert caused by the discharge of heated water from Florida Power and Light Company's power plant. This problem is commonly referred to as thermal pollution.

Thermal pollution is an increase in the temperature of water that adversely affects organisms living in it. Although thermal pollution may result from both natural causes (excessive heating by the summer sun) and human actions, the latter are far more significant. What are the sources of thermal pollution?

Many industries remove water from lakes and streams to cool equipment or products. Among the most important are electric utilities, steel manufacturing, and chemical plants.

Electrical generating plants in the United States withdraw and use 752 billion liters (195 billion gallons) per day, nearly half of the total U.S. water withdrawal. Where is this water used? As shown in Figure 11.8, water plays two

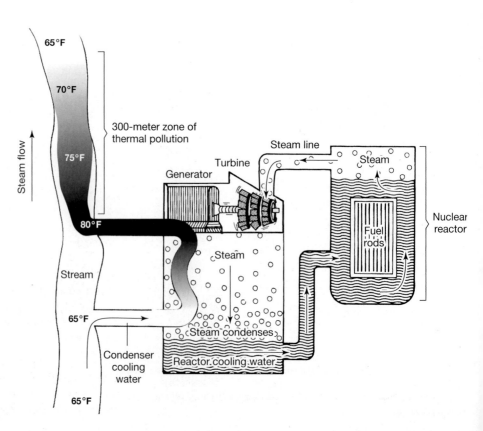

FIGURE 11.8 Nuclear reactor cooling system. Note the 300-meter (1,000-foot) zone of thermal pollution caused by the discharge of cooling water.

important roles in coal and nuclear power plants. As shown on the right side of the drawing, water is heated inside a nuclear power plant facility to produce steam. Steam, in turn, is used to propel the blades of a turbine. The mechanical energy is converted to electrical energy by a generator. The generator produces electricity. Steam leaving the turbine is often cooled and used again. To do this, the steam is passed over coils that carry cold water drawn from a nearby stream or lake. Cooling the steam converts it to water. In the process, heat once in the steam is transferred to the cooling water, raising its temperature by 11°C (20°F). The heated cooling water may then be discharged into a lake or stream. This warms up one area of that lake or river, creating a **thermal plume**—a heated body of water. Such plumes may extend 1 kilometer (3,280 feet) or more from the point of discharge (Figure 11.9).

Biological Effects of Thermal Pollution   Thermal pollution has many adverse effects on aquatic ecosystems. Let us consider some of them.

Reduction in Dissolved Oxygen   When water is warmed, its capacity to dissolve oxygen falls. For example, at 0°C (32°F), water that has been thoroughly mixed with oxygen has an oxygen content of 14.6 ppm, but at 40°C (104°F), it contains only 6.6 ppm. Unfortunately, as water temperature increases, oxygen requirements for fish increase. Thus, rising temperature decreases oxygen concentration in waters and increases a fish's demand for oxygen.

Fish vary in their need for oxygen. Carp, for example, require concentrations of only 0.5 ppm of oxygen at 0.6°C

(33°F). Cold-water fish such as trout and salmon need about 6 ppm to survive, so they cannot tolerate the warm water in a thermal plume. If they remain in the plume, they will die from oxygen starvation.

Interference with Reproduction   Fish are **cold-blooded animals** whose body temperatures and activity levels vary with the temperature of the external environment. In addition, most fish are extremely sensitive to slight changes in water temperature. Many kinds of fish are instinctively "tuned" to certain thermal signals that trigger such activities as nest building, spawning, and migration. Changes in temperature can disrupt these activities and may also kill eggs.

Increased Vulnerability to Disease   Another problem with thermal pollution is that it may increase the vulnerability of fish and other aquatic organisms such as turtles to disease organisms. The ability of some bacteria such as *Chondrococcus* to penetrate the body of a fish is poor at 16°C (60°F), but it gradually increases with rising water temperatures.

Direct Mortality   The body temperature of a fish is determined by the temperature of the water in which it lives. A lake trout will perish if the water temperature is much higher than 10°C (50°F). Because the discharge of cooling water may raise stream or lake temperatures 11°C (20°F), cold-water fish in the thermal plume will die from shock if they cannot quickly escape to cooler surrounding waters.

Invasion of Destructive Organisms   Thermal pollution may permit the invasion of highly destructive organisms that tolerate warm water. A good example is the invasion of shipworms (highly specialized relatives of clams and oysters) into New Jersey's Oyster Creek where heated water discharged from a power plant warms the creek. Invading shipworms burrow into wooden dock pilings and the hulls of boats, creating considerable damage.

Undesirable Changes in Algal Populations   Shifts in water temperature can also alter the types of algae that live in a body of water. Three major groups of algae live in freshwater ecosystems: diatoms, green algae, and cyanobacteria, once called blue-green algae (Figure 11.10). Each has a distinct range of tolerance for water temperature. Diatoms prefer the coolest water (14°C or 58°F), green algae like it warmer (32°C or 90°F), and cyanobacteria like it even hotter, 40°C (104°F). The most valuable algae, as far as fish and human food chains are concerned, are the diatoms. Those that prefer the warmest water—the cyanobacteria—are the least desirable as aquatic animal food. Many are too large to be eaten by small crustaceans such as water fleas. In addition, cyanobacteria give off toxic substances and cause the multiple problems already discussed earlier (in the section on inorganic nutrient pollution).

Destruction of Organisms in Cooling Water   Warm waters released from power plants and factories adversely affect fish and other organisms. Many organisms, such as plankton, insect larvae, and small fish, are sucked into the condenser

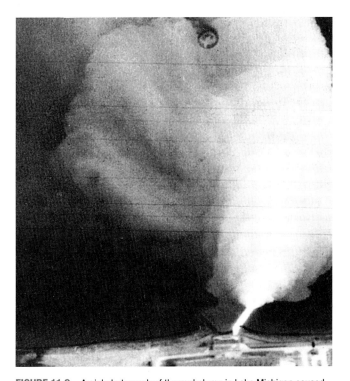

FIGURE 11.9   Aerial photograph of thermal plume in Lake Michigan caused by the discharge of warmed-up cooling water from the Point Beach nuclear plant on the Wisconsin shore. The photograph was taken with infrared film, which is sensitive to heat.

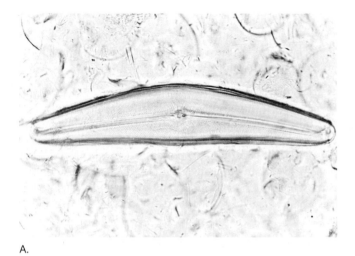

A.

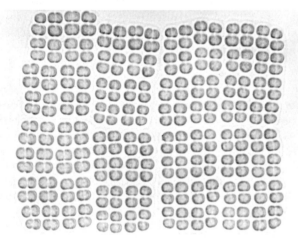

B.

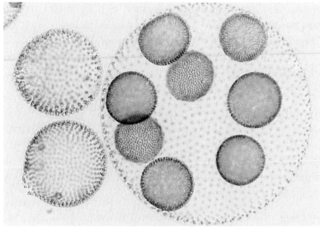

C.

FIGURE 11.10 Aquatic microorganisms. (A) Photomicrograph of a diatom. This alga prefers a water temperature of about 14°C (58°F). It is an important producer link in fish food chains. (B) Photomicrograph of a green alga. This is *Volvox*. It prefers water temperatures of about 32°C (90°F). (C) Photomicrograph of a blue-green alga. This colonial form prefers water temperatures of about 40°C (104°F). An abundance of blue-green algae may be a sign of thermal pollution. In general, blue-green algae are not important producer links in fish food chains. Moreover, some species may be toxic to fish and other aquatic animals, as well as to humans.

along with the cooling water and are destroyed immediately by thermal shock, as well as by water turbulence and pressure.

Controlling Thermal Pollution Power plants can reduce or eliminate thermal pollution by installing cooling towers, as shown in Figures 11.11 and 11.12. **Cooling towers** are devices that transfer heat from the water to the atmosphere. They cool the water before it is released back into water ways. Cooling towers on some power plants, especially nuclear power plants, are mammoth structures about 30 stories tall (Figure 11.11). They are large enough at the base to cover a football field.

Power companies may install one of two types of cooling towers, wet and dry. In the **wet cooling tower,** shown in Figure 11.12, the heated water is piped to the top of the tower. The water flows downward over a series of plates. As it does, it meets cool air flowing in from the bottom of the tower, powered by a fan. When the hot water flowing down through the interior of the tower mixes with cooler air rising through the structure, the water is cooled. The cooled water may then be reused, or it may be discharged into a stream, lake, or ocean from which it was drawn or directed into an artificial lagoon for further cooling before being released into its original source. (Smaller cooling towers are used in factories.)

Dry cooling towers are very similar to wet cooling towers, except the water flows downward through the interior of the structure through pipes. These towers achieve the same result without evaporative losses.

Although cooling towers reduce the thermal pollution, they can create some problems. First, wet cooling towers generate a considerable amount of fog on cold winter days in colder regions. The fog may come into contact with solid surfaces such as roads, forming a thin layer of ice. Ice, in turn, can be a hazard to motorists. Fog may also obstruct visibility. Second, the water that evaporates from wet cooling towers is lost to the aquatic ecosystems (rivers and lakes) from which it came. Third, toxic chemicals such as chlorine are used to prevent bacteria and other organisms from growing in the pipes in cooling towers. These chemicals can pollute waterways, killing other organisms. Fourth, although the tall towers are remarkable engineering accomplishments, they are 121-meter-tall (400-foot) masses of concrete and steel that dominate the skyline. Many people find them to be visually obtrusive. Fifth, the towers are very expensive. Such expenditures, of course, are eventually passed along from the utility to the consumer in the form of increased electrical rates, adding perhaps 1% to a customer's annual bills. This cost, say environmentalists, seems reasonable when weighed against the damage caused by thermal pollution.

Beneficial Effects of Heated Water Artificially heated water is usually harmful to aquatic ecosystems. Under certain

GO GREEN!

Use energy efficiently. Energy conservation in the home reduces your demand for electricity, reducing thermal pollution.

FIGURE 11.11   Wet cooling towers in operation. Note water vapor being discharged into the air by the towers.

circumstances, however, it can be beneficial. Consider some examples:

1. In Eugene, OR, heated water from a paper mill is sprayed onto fruit trees to prevent frost damage.

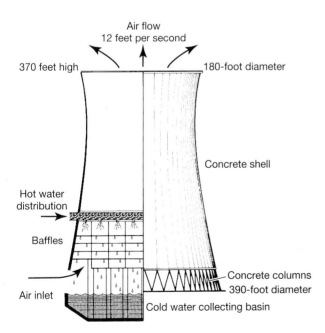

Air flow
12 feet per second

370 feet high   180-foot diameter

Concrete shell

Hot water
distribution

Baffles

Concrete columns
390-foot diameter

Air inlet

Cold water collecting basin

FIGURE 11.12   Operation of a wet cooling tower. Heat from the cooling water is removed by evaporation on direct contact with air rising up through the hollow concrete shell. Its large size is necessary to provide sufficient surface area and draft to cool thousands of liters of water each day. The distinctive shape channels the air flow and provides great structural strength with a minimum of material.

2. In Vineyard Haven, MA, heated cooling water is used by a hatchery to accelerate the growth of lobsters. The time required to produce a marketable lobster has been reduced from eight years to two years.

3. Researchers in Georgia have found that thermal pollution of the Savannah River resulted in an increased rate of growth in black bass. The heated water also attracts fish to the plume area, resulting in record catches for anglers.

Some scientists and engineers have suggested that warm water from various factories and power plants be combined with the effluent of domestic sewage treatment plants and emptied into specially built ponds stocked with fast-growing fishes that are accustomed to warm waters—for example, the Asiatic milkfish. Thus, both the nutrient and thermal pollutants could be put to constructive use: producing food. During the winter in the northern states, the warm water would prevent ice from forming and permit the milkfish to feed and grow year-round. Furthermore, because the fish are plant eaters, not only would the nutrients be efficiently converted into fillets, but the weed problem commonly associated with eutrophic ponds would be held in check.

Warm coolant water could also be used to heat homes or other buildings. Because pipes required for conducting the heated water are very expensive, however, the power plants and the homes that are to use the water must be designed with this heating scheme in mind.

Disease-Producing Organisms

Water contaminated with infectious microorganisms is responsible for more cases of human illness worldwide than any other environmental factor. Diseases include cholera, typhoid fever, dysentery, polio, and infectious hepatitis. Each one is caused by a specific microorganism that is transmitted by water polluted with human or animal wastes. Humans contract infectious diseases when they drink, swim in, or bathe in these waters. Fortunately, the death rate from these diseases has dropped dramatically in the past century in many countries. According to a report by the Natural Resources Defense Council, however, bacteria, viruses, and protozoa in U.S. drinking water cause 900,000 Americans to become ill every year. Even worse, these contaminants kill about 900 people annually—usually the very young, the elderly, and those people already weakened from surgery or diseases such as AIDS, cancer, or pneumonia.

One microorganism that has been causing problems in recent years in parts of the United States is *Pfiesteria* (fee-steer´ee-uh). Although most species are not harmful, one species of this single-celled microscopic organism causes open sores (lesions) and massive fish kills in the coastal waters of the eastern United States. Outbreaks have occurred from Delaware to North Carolina (Figure 11.13). Found in tidal rivers and estuaries, this organism feeds on algae and bacteria. When in the presence of fish, and in particular schooling fish, *Pfiesteria* changes form and begins to release a powerful toxic chemical that stuns the fish, causing them to become sluggish. Other toxicants produced by the organism are believed to cause fishes' skin to break down, resulting in open sores or lesions. *Pfiesteria* then begins to feed on tissues and blood of fish.

Pfiesteria is not an infectious agent, like other bacteria and viruses. It is not transmitted from one fish to another. In addition, it does not kill fish itself. Rather, scientists believe that fish deaths result from the toxic chemicals it produces or the open sores, which become infected by other harmful organisms.

Pfiesteria is responsible for many fish kills, including some massive ones in tributaries of Maryland's Chesapeake Bay and in the brackish coastal waters of North Carolina, where millions of fish have perished. There's also some evidence of human health damage as a result of exposure to the organism. Studies suggest that memory loss, confusion, and a variety of other respiratory, skin, and gastrointestinal problems may result from exposure to waters containing toxic forms of *Pfiesteria*. To date, there's no evidence that eating fish or shellfish exposed to this organism causes any health problems in people.

Although *Pfiesteria* outbreaks occur in warm, brackish, poorly flushed waters when schools of fish are present, there is growing evidence to suggest that certain pollutants may play a contributing role. Elevated levels of nitrogen and phosphorus, in particular, may cause outbreaks, either by stimulating the growth of algae upon which *Pfiesteria* feeds or stimulating the organism's growth more directly. As noted earlier, these pollutants, which are common in coastal waters, come from a variety of sources, including sewage treatment plants, septic tanks, suburban runoff, and farms.

Protecting Ourselves from Pathogens in Water Supplies

Protecting ourselves from disease-producing organisms requires a variety of measures. Good sewage treatment is essential. As discussed later in the chapter, sewage treatment is designed to remove pollutants and kills potentially infectious agents contained in wastewater, which, after treatment, is often dumped into drinking water supplies.

Purification of drinking water also is essential to protecting humans from disease. In many countries, the use and reuse of water from a single river or stream by communities along the course of the waterway is very intensive. If you live in a densely populated region, for instance, the last glass of water you drank may already have passed through the bodies of eight other people living upstream from you. Of course, it was cleansed over and over again after each use. As the world population grows, water use and reuse will grow. Safe reuse will depend on extremely effective fail-safe water treatment methods to reduce harmful microorganisms to very low levels.

Drinking water is filtered at water treatment plants, often through elaborate sand filters. It is then chlorinated to kill potentially harmful microorganisms. Water suppliers take frequent samples of the drinking water they distribute to ensure that disease-causing bacteria are held to an absolute minimum, as required by the Safe Drinking Water Act (1974), which will be discussed shortly.

Because so many species of disease-causing bacteria, or **pathogens**, can be ingested through drinking water, it is physically impossible to test for all of them. Instead, the law requires sampling of **coliform bacteria**. Coliform bacteria, an extremely common form, are found in the intestines of humans and other warm-blooded animals. Most types are harmless.[1] Domestic

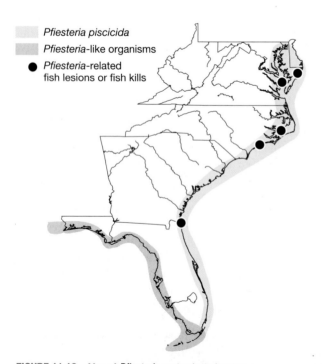

FIGURE 11.13 Map of *Pfiesteria*-contaminated waters.

Pfiesteria piscicida
Pfiesteria-like organisms
● *Pfiesteria*-related fish lesions or fish kills

---

[1]Some strains of coliform bacteria can be very harmful, causing diarrhea and death. In recent years, coliform bacteria have shown up on lettuce, tomatoes, and spinach in the United States, causing serious health problems in many who ingested the contaminated foods.

sewage contains 2 to 3 million coliform bacteria per 100 milliliters due to the high coliform content of human feces. Waterborne disease organisms also multiply in the intestines of infected individuals and can be spread to others through untreated or partially treated sewage released into surface waters. Coliform bacteria in drinking water therefore are an indicator of levels of potentially harmful bacteria and viruses. A low coliform count suggests a low level of harmful bacteria. A high coliform count suggests that a large number of pathogenic bacteria could be present. If water has more than 2 coliform bacteria per 100 milliliters, it is considered unsafe to drink. The city's water treatment plant then has to increase **chlorination** to destroy the bacteria. An alternative option might be to shift to another available water source, such as a well or lake. When the coliform count of a lake or stream exceeds 200 per 100 milliliters, it is considered unsafe for swimming, and the beaches are closed (Table 11.3). For a discussion of the effects of another biological pollutant, see Case Study 11.1, which discusses zebra mussels.

Studies suggest that, as important as monitoring coliform bacteria is to protecting public health, it is not foolproof. Swimmers, in fact, have developed gastrointestinal disorders and flulike symptoms after swimming in waters deemed safe by virtue of low coliform levels. Further investigation suggested that the culprit was a group of noncoliform bacteria, known as fecal streptococcal bacteria. As a result, the U.S. Environmental Protection Agency (EPA) suggests that waters be monitored for coliform bacteria and fecal streptococcal bacteria to ensure safety.

Toxic Organic Compounds

Chemical substances are divided into two groups: inorganic and organic. The inorganic compounds such as nitrates and phosphates are those that generally contain no carbon atoms. They are usually small-molecule compounds, including sodium chloride, water, and oxygen. In contrast, organic compounds are made primarily of carbon and hydrogen. Over a million organic compounds are known to science. Organic compounds such as protein, fats, and carbohydrates are important components of living organisms. Our interest here, however, is with a group of toxic organic compounds that are synthesized in chemical factories and sometimes released into groundwater and surface water.

Since the 1950s, hundreds of thousands of such synthetic organic compounds have been produced. They are used to make drugs, plastics, solvents, pesticides, and synthetic fibers. Unlike naturally occurring organic compounds, many of the synthetic organics resist decomposition by bacteria, sun, air, or water. As a result, once they are discharged into a body of water, they may persist as pollutants for several years, sometimes decades. Good examples of such persistent organics are the pesticide DDT and the industrial chemicals called polychlorinated biphenyls (PCBs). These and other toxic chemicals can enter the food chain or may be present in the water we drink or swim in.

Toxic chemicals are also released by plastic containers, such as drinking water bottles and baby bottles, and by steel cans (which are lined with a thin layer of plastic). These chemicals, known as **plasticizers**, are added to plastic to make it less brittle. Two plasticizers, bisphenol A and a group of chemicals known

TABLE 11.3	Diseases Transmitted by Water Contaminated with Fecal Matter		
Disease	**Organism**	**Symptoms**	**Comments**
Typhoid fever	Bacterium	Vomiting, diarrhea, fever, intestinal ulcers, reddish spots on skin; may be fatal	500 cases in the United States annually
Cholera	Bacterium	Vomiting, diarrhea, water dehydration; may be fatal	Rare in the United States
Traveler's diarrhea	Amoeba	Diarrhea, vomiting	Not uncommon
Amoebic dysentery	Amoeba	Diarrhea, chills, fever, abdominal pain; death may occur	May be contracted by eating infected oysters and clams
Infectious hepatitis	Virus	Headache, fever, loss of appetite, enlarged liver	May be contracted by eating infected oysters and clams
Polio	Virus	Headache, fever, sore throat, weakness, often permanent paralysis; may be fatal	Rare in the United States since the use of the polio vaccine

CASE STUDY 11.1   THE ZEBRA MUSSEL: A WATER CONTAMINANT FROM EUROPE

The waters of North America are being invaded by a tiny but dangerous creature. It is not a fish or eel parasite, but rather a small mollusk known as the zebra mussel, so named because of the black-and-white striped pattern on many of their shells. Arriving in North America in 1988, as well as scientists can tell, the zebra mussel first appeared in Lake St. Clair, a small body of water connecting Lake Erie and Lake Huron. By 2000, this tiny mussel had spread widely throughout U.S. and Canadian waters, and continues to move (Figure 1).

Zebra mussels are freshwater mollusks that are about the size of a man's fingernail but can grow to a maximum length of about 5 centimeters (2 inches). They live for approximately four or five years. Females can produce up to a million eggs in a single spawning season. The fertilized eggs develop into free-swimming larvae, which are carried by water currents from their origin to other sites, where they soon find a point of attachment—a rock, pier, dock, boat, or water inlet.

Because this tiny creature is sedentary as an adult, its rapid spread seems amazing. Although the microscopic larvae are capable of swimming away from their parents, their dispersal is greatly facilitated by waves and water currents. The larvae may also be spread inadvertently by boaters and anglers while the larvae are in bait buckets, live wells (water containers in boats for holding bait), or the cooling passages of outboard motors. Another method of dispersal is transport by air: The tiny larvae ride in small droplets of water on the feathers of migrating waterbirds, such as ducks, geese, and herons. Adults may even be moved about while attached to the backs of turtles and crayfish or to the hulls of speedboats, barges, freighters, kayaks, and canoes. Although they disperse by floating in currents as larvae, the main means of transport appears to be boats. Adults attached to ships, boats, and barges are dislodged. Wherever they land, they set up new colonies.

Evidence of their ability to spread rapidly in North America became evident in 1993, five years after they were first observed in Lake St. Clair, when a weary rescue worker from Hannibal, MO, happened to look at some muddy tree trunks exposed after the floodwaters receded along the banks of the Mississippi River. What he saw were hundreds of tiny, brown-striped "clams," holding tightly to the trees. It was a long way from Lake St. Clair!

Their invasive properties were clearly known to European scientists. From their original home in the Caspian Sea, the zebra mussel spread through all of Europe over the past century, creating an incredible amount of economic and ecological damage. How the zebra mussel got to North America is not known. Its appearance in the Canadian waters of Lake St. Clair in 1988, however, is believed to be linked to shipping. The zebra mussel may have "stowed away" in the ballast water of a trans-Atlantic freighter. (Ballast water is carried in special tanks of an empty ship to steady it when at sea. After cargo has been loaded, the water is pumped out.)

Since its first appearance, the zebra mussel has spread relentlessly through many of the lakes and rivers of the United States. It could spread throughout waterways along the entire East Coast of the United States. It is now found throughout the Great Lakes and along the entire length of the Mississippi River and all its major tributaries. Some scientists believe that eventually it will be found in lakes and rivers throughout the entire United States east of the

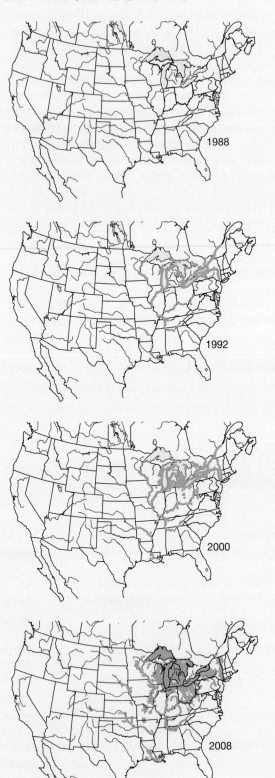

FIGURE 1   Maps showing the spread of zebra mussels in the United States and Canada from 1988 to 2008.

Rockies as well as in southern Canada. In 1999, it was spotted on the Missouri River near Sioux City, IA.

Zebra mussels feed by filtering green algae and organic material from the water. As you may recall, algae form the base of aquatic food webs that contain species valuable to humans including fish, waterfowl, eagles, otter, and mink. Some scientists and conservationists believe that large colonies of zebra mussels pose a serious threat to these and other species. Some sport fish such as walleyes and lake trout deposit eggs mainly on lake bottoms that are covered with boulders. However, because the zebra mussel frequently colonizes such surfaces, the reproductive success of these fish may be jeopardized. This would be a serious financial blow to the walleye fishery, which is worth almost $1 billion in Lake Erie alone.

Zebra mussels often attach themselves to the inner surface of water intake pipes of municipal water systems, power plants, factories, and irrigation systems. In 1990, this caused the entire city of Monroe, MI, to be without water for two days. At about the same time, Bill Kovalak, biologist with the Detroit Edison power plant, was startled when he found a zebra population density of 700,000 mussels per square meter blocking the company's intake pipes. Detroit Edison had to spend $500,000 to scrape them away. U.S. and Canadian water treatment plants along the shores of the Great Lakes will probably spend billions of dollars to control the problem.

Zebra mussels have even been known to grow so heavily on their substrate that they can sink navigational bouys. Continued attachment to steel and concrete can cause corrosion and can lead to structural failure.

Scientists are working on a variety of strategies to control zebra mussels and stop the invasion. One major method for controlling the populations of this European invader is the manipulation of its reproductive activity. Male mussels are normally stimulated to release sperm by the scents they get from the explosive growth (bloom) of billions of green algae. The female mussels, in turn, are triggered to produce eggs when they are exposed to mussel sperm in the water around them. This type of breeding behavior ensures that the young larvae have a superabundant supply of algal food. Jeffrey L. Ram of Wayne State University and his coworkers are trying to develop a false scent that would trigger the male mussels to release sperm in the absence of an algal bloom. The sperm would stimulate the females to release eggs. If successful, the resulting larvae would face mass starvation. Eventually, after repeated releases of such pseudoscents, the pest population might be brought under control.

Susan Fisher, a biologist at Ohio State University, recently discovered that a small amount of potassium released in the water prevents the zebra mussel from settling down and fastening itself to a hard surface. She suggests, therefore, that potassium could be constantly released at the openings of water intake pipes to prevent the mussels from forming colonies and blocking flow. Marine paint manufacturers have become interested in adding potassium to their paints to prevent zebra mussels from fouling boat hulls, buoys, and piers.

High-tech methods of control are also being investigated. They include acoustics to kill larvae and the use of ultraviolet light and low-intensity electrical fields to repel them from intakes. Zebra mussels are consumed by natural predators, too, such as diving ducks, catfish, drum (a type of fish), sturgeon, and raccoons. However, studies suggest that these natural controls are currently no match for the reproductive capacity of the zebra mussel. There are simply not enough predators to make a dent in the population.

Surprisingly, not all of the environmental effects of this European invader have been bad. The zebra mussel has been found effective in clearing up water that for years had been the color of pea soup because of successive algal blooms. A single zebra mussel can filter 1 liter of water every 24 hours. A colony of 5 million could filter 1 million liters per day. Improvement in water clarity can be dramatic. A good example is the transformation in Lake Erie. Charles O'Neal, biologist at the State University of New York, commented, "The zebra mussels are like 'vacuum cleaners.' They suck all the nutrients out of the water. The West Basin of Lake Erie used to be green. You couldn't see your fingers. But now you can see down 30 feet!" Water clarity improves lake aesthetics and lakeside property values. It is a delight to swimmers and boaters as well. It has also increased sunlight penetration into previously murky waters, causing bottom-dwelling plants to thrive. They serve as a source of shelter for many smaller fish.

Zebra mussels also filter out considerable amounts of heavy metals such as lead, cadmium, and mercury. In so doing, they may lessen the degree of serious illness that humans may suffer from drinking metal-contaminated water or from eating fish (which can accumulate metals in their flesh).

Hardest hit by the zebra mussel are native mussel species. Already eliminated from the western basin of Lake Erie by zebra mussels, native species are simply outcompeted by their prolific European relative. Loss of plankton and dissolved nutrients caused by the thick masses of zebra mussels that often grow right on native species causes the latter to starve to death.

This case study on the zebra mussel is a classic example of how a natural ecological system, in balance for millennia, can be severely disrupted by the accidental invasion of an exotic organism. For better or for worse, despite strenuous attempts to control this pest, it seems likely that the zebra mussel will spread to most lakes and streams east of the Rockies. They will be here throughout your lifetime, and maybe forever!

as phthalates, are widely used and are present in all of us. For more on them, see A Closer Look: The Hazards of Plasticizers.

The toxicity of synthetic organic compounds is largely a result of their ability to disrupt normal enzyme function in living cells. Enzymes are proteins in the cells of organisms that speed up the rate of important chemical reactions. Without enzymes, we would perish.

For years, toxic organic compounds have been released by factories directly into streams and other surface waters. Factories have also poured their toxic organic wastes into surface impoundments, ponds designed to retain organics where they evaporate. These substances then entered the atmosphere, only to rain down on the land downwind from the site. Some of this waste, of course, percolated into the ground, contaminating groundwater. Some companies have even injected toxic organics and other wastes, such as arsenic compounds, cyanides, and radioactive materials (all potentially harmful to humans) into deep wells in the ground that range from 90 meters (300 feet) to over 3.3 kilometers (2 miles) deep. In theory, the wastes are supposed to remain trapped underground, out of harm's way.

Deep-well injection is surprisingly common in some parts of the United States—and becoming more common as time passes.

A CLOSER LOOK 11.1   The Hazards of Plasticizers

In April 2008, Canada's health minister Tony Clement announced that his government would take steps to limit exposure of infants up to 18 months of age to a common chemical, a plasticizer known as bisphenol A (BPA). This chemical is added to hard plastics (polycarbonate) used to make baby bottles and is also found in polycarbonate plastics that line cans of infant formula and other food products. The health minister soon took steps to ban the importation, sale and advertising of polycarbonate baby bottles in Canada.

This announcement was based on a draft report by Health Canada, which concluded that bisphenol A has the potential to endanger individuals exposed early in life. This report and a similar report issued in April 2008 by the U.S. National Toxicology Program noted that scientific evidence, mostly from studies of laboratory animals, raises some concern for the ill effects of bisphenol A. Low-level exposure of fetuses, infants, and children, the report concludes, could cause behavioral changes in infants as well as adverse changes in their brains, prostates, and mammary glands. Prostate and breast cancer are two possible health outcomes of early exposure. Bisphenol A, the report notes, could also reduce the age at which girls reach puberty.

The report concluded that exposure of pregnant women to low levels of bisphenol A during pregnancy probably posed no risk of birth defects or decreased birth weight. While exposure of adults to high levels—for example, in the workplace—does appear to adversely affect reproduction, exposure to low levels, which are common in everyday life, is considered to present no risk to fertility.

In response to the Canadian report, the world's largest retailer, Wal-Mart, announced in April 2008 that it would immediately halt the sale of products such as baby bottles, sippy cups, pacifiers, food containers, and water bottles that contain BPA in its Canadian stores. Wal-Mart also announced that in early 2009 it would stop selling baby bottles and other products made with polycarbonate plastic containing bisphenol A in its stores in the United States.

Other retailers, including Target and Babies "R" Us, are also removing plastic products containing bisphenol A and/or carrying products made without this potentially harmful chemical, including glass bottles and BPA-free plastic bottles.

This may mark the beginning of even more bans. Studies by Shanna Swan of the University of Rochester School of Medicine in Rochester, NY have shown that exposure of human embryos to moderate to high levels of four common chemicals added to plastics and numerous other products may cause abnormalities in the reproductive organs of newborn boys. These chemicals belong to a group of chemicals known as phthalates and are found in perfumes, candle scents, nail polish, plastic containers, flooring, paints, adhesives, and other products.

The researcher found that newborns of mothers who had moderate to high levels in their blood were more likely to exhibit reproductive organ abnormalities, including smaller penises and undescended testes—that is, testes that fail to enter the scrotum during fetal development. Signs of demasculinzation such as these have been observed in laboratory animals exposed to phthalates as well.

Scientists believe that phthalates in the bloodstream of mothers may reduce the production of the male sex hormone such as testosterone by human fetal testes. This, in turn, leads to abnormal development of the sex organs. The effects, they postulate, could lead to further detrimental effects later in life, including lower sperm production in adulthood and an increased risk of testicular cancer.

Most of these wells exist in the states of Texas, Oklahoma, Arkansas, Louisiana, and New Mexico. Unfortunately, pollution dispersal from these wells is not actively monitored.

There is good evidence to suggest that the deep-well injection technique may actually increase the frequency of earthquakes in some areas. Scientists have demonstrated a well-defined correlation between the volume of waste injected by the U.S. Army's chemical plant near Denver, CO, and the frequency of earthquakes in the immediate region.

Although deep **injection wells** in the United States are supposed to contain wastes for 10,000 years, not all critics believe that they will perform as hoped. Critics view this technique as a means of "sweeping pollution under the rug" and fear that it may be a stopgap measure that comes back to haunt future generations.

Groundwater Contamination Let's look at groundwater contamination by toxic organic chemicals. Bear in mind that although we'll be examining toxic organics, all forms of water pollution can contaminate groundwater.

Groundwater is a massive resource. Within only 0.5 mile of the earth's surface, this vast water resource has four times the volume of the Great Lakes! At present, Americans pump more than 300 billion liters (80 billion gallons) of groundwater each day and consume more than 24 trillion liters (6 trillion gallons) yearly. This underground resource serves as drinking water for 95% of the nation's rural households and for 35 of our nation's largest cities.

Unlike a flowing river, groundwater has virtually no natural cleansing mechanisms. As a result, once it becomes contaminated, it may remain so for thousands of years.

The toxic organic compounds that pollute groundwater come from a variety of sources, as shown in Figure 11.14. Two sources of groundwater contamination are municipal and industrial landfills (sites where waste is buried). Groundwater can also be polluted by toxic chemicals percolating down from the contaminated grounds of industrial facilities, evaporation ponds containing industrial wastes, and agricultural and mining sites. Deep injection wells contaminate groundwater with industrial chemicals, while septic tanks in rural and suburban areas release household toxic organics. Leaky underground gasoline storage tanks at service stations can also contaminate our groundwater supplies.

Because water flows so slowly through most aquifers, it may take decades for a pollutant to move 1.6 to 3.2 kilometers (1 to 2 miles) from the site of contamination, but once groundwater has been contaminated, it is extremely difficult to cleanse.

Adverse Effects on Health As better instruments for detecting toxic organics become available, more and more

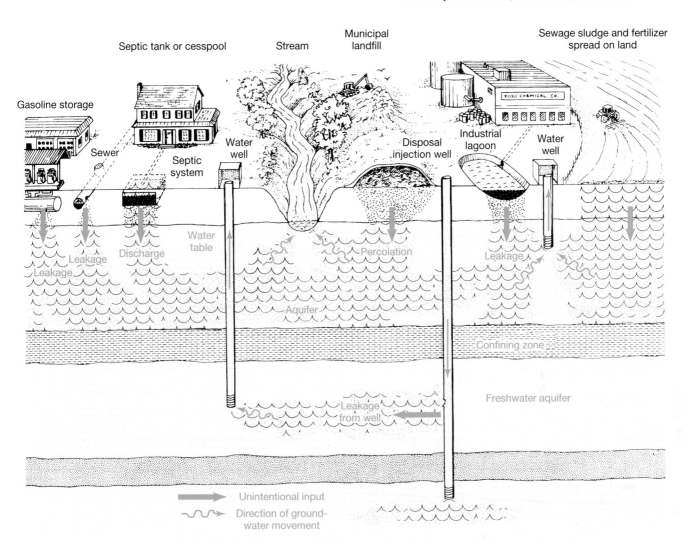

FIGURE 11.14 Sources of groundwater pollution.

toxic chemicals are being found in our nation's groundwater. The question naturally arises: "If I drink contaminated water, will I eventually come down with some serious illness?" Unfortunately, the answer is not clear. Health officials simply cannot say that if you drink x liters of water contaminated with y chemical at a concentration of z ppm over a period of many years, you will develop an illness such as cancer. However, numerous studies have linked groundwater contamination by **toxic organic chemicals** to serious health effects (Table 11.4).

It should be emphasized that polluted well water rarely contains only a single organic contaminant. Usually a given sample contains several. This means that it is possible for two or more organic compounds to interact, producing a synergistic effect on the body—an effect that is greater than the sum of the effects of each pollutant acting separately. Medical researchers have found, for example, that the industrial solvent TCE (trichloroethane) causes greater kidney damage when the individual is simultaneously exposed to a group of toxic organic chemicals, the PCBs (Table 11.5).

Another question that arises is whether a fetus can be adversely affected by toxic organic compounds ingested by the mother. The answer is yes, at least in some cases. For example,

studies show that PCBs are passed from the mother to the fetus. They can cause delayed development, low birth weight, retarded reflexes, and impaired memory.

Despite greater awareness of the problem and stricter regulations, groundwater contamination continues today from the improper and sometimes illegal disposal of hazardous chemicals. Moreover, similar, if not worse, problems exist in other countries. In many eastern European nations of the former Soviet Union, for example, numerous industrial sites were badly contaminated by hazardous materials. Improper disposal of wastes at these facilities and off site was common. Although statistics on the extent of the problem are not available, the problem is considered very serious.

Reducing the threat from toxic organic pollutants requires many different actions, many of them preventive. Preventing groundwater from becoming contaminated requires proper disposal or destruction (for example, incineration) of wastes. Bans on evaporation ponds and on burial of toxic wastes in landfills are also vital to meeting this goal. Better yet are pollution prevention methods, such as hazardous-waste recycling and substitution of nonhazardous materials in chemical processes for hazardous ones, as discussed in Chapter 17. Such measures eliminate toxic-waste production altogether and are now being

TABLE 11.4 Effects of Some Toxic Organic Chemicals on the Health of Occupationally Exposed Workers

Chemical	Exposure	Effects*
Carbon tetrachloride	Inhalation; absorption through skin	Liver and kidney damage
		Vomiting
		Abdominal pain
		Diarrhea
		Jaundice
		Red and white blood cells in urine
		Coma
		Death
Vinyl chloride	Inhalation	Chromosome abnormalities
		Increased spontaneous abortion (in the opinion of exposed workers)
Benzene	Inhalation	Prolonged menstrual bleeding
		Leukemia (blood cancer)

*All of these chemicals have caused cancer in laboratory rats and mice.

Source: Council on Environmental Quality, *Contamination of Ground Water by Toxic Organic Chemicals* (Washington, DC: U.S. Government Printing Office, 1981).

pursued vigorously by many businesses in many countries, including Canada and the United States.

Heavy-Metal Pollution

Heavy metals are highly toxic elements such as lead and mercury. They come from a variety of sources, including active and inactive mines. Abandoned or poorly maintained operating underground mines, for instance, may release water containing heavy metals. Mine tailings, the waste material removed from the mine and sometimes discarded around the mine, also contain heavy metals such as copper, zinc, and cadmium. These metals may be washed into surface waters during heavy rains. Heavy metals in surface waters, such as mercury, also come from the gaseous emissions of coal-fired power plants, garbage incinerators, and other industrial facilities. These metals may be washed from the sky in rain or snow, falling on land and water. Those that land on the Earth may be washed into waterways. Several key point sources also release heavy metals directly into waterways. They include metal-processing plants, dye-making firms, and paper mills.

Heavy metals in drinking water may also arise from other sources. Lead, for instance, may be leached from lead pipes in homes built before 1930 and from lead solder often used for joining copper pipes in newer homes, offices, and other facilities.

The problem with heavy metals is that, unlike organic pollutants, they are not broken down by bacteria. Consequently, they may persist in the water or bottom sediments for many years and eventually enter human food chains.

Metals are also poisonous. Their toxicity results from the fact that many heavy metals interfere with the normal function of enzymes—the proteins that facilitate many life-sustaining functions in our cells. Therefore, ingestion of metals through contaminated food and water may cause serious illness. **Lead poisoning**, for example, may result in a broad spectrum of effects, including decreased learning ability, gastric upsets, convulsions, coma, and death. Lead poisoning also causes mental retardation and stunted growth in children.

At present, the greatest concern among health officials and scientists is the subtle adverse effect of relatively low concentrations of lead in drinking water on embryonic development and on learning and memory in children. Children under the age of ten are especially vulnerable. The EPA estimates that the intellectual capacity of 143,500 American children is reduced by up to five IQ points because of lead-contaminated tap water. Unfortunately, lead in drinking water cannot be seen, smelled, or tasted. Consequently, its presence is not suspected until some time after the first symptoms of poisoning appear.

TABLE 11.5	Some Toxic Chemicals in the Great Lakes: Use, Source, and Effects on Human Health			
Name	Use	Probable Source	Found In	Characteristics/ Health Effects
Asbestos in taconite tailings	No current use.	By-product of iron ore mining. Secured on-land disposal ordered by court.	Lake Superior	Airborne effects may include asbestosis and lung cancer; water-borne effects not known, but cancer is implied.
DDT, chlordane, dieldrin, aldrin	Pesticides used widely in Great Lakes region to control insects and rodents. DDT banned in 1971; others now restricted.	Residues from previous widespread use; runoff from agricultural and forested areas; leaching from improper waste disposal sites; and atmospheric deposition.	All five Great Lakes	Bioaccumulation in fish, wildlife, and humans. Persistent in the environment. Long-range effects include reproductive disorders in wildlife; suspected cause of cancer in humans.
Heavy metals (mercury, lead, arsenic, cadmium, copper, chromium, iron, selenium, and zinc)	Wide variety of industrial uses, from antiknock agent in gasoline to paints, pipes, pesticides, glass, and electroplating.	Industrial discharges; medical profession wastes via municipal discharges; agricultural runoff; disposal of waste products; mine tailings; urban nonpoint sources.	Lake Superior, Lake Ontario, Lake Huron, and Lake Erie	Excessive levels of heavy metals bioaccumulate in fish and wildlife. Human consumption of contaminated food may cause a variety of health problems. See Table 11.6 for details.
PAHs (poly-aromatic hydrocarbons)	Variety of industrial uses.	Industrial oil and grease discharges; by-product of all types of combustion; urban nonpoint sources; smelting.	All five Great Lakes	Persistent in the environment. Can induce cancer and cause chromosome damage in fish, wildlife, and humans.
PCBs (polychlori-nated biphenyls)	Insulation for electrical capacitors, transformers; plasticizer, carbonless copy paper, wide industrial use. Total ban except by special EPA permit in July 1979.	Industrial discharges; municipal sewage treatment plant discharges; harbor sedi-ments; low-temperature incineration of wastes; atmospheric deposition.	All five Great Lakes	Bioaccumulation in fish, wildlife, and humans. Persistent in the environment. Test monkeys developed reproductive failures and skin and gastrointestinal disorders. Probable human carcinogen.

(Continued)

TABLE 11.5	Continued			
Name	Use	Probable Source	Found In	Characteristics/ Health Effects
Dioxins	No known technical use.	Microcontaminants in chlorophenols and banned pesticide 2,4,5,-T. Also bleach kraft paper process and atmospheric deposition.	All five Great Lakes	Bioaccumulation in fish. Probable human carcinogen. Cause of birth defects and reproductive disorders in wildlife.

The EPA estimates that lead in drinking water is responsible for at least 680,000 cases of high blood pressure in adult men in the United States. In addition, about 560,000 children have unacceptably high lead levels in their blood, a considerable amount of which probably had its source in lead-tainted water. At least 350,000 Americans are drinking kitchen tap water with lead concentrations higher than deemed safe by the EPA.

Another heavy metal of grave importance is **mercury**. Mercury in waterways comes from direct industrial discharges and also from rain and snow. The latter comes from coal-fired power plants and incinerators that burn municipal garbage containing mercury batteries. Elemental mercury is fairly innocuous in aquatic ecosystems, but bacteria in sediments convert it to methyl mercury, a toxic form. Methyl mercury accumulates in body tissues and increases in concentration as one goes higher in the food chain. In lakes in the northern United States, concentrations in pike, a predatory fish, are 225,000 times higher than they are in the water. This phenomenon is known as **biological magnification** or **biomagnification**.

Mercury is found in high concentrations in upper-level organisms, including people, in many parts of the world. Three-quarters of the U.S. states (virtually all that have examined the issue) have posted guidelines for eating fish contaminated by mercury. In Michigan, health officials recommend that people eat no more than one meal a week of various fish taken from the state's inland lakes. The U.S Food and Drug Administration has issued guidelines for the consumption of oceangoing fish such as shark and swordfish by pregnant women and women of childbearing age who may become pregnant. Although health studies of the effects of low levels of mercury are not conclusive, some researchers warn that mercury may have significant developmental effects, including reduced cognitive abilities. Mercury can also affect wild species, especially fish-eating birds, such as loons, eagles, merganser ducks, herons, osprey, and kingfisher. All of these species generally have high levels of mercury in their body tissues. In loons, high mercury levels are suspected of reducing the birds' reproductive success and growth, which could lead to increased death rate and decreased birth rate, resulting in a decline in populations. A summary of the effects of four heavy metals on human health is presented in Table 11.6.

Another source of potentially toxic heavy metals is agriculture. As noted in Chapter 9, irrigation water can remove toxic metals from the soil. Not long ago, officials at the Kesterson Wildlife Refuge in California were mystified by an extensive die-off of fish and waterfowl. In addition, many birds were hatched with deformities, such as crossed bills, which made feeding impossible. Furthermore, an abnormally high number of embryos died before hatching. Chemical analysis of the water in the artificially made marshes revealed very high levels of the heavy metal selenium (seh-lean´-ee-um). The selenium was traced to runoff from irrigation water from nearby croplands. Irrigation water leaches selenium from soils that naturally contain high concentrations of the potentially toxic element.

GO GREEN!

Recycle old batteries, or properly dispose of mercury-containing batteries such as watch batteries. Whatever you do, don't throw them in the trash, especially if your city's garbage is incinerated. This will release mercury into the environment.

Controlling Heavy-Metal Pollution Reducing our exposure to heavy metal, as with other pollutants, requires a variety of measures that reduce discharges into the air and water or eliminate them entirely. Interestingly, conventional municipal sewage treatment plants that receive wastes from homes and many businesses do not efficiently remove metals from the wastes. In fact, some metals may be toxic to the very bacteria the plant relies on to digest organic materials. It is therefore important that the amount of metal contamination in the incoming waste be reduced to a minimum. In the United States, local, state, and federal regulations require industries to pretreat their metal-laden waste (to remove toxic metals) before sending it on to a municipal sewage treatment plant. Under ideal circumstances, most of the metals are removed and transported to a certified hazardous-waste dump. Separating

TABLE 11.6	Effects of Four Heavy Metals on Human Health

Mercury	Arsenic
Fatigue	Headache
Headache	Dizziness
Irritability	Fatigue
Loss of coordination	Vomiting
Numbness of hands and feet	Diarrhea
	Abdominal pains
Shortening of attention span	Muscular pains
	Blood in urine
Memory loss	Anemia
Kidney damage	General paralysis
Death	Heart malfunction
Lead	Coma
Intestinal colic	Death
Irritability	**Cadmium**
Reduced resistance to infectious diseases	Degenerative bone disease
Anemia	Severe crippling
Blood in urine	High blood pressure
Brain damage	Heart malfunction
Partial paralysis	
Mental retardation	

industrial waste from municipal waste and requiring on-site treatment of industrial wastes have gone a long way in helping reduce heavy-metal contamination of rivers and other surface waters. Despite progress, improvements are still necessary.

Tighter controls on air pollution emissions from factories and power plants have assisted, too, as many heavy metals in the environment originate as air pollutants, which may rain down on the land and water. This process, known as **cross-media contamination**, suggests the need for a holistic approach to controlling this and other forms of pollution. In our homes, changes in the type of pipe used to supply water and changes in the type of solder used to attach pipe have helped reduce exposure.

Oxygen-Demanding Organic Wastes

Most of us know of a friend or a child who bought a goldfish and then found it floating belly up in its bowl a few weeks later. In many cases, the demise of the new pet was not due to neglect, but rather too much food provided by an overzealous owner. Fish food consists of **biodegradable** organic matter. The excess is consumed by bacteria in the water, and as they consume the excess, they deplete oxygen, reducing concentrations in the fish tank. (The decomposition of organic matter, as you will soon see, often requires oxygen.) The pet fish was actually asphyxiated because of a lack of dissolved oxygen.

Organic matter may accumulate in aquatic environments, as, for example, when an autumn leaf fall blankets a woodland stream or when slaughterhouse waste is discharged into a stream. Streams contain a wide assortment of bacteria that decompose this and other organic material such as sewage or animal wastes. These microorganisms are part of a self-purifying mechanism of streams.

The process by which such organic material is eventually decomposed by bacterial action may be summarized as follows:

$$\begin{array}{ccl} \text{high-energy organic} & : & \text{low-energy carbon} \\ \text{molecules (fats} & & \text{dioxide} + \text{energy} \\ \text{carbohydrates, and} & & \text{(used by bacteria} \\ \text{proteins)} + \text{oxygen} & & \text{to sustain life)} \\ & & + \text{water} + \text{nitrate} \\ & & \text{ions (NO}_3^{-2}) \\ & & + \text{phosphate ions} \\ & & \text{(PO}_4^{-3}) + \text{sulfate} \\ & & \text{ions (SO}_4^{-2}) \end{array}$$

Note from this equation that oxygen in the water is essential to the process. In goldfish bowls and aquatic ecosystems, the bacteria actively compete with other oxygen-demanding aquatic organisms (fish, crustaceans, insect larvae, and so on). If sufficient organic material is present in the water, and if other conditions such as water temperature are favorable, the oxygen-consuming bacteria multiply rapidly. Levels of dissolved oxygen fall as the bacterial population increases. Levels may plummet from 10 ppm to less than 3 ppm, to the detriment of other aquatic organisms.

Because their decomposition consumes oxygen in aquatic ecosystems, naturally occurring organic wastes and human-produced wastes such as sewage are called **oxygen-demanding organic wastes**. The federal government maintains a network of stream-monitoring sites at which dissolved oxygen levels are systematically checked. Of the thousands of measurements taken in the past few years, fewer than 5% were below 5 ppm of dissolved oxygen—the minimal level required for quality fish populations.

Water quality chemists measure the concentration of oxygen-demanding organic matter through a simple technique in which they mix water samples with a certain amount of oxygen and then measure the decline in oxygen over time. The amount of oxygen consumed by bacteria depends on the amount of organic material in the sample. The more organic material, the greater the oxygen consumption.

The organic content of water is not expressed directly, but rather as the **biological oxygen demand (BOD)**. More recently, water quality personnel have switched to the term **biochemical oxygen demand**. Thus it is customary to speak of the BOD of human sewage, of slaughterhouse wastes, and so on, since this material supports bacteria that require or "demand" oxygen.

If you live in a city or town served by a sewer system, every time you flush your toilet, you are making it a bit tougher for game fish in the stream or lake near your home to survive, for there are about 250 ppm BOD in the wastewater going down the pipe. Many of the wastes from canneries, cheese factories, dairies, bakeries, and meat-packing plants have BOD levels ranging from 5,000 to 15,000 ppm.

It is important to point out that oxygen-depletion depends on the amount of organic matter added to a body of water. Small amounts of organic matter, which occur naturally, generally cause no problem. Bacteria and other microorganisms remove the organic material, but oxygen levels are replenished naturally. Oxygen levels decline significantly, however, when large amounts of organic matter are dumped into a water body. In other words, it is only when the natural assimilation capacity of a lake or stream is exceeded that severe oxygen depletion and damage occur.

As you will soon see, a river's organic assimilation capacity is determined by the rate of aeration—how fast oxygen is replenished. Cold, turbulent water contains lots of oxygen and replenishes it more quickly than a warm, slow-moving river. Rivers most vulnerable to oxygen depletion are the warm, slow-flowing, stagnant waterways.

Effect of a High BOD on Aquatic Species   High BOD wastes from sewage treatment plants and other facilities such as pulp mills and slaughterhouses can have a devastating effect on aquatic organisms. Results of studies of the kinds and numbers of organisms occurring immediately above and at several sites below the point of sewage discharge are shown in Figure 11.15. At point *A,* just above the outfall, the river is characteristic of an unpolluted stream. The high levels of dissolved oxygen (8 ppm) and the abundant food in the form of mayfly and caddis fly larvae make possible the survival of highly prized fish such as bass and trout. However, at point *B,* in the **zone of decline** immediately below the outfall, dissolved oxygen levels drop rapidly because of the high organic component of the waste. In some streams, the dissolved oxygen may drop to 3 ppm or less, which is insufficient to support the oxygen requirements of more-desirable fish. Instead, only less-desirable fish, such as carp and bullheads, which have low oxygen requirements, can survive. The larvae of mayflies, stoneflies, and caddis flies, which require higher oxygen levels, are virtually absent, too. The dissolved oxygen concentration is so drastically reduced in the **damage zone,** from *C* to *D,* that even carp and bullheads cannot survive.

The most typical bottom-dwelling animals in the damage zone are reddish **sludge worms,** of which there may be 180,000 per square meter of stream bottom; bloodworms; and the red rat-tailed maggot (Figure 11.15). These animals are sometimes used as index organisms; their occurrence indicates that a particular stretch of stream is highly contaminated with organic waste.

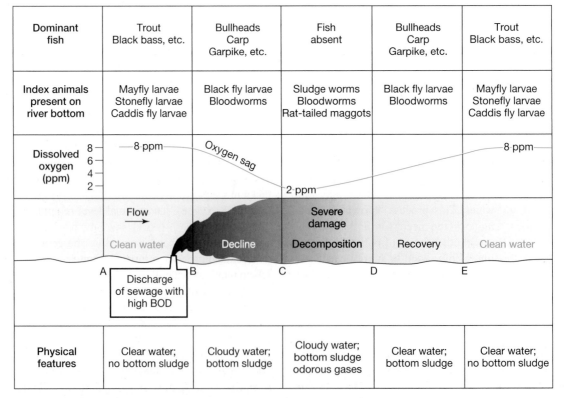

Dominant fish	Trout Black bass, etc.	Bullheads Carp Garpike, etc.	Fish absent	Bullheads Carp Garpike, etc.	Trout Black bass, etc.
Index animals present on river bottom	Mayfly larvae Stonefly larvae Caddis fly larvae	Black fly larvae Bloodworms	Sludge worms Bloodworms Rat-tailed maggots	Black fly larvae Bloodworms	Mayfly larvae Stonefly larvae Caddis fly larvae
Dissolved oxygen (ppm)	8 ppm	Oxygen sag	2 ppm		8 ppm
	Flow → Clean water	Decline	Severe damage Decomposition	Recovery	Clean water
	A	B	C D	E	
	Discharge of sewage with high BOD				
Physical features	Clear water; no bottom sludge	Cloudy water; bottom sludge	Cloudy water; bottom sludge odorous gases	Clear water; bottom sludge	Clear water; no bottom sludge

FIGURE 11.15   Effect of sewage with a high BOD level on the amount of dissolved oxygen and the type of aquatic organisms in the stream.

Beginning at point *D* in the **recovery zone**, the amount of oxygen removed by the sewage bacteria is more than counterbalanced by the oxygen entering the stream from the atmosphere because of wind action or photosynthesis of algae and stream-dwelling plants. As a result, the dissolved oxygen level rises, permitting the occurrence of carp and garpike. Finally, still farther downstream, at point *E*, most of the organic material discharged from the sewage plant has been decomposed; the level of dissolved oxygen rises to its original value. Fish and other organisms supported by the stream above the point of discharge can survive in the water below point *E*.

The characteristic dip of the oxygen curve at points *B* and *C* is known as the **oxygen sag**. The slope of the dissolved oxygen curve, which is highly variable, depends on the BOD of the sewage, the rate at which oxygen enters the stream, the water temperature, and the water velocity. If additional sources of pollution are present, recovery may be impossible.

The oxygen sag curve is not just of scientific interest. It is very practical information that is used to set BOD discharge standards for wastewater treatment plants. A wastewater treatment plant on a stream that is more prone to oxygen depletion will have a lower BOD discharge permit standard to maintain a minimum dissolved oxygen of 5 ppm in the receiving water than one that is swifter moving and better aerated. The discharge permit procedure, of course, is designed to protect aquatic life in the lake or stream into which treated sewage is released. The control of oxygen-demanding organic wastes is discussed in section 11.3.

GO GREEN!

When the time comes to buy your first home, consider installing a composting toilet to reduce waste output.

Hormone Disruptors and Drugs

Another form of water pollution that is gaining recognition these days is a group known as **hormone disruptors**. These chemical pollutants alter the hormonal system of animals that ingest them, affecting reproduction and other vital functions in humans and wildlife. The effects can be dramatic and devastating. For instance, pulp mills, where trees are ground up to make paper, release an estrogen-like compound (genistein) in significant amounts into nearby lakes and rivers. Studies suggest that this compound—and others like it—are responsible for dramatic disruptions in the reproduction of fishes in the Great Lakes and in streams that flow into the lake. How do estrogenlike compounds affect fish? When exposed to these compounds, male fish develop both male and female reproductive organs and are unable to reproduce. Unfortunately, these chemicals are not removed by water treatment facilities used at paper mills.

Paper mills also release a group of potentially harmful chemicals known as dioxins. At high levels, dioxins are thought to be carcinogenic. At low levels, they may suppress the immune system of animals, including humans. Fortunately, new methods of papermaking, known as elemental chlorine-free pulp processes, have been developed that dramatically lower dioxin production. Unfortunately, very little paper is currently made this way.

Hormone disruptors can affect species many miles from the source. Male cod (a species of fish) that live in the open ocean far from sources of human-made pollution, for example, produce an egg yolk protein normally made only in females. In female fish, the protein is produced by the liver and deposited in their eggs. The egg yolk protein is also abnormally produced in male cod and other fishes inhabiting waters polluted by certain pollutants from sewage treatment plants and paper mills. When ingested, these chemical pollutants mimic the female sex hormone estrogen, causing feminization. Exposure to such pollutants may also interrupt male sexual development and fertility.

Researcher Alexander P. Scott of the Centre for Environment, Fisheries, and Aquaculture in Weymouth, England, wanted to see whether male cod that spend their entire life at sea far from sources of pollution suffered similar anomalies. In his study, he found the egg yolk protein but only in larger, older male cod.

The presence of the protein in fish isolated from pollution sources suggests that the pollutants may be working their way through the food chain. Scott believes that these chemicals are deposited in the sediments of the deep ocean far from sources and are ingested by bottom-dwelling eels and other fish upon which cod feed. As the cod grow older, concentrations of the estrogen-mimicking chemicals build up in their tissues, including the liver, where they stimulate the production of the egg yolk protein. Such exposure could reduce reproduction in this commercially important food source.

Other chemicals showing up in waterways of the world are synthetic female sex hormones derived from birth-control pills. Studies suggest that these substances, excreted in the urine of women, appear to cause a shift in the gender of male fish. Male carp and walleyes taken from waters near the effluent of a sewage treatment plant, for instance, were found to be producing large quantities of egg yolk protein, which is normally produced only by females.

Laboratory studies show that two female sex hormones from birth-control pills cause some male fish to exhibit both male and female reproductive tissues. Such fish often fail to spawn; they've effectively become sterile or nonreproductive. Scientists are now only beginning to consider the effects on humans of consuming these chemicals.

Contraceptive patches worn by women and flushed down toilets may also cause significant biological effects on wildlife, particularly fish and other aquatic organisms that live in rivers, streams, and lakes. When flushed, the patches still contain substantial amounts of the hormone. Even a small number flushed into a municipal sewage system could impair fish reproduction downstream from the plant, according to researchers.

Prescription and nonprescription drugs such as antibiotics, painkillers, blood pressure medications, and antidepressants

GO GREEN!

Ladies, if you wear contraceptive patches, discard them in the trash. Don't flush them down the toilet.

are also polluting waterways. Trace amounts of many common drugs have been turning up in surface waters since the 1970s. Where do they come from? These drugs come from the urine of humans released into surface waters from sewage treatment plants. They also come from livestock.

Although levels of these drugs are relatively low, studies show that some of these drugs may, in combination, maim and kill tiny crustaceans—vital components of aquatic food chains. In one study, exposure of *Daphnia* (a type of crustacean) to a cholesterol-lowering drug (clofibric acid) and an antidepressant (fluoxetine) simultaneously for six days at concentrations typically found in the environment killed most of the *Daphnia*. When levels of the cholesterol-lowering drug were reduced, mortality declined, but some of the *Daphnia* offspring developed serious physical abnormalities.

Tests with five common antibiotics using *Daphnia* showed similar results: Individually, the antibiotics had no measurable effect. However, when combined in concentrations presumably representative of levels in surface waters, three of the antibiotics resulted in the production of generations of *Daphnia* with a much higher percentage of males.

Mixtures of common medications found in rivers, lakes, and other surface waters have also been shown to adversely affect a number of fish species. However, the long-term impact of possible changes on their food sources is yet to be assessed.

11.3 Sewage Treatment and Disposal

Cities and towns produce enormous quantities of water pollution from nonpoint and point sources. In many parts of the world, especially the more-developed nations such as those of Europe and North America, point sources such as homes and offices discharge their wastes into underground pipes. These pipes, in turn, transport the wastes to **sewage treatment plants**, facilities that process the wastes, removing many of the pollutants. Sewage treatment plants receive numerous types of water pollutants, including sediment, infectious organisms, detergents (containing inorganic nutrients), human excrement, drugs excreted in human urine, heavy metals, and toxic organic chemicals (Table 11.7). Many of the pollutants are removed by sewage treatment plants before the waste is discharged into a lake, stream, or ocean. These plants have done much to help clean up the waters of the United States, Canada, and other more-developed nations. (For a discussion of the successes and hidden problems that remain, see Case Study 11.2.)

TABLE 11.7	Water Pollution: Sources, Effects, and Control		
Contaminant	**Source**	**Effects**	**Control**
Oxygen-demanding waste	Soil erosion Autumn leaf fall Fish kills Human sewage Domestic garbage Remains of plants and animals Runoff from urban areas during storms Industrial wastes (slaughterhouses, canneries, cheese factories, distilleries, creameries, and oil refineries)	Bacteria that decompose the organic matter deplete the stream of oxygen Game fish replaced by less desirable fish Valuable food for game fish (mayflies, etc.) destroyed Foul odors develop	Reduce runoff from barnyards Reduce BOD of sewage with modern secondary sewage treatment plants Reduce runoff from feedlots with catch basins
Disease-producing organisms	Human and animal wastes Contaminated aquatic foods (clams, oysters)	High incidence of waterborne diseases such as cholera, typhoid fever, dysentery, polio, infectious hepatitis, fever, nausea, and diarrhea	Reduce runoff from barnyards and feedlots More effective sewage treatment Proper disinfection of drinking water

(Continued)

| TABLE 11.7 | Continued | | |

Contaminant	Source	Effects	Control
Nutrients (phosphates and nitrates)	Soil erosion Food-processing industries Runoff from barnyards, feedlots, and farmlands Untreated sewage Industrial wastes Exhaust of trucks, cars, trains, and buses	Eutrophication May cause methemoglobinemia in infants Decreased recreational and aesthetic values	Tertiary sewage treatment Reduce use of commercial fertilizers Change detergent formula Control soil erosion with strip cropping, contour plowing, and cover cropping Control feedlot runoff with catch basins
Sediment	Soil erosion from farmland, strip-mined land, logged-off areas, and construction sites (roads, homes, and airports)	Fills in reservoirs Clogs irrigation canals Increases probability of floods Impedes progress of barges Interferes with photosynthesis, reducing dissolved oxygen levels Destroys fresh water mussels (clams) Causes fish mortality due to asphyxiation Destroys spawning sites of game fish Necessitates expensive filtration of drinking water	Employ erosion-control practices on farms such as contour plowing, cover cropping, and shelterbelting Employ erosion-control practices at construction sites: sodding, use of catch basins, etc. Use mulching and jute matting on seeded road banks Establish temporary cover, such as rye and millet, at construction sites
Heat	Midsummer heating of shallow water by the sun Discharge of warm water from electrical power, steel, and chemical plants	Disrupts structure of aquatic ecosystems Causes shift from desirable to undesirable species of algae and fish Kills cold-water fish such as salmon and trout Blocks spawning migrations of salmon Interferes with fish reproduction Increases susceptibility of fish to diseases and to the toxic effects of heavy metals such as zinc and copper	Reduce the nation's energy demands for electricity Use closed cooling systems exclusively Instead of discharging heated water to streams, use it to heat homes, extend growing seasons on cropland, increase growth rate of food fish and lobsters, and prevent frost damage to orchards

Sewage Treatment Methods

Since 1880, when the first sewage treatment plant in the United States was built in Memphis, TN, more than 16,000 have been constructed, according to the EPA. They serve 70% of the U.S. population. (Most of the rest of the population is served by septic tanks, discussed shortly.) Domestic sewage treatment plants fall into one of three categories: primary (rudimentary and relatively inexpensive), secondary (more effective and more costly), or tertiary (most effective but the most expensive). Before you study each system, we recommend that you take a look at Figure 11.16. It summarizes the outcomes of primary, secondary, and tertiary treatment of sewage. Be sure to take note of the white squares, which highlight the major "accomplishments" of each stage.

Primary Treatment The wastewater entering a sewage treatment plant contains solid and liquid wastes, but human solid waste (feces) is highly liquefied by the time it arrives at the plant. (Organic fecal matter is suspended in the incoming wastewater.) In addition to suspended solids, the incoming wastewater carries solid materials such as Band-Aids, sediment from storm water (discussed later), and toys flushed down the toilet by children. **Primary sewage treatment** is mainly a physical process to remove solids from the wastewater, as shown in Figure 11.17. Wastewater enters the plant and travels through a screen that removes large objects (gravel, garbage, leaves, and so forth). The stream of wastewater then passes into **settling tanks**, which are often called clarifiers, where suspended organic solids settle to the bottom. In primary-treatment-only plants, the fluid that remains is then chlorinated to destroy disease-causing organisms and discharged into lakes or streams.

The solids that have accumulated at the bottom of the settling tank are pumped to a **sludge digester**, where millions of bacteria "feed" on the organic waste, breaking it down in the absence of oxygen. One decomposition product, methane gas, is frequently captured and burned on site to heat the digester to the temperature required for most effective bacterial action. In some plants, the methane is burned to produce electricity that can be used on site.

Primary treatment removes about 60% of the suspended solids and about 33% of the oxygen-demanding waste (BOD). Although primary treatment makes sewage look a lot better, it leaves a substantial amount of organic material, nitrates, phosphates, and bacteria, some of which may cause human disease. Fortunately, with the aid of cost-sharing grants from the federal government, thousands of American cities have been able to replace their primary plants with secondary plants.

Secondary Treatment Virtually 100% of the municipal sewage in the United States receives secondary treatment (Figure 11.18).[2] Primarily biological in nature, **secondary sewage treatment** facilities rely on aerobic bacteria to break down degradable organic materials. Secondary treatment may also remove ammonia by biological oxidation to nitrates. The two major methods available for secondary treatment are (1) the activated sludge process and (2) the trickling filter.

Activated Sludge Process In a sewage treatment plant with a secondary treatment capacity, fluid from the first settling tank (clarifier) is piped to another tank, the aeration tank, in which air is bubbled to provide a maximal supply of oxygen (Figure 11.18). In the aeration tank, aerobic (oxygen-using) bacteria decompose the organic compounds, and because there is plenty of oxygen, they do so at a fairly rapid rate. BOD drops rapidly because of the presence of oxygen and aerobic bacteria. The mix is then transferred to a settling tank. Here the remaining suspended organic matter and bacteria settle to the bottom, forming sludge. Most of the sludge from this process and also from primary treatment stages is then drained into a device known as an anaerobic digester, labeled "sludge digester" in Figure 11.18. Here it is further broken down in the absence of oxygen. (The term **activated sludge** refers to the highly concentrated mix of aerobic bacteria and organic matter within the aeration chamber.) As illustrated, some sludge is recycled back to the aeration tank where it provides a "seed" population of bacteria to act on the incoming waste. What's left at the anaerobic digester is dried and either incinerated, landfilled, or used as fertilizer. The liquid that accumulates at the top of the second sedimentation tank is eventually chlorinated to kill bacteria and then discharged.

Secondary treatment reduces the BOD by 90% and removes 90% of the suspended solids. However, 50% of the nitrogen compounds and 70% of the phosphorus compounds (the chemical culprits responsible for eutrophication) still remain. To remove them, tertiary treatment is required.

GO GREEN!

Take shorter showers and install a water-efficient showerhead. These measures reduce water consumption, save energy, and reduce the amount of wastewater flowing to treatment plants, reducing energy use at such plants and the use of chemicals such as chlorine needed to treat the effluent.

Trickling Filter In the **trickling filter** process, the sewage is sprayed by the arms of a slowly rotating sprinkler onto a filter bed made up of stones or large chunks of bark (Figure 11.19). The filter bed may be about 2 meters (6 feet) thick and up to 60 meters (200 feet) in diameter. The stones or bark are coated with a slime of bacteria that has accumulated during the operation of the filter. The sewage, containing a load of dissolved organic compounds, trickles down through the stones, and the bacteria consume the organics. Leftover solids are piped to a settling tank and later transferred to a sludge digester. About 80 to 85% of the dissolved organic compounds are removed by the trickling filter system. However, the wastewater still contains a high level of nutrients, such as phosphates, ammonia (NH_3), and nitrates, which could cause eutrophication of the lakes and streams into which they are discharged. The removal of these materials depends on still another process: tertiary treatment.

---

[2] Of the 16,000 sewage treatment plants in the United States, all but 68 are secondary treatment facilities. Those that are only primary treatment tend to be small.

CASE STUDY 11.2   INVISIBLE THREAT: TOXIC CHEMICALS
IN THE GREAT LAKES

In the 1960s, the Great Lakes were plagued by massive pollution primarily from sewage treatment plants and factories. Symptoms of gross eutrophication abounded: turbid bays, weed-choked shallows, floating mats of algae, rotting fish that fouled beaches. Fortunately, those obvious signs of pollution gradually receded, thanks to a $9 billion Canadian–American investment in modern municipal and industrial wastewater treatment systems along the perimeter of the Great Lakes. Today, their waters are, for the most part, clear and sparkling.

Appearances can be deceiving, however. An invisible threat is present in these waters—posed by a chemical "broth" of toxic chemicals (Table 11.5). In fact, a joint study by American and Canadian scientists has indicated that human exposure (40 million people) to toxic pollutants in the Great Lakes region is greater than in any other area on the North American continent.

One of the symptoms of the pervasive toxic contamination of this 162,500-square-kilometer (65,000-square-mile) expanse of fresh water is the appearance of skin lesions and cancers in fish. For example, liver cancers are frequently found in bullheads taken from Ohio's Cuyahoga River where it flows into Lake Erie. Bottom-feeding fish such as carp, suckers, and catfish are showing an increasing frequency of cancer—probably because they are ingesting toxic chemicals that accumulated in and are now being released from sediments. In the most contaminated tributaries of the Great Lakes, nine of every ten fish may have some form of cancer!

When the bodies of these fish are ground up and analyzed, they are usually found to contain relatively high levels of toxic chemicals, including mercury and organic pesticides; the older and bigger the fish, the higher is the concentration of contaminants. As a result, health officials have issued fish consumption advisories that instruct people to restrict their consumption of salmon, lake trout, and certain other species of Great Lakes fish. Fetuses as well as young children are especially vulnerable to the toxicants.

Two questions naturally arise: What did these chemicals come from? And how did they get into the Great Lakes? Consider the case of a chemical known as toxaphene. This organic pesticide has been found in the tissues of fish taken from a lake on Isle Royale, an island in Lake Superior. Since toxaphene has never been used on Isle Royale, the only possible mode of entry to the island lake was by "fallout" from the atmosphere. Scientists believe that the toxaphene comes from southern states including Texas, where it has been used to control an insect that attacks cotton, the cotton boll weevil. Northward-blowing winds carry the toxaphene to the lake. Other pesticides, including DDT and chlordane, as well as PCBs and heavy metals such as lead and zinc, apparently also enter the Great Lakes in substantial amounts via atmospheric deposition. Other modes of entry for the more than 450 chemical contaminants of the Great Lakes ecosystem include (1) industrial wastewater discharge, (2) leaching from industrial waste storage lagoons, (3) municipal waste discharge, (4) agricultural runoff, (5) urban runoff, (6) mining site runoff, and (7) release from bottom sediments.

The health effects of many of these toxic substances, which are ingested in contaminated fish and drinking water, are only beginning to be understood. After all, it is one thing to feed laboratory animals high levels of a given toxic chemical and then observe the harmful effects, and quite another to determine precisely what will happen to human beings who ingest a few parts per trillion of numerous toxic chemicals every day for 30 years. And since these contaminants in fish flesh or drinking water are not only invisible but also tasteless and odorless, it is extremely difficult for a regulatory agency to convince legislative bodies and the public that something should be done about them.

In 1990, the United States and Canada formed the **International Joint Commission** to identify mutually important problems. As the water pollution problem in the Great Lakes became increasingly serious, the two nations entered into a **Great Lakes Water Quality Agreement.** Under the terms of this agreement, each nation agreed to monitor the Great Lakes water under its jurisdiction, to identify areas of concern (hot spots), and to explore ecologically sound methods to control the pollutants. The United States and Canada identified about 50 toxic hot spots that needed immediate attention. One of them is the Detroit River, which is a tributary of Lake Erie. This waterway serves as a sewer for both municipal and industrial waste from scores of nearby cities. Even though this waste receives conventional treatment, the river contains a diverse mix of toxicants, including heavy metals and toxic organic compounds. Hundreds of contaminants are present in the bottom sediments of this stream. Other areas of concern include the harbors of Milwaukee (WI), Gary (IN), Muskegon (MI), Cleveland (OH), Toronto (ON), and Rochester (NY), as well as the mouths of many great rivers.

Under the terms of the Water Quality Agreement, the United States and Canada are using an ecosystem approach to deal with these focal points of contamination. The Remedial Action Plans drawn up so far are based on the premise that problems at a hot spot must be considered in their total ecological context before appropriate solutions can be developed. An important component of Remedial Action Plans is the examination and regulation of the modes of entry of each toxic chemical. For example, if pesticide contamination of the Buffalo River originates in agricultural runoff, farmers in the Buffalo River drainage will be urged to reduce their use of chemicals in pest control and to adopt more effective strategies to prevent soil erosion, such as contour farming and conservation tillage.

If the Remedial Action Plans developed by the United States and Canada for the detoxification of the Great Lakes are to succeed, the lifestyles of many people living in this region may have to change. The use of low-phosphate detergents and more conservative application of lawn fertilizer could help. Moreover, great commitment and cooperation will be required from the industrial sector, environmental agencies, and government at all levels. Only in this way will the "invisible" chemical threat to the 40 million people of the Great Lakes really be made to disappear.

Pollutant	Treatment		
	Primary	Secondary	Tertiary
Solids	Solids removed	Little removed	Little removed
Harmful bacteria	Bacteria removed	Little removed	Little removed
Dissolved organics	Little removed	Dissolved organics removed	Little removed
Harmful viruses	Little removed	Viruses removed	Little removed
Phosphorus	Little removed	Little removed	Phosphorus removed
Nitrogen	Little removed	Little removed	Nitrogen removed

FIGURE 11.16 Relative effects of treatment on the removal of pollutants from sewage.

Tertiary Treatment The most advanced form of sewage treatment is **tertiary sewage treatment**. As shown in Figure 11.16, tertiary treatment is designed to remove most of the remaining pollutants, notably the nitrates and phosphates. The water quality of our streams and lakes would be considerably better if all sewage underwent tertiary

treatment. Unfortunately, these plants are twice as expensive to build as secondary sewage plants and four times as expensive to operate. As a result, tertiary treatment is not used unless it is necessary to maintain a high level of purity in the receiving body of water. Currently, about 50% of the sewage is treated in this way in the United States. Some cities and towns are also exploring other, less expensive, methods of providing tertiary treatment—for example, diverting waste after secondary treatment into ponds where water hyacinths grow. These aquatic plants remove many of the pollutants at a fraction of the cost of conventional tertiary treatment facilities.

Managing Stormwater Runoff

In newer cities, stormwater, rain, and snowmelt running off of streets and sewage from homes are typically carried in separate sets of pipes under the streets. That way, water running off parking lots, streets, rooftops, and lawns is kept separate from sewage. (It's delivered directly into streams, while sewage is piped to sewage treatment plants.)

As noted earlier, in older cities, however, stormwater runoff, snowmelt, and rain frequently empty into the same underground pipes that transport domestic sewage from homes and businesses to sewage treatment plants. **Combined sewer** systems such as this cause few problems most of the time. Sewage treatment plants can handle the inflow. When heavy rain occurs, however, sewage treatment plants are often overwhelmed by the additional flows. To prevent the tanks at the plant from overflowing, excess stormwater containing raw sewage is shunted directly into streams and lakes. Wastewater diversion during periods of heavy rainfall, in fact, is a major reason why many rivers do not meet federal standards established by the Clean Water Act.

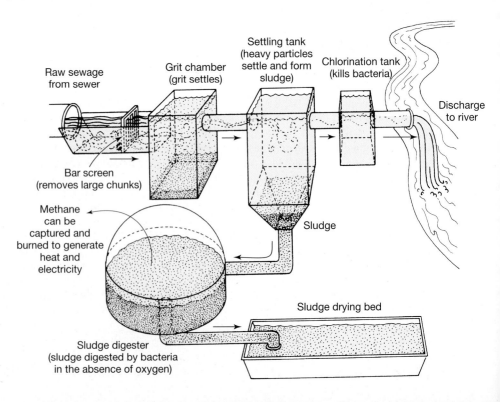

FIGURE 11.17 Primary sewage treatment.

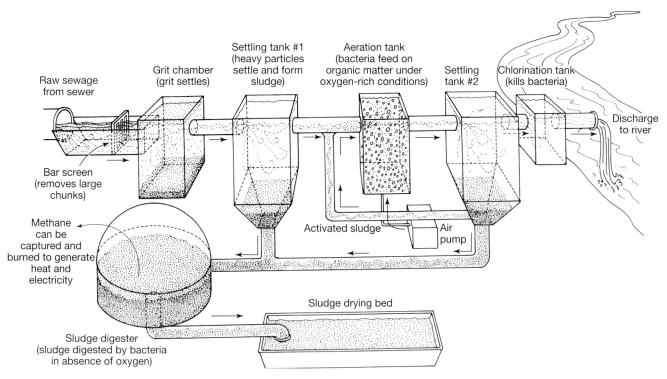

FIGURE 11.18   Primary and secondary sewage treatment.

The Clean Water Act, discussed in more detail shortly, requires cities and towns to create **stormwater management programs** to protect surface waters of the nation from pollutants carried in stormwater. These programs must seek to identify and eliminate illegal discharges of pollutants into storm drain systems—for example, companies that are illegally dumping waste into a city or town's storm sewers. They also must reduce effects of development by private and public entities projects (new housing or road construction, for example) on runoff and erosion, to reduce nonpoint water pollution. And they must help citizens learn more about nonpoint water pollution and ways they can help minimize problems. Stormwater management programs also require the cities and counties to periodically monitor water quality, identify problem areas that need addressing, and report their findings to state and federal agencies.

FIGURE 11.19   Secondary sewage treatment. This rotary trickling filter device handles 4 million gallons of wastewater from Sacramento, CA, daily. After chlorination, the effluent is discharged into the American River.

Septic Tanks

Many rural and (a few) suburban families in the United States and other countries have backyard septic tank systems that process household wastes (Figure 11.20). **Septic tanks** are underground sewage containers made of concrete or plastic into which all household wastewater flows. Solids settle to the bottom of the tank and form sludge. The fluids flow from the tank into a system of perforated pipes buried underground above a bed of crushed rock. This creates a drainage field, known as **leach field**. The liquid waste passes through holes in the pipes in the leach field and slowly percolates through the soil. The soil acts as a natural filter, removing bacteria, some viruses, and suspended materials. Phosphate binds chemically to the soil particles. Organic material in the leach field is decomposed by soil bacteria. The sludge that accumulates at the bottom of the septic tank also undergoes bacterial decay but is also pumped out every three to four years, especially if hard-to-digest materials like carrot peelings and toxic household cleaners are flushed down the drain. The sludge is then delivered to a sewage treatment plant or applied to farm fields.

Septic systems are an effective means of treating household wastes, but they do have drawbacks. For example, they cannot be used if the water table is high enough that they pollute the groundwater, as it is in the southern United States. They cannot be used if the soil is relatively impermeable, as there would be no place for the leachate to go. If septic systems are overtaxed and not regularly pumped, their leach fields may become clogged with organic material. As a result, the partially decomposed waste may rise to the surface, causing visual pollution and generating foul odors. Septic systems also have a limited life span. Leach fields clog with undecomposed organic matter and must be replaced. Septic systems are the most frequent cause of bacterial contamination of groundwater, especially in areas where the population density is high. Septic systems servicing homes around lakes can also release substantial amounts of waste into the groundwater, which interfaces with surface waters. In such instances, nitrates and phosphates from the tanks can pollute surface waters, causing algal blooms.

Septic tank performance can be improved by using nontoxic cleaning agents and composting kitchen wastes, especially those that don't break down easily, such as carrot peels and rinds from citrus fruit. It can also be improved by filtering the liquid before it leaves the tank and enters the leach field. The performance and longevity of the leach field can also be enhanced by installing a small pump to force liquid from the septic tank through the field. This results in a more even distribution of liquid waste. Many leach fields fail when the first 10 meters (30 feet) of the system clog. The rest of the leach field is still in fine working order.

Alternative Treatment Methods

Although sewage treatment plants like those described in this section are the norm in most developed nations of the world, many countries are experimenting with alternative biological treatment facilities that are, quite often, cheaper to build, much cheaper to maintain and operate, and more in harmony with nature.

Holding Ponds, Indoor Biological Treatment Facilities, and Other Technologies In some cities and towns, sewage flows into a series of specially built ponds or marshes. In these facilities, various species of aquatic plants such as algae, cattail, water lilies, water hyacinths, and duckweed break down the waste. The tiny duckweed is a floating plant smaller than your little fingernail. In some ponds, duckweed grows profusely enough to form a solid green "living blanket" over the water. All these plants, of course, grow rapidly because they have access to an abundant supply of nutrients. Not only does duckweed absorb dissolved organics directly from the water, but it also can be harvested and fed to livestock or even used as human food. In Thailand and Burma, residents have long consumed duckweed, a highly nutritious plant that has six times the protein content of a soybean field of equal area! (The vegetation should be analyzed for the presence of heavy metals before being eaten.) Such sewage **holding ponds** may even serve as wildlife habitats. Interestingly, the effluent from carefully built and well-thought-out sewage treatment ponds can be much cleaner than that coming from primary and secondary treatment facilities.

Indoor facilities are also being built. These generally consist of greenhouses that contain a number of tanks, each containing aquatic plants, microorganisms, and animals that degrade the waste material (Figure 11.21).

Sewage from homes can also be treated in specially built wetlands (Figure 11.22). One of the most popular designs is submerged wetlands. They consist of lined depressions filled with crushed rock or pumice and then covered with dirt. Sewage from the house empties into the system underground, so there is no possible means of human contact. Organic matter is then decomposed by bacteria in the rock bed. Plants growing in the soil absorb moisture and nutrients from the sewage, too. Effluent from such systems can be quite pure.

In areas where the topsoil is too shallow for septic tanks, many homeowners use composting toilets (Figure 11.23). This technology, invented in Scandinavia, has improved dramatically, and numerous models are now available for home use. In many composting toilets, wastes are deposited into a chamber just below the "throne," as in Figure 11.23. A handful of sawdust is used to cover the waste. Water in urine and in the feces evaporates through a vent pipe, and the organic matter is converted into a rich, organic matter called humus. It can then be buried in flower gardens or around trees. In other composting toilets, the tank that receives the feces is much larger and is located in a room (usually the basement) below the toilets. Here waste from several toilets is collected. Some composting toilets use small amounts (less than 1 pint or .47 liters) of water to flush.

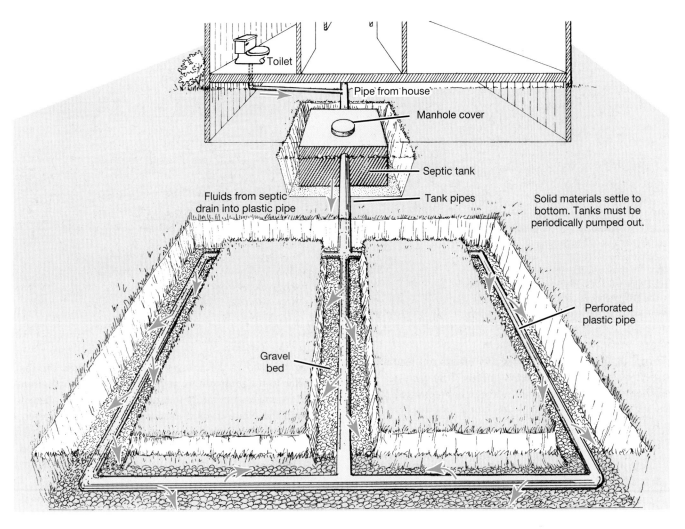

A. Septic tank and drain field

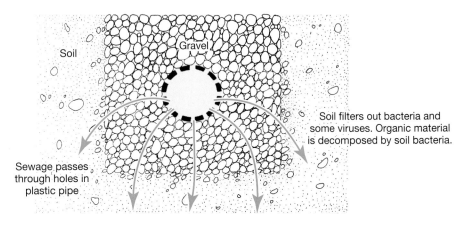

B. Cross section of drain pipe

FIGURE 11.20   The septic tank and drain field system of sewage disposal.

Sewage Sludge: A Resource in Disguise?

For many years, sewage sludge has been buried in landfills or burned to get rid of it. Recently, however, many people have begun to see sludge as the valuable resource it is, rather than as a waste product that must be disposed of. They recognize that sludge has many potential uses. It can be used as a fuel, livestock feed supplement, soil conditioner, and fertilizer. It has even been used to make bricks to build experimental homes and other buildings. The wide use of sludge depends on efforts

FIGURE 11.21   Greenhouse containing an artificial ecosystem constructed from plants, animals, and microorganisms that remove virtually all of the wastes from sewage. Such systems require few mechanical parts and are cheaper to operate than traditional sewage treatment plants. They also purify water without costly and toxic chemicals used in traditional sewage treatment plants.

to prevent its contamination, especially efforts to ensure that industrial wastes containing heavy metals from factories are not commingled with municipal sewage or are pretreated to remove such pollutants. Let's consider some of the alternative uses for this material:

1. *Fuel.* During primary treatment of municipal sewage, much of the organic material in sewage is sent to a digester, where it is decomposed by bacteria in the absence of oxygen. During this process, an organic fuel called **biogas** is released. Biogas is mainly methane, the principal component of natural gas, a common household and industrial fuel. As noted earlier, biogas

may be used at the sewage plant itself to heat the digester to the temperature required for proper operation and to power generators that provide electricity needed to run the plant. Even after removal from the digester, the treated sludge still has some fuel value left. If dried, it can be burned in an incinerator. The heat can then be recovered and used in industrial processes or to warm buildings.

2. *Livestock feed.* Sludge has a fair content of nutritional proteins and fats and with proper treatment can be converted into a tasty feed supplement for cattle, pigs, and chickens.

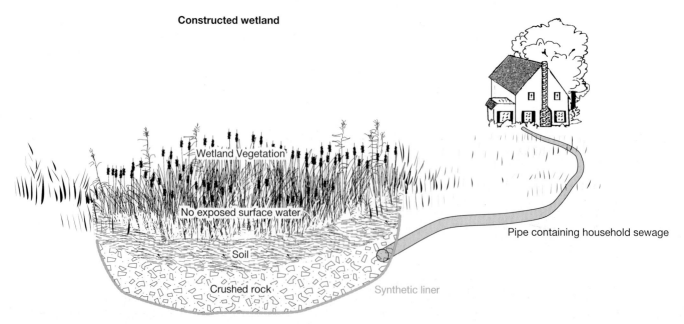

Constructed wetland

Wetland Vegetation

No exposed surface water

Soil

Crushed rock

Synthetic liner

Pipe containing household sewage

FIGURE 11.22   Artificial wetland used to treat sewage from a house. Community-wide systems are also being used successfully in the United States and other countries.

FIGURE 11.23 Composting toilet. This SunMar composting toilet reduces toilet paper and feces to a fluffy organic matter rapidly and without odors. In this system, the tank that holds the waste is located in the basement. A vent pipe allows moisture from urine and odors to escape.

3. *Soil conditioner and fertilizer.* Sludge has considerable value as a fertilizer and soil conditioner because it improves the soil's ability to retain nutrients, reduces erodibility, and promotes the soil's ability to hold oxygen and moisture (Figure 11.24). In the United States, nearly 7 million tons of sludge (dry weight) is produced each year by sewage treatment plants; about 53% of the sewage sludge is returned to the soil or applied to the land. While still watery, sewage sludge may be sprayed directly on land by means of tank trucks or conventional irrigation systems. In Lubbock, TX, and Muskegon County, MI, this method is used extensively. Phosphates and nitrates in the sludge are absorbed by the roots of crop plants, promoting growth and increasing yields. Any disease-causing microorganisms in the wastewater can be destroyed by preheating the waste before disposal. The organic material in the wastewater improves soil structure and increases its ability to absorb moisture and resist erosion. Interestingly, the Metropolitan Sewage District of Milwaukee packages much of its sludge and markets it under the name Milorganite. Much of it is bought for use on lawns and gardens. At least 500 farms in the Milwaukee area regularly receive applications of Milwaukee sewage plant sludge.

4. *Recharge aquifers.* Sewage sludge can also be used to recharge aquifers. For many years, Nassau County in Long Island, NY, dumped its sewage in the Atlantic Ocean. This practice is now illegal. As a result, the county sprays its treated sewage on land that overlies a severely depleted sandstone-limestone aquifer. As the wastewater percolates down through the soil, the impurities are gradually removed, and the aquifer is recharged. The citizens of Nassau County are now taking showers and brewing their breakfast coffee with water that once suspended human waste—a remarkable example of the recycling of a precious resource.

5. *Building materials.* In Washington, DC, stands one of the world's most unusual buildings: It is made in part of sewage sludge! The existence of this construction marvel is based on a technique developed by scientists at the University of Maryland. They succeeded in producing bricks from a mixture of sludge, clay, and slate. These so-called biobricks have no odors and look like ordinary bricks. The 750-square-meter (8,300-square-foot) sludge building was constructed by Washington's Suburban Sanitary Commission out of 20,000 biobricks. The widespread use of biobricks could have multiple environmental benefits, such as reducing the cost of sewage disposal, slowing the rate at which landfills are filling up, and reducing both the soil erosion and the visual pollution caused by the mining of clay.

FIGURE 11.24 Application of sludge fertilizer on a Wisconsin farm.

11.4 Legislating Water Pollution Control

So far, we have examined the major forms of water pollution and discussed how pollution can be reduced through pollution-control devices or, better yet, eliminated through preventive measures, the sustainable option. Water pollution control efforts in the United States and other nations have been inspired by numerous laws and regulations. In the United States, the **Federal Water Pollution Control Act (FWPCA)** of 1972 is one of the most important environmental laws passed by Congress. The ambitious goal of the FWPCA was to make the nation's waters fishable and swimmable by 1985. Toward this end, the law classifies surface waters according to their designated use: (1) drinking water, (2) swimming and fishing, and (3) transportation and agriculture. Most of the nation's surface waters fall into the swimming and fishing category, and pollution controls applied to facilities that discharge into these waters are meant to make such use possible. The nation has 120,000 miles of streams whose designated use is as a supply for drinking water. The transportation and/or agriculture category contains 32,000 miles of streams. The quality of these waters can be somewhat less than those in the swimming and fishing category.

The FWPCA also established minimal water-quality standards for the nation's lakes, rivers, and streams and deadlines for industries and cities to reduce or eliminate their waste discharges. All U.S. cities were required to provide at least secondary sewage treatment. The law established fines of up to $50,000 per day and jail sentences of up to two years for violators.

In 1977, the FWPCA was amended under the name of the **Clean Water Act**. Amendments were also passed in 1981 and 1987. The 1987 amendments required all municipalities to have secondary sewage treatment plants in operation by July 1, 1988. The construction, operation, and maintenance of such plants were supported with $45 billion in federal funding and $15 billion in state and local funding during the period 1972–1986. An expenditure of $18 billion in additional federal funds was authorized by the 1987 amendments for the period 1987–1996.

The Clean Water Act and its amendments require water providers to drastically reduce or even eliminate chemical pollutants from drinking water that might cause serious human illness or death. How effective has the Clean Water Act been?

The most inclusive information on water quality in the United States to date has been provided by the monitoring program of the U.S. Geological Survey. It is called the **National Ambient Stream Quality Accounting Network (NASQAN)**. It includes 501 monitoring stations located on streams distributed throughout the country. In 1995, NASQAN reported that 4% of streams monitored exceeded the standard for phosphorus, and 1% of streams had unacceptable levels of dissolved oxygen, cadmium, and lead. U.S. streams were not so well off when it came to other pollutants. For example, 35% of all streams monitored exceeded the EPA standard for coliform bacteria, up from 25% four years earlier. What many cities have found is that water quality has not improved or has even declined,

despite the construction of expensive treatment plants. Studies have shown that the reason for this is the proliferation of nonpoint sources. Although cities and towns have made remarkable progress in reducing pollution from sewage treatment plants, nonpoint sources have increased as populations have expanded. Nonpoint water pollution has proved to be difficult to control, for reasons discussed earlier in the chapter. Nonpoint pollution has therefore thwarted progress in cleaning up our lakes and rivers.

The Clean Water Act (1987) authorized the expenditure of millions of dollars of federal money by the states to control nonpoint water pollution. In the first five years of this program, the U.S. government spent $270 million. To receive the money, the states were required to match 40% of federal grants with their own money. One of the most popular approaches has been an "effluent trading policy" in which money has been spent to tighten controls on point sources to offset nonpoint pollution entering lakes and rivers. Over the years, hundreds of successful programs have been launched, but much more is needed to eliminate or greatly reduce this growing and serious problem.

The Clean Water Act is currently subject to intense debate, and numerous efforts have been made to weaken it by the Congress during the Bush Administration. Another force that is threatening the law is the current shift from federal to state control. At the state level, clean-water laws are being weakened by industries that oppose the legislation.

Another key piece of legislation aimed at protecting water is the **Safe Drinking Water Act** of 1974. This law, as its name implies, is designed to protect the quality of drinking water. It establishes the EPA as the main regulator of drinking water quality, overseeing state and local governments and other water suppliers. The Safe Drinking Water Act called on the EPA to establish regulations for pollutants in drinking water—that is, standards that tell us how much can be present in our water supplies. It also instructed the EPA to establish programs to protect groundwater drinking water supplies. Implementation of the 1974 law was slow, so in 1986, the U.S. Congress passed a sweeping set of amendments to accelerate the pace at which the EPA was moving. The amendments called for the establishment of regulations for 89 contaminants by 1989. It also called on the EPA to establish regulations to ensure that drinking water from public supplies was filtered and disinfected. It banned the use of lead pipes and lead solder in plumbing systems and established a program to protect areas around wells supplying public drinking water.

Many of these goals have been met. Currently, nearly 80 National Primary Drinking Water Standards have been set. The EPA has also established standards for other pollutants, although compliance with these standards is voluntary.

In 1996, the act was once again amended. The newest amendment established a fund to help communities upgrade outdated drinking water filtration and purification systems. It established a reporting system, too, calling on all water suppliers who serve more than 10,000 customers to report violations of standards and to pinpoint sources of water pollution in their system, which then becomes public knowledge. The amendments also required the EPA to use sound science when establishing drinking water standards and to base decisions on risk and benefit calculations.

Another law that has helped in the battle against water pollution is the **Emergency Planning and Community Right-to-Know Act**, passed by the U.S. Congress in 1986. This law established a nationwide **Toxics Release Inventory (TRI)**, a pollution accounting system. Under the law, major industrial facilities are required to publish data on the levels of pollutants that they have discharged into the air, water, and land or transferred to other sites for incineration, recycling, and disposal. These reports are open to the public via the Internet. TRI data, sometimes showing the release of huge amounts of toxic substances, has been used by local and national activists from national environmental groups to pressure companies to reduce toxic emissions. Many companies took steps to reduce their emissions because of such pressure. Others were astounded at the amount of pollution and the cost of getting rid of it and took steps to reduce emissions for financial reasons.

Watershed Management Plans

With nonpoint water pollution becoming such a major problem and countering many gains in controlling point sources, cities and towns have begun to explore other avenues to protect water quality. One approach is **watershed management**, better management of entire watersheds. Watershed management involves actions by many individuals living and working within a watershed, among them farmers, homeowners, gardeners, city park managers, city street crews, and boaters. Watershed management involves many steps by many individuals with several key goals: to protect, even increase, vegetative cover; reduce impervious surfaces; and reduce nonpoint pollution sources in a watershed. To get an idea of the depth and breadth of this approach, consider a few examples.

In some cities and towns, officials are helping to establish vegetative **buffer zones** along streams, rivers, and lakes. Buffer zones help reduce surface flow into streams, cutting down on flooding and sediment pollution. Vegetative buffer zones may also help remove nitrates, phosphates, and organic pollutants from surface runoff. Naturally occurring bacteria and other microorganisms in the zone decompose or incorporate organic and inorganic nutrients, respectively. Toxic metals are physically filtered out.

In other areas, officials are setting aside more open space, ranging from public parks to undeveloped wildlife habitat. In Boston, officials have preserved land along the Charles River, which runs into Boston Harbor. Consisting of parks and wetlands, this vegetative zone not only preserves the aesthetic qualities of the area and provides recreational opportunities, it also helps protect wildlife habitat. The protected areas reduce surface runoff, thus preventing flooding, erosion, and sediment deposition in the river.

Some cities and towns divert storm water into retention ponds, artificial structures that help reduce flooding by absorbing massive water flows from impervious surfaces such as streets, driveways, parking lots, and rooftops. Water is then allowed to flow into streams at more reasonable rates or soak into the ground, replenishing groundwater.

The City of Austin, TX, promotes responsible actions on the part of citizens through educational programs. Citizens are asked to wash cars on their lawns rather than their driveways, so soaps don't run down into the sewer and into local waterways. Citizens are asked to apply fertilizer and pesticides sparingly and never before a storm, or to use natural, nontoxic alternatives. City officials offer advice on disposal of chemical wastes, including motor oil and antifreeze. They even install "No Dumping, Drains to Creek" frog markers on storm drains to remind citizens that a sewer is not a place to dispose of paints, oils, and other fluids. With boating a popular recreational activity, they promote cleaner, more environmentally sound ways of enjoying the sport. They, for instance, encourage boaters not to throw litter overboard, especially fishing line and plastic six-pack rings. They urge boaters to scrub their boats with a brush and not use soap, or to use phosphate-free and nontoxic soaps.

Many cities and towns have also written formal **watershed protection plans**. They are usually derived after months of work by committees composed of citizens, environmentalists, government officials, businesses, farmers, and other local stakeholders (people who have a stake in the outcome and will be affected by recommendations of the plan). A watershed protection plan outlines ways to protect watersheds for multiple reasons—maintenance of water quality being one of the key ones. Citizens and public officials from Frederick County, MD, worked with the Center for Watershed Protection, a nonprofit group, to develop an extensive set of recommendations to ensure that future development helps to protect existing watersheds. For new development, the plan calls for shorter, narrower streets, fewer and smaller cul-de-sacs, and smaller parking lots to reduce impervious surfaces. It also calls for measures to reduce storm water flow into streams and other surface waters. And it recommends establishment of more community open space, increased vegetated buffers, limited clearing and grading of sites, and ways to enhance native vegetation.

Important as they are, watershed management plans often fail to produce the anticipated results, according to an analysis by the Center for Watershed Management. The reasons for their failure are many. But in general, most are overly ambitious and prescriptive—that is, they outline what needs to be done, but they don't result in the formulation of regulations or mechanisms to fund them. According to the center, many of these plans simply end up on the shelves of city officials, gathering dust with many other plans that have been produced by similar processes. Unless city officials and developers are required to take action and there is money to support actions, it won't happen. That said, we mustn't forget that watershed management programs do note that many existing laws, zoning regulations, and ordinances actually work against watershed protection, permitting and sometimes encouraging activities that increase impervious surfaces, increase erosion, and decrease vegetative surfaces. The true value of watershed management programs, however, will only be achieved when plans are translated into new laws, zoning regulations, and ordinances, and when there are agencies to monitor watershed development. To be successful, these plans will need to produce desired, long-term outcomes of protecting streams and other resources from degradation.

11.5 Pollution of Oceans

Despite the many benefits humans derive from the seas, we have generally viewed oceans as vast and bottomless dumping grounds for domestic, municipal, and industrial garbage and other wastes. Disposal of waste materials in the ocean was convenient and economical, and appeared safe. Since the 1970s, however, increases in scientific information and public awareness of ocean pollution have stimulated important legislation and regulations designed to protect our oceans. The Federal Water Pollution Control Act (FWPCA), discussed previously, sets water quality standards and regulates the discharge of pollution into U.S. waters, many of which flow into the oceans. In addition, the U.S. Congress passed laws that ban or regulate disposal of waste into the ocean. The enforcement of the laws was assigned primarily to the U.S. Coast Guard and the U.S. Army Corps of Engineers. Many other countries took actions as well, as did the international community. The United Nations (UN), for example, set up the International Maritime Organization (IMO), an agency that has developed a number of programs to reduce pollution of the world's oceans. Another UN agency, the International Seabed Authority, was established to regulate mining in international waters. These and other attempts to protect ocean resources are encouraging. However, conflicts over pollution and ocean resources continue today.

Sewage

For decades the neritic zone bordering America's coastline has been used as a dumping ground for sewage sludge, industrial wastes, and even household garbage. The harmful effects are many, as illustrated by the problems at the New York Bight.

The bight is a relatively shallow area over the continental shelf opposite New York Bay. The site was used as a dumping ground for sewage sludge and other wastes for more than 60 years (Figure 11.25). In the early 1970s, for example, more than 7 billion liters (1.8 billion gallons) of sewage treatment plant effluents were dumped into the bight through 130 discharge pipes each year from the New York metropolitan area. More than 16% of this sewage had received no treatment whatsoever. In addition, raw sewage from 23 New Jersey towns was disposed of into the bight. As a result, the bottom of the bight became blanketed with a layer of black sludge covering 105 square kilometers (40 square miles). This "blanket" is contaminated with a large variety of pollutants, from toxic metals and organic compounds (PCBs) to pathogenic viruses and bacteria.

The long-continued dumping of sludge and raw sewage in the New York Bight had many adverse effects on the marine ecosystem:

1. Because of the high BOD of much of the waste, the concentration of dissolved oxygen in the region was often

FIGURE 11.25   Barge hauling refuse from New York City to dumping grounds.

less than 2 ppm. Populations of microscopic algae and crustaceans fell sharply or disappeared altogether. This caused a decline in many commercially valuable species of plankton-eating fish.

2. Some fish suffered from black gill disease, characterized by abnormally dark gill membranes and reduced respiratory function.

3. Toxic metals such as nickel, chromium, and lead reached unusually high levels in some fish.

4. A considerable number of harmful mutations resulting from chromosome damage were observed in young mackerel. Moreover, clam and oyster beds were so highly contaminated with disease-causing microorganisms that they were unfit for human consumption. Apparently the bacteria and viruses were transported from the waste to the clam and oyster beds by shoreward-moving currents.

In 1986, the EPA directed New York City to dump its sewage sludge at a newly designated site at the edge of the continental shelf—170 kilometers (106 miles) from the coast. The city began dumping its sludge at the new site in 1987. In 1988, however, the U.S. Congress passed the **Ocean Dumping Act**, which banned all ocean garbage dumping (by January 1, 1992). The law established severe fines on any community that violated the ban. Today, nearly all ocean dumping that occurs in U.S. waters consists of dredged materials (sediment from harbors, for example); however, other countries still dump sewage sludge and nontoxic industrial wastes into the ocean.

Dredge Spoils

Eighty percent of the waste that has been dumped into U.S. coastal waters is dredge spoil. **Dredge spoil** is sediment (sand, silt, clay, and gravel) scooped from harbor and river bottoms to deepen channels for navigation. More than 400 million cubic yards are dredged annually from U.S. channels and harbors—the equivalent of a four-lane highway 6 meters (20 feet) deep, beginning in New York and ending in Los Angeles. This waste poses an enormous disposal problem. Annually, about 15% of this material, or 60 million cubic yards, is disposed of in the ocean; where to dump can be a contentious decision. The urgently needed dredging of Baltimore Harbor was postponed for 15 years because no agreement could be reached on where to dump the spoil. Much of the spoil generated in the mid-1980s was dumped at 70 different ocean sites.

Unfortunately, about 1 in every 3 tons of dredge spoil is contaminated with both urban and industrial waste, as well as with pollutants resulting from urban and agricultural runoff. These contaminants (PCBs, heavy metals, and so on) eventually enter marine food webs and may harm not only ocean life but humans as well. Under the terms of the **Marine Protection, Research, and Sanctuary Act**, the U.S. Army Corps of Engineers, which does most of the dredging, was charged with finding suitable disposal sites beyond the continental shelf. At such sites, the water is deep enough that most of the pollutants should be greatly diluted, minimizing their adverse effects on the marine ecosystem.

Plastic Pollution

The world's oceans are seriously polluted with plastic—about 10 million tons per year, according to Greenpeace, a nonprofit environmental organization. The enormous volume of plastic debris riding the waves is not appreciated until one hikes along a beach. During a three-hour cleanup of a 260-kilometer (157-mile) stretch of the Texas coast, the following plastic objects were removed: 31,800 bags, 30,000 bottles, 29,000 lids, 7,500 milk jugs, 15,600 six-pack rings, 2,000 disposable diapers, and 1,000 tampon applicators. Even remote islands are not immune from an accumulation of plastic and other litter. The Ducie Atoll in the South Pacific, an uninhabited island 300 miles from the nearest inhabited island and 3,000 miles from the nearest continent, was found littered with human trash in 1991. In a 2.4-kilometer (1.5-mile) stretch of beach, 950 miscellaneous pieces of trash were collected, much of it plastic. Although seemingly harmless, such materials kill 1 million to 2 million seabirds and more than 100,000 whales, porpoises, and seals every year, according to Greenpeace (Figure 11.26). The gut of a sea turtle found dead in Hawaii was jammed with a variety of lethal objects, including golf tees, bottle caps, bags, and imitation flowers. Hundreds of seabirds, salmon, and marine mammals die when they become entangled in discarded fish nets.

Plastic causes wildlife mortality is several ways: After being swallowed, it can be neither digested nor voided, so death is

FIGURE 11.26   Western gull on a California beach, where it is threatened with death by strangulation by a plastic six-pack frame.

caused by blockage of the digestive tract. Plastic entanglement may cause death by drowning. Plastic entanglement also may cause starvation because it prevents marine birds and mammals from searching for or swallowing food.

The sources of these lethal plastic pollutants are many. Every industrialized society lives in a plastic world. Manufacturers in the United States alone annually produce more than 6 million metric tons of plastic. Some of this plastic is discarded into streams by humans and then carried downstream to the ocean or dumped directly into the ocean from fishing boats and other commercial or recreational vessels or from garbage barges. The National Academy of Sciences once reported that more than 5 million plastic containers are tossed overboard from oceangoing vessels every day.

Most plastic cannot be broken down by bacteria. In other words, plastic is nonbiodegradable. Another problem is that most plastic items are quite buoyant. These characteristics make it possible for a golf tee from a Seattle golf course to be ingested by a seabird in the South Pacific.

The amount of plastic floating and bobbing on the global seas will certainly increase. After all, population continues to grow, and therefore so will demand for products. Commercial fishers lose more than 136,000 metric tons of plastic lines and nets annually. In the North Pacific alone, fishermen set out more than 32,000 kilometers (20,000 miles) of plastic nets nightly. Within a year, more than 4,800 kilometers (3,000 miles) of netting is lost, forming a considerable threat to marine life.

Control of Plastic Pollution How can plastic pollution be controlled? At present, dumping in the ocean is regulated by the **London Convention** (1972), an agreement signed by more than 85 nations. Additions to the agreement regulate disposal from all trash-hauling ships. In the United States, this convention was implemented by the Ocean Dumping Act.

In 1996, much stricter guidelines were drawn up. This new agreement, known as the Protocol, replaces the London Convention of 1972. Although the purpose of the Protocol is similar to that of the London Convention, the Protocol is more restrictive. The agreement calls for a "precautionary approach," which implies that all dumping is prohibited unless explicitly permitted. It also prohibits the incineration of wastes at sea and the export of wastes to other countries where they will be dumped or incinerated at sea. A transitional period of five years was created to allow countries to switch to the new and stricter guidelines.

Another international law that controls plastic pollution is the 1973 **Marine Pollution Convention**, or MARPOL Act. Annex V of the act bans the dumping of plastic by all ships other than trash ships. In 1987, after several years of deliberation, Annex V was eventually ratified by the United States. As a result, beginning in December 31, 1988, it became illegal for the ships of the 15 signatory nations and for the ships of any nation plying their waters to dump plastic at sea. Since the 15 signatory nations accounted for over 50% of the gross tonnage of the world's commercial ships, this agreement should help reduce plastic pollution. The Coast Guard enforces Annex V for the United States.

Another way to reduce this problem is to recycle more plastic materials. Some progress is being made. For example, in the United States, plastic soft-drink bottles are reprocessed into paintbrushes, stuffing, and industrial straps. Some discarded plastic is recycled into building materials, such as deck "lumber" and carpeting.

Perhaps the ultimate solution to plastic pollution is at the source of the plastic—the manufacturing process. Some manufacturers, for example, have recently developed a type of photodegradable plastic that will disintegrate when exposed to ultraviolet light from the sun. At least 11 U.S. states have laws that require photodegradable plastic in some products, such as the rings for soft drink and beer six-packs. A few manufacturers in the United States, Canada, and Italy are now producing biodegradable plastic bags. Should photo- or biodegradable plastics come into mass production, this potent threat to marine life may gradually disappear, assuming the breakdown products are not harmful. In recent years, some companies have begun to produce biodegradable plastic made from chemicals extracted from corn. Some cities and states have banned the use of plastic shopping bags altogether.

Oil Pollution

Oil has always polluted the sea, seeping naturally through cracks in the ocean floor. This naturally occurring oil, however, is of little concern, because the sources are widely distributed and contribute only 9% to the total oil input annually. The concern for most conservationists is oil from human activities, which accounts for the remaining 91% of the annual input (Figure 11.27).

Oil Tanker Spills One of the most dramatic and newsworthy sources of oil in the oceans is tanker spills, contributing about 5% of the oil that enters the ocean each year (Figure 11.28). Although their contribution is small, oil spills can have a major impact because oil tankers usually run aground close to shore, often biologically sensitive coastal shorelines.

One of the largest tanker spills in human history occurred on March 17, 1978, when the *Amoco Cadiz* ran aground 2 kilometers (1.2 miles) off the coast of France. Efforts to stop or contain the spill were unsuccessful. The ship broke apart, and high winds and violent seas made it impossible to transfer oil to other tankers. As a result, the entire cargo of 228 million liters (60 million gallons) of crude oil was released from the tanker. Oil polluted the waters and 198 kilometers (124 miles) of coastline. Fish and seabirds died in the thousands, and oil ruined the beauty of the coastal area, affecting the economy of the shore-based villages. A number of citizens' lawsuits were filed against the Amoco Oil Company, owners of the tanker. In 1988, almost ten years after the spill,

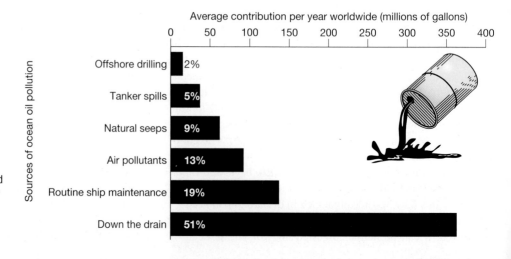

FIGURE 11.27 Sources of oil pollution in the ocean. Most of the oil comes from inland sources via drain and sewer pipes and river runoff. Other human-related sources are tanker spills, routine ship maintenance, air pollution, and offshore oil well accidents.

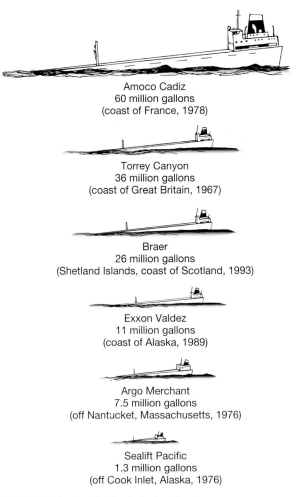

Amoco Cadiz
60 million gallons
(coast of France, 1978)

Torrey Canyon
36 million gallons
(coast of Great Britain, 1967)

Braer
26 million gallons
(Shetland Islands, coast of Scotland, 1993)

Exxon Valdez
11 million gallons
(coast of Alaska, 1989)

Argo Merchant
7.5 million gallons
(off Nantucket, Massachusetts, 1976)

Sealift Pacific
1.3 million gallons
(off Cook Inlet, Alaska, 1976)

FIGURE 11.28 Sample of major tanker oil spills.

FIGURE 11.29 The oil tanker *Exxon Valdez* ran aground in Prince William Sound, AK, March 24, 1989.

the company was ordered by the courts to pay millions of dollars in compensation.

On March 24, 1989, citizens of the United States were horrified by news of a 40-million-liter (11-million-gallon) oil spill in Alaska's Prince William Sound near the port of Valdez (Figure 11.29). The supertanker *Exxon Valdez* ran aground on a reef in the sound, releasing much of its oil into the pristine, biologically rich waters. The oil slick spread quickly and, by the end of the summer, had polluted 2,300 kilometers (1,400 miles) of shoreline (Figure 11.30). Thousands of seabirds and otters perished in the oil, which formed a layer 1 meter (3.3 feet) thick in some places (Figure 11.31). Making matters worse, the spill occurred only two weeks before migrating flocks of waterfowl arrived. Many birds spend the summer in the waters of Prince William Sound; others merely stop there to feed and rest on their way to the Arctic tundra, where they breed.

The *Valdez* spill was not the largest in history but probably will go down as one of the most costly, economically and environmentally. The damage can be attributed to four key factors. First, the spill occurred in the relatively protected waters close to land. Second, cleanup was delayed for several days. The special cleanup force stationed at Valdez had been all but abandoned by the oil companies operating there. Consequently, too

few oil skimmers were close at hand. Third, the waters of the sound are cold, which retarded the biological degradation of the oil. Fourth, the waters were extraordinarily rich in sea life.

One of the most recent oil spills occurred in San Francisco. A container ship bound for South Korea released 224,000 liters (58,000 gallons) of heavy fuel oil into San Francisco bay after it struck a tower supporting the San Francisco–Oakland Bay Bridge in heavy fog on the morning of November 7, 2007. Oil quickly spread throughout the bay and forty miles up the California coastline, forcing officials to close nearly a dozen beaches in San Francisco and Marin County. The governor immediately suspended fishing for human consumption. Area crab fishermen voted for an immediate delay of the official harvesting season, which was scheduled to begin that week.

Although it is too early to determine the long-term ecological impact of the spill, the oil killed at least 2,000 sea birds, including brown pelicans, a federally endangered species, and marbeled murrelets, listed as a threatened species under federal law and an endangered species under California law.

Ecologists are concerned the spill could affect fish and the fishing industry for years to come. Herring, the Bay's only commercially fished species, spawns in its waters during the time of the spill. The spill could also threaten steelhead and Chinook salmon that migrate through the waters of their bay on their way to spawning grounds in two major rivers that empty into the bay. Scientists also worry that it could affect longfin smelt, a fish species whose population reached an all-time low in 2007.

Scientists are also concerned about the effect of oil on plant life. Oil that accumulated on eelgrass, an underwater grass in tidal marshes, could damage fish that inhabit these areas and birds that feed in them.

FIGURE 11.30   An Exxon worker cleans oil-covered rocks with a hot-water spray in the aftermath of the massive oil spill.

FIGURE 11.31   An oil-soaked cormorant registering its protest against the atrocious oil spill off Alaska's coast in the once-pristine waters of Prince William Sound. Thousands of birds and mammals died. Volunteers cleaned many birds and mammals that were caught in the oil, but the animals' prospects for survival were dim.

Routine Ship Maintenance   Nearly 20% of the oil released into the ocean each year comes from routine operations and ship maintenance such as loading and discharging oil, tank cleaning, oil ballast discharge, and other operations. Millions of these routine operations spill just a few liters each, but together amount to millions of liters of oil.

Offshore Oil Well Accidents   Oil is also released during the operation of offshore oil wells and by accidents. In 1969, a major oil well off the coast of Santa Barbara, CA, accidentally released thousands of gallons of oil because of a faulty drilling technique. However, the spill was a mere grease spot compared to the 140-million-gallon oil well blowout that occurred in the Bay of Campeche off Mexico's east coast in 1979. Oil escaped from the well for several months, threatening marine life along the Texas shore several hundred kilometers to the north. The world's largest oil spill occurred as a result of fighting in the Persian Gulf War during 1991. The spill involved a combination of wells, terminals, and tankers that amounted to a 926-million-liter (240-million-gallon) disaster in the Persian Gulf and surrounding area.

Offshore drilling accidents contribute less oil than tanker spills on an annual basis but, as in these examples, can be equally, if not more, devastating. American oil companies alone drill about 1,300 offshore oil wells a year (Figure 11.32). Four thousand new offshore wells are drilled worldwide each year. Careful controls on these wells are important to reduce the likelihood of accidents.

Land Sources of Oil Pollution   Surprisingly, over half of the oil released into the ocean each year comes from terrestrial sources in both inland and coastal communities. The major sources are service stations, motor vehicles, factories, and runoff from parking lots and highways. This oil finds its way to the ocean via storm and sewage drains and as runoff in rivers and streams.

For many years, ordinary citizens changing their own oil dumped it into sewers that empty into rivers and eventually flow to the sea. Although three or five quarts of oil doesn't seem like much, billions of quarts released this way add up.

Air Pollution   Airborne hydrocarbons released from factories, service stations, and vehicles also rain down on our oceans. You may be contributing to this problem yourself. Most of us are. How? When you fill up your gas tank at a self-service pump, notice the pungent odor of evaporating

FIGURE 11.32 Offshore oil platforms. These oil wells in the Gulf of Mexico are just a handful of the many rigs operating near sensitive wetlands and coastlines.

gasoline. Obviously, not all of the gasoline goes into your tank. Some of it escapes into the air. Some unburned gas also escapes from the exhaust pipe of your car and becomes airborne. Evaporation of petroleum also occurs at thousands of industrial plants throughout the world. Combined, oceangoing vessels burn about 1.5 million metric tons of diesel fuel per day, spewing airborne effluent. As just noted, these pollutants wash from the sky, polluting the oceans with at least 20 million metric tons of airborne petroleum hydrocarbons annually, contributing 13% of the oil input.

Adverse Effects of Oil Pollution   Precisely how a given oil spill will affect marine life is difficult to predict. The effects depend on a number of factors, such as the amount and type of oil (crude or refined) and proximity of the spill to biologically sensitive areas. Season of the year and weather, as well as the ocean currents and the wind velocity, also determine the effects.

Oil spills have their greatest impact when they occur close to shore, which as noted earlier, is the most common site of tanker spills. Over the years, the heaviest influx of oil occurs in the neritic zone near the continental margins—the zone where virtually all of our shellfish (oysters, lobsters, and shrimp) and over half of our commercial fish crop are produced (Figure 11.33). Oil can kill or contaminate sea life. Shellfish beds can become contaminated with toxic chemicals from oil, causing millions of dollars in lost revenue to local residents who make their living from harvesting the sea. Many seabirds perish in oil. The number of seabirds killed annually worldwide is enormous. In a single winter, more than 250,000

murres, eiders, and puffins were destroyed by oil pollution. In 1988, thousands of these birds were destroyed by oil spills in the North Sea.

Oil spills can have less obvious effects, too. For example, several of the hydrocarbons in crude oil mimic chemicals used by marine animals to guide them during mating, feeding, homing, and migrating. Flooding the ocean with pseudosignals from oil spills might alter the behavior of marine animals, disrupting vital life functions.

Crude oil is not just a single compound but a complex mixture of dozens of different hydrocarbons such as benzopyrene, an acknowledged cancer-inducing chemical, or carcinogen. These carcinogens may be concentrated in marine organisms such as shrimp, lobsters, and fish, and may eventually be consumed by humans.

Oil spills also devastate beaches and the recreation industry. Cleanups can be extremely costly.

Control of Oil Pollution   In the U.S, the **Oil Pollution Act of 1990** was passed in response to public outcry over the devastation caused by the *Exxon Valdez* grounding. The law established provisions for new tankers' structural design, including double-hulled construction; the retirement or phaseout of existing tanker fleets; improved spill response strategies; and more rigorous inspection systems. It also more clearly defined financial responsibility, compensation, and liability for oil spills. With international cooperation, the number of oil spills from tankers has been reduced. But tighter controls and improved enforcement methods are still needed.

Controlling oil pollution after a spill has occurred can involve several strategies:

1. *Physical cleanup.* Oil that washes ashore may be cleaned up manually or with machines. In the *Valdez* spill, for example, workers sopped up oil with absorbent pads. Other workers scoured the beaches with hot water, washing the oil back into the sound, where it was picked up by oil skimmers, vacuum-type devices that skim the oil off the surface and empty it into barges. Early in the spill, planes dropped absorbent material on slicks, which were later picked up by boats. In many spills, straw has been used to sop up oil that has washed ashore.

2. *Dispersion.* While authorities first attempt to clean up as much oil as possible after a spill, chemical dispersants are used to break up the spill if the weather is too rough or the oil slick is approaching ecologically sensitive areas, such as coastal wetlands and coral reefs. Chemical **dispersants** consist of solvents and another group of chemicals known as surfactants. Together, like a detergent, they cut through oil, breaking it into tiny droplets that readily mix with water. This, in turn, allows the oil slick to disperse much more quickly, preventing damage. Unfortunately, recent studies by Baruch Rinkevich of the National Institute of Oceanography in Haifa, Israel, and his colleagues show that chemicals used to disperse oil spills can damage nearby coral more than the oil itself.

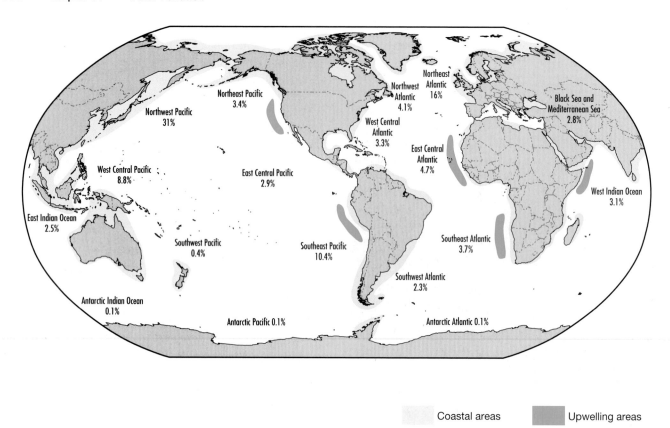

FIGURE 11.33 Distribution of the world's fisheries. Coastal areas and upwelling areas together supply over 99% of world fish production. The deep ocean forms 90% of the ocean area but accounts for only 1% of the fish catch, if upwelling areas are excluded.

3. *Decomposition of oil by bacteria.* Two Israeli scientists developed a technique that uses bacteria to break down oil. Theoretically, an oil slick could be "seeded" by helicopter with the bacterial powder, accelerating the rate of decomposition and ultimate oil slick breakup. In the process of breaking down the oil, the bacterial population multiplies rapidly, to the point where it can be used as protein feed for livestock. The Israeli scientists estimate that hundreds of tons of animal food could be obtained from a spill. In the *Valdez* spill, Exxon applied oil-degrading bacteria to some beaches and found that the bacteria accelerated the destruction of oil.

4. *In situ burning.* **In situ burning** is a technique that can significantly reduce the amount of oil on the water, thereby minimizing adverse effects to the local environment. The controlled burn can be economical in certain instances, rapidly removing oil from the environment and reducing the need for a large amount of equipment and labor. The advantages of this method must be evaluated on a case-by-case basis. In situ burning may not be advisable in some situations. Disadvantages include the generation of highly visible smoke, short-lived air pollution that increases health risks to human and animal populations downstream, localized temperature elevations that can harm or kill wildlife, and unknown long-term effects. The *New*

Carissa tanker grounding off the coast of Coos Bay, OR, in 1999 was an unusual event in which mitigation to avoid an oil spill included burning the ship's oil on board and sinking the ship out at sea.

In most oil spills, a combination of these techniques is used. Appropriate methods should be chosen based on the unique conditions of each site. All of these methods are time-consuming and costly. Many of them are rather ineffective. Oil washed off the beaches near Valdez, for example, was spread back on by the rising tides. Cleanup methods also generate substantial amounts of waste that must be disposed of safely to avoid creating problems elsewhere. Given their shortcomings, these methods are clearly no substitute for prevention. To remind the public of the need for prevention, one environmental group sells T-shirts with the motto, "An ounce of prevention is worth 11 million gallons of cure."

The fact remains that the majority of oil and other sources of ocean pollution is generated from the result of small but numerous individual land activities. Perhaps the best strategy for ocean pollution control is one that emphasizes and increases public education and awareness of everyday activities that contribute to this insidious form of ocean resource degradation. Educate yourself, and then share your knowledge with others. The following list is a good beginning:

• Never pour oil, engine fluids, cleaners, or household chemicals down storm drains or sinks.

- Find government and industry-sponsored oil collection and recycling programs in your community, and use them.
- Repair automobile and boat engine leaks immediately.

11.6 A World View of Water Pollution

Water pollution cleanup in the more-developed countries has been moderate and, in many nations, is probably on a par with progress in the United States. The water pollution problems in the less-developed countries in general are much worse than in the United States. There are several reasons for this: (1) lack of properly educated and technically trained personnel, (2) lack of funding for construction of waste treatment plants, (3) lack of tough pollution-control legislation, and (4) lack of enforcement of such laws, if indeed they do exist. South American countries, for the most part, have relatively safe drinking water. However, many streams are seriously polluted with runoff from lead, zinc, and silver mines. As deforestation intensifies on this continent, river contamination with pesticides, fertilizer, and sediment is expected to increase accordingly. In Mexico, the drinking water has such high counts of infectious bacteria that American college students studying there have been advised to boil the water before drinking it, lest they come down with diarrhea, fever, chills, and nausea, a complex of symptoms dubbed Montezuma's revenge. The scarcity of safe drinking water is even greater in Africa. For example, in rural Guinea, only 1 of every 50 people has access to it! Less than 10% of the rural populations of Madagascar, Mali, Sierra Leone, and Zaire have good drinking water available. In Pakistan, most of the human diseases, such as typhoid, diarrhea, dysentery, and infectious hepatitis, are caused by microorganisms that have contaminated public water supplies. In India, the Yamuna River receives 200 million liters (54 million gallons) of untreated sewage from New Delhi every day. As a result, the coliform count in this stream is an almost unbelievable 24 million per 100 milliliters. (Recall that 200 per 100 milliliters is the standard for swimmable waters in the United States.) In Malaysia, 42 major rivers have been declared "ecological disasters," since they are virtually unable to support desirable aquatic life. Seventy percent of some stretches of Manila's Pasig River consists of untreated sewage.

In 1980, the UN launched the **International Drinking Water and Sanitation Decade** (1981–1990). The major objective of this program was to make all nations, especially the less-developed countries, acutely aware of the importance of safe drinking water in the fight against disease. The ambitious goal of the program was to supply an additional 500,000 people with drinkable water every day of the decade! Unfortunately, this goal was not attained, but in the United States and other countries, efforts are still under way to improve water quality—a goal that remains elusive today, in large part because of the continued increase in world population, resource demand, and industrial output. These three forces make it more important than ever to seek sustainable solutions—measures like erosion control and pollution prevention—to avoid creating problems in the first place.

Summary of Key Concepts

1. Water pollution can be defined as any contamination of water that lessens its value to humans and nature and may cause disease and death.

2. Two broad classes of water pollution are point source and nonpoint source pollution.

3. Point pollution has its origin in specific, well-defined sources such as the discharge pipes of sewage treatment plants and factories. Nonpoint pollution stems from widespread sources such as the runoff from agricultural lands or urban areas. Point sources are easier to control than nonpoint sources because they are so identifiable and are the source of concentrated releases of pollutants.

4. Pollution can also be classified by type. The major categories include sediment, inorganic nutrients, thermal pollution, disease-producing organisms, toxic organic compounds, heavy metals, and oxygen-demanding organic wastes.

5. Sediment comes from natural soil erosion and from erosion occurring on improperly managed and maintained farms, forests, construction sites, and the like.

6. Sediment fills up reservoirs, damages hydroelectric plants, clogs irrigation canals, interferes with barge traffic along major rivers such as the Mississippi, destroys fish spawning grounds, reduces the photosynthetic activity of aquatic plants, and necessitates the costly filtration of water.

7. Better land management can help reduce sediment pollution, with numerous biological and economic benefits.

8. Inorganic nutrients include the phosphates and nitrates that arise from a variety of natural and human sources, such as septic tanks, sewage treatment plants, and agricultural runoff.

9. Nitrates and phosphates contribute to a phenomenon called natural eutrophication, a slow process of nutrient buildup in lakes that occurs over a period of thousands of years. Human activities can accelerate the buildup of nutrients in waterways, leading to cultural or accelerated eutrophication of lakes, and may age a lake 25,000 years in only 25 years.

10. Many factories and power plants use surface waters to cool various processes. This heated water is often released back into the source from whence it came, creating thermal pollution.

11. Thermal pollution has multiple adverse effects. It kills fish because it lowers the concentration of dissolved oxygen. It also kills fish directly if the temperature of the water climbs above the range of tolerance. Thermal pollution causes a shift in populations of microorganisms from the beneficial diatoms to undesirable blue-green algae. It also interrupts migration and enhances disease-causing organisms. Thermally enriched water, however, may be put to a variety of beneficial uses such as aquaculture.

12. Surface waters and groundwater may become contaminated with disease-causing organisms, especially bacteria and viruses. Because it is costly to measure the many possible pathogens, water quality officials monitor fecal contamination and the potential for the presence of pathogens by measuring coliform bacteria, common but largely harmless bacteria found in the human intestine and in feces. A high coliform count indicates that the water

sample may contain high levels of other microorganisms from human or livestock feces, some of which may be capable of causing disease. Research, however, shows that fecal coliform levels may not be a sufficient indicator; the EPA is now promoting the use of an additional bacterium, fecal streptococcal bacteria, to monitor the safety of water.

13. Human society depends on a large number of potentially toxic organic compounds such as pesticides. These compounds can contaminate ground and surface waters. Groundwater contamination is especially troublesome, because so many people throughout the world depend on groundwater for drinking water and because, once contaminated, aquifers remain polluted. They aren't naturally purified like rivers.

14. Sources of groundwater pollution from toxic organic compounds are (1) industrial landfills and lagoons, (2) municipal landfills, (3) septic tanks, (4) underground storage tanks for gasoline and other chemicals, (5) contaminated industrial sites, and (6) injection wells.

15. Toxic organic compounds cause many problems, depending on the chemical. Health effects include skin and eye irritation; brain and spinal cord damage; interference with normal kidney, liver, and lung function; cancer; and genetic mutations.

16. Plasticizers, widely used toxic organic chemicals, are also present in drinking water that has been stored in plastic water bottles and in food and beverages in plastic and metal containers. These chemicals have a wide range of effects, but one of the most significant is their effect on sexual development in humans and aquatic species, especially fish.

17. Drugs such as blood pressure medication, cholesterol-lowering medications, heart medication, and hormones in birth-control pills are also ending up in our waterways and may have a profound effect on fish and other aquatic organisms. These chemicals are generally not removed by sewage treatment plants.

18. Heavy metals such as mercury and lead come from many different sources, including coal-fired power plants and garbage incinerators. Most are toxic because they interfere with normal enzyme function.

19. Lead is one of the most troublesome of all heavy metals. It comes from solder used to join copper pipes and from old lead pipes in our homes, among other sources. Fortunately, newer types of pipe (copper and PVC) and new lead-free solder are being used in new construction.

20. Lead impairs mental development permanently. The EPA estimates that the IQ of more than 140,000 American children has been reduced by five points because of lead in their drinking water.

21. Lead in drinking water is responsible for at least 680,000 cases of high blood pressure in American males.

22. Oxygen-demanding organic wastes come from (1) fruit- and vegetable-processing industries, (2) cheese factories, (3) creameries, (4) distilleries, (5) pulp and paper plants, (6) slaughterhouses, (7) bakeries, and (8) natural sources.

23. The discharge of organic waste reduces the dissolved oxygen in streams and lakes and is lethal to many species of fish.

24. Sewage treatment plants primarily remove sediment, inorganic nutrients, and oxygen-demanding organic wastes.

25. Primary treatment of sewage is mainly a physical process in which solids are removed by sedimentation.

26. Secondary treatment of sewage is primarily a biological process in which organic wastes are decomposed by bacterial action. This process removes organic material and some of the nitrogen and phosphate, the inorganic nutrients.

27. The bacterial decomposition of organic waste during secondary treatment can be accomplished either by the activated sludge process or by trickling filters.

28. Tertiary sewage treatment, which is rather expensive, removes much of the nitrogen and phosphorus from the waste.

29. State and national laws promote measures to remove pollution from waste streams and control nonpoint water pollution.

30. In the United States, the most important federal water pollution control law is the FWPCA of 1972 (now called the Clean Water Act) and its amendments (1977, 1981, 1987). Many other countries have similar laws patterned after this one. The Safe Drinking Water Act helps ensure that water sent to our homes is safe for consumption.

31. Toxic release inventory data have also been useful in the United States in pressuring major corporations to reduce their emissions of toxic substances into the air, water, and the soil.

32. Water pollution controls help reduce pollution, but in many instances, water quality in and around major metropolitan areas has not improved because of an increase in nonpoint water pollutants resulting from the expansion of populations and resultant land development. To address this problem, many cities and towns have begun to adopt watershed management.

33. Watershed management involves steps taken by a wide variety of people, including government officials, homeowners, gardeners, city park officials, and farmers. They are designed to reduce the sources of pollution in a watershed and minimize the disturbance of vegetation to reduce surface runoff and hence reduce the flow of pollutants into streams and lakes. Storm water retention ponds and buffer zones are two of many measures used in watershed management.

34. Water pollution is common in all countries, rich and poor. Although water pollution control efforts in more-developed nations have been remarkable, there is much to be done to prevent further deterioration. In less-developed countries, the challenge is even greater because of a lack of funds and technical expertise.

35. For decades, the neritic zone bordering our nation's coasts has served as a dumping ground for raw domestic sewage, sewage sludge, medical wastes, industrial wastes, and dredge spoils.

36. Waste dumping at the New York Bight, one of many dumping grounds, reduced levels of dissolved oxygen, caused declines in plankton and plankton-dependent fish, resulted in disease epidemics among fish, contaminated fish with toxic metals, and resulted in high rates of mutations and cancer in fish.

37. Plastic pollution causes wildlife mortality by blockage of digestive tracts, entangling-induced drowning, and entangling-induced starvation.

38. A U.S. law that helps control plastic pollution is the Ocean Dumping Act, which bans plastic dumping from trash ships.

39. Today, nearly all ocean dumping that occurs in U.S. waters consists of dredged materials; however, other countries still dump sewage sludge and nontoxic industrial wastes into the sea.

40. The main sources of oil in the marine environment are (1) natural seeps, (2) oil well blowouts, (3) tanker spills, (4) routine tanker and ocean vessel maintenance, (5) inland disposal via river and pipeline runoff, and (6) air pollution.

41. The often dramatic and highly publicized oil tanker spills contribute only 5% of the total oil inputs to the ocean each year.

42. The majority of ocean oil pollution is generated from the small but numerous individual inland activities.

43. Oil pollution adversely affects the marine ecosystem by (1) reducing photosynthetic rates in marine algae; (2) concentrating chlorinated hydrocarbons such as pesticides; (3) contaminating human food chains with carcinogens such as benzopyrene; (4) disrupting chemical communication in marine organisms, which adversely affects such activities as feeding, reproduction, and escape from predators; (5) killing animals; and (6) causing long-term effects such as cancers due to chronic exposure to low levels of oil.

44. The Oil Pollution Act of 1990 requires (1) improved structural designs for new tankers; (2) phaseout of the existing tanker fleet; (3) financial responsibility, compensation, and liability for spills; and (4) improved spill response strategies and inspection systems.

45. Strategies for cleanup of an ocean oil spill include (1) physical cleanup, (2) dispersion of oils by chemical agents, (3) decomposition by oil-eating bacteria, and (4) in situ burning.

46. The best strategy for ocean pollution control may be one that emphasizes and increases public education and awareness of everyday activities that contribute to this insidious form of ocean resource degradation.

Key Words and Phrases

Accelerated Eutrophication	Coliform Bacteria
Activated Sludge	Combined Sewers
Algal Bloom	Cooling Tower
Biochemical Oxygen Demand (BOD)	Cross-Media Contamination
	Cultural Eutrophication
Biodegradable	Damage Zone
Biogas	Detergent
Biological Magnification	Dispersant
Biological Oxygen Demand (BOD)	Dissolved Oxygen
	Dredge Spoil
Biomass	Dry Cooling Tower
Blue-Green Algae	Emergency Planning and Community Right-to-Know Act
Buffer Zone	
Chlorination	
Clean Water Act	Eutrophic Lake
Cold-Blooded Animal	Eutrophication

Federal Water Pollution Control Act	Oligotrophic Lake
	Output Control
Feedlot	Oxygen Sag
Great Lakes Water Quality Agreement	Oxygen-Demanding Organic Waste
Groundwater	Pathogen
Heavy Metals	Plasticizer
Holding Ponds	Point Source Water Pollution
Hormone Disruptors	Pollution-Control Devices
Hydroseeder	Pollution Prevention
Infectious Hepatitis	Primary Sewage Treatment
Injection Well	Recovery Zone (of a River)
Inorganic Nutrient Pollutants	Safe Water Drinking Act
Input Control	Secondary Sewage Treatment
In Situ Burning	Sediment
International Drinking Water and Sanitation Decade	Septic Tank
	Settling Tank
International Joint Commission	Sewage
	Sewage Treatment Plant
Landfill	Sludge
Leach Field	Sludge Digester
Lead Poisoning	Sludge Worms
Limiting Factor	Stormwater Management Program
London Convention	
Marine Pollution Convention	Tertiary Sewage Treatment
Marine Protection, Research, and Sanctuary Act	Thermal Plume
	Thermal Pollution
Mercury	Throughput Control
Mesotrophic Lake	Toxic Organic Chemicals
National Ambient Stream Quality Accounting Network (NASQAN)	Toxics Release Inventory (TRI)
	Trickling Filter
	Water Pollution
Natural Eutrophication	Watershed Management
Nonpoint Source Water Pollution	Watershed Protection Plans
	Wet Cooling Tower
Ocean Dumping Act	Zone of Decline
Oil Pollution Act of 1990	

Critical Thinking and Discussion Questions

1. Define water pollution.

2. What is the difference between point source and nonpoint source water pollution? Give some examples of each. Which ones are easiest to address? Why?

3. List the seven basic types of water pollution.

4. List five differences between oligotrophic and eutrophic lakes.

5. Distinguish between natural and cultural eutrophication.

6. Because photosynthetic levels are high in eutrophic lakes, you might suppose that the water would contain a relatively large amount of dissolved oxygen. Does it? If not, why not? Discuss your answer.

7. Using your critical thinking skills, discuss the following statement: "Eutrophication is a natural process, so there's

no need to worry about human pollutants that contribute to this phenomenon."

8. List five adverse effects of algal blooms.

9. Describe the effects of a high level of BOD waste on the aquatic life of a stream.

10. Using your critical thinking skills and your knowledge of water pollution and ecology, discuss the following statement: "Lakes and rivers have naturally occurring bacteria that can cleanse water, removing organic wastes. Pollution from sewage treatment plants can therefore be released into them without harm."

11. Roughly what levels of dissolved oxygen does a particular stretch of stream have if one of the dominant organisms found in it is (a) a carp, (b) a trout, (c) a sludge worm, (d) a mayfly larva?

12. What does the term *biological oxygen demand* mean?

13. Some manufacturers believe that thermal pollution should really be called thermal enrichment. Do they have a case for such a change in terminology? Discuss your answer.

14. Why are coliform bacteria levels monitored in lakes and streams?

15. Describe four important methods for controlling sediment pollution.

16. Summarize the main benefits derived from (a) primary sewage treatment, (b) secondary sewage treatment, and (c) tertiary sewage treatment.

17. Using your knowledge of ecology and water pollution, discuss the following statement: "Groundwater contamination is more serious than contamination of surface waters."

18. What is a plasticizer? Give some examples. What impacts do they have? How are we exposed to them? Are they harmful to aquatic organisms? Why or why not?

19. Name six heavy metals that are toxic to humans when ingested.

20. What are the possible sources of lead in drinking water in a home?

21. Discuss the effect of plastic pollution on marine life.

22. List five sources of the oil that pollutes the oceans, and explain the contribution of each to the total oil input each year.

23. Discuss five adverse effects of oil on marine organisms.

24. Where should the United States focus efforts on controlling ocean pollution? Why?

25. What can you do to reduce the amount of ocean pollution?

Suggested Readings

Bai, X., and P. Shi. 2006. Pollution Control in China's Huai Basin: What Lessons for Sustainability? *Environment* 48(7): 22–38. An in-depth study of water pollution and water pollution control in the most rapidly industrializing nation in the world.

Canby, T. Y. 1991. After the Storm. *National Geographic,* August: 2–33. Account of the devastation on the ground, in the water, and in the air from the largest oil spill in the

world's history as a result of the fighting in the Persian Gulf War.

Carey, J. 1996. Lessons from Loons. *National Wildlife* 34(5): 12–19. An excellent look at the effects of water pollutants on wildlife.

Cheremisinoff, P. N. 1993. *Water Treatment and Waste Recovery: Advanced Technology and Applications.* Englewood Cliffs, NJ: Prentice Hall. Authoritative survey of the methods of water purification.

Doppelt, B., et al. 1993. *Entering the Watershed: A New Approach to Save America's River Ecosystems.* Washington, DC: Island Press. Good overview of watershed protection.

Frankl, E. 1995. *Ocean Environmental Management: A Primer on the Role of the Oceans and How to Maintain Their Contributions to Life on Earth.* Englewood Cliffs, NJ: Prentice Hall. Overview of current environmental problems relating to ocean resources and discussions of alternatives for sustainable use.

Garelik, G. 1996. Russia's Legacy of Death. *National Wildlife* 34(4): 36–41. A telling story of water pollution and other environmental atrocities caused by lax environmental enforcement.

Gleick, P. H. 2000. *The World's Water 2000–2001.* Washington, DC: Island Press. A vast amount of information on the world's water resources, including waterborne diseases.

Hearne, S. A. 1996. Tracking Toxics: Chemical Use and the Public's Right to Know. *Environment* 38(6): 4–9, 28–34. A description of an effective tool in tracking water pollution.

Klucas, G. 2004. Leadville: The Struggle to Revive an American Town. Washington, DC: Island Press. A look at the efforts to clean up and revive a town after years of mining and pollution.

Krimsky, S. 2001. Hormone Disruptors: A Clue to Understanding the Environmental Causes of Disease. *Environment* 43(5): 22–31. An important article about an important issue.

Platt, R. H., P. K. Barten, and M. J. Pfeffer. 2000. A Full, Clean Glass? Managing New York City's Watersheds. *Environment* 42(5): 8–20. Useful case study for those interested in learning more about watershed management.

Platt, R. H. 2006. Urban Watershed Management: Sustainability, One Stream at a Time. *Environment* 48(4): 26–42. A must read for anyone interested in watershed management.

Riley, A. L. 1998. *Restoring Streams in Cities: A Guide for Planners and Citizens.* Washington, DC: Island Press. Examination of a host of sustainable approaches to protecting watersheds and rivers.

Sampat, P. 2001. Uncovering Groundwater Pollution. In *State of the World 2001,* ed. L. Starke. New York: W. W. Norton. Valuable information on the state of groundwater pollution in the United States.

Satterthwaite, D., and G. McGranahan. 2007. Providing Clean Water and Sanitation. In *State of the World 2007,* ed. L. Starke. New York: W. W. Norton. A look at the many challenges of providing water and treating waste in growing urban areas throughout the world.

Thompson, J. W., and K. Sorvig. 2000. *Sustainable Landscape Construction: A Guide to Green Building Outdoors.* Washington, DC: Island Press. Invaluable advice on landscape design in cities and towns to help protect watersheds.

United Nations Environment Programme (UNEP). 1995. *The Pollution of Lakes and Reservoirs.* Nairobi, Kenya: UNEP. Documentation of pollution issues of the major lakes and reservoirs of the world and ways to address them.

UNEP. 1995. *Water Quality of the World River Basins.* Nairobi, Kenya: UNEP. Summary of data on water quality from 82 major river basins around the world.

UNEP/GEMS. 1996. *Groundwater: A Threatened Resource.* Nairobi, Kenya: United Nations Environment Programme. Explanation of global groundwater pollution.

Weber, M. L. 2002. *From Abundance to Scarcity: A History of U.S. Marine Fisheries Policy.* Washington, DC: Island Press. An in-depth look at U.S. policy related to marine fisheries.

Web Explorations

Online resources for this chapter are on the World Wide Web at: **http://www.prenhall.com/chiras** *(click on the Table of Contents link and then select Chapter 11).*

FISHERIES CONSERVATION

Worldwide, there are at least 27,000 species of fish, making them the most abundant of the vertebrate animals. Over the past 200 years, however, human activities have adversely affected aquatic ecosystems and fish habitat. In the past few decades in particular, these activities have increased. The result of this acceleration in human pressure has been a decline in the number and range of many fish species considered commercially and recreationally important to society.

Historically, management of both freshwater and marine fisheries has emphasized the manipulation of fish populations and their environment for the sole purpose of increasing sport and commercial fish harvests. Indeed, this is an important goal, justified by the ever-increasing demands for food and recreation of a growing world population. As evidenced by the continued decline of many fish populations worldwide, however, fisheries management has yet to attain strategies that produce sustainable fish harvests for many species. Understandably, this ambitious goal will take time. To return the health of our declining fish stocks will also require extensive commitment and cooperation on national and international levels. Most experts agree that the health of fisheries depends on the protection and restoration of aquatic habitats.

To protect and restore aquatic ecosystems for fish survival, we must understand the complex interrelationships between fish and their habitats. The fisheries manager must be able to accurately assess fish responses to alterations of their habitat, whether caused by natural events or human activities, to make intelligent management decisions. In this chapter, we discuss freshwater and marine fisheries, environmental limitations to freshwater fish productivity, problems facing the marine fishing industry, management options for sustainable freshwater and marine fisheries, and aquaculture.

Before we begin, however, we should pause for a moment to define fishery. The term **fishery** is extremely broad. It commonly refers to a place where fish or other aquatic species such as shrimp and sponges are caught. We could, for example, speak about a particular marine fishery. The term is also used to define the species caught. It is even used to describe the occupation or industry—and technologies—involved in catching fish and other aquatic species. The most common usages have to do with place and species.

In this chapter, we'll focus on freshwater and marine fisheries. According to the Food and Agriculture Organization, about 26% of the global commercial fish harvest in 2004 was taken in inland fresh waters, while the remaining 74% came from marine waters. As you shall soon see, both pose significant problems. You'll also see that the tools of freshwater fisheries managers are more numerous than those of marine fisheries managers. Moreover, it is more difficult to assess populations of fish and other commercial species in the marine environment than in freshwater ecosystems. As a result, marine fisheries managers work with more uncertainty and face more pressure from commercial fishing interests, who have a much greater financial commitment than the average recreational angler.

12.1 Freshwater Fisheries

Most freshwater fisheries are managed for recreational anglers. The freshwater game fish sought after by our nation's anglers include catfish, northern pike, muskellunge, pickerel, trout, bass, bluegill, sunfish, crappie, perch, and walleye. Although most salmon are caught in marine commercial harvests, they are also an important freshwater fish, along with other anadromous fish. **Anadromous fish** are species that begin their lives in freshwater but then travel to and mature in the sea. After maturation, they return to their native stream to reproduce and die (Figure 12.1). Nonanadromous fish are called **resident fish** and live their entire life in freshwater, sometimes confined to a lake, a particular section of a stream, or a single tributary.

Habitat Requirements

All anadromous and resident fish species need relatively unaltered, or pristine, freshwater habitats during part or all of their life cycles. This ensures optimum growth and survival. The life cycle includes various stages such as spawning, incubation, rearing, and migration. However, the successful completion of each of these stages also depends on many other environmental conditions, including proper water temperature, water depth, and for riverine species, stream flow. But that's not all. Reproduction and survival are also affected by turbidity, levels of dissolved oxygen and salinity, the substrate (that is, the condition of stream and lake bottoms), the availability of cover for protection, and food supply. Let's consider each of these factors.

Temperature   Stream temperature affects many stages of the life cycle of fish. Temperatures that are too cold or too warm can delay **upstream migration** of fish. Temperature also affects **spawning** (egg fertilization and deposition). Each native fish species has a specific time of year and temperature range in which spawning occurs. These factors ensure the maximum survival rate of their offspring. If water temperatures are too cold or too warm during incubation, embryos will not develop properly and will have a low chance of survival. Specific temperatures are also required for normal behavior and optimal growth of juvenile fish during the rearing stage. Growth is retarded when temperatures become too low or too high. Many native fish species will move up- or downstream in response to adverse temperatures.

Water Depth and Velocity   Successful upstream migration depends on adequate water depths and velocities. As a general rule, larger fish require a greater minimum depth than smaller fish. They can also tolerate stronger stream flows. During upstream migration, for example, most salmon require a water depth of at least 0.24 meter (9.5 inches). Moreover, they can tolerate a flow velocity of up to 2.4 meters per second (7.9 feet per second). Trout, which are smaller, need only a depth of 0.12 meter (4.7 inches). They

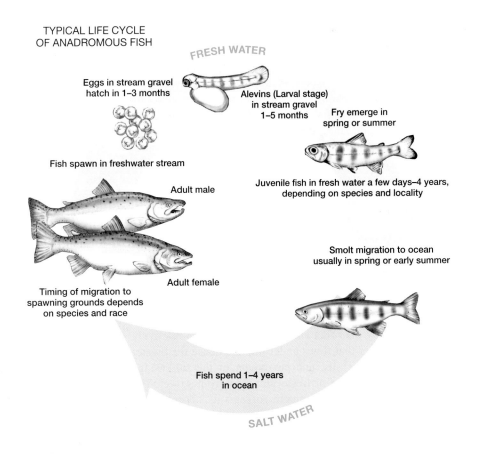

TYPICAL LIFE CYCLE
OF ANADROMOUS FISH

FRESH WATER

Eggs in stream gravel hatch in 1–3 months

Alevins (Larval stage) in stream gravel 1–5 months

Fry emerge in spring or summer

Fish spawn in freshwater stream

Adult male

Juvenile fish in fresh water a few days–4 years, depending on species and locality

Adult female

Smolt migration to ocean usually in spring or early summer

Timing of migration to spawning grounds depends on species and race

Fish spend 1–4 years in ocean

SALT WATER

FIGURE 12.1   Typical life cycle of anadromous fish.

can tolerate a velocity of only 1.22 meters per second (4.0 feet per second).

Stream depth also affects spawning habitat. When stream level is high, spawning habitat tends to increase. During periods of lower levels, spawning habitat decreases. Spawning habitat also declines when velocity of water flow through a stream becomes too high. (It can impair successful spawning.) Juvenile fish need space for rearing—which is, in part, a function of water depth and stream flow.

Turbidity As you learned in Chapter 9, turbidity is a measure of suspended sediment in water. Researchers have found that turbidity affects a variety of fish behaviors. Migrating fish, for example, may avoid or delay their migration when turbidity is excessive. Researchers have also found that adults and juvenile fish can tolerate temporary episodes of high turbidity levels, for example, during rainstorms or during spring runoff. However, fish generally avoid streams that have chronically high levels of suspended sediments.

Dissolved Oxygen Streams are usually very well aerated and, therefore, well oxygenated. As noted in Chapter 9, aeration and oxygen content are influenced by a variety of factors, such as the turbulence created by the flow of water over rocks. Stream aeration is also influenced by the depth of streams relative to surface area. All things being equal, the waters of shallow, fast-moving streams have much higher oxygen levels than those of deep, slow-moving streams. As a general rule, because streams are well aerated, the production of oxygen by photosynthetic organisms is not nearly as important as it is in a pond or lake. In addition, because stream water tends to be more thoroughly mixed than lake water, oxygen depletion occurs less frequently than in lakes. However, stream fish are very sensitive to even slight reductions of oxygen levels. For example, a reduction in dissolved oxygen to below 5 parts per million (ppm) decreases the swimming speed of adult fish during migration. This reduction also limits growth rate and food conversion efficiency of juvenile fish. When oxygen levels fall to the range of 1 to 2 ppm, they are generally inadequate for aquatic life. This condition is called **hypoxia.** During such times, slow-moving fish or fish that are, for one reason or another, unable to escape will perish, as will nonmobile aquatic life. **Anoxia,** where no oxygen is available, occurs when dissolved oxygen levels fall to 0.5 ppm.

Salinity As noted in Chapter 9, the concentration of salts, or salinity, is especially important for species that live in the estuarine zone and coastal marshes. Salt levels can limit the movement of marine and freshwater organisms through these areas. As noted in Chapter 9, estuaries are a transition zone where a river or rivers enter the ocean. They are characterized by tidal flux and a moving wedge of relatively dense, salty water that flows under the less dense freshwater from the river. The size and movement of the salt wedge depends on the volume of freshwater flow from a river. The salt wedge can affect aquatic organisms, commercial fishing, and even drinking-water supplies of an adjacent

city. For example, if river flow is high, the salinity level in parts of the estuary can drop, adversely affecting shellfish beds.

Substrate For successful spawning of many species, the bed materials of a stream or river must consist of gravel particles of a suitable diameter. Spawning fish use the gravel to make a "nest," called a **redd,** for their eggs. Most eggs are deposited in spawning gravels ranging from 1.3 to 15 centimeters (0.5 to 6 inches) in diameter, with the largest proportion in the range of 1.3 to 3.8 centimeters (0.5 to 1.5 inches). In general, larger fish are able to use larger spawning gravels.

Gravel beds make good spawning grounds because the pores in the gravel itself allow water to pass through. This, in turn, provides the embryos with much-needed oxygen. Pores also permit removal of waste materials. Unfortunately, fine sediment in streams can be deposited on spawning gravels. If this sediment layer becomes thick enough, developing embryos will die from suffocation.

Cover Cover is vital to the survival and reproduction of fish. It is provided by riparian vegetation as well as certain features of streambanks and streambeds. Overhanging vegetation, undercut banks, aquatic vegetation, large woody debris, boulders, and deepwater pools all provide adult and juvenile fish with shade, resting areas, and protection from predation (see Figure 9.17).

Food Supply How fast fish grow and how abundant they are also depend on food. As noted in Chapter 9, the productivity of streams—that is, their ability to supply nutrients and energy—varies from one stream to the next. It also varies between sections, or reaches, of any one stream. In general, the more organic matter (dead or living) in and adjacent to a stream, the larger the aquatic and terrestrial invertebrate populations that feed on such materials. The larger the aquatic and terrestrial invertebrate populations, the higher the numbers of adult and juvenile fish.

Organic matter in streams comes from many sources. Trees, shrubs, grasses, and other plants that live on the stream banks, for example, contribute leaves, needles, and twigs. Some, especially trees, may even contribute large woody debris. Aquatic vegetation—plants living in or on the water—also contribute organic matter that "feeds" streams. Dissolved organic matter may come from sources such as sewage treatment plants or nearby cattle or pig operations. Within a stream, you will find this matter in various stages of decomposition.

12.2 Environmental Limitations to the Reproductive Potential of Freshwater Fish

Like most organisms, freshwater fish have a great ability to reproduce. For example, a 16-kilogram (35-pound) female muskellunge may produce 225,000 eggs during a single breeding season. Some bass nests in Michigan lakes and streams

TABLE 12.1	Reproductive Characteristics of Major Species of Fish			
Common Name	Reproductive Age (or Length)	Spawning Time	Number of Eggs	Type of Reproduction
Bass, largemouth	2 yr.	Spring	2,000–100,000	Nest
Bass, smallmouth	2 yr.	Spring	2,000–20,800	Nest
Bluegill	1 yr.	May–August	2,300–67,000	Community nest
Carp	12 in.	Spring	790,000–2,000,000	Eggs scattered
Channel catfish	12 in.	Spring	2,500–70,000	Nests
Muskellunge	3–4 yr.	Spring	10,000–265,000	Eggs scattered
Northern pike	2–3 yr.	Spring	2,000–600,000	Eggs scattered in marsh
Salmon, chinook	4–5 yr.	Fall	3,000–4,000	Eggs buried in gravel
Salmon, coho	3–4 yr.	Fall	3,000–4,000	Eggs buried in gravel
Trout, brook	2 yr.	Fall	25–5,600	Eggs buried in gravel
Trout, brown	3 yr.	Fall	200–6,000	Eggs buried in gravel
Trout, lake	5–7 yr.	Winter	6,000	Eggs scattered over gravel
Trout, rainbow	3 yr.	Spring	500–9,000	Eggs buried in gravel
Walleye	3 yr.	Spring	35,300–615,000	Eggs scattered

Source: Illinois Department of Conservation, *What Fish Is This?* June 1986.

contain more than 4,000 young per nest. In some species, such as bluegills, nests are built very close together. Such gregarious breeding behavior promotes reproductive success, because many individuals can share an optimal spawning habitat. The reproductive characteristics of some species of fish are summarized in Table 12.1.

Although fish produce many eggs and offspring, many of them die as a result of **environmental limitations,** natural and human-induced changes in their environment that hinder survival and reproduction. Were it not for these limitations, lakes and rivers would be choked with fish. Tagging studies have revealed that roughly 70% of a given fish population dies each year. Thus, for every million young of a given species that hatch, 300,000 will be alive at the end of the first year. By the end of the second year,

only 90,000 will be alive. By the end of the tenth year, only 6 fish will survive. Environmental limitations—natural and human-induced—are summarized in Table 12.2. Let's take a look at them.

Natural Limitations

Storms and Soil Mass Movements   A major storm can have serious effects on stream channels and fish habitat. For example, flooding and high-velocity stream flow that occur as a result of high rainfall and surface runoff can alter the distribution of pools and riffles. They can also increase sediment levels in streams, causing stream turbidity to increase. High-water flows can also cause stream banks to erode, damaging fish habitat. Such events can disrupt

TABLE 12.2   Major Factors That Limit the Reproductive Potential of Fish

Cause of Limitation	Effects on Fish and Their Habitats
Natural*	
1. Major storms	1. Altered pool and riffle distribution and structure, increased turbidity and sedimentation, stream bank erosion, siltation of spawning gravels, burial of food sources, disturbance to rearing areas, blockage of fish passage
2. Soil mass movements (mudslides, avalanches, slumps, debris flows)	2. Damming and obstruction of stream channel, increased sediment load and turbidity, localized flooding, siltation of spawning gravels, burial of food sources, disturbance of rearing areas, blockage of fish passage
3. Animal activities (beavers, ungulates)	3. Diversion and ponding of water, localized temperature increases, flooding of rearing areas, siltation of spawning gravels, blockage of fish passage at low flows, trampling and browsing of riparian vegetation, reduction of cover and nutrient sources, alteration of stream temperatures, increased soil erosion, bank instability
4. Natural barriers (waterfalls, debris jams, excessive water velocities)	4. Interference with upstream migration, delay of spawning activities
5. Vegetation disturbances (windthrow, wildfire, insects, disease)	5. Reduction of cover, alteration of stream temperatures, decrease in nutrient and energy sources, bank instability, increase in erosion and sedimentation, increase in woody-debris jams
6. Predation by native animals	6. Fish mortality
7. Winterkill	7. Fish mortality
Human-Induced	
1. Water pollution (eutrophication, sedimentation, acid deposition)	1. Excessive growth of aquatic vegetation, oxygen depletion, increase in sediment load and turbidity, acidification of water, alteration in chemical composition of water, interference with reproductive and feeding behavior, siltation of spawning beds, burial of food sources, decrease in spawning success, asphyxiation and mortality of fish at all stages of development
2. Alteration of stream temperatures	2. Disruptions in timing of migration and spawning, abnormal growth rates, increase in disease outbreaks, abnormal behavior patterns
3. Predation and competition by exotics	3. Effects dependent on the behavior of the introduced species, possibly including loss of aquatic vegetation and food sources, loss of spawning grounds, increased turbidity, decreased dissolved-oxygen levels, decreased spawning success due to predation of eggs, disease outbreaks
4. Predation by humans	4. Increased fish mortality and injury, population decline or elimination
5. Dams	5. Altered water levels and velocities, loss of rearing habitat, increased mortality from nitrogen intoxication and contact with turbines, population decline, delay and blockage of fish passage, spawning delay
6. Resource extraction (logging, grazing, mining)	6. Loss of riparian vegetation, alteration of stream temperatures, modifications of stream channel structure, abnormal fluctuations in water depth and flow, chemical pollution, siltation of spawning gravels, increase in sediment loads and turbidity, stream bank erosion, decrease in dissolved-oxygen levels, decrease in spawning success, population decline
7. Channelization	7. Alteration of stream flows and depths, loss of spawning and rearing habitat, increased sediment load
8. Human recreation (swimming, boating, hiking, camping, horseback riding, mountain biking)	8. Riparian vegetation trampling, streambank erosion, soil compaction, decreased water quality, loss of cover, alteration in stream temperatures, loss of nutrient and energy sources

* Note that these natural events, in the long term, may in some cases actually increase productivity and improve habitat quality.

spawning, smother spawning beds with silt, and deposit fine sediment over food sources, such as bottom-dwelling flora and fauna. They can also damage side-channel rearing areas and block fish migration when woody debris that washes into streams during floods creates log jams.

Mudslides, avalanches, slumps, and debris flows can have similar effects on fish habitat. Sediment and boulders from a mudslide, for instance, may be transported into streams and can dam stream channels. Heavy sediment loads can increase turbidity.

Animal Activities   Beaver dams are common on relatively undisturbed, lower-order streams (Order 4 or less) in the forested regions of North America. Construction of these dams alters fish habitat by diverting and ponding stream water, increasing local water temperatures, blocking fish passage at low flows, flooding side-channel rearing areas, and causing siltation of spawning areas. However, in some cases, the benefits to fish habitat provided by these dams may outweigh the costs. For example, the flooding caused by beaver dams can provide important rearing and overwintering habitat. The silt that accumulates in the ponds behind the dams, with subsequent warmer water temperatures, can result in an increase in local biological productivity. And finally, beaver ponds may act as sediment traps, improving water quality in downstream regions.

Deer, elk, and moose browsing also affects aquatic ecosystems. For example, heavy use of an area by these species may result in trampling of riparian areas. Heavy browsing by these species may reduce the amount of under-story shrubs, grasses, and wetland plants growing along the shores of streams and lakes. The loss of this vegetation may reduce the quality of fish habitat, for example, by reducing cover and important nutrient and energy sources. Reductions in vegetation may also increase soil erosion and make the banks of rivers and lakes less stable.

Natural Barriers   Waterfalls, excessive water velocities, and debris jams can impede the upstream migration of fish. Because some species of fish spawn in "upstream habitats," these factors may have an impact on spawning. Researchers have found that ideal leaping conditions for fish occur when the depth of a pool below a waterfall is 1.25 times the height of the falls (Figure 12.2A, B). Of course, the success of any one fish in passing a given barrier also depends on other factors, including the swimming speed the fish can attain and its jumping ability, as well as the horizontal and vertical distances to be jumped and the steepness of the incline (Figure 12.2C, D).

Fish may be unable to continue their upstream migration if water velocities are too high. For example, large fish such as salmon and steelhead can swim for extended periods of time in water velocities of 2.4 meters per second (7.9 feet per second) or less. They can swim for a few minutes at a time in water velocities of 3 to 4 meters per second (9.8 to 13.1 feet per second) or less. However, they are unable to navigate in velocities above 4 meters per second (13.1 feet per second).

Some woody-debris jams can delay or prevent upstream migration. Removal of the jam, however, must be done with care to prevent siltation of active spawning and rearing areas downstream and harmful alterations of water levels and velocities.

Vegetation Disturbances   We have previously established the crucial role played by riparian vegetation in maintaining high-quality fish habitat. We have pointed out that some species, such as deer, elk, and moose, can alter riparian habitat. However, major disturbances such as wildfires, insects, and diseases also can destroy riparian vegetation. These natural forces can also damage fish habitat. They may, for instance, reduce cover, alter stream temperatures, or decrease litterfall. They may also increase soil erosion, which increases sediment deposition in streams. They can reduce the stability of stream banks, making them more prone to erosion, and they can increase jams created by woody debris.

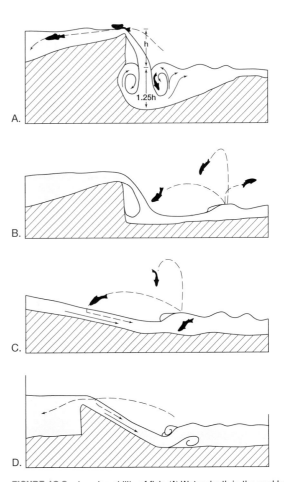

FIGURE 12.2   Leaping ability of fish. (A) Water depth in the pool below the falls is 1.25 times the distance (h) from the top of the falls to the water surface of the pool. Fish use the upward momentum of the standing wave, formed close to the waterfall, to leap over the falls. (B) The height of the waterfall is equal to that in A, but the pool is shallower. The standing wave is formed too far from the falls to be of any use to the fish. (C) An incline too steep and too long for fish to negotiate. Fish can be thrown from the standing wave at the foot of the incline. (D) An incline steeper than in C but shorter in length. Some fish, depending on their leaping ability and the amount of energy in the standing wave, may be able to pass over this barrier.

Predation by Native Animals Fish are subjected to intense predatory pressure from other fish. They are also preyed upon by reptiles, birds, and mammals. For instance, a 6-inch muskellunge, a predatory fish, can consume 15 minnows a day! A walleye, another predatory fish, will eat 3,000 young fish by the time it is three years old. When other food is scarce, many fish resort to cannibalism—eating one another. Large fish will consume smaller fish of their own kind, and smaller fish species may eat the eggs of larger species.

Wading birds and other waterfowl, such as egrets, herons, and ducks, also consume large numbers of fish. A single merganser (a diving duck) may eat more than 35,000 fish annually. Bear, otter, fisher, and mink prey extensively on fish, especially during droughts, when water levels are low and the fish are easily caught. The Alaskan brown bear can easily eat 15 large salmon a day.

Winterkill Another environmental factor that can reduce fish populations is **winterkill.** During the long winters of the northern states, an icy barrier seals off lakes from atmospheric oxygen. As long as the ice remains clear of snow, sufficient sunlight may penetrate the ice to sustain photosynthesis (Figure 12.3). As a result, oxygen levels remain adequate. Snow, however, forms an opaque barrier over the ice, reducing the amount of sunlight penetrating the water, which reduces photosynthesis. If photosynthesis declines, so do oxygen levels. The result? Heavy fish kills, especially if the lake is fertile and shallow. The decay of dead vegetation worsens the problem. As winter progresses, oxygen levels may drop to 5 ppm, at which point, many of the more sensitive fish die; the more resistant fish, such as carp and bullheads, may die later if levels drop to about 2 to 3 ppm.

Human-Induced Limitations

Water Pollution More than 30 million fish are killed by water pollution in the United States each year. In Chapter 11, we described fish kills caused by industrial and municipal pollutants. As shown in Table 12.3, many different pollutants released into waterways from point and nonpoint sources can cause fish kills. Here we describe mortality caused by eutrophication, sedimentation, and acid deposition.

Eutrophication As you may recall from Chapter 11, the enrichment of a body of water with nutrients that promote

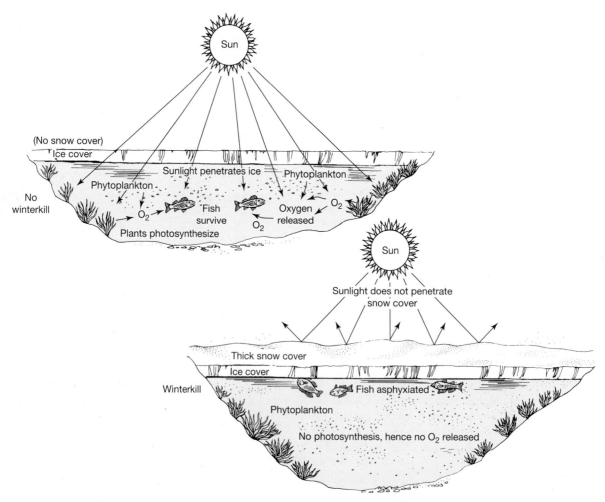

FIGURE 12.3 Winterkill of fish.

TABLE 12.3	Fish Kills Caused by Water Pollution in Ohio, 1984			
Date	County	Name of Water Body	Number Killed	Suspected Pollutant
9/2	Montgomery	Great Miami River	158,234	Sewage and corn syrup
7/17	Daske	Greenville Creek	122,057	Hog manure
6/15	Fulton	Brush Creek	79,110	Ammonium nitrate
4/16	Crawford	Broken Sword Creek	57,237	Nitrogen fertilizer
9/1	Butler	Four-Mile Creek	41,335	Sewage
3/26	Columbiana	Beaver Creek	26,986	Gasoline
9/10	Coshocton	White Eye Creek	13,170	Cow manure
6/27	Morgan	Bell Creek	3,274	Cleaning chemicals
5/3	Marion	Riffle Creek	2,905	Herbicides

Source: *Water Pollution, Fish Kill, and Stream Litter Investigation, 1984.* Columbus, OH: Department of Natural Resources, 1985.

the excessive growth of aquatic vegetation is known as **eutrophication.** As noted in that chapter, dense algal blooms form in the surface waters, preventing sunlight from reaching the billions of algae at lower depths. This, in turn, results in a decline in oxygen production in deeper waters.

In the fall, when sunlight declines, algae and submerged vegetation die and sink to the bottom of lakes in temperate regions, forming a dense organic ooze. Billions of aerobic (oxygen-requiring) bacteria in the bottom sediment decompose this organic material. As they do, they consume the oxygen dissolved in the deep waters, the hypolimnion. As a result, the oxygen concentration in the hypolimnion may fall rapidly from 7 to 2 ppm or less.

As a water body becomes polluted with oxygen-demanding organic material, such as human sewage or the waste from slaughterhouses, pulp mills, and canneries, oxygen levels in the water may fall drastically. This reduction may trigger massive fish kills. Eventually the dead fish float to shore, decompose, begin to smell, and attract flies.

Sedimentation Many tons of soil are washed into lakes and streams by runoff as a result of improper land practices on farms, around mines, or on urban construction sites. Sediment depresses the photosynthetic activity of aquatic plants because it reduces sunlight penetration of the water. This, in turn, causes levels of dissolved oxygen to drop, often sharply. A dramatic decline in oxygen is especially stressful to fish such as trout and salmon, which require a minimum of 5 ppm of dissolved oxygen.

Turbidity also affects fish directly. Although fish may tolerate turbidities of up to 100,000 ppm for brief periods, concentrations of 100 to 200 ppm are harmful if they persist for any length of time. Thousands of fish die annually in American lakes and streams from asphyxiation caused by silt-clogged gills (Figure 12.4).

FIGURE 12.4 Fish kill caused by sediment. Many fish suffocate each year when sediment clogs their gills following heavy rainfall that erodes soil from unprotected land.

Besides reducing oxygen levels and clogging gills, **suspended sediment** also interferes with the reproductive behavior of fish. As you may know, fish depend on visual cues provided by the gravel and sand of the stream or lake bed as well as on the color, shape, and behavior of the sex partner. Sediment also smothers spawning beds, as noted earlier in the chapter. In fact, vast beds of inshore aquatic vegetation that were once important spawning beds for Great Lakes fish have been smothered by sediment eroded from

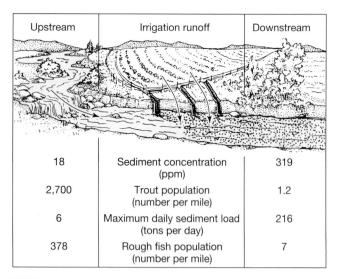

Upstream	Irrigation runoff	Downstream
18	Sediment concentration (ppm)	319
2,700	Trout population (number per mile)	1.2
6	Maximum daily sediment load (tons per day)	216
378	Rough fish population (number per mile)	7

FIGURE 12.5 Effect of sediment from irrigation runoff on a Montana trout stream.

the land. Mud may also cover fertilized eggs and reduce hatching success. A study of trout reproduction in Bluewater Creek, MT, for instance, showed that egg-hatching success was highest (up to 97%) in areas where **siltation** was minimal (Figure 12.5).

Silt also affects food supplies. For example, the larvae of aquatic insects, such as mayflies and stoneflies, are favored fish foods. However, they may be destroyed by silt. Moreover, mud sharply reduces the visual range of predatory fish, including bass, pike, and muskellunge, in their search for the smaller fish on which they feed.

Acid Deposition Can the burning of coal in an Ohio steel plant cause the death of trout high in the Adirondack Mountains of New York? Although it seems highly improbable, the answer is yes. The sulfur dioxide gas (SO_2) released from the smokestacks undergoes chemical reactions with oxygen to form sulfuric acid and water. Prevailing winds carry the acid droplets northeastward in clouds to a point high over the Adirondacks,

the United States' most acid-sensitive region. Rain (or snow) then washes the acid into the lakes.

Normal, unpolluted rain has a pH of about 5.6—slightly acid, because carbon dioxide is dissolved in it to form carbonic acid. However, much of the acid rain in the eastern states has a pH of 4 or below—almost 100 times as acid as normal rain.

The effects of **acid rain** were not fully appreciated until the 1970s. Biologists have found that when the pH of lake water falls as low as 5, fish begin to die. The pH of many Adirondack lakes is below 5. Under such conditions, lake trout become deformed, and embryos suffer high mortality.

The acidic water from rain and snowmelt also affects the chemical composition of the water. As the water drains off the land, the acids dissolve toxic metals from the soil and carry them into lakes and streams. Aluminum leached from the soil, for example, causes mucus to build up on the gills of fish. As a result, the fish suffocate. Aluminum is especially troublesome in the spring when the snows begin to melt, sending torrents of acidic water across the soil and eventually into lakes and streams.

According to one survey, many of the Adirondack lakes became devoid of fish. The elimination of a species of fish from a lake may not occur suddenly. Instead, it may develop gradually over a period of years, owing to the inability of fish to spawn successfully. As an example, the effects of increasing acidity on fish populations in another lake, George Lake in Ontario, are shown in Figure 12.6. (For more on acid deposition, see Chapter 19.)

Two scientific studies published in 2003 and based on the same data set indicated that the lakes in the Adirondack Mountains were showing signs of slow recovery from acidification. Roughly 60% of lakes surveyed in the Adirondacks displayed a significant increasing trend in acid-neutralizing capacity. a key indicator of recovery because it measures a watershed's ability to counteract the acid pollutants in rainfall. The research demonstrates that the region-wide cuts in SO_2 emissions mandated by the Clean Air Act helped lakes recover over a wide area. Emissions of SO_2 declined 38% between 1973 and 2003. Although most lakes were improving, 40%

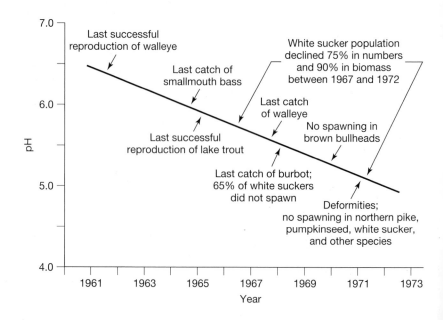

FIGURE 12.6 Effect of increasing acidity (decreasing pH) of water on fish populations in George Lake, Ontario.

still showed no change or continued to lose acid-neutralizing capacity. The scientists concluded that more needs to be done.

Alteration of Stream Temperatures Abnormal fluctuations in stream temperatures can harm fish populations, a subject described in Chapter 11. As you may recall, they do so by disrupting the timing of migration and spawning, retarding or accelerating growth stages, increasing disease outbreaks, and altering behavior patterns. A variety of human activities adversely affect stream temperatures and fish populations. Forestry, mining, and cattle ranching, for example, all reduce stream bank vegetation, although regulations now exist that help to reduce this practice. Water temperatures are altered when water is removed from streams and then returned after irrigation of agricultural lands and when water is released from deep reservoirs for flood control or power generation. Cooling towers associated with nuclear and coal-fired power plants often discharge warm water into rivers and streams.

Predation, Competition, and Habitat Modification by Exotics The U.S. Fish and Wildlife Service is increasingly concerned with the "biological pollution" of native fish populations with **exotic (nonnative) species.** Over the years, foreign fish have been deliberately introduced by fisheries biologists to provide a desirable game or food fish—or to aid in controlling an environmental problem. Many other exotic fish species have been introduced accidentally. Several intentionally and accidentally introduced species have adversely affected native fish, altering their reproduction, growth, and survival.

The most destructive exotic fish ever intentionally brought to the United States is the **European carp** (Figure 12.7). It was originally introduced to California (1872), the Great Lakes (1873), and Washington, DC (1877). Hundreds of requests

from all over the nation came to the US Bureau of Fisheries to stock this so-called "wonder fish" in order to provide Americans with a valuable source of food.

The carp proved to be an extraordinarily adaptable fish. The introduced populations grew rapidly, following the characteristic S-shaped growth curve. Only 20 years after the introduction of European carp to Lake Erie, fishers were able to harvest 1.6 million kilograms (3.6 million pounds) in a single year. However, as their numbers increased, researchers discovered that carp uprooted aquatic vegetation during bottom-feeding. This impact has been exceedingly harmful to game fish populations, as it has (1) destroyed their spawning grounds, (2) reduced their food supplies, and (3) reduced levels of dissolved oxygen as a result of interference with photosynthesis caused by a muddying of the waters.

One of the most recent import tragedies is the **river ruff,** another European species. The river ruff may have been accidentally introduced into Lake Superior in 1987 from a cargo ship that docked at Duluth, MN. The ruff is a potential threat to the multi-million-dollar commercial fishery of the Great Lakes because it is a ravenous consumer of the eggs of the whitefish and other species. Since the ruff becomes sexually mature when only one year old, it has a reproductive edge on most species of native fish. In 1989, only two years after its introduction, the ruff's population had increased sharply in the harbor of Duluth. To control this problem, fisheries biologists planned to reduce the number of muskies and pike that anglers could take in Duluth Harbor. They hope these fish will prey on the ruff.

A nonnative or exotic fish species does not have to be introduced from another country to cause troubles. When intentionally or accidentally transplanted outside of its original lake or stream, a native fish species from one area of the country becomes a nonnative species in another area. These introduced species can compete with native fish species in their new home or prey on them, sometimes severely reducing populations (see Case Study 12.1).

Predation by Humans Fishing has been a major factor in the decline of many of North America's freshwater commercial and game fish. For example, lake sturgeon, *Acipenser fulvescens,* were eliminated from Lake Erie as a result of overexploitation by sport and commercial fisheries—an extreme example of **human predation.** A classic example of fishing pressure is offered by anglers on the opening weekend of the trout season. During their enthusiastic quest for the king of American game fish, in some areas they stand shoulder to shoulder along stream margins.

Fishing is popular; at least one in ten Americans fish each year. In 2006, about 30 million U.S. residents 16 years old and older fished, according to the U.S. Fish and Wildlife Service.

Dams Dam building on rivers that eventually drain into the Pacific Ocean has had devastating effects on many anadromous fish species. For example, in the Columbia River on the Washington–Oregon border, many young salmon on their way out to sea are killed by **nitrogen intoxication** resulting from the high levels of nitrogen in the turbulent waters immediately below dams.

FIGURE 12.7 The carp, an exotic introduced into American waters from Europe during the late 19th century. Note the mouth, specialized for bottom-feeding. This feeding habit causes this species to muddy waters and spoil habitat for game fish.

CASE STUDY 12.1   THE SEA LAMPREY—SCOURGE OF THE GREAT LAKES

Imagine a predator so efficient that it could destroy 97% of the lake trout population of the Great Lakes in only 21 years! That predator is the *sea lamprey*—an olive-gray, blood-sucking killer that completed its invasion of the Great Lakes by 1950. The lamprey is a primitive, jawless vertebrate with a slender, eel-like body. The muscular funnel around its circular mouth enables it to attach firmly to its prey (Figures 1 and 2). It moves its pistonlike tongue, armed with numerous hard, rasping teeth, back and forth through the lake trout's tissues, tearing flesh and blood vessels and causing severe bleeding. An anticoagulant prevents the blood from clotting. After gorging itself on a meal of blood and body fluid, the predator may drop off its host and permit it to swim weakly away. The trout may die from the direct predatory attack, or it may eventually succumb to bacterial and fungal infections that become established in the open wounds. During its short adult life of about 15 months, the average lamprey kills about 18 kilograms (40 pounds) of trout, salmon, and other Great Lakes fish. Even if a lake trout survives, the ugly scar left on its body would scarcely be admired by the grocery-buying homemaker.

The Great Lakes lampreys spend their entire life cycle in freshwater. When sexually mature, the adults swim up tributary streams to mate, spawn, and die. After hatching from the eggs, the larval lampreys are needle-thin and about 3 millimeters (1/8 inch) long. They drift downstream until they come to a muddy bottom. Then they burrow tail-first into the mud, allowing only their heads to remain exposed to the current. During this time, they feed on small algae, insects, worms, and crustaceans. Several years later, when they have grown to the size of a pencil, they acquire the muscular funnel and rasping tongue of the adult, emerge from their burrows, and swim into the open waters of the lake to prey on fish (Figure 3).

The lamprey originally occurred in the shallow waters off the Atlantic seaboard from Florida to Labrador, in the waters of the St. Lawrence River, and in Lake Ontario, at the eastern end of the Great Lakes chain. For many centuries, the westward extension of the lamprey's range into Lake Erie was blocked by Niagara Falls. However, in 1833 the Welland Canal was constructed to promote commercial shipping. Unfortunately, however, the canal also provided the lamprey with an invasion channel to Lake Erie (Figure 4).

The lamprey's colonization of Lake Erie was a slow process, probably because of a lack of suitable tributary spawning streams. However, once it invaded Lake Huron, it spread rapidly into Lake Michigan and Lake Superior. By 1950, the lamprey was found in the western end of Lake Superior and had completed its Great Lakes invasion (Figure 4). Its predatory activity soon threatened the multi-million-dollar Great Lakes trout-fishing industry with total collapse. The annual catch declined from 4,545 metric tons (5,000 tons) in 1940 to 152 metric tons (166 tons) in 1961, a 97% reduction in only 21 years. Idle nets rotted along the waterfront. Veteran fishermen, too old to acquire new skills, went on relief. Many of the younger men moved to Minneapolis, Milwaukee, Chicago, and Detroit in search of work.

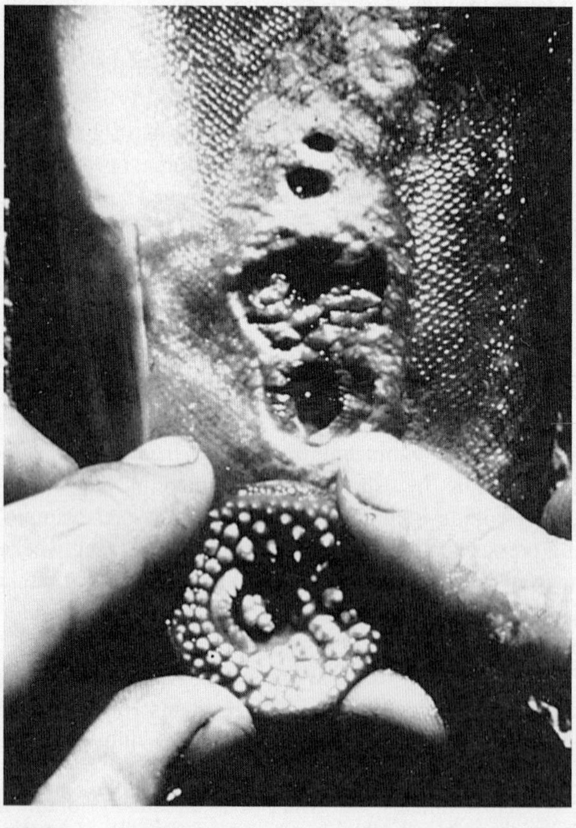

FIGURE 1   Close-up of the muscular funnel, mouth, and rasping "tongue" of the sea lamprey. The end of the pistonlike tongue is visible inside the circular mouth in the center of the funnel. Note the wounds caused by the lamprey on the lake trout.

FIGURE 2   Lamprey adhering to netted lake trout.

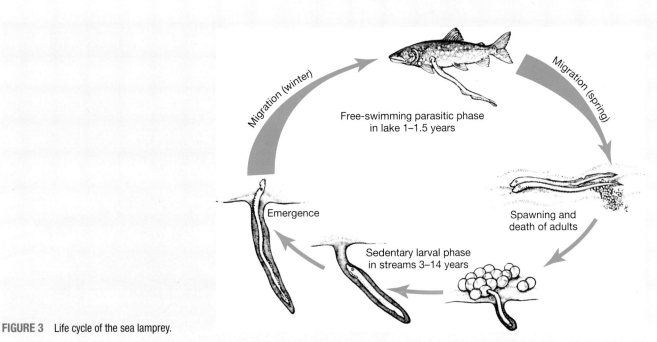

FIGURE 3 Life cycle of the sea lamprey.

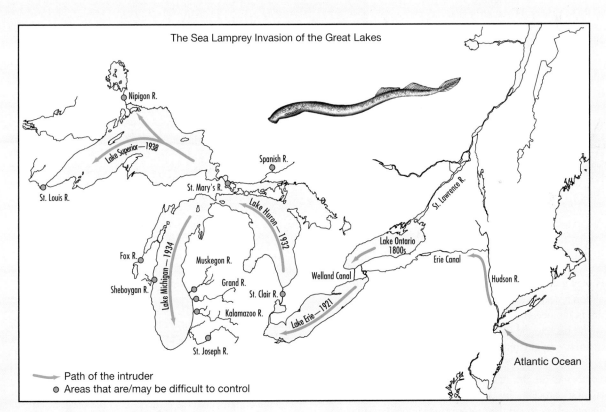

FIGURE 4 Sea lamprey invasion of the Great Lakes.

In 1955, the Great Lakes Fishery Commission was formed by treaty between the United States and Canada to control the sea lamprey. Members of the commission represented all the Great Lakes states and the province of Ontario. A variety of control strategies were tried. Adult lampreys were netted and seined. They were even shocked with electric "fences" as they tried to move up their spawning streams (Figure 5). These methods had only limited success, however. It was decided, therefore, to try chemical control—to use a **lampricide.** From 1951 to 1959, more than 6,000 compounds were tested as potential lamprey killers. Eventually an obscure poison, known as TFM, was selected. TFM is essentially nontoxic to humans as well as to game fish and their food organisms, such as minnows and aquatic insects. But it is

(continued)

FIGURE 5 Electric "fence" placed across Michigan stream near its entrance to Lake Michigan. As the lampreys swim through the fence, they are stunned by an electric charge and float to the surface, where they are easily removed.

lethal to lampreys. Larvae treated with low levels of TFM pop out of their burrows and quickly die from hemorrhage.

By 1960, the Great Lakes Fishery Commission had treated all of the lamprey-infested tributaries of Lake Superior with TFM. Only two years later, this chemical had reduced the number of spawning lampreys in these streams by 85%. In the ensuing years, the lampricide was also used on tributary streams of Lake Michigan (1963), Lake Huron (1970), and Lake Ontario (1972), with equally effective results. TFM currently remains the primary lampricide in the Great Lakes area.

Today the sea lamprey control program is formulated and implemented by the Great Lakes Fishery Commission, in partnership with the U.S. Fish and Wildlife Service, Department of Fisheries and Oceans Canada, U.S. Army Corps of Engineers, and U.S. Geological Survey. This control program has contributed significantly to the maintenance of the $4 billion Great Lakes sport and commercial fisheries. Infested areas must be treated every three to five years with lampricides to control sea lamprey populations. Fishery biologists and technicians conduct surveys for sea lamprey larvae in hundreds of Great Lakes streams each year.

(Nitrogen is introduced to the water from the air.) Additional fish perish as they are drawn into the dams' giant turbines. Only 10% of the salmon fry ultimately reach the ocean. After spending one to four years in the ocean, where they become sexually mature, the salmon swim back to the Pacific Coast. The fish then ascend their native streams, recognizing them by their distinctive smell. (The hatching fish apparently "learn" how their native stream smells and remember it for the rest of their lives.) The psychological or behavioral process by which migratory fish assimilate environmental clues to aid their return to their stream of origin as adults is called **imprinting.** On their upstream migration to spawning grounds, the adult salmon must get around the dams—or die without reproducing. To accommodate migrating fish, many dams have been retrofitted with **fish ladders** (Figure 12.8) and other fish passage systems. These devices were built into dams to allow fish a way to bypass a dam. However, many systems are ineffective and have been redesigned and upgraded.

Dams not only block fish from migrating to and from the sea, they also alter the behavior of anadromous fish and their habitat by changing the natural dynamics of riverine

ecosystems—in particular, by causing abnormal fluctuations in water levels and water flow rates.

Resource Extraction   A variety of human land-use activities also affect fish habitat. For instance, logging, road construction, mining, and livestock grazing near rivers or within a watershed—if not carried out properly—can all cause an increase in soil erosion. This, in turn, may result in greater sediment loads in streams. Increased sediment flows in streams can have many adverse effects, as noted earlier in the chapter.

Logging   Harvesting and log transport methods have changed over the life span of the timber industry on the West Coast, but each practice has damaged natural river ecosystem processes and native fish habitat. In the late 1800s and early 1900s, for example, logging companies cleared instream debris, boulders, and riparian vegetation from streams. Companies built **splash dams** to make it easier to float logs downstream to sawmills. In particular, water was released from the splash dams when logs were ready for transport. These high-velocity surges of logs and water gouged and

FIGURE 12.8   King (chinook) salmon leaping up the top step of a fish ladder at the Red Bluff Diversion Dam during a late spring spawning run. This fish ladder enables salmon to move upstream beyond the dam and eventually spawn in California's Sacramento River.

Channelization   Rivers and streams the world over have undergone a myriad of hydrological alterations over the years to control floods, drain wetlands, and divert waters for cities and farms. Most of these projects have resulted in significant losses or alterations of natural fish habitats.

Recreation   Other forms of human recreation, in addition to fishing, involve the use of rivers, lakes, and riparian areas. Future human population increases will no doubt increase the demand in these areas for recreational activities such as swimming, boating, hiking, camping, horseback riding, and mountain biking. Overuse can have detrimental effects on riparian and aquatic habitats, including vegetation trampling, soil compaction, stream bank erosion, and water pollution.

12.3   Sustainable Freshwater Fisheries Management

The fisheries manager has a challenging job. One reason the manager's job is so demanding is that each fishery, whether in a lake, stream, or reservoir, is a unique system and needs to be managed according to its particular needs. Another reason the manager's job is taxing is that the successful production of just a single species depends on the interaction of numerous chemical, physical, and biological factors. Among these are lake (or stream) area, bottom, and depth; water currents; length of growing season; water temperature; levels of dissolved oxygen; water acidity or alkalinity; water pollution; water fertility (dissolved nutrients); shelter (cover); food availability; predator–prey relationships; reproductive potential; mortality rates; fishing pressure; fishing regulations; species composition; population size; population age structure; and growth rates. In addition, the fisheries manager must determine a suitable method to sustain fish populations on public lands that are simultaneously managed for other resource uses (logging, grazing, mining, recreation). Additionally, a single river or stream ecosystem can flow through a large geographic area that includes both private and public lands. Applying consistent management practices throughout an entire watershed may be difficult when public and private land owners have different objectives.

Over the past 200 years, humans have inadvertently degraded and destroyed considerable amounts of fish habitat; there is no quick fix. To create sustainable freshwater fisheries, resource managers must be able to restore these systems. Some of the possibilities and difficulties faced by fisheries managers in attempting to restore fish populations are described in Case Study 12.2.

Clearly, the task of a fisheries manager is not easy. He or she must be able to develop and administer programs that regulate human activities in watersheds, protect land and water, and restore aquatic ecosystems, all of which lead to the long-term health of aquatic ecosystems. With the exception of a few successes and the strategies acceptable in last-resort situations, much evidence indicates that most artificial biotic manipulations designed to enhance or restore fish populations are, in general, poor management options. They create too many risks, negative effects, and are primarily short-term in nature. Resource managers do agree on three key elements of sustainable freshwater fishery management: (1) population enhancement such as artificial propagation and

eroded stream banks, increasing the sediment load downstream. Streambed gravels were scoured out, and with them, spawning beds and developing embryos. Fluctuations in streamwater levels as these dams were opened and closed and secondary, unnatural hydrologic changes wreaked havoc on fish habitat, diversity, and productivity.

Logging continues to damage fish habitat today, despite laws and improved practices. Clear-cutting, road building, planting, thinning, burning, chemical application, mechanical site preparation, and milling practices degrade fish habitat in a variety of ways. They do so by altering sediment and nutrient delivery rates and changing water temperature levels, dissolved oxygen levels, and overall water quality.

Grazing   Cattle graze heavily on riparian vegetation, if available, because of its succulence and variety. Unfortunately, however, after several years, the heavy grazing pressure can eliminate plant cover, trigger erosion, and result in many harmful effects on the stream's **carrying capacity** for fish. (Carrying capacity of a stream is the number of fish the stream can support indefinitely.) A study of a Montana stream showed that the portions with ungrazed banks had 27% more fish over 15 centimeters (6 inches) long than the grazed sections.

Mining   Miners were once able to dredge, straighten, and pollute streams and strip vegetation from river channels on public lands without constraint. In most areas today, however, these practices are outlawed or significantly reduced, thanks to laws and regulations. For the duration of the mining operation, strict water quality standards must be met. Unfortunately, monitoring and enforcement of these regulations are often lacking, owing to personnel limitations.

Mining companies are also often required to implement approved restoration plans to fix any damage they cause. A reclamation performance bond must be posted before mining permission is granted. (That is, companies must put up money for restoration, so that if they go out of business, the money to fix any problems is available.)

CASE STUDY 12.2   REBUILDING FISH AND WILDLIFE POPULATIONS ON THE COLUMBIA RIVER DRAINAGE SYSTEM

Estimates based on tribal accounts and historical catch records—stretching back to canneries operating before the industrial era—on the Columbia River indicate that somewhere between 10 million and 16 million adult salmon and steelhead returned every year to the mouth of this river—the starting point for their long upstream migration to their spawning grounds. Fish harvests by commercial interests and anglers were bountiful. Some of the early anglers spoke of seeing salmon so numerous during their spawning runs that "you could almost walk across the stream on their backs." The brightness of the sun glinting off the backs of the silvery mass of migrating adult salmon was the inspiration for the name of White Salmon, a city located on the Washington side of the Columbia River, about 60 miles east of Portland. Huge chinook salmon, some weighing over 27 kilograms (60 pounds), were so big and fat they were referred to as "hogs" (Figure 1).

Unfortunately, such sights are long gone. Populations of all the salmon species native to the Columbia River system have declined dramatically in the past few decades. Both the sockeye and the chinook salmon of the Snake River in Idaho are listed as endangered. Moreover, the native coho salmon of the upper Columbia and Snake Rivers have disappeared entirely.

By any measure, the salmon situation in the Columbia River Basin is grim. In 1992, for example, only a single male Snake River sockeye completed the 900-mile upstream migration to its spawning grounds in Redfish Lake in central Idaho. Fisheries biologists dubbed him "Lonesome Larry." His sperm was removed and refrigerated, and has been used in artificial propagation of the species. In recent years, only about 1 million adult salmon and steelhead trout return to the Columbia River Basin system—less than 10% of the region's historic runs.

What caused the dramatic decline in the Columbia River Basin salmon and steelhead? One factor was the sediment pollution resulting from the construction of logging roads near stream margins. Another was thermal pollution caused by the removal of trees along streams and river banks—trees that used to shade the streams and keep their waters cool. Temperatures of 20°C (68°F) or above effectively block the upstream movement of salmon. However, when water temperatures drop below 20°C, migration is quickly resumed. Still another factor is urbanization, with its attendant housing developments, malls, and parking lots. The water runoff from thousands of acres of asphalt and concrete forms a "witch's broth" of pollutants, such as oil, grease, salt, and heavy metals. They have contaminated nearby salmon and steelhead streams. Overfishing has also been highly detrimental.

The most important factor in the decline of salmon and steelhead, however, was the construction of the vast hydropower system along the Columbia River and its tributaries. This huge complex—the greatest hydropower system in the world—consists of 66 major hydroelectric dams, including the Grand Coulee, which is 107 meters (350 feet) tall (Figure 2). Such structures constitute a daunting array of concrete barriers to successful upstream migration by adult salmon and steelhead. Juvenile fish that try to move downstream are often drawn into the giant turbines and die. Only about 10% of the juveniles make it to the ocean every year.

FIGURE 1   Chinook (king) salmon.

For many years, attempts by state or federal agencies to rebuild the salmon populations of the Columbia River and its tributaries were best described as piecemeal, haphazard, and ineffectual. In

FIGURE 2   Bonneville Dam on the Columbia River.

1980, however, Congress passed the Northwest Power Act. The agency created by this act, known as the **Northwest Power Planning Council (NPPC),** was formed by an interstate compact involving Washington, Oregon, Montana, and Idaho. Provision was made for influential input to the NPPC from federal fish and wildlife managers, the four northwestern states, Native American tribes (who have asserted their salmon fishing rights), and public and private utilities.

The Northwest Power Act directs the NPPC to carry out three responsibilities: (1) prepare "a regional conservation and electric power plan" [Section 4(d)(1)]; (2) "inform the Pacific Northwest public of major regional power issues" [Section 4(g)(1)(A)]; and (3) "develop . . . a program to protect, mitigate, and enhance fish and wildlife, including related spawning grounds and habitat, on the Columbia River and its tributaries" [Section 4(h)(1)(A)].

The NPPC built several new hatcheries to augment the 100 state and federal hatcheries already in existence. Paid for by citizens—that is, power consumers supplied by the Bonneville Power Administration—this program for the artificial propagation of salmon may seem admirable. However, as pointed out in the text, artificial propagation can be iffy. Care must be exercised so that the natural diversity of native populations of salmon is not overwhelmed by genetically inferior, hatchery-reared fish, and so that disease is not introduced into the wild. The NPPC, in its Columbia River Basin Fish and Wildlife Program, has cited concerns over **genetic erosion.** Its studies show that the offspring of a hatchery fish bred with a native parent have a lesser chance of survival than pure native offspring.

The NPPC's program for mitigating the destructive effects of the hydropower system on salmon also included measures to improve fish ladders (Figure 3) on older dams constructed by the U.S. Army Corps of Engineers. To prevent the future construction

of hydropower projects that would be destructive to salmon-spawning habitat, the NPPC has declared 72,000 kilometers (44,000 miles) of salmon streams off-limits to future dam builders.

Another measure investigated by the NPPC is the permanent drawdowns of mainstream reservoirs, so that the stream channel would assume once more its predam character. A report prepared for the NPPC by nine independent scientists suggested that this hydromodification would need to be permanent in order to restore the riverine habitat for fish and other wildlife. Such a strategy would greatly accelerate the speed of water flow, reducing the migration to the Pacific Ocean from 30 days with dams and reservoirs in place to 10 days. This would reduce the amount of time young salmon would be exposed to pollution and predation. Many fisheries experts consider **hydromodification** the most important component of the NPPC's ambitious program to rebuild the fish

FIGURE 3   Fish ladder on the Bonneville Dam, which permits salmon to swim upstream to their spawning grounds.

(continued)

and wildlife populations of the Columbia River Basin. This strategy has not been universally accepted, however. Opponents of drawdowns worry about the potential impact on hydropower generation and the potential for higher electricity rates. It could, they also point out, reduce supplies of irrigation water. Loss of hydropower and reduced reservoir levels could lead to a loss of jobs. In addition, opponents argue that the idea should be proved effective or at least endorsed by scientists before being implemented.

In contrast, proponents of drawdowns believe that fish and wildlife habitat cannot be restored without reverting to natural river dynamics and that salmon need a free-flowing river, not a chain of slow-moving lakes. Proponents argue that drawdowns should be implemented despite the potential for higher electricity prices.

Even more controversial than drawdowns is the breaching (destruction) of four dams on the Snake River, a tributary of the Columbia River. This has been touted by proponents as the best strategy to reverse the rapid decline of salmon stocks on the Columbia River's largest tributary. Because all four of the Snake River's salmon and steelhead populations are listed as endangered under the Endangered Species Act, the National Marine Fisheries Service (NMFS) has been charged with developing and implementing plans to reverse the declines in the river's salmon and steelhead populations.

Opponents to dam breaching argue that dam removal would likely cost more than $1 billion, reduce the region's power-generating capacity, eliminate an important transportation artery for neighboring farmers shipping their products to markets via Snake River barges, and have an uncertain impact on salmon and steelhead populations.

A 1999 NMFS plan put dam breaching on the back burner, and instead called for immediate improvements to habitat, hatchery, and harvesting strategies. With opposition to dam breaching running strong among the region's political leaders, supporters of the NMFS's plan feel that focusing on immediate, less controversial strategies will further salmon recovery efforts more than waiting the decade or so likely required to overcome political opposition to dam removal plans.

A serious obstacle to successful completion of either NPPC or NMFS efforts is the complex nature of the many stakeholders in the Columbia River and its fish populations. At least this is the view of Kai N. Lee, former member of the NPPC and director and professor of environmental studies at Williams College in Williamstown, MA. As Lee has noted, the Columbia River Basin Fish and Wildlife Program is either implemented or significantly influenced by 11 state and federal agencies, 13 Native American tribes, over 10 public and private utilities that own and operate major hydroelectric projects in the Columbia drainage, and numerous organized interests ranging from agricultural groups anxious to protect water rights to fly fishers impatient for the return of native fish stocks. Lee commented, "If the river is to revive in any sustainable sense, it will have to be managed with a stability, a durability, and an awareness of biology which is exceedingly rare in human affairs."

stocking; (2) human management, for example, regulating the numbers of fish caught; and (3) habitat management (maintaining adequate habitats) and restoration. Let's consider each topic, examining the pros and cons of each one.

Population Enhancement Techniques

Artificial Propagation and Stocking

In the early history of fish management, it seemed logical to biologists and anglers that if human beings could supplement the natural reproduction of a given fish species by artificial methods and introduce those artificially propagated fish into lakes and rivers, fish populations would be augmented and the angling success of fishers virtually assured. Since then, hundreds of programs have been initiated, and billions upon billions of fish have been released.

Problems After intensive studies of population dynamics and reviews of numerous projects, it has become apparent that **artificial propagation** (also referred to as **captive breeding**) is more often a failure than a success. Moreover, the cost of artificial propagation in terms of facilities, maintenance, staff, and the rearing and eventual distribution of the young fish is almost prohibitive.

Artificial propagation is not well regarded by some fisheries biologists for several other reasons as well. Hatchery-bred fish are reared under more stressful environmental conditions and often are less well suited to the streams into which they are released. These fish tend to be physically weaker, generally more susceptible to parasites and diseases, and prone to more severe manifestations of a particular disease than wild or native fish. Therefore, artificial propagation and **stocking** of lakes or streams with these fish carries a high risk for the spread of disease to wild (transplanted but now self-sustaining) and native fish populations. In addition, captive breeding of *native* fish to enhance populations in their native stream, carried out for more than one generation, puts the population at risk for inbreeding and loss of genetic diversity.

Whirling Disease An extremely pressing problem in the intermountain West is **whirling disease,** which is currently infecting wild trout populations. Whirling disease is caused by the fish parasite *Myxobolus cerebralis.* Researchers believe that it was unintentionally introduced in 1955 from Europe to fish hatcheries on the East Coast of the United States. The mud bottoms of some of these hatchery rearing ponds presented an ideal habitat for the *Tubifex* worm, another host of the parasite (Figure 12.9). The transfer of the infected hatchery fish may have spread the disease to wild and native fish species in other regions of the country. Rainbow trout, especially those less than two years old, appear to be the most susceptible to whirling disease. Visible symptoms of the disease are whirling behavior, black tail, and cranial and skeletal deformity. There is no treatment for this disease; infected hatchery fish must be destroyed to prevent further spread.

In the upper Colorado River drainage, wild, young rainbow trout are severely infected. A 1994 study of an 8-kilometer (5-mile) reach of the river confirmed what may be interpreted as a catastrophic decline in population. Between 1991 and 1994, the rainbow trout population decreased so rapidly in this section of river that only a few fish remained that were less than 30 centimeters (12 inches) long. The lack of natural recruitment of young fish over several seasons suggested that the local population could collapse.

Whirling disease is also affecting an extremely popular trout-fishing destination, the Madison River in western Montana. After years of a stable wild-trout fishery, 1991 population

Life Cycle of Whirling Disease

Whirling Disease poses a serious threat to trout populations in 25 states.
To prevent the spread of this disease it is helpful to understand its life cycle.

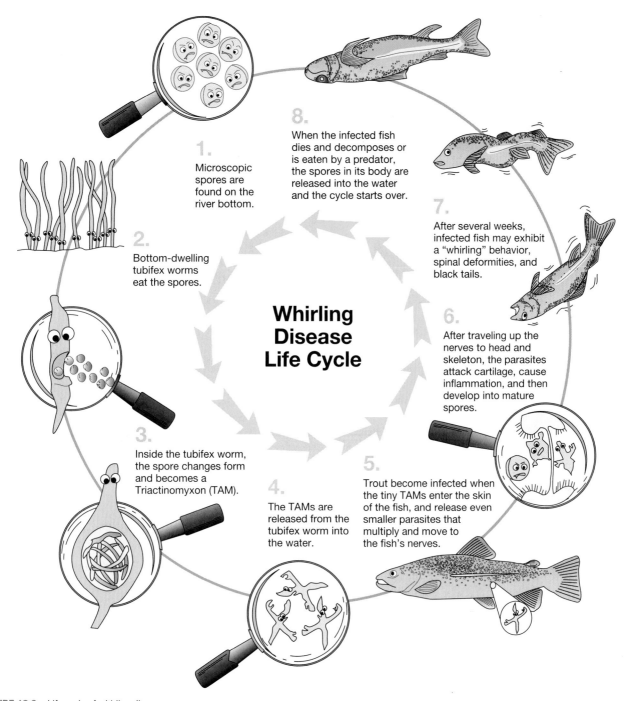

1.
Microscopic spores are found on the river bottom.

2.
Bottom-dwelling tubifex worms eat the spores.

3.
Inside the tubifex worm, the spore changes form and becomes a Triactinomyxon (TAM).

4.
The TAMs are released from the tubifex worm into the water.

5.
Trout become infected when the tiny TAMs enter the skin of the fish, and release even smaller parasites that multiply and move to the fish's nerves.

6.
After traveling up the nerves to head and skeleton, the parasites attack cartilage, cause inflammation, and then develop into mature spores.

7.
After several weeks, infected fish may exhibit a "whirling" behavior, spinal deformities, and black tails.

8.
When the infected fish dies and decomposes or is eaten by a predator, the spores in its body are released into the water and the cycle starts over.

Whirling Disease Life Cycle

FIGURE 12.9   Life cycle of whirling disease.

estimates in a section of the river indicated a significant decline in the numbers of wild rainbow trout. In 1993, a second section of the river also experienced an abrupt decline, and by 1994, both sections reported a 90% decrease in wild-trout numbers, based on the averages reported in the 1970s and 1980s. Tissue analyses of fish specimens from these sections confirmed whirling disease. Further sampling in the upper 88 kilometers

(55 miles) of the river indicated that 75% of the young trout examined were infected. In 1995, population estimates showed further declines in the Madison River trout numbers, and whirling disease was discovered in other Montana waters.

Whirling disease is currently found in wild fish and fish hatcheries in 25 states. Researchers are trying to determine whether unique conditions exist that may put wild-trout populations in this region at risk. An increased understanding of the disease will also aid in the development of management strategies that will prevent the spread of whirling disease to uninfected waters.

Values   Without the artificial propagation of trout, the thrill of hooking one of those fish would soon be nothing but a memory for most anglers. For instance, in Virginia, about 850,000 catchable trout are stocked annually in 185 streams and 20 lakes. Currently, federal and state fish hatcheries rear brook, brown, cutthroat, rainbow, and lake trout. In mountainous areas of Wyoming and Colorado, trout fingerlings may be stocked by means of aerial drops. In Colorado, nearly all of the trout stocked in its rivers are caught in just a few months!

Captive breeding and fish stocking can be valuable as a short-term measure in some cases. Fish stocking, for example, may be used to reestablish fish populations that have been destroyed by predators, drought, pollution, disease, or some other environmental factor. It can also be used for other reasons. For example, fisheries biologists may stock southern farm ponds with tilapia to get rid of excess aquatic vegetation. They may stock reservoirs with predatory species to reduce the number of bluegills. Reservoirs may be stocked with rainbow, brook, or brown trout if the water temperature is suitable. Largemouth bass may be stocked in reservoirs and farm ponds where water gets too warm for trout. However, this technique should be saved as a last resort to restore native species to their original habitat.

The official "trout policy" of federal hatcheries is to propagate to fill the following needs: (1) to stock trout in suitable waters in which they do not occur (newly created reservoirs or waters from which competitive nongame fish have been removed); (2) to stock trout in waters where conditions for growth are good but where natural spawning sites are inadequate (in which case growth is usually rapid, but streams must be restocked at intervals of one to three years); and (3) to stock trout in waters where fishing pressure is heavy but there is no natural production.

In all three instances, trout released into streams and lakes are of catchable size. Most of them are caught the same season they are "planted," and this strategy is known as **put-and-take stocking.** Put-and-take stocking is common in urban areas. For instance, thousands of legal-sized rainbow trout are planted annually in lakes in the metropolitan area of Denver, Colorado. Several U.S. presidents who were sportsmen, including Dwight Eisenhower and Lyndon Johnson, boasted about the fishing potential of a particular trout stream on the basis of the lunkers they hooked only minutes after strategic stocking by publicity-sensitive conservation officials!

Introductions   An **introduction** is the stocking of a nonnative fish. Although this practice was popular in the United States in the past, it can have some detrimental affects on the existing biological community, as noted earlier in the chapter. If, for example, the nonnative fish species becomes established, it can compete for resources with native fish. Introduced species can hybridize with genetically similar native species, and they can prey on native fish species or their eggs. They can also alter the community structure. These effects can individually or collectively lead to the decline or extinction of native species. The negative effects from introductions of the European carp and river ruff were discussed earlier.

Some introduced species can survive without causing a major disruption by finding an ecological niche not fully utilized by the existing fish community. Coho salmon and chinook salmon, native to many drainages to the Pacific Ocean from Alaska to California and in Japan, have been successfully introduced into the Great Lakes. Many lakes in Minnesota, Michigan, New York, and other states that previously had no walleye or muskie fishing now produce trophy-sized specimens, thanks to introductions from other waters in this country. New reservoirs are frequently stocked with species that were not originally found at the site. The great success of warm-water sport fishing in California has been possible largely because of introductions. Twenty-one of 24 warm-water species in California were introduced from states east of the Rocky Mountains, mostly in the late 1800s (Table 12.4).

The introduction of the European brown trout more than 100 years ago is also considered highly successful in some areas of the country. This species has established itself in waters either too warm or too badly polluted for native trout. As a result, it has provided thrills for the anglers even in urban areas. The brown trout is able to survive in the relatively warm, somewhat muddy waters of Lowes Creek, for example, which is only a stone's throw from the city limits of Eau Claire, WI, a bustling city of 58,000.

Vibert Boxes   Local angler groups can greatly increase trout populations in their favorite streams by using a simple but ingenious device called a **Vibert box** (Figure 12.10). This is a plastic box with slots on all sides to permit the free flow of stream water. About 100 trout eggs are placed in the box, which is then planted in the gravel bed of the stream. The Vibert box (1) permits the eggs to develop under natural conditions, (2) protects the eggs from predation, and (3) is inexpensive. (One trout-fishing club planted 50,000 brown trout eggs by this method at a cost of only $300. Hatchery production methods would have been ten times as expensive.) Ninety percent of the eggs in a Vibert box hatch, compared with natural hatching success of only 15%. The newly hatched fish (fry) are immediately conditioned to their environment of dissolved oxygen, water temperature, water chemistry, and stream flow. They therefore have greater ability to resist environmental stress than hatchery-reared stock.

Reservoir Stocking   Few states are blessed with the number of natural lakes found in Minnesota (22,000) or Wisconsin (8,000). However, many benefit from artificial lakes created when rivers were dammed to produce hydropower or to control flooding or provide recreational opportunities. These measures have resulted in the formation of thousands of artificial lakes, or **reservoirs,** throughout the United States, especially in the South and West. Interestingly, today reservoirs account for more than 25% of all freshwater fishing.

As you can imagine, the temperature, chemistry, and biology of the water in a reservoir are quite different from those of the river above the dam. The waters of newly created reservoirs dissolve nutrients from the newly submerged soils, which is of great significance to fish management. The sudden increase in water fertility that results is called a **nutrient flush.** In addition to soil nutrients, reservoirs are enriched by various organisms such as grass, shrubs, insects, worms, and mice from the flooded land. These organisms decompose and release additional nutrients into the waters of reservoirs.

TABLE 12.4 Some Native Fish Introductions to California's Inland Waters

Species	Year	Source	Introduction Site
Smallmouth bass	1874	Lake Champlain, VT	Napa River
Channel catfish	1874	Mississippi River	San Joaquin River
Largemouth bass	1879	Eastern United States	Crystal Spring Reservoir (San Mateo County)
Yellow perch	1891	Illinois	Feather River (Butte County)
Lake trout	1894	Michigan	Lake Tahoe
Bluegill	1908	Illinois	Placer and Orange Counties
White bass	1965	Nebraska	Lake Nacimiento (San Luis Obispo County)
Blue catfish	1969	Arkansas	Lake Jennings (San Diego County)

Nutrients from these and other sources are "channeled" upward through the aquatic food chain. As a result, fish growth and reproduction are usually excellent. In one impoundment in Kentucky, anglers caught twice as many fish as they had at the same location prior to the construction of the reservoir.

Stocking Reservoirs Fishing success may improve dramatically if reservoirs are stocked. For example, it is possible to catch trophy-sized muskies in Virginia's Smith Mountain Lake thanks to stocking. Fish stocking is most effective in four situations: (1) in newly formed impoundments; (2) when introducing predators (bass) to control an overpopulation of stunted prey

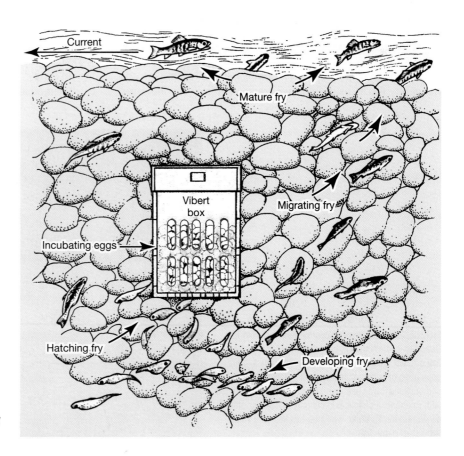

FIGURE 12.10 Stream-planted Vibert box. View of section through gravel on stream bottom. All stages of fry development are shown.

(bluegill); (3) when compensating for the severe reproductive failure of a game species; and (4) when stocking a forage fish (bluegill, threadfin shad) to provide food for desirable predatory species (spotted bass, northern pike).

If the reservoir is deep enough, it becomes thermally stratified during the summer, complete with epilimnion, thermocline, and hypolimnion. Such an impoundment is referred to as a **two-story reservoir.** At least 30 states stock their two-story reservoirs with both warm- and cold-water species: bass, catfish, and bluegill for the warm epilimnion and rainbow, brown, and lake trout for the cold hypolimnion.

Translocations Replacing or enhancing native fish populations that have disappeared from or are diminished in their original water body can be accomplished with little threat to the existing species through **translocation** from another water body. The translocation process involves obtaining large numbers of fertilized eggs directly from spawning adults of the same species that are well established in an ecologically similar water body. The eggs are then immediately transferred to and placed in the original water body, thus creating a new population. The eggs may also be hatched in a captive breeding station (hatchery) and later introduced as young into the original water body. However, this second option is the least desirable, owing to the aforementioned risks of stress, disease, and loss of genetic integrity. Before all translocations, however, the root cause(s) of the decline or extinction of the target species in the original water body must be identified and ameliorated. Bypassing this first, important step will undermine the success of the translocation.

Removing Undesirable Fish Because of their destructiveness to game fish, large populations of **nongame fish** (gar, alewife, threadfin shad, common carp, quillback, spotted sucker, brook stickleback, mottled sculpin) and "stunted" **panfish** (black bullhead, white perch, rock bass, bluegill, black crappie, warmouth, redbreast sunfish, orange-spotted sunfish) are frequently the focus of intensive eradication projects. However, eradication of any of these species from any body of water is enormously difficult.

Various control methods under study involve chemicals, seining (netting), commercial fishing, manipulation of water levels, and fish-spawning control. Before state or federal biologists use a specific chemical, it must first be registered with the U.S. Department of Agriculture and approved by state health and pollution agencies and by the Federal Committee on Pest Control.

Rotenone, a chemical derived from the roots of an Asiatic legume, kills fish at a concentration of only 1 ppm within minutes at a water temperature of 21°C (70°F). Unfortunately, poisoning with rotenone is nonselective, resulting in the indiscriminate death of many species (Figure 12.11). The chemical control agent **antimycin** kills carp more readily than it does most other fish and does not appear to be deleterious to invertebrates. The long-term effectiveness of these techniques is questionable, however. In 1994 a group of researchers in Wisconsin reviewed 250 fish-control projects located in 36 states and three counties. The study results revealed that less than 50% of the projects were considered successful, suggesting that improvements are necessary. The researchers concluded that many fish-control projects are initiated without thorough investigation into the causes of the overpopulation of the undesirable species. Many times the underlying cause of the problem is habitat degradation, decreased water quality, or overexploitation of these species' predators by a fishery. Therefore, chemical or physical removal techniques are only short-term treatments for the symptom rather than for the cause of the problem.

Controlling Oxygen Depletion in Winter Various methods are available to reduce winterkill of fish caused by oxygen depletion: (1) If the lake is small, the opaque snow blanket may be removed with plows. This will permit sunlight to penetrate to aquatic vegetation so that photosynthesis can occur and the water can be oxygenated. (2) Dynamite may be used to blast holes in the frozen lake to expose surface waters to atmospheric oxygen. (3) Oxygen can be introduced through ice borings by motorized aerators.

FIGURE 12.11 Chemical control of undesirable fish. Dead fish by the thousands float belly-up in a small bay of Clear Lake near Watkins, MN. The lake's entire population of undesirable (rough) fish, including carp and bullheads, was destroyed after the lake was treated with rotenone, a chemical lethal only to fish and other gill-breathers. This lake was later stocked with valuable species of game fish.

Selectively Breeding Superior Fish Fisheries biologists in Wisconsin have crossed northern pike with muskellunge to develop a hybrid known as a tiger muskie. The tiger muskie has been successfully stocked in reservoirs. In these artificial lakes, they fare much better than either of the parental species. Larger and higher-quality fish are being developed at the Federal Fish Farming Station at Stuttgart, AR. A rapidly growing hybrid catfish, for instance, has been produced by crossing a channel catfish with a blue catfish. When two years old, the hybrids weigh 32% more than similarly aged blue catfish and 41% more than channel catfish of the same age. Another successful hybrid in the South is the hybrid striped bass, which has been widely stocked and exhibits rapid growth and good environmental tolerance.

Because it is essentially impossible to predict the effects of a "new" fish on a habitat or native organisms, selectively breeding superior fish is highly controversial. Introducing genetically engineered or **hybrid fish** species should be done on a very limited basis, with considerable forethought of potential consequences, and only in artificial impoundments or small, "closed" water bodies. Even then, it may be impossible to prevent people from catching the fish and releasing them into other waters.

Protective Legislation

Fish populations are also controlled by regulations that limit the take—the size and number of fish an angler can take home. Similarly, some species or fishing areas are regulated by **"catch and release only" restrictions.** In such instances, as the name implies, anglers must release all the fish they catch. Careful measures are required to minimize injury to such fish. **Closed seasons** also are used to protect species at critical times (Table 12.5). Fisheries biologists have long recognized that when female bass or walleye are taken when swollen with eggs, anglers are removing much more than a single adult. They are also removing hundreds of future young fish. Over the years, governments have outlawed certain fishing techniques, such as seining, poisoning, dynamiting, spearing, and using multiple-hook lines.

In recent years, fisheries biologists have been experimenting with more liberalized regulations on many species of warm-water fish. In many states, size limits on panfish (sunfish, bluegills, rock bass, and crappies) have been lifted, permitting fish of any size, from runts to giants, to be taken. In contrast, **minimum size limits** have been placed on predatory species such as bass, pike, walleyes, and muskies. The main objective of these regulations is to ensure the presence of large predators that can control populations of panfish. Such regulations provide more opportunities for anglers to land a lunker bass or pike. However, researchers have shown that size limit regulations do not affect the survival rate of northern pike (Figure 12.12). The effects of **creel (catch) limits,** varying fish methods and gear, open and closed seasons, and winter fishing on fish populations are continually being evaluated.

The most ecologically sound and hence most effective regulations are those tailored to a given body of water. Such regulations are formulated on a lake-by-lake or stream-by-stream basis. Unfortunately, however, their administration and enforcement are often difficult.

Habitat Management and Restoration

Boosting fish populations and limits on fishing help maintain fish populations. Fisheries biologists also employ another approach: alterations of habitat in artificial and natural surface waters. Consider first some measures that are used in reservoirs to enhance freshwater fisheries.

Preventing Harmful Drawdowns The water level of a reservoir can be purposely lowered or raised by the dam operator

TABLE 12.5	Typical Fishing Regulations for a Northeastern State		
Species	Season	Daily Limit	Minimum Size
Largemouth bass	May 7–March 1	5	None
Bluegill, sunfish, crappie, perch	Open all year	50 in all	None
Catfish	Open all year	10	None
Muskellunge	May 28–Nov. 30	1	32 in.
Northern pike	May 7–March 1	5	None
Walleye	May 7–March 1	5	None
Lake trout	Jan. 2–Sept. 30	2	17 in.
Trout (other than lake trout)	May 7–Sept. 30	3 in all	Brook trout—10 in.
			Brown trout—13 in.
			Rainbow trout—6 in.

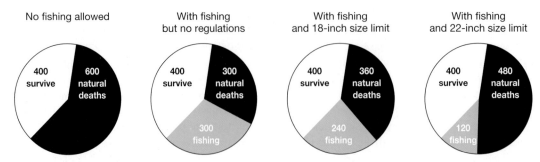

FIGURE 12.12 Impact of size limits on the number of northern pike that die during a given year. These size limits have little, if any, effect. Even if no fishing at all is permitted, the mortality of a given northern pike population will be the same.

to control floods or generate hydropower. Unfortunately, the fluctuating water level may seriously affect fish populations. For example, a **drawdown** of just a few meters shortly after the lake trout have spawned could leave their eggs high and dry. The entire spawn could be destroyed. If a drawdown occurs shortly before spawning, the lake trout are forced to spawn in an area where many of the eggs might be consumed by other fish, such as bullheads. Thus, to realize the reservoir's fish production potential, the fish manager needs the cooperation of the owner and operator of the dam and reservoir.

Beneficial Drawdowns Drawdowns can also benefit the fisheries of a reservoir. For example, in late summer, fish managers may request dam operators to draw down 10% to 80% of the water (Figure 12.13). The objectives include (1) aeration of the bottom muck, (2) acceleration of the bacterial decomposition of organic material and the release of nutrients, (3) restriction of forage fish to a small area where they can be more easily caught and eaten by predatory game fish, and (4) facilitation of nongame fish removal.

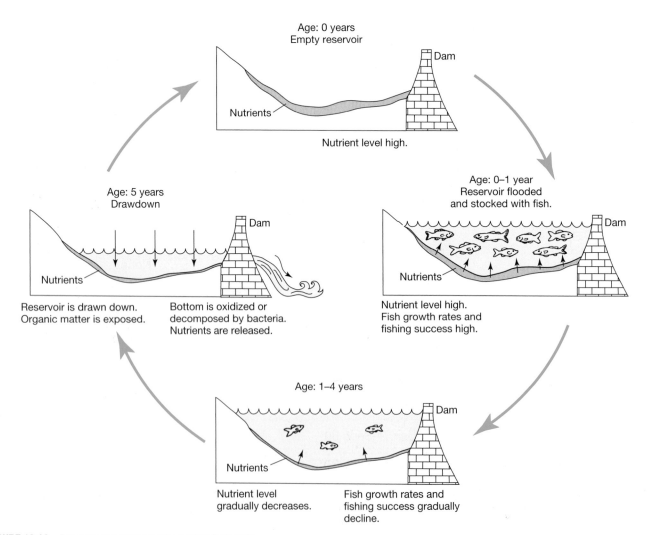

FIGURE 12.13 Reservoir drawdown and the nutrient flush cycle.

Some time after drawdown, the reservoir is refilled and re-stocked with game species. Due to the nutrient flush effect, a large quantity of food (diatoms, crustaceans, insects, minnows, and so on) become available to the fish of this artificial ecosystem. As a result, both growth and reproduction are enhanced. For example, in the 11,200-hectare (28,000-acre) reservoir in Beaver Creek, AR, the average weight of pike increased 2 kilograms (4.4 pounds) annually during the first three years after drawdown. In an impoundment on the Rough River in Kentucky, the average weight of catfish progressively increased through the fourth year after drawdown, at which time it was 473% greater than during the years before impoundment! Unfortunately, however, in most reservoirs, the positive effects of the nutrient flush diminish by the third or fourth year. At this time, therefore, another drawdown is required.

Habitat protection—preventing streams and lakes from being damaged—and restoration (described next) are also essential to successful fisheries management. Habitat protection requires efforts to limit or prevent pollution from entering surface waters. Many such measures were discussed in the previous chapter. Because many pollutants come from nonpoint sources, watershed management and protection also are essential. This topic is also discussed in Chapter 11.

Habitat restoration—repairing damage to water bodies—involves measures to actively repair damage to streams,

lakes, and rivers. The success of habitat restoration efforts hinges, in large part, on an effective management plan that includes routine, long-term monitoring of vegetation in watersheds, especially along the banks of rivers and lakes. It also requires monitoring of water quality and fish productivity. Managers must be able to evaluate changes in the habitat over time and adapt necessary strategies that address problems as they arise. Unfortunately, comprehensive programs of this sort are rare because of the financial and time obligations they require.

The first step in habitat management is to identify and mitigate the existing factors that are limiting production. Native fish populations may increase considerably when the carrying capacity of their environment is raised; that is, when adequate water flows, depths, and temperatures, space, substrate, cover, and food supplies are provided, fish populations may make a dramatic recovery. Various strategies can be used to improve the stream habitat for fish: adding spawning gravels, revegetating stream banks and lake edges, restoring the historic meanders of rivers, creating pools and riffles, and placing instream woody debris. Some methods of stream improvement are shown in Figure 12.14.

A series of **brush shelters** may be anchored along the inner margin of a lake's littoral zone with great effectiveness. In the winter season, brush piles can be set up on the ice cover in

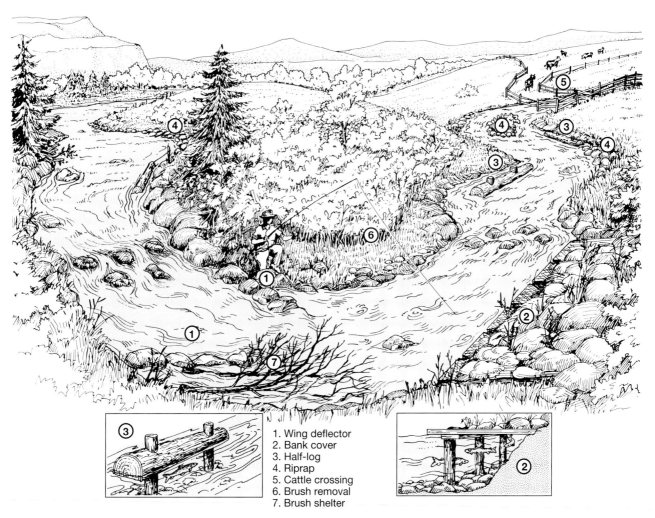

1. Wing deflector
2. Bank cover
3. Half-log
4. Riprap
5. Cattle crossing
6. Brush removal
7. Brush shelter

FIGURE 12.14   Stream habitat improvements.

strategic areas and weighted with bags of stones. When the ice melts, the brush shelters will sink to the lake bottom, providing important habitat.

In the case of heavy grazing, sometimes all that is necessary is to eliminate the culprits. Research by the U.S. Forest Service has shown that if the sides of a severely grazed stream are fenced off from cattle for five to ten years, the stream's original ability to support large trout populations will be restored, as shown in Figure 12.15.

Fish populations may be increased by improving natural **spawning sites** and by providing artificial spawning surfaces where suitable natural ones are lacking.

1. **Sites for bass.** Sand or small gravel can be spread on the muddy bottoms of lakes and streams to provide spawning sites for bass. Nylon mats can be used as an artificial spawning surface for largemouth bass. In one experiment, 90 mats were tested in several ponds. Over a two-year period, the bass spawned on nearly 75% of the mats.

2. **Sites for northern pike.** Intensive study of the breeding behavior of the northern pike has shown that the shallow, marshy fringes of the littoral zone are the preferred spawning habitat. Regrettably, in recent years, suitable spawning sites for this species have been greatly reduced as a result of real estate development, marina construction, and industrial expansion. State fish and game departments are attempting to rectify the situation. Wisconsin, Iowa, and Minnesota, for instance, have acquired thousands of acres of marshes to provide suitable breeding habitats for pike.

3. **Sites for lake trout.** Hatchery-reared lake trout have been stocked in the Great Lakes for several decades. However, only a small percentage of the trout successfully spawn. Ross Horrall, a researcher from the University of Wisconsin–Madison, located in the late 1970s traditional lake trout spawning reefs (mounds of submerged rocks) that were successfully used long before the sea lamprey decimated this species. One of these reefs was located near the Apostle Islands in Lake Superior. The Horrall research team placed 273,000 fertilized lake trout eggs in huge **Astroturf "sandwiches"** (Astroturf is the plastic "grass" used on indoor football fields). The sandwiches stabilize the eggs, protect them from sediment and predators, and make retrieval easy (Figure 12.16). After a seven-month incubation period, the Horrall team found that 88% of the eggs had hatched.

Often a compounding effect is associated with restoration activities. Each habitat management method will have a positive

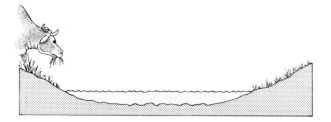

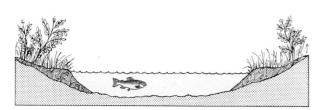

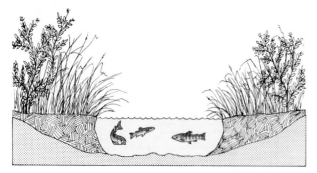

FIGURE 12.15 Schematic representation of stream conditions after heavy grazing and recovery. With heavy grazing (top), stream banks are grazed and trampled, leading to increased stream width, shallow water, and poor habitat for trout. Shallow water may be warmed enough by direct sunlight to limit a trout fishery. After two to three years without grazing (middle), vegetation begins recovering as stream banks again develop their structure. The habitat is improving for trout as food, cover, and spawning conditions become more favorable. Conditions after five to ten years without grazing (bottom) offer excellent trout habitat. Overhanging banks and deeper water have been restored as the vegetation recovered. Sedimentation also is reduced significantly.

FIGURE 12.16 Astroturf egg "sandwich." Ross Horrall, a fisheries biologist at the University of Wisconsin–Madison, displays an Astroturf sandwich, a device that holds and protects lake trout eggs during incubation.

effect that, in turn, may result in other positive effects. For example, suppose that cover is increased by the introduction of brush shelters (Figure 12.14). The cover protects the fish from predation by mink, otters, bears, fish hawks, and herons. As a result, fish become more numerous. Shelters have an additional positive effect. As shown in Figure 12.14, the twigs and branches of the shelter serve as attachment sites for insect larvae, snails, and crustaceans. Those organisms provide food for the growing fish populations. As a consequence, the growth rates increase, resulting in bigger fish. Such shelters also provide shade where fish may retreat during the heat of the day. The cooling effect on the water enables it to dissolve more oxygen. The increased levels of oxygen enable the fish to swim faster. Figure 12.17 shows the pathways of effect for several other stream management methods; among these effects are cover, an increased water flow, and improved water fertility. For a summary of **stream improvements** for fish, see Table 12.6.

Priority for restoration efforts and funding should be given to the protection of remaining high-quality sites and to the restoration

GO GREEN!

Join or start a local stream restoration group or a local Adopt-a-Stream program. These volunteer groups monitor their local streams and help with stream cleanup, planting of riparian vegetation, and other stream restoration activities.

of sites that are, at worst, moderately degraded yet still contain valuable, threatened, or endangered fish species. Restoration of habitats that are severely degraded or have lost most of their valuable fish may be prohibitively expensive and may have a much lower chance for success.

12.4 Marine Fisheries

In 2004, the oceans provided approximately 104 million metric tons of food—mostly fish such as tuna, shellfish such as clams, and aquaculture—out of a total of 140.5 million metric tons for marine and freshwater fisheries, according to the FAO Fisheries and Aquaculture Department. China remains by far the largest producer, with reported fisheries production of 47.5 million metric tons in 2004 (Figure 12.18). Most of the species that form the basis of the world's marine fishing industry are harvested near shorelines in the areas of the continental shelf and margin. With production totaling about 10.7 million metric tons in 2004, the Peruvian anchoveta leads by far the ranking of the ten most caught marine species (Figure 12.19).

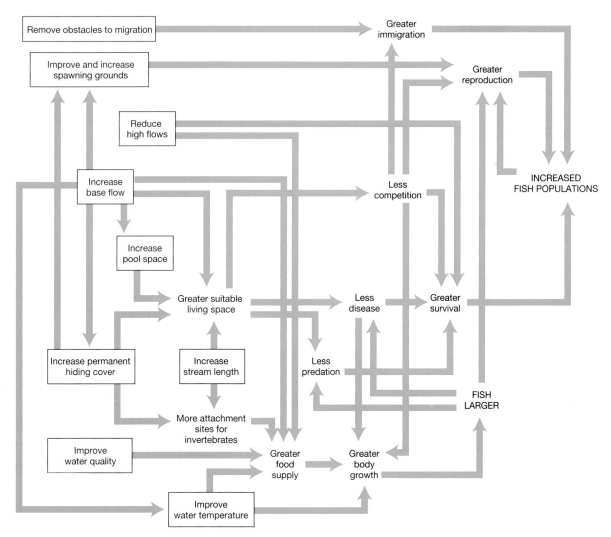

FIGURE 12.17 "Chain responses" resulting from the application of various habitat improvement methods to streams. The methods are shown in boxes.

TABLE 12.6   Stream Habitat Improvements

Problem	Habitat Improvement	Result
Not enough shelter or living space	Wing deflector	Channel deepens; pools form.
	Bank cover	Cover increases; predation and competition decrease.
	Half log	
Stream overgrown with trees and shrubs	Brush removal	Sunlight reaches stream; more food is produced.
Erosion of stream banks	Riprap	Banks stabilize; water clears; channel deepens.
	Cattle crossing	
	Fencing	
Poor spawning success	Wing deflector	Silt is scoured from gravel beds.
Water too warm	Narrow and deepen channel	Water temperature falls.
Too much predation; lack of food	Brush shelter	Brush provides cover from predators and habitat for food organisms.

Marine Commercial Fishing Techniques

Methods of Locating Fish   In recent years, some highly sophisticated methods have been developed for locating commercial stocks of marine fish:

1. **Sonar,** or echo-sounding systems, which use sound waves, help to locate fish schools and determine their relative abundance.

2. **Moored buoys** equipped with sonar or other devices detect the presence of fish schools swimming nearby and then radio the information to fishing vessels.

3. **Color enhancement** by either photographic or electronic means can detect color differences in the ocean that are undetectable with the unaided eye. These images provide information on the location of fish.

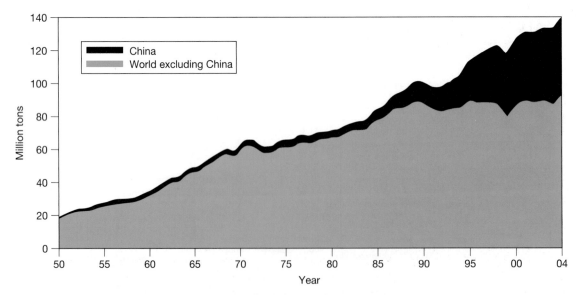

FIGURE 12.18   Total global fisheries production from 1950 to 2004. Out of a total of 140.5 million metric tons in 2004, 104.1 million metric tons of food came from oceans and 36.4 million metric tons came from inland freshwater fisheries. China was the largest producer with 47.5 million metric tons of production in 2004. Aquaculture, both freshwater and marine, accounted for 45.5 million metric tons of the total 140.5 million metric tons in 2004. (Source: FAO)

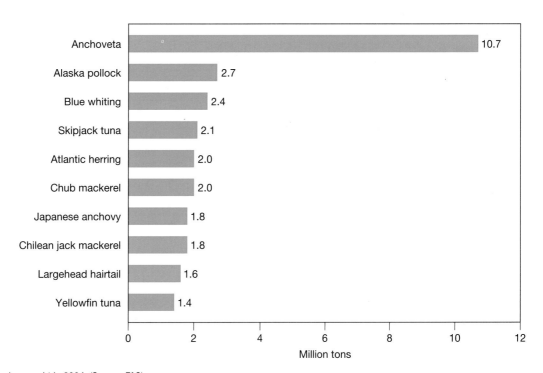

FIGURE 12.19 Top ten marine species caught in 2004. (Source: FAO)

4. **Infrared sensors** borne by airplanes or spacecraft sense the ocean's temperature, which can determine fish movement. Temperature detection is valuable because some species of commercially valuable fish, such as tuna, have highly specific water temperature preferences.

5. **Ultraviolet sensors** can detect the presence on the ocean surface of oils given off by fish as they swim through the water. Since the oil rapidly dissipates when exposed to air, its occurrence indicates the relatively recent presence of a school of fish.

6. Airborne **electronic image intensifiers** can detect the flashes of light (bioluminescence) given off at night by microscopic marine organisms when disturbed by passing fish. The faint light is intensified 55,000 times and then projected on a television monitor.

Methods of Harvesting Fish Commercial fishing fleets use both passive and active fishing gear. **Passive fishing gear** includes methods that are stationary and, therefore, depend on the fish or shellfish to come into contact with them. **Active fishing gear** includes methods that are mobile and are able to seek out and capture the fish or shellfish.

Passive Gear

1. **Pots and cages.** Pots and cages are used for catching crustaceans such as crabs and lobsters (Figure 12.20).

2. **Set nets, pound nets, and traps.** These nets are situated in such a manner that fish can easily swim in but can escape only with extreme effort.

3. **Hooks and lines.** A line with numerous baited hooks is set out to the desired depth. A long line can be over 1 kilometer (0.62 mile) in length and contain more than 400 hooks (Figure 12.21).

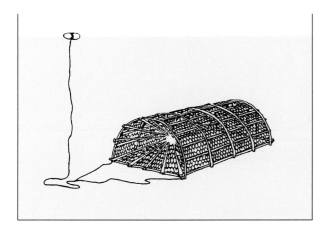

FIGURE 12.20 Example of a pot used for crabs or lobsters.

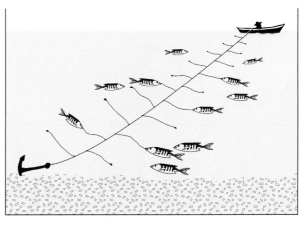

FIGURE 12.21 Hook and line.

4. Gill net. The fishing vessel bearing a gill net moves into a promising area shortly before nightfall. Up to 4.8 kilometers (3 miles) of net may be laid at right angles to the incoming or outgoing tide or a current. The net is buoyed up by floats at the top and is weighted at the bottom with lead sinkers, thus forming a wall across the path of targeted fish stocks (Figure 12.22). The nets can be anchored (set nets) or allowed to float freely (drift nets). As schooling species try to pass through the net, their gills become entangled. Unable to swim so water can continue to pass over their gills, the fish suffocate and die from lack of oxygen. The mesh is designed so that it catches only fish above a minimum size. The catch is hauled into the boat at dawn.

Active Gear

1. Purse seine. More fish are caught by purse seines than by any other type of net (Figure 12.23). The nets are played out in circles around schools of fish. When a school is completely encircled, the bottom of the net is drawn in to prevent fish from escaping. The purse line at the net bottom works like the drawstring on an old-fashioned purse. A large purse seine may be almost 1.6 kilometers (1 mile) long, have 16,000 floats, and weigh 13.6 metric tons (15 tons).

2. Trawl net. The trawl is a baglike net that is pulled by the fishing vessel. Most trawls are drawn over the ocean bottom and are used to harvest cod, flounder, haddock, ocean perch, and shrimp (Figure 12.24).

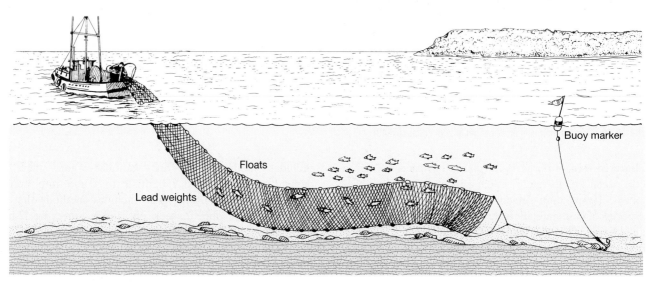

FIGURE 12.22   Gill nets being used on the ocean bottom. These nets may also be suspended from the surface by floats. Fish become entangled in the net as they try to swim through it.

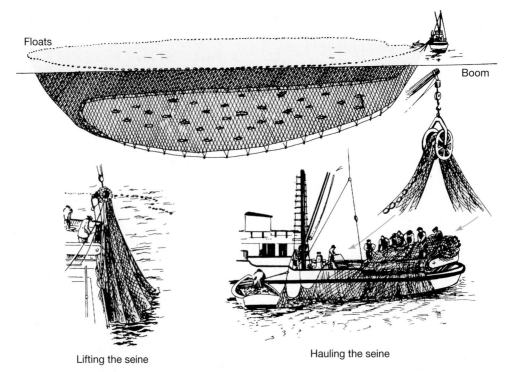

FIGURE 12.23   Purse-seining for salmon: equipment and techniques.

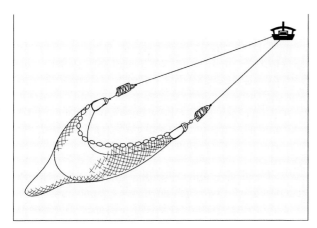

FIGURE 12.24 Trawl net (otter type).

3. **Drift net.** In theory, drift netting is a passive technique, but the enormous size of some nets (sometimes called "walls of death") makes it impossible for many marine organisms to avoid them. Therefore, they are considered by some as an active technology.

12.5 Problems Facing Marine Fisheries

According to the Food and Agriculture Organization (FAO), the overall state of exploitation of the world's marine fishery resources has tended to remain relatively stable. During the past 10 to 15 years, the proportion of overexploited and depleted stocks has remained unchanged, after showing a marked increase during the 1970s and 1980s. It is estimated that in 2005, as in recent years, about one-quarter of the stock groups monitored by FAO were underexploited or moderately exploited and could perhaps produce more, whereas about half of the stocks were fully exploited and therefore producing catches that were at, or close to, their maximum sustainable limits, with no room for further expansion. The remaining stocks (about one-quarter) were overexploited, depleted, or recovering from depletion; that is, they were yielding less than their maximum potential, owing to excess fishing pressure.

With more than 75% of world fish stocks reported as already fully exploited or overexploited (or depleted and recovering from depletion), the maximum wild-capture fisheries potential from the world's oceans has probably been reached. More cautious and effective fisheries management is needed to rebuild depleted stocks and prevent the decline of those being exploited at or close to their maximum potential. The decline in economically important fish species has resulted from several factors. One of the most important is the degradation and loss of critical habitat in coastal wetlands, estuaries, and bays—habitat that is vital for spawning, feeding, and rearing. As you may recall from Chapter 11, the loss of habitat has resulted from residential, commercial, and industrial development. Pollution from various human activities also has played a role in the degradation of critical habitat and the accompanying decline of marine species. In this chapter, we examine two additional factors: overfishing and bycatch. Both threaten the long-term health of the marine environment.

Overfishing

Overfishing is a term conservationists and scientists use to describe harvests of aquatic organisms that exceed the capacity of fish populations to sustain themselves. A large part of the overfishing problem stems from the fact that the open ocean (all areas except the 322-kilometer, or 200-mile, strip extending out from each nation's shoreline) is the world's biggest **commons,** a public area over which no nation has sovereignty. The tragedy of such a commons, however, as pointed out by professor Garrett Hardin of the University of California at Santa Barbara, is that its resources are often overexploited. After all, with no restriction on the take, it would seem foolish for a country with a marine fishing fleet not to get what it can as long as possible. This attitude has dominated the fishing industries of seaboard nations for years.

Unfortunately, such attitudes often lead to the depletion of the commons. A classic example is the collapse of a once-

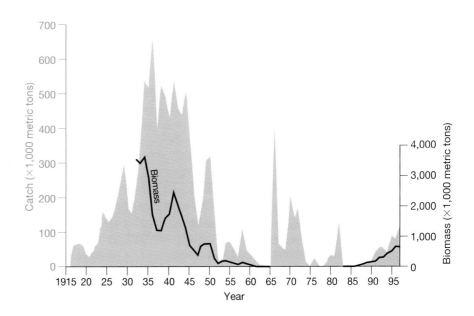

FIGURE 12.25 Pacific sardine catch, 1916–1997.

major Pacific sardine fishery caused by intense human predation. In 1936–1937, the Pacific Coast sardine industry reached its peak. About 660,000 metric tons (727,000 tons) were netted (Figure 12.25), and sardine biomass was about 3 million metric tons (3.3 million tons). Biomass or fish weight is one way of expressing fish abundance or population size. This industry ranked first in the nation in metric tons harvested and third in value of catch. It grossed $10 million annually. The fish were used in many ways, from canned sardines to fish bait, pet food, and fertilizer. Unfortunately, the industry's short-term prosperity depended on overexploitation. And as harvests began to decline, the fishing fleet was enlarged to make up for decreases in the harvest per boat. The industry rejected regulations based on the advice of fisheries scientists. After World War II, however, the sardine population and catches declined. The Washington–Oregon fishery collapsed in 1947–1948. In 1951, the San Francisco fleet returned with only 72 metric tons, well less than 1% of its take only a decade earlier, and the fishery closed down. In the late 1970s, Pacific sardine biomass declined to low levels (a few thousand metric tons). (However, since 1986, sardine populations have increased by 30% to 40% annually. Commercial fishing has resumed but this time with quotas to prevent the fisheries from experiencing the previous problems. In 1997, biomass was about 600,000 metric tons.)

Unfortunately, the sardine fishery is only one among many marine fisheries that have been severely depleted. According to FAO, most of the stocks of the top ten species (Figure 12.19), which account in total for about 30% of the world capture fisheries production in terms of quantity, are fully exploited or overexploited and therefore cannot be expected to produce major increases in catches. Populations of many species, including such highly desirable food fish as Pacific perch, Spanish mackerel, grouper, flounder, bluefin tuna, Chilean sea bass, Atlantic salmon and cod, orange roughy, rockfish (Pacific red snapper), and swordfish, have been critically depleted or are overfished. Figure 12.26 illustrates many of the exhausted fish stocks in the North Atlantic alone.

Overfishing of many species also leads to another problem, sometimes referred to as "fishing down the food web." That is,

as commercial fishing interests deplete large, long-lived predatory fish, such as cod and salmon, they frequently move lower on the food chain—that is, they turn to species that feed on plankton. However, these smaller, less valuable planktivorous species of fish, crustaceans, and shellfish are often the food supply of the larger, more commercially valuable predatory fish. By depleting their food supply, commercial fishing interests—and hungry consumers—make it more difficult for predatory fish to recover after fishing pressure on them has been reduced.

Economics and legislation also play a role in overfishing. As noted previously, each nation has economic jurisdiction over a 322-kilometer (200-mile) belt of sea space bordering its coasts, an area termed an **Exclusive Economic Zone (EEZ).** Scientists estimate that 95% of the world's living marine resources are contained within these EEZs. The intent of the EEZs was to reduce competition from foreign fleets and create a stronger incentive for each nation to manage its marine resources on a long-term, sustained-yield basis. What occurred, however, was a frenzied rush to expand national fishing fleets to take advantage of these new "noncompetitive" fishing opportunities. Governments subsidized construction of fishing vessels and fish-processing facilities, and industry developed more powerful vessels and more effective fishing gear for finding and harvesting fish. The result has been overcapacity, overinvestment, and overexploitation. Because the fishing industry employs and feeds a large number of people and is an important source of export earnings, governments continue to compensate losses through **subsidies.** In the presence of government subsidies, the fair market price does not accurately reflect the value or the scarcity of our marine resources.

Bycatch and Discards

Bycatch is the term used to describe captured marine organisms that are not the target species. **Discards** are the bycatch that are thrown back for various reasons, including nontarget species, juvenile fish, endangered species, wrong size, inferior quality, or surplus to quotas. Bycatch also includes the capture of

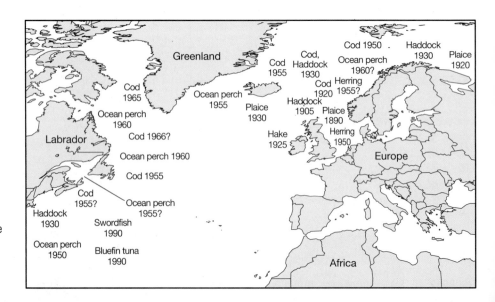

FIGURE 12.26 Many of the exhausted fish stocks in the North Atlantic. The depletion of the plaice (flounder) fishery in the North Sea became obvious as early as 1890. Since that time, many other fisheries have declined sharply. Note the depletion of the cod, ocean perch, and haddock fisheries off Labrador and Newfoundland.

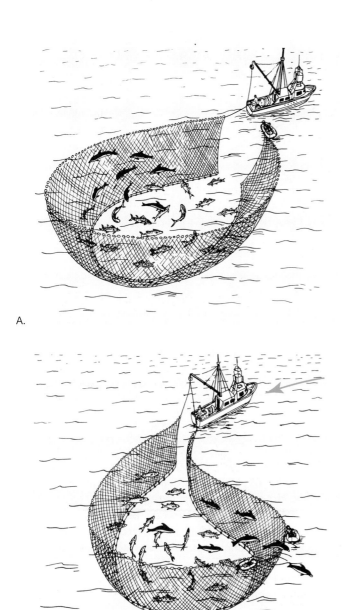

A.

B.

FIGURE 12.27   Dolphin bycatch. (A) Tuna fishers accidentally net dolphins along with tuna. This happens rather frequently, because dolphins swim underneath tuna schools. Fishers are now legally obligated to rescue as many dolphins as possible. (B) To permit the escape of the dolphins, the fishers first draw their net into an elongated shape. Then fishers on rafts help "spill" the dolphins over the end of the net into the open sea.

oceanic birds and marine mammals, the most famous example being dolphins caught in purse seine nets intended for tuna (Figure 12.27). Once thought to occur on a limited basis and otherwise considered an unavoidable aspect of the extensive use of nonselective fishing gear, bycatch is now recognized as a serious problem that may have far-reaching effects for the entire marine ecosystem.

Estimates for annual global discards in commercial fisheries are 7.3 million metric tons, about 8.5% of the weight of the total marine catch worldwide. The shrimp and prawn trawler fisheries have been singled out as the largest contribu-

tors to this problem, accounting for more than 27% of the world's bycatch discards. In recent years, discards have been reduced substantially, mainly due to a reduction in unwanted bycatch and increased utilization of catches. Bycatch reduction is largely a result of using more selective fishing gear, introduction of bycatch and discard regulations, and improved enforcement of regulatory measures. Increased utilization of catches for human or animal food results from improved processing technologies and expanding market opportunities for lower-value catch.

There is still a great need for the development of better bycatch management plans and promotion of best practices for bycatch reduction and mitigation of incidental catches. The bycatch organisms hauled aboard the vessel that are still alive sustain trauma from exposure and handling, resulting in death or injury. Most of the bycatch is dead by the time it is thrown overboard. Organisms that are thrown overboard in an injured or weakened state become easy prey for hungry predators or have an otherwise poor chance of survival. The incidental capture and death of large numbers of juveniles of commercially and recreationally important species may reduce the potential biomass and yield of their stocks. Bycatch of endangered or threatened species threatens the recovery of these species. The bycatch of species that are not of direct commercial value also may represent a serious problem, as these species are very likely an important food source for commercially valued or endangered fish. So, not only can bycatch affect a single species, it can cause a breakdown in the structure of entire biological communities.

12.6 Sustainable Marine Fisheries Management

Under the current marine management schemes, the overriding assumption appears to be that, if the fishing pressure is reduced, depleted fish stocks will bounce back. However, for the return of a healthy, naturally productive environment and a sustainable world fishery industry, many experts recommend a more comprehensive strategy. This strategy calls for the protection of crucial habitat and controls on pollution, in addition to quotas and limits on bycatch. Although a majority of the world's governments, scientists, and fisheries recognize the problems and acknowledge an urgent need to take action, this will be a formidable task requiring international cooperation and consensus.

Toward the goal of sustainable fish harvests, several management options have been suggested and debated over the years. Some strategies focus on regulations that limit harvests, others address the government subsidy issue, and still others aim at pollution cleanup and habitat restoration. As just noted, the effort ultimately must include a combination of all approaches.

Optimum Yield

To set quotas that ensure sustainable catches, many aquatic biologists believe, scientists must obtain scientifically sound estimates of the optimum yield of marine fisheries. In addition, they note that it is crucial that a strict monitoring system be in place to discourage fishers from the temptation to exceed quotas. The scientific research necessary to set quotas and the workforce necessary to monitor catches will require considerable financial resources and technical support.

Most fisheries management has been operating under the principle of **maximum sustained yield (MSY).** MSY, established for each species, is the greatest fish or shellfish catch (in terms of number or weight) that can occur each season without affecting the reproductive capacity of that species. We know from the preceding discussion that, for several reasons, fisheries management is not meeting this goal. Why?

Maximum sustained yield is established through calculations based on factors such as spawning mass, annual recruitment (the number of fish that come of reproductive age each year), annual growth in biomass, nonfishing mortality (natural death, predation), and fishing mortality. Unfortunately, the calculation does not take into account other important factors such as pollution, habitat destruction, and declines in other marine organisms (for example, food organisms). All of these factors affect fish or shellfish breeding, feeding, and rearing success. In addition, the benefits and values that a fishery industry provides to a particular culture are not considered. In summary, then, maximum sustained yield is not really a sustainable yield.

A more holistic view of fisheries management—and one some experts believe will achieve the requirement set forth in the 1976 **Magnuson Fishery Conservation and Management Act** for sustainable yields of important commercial fish and shellfish—is the principle of **optimum yield.** Optimum yield management considers the biological, economic, social, and political values of a given fishery in order to maximize benefits to society. It recognizes that reductions in annual catch alone will not improve the health of fish and shellfish stocks or generate greater social, economic, and nutritional benefits for society.

The optimum yield of a particular fishery must be considered on a case-by-case basis. Establishing it involves an interdisciplinary team of scientists, managers, industry representatives, fishers, and laypeople. Take a moment to examine Figure 12.28 to better understand the concept of optimum yield. Yield is shown on the vertical axis. Use is plotted along the horizontal axis, starting with no use. Optimum and maximum harvest are indicated by the letters O and M, respectively.

This graph shows the intrinsic yield of a fishery. When no fishing takes place, populations provide maximum benefit to the ecosystem's health. As harvesting increases, the population's contribution to the ecosystem declines.

The extrinsic yield of a fishery, also shown on the graph, is its sociocultural, economic, and nutritional benefits to human society. As the harvest increases, the extrinsic yield of a fishery increases. A maximum extrinsic yield—maximum benefit to human society—is reached (at point M in Figure 12.28), after which the benefits to society decrease along with ecosystem health.

The chosen optimum level of use established for any given fishery depends on the goals and resource objectives

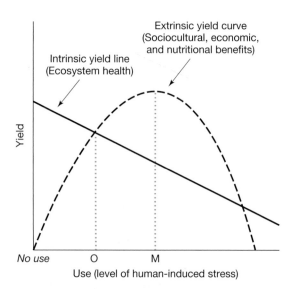

FIGURE 12.28   Conceptual illustration of optimum yield related to fisheries management. The intrinsic yield (health) of an ecosystem declines from its pristine state as the extrinsic yield (sociocultural, economic, and nutritional benefits to a society) of a fishery increases. At point *M*, the maximum extrinsic yield is reached, after which benefits to society decrease along with intrinsic yield. Maintaining the maximum extrinsic yield comes at a high cost to ecosystem health that may eventually outweigh the benefits to society. In theory, point *O* is the optimum yield, which allows the greatest extrinsic yield to society with the smallest loss to ecosystem health (intrinsic yield). Note that optimum yield will be less than the yield necessary to maximize benefits to society.

of the associated society. Point *M*, where extrinsic yield is the highest, might be chosen as the optimum level. However, when maximizing extrinsic yield, costs to the health of the ecosystem outweigh the benefits to society. Ideally, an optimum yield allows the greatest extrinsic yield to society with minimum loss of intrinsic yield to the ecosystem (point *O* in Figure 12.28). In other words, the optimum yield is always less than the yield that would give maximum benefits to society.

Regulations and Economic Incentives

Over the years, scientists, government officials, and others have proposed a variety of methods—some of which have been implemented—to address the problems in the world's marine fishing industry. Let's examine them:

1. **Reducing bycatch.** Some individuals believe that bycatch could be reduced significantly by banning some types of fishing gear, such as trawl nets and drift nets, in certain waters. Economic incentives may need to be created to develop more selective gear and fishing practices—that is, gear that captures only the target species.

2. **Decreasing fish catch.** Some studies have estimated that sustainable harvests can be attained only by decreasing the intensity of fishing 30% to 50%—which would require a considerable downsizing of the global fishing fleet. **Downsizing** could be accomplished by direct buyouts (governments buy fishing companies and retire the fleets to protect existing stocks and allow for recovery). Downsizing can also be achieved by

reducing subsidies to fishing interests to encourage the retirement of old vessels and discourage the entry of new vessels.

3. **Implementing a harvest tax.** A harvest tax or user fee for fishing could be imposed.

4. **Changing the quota system.** The current quota system encourages waste by forcing fishing vessels to rush and catch as much as they can in as little time as possible before the "blanket" quota is reached. This "free-for-all" fishing causes companies to invest in fishing equipment and employ practices that are efficient but not necessarily environmentally friendly. If more time were allowed, friendly methods might be used—for example, methods that reduce bycatch. However, with less-stringent time constraints on their catch limit, fishers may actually exacerbate the bycatch problem by discarding smaller fish caught earlier in return for larger, higher-valued fish caught later.

5. **Implementing individual transferable quotas. Individual transferable quotas (ITQs),** which give fishers a guaranteed portion of the catch, are currently being tried in various parts of the world. Evaluations of this strategy have had mixed reviews, however. ITQs essentially transform a common property resource into a private one; fishers have an economic incentive to protect the resource in order to maintain the same quota in subsequent years. A drawback of this system may be that ITQs can be bought and sold, and over time a small group of fishing fleet operators (the largest and wealthiest) may accumulate quotas, eliminating smaller operators.

6. **Regulating harvest beyond EEZs.** The problems of open access in international waters (beyond the EEZs) may be mitigated by the following regulations:
   - Establish size limits on fish that may be harvested.
   - Limit the number (or poundage) of a given species of fish that may be taken.
   - Restrict the number of times a particular vessel can fish.
   - Restrict the number of boats that can fish in a particular region.
   - Limit the number of fishing vessels that may be built during a given year.

All nations exploiting fish stocks in international waters would have to comply with mutually accepted restrictions to ensure a continued optimum yield year after year.

Precautionary Approach to Fisheries

Agreements on harvesting fish in a commons are usually difficult, if not impossible, to attain. However, in December 1995, a UN-sponsored conference resulted in an international agreement that calls for a more conservative, or precautionary, approach to fisheries management. This agreement became part of the 1982 United Nations Convention on the Law of the Sea. The **precautionary approach** requires increased monitoring, inspecting, and reporting in an effort to protect fish stocks *before* they show signs of decline instead of managing by reacting to the decline itself. The principles of the precautionary approach are outlined in Table 12.7.

Marine Protected Areas

One of the keys to protecting the world's marine fisheries is the establishment of **marine protected areas.** Like wildlife refuges for terrestrial species, a subject you'll read about in Chapters 15 and 16, marine preserves are areas off limits to many human activities, especially commercial fishing. As such, they provide vital habitat for fish and other species, permitting them to survive without human interference. In recent years, scientists are finding that these preserves do

TABLE 12.7 Precautionary Approach to Fisheries	
Principle*	Management Objective
1. Fisheries stocks must be maintained at levels of abundance that are not substantially below their range of natural fluctuation.	1. Reduce fish stocks to no less than 75%–80% of their unfished level. (Currently, fish stocks are managed at maintenance levels below 1/2 their original abundance.)
2. New fishing gear and techniques must be evaluated before being introduced to a fishery.	2. Fishing gear types would be widely tested before they are used on a commercial scale. Technologies that result in excessive levels of bycatch or substantial disturbance to the habitat would not be allowed unless modifications are made to reduce these effects to minimal levels.
3. Closed areas must be established to protect the marine habitat.	3. Set aside large areas where fishing gear types that are destructive to the sea bottom are banned, protecting valuable habitat or allowing the recovery of a damaged habitat.

*These three principles are mutually supportive. Because abundant stocks require much less fishing effort for the same level of catch, there is less incentive to use destructive or nonselective fishing methods. Consequently, large parts of the habitat can be closed with only minimal impact on the fishery.

(Source: Modified from M. Earle, "The Politics of Overfishing." *Ecologist* 25 (1995): 70.)

more than protect small pockets of habitat and the species that live there. They also seed surrounding areas. That is, excess fish and other commercially important species in these areas migrate into depleted areas, helping restore commercial stocks.

Many countries, including the United States, have set aside marine preserves. However, many of these countries, especially poorer countries in the less-developed world, are finding it difficult to regulate human activities in these sanctuaries. Money for adequate monitoring and enforcement typically is unavailable. That's unfortunate, too, because many of these countries—especially the tourist-dependent nations of the Caribbean—also depend on revenues from scuba divers and snorkelers. Coral reefs in these countries are in bad shape because of overfishing, pollution, global warming, and a host of other factors.

In 1972, the U.S. Congress passed the Marine Protection Research and Sanctuaries Act, also known as the Ocean Dumping Act. Besides regulating ocean dumping, a subject covered in Chapter 11, this law established marine sanctuaries off the U.S. coastlines. Since 1972, 13 national marine sanctuaries and the Northwestern Hawaiian Islands Marine National Monument have been established. One of the largest sanctuaries is off the coast of the Florida Keys, the long string of islands off the southernmost tip of Florida. When used in conjunction with other management tools described in this section, marine sanctuaries are vital component of a sustainable strategy.

Marine Habitat Restoration: Constructing Artificial Reefs

In addition to the ideas presented so far, marine fisheries biologists are currently exploring the potential of creating artificial reefs to provide food and cover for sport and commercial fish species. Artificial reefs are especially helpful in raising the carrying capacity of flat, sandy coastal plains. An artificial reef is a human-made, underwater structure, typically built for the purpose of promoting marine life in areas of generally featureless bottoms.

Marine scientists at the State University of New York at Stony Brook constructed a reef from blocks of compacted sludge and fly ash—materials that create major disposal problems. These researchers formed a 454-metric ton (500-ton) reef by dumping 18,000 blocks in the Atlantic Ocean 4 kilometers (2.5 miles) south of Saltaire, Long Island. It is hoped that such reefs will enhance sport and commercial fishing in an area of high human population density and will not release toxins that contaminate the very fish they support. It could help solve an increasingly serious waste disposal problem for operators of coal-fueled power plants.

A series of reefs have been constructed along the New Jersey coast from Sandy Hook to Cape May by the New Jersey Department of Environmental Protection. Construction materials range from discarded tires and stainless-steel drums

FIGURE 12.29   Explosion and sinking of a cargo ship that had previously been used to smuggle drugs into South Florida, July 31, 2001, off the coast of Miami, FL. The U.S. Customs Service, working with Miami–Dade County and the Atlantic Gamefish Foundation, sunk three cargo ships in order to make an artificial reef. The ships were sunk 6.4 kilometers (four miles) east of Key Biscayne in 40 meters (130 feet) of water to create the "U.S. Customs Reef."

to concrete bridge rubble and purposely sunken barges. More than 36,000 bald tires were placed on the Garden State Reef in one year alone. The reefs provide excellent recreational and commercial fishing for such highly regarded species as sea bass, cod, bluefish, mackerel, and tuna (see Figure 12.29 for another human-made reef in Florida).

Choosing Seafood Wisely

When the Monterey Bay Aquarium, one of the world's most prestigious aquariums, buys seafood for its restaurant, it wants to support sustainable fisheries so that there will be plenty of fish for the future. Its seafood choices are based on the best available information for the region (the West Coast), including data from the U.S. National Marine Fisheries Service, the Food and Agriculture Organization, and the Australian Bureau of Rural Sciences.

The aquarium has come up with sustainable-fish eating guides, called Seafood Watch Guides, for the different regions of the United States, as well as a national guide, as shown in Table 12.8. The Seafood Watch Guide puts fish into three categories: best choices, good alternatives, and avoid. Individuals can use this chart when ordering food to cook at home or in a restaurant. If you do so, you play an important role in supporting sustainable fisheries.

GO GREEN!

When buying fish to cook or ordering fish in a restaurant, use the Seafood Watch guide in Table 12.8, or download a pocket guide for your region at the Monterey Bay Aquarium Web site, www.mbayaq.org.

12.7 Aquaculture

For thousands of years, human beings obtained their food by hunting animals and gathering eggs, fruits, berries, nuts, and seeds—a rather inefficient process that could support only a few million people the world over. Eventually, humans invented agriculture—the controlled rearing of plants and animals—a process that has

| TABLE 12.8 | Seafood Watch Guide of the Monterey Bay Aquarium (2008). (For more detailed information, see the Monterey Bay Aquarium web site at www.mbayaq.org.) |

Best Choices (Best Choices are abundant, well-managed and caught or farmed in environmentally friendly ways.)	Good Alternatives (These are an option, but there are concerns with how they're caught or farmed or with the health of their habitat due to other human impacts.)	Avoid (These are caught or farmed in ways that damage the environment or harm other marine life.)
Arctic Char (farmed)	Basa, swai (farmed)	Chilean sea bass/toothfish
Barramundi (U.S. farmed)	Clams (wild)	Cod: Atlantic
Catfish (U.S. farmed)	Cod: Pacific (trawled)	Crab: king (imported)
Clams (farmed)	Crab: blue, king (U.S.), snow	Flounders, soles (Atlantic)
Cod: Pacific (Alaska longline)	Crab: imitation/surimi	Groupers
Crab: Dungeness Stone	Flounders, soles (Pacific)	Halibut: Atlantic
Halibut: Pacific	Herring: Atlantic/sardines	Lobster: spiny (Caribbean imported)
Lobster: Spiny (U.S.)	Lobster: American/Maine	Mahi mahi/dolphinfish (imported)
Mussels (farmed)	Mahi mahi/dolphinfish (U.S.)	Marlin: blue, striped
Oysters (farmed)	Oysters (wild)	Monkfish
Pollock (Alaska wild)	Scallops: sea	Orange roughy
Salmon (Alaska wild)	Shrimp (U.S. farmed or wild)	Rockfish (Pacific)
Scallops: Bay (farmed)	Squid	Salmon (farmed, including Atlantic)
Striped Bass (farmed or wild)	Swordfish (U.S. longline)	Sharks
Sturgeon, Caviar (farmed)	Tuna: bigeye, yellowfin (troll/pole)	Shrimp (imported farmed or wild)
Tilapia (U.S. farmed)	Tuna: canned light, canned white/albacore	Snapper: red
Trout: Albacore (U.S., British Columbia troll/poll)		Sturgeon, caviar (imported wild)
Tuna: Skipjack (troll/poll)		Swordfish (imported)
		Tuna: albacore, bigeye, yellowfin (longline)
		Tuna: bluefin

been able to feed many more people. In a similar fashion, humans have for many years harvested fish from the sea by a relatively inefficient hunting-and-gathering technique: Fishing vessels move to fishing grounds where the harvest is unpredictable and then transport the catch back to market. The controlled culturing of fish and other aquatic food organisms—called **aquaculture**—is potentially much more efficient. **Freshwater aquaculture** is practiced in inland ponds. **Marine aquaculture** (also called **mariculture**) is practiced in shallow bays or estuaries.

Aquaculture has great potential to help meet the nutritional needs of a growing global population and to relieve pressure on stressed ocean fisheries. In 2004, aquaculture produced 45.5 million metric tons of fish and shellfish, according to the Food and Agriculture Organization. This translates to 32.4% of world total fisheries production. China is reported to have accounted for nearly 70% of the quantity and over half the global value of world aquaculture production. India follows at a distant second in aquaculture production with 2.5 million metric tons, and the United States ranks tenth with slightly more than 0.6 million metric tons. Aquaculture continues to grow faster than all other animal-food-producing sectors. Worldwide, the sector has grown at an average rate of 8.8% per year since 1970, compared with only 1.2% for capture fisheries and 2.8% for terrestrial meat production systems over the same period.

Worldwide, over 200 species of fish and shellfish are currently being grown in aquaculture, as are seaweed and numerous species of crustaceans. Important farmed marine organisms include shrimp, salmon, and bivalves (such as oysters, clams, and mussels). Freshwater-farmed fish include carp, tilapia, and catfish, making up slightly more than 60% of global aquaculture production.

Production Methods

Mollusks (for example, oysters) are often grown in shallow, near-shore environments on rafts, in trays, or on other structures to which the mollusks attach (Figure 12.30). Crustaceans have

traditionally been reared in saltwater ponds in coastal environments, although more recently they are also being raised inland to expand production. Most marine fish are produced in near-shore floating cages. For anadromous fish, such as salmon, that hatch and develop in freshwater and mature in saltwater, both freshwater tanks and saltwater cages are used. Freshwater fish are commonly raised inland in earthen ponds or concrete pools.

Food sources used in aquaculture production are important considerations for producers and greatly determine the ecological effects of the particular system. Low-trophic-level organisms, such as mollusks, rely on plankton and organic material provided by nutrient-rich seawater circulating through the production system. Midlevel herbivores and omnivores are most often maintained simply by providing nutrients to the naturally occurring aquatic organisms on which the fish feed. The high-trophic-level carnivores are reared on nutrient-rich diets requiring large amounts of fish meal and fish oil derived primarily from ocean catches. The rising costs of feed inputs, particularly of fish meal and fish oil, have helped to focus research efforts on developing lower-cost feed formulations with greater portions of vegetable protein.

Ecological Impact

The ecological effects of aquaculture, as with terrestrial agriculture, vary depending on the intensity of production, the organism being produced, and the site of production. The factor that perhaps most determines the effect of an aquaculture system on the surrounding environment is the trophic level of the organism being produced. Mollusks, tilapia, and carp are low-trophic-level organisms. They are typically raised in **extensive aquaculture systems** that rely on naturally occurring food sources. These systems typically produce relatively small amounts of waste that are easily assimilated into the environment. In addition, fish farming at this level has traditionally been carried out on a small scale and at low stocking densities. It has traditionally been integrated within complex polyculture systems—that is, systems in which

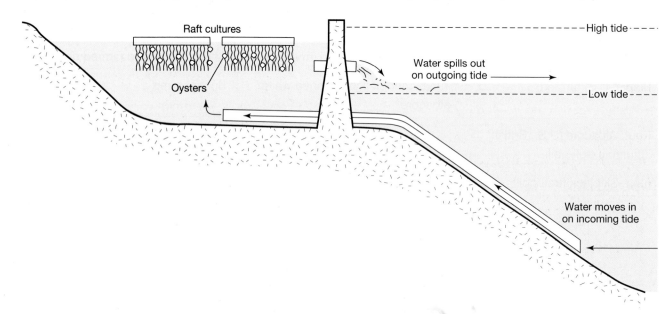

FIGURE 12.30 Oyster culture. Deep, nutrient-rich water of the incoming tide enhances the growth of algae in the artificial impoundment. Because the oysters either feed on the algae directly or feed on crustaceans that feed on the algae, oyster production on these inshore rafts is increased significantly.

many other species are produced, including fish and plant species. In such systems, nutrient-rich aquaculture effluent often is applied to surrounding crops. Although such production methods are still quite common throughout the world, greater market demand has encouraged their replacement by more **intensive aquaculture systems.** These systems, like modern agriculture, tend to focus on one species. They typically employ higher stocking densities, generate greater amounts of waste, and are less likely to be involved in tight nutrient-cycling systems.

Aquaculture systems involving high-trophic-level species (species that feed higher on the food chain) are often intensive, high-input farms and are rapidly increasing in number. Global production of salmon and shrimp, for example, has increased six- to sevenfold since the 1980s. To feed fish and shrimp, growers typically rely heavily on wild-caught ocean fish. For example, about 3 metric tons of wild-caught fish are required to produce 1 metric ton of farmed salmon or shrimp. Ocean stocks of small open-ocean (or pelagic) fish, such as anchovies, sardines, and herring, are particularly affected. Some fisheries experts question the sustainability of the intensive production of high-trophic-level species.

Aquaculture of all species, regardless of their trophic level, can have other serious ecological effects. These include soil and water contamination, loss of natural ecosystems, and biological contamination of native species by escaped farmed species. Levels of waste discharge that exceed the ecosystem's ability to assimilate it can pollute surrounding water and soils. Researchers have found that high stocking rates of farmed salmon in the North Atlantic have resulted in potentially damaging discharges of nitrogen and phosphorus into coastal waters. Inland saltwater ponds used to produce shrimp have contaminated the soil and water with salt in China, Thailand, and Indonesia.

The conversion of crucial ecosystems to aquaculture systems also is a common problem. In Asia, important coastal mangroves have been cleared to make way for shrimp and milkfish farms. The loss of mangroves can have serious implications, because mangroves provide habitat for many plants and animals. Shrimp farming alone may account for as much as 10% of the total global loss of coastal mangroves. Conversion of coastal areas to aquaculture can also deplete natural fisheries, because coastal ecosystems play a key role in the life cycle of many marine fishes. Researchers estimate that as much as 0.4 kilogram of fish and shrimp is lost from ocean fisheries due to habitat conversion for every kilogram of pond-raised shrimp produced in Thailand.

"Biological contamination" of native species by farmed species has also proven to be a widespread problem. Escaped, farm-raised Atlantic salmon account for as much as 40% of the salmon catch in the North Atlantic region, and more than 250,000 farmed salmon have escaped into the North Pacific Ocean in the past two decades. Escaped salmon may hybridize with native stocks, irreversibly altering the genetic makeup of wild stocks, many of which are already endangered. A study published in 2006 in the *Proceedings of the National Academy of Sciences USA* found that sea lice—a fish parasite—from salmon farms along the British Columbia coast can kill up to 95% of the wild juvenile salmon that pass near them heading out to sea. Biological pollution also includes the spread of diseases from farmed populations to wild populations. Whitespot and yellowhead viruses have spread from domestic to wild

shrimp stocks in Asia and the United States, causing tremendous economic losses to shrimp farmers and fishers alike.

To ameliorate some of these negative environmental effects of the aquaculture industry, researchers recommend decreasing the production of high-trophic-level fish, reducing levels of fish meal and fish oil in feed, increasing the use of integrated polyculture systems, and improving aquaculture practices. Increasing production of herbivores and omnivores, or farming lower down on the food chain, could reduce demand for fish meal and fish oil. Integrated polyculture systems, where waste effluent is spread as fertilizer on adjacent cropland and provided to other aquaculture species, can increase production efficiencies and can reduce pollution and input requirements. Careful site planning can greatly mitigate damage to the surrounding environment. To provide a lasting source of protein for the world's growing population, such recommendations will need to be rapidly adopted to keep pace with the expanding aquaculture industry around the world.

Summary of Key Concepts

1. Anadromous fish begin life in freshwater, travel to the sea to mature, and return to their native stream to reproduce and die.

2. The successful completion of each stage in the life cycle of a fish depends on one or more of the following environmental conditions of the freshwater habitat: proper water temperature, depth, velocity, turbidity, and level of dissolved oxygen; substrate; cover; and food supply.

3. Fish respond physiologically and behaviorally to degraded habitat caused by long-term or chronic alterations to stream and lake ecosystems.

4. Roughly 70% of a given fish population dies each year as a result of environmental limitations.

5. The major environmental limitations that cause mortality in fish populations can be divided into two categories: natural and human induced. Natural limitations include major storms, soil mass movements, animal activities, natural barriers, vegetation disturbances, native-animal predation, and winterkill. Human-induced limitations include water pollution, stream temperature alterations, predation and competition by exotics, human predation, dams, resource extraction, channelization, and recreation.

6. Predation by the sea lamprey caused the annual lake trout harvest in the Great Lakes to drop from 4,545 metric tons in 1940 to 152 metric tons in 1961—a 97% reduction in only 21 years.

7. Lamprey populations in the Great Lakes have been controlled by treating their spawning streams with lampricides.

8. Severe winterkills of fish caused by oxygen starvation occur in shallow lakes when the snow cover prevents aquatic plants from getting sufficient sunlight for photosynthesis.

9. The 1980 Northwest Power Act directs the Northwest Power Planning Council to (1) prepare "a regional conservation and electric power plan"; (2) "inform the Pacific Northwest public of major regional power issues"; and (3) "develop...a program to protect, mitigate, and enhance fish and wildlife, including related spawning grounds and habitat, on the Columbia River and its tributaries."

10. Three components for sustaining freshwater fisheries are biota management (such as propagation and stocking), habitat management (maintaining adequate habitats), and human management (regulating the numbers of fish caught).

11. Controversial biotic manipulation techniques for enhancing or restoring fish populations include artificial propagation and stocking, introductions, chemical or physical removal of undesirable fish and predators, and selective breeding of superior fish.

12. Hatchery-reared fish tend to be physically weaker, more susceptible to diseases and parasites, more likely to contract severe manifestations of a particular disease, and less suited to the streams into which they are released than their wild or native counterparts.

13. Whirling disease, caused by a fish parasite thought to have been introduced from Europe, has infected several trout streams in 25 states in the United States.

14. Federal and state fish hatcheries stock trout in suitable waters in which they do not occur, in waters where conditions for growth are good but where natural spawning sites are inadequate, and in waters where fishing pressure is heavy but there is no natural production.

15. Carp populations have been brought under partial control in some regions by periodic seining and the use of selective chemicals such as antimycin.

16. Many fish-control projects are only short-term treatments for the symptom rather than addressing the cause of the problem. Thorough research into the causes for the overpopulation of the undesirable species must be an initial step in the development of long-term solutions.

17. Levels of dissolved oxygen in snow-covered northern lakes can be increased by snow removal, opening the ice cover with dynamite, and using motorized aerators.

18. Two-story reservoirs are stocked with warm-water species in the epilimnion and cold-water species in the hypolimnion.

19. The water of a newly created reservoir receives a nutrient flush that greatly increases the abundance of fish food.

20. Replacing or enhancing fish populations that have disappeared or diminished from their original water body can be accomplished with little threat to the existing species through translocations.

21. Legislation that has been utilized to protect fish populations includes catch (size and number) limits, catch-and-release-only restrictions, closed seasons, and outlawing destructive fishing techniques.

22. Habitat protection and restoration, supported by legislation to control fishing pressure and potentially harmful population enhancement techniques, are the principal long-term measures by which natural fish populations can be sustained.

23. Native fish populations may increase considerably when the carrying capacity of their environment is raised—that is, by providing adequate water flows, depths, and temperatures; space; substrate; cover; and food supplies.

24. Priority for restoration efforts and funding should be given to the protection of remaining high-quality sites and the restoration of sites that are, at worst, moderately degraded yet still contain valuable, threatened, or endangered fish species.

25. Among the techniques and instruments employed by commercial fishers to locate marine fish are sonar or echo-sounding systems, moored buoys, color enhancement of aerial photos, infrared sensors, ultraviolet sensors, and electronic image intensifiers.

26. Fish and shellfish are harvested with passive and active fishing gear. Passive gear includes pots and cages; set nets, pound nets, and traps; hooks and lines; and gill nets. Active gear includes purse seine nets, trawl nets, and drift nets.

27. With more than 75% of world fish stocks reported as already fully exploited or overexploited (or depleted and recovering from depletion), the maximum wild-capture fisheries potential from the world's oceans has probably been reached. More cautious and effective fisheries management is needed to rebuild depleted stocks and prevent the decline of those being exploited at or close to their maximum potential.

28. Overfishing is the result of a combination of factors, including fishing in a "commons," government subsidies, economic and political forces, advances in fishing gear technology, and poor enforcement of harvest regulations.

29. Among the fisheries that have been depleted due to overfishing are the Pacific sardine fishery, northwestern Atlantic cod and herring fisheries, northwestern Pacific salmon fishery, and eastern Atlantic bluefin tuna fishery.

30. Bycatch is the collective term used to describe captured marine organisms that are not the target species of the fishery. Discards are the bycatch that are thrown back because they are either a nontarget species, juvenile fish, endangered species, improper size, inferior quality, or surplus to quotas.

31. Overfishing, bycatch, and habitat loss and degradation not only result in population and productivity declines of commercially important marine organisms but also disturb nontarget species, disrupt predator–prey relationships, alter marine community structures, and reduce genetic diversity.

32. Recognition and concern for the state of the marine ecosystem is increasing throughout the international community. Several management options for sustainable fisheries have been proposed or are being initiated, including the optimum-yield principle, economic incentives for downsizing the global fishing fleet and developing selective fishing gear, individual transferable quotas, stricter harvesting regulations, the precautionary approach, and schemes for habitat enhancement and restoration.

33. Individuals play an important role in supporting or not supporting sustainable fisheries by the choices we make in purchasing seafood.

34. The practice of aquaculture is increasing and holds promise as a protein source for a growing global population. However, if not managed correctly, aquaculture systems can destroy wetlands, pollute water, diminish ocean stocks of small open-ocean fish, and cause biological contamination of native species by escaped farmed species.

35. To make the aquaculture industry more sustainable, researchers recommend decreasing the production of high-trophic-level fish, reducing levels of fish meal and fish oil in feed formulations, increasing the use of integrated polyculture systems, and improving aquaculture practices.

Key Words and Phrases

Acid Rain
Active Fishing Gear
Anadromous Fish
Anoxia
Antimycin
Aquaculture
Artificial Propagation
Astroturf Sandwich
Brush Shelters
Bycatch
Captive Breeding
Carrying Capacity
Catch-and-Release-Only
  Restrictions
Closed Seasons
Color Enhancement
Commons
Creel (Catch) Limits
Discards
Downsizing
Drawdown
Electronic Image Intensifiers
Environmental Limitations
European Carp
Eutrophication
Exclusive Economic
  Zone (EEZ)
Exotic (Nonnative) Species
Extensive Aquaculture
  System
Fishery
Fish Ladders
Freshwater Aquaculture
Genetic Erosion
Habitat Protection
Habitat Restoration
Human Predation
Hybrid Fish
Hydromodification
Hypoxia
Imprinting
Individual Transferable
  Quotas (ITQs)
Infrared Sensors
Intensive Aquaculture System

Introduction
Lampricide
Magnuson Fishery
  Conservation and
  Management Act
Marine Aquaculture
  (Mariculture)
Marine Protected Area
Maximum Sustained Yield
  (MSY)
Minimum Size Limits
Moored Buoys
Nitrogen Intoxication
Nongame Fish
Northwest Power Planning
  Council (NPPC)
Nutrient Flush
Optimum Yield
Overfishing
Panfish
Passive Fishing Gear
Precautionary Approach
Put-and-Take Stocking
Redd
Reservoir
Resident Fish
River Ruff
Rotenone
Sea Lamprey
Siltation
Sonar
Spawning
Spawning Sites
Splash Dams
Stocking
Stream Improvements
Subsidies
Suspended Sediment
Translocation
Two-Story Reservoir
Ultraviolet Sensors
Upstream Migration
Vibert Box
Whirling Disease
Winterkill

Critical Thinking and Discussion Questions

1. How does water temperature affect fish behavior?

2. Discuss the importance of streambed materials (substrate) on fish spawning.

3. What factors limit food and energy supplies in the stream environment?

4. List three human-induced limitations to fish productivity, and describe their effects on fish habitat.

5. What is the source of acid rain? What is its effect on fish?

6. What causes the winterkill of fish in northern lakes? How can it be prevented?

7. Describe three activities of carp that are harmful to game fish populations.

8. What would you suggest as a solution for the increasing angler pressure on freshwater fish?

9. How have dams affected the survival of anadromous fish?

10. Discuss the pros and cons of artificial propagation and stocking.

11. What is the suspected cause of the spread of whirling disease to the intermountain West? What are the visible symptoms? Discuss some possible strategies to stop the further spread of this disease.

12. What three general strategies are available to fisheries managers to protect and restore freshwater fish populations?

13. How can the stream habitat be improved for fish?

14. Discuss the following statement: "Predator control is an effective method for increasing game fish populations in the United States." Is it valid? Why or why not?

15. Describe the life cycle of the sea lamprey.

16. In what situation would you use translocation as a management option?

17. Describe efforts to control the sea lamprey in the Great Lakes.

18. Discuss the use of a Vibert box.

19. Why would a fish manager loosen regulations on the size and take of panfish?

20. Briefly list four benefits that may be derived from the drawdown of reservoirs.

21. Describe the types of protective legislation used to control freshwater fish populations.

22. Describe two methods used by commercial marine fishers to locate fish.

23. Why has the bycatch problem become an international concern?

24. Discuss the negative environmental effects caused by modern fishing gear.

25. How has the establishment of EEZs enhanced marine fishing pressures?

26. List two methods proposed to address the problems of the world's marine fishing industry. Explain how each would reduce the overfishing or bycatch problem.

27. Explain the theoretical concept underlying the optimum-yield principle.

28. What is meant by the "precautionary approach" to fisheries management?

29. Do you feel artificial reefs composed of human "garbage" are viable options for habitat restoration or enhancement? Explain.

30. Discuss the pros and cons of aquaculture.

Suggested Readings

Cicin-Stain, B., and R. W. Knecht. 2000. *The Future of U.S. Ocean Policy.* Washington, DC: Island Press. In-depth analysis of the evolution of U.S. ocean policy and a discussion of current issues.

Crivelli, A. J. 1995. Are Fish Introductions a Threat to Endemic Freshwater Fishes in the Northern Mediterranean Region? *Biological Conservation* 72: 311–319. Discussion of the controversies surrounding fish introductions.

Driscoll, C. T., K. M. Driscoll, K. M. Roy, and M. J. Mitchell. 2003. Chemical Response of Lakes in the Adirondack Region of New York to Declines in Acidic Deposition. *Environ. Sci. & Technol.* 37:2036–2042. A study, discussed in this chapter, showing declines in acid deposition in the Adirondack Lakes.

Fairlie, S., M. Hagler, and B. O'Riordan. 1995. The Politics of Overfishing. *Ecologist* 25:46–73. Interesting explanation and discussion of the underlying reasons for the overfishing problem.

Fisheries Statistics Division. 2008. *Fisheries of the United States, 2007.* Washington, DC: U.S. Department of Commerce, National Oceanic and Atmospheric Administration, U.S. Government Printing Office. <http://www.st.nmfs.noaa.gov/st1/fus/fus07/index.html>. Annual report detailing the status and trends of the U.S. commercial and recreational fishing industries.

Food and Agriculture Organization (FAO), Fisheries and Aquaculture Department. 2007. *The State of World Fisheries and Aquaculture 2006.* Rome, Italy: FAO <http://www.fao.org/docrep/009/A0699e/A0699e00.htm>. A report published every two years with the purpose of giving policy makers, civil society, and those whose livelihoods depend on the sector a comprehensive, objective, and global view of capture fisheries and aquaculture.

Frankl, E. 1995. *Ocean Environmental Management: A Primer on the Role of the Oceans and How to Maintain Their Contributions to Life on Earth.* Englewood Cliffs, NJ: Prentice Hall. Overview of current environmental problems relating to ocean resources and suggestions for sustainable use.

Heinz, H. J. III. 2000. *Fishing Grounds: Defining a New Era for American Fisheries Management.* Washington, DC: Island Press. Comprehensive assessment of many issues related to fisheries management in the United States.

Hobbie, J., ed. 2000. *Estuarine Science: A Synthetic Approach to Research and Practice.* Washington, DC: Island Press. Essays, written by leading scientists, summarizing the current state of knowledge about estuaries that is important for proper management.

Iudicello, S., M. Weber, and R. Wieland. 1999. *Fish, Markets, and Fishermen: The Economics of Overfishing.* Washington, DC: Island Press. Describes how current policies lead to overfishing and depletion of fish stocks.

Klein, L. R., S. R. Clayton, J. R. Alldredge, and P. Goodwin. 2007. Long-Term Monitoring and Evaluation of the Lower Red River Meadow Restoration Project, Idaho, U.S.A. *Restoration Ecology* 15:223–239. Excellent long-term monitoring study of a river restoration project in Idaho, which evaluates the effectiveness in shifting a degraded stream ecosystem onto a path of ecological recovery.

Kohler, C. C., and W. A. Hubert. 1999. *Inland Fisheries Management in North America,* 2nd ed. Bethesda, MD: American Fisheries Society. A mandatory reading assignment for current and future fisheries managers.

Malvestuto, S. P., and M. D. Hudgins. 1996. Optimum Yield for Recreational Fisheries Management. *Fisheries* 21: 6–17. Detailed explanation of the optimum yield management concept.

Mann, C. C., and M. L. Plummer. 1999. Can Science Rescue Salmon? *Science* 289:716–718. Excellent discussion of the Snake River salmon issue in the Pacific Northwest.

Meronek, T. G., et al. 1996. A Review of Fish Control Projects. *No. Amer. J. Fisheries Mgmt.* 16:63–74. Literature review assessing the "success" of 250 fish-control projects.

Moyle, P. B., and J. J. Cech Jr. 2004. *Fishes: An Introduction to Ichthyology,* 5th ed. Englewood Cliffs, NJ: Prentice Hall. A comprehensive source of information on the structure and physiology, evolution and taxonomy, zoogeography, and ecology and conservation of fishes.

Naylor, R. L., et al. 2000. Effect of Aquaculture on World Fish Supplies. *Nature* 405:1017–1024. Excellent article on the pros and cons of aquaculture.

Naylor, R. L., J. Eagle, and W. L. Smith. 2003. Salmon Aquaculture in the Pacific Northwest: A Global Industry with Local Impacts. *Environment* 45(8): 18–39. Detailed look at salmon aquaculture and its many effects.

Nehring, R. B., and P. G. Walker. 1996. Whirling Disease in the Wild: The New Reality in the Intermountain West. *Fisheries* 21:28–30. Interesting account of the history and extent of the whirling disease threat to trout in the rivers of the intermountain West.

Ofiara, D. D., and J. J. Seneca. 2001. *Economic Losses from Marine Pollution: A Handbook for Assessment.* Washington, DC: Island Press. Great insights into the economic impact of pollution on marine fisheries.

Sobel, J., and C. Dahlgren. 2004. *Marine Reserves: A Guide to Science, Design, and Use.* Washington, DC: Island Press. An in-depth look at marine reserves and their importance to fisheries management.

Weber, M. L. 2002. *From Abundance to Scarcity: A History of U.S. Marine Fisheries Policy.* Washington, DC: Island Press. An in-depth look at U.S. policy related to marine fisheries.

Young, O. R. 2003. Taking Stock: Management Pitfalls in Fisheries Science. *Environment* 45(3): 24–33. A critical look at fisheries management; important reading for anyone interested in pursuing this career.

Web Explorations

Online resources for this chapter are on the World Wide Web at: **http://www.prenhall.com/chiras** *(click on the Table of Contents link and then select Chapter 12).*

CHAPTER 13

RANGELAND MANAGEMENT

Rangelands are areas of the world that are a source of forage such as grasses and shrubs for free-ranging native and domestic animals. Rangelands may be a source of wood products, water, minerals, and energy. They support wildlife and provide a number of recreational opportunities. They also produce intangible benefits such as natural beauty, open space, and wilderness that satisfy important societal values. They even store carbon and thus reduce atmospheric greenhouse gases. Most rangelands are unsuitable for cultivation because of physical limitations, such as low precipitation, rough topography, poor drainage, or cold temperatures.

Worldwide, rangelands occupy about half of the terrestrial surface (that is, if you exclude permanently frozen lands). The United States is home to about 312 million hectares (770 million acres) of rangelands. U.S. rangelands are a diverse group of ecosystems such as the wet grasslands of Florida, the desert floor of California, and the mountain meadows of Utah. More than 99% of the nation's rangeland is west of the Mississippi River.

This chapter explores rangelands—their ecology, history, distribution, and condition. It concludes with a look at ways to manage these important lands more successfully.

13.1 Ecology of Rangelands

Types of Rangelands

Seven major biomes provide most of the world's grazing lands: grasslands, tropical savannas, tundra, desert shrubs, shrub woodlands, temperate forests, and tropical forests. **Grasslands,** as used here, occur in temperate regions. They are made up of a mixture of grasses and some forbs (broad-leaved flowering plants), and are typically free of trees or shrubs. Grasslands usually develop in areas with 25 to 75 centimeters (10 to 30 inches) of annual rainfall. They occur on every continent in the world and represent some of the world's richest and most productive grazing areas (Figure 13.1). (Grasslands are called *pampas* in Argentina, *pastures* in Europe, *steppes* in Eurasia, and *veldt* in South Africa.)

At one time, great herds of large, wild herbivores roamed the world's grasslands. Today, most of these wild animals have been killed or driven away. In wetter areas, grasslands have been converted to cropland because of their deep, fertile soils. In the areas too dry to grow crops, grasslands are now used primarily for livestock. Because grazed rangelands have been dramatically altered under pressure from livestock, there is a growing movement to restore and preserve some of the native grassland areas around the world (see A Closer Look 13.1).

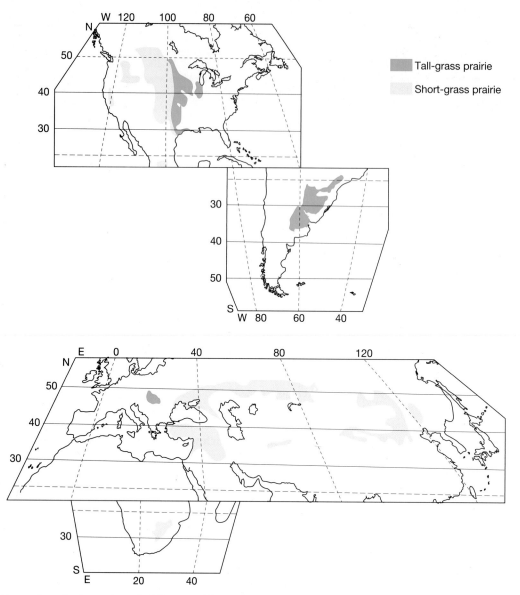

FIGURE 13.1 World map of grassland ecosystems: the tall-grass prairie and the short-grass prairie.

Tropical savannas, essentially tropical grasslands, occur mostly in Africa (Figure 13.2) and do not occur in the United States. They are characterized by a mixture of grasses, shrubs, and scattered trees. Savannas are called *campos* in Brazil and *llanos* in Venezuela. The crop-growing potential of tropical grasslands is usually limited by water availability or poor soil.

Tundra vegetation can be found in very cold areas in the Arctic or as meadows at high mountain elevations. The Arctic tundra is used mostly by wild animals, whereas the alpine tundra is used by both livestock and wild animals. Temperatures are too cold and soils too poor for tundra to be converted to crop production.

Desert shrublands make up the largest area of the world's rangelands. They are characterized by an arid climate (usually less than 25 centimeters or 10 inches of rainfall), poorly developed soils, and sparse vegetation dominated by low-growing shrubs usually less than 2 meters (6.6 feet) in height (Figure 13.3). All arid regions of the world support some desert shrubland.

FIGURE 13.2 Rhinoceros grazing on open tropical savanna in Tanzania.

A CLOSER LOOK 13.1	Prairie Restoration and the National Grasslands Story

The grasslands of the Great Plains form the largest single expanse of true grass range in the world, extending from southern Canada to Texas. They include tall-grass prairies and short-grass prairies. Prairies, especially tall-grass prairies, which grow in moister areas than do short-grass prairies, have helped produce humus- and nutrient-rich soils. Before the European settlement in North America, these two types of prairie could be found in the Great Plains and what are now southern Alberta, the northern border region of Montana, eastern Washington State, and western Idaho (Figure 13.1). These prairies supported abundant buffalo herds, elk, and other wildlife. They were also home to over 30 Native American tribes, including Sitting Bull's Hunkpapa Sioux, Apache, Arapaho, Blackfeet, Cheyenne, Chippewa, and Pawnee.

Since the time of European settlement, however, many of these areas have been overgrazed and plowed for growing crops. The settlers discovered that while these vast grasslands were productive in wet years, they were also subject to serious drought and bitter winters. In the Great Plains, land that should never have been plowed yielded its topsoil to incessant dry winds, resulting in the Dust Bowl in the 1930s (see Chapter 7 for more details). Today, most of these prairies are farmed, and some are used for grazing. However, since the birth of the **National Grasslands** in 1960, 20 native grassland areas have been revegetated and restored to conserve the natural resources of the prairie ecosystem.

The 20 National Grasslands, totaling almost 1.6 million hectares (4 million acres), are important lands managed for sustainable multiple uses as part of the U.S. National Forest System. They have made important contributions to conserving grassland ecosystems while producing a variety of goods and services that, in turn, have helped to maintain rural economies and lifestyles. Wildlife, including many declining, threatened, or endangered species, thrives in reborn habitats. With restored prairie vegetation,

once-wounded soil rebuilds itself. Construction of livestock ponds allows cattle-grazing while simultaneously expanding the range of many wildlife species by providing water where none existed before. Private farmlands within the National Grassland boundary add diversity to the prairie habitat. The National Grasslands also provide diverse recreational uses, such as mountain biking, hiking, hunting, fishing, photography, birding, and sightseeing.

Although most of the National Grasslands are located in Great Plains states, three are located in the Great Basin states of California, Oregon, and Idaho. The Little Missouri National Grassland in North Dakota is the biggest, with 416,215 hectares (1,028,051 acres). The Rita Blanca National Grassland, which includes 31,362 hectares (77,463 acres) in Texas and 6,421 hectares (15,860 acres) in Oklahoma, has wildlife that varies as much as does the climate over the wide expanse of country.

Restoration and preservation of native prairies have taken place thanks to the work of other organizations and foundations. The largest is **The Nature Conservancy,** whose mission is to preserve plants, animals, and natural communities by protecting the lands and waters they need to survive. The Nature Conservancy has helped to protect more than 6 million hectares (15 million acres) of habitat in the United States and more than 47 million hectares (117 million acres) in Africa; North, Central, and South America; the Caribbean; and the Asian Pacific. The organization currently manages more than 1,400 preserves in the United States, including numerous native prairies. For example, the Oregon chapter of the Conservancy purchased 11,000 hectares in northeast Oregon, part of the largest expanse of bunchgrass prairie remaining in North America. In addition to being home to one of the highest concentrations of nesting birds of prey in the country, the Zumwalt Prairie Preserve also provides habitat for elk, mule deer, bighorn sheep, bobcat, and the endangered Snake River steelhead.

Some are productive and of major grazing value, whereas others occur in areas of such low precipitation that they are little used.

Shrub woodlands usually occur in about the same annual rainfall belt as grasslands. However, in these areas, the dominant vegetation is low-growing trees (usually less than 10 meters or 33 feet) and dense shrubs. Considerable grazing opportunities

may also exist in **temperate forests**—naturally occurring open stands (areas with plenty of open sunny land for grasses to grow). Timber cutting may also open up the forest, providing opportunities for grass to grow. **Tropical forests** provide little herbaceous vegetation under their dense canopies and are of little relative importance for grazing.

FIGURE 13.3   Cattle grazing on desert shrubland in Nevada.

Characteristics of Rangeland Vegetation

The vegetation on rangelands is mostly grasses, grasslike plants (sedges and rushes), forbs, and shrubs (Figure 13.4). These plants are known as **forage.** They supply food and energy for domesticated animals such as cattle, sheep, and horses, and for wild animals such as deer, giraffes, and wildebeests. Some of these animals graze on grass, whereas others browse on leaves, twigs, and shoots. Humans, in turn, acquire food and energy from cattle and sheep in the form of veal, beef, mutton, or lamb.

Grasses and other rangeland vegetation are ideal forage because a leaftip can be nibbled off without affecting growth. As long as the **basal zone** (the lowermost portion of the leaf) remains intact, the plant will survive and continue to produce more food. In a short time, the leaf can grow to its original length. In fact, the grass leaf can be grazed again and again without adverse effects, as long as the plant has some time to recover. Grasses can therefore provide a continuous food reservoir for grazing animals.

Range ecologists generally regard the upper 50% of the grass shoot (stem and leaves) as a "surplus" that can be safely eaten by livestock or wild herbivores (deer, antelope, elk, and so on) without damaging the plant. The lower 50%, known as the **metabolic reserve,** is necessary for the plant's survival (Figure 13.5A). This provides the minimum amount of photosynthesis needed to manufacture foods for the roots. Many root systems extend to a depth of 2 meters (6 feet) or more. Large amounts of nutrients are stored in them. This nutrient reserve enables range grasses to survive drought as well as brief periods of overgrazing. When a

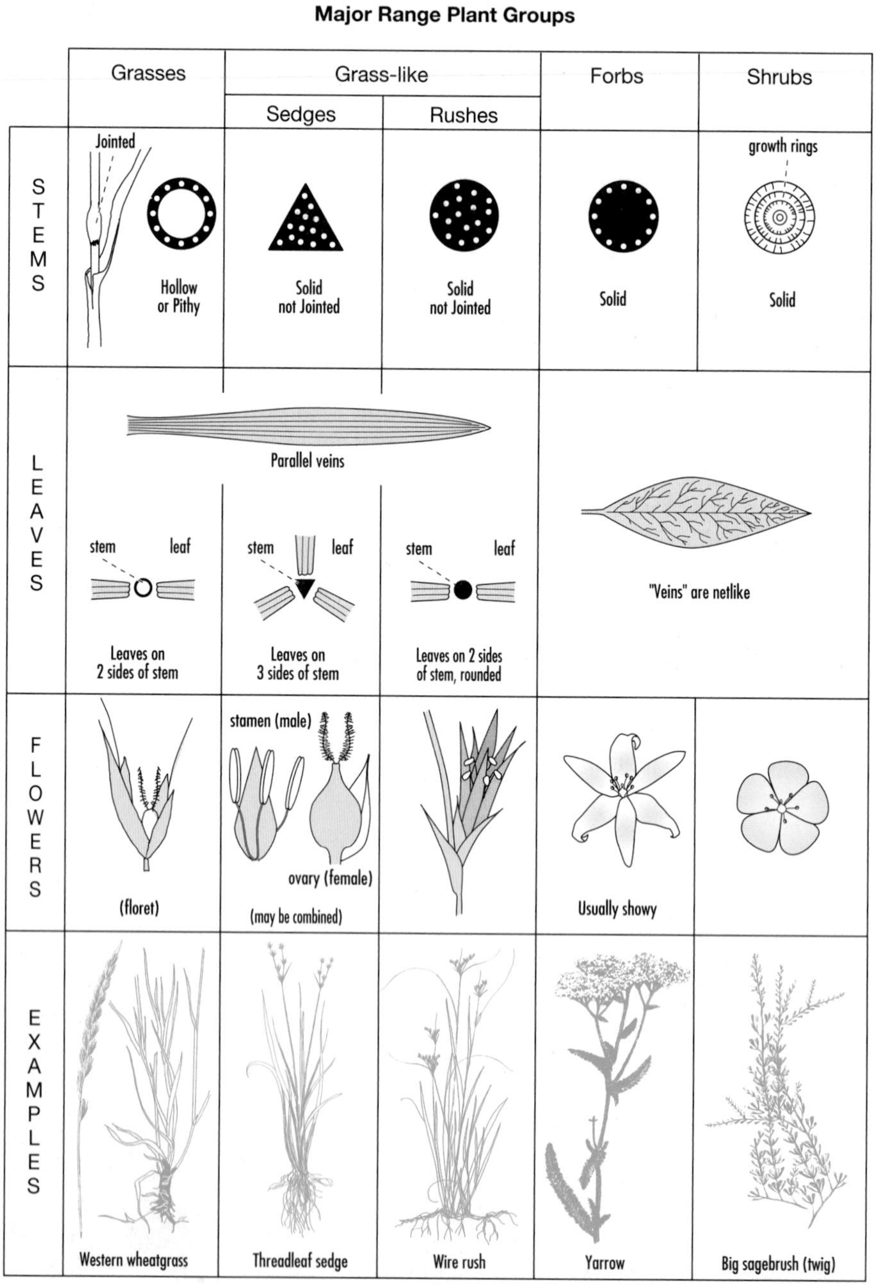

Major Range Plant Groups

| | Grasses | Grass-like | | Forbs | Shrubs |
		Sedges	Rushes		
STEMS	Jointed / Hollow or Pithy	Solid not Jointed	Solid not Jointed	Solid	growth rings / Solid
LEAVES	Parallel veins / stem leaf / Leaves on 2 sides of stem	stem leaf / Leaves on 3 sides of stem	stem leaf / Leaves on 2 sides of stem, rounded	"Veins" are netlike	
FLOWERS	(floret)	stamen (male) / ovary (female) (may be combined)		Usually showy	
EXAMPLES	Western wheatgrass	Threadleaf sedge	Wire rush	Yarrow	Big sagebrush (twig)

FIGURE 13.4 Characteristics of the major groups of range plants.

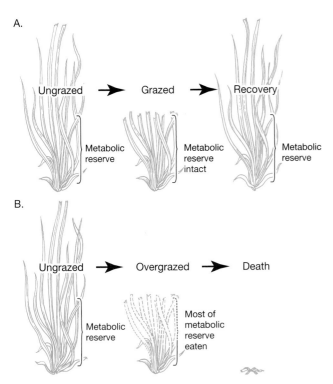

FIGURE 13.5 Impact of metabolic reserve. (A) Grasses can be grazed without harm as long as the metabolic reserve is left intact. (B) With overgrazing, most of the metabolic reserve is eaten, and the plant dies.

range is overgrazed for longer periods, herbivores destroy the metabolic reserve by frequently clipping the grass to the ground (Figure 13.5B). This starves and kills the root system, leaving the land vulnerable to erosion; hence the need for recovery between grazings.

Range ecologists and ranchers frequently classify the various range plants into three categories with respect to the dynamics of plant succession: decreasers, increasers, and invaders. **Decreasers** are highly nutritious, extremely palatable plants that generally decrease under even moderate grazing pressure (Figure 13.6). Representative decreaser species are big bluestem, little bluestem, wheatgrass, and buffalo grass. Additional examples are shown in Figure 13.6.

Increasers are generally less palatable but are still highly nutritious climax species that tend to increase (at least temporarily) when a range is heavily grazed (Figure 13.7). Apparently this increase is the result of reduced competition from the decreasers. When severe grazing pressure continues over a long period, even the increasers begin to decline, apparently unable to withstand trampling by hoofs of grazing animals. They are replaced by invaders.

The **invaders,** such as ragweed, cactus, and thistle, are generally considered to be undesirable weed species. They are low in nutritional value and are not very desirable for grazing animals (Figure 13.8). Some may be poisonous. The invader known as downy chess has sharp seeds that can harm animals by lodging in their throats or piercing their skin. Invaders frequently are perennial plants (which grow year after year from the same root system) and thrive under sunlight intensities much higher than the 1% to 2% of full sunlight occurring in a dense stand of climax grasses. Because their roots are taproots

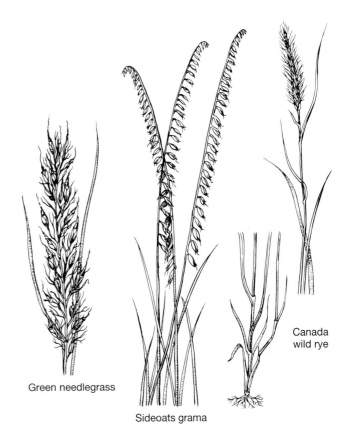

FIGURE 13.6 Decreasers. These plants are usually the plants that grazing animals prefer. They are highly nutritious and palatable, and decrease under heavy grazing pressure.

instead of the dense fibrous roots of a typical grass, these invaders are not very effective at binding the soil.

A range in excellent condition has a high percentage of decreasers and almost no invaders. Conversely, a range in poor condition has a low percentage of high-forage-value decreasers and a large percentage of low-forage-value invaders (Figure 13.9).

Rangeland Carrying Capacity

The **carrying capacity** of a habitat is the size of the population of a species that can be sustained in that habitat. Grazing carrying capacity is the maximum number of animals (or quantity of herbivore biomass) that can graze each year on a given area of range, for a specific number of days, without causing a downward trend in forage production, forage quality, or soil quality. Grazing carrying capacity is affected by annual climatic conditions, previous grazing use, kinds of grazing animals and how long they graze a particular area, and soil type. It varies from site to site and from season to season for each species. The carrying capacity or number of grazing animals for dry years should not be exceeded if a particular grazing site is to maintain a sustainable herd. The reason is that, in dry years, the carrying capacity may be half that of a normal year.

The forage potential of a given range is usually described in terms of **animal unit months (AUMs).** An AUM is the amount of forage needed to keep a 1,000-pound (454-kilogram) animal

Nebraska sedge

Blue grama

Needle-and-thread

FIGURE 13.7 Increasers. These plants increase in number (at least temporarily) under grazing pressure and replace the decreasers. Although they are fairly nutritious, they are less palatable and less preferred by cattle.

TABLE 13.1	Animal Unit Equivalents*
Number of Animals	Animal Unit Month (AUM)
1 steer	1.0
5 sheep	1.0
5 goats	1.0
4 deer	1.0
1 horse	1.25
1 bull	1.25
1 elk	0.67

*One steer is considered equal to five sheep grazing the same habitat.

to be taken into account. For example, when a range is in excellent condition, 0.4 hectare (1 acre) may equal 1 AUM. However, on poor range, 2 hectares (5 acres) may be needed to equal 1 AUM.

Effects of Human Activities and Overgrazing on Rangelands

When people plow native rangelands and plant them with crops (called **sodbusting**), they destroy the rangelands' ecological stability. When the vegetation is grasslands, the grasses are replaced with crops whose root systems are less effective at holding the soil in place. Without appropriate soil conservation practices, the soil may become vulnerable to wind or water erosion. A classic example is the Dust Bowl of the 1930s, which was associated with massive wind erosion on the Great

well fed for one month. Accepted animal unit equivalents are listed in Table 13.1. Notice that one steer is equivalent to five sheep or five goats grazing the same habitat. This means that five sheep or five goats eat as much as one steer. In addition, a 1,000-pound horse typically eats 25% more than a 1,000-pound steer. When considering AUMs, range condition needs

Canada thistle

Salsify

Curly cup gumweed

Downy chess

FIGURE 13.8 Invaders. These weeds replace the increasers if the range continues to be grazed heavily. Some of the invaders, such as military grass (downy chess), have sharp seeds that can harm livestock by lodging in their throat or piercing their skin. Thistles have sharp spines that make them useless as forage.

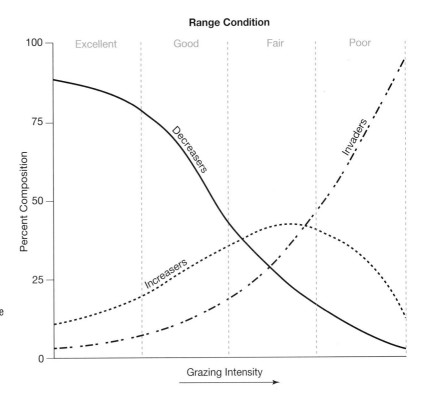

FIGURE 13.9 Intensity of grazing versus range condition and the relative proportion of decreasers, increasers, and invaders. Figure is taken from L. A. Stoddart, A. D. Smith, and T. W. Box, *Range Management*, 3rd ed. (New York: McGraw-Hill, 1975), p. 191, which in turn was adapted from data from P. L. Sims and D. D. Dwyer, "Pattern of Retrogression of Native Vegetation in North Central Oklahoma," *Journal of Range Management* 18 (1965): 20–25.

Plains (see Chapter 7). A more recent example is the Palouse region in eastern Washington State. Once a rich, rolling prairie, it now produces high yields of rain-fed wheat, barley, peas, and lentils. Because of a combination of steep hills and lack of proper soil conservation practices, the Palouse region is one of the most erodible areas in the United States.

Rangelands not converted to croplands can be properly grazed by domesticated livestock and wild animals. In fact, light to moderate grazing is believed to be necessary for the health of rangelands, especially grasslands. Studies show that normal grazing promotes healthier root systems and vigorous leaf growth. It also promotes nutrient cycling and the buildup of soil organic matter and hinders soil erosion. Too much of a decrease in the number of grazers causes replacement of grasses by forbs and woody shrubs. An excessive increase in grazers causes overgrazing.

Overgrazing is continued heavy grazing that exceeds the carrying capacity of the community, resulting in deterioration of the range. Although most overgrazing is caused by too many livestock feeding too long in a particular area, large numbers of wild herbivores also can overgraze rangeland in long dry periods. Overgrazing changes a plant community's makeup and decreases its productive potential. More important, it can expose soil to water and wind, increasing its erosion potential.

Overgrazing can sometimes be so devastating that all vegetation disappears and the land becomes barren and highly prone to erosion. When combined with drought, severe overgrazing can cause desertification (see A Closer Look 13.2).

Undergrazing also can damage rangeland by allowing too much leaf and stem to get old, which in turn reduces the growth of grasses and increases the growth of forbs and woody shrubs. It can even lead to an increase in soil erosion and degradation. Thus, rangeland declines as a source of food for livestock and wild herbivores.

Effect of Drought on Range Forage

Drought is one of the greatest environmental problems encountered by ranchers. They can, to some degree, control rodents, poisonous plants, brush and weeds, predators, insects, and unfavorable soil chemistry, but there is absolutely nothing they can do to control drought. They can adjust to it, but that is all. Moreover, drought is unpredictable. A severe drought results in a drastic deterioration of the rangeland plant community, even when grazing is light. In the Snake River region of southern Idaho, during the height of the Dust Bowl era in 1934, extended drought caused an 84% reduction in plant cover in ungrazed areas.

Once a dry spell has ended, range recovery may be fast or slow, depending on the amount of precipitation. Thus, a lightly grazed Montana pasture required eight years to return to good condition after a severe drought because of limited rainfall. In contrast, a Kansas range recovered very rapidly when rainfall was adequate (Figure 13.10).

13.2 A Brief History of Range Use in the United States

Before European settlers arrived on the Great Plains of North America, the land was populated by a number of large herbivores—bison, pronghorn antelope, and elk. At one time, researchers believe, as many as 34 million head of these species may have roamed the range. Despite this massive number of animals, overgrazing was uncommon. These grazers selectively grazed the land—taking the most edible plants—and moved about, migrating throughout the range. Moreover, their numbers were kept at the carrying capacity of the land because of competition and predation within and between species. Native Americans used their meat and hides for food, clothing, and shelter.

A CLOSER LOOK 13.2   Causes of Desertification

Desertification has been defined in many ways. The United Nations Desertification Prevention Treaty defines desertification as land degradation caused by natural factors (climatic changes) and artificial factors (human-induced activities) in dry, semidry, or dry–semihumid areas. The climate factor refers to large-scale global changes in climate in a dry area. Human-induced activities include overgrazing, overcutting, cultivation of unsuitable land, and inadequate irrigation. These destructive activities are intensified by high population density, poverty, and poor land management. The bottom line is that desertification reduces the land's productive potential and diminishes its capacity to provide benefits to humans and other life on the planet.

Desertification is characterized by devegetation, brush invasion, groundwater depletion, salination, and severe erosion. Areas probably least affected by desertification are the naturally barren deserts of the world that are little used by humans. However, sparsely vegetated grazing lands adjacent to these deserts have, in many places, been severely affected.

Because of the different definitions of desertification, estimates of the degree of desertification vary. A recent detailed study on soil degradation worldwide found that some 680 million hectares (1680 acres) of land have been degraded since 1945 due to overgrazing, most of which has occurred in Africa and Asia. This study is discussed in more detail later, in the section on range condition.

Desertification is probably most severe in the **Sahel,** or Sahelian zone, just south of the Sahara Desert in Africa (Figure 1). The combination of long droughts in the period from 1969 to 1973 and again during the 1980s, overgrazing of common lands, over-harvesting of fuelwood, and expanded cultivation on marginal lands reduced plant cover and left the soil bare in places. Hot, dry winds picked up the soil and moved it toward the Sahara Desert, resulting in an apparent advance southward of the desert. During the 1969–1973 drought, grain crops failed, and cattle starved or had to be sold due to a lack of available forage. In 1973 alone, an estimated 100,000 people died of starvation and disease, and 5 million cattle perished.

Desertification also occurs in the United States, mainly in the Southwest. For example, the Sonoran and Chihuahuan Deserts of the Southwest are perhaps a million years old, yet in places, they have become even more barren during the past 100 years. Their animal populations have diminished. Valuable grasses have declined. Invader species such as Russian thistle have multiplied. The original floodplain vegetation has changed beyond recognition in the Santa Cruz River Valley of Arizona.

Although opinions differ about the steps needed to prevent and reverse desertification, scientists believe that this process can be stopped and desertified regions rehabilitated. They agree on the critical role that plants play in stabilizing the soil but differ somewhat on which soil, water, rangeland, and forest management strategies are most appropriate. The United Nations Environment Programme estimates that the global cost would be close to $150 billion. However, the enormous expense of the rehabilitation effort would be more than offset by income from the increased agricultural productivity that would result.

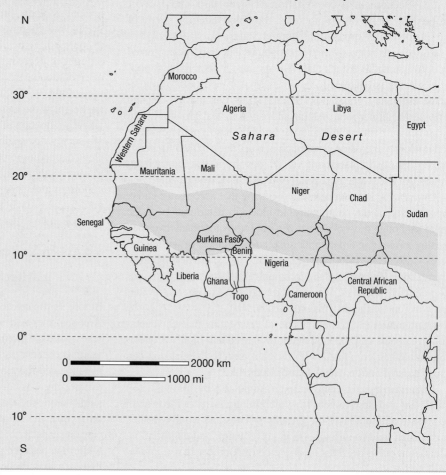

FIGURE 1   The Sahel or Sahelian zone (shown in color). This zone has undergone periodic droughts throughout past decades.

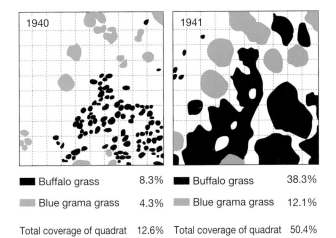

Buffalo grass   8.3%   Buffalo grass   38.3%

Blue grama grass   4.3%   Blue grama grass   12.1%

Total coverage of quadrat   12.6%   Total coverage of quadrat   50.4%

FIGURE 13.10 Recovery of range from drought following a period of adequate rainfall. The basal cover of range vegetation near Ness City, KS, had been reduced to only 12.6% of the area by autumn of a drought year. Overgrazing had contributed to the deterioration. However, within one year after the return of adequate rainfall, range grasses responded sufficiently to cover 50.4% of the sample plot.

Introduction of Domesticated Animals

In the 16th and 17th centuries, European colonists, mostly Spaniards, immigrated to the New World and brought a domesticated-animal-based culture with them. Many wild herbivores were killed or driven from rangelands to reduce their competition with livestock. By 1800, cattle, sheep, and horses were a significant part of rangelands in the western United States and Canada. The cattle industry's phenomenal expansion during the 1800s was responsible for much of the settlement of the West. For this we can thank thousands of courageous, hardworking ranchers and farmers. However, in their zeal, many abused our once-bountiful grassland and caused widespread destruction that is still evident today.

On the range, cattle and sheep replaced the bison, which was nearly driven to extinction by the end of the 19th century. Between 1870 and 1890, cattle numbers jumped from about 5 million to almost 27 million. Sheep numbers rose from about half a million in 1850 to more than 20 million in 1890. As the numbers of domesticated grazing animals increased, the quality and carrying capacities of rangelands declined.

The problem with cattle is not so much the cattle but their owners. Cattle ranchers allowed their cattle to overgraze the ranges in many areas. If one person did not exploit the forage of an area, then a neighbor would. Some called this the "tragedy of the commons," first described in Chapter 12 (Figure 13.11). Where a given pasture could support 25 cattle, many grazed 100. Big bluestem, bluegrass, and buffalo grass were chewed off at the roots. Once the grass plant's metabolic reserve had been eliminated, the root system withered and died. Many of these livestock owners were so obsessed with large numbers of range animals that they ignored the fact that four head of livestock, sick and scrawny from undernourishment, would sell for less than one head in prime condition after grazing on good forage.

Public Land Distribution and Abuse

The **Homestead Act of 1862** accelerated the development of the West. It provided 160-acre (64-hectare) parcels of publicly owned land to anyone who would settle in the West and live on the land for five years. In the eastern United States, a farm of 160 acres was plenty of land for a family to make a living. In the West, that was not true. Land in the arid West that should never have been cultivated was plowed and damaged by homesteaders trying to make a living.

In 1878, **John Wesley Powell,** an explorer who had studied the soil, water resources, plants, and animals of much of the arid West, submitted his *Report on the Lands of the Arid Region* to the U.S. Interior Department. In it he warned that the prevailing methods of distributing land and practicing agriculture would not work in the arid West. He recommended that the land was suitable for grazing sheep and cattle, not for growing crops. He felt that a single-family ranch should be 2,500 acres (1,000 hectares) or larger and that the arid range could support far fewer cattle than cattle ranchers were raising. He also believed that if the arid land were to be farmed, farmers would have to be assured of irrigation water from building of dams or diverting of rivers, and that a single farm should be no more than 80 acres (32 hectares). Powell's report was ignored by government leaders. If his suggestions had been taken to heart, the health of our current rangelands would most likely be better today.

In 1873 barbed wire was developed, which allowed farmers, sometimes called sodbusters, to build fences to keep cattle from trampling their crops. Barbed wire was also used to limit grazing and keep other livestock from trespassing. Farmers and ranchers fought and even killed each other because angry ranchers, wanting the land for grazing, cut and pulled down the farmers' fences. This dispute between farmers and ranchers is depicted in the famous western movie *Shane.*

By 1900, many rangelands had been degraded because of heavy grazing by cattle, sheep, and horses for more than 50 years. Other areas of rangeland had been degraded by plowing because they were too dry to support crops without irrigation. In 1905, the **U.S. Forest Service (USFS)** was formed and began to restrict livestock numbers and grazing seasons in national forests.

Although the condition of some rangelands in national forests began to improve, the remaining millions of hectares of public lands that were not controlled by grazing statutes continued to experience severe abuse from uncontrolled grazing. Finally, in 1932, the seriousness of the range problem prompted Congress to request the USFS to survey range conditions. The survey showed that rangeland productivity in the United States had been reduced by 50%. On some Utah ranges, intense grazing had reduced rice grass (a valuable winter forage species when most rangeland is covered with snow) by 90%. The survey further revealed that the extensive removal of grass cover had resulted in erosion on 80% of the range.

Taylor Grazing Control Act and Other Laws

In 1934, as a direct consequence of the Forest Service report, Congress enacted the **Taylor Grazing Control Act.** This was the successful culmination of a long struggle on the part of conservationists to place ailing public rangelands under federal control. This act

FIGURE 13.11 Cartoon illustrating the "tragedy of the commons." Any individual user of a commons, or a publicly owned rangeland, can profit by grazing as many of his or her cattle as possible, because the forage is free. But if all members of the community do the same thing, the commons will quickly be overgrazed, becoming useless for feeding animals and, thus, humans in the future.

had three major objectives: (1) to halt overgrazing (Figure 13.12) and soil deterioration, (2) to improve and maintain ranges, and (3) to stabilize the rangeland economy. Although much grazing land was under private ownership, the major focus of the Taylor Grazing Control Act was on rangeland owned by the public, which had been seriously abused by western ranchers. Under the provisions of the Taylor Grazing Control Act, the public range was divided into operational units called **grazing districts** and managed by the newly established Grazing Service. Unfortunately, the Grazing Service, which became the **Bureau of Land Management (BLM)** in 1946, was not effective in administering this management system.

In 1976, Congress passed the **Federal Land Policy and Management Act,** which consolidated legislation related to public land management and gave the BLM authority to manage all public rangelands not in national forests or national parks. The BLM is responsible for preventing overgrazing and seeing that damaged rangelands recover. Also in 1976, Congress passed the **National Forest Management Act,** directing the USFS to develop and maintain a comprehensive inventory of all National Forest Service lands (including rangelands).

FIGURE 13.12 A severely overgrazed range. Note the denuded land and the scrawny cattle.

In 1978, the **Public Rangelands Improvement Act** was passed; among its provisions are policies to manage and improve the condition of public rangelands. These three major pieces of legislation provide BLM and the USFS with the framework for managing, inventorying, and improving U.S. public rangelands.

In 1985, Congress passed the **Food Security Act** (known as the Farm Bill). One of the programs of the Farm Bill is called the **Conservation Reserve Program (CRP),** which calls for the removal of 18 million hectares (45 million acres) of highly erodible cropland from the cropland base. This land is to be planted back to grassland (and some trees) to control erosion. Under the terms of this program, the farmer makes a contract with the U.S. Department of Agriculture (USDA) to withdraw erodible farmland from crop production for ten years and to establish vegetative cover (most often grass) on this land to stabilize the soil. The USDA, in turn, makes "rental" payments to the farmer during this period. Not only are the grass plantings valuable in checking erosion, they also are useful in providing food and cover for wildlife. For more on CRP, see Chapter 7.

13.3 Rangeland Resources and Condition

Rangeland Resources

Worldwide, rangelands occupy about half of the planet's ice-free land surface (Figure 13.13) and provide about 80% of the feed for domestic livestock. Most rangelands are in semiarid areas too dry for rain-fed crop production. About 42% of the world's rangelands are used for grazing livestock. The remaining 58% are too cold, dry, or remote from population centers to be utilized. However, with good management, some of this land could be grazed by livestock.

Globally, more than 200 million people use rangelands for some form of livestock (pastoral) production. More than 15% are nomadic and pastoral people who are completely dependent on livestock-grazing on rangelands. Most pastoral people live in the less-developed countries in Africa and Asia. However, pastoralism can be found in industrialized countries such as Australia, where rangelands cover more than 70% of

the country. In Australia, pastoralism consists of mostly sheep- and cattle-grazing and uses 60% of the nation's rangelands.

Rangelands occupy about 29% of the total land area in the United States. Most of this is short-grass prairies in the arid and semiarid western half of the country (Figure 13.14). More than half of these rangelands in the United States are privately owned, 43% are owned by the federal government, and the remainder are owned by state and local governments. Privately owned rangelands, compared with publicly owned rangelands, have relatively few restrictions placed on them by the government and are usually managed for the single use of grazing. The federal government (mostly the BLM and USFS) manages public rangelands according to the principle of **multiple use.** This means that, besides grazing, these lands are used for wildlife conservation, recreation, mining, energy resource development, soil conservation, and watershed protection.

About 2% of U.S. ranchers are issued permits by the BLM or USFS to graze their herds on public rangeland. How much ranchers should pay for public **grazing permits** has been a controversial issue between ranchers and environmentalists for many years. (For details, see A Closer Look 13.3.)

About 75% of U.S. public and privately owned rangeland is grazed by livestock at some time during each year. According to one estimate, however, America's rangelands have supplied only about 16% of the total amount of the forage that cattle consume before slaughter. The remaining 84% is provided by crops such as alfalfa and corn. This surprising statistic is explained by the fact that most cattle graze on the range only until they are mature. They are then shipped to **feedlots** (enclosed pens), where they are crowded together and fed with grain and special feeds to fatten them up for slaughter (Figure 13.15). Some feedlots in Nebraska hold as many as 10,000 cattle at a time.

The overriding importance of feedlots and the relatively minor role of rangelands in livestock nutrition is being modified somewhat because of America's awareness of the role that high-fat, high-cholesterol meat plays in heart disease and stroke. These health concerns have resulted in a decrease in annual per capita beef consumption from almost 39 kilograms (85 pounds) in 1975 to only 30 kilograms (66 pounds) in 1995, where this lower figure

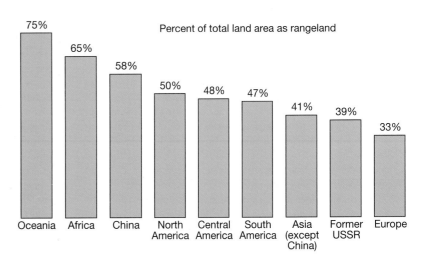

FIGURE 13.13 Distribution of rangelands in the world. Data from UN Food and Agriculture Organization.

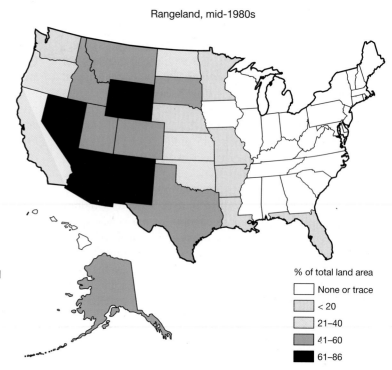

Rangeland, mid-1980s

% of total land area

	None or trace
	< 20
	21–40
	41–60
	61–86

FIGURE 13.14   Rangeland as a percentage of total area of each state in the mid-1980s. Note that the western United States is dominated by rangeland. Today the area in rangelands has decreased slightly. For example, privately owned rangelands decreased from 168 million hectares (415 million acres) in 1982 to 164 million hectares (405 million acres) in 2003, according to USDA-NRCS National Resources Inventory data. Figure is reprinted with permission from D. A. Castillon, *Conservation of Natural Resources* (Dubuque, IA: William C. Brown Publishers, 1992), p. 216.

A CLOSER LOOK 13.3 — Range Wars: Ranchers Versus Environmentalists

Cattle-grazing in the West began in the 1500s, when Spanish long-horns were brought in from Mexico. The peak grazing year was 1884, when almost 40 million cattle munched the grasses of the open range. Sometimes the ranchers tried to flout federal laws. In 1896 some even tried to graze their livestock in Yosemite National Park. The U.S. Cavalry took charge and drove 189,000 sheep, 1,000 cattle, and 300 horses from the park.

Today, cattle-grazing is legal in 45 national parks, including such celebrated ones as Grand Canyon, Rocky Mountain, Grand Teton, and Mesa Verde. Federal law even allows cattle-grazing in 150 National Wildlife Refuges, as well as in designated wilderness areas such as the Big Blue Wilderness in Colorado and the Gila Wilderness of New Mexico. Additional grazing occurs in many national forests. However, the great majority of livestock graze on lands administered by the BLM. About 80% of the 110 million hectares (270 million acres) of public land in the West is now being used for grazing.

In recent years, ranchers bought permits to graze 2 million cattle and 2.3 million sheep annually on public rangelands. The cost of a grazing permit for cattle in 2004 was $1.43 per AUM—far less than a permit to graze on private rangeland, which cost between $8 and $23 per AUM the same year. Critics point out that U.S. taxpayers, who own the public grazing lands, are therefore subsidizing ranchers who hold federal grazing permits. They further assert that low fees contribute to budget shortfalls because federal fees are much lower than lease rates for private grazing land and do not cover the costs of range management.

Some environmental groups, such as the Sierra Club, the Audubon Society, and the National Wildlife Federation, are very concerned about grazing in western states. Not only are these ranchers (and their livestock) feeding at the public trough, they complain, but the livestock are destroying grassland ecosystems in the process. Consider streamside habitat, for example. Riparian habitat provides food, shelter, and breeding sites for 75% of the rangeland's wildlife. Unfortunately, cattle and other livestock also prefer to feed in these areas. Heavy grazing pressure is destroying these areas. A recent report of the Environmental Protection Agency (EPA) stated, "Riparian [streamside] areas throughout much of the West are in the worst condition in history."

Environmentalists believe that federal agencies tend to issue too many grazing permits. The result, they say, is that public rangelands are inevitably being overgrazed and degraded. Overgrazing results in a shift in plant life, as noted in this chapter. Valuable forage grasses are replaced by worthless weeds. Not only does this decrease the ecological vitality of these regions, it is the most important cause of grassland plant extinction in America. Although livestock and wildlife compete for the same food supplies, wildlife usually loses out in this competition. A good example is the situation in the Burns BLM district in Oregon, where wild animals such as grouse, deer, and antelope consume only 3% of the total plant food available.

Conservationists argue that higher fees would provide more money for improvements in range condition, wildlife conservation, and watershed management. They contend that any rancher with a permit who cannot remain in business without government subsidies should not be in the ranching business.

Ranchers with permits are opposed to higher grazing fees because, for most of them, profit margins are already razor thin. They argue that most federal land is steep or otherwise difficult land on which to manage livestock, whereas private land usually is more productive and manageable. Most public rangelands remained in the federal domain because homesteaders found other areas more attractive. They say overgrazing on public rangeland is also caused by increasing numbers of elk, deer, and other wild animals.

Private grazing leases are for single use, whereas public grazing leases have restrictions for multiple use. In addition, the restrictions imposed on federal grazing permits are more severe than those typically imposed on private grazing leases. These restrictions impose extra costs on federal permit holders. For example, federal grazing permits include requirements for range improvements and maintenance (such as fences) paid for by the permit holder. They may also place forage limitations and restrictions that prohibit grazing in some areas to protect water-

sheds and wildlife habitat. Almost all grazing permittees also incurred extra costs when they acquired their ranches, because they had to pay for the land's increased value gained from having a public land grazing permit assigned.

Some people argue that if public grazing fees are raised to the level of prevailing private grazing fees, many ranchers dependent on public rangelands will go out of business. In turn, rural communities where they buy their supplies will decline.

The grazing permit fee has become a perennial public-policy issue. Maybe the debate should focus instead on the bigger issue: the uses of federal lands and their resources. With improved co-operation among all parties and organizations, these issues should be resolved.

---

has remained with almost no change to 2004, according to data from the USDA Economic Research Service. Studies have shown that the longer cattle are grazed on rangelands, the lower the fat content of their meat. Ever sensitive to consumer demands, much of the livestock industry is now keeping cattle on the range until they weigh 300 kilograms (660 pounds)—twice the weight at which they were once shipped to feedlots.

A few farsighted ranchers are replacing some of the traditional strains of cattle, such as the shorthorn and Hereford, with the leaner-meat longhorn—a favorite of the young livestock industry in the 19th century. Even more interesting, the American Water Buffalo Association, established in 1986, promotes the scientific study of raising, breeding, and marketing water buffalo (oxlike animals from Southeast Asia). Water buffalo cows and bulls are now for sale in Florida, and there is a market for water buffalo meat in the Washington, DC, area. Water buffalo steaks are tasty and lean, and should satisfy fat- and cholesterol-conscious Americans.

Rangeland Condition

U.S. Rangelands The term **range condition** may be defined as the current state of vegetation of a range in relation to the potential natural plant community. In a sense, it is an estimate of how close a particular rangeland is to its productive potential.

Interestingly, the three major U.S. federal land management and advisory agencies—the USFS, BLM, and Natural Resource Conservation Service (NRCS)—use somewhat different definitions and methodologies to determine range condition. In addition, methodologies and concepts within agencies have changed somewhat over time. Despite their differences, these agencies have traditionally separated range condition into four (or five) classes: excellent, good, fair, and poor. (Sometimes a fifth class of "very poor" or "depleted" has been used.) Although the agencies' terminologies have differed somewhat—and still do—as to what each class represents, their classifications have reflected comparisons between a site's existing vegetation (that is, the plant

FIGURE 13.15 Aerial view of a large western feedlot. Cattle are crowded together, and food is provided so that they "fatten up" before being slaughtered.

FIGURE 13.16 This range near Miles City, MT, is classified as being in good condition. Because of the method for comparing range conditions, a range in the arid sagebrush region of Wyoming may be in excellent condition and still be much less productive than a range in good condition in the semiarid short-grass region of Nebraska.

species present and their relative amounts) and what the site could potentially support if natural plant succession had progressed unimpeded through time. A rating of excellent means the existing vegetation closely resembles its natural potential, whereas a poor rating means the existing vegetation is very dissimilar to its natural potential (Figure 13.16). Thus, although comparing range condition data from these agencies is difficult, examining past data can still give us an idea of the general trend of the condition of U.S. rangelands.

Public rangelands in the United States were severely abused in the late 1800s and early 1900s due to improper livestock grazing. In some places, soil and vegetation are still recovering from these past abuses. Although better grazing management practices are needed in many areas, U.S. rangelands today are in their best condition in the past 100 years (with some exceptions). For example, BLM data on trends in range condition on the bureau's lands indicate that acreage in the combined category of excellent and good has more than doubled between 1936 and 1992, and acreage rated poor has decreased by almost two-thirds (Figure 13.17). NRCS data for private rangelands indicate similar improvement during a briefer period (Figure 13.18).

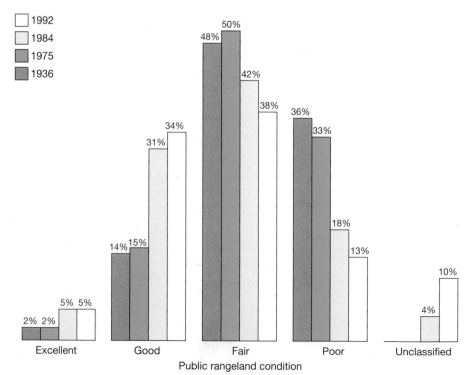

FIGURE 13.17 Trends in range condition on lands administered by the Bureau of Land Management, 1936–1992. Data from BLM and USFS.

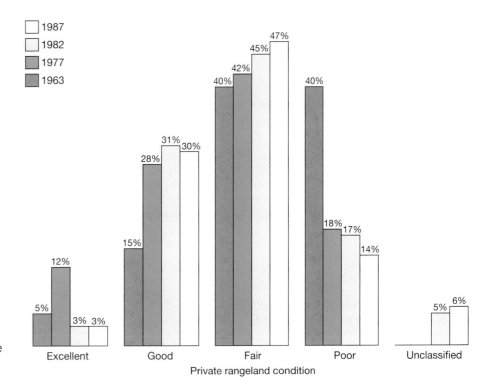

FIGURE 13.18 Trends in the condition of private rangelands, 1963–1987. Data from the Natural Resource Conservation Service.

Despite this improvement, a large body of evidence still indicates that significant areas of U.S. rangelands are being overgrazed and need improvement. Let us not forget, we still have large areas of fair and poor rangelands, both public and private. In all fairness, though, some rangelands classified as fair or poor may be in satisfactory condition under current management practices. Here's why: Remember that each rangeland site is rated relative to its potential natural plant community. A site being managed for multiple uses (such as wildlife habitat, camping, hiking, and livestock grazing) may be most suited to vegetation that is very different from its natural potential composition. For example, deer forage is usually maximized on a site when shrubs and forbs, rather than perennial grasses, are plentiful. But if such a site's natural potential were abundant perennial grasses, its existing vegetation (of abundant shrubs and forbs) would be judged dissimilar from its potential, and the site would get a lower condition class rating, such as fair or poor. Yet the site may currently support the desired plant species in the desired relative amounts.

Rangelands Outside the United States Partial surveys indicate that rangelands in Canada have followed a similar upward trend as in the United States. Australia's rangelands have generally shown an improvement in recent decades, although some areas have certainly been overgrazed and degraded. Most of the rangelands in Europe are in excellent or good condition and are among the most productive in the world. This cannot be said for most rangelands in other parts of the world. Limited data from partial surveys in Africa, Asia, and South America show that many of these regions' rangelands have been degraded to some degree.

Rangeland deterioration is probably most visible in Africa and Asia, which together contain nearly half the world's rangelands. A three-year UN study, *Global Assessment of Soil Degradation*, involving more than 250 scientists throughout the world, found that overgrazing has degraded some 680 million hectares since midcentury, of which almost 65% are in Africa and Asia (Table 13.2). This study suggests that 20% of the world's pasture and range is losing productivity and will continue to do so unless herd sizes are reduced or more-sustainable livestock practices are put in place.

The nine countries in the arid Sahel region of West Africa, where livestock numbers commonly exceed rangeland carrying capacity by 50% to 100%, have reported severe overgrazing and an increase in desertification. The northern African countries of Morocco, Algeria, Tunisia, Libya, and Egypt are

TABLE 13.2	Worldwide Land Degradation by Overgrazing, 1945–1991
Region	Land Overgrazed (million hectares)
Africa	243
Asia	197
Europe	50
North and Central America	38
Oceania	83
South America	68

experiencing overgrazing and decreased forage production. Sudan shows signs of a rapidly deteriorating range and increasing desertification. Zambia has a serious overgrazing problem, with livestock numbers far exceeding the rangeland's carrying capacity.

In China, about one-fourth of the rangeland has been severely degraded, most of it in the northern part of the country. Much of the rangeland in the Middle East, particularly in parts of Iran, Iraq, Jordan, Oman, Pakistan, and Syria, is in poor condition. In India, much of the rangeland is degraded or overgrazed, although some is improving through rangeland improvement programs. Many rangelands in the South American countries of Brazil, Argentina, Uruguay, and Paraguay have been degraded from overstocking.

13.4 Range Management

Range management is an interdisciplinary field. That is, it requires information from a variety of scientific disciplines, such as soil and plant sciences, animal and wildlife sciences, forestry, hydrology, economics, and other related fields. The major goal of range management is to maximize livestock or wild-herbivore productivity while protecting the long-term health of rangeland.

Like other areas of resource management, the best strategy in range management is prevention: keeping rangelands from deteriorating. One of the first steps in good range management is determining the carrying capacity of the rangeland sites to be managed. This estimate is then used to manage the number and kinds of animals grazing rangeland sites to avoid overgrazing. To protect rangeland, many ranchers and rangeland managers also rotate their livestock from one site to another to prevent overgrazing. The task of managing rangeland is made more difficult by the fact that other rangeland species, such as grasshoppers and jackrabbits, can compete with livestock for forage, decreasing the carrying capacity of a particular pasture or open range. Predators such as coyotes may also play a role in determining the carrying capacity of a range, for reasons explained shortly. Several range management options are available to increase the carrying capacity of rangeland, among them stock manipulation, artificial seeding, and the control of plant and animal pests.

Stock Manipulation

Distribution of Water and Salt Cattle and other livestock tend to concentrate in certain areas, for example, in wet meadows and along stream margins, where forage is most abundant and most nutritious. They also tend to avoid other locations, such as ridges and slopes, where forage is less abundant. As a result, part of a given range may be severely overgrazed while another part is ignored.

To prevent this problem, ranchers often take steps to ensure that livestock graze their land more uniformly. Grazing can be directly controlled by barbed-wire fencing and herding, that is, fencing off fields and herding livestock from one location to another. Both methods are rather costly.

Ranchers also use indirect methods that are much less expensive but highly effective. For instance, they can strategically locate water holes and salt blocks throughout the range. Because cattle and sheep normally congregate around water sources, the salt blocks should be placed between 0.4 and 0.8 kilometer (0.25 and 0.5 mile) from the nearest water source, although this distance can vary depending on length of grazing time and other factors. Salt blocks should also be located in ungrazed areas on ridges, gentle slopes, or openings in brush or forest to induce livestock to frequent areas they would normally avoid (Figure 13.19).

Salt is essential to the vigorous health of range animals. Within three weeks after having been deprived of salt, cattle develop an unusual craving for it. When salt deprivation continues, the animals lose their appetite, become emaciated and weak, and may collapse. On ranges where the soil is naturally high in phosphate and sulfate salts, livestock may partially satisfy their salt requirements by grazing on salt-absorbing vegetation.

Grazing Systems Some ranchers allow their livestock to graze in one field year-round, an approach known as a continuous-grazing system. However, many ranchers rotate livestock from one field to another. Over the years, ranchers and rangeland managers have developed eight grazing systems: (1) continuous, (2) deferred-rotation, (3) rest-rotation, (4) short-duration, (5) Merrill three-herd/four-pasture, (6) high-intensity/low-frequency, (7) best-pasture, and (8) season-suitability grazing. In selecting an appropriate grazing system, ranchers need to consider many factors, such as the climate, topography, vegetation, and the kind or kinds of livestock to be grazed. They also take into account wildlife needs, watershed

FIGURE 13.19 Salt block for cattle. Proper placement of salt, either alone or in mixture with some other supplement, can help obtain the desired distribution of grazing animals. Livestock usually go from water to grazing and then to salt. When salt is placed away from watering points, livestock can be enticed to use grazing areas otherwise avoided.

protection, and labor requirements of the grazing system. But that's not all; they must also consider whether fencing is required, as well as the availability of water. While students who go on to study range management will be exposed to the eight systems, we will discuss only a few grazing systems in this chapter.

Allowing livestock to graze in a pasture continuously throughout the grazing season is called **continuous grazing.** One problem with continuous grazing is that livestock have preferred areas of grazing that will often receive excessive use, resulting in a mosaic of overgrazed and undergrazed patches in an area. In preferred grazing areas, the more palatable (and usually more nutritious) range plants can be so seriously overgrazed that they lose vitality and nutritional value, or become replaced by less nutritious and less palatable plants.

The first specialized grazing system developed in the United States was **deferred-rotation grazing.** This system is a type of crop rotation, the "crop" being cattle or other livestock. The purpose is to temporarily withdraw fields from grazing by moving livestock. The main features of this system are presented in Figure 13.20. As illustrated, in this system, a ranch is divided into three pastures, A, B, and C. In year one, pasture B is grazed first, and then cattle are shifted to pasture C. Pasture A is grazed last, but only after the grasses have matured—that is, after the plants have gone to seed and dropped their seeds. In year two, pasture C is grazed first, followed by pasture B. Pasture A is grazed last—again, only after the seeds have matured and dropped to the ground. In years three and four, pasture B is deferred—that is, grazed last, only after seed production has been completed. In years five and six, pasture C is deferred. In summary, with this system, grazing in each pasture is deferred for two successive years within a six-year period.

During years one and two, forage plants in pasture A are allowed to reach maturity and drop their seeds before livestock are permitted to graze on them late in the season. Even though the grasses become rather dry at this time, they still are highly nutritious. A certain amount of grazing after the seeds have been produced may be advantageous to the pasture, because foraging cattle scatter the seeds and trample them underfoot, which forces the seeds into the ground and enhances germination.

Deferred-rotation grazing has many benefits. Most important is that it helps to ensure long-term health of the second crop: the forage. This system increases the size, density, and weight of forage. It also increases its vitality, reproductive capacity, and nutritional value.

Short-duration grazing was developed in Zimbabwe by Allan Savory in the 1960s and introduced in the United States in the 1970s. Savory has since made refinements to this system, which is now called **holistic resource management.** In this system, cattle or other livestock are allowed to intensely graze a field. Because they are in a field for only a short time and many animals are competing for the same food, they eat rather indiscriminately. That is, they eat the more palatable as well as the less palatable plant species. When the field is thoroughly grazed, the livestock are moved out, allowing the plants to grow back. This system, which has gained popularity in specific areas of the United States, involves a high stock density (number of animals per unit area). Under good management, it is claimed that labor costs are reduced, individual animal performance is increased, and range condition is improved. Although holistic resource management has been advocated for all rangeland types around the world, there is not enough long-term research proving its effectiveness. For a complete discussion of Savory's views on grazing management, refer to Savory's book listed in Suggested Readings at the end of this chapter.

Artificial Seeding (Reseeding)

Ranchers can improve their rangelands by periodic seeding, also called **reseeding.** Reseeding helps restore severely degraded rangeland and increase its carrying capacity. Seeds may be broadcast by hand or airplane. However, unless some provision is made for covering the seeds, reseeding will probably fail. Uncovered

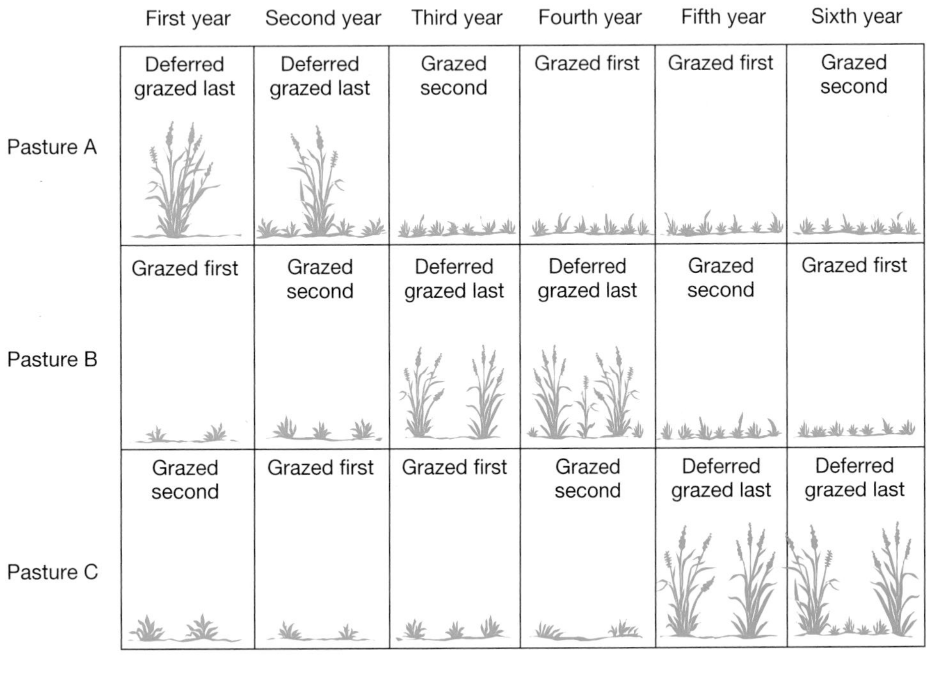

	First year	Second year	Third year	Fourth year	Fifth year	Sixth year
Pasture A	Deferred grazed last	Deferred grazed last	Grazed second	Grazed first	Grazed first	Grazed second
Pasture B	Grazed first	Grazed second	Deferred grazed last	Deferred grazed last	Grazed second	Grazed first
Pasture C	Grazed second	Grazed first	Grazed first	Grazed second	Deferred grazed last	Deferred grazed last

FIGURE 13.20 Deferred-rotation grazing.

seeds may be blown away by strong winds, killed by severe winter cold, eaten by birds and rodents, or washed away during heavy rainstorms. Ranchers can drive a herd of cattle over the area to trample the seeds into the ground. If seeding is done in recently burned-over timberland, the loose covering of ashes may ensure the success of reseeding. (In very intensely burned areas, though, the soil can be too hard to permit seeds to penetrate.) Similarly, if reseeding is synchronized with autumn leaf fall, the leaf litter may provide sufficient seed cover for successful germination.

Aerial broadcasting is the only feasible method in rugged, mountainous regions. It can be highly effective. For example, several years ago, a burn in a fir and pine stand in the Cabinet National Forest of Montana was seeded by airplane. Two years later, the area was cloaked with a dense stand of timothy grass and Kentucky bluegrass. This vegetation provided not only excellent protection from erosion but also 2,100 kilograms of food per hectare (1,900 pounds per acre) for grazing animals.

In general, ranges that have been properly reseeded support more livestock in better condition over a longer period than equivalent ranges that have not been reseeded. Many reseeded areas in the West, for example, have been grazed for 15 successive years and still produce 3 to 20 times as much forage as they did before seeding. A classic example of how artificial reseeding can improve rangeland is provided by a 200-hectare (500-acre) plot in the Fish Lake National Forest in Utah. Before seeding, this area supported a vigorous cover of sagebrush and rabbit brush, which provided forage for only 8 head of cattle. However, only three years after it was seeded to wheatgrasses (Figure 13.21) and bromes, it was able to sustain 100 head.

Control of Rangeland Pests

Another important aspect of rangeland management is the control of plants, such as weeds and woody vegetation, that compete with range grasses; the control of herbivores (grasshoppers, jackrabbits, rodents) that compete with livestock for food; and the control of predators, particularly the coyote.

Control of Plant Pests Ranchers are periodically confronted with invasion of their grazing land by woody, low-value shrubs, such as mesquite, sagebrush, and juniper. These and other species compete with range grasses for soil moisture, nutrients, and sunlight (Figure 13.22). Invasion of these shrubs is especially likely if the land is overgrazed. Unless effective control methods are established, the aggressive spread of these species may seriously lower the livestock-carrying capacity of the range. Consider the mesquite.

FIGURE 13.21 Crested wheatgrass, an exotic bunch grass that was introduced from Russia. It thrives in the northern plains states, where summers are cool.

The **mesquite** is a thorny desert shrub with small, leathery leaves. Like all members of the pea family, it produces large, pulpy seed pods. The extensive root system may extend to a depth of 15 meters (50 feet).

Of all the woody plant invaders of southwestern grasslands, mesquite ranks first in distribution, abundance, and aggressive encroachment on rangeland. Plant ecologists believe that, for millennia, mesquite invasion was prevented by periodic fires ignited by lightning strikes or by Native Americans who lived on the Great Plains (as an aid to hunting). Grasses were consumed in the fires along with mesquite. Many grass species, however, can mature and produce seeds in two years, whereas mesquite requires a longer period. For this reason, recurrent fires can control mesquite growth.

Ecologists believe that the grassland biome's climax vegetation, with its deep, fibrous root systems, can compete successfully with mesquite for sunlight and limited soil moisture. However, when the white settlers drove off the Indians, they introduced cattle and sheep by the millions and instituted new methods to control fire. As a result, the main factor responsible for confining mesquite vanished. Overgrazing caused the climax species (decreasers) to decline. The decreasers, in turn, were replaced by increasers, which were then replaced by invaders. The first invaders (with shallow taproots) were inferior to mesquite, and the mesquite gradually took over. The invasion of the mesquite was also facilitated by cattle. The late-summer-maturing mesquite pods, some up to 20 centimeters (8 inches) long, provide cattle with nutritious food. Although the cattle digest the pulp, seeds

FIGURE 13.22 Mesquite frames an expanse of Red Rock Canyon National Conservation Area in the Mojave Desert of Nevada.

generally pass through the digestive system and are voided with their feces, frequently at a considerable distance from the parent plant. The seeds, still viable and well fertilized, show a surprisingly high germination rate.

Dense, mature mesquite stands or other invasive plants can be regulated through **controlled burning** (Figure 13.23). In this way, the rancher can duplicate the natural control by fire that operated for thousands of years before the arrival of the white settlers. Like natural fires, controlled burning encourages the growth of valuable forage plants such as grama, bluestem, and buffalo grasses. It can also benefit many wildlife species. Furthermore, controlled burning can result in a more uniform distribution of grazing livestock. Studies also show that cattle "bulk up" much more quickly on control-burned ranges than on unburned ranges. This rapid weight gain is due not only to the increase in the protein, phosphorus, and moisture content of the forage plants after a burn, but also to the plants' enhanced palatability and digestibility. Despite its many benefits, controlled burning in the United States has been restricted in recent years because of concern about air pollution and fires.

Grubbing the mesquite plants out of the ground (pulling them out with chains and heavy equipment) or plowing them up is also effective—but very costly. In the case of plowing, the whole area would have to be carefully reseeded with nutritious forage grasses.

The control of extensive acreages of mesquite can be accomplished by the aerial spraying of herbicides. Great care must be taken, however, to prevent damage to valuable plants and animals. As a rule, herbicides are not used very often in the United States to control range vegetation, because they are expensive and may pose threats to human health.

Biological control of unwanted range plants includes the introduction of goats, camels, and predatory insects. Goats, in particular, are used in many countries to control brush and other weedy species. Careful and complete ecological study should precede the introduction of any biological agent—that is, any species that will very like remain on the range—to be sure that the animal (or plant) concerned will not displace the desirable native species. Some range managers control unwanted vegetation with periodic, short-term trampling by large numbers of livestock.

Control of Herbivores Insects can more severely overgraze ranges than can livestock. Although range caterpillars, black-grass bugs, Mormon crickets, and harvester ants cause the greatest damage, the grasshopper is the number one insect in influencing range vegetation. Of the more than 100 species of **grasshoppers** collected in western range vegetation, the most destructive and widely distributed are the lesser migratory grasshoppers (Figure 13.24). During periods of peak abundance, they may gather in swarms and migrate several hundred kilometers.

When the weather is wet, grasshopper populations remain small. The rancher may not even be aware of their presence. In severe drought, grasshopper populations rapidly increase until it is almost impossible to take a single step through a pasture without flushing several of them. During a peak year, grasshoppers may so deplete forage that livestock must move to other pastures or starve. During a severe drought, grasshopper density

FIGURE 13.23 Controlled burning of a Montana range.

FIGURE 13.24   Grasshopper outbreak. An insect specialist examines grasshoppers in his sweep net. By sweeping the range grasses a few times with his net, he can determine the severity of the outbreak. Appropriate control measures can then be applied.

may reach more than 30 per square meter; the insects may consume up to 99% of all vegetation.

One interesting facet of the grasshopper–rangeland relationship is that these insects are much more numerous in overgrazed ranges than in moderately grazed fields. This is because most grasshopper species prefer ranges with little grass and a high forb (broad-leaved flowering plants) component. A study in southern Arizona found that the grasshopper population was 450,000 per hectare (180,000 per acre) on overgrazed lands, compared with only 50,000 per hectare (20,000 per acre) on range in average condition—a 9:1 differential. Therefore, a possibly effective method of controlling grasshopper plagues may be ensuring that the range is not subjected to excessive grazing. Again, prevention is the best medicine!

During a drought, **jackrabbits** compete aggressively with cattle and sheep for high-quality forage (Figure 13.25). Between 75 and 150 jackrabbits consume as much forage as one cow, and 15 to 30 eat as much as one sheep. Studies show that jackrabbits consume about 6.5% of their body weight per day. Such heavy intake is more than three times the daily rate of most ruminants on range forages. A dense jackrabbit population, therefore, exerts heavy grazing pressure on vegetation in a range. This, in turn, may prevent the reestablishment of highly nutritious decreaser species while favoring the intrusion of lower-value increasers as well as invaders such as cactus and thistle.

The rangeland pest, whether grasshopper, jackrabbit, or even prairie dog, becomes a serious problem only during population peaks, and these peaks usually coincide with rangeland deterioration. It should be emphasized that these pests do not cause the initial depletion of the pasture. They are a symptom rather than a cause of range deterioration. We can compare range abuse to a wound. The "wound" was initially inflicted by excessive stocking, and the ensuing pest buildup merely irritated the wound and prevented it from healing properly.

Just why a range in good condition (well stocked with climax plants) should be an unsuitable habitat for certain rabbits and rodents has never been fully explained. Perhaps the tall vegetation obstructs the vision of these relatively defenseless animals and makes them more vulnerable to predators. In any event, most rangeland experts agree that shooting, trapping, and poisoning campaigns are only stopgap procedures and are rarely worth the cost. Rabbits and rodents have high reproduction rates, and their numbers can usually recover in a short

FIGURE 13.25   The jackrabbit competes with livestock for forage. However, it becomes a serious pest only when the range has been overgrazed, as in this photograph of the Santa Rita Experimental Range in Arizona.

time. The best long-term solution to the pest problem seems to be vegetation management, which often is simply a four-strand barbed-wire fence to keep excess cattle off the deteriorating range, coupled with a rotational grazing plan.

Control of Predators: The Coyote   Predators such as coyotes, black bears, golden eagles, bobcats, foxes, and mountain lions can have a considerable influence on range livestock. Of all rangeland predators, the **coyote** (Figure 13.26) poses the most serious problem, mainly to sheep. In one study on losses of sheep to predators in the Great Basin, coyotes accounted for 90% of the losses, bobcats for 2%, and undetermined predators

FIGURE 13.26   The coyote, a stealthy rangeland predator that has been accused, justly or unjustly, of killing many thousands of sheep annually.

for 8%. The wily "brush wolf" has become a thorn in the side of sheep ranchers, with some reporting sheep losses of more than 20%. Studies show that losses of lamb to predators are much lower for herded sheep than for unherded sheep. However, the number of sheepherders has been declining in the past four decades.

When a rancher destroys a coyote that has been killing, say, 20 sheep per year, simple arithmetic might suggest that this rancher will be 20 sheep richer each year thereafter. However, nature is not quite that simple and straightforward. For one thing, the coyote feeds on animals other than sheep. Fifty percent of its diet is composed of range grass-consuming jackrabbits and rodents. Therefore, the value of the few sheep saved by destroying a coyote may be less than the value of the forage consumed by the hundreds of rodents and rabbits that the coyote would have removed from the rangeland community, had it been allowed to live. Nevertheless, many ranchers appear irrevocably committed to predator control as a range management tool. In fact, where coyotes have become numerous, sheep ranchers have waged all-out extermination campaigns—poisoning, trapping, shooting, and even pursuing them in their dens.

Predator control on grazing lands became a federal government responsibility by an act of Congress in 1931 called the Animal Damage Control (ADC) Act. Under the federal ADC program, in 1994 alone, more than 100,000 predators, mostly coyotes, were destroyed. In light of the latest predation data, many experts believe that coyote control should focus on the relatively few ranches where sheep kills have actually occurred. A widespread, nonselective campaign to destroy all coyotes is expensive, time-consuming, and highly unwarranted. The pros and cons of various methods of coyote control are discussed in Case Study 13.1.

CASE STUDY 13.1   METHODS OF COYOTE CONTROL

Coyote populations can be controlled in a variety of ways. Some methods for controlling coyote populations are nonlethal. These include the use of sheepherders, guard dogs or cattle, birth control, and chemical repellents. Others are lethal. They include poisoning, trapping, shooting, and den hunting. Most of these control strategies are discussed in this case study.

Guard Dogs

For several centuries, European and Asian sheepherders successfully used guard dogs to protect sheep from predatory coyotes. Some of the more common breeds used in the United States originated in Europe and Asia. They include Great Pyrenees (from France), Akbash and Anatolian shepherd (from Turkey), Komondor (from Hungary), Maremma (from Italy), and Tibetan mastiff (from Tibet). Since 1978, sheep ranchers in 31 states have been using these dogs to protect their flocks (Figure 1). When the dogs are placed in sheep flocks while still pups, they soon consider themselves a natural part of the flock. When a coyote approaches a

sheep, a mature dog's protective instincts kick in. The dog reacts quickly, rushing fiercely at the intruder, causing it to flee.

The guard dog system is highly effective and can be a financial boon to the sheep rancher. For instance, a survey found that one of every three ranchers who had experienced heavy sheep losses from marauding coyotes reported not a single attack once guard dogs were used. Environmentalists firmly support the guard dog system. Nevertheless, many ranchers do not use guard dogs, either because they are unwilling to try a new approach to coyote control or because they mistakenly believe the dogs themselves will kill some sheep.

Guard Cattle

A recently developed predator-control strategy involves the intermingling of lambs and calves in pens for one month. During this period, the animals develop a strong attachment for each other. When released on the open range, the cattle protect the sheep from predators by kicking and butting the predators. According to a report by the USDA, this method can sharply reduce sheep kills by coyotes.

(continued)

FIGURE 1   A Great Pyrenees guard dog watching over a flock of sheep in New Zealand.

Birth-Control Chemicals

In an effort to limit a coyote population, carcasses of livestock can be laced with birth-control chemicals that reduce the reproductive ability of coyotes that consume the meat. Theoretically, coyote populations should then decline. Unfortunately, however, such a decline would probably be only temporary. The reason? Like many other species of wildlife, coyotes have tremendous reproductive resilience. When a population declines in a given year, the number of young the next year generally increases. The coyote is so resistant that, to eradicate it completely, 75% of the population would have to be destroyed year after year for half a century!

Chemical Repellents

The use of chemical repellents, still in the experimental stage, involves the injection of lithium chloride—a bad-tasting, nausea-inducing chemical—into the carcasses of dead sheep. The carcasses are then left out for coyotes to feed on. When a coyote eats the tainted flesh, it becomes very sick and may thereafter avoid coming into contact with live sheep.

Lethal Poison: Compound 1080

The use of sodium monofluoroacetate, popularly known as Compound 1080, for coyote control has been highly controversial. The general public, environmentalists, and organizations including Defenders of Wildlife, the National Audubon Society, and the National Wildlife Federation have strongly opposed its use. For one thing, it is extremely toxic; 28 grams (1 ounce) is sufficient to kill 20,000 coyotes.

One major complaint concerning its use has been the unintentional but fatal poisoning of nontarget species such as golden eagles and bobcats that accidentally eat Compound 1080–laced bait intended for coyotes or even the carcasses of poisoned coyotes. Some wildlife biologists estimate that while the chemical was being used in the 1960s, about 9,000 bobcats were accidentally poisoned per year.

In 1972, the use of Compound 1080 on all federal lands and by federal agencies anywhere was banned by an executive order of then-president Richard Nixon. A short time later, the EPA banned its use by state agencies and private individuals as well.

Sheep ranchers complained bitterly to their congressional representatives, saying the ban on Compound 1080 deprived them of the most effective weapon in their coyote-control arsenal. In 1985, the EPA yielded to pressure from the Reagan administration and approved the use of the poison in special collars that are worn around the necks of sheep. Since coyotes frequently lunge for the neck of their prey, Compound 1080–containing collars would seem to be an effective way to control these predators.

Is the political struggle to use Compound 1080 really worth the effort? The answer, ironically, is no. In one highly regarded study, sheep losses from predators and other causes before the use of Compound 1080 (1940–1949) were compared with losses from 1950 to 1970, when it was widely used. The study found no detectable difference in sheep losses from predation with or without Compound 1080. The present policy is for the USDA to restrict predator-control activities to ranchers who suffer substantial, legitimate, known predator losses.

Congressman Peter DeFazio (Oregon-Dem) has been advocating for the elimination of Compound 1080 since 2004. In 2005, DeFazio introduced legislation in the U.S. House of Representatives that would ban this lethal poison, but Congress failed to act on the legislation. Since then, DeFazio has written numerous letters to the EPA and other federal agencies, urging them to utilize their authority to ban Compound 1080 because of its deadly nature. DeFazio's most recent action, the introduction of H.R. 4775, the Compound 1080 and M-44 Elimination Act, would halt the legal use of these poisons once and for all.

Summary of Key Concepts

1. Rangelands are areas of the world that are a source of forage (such as grasses and shrubs) for free-ranging native and domestic animals, as well as a source of wood products, water, energy, wildlife, minerals, and recreational opportunities.

2. Worldwide, rangelands occupy about half of the planet's ice-free land surface and provide about 80% of the feed for domestic livestock. Rangelands occupy about 29% of the total land area in the United States.

3. Seven major categories of vegetation provide most of the world's grazing lands: grasslands, tropical savannas, tundra, desert shrubs, shrub woodlands, temperate forests, and tropical forests.

4. The upper 50% of the grass shoot is a surplus that can be safely eaten by livestock; the lower 50%, known as the metabolic reserve, is necessary for the plant's survival.

5. Range ecologists and ranchers frequently classify the various range plants into three categories with respect to the dynamics of plant succession: decreasers, increasers, and invaders.

6. Grazing carrying capacity is the maximum number of animals that can graze each year on a given area of range, for a specific number of days, without causing a downward trend in forage production, forage quality, or soil quality.

7. Overgrazing is defined as continued heavy grazing that exceeds the carrying capacity of the community and results in a deteriorated range.

8. Desertification is land degradation caused by climatic changes and human-induced activities, such as overgrazing, overcutting, cultivation of unsuitable land, or inadequate irrigation, in dry, semidry, or dry–semihumid areas.

9. The Homestead Act of 1862 offered 160 acres (64 hectares) of public land to anyone who would settle the land and reside on it for five years.

10. In 1905, the U.S. Forest Service (USFS) was formed and began to restrict livestock numbers and grazing seasons in U.S. national forests.

11. The major objective of the Taylor Grazing Control Act of 1934 was to improve the quality of U.S. rangelands.

12. About 2% of U.S. ranchers are issued grazing permits (subsidized by American taxpayers) by the Bureau of Land Management (BLM) or USFS to graze their herds on public rangeland.

13. The term *range condition* is defined as the current state of vegetation on a range site in relation to the potential natural plant community.

14. Public rangelands in the United States were severely abused in the late 1800s and early 1900s due to improper livestock grazing.

15. Although better grazing management practices are still needed in many areas, today U.S. rangelands are in their best condition of the past 100 years.

16. Many of the rangelands in Africa, Asia, and South America have been degraded to some degree.

17. Livestock may be better distributed to prevent overgrazing by the strategic distribution of salt and water.

18. Various grazing systems, such as continuous grazing, deferred-rotation grazing, and short-duration grazing (also called holistic resource management), are commonly used in the United States and other parts of the world.

19. Rangelands in poor condition may benefit from reseeding.

20. An important aspect of rangeland management is the control of plant pests, such as weeds and woody vegetation, that compete with range grasses, as well as the control of animal pests (grasshoppers, jackrabbits, rodents) that compete with livestock for food.

21. Various methods have been used to control coyote predation of sheep, including guard dogs, guard cattle, birth-control chemicals, chemical repellants, lethal poisons such as Compound 1080, trapping, shooting, and den hunting.

Key Words and Phrases

Animal Unit Months (AUMs)
Basal Zone
Bureau of Land Management (BLM)
Carrying Capacity
Conservation Reserve Program (CRP)
Continuous Grazing
Controlled Burning
Coyote
Decreasers
Deferred-Rotation Grazing
Desert Shrublands
Desertification
Federal Land Policy and Management Act
Feedlots
Food Security Act
Forage
Grasshoppers
Grasslands
Grazing District
Grazing Permit
Holistic Resource Management
Homestead Act of 1862
Increasers
Invaders
Jackrabbits
Mesquite
Metabolic Reserve
Multiple Use
National Forest Management Act
National Grasslands
The Nature Conservancy
Overgrazing
Powell, John Wesley
Public Rangelands Improvement Act
Range Condition
Rangelands
Reseeding
Sahel
Short-Duration Grazing
Shrub Woodlands
Sodbusting
Taylor Grazing Control Act
Temperate Forests
Tropical Forests
Tropical Savannas
Tundra
Undergrazing
U.S. Forest Service (USFS)

Critical Thinking and Discussion Questions

1. What are the differences between grasslands, tropical savannas, and desert shrublands?

2. What unique growth characteristics of grasses enable them to survive despite moderate grazing pressure?

3. What is the relationship among the three groups of rangeland plants called decreasers, increasers, and invaders?

4. Briefly list three major objectives of the Taylor Grazing Control Act of 1934.

5. Name the two federal agencies that administer most of the public rangelands in the United States. When were they established?

6. Should fees for grazing on public rangelands in the United States be increased? Why or why not?

7. What is the function of the feedlot?

8. Describe the deferred-rotation grazing system.

9. Briefly describe some of the positive effects of rangeland burning.

10. What precautions must be taken when reseeding a range by broadcasting?

11. Why has mesquite become a serious rangeland pest?

12. What would be an effective strategy for controlling grasshoppers on rangeland?

13. Describe the different methods, both lethal and nonlethal, of coyote control.

14. Define *range condition*. Describe the condition of U.S. rangelands today.

15. What are the major causes of desertification in Africa?

16. What are rangelands? Where are significant areas of rangelands in the world?

17. What are the benefits of restoring native prairies?

Suggested Readings

Behan, R. W. 2001. *Plundered Promise: Capitalism, Politics, and the Fate of Federal Lands.* Washington, DC: Island Press. An in-depth view of the history of public land management and what is wrong with it.

Council for Agricultural Science and Technology (CAST). 1996. *Grazing on Public Lands.* Task Force Report No. 129. Ames, IA: CAST. Discussion and scientific information concerning livestock grazing on public lands in the western United States.

Dagget, D. 2000. *Beyond the Rangeland Conflict: Toward a West That Works.* Reno: University of Nevada Press. Ten real-world examples of environmentalists and ranchers working together to deal with issues of open space, endangered species, functional ecosystems, environmental restoration, holistic management, and conflict resolution.

Heady, H. F., and R. D. Child. 1999. *Rangeland Ecology and Management.* Boulder, CO: Westview Press. Rangeland management textbook that focuses on the ecology of rangeland grazing, practical management of animals, and vegetational manipulation.

Higgins, K. F., A. D. Kruse, and J. L. Piehl. 1989. *Effects of Fire in the Northern Great Plains.* Extension Circular 761. Brookings: South Dakota State University. Excellent discussion of the effects of fire on the grassland ecosystem of the northern Great Plains, with special emphasis on the use of fire for wildlife management.

Holechek, J. L., R. D. Pieper, and C. H. Herbel. 2004. *Range Management: Principles and Practices,* 5th ed. Upper Saddle River, NJ: Prentice Hall. Solid fundamental textbook covering almost all aspects of range science and management.

Kemmis, D. 2001. *This Sovereign Land: A New Vision for Governing the West.* Washington, DC: Island Press. A radical new approach to managing resources (collaborative management) regionally.

Knapp, A. K., J. M. Briggs, D. C. Hartnett, and S. L. Collins. 1998. *Grassland Dynamics: Long-Term Ecological Research in Tallgrass Prairie.* New York: Oxford University Press. Comprehensive description of the long-term research studies on the ecology of the tall-grass prairie, focusing on the Konza Prairie in Kansas.

Mitchell, J. E. 2000. *Rangeland Resource Trends in the United States.* Fort Collins, CO: USDA Forest Service. Report documenting trends in America's rangelands as required by the Renewable Resources Planning Act of 1974.

National Research Council. 1994. *Rangeland Health: New Methods to Classify, Inventory, and Monitor Rangelands.* Washington, DC: National Academy Press. Examination of methods used to inventory, classify, and monitor rangelands, with recommendations for evaluating the ecological health of U.S. rangeland ecosystems.

Pollan, M. 2006. *The Omnivore's Dilemma: A Natural History of Four Meals.* New York: Penguin. Tracing from source to table the food chains that sustain us, the author develops a portrait of the American diet (partly based on high fructose corn syrup and too much meat) and the profound implications our food choices have for the health of our species and the future of our planet.

Savory, A., and J. Butterfield. 1998. *Holistic Management.* Washington, DC: Island Press. Good discussion of Savory's grazing system, called holistic resource management, and how it can be applied.

Shenk, T. M., and A. B. Franklin, eds. 2001. *Modeling in Natural Resource Management: Development, Interpretation, and Application.* Washington, DC: Island Press. Information about one of the most useful tools in resource management: computer modeling.

Stegner, W. 1954. *Beyond the Hundredth Meridian.* New York: Penguin Books. A fascinating look at the American West in the late 1800s and at John Wesley Powell, who explored it and warned against the dangers of settling it.

Stoddart, L. A., A. D. Smith, and T. W. Box. 1975. *Range Management,* 3rd ed. New York: McGraw-Hill. A classic on the characteristics, ecology, and management of rangelands.

Stubbendieck, J., S. L. Hatch, and C. H. Butterfield. 1997. *North American Range Plants,* 5th ed. Lincoln: University of Nebraska Press. Comprehensive treatment of the habitats and forage values of American range grasses, with excellent drawings of each species.

Vincent, C. H. 2006. *Grazing Fees: An Overview and Current Issues.* Washington, DC: Congressional Research Service, Library of Congress. Excellent review of issues related to private versus public grazing fees.

White, C. 2008. *Revolution on the Range: The Rise of a New Ranch in the American West.* Washington, DC: Island Press. Discussion of how ranchers, working with environmentalists (originally a source of conflict and friction), create a common alliance to make the most of the land without destroying it.

Web Explorations

Online resources for this chapter are on the World Wide Web at: **http://www.prenhall.com/chiras** *(click on the Table of Contents link and then select Chapter 13).*

FOREST MANAGEMENT

Our nation's forests range from the virgin stands of hemlock and Douglas fir in Alaska and the Pacific Northwest, to second-growth oak and hickory in the Mideast and East, to plantations of pine in the South. As shown in Figure 14.1, six major forest regions can be found in the conterminous United States (the lower 48). The occurrence of a particular forest region is the biological expression of the prevailing environmental conditions, including rainfall, temperature, and soil.

Healthy forest ecosystems are ecological life-support systems. Forests provide a full suite of goods and services that are vital to humans' and other animals' health and livelihood, natural assets we call **ecosystem services.** Many of these goods and services are traditionally viewed as free benefits to society. Examples of such "public goods" include wildlife habitat and diversity, watershed services, carbon storage, and scenic landscapes. Lacking a formal market, these natural assets are traditionally absent from society's economic balance sheet, with their contributions often overlooked in public, corporate, and individual decision making. Recognizing forest ecosystems as natural assets with economic and social value can help promote conservation and more responsible decision making.

In this chapter, we'll examine forests and forest management. We'll also explore wilderness and national parks and conclude with a look at tropical rain forests.

14.1 Forest Ownership

Given all of the media coverage of forests, it may be surprising to find out that the United States has more trees now than it did in 1920 on approximately the same amount of forestland. The country also has the largest legally protected wilderness system in the world, while at the same time sustaining a highly productive and efficient wood products industry. Many changes have occurred over the years. Before the 1920s, forests were generally logged and abandoned. Today, foresters replant harvested forestlands. According to a 2000 report by the **U.S. Forest Service (USFS),** an agency of the U.S. Department of Agriculture, tree planting in the United States totals approximately 1 million hectares (2.5 million acres) annually. This annual planting covers an area roughly equal to the size of the state of Connecticut. The forest industry plants 45%, nonindustrial private owners plant 42%, the National Forest System plants 6%, and other government and industries plant 7%.

Forestland is defined by the USFS as land that is at least 10% covered by forest trees of any size. According to the Forest Service, in 2006 the United States had almost 302 million hectares (745 million acres) of forests, covering 33% of the total land area of the United States. America's forests are owned by private individuals (54%), public agencies (37%), and private industries (9%). Since the early 1900s, total forestland area in the United States has remained static at around 300 million hectares, much lower (29%) than estimated upon European settlement.

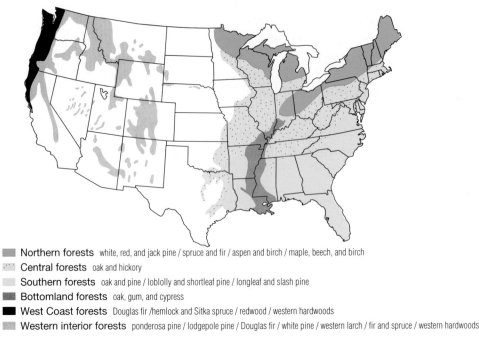

Northern forests white, red, and jack pine / spruce and fir / aspen and birch / maple, beech, and birch
Central forests oak and hickory
Southern forests oak and pine / loblolly and shortleaf pine / longleaf and slash pine
Bottomland forests oak, gum, and cypress
West Coast forests Douglas fir /hemlock and Sitka spruce / redwood / western hardwoods
Western interior forests ponderosa pine / lodgepole pine / Douglas fir / white pine / western larch / fir and spruce / western hardwoods

FIGURE 14.1   Distribution of the six major forest regions in the conterminous United States.

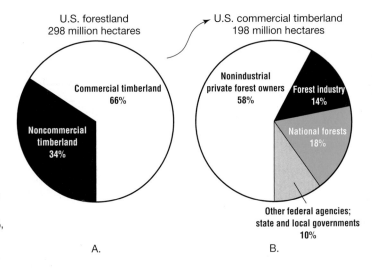

FIGURE 14.2   Classification of U.S. forestland. (A) About two-thirds of U.S. forestland is commercial timberland, while one-third is noncommercial. (B) Commercial timberland is classified by type of owner, with 58% owned by nonindustrial private forest owners (private owners other than the timber industry). The forest industry owns 14%, the National Forest System has 18%, and other federal agencies and local and state governments own the remaining 10%.

About two-thirds of U.S. forestland (198 million hectares or 490 million acres) is classified by the USFS as commercial timberland, as illustrated in Figure 14.2A. The remaining 34% is noncommercial.

Commercial timberland is considered of high enough quality to be used by the timber industry. As shown in Figure 14.2B, of the commercial timberland, 58% is owned by private individuals—that is, owners other than the timber industry, such as farmers and owners of estates. Their major source of income comes from activities other than producing timber. These individuals are classified as **nonindustrial private forest owners.**

Also shown in Figure 14.2B, 14% of the commercial timberland is owned by the forest industry, which produces a variety of wood products. More than half of the forest industry's land is located in the South, an important region in the nation's timber economy. These companies range from small open sawmills with a few thousand hectares of land to multinational

conglomerates operating several dozen mills with holdings covering millions of hectares of their forestland.

Also shown in Figure 14.2B, 18% of the nation's commercial timberland is located in the National Forest System, publicly owned land. Other federal agencies, such as the National Park Service, and local and state governments own the remaining 10%.

14.2   The U.S. Forest Service

Four federal bureaus are charged with administering and managing U.S. forests. First is the Natural Resource Conservation Service, which is concerned with farm management–associated forests. Second is the Tennessee Valley Authority, which is charged with timberland management near numerous reservoirs along the Tennessee River and its tributaries. Third is the

Fish and Wildlife Service, which is interested in improving the forest habitat for wildlife and fish. The fourth is the U.S. Forest Service, which has the primary responsibility for managing the nation's forests to promote the greatest good for the most people over the long run.

The USFS was established in 1905. President Theodore Roosevelt appointed **Gifford Pinchot** to be its first chief forester (Figure 14.3). A forestry professor at Yale University, Pinchot promoted the use of several forest management methods he had learned in Europe. Pinchot was a zealous crusader for conservation, which he defined as the wise use of natural resources.

The Forest Service divides its attention among three major areas: (1) administering and protecting the national forests; (2) researching forest, watershed, range, and recreation management, wildlife habitat improvement, forest product development, and fire and pest control; and (3) cooperating with the state and private forest owners in the 50 states, Puerto Rico, and the Virgin Islands to promote sound forest management.

The Forest Service protects and manages 155 national forests and 20 national grasslands (Figure 14.4), embracing 77 million hectares (191 million acres). About 18% of this land is protected as wilderness areas. The remaining 82% is managed by the Forest Service according to the principles of multiple use and sustained yield.

Multiple Uses

Multiple-use management requires meeting a number of different needs, including timbering, grazing, agriculture, mining, oil and gas leasing, hunting and fishing, recreation, soil conservation, wildlife protection, and watershed management. A primary objective of the Forest Service is to make the greatest number of forest resources available to the greatest number of Americans, a goal mandated by the **Multiple**

Use–Sustained Yield Act of 1960. The multiple-use management of forests looks simple on paper. In operation, however, it is an extremely complex ecological problem. For example, the Forest Service is frequently forced to use a given forest primarily for one purpose, thus sacrificing its potential use for others. A forest cannot be all things to all people. If a stand of Douglas fir, for example, is developed for high-quality timber, clear-cutting may be the best way to harvest it. However, the wholesale removal of timber may impair natural flood and erosion control and may eliminate wildlife and recreational opportunities.

Sound multiple-use management must weigh the needs of many people, and these needs vary. Thus, timber production may have top priority in the Douglas fir and western hemlock stands of Washington and Oregon, but in the low-value second-growth forests of populous New York, where many city dwellers go for a dose of wilderness tonic, recreational values have high priority.

Forests as a Source of Wood Products From early colonial days, when the straight, sturdy trunks of New England spruce and white pine were fashioned into masts for the Royal Navy, to the present, almost three centuries later, the nation's forests have been the source of a variety of valuable products. Today commercial forests in the United States provide the raw materials for more than 10,000 products. They support an industry that ranks among the top ten employers in 40 of the 50 states.

The global average consumption of wood per person is about 0.7 cubic meter (almost 25 cubic feet) per year. The United States uses more wood per capita than any nation on Earth—about 2 cubic meters (81 cubic feet) of lumber per person per year. (A considerable amount is imported from Canada and Scandinavia.) Americans eat, sleep, work, and play in a world of wood. Whether in the form of toothpicks, telephone poles, photographic film, maple syrup, paper, or structural timbers, we depend heavily on wood and wood-derived products,

FIGURE 14.3 Theodore Roosevelt and Gifford Pinchot (to the left of Roosevelt) standing at the base of a giant redwood called Old Grizzly. Pinchot was the first chief of the U.S. Forest Service.

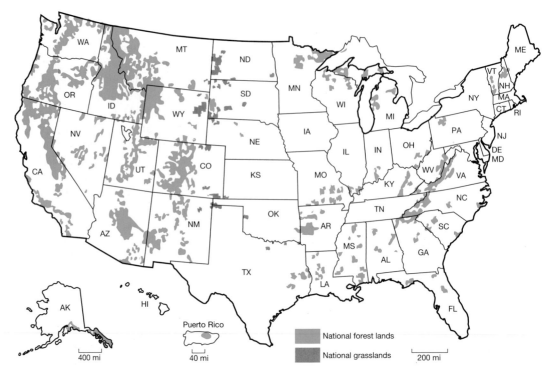

FIGURE 14.4 Location of the national forests and national grasslands of the United States.

GO GREEN!

Recycle paper as well as other recyclables. Making new paper from old paper saves trees, uses 30% to 55% less energy than making paper from trees, and reduces related air pollution by 95%.

---

as illustrated in Figure 14.5. Globally, about 50% of all wood harvested is used for fuel for direct combustion. Wood fuel can be used for cooking and heating (Figure 14.6) and occasionally for fueling steam engines and steam turbines that generate electricity.

Forests in Flood and Erosion Control Forests are more than a source of valuable wood products. They also provide valuable ecological services. For example, forest vegetation reduces flooding and soil erosion. This was demonstrated in Davis County, UT, on the eastern edge of the Great Salt Lake, a region frequently plagued by flash floods. Forest Service investigators discovered that much of a flood-triggering runoff originated from areas that had been depleted of vegetation. These denuded parts of the watershed had been burned, overgrazed, or plowed up and converted into marginal croplands. In some areas, the runoff waters carved gullies 21 meters (70 feet) deep. During one rainy period, runoff was 160 times greater on an abused plot than on a nearby undisturbed one. With the aid of bulldozers, the gullies were filled in, slopes were contoured, and the bare soil was carefully prepared as a seedbed and planted with rapidly growing shrubs and trees. Only 11 years later, severe August rainstorms put the rehabilitated watershed to the test. An investigation revealed that fully 94% of the rainfall was retained by the newly forested area. Moreover, soil erosion was reduced from the pretreatment figure of 80 metric tons per hectare (36 tons per acre) to a mere trace.

Forests as Rangelands In addition to timber and ecological services, U.S. forests frequently include considerable areas of high-quality livestock forage. Thus, of the 77 million hectares (191 million acres) of national forests and national grasslands, 40.5 million hectares (100 million acres) provide forage for 6 million cattle and sheep belonging to 19,000 farmers and ranchers. (Most of this is in the West. In the lake and central states, most forest grazing occurs on farm woodlots.) Ranchers pay fees for the privilege of grazing their livestock in national forests, as discussed in Chapter 13.

Forests as Wildlife Habitat The U.S. national forests, as well as many private woodlands, provide excellent wildlife habitat. More than 60% of the elk in the Rocky Mountain region find food, cover, and shelter in national forests.

The Forest Service tries to manage the national forests to provide the best possible wildlife habitat. Sometimes the best management involves measures to increase the amount of forest edge. Forest edge is habitat that occurs between the forest and adjoining fields, meadows, and marshes. It frequently includes a considerable number of shrubs that provide food and cover for wildlife. Development of such habitat can be integrated with timber harvesting and the construction of fire breaks and logging roads, without causing continuous habitat fragmentation. Because food and cover for elk and deer are more abundant in early stages of a succession than in climax stages, setting back the succession by periodic controlled burns may be very beneficial to them.

Sustained Yield

Those who are in the lumber business today are a different breed from the cut-out-and-get-out loggers of the late nineteenth and early twentieth centuries. After studying German silvicultural

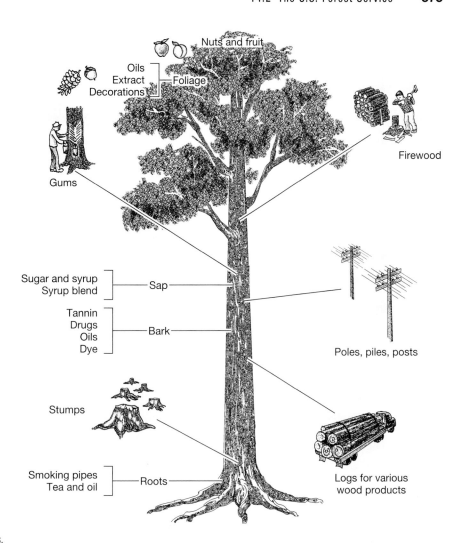

FIGURE 14.5 Useful wood and wood-derived products.

FIGURE 14.6 A child of Bali, Indonesia, carrying firewood, which has been cut in the nearby woods and will be used for cooking.

techniques, American foresters learned that a forest can be managed in such a way that a modest timber crop can be harvested indefinitely, year after year, if annual decrements are counterbalanced by annual growth. This is the **sustained-yield concept.** Under the terms of the Multiple Use–Sustained Yield Act of 1960,

foresters have a mandate from Congress to employ the principle of sustained yield in their management of national forests.

If a forest is managed for sustained yield, the wood produced in a given year should equal the volume removed. Let us see how this might work with clear-cutting, a harvest method in which all

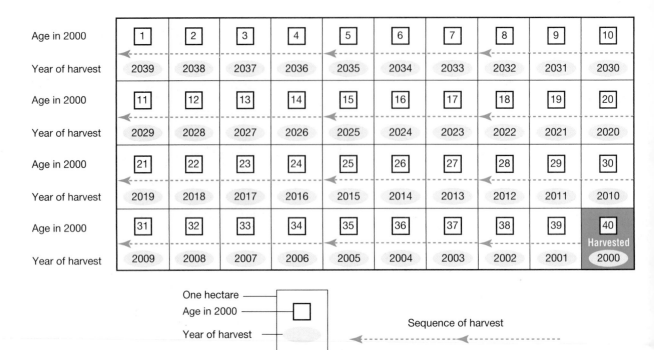

FIGURE 14.7   A 40-year rotation harvest in a 40-hectare (100-acre) forest.

the trees in a given area are cut. Suppose that a Georgia farmer owns 40 hectares of pine woods and that the trees are harvested when they are 40 years old. If this pine stand has a normal distribution, as foresters say, this stand would have 40 age classes, ages 1 to 40, each covering 1 hectare (Figure 14.7). One 40-year-old hectare of forest could be harvested each year. This operation could be carried out indefinitely as long as the clear-cut hectare is properly reseeded or replanted.

The length of the cutting cycle, or rotation, depends on the species of tree and on its intended commercial use. For aspen and birch to be used as pulpwood, the cutting cycle varies from 10 to 30 years; for pine pulpwood, it is 15 to 20 years. In contrast, the rotation for Douglas fir to be used as lumber may be up to 100 years.

14.3   Harvesting Trees

Preparation for Harvest

Once a forest matures, it can be harvested. Before harvesting a forested area, foresters estimate the volume and grade of standing timber on the site. Such an on-site survey is called a **cruise.** Individual trees to be cut as well as the boundaries of areas to be cut are marked. Data from the cruise are used by the forest manager to prepare a detailed **logging plan.**

In some states, including California and Massachusetts, the plan must be submitted to a state board for approval. A typical plan might include the following items:

1. A map showing the stand to be logged.
2. A map that shows the location, distribution, age, and volume of the species to be logged. The distribution of species is shown on areas that are to be selectively cut.

3. Harvesting method (clear-cutting, strip cutting, or selective cutting) to be employed.
4. The most suitable access roads.
5. An estimate of the amount of time required to complete the logging operation.
6. The cost of the operation and the probable gross income from the sale of the saw timber and/or pulpwood.

Harvest Methods

Managing a stand of trees requires serious planning and involves a silvicultural system. A **silvicultural system** is a long-range harvest and management program designed to optimize the growth, regeneration, and administrative management of a particular forest stand, usually with the goal of obtaining a perpetual and steady supply of timber. Such a stand is managed on a sustained-yield basis.

Silvicultural systems are generally classified by the method used to harvest and regenerate the stand. Several harvest methods are available to timber companies. The choice of a particular harvest method depends on many factors, both biological and economic. Harvest methods may be grouped under two categories: even-aged and uneven-aged stands. Clear-cutting, seed-tree, shelterwood, and coppice methods are used to produce even-aged stands. Selective cutting and strip cutting are used to produce uneven-aged stands.

Even-Aged Stand Methods   The most widely used method of timber harvesting is **clear-cutting.** The clear-cutting method, which is the standard logging practice in the Northwest and other areas in both private and public forests, is employed on **even-aged stands** and stands composed of one or two tree species. Clear-cutting is advisable only for trees whose seedlings thrive in full sunlight.

Even-aged stands of trees are maintained at about the same age and size and harvested all at once (clear-cut). Growing a single-species stand of trees is called **monoculture,** which is a controversial practice (see A Closer Look 14.1).

The **Douglas fir** on the Pacific Coast is harvested by clear-cutting. Perhaps the most valuable timber species in the world, Douglas fir from this region has been exported to Europe, where it has proved superior to native species. Some Douglas fir trees in Washington and Oregon are more than 60 meters (200 feet) tall and are over 1,000 years old. A Douglas

A CLOSER LOOK 14.1 — The Monoculture Controversy

Monoculture is the practice of growing trees much as the farmer grows a crop of corn or oats. It involves planting and raising a single-species stand of trees, with all individuals in the stand being of the same age and size (Figure 1). Monoculture is practiced on many tree farms and some public lands around the world. A **tree farm** is a private land area used to grow timber for profit. Tree farms make up 2% to 3% of the world's forest area. There are more than 74,000 tree farmers in the United States who in aggregate manage 38.5 million hectares (95 million acres) of woodland.

Monoculture is advocated by some forest managers, forest products corporations, and economists, but it has been criticized by many professional foresters and ecologists. Since it is controversial, it is instructive to list the major points both for and against monoculture.

FIGURE 1 A monoculture. This pine plantation is located in the Tennessee River Valley in the southern coniferous forest region. These trees will be used as a source of pulpwood for the paper industry. Some trees are also big enough to make lumber or panels (oriented strand board). Today, small trees make most products, due to advances in engineered wood products.

ARGUMENTS FOR MONOCULTURE

1. It is an efficient method of growing and harvesting a large volume of timber. Because growth is rapid, harvesting (by clear-cutting) can be done on a relatively short rotation.
2. It increases short-term profits for timber companies and private landowners.
3. Planting a forest monoculture is the quickest way to reforest degraded land and prevent soil erosion and desertification.
4. It is amenable to the intensive application (frequently by air) of fertilizers, herbicides, fungicides, and insecticides.
5. It makes possible maximal use of such recent technological developments as machine seeders, tree-planting machines, the tree monkey (which climbs and prunes trees simultaneously), the chip harvester, the one-man logger, and the crusher, which can clear 240 hectares (600 acres) of forested land in one month.
6. It makes possible the establishment of sun-loving (shade-intolerant) seedlings of such valuable species as Douglas fir, redwood, longleaf pines, ponderosa pine, yellow poplar, red oak, cherry, and black walnut.
7. It can help increase wood production to meet growing demands.
8. High yields from forest monocultures can reduce pressure to clear large areas of old-growth forests.

ARGUMENTS AGAINST MONOCULTURE

1. A forest under monoculture is an artificial, simplified ecosystem. As such, it lacks the built-in balancing mechanisms found in the more complex natural ecosystem represented by the multiage, multispecies forest.
2. Although monoculture admittedly grows wood faster, the wood is inferior to the slower-growing wood in a natural forest.
3. The intensive use of fertilizers and pesticides can pollute aquatic ecosystems. Pesticide contamination of food chains may have adverse effects on wildlife and humans (Chapter 8). Continuous use of insecticides can result in the development of insecticide-resistant strains of forest insects.
4. Monoculture depends on the intensive use of energy derived from fossil fuels. This energy may be used directly, as in the consumption of gasoline by tree planters, pruners, chain saws, helicopters, and airplanes (used in seeding and in applying fertilizers and insecticides), or indirectly, in the manufacture of the heavy forest-planting and harvesting machinery or in the production of fertilizers and pesticides.

(continued)

5. The single-species, single-age forest primarily serves one function: wood production. But other functions, such as erosion and flood control, maintenance of wildlife habitat and diversity, scenic beauty, and recreational opportunities, are often better served by the naturally developed multispecies, multiage forest.

6. Because of the relative scarcity of moisture-absorbing organic material on the floor of the monoculture forest, the forest floor tends to be drier and warmer than that of the natural forest. As a result, the monoculture forest is more susceptible to fire.

7. A forest monotype is very vulnerable to destructive outbreaks of insects and disease organisms.

8. Eventually, of course, the monoculture forest will be clear-cut, a harvesting method that can result in environmental problems.

One solution to the monoculture controversy is to limit forest monocultures on public lands and regulate them on private lands to prevent soil erosion and water pollution from runoff of sediment, pesticides, and fertilizers. What do you think?

fir, unlike a beech or a maple, is not a climax tree and is not shade tolerant as a seedling. Its seeds do not germinate in the shade of the forest floor. Therefore, the species is not amenable to selective cutting. If it were, its place in the forest would rapidly be appropriated by shade-tolerant species. In addition, a 30-meter (100-foot) Douglas fir weighing several tons is difficult to remove without badly bruising and killing younger growth.

With the clear-cutting technique, an entire stand of trees, usually ranging in size from 16 to 49 hectares (40 to 120 acres), is removed, leaving an unsightly scar in the midst of the forest. In national forests, the maximal size of a cut is 16 hectares (40 acres). Because many such blocks may be removed, a clear-cut forest may resemble a giant green-and-brown checkerboard when viewed from the air (Figure 14.8). In addition to its use on Douglas fir in Oregon and Washington, the clear-cutting method has been used effectively in harvesting even-aged stands of southern pine; aspen forests in northern Minnesota, Wisconsin, and Michigan; and coniferous forests in the West.

Clear-cutting is performed on **rotation.** A rotation is the cycle between planting and harvesting. If saw timber is wanted, the rotation may be 100 years. The reason for this relatively long rotation is that the trees must be quite mature before the wood has the desirable density and durability required for dimensional lumber used in construction. In contrast, if pulpwood is desired, a rotation of only 30 years is satisfactory. At that age, pulpwood species such as pine, aspen, and birch have optimal characteristics. Moreover, if birch and aspen get much older, they become highly susceptible to diseases and insect attacks. Rotations of 100 years for saw timber and 30 years for pulpwood are also the most efficient from the standpoint of harvest volume. That's because the trees are harvested before their growth rates sharply decline.

Large-scale clear-cutting is controversial in many areas, especially the United States, Canada, Australia, and the tropical forests of Latin America, Indonesia, Asia, and Africa. The clear-cutting practices of the USFS became a storm center of controversy in 1971. Much of the criticism focused on the ponderosa pine logging in the Bitterroot National Forest of Montana.

The clear-cutting practices of private industry in the United States have also been the subject of criticism. Consider the Pacific Lumber Company of California, for example. In 1988, it began to clear-cut magnificent, centuries-old redwoods (Figure 14.9) near Eureka at an accelerated pace, allegedly to pay off junk bonds (bonds issued with little or no collateral) that paid for the takeover

FIGURE 14.8   Checkerboard clear-cut area in Olympic National Forest, Washington.

FIGURE 14.9 Redwood trees in Redwood National Park, northern California. The coast redwood *(Sequoia semper-virens)* towers over all other trees in the world, with some reaching more than 113 meters (370 feet) in height.

of a company with a long track record of sustainable harvest. Since one 500-year-old redwood had a market value of more than $50,000, the accelerated clear-cutting made financial sense to Pacific Lumber. But it did not make aesthetic or ecological sense to an environmentalist group called the Coalition to Save the Redwoods. The group's members protested the clear-cut vigorously, even to the point of staging a sit-in on pulley-suspended platforms high up in the doomed trees. The advantages and disadvantages of clear-cutting are listed in Table 14.1.

Other even-aged methods are designed to overcome some of the problems with natural regeneration inherent in clear-cutting. For example, the **seed-tree method** is a silvicultural system in which all timber is removed in one cut, except for a scattered number of mature trees. They're left to provide a source of seed for the new stand. These trees need to be spaced to ensure uniform seed distribution. The seed-tree method is most suitable in situations where intensive site preparation is feasible and the trees are reasonably protected from the wind and are stable.

For trees whose seedlings do not germinate well in open conditions, foresters may use the **shelterwood method,** illustrated in Figure 14.10. As shown in the figure, foresters using this technique leave seed trees in numbers sufficient to provide protection and shade for the new tree seedlings. In a typical shelterwood cut, the first major harvest leaves enough mature trees to shade 30% to 80% of the ground, depending on species and local conditions.

Once the seedlings become firmly established (usually after several years), the remaining trees are completely cut down so that they do not reduce the growth of the saplings (young trees).

The **coppice method** depends on vegetative regeneration by stump sprouts rather than seeds, as in previous methods. Coppice stands are usually harvested by clear-cutting of even-aged forests. The coppice method is typically used for species, such as aspen and oak, that sprout vigorously and have sprouts with the potential to reach commercial size.

Uneven-Aged Stand Methods   Clear-cutting will not work in timber stands composed of **uneven-aged trees** (different ages and sizes) or in mixed stands composed in part of valuable timber species and in part of commercially unattractive species. Under such conditions, trees are harvested by **selective cutting,** a hunt-and-pick method in which mature trees of quality species are harvested at repeated intervals. Trees are marked in advance with spray paint or some other method. Deformed trees and "trash" species are removed to upgrade the stand.

The selective-cutting method can be used to harvest a single species from a forest, such as maple, beech, and hemlock. Like other species for which this technique is applied, their seedlings germinate well in the shade of the forest floor. Selective cutting has been used extensively in mixed coniferous–hardwood stands and in deciduous forests (oak, hickory, butternut, and walnut) (Figure 14.11). It is more costly and

TABLE 14.1	Advantages and Disadvantages of Clear-Cutting

Advantages	Disadvantages
1. Clear-cutting is the quickest and simplest method of harvesting, requiring less skill and planning than other harvesting methods. It also reduces road building.	1. Clear-cutting accelerates surface runoff and increases erosion on sloping land because the trees that protected the soil from rain and wind are removed. This erosion of topsoil often leads to siltation and sedimentation of stream channels, which may have downstream effects on dams.
2. A few years after the area has been clear-cut, sun-loving shrubs and saplings usually become established on the logged-off site, providing cover, food, and breeding sites for a great variety of wildlife such as rabbits, grouse, deer, and many songbirds.	2. It replaces an old-growth forest that has a diverse, uneven-aged stand of trees with a monoculture more vulnerable to attack from disease, insects, and fire. In addition, high-quality, old-growth timber is replaced with faster-growing, lower-quality timber.
3. It is the best way by which forests of highly desirable species, such as Douglas fir, can be regenerated.	3. It promotes the blowdown of trees. In a solid stand, most of the trees are protected from a windstorm. However, when a forest is clear-cut, the trees bordering the open areas are left unprotected.
4. It is the only effective method for controlling some disease and insect outbreaks. To save the life of an infected stand, the forester uses the surgery of clear-cutting, just as a surgeon amputates infected limbs to save the life of the patient.	4. It reduces biological diversity and greatly diminishes the carrying capacity (for some species) and habitat quality of an area, at least temporarily. How many grouse or deer can be supported by a bare patch of ground?
5. It usually gives the maximum economic return.	5. It reduces the recreational value of a forest and destroys the scenic beauty of a region, converting it into ugly, desolate scars.
6. It increases the volume of timber harvested per hectare and shortens the time needed to establish a new stand of trees. It often permits reforesting with genetically improved stock.	6. It creates a fire hazard. A large amount of debris (loose bark, branches, sawdust, broken logs) is often left behind after a forest is clear-cut. Such slash could easily be ignited by lightning and start a wildfire.

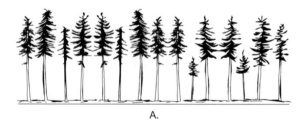

A.

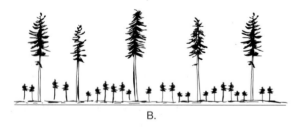

B.

C.

FIGURE 14.10 Typical shelterwood method. (A) This is a mature tree stand before harvest. (B) The first major cut leaves enough trees as a temporary shelterwood overstory to provide 40% crown cover. (C) After tree saplings have become established, the remaining mature shelterwood is cut down.

time-consuming than clear-cutting but has many advantages over the latter method. These advantages include the following:

1. It minimizes environmental abuses such as land scarring, accelerated runoff, soil erosion, and destruction of wildlife habitat.
2. It reduces blowdown.
3. It decreases the fire hazard because it reduces the volume of slash left after the harvest.
4. It results in a high rate of natural reproduction.

Selective cutting on a given stand may be performed on a ten-year rotation. As a rule, the harvest removes a small number of trees per hectare. The entire stand is never completely cut. Trees may be cut individually or in small groups. Sustained-yield management is practiced—the volume of wood harvested during a given year being equal to the volume grown since the previous cutting. Trees cut down are eventually replaced by natural reproduction, an ongoing process because openings are continually created.

The **strip-cutting** method of harvesting timber has been used effectively in the forests of the northeastern United States as well as in tropical rain forests of Central and South America. In our discussion of erosion control on farmland (Chapter 7), strip cropping on the contour was mentioned as a highly desirable technique. Strip cutting is somewhat similar. It is usually used in hilly terrain where clear-cutting might result in excessive soil erosion and therefore massive pollution of streams just below the logging site.

During strip cutting, loggers remove narrow strips of forest, as shown in Figure 14.12. A typical technique involves cutting a strip about 80 meters (250 feet) wide. Residual forested strips

FIGURE 14.11 Selective cutting of an unevenly aged northern hardwood–hemlock stand. (A) Before selective cutting. Trees to be felled are those with a line drawn through the trunk. (B) Same stand ten years after selective cutting.

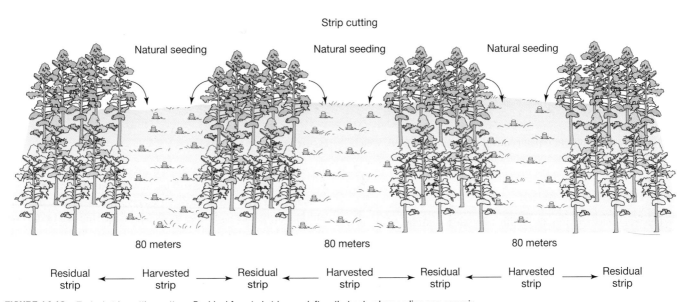

FIGURE 14.12 Typical strip-cutting pattern. Residual forested strips are left so that natural reseeding can occur in the harvested strips.

are left between the cut strips to serve as seed sources. Strip cutting has been successfully used for southern pines.

Strip cutting has several advantages over clear-cutting: (1) It minimizes the loss of soil nutrients from the forest by minimizing soil erosion. (2) It curbs the pollution of mountain streams with sediment, thus preventing the destruction of spawning sites for trout and other species. (3) It minimizes the visual impact associated with much larger clear-cut areas. (4) It permits more effective **reforestation** by natural mechanisms.

The Logging Operation

During logging, trees are cut down (felled), limbs are removed, and the trunks are cut into logs. The logs are then dragged across the forest floor or airlifted to a central point, where they are loaded onto trucks or train cars. The wood is then transported to a pulp-processing plant or sawmill.

Using a chain saw, a logger can cut about 14 cubic meters (500 cubic feet or 4 cords) of saw timber or pulpwood per day.

Large companies frequently use feller-bunchers and whole-tree chippers. A **feller-buncher** is a large logging machine with an attachment that cuts trees in place (Figure 14.13). It consists of a standard heavy-equipment base with a tree-grabbing device furnished with a circular saw or a shear—a pinching device designed to cut small trees off at the base. The machine then places the cut tree on a stack suitable for a skidder or forwarder, or other means of transport for further processing, such as delimbing or chipping.

A whole-tree chipper, used by the pulpwood companies, can chew up an entire pine trunk with a 50-centimeter (20-inch) diameter, branches and all, into thousands of small wood chips in less than a minute. The chips are then blown into a waiting truck or trailer and hauled to a paper mill. The chipper has two advantages: (1) it enables the forester to utilize much more of a tree's biomass than the 60% harvested by conventional methods; and (2) it leaves the forest floor relatively clean, facilitating the growth of new trees and reducing the fire hazard posed by accumulated slash from conventional logging operations. Unfortunately, the removal

FIGURE 14.13 A feller-buncher. A feller-buncher is a logging machine that cuts trees in place and then piles them.

of slash reduces nutrient replenishment and soil formation, which may come back to haunt companies in the long term.

14.4 Reforestation

Whenever timber is removed, either by clear-cutting, strip cutting, or selective cutting, the denuded area must be reforested to ensure a sustained yield. This may be done by natural or artificial methods. Similarly, any forested land that has been destroyed by fire, insects, disease, hurricanes, or strip mining also should be reforested, even though timber may not be its ultimate primary use. Reforestation helps reduce erosion and restores wildlife habitat and recreational area. In recent decades, millions of hectares in the United States have been either seeded or planted.

Natural Reseeding

As noted earlier, in some clear-cutting operations, a few mature, wind-firm trees are left intact as a seed source within the otherwise logged-off site. Scattered by wind and, to a lesser degree, by birds, rodents, and runoff water, the seeds are eventually dispersed throughout the denuded area. **Natural reseeding** is usually not completely adequate, however. One reason is that some tree species, such as loblolly pine, may have only one good seed-producing year every two to five years. (In a good year, seed production may be ten times greater than in a poor year.) Another reason is that the dispersed seeds must reach bare ground to develop properly so that the seedlings can absorb moisture and nutrients from the soil. If they fall on bark, logs, or stones, survival is greatly reduced. Seeds may also be consumed by rodents, birds, or other animals. Because of these drawbacks, natural reseeding is usually supplemented by aerial, hand, or machine seeding.

Seeding by Foresters

In rugged terrain, the best method is **aerial seeding.** Seeds are sown from airplanes or helicopters flying slowly just above the treetops. A helicopter can seed 1,000 hectares (2,500 acres) per day. Unfortunately, many of these seeds fall on infertile soil or are consumed by birds, mice, and squirrels. To minimize losses to animals, the seeds are frequently coated with a toxic deterrent, interfering with digestion or making the seeds unpalatable. Except in the case of unusually small-seeded trees, such as hemlock and spruce, rodent eradication is virtually a prerequisite to successful seeding.

If a logged-off site is flat, power-driven seeding machines may be used. Widely used in the southeastern United States and elsewhere, these machines plant up to 3.3 hectares (8 acres) per day. They simultaneously fertilize the soil and apply a herbicide to prevent weed encroachment.

Planting

In addition to bird and rodent problems, a major disadvantage of seeding is the high number of first-year seedlings killed by frost, drought, hot weather, insects, and autumn leaf fall. As a result, seeding, even by artificial methods, is less successful than planting young trees from seedlings grown in nurseries or greenhouses. Moreover, no rodent control is needed. Successful growth of these trees requires a good seed source, good nursery or greenhouse practices, and good planting practices. In the South and in the Great Lakes states, trees can be planted at a rate of 150 per worker-hour. On flat land, three workers, a tractor, and a planting machine can set 1,000 to 2,000 trees per hour.

Weed control is often critical in the first few years of tree seedling planting because aggressive weeds give the seedlings stiff competition for sunlight and moisture. In the first few years, weeds can be controlled with mowing, weed barriers, and herbicides. For protection from deer and rabbit browsing, seedlings can be encased in semitransparent plastic tubes.

Developing Genetically Superior Trees

Forest managers have spent a great deal of time and money developing genetically superior trees for replanting forests. Crossing two species of trees, known as **hybridization,** may result in offspring that combine the best traits of the parents. For example, in northern California, plantations of Jeffrey pine were formerly very vulnerable to the attacks of the pine weevil. Economic damage was severe. The problem has been somewhat reduced, however, by crossing the cold-resistant Jeffrey pine with the weevil-resistant Coulters pine. The resultant hybrids are resistant to both cold and the pine weevil. Some other possibilities through genetic engineering are discussed in A Closer Look 14.2.

Tree stocks can also be improved by older genetic techniques (for example, selective breeding). In **selective breeding,** one organism with a desired trait is bred to another, in hopes of passing the genes on to the offspring. Research shows that in southern pine stands, selective-breeding techniques may increase wood volume by 10%, straightness of the trunk by 9%, wood density by 5%, and resistance to rust (a fungal disease) by 4%. The economic gains resulting from selective breeding may be considerable. A forest-breeding program in California, for example, resulted in an increased return of $68 per hectare ($27 per acre).

Another way to produce better trees is through **seed orchards,** which produce large quantities of high-quality seeds. To begin, the tops of young trees with superior traits are removed. These can be planted in the ground and are called rootstocks. Small branches are then cut from other trees of high quality and grafted onto the rootstocks. The tree that develops from this graft is then crossed with still other commercially valuable trees. The seeds from this cross may then be saved for planting in commercial forests.

Another method to produce better trees is through **tissue culture.** In this technique, seeds are collected from superior trees and grown into seedlings in glass dishes. Tiny shoots from these seedlings are then cut off, chopped up, and placed in nutritive solutions. The cells grow into baby trees complete with roots, shoots, and leaf buds. Because all these individuals are derived from parts of the same parent tree, they have identical hereditary material and are called **clones.**

A CLOSER LOOK 14.2 — Genetic Engineering: The Key to Tomorrow's Superforests?

The rapidly developing field of **genetic engineering,** described briefly in Chapter 5, holds promise for dramatic improvements in our nation's forests. Hereditary traits are determined by genes. These genes, in turn, are present in the DNA molecules of the chromosomes. In one technique, known as direct DNA transfer, the gene responsible for a given trait, such as rapid growth, can be transferred directly to the DNA of single, isolated tree cells. Each of these cells may then develop into a tree with the ability to grow at an accelerated rate.

What does the future hold for genetically engineered (GE) trees? On the basis of recent research in the United States, Brazil, New Zealand, South Africa, and several European countries, the following outcomes already are, or soon may be, attainable:

1. GE trees that will be distasteful to potentially destructive browsing herbivores, such as deer.
2. Spruce and hardwood (oak, maple) planting stock with a whole series of GE traits such as accelerated growth, reduced need for fertilizer, improved ability to compete with weeds, and higher survival rates.
3. Pulpwood species (spruce, aspen, birch, and loblolly pine) with GE traits that make them much more valuable to the paper industry.
4. GE trees that produce natural insecticides to ward off insects, as well as natural herbicides to reduce or eliminate weeds that compete with them for soil moisture and nutrients.

There are many scientists and citizens who want genetically engineered trees to be safety tested before these new-technology trees are released. Since GE trees have not been proven to be safe, many believe that the engineering of trees is strictly about speculative science and economic return. Some scientists argue that the incentive to develop faster-growing GE tree plantations to feed the rising global demand for timber may lead to the clearing of forests to make room for more economically valuable plantations, with serious consequences for forest biological diversity, forest-dependent communities, and the climate.

If commercialized, GE trees will invariably result in contamination of native forests with GE tree traits, such as the ability to kill insects. In a survey documented in a 2005 report on GE trees prepared by the UN Food and Agriculture Organization, over half of GE tree scientists surveyed reported unintended contamination of native ecosystems and plants by GE trees as a major concern. For example, contamination of native forests by reduced-lignin GE trees could lead to serious forest health crises. Lignin is an important structural polymer that is also significantly responsible for the high levels of insect and disease resistance in trees.

Whether or not to plant GE trees is an important issue. What do you think about the release of GE trees into the environment?

14.5   Control of Forest Pests

Growing trees is not as easy as it may sound. Many factors affect tree growth and mortality. The most serious agents of mortality and growth loss for most forests are disease and insect pests (Figure 14.14). Let's look at diseases first.

Diseases

When you think about the cause of many common human diseases that lead to death, you probably think about viruses and bacteria. For forest trees, the most numerous and important diseases are caused by fungi (Table 14.2). These diseases vary in the species and parts of tree affected, in symptoms, and in the type of damage they cause. For example, fungi can cause decay in wood, most of which occurs in the tree's central core of dead wood, or "heartwood." Such heartwood decay, called **heart rot,** lowers the wood's quality or makes it totally unmarketable. Heartwood decay has the greatest effect on the growth of forest trees of all destructive agents. Many different species of fungi cause heart rot. (Heart rot fungi are also beneficial, however, as important agents in the decay of fallen logs, dead stubs, and slash, the recycling of elements, and the removal of flammable debris.) The remaining disease damage can be attributed primarily to white pine blister rust (Figure 14.15), dwarf mistletoe, and Dutch elm disease. The most injurious diseases are exotics, accidentally introduced into the United States, that have suddenly been released from environmental factors that ordinarily kept them in check in their native habitat.

Insects

Under the authority of the **Forest Pest Control Act of 1947,** surveys are conducted annually in both private and public forests to detect pest population buildups so that they can be arrested before they reach disastrous levels. Insects account for about 20% of all timber destroyed, ranking second to diseases as agents of forest damage. Each species of tree has its own unique assemblage of insect pests; for example, an oak tree may be eaten by more than 100 species. No part of a tree is spared (Table 14.3). A healthy tree can withstand the nibbling of insects, but if trees are already stressed because of drought, crowding, pollution, or a combination of the three, they may be seriously damaged by insects.

Bark Beetles   **Bark beetles** are among the most devastating insects of forests in the world. Adult beetles attack a tree by boring through the bark and then tunneling out egg chambers and galleries with their powerful jaws. The tiny grubs that hatch from the eggs consume the soft inner bark. If sufficiently numerous (1,000 per large tree), they may actually girdle the tree and kill it within a month. The bark beetle group includes a large number of destructive species.

Pine beetles kill trees by transmitting a fungus that develops inside the tree. The fungus plugs up the vessels that transport water to the leaves, branches, trunk, and roots. The western pine beetle killed 59 million cubic meters (more than 2 billion cubic feet) of ponderosa pine along the Pacific Coast between 1917 and 1943. The mountain pine beetle has done extensive damage to sugar pine, western white pine, and lodgepole pine in California. The southern pine beetle, the most damaging insect in the southeastern United States, is responsible for killing large areas of southern pine.

Pine beetles are a natural member of the forest ecosystem, but human activities can cause them to proliferate. Overharvesting of trees, for example, often results in thick regrowth. These trees compete for water. In drought years, they are unable to acquire enough water and are less able to ward off insects like bark beetles. Thinning trees, therefore, effectively controls the beetles.

On a long-term basis, perhaps the best method of control is to prevent infestations from occurring in the first place. This may be done by using **sanitation techniques:** burning all potential bark beetle breeding sites, such as senile and windblown trees, and burning or removing logging-accumulated slash and debris and the broken-off trunks of lightning- and fire-killed trees. (In removing those trunks, however, the forester is also

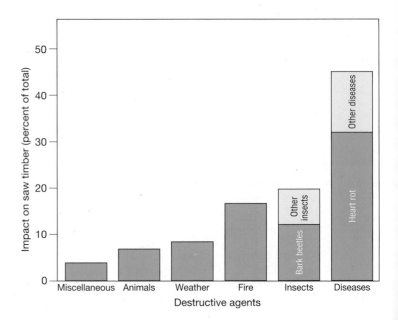

FIGURE 14.14   Annual impact of destructive agents on saw timber. Saw timber or sawlogs are logs of sufficient size and quality to produce lumber or veneer.

TABLE 14.2 Common Fungus Diseases of Forest Trees: Damage and Control

Common Name	Host	Symptoms	Control
Root and butt rot	Species of pine: mature stands, and plantations 15–25 years old	Decline in vigor; needles short, become yellow and then die; cones appear prematurely; fruiting bodies of fungus at base of tree or on ground surface arising from roots	Avoid planting white or red pine on poorly drained or high-pH soils
Western red rot	Ponderosa pine	None, except presence of decay and red discoloration in heartwood	Periodic pruning of dead branches on crop trees
Brown spot	Longleaf pine	Small spots on needles, causing needle dieback; retards growth of seedlings	Controlled burning in third year of seedling's life or spraying with prescribed chemicals
Oak wilt	Oak species, particularly those of the black oak group	Leaves crinkle and become pale green, later turning brown or bronze; mature leaves shed at any symptom stage; lower branches affected last; trees usually die after the summer symptoms appear	Some promise of control obtained by cutting or poisoning healthy oaks for 50–100 ft. around spot infections Mechanical or chemical root barriers
Dutch elm disease, associated with attacks by scolytid beetles	American elm	Progressive dwarfing and yellowing of leaves, accompanied by various degrees of defoliation; followed by death of branches of entire tree	Mainly through destruction of infected trees and protection against bark beetles by insecticidal sprays and maintenance of tree vigor by appropriate pruning and feeding practices Lignasan injection

removing the potential nesting cavities of beetle-eating woodpeckers and is removing nutrients that would be returned to the forest floor, so a cost–benefit analysis would have to be made.) Insecticides can be applied by planes that fly low over the infected area.

In a forest infested with beetles, the spread of beetles can be stopped if infected trees are cut down and then sawn into small lengths. The logs are piled and then covered with clear plastic. A pesticide that vaporizes when the sun heats the interior of this artificial greenhouse is placed under the plastic.

Integrated Pest Management In the late 1970s, certain environmental groups became concerned over the large amounts of chemical pesticides used by the Forest Service to control insect outbreaks. This criticism reached its peak when the Forest Service used DDT (banned except for emergencies) to control an explosion of tussock moths in conifer stands of the Northwest. Ecologists considered this strategy particularly questionable because ordinarily the tussock moth population is held in check by a death-causing virus. Eventually, the Forest Service, stung by criticism from many quarters, decided to adopt the **integrated pest**

FIGURE 14.15 Spraying of a young western white pine stand in the St. Joe National Forest in Idaho with phytoactin, a chemical that kills blister rust.

TABLE 14.3	Common Insect Pests of Forests: Damage and Control				
				Control Method	
Name	Principal Species Affected	Type of Damage	Prevention	Direct Control	Control Season
Bark beetle	Pines	Girdles tree by killing cambium layer under bark	Salvage green blown-down timber	Feel, peel, and burn bark; salvage infested logs; burn slabs	Spring, summer, fall
Gypsy moth	Oaks, birch, aspen	Defoliation		Spraying insecticide	May–June
Sawfly	Eastern and southern pines and tamarack	Reduces growth by defoliation; epidemics kill stands	Cut mature and overly mature stands	Aerial insecticide spray	Early summer in period of greatest activity
Spruce budworm	True firs, Douglas fir	Reduces growth by defoliation; kills older trees extensively	Cut mature and overly mature stands	Aerial insecticide spray	Early summer as insects emerge
White pine weevil	Eastern white pine, Norway spruce	Kills leaders, causes forked and crooked boles	Maintain shade where possible to 20 ft.	Hand spray upper stems and new growth with sodium arsenate or cut and burn infected tips; leave only best side branch on first lower whorl	Early summer as insects emerge

A CLOSER LOOK 14.3   Controlling Insect Outbreaks with Heterotypes

Insect pest populations sometimes build up to impressive numbers. Charles Kendeigh, an ecologist at the University of Illinois, studied a severe outbreak of spruce budworm in Ontario. The larvae, which were feeding on coniferous foliage, occurred in such vast numbers that their excreta sounded like drizzling rain as it fell on the forest floor. Another example is an explosion of forest tent caterpillars that virtually destroyed 11,000 hectares (26,850 acres) of water tupelo (a type of tree) in the region of Mobile Bay, AL. Aerial photographs revealed almost complete defoliation of extensive stands of once-healthy trees.

Is it possible to control such outbreaks without using expensive insecticides that may have adverse effects on the forest ecosystem? Most experts say yes. One way is to establish forest heterotypes, that is, forests with a more diverse population of trees. This strategy was suggested by Kenneth Watt, an ecologist at the University of California–Davis. How does it work?

To understand, let's backtrack a little bit. When monoculture is practiced in a large area, populations of forest pests tend to fluctuate widely. At times, their populations can reach tremendous numbers. Over an evolutionary period of millions of years, certain species of insects have specialized to feed on a particular species of tree. Spruce budworms, for example, feed on balsam fir, larch

sawflies feed on larch, gypsy moths dine on oak, and pine bark beetles are specialized to feed on pine. Many pests feed on a particular kind of tree at a particular stage (seed, seedling, sapling, or mature tree) in its life cycle.

In natural forest ecosystems, the particular food tree to which a species of insect has become adapted may be widely dispersed. An insect such as a pine bark beetle, for example, might have to creep, crawl, or fly a considerable distance from one food tree to another. In the process, of course, such an insect becomes vulnerable to predation, windstorms and rainstorms, fire, and other mortality factors. Even if it survives, it may eventually starve to death before it is able to find another food tree. Conversely, in an artificial forest ecosystem—a monotype with one species covering many hectares—the pine bark beetle is surrounded not only by the right species of food tree, but also those of the precise age that the beetles prefer.

Interestingly, the severity of forest insect outbreaks can be substantially reduced simply by breaking up large contiguous stands of single-age, single-species trees into small, isolated stands interspersed with trees of different kinds and ages—that is, a heterotype. It's a simple and effective way of reducing our dependence on pesticides.

---

management (IPM) policy, a concept described more fully in Chapter 8.

With IPM, methodologies from several disciplines are integrated in an ecological manner to control a pest or pests. The goal is to control pests and reduce damage without damaging the environment. Among the IPM strategies now used by the Forest Service are the following:

1. Use of biological control agents, such as natural predators and parasites.

2. Use of selective cutting methods rather than clear-cutting methods (because even-aged trees are highly vulnerable to insect attacks).

3. Removal of bark-damaged trees because they serve as focal points of insect infestation.

4. Growing heterotypes rather than monotypes whenever feasible (see A Closer Look 14.3).

14.6 Fire Management

Wildfires

In the early part of the 20th century, there were single years when more than 100,000 forest **wildfires** occurred in the United States. They burned more than 27,500 square kilometers (10,450 square miles) of forest, an area the size of Maryland. Because of better **fire control** techniques, the total acreage of destroyed timber in the United States has been reduced in re-

cent years. Nevertheless, occasional serious wildfire outbreaks do occur, especially during periods of extended drought. Consider 1988, for example. In that year, the most severe drought in half a century transformed much of the nation's timber into kindling wood. Lightning strikes set the kindling ablaze. By midsummer, scores of forest fires were raging in Alaska, Idaho, California, Oregon, Colorado, Utah, Wyoming, and Wisconsin. The Forest Service called the summer of 1988 the worst fire season in 30 years. The fires charred at least 1.5 million hectares (3.7 million acres), an area larger than Connecticut. The largest burn in Colorado's history destroyed more than 7,300 hectares (18,000 acres) of prime habitat for elk and deer. The Black Hills blaze in South Dakota scorched 7,000 hectares (16,500 acres), forced the evacuation of 1,000 campers, and smoke-shrouded the presidential shrine at Mount Rushmore.

Most of the public's attention, however, was focused on the devastation in **Yellowstone National Park.** Fire boss Fred Roach, who had battled such blazes for two decades, told reporters that he had never seen anything as awesome as this! Twenty-five hundred army personnel were flown to the scene to help contain the fire. President Ronald Reagan became so concerned that he sent Agriculture Secretary Richard Lyng, Interior Secretary Donald Hodel, and Deputy Defense Secretary William O. Taft to Yellowstone to assess the destruction. Eventually, in early September, as a result of cooler weather, snow, reduced wind velocities, and the heroic efforts of thousands of firefighters, the fires went out. In the end, eight separate fires burned about 45% of the park's

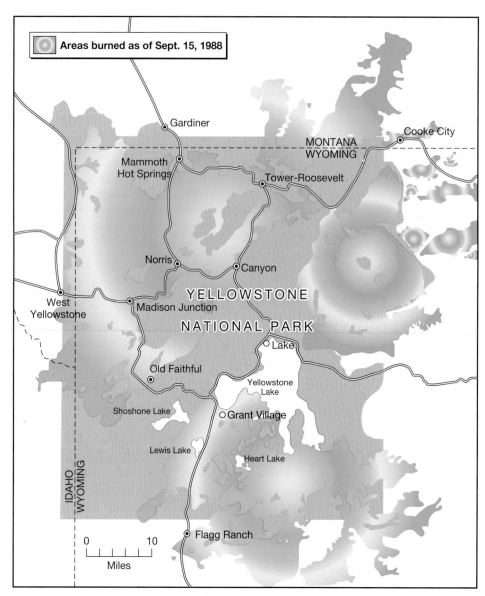

FIGURE 14.16 Areas burned inside and outside of Yellowstone National Park after the 1988 fires. About 45% of the park's 890,000 hectares (2.2 million acres) burned.

890,000 hectares (2.2 million acres) (Figure 14.16). But in nearly half of the burned area, the trees were not killed. Often not discussed is the fact that more than half of the burned area resulted from three human-caused fires against which suppression action was taken from the very beginning. In addition, two other fires burned into the park from adjacent lands.

Since the Yellowstone event, fires have continued to ravage the West, in large part because of a prolonged drought (possibly caused by global warming) and previous fire prevention programs that protected forests yet permitted dead timber and debris to build up on the forest floor, which greatly increases the severity of fires. In 2002, for example, several thousand forest fires ravaged the state of Colorado. One burned 55,900 hectares (138,000 acres) just south of the senior author's home in Evergreen. In 2003, huge fires burned in California, destroying 394,000 hectares (750,000 acres).

About 15% of our nation's forest fires are caused naturally (for example, by lightning). The rest are caused by people, either deliberately or accidentally through carelessness.

Firefighting

The actual suppression or attack pattern used by firefighters varies greatly, depending on the size of the fire, terrain, type of fire, wind direction, location of roads, availability of water, and relative humidity. A variety of fire suppression methods are used, including firebreaks, backfires, and fire-retarding chemicals (Figure 14.17). Firefighters often use **infrared systems**—computers, satellites, satellite dishes, and video cameras—to pinpoint the location of wildfires and their movement. Because such movement can be highly erratic and could trap unsuspecting firefighters, infrared systems have the potential to save lives and houses in the path of fires.

In rugged mountainous country, **smoke jumpers** may parachute into the burn area. The national forests of mountainous regions in Washington, Oregon, California, Montana, Idaho, and New Mexico are protected by USFS smoke jumpers. With their help, many remote blazes that important role in maintaining many economically important timber stands. Examples include the old-growth, even-aged stands of Douglas fir in the Northwest, red pine in Minnesota, and white pine in Pennsylvania and New Hampshire. Certain less valuable species, such as pitch pine on sandy soils near the mid-Atlantic Coast and jack pine in the Great Lake states, are also considered to be fire types, or **fire climaxes**—that is, the health of the forest depends on periodic fires, for reasons explained shortly.

Today foresters in many countries use controlled or prescribed burns to improve the quality of timber, livestock forage, and wildlife habitat (Figure 14.18). A **controlled burn** is a fire that is purposely ignited by highly trained foresters for specific purposes. It has a low flame and moves slowly along the forest floor. Forest managers must be extremely cautious when performing controlled burns. They must be certain that the wood is not too dry and that the wind is neither too strong nor headed in the wrong direction. A huge forest fire that broke out in New Mexico in 2002 and burned part of a nearby town was a prescribed burn that got out of control as a result of strong winds.

About 0.8 million hectares (2 million acres) of pine stands in the South are control-burned annually. At the Alpha Experimental Range in Georgia, controlled burning is conducted only in the afternoon, under damp and relatively cool conditions. The fire is usually extinguished by night. Any pine stand 2 to 4 meters (6.5 to 13 feet) high can be control-burned because trees of this size are protected from flames by their corklike bark. The highly resistant longleaf pine, whose terminal bud is protected by a group of long needles, can be burned without ill effect when the seedlings are only 15 centimeters (6 inches) high.

In the longleaf slash pine stands of the South, controlled burning has many benefits: it (1) reduces the crown-fire hazard by removing highly combustible litter; (2) prepares the forest soil as a seedbed; (3) increases the growth and quality

FIGURE 14.17 Firefighters setting a backfire to create a containment line around a huge wildfire in California. By burning off brush and grasses in advance of a fire, firefighters can slow a fire or stop it from spreading.

in 1930 might have burned out of control may now be extinguished within hours.

Use of Controlled Fires

Since the 1930s, the Forest Service has used the symbol of Smokey Bear to alert the American public to the highly destructive effects of forest fires. In recent years, however, the service has come to realize that some wildfires are ecologically beneficial. In fact, it is now believed that fire plays an

FIGURE 14.18 A controlled burn in a forest in Nepal. Note the low flame.

of livestock forage; (4) retards forest succession, which would lead to a low-value scrub oak climax, and maintains the high-timber-value pine subclimax; (5) promotes legume establishment and resultant soil enrichment; (6) increases the amount of soluble mineral ash (phosphorus and potassium) available to the forest plants; (7) stimulates the activity of soil microorganisms; (8) controls the brown-spot needle blight, a fungus that is highly destructive to longleaf pine seedlings; and (9) improves food and cover for wildlife, including quail.

The "Let-It-Burn" or "Prescribed Natural Fire" Policy

In 1972, the National Park Service established a policy that utilizes wildfires in 17 national parks, including Grand Teton (Wyoming), Rocky Mountain (Colorado), Sequoia (California), Yosemite (California), and Yellowstone (Wyoming). Under this policy, wildfires are permitted to burn under careful surveillance. This **"let it burn"** (or let-burn) **policy** has also been implemented by the USFS in wilderness areas in national forests. Since the phrase "let it burn" more appropriately describes what the fire does and not what the management strategy entails, many people prefer to call the policy **"prescribed natural fire"** to reflect the monitoring and decision processes associated with these naturally ignited fires. This policy is quite a turnabout from the Park Service's policy between 1920 and 1972, when wildfires were considered the forest's prime evil, and all were to be fought. Periodic natural fires help revitalize old-growth forests by burning up dead trees and debris and killing off diseased trees. In turn, this encourages new plant growth.

The massive Yellowstone fires in 1988, the most severe in the park's 112-year history, forced the National Park Service and Forest Service to reevaluate their prescribed natural fire policy. They admit, for example, that when the present fire management plans were established in the early 1970s, no provision was made for a firestorm of the 1988 variety. After all, data from fire-scarred trunks that are centuries old indicate that a conflagration on such a scale would occur only once in 150 to 200 years (Figure 14.19)! In the aftermath of the Yellowstone fires, then-Interior Secretary Donald Hodel suggested that the prescribed natural fire policy be scrapped. However, after reappraisal, it was decided that the policy would be retained except under conditions of extreme drought like that of 1988. Still, some believe that the natural fire policy procedures are now too conservative. Greatly expanded checklists in too many instances result in a decision to suppress a fire.

Even if the natural fire policy had not been in place, it is estimated that 25% to 30% of Yellowstone National Park would have burned. Today, more than 20 years after the big fires, forests are reborn and wildlife blooms in the part of Yellowstone once described as destroyed. Many ecologists and biologists now believe that the unusually large fires of 1988 are in the normal range of natural disturbance for a lodgepole pine ecosystem and that these fires will lead to a significant increase in the number of Yellowstone's plant species.

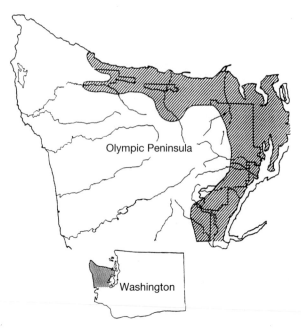

FIGURE 14.19 Estimated area of a single fire or series of fires around 1700 in the eastern Olympic Mountains of Washington. This fire (or fires) burned from 1.2 million to 4 million hectares (3 million to 10 million acres) of forestland. Although we really do not know, such widespread fires were probably weather-driven by large lightning storms rather than fuel-driven. Other large fires in the western Olympics appear to have occurred around 1230 and again in 1480.

14.7 Meeting Future Timber Demands Sustainably

Forests cover 30% of the planet's total land area. The total forested area in 2005 was just under 4 billion hectares (9.9 billion acres), at least one-third less than before the dawn of agriculture, some 10,000 years ago. The ten most forest-rich countries, which account for two-thirds of the total forested area, are (in order of most to least forested area) Russia, Brazil, Canada, the United States, China, Australia, Democratic Republic of Congo, Indonesia, Peru, and India. About 84% of the world's forests are publicly owned, but private ownership is on the rise.

Total forest area in the world continues to decrease, but the rate of net loss is slowing (Figure 14.20), according to the UN Food and Agriculture Organization. Deforestation, mainly conversion of forests to agricultural land, continues at an alarming rate—about 13 million hectares (32 million acres) per year. Simultaneous forest planting, landscape restoration, and natural expansion of forests have significantly reduced the net loss of forest areas. The net change in forest area from 2000 to 2005 is estimated at −7.3 million hectares (−18.0 million acres) per year, down from −8.9 million hectares (−22.0 million acres) per year from 1990 to 2000. Africa and South America continued to have the largest net loss of forests (Figure 14.20). North and Central America as well as Oceania also had a net loss of forests. Europe continued to expand, while Asia, which had a net loss in the 1990s, reported a net gain of forests between 2000 and 2005, mainly due to China's large-scale afforestation projects.

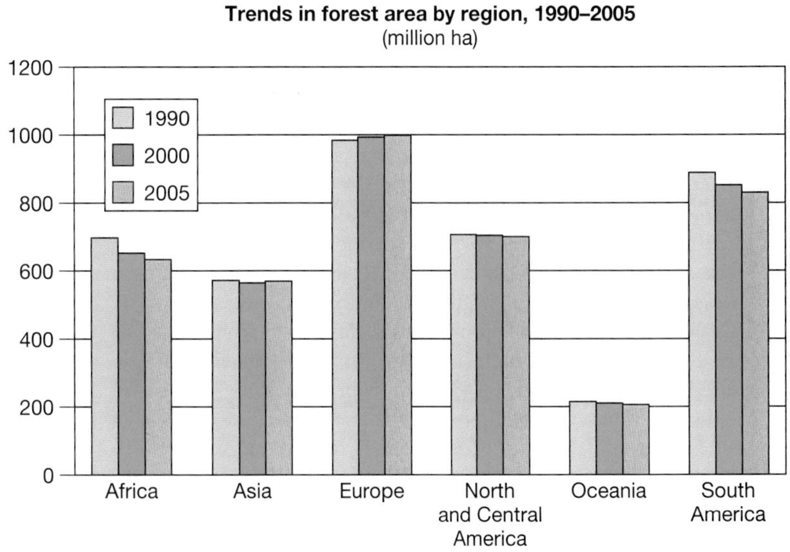

Trends in forest area by region, 1990–2005
(million ha)

FIGURE 14.20 Worldwide trends in forest area by region, 1990–2005. (Source: UN Food and Agriculture Organization)

According to the UN Environment Programme, global removal of wood for timber and fuel amounted to 3.1 billion cubic meters (109.5 billion cubic feet) in 2005. With the anticipated growth in the human population, global demand for wood is expected to double by 2050. Some conservationists question whether this rate of increase in forest harvesting can continue. In fact, some people believe that the world will face conditions of "timber famine" within the next 20 years. Although this dire prediction has been made many times before, it has not been fulfilled. The increasing demand for wood for such uses as fuel, paper, and home building has always been satisfied by technological developments, exploitation of areas of previously unexploited forest, and utilization of tree species previously considered unusable. However, because of increasing world population, deforestation in less-developed countries, and the establishment of new plantations lagging far behind the rate of forest depletion in developing countries, this prophecy may finally come true.

Turning our attention to the United States, by 2030 Americans may be using 50% more timber than they do now. Theoretically, there are several ways in which the increasing demand for timber may be satisfied. Unfortunately, some of the solutions are fraught with economic or ecological problems. Among them are the following ideas:

1. Upgrade and extend forest management in all forests, public and private, large and small. Where can the greatest gains be made? Primarily on small, privately owned woodlots. Of the 4.5 million small forests in the United States, over 50% are smaller than 12 hectares (30 acres) each. Yet these small-forest owners control over three times as much timberland as the USFS. Until now, few small-forest owners have cut trees on their land for commercial sale. With appropriate management, many of these small forestland holdings could yield valuable forest products.

2. Make more effective use of wood residues and weed species of trees. In a given year, more than 30 million cubic meters (1 billion cubic feet) are left as unused residue. (Of course, eventually this material will decompose, and the resulting nutrients will enrich the soil for the growth of future generations of trees.) Another 30 billion cubic meters (1 billion cubic feet) of wood are left as milling residue. Both types of residue could be used to manufacture wood products (see A Closer Look 14.4). It may well be, however, that the value of such residues and weed species as a source of fuel will exceed their value as wood products.

3. Develop superior (faster-growing, better-grained, and disease-, insect-, fire-, and drought-resistant) trees through the techniques of grafting, hybridization, and genetic engineering.

4. Increase the use of wood substitutes. Plastic and other materials might be substituted for wood in packaging. The fibrous waste left over after the juice has been extracted from sugarcane can be used to manufacture paper. It must be emphasized, however, that the manufacture of some of these substitute materials depends on the intensive use of energy from oil fuels—a source that is being depleted. Moreover, the extensive use of plastics is fraught with problems. Plastics do not decompose easily, so they pose a serious waste disposal dilemma.

5. Increase imports.

6. Recycle paper to reduce demand.

7. Reduce demand by building smaller houses or reducing paper packaging.

Sustainably Harvested Forests and Certified Lumber

To minimize the impact of the wood harvesting, many homebuilders use lumber and wood products obtained from sustainably managed and harvested forests. How do they know a forest has been sustainably managed? In 1994, the Forest Stewardship Council (FSC), headquartered in Oaxaca, Mexico, launched an international effort to promote global sustainable wood production.

A CLOSER LOOK 14.4 Forest Conservation by Efficient Utilization

During the cut-it-and-get-out logging operations of the 1890s, lumbermen were interested only in logs. The rest of the tree—the stump, limbs, branches, and foliage—was left in the forest, where it frequently served as tinder for a catastrophic fire. (Of course, much of this biomass decomposed, and the nutrients were released into the soil for new growth.) Further waste occurred at the sawmill, where square timbers were fashioned from round logs. Slabs, trimmings, bark, and sawdust were hauled to the refuse dump and burned. Today, to help meet the growing demand for wood and wood products, it is necessary to practice even more conservation after trees are harvested and removed from the forest.

A definite trend toward more efficient utilization of our timber resources by the U.S. wood and wood products industry is under way. Whereas early in the timber industry, wood had only two primary uses, as lumber or as fuel, a number of ingenious methods have been developed for utilizing almost every part of the tree,

including the bark, thanks to the efforts of the Forest Products Laboratory and similar research centers.

The forest industry has become much more diversified. Whereas in 1890, almost 95% of the forest harvest was converted into lumber, today only about 30% of the harvest consumed by Americans is fashioned into boards and timbers. In the past few years, techniques have been developed for making extremely useful products from scrap boards, shavings, wood chips, bark, and sawdust. For example, years ago, extremely short boards went to the scrap heap. Today, however, thanks to superior waterproof glues, such boards can be joined to form structural beams of almost unlimited length. Much wood that was once considered worthless is now converted into thousands of small chips. These chips are then compressed into sturdy and durable chipboard or hardboard. The forest industry has also been increasing its use of wood wastes and mill residues for energy production. Some uses include firing boilers in pulp and paper mills for process heat and providing energy for drying lumber.

This independent, nonprofit organization, free of business and government affiliations, created a set of standards for forest management that promote sustainable timber production. Their standards include region-specific guidelines to ensure proper management in a wide variety of forest types, trees species, and climates. The FSC does not inspect forests themselves, but rather relies on nine accredited international organizations. They inspect and certify timber management practices of companies harvesting wood throughout the world on tree farms, plantations, and natural forests. FSC standards also ensure that forests are managed for values other than timber production—for example, wildlife habitat. In addition, FSC certification guarantees that wood is produced in ways that respect the rights of indigenous peoples (for example, in tropical regions) and maintain the economic and social well-being of local communities. Wood is tracked from forest to mill to building supply outlet so that buyers are sure they are getting what they pay for. Products bearing the FSC logo, which guarantees that the wood comes from a certified well-managed forest, are available across the world from a variety of mills, manufacturers, and distributors. The FSC has certified almost 103 million hectares of forestland in 81 countries. Although the FSC has been criticized by U.S. activists for a variety of reasons, such as permitting the use of herbicides to control weedy species when replanting land after harvest, its certification is generally considered to be one of the most rigorous of all certifying organizations. Also, other organizations and programs with objectives similar to those of the FSC aim to ensure that forest practices are sustainable.

The success of the FSC certification program depends not just on independent examiners but also on the availability of retail outlets—stores that stock certified lumber. Several companies have taken a lead in developing this market,

including EcoTimber of Berkeley, CA. EcoTimber began operations in 1992 and has grown to a multimillion-dollar business over the years, offering certified lumber as well as reclaimed and salvaged lumber. Late in 1999, Home Depot, which supplies about 10% of America's lumber, announced that it would discontinue the sale of old-growth timber in its stores across the nation by 2002 while phasing in lumber and wood products from FSC-certified forests. In the summer of 2000, Lowe's, its major competitor, announced plans to aggressively phase out wood products from endangered forests while supporting FSC-certified lumber and wood products. Although both have fallen far short of their goal, their continuing commitment to certified lumber could dramatically increase the availability of sustainably produced lumber and wood products and spur more timber companies to join the burgeoning movement. Certified wood offers buyers a sense of satisfaction that their homes are being built from wood that has been produced in ways that are gentle on the Earth. Because of this, certified lumber typically costs more than conventionally produced lumber.

Wade Mosby, vice president of marketing for Collins Pine, one of the largest suppliers of FSC-certified lumber in the United States, notes that its forests are managed on 140-year rotations and its operations produce far fewer impacts. Although wood from Collins Pine costs more, trees the company harvests tend to be of higher quality, with clearer, tighter grains. The slight additional cost, Mosby argues, is "more than made up for in quality and performance." Longer harvest rotations, notes Dan Imhoff, author of *Building With Vision*, "yield older, straighter wood. Straighter wood means fewer on-site rejections and a tighter building envelope overall. Using lumber that doesn't warp or twist lessens the chance of gaps developing in the framing structure, which can compromise energy efficiency and comfort as well as the durability of a building."

GO GREEN!

When remodeling a house, buy certified wood and lumber. If building or buying a green-built home, look for certified wood and lumber—that is, framing lumber, plywood, finish products (trim), and particleboard.

14.8 Preserving Wilderness

From time to time in U.S. history, farsighted conservationists have urged that the federal government set aside large acreages of pristine forest as wilderness preserves. One of the most influential of these was **John Muir.** An immigrant from Scotland, he fell in love with the forests of America. He tramped through them with endless fascination. In 1892, he founded the **Sierra Club**—one of the nation's most active environmental organizations. A magnificent redwood stand just outside San Francisco was named the John Muir National Monument in his honor.

For several decades, Muir's crusade met only apathy. Truly, his seemed to be a "voice crying in the wilderness." However, midway through the twentieth century, a surge of environmental awareness and responsibility moved the American public. This awareness was fostered by the writings of such people as Aldo Leopold and Rachel Carson. At long last, the U.S. Congress passed the **Wilderness Act of 1964.** The purpose of this act was to preserve primitive areas in their natural state. The act allowed for the designation of primitive areas in national forests, national parks, and national wildlife refuges. Together, they form the **National Wilderness Preservation System.** Congress defined its concept of **wilderness** in the following words:

A wilderness, in contrast with areas where man and his own works dominate the landscape, is hereby recognized as an area where the earth and its community of life are untrammeled by humans, where humans themselves are visitors who do not remain.

Congress described a wilderness as having four major characteristics:

(1) It generally appears to have been affected primarily by the forces of nature, with the imprint of the work of humans substantially unnoticeable. (2) It has outstanding opportunities for solitude or a primitive type of recreation. (3) It has at least 5,000 acres of land, except in the case of islands, which might have a smaller area. (4) It may also contain ecological, geological, or other features of scientific, educational, scenic, or historical value.

Access to an official area within the National Wilderness Preservation System can be made only by trail or canoe. All motorized vehicles or boats are prohibited, as are roads, buildings, and any commercial activities. Grazing, timber cutting, mining, and drilling for oil also are prohibited except to those parties who had filed claims by 1983. Human recreation is restricted to such quiet forms as hiking, camping, bird watching, studying rock formations, identifying wildflowers, canoeing, and sportfishing. Quiet forms of human recreation are sometimes referred to as low-impact recreation, whereas the use of ATVs (all-terrain vehicles) is a high-impact form of recreation.

The 1964 act designated 54 wilderness areas on 3.6 million hectares (9 million acres) of Forest Service land. Congress started expanding the wilderness system in 1968. As of 2008, the National Wilderness Preservation System includes 704 wilderness areas (Figure 14.21), totaling 43.5 million hectares (107.5 million acres), of which 54% is in Alaska. They are

FIGURE 14.21 Distribution of wilderness areas in the National Wilderness Preservation System in the United States.

managed by four government agencies: the National Park Service (42%), the U.S. Forest Service (33%), the Fish and Wildlife Service (20%), and the Bureau of Land Management (5%).

Why is our wilderness system important? Today, when urban dwellers from Seattle to Miami and from San Diego to Boston are being crowded together shoulder to shoulder, when urban air is polluted with industrial gases and automobile fumes, when drinking water tastes of chlorine, and when noise assaults the ears in so many areas, it is reassuring to know that somewhere in the great forests of America is wilderness. And once you get there, you can hike or canoe for miles, free of modern civilization.

Of the 77 million hectares (191 million acres) of national forest in the United States, about 40 million hectares (51.5%) are roadless—they contain no roads whatsoever. About 13 million hectares of the roadless forests have been included in the National Wilderness Preservation System. However, an ongoing controversy has been raging concerning the designation of the remaining 27 million roadless hectares. In 2003, then–Secretary of the Interior Gale Norton issued a "no more wilderness" decree on land under the Bureau of Land Management (an agency under the Interior Department's control). BLM has millions of acres of roadless area that could become part of the National Wilderness Preservation System. Besides declaring her objection to further wilderness designation, Norton began opening up potential wilderness areas for development—for mining, oil drilling, and other activities—so that they would become ineligible for wilderness designation. In 2003, for instance, Norton opened up 2.6 million acres of potential wilderness in Utah for development—cutting a deal for business behind closed doors, say environmental critics. (For more details, see A Closer Look 14.5.) Environmentalists, led by organizations such as the Sierra Club, the National Audubon Society, and the Wilderness Society, would like most or all of the roadless acreage to be included in the National Wilderness Preservation System. In the name of progress, resource developers lobby elected officials and government agencies to build roads in these areas to prevent them from being designated wilderness in the future. They have mounted extensive lobbying campaigns to prevent further wilderness designation.

Under the terms of the Wilderness Act, obtrusive activities such as logging, mining, and the use of automobiles, motorboats, and snowmobiles are prohibited in designated wilderness. In some areas, even aircraft are not permitted to descend below a certain altitude. In an officially designated wilderness area, every attempt is made to permit the forest ecosystem to operate without human interference. If an overly mature pine riddled with bark beetle blows down during a windstorm, it remains where it falls. Barring a major catastrophe, such as a crown fire—that is, a fire that spreads rapidly from the top or crown of a tree to another—the official policy is to let nature take its course. As a result, these areas have considerable scientific as well as recreational value. In such an area, university researchers can make observations, collect data, and formulate hypotheses, and

college students can acquire valuable field experiences in geology, entomology, mammalogy, ornithology, ecology, soil science, natural history, game management, and forestry, as well as other fields.

14.9 Protecting Natural Resources: National Parks

Much of the forestland of the United States is located in **national parks** administered by the **National Park Service.** In addition to the preservation and appropriate use of these woodlands, the national park system is committed to protecting other national resources such as wildlife and scenic beauty.

Brief History

The hunters and trappers who brought back game and furs from the rugged wilderness country of northwestern Wyoming also brought back tales of wondrous sights and sounds: exploding geysers that shot a stream of water 200 feet in the air; an awesome gorge almost a quarter-mile deep; dozens of plunging waterfalls; multicolored springs that bubbled water hot enough to heat your morning coffee; vast forests of spruce, fir, and pine; snow-crowned mountain peaks; and huge herds of antelope, elk, and deer. These accounts seemed almost unbelievable. However, U.S. Army General Henry Washburn led an expedition to Wyoming and found them to be true. Interest in the region grew rapidly. Ultimately, a Montana judge named Cornelius Hedges suggested the area be converted into a national park so that all Americans could thrill to the sights and sounds of this natural wonderland. On March 1, 1872, President Ulysses S. Grant signed legislation that made Yellowstone the first national park in the United States—and in the entire world.

The national park concept seemed so attractive to Congress that several more parks were added in the 1890s, including Yosemite (Figure 14.22), Sequoia, and Mount Rainier. In 1916, the National Park Service was established as a bureau within the Department of the Interior. Congress gave the new agency the following mandates:

1. National parks are to be set apart for the use, observation, health, and pleasure of the people.

2. They are to be maintained in absolutely unimpaired form for the use of present and future generations.

3. The national interest must dictate all decisions affecting public or private enterprise within the parks.

Today the National Park System comprises 391 diverse units—that is, individual holdings. Besides national parks, they include national monuments, national historical sites, national recreation areas, national memorials, the White House, and more (Table 14.4). This system embraces more than 34 million hectares (more than 84 million acres). Units of the National Park System are located in every state, as well as in the District of Columbia, American Samoa, Guam, Puerto Rico, the

A CLOSER LOOK 14.5   The Wilderness Controversy

About 43.5 million hectares (107.5 million acres), or 4.6% of the total land area of the United States, has been officially designated as part of the National Wilderness Preservation System. However, 54% of this wilderness is in Alaska. In the contiguous 48 states, less than 2% of the land area is so designated. Government agencies, in reviewing remaining roadless areas in public lands, found that an additional 5% could qualify for wilderness designation. A spirited controversy now rages concerning the advisability of expanding the National Wilderness Preservation System. Conservationists say we don't have enough wilderness areas, whereas many ranchers and officials of the timber, mining, and energy industries operating on public lands say we already have too much. Let's examine arguments for and against wilderness expansion.

ARGUMENTS FOR EXPANSION

1. Access to wilderness provides temporary escape from the hustle and bustle of modern society. The awesome grandeur of the wilderness replenishes the human spirit.
2. Wilderness is an insurance policy that protects natural ecosystems against the type of disruption and degradation that occurs in areas not so designated.
3. Wilderness preserves biological diversity. It prevents the extinction of endangered species of plants and animals. It provides essential habitat for mountain lion, bear, moose, elk, bighorn sheep, and deer, and it protects the spawning beds of salmon and trout.
4. It preserves the inherent right of all creatures, both human and nonhuman, to survive.

5. It supports thriving outdoor gear, camping supply, and tourist industries.
6. Supplies of cattle forage, timber, copper, oil, and natural gas that are available in nonwilderness areas are completely adequate to support our economy and ensure our national security. In addition, potentially new wilderness areas (in the 5% of public lands that qualify) would not add much to the country's timber, mineral, and energy resources.

ARGUMENTS AGAINST EXPANSION

1. The current interest in wilderness is a fad and will soon dissipate.
2. Wilderness use by visitors is only a fraction of the visitor use in our national parks.
3. Wilderness is just a playground for the wealthy members of our society.
4. The designation of wilderness lands has caused a substantial employment decline in the cattle, logging, and mining industries.
5. Resources like grasslands, timber, coal, copper, oil, and natural gas should not be locked up in wilderness areas. They should be fully utilized to promote our nation's economic health and security. After all, the original vegetation quickly becomes reestablished soon after grazing, logging, and mining have ceased.
6. All of our public lands, including so-called wilderness, should be managed according to the multiple-use concept.

After reflecting on the arguments pro and con, do you think our nation's wilderness system should be expanded?

Mariana Islands, and the Virgin Islands (Figure 14.23). Let's briefly examine some of the unit categories:

1. *National parks.* National parks are usually extensive acreages whose major function is the preservation of scenic grandeur, wilderness, and wildlife. Examples include Yellowstone, Yosemite, Glacier, and Grand Teton National Parks. (Worldwide, there are more than 1,000 national parks today in more than 120 countries.)
2. *National monuments.* These units range in size from the Statue of Liberty in New York Harbor to the 800,000-hectare (2-million-acre) Death Valley National Monument in east-central California. Death Valley's Badwater region, at 85 meters (280 feet) below sea level, has the lowest elevation on the North American continent.
3. *National recreation areas.* National recreation areas include rural and urban sites set aside for their recreation potential—fishing, boating, camping, hiking, and picnicking. Hells Canyon National Recreation Area, for example, is in Oregon and Idaho. It's a deep gorge cut by the Salmon River and is known for unsurpassed scenic grandeur and sizable trout populations. Two examples of national recreation areas that are located in heavily populated urban

areas, bringing parks closer to people, are the Golden Gate National Recreation Area, near San Francisco, and the Gateway National Recreation Area, near New York City.

4. *National lakeshore and seashore areas.* Many coastlines are also part of the National Park System. One of the most popular national seashores is Padre Island, off the cost of Texas. It provides more than 109 kilometers (68 miles) of white sand beaches along the Gulf of Mexico. Among the varied recreational activities provided are bird watching, swimming, surfing, sailing, and horseback riding. Padre Island is inundated with college students every year during their spring break.
5. *National historical sites.* These units include both national historic sites and national historical parks. The major function of these sites is to commemorate some significant event in the history of the nation. For example, Valley Forge National Historical Park was established when President Gerald Ford signed a bill on July 4, 1976, to celebrate the bicentennial of the American Declaration of Independence. It includes the house where George Washington lived in 1760 as well as the site of the colonial encampment in 1777–1778.

FIGURE 14.22 Half Dome in Yosemite National Park in California. This scenic vista and others attract hundreds of thousands of visitors each year.

How Is Land for National Parks Acquired?

Early in the history of the National Park System, new parks were established on lands already owned by the federal government. These parks were usually quite large and were established to preserve scenic beauty, wilderness, and wildlife. However, in the past few decades, most units added to the National Park System have been purchased from private individuals, have been located east of the Mississippi River, are relatively small, and serve as monuments, historical sites, and recreational areas. Money for such purchases was authored in 1965 when the U.S. Congress passed the **Land and Water Conservation Act.** More than 128 units have been purchased in this way, at a cost of more than $2 billion.

Sometimes land that is incorporated in the National Park System is donated by private individuals, corporations, or even states. For example, in 1984 the state of New York donated 5 hectares (12.5 acres) of an attractive beach to the Fire Island National Seashore. The wealthy Rockefeller family donated the major portion of what is now Acadia National Park in Maine. They also donated the exquisite ocean beaches of what is now the Virgin Islands National Park on St. John Island.

Preservation of Natural Beauty

The national parks preserve the natural beauty of the nation's mountains, forests, canyons, waterfalls, wetlands, lakes, and streams. However, the concept underlying the National Park System has changed considerably from that visualized by Congress when it established Yellowstone Park in 1872. In that era, the concept was that lands should be left alone, protected completely from interference by humans. However, in those days, knowledge of ecology and the principles of biological succession and carrying capacity of habitat was quite limited.

Today, scientists know considerably more. They realize that if the National Park Service is to preserve plant and animal communities in their original form, some human manipulation of succession and carrying capacity is essential. Of course, such manipulation must be based on research that is sound and comprehensive.

TABLE 14.4	Categories of the National Park System (not all categories are listed). As of the year 2006, there were a total of 391 units in the National Park System.
International historic sites	National parks
National battlefields	National parkways
National battlefield parks	National preserves
National battlefield sites	National recreation areas
National historic sites	National reserves
National historical parks	National rivers
National lakeshores	National scenic trails
National memorials	National seashores
National military parks	National wild and scenic rivers
National monuments	Other designations

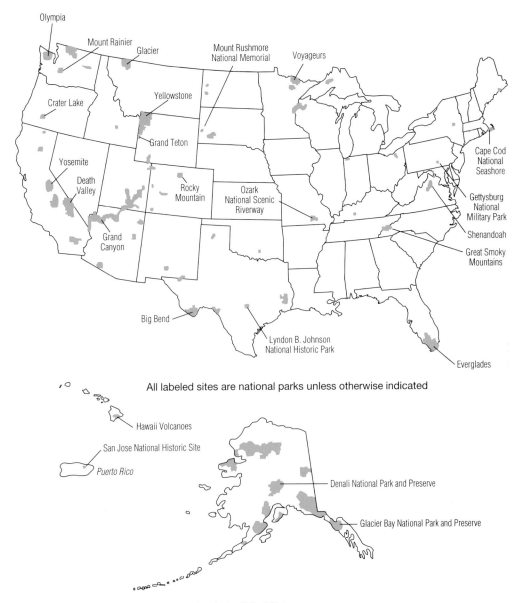

FIGURE 14.23 Major national parks and other national designations in the United States.

Overcrowding: A Major Stress on the National Park System

In recent years, the National Park System has been subjected to increasing stress. Much of it has been caused by attempts of private interests to engage in logging, mining, oil drilling, grazing, urban development, and other activities in nearby areas. Such human activities can threaten wildlife and recreational values of the park system. However, one of the most serious problems, ironically, has resulted from the parks' popularity. For example, consider the following statistics.

Since 1916, when the National Park Service was established, the annual number of visitors has soared from a mere 400,000 to more than 272 million in 2006. In fact, today there are more annual visits to the national park system than there are people living in the United States! By 2020, annual visits to the national park system are expected to at least double.

Nowhere is overcrowding more noticeable than in California's Yosemite National Park. In midsummer throughout the 1990s, Yosemite was inundated with tens of thousands of visitors. It suffered from all the environmental ills of a major city, such as traffic jams and pollution of air, water, and land. Motorists to Yosemite cursed as they leaned on their horns in "beep and creep" traffic. Pollutants from thousands of exhaust pipes caused vile odors. What's more, eventually they contributed to a haze that sullied the beauty of Yosemite Falls and the mighty granite dome of El Capitan. Sewage systems were overtaxed. Soiled paper, candy wrappers, pop cans, paper plates, and other assorted litter desecrated once-lovely campsites. Lines of tourists waited impatiently at checkout counters in Yosemite Mall—a virtual modern urban shopping center located in what was once a wilderness. In a historic effort to restore the park, officials announced on November 4, 1997, a proposed ban on cars by 2001, requiring visitors to walk, bike, or take buses to get around Yosemite Valley. However, as of 2008, the ban had still not been implemented. The use of shuttle buses is encouraged, where available. Also, some roads are closed at certain times of the year. Today nearly 4 million people visit the park.

Many other national parks suffer from similar problems. In Yellowstone National Park in Wyoming (Figure 14.24), the list of modern developments in the 5% of the park most frequently visited by tourists includes 1,207 kilometers (750 miles) of roads, 2,100 permanent buildings, seven amphitheaters, 24 water systems, 30 sewer systems, 150 kilometers (93 miles) of electrical transmission lines, numerous garbage dumps, 54 picnic areas, 17,000 signs, and hotel or cabin accommodations for 8,586 people per night. Other parks that are often overcrowded in the summer include Mount Rainier, Sequoia, Crater Lake, and Rocky Mountain.

Like any urban center, the most popular national parks experience frequent crimes, including sales of illicit drugs, sexual assaults, and wallet snatching. Overzealous tourists often make off with illegal souvenirs such as endangered species of cacti and alpine flowers as well as redwood bark and slivers of petrified wood. Even in the backcountry of the parks, people pollution is in evidence. For example, in Rocky Mountain National Park, backpackers caused so much trampling of wildflowers and erosion that some wilderness trails had to be paved over with asphalt.

Blueprint for the Future of the National Park System

Conservation and environmental organizations are concerned with the ongoing pressures on the National Park System. They have suggested ways of managing these problems in the next few decades:

1. *Restrict commercial activity.* All logging, mining, oil drilling, and grazing now permitted on federal lands adjacent to any national park should be banned. All commercial ventures, including restaurants, hotels, and camping equipment sales, should be reduced to a minimum.

2. *Expand the park system.* To alleviate crowding and resultant pollution and deterioration of the present parks, the system should be greatly expanded. The Wilderness Society proposes the enlargement of many of the 391 units in the National Park System. In addition, it suggests that many new parks be added in the near future.

3. *Control traffic congestion and pollution.* Bar all private motor vehicles from each park's interior. All transportation within each park should be performed exclusively by federal shuttle buses (if needed), powered preferably by electric motors, which are not only quiet but nonpolluting as well.

4. *Reduce crowding.* Crowds will be reduced if the system is expanded. However, another strategy must be employed: greatly restrict the number of visitors by issuing a restricted number of entry permits. For example, a given park might issue only 1,000 permits per day, to a maximum of 180,000 during the six months the park is open to the public.

5. *Greatly increase park budgets.* The federal money available to the National Park System should be boosted considerably. A park's nature education facility provides exciting and important information on the kinds of biological communities in the park and how their ecological systems work. These educational centers are in urgent need of improved equipment as well as additional staff. Money should also be available to provide summer research grants to college students who are interested in pursuing a career in wildlife biology, ecology, nature education, or park administration. More funds are required so that deteriorating roads, shelters, and visitors' centers can be upgraded. A minimum of $2 billion is urgently needed for road repair alone. And, of course, billions of dollars will be required in the near future to acquire land for the proposed expansion of the National Park System.

FIGURE 14.24 Traffic jam caused by animal viewing on a road in Yellowstone National Park.

14.10 Reversing Tropical Deforestation

One of the world's most endangered forests is the tropical rain forest. By definition, a **tropical rain forest** is a forest with 200 centimeters (79 inches) of annual rainfall spread evenly enough through the year to support broad-leaved evergreen trees. Trees are typically arrayed in several irregular canopy layers. They grow dense enough to capture more than 90% of the sunlight before it reaches the ground. The ground is so dark where the sunlight does not hit that a flashlight is needed to closely examine the insects.

Tropical rain forests form a band around the equator of about 7.8 million square kilometers (3 million square miles)—roughly the area of the United States (Figure 14.25). They are found in Latin America, Asia, and Africa. More than half of the world's intact tropical forests are found in three countries: Brazil, the Democratic Republic of the Congo (formerly Zaire), and Indonesia. The largest continuous rain forest lies in South America, where about 6.9 million square kilometers (2.7 million square miles) of forest cover the Amazon Basin.

For many years, rain forests have been cleared to make way for roads, ranches, farms, cities, towns, mines, factories, and reservoirs. Researchers disagree about how much tropical forested land has been and is being logged. Most believe that humans have already cut about 50% of the world's closed-canopy

tropical rain forests, with much of that loss in the past 40 to 50 years. Tropical deforestation continues today. Each year, about 10.5 million hectares (25.9 million acres), an area roughly the size of Tennessee, are cut down or altered by human activity. Figure 14.26 illustrates tropical deforestation rates between 2000 and 2005 in various nations. By some estimates, present rates of removal will leave only scattered remnants of tropical rain forests by 2025, except for parts of the Amazon Basin and Central Africa.

Some have called tropical rain forests a green hell, others a green cathedral. Whatever the name, the world's tropical rain forests, which once covered much of the Earth like a thick green blanket, are rapidly declining. You might say, "So what? After all, what is a tropical forest good for? Snakes and squawking parrots and monkeys, maybe. But what does it have to offer me?" The answer is, plenty. From the medicine you took for your last illness to the rubber tires on your car, tropical forests have a lot to do with you. And they have significance for the well-being of nations as well as individuals.

Value of Tropical Forests

Tropical forests (of which two-thirds are rain forests) contain almost half of the world's growing wood. More than 1 billion cubic meters (35 billion cubic feet) of wood are removed from the world's tropical forests for human use each year. Half of the world's annual harvest of hardwood is supplied by tropical forests. Tropical nations receive billions of dollars annually for the export of wood from the forests. Brazil, for example, exports much of its wood to the United States.

Rice production and irrigation farming in the tropics depend on water supplies slowly released from tropical forests. Tropical forests produce food products such as coffee, cocoa, nuts, tropical fruits, spices, and sweeteners. They also produce other materials, such as gums, dyes, waxes,

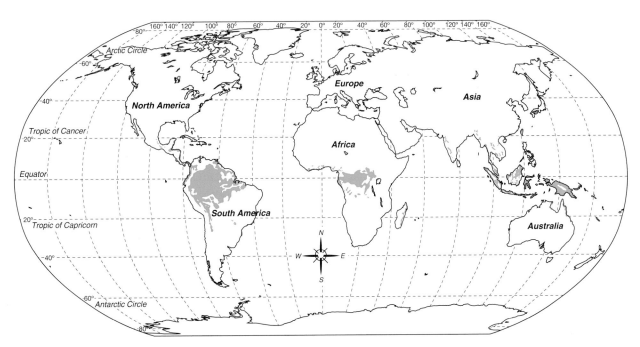

FIGURE 14.25 Tropical rain forests of the world.

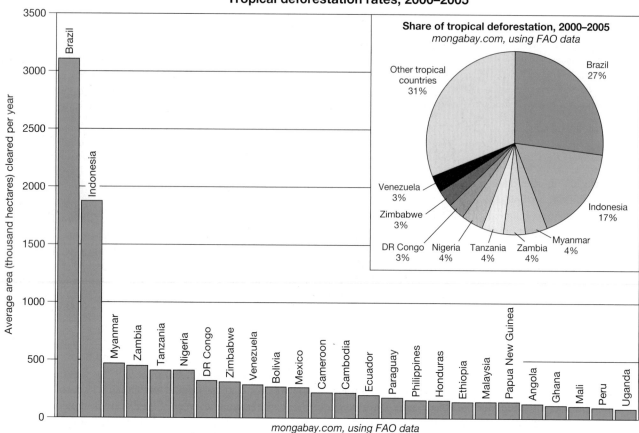

FIGURE 14.26 Tropical deforestation rates in various nations, 2000–2005. (Source: UN Food and Agriculture Organization and mongabay.com)

resins, and oils that are used in ice cream, shampoo, deodorant, sunscreen lotion, tires, shoes, and many other products.

Tropical forests provide habitat for an estimated 3 million to 7 million species of plants and animals—the greatest variety of life occurring anywhere on this planet. They are home to at least half of all the Earth's species of plant and animal life. On a global basis, the tropical forest biome has unsurpassed scientific and educational value. For example, the raw material for productive studies of plant and animal ecology is almost unlimited.

Many modern medicines, such as those that control malaria (quinine), relieve mental stress (reserpine), or show promise to cure cancer (vincristine), are derived from plants that grow in tropical forests. Humans have examined only a very small portion of the millions of plant species found in these forests as potential sources of medicines. In other words, they could offer hundreds of additional drugs to fight an assortment of illnesses.

Causes of Deforestation

The main direct causes of tropical deforestation are permanent and slash-and-burn agriculture, fires, livestock grazing, fuelwood gathering, and commercial logging (Figure 14.27). Three root causes are continuing population growth, continuing economic activity in tropical countries, and poverty.

Agriculture Although permanent conversion to agriculture occurs, generally forest clearers use slash-and-burn techniques, but on a much larger scale than as traditionally practiced. **Slash-and-burn agriculture,** also called **shifting agriculture,** is a farming system in which plots of forested land are cleared and then farmed for a few years until the soil loses fertility. The land is then abandoned to be eventually reclaimed by the natural forest vegetation. This regeneration or fallow period has been reduced today to 3 to 5 years in many areas when it needs to be much longer, from 10 to 25 years, to regenerate the forest and restore soil fertility. In theory, at least, this system appears to be ecologically sound, as long as clear-cuts are kept small and the fallow period is long enough. If the clear-cut areas are too large, the effects can be devastating. In practice, because of the increasing food demands imposed by increased population growth, the fallow period is often too short for soil fertility to be restored.

Fire Fire is used as a tool in slash-and-burn farming to dispose of felled trees and to prevent the invasion of weeds and shrubs into the new cropland. Millions of hectares of tropical forests have been destroyed when such fires have raged out of control. Forest destruction by fire in Brazil's Amazonian rain forest is accelerating.

Cattle Ranching and Biofuel Farming To raise the cattle to supply the U.S. demand for hamburgers, hot dogs, and luncheon meat, cattle ranchers in Central America are clearing

FIGURE 14.27 Example of tropical deforestation. This aerial view shows a large soybean field eating into a tropical rain forest in Brazil.

away tropical forests and converting the newly opened areas to grazing lands. Next time you bite into a hamburger, you should momentarily reflect on the fact that in a very small but very real way, you may be contributing to one of this hemisphere's most serious environmental problems.

Areas of tropical rain forests are also being cleared for farming crops for biofuel use. For example, in recent years, vast areas of tropial rain forests have been cleared across Asia for palm oil plantations. Palm oil is used as one of the raw materials in biodiesel production, a fuel that is derived from vegetable oils or animal fats. The crop is also used in food products, engine lubricants, and cosmetics. This conversion has reduced biodiversity, increased vulnerability to catastrophic fires, and affected local communities dependent on services and products provided by forest ecosystems.

Gathering Fuelwood   More than 1 billion cubic meters (35 billion cubic feet) of wood are harvested for fuel in the tropics. This wood is either used directly for firewood or converted to charcoal. In Haiti the production of charcoal is big business. Unfortunately, the resulting forest destruction there is also immense.

Industrial Logging   Commercial logging in tropical regions is frequently wasteful and inefficient. For example, during a selective logging operation in Malaysia, more than 55% of the uncut trees were either severely damaged or destroyed. Until recently, commercial logging employed selective cutting. The only trees taken were mature specimens of the most valuable species, such as teak, mahogany, and rosewood. Today, however, any tree has potential value if it can be reduced to wood chips. As a result, clear-cutting is being used more widely. Unfortunately, most clear-cut areas do not become reforested by natural means. They must be replanted by people. Frequently, however, forestry personnel either are not knowledgeable about effective reforestation or simply refuse to

use it. Forests that are replanted tend to be planted in monoculture. In many instances, land-starved peasants move onto deforested land and attempt to eke out an existence using slash-and-burn agriculture.

Effects of Deforestation

Firewood Scarcity   Because of the severe firewood shortages in India, Haiti, and Nepal, hot meals are almost a thing of the past. Eric Eckholm, a former research ecologist with the Worldwatch Institute in Washington, has pinpointed the problem for many people living in these and other less developed countries: "For one-third of the world's people, the energy crisis does not mean…high prices of petroleum.…It means something much more basic—the daily scramble to find the wood needed to cook dinner." As a result of increasing wood scarcity, the poor are now shifting from wood fuel to animal dung. Unfortunately, however, dung also is valuable as fertilizer. If the cow dung being burned as fuel in Africa, Asia, and the Near East were used as fertilizer instead, a lot more grain could be produced.

Climatic Changes   Tropical forests influence the weather and the climate in important ways that scientists do not as yet fully understand. For example, the Amazon rain forest receives roughly twice as much rainfall as can be accounted for by moisture moving in from the ocean. How can this be so? Much of the rain that falls on the land evaporates and simply rains down on other areas downwind. Put another way, the Amazon rain forest generates much of the rain that supports the forest. Thus, by means of transpiration and evaporation followed by precipitation, water is kept in circulation in these densely forested areas.

Deforestation could alter this vital recycling system. When land is cleared of trees, more water flows into streams, so less water is available for evaporation. Thus, scientists who have studied the issue using computer models suggest that continued deforestation could eventually render the forest too dry to support tree growth. When a tropical forest is removed, the radiation of heat from the denuded area is greatly increased. As a result, the local climate and even the global climate may be adversely affected. The burning of wood releases carbon dioxide into the atmosphere. Trees also absorb carbon dioxide, so their loss accounts for an additional increase in global carbon dioxide levels. The result of the increased heat radiation and increased levels of carbon dioxide may contribute to a progressive warming of the Earth (Chapter 19).

Loss of Gene Pools and Extinction of Species Tropical forests contain the parent species from which many of our present agricultural crops were derived. Many agricultural scientists believe that these species may be needed as genetic reservoirs from which new varieties of disease- and pest-resistant crops can be developed. Unfortunately, as rain forests are destroyed, we lose these species.

Tropical deforestation also threatens the survival of many birds and mammals. Species diversity is greater in this biome than anywhere else on Earth. That is, tropical rain forests possess more species of animals and plants than any other ecosystem. Extinction of such species is occurring at alarming rates.

Saving the Tropical Forests

How can the tropical forests be saved? To answer this question, we must ask another: How can people living in tropical countries achieve a decent standard of living from the land without destroying it? Many people believe that forest protection should be tied to economic development, notably sustainable economic development, a subject described in Chapter 2. Protection and economic benefit can go hand in hand. For example, tree-planting programs have been shown to be most successful when local people not only are involved in their planning and implementation but also own the land or are given ownership of any trees grown on commonly owned land. With ownership comes the needed incentive for local villagers to plant and protect trees for their own use and for sale. It may even help to privatize a small portion of the tropical forests, to give the protection of ownership or contract control. Where a government cannot afford to protect a forest at risk, a well-regulated timber company, given the right long-term incentives, might provide good management and reforestation. National forests and wildlife preserves should certainly be kept and protected. The World Bank recommends that countries with rain forests convert 15% of their forests to national parks and preserves.

Some ecologists believe that the best method of harvesting timber in rain forests profitably with little loss of biodiversity is strip cutting or logging. As you may recall, with this method, a strip is harvested on the contour of a slope, and logs are hauled out on a road along the upper edge of the strip. The area is then left for a few years until seedlings begin to grow in the harvested strip. Another strip is clear-cut, but this time above the road. One advantage of strip cutting is that nutrient runoff from the freshly harvested second strip is used by the regenerating vegetation downslope in the first strip. In addition, the recently cut second strip catches the seeds rolling down from the mature forest above. These benefits do not occur in a clear-cut system.

Agroforestry shows promise of increasing the production of food and fuelwood in the tropics. **Agroforestry** is the practice of growing trees among crops or on pastureland. Trees can supply timber, fuelwood, fruits, nuts, and food for livestock. They also provide fertilizer for crops, reduce soil erosion and weeds, and protect crops from winds. Agroforestry programs are already operating successfully in areas of many tropical countries in Asia, Africa, and Latin America.

The extraction of nontimber products from rain forests has been shown to yield financial returns similar to those from logging and farming in Peru and Brazil. In the Amazon, half a million rubber tappers, called *seringueiros,* make their living today from rubber, Brazil nuts, tonka beans, palm hearts, and other wild products. They also hunt, fish, and practice small-scale agriculture in forest clearings. The tappers preserve the rain forests because they depend on the forests' biological diversity for their livelihood. They attempt to harvest resources within the forests' capacity to sustainably regenerate them. In 1987, the Brazilian government established *seringueiro* extractive reserves on government land with 30-year renewable leases and a prohibition on clear-cutting timber.

So far, we have given some helpful but modest strategies. Governments and local people will need to be persuaded to adopt such innovations as agroforestry, strip cropping, and extractive reserves. As stated earlier, private ownership of some kind should help. Just as important, **sustainable development** will depend on education and social change.

Countries with rain forests could provide grants (with United Nations financial assistance) to support research and education on the ecology and management of their tropical forests. Science is already demonstrating its potential to preserve tropical forests through its advances in increasing tree yields on forested land and crop yields on agricultural land (so that less forestland has to be cleared for crop production). The capability and knowledge also exist to increase the efficiency of forest product harvest and use.

Many people believe that the more-developed countries—like the United States—must help because the loss of species anywhere in the world diminishes wealth everywhere. More-developed countries already provide and need to continue providing funds to help plan and implement successful strategies. The funds should be used to provide financial incentives to villagers and village organizations for reforesting areas from which trees have been removed for use as fuel. Incentives could be used to encourage sustainable management of fuel plantations—harvesting that is on a sustained-yield basis. Funds could also provide incentives to improve management practices in natural forests to increase yields and reduce forest degradation. Helping raise the standard of living of the people of tropical less-developed countries is also important because people who worry little about food, clothing, and shelter can afford to care about forest preservation.

Summary of Key Concepts

1. The national forests of the United States are administered and managed by four federal agencies: the U.S. Forest Service, the Natural Resource Conservation Service, the Tennessee Valley Authority, and the Fish and Wildlife Service.

2. President Theodore Roosevelt appointed Gifford Pinchot to be the first chief forester of the newly created U.S. Forest Service in 1905.

3. Under the terms of the Multiple Use–Sustained Yield Act of 1960, national forests must be managed for many uses in addition to timber harvesting. Among them are flood and erosion control, grazing land, wildlife habitat, biomass fuel, mining, oil and gas leasing, and scientific, educational, wilderness, and recreational uses.

4. Sustained-yield management ensures that annual forest harvests are balanced by annual timber growth. It also dictates that whenever timber is removed, the denuded area must be reforested, either by natural or by artificial means.

5. As of 2008, the National Wilderness Preservation System includes 704 primitive areas. They are managed by the National Park Service (42%), U.S. Forest Service (33%), Fish and Wildlife Service (20%), and Bureau of Land Management (5%).

6. Much of the nation's forestland is located in national parks and administered by the National Park Service. As of 2006, the National Park Service administered 391 diverse units, such as national parks, national monuments, national historical sites, national recreation areas, and national lakes and seashores.

7. The national park system endures many stresses today, including the attempts of private interests to engage in logging, mining, oil drilling, grazing, and urban development in nearby areas, as well as overcrowding by visitors in summertime.

8. A logging plan includes the following information: (1) a map of the forest on which the location, species, age, and volume of the species to be logged are indicated; (2) a discussion of the harvesting method; (3) selection of haul roads; (4) consideration of the scientific, recreational, and wildlife functions of the forest; and (5) an estimate of gross income from the operation.

9. Harvest methods may be grouped under two categories: even-aged and uneven-aged stand methods. Clear-cutting, seed tree, shelterwood, and coppice methods are used to produce even-aged stands. Selective cutting and strip cutting are used to produce uneven-aged stands.

10. Clear-cutting is cutting the entire timber stand in a designated area.

11. Opponents of clear-cutting criticize it for the following reasons: (1) It accelerates surface runoff, flooding, and soil erosion. (2) Wildlife-carrying capacity is sharply reduced, at least temporarily. (3) Scenic beauty is destroyed. (4) The fire hazard is increased. (5) Tree blowdown is facilitated.

12. Supporters of clear-cutting claim it has many advantages: (1) It is the only way to regenerate forests in some areas. (2) It can help to control certain disease and insect outbreaks. (3) It is cheaper than selective cutting. (4) Timber grows faster on clear-cut areas than on areas that are selectively cut.

13. The seed-tree method is a silvicultural system in which all timber is removed in one cut, except for a scattered number of mature trees left to provide a source of seeds for the new stand.

14. The shelterwood method is a silvicultural system in which the mature timber is removed, leaving enough trees standing to provide shade and protection for new seedlings.

15. Strip cutting is the harvesting of wide strips (for example, 80-meter-wide strips), allowing natural regrowth.

16. Selective cutting is the harvesting of scattered trees or small groups of trees, usually in mixed-species stands of unevenly aged trees. Selective cutting is more costly and time-consuming than clear-cutting but reduces land scarring, runoff, soil erosion, and wildlife depletion.

17. Arguments in favor of monoculture practiced on tree farms and in national forests say that it (1) takes optimal advantage of fertilizers and pesticides; (2) is rapid and efficient; (3) increases short-term profits; (4) makes maximal use of modern tree planting and harvesting methods; (5) is needed to satisfy America's growing demand for lumber and wood products; and (6) is appropriate for the sun-loving seedlings of timber-valuable species such as Douglas fir, lodgepole pine, and ponderosa pine.

18. The following arguments have been offered against monoculture: (1) It is an artificial system made by humans that lacks the built-in balancing mechanisms of a multispecies, multiage forest. (2) The fast-growing wood is inferior. (3) Intensive fertilization causes the eutrophication of lakes and streams. (4) Intensive use of pesticides may contaminate wildlife and human food chains. (5) It requires intensive inputs of costly energy derived from fossil fuels. (6) It is more susceptible to fires than the natural forest. (7) It is highly susceptible to outbreaks of destructive diseases and insect infestations.

19. A tree farm is a private land area used to grow trees for profit under sound resource management principles.

20. A logging operation includes felling trees, removing them from the forest, and transporting them to a sawmill or pulp mill. If trees are to be used as pulpwood, they are often harvested with feller-bunchers and whole-tree chippers.

21. Genetic engineering holds promise for improvements in the nation's forests by providing trees with desirable traits such as accelerated growth, reduced need for fertilizer, and built-in chemical resistance against weeds and insects. However, since GE trees have not been proven to be safe, many believe that the engineering of trees is strictly about speculative science and economic return.

22. Ingenious methods have been developed to use almost every part of a tree. Useful products are being made from scrap boards, shavings, wood chips, and bark.

23. Major agents of forest destruction are disease (such as heart rot fungus), insect pests (such as bark beetles), and fire.

24. Insect pests can be partially controlled by using insecticides; clear-cutting the infested area; using biological control agents such as viruses, parasites, and predators; and replacing monotypes with heterotypes.

25. Fires are controlled with the use of backfires, fire lanes, fire-retardant chemicals released from planes, and water pumped from tank trucks or released from planes.

26. The idea that every forest fire is destructive has been replaced by the concept that occasional fires are a natural and beneficial feature of the forest ecosystem and are needed to maintain the forest's health.

27. Controlled fires, purposely started by highly trained personnel, serve many functions: (1) reducing the crownfire hazard, (2) preparing the soil as a seedbed, (3) maintaining subclimax species that are valuable as timber and as a wildlife habitat, (4) controlling insect and disease outbreaks, (5) stimulating soil microbial activity, and (6) increasing the quality of livestock forage.

28. By 2030, Americans may be using 50% more timber than they do now. This increased demand may be met by upgrading small private forests; more effectively controlling the agents of forest destruction (diseases, insects, and harmful wildfires); developing superior (faster-growing, bettergrained, and insect-, disease-, and drought-resistant) trees through the techniques of grafting, hybridization, and gene splicing; increasing the use of wood substitutes; increasing imports; recycling paper; and reducing wood use.

29. Tropical forests have multiple values: (1) erosion and flood control, (2) timber and forest products, (3) medicines, (4) wood fuel, (5) outdoor laboratories for scientific research, (6) the habitat for at least half of all the Earth's species of plant and animal life, and (7) recreational functions.

30. Tropical forests have been greatly reduced because of slash-and-burn agriculture, conversion to grazing land, industrial logging, and fuelwood gathering.

31. The effects of deforestation include firewood shortages, climatic changes, loss of gene pools, and extinction of species.

32. The issue of saving tropical forests boils down to the following question: How can people living in these tropical areas achieve a decent living from the land without destroying it? Some modest strategies include private ownership of land or of trees, agroforestry, strip cropping, extractive reserves, research and education, and donations from more-developed countries.

Key Words and Phrases

Aerial Seeding	Land and Water Conservation
Agroforestry	Act
Bark Beetle	"Let It Burn" Policy
Clear-Cutting	Logging Plan
Clone	Monoculture
Coppice Method	Muir, John
Controlled Burn	Multiple-Use Management
Cruise	Multiple Use–Sustained Yield
Douglas Fir	Act of 1960
Ecosystem Services	National Park Service
Even-Aged Stands	National Parks
Feller-Buncher	National Wilderness
Fire Climax	Preservation System
Fire Control	Natural Reseeding
Forest Pest Control Act of	Nonindustrial Private Forest
1947	Owners
Genetic Engineering	Pinchot, Gifford
Heart Rot	"Prescribed Natural Fire"
Heterotype	Reforestation
Hybridization	Rotation
Infrared System	Sanitation Techniques
Integrated Pest Management	Seed Orchards
(IPM)	Seed-Tree Method

Selected Breeding	Sustained-Yield Concept
Selective Cutting	Tissue Culture
Shelterwood Method	Tree Farm
Shifting Agriculture	Tropical Rain Forest
Sierra Club	Uneven-Aged Trees
Silvicultural System	U.S. Forest Service (USFS)
Slash-and-Burn Agriculture	Wilderness
Smoke Jumpers	Wilderness Act of 1964
Strip Cutting	Wildfire
Sustainable Development	Yellowstone National Park

Critical Thinking and Discussion Questions

1. Discuss the economic importance of our nation's forests.

2. Describe the functions of our nation's forests in terms of (a) flood and erosion control, (b) rangeland, (c) wildlife habitat, (d) wilderness areas, and (e) source of fuel.

3. What is meant by the term *sustained yield*?

4. Discuss the following statement: "Clear-cutting is the best harvesting method for our nation's forests."

5. Describe six advantages and six disadvantages of monoculture.

6. What type of forest is most suitable for clear-cutting? For strip cutting? For selective cutting?

7. List four factors that make reforestation by seeding difficult.

8. What are the pros and cons of genetic engineering for forest ecosystems in the 21st century?

9. What might be the reasons that so many forest pests in the United States are exotics?

10. Discuss the factors that make heterotypes resistant to violent outbreaks of insect pests.

11. Suppose a forest fire breaks out in the Rockies. Wind velocities are high. Discuss methods that might be used to bring this fire under control.

12. List five effects of controlled burns in the longleaf pine stands of the South.

13. Describe several methods by which our nation's future timber needs might be satisfied.

14. Why is it important to have wilderness areas?

15. The National Park Service was established in 1916. It administers close to 400 units, including parks, monuments, historical sites, recreation areas, and lakes and seashores. A blueprint for the future should include expansion of the system, control of traffic congestion, restriction of commercial activity, reduction of crowding, and major increases in budget. Why? Discuss your answer.

16. Suppose that tomorrow all of the tropical forests in the world were suddenly destroyed. Discuss the aftermath of such a catastrophe. What might be the long-range climatic effects? The biological effects? The social and economic effects? The effects on world peace?

17. What are some strategies for saving the tropical forests from deforestation? Discuss such strategies.

Suggested Readings

Behan, R. W. 2001. *Plundered Promise: Capitalism, Politics, and the Fate of Federal Lands.* Washington, DC: Island Press. An in-depth view of the history of public land management and what is wrong with it.

Brick, P., D. Snow, and S. Van de Wetering. 2000. *Across the Great Divide: Explorations in Collaborative Conservation and the American West.* Washington, DC: Island Press. Exploration of a new movement in conservation: collaborative conservation. A must-read for anyone interested in a career in environmental conservation or natural resource conservation.

Brown, C. D. 1993. Mapping Old Growth. *Audubon* 95(3): 130–133. Well-written account of how environmental groups determined the amount of old-growth acreage in some of the national forests in the Northwest.

Burton, L. D. 2007. *Introduction to Forestry Science,* Albany, NY: Delmar Publishers. Solid introductory text on forests and forestry.

Carvalho, G. O., et al. 2002. Frontier Expansion in the Amazon: Balancing Development and Sustainability. *Environment* 44(3): 34–45. A detailed look at Brazil's ongoing policy of tropical rain forest development, which could lead to huge losses.

Food and Agriculture Organization. 2006. *Global Forest Resources Assessment 2005: Progress Towards Sustainable Forest Management.* FAO Forestry Paper 147. Rome, Italy: UN-FAO. FAO's Forestry Department gives excellent, up-to-date information on the status and trends of world forests.

Jensen, D., and G. Draffan. 2003. *Strangely Like War: The Global Assault on Forests.* White River Junction, VT: Chelsea Green. A popular book with an important, somewhat alarming message.

Jordan, C. F. 1980. Amazon Rain Forests. *American Scientist* 70: 394–401. Good discussion of strip cutting as a sustainable industry in the Amazon.

Kemmis, D. 2001. *This Sovereign Land: A New Vision for Governing the West.* Washington, DC: Island Press. A radical new approach to managing natural resources regionally.

Kerasote, T., ed. 2001. *Return of the Wild: The Future of Our Natural Lands.* Washington, DC: Island Press. A collection of writings about the value of wilderness, current threats, and what can be done to truly protect it.

Kimmins, J. P. 2004. *Forest Ecology: A Foundation for Sustainable Forest Management and Environmental Ethics in Forestry,* 3rd ed. Upper Saddle River, NJ: Prentice Hall. Comprehensive and in-depth treatment of forest ecology and management.

Kohm, K. A., and J. F. Franklin, eds. 1997. *Creating a Forestry for the 21st Century.* Washington, DC: Island Press. Authoritative examination of the current state of forestry and its relation to the emerging field of ecosystem management.

Lambin, E. F., and H. J. Geist. 2003. Regional Differences in Tropical Deforestation. *Environment* 45(6): 22–36. Examination of the many causes of tropical deforestation.

McComb, B. C. 2008. *Wildlife Habitat Management: Concepts and Applications in Forestry.* Boca Raton: CRC Press. Integration of silvicultural and forest-planning principles with principles of habitat ecology and conservation biology.

Miller, C. 2001. *Gifford Pinchot and the Making of Modern Environmentalism.* Washington, DC: Island Press. Biography of a man who profoundly influenced resource conservation.

Mitchell, J. G. 1997. Our National Forests. *National Geographic* 191(3): 58–87. Excellent account and pictorial of the management and health of U.S. national forests.

Parfit, M. 1996. The Essential Element of Fire. *National Geographic* 190(3): 116–139. Wonderful article and pictorial about the blessings and curses of forest fires.

Perry, D. A., R. Oren, and S. C. Hart. 2008. *Forest Ecosystems.* Baltimore, MD: Johns Hopkins University Press. Straightforward language and clear diagrams emphasizing ecosystem processes and applications in forest management.

Repetto, R. 1998. *The Forest for the Trees: Government Policies and the Misuse of Forest Resources.* New York: Basic Books. Discussion of the manner in which governments that are committed in principle to conservation are aggravating the losses of their forests through ill-conceived policies.

Stewart, D. 2003. Growing Greener Forests. *National Wildlife* 41(6): 37–42. A must-read for anyone interested in learning more about sustainably harvested forests and certified lumber.

Wilson, E. O. 1992. *The Diversity of Life.* New York: W. W. Norton. Fascinating, well-written book, especially the sections on tropical forests.

Young, R. A., and R. L. Giese, eds. 2002. *Introduction to Forest Ecosystem Science and Management,* 3rd ed. New York: John Wiley & Sons. Good general introduction to the science and practice of forestry.

 ## Web Explorations

Online resources for this chapter are on the World Wide Web at: **http://www.prenhall.com/chiras** *(click on the Table of Contents link and then select Chapter 14).*

PLANT AND ANIMAL EXTINCTION

Many of us live our lives seemingly apart from nature, and for good reason. In the cities and suburbs where most of us live, there's often little evidence of nature. As a result, to many people, nature seems abstract and inconsequential. But nothing could be further from the truth. Wild plants and animals—and the ecosystems they inhabit—enrich our lives in numerous ways (Figure 15.1).

Nature is a source of recreation and sheer beauty. It provides inspiration and solitude. Beyond these benefits, natural systems provide many practical benefits. Researchers have discovered chemicals in plants and other organisms that are now manufactured to produce antibiotics, anticancer drugs, and a host of other medicines to combat heart disease, malaria, and others. Today one of every two prescription and nonprescription drugs originates from a wild plant. These drugs not only reduce human suffering, but also contribute significantly to our economy, resulting in billions of dollars' worth of revenue to pharmaceutical companies and retailers.

The economies of the United States and many other nations derive further economic benefits from nature. The hordes of hunters, anglers, birders, hikers, campers, and river runners annually spend billions of dollars on travel, lodging, food, and supplies. And nowadays, many people are flocking from richer nations like Canada, the United States, England, and Germany to view wildlife in remote corners of the world, spending billions of dollars on **ecotourism,** defined here as a form of tourism focused on adventures in the wild with the intent of doing little harm in the process.

Wild species are also an important reservoir of genes that aid agricultural scientists looking for ways to improve food crops and domestic livestock. Plants and animals provide food for many of the world's people, from the Chicago stockbroker who sips white wine after a seafood dinner in an elegant lakeside restaurant to the hunter who, crouched by a fire in the grasslands of Africa, tears meat from the bones of the gazelle he has killed and washes it down with water from a gourd.

Nature also provides many ecological services free of charge. Many species of birds, for instance, help to control pests in forests and on farms. Vegetation recycles water, reduces flooding, and controls erosion. Plants and algae provide oxygen and remove carbon dioxide from the air. Plants also help to maintain climate, and they and others recycle nutrients essential to the survival of humans and all other species. Just because these services come without a price tag, they are not without value. In fact, quite the opposite is true. Nature's services are priceless. The services that we have lost, including the flood-control function of wetlands in many river basins, cost society billions of dollars per year.

Clearly, nature serves us well. And although we humans have seemingly isolated ourselves from nature in cities and suburbs, we have not emancipated ourselves from the bonds that link us to the environment. Few people recognize that nature is the **biological infrastructure** of

FIGURE 15.1 Courtship dance of the whooping crane, an endangered species of North America.

society. That is, it is the source of all the resources that make our lives and our economy possible.

Despite these important benefits, humans are systematically destroying the Earth's life support systems. Economist Herman Daly argues that most nations are "treating the Earth as if it were a corporation in liquidation." This chapter looks at evidence of one aspect of the Earth's impending liquidation: plant and animal extinction. It examines the causes of extinction and ways to prevent it. It also provides background information on plant and wildlife populations to help you better understand the natural forces at work in a population—information that is essential to understanding ways to preserve our vanishing biological resources.

GO GREEN!

Join one of the many excellent national nonprofit organizations that fight to protect wildlife, such as Defenders of Wildlife, Audubon, National Wildlife, and the Nature Conservancy. Local organizations often do valuable work and operate on shoestring budgets, so they can benefit from your support as well.

15.1 Extinction: Eroding Earth's Biological Diversity

Extinction is the disappearance of a species. Extinction results from natural and human forces, discussed in this chapter.

Biologists estimate that 500 million species have made Earth their home since life began over 3.5 billion years ago. To date, biologists have catalogued approximately 1.8 million species. Recent estimates put the number of species alive today at well over 30 million, perhaps as high as 80 million. At least half, perhaps as many as two-thirds, of those species live in the rich tropical rain forests of the world.

Despite the impressive number of fellow travelers on spaceship Earth, at least 90% of the species that once lived on this planet have vanished—become extinct. Why have so many species gone extinct?

Biologists believe that many organisms such as the dinosaur became extinct because they were unable to adjust to changing environmental conditions. Many of these species left behind only fossil remains. That is, their line disappeared entirely. In other instances, changing environmental conditions resulted in the evolution of new species from existing organisms. The original species became extinct, but their descendants remain today, carrying on the genetic line.

What caused the environmental conditions to change for dinosaurs? No one knows for sure, but theories abound. One popular theory that explains the disappearance of the dinosaurs some 65 million years ago says a giant asteroid struck the Earth. This produced a cloud of dust that shrouded the Earth for many years and brought about a temporary but biologically devastating cooling in global temperature. Other scientists believe that intense global volcanic activity may have produced conditions hazardous to dinosaurs' existence. Gases from volcanoes may have formed harmful acids that fell on the land and waters, wiping out plants and animals. And dust from the towering volcanoes may have blocked out the sun, shading Earth and spawning a period of bitter cold intolerable to these magnificent creatures.

Changing conditions today, brought about by human actions, continue to drive species to extinction (Figure 15.2). Many biologists warn that we are on the brink of another mass extinction. Today, scientists estimate that one vertebrate (backboned) species disappears every nine months. Add to that the plants, microorganisms, and invertebrates that disappear from the face of the Earth, and according to Harvard's eminent biologist E. O. Wilson and others, the rate of extinction is an alarming 100 species or so every day!

FIGURE 15.2 A handful of the planet's endangered species. Top: key deer, black rhinoceros; center: Florida panther, golden lion, tamarin; bottom: gorilla, manatee. The rate of animal extinctions is increasing.

Many species may disappear right under our noses with hardly a person being aware of it. Throughout the world, for instance, scientists have reported an alarming disappearance of amphibians. Many species of frogs, toads, and salamanders either are experiencing extremely steep declines in population size or have vanished altogether. Amphibians are disappearing from a wide variety of habitats—from the jungles of Brazil to the suburbs of Kansas City. The cause is still unknown, but evidence suggests that pesticides may be the main culprit.

(See A Closer Look 15.1: Pesticide Drift and the Worldwide Disappearance of Amphibians.)

With continued population growth and economic development, which is occurring at a rapid rate in the biologically rich tropics of Central and South America, biologists fear that the rate of extinction could climb even more. Biologist E. O. Wilson contends that the consequences of widespread loss of species are potentially far more devastating than fossil fuel depletion, economic collapse, and limited nuclear

A CLOSER LOOK 15.1 **Pesticide Drift and the Worldwide Disappearance of Amphibians**

Pesticides applied to crops in the Central Valley of California make their way into high-elevation lakes and streams in the Sierra Nevada, mountains that border the rich agricultural region, killing some species of frogs and toads, according to a study presented at the Society of Environmental Toxicology and Chemistry in November 2005. In an attempt to control damage by insects and other pests, pesticides are sprayed on crops by plane or by tractor-drawn spray applicators. Some pesticides are carried away immediately by the prevailing winds and are deposited on nontarget downwind sites. In addition, pesticides that land on the crops initially may later become airborne through evaporation. They, too, may be deposited in downwind sites.

The migration of pesticides from the site of application, known as pesticide drift, can expose neighboring human and wildlife populations to potentially dangerous levels of toxic chemicals. To test the effects of pesticides on amphibians in the

Sierra Nevada, researcher Donald W. Sparling, a wildlife toxicologist at Southern Illinois University in Carbondale, IL, exposed amphibian eggs collected from ponds in the mountains to endosulfan. Endosulfan is commonly used in Central Valley agricultural production. Sparling found that endosulfan killed half of the foothills yellow-legged frog eggs when exposed to levels commonly measured in mountain ponds and lakes. Even lower concentrations caused increased death rates in this frog and one species of toad.

Researchers have long been concerned about a massive die-off of amphibians throughout the world and the potential impact it could have on aquatic ecosystems. Scientist have also examined a number of potential reasons for the die-offs, including ozone depletion and environmental pollution. This research suggests that pesticides may play a significant role in this phenomenon.

war. The erosion of Earth's rich biological diversity, says Wilson, will take millions of years to repair. It is "the folly [for which] our descendants are least likely to forgive us."

Preserving biological diversity has become one of the most important and pressing environmental concerns of modern times.

15.2 Causes of Extinction

To prevent species from going extinct, we must first understand the factors that cause extinction. Let's begin by focusing on anthropogenic causes—that is, human actions. "If a species becomes extinct," wrote David Day, author and conservationist, "its world will never come into being again. It will vanish like an exploding star. And for this we hold direct responsibility." What Day means is that most of the species that have become extinct since the beginning of the Industrial Revolution have done so because they have been unable to adapt to changes in the environment caused by another species, *Homo sapiens.*

Plants and animals become extinct for several reasons. Most of them vanish because their habitat is destroyed or altered beyond habitability. Therefore, habitat destruction or alteration is the number one cause of species extinction today. Other species disappear because commercial interests overharvested them to supply food or other valuable resources for human society. Still others succumb to **alien species** that have been accidentally or intentionally introduced into their habitat. Alien species may kill a native species outright or simply outcompete it. Sport-hunting has contributed to the decline of some species, as have pest- and predator-control programs. Pollution and the pet business have also contributed to the steady decline in the world's species. In some cases, several factors combine to drive a species to extinction. This section describes the three main causes of extinction: habitat destruction and alteration, hunting (commercial and sport), and the introduction of alien species.

Habitat Destruction and Alteration

Habitat is the area a population occupies. It provides food, water, shelter, and many other requirements of life. Destroying or altering that habitat can have profound effects on the plant and animal species living there.

Habitat is altered and destroyed by many human activities: expanding urban areas, farming, logging, mining, and the construction of highways, railroads, pipelines, and dams. Today, these and other activities are growing at a rapid rate. In the past 50 years, the destruction or severe alteration of habitat has claimed innumerable species and threatens millions of others. In the United States, for instance, 500 species are known to have become extinct since Europeans arrived. Currently, approximately 598 animal and nearly 600 plant species are listed as endangered (in imminent danger of extinction) in the United States. Another 146 animal species and subspecies and 146 plant species are listed as threatened (likely to become extinct), bringing the total of endangered and threatened species to over 1,500 (see Table 15.2 on page 424). Nearly 300 species are awaiting listing. The story of one species, the snail darter, is presented in Case Study 15.1.

CASE STUDY 15.1   DAM VERSUS DARTER: A CLASSIC CONFRONTATION

In 1973, David Etnier, a fish expert at the University of Tennessee, discovered a new species of fish in the Little Tennessee River. The tiny fish, only 75 millimeters (3 inches) long, was given the name snail darter (*Percina tanasi*). According to Etnier, the entire world population of this species, numbering about 1,400 individuals, was confined to a 2.5-kilometer (1.5-mile) stretch of the Little Tennessee River. The species qualified as endangered under the criteria of the Endangered Species Act, discussed later. It was most unfortunate, however, that the fish's habitat would soon be destroyed because the Tennessee Valley Authority (TVA), a federal agency, was constructing the $116 million Tellico Dam on the Little Tennessee, a short distance from where the darter was discovered. The resulting reservoir, impounded behind the dam, would replace the shallow, fairly rapidly flowing water essential for snail darter survival with deep, quiet water—a completely different aquatic habitat.

The discovery of the tiny fish set the stage for a confrontation between technology and a powerful federal agency (the TVA) on the one hand, and a vanishing species, supported by a few dedicated environmentalists, biologists, and nature lovers, on the other. In a letter to us, Etnier described the confrontation:

The story is an interesting one that involves a small fish and a group of people with virtually no resources attempting to force a governmental agency to comply with federal law. The number of participants has been immense, on both sides. Our side fought with money from the sale of snail darter T-shirts. TVA's efforts to thwart us have probably cost that agency well in excess of a million dollars in lobbying and expenses associated with their staff of lawyers, biologists, and administrators working on the case. Virtually every newspaper, press service, and TV network has devoted some time or space to the issue and magazines such as *People, The New Yorker, Time,* etc., have carried lengthy articles.

Eventually the case was brought to the courts. In early 1977, a federal appeals court ruled that the TVA would have to terminate construction of the Tellico Dam. But the pro-dam people did not give up easily. In April 1978, no less an official than U.S. Attorney General Griffin Bell asked the U.S. Supreme Court to spare the dam and scrap the darter. But to no avail. In June 1978, our nation's highest court ruled in favor of the fish.

However, legislators began pondering the question, "When we formulated and enacted the Endangered Species Act, were we

(continued)

really concerned about saving tiny fish, or were we thinking of eagles and moose?" As one scientist wrote, "Congressmen are now finding themselves confronted with a Pandora's box containing infinite numbers of creeping things they never dreamed existed."

Nevertheless, in 1978, when the Tellico Dam was 90% finished, the U.S. Supreme Court ruled that the project had to be stopped because it violated the provisions of the Endangered Species Act. The court ruled that the act made any federal construction project illegal if it jeopardizes the survival of any organism formally classified as endangered under the provisions of the act.

A storm of controversy concerning the act raged in Congress for many months. Verbal battles were waged. Lobbyists pressured members of Congress. The construction industry naturally was interested in weakening the Endangered Species Act so that federal projects could be exempted. Environmentalists, by contrast, were steadfastly opposed to any modification of the act that would lessen its influence in preserving endangered organisms.

Eventually the act was amended. Under the amendment, any requests for exemptions from the act were to be considered by a special high-level review committee. In 1979, the committee ruled to block any further construction of the Tellico Dam on the grounds that the economic benefits resulting from its construction did not justify its costs—in addition to threatening the survival of the snail darter. However, supporters with a financial interest in such multi-million-dollar projects as dams, levees, reservoirs, and highways refused to concede defeat. They succeeded in convincing legislators to amend a public works bill that would permit the completion of the Tellico Dam.

The pro-darter people were understandably dismayed. However, they did the best they could, under the circumstances, to preserve the fish. The entire snail darter population was removed from the Little Tennessee River and introduced into the Hiwassee River nearby.

The Hiwassee River population survived the transplant and appears to be doing well in its new home. Recently, scientists have discovered snail darter populations in four additional creeks in Tennessee and Alabama.

As opponents projected, however, the TVA's Tellico Dam and reservoir have not fared as well. Built to stimulate industrial development and increase recreational fishing, the dam appears something of a failure. So far it has brought no industrial development to the region, and the projected recreation fishery is generously described as "average." Many critics think of it as a "pork barrel" project—built to please local and state politicians at a huge federal expense.

In many places, only tiny patches of natural habitat, called **ecological islands,** now exist amid a sea of crops, pasturelands, towns, and housing developments (Figure 15.3). In the eastern United States, for example, small, scattered patches of deciduous forest are all that remain of once lush and vast old-growth forests. **Habitat fragmentation,** the progressive destruction of habitat that results in the formation of isolated ecological islands, reduces populations of wild plants and animals, making them vulnerable to extinction. Chandler Robbins of the U.S. Fish and Wildlife Service found that fragmentation of the forests in Maryland, New York, and Pennsylvania has drastically reduced the number of birds that once bred in them. The long-distance migrants such as vireos, tanagers, and orioles have been most severely affected.

Princeton biologist Robert MacArthur studied species diversity on naturally occurring islands and found that, all other things being equal, the smaller the island, the fewer species it supports. Other scientists found that this conclusion is also valid for terrestrially based ecological islands cut off from similar habitat by human activities. Thus, a 10-hectare plot of

FIGURE 15.3   Aerial photo showing very little of the land in its original state (forest). Farms, towns, roads, and other forms of development often leave only small islands of natural habitat, which support far fewer species than larger, undisturbed tracts of land.

deciduous forest cut from a continuous forest supports far fewer species per hectare than a 10,000-hectare plot.

At least three reasons explain the reduced species diversity on ecological islands. First, small plots may not contain enough room and food for some species. To access seasonally abundant and widely dispersed foods, such as berries, grizzlies must travel great distances. In less productive habitat in the Rocky Mountains, for instance, a female grizzly bear requires a 500-square-kilometer (193-square-mile) range, and a male requires up to 1,000 square kilometers (386 square miles) to find the food it needs, according to the Eastern Slopes Grizzly Bear Project. Anything smaller is inadequate.

Second, small habitats may reduce the number of organisms in a given population below the critical size needed to reproduce. For instance, the now-extinct passenger pigeon once roamed in flocks that contained many millions of birds (Figure 15.4). Commercial hunters eliminated the huge flocks in mass slaughters to provide food for urban dwellers (Case Study 15.2). Excessive hunting accompanied by heavy deforestation spelled doom for the bird. By 1878, only 2,000 birds remained, in flocks too small to reproduce successfully. The population had fallen below the **critical population size,** the size needed to support a breeding population. It could never recover. Habitat destruction has the same effect on other species.

Third, tiny habitats may promote extensive inbreeding—that is, mating with close relatives. Inbreeding often results in inferior offspring.

A related problem that occurs when a population reaches a low level is that smaller populations have less genetic diversity. **Genetic diversity** is the amount of genetic variation in a population. **Genetic variation,** differences in the genetic makeup of the members of the population, results in slight variations in the structure, function, and behavior of organisms. Genetic variation means that not all members of a population are identical. For example, in a population of plants, some may be better able to survive drought or may produce more seeds. The greater the genetic diversity, the more likely it is for a species to survive and reproduce when environmental conditions change. When a population is reduced, its genetic diversity declines, and its options for survival may decline, too. In the late 1800s, for example, the northern elephant seal was hunted extensively, nearly to the brink of extinction. When hunting pressure stopped, the 20 or so remaining seals bred, which eventually resulted in all 150,000 of the modern descendants. Needless to say, they're all fairly similar genetically. This, in turn, makes them more vulnerable to environmental change.

Habitat fragmentation leads to **faunal collapse,** a decrease in animal species. Human activities may undermine the survival of species living in fragmented habitat. Chemical contaminants in the air, water, and soil, for example, may hinder reproduction or kill organisms outright. Global warming may alter the distribution of plants and eliminate the animal species dependent on them. Increased ultraviolet radiation resulting from ozone depletion may destroy other plants or increase rates of cancer and mutations among animals. "The cumulative effects of such changes," notes Edward C. Wolf of the World-Watch Institute, "can alter ecosystems in ways that increase the vulnerability of plants and animal species to extinction."

To protect biological diversity, human society has embarked on an ambitious program of park designation. Today, 425 million hectares (174 million acres) of land are protected the world over. However, park protection is woefully inadequate. By some estimates, three times as much parkland as currently exists would be required to protect samples of each of Earth's major ecological

FIGURE 15.4 Shooting "wild pigeons" in Iowa, an illustration copied from *Leslie's Illustrated Newspaper* of September 21, 1867. Note the gunner firing into the densely massed birds. Over 100 birds are resting on the bare branches of the oak in the background.

CASE STUDY 15.2   THE PASSENGER PIGEON: THE MANY CAUSES OF EXTINCTION

The passenger pigeon (*Ectopistes migratorius*) was once the most abundant bird on Earth. Early in the 19th century, the renowned ornithologist Alexander Wilson observed a migrating flock that streamed past him for several hours. Wilson estimated the single flock to be 1 mile wide and 240 miles long and composed of about 2 billion birds. (The population of this flock was roughly ten times the total North American waterfowl population today.) Yet not one passenger pigeon is left.

What factors contributed to the passenger pigeon's extinction? First, many potential nest and food trees were chopped down or burned to make room for farms and settlements. The pigeon fed extensively on beechnuts and acorns; the single flock observed by Wilson could have consumed 17 million bushels per day.

Second, disease may have taken a severe toll. The breeding birds were susceptible to infectious disease epidemics because they nested in dense colonies.

Third, many pigeons may have been destroyed by severe storms during the long migrations between the North American breeding grounds and the Central and South American wintering region. Cleveland Bent cites a record of an immense flock of young passenger pigeons that descended to the surface of Crooked Lake, MI, after becoming bewildered by a dense fog. Thousands drowned and lay a foot deep along the shore for miles.

Fourth, their low biotic potential may have been a factor in their extinction. Although many birds, such as robins, lay 4 to 6 eggs per clutch, and ducks, quail, and pheasants lay 8 to 12 eggs, the female pigeon produced only a single egg per nesting.

Fifth, the reduction of the flocks to scattered remnants possibly deprived the birds of the social stimulus requisite for mating and nesting.

Sixth, one of the most important factors in the bird's demise was intense pressure from market hunters. Commercial hunters slaughtered the birds in their nests. Every imaginable instrument of destruction was employed, including guns, dynamite, clubs, nets, fire, and traps. Over 1,300 densely massed birds were caught in one pass of the net. Pigeons were burned and smoked out of their nesting trees. Migrating flocks were riddled with buckshot. Over 16 tons of shot were sold to pigeon hunters in one small Wisconsin village in a single year. Pigeon flesh was considered both a delectable and a fashionable dish in the plush restaurants of Chicago, Boston, and New York. Sold for two cents per bird, almost 15 million pigeons were shipped from a single nesting area at Petoskey, MI, in 1861.

The last wild pigeon was shot in 1900. Martha, the last captive survivor, died in the Cincinnati Zoo on September 1, 1914, at the age of 29 (Figure 15.5).

FIGURE 15.5   The last living passenger pigeon, Martha. When Martha died in the Cincinnati Zoo on September 1, 1914, a unique organism was removed from the human ecosystem forever.

zones. Obviously, as the previous paragraph pointed out, strict measures to control pollution, especially greenhouse gas emissions and pesticides, are needed to enhance the survival of species living in these protected regions.

Recent studies have shown that many national parks in the United States, long seen as the last hope for America's vanishing wildlife, inadequately protect species diversity. Ecologist William Newmark studied the loss of mammal species in national parks and found an alarming drop in the number of species in all but the largest parks (Table 15.1). Bryce Canyon National Park, one of the smallest, lost 36% of its species. Yosemite, nearly 20 times larger than Bryce, lost 25%. Only the mammoth parks like Yellowstone suffered minor losses.

What lessons can we draw from these observations? Many parks are simply too small to support the diverse array of species that lived in the area before it became a park. If a park is cut off from neighboring areas by development in surrounding tracts, the park becomes an ecological island too small to support the diversity it once enjoyed. What is more, researchers point out, the faunal collapse in the world's parks may be continuing today.

Nowhere is the problem of extinction as critical as in the tropical rain forests, which contain an estimated one-half to two-thirds of the world's species. Widespread deforestation in the tropics is wreaking havoc on native species. In Brazil, for example, 50,000 square kilometers (19,300 square miles) of rain forest are lost each year. Throughout the world, large human populations live near estuaries, bays, and other coastal wetlands. Roads, highways, cities, homes, and airports now occupy space

TABLE 15.1 Faunal Collapse in America's National Parks

Park	Area (Square Kilometers)	Share of Original Species Lost (%)
Bryce Canyon	144	36
Lassen Volcano	426	43
Zion	588	36
Crater Lake	641	31
Mount Rainier	976	32
Rocky Mountain	1,049	31
Yosemite	2,083	25
Sequoia–Kings Canyon	3,389	23
Glacier–Waterton	4,627	7
Grand Teton–Yellowstone	10,328	4
Kootenay–Banff–Jasper–Yoho	20,736	0

Source: Based on W. D. Newmark. "A Land-Bridge Island Perspective on Mammalian Extinctions in Western North American Parks," *Nature,* Jan. 29, 1987.

that once supported wetlands and an abundance of plants and animals. Inland wetlands have not fared any better. From Florida to Wisconsin, farmers have drained swampland and plowed it under. Thus, many of the world's coastal and inland wetlands have already been destroyed. In the Philippines, for instance, 50% of the mangrove wetlands have been filled in or dredged. In southern California, 90% of the salt marsh wetlands have met a similar fate. In the United States as a whole, more than half of the wetlands have vanished. Although the rate of loss has decreased dramatically in recent years (partly because of new laws and regulations, but also because many of the prime wetlands have already been destroyed), we still continue to lose wetlands.

Wetlands provide many benefits to people and the species that live in them, many of which have been discussed in previous chapters. (See Table 9.1 for a summary of them.) As the benefits of wetlands are many, so also are the dangers of destroying them. At least half of the biological production of the world's oceans occurs in coastal wetlands and estuaries. Of the world's commercially important marine fish, 60% to 80% either spend time in estuaries or depend on them for food. And 60% of the fish caught in the ocean by commercial fishermen depend on the estuarine zone—the mouths of rivers and coastal wetlands—at some point in their life cycle. By filling in wetlands and polluting them with wastes from our homes and factories, we do a disservice to the other species that share this planet with us, and to ourselves.

GO GREEN!

Join a local conservation group involved in habitat restoration—planting trees or rehabilitating streams—in your area. Ask your professor for suggestions. If you can't find a group, start one yourself.

The loss of wetlands increases the frequency and severity of floods. In seven states in the upper Mississippi River watershed, nearly 80% of the wetlands have been destroyed. Scientists calculate that if only a portion of those wetlands had been left intact, they could have contained the devastating floodwaters that caused nearly $16 billion in damage in the flood of 1993. They could have also helped reduce devastating flooding in 2008. Loss of wetlands worsens flooding and results in poorer water quality, necessitating more costly water treatment.

Filled wetlands also make poor sites for building, as many a flooded homeowner will tell you. Not only are they prone to flooding, because they're located in floodplains, but homes built on filled wetlands also tend to sink.

Sir Edmund Hillary once noted that environmental problems are really social problems. They begin with people as the cause and end with people as the victims. Deforestation and wetland destruction show this relationship very clearly. Unfortunately, millions of species suffer along with us.

Hunting for Profit and Sport

Commercial hunting has a long history of causing species extinctions and near extinctions. The passenger pigeon, mentioned earlier, became extinct in the early 1900s because of widespread clubbing and shooting by commercial interests—in addition to habitat destruction. The great auk, a large

FIGURE 15.6 The snowy egret. This bird was nearly hunted to extinction in the early 1900s to provide plumes for ladies' hats. Fortunately, a prohibition on hunting allowed this elegant bird species to recover.

penguin-like bird that once lived along the North Atlantic coast, became extinct in 1884 as a result of overharvesting. Sailors killed the birds for meat. The heath hen, a bird similar to the prairie chicken, once lived in an area that stretched from New England to Virginia but was severely depleted by commercial hunters.

At the turn of the 20th century, women flocked to stores to buy fashionable hats adorned with the elegant plumes of the snowy egret. To meet the demand, hunters gunned the birds down unmercifully in Florida, nearly wiping out the population (Figure 15.6). Fortunately, however, when restrictive laws were passed to protect the bird early in the 20th century, the population rebounded.

The bison (commonly known as the buffalo), whose herds once blackened the prairies, remains today in tiny remnant populations. Commercial hunting and habitat destruction caused its decline and contributed to the decline of Native American populations, which were highly dependent on the bison, as well (Figure 15.7). Overzealous whalers drove many species of whale, among them the blue whale and the humpback whale, to the brink of extinction. Protected today, some whale species still remain in jeopardy. Whales are being killed when struck by cargo ships, for instance. Today, numerous populations of some commercially important fish species have fallen to dangerously low levels because of overharvesting (Chapter 12).

Not all species are harvested for their food. The big cats of Africa, for instance, have been severely depleted by commercial hunters to provide furs for fashionable coats. Despite protective laws, poachers still hunt jaguars, cheetahs, tigers, and other furbearers.

FIGURE 15.7 The bison, a multiple-use species. It formed an important base for the culture of the Great Plains Indians. Commercial hunters destroyed the wild herds of bison, killing them for meat and sport, eliminating an important resource of Native Americans.

Several species of African rhinoceros also have been hunted to near extinction. Although the law protects them, rhinos are still killed by poachers for their horns, which are sold to countries such as Yemen and China. The Yemenis, wealthy from oil, carve the horns into dagger handles for businessmen. The Chinese grind the horns to produce an alleged aphrodisiac and a fever-depressing drug that is reportedly useless.

Sport hunting may contribute to the decline of animal populations if their populations are not well managed. For the most part, however, hunting has helped wildlife populations. The United States has over 400 national wildlife refuges and thousands of state wildlife areas all paid for, in large part, by the sale of duck stamps and hunting licenses. Regulations on hunting have helped protect many species from overhunting. Hunting also helps control the population size of deer and other game species, because their natural predators have long been eliminated from most regions.

Introduction of Alien Species

Humans have all too often introduced alien species of animals and plants only to discover, too late, that the anticipated benefits failed to materialize or were greatly offset by the negative effects. A classic example is the introduction of the water hyacinth into Florida. This South American flowering plant was brought in to adorn private ponds, but it was accidentally released into the waterways of Florida. It spread wildly throughout the southern states, clogging rivers and lakes, killing native plants, and making navigation impossible. Several southern states now spend millions of dollars each year to clear waterways of this fast-growing species. Another example is the mongoose, imported from India into the islands of Hawaii and Puerto Rico. A fierce, quick-moving, weasel-like predator, the mongoose was brought to these islands to control rats, which caused extensive damage to sugar cane. Unfortunately, the people in charge of this introduction had not studied the animal very carefully. Shortly after the mongoose arrived, they found that it hunts primarily during the day. The rat, on the other hand, is nocturnal. As a result, mongoose–rat encounters were rare, and few rats were killed. Unfortunately, the mongoose soon began to prey on ground-nesting birds. Some, including the Newell's shearwater and the dark-rumped petrel, were eradicated from the island of Molokai. Others, such as the Hawaiian goose, were driven to the brink of extinction.

As this example shows, islands are extremely vulnerable to alien species. The reason is that native species are often ill equipped to cope with introduced species, especially predators, and the habitat is too limited for them to escape the pressure exerted by the newcomers. The Hawaiian Islands have been particularly hard hit by alien species. Before humans settled there, the islands lacked any natural mammalian predators. Many bird species that had lived on these volcanic islands for hundreds of years had lost their ability to fly. Of what use are wings if there's plenty of food and there are no natural predators? Flightless birds have no other choice but to nest on the ground. After humans settled the islands, their dogs, pigs, and goats decimated the populations of the flight-less birds. The birds could easily be clubbed to death, and pigs raided their nests.

Alien species are sometimes introduced intentionally but may also be introduced accidentally. Dutch elm disease was introduced accidentally into the United States, as was the West Nile virus. Dutch elm disease destroyed nearly all of the nation's majestic elms. West Nile, accidentally introduced in 1999, is spreading throughout the United States, killing birds such as crows, ravens, magpies, and bluejays by the thousands. It has also infected thousands of people, creating no symptoms in some instances or mild flulike symptoms that last a week or less in others. In some cases, however, the virus causes infections of the brain and can kill its victims.

Alien species are sometimes assimilated into a new ecosystem without a wrinkle in the ecological fabric. In other instances, they perish because their new environment lacks essential resources or because environmental conditions differ too much from their native habitat. Many hardy species, however, tend to thrive in new environments. The zebra mussel that is spreading through the waterways of the United States and Canada, which was discussed in Chapter 10, is a good example. Without predators, competitors, disease, or parasites, such species proliferate, interfering with native plants and wildlife—literally reweaving the ecological fabric, turning a rich and varied cloth into a threadbare, often colorless one.

For a discussion of ways in which geographic information systems (GIS) and remote sensing are being used to monitor and control noxious alien weeds in the Rogue River National Forest, see the GIS and Remote Sensing box.

The Many Causes of Extinction

As noted earlier, habitat destruction, commercial hunting, and the introduction of alien species are only three of a handful of factors that influence populations. Pollution, the pet trade, pest control, and predator control can have dramatic effects on wildlife populations even in remote areas. As an example, scientists have recently discovered that 4% of polar bears are contaminated with PCBs, a toxic chemical used to clean nuclear reactors in Russian submarines. This toxicant passes up through the food chain, and today about 4% of all polar bears are unable to successfully breed because of high levels of PCBs in their bodies.

As the human population grows and as more and more countries become industrialized, pollution may take an even larger toll on wildlife populations. Especially harmful to plants and animals will be the changes in weather and ultraviolet light penetration caused by an increase in carbon dioxide and the destruction of the ozone layer, respectively (Chapters 19 and 20).

It should also be emphasized that many species become extinct as the result of not just a single action, but many. The California condor once soared above much of the southern United States from California to Florida. This magnificent bird, which lived for 45 to 80 years, succumbed to habitat destruction, ingestion of lead in its scavenged prey (pollution), and pesticides (pest control). The bald eagle, the symbol of our great nation, was also pushed to the brink of extinction as a

GIS AND REMOTE SENSING

MAPPING NOXIOUS WEEDS WITH GIS

GIS and remote sensing technologies are helping state and federal land managers combat an invasion of weeds that is damaging and displacing native vegetation all across the country. Many of these species have come inadvertently from other countries. Seeds reach North America's shores in seed stock or on livestock. In a new land without natural controls, such as diseases, the newcomers frequently outcompete native plants, spreading widely and causing considerable economic damage.

A weed is defined as a plant that is useless, undesirable, or detrimental. Weed species tend to restrict or interfere with the use of the land. Examples of important weed species in the United States are ragweed, vetch, chicory, kudzu, sesbania, Russian thistle, black mustard, lamb's-quarter, crabgrass, horseweed, wild buckwheat, wild carrot, bull thistle, wild parsnip, Canada thistle, leafy spurge, quack grass, Johnson grass, horsetail, St.-John's-wort, horse nettle, sheep sorrel, wild rose, and sweet clover.

Weeds are commonly classified as either common or noxious. Common weeds are species of weed plants that are readily controlled by ordinary good-farming practices. Noxious weeds are those that are difficult to control because of an extensive perennial root system, effective means of propagation, and adaptability (meaning they do well under a variety of conditions). These and other features make them hardy and invasive.

The U.S. Department of Agriculture (USDA) defines noxious weeds as species of plants that cause disease or injure crops, livestock, or land, and thus are detrimental to agriculture, commerce, or public health. An example of a very invasive weed species that is causing ecological havoc in much of the United States is a stunningly beautiful wetland plant known as purple loosestrife. This plant aggressively displaces native plants, disrupts fish and wildlife populations, and adversely affects agriculture and public recreation.

Noxious weeds have become well established on much public and private land. Because of this, federal and state agencies have implemented numerous programs to control the spread of these foreign invaders and eliminate them wherever possible. Until recently, most efforts involved field surveys that required visual assessment of weed infestation. Maps were then prepared to show the extent of weed growth. Such techniques were crude and inaccurate, depending on a subjective analysis by surveyors.

Today, GIS and remote sensing technologies are helping state and federal agencies understand the extent of infestation and mount strategies to control them. One example of this work is the management of the Rogue River National Forest (RRNF) in southern Oregon and northern California by the U.S. Forest Service (USFS).

In the past, USFS staff of the RRNF identified, located, and treated weedy species that were of particular interest to their land management objectives as the problem arose or was discovered. The approximate locations of infestations were marked on forest district maps, with only sparse recording of data describing the plant, its location, and density. Field workers often reported their findings quite differently. The result was a somewhat confusing, often inefficient way of plotting weed infestation.

Knowing the potential benefits of GIS and digital mapping techniques to their land management mission, the RRNF worked with a private company to develop a standardized noxious-weed database to assist them in developing efficient, effective control strategies for noxious weeds. Because they needed to locate and map weed infestations with great accuracy, Forest Service workers and consultants decided to use GPS technology in their field mapping. GPS stands for Global Positioning System. It uses military navigation satellites orbiting Earth to triangulate the position of GPS receivers on the ground. In this case, workers used handheld GPS receivers to pinpoint infestation.

An advantage of the computer technology aspect of GPS is that the receiver system can be programmed to input data in specific formats in a very standardized and user-friendly way. These programs, called *data dictionaries,* allow workers to enter data on location and other features (density of weed species) in a digital format that is readily transferable to a GIS. Digital photographs were taken at each site as well.

Standardizing data options helped workers establish a uniform system of reporting. After each day of mapping, field workers would download the GPS data, which was then exported to their GIS for integration with their existing database and map analysis system. Aerial photos of the forest already existed and were used to identify the locations of large noxious weed populations.

GIS, GPS, and remote sensing helped the RRNF develop a Noxious Weed Strategy. For the first time in a decade, the RRNF is battling noxious weeds with organized, accurate data. Workers are using integrated weed management strategy that minimizes the use of potentially harmful herbicides.

A few years ago, a television news reporter interviewed a bulldozer operator regarding the imminent destruction of a prairie dog colony in Denver, CO, to make room for a new subdivision. The bulldozer operator rationalized his action by saying it was "survival of the fittest." Another construction worker shrugged his shoulders and said, "What's a few prairie dogs? They're in the way of progress."

Some animal rights advocates were also interviewed and argued that it was wrong to destroy the prairie dogs. At the very least, they said, the animals should be humanely trapped and transplanted elsewhere, even if it cost a little money.

This scenario is typical of what happens when humans and wildlife conflict. It hinges on an issue of rights. What rights do other species have, if any? Do human rights supersede the rights of all other species?

A survey of the thinking on this subject reveals a wide range of ideas. Some people think humans reign supreme. Our rights trump all others. Others argue that the many species that share this planet with us have value and that humans ought to respect that. Part of that value lies in their usefulness to us. They may provide opportunities for hunters or anglers or bird-watchers, who provide valuable revenue to local and state economies. But just saying we ought to respect other species isn't enough to ensure that we will. We ought to exercise more and eat better and respect the feelings of others, but that doesn't mean we will.

Some individuals believe that living and nonliving things have moral standing in their own right. That is to say, they have a right to exist, regardless of their value to us—a viewpoint with which many people are uncomfortable. When proponents of this point of view articulate it, they are often met with disbelief.

In his book, *The Rights of Nature,* historian Rod Nash points out that incredulity met the first proposals to grant independence to American colonists, to free the slaves, to respect Native American rights, and to permit women to vote. Historically, according to Nash, certain groups of people have benefited from the denial of rights to other groups or to nature. Today, in fact, many people see the idea of the rights of nature as a threat to human prosperity and progress.

If you poll environmentalists, you will find diverging views on this issue. To many, it is right to protect and wrong to abuse nature, because the abuse could harm people and their way of life. It denies us the goods and services of nature. At the other end of the spectrum are those who maintain that nature has intrinsic value and intrinsic rights irrespective of human needs. In other words, other species possess a right to exist just as humans do, and the same rights that we attach to people in general should be attached to nature's cornucopia of living things.

Of course, as Nash points out, nature does not demand rights as people do. Wolves and redwoods don't petition for their rights. They can't. Thus, it is up to people to act as moral agents for nature. That is, we humans have a responsibility to articulate and defend the rights of the other occupants of the planet.

This seemingly radical notion of the rights of nature is an extension of the idea of liberty to nature, Nash points out. Liberty is an American tradition. The U.S. Constitution proclaims that we have certain inalienable rights: life, liberty, and the pursuit of happiness. Advocates of the rights of nature simply want to extend these American ideals to other living things.

Radical?

Believe it or not, these ideas are not new. Greek and Roman philosophers spoke of natural rights, called *jus naturale.* Jus naturale held that people had rights based solely on the fact that they existed. The Romans found it logical to assume that other animal species had rights, too, which they called *jus animalium.*

But after the decline of Greece and Rome, Nash points out, nature did not fare so well. Increasingly, people assumed, and still do assume, that nature, animals included, has no rights, and that nonhuman beings exist to serve human beings. This relationship of people to nature emphasized expediency and utility.

Several centuries later, the rights of animals once again came to the forefront, this time in England, over the issue of vivisection—the dissection of living animals. René Descartes (1596–1650) was called on to support vivisection because he believed that animals could not feel pain. Moreover, he argued, animals do not think and therefore cannot be harmed. Others, of course, disagreed.

In 1641, the Massachusetts Bay Colony passed the first law respecting the rights of domestic animals. The law read, in part, "No man shall exercise any Tirranny or Crueltie toward any bruite Creature which are usually kept for man's use."

John Locke, the 17th-century English philosopher, became an important source of thinking on animal rights. He argued that people have certain natural (inalienable) rights by simple virtue of our existence. For example, we share a natural right to continue existing. Interestingly, however, Locke did not assert that nature or animals have natural rights. He argued against cruelty toward animals for the way it affects people, maintaining that cruelty to animals would harden people toward people.

In the early history of the English humane movement, though, it was argued that animals are part of God's creation, so people have the responsibility for being good trustees on God's behalf. All nature exists because of and for the glory of God, the Creator. Proponents of this view argued that God cares as much about the welfare of the most insignificant being as about human beings.

Additional support for the rights of nature came from the philosophy of animism, which holds that a single and

(continued)

continuous force permeates all beings and things. The philosopher Benedict de Spinoza (1632–1677) put forth the notion that every being or object—wolf, maple tree, human, rock, star—is a temporary manifestation of a common God-created substance. When a person dies, the matter in his or her body becomes something else: soil and food for a plant, which might nourish a deer and, in turn, a wolf or another person.

Spinoza's understanding of these interrelationships made it possible for him to place ultimate ethical value on the whole rather than on any single transitory part, such as a human life. To Spinoza, there were no higher and lower organisms; his idea of community held no bounds either. A tree or a rock had as much value and right to exist as a person.

Alexander Pope (1688–1744), the gifted British poet, summarized thousands of pages of animist philosophy when he wrote that living things "Are all but parts of one stupendous whole, whose body Nature is, and God the soul."

Things gradually got better for animals and for nature as people began to see that animals can think and feel pain. In 1789, England's Jeremy Bentham, who believed that animals can feel pain, said that pain is bad and pleasure good. He understood how the maximization of happiness could be extended from colonists to slaves to nonhuman beings. He rejected the ability to reason or to talk as an ethical dividing line between people and other forms of life. The question, he said, is not "Can they reason?" or "Can they talk?" but "Can they suffer?" Edward Nicholson, a librarian at the London Institution, wrote in 1879 that saying animals have no powers of reason is inconsistent with common observations of household pets. Granted, an animal's "functions of mind are fewer and its feeling more limited than that of a man, but so are those of a human idiot." And, he maintained, no ethical person proposes to deny such a person's rights to life and liberty. Nicholson argued that animals are capable of experiencing pain and pleasure and so have the "same abstract rights of Life and Personal Liberty" as humans.

Limited as it is, this humanitarian view of animal rights is the fundamental building block that is leading to a growing environmental ethic.

One can see the extension of human ethics to the environment in the writings of Henry David Thoreau. Thoreau referred to nature and its creatures as his "society," thus transcending the usual human connotation of the word. "What we call wildness," he wrote in 1859, "is a civilization other than our own." He regarded sunfish, plants, skunks, even stars as fellows and neighbors. "If some are prosecuted for abusing children, others deserve to be prosecuted for maltreating the face of nature committed to their care."

This kind of thinking is prevalent today. Michael Fox of the Humane Society of the United States has said, "If a human has a natural right by virtue of his very being to be free, then surely this right should be accorded to all other living creatures."

"Firms that destroy the integrity of an ecological system are the same as individuals who make cash withdrawals from a 7-Eleven with a shotgun," said Tom O'Leary, not an environmental radical, but president of a Seattle energy resources company. His sentiments are echoed in other high places.

What can one conclude from these observations?

First, there is no basis for the rights of nature except that which we assign. The whole idea of rights is a human construct. We call the shots. We can deny rights to other species, or we can assign them on the basis of existence, pain, thinking ability, or God spirit. They can be partial rights or complete rights or something in between. The choice is ours.

What do rights imply? Again, that's up to us. They can be sentiment without action, or they can call for the complete negation of humanity. We believe that the goal should be to find some place on the continuum that works for us and for the planet and its millions of creatures.

Source: Adapted with permission from R. F. Nash, *The Rights of Nature* (Madison: University of Wisconsin Press, 1989).

result of numerous factors, including habitat destruction, shooting, and pesticides. Fortunately, protective measures have helped the species recover. (Some of the ethical considerations that arise from extinctions due to human pressures are discussed in Ethics in Resource Conservation 15.1.)

GO GREEN!

Use energy and other resources, such as water, efficiently. Energy efficiency measures help reduce pollution and habitat destruction, both of which are beneficial to wildlife. Water conservation helps to ensure adequate water flow in local streams and rivers, vital for fish and other species.

Traits of Vulnerable Species

Making matters more complicated, some species have attributes that make them more vulnerable to extinction than others. Some of the major characteristics of concern are specialization, low biotic potential, and nonadaptive behavior.

Specialization **Specialists** are organisms that have rather narrow requirements for reproduction and survival. Because of this, they are extremely vulnerable to extinction.

A good example of a specialist that is highly vulnerable to extinction is a bird known as Kirtland's warbler (Figure 15.8). A tiny bird with a powerful song, Kirtland's warbler has an extremely small breeding range. It is found in 13 counties in Michigan. Within this area, the warbler lives only in jack pine habitat, among trees that are 6 to 15 years old and 2 to 7 meters (6 to 20 feet) high.

FIGURE 15.8 Kirtland's warbler at nest. Nests are located in the protective lower branches of jack pines that are 6 to 20 feet tall.

Although it is a ground nester, the bird's survival depends on trees of this size because their branches extend to the ground, providing protection while the birds are nesting and at other times as well. In younger trees, the lower branches do not provide adequate nesting cover; in pines older than 15 years, the bottom branches become shaded out and die and are no longer suitable for nesting. Before humans intervened, pine growth was held in check by naturally occurring forest fires. Fires wiped out sections of the forest and created new growth, which in a few years provided trees of a suitable age and size. Well-intentioned forest fire protection, however, upset the natural cycle of renewal and caused the forests to age, wiping out the warbler's nesting ground.

Today, the USFS mimics nature by periodically burning sections of the forest in prescribed burns (Chapter 14).

Prescribed burns are small, well-managed fires that, in this case, are used to destroy the mature pines and cause the jack pine tree cones to pop open, releasing seeds. The new seeds grow in the burned patches, ensuring a constant supply of trees of the right age for nesting. The Forest Service and other state and federal agencies are also planting new areas. Moreover, wildlife officials are trapping cowbirds, a bird that poses a threat to warblers. The cowbird lays its eggs in warbler nests. Unaware of the addition, the warbler raises the cowbird young at the expense of its own young.

Clear-cutting, burning, replanting, and cowbird control are now working wonders. The Kirtland's warbler population reached approximately 1,600 in 1998, up from 300 in 1974, according to the U.S. Fish and Wildlife Service. By 2002, its population had increased to 2,100. In 2007, the population was estimated to be around 4,500, the highest recorded since the first census was taken in 1951, according to the U.S. Fish and Wildlife Service. Most of the increase has occurred in areas specifically planted for warblers. This effort not only provides habitat for the warbler, it also provides habitat for a variety of other species of songbirds, plants, and mammals and commercial forest products.

Despite success in the United States, Kirtland's warbler faces another danger: deforestation of its wintering ground in the Caribbean. Without it, the bird remains in peril, a fact that illustrates the importance of international cooperation in protecting a species.

The restricted habitat of some species makes them targets for extinction. So do restricted diets. China's panda, for instance, eats the leaves of certain species of bamboo and little else. If bamboo is destroyed, the panda will vanish (Figure 15.9).

Another example of an endangered specialist is the graceful hawk known as the Florida Everglades kite, now called the snail kite, one of the rarest species of birds in the United States. This graceful bird lives in central and south Florida and in parts of the panhandle of Florida. An important factor contributing to its

FIGURE 15.9 The panda, a specialist that feeds only on one species of bamboo, shown here. Destruction of the bear's only food source threatens to wipe out this magnificent creature. Incidentally, the species of bamboo used for flooring and other building materials (scaffolding) and to provide human food (bamboo shoots) is a different species, not consumed by pandas.

falling numbers has been its highly specialized diet. The kite, so named because it hovers like a kite above the swamplands where it feeds, subsists almost exclusively on apple snails. The snails are dependent on emergent vegetation (aquatic plants that protrude above the surface) such as spike rush, sawgrass, and cattails. These plants enable the snail to climb near the surface to feed, breathe, and lay eggs. The kites pick them off the vegetation and then extract the snails from their shells.

Snail kite populations began to decline in the 1940s as farmers and real estate developers drained and filled in marshland, in the process wiping out the snail's once vast habitat. To date, the Everglades habitat has been cut in half by these and other activities. Further declines have resulted from rising demand for water for irrigation and domestic uses, which lowers water levels in remaining wetlands and reduces snail habitat. Pollution of the waters from dairy and vegetable farms has resulted in a massive die-off of snails, too. In 1972, the snail kite population had plummeted to 65 birds. Since then, the numbers have climbed nicely. In 1999, researchers counted 3,577 kites in Florida, but their numbers suddenly went into decline. By the summer of 2005, there were only about 1,300 in the state. Although the population has declined, it is currently considered to be stable. (The species is well represented in Mexico and South America, too.)

Generalists are organisms that occupy a variety of habitats and eat a number of different foods. Because of their ecological versatility, they generally fare better alongside growing human populations. If their habitat is destroyed, they can move elsewhere. If they lose a food supply, they can shift to another. The coyote of North America is an excellent example. This marvelously adaptable canid is expanding throughout much of the United States, especially the Northeast, filling the empty niche created by the extinction of certain subspecies of wolves. You can even find them occasionally wandering the streets of some major cities in the wee hours of the morning.

Low Reproductive Rates Some animal species are extremely vulnerable to environmental stress, such as storms, drought, and disease, because of their low reproductive rates. The female polar bear, for example, breeds only once every three years and then gives birth to only two cubs. The female California condor lays only a single egg every other year. The problem is further complicated by the fact that condors require six to seven years to reach reproductive age. The slow-moving orangutan of tropical rainforests of Borneo and Sumatra breeds once every seven years. These and other species like them may have a difficult time coping with human-induced stress, that is, human factors that change their environment.

Nonadaptive Behavior The Carolina parakeet, the only parrot native to the United States, became extinct in 1914, when the last survivor died in a zoo. Fruit farmers hunted the parakeet extensively because the birds descended on their orchards in huge flocks, ravaging the trees. However, this exquisite red, yellow, and green "paint pot" might still be with us if it were not for one peculiar trait: When one member of a flock was shot, the remaining birds would hover above it, becoming easy targets for gunners.

Of more recent interest is the red-headed woodpecker, which ranges over two-thirds of the United States. The woodpecker's population has declined in the past few decades in part because the bird has a curious tendency to fly along highways directly ahead of automobiles. Unfortunately, the latter usually win the fatal race.

15.3 Methods of Preventing Extinction

Three major methods are currently used to protect wildlife and plants, not just rare and endangered species, but all species. They are (1) the **zoo–botanical garden approach,** (2) the **species approach,** and (3) the **ecosystem approach.**

The Zoo–Botanical Garden Approach

Were it not for botanical gardens, several plants species would be extinct. Nurtured in climate-controlled facilities, trees that are extinct in the wild today hang on by a thread. Many animal species face a similar future. Pere David's deer, originally from China, continues its existence in zoos throughout the world. California condors have been given a boost thanks to efforts of the Los Angeles Zoo and the San Diego Wild Animal Park. Once widely dispersed over much of the southern United States, the condor had suffered enormous losses. By the early 1980s, there were only 21 condors left in the wild and in zoos. Wildlife officials took steps to solve the problem. They trapped the remaining condors and sent them to the two zoos and began a captive breeding program. By late 1995, the condor population had increased to 105 birds.

In January 1992 wildlife officials began releasing captive-raised condors into the wild in carefully selected sites in California. Their hope was that the birds would establish wild breeding populations and expand to other habitat. Additional birds were introduced into habitat near the Grand Canyon in Arizona.

Much of the initial work done to protect condors depended on state and federal government officials working jointly with zoos. Some time later, the Peregrine Fund, a nonprofit organization based in Boise, ID, became a partner in the costly efforts to save the condor. Thanks to the efforts of these organizations, as of March 2007, there were approximately 279 California condors in the world, about 130 of which were living in the wild in California, Arizona, and Mexico. The condor, still rare by any standard, is facing a much brighter future.

However important they are in preserving the rich biological diversity of the world, zoos and botanical gardens have some major drawbacks. First, they are a last-ditch effort, saving species before their final extinction. Second, many organisms do not do well in captivity. They may not breed, or they may succumb to disease. Animal species accustomed to roaming over many square kilometers may become bored and restless when confined to a few square meters. Some refuse to care for their young. And caring for the offspring of these temperamental creatures is often costly and time-consuming. Third, captive-raised animals may be difficult to release into the wild.

Most zoos have taken important steps to mimic species' habitat and have provided areas that allow animals to range more widely. Their reward: healthier, more productive animals that may help save some of our endangered species.

While harboring and breeding endangered animals in zoos may help save species in the short run, this approach is of limited value—a bit like saving a few of Renoir's and Monet's paintings for the sake of art while allowing the rest to be destroyed. In the long run, however, saving species requires a more permanent solution: rebuilding wild populations in protected habitats. Realizing the importance of habitat, many zoos have taken on this role, too. The San Diego Zoo's condors, as noted earlier, have bred successfully and are now being released into the wild in small numbers. Numerous zoos throughout the world are also cooperating in a program to breed golden lion tamarins that will be released into protected jungles in South America. In recent years, many progressive zoos have begun taking an active role in protecting habitat.

The Species Approach

Most work to save plants and animals has focused on the protection of those that are on the brink of extinction. Species not yet reduced to the critical level need to be protected, too. One of the best approaches to protect a species, whether its numbers have fallen precipitously or not, is through species management programs. In such instances, scientists carefully study the niches of species they want to protect, to determine their habitat, food sources, and other requirements. They then design programs to enhance or expand the resources that the species needs to survive and prosper. This may require that human activities be altered. For instance, a species protection program may call for a ban on habitat destruction or controls on harmful pollutants. In other instances, it may require measures that improve the habitat, such as prescribed burning and predator control. Stream improvements that protect spawning grounds and provide shelter may be needed to protect fish populations. Today, the California sea otter and the bison owe their survival to such actions.

One problem with this method is that it tends to overlook the needs of other species. By narrowly concentrating on one species, society may overlook other species that, in the long run, are more valuable. It is quite possible, for example, that a species that might be of more value to society would be neglected in a program aimed at protecting one that happens to be visually more appealing. For example, many people who would enthusiastically support protecting the grizzly bear would balk or laugh at similar efforts in behalf of the furbish lousewort or some insect of far greater value to society.

Also, a narrow approach overlooks the fact that species are part of a complex ecosystem. To protect a species requires protection of its ecosystem and all the members of it, the subject of the next section.

The Ecosystem Approach

Perhaps the most significant outcome of the science of ecology is the concept of the ecosystem—an interacting and interdependent network of biotic and abiotic factors. Scientists recognize that to protect the Earth's species, we must preserve its ecosystems—a step that will benefit us as well. The ecosystem approach to species protection is perhaps the most effective and least costly means of saving plants, animals, and microorganisms. The concept of ecosystem management was introduced in Chapter 1. We've talked about a form of ecosystem management in Chapters 10 and 11 under the topic of watershed management. One of the key components of ecosystem management is habitat protection.

Habitat Protection By setting aside large areas of habitat that are populated with a sufficient number of species and letting nature take its course, biologists believe we can prevent the steady decline in species diversity. This important measure, however, requires an understanding of the habitat requirements of species in need of protection and the species they depend on. It also requires an understanding of where the most critical wildlife habitats are.

The drawbacks to habitat protection, biologists are now finding, are significant. First, as discussed earlier in this chapter, unless the habitat is extremely large or connected to another natural habitat that is also relatively undisturbed, many species will be lost. A recent study of the tropical rain forests showed that small plots, 1 hectare (2.5 acres), lose all of their primates and other mammals and about half of their birds. Slightly larger plots of 10 hectares do not fare much better. Even 100-hectare plots lose nearly half of their bee species and a few of their primates.

The ecosystem approach works best if special care is taken to set aside enough habitat to retain the original biological diversity. These are called **core preserves.** It also requires protection and careful management of surrounding lands, called buffer zones. A **buffer zone** is a region around a preserve in which limited human activity is permitted. The buffer, as its name implies, protects the core reserve from outside influences. In the tropics, many countries have set aside huge tracts of land for protection. Called **extractive reserves,** these lands are also used by indigenous people for harvesting rubber, fruits, nuts, and other forest products from the forest. Such activities are thought to have little, if any, impact on the forests themselves and therefore serve human needs for food and other materials while protecting native species. Although extractive reserves are a good idea, scientists are finding vast areas of tropical rain forest that are practically devoid of animal life, having been overhunted by native peoples.

Another important element of ecosystem management is the use of **wildlife corridors**—connecting corridors that link similar habitats. Although not a panacea, wildlife corridors permit animals to escape predatory pressure and to expand into new territory. They increase food options and permit an intermixing of populations to preserve genetic diversity in neighboring populations.

Several extremely ambitious projects of this nature are now under way. One project, if successful, will create a series of connecting corridors in Florida that permit the endangered Florida panther freer access to its historical range. Another project in the midwestern United States hopes to eventually create a 362,000-square-kilometer (140,000-square-mile) bison preserve across 10 Great Plains states. In China, officials concerned about the endangered panda bear are expanding preserves, creating additional ones, and connecting them by corridors.

Combined with human population stabilization and reductions in pollution, habitat protection remains one of the sustainable answers to the problem of vanishing species. It may even surprise readers to learn that some parks, including New York City's massive Central Park, play a key role in protecting species. In this case, many birds use the park during the annual migration (Figure 15.10). It provides an important resting spot. Efforts are now under way by nonprofit conservation organizations and government entities to identify and protect such areas, known simply as Important Bird Areas.

FIGURE 15.10   Central Park, New York City. Located in one of the most heavily populated cities in the world, Central Park is one of thousands of designated Important Bird Areas throughout the world. IBAs provide places for birds to rest and feed during migration and provide nesting sites as well for birds that spend their summers in the north.

Important Bird Areas (IBAs) refer to any habitat that is valuable to birds. An IBA may be a great nesting area or a convenient resting spot for migrants. The first IBA program emerged in Europe in 1985. Established by an international conservation organization, BirdLife International, the IBA program has spread to over 100 countries in Asia, Africa, the Middle East, and the Americas. Canada has an extremely active program with dozens of IBAs spread throughout its vast territory. Today, more than 8,000 sites in 178 countries have been identified as IBAs, accoridng to BirdLife International. Hundreds of these sites and millions of acres have received better protection as a result of this program. In the United States, at this writing (July 2008), there are more than 2,100 IBAs in 41 states.

Interestingly, IBAs are almost always private ventures, involving nonprofit organizations such as the Audubon Society and private landowners. Individuals nominate sites that are essential to the long-term survival and reproduction of naturally occurring bird species. In Canada, submissions are made to Bird Studies Canada, located in Port Rowan, ON. If the land is selected, it is added to the list. Nonprofit organizations and other interested parties in the region work together to determine the type of protection and management as well as those who will assume responsibility for the site. Conservation plans are drawn up in many cases.

Recent studies have also shown that many important wildlife areas in the United States are on private property. For example, vast government-owned tracts of land in the West, while important, may not be nearly as species-rich as private holdings. These biologically rich areas need to be protected to ensure the survival of many species. Some conservation groups such as the Nature Conservancy purchase such lands. Individual landowners are often inspired on their own to protect their lands, but in many cases, some form of financial incentive is needed to ensure protection. Some state governments, for instance, pay for **conservation easements.** They give the farmer money to protect the land, paying him or her for the development rights. The farmer or rancher still owns the land and can pass it on to his or her heirs, but the land cannot be developed. The development

rights are owned by the state. Thus, the farmer or rancher sets the land aside in perpetuity.

Habitat Restoration   Another element of the long-range plan is restoring damaged lands. In the Amazon Basin, for instance, at least 15 million to 17 million hectares (37 to 42 million acres) of forest have been converted to cropland and pasture. Approximately half of this land has already been abandoned because the poor soil of tropical forests lasts only four to eight years when planted in crops or grazed by livestock. Refurbishing this land and other areas like it throughout the world could greatly slow the rate of loss of biological diversity.

But forest regeneration may be lengthy. The larger the disturbance, the slower is the recovery. A large clear-cut in the tropics may take 150 years to fully restore itself. Already a number of studies suggest that humans can accelerate forest regrowth and that careful work can reestablish a forest's full ecological diversity. But the costs could be exorbitant. Severe erosion must be stopped to prevent the soils from vanishing. Native species will have to be reintroduced to start natural succession.

In 1985, Rajiv Gandhi, then India's prime minister, established a program of tropical reforestation. Nearly 100% of India's large, intact forests have been cut down. The government hopes to replant 5 million hectares (12 million acres) a year but will limit its plantings to a few species of trees that can be used as food for livestock and fuel to supply the needs of India's rural poor. This effort could help reduce soil erosion and rural despair and would benefit wildlife, but it will not come close to restoring the full ecological diversity of the natural forests.

Some observers point out that India could easily broaden its project to restore some of its abandoned land to natural forest. Research in a variety of locations indicates that native populations can restore forests to near-natural conditions and that these forests can provide food for rural populations on a sustainable basis. For example, researchers in Mexico found that descendants of the Mayans protected and cultivated forests containing a variety of fruit- and nut-bearing trees. These forests, while not identical to native forest, were in many ways similar and supported a variety of species. Similar practices have been observed in Brazil, Colombia, Java, Sumatra, Tanzania, and Venezuela.

Habitat protection and restoration are occurring throughout the world. In the United States, much of our grassland biome has been plowed under for farms. Much of the deciduous forest biome has been leveled for pastures, farmland, towns, and cities. Conservation organizations, such as the Nature Conservancy, have taken an active role in setting land aside to protect plants and animals. And efforts are being made to restore native prairie, although on a small scale. Scientists believe that native prairie vegetation may prove to be one of the best long-term rotation crops for farmland that has been severely compacted, eroded, and depleted of its nutrients.

Protecting Keystone Species

In a stone archway, one stone located in the middle of the arch, the keystone, holds the others in place. In ecosystems, scientists are finding that single species have extraordinary influence on the well-being of others. These are called keystone species. Technically, a **keystone species** is one that, if lost, results in the loss of many others. It may, for instance, be a key food source, as in the case of the fig trees in tropical rain forests. They provide a steady source of food during times when other fruits are not available for monkeys and birds. (Most fruits are available only nine months a year; the fig produces fruit year round.) Or a keystone species' home may serve as a home to others. In the southeastern

FIGURE 15.11   Gopher tortoise. This animal lives in Florida, the extreme southern tip of South Carolina, Georgia, Alabama, and Louisiana, and is one of many keystone species, organisms whose existence is vital to the well-being of many other species in the ecosystems in which they live. Its burrows are home to at least 37 other species. It is listed as a threatened species throughout much of its habitat.

United States, the gopher tortoise digs burrows in the sandy soil that many other species live in or seek shelter in (Figure 15.11). So important is the gopher tortoise that if it vanishes, at least 37 other species will disappear, in part because of the loss of shelter.

Understanding ecosystems and the importance of various species is therefore very important to proper management. Many little-known species may turn out to be extremely important for the ecological integrity of a region. Special efforts must be made to identify and protect such species. Unfortunately, most conservation money is spent on nonkeystone species.

Improving Wildlife Management and Living Sustainably

As noted earlier, wildlife managers take steps to improve habitat and protect individual species. More and more managers are practicing ecosystem management. Efforts to better regulate commercial harvesting of fish and other species are also badly needed. Good scientific knowledge of the ecology, population dynamics, and sustainable harvest levels of commercially important species is vital to this effort, as are strong means of enforcing quotas. Cooperation on the part of individuals and companies that harvest fish and other species would go a long way, too.

Individuals, businesses, and governments can also contribute by creating a more ecologically sound society. Efforts to increase recycling, to use resources with much greater efficiency, and to tap into renewable energy supplies all reduce the pressure on natural systems. Population stabilization and growth management also must occur. Only by scaling back human impact can we ensure the well-being of the millions of species that share this planet with us.

15.4   Endangered Species Act

The United States has long been a leader in protecting endangered species. America's first concerted effort to protect endangered species came in 1973, when the U.S. Congress passed the **Endangered Species Act.** This monumental and still-controversial act has helped to thwart the loss of species in the United States and abroad. It has also become a model for other countries.

The act requires the U.S. Fish and Wildlife Service to identify species that are **endangered,** that is, in imminent danger of going extinct, or **threatened,** that is, likely to become endangered in the foreseeable future. These classifications are based on the population size of species under consideration and the rate of decline of their population. In 2008, for instance, the Interior Department declared the polar bear a threatened species under the Endangered Species Act. Declining population and melting Arctic sea ice (loss of habitat) were the main reasons for this decision, making the polar bear the first species to be protected because of the threats posed by global warming. Table 15.2 shows the number of species officially listed as endangered in the United States and in foreign countries.

After a U.S. species is listed as endangered, it is afforded full legal protection under the act. It cannot legally be hunted, killed, or harassed. Individuals cannot be exported. Violators can be fined up to $20,000 and can be imprisoned for one year.

TABLE 15.2 Endangered and Threatened Species (2008)

Item	Mammals	Birds	Reptiles	Amphibians	Fishes	Snails	Clams	Crustaceans	Insects	Arachnids	Plants
Total listings	358	275	119	32	151	76	72	22	61	12	747
Endangered species, total	325	254	79	21	85	65	64	19	51	12	599
United States	69	75	13	13	74	64	62	19	47	12	598
Foreign[†]	256	179	66	8	11	1	2	—	4	—	1
Threatened species, total	33	21	40	11	66	11	8	3	10	—	148
United States	13	15	24	10	65	11	8	3	10	—	146
Foreign[†]	20	6	16	1	1	—[*]	—	—	—	—	2

[*]Dash represents zero.
[†]Species outside the United States and outlying areas as determined by the Fish and Wildlife Service.

The act also bans the importation of endangered species or their products from outside the United States. Recognizing the importance of habitat protection, Congress also directed the Department of Interior's Fish and Wildlife Service to identify the habitats of endangered species. Money was provided for habitat purchase.

The Endangered Species Act also promotes the protection of the habitat of endangered species by prohibiting federal projects (dams and highways, for instance) or federally funded projects in areas deemed to be critical to the survival of endangered species. Since the act was passed, thousands of projects have been modified to protect endangered species, with little or no problem. The exception to the rule is the TVA's controversial Tellico Dam, whose construction was temporarily stopped when scientists discovered the tiny snail darter in the stream to be dammed. The ensuing controversy was described in Case Study 15.1. Logging of old-growth forest in the Pacific Northwest was curbed (not stopped) because of concerns over the spotted owl, which was listed as an endangered species in 1989.

The Endangered Species Act is considered by some to be a very effective tool in protecting species from extinction. Had the law not been enacted, a number of species might not have survived. Russell Peterson, formerly president of the National Audubon Society, agrees in part. He notes that the Endangered Species Act has been reasonably successful in protecting endangered species, but that it lacks the funding needed to restore and set aside habitat to protect species in the United States. Furthermore, he contends, the act "fails to address the threat of extinction where it is greatest: In less developed countries, that threat is now being addressed by numerous agencies and private organizations." However, the specter of species extinction will not go away easily, and much needs to be done here and abroad to protect the world's rich diversity.

Despite its successes, the Endangered Species Act has long been under attack by pro-development interests. In recent years, they have tried to make economics a factor in classifying a species. They argue that if protecting an endangered species could result in economic hardship, the listing should be denied. Alternatively, when development of private property is prevented because of the presence of an endangered species, some argue that the private landholder (such as a developer) should be reimbursed for the legal "taking" of his or her property. Attempts are also under way to reduce or eliminate the protection of internationally endangered species.

The federal government, under the Clinton administration, also reacted to criticism of the Endangered Species Act. A previously rarely used provision of the act permits private landowners, corporations, state or local government, or other nonfederal landowners who wish to conduct activities on their land that negatively affect a threatened or endangered species to do so as long as they have a permit known as an **incidental take permit.** These permits essentially allow for the potential loss of some endangered species. To obtain a permit, however, the interested party must submit a Habitat Conservation Plan to the Secretary of the Interior or the Secretary of Commerce. **Habitat Conservation Plans** are documents that outline what parties will do to ensure that their activities won't affect the survival or recovery of an endangered or threatened species. (Some plants or animals may be lost, but the species will be protected by the habitat conservation plan. At least that's the theory.) They also outline ways that any adverse effects would be mitigated or offset. If these conditions are met, they are approved.

As of July 2008, over 430 Habitat Conservation Plans had been approved, up from 300 in 2000, covering tens of millions of acres and protecting more than 200 endangered or threatened species. Although this sounds great, some critics say that the plans have serious shortcomings. Although activities outlined in the plan could have benefits, submission and approval of a plan does not mean that the permittee will necessarily carry through on it. Critics such as the National Wildlife

Federation also point out that the plans authorize activities that destroy significant amounts of habitat of endangered or threatened species. They argue that the U.S. Fish and Wildlife Service doesn't adequately assess species recovery needs or doesn't assess the habitat destruction that is occurring elsewhere in a species range. Approval of a plan without a full understanding of these factors could result in significant losses. The American Lands Alliance, a nonprofit group in Portland, OR, argues that most forest Habitat Conservation Plans in the West fail to offset the losses of habitat and do little, if anything, to restore degraded habitat. The Alliance says the federal government is, in essence, giving exemptions to wood products companies and other major landowners to build roads, log, and damage habitats of threatened or endangered species on millions of acres of forestland for commercial gain. One of the many problems they cite is the failure of plans to provide habitat before eliminating existing habitat. As a result, many species fail to colonize the new habitat, which is established after the original habitat is destroyed. In some instances, late-succession forests are replaced with new forests that will eventually grow into buffer strips along rivers and lakes. Most often, however, they say the plans fail to provide any mitigation habitat.

"Agencies entrusted to protect animals have allowed driving on Florida beaches where threatened sea turtles nest, the electrocution of rare birds on security fences at California prisons, and the killing of protected salmon in one of southwest Washington's last undammed rivers," according to Robert McClure and Lisa Stiffler, reporters for the *Seattle Post-Intelligencer.* In summary, many of the nation's habitat plans appear to have serious shortcomings that tip the scales in favor of development over endangered species.

Another criticism is that the Clinton administration gave assurances to those whose plans were approved that there would be no additional conservation efforts besides those required by their plans for up to 100 years. This guarantee, sometimes called the "no surprises" policy, means that no changes in the management plan may be required even if further information suggests their need. Many changes are needed to make them work. You can learn about them by visiting our Web site, http://www.prenhall.com/chiras, for links to other sites where the issue is discussed in more detail.

In response to criticism of the act, the National Research Council, which provides scientific advice to Congress, established a committee of scientists to examine the issues and make recommendations. They outlined numerous changes in the act to make it more scientifically sound and economically responsible. For example, they called for faster development of recovery plans. Such plans, they said, should spell out which human activities in and around protected areas are likely to harm recovery and which ones are not—a step that would allow for better economic planning.

The committee also recommended that a core of survival habitat be established as an emergency, stopgap measure when a species is first listed as endangered. This habitat would be able to support the population for 25 to 50 years. After more careful study, scientists could determine the exact dimensions of the critical habitat needed for the species to recover. This might result in either a downsizing or an increase in the protected habitat.

Convention on International Trade in Endangered Species

Protecting wildlife requires efforts by all nations. Many have passed laws similar to the United States' Endangered Species Act. Despite this, species are still vanishing at a remarkable rate. One reason is the legal and illegal trade of plants and animals. The sale of animal skins, live animals, and plants is worth billions of dollars to those who engage in it, both legally and illegally. Recognizing that threat posed by trade, the United Nations Environment Programme, the World Conservation Union, and the World Wide Fund for Nature launched an effort to ban international trade in endangered or threatened species. The result, a treaty known as the **Convention on International Trade in Endangered Species of Wild Fauna and Flora (CITES),** went into effect on July 1, 1975. The treaty is currently endorsed by 150 nations. It bans the hunting, capture, or sale of endangered and threatened species. CITES, the organization that administers the agreement, and the UNCITES secretariate, keeps a list of species that are currently threatened by extinction as a result of international trade. This list is known simply as Appendix 1.

Although efforts by the CITES's nations are ongoing, illegal trade continues. Traders falsify documents so that it appears as if species being shipped have been taken from areas where they are relatively abundant. Species are often mislabeled. Wildlife officials are limited in number and usually underpaid and can be bribed to look the other way. Enforcement is often lax. Penalties are mild.

As you have learned in this chapter, protecting species, especially those that are already threatened or endangered, will require a mix of actions by citizens, businesspeople, and government officials. The task may not seem important to those who fail to understand the importance of ecosystems to human health and well-being.

Summary of Key Concepts

1. Wildlife and plants enrich our lives in many ways. Beyond the countless aesthetic benefits are purely economic ones derived from the sale of medicines, ointments, and foods developed from and sometimes extracted from natural ecosystems. Sports enthusiasts and nature lovers spend billions to hunt, fish, and photograph nature's rich offering. Wild plants and animals provide food and innumerable ecological benefits, such as flood protection, erosion control, and nutrient recycling.

2. An estimated 30 million to 80 million species share this planet with us, but this rich and varied biological world is fast disappearing. Currently, estimates suggest that approximately 100 species go extinct every day. Most species are lost from the tropics.

3. Preserving biological diversity has become one of the most important and pressing environmental concerns of our time and is essential for creating a sustainable society.

4. Most of the species that have become extinct since the beginning of the Industrial Revolution have done so not because they have been unable to adapt to natural changes

in the conditions on Earth, but rather because they have been unable to adapt to changes wrought by humans.

5. The major causes of extinction are (1) habitat alteration and destruction, (2) commercial overhunting and overharvesting, (3) the introduction of alien species, (4) pest and predator control, (5) pollution, and (6) the pet trade. Controlling these causes, especially the first three, could greatly improve the chances of survival for many of the world's vanishing species.

6. In many places, only tiny islands of natural habitat, called ecological islands, now exist amid a sea of crops, pasturelands, towns, and housing developments. The smaller the island, the fewer species it can support. Thus, the progressive fragmentation of Earth's biomes is gradually eroding its biological diversity. Ecologists are especially concerned with the damage now occurring in tropical rain forests.

7. Biologists are also concerned about the impact of destroying wetlands. Already many of the world's wetlands have been destroyed by human development. The wetlands provide many direct and indirect benefits to human society and are an important habitat in need of protection.

8. Commercial hunting has a long history of causing species extinctions and near-extinctions. Sport hunting may contribute to the decline in animal populations if their populations are not well managed. For the most part, however, hunters do more good than harm.

9. Humans have frequently introduced alien species of animals and plants only to discover, too late, that the benefits never really materialize. In some cases, the introductions backlashed as the alien species proliferated and displaced native plants and animals. Islands have been particularly hard hit by the introduction of alien species.

10. Many species become extinct not as a result of a single action but of many. And some species have traits that make them more vulnerable than others, such as specialization, low biotic potential, and nonadaptive behavior.

11. Three major methods are now used to protect plants and wildlife: (1) the zoo–botanical garden approach, (2) the species approach, and (3) the ecosystem approach.

12. The zoo–botanical garden approach was once narrowly confined to raising endangered species in captivity, but has been expanded to include programs in which endangered species that are bred in captivity are released in the wild.

13. The species approach involves plans developed to protect individual species by controlling human activities that threaten them and by improving habitat.

14. The ecosystem approach is perhaps the most effective and least costly means of saving endangered species. By setting aside large areas that are populated with a sufficient number of each natural species and letting nature take its course, biologists believe, we can reduce species extinction. Recent studies indicate, however, that large tracts must often be set aside to preserve species diversity.

15. The ecosystem approach also entails plans to restore damaged lands and waters, to better manage entire ecosystems, to establish preserves surrounded by buffer zones, and to create wildlife corridors.

16. Extractive reserves are also a valuable tool in protecting wildlife habitat. Extractive reserves allow humans access to natural areas to sustainably harvest nuts, fruits, and other plant products. Such uses can safeguard wildlife and plant populations.

17. Keystone species are often key components of their environment, and their loss can result in the loss of many other species. Protecting keystone species is obviously very important to the health and welfare of the world's ecosystems.

18. The Endangered Species Act of 1973 represents the United States' first concerted effort to protect endangered species. The act requires the U.S. Fish and Wildlife Service to identify endangered or threatened species and protect their habitat through a number of means. It also bans the importation of endangered or threatened species and has helped save many species from extinction.

19. The Convention on the International Trade in Endangered Species is another valuable tool. This international agreement bans the import of endangered species, although it is not uniformly or vigorously enforced in all countries.

Key Words and Phrases

Adaptive Management
Alien Species
Biological Infrastructure
Biotic Impoverishment
Buffer Zone
Conservation Easement
Convention on International
  Trade in Endangered
  Species of Wild Fauna and
  Flora (CITES)
Core Preserves
Critical Population Size
Ecological Island
Ecosystem Approach to
  Species Protection
Ecosystem Management
Ecotourism
Endangered Species
Endangered Species Act
Extinction
Extractive Reserve
Faunal Collapse
Generalists
Genetic Diversity
Genetic Variation
Habitat
Habitat Conservation Plan
Habitat Fragmentation
Important Bird
  Areas (IBAs)
Incidental Take Permit
Keystone Species
Prescribed Burn
Specialists
Specialization
Species Approach to Species
  Protection
Threatened Species
Wildlife Corridor
Zoo–Botanical Garden
  Approach to Species
  Protection

Critical Thinking and Discussion Questions

1. Suppose that a classmate of yours said, "I couldn't care less about the extinction of some weed or bug in Africa." How would you respond?

2. Describe the value of plants and animals to modern society.

3. Discuss the major hypotheses that attempt to explain why so many species have become extinct since life began on Earth.

4. Using your critical thinking skills, analyze the following statement: "Extinction is a natural phenomenon, so we really shouldn't worry about it."

5. List the causes of extinction. Which ones are the most important?

6. What is habitat fragmentation, and why is it so harmful to species diversity?

7. Suppose that a population of 100 passenger pigeons was discovered in Illinois. Would you expect this population to be able to survive, knowing what you do about the pigeon's reproductive requirements?

8. Give some examples in which commercial hunting has driven a species to extinction or near extinction.

9. Why are alien species such a threat to native populations?

10. Some species are vulnerable to extinction. What makes them so?

11. Suppose that fire was excluded from the Kirtland warbler's habitat. Would this promote or hinder the species' survival? Explain your answer.

12. In what ways can zoos help preserve endangered plants and animals?

13. What is meant by the ecosystem approach to species protection? Give some examples.

14. How would you respond to these questions put to you by a frustrated citizen: (a) "Who's more important, humans or alligators?" (b) "Wilderness is something I'll never be able to visit and don't care to, so why should we save it?"

15. Using your critical thinking skills and your knowledge of ecology, debate the following statement: "The Endangered Species Act is an impediment to progress and should be severely watered down. Environmental protection will only harm our economy."

16. Debate the following statement: "Animals and plants have a right to live and prosper, and human activities should be curbed to protect them."

Suggested Readings

Brandon, K., K. H. Redford, and S. E. Sanderson, eds. 1998. *Parks in Peril: People, Politics, and Protected Areas*. Washington, DC: Island Press. Analysis of trends in park management and implications for protection of biodiversity.

Chester, C. C. 1996. Controversy over Yellowstone's Biological Resources. *Environment* 38(8): 10–15, 34–36. Interesting insights into the value of biological resources.

Cox, G. W. 1999. *Alien Species in North America and Hawaii*. Washington, DC: Island Press. Valuable resource on invasive species.

Cubie, D. 2007. A Rare Species Gets a Second Chance. *National Wildlife* 45(1): 12–13. Chronicle of the successful efforts to save a species and reintroduce it into the wild.

Dupree, J. 2007. Coral Crisis. *National Wildlife* 45(4): 22–30. An in-depth look at the ongoing destruction of coral reefs due to climate change and other factors.

Dupree, J. 2007. Cat on a Collision Course. *National Wildlife* 45(1): 22–30. Description of efforts to save a species despite continual loss of habitat.

DiSilvestro, R. 1996. What's Killing the Swainson's Hawk? *International Wildlife* 26(3): 38–43. An examination of the effects of pesticide use on migratory birds.

DiSilvestro, R. 2007. Fair Funding for Endangered Species. *National Wildlife* 45(5): 22–30. Description of the importance of adequate funding to protect endangered species.

Freese, C. H. 1998. *Wild Species as Commodities: Managing Markets and Ecosystems for Sustainability*. Washington, DC: Island Press. Valuable reading.

Glick, D. 2007. On Thin Ice. *National Wildlife* 45(1): 22–30. A look at the many threats to polar bears, especially global warming.

Laycock, G. 1996. *The Alien Animals*. Garden City, NJ: Natural History Press. Superb account of the folly of introducing alien species into the United States.

Margolis, R., and N. Salafsky. 1998. *Measures of Success: Designing, Managing, and Monitoring Conservation and Development Projects*. Washington, DC: Island Press. A guide for development projects that protect the environment.

Mattoon, A. 2000. Amphibia Fading. *World-Watch* 13(4): 12–23. Disturbing article about the worldwide disappearance of amphibians.

Monks, V. 1996. The Beauty of Wetlands. *National Wildlife* 34(4): 20–27. An excellent article with new insights into wetlands and the value of protecting them.

Mooney, H. A., and R. J. Hobbs. 2000. *Invasive Species in a Changing World*. Washington, DC: Island Press. Description of how changing patterns of global commerce result in the spread of alien species, and how climate change and other factors influence this spread.

Noss, R. E., and A. Y. Cooperrider. 1994. *Saving Nature's Legacy: Protecting and Restoring Biodiversity*. Washington, DC: Island Press. The case for protecting biodiversity and a description of how it can be done.

Perrine, D. 2008. Rescuing the Reefs. *National Wildlife* 46(1): 30A–30G. Description of measures in Belize to protect its marine fishes, an effort that could serve as a model for other Caribbean nations.

Plotkin, M., and X. Famolare, eds. 1992. *Sustainable Harvest and Marketing of Rain Forest Products*. Washington, DC: Island Press. Outline of ways of sustainably harvesting tropical rain forest products while protecting native species.

Posey, D. A. 1996. Protecting Indigenous Peoples' Rights to Biodiversity. *Environment* 38(8): 6–9, 37–45. A look at the role of indigenous people in protecting biological resources.

Quammen, D. 1997. *Song of the Dodo: Island Biogeography in an Age of Extinctions*. New York: Touchstone. A massive book with much to offer on ecological science and extinction.

Rosenthal, D. 1996. Showdown in Zimbawe. *International Wildlife* 26(6): 28–35. A look at what people can do to save wildlife.

Ryan, J. C. 1992. Conserving Biological Diversity. In *State of the World 1992*, ed. L. Starke. New York: W. W. Norton. Description of many ways to protect biological diversity.

Sachs, J. S. 2004. Poisoning the Imperiled. *National Wildlife* 42(1): 22–29. A look at endangered species whose future is threatened by pesticides.

Sharp, E. Alien Invasion: A Great Lakes Dilemma. *National Wildlife* 45(5): 40–47. A look at the effect of 180 alien species introduced into the Great Lakes in the past 200 years.

Sunquist, F. 1988. Zeroing in on Keystone Species. *International Wildlife* 18(5): 18–23. Good reference on keystone species.

Temple, S. A. 1998. Easing the Travails of Migratory Birds. *Environment* 40(1): 6–9, 28–32. Description of the need for coordinated efforts among countries to protect migratory bird species.

Tudge, C. 1992. *Last Animals at the Zoo: How Mass Extinction Can Be Stopped.* Washington, DC: Island Press. Description of how captive breeding programs and restoration of natural habitat can be used to save endangered animals from extinction.

Tuxil, J. 1999. Appreciating the Benefits of Plant Biodiversity. In *State of the World 1999*, ed. L. Starke. New York: W. W. Norton. Great information on the value of plants.

Wilcove, D. 1990. Empty Skies. *Nature Conservancy* 40(1): 4–13. Excellent overview of factors causing the decline of songbirds in the United States.

Youth, H. 2003. Watching Birds Disappear. In *State of the World 2003*, ed. L. Starke. New York: W. W. Norton. Excellent look at the many factors that cause the decline of birds worldwide.

 Web Explorations

Online resources for this chapter are on the World Wide Web at: **http://www.prenhall.com/chiras** *(click on the Table of Contents link and then select Chapter 15).*

16

WILDLIFE MANAGEMENT

Wildlife management may be defined as the planned use, protection, and control of wildlife by the application of ecological principles. Carried out by wildlife managers in state and federal agencies, as well as by private individuals, wildlife management often seeks to achieve its goals through habitat protection or manipulation of habitat to enhance wildlife populations, as you'll see in this chapter. One major function of wildlife management is to protect endangered species, as described in the previous chapter. But wildlife management has other important functions. For example, it protects and enhances wildlife for **consumptive uses**, such as **hunting**, as well as **nonconsumptive uses**, such as wildlife photography and observation. For much of the past century, wildlife managers have emphasized the management of wildlife for the hunter. In recent years, however, more attention has been directed toward management for nonconsumptive uses such as wildlife photography and bird-watching.

Under some conditions, wildlife, though ordinarily a desirable resource, can be harmful to society. Animals as varied as birds and elk can damage agricultural crops, natural habitat, lawns, and gardens, as well as other organisms. They can even pose a danger to people. In such cases, the wildlife manager must develop strategies to control the destructive animals or populations.

Few areas, if any, are off-limits to human activity—areas that we deliberately choose not to enter so that the species living there can flourish unmolested by humans. Many humans think that we own and have the right to dominate every square inch of the Earth. Invading wildlife habitat, such as paving it or turning it into farms, golf courses, housing developments, or parks, is not the only way that an area can become untenable as habitat. Anything that makes it unattractive or unavailable to a given species can cause habitat loss.

It is apparent that the effective management of our nation's wildlife is highly challenging and demands a great range of knowledge and skills. The involvement of well-trained professional wildlife biologists in decision making and action in many diverse arenas is essential to the wise use, protection, and control of our nation's wildlife resources.

16.1 Wildlife

What Is Wildlife?

The term **wildlife**, in its most comprehensive sense, includes all animals on Earth that have not been domesticated by humans. Even insects fall into this category. However, for most professionals in wildlife management, the term is largely restricted to wild vertebrate animals—that is, animals with backbones, including birds, mammals, amphibians, and reptiles. Wildlife includes many game and nongame species. For our purposes, **game animals** are species that are harvested for sport, whereas **nongame animals** are, by subtraction, the majority of species that are not so harvested. Nongame species include songbirds, many rodent species, and most species of amphibians and reptiles. In reality, game animals are designated by legislation. For example, states designate bear and mountain lions as game animals. But the states are not uniform. In some states, mourning doves are considered game birds by law; in others, hunting them is not legal, and they are considered nongame birds.

Wildlife Habitat

Habitat is the general environment in which an organism lives—its natural home. It provides the essentials for survival: cover, food, water, and breeding sites (den, nest, or burrow).

Cover   **Cover** protects animals from adverse weather. Good examples are the dense cedar swamps that protect whitetail deer herds from winter winds and drifting snow. Even the leafy canopies of backyard apple trees are habitat that shields nestling robins from the heat of the midday sun. Cover may also protect wild animals from predators (Figure 16.1). Good examples are a thicket into which a cottontail plunges when eluding a fox and the marsh grasses that conceal a teal from a hawk. Even water may serve as cover—for example, for muskrats and beavers—because it provides relative security from landbound predators including wolves and humans.

Food   Within a single species, **food** preference often varies widely. It also varies in individuals, depending on the health and age of the animal, season, habitat, and food availability.

Birds and mammals spend a great deal of time searching for food. An animal's access to food is influenced by many factors. These include its population density, the weather, and the condition of its habitat (Figure 16.2). Occasionally, when a food source is abundant, an animal will exploit a source even though it is not usually a dietary item. Consider some examples: Even though the green-winged teal primarily consumes vegetation (90% of its diet is plant matter), this duck avidly consumes the maggoty flesh of rotting Pacific salmon when present. Although the lesser scaup (duck) is not normally a scavenger, the stomachs of these ducks feeding at the mouth of a sewer have been found to be filled with slaughterhouse debris and cow hair (as well as rubber bands and paper). A house wren, normally an insect eater, feeds nestlings large quantities of newly hatched trout from an adjacent hatchery.

Animals that consume a great variety of foods are **euryphagous**. The opossum is a good example. It consumes blackberries, corn, apples, earthworms, insects, frogs, snakes, lizards, newly hatched turtles, bird eggs, young mice, and even bats. This euryphagous animal is well adapted to survive because it will eat so many different foods. If one is in short supply, usually it's no problem. The opossum has plenty of additional options.

A **stenophagous** animal, in contrast, maintains a specialized, or limited, diet. Such species are more vulnerable to starvation when their usual foods are scarce. For example, an early freeze that kills off insects frequently causes many swifts and swallows to die of starvation.

Water   Roughly 65% to 80% of wild animal weight is **water**. Water serves many functions. As a major blood constituent, it transports nutrients, hormones, enzymes, and respiratory gases. It also transports wastes from the cells of the body to the organs of excretion and helps to distribute body heat.

FIGURE 16.1   A red fox peering from behind a rock in Kettle Moraine State Forest in Wisconsin. To survive, this species requires a habitat that provides cover, food, water, and adequate breeding sites.

FIGURE 16.2   Parent bluebird bringing a bill full of insects to hungry young. A young bluebird can consume half its weight in insects in one day. Nesting boxes provide additional protection for the young, helping boost the survival rate of this species.

Animals can survive for weeks without food but only a few days without water. In the 19th century, buffalo herds living in the arid western grasslands of the United States traveled many kilometers to find water holes. Mourning doves may fly 50 kilometers (30 miles) from their nest site to a watering place. Dove and quail populations have been increased in the Southwest by the installation of "guzzlers"—devices that collect rainwater (Figure 16.3). Birds and mammals may get their water from dew or may drink it as it drips from foliage and tree trunks after a shower. During the northern winter, when liquid water is scarce, house sparrows and starlings will eat snow. Desert carnivores, such as the rattlesnake, fox, and bobcat, may obtain water from the blood of their prey. Another desert animal, the kangaroo rat, may not need to ingest water during its entire life! It can use the water formed during cellular energy production.

The Edge Effect

The habitat essentials (cover, food, water, and breeding sites) for a given species are rarely all found in a single type of plant community. Usually an animal must rely on two or more plant communities to satisfy its needs. For deer, a clump of evergreens provides cover from storms and predators, a forest margin provides adequate food, and a dense thicket of shrubs provides a fawning site. The region where two different ecosystems, such as marsh and oak woods, come together is called an **ecotone**. Ecologists also refer to it as an **edge**. As a general rule, the greater the amount of edge, the greater the population densities of many species. Take quail, for example, in an area composed of grassland, shrubs, cornfields, and woods. The vegetation shown in Figure 16.4A, with minimal edge, supports only one covey (flock) of quail. In contrast, the vegetation in Figure 16.4B, with much more edge, supports nine coveys, even though the total area of each vegetational type is the same in both cases. Although increased edge is advantageous to many species, it is not always an advantage. For example, it can lead to more nest predation on birds that nest in forests.

Thus, creating edge is a beneficial wildlife management practice for certain game species such as whitetail deer and ruffed grouse, but it may cause a decline in interior species such as the spotted owl and wood thrush. Creating too much edge can produce **habitat fragmentation**, which is usually caused by humans when native vegetation is cleared for human activities such as agriculture, rural development, or urbanization. Habitats that were once continuous become divided into separate fragments and can become small islands isolated from each other by cropland, pasture, pavement, or even barren land. These small fragments are unfavorable for species that require interior habitat.

Corridors

As noted in Chapter 15, **corridors** allow wildlife populations to expand habitat availability. A terrestrial corridor is a narrow strip of land that often differs from the surrounding areas—usually in the type of vegetation found in it. A good example of a corridor is a narrow strip of trees that extends across a large meadow and connects extensive patches of forest on either side of that meadow. A corridor can also be a narrow stretch of grassland running through a forested region. Corridors can be undeveloped areas that connect wildlife habitat within a more developed landscape.

FIGURE 16.3 Wildlife habitat improvement. Chukar partridge (introduced from Asia) have been attracted by a "guzzler," a watering device used in desert country for wildlife.

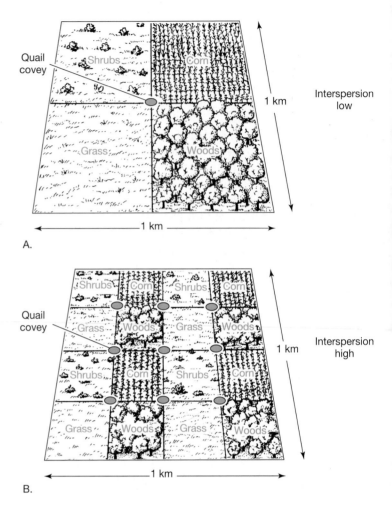

FIGURE 16.4 Impact of edge on population density. (A) This square-kilometer area with minimal interspersion of four habitat types can support only one covey (flock) of quail. (B) The same square kilometer with much greater interspersion of four habitat types can support nine coveys of quail.

Disrupting corridors can have serious consequences for the movement of some animals. For example, in prairie or agricultural regions, wooded corridors next to rivers may be important for migratory birds and other animals.

Home Range

Ecologists define a **home range** as the area over which an animal habitually travels while engaged in its usual activities. The size of a home range can be determined by marking, releasing, and recapturing an animal. Animals can be fed dyed foods that result in colored feces. The home range can then be determined by studying the distribution of droppings. Birds can be individually marked with colored leg bands or spray paint. Small mammals can have their ears notched or toes clipped. Large animals (buffalo and elk) can be tattooed or marked with plastic collars so that visual identification is possible at a distance. Today, a wide variety of animals can be tracked by radio waves. Animals are captured, fitted with a radio transmitter, and then released (Figure 16.5). Scientists can then track the animal by remote sensor.

Herbivores usually have smaller home ranges than carnivores. A plant-eating moose, for example, may have a home range of only 40 hectares (100 acres or 0.4 square kilometer). The omnivorous grizzly bear, in contrast, requires 52 square kilometers (20 square miles). Timber wolves, which are carnivores, have home ranges of at least 100 square kilometers (39 square miles).

Territory

A **territory** is defined as any defended area. Territories are usually defended against individuals of the same species. Many kinds of birds use threat displays (gaping, crouching, fanning the tail) and/or songs to defend territory, rather than fighting. This cuts down on potential injuries.

Territorialism in birds may serve many functions. It can ensure adequate food, establish and maintain the pair bond, spread birds out to control infectious disease, reduce interference with breeding (mating, nest building, incubation), and reduce predation because the territorial birds become familiar with refuge sites.

The size of a bird's territory varies widely, from 0.3 square meter (3.3 square feet) for the black-headed gull to 9,300 hectares (23,000 acres) for the golden eagle (Figure 16.6). The majority of songbirds (like the robin) have a territory size of 0.1 to 0.3 hectare (0.25 to 0.75 acre).

16.2 Types of Animal Movements

For the greater part of their lives, birds and mammals occupy a relatively small area represented by their home ranges or territories. However, under certain conditions, many species move considerable distances from the original home ranges or territories. Such movements promote the survival of the species. Three basic types of movements are dispersal of the young, mass emigration, and migration.

FIGURE 16.5 A female mallard fitted with a radio transmitter. After the bird is released, the radio waves generated by the transmitter will be picked up by the receiving set shown in the foreground. This is part of a University of Minnesota project conducted at the Cedar Creek Wildlife Area in Minnesota.

Dispersal of the Young

The phenomenon of **dispersal of young** occurs with many birds (for example, gulls, herons, egrets, grouse, eagles, and owls) and mammals (for example, muskrats, fox squirrels, and gray squirrels). In a pine–oak forest in central Pennsylvania, for example, up to half of the juvenile ruffed grouse leave their nesting areas, some traveling up to 12 kilometers (7.5 miles). Young bald eagles in Florida move north immediately after fledging, some arriving 2,400 kilometers (1,500 miles) away in Maine and Canada (Figure 16.7). Up to 40% of a wintering muskrat population may disperse in spring. Muskrats are primarily young animals that have been ejected by the established, more aggressive adults. Such dispersals control population densities. Many of the dispersed young move into marginal habitats, where they incur heavy mortality from predation and accidents.

Mass Emigration

Mass emigrations frequently occur when a population has peaked because of extremely favorable conditions (such as abundant food), followed by unfavorable conditions (such as a lack of food). In such instances, the alternatives to starvation are summer dormancy, hibernation, or emigration. Snowy owl emigrations into the United States from the Canadian tundra are correlated with the population crash of their lemming prey. Ornithologists recorded 13,502 snowy owls during the 1945–1946 emigration, which extended as far south as Oregon, Illinois, and Maryland. Twenty-four were observed over the Atlantic. Some were even seen in Bermuda. It is believed that very few of these owls lived long enough to make the return flight to the Canadian tundra the following spring. Many were shot illegally and wound up stuffed with cotton on someone's mantelpiece.

Migration

Two types of migration occur with wildlife species. **Latitudinal migration** (north–south movements) involves changes in latitude, whereas **altitudinal migration** involves changes in elevation.

Latitudinal Migration   Winter bird densities in the southern United States are high because many birds that breed in more northern latitudes temporarily join the permanent residents (Figure 16.8). Foods such as insects, fruits, and seeds are more available in the south than in the snow-covered lands to the north. In spring, however, the increasing day length eventually triggers hormonal secretions that stimulate migration. Presumably the northern habitats have a higher carrying capacity for the migrants and their future offspring. In far northern latitudes during the summer, such as the tundra of northern Alaska, migrants encounter much more daylight in one 24-hour cycle for feeding their young. Biologists believe that the exploitation of two different habitats (winter and summer) may ensure a more balanced supply of vitamins and minerals.

Altitudinal Migration   Latitudinal migrants move thousands of kilometers to find warmth and food. In contrast, altitudinal migrants achieve the same result simply by moving a few kilometers down the mountainside. The elk herds of the Rocky Mountains, for example, ascend the mountains in spring, keeping pace with the receding snow line, and spend the summer at the relatively cool upper levels. When the first snows cover their food supplies, the elk move down to the valleys for the winter. Herds of bighorn sheep make similar migrations.

16.3   Mortality Factors

In the previous chapter, we saw that the population level of any species at a given time is the expression of two opposing forces—the biotic potential, which tends to push the population up, and the environmental resistance or mortality factors, which tend to force it down. Let's examine a variety of mortality factors that affect populations of deer and waterfowl.

Mortality Factors Affecting Deer

Starvation   Winter is a critical season for deer in the northern states, because available food is extremely limited. Herbs, mosses, fungi, seedlings, and stump sprouts are often covered by snow. Under such conditions, the only available plant

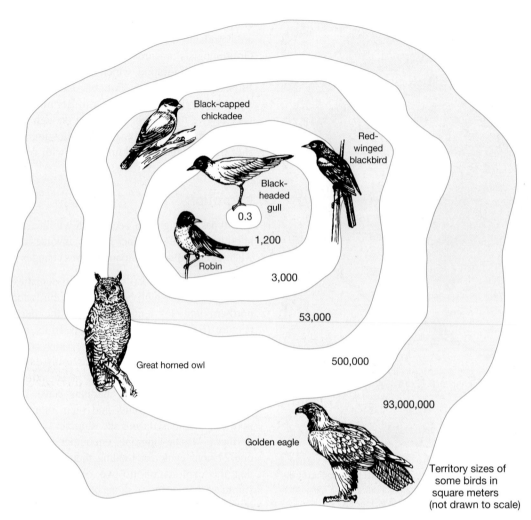

FIGURE 16.6 Territorial sizes of several species of birds, ranging from 0.3 square meter for the black-headed gull to 93 million square meters for the golden eagle.

materials are buds, twigs, and the foliage of conifers, such as white cedar and pine. If the deer populations grow too large, they will consume all the available browse up to the height they can reach when rearing up on their hind legs (Figure 16.9). As a result, a conspicuous **browse line** will form at a height of about 1.5 meters (5 feet), a definite warning to the wildlife biologist that the deer herds have overtaxed their food supplies.

Heavy snowfall in the Rocky Mountains may confine mule deer to 10% of their normal winter range. Under these conditions, the only available food is on the sunny south slopes, where snow melts rapidly. Of course, as deer crowd onto these slopes, food supplies can be depleted. A wildlife biologist once counted the carcasses of 381 starved deer in only 5 square kilometers (2 square miles) of a heavily used range in the Colorado Rockies.

When deer and elk begin to die of starvation, many people clamor for emergency feeding programs. However, most wildlife biologists believe that emergency feeding of starving deer is not sound management. They argue that it permits the survival of deer whose future offspring will exert even greater demands on the available natural browse, thus aggravating the problem in subsequent years. Artificial feeding also may facilitate the spread of **disease** by promoting concentrations of highly susceptible animals. Moreover, it is expensive.

Predation A number of predators, including wolves, cougars, bobcats, coyotes, and dogs, feed on deer. In Superior National Forest (Minnesota), wolves have been known to kill 15 deer per 2.6 square kilometers (1 square mile) annually. One year, the wolves killed 6,000 (17%) of the 37,000 deer in the forest. Because hunting pressure in the remote backwoods country of Superior National Forest is extremely light, accounting for only 0.65 deer per 2.6 square kilometers, wolf predation theoretically might help control a deer herd that is often on the verge of exceeding the carrying capacity of the range. Because only around 2,900 wolves inhabit the entire state of Minnesota (as of 2008) and the population has remained stable for nearly ten years, according to Minnesota's Department of Natural Resources, the impact of wolf predation on deer in that state is probably negligible. In fact, wildlife biologists believe that wolves improve the deer herd and do not compete for the same deer that hunters want. A 1971 study conducted in Minnesota by the U.S. Forest Service found that the vast majority of deer killed by wolves were at least five years old and in poor health, while the majority of deer killed by hunters were two years old or less and in good health. In addition, predator species populations are delicately balanced with their prey species. When deer populations decline, the wolf population declines soon after.

FIGURE 16.7 Juvenile dispersal of young bald eagles that were leg-banded as nestlings in western Florida. Many of these birds moved up the coast, some as far as 2,400 kilometers (1,500 miles) away! The function of this dispersal is unknown. After spending their first summer in the northern states, they return south to their breeding grounds.

2,400 km (1,500 miles)

1,600 km (1,000 miles)

800 km (500 miles)

Nesting area

● Recovery of banded eagle

Although one cougar may kill 50 or more deer annually, cougars are unimportant as regulators of deer populations because of their scarcity, except in localized areas of the Southwest. Free-ranging domestic dogs can be a problem for deer, especially does and young fawns, but mostly in local cases.

Hunting   A major source of mortality for some deer populations, especially the white-tailed deer, is legal sport hunting. Illegal hunting and poaching can be important in some areas of the United States, but estimates are hard to acquire. Because deer populations can reproduce rapidly, effective control by natural predators and human predators (hunting) is necessary so that populations do not exceed the carrying capacity of their habitat. However, some people and conservation groups oppose hunting and would like to see it banned or reduced. The hunting controversy is discussed in A Closer Look 16.1.

Disease   Disease can be a concern for deer populations. For example, white-tailed deer are susceptible to outbreaks of epizootic hemorrhagic disease. This disease is caused by a virus. Outbreaks are common in the southeastern United States and sporadically in other regions and Canada. Outbreaks are usually associated with high densities of the white-tailed deer population. There is no effective control of this disease.

Two other important diseases today are chronic wasting disease and West Nile virus. The former is a transmissible neurological disease of deer and elk that produces small lesions in brains of infected animals. West Nile virus mainly infects birds but is known to infect humans, horses, dogs, cats, bats, chipmunks, skunks, squirrels, and domestic rabbits. The main route of infection is through the bite of an infected mosquito.

Accidents   Collisions between automobiles and deer can be a significant problem in the United States. For instance, according to the Insurance Institute for Highway Safety's Web site, roughly 1.5 million deer-related accidents occur each year in the United States, causing more than 150 human fatalities, tens of thousands of injuries, and more than $1 billion in property damage. Few deer survive such encounters. Erie Insurance, which reports using ten years' worth of deer accident claims to perform an in-depth annual analysis of deer–vehicle collisions, found that the frequency of claims for these accidents was highest in West Virginia, New York, Pennsylvania, and Virginia.

Mortality Factors Affecting Waterfowl

The major mortality factors affecting waterfowl populations include loss of habitat, oil and chemical pollution, hunting and lead poisoning, disease, and acid rain (Figure 16.10).

Loss of Habitat   Since habitat provides cover, food, water, and breeding sites, its loss is a major threat to migratory birds.

FIGURE 16.8 Latitudinal migration of the Arctic tern. Only the southward movement is shown. Note that many of these birds nest in the Arctic and winter in the Antarctic. During their southward migration, some of these birds will cross the Atlantic Ocean twice and will complete an annual migration of about 40,000 kilometers (25,000 miles)—the longest migration of any organism in the world.

● Breeding areas
▲ Wintering areas

Ducks and other waterfowl breed in areas with wetlands, small ponds, and lakes. As pointed out in Chapter 9, more than half of our nation's original inland and coastal wetlands have been lost to farming and other development. Wetland losses seriously affect thousands of waterfowl, such as Canada geese, mallards, black ducks, wood ducks, and teal, which use U.S. wetlands as wintering grounds, especially on the Atlantic, Gulf, and Pacific Coasts, in the lower Mississippi River Valley, and in California.

GO GREEN!

If you are a coffee drinker, purchase only fair-trade, shade-grown coffee and cacao. Much of the coffee we drink comes from Central and South America. Traditional plantations grow coffee under a canopy of trees, which provides habitat needed by overwintering birds. Increasingly, however, land is being cleared to grow coffee in full sun; this habitat loss contributes to the decline in neotropical migratory birds. You can help preserve critical winter habitats for birds such as the ruby-throated hummingbird, Baltimore oriole, Kentucky warbler, and swallow-tailed kite by purchasing only shade-grown coffee and cacao.

The most productive "duck factory" on the North American continent is located in the grassland biome of Manitoba, Saskatchewan, Alberta, the Dakotas, western Minnesota, and northwestern Iowa (Figure 16.11). This region produces more than half of this continent's waterfowl. The ducks are raised primarily in tiny **potholes**, 0.5 to 1 hectare (1.25 to 2.5 acres) in size, where all their requirements for food, cover, water, and nesting sites are usually met. The density of the potholes may reach 50 per square kilometer (125 per square mile). An estimated 10 million potholes once existed in the prairie provinces of Canada alone.

Unfortunately, thousands of potholes in the United States have been drained by farmers, seriously threatening waterfowl populations. Much of this drainage has been subsidized by the U.S. farm bills promoting production of more corn, wheat, and soybeans. In Iowa alone, the number of potholes has been reduced drastically since the late 1930s, in turn causing a large decline in duck populations. In response, the U.S. Department of Agriculture's (USDA's) Natural Resource Conservation Service through its Wetlands Reserve Program, the USDA's Forest Service through its Prairie Pothole Joint Venture, and some states' departments of natural resources have restored millions of hectares of wetlands since the mid-1980s, helping to offset the this decline.

FIGURE 16.9   A deer rearing up on its hind legs for food in a Michigan forest. This sight is a possible sign that the deer herd is overtaxing the carrying capacity of the range.

Drought can be as destructive to wetlands as agricultural drainage. For example, about 1 million potholes in the duck factory of Canada's prairie provinces and the Dakotas, Iowa, and Minnesota went bone dry during the drought of 1988. This was the most severe dry spell since the Dust Bowl days of the 1930s. As a result of the sharply reduced acreage of prime breeding habitat, the total North American duck population in autumn, just before the hunting season, was about 66 million—8 million less than in 1987, representing the second-lowest population ever recorded.

Carp also are a notorious destroyer of waterfowl habitat. This type of fish can eradicate dense growths of favored duck foods such as sago pondweed, water milfoil, and coontail. Lake Koshkonong in southern Wisconsin was once almost blanketed with rafts of canvasback ducks, which consumed the abundant wild celery buds and pondweed nuts. Late in the 19th century, however, carp were introduced to the lake. In a brief time, the fish uprooted the choice waterfowl food plants, and the thrilling panoramas of ducks quickly vanished. To make matters worse for the canvasbacks, young carp compete directly with ducklings for the protein-rich crustaceans so essential for growth and development.

Oil and Chemical Pollution   Every year at least half a million water birds die worldwide from encounters with spilled oil, according to the International Bird Rescue Research Center in Calfornia (Figure 16.12). On occasion, rescue teams arrive on the scene in time to scrub the birds' feathers clean and prevent calamity. The timing of the spill greatly affects the mortality. On January 2, 1988, a storage tank collapsed near Pittsburgh, PA, and released a tidal wave of diesel fuel into the Monongahela River. The 22.5-kilometer-long (14-mile-long) slick caused some waterfowl to die. The death toll would have been much higher had the spill occurred during the spring, when thousands of ducks move up the Monongahela during their migration.

Why does oil kill ducks? There are several reasons. First, oil mats the feathers of waterfowl and reduces their ability to keep warm in ice-cold water. Death then results because of rapidly dropping body temperature. So dangerous is oil on a bird's

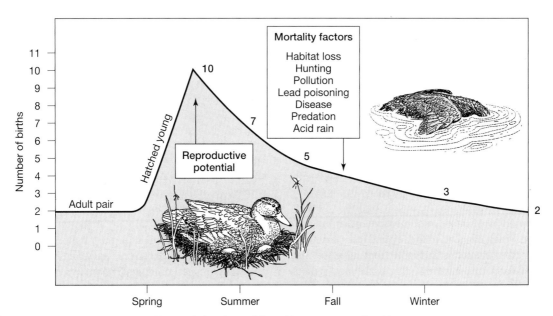

FIGURE 16.10   Survival curve for a mallard family. The population of two adults and ten young was reduced by a variety of mortality factors to only two birds by the beginning of the next year's breeding season.

A CLOSER LOOK 16.1   The Hunting Controversy

The hunter spots an elk herd edging into a clearing in the Colorado Rockies. He raises his rifle, picks out a big bull with a magnificent rack, squints down the sights, and squeezes the trigger. The crack of the rifle shot echoes through the valley. The bull staggers and slumps to the ground, blood spurting from the wound. He twitches his legs a few times and then lies still. Once again, the human predator has made a kill.

In 2006, about 13 million Americans hunted some form of wildlife. About 91% of the hunters in the United States are males. About 28% of them are young, ranging from 16 to 34 years of age. Less than one in every two adult hunters has attended college. Their occupations are highly diverse. Almost every walk of life is represented, from farmers to surgeons, from factory workers to business executives. Many are still in college. Regardless of their occupation, however, most American hunters grew up in a rural or semirural environment and received hunting instructions from their fathers when they were teenagers.

In the rural environment of the past century, hunting had much greater acceptance than it does today. Probably this was because the slaughter of livestock for food was a common farm activity. As a result, the death of animals was accepted as necessary for human survival.

Today, however, our nation is highly urbanized. Most Americans have never seen a farmer butcher a chicken, pig, or cow. Many consider the killing of wild game unnecessary and cruel, and about 50% of Americans today oppose hunting. They believe hunting causes animals to suffer unfairly. They also draw a distinction between the behavior of hunters and wild predators. For instance, wolves, cougars, and other wild predators usually take prey that is weak, sick, crippled, aged, or suffering from disease. In contrast, the human hunter prefers to bag trophy-sized animals in prime condition. This may weaken the species.

Predators also have other effects on ecosystems that hunters can't duplicate. After much controversy, wolves were reintroduced into Yellowstone National Park in 1996. Since their introduction, wolves have increased biodiversity in the park. One way is by trimming the coyote population, which grew excessively in the wolf's absence. (Wolves were eliminated from the park and surrounding areas around 1920, and then coyotes filled their niche.) As coyotes flourished, red fox populations declined. (Coyotes and red fox compete for resources.) Now that wolves are back and the coyote population is on the decline, red fox have returned to areas where

they had been eliminated. Wolves have also had a beneficial effect on beaver populations. How? Over the years, since the elimination of the wolf, elk populations have skyrocketed in the park. Elk feed on aspen, which is also a food source for beavers. Beavers declined as a result of this and other factors. Now that wolves are driving elk from aspen groves (because wolves prey on elk in aspen, and now elk are learning to avoid aspen groves), the groves are recovering, and so are the beavers. As beavers dam streams, they create habitat for a variety of other species, notably songbirds.

Many people opposed to hunting believe that state game commissions set hunting quotas not to keep wildlife in balance, but to satisfy the desires of hunters.

A number of citizens' groups have been organized to outlaw hunting. Among them are the Friends of Animals, based in New York City, and the Humane Society of the United States, with headquarters in Washington, DC. A common view held by such groups is aptly expressed by Joseph Wood Krutch, a nature writer from New York City: "When a man wantonly destroys one of the works of man we call him Vandal. When he wantonly destroys one of the works of God we call him Sportsman."

Hunters believe that, since we have eliminated most of the natural predators of deer and other large game animals, properly regulated hunting by humans is necessary. Without effective control by natural and human predators, these game animals will exceed the carrying capacity of their habitat. Left unchecked, this could result in a dramatic destruction of the landscape. Hunting helps prevent this problem. Sport hunters and hunting groups further argue that hunting is enjoyable recreation and brings in income to local economies—to local gun shops, restaurants, and hotels, for example. Taxes on firearms and ammunition and sales of hunting licenses provide states with funds for wildlife management, habitat work, and wildlife research.

Hunters have also been supported by many local sportsmen's groups as well as state and national organizations. Nationwide support for hunters is given by the National Rifle Association, the National Wildlife Federation, and the Wildlife Management Institute. Their pro-hunting philosophy is well represented by N. Adams in an article entitled "Hunting: An American Tradition." Adams writes, "You show me a person who doesn't directly or indirectly kill on a regular basis, and I'll show you a bleached, well-weathered pile of human remains. Every living creature takes life to stay alive, and if it doesn't, it quickly starves and dies a slow agonizing death."

feathers that an oil-soaked area the size of a quarter is sufficient to kill some waterfowl. Another reason for deaths is that waterfowl with oil-matted feathers may starve to death because they lose their ability to swim or fly in search of food. Finally, waterfowl may swallow toxic oil while feeding, drinking, or preening feathers, which can lead to kidney and liver failure.

Pesticides and other chemicals in runoff from farmland also pollute wetlands, endangering waterfowl. One of the most famous examples occurred in the Kesterson National Wildlife Refuge, which sits adjacent to California's heavily farmed San Joaquin Valley. Toxic selenium leached from the soil by irrigation water has caused birth defects in young birds hatched in the refuge.

Hunting and Lead Poisoning   About 14 million ducks are killed each year by hunters in the United States and Canada, of which 12.7 million are in the United States. Today there are an estimated 1.5 million waterfowl hunters in the United States, according to the nonprofit organization Ducks Unlimited. At one time, hunters deposited more than 3,000 tons of lead shot in our nation's lakes, rivers, and marshes annually. Because one shotgun shell contains 280 pellets of Number 6 shot, and the average hunter needs six shots to kill one duck, about 1,400 pellets were deposited on waterfowl habitat for each bird taken. In one study, researchers counted 150,000 pellets per hectare (60,000 per acre) in the San Joaquin River marshes

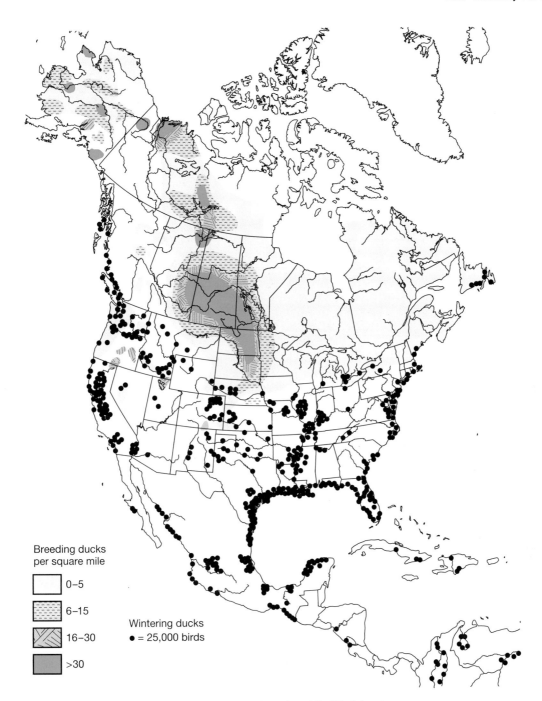

FIGURE 16.11 Distribution of North American breeding and wintering ducks. More than 50% of North American ducks are produced in the "duck factories" of Manitoba, Saskatchewan, Alberta, the Dakotas, western Minnesota, and northwestern Iowa. The major wintering areas are the Atlantic, Gulf, and Pacific Coasts, the lower Mississippi River Valley, and California. Because many ducks that migrate through the United States winter in Mexico, effective waterfowl management depends on the cooperative actions of Canada, the United States, and Mexico.

Breeding ducks
per square mile

- 0–5
- 6–15
- 16–30
- >30

Wintering ducks
● = 25,000 birds

of California; 300,000 per hectare (120,000 per acre) were found on the bottom of Wisconsin's Lake Puckaway.

Until recently, **lead poisoning** from accidentally ingested spent shot killed an estimated 2% to 3% of U.S. waterfowl each year, an amount that nearly equals the combined duck production of North and South Dakota. (Secondary poisoning of bald eagles may occur when they eat the carcasses of lead-poisoned ducks.) The heaviest duck mortality from lead poisoning has occurred along the Mississippi flyway, especially in Illinois, Indiana, Missouri, and Arkansas.

To reduce this problem, the U.S. Fish and Wildlife Service began phasing out lead shot and replacing it with steel shot in 1987. Beginning with the 1991 season, lead shot was banned in the United States for waterfowl hunting. Starting with the 1999 season, lead shot was declared illegal in Canada. The advantage of steel shot is that it saves the lives of more than 2 million ducks per year. The disadvantages of steel shot are that it is more expensive than lead shot, does not kill at a distance of more than 40 meters (125 feet), and ruins outdated shotgun barrels made of soft steel.

FIGURE 16.12 Waterfowl mortality caused by an oil spill. Oil-coated ducks pile up on the shore of the Mississippi River near Spring Lake, MN. They were victims of a combined petroleum and soybean oil spill that killed 20,000 ducks.

Disease Waterfowl are subject to a wide variety of diseases. **Botulism**, a disease caused by the toxic metabolic wastes of the anaerobic bacterium *Clostridium botulinum,* can cause severe disease outbreaks in waterfowl. Although most prevalent in the West, it has been recorded from Canada to Mexico and from California to New Jersey. During the summer of 1910, this microscopic organism was responsible for millions of waterfowl deaths. Even today, botulism may kill 100,000 waterfowl a year in California and Utah (Figure 16.13).

Clostridium thrives in stagnant alkaline mudflats, where there is an abundance of trapped organic material (such as dead aquatic vegetation) and water temperatures are high. These conditions are most likely to occur in the late summer during extended drought. Ducks become ill after eating contaminated organic material (decomposing plants and animals) or maggots and other insects that harbor the bacteria. Apparently, insect larvae are a specialized microhabitat for the bacteria. After being absorbed by the bloodstream of the waterfowl, the toxin produced by the bacteria kills the birds by paralyzing their breathing muscles. Sick waterfowl cannot hold their necks erect, so the disease is also called "limber neck."

Other common diseases in waterfowl are avian cholera (a bacterial disease) and aspergillosis (a fungal disease). A viral disease that primarily attacks waterfowl is duck virus enteritis.

Acid Rain Another problem for waterfowl is acid rain in the United States and Canada (Chapter 19). The ultimate effect of decreasing pH on these birds varies with the feeding habits of the species and the severity of the acidification, but acid rain has clearly led to adverse reproductive effects for some water birds arising from shifts in habitat selection or diet. Several common duck species, including common goldeneye, hooded merganser, ring-necked duck, and American black ducks, which utilize small bodies of water as preferred nesting and brood-rearing habitat, are threatened by the vulnerability of these water bodies to acidification.

The black duck is highly prized by hunters, especially along the Atlantic Coast. It breeds in the northeastern United States and southeastern Canada, a region plagued by **acid deposition**. The mayfly, a common aquatic insect, is a valuable food for young black ducks. Unfortunately, however, mayfly populations have been sharply reduced in acidified lakes (Figure 16.14). Biologists wonder whether future generations of young black ducks will be able to shift to alternative food sources. It is possible that such substitute food will be either unpalatable or too difficult to catch in sufficient quantities. In addition, acidity may also kill alternative food sources, causing even greater problems for the black duck.

16.4 Wildlife Management

We have defined wildlife management as the planned use, protection, and control of wildlife, using sound ecological principles. One basic approach to the management of wildlife is the acquisition and development of quality wildlife habitat.

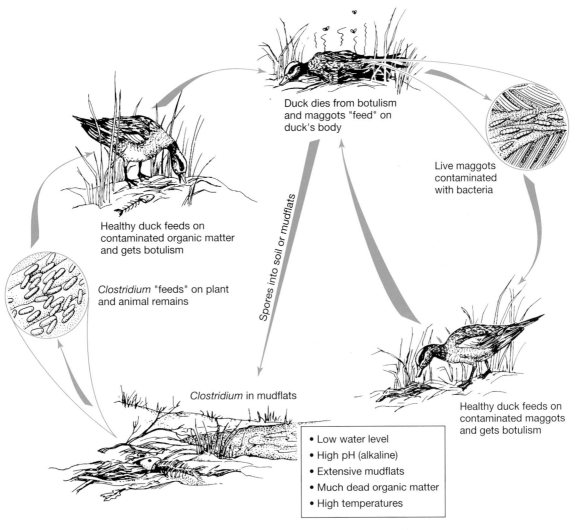

Duck dies from botulism and maggots "feed" on duck's body

Live maggots contaminated with bacteria

Healthy duck feeds on contaminated organic matter and gets botulism

Clostridium "feeds" on plant and animal remains

Spores into soil or mudflats

Clostridium in mudflats

Healthy duck feeds on contaminated maggots and gets botulism

- Low water level
- High pH (alkaline)
- Extensive mudflats
- Much dead organic matter
- High temperatures

FIGURE 16.13   Cycle of events leading to a botulism outbreak.

Acquiring and Developing Habitat for Terrestrial Wildlife

Currently, the best prospect for increasing wildlife populations is to increase the amount and quality of habitat. Many wildlife biologists consider habitat development to be indispensable.

The best management programs are meaningless without habitat protection. For waterfowl, wetlands are vitally important. As pointed out in Chapter 9, wetland losses have slowed considerably in recent years, thanks to federal, state, and private efforts—mostly efforts to purchase wetlands for waterfowl. Federal funds for this purpose come from an excise tax on the sale of hunting equipment. This tax was established by the 1937 Federal Aid in Wildlife Restoration Act, popularly known as the **Pittman–Robertson Act**. The tax was originally set at 10% but was later increased to 11%. Revenues from the federal tax on guns and other hunting equipment are distributed to the states for habitat purchase and for wildlife management and research. In addition to the Pittman–Robertson excise tax, other monies are derived from the sale of federal **duck stamps**, which must be purchased by waterfowl hunters at the beginning of each season. Unfortunately, federal funds have been insufficient. In 1961, the U.S. Congress passed the Wetlands Loan Act, which has enabled the U.S. Fish and Wildlife Service to borrow funds from federal sources for wetland purchases.

Since federal and state budgets for the development of parks and wildlife sanctuaries are tightening, private lands play an increasingly important role in supporting wildlife. Among the organizations that have had great success in acquiring private land for the benefit of wildlife are the **Nature Conservancy** and the **National Audubon Society**.

The Nature Conservancy does not make a lot of headlines. Quietly and effectively, it goes about its business of purchasing or leasing essential habitat to preserve plants, animals, and natural communities. It has helped protect more than 6 million hectares (15 million acres) of habitat in the United States and more than 47 million hectares (117 million acres) in Africa, the Caribbean, the Asian Pacific, and North, Central, and South America. With the aid of a huge army of volunteers, it manages hundreds of wildlife sanctuaries in the United States. The National Audubon Society maintains more than 100 wildlife sanctuaries and nature centers throughout the United States. Its high-priority campaigns include preserving wetlands and endangered forests, protecting corridors for migratory birds, and conserving marine wildlife.

FIGURE 16.14   Regions with acid waters and breeding areas of the black duck population of eastern North America. Where the breeding areas broadly overlap with regions that have acid waters, an important food supply is imperiled. The acid conditions have caused a sharp decline in the population of mayflies, an aquatic insect that is an important source of protein for young black ducks.

👣 GO GREEN!

You can make your yard a stopover for migratory birds by providing them with some essentials: water (even just a shallow dish), cover (evergreens and brush piles), and berry-producing trees and shrubs (such as dogwoods, American holly, viburnums, American highbush cranberry, and serviceberry). To attract the insects that birds eat, mulch your leaves in a sheltered corner, and don't deadhead flowers such as coneflowers, marigolds, cosmos, and sunflowers. Birds will feed on the seeds. You may also choose to put out a variety of seed feeders during migratory periods, along with cut or dried fruit.

Developing Habitat on the Farm and in Our Backyards

More than 85% of the hunting land in the United States is privately owned or controlled. Private farms, ranches, and woodlots, in fact, produce most of our nation's grouse, quail, doves, pheasants, and rabbits. Therefore, the biggest contribution to

an abundant and varied game resource (as in the case of forest development) can be made by the private citizen. Fortunately, many soil and water conservation practices (Chapter 7), such as shelterbelting and conservation tillage, also improve the habitat for wildlife.

During the Dust Bowl era of the 1930s, the USDA planted more than 30,000 kilometers (18,000 miles) of **shelterbelts** in the Great Plains. These narrow belts of trees on farms from the Canadian border down to Texas were planted primarily to control soil erosion by wind. However, they also provided food, cover, and breeding sites for dozens of species of nongame birds such as thrushes and warblers, as well as game such as grouse, quail, pheasants, squirrels, rabbits, and deer.

No-till has been discussed (Chapter 7) as a superb method for the control of soil erosion. Residues from the harvested crops are left on the land. The new crop is planted the following spring with specially designed seeders that penetrate through the residue to the undisturbed soil below. The acreage of farmland being farmed in this way is slowly increasing throughout the United States and in other parts of the world. No-till has benefited a great variety of wild animals, from pheasants in Nebraska to prairie chickens in Texas. Other ground-nesting species, such as quail, partridge, and early-nesting waterfowl, also benefit, since the stubble and other crop residues provide cover, breeding sites, and some food in the form of waste grain.

In 1985, Congress passed the **Food Security Act** (or Farm Bill). An extremely important provision of the act with regard to wildlife habitat improvement is the **Conservation Reserve Program (CRP)**, which started in 1986 and continues today. As noted in Chapter 7, this provision enabled the USDA to make contracts with farmers to control soil erosion. In essence, the farmers received payments for discontinuing farming on highly erodible land. Instead of raising crops, the farmers planted grasses or trees. The program is responsible for setting aside 14.9 million hectares (36.8 million acres) of CRP land, which has greatly increased wildlife habitat.

The 1985 Farm Bill (and most recently the 2008 Farm Bill) created other USDA programs important to wetlands and wildlife protection. For example, the **"Swampbuster" provision** protects wetlands by denying eligibility for other USDA programs to growers who drain and farm certain wetlands. The **Wetlands Reserve Program** is a voluntary program offering landowners the opportunity to protect, restore, and enhance wetlands and their values (including optimum wildlife habitat) on their property. Landowners receive financial incentives to enhance wetlands in exchange for retiring marginal agricultural land. This program offers landowners an opportunity to establish long-term conservation and wildlife practices and protection beyond what can be obtained through any other USDA program. The **Wildlife Habitat Incentives Program** also provides financial incentives to develop habitat for fish and wildlife on private lands. Participants agree to implement a wildlife habitat development plan, and the USDA agrees to provide cost-share assistance for the initial implementation of wildlife habitat development practices. This agreement generally lasts a minimum of ten years from the date the contract is signed.

In addition to all of these USDA programs, the U.S. Fish and Wildlife Service has a **Partners for Fish and Wildlife Program**, formerly named the Partners for Wildlife Program, which offers technical and financial assistance to private landowners who voluntarily restore wetlands and other fish and wildlife habitats on their land. The program emphasizes the reestablishment of native vegetation and ecological communities for the benefit of fish and wildlife in concert with the needs and desires of private landowners. Since the program began in 1987, these partnerships have generated significant habitat restoration on private lands, primarily focused on the restoration of wetlands, native grasslands, stream banks, riparian areas, and in-stream aquatic habitats.

GO GREEN!

If you love gardening and connecting with people in your community, the National Wildlife Federation can help you certify a Community Wildlife Habitat at www.nwf.org/community. A Community Wildlife Habitat is a community that provides habitat for wildlife throughout the community—in individual backyards, on school grounds, and in public areas such as parks, community gardens, places of worship, and businesses. It is a place where the residents make it a priority to provide habitat for wildlife by providing the four basic elements that all wildlife need: food, water, cover, and places to raise young.

Backyard Wildlife Habitat

While many people think about wildlife habitat as land outside of their own personal space—that is, farmland or **wildlife refuges** owned by state or federal governments—other possibilities exist. For example, homeowners, even apartment dwellers, can provide habitat in their own backyards. Rural, suburban, and even urban backyards can provide significant habitat for birds, butterflies, and other forms of wildlife. By planting species that provide food, shelter, and cover—and by providing a source of water such as a small pond or a birdbath—individuals can help improve the prospects for local wildlife. In urban settings, rooftop gardens can provide habitat for birds. Flowers and a birdbath on an apartment balcony can provide a small bit of green in an urban setting.

In 1973, the National Wildlife Federation began a Backyard Wildlife Habitat Program to promote the concept. Today, the organization provides information to help individuals create wildlife habitat, and there are currently more than 36,000 certified backyard wildlife habitat properties in the United States. Wildlife habitat has also been created around churches, in public parks, on school grounds, and near government facilities. The senior author, who lives in a rural setting, has planted trees in his meadow that provide shelter for birds and numerous bushes that produce berries for birds.

Backyard wildlife habitat not only helps wild species survive better in an increasingly developed world, it also provides people who might otherwise not be able to experience the wild world an opportunity to view wildlife just outside the kitchen window.

Manipulating Ecological Succession

In our discussion of **succession** in Chapter 3, we said that both plant and animal communities change as the physical environment changes. Thus, different organisms occupy a region as it goes through succession. Ecologist Raymond Dasmann has classified a number of species according to the successional stage: climax species (bighorn sheep, caribou, and grizzly bear); mid-successional species (antelope, elk, moose, deer, and ruffed grouse); and early-successional species (quail, dove, rabbit, and pheasant).

Wildlife biologists can regulate the abundance of these species by manipulating ecological succession. Thus, they can permit a succession to proceed on its natural course to a climax or, by employing such artificial devices as controlled burning (Figure 16.15), controlled flooding, plowing, and logging, can retard the succession or even return it to the pioneer stage.

The **early-successional species**, such as the rabbit, quail, and dove, depend heavily on major disturbance of ecological succession by humans. These species prefer the weedy

FIGURE 16.15 Wildlife biologists developing suitable habitat for prairie chickens in Minnesota. The controlled burn will prevent plant succession and the growth of shrubs and trees. As a result, the grassland habitat, which the prairie chickens require, will be maintained.

pioneer plants that invade an area denuded by human activity. Such vegetation may become established when farmland is abandoned and a pioneer community of invading weeds and shrubs becomes established.

Because **climax-associated species** such as caribou, bighorn sheep, and grizzly bear flourish only in relatively undisturbed climax communities, their survival depends largely on the establishment of state or national refuges and wilderness areas. Without such protected "islands" in the "oceans" of successional disturbance caused by humans, these climax-associated species will decline and become extinct.

As an example of manipulating ecological succession to encourage the population growth of a desired species, let's look at improving habitat for the ruffed grouse. The ruffed grouse is a highly prized game bird in the northern United States. It is named after the feathered ruff, or collar, it displays during courtship. When flushed by the hunter, it bursts from cover with a thunderous whir. A typical **mid-successional species**, grouse occupy a given region only temporarily.

Gordon Gullion, the nation's leading grouse expert, has found that young aspen stands provide virtually all of the food, cover, and breeding requirements of the grouse in the Great Lakes forests. Winter aspen buds provide an abundance of high-quality nourishment during a season when other foods are scarce. Note in Figure 16.16 that the highest population density, 81 birds per square kilometer (210 birds per square mile), occurs in aspen forests that are 12 to 25 years old. However, as the succession proceeds, other hardwoods, such as oak, gradually replace the aspen and the grouse population declines. Gullion found that the carrying capacity of aspen forests in northern Minnesota can be increased by more than 600% when blocks of aspen are clear-cut at 10, 20, and 30 years of age. This harvesting pattern gives grouse access to aspen stands of various ages and staves off the invasion of other tree species.

Managing Habitat for Waterfowl

Waterfowl habitat can be improved by creating openings in dense marsh vegetation, constructing **artificial ponds** and islands, developing artificial nests and **artificial nest sites**, and establishing waterfowl refuges.

Creating Openings in Marshes Although waterfowl require cover for protection from weather and predators, they also need channels and openings through which they can paddle or waddle between nesting sites and feeding areas and between feeding and loafing areas. Channels also provide areas where the birds can feed. These essential openings result from natural causes such as hurricanes and lightning-triggered fires, or they may be made by humans.

Constructing Artificial Ponds Where sloughs and potholes are scarce, waterfowl habitat can be improved by constructing ponds. Between 1936 and 1994, the USDA helped farmers build more than 4 million farm ponds. Roughly two-thirds of those ponds are used by waterfowl, either as nesting, feeding, and loafing areas for resident birds or as resting areas where migrating waterfowl can touch down for a brief respite before resuming their strenuous journey. Farmers can increase the carrying capacity of these ponds by erecting artificial nest boxes for mallards and wood ducks; by dumping piles of rocks or anchoring logs and bales of hay in the open water, where birds can preen and sun; and by seeding the pond with choice duck-food plants.

Constructing Artificial Islands In recent years, society has destroyed much valuable wildlife habitat. As a result, wildlife populations have frequently been decimated. And for years it has been assumed, by at least a few biologists, that virtually all human-induced changes of the natural environment were detrimental to wildlife. The error of this type of thinking is shown by the effect of **artificial islands** in boosting waterbird

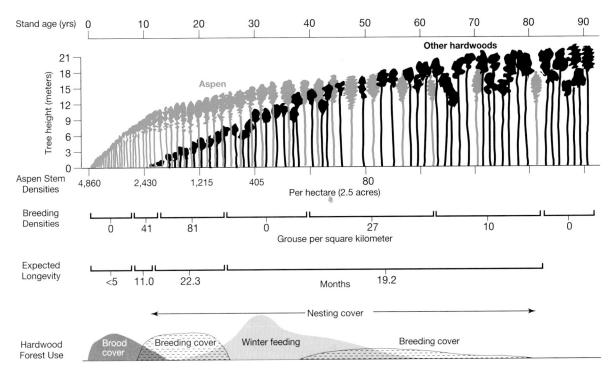

FIGURE 16.16 Forest succession and grouse habitat. Note that the carrying capacity of the forest for ruffed grouse changes as the forest succession proceeds. Use of the forest as brood and breeding cover is high during the early successional stages, when young aspen dominate the woods. When the aspen have attained an age of about 30 years, the grouse no longer use the forest for breeding purposes but use it primarily as a source of winter food. Grouse feed extensively on winter aspen buds. By the time the stand is 85 years old, the aspen have been completely replaced by other hardwood species that are of no value to grouse. The highest grouse densities (81 per square kilometer) occur in a forest that is 12 to 25 years old and is dominated by aspen.

FIGURE 16.17 Artificial nesting islands for Canada geese. This view shows the Westside Pond in Canyon Ferry Lake near Townsend, MT. Fill material was used to create about 62 artificial nesting islands. Note the Canada geese nesting on the island in the foreground.

populations. For example, in 1977, the Bureau of Reclamation (frequently assailed by environmentalists because of its obsession with big dam construction) used fill material to form 62 artificial nesting islands for Canada geese in Canyon Ferry Lake near Townsend, MT (Figure 16.17). Since the construction of these artificial islands, biologist Robert Eng of Montana State University has noted a threefold increase in Canada goose production in the region.

For a number of years, the U.S. Army Corps of Engineers has constructed coastal "dredge islands" from the sand, mud, and shells it removes during dredging operations. More than 2,000 of these islands are scattered along the nation's coast from Long Island, NY, to Brownsville, TX. Some have also been formed in the Mississippi River, in the Great Lakes, and along the Pacific Coast. They are usually located far enough from shore to afford protection from such predators as raccoons, foxes, and free-running dogs. Moreover, because the dredge islands are frequently about 2 to 3 meters (6.6 to 9.8 feet) high, they are not often flooded during high tides, as are many of the low-lying natural islands near shore. Sidney Island, now administered by the National Audubon Society, was formed from spoil resulting from the dredging of the ship channel at Orange, TX. Clouds of herons and ibises leave their nests and circle above when visitors set foot on the island. According to Audubon Society counts, Sidney Island is home to 20,000 egrets, almost 8,000 herons, 2,000 ibises, 1,400 roseate spoonbills, and 380 cormorants. Several of these species have declined in other areas.

Developing Artificial Nests and Nest Sites Through the process of natural selection operating over millions of years, each species of waterfowl has evolved its own unique instinct for nest site selection and nest construction. It would appear almost impertinent, therefore, for humans to attempt to improve on nature by constructing artificial nests and sites for waterfowl. However, wildlife biologists have done precisely this, with encouraging results (Figure 16.18). These artificial nests promote reproduction and may be even more effective than natural nests in minimizing mortality caused by mowing machines, predators, and nest site competitors.

Establishing National Wildlife Refuges Theodore Roosevelt launched America's system of national wildlife refuges in 1903. He established the Pelican Island Refuge in Florida's Indian River to protect the brown pelican. From this modest beginning, the federal refuge system has grown to 548 individual refuges and 37 wetland management districts, covering more than 39 million hectares (97 million acres), as of 2008. National Wildlife Refuges are home to more than 700 species of birds, 220 species of mammals, 250 reptile and amphibian species, and more than 200 species of fish. Every state has at least one refuge, and most of the refuges were established primarily for use by waterfowl. In 1934, Congress passed the **Migratory Bird Hunting Stamp Act**, which provides funds to acquire, maintain, and develop waterfowl refuges through the sale of duck stamps.

National wildlife refuges provide over 1.2 billion waterfowl-use days. (One waterfowl-use day is one day's use by one duck, coot, swan, or goose.) These refuges produce over 500,000 ducklings each year. The Tule Lake (California) and Agassiz (Minnesota) Refuges each produce 30,000 ducks yearly, and the Malheur (Oregon) Refuge produces 40,000 annually. Huge concentrations of ducks and geese use many of the refuges during the fall migration (Figure 16.19). For example, in the Klamath Basin Refuge on the California–Oregon border, where considerable acreages of wheat and barley are grown exclusively as waterfowl food, a peak of 3.4 million ducks and geese has been

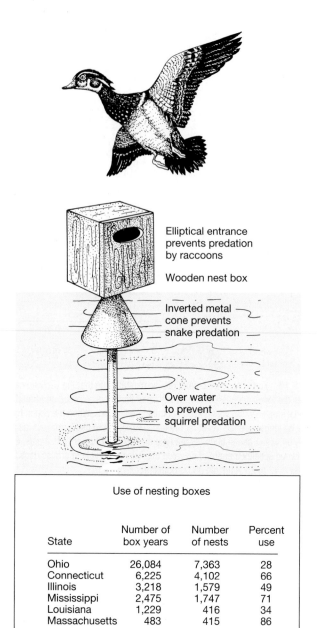

Use of nesting boxes			
State	Number of box years	Number of nests	Percent use
Ohio	26,084	7,363	28
Connecticut	6,225	4,102	66
Illinois	3,218	1,579	49
Mississippi	2,475	1,747	71
Louisiana	1,229	416	34
Massachusetts	483	415	86

FIGURE 16.18 Design and use of nesting boxes. Construction of nesting boxes for wood ducks has been an important factor in the population increase of this beautiful species.

recorded. Nearly 150,000 Canada geese have stopped over at the Horicon National Wildlife Refuge in southern Wisconsin—the greatest concentration of this species ever recorded in the United States. The geese fatten up or "refuel" at these stopover refuges before continuing their migration (Figure 16.20).

Many refuges have the specific mission of preserving one or more of the nation's endangered species. For example, the Aransas Wildlife Refuge in southern Texas provides a safe wintering ground for the whooping crane, while the Red Rocks Lake Refuge in Montana serves as a secure nesting area for the trumpeter swan. Some refuges are devoted primarily to saving endangered species of plants. In 1980, for example, the Antioch Dunes Refuge was established in California to ensure the survival of the Antioch Dunes evening primrose and

FIGURE 16.19 Sky darkened with thousands of pintail ducks at the Sacramento National Wildlife Refuge in California. As of 2008, the National Wildlife Refuge System includes 548 individual refuges and 37 wetland management districts, covering more than 39 million hectares (97 million acres).

the Contra Costa wallflower. Although the national wildlife refuges have the primary mission of conserving wildlife, the U.S. Fish and Wildlife Service, which administers the refuge network, does permit a limited amount of farming, logging, sheep and cattle grazing, and even oil drilling and mining, as long as the wildlife is not jeopardized in the process. Unfortunately, on some refuges, these extracurricular activities have destroyed some feeding areas and breeding grounds. Because of this, a number of environmental groups, including the Audubon Society, the Sierra Club, and Ducks Unlimited, have urged Congress to enact legislation that would ban such activities permanently.

The history of the Everglades, in Florida, and the subsequent creation of **Everglades National Park**, which has some qualities of a refuge, are discussed in Case Study 16.1.

16.5 Regulating Populations

In addition to habitat acquisition and development, wildlife managers use another basic approach to accomplish their objective: the regulation of populations. Most of our discussion will be concerned with deer and waterfowl.

Controlling the Harvest of Game Populations

Game managers may use hunting to control the populations of well-established species. Populations of upland game birds (quail, pheasants, grouse) and small mammals (rabbits, squirrels) usually recover rapidly after being reduced by hunting. Such populations are said to be **resilient**. Such resilience is due in part to their high biotic potential. For example, a hen partridge may lay 20 eggs per clutch; a doe rabbit may have six young per litter and rear several litters per year. Even deer have resilient populations. Under optimal conditions, a deer herd of six individuals can build up to 1,000 head in only 15 years!

In all species of animals, mortality factors, or environmental resistance, counteract the biotic potential. As a consequence, the population of a species remains about the same from year to year. Wildlife biologists distinguish between two types of mortality—additive and compensatory.

Additive mortality simply adds to the deaths caused by other factors (Figure 16.21A). For example, if predation accounts for 15% mortality of a population and an ice storm accounts for 15%, then the total mortality for the year is 30%. If, in the next year, predation takes 20% and an ice storm takes

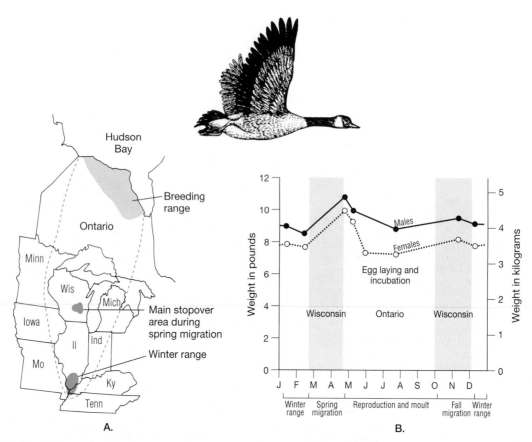

FIGURE 16.20 The value of "stopover" refuges in Wisconsin for the Mississippi River Valley population of Canada geese. These geese fly 725 kilometers (450 miles) nonstop to stopover refuges like Horicon Marsh in southeastern Wisconsin. There they find an abundance of food, which enables them to acquire the fat needed to fuel their 1,375-kilometer (850-mile) nonstop flight to their breeding ground in northern Ontario.

25%, for a total mortality of 45%, the effects of the two factors are said to be additive.

Compensatory mortality denotes the replacement of one kind of mortality with another kind of mortality in animal populations (Figure 16.21B). Compensatory mortality occurs if increased mortality from, say, hunting results in decreased mortality from other factors, such as starvation, disease, and predation, so that total mortality remains the same. Note that with or without hunting in a properly managed ecosystem, the total mortality from all causes remains the same. For example, about 70% of a northern bobwhite quail population will die in a normal year, regardless of whether hunting mortality occurs. Furthermore, a **harvestable surplus** of animals, within limits, can be safely removed. A harvestable surplus is that portion of the population that can be taken by humans (hunters) without adversely affecting subsequent populations of that animal.

The number of animals taken by hunters depends on hunting regulations. The hunting kill can be adjusted by extending or shortening the hunting season, increasing or decreasing bag limits, permitting the kill of both sexes or limiting it to males, and regulating the use of weapons (bow, rifle, or shotgun). Wildlife managers should set hunting season regulations by annually surveying the land under their management to determine habitat conditions and numbers, reproduction rates, and age structures of game animals.

Regulating the Deer Harvest

Several decades ago, when deer were relatively scarce, state legislatures restricted hunting by closing or shortening the season, by timing the seasons to ensure an absence of tracking snow, by restricting firearms to shotguns, and by restricting the kill to one per hunter. The doe was afforded special status. The deer herd buildup was further promoted by winter feeding, introductions, predator control, and the establishment of refuges.

In response to these measures, as well as to the great abundance of edge and food available in the wake of extensive fires and logging, the herd increased rapidly—too rapidly. In only 13 years, the whitetail population in 45 states increased from 3.2 million (1937) to 5.1 million (1949). It soon exceeded the range's carrying capacity. Browse lines appeared, winter starvation became commonplace, and the range rapidly deteriorated.

Many state game departments advised legislators to reverse the trend by liberalizing hunting regulations. After much prodding from wildlife biologists, herd reduction was implemented by opening and extending seasons, timing the season to coincide with the occurrence of tracking snow, legalizing the use of rifles, lifting the ban on does, establishing bow seasons, and removing bounties on predators. Roads were built to facilitate hunter access in the backcountry.

THE EVERGLADES: WATER TROUBLES IN A WILDLIFE PARADISE

The Everglades is a gigantic, 13,000-square-kilometer (5,000-square-mile) sea of grass that extends south from Lake Okeechobee almost to the tip of the Florida peninsula (Figure 1). This magnificent wetlands is roughly 64.4 kilometers (40 miles) wide and more than 161 kilometers (100 miles) long. The characteristic vegetation is saw grass, some of which grows 3.6 meters (12 feet) tall. The Everglades is dotted with raised oblong areas, called tree islands, that are vegetated with willow and myrtle. Wildlife is highly varied and of great interest, including huge turtles, alligators, and crocodiles, as well as deer, cougars, and long-legged wading birds. These birds include the wood stork, great egret, tricolored heron, great blue heron, white ibis, and roseate spoonbill. In the 1930s, more than 200,000 of these birds nested in the Everglades.

For many centuries, water moved slowly through this great swamp from its source in Lake Okeechobee. The water flowed southward through two grassland rivers, the Shark River Slough, which empties into the Gulf of Mexico, and the Taylor Slough, which flows into Florida Bay. Water movement was so slow it took almost a year to move from Lake Okeechobee to the ocean. During the rainy season the sloughs would overflow, and large numbers of fish would "spill out" into the grasslands. The result was a food bonanza for the huge population of wading birds and other fish-eating wildlife.

Such was the water flow pattern in the Everglades for millennia. Then, in 1905, Governor Napoleon Bonaparte Broward instituted a massive campaign to convert the northern part of the Everglades to agricultural and urban use. Since that time, many thousands of acres have been drained, diked, and channeled to control flooding and to provide sites for crop production as well as housing developments. Much of the work was done by the U.S. Army Corps of Engineers. American taxpayers footed the $30 million bill for this grandiose adventure in swampland "plumbing."

Regardless of the heightened concerns of professional ecologists and such groups as the Audubon Society, the project continued. Additional water was diverted from the Everglades by the South Florida Water Management District to supply the city of Miami. Because of these and other projects, the natural rain-driven sheet flow of water through the Everglades was severely disrupted. In effect, the entire northern half of the Everglades was converted into a series of diked, human-made pools connected by canals.

These Everglade plumbing projects have caused a lowering of the water table. During the dry season, the parched grasses are highly vulnerable to fire. Encroachment of ocean water has resulted in an increase in salinity. The original natural condition of the vast Everglades ecosystem has been greatly altered. Wildlife populations have suffered. Indeed, the numbers of some species are so low that the U.S. Fish and Wildlife Service has placed them on its official list of endangered species. They include the American crocodile, wood stork, snail kite, Cape Sable seaside sparrow, Florida panther, and manatee—all native to the Everglades. At the same time, the populations of some foreign species of plants have exploded under the artificially altered conditions. Back in the 1920s, then-Governor Broward introduced the Australian eucalyptus. The purpose was to help dry out the Everglades so that sugarcane and other crops could be raised. The plant thrived where water levels had been reduced by drainage. Today this shrubby plant infests more than 40,000 hectares (100,000 acres). It forms dense stands and steals moisture and nutrients from more desirable vegetation that had served as good nesting and feeding habitat for wildlife.

Environmentalists argued for decades about the need to restore the biological health of the Everglades. They got a big boost in 1947, when Congress created Everglades National Park in the southern portion of the marsh. (Today, the boundaries of Everglades National Park protect only the southern one-fifth of the historic Everglades ecosystem.) Unfortunately, however, by that time, the northern Everglades had been desecrated by powerful agribusiness and real estate interests. A maze of canals, dikes, dams, and levees had replaced the natural saw grass ecosystem. However, in the 1980s, an emerging consensus developed on two critical points: First, the original Everglades

FIGURE 1 The Florida Everglades. A large portion of the state of Florida was once covered by the Everglades (shown in light blue), but urban development and drainage have severely damaged this ecosystem. Today the Everglade ecosystem is only half this size. Because the Everglades ecosystem is so important to the people and wildlife of Florida, the Comprehensive Everglades Restoration Plan became law in 2000, commiting nearly $8 billion to restore and protect these water resources.

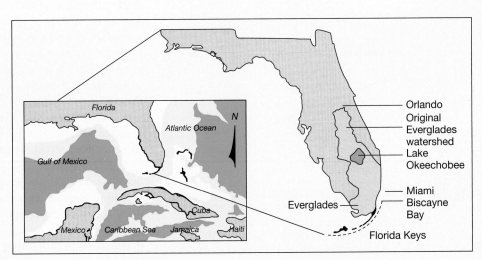

(continued)

ecosystem had to be restored (Figure 1). Second, this could be done only by reestablishing the natural, rain-driven sheet flow of the water from Lake Okeechobee southward through the marsh. Among the parties endorsing this policy were the National Park Service, the U.S. Fish and Wildlife Service, the National Audubon Society, concerned conservationists nationwide, and even the U.S. Army Corps of Engineers and the South Florida Water Management District.

This objective was given impetus by the Everglades National Park Protection and Expansion Act, which Congress passed in 1989. It provided for a 43,000-hectare (107,000-acre) addition to the park—today covering about 6,000 square kilometers (2,300 square miles)—that was absolutely essential for reestablishing the original sheet flow of water from the Shark River and Taylor sloughs. The gigantic restoration project received additional assistance in 1996, when Congress passed a controversial act limiting the activities of the sugarcane industry, which allegedly damages the fragile Everglades with its industrial waste and refinery by-products. Although these two acts aided in restoring the natural flow to parts of the Everglades, many believed that they would only temporarily prolong its life.

On December 11, 2000, the Water Resources Development Act of 2000 was signed into law by President Bill Clinton. Part of the act contained the **Comprehensive Everglades Restoration Plan,** a $7.8 billion federal–state project intended to restore the natural flow of water in the national park. The goal of the comprehensive plan is to restore, protect, and preserve the defining ecological features of the original Everglades and other parts of south Florida, by capturing freshwater that now flows unused to the ocean and the gulf and redirecting it to areas that need it most. The majority of the water will be devoted to environmental restoration, reviving a dying ecosystem. The remaining water will benefit cities and farmers by enhancing water supplies for south Florida. Led by the Army Corps of Engineers, the project is to span 30 years.

The plan is supposed to benefit Everglades National Park directly in several ways. Most importantly, it greatly improves the quality, quantity, timing, and distribution of flows into the park. In addition, more than 386.4 kilometers (240 miles) of canals and levees within the Everglades are to be removed to reestablish the natural sheet flow of water through the park. That change will support the return of the large nesting rookeries of wading birds to the park and the recovery of several endangered species.

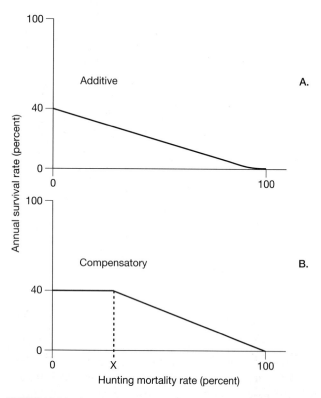

FIGURE 16.21 Comparison of the influence of hunting mortality on annual survival rates of a theoretical bird population under additive and compensatory mortality conditions. (A) With additive mortality, an annual survival rate of 40% declines linearly beginning with the death of the first bird shot during hunting season; that is, the survival rate diminishes whenever a bird is killed. (B) With compensatory mortality, the annual survival rate of 40% is unaffected by hunting mortality until a certain point (*X*) is reached, after which hunting mortality becomes additive.

The overpopulation problem is far from solved. Despite liberalized laws, hunters rarely harvest more than 10% of herds. One reason is the deer's secretive behavior: The animals rarely emerge from protective cover during daylight hours of the hunting season. Hunters sometimes do not see a single deer even in an area with large populations. A 10% annual harvest is simply not enough to appreciably check herd increase. In some states, at least one-third of the autumn deer herd could be taken during the hunting season year after year without affecting deer herd size.

Deer populations frequently vary from region to region within a state, often being highest in semiwooded agricultural regions and low in climax forests and urban areas. (Today, however, it is not uncommon to see deer on the outskirts of major cities such as Milwaukee and Chicago.) Therefore, a state may be divided into several zones, each with its own set of regulations. One state has had 60 zones. In zones where herds are small, the season may be closed completely or may be open to bow hunters only. In overpopulated zones, the season may be opened on bucks, does, and even fawns.

In some states, wildlife managers are not permitted to practice what they preach because their technical knowledge in game management is far in advance of a receptive public or political climate. Too often, the framers of our hunting laws yield to pressures exerted by hunters and resort owners, to whom the essence of game management is "more deer." Only when regulations are formulated in accordance with the advice of professionally staffed conservation departments will hunting regulations serve as an effective management tool. Some of the ethical considerations in regulating deer populations are discussed in Ethics in Resource Conservation 16.1.

ETHICS IN RESOURCE CONSERVATION 16.1   TO KILL OR NOT TO KILL?

Few areas of natural resource management are as heavily infused with (and confused by) ethical issues as wildlife management. One of the major issues that crops up from time to time has to do with managing wild species, especially deer and elk.

Individuals trained in wildlife management generally arrive at management decisions based on scientific facts. Basically, wildlife managers assess population size and growth rate, as well as habitat conditions, and then determine the level of hunting needed to keep populations within the local or regional carrying capacity. Their goal is to keep deer and other wildlife from eating themselves out of house and home.

One thing that is rarely realized by outsiders is that the scientific approach to wildlife management is ultimately geared toward ecosystem protection. In other words, to the wildlife manager, the ecosystem is of paramount importance. He or she manages wildlife populations in order to protect the entire system upon which all populations depend.

Critics of wildlife management professionals usually come at the issue from an entirely different vantage point. Their criticism of wildlife management, often presented in scientific rationale, is frequently underpinned by a profound belief in animal rights—that is, the right of individual species to flourish without human intervention. Some people call this the Bambi syndrome, a derogatory appellation meant to imply that people are emotionally involved, not intellectually engaged, in the process of managing our wild species.

Varying from passionate zealots to ordinary citizens, animal rights advocates often pit themselves against wildlife managers, who are seen as cruel and manipulative. Time and again, conflicts between these two groups generate more heat than light. Why?

We believe that the chief reason for the degree of conflict is that the two groups are talking on different planes. Wildlife managers talk about the science of population management and ecosystem management, or so it seems, but

animal rights advocates are talking moral issues. The former argue that we need to control wildlife populations, such as deer and elk; the latter argue that killing animals is inhumane. The former group promotes human control over nature. Animal rights advocates say this outlook is arrogant and morally repugnant.

But in the debates that ensue, infrequent are the times when both groups talk on the same level. Rare is the individual who can dissect the arguments, teasing out moral or ethical concerns from scientific ones.

Animal rights advocates say it is wrong to kill deer or wild horses that are overrunning their habitat. Wildlife management officials argue that it is absolutely essential for the health and long-term future of the land and other species as well.

An important point is often missed in the acrimonious debates: Wildlife managers do have an ethical position. They have aligned themselves on the side of ecosystems. They have taken the moral stance that ecosystem health is more important than an individual species' right to be free from human manipulation. You could argue that the ecosystem ethic (for lack of a better term) is really a scientific position, but it is not. It is ethics based on science.

Can there be any reconciliation? Not until both parties begin to sort out ethical from scientific considerations and agree to talk a similar language. As a wildlife manager, you may want to meet with people who oppose your actions and lay out your ethical position. It may help to outline the argument for the primacy of ecosystems in scientific terms as well as ethical terms. In addition, it may be helpful to note that you care about humane treatment. Although a few animals may be killed to control population size, the benefit to the rest of the population and to the ecosystem is substantial.

With ethical positions laid out, you can debate scientific discrepancies and make real headway. You may find legitimate points that bear consideration or that may force you to alter your plans. Even then, differing ethical viewpoints may make reconciliation impossible.

Regulating Destructive Deer Populations

Deer can become pests in rural, suburban, and urban environments. Farms, in particular, provide sources of food for deer. Deer frequently cause excessive losses of soybeans and other crops. Wisconsin provides a good example of crop depredation by deer.

In 1994, the Wisconsin white-tailed deer population stood at more than 1 million. This was a population increase of 162% for the state as a whole relative to the deer population in 1962. However, when agricultural lands alone are considered, the increase was a dramatic 488%. And these farming country deer were taking a huge $37 million annual bite out of Wisconsin agricultural production. Fields of corn and soybeans

were damaged when deer invaded them with crops still in the seedling stage. In response, the Wisconsin Department of Natural Resources liberalized its hunting regulations in areas where deer-inflicted crop damage was high. For example, during Wisconsin's nine-day hunting season in 1993, about 255,000 deer were taken—slightly over 25% of the total herd. Despite the huge annual harvests, however, deer damage to Wisconsin fields and orchards continues to rise. Wisconsin's white-tailed deer population stood at 1.64 million going into the fall 2004 deer-hunting season. The Wisconsin Department of Natural Resources has liberalized its hunting rules and hours even more in hopes that hunters will curb the growth in the deer population that Wisconsin has been experiencing.

Regulating Other Wildlife Pests

Deer are not the only wild species that can become pests. As human populations push into surrounding rural areas, many other species can become bothersome to suburban residents. Even in urban settings, wildlife can become a nuisance. One main reason is that our cities and suburbs keep expanding in the countryside, leaving less wildlife habitat and increasing human and wildlife conflicts. Canada geese, for instance, invade parks, golf courses, and playing fields in suburban and urban settings, feeding off lush grasses. In some areas, geese may stay all year, rather than following their customary annual north–south migration. Geese delight many urban and suburban residents, but they litter parks, golf courses, and other areas with feces, making them unpleasant for human use. They can also be quite noisy.

Pigeons or rock doves, also a wild species, have made cities their home. In huge flocks, they can be quite troublesome. Over the years, cities have poisoned and have captured and killed pigeons by the thousands to clean up public places where the birds often congregate en masse. In some cities, wildlife officials have introduced the endangered peregrine falcon. This bird makes its home high among the canyons of concrete and steel, preying on pigeons.

Foxes, skunks, coyotes, and raccoons can also invade urban and suburban areas, setting up homes under porches or other areas. Beavers have even been removed from urban and suburban areas, where their dams have plugged ditches, culverts, and water-control structures, in turn flooding lawns, fields, and roads. In drought years, even larger types of wildlife invade human neighborhoods. Bears, for instance, frequently venture into suburban areas in Colorado during drought years when food supplies are low.

Wildlife incursions are handled in various ways. Bears, for instance, are typically tranquilized and then relocated, although some are killed outright. Foxes and other smaller mammals are captured alive or poisoned. Large flocks of geese may be captured to thin them. Feeding is discouraged as well.

Regulating the Waterfowl Harvest

The mallard is one of the most highly prized ducks sought by hunters on the North American continent. Each autumn, 12 million to 18 million mallards wing their way south from their breeding grounds in southern Canada and the northern United States. And each autumn, several million duck hunters crouch in their blinds, hoping their lifelike decoys will lure these swift-flying birds to within shooting distance.

But how many mallards may the hunter take each day? One? Three? Five? Moreover, how many days may he or she hunt? The answers to these questions may vary from one state to another, and from year to year as well. Let's find out how, and on what basis, such waterfowl hunting laws are made.

Much of the information on which waterfowl-hunting regulations are based is derived from population censuses conducted on the breeding grounds several months before the fall migration. Regulations are also based on population data taken in the major migratory pathways of the ducks, which are determined from studies of banded birds.

As of 2004, approximately 60 million birds had been banded across the United States and Canada under the North American Banding Program. About 4 million of these bands had been recovered. On average, about 1.1 million birds are banded every year. Game birds (mostly waterfowl) make up only 31% of birds banded but account for 72% of band recoveries. In 2001 alone, almost 355,000 waterfowl (ducks, geese, and swans) were leg-banded in the United States and Canada with serially numbered bands. These banding operations are conducted by the U.S. Fish and Wildlife Service and the Canadian Wildlife Service. Records of the species, age, sex, weight, date, and banding locality are computerized. In 2001, nearly 89,000 waterfowl bands were recovered. Although most recoveries are made by hunters, a considerable number are also recovered by bird-watchers and amateur naturalists who retrieve bands from birds killed by storms, pollution, predation, and disease.

Analysis of recovery data gives waterfowl biologists information concerning growth rate, life span, and mortality, as well as the length, speed, and route of migration. For example, from such banding studies, we now know that some snow geese may travel 3,200 kilometers (2,000 miles) nonstop from James Bay, Canada, to the Texas coast in only two days!

From the practical standpoint of waterfowl population management, the most significant information derived from banding studies is that waterfowl, such as mallards that breed in Canada and the northern states, migrate along four (rather poorly defined) **flyways** on their way to southern wintering grounds. Known as the Pacific, Central, Mississippi, and Atlantic Flyways, they have served as administrative units in the development of hunting regulations (Figure 16.22). These flyways have corridors that connect them (Figure 16.23). For example, many of the mallards that nest in the prairie provinces of Canada begin their fall migration by moving south along the Central Flyway into South Dakota. Eventually, however, they swerve southeastward, joining the Mississippi Flyway in Minnesota and Illinois, where they continue to their wintering grounds along the Gulf of Mexico.

Each of the four flyways is administered by a flyway council made up of directors of the various state conservation departments or their representatives. The flyway councils meet in August to frame regulations for the fall hunting season based on the size and distribution of the waterfowl population expected to be moving through their respective flyways. The councils then submit their recommendation to the U.S. Fish and Wildlife Service. The service also receives suggestions from various national conservation organizations, such as the National Audubon Society, Ducks Unlimited, and the Wildlife Management Institute. Armed with all this input, the service decides on regulations that it considers appropriate.

Let's examine a specific example of how waterfowl hunting regulations are tailored to the population level of the target species. The mallard is North America's most abundant and familiar duck. In 1962, the estimated mallard population on the North American continent was extremely low—about 7.6 million. Highly restrictive hunting regulations were therefore established. In the state of Washington, for example, the hunting season in 1962 was shortened to 75 days, and the daily bag limit was reduced to four. As a result, the mallard harvest in that state in 1962 was only

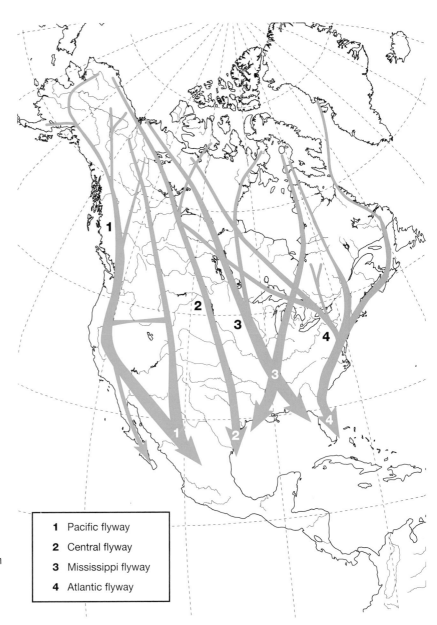

FIGURE 16.22   Major waterfowl flyways. Recovery data on many thousands of waterfowl led biologists to believe in the mid-20th century that the birds migrated along four major flyways: (1) Pacific, (2) Central, (3) Mississippi, and (4) Atlantic.

1 Pacific flyway
2 Central flyway
3 Mississippi flyway
4 Atlantic flyway

177,000 birds. Such restrictive laws were continued. The mallard population responded to this protection. By 1970 it had reached 11.6 million on the North American continent. Since then, its North American population has been generally increasing and sits at more than 13 million today, according to the National Audubon Society.

Regulations on the mallard harvest were, therefore, liberalized. Washington extended its season to 93 days and increased the bag limit to six birds. The result was a kill of 311,000 birds, 76% greater than in 1962. True, liberal hunting regulations (bag limits of three to seven per day and long seasons) increase the number of mallards shot compared with the number shot in restrictive seasons. However, a study done in the late 1970s analyzing band recoveries found that the survival rates for mallards, from September 1 to August 31 of the following year, were no different whether hunting regulations were liberal or restrictive. This indicates that other forms of mortality compensate for hunting mortality.

It is important to have population studies of mallards and other waterfowl that analyze whether a specific mortality factor like hunting is additive or compensatory. Such information is vital for determining bag limits and hunting season lengths. For example, if the hunting mortality is additive, then restrictive regulations should be adopted to help depressed populations recover. However, if it is compensatory, then liberal regulations (up to a point) may be adopted.

Regulating Destructive Waterfowl Populations

Waterfowl can damage farmlands. The most serious example of wildlife-inflicted crop damage involves grain consumption by waterfowl in Canada. Losses to Canadian grain farmers may reach 380,000 metric tons per year. Financial setbacks have approached $40 million annually, although the annual average is about $14 million.

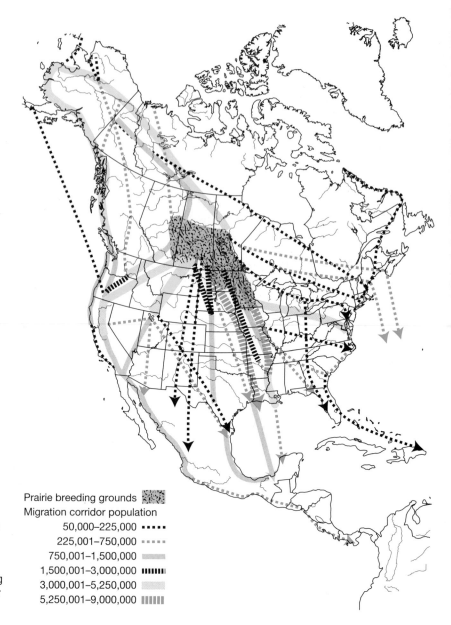

FIGURE 16.23   Duck migration corridors. Wildlife biologists now know that the "flyway" concept was an oversimplification of an extremely complex migration phenomenon. The routes of many species of waterfowl may actually cross each other, as shown here. Moreover, many waterfowl may fly east or west during migration, rather than north or south, as suggested by the flyway concept. Frank Bellrose of the Illinois Natural History Survey identified migration corridors for southward-flying ducks and was able to estimate the approximate number of ducks using a particular corridor.

Prairie breeding grounds
Migration corridor population
50,000–225,000
225,001–750,000
750,001–1,500,000
1,500,001–3,000,000
3,000,001–5,250,000
5,250,001–9,000,000

Although the long-term solution to crop destruction by waterfowl in Canada may involve some type of habitat modification, either in the cropped fields or in the nearby wetlands, wildlife managers are not even close to solving the problem. However, several management strategies that are currently available lessen the damage. A variety of noisemakers, including propane gas exploders (Figure 16.24), are used to scare waterfowl away from their "Wheaties."

Another practice is planting a less commercially valuable, but more attractive, waterfowl "lure crop", such as millet, to attract ducks and geese away from more valuable cash crops, such as wheat. Lure crops can be effective in drawing birds from nearby fields, thus minimizing crop losses on farmlands. Sometimes feeding or bait stations are even deployed to attract ducks away from crops.

The Canadian government offers farmers insurance programs that partially offset the financial losses caused by

marauding waterfowl. It's an interesting dilemma, to say the least. On the one hand, farming destroys waterfowl habitat when wetlands are converted to cropland. On the other hand, waterfowl destroy the crops that are planted on the original wetlands.

Canada geese have recently established breeding populations in some areas of northern Illinois. Unfortunately, however, huge flocks of these birds have become a serious nuisance on golf courses. Not only do they damage the greens, they cause problems at water hazards as well. In recent years, hundreds of these geese have been trapped alive and transplanted to rural areas in southern Illinois, far from the golfing scene.

Wildlife Diseases Affecting Humans

Wildlife populations can be carriers of some diseases that can spread to humans. Tularemia (mostly transmitted by ticks or

FIGURE 16.24 A propane gas exploder in a field of young brussel sprouts in California. Such noisemakers have been used to scare birds and other wildlife from farm fields. Such strategies often lose their effectiveness when the animals get used to the disturbance and continue eating the crops.

other arthropods) and sylvatic plague (transmitted by wild rodents) are two wildlife diseases of concern to humans, but rabies strikes the most fear in humans.

Rabies is a disease caused by a virus that lives in the saliva of the host (carrier) animal. The virus is transmitted to humans by the bite of an infected animal. It moves along nerve cell pathways and eventually reaches the brain. Unless the infected person is vaccinated against the virus, he or she will sink into a coma and die. Rabid animals will snap and bite at animals and humans. Many different species of wild animals can harbor the rabies virus, including skunks, bats, raccoons, and foxes. Extreme measures, such as poisoning and shooting, must sometimes be used to control rabid animals. Many years ago, strychnine-treated baits were used to control rabid foxes in Kentucky. Unfortunately, however, in addition to destroying 65 foxes, the poison killed 135 dogs. Moreover, the control measure was prohibitively expensive, costing $208 for every poisoned fox. More recently, an effective oral vaccine bait has been developed. It is dropped from airplanes over rabies-prone regions and confers immunity to the carnivores that consume it.

16.6 Nongame Wildlife

In recent years, state and federal wildlife agencies have developed a new focus of attention: nongame management. This is certainly appropriate, since wildlife watchers in the United States spent $45.7 billion in 2006, according to the U.S. Fish and Wildlife Sevice. Their activities range from whale watching to winter bird feeding to wildlife photography (Figure 16.25). The expenditure breakdown for these wildlife-watching activities in 2006 is as follows: 28% of all

expenditures for trip-related expenses (lodging, food, and transportation), 51% for equipment, and 21% for other expenses, such as magazines, membership dues to wildlife-related organizations, and land leasing and ownership for the purpose of wildlife watching.

The status of nongame populations, such as those of amphibians and reptiles, is not well known. In contrast, population trends in birds have been followed closely for many years through annual surveys conducted by the U.S. Fish and Wildlife Service and the National Audubon Society. These surveys have shown that the populations of most nongame birds are mixed—some are increasing, and some are decreasing. For example, the annual Christmas bird censuses conducted by the National Audubon Society have shown that hawk populations are on the rebound after being decimated by pesticide contamination of their food chains in the 1950s and 1960s. However, populations of many songbirds— vireos, tanagers, and flycatchers—that winter in Central and South America are declining. Many biologists attribute this to the destruction of habitat in the United States and tropical rain forests (Chapter 14), the wintering range of these species.

Several species of fish-eating birds of the Great Lakes— for example, cormorants, terns, and gulls—have suffered dramatic population declines in the past two decades. The suspected cause is their consumption of fish contaminated with toxic chemicals such as PCBs and dioxin. Female birds transmit the poisons to their eggs. A survey conducted by John Giesy, professor of wildlife management at Michigan State University, revealed toxic chemicals in the eggs of fish-eating birds throughout the Great Lakes. Many of the poisoned eggs never hatch. Many of the chicks that do emerge from the egg cannot survive, owing to gross deformities. The abnormalities are crossed bills, club feet, dwarfed wings, and lack of eyes and skull. The incidence of these deformities has increased 30-fold in the past 40 years.

The restoration of the populations of these threatened species poses a formidable and ongoing challenge to state and

FIGURE 16.25 The beautiful osprey, an example of nongame wildlife that people love to see. Here an osprey is flying with a fish caught in the Pacific Ocean.

federal wildlife managers. The eventual solution to the problem, of course, depends on cleanup of the chemical pollution of the Great Lakes.

Summary of Key Concepts

1. Wildlife management may be defined as the planned use, protection, and control of wildlife by the application of ecological principles.

2. The habitat of a wild animal provides certain essentials, chiefly shelter, food, water, and breeding sites.

3. Wildlife population densities tend to be higher in areas where there is a large amount of interspersion of plant communities, or edge.

4. A home range is the area over which an animal habitually travels while engaged in its usual activities.

5. A territory is any area that is defended. It is usually smaller than the home range.

6. Four types of movements in wild animals are dispersal of young, mass emigration, latitudinal migration, and altitudinal migration.

7. The movements of animals promote survival by providing supplies of food and water and a hospitable climate.

8. Additive mortality simply adds to the deaths caused by other factors, whereas compensatory mortality is the concept that one kind of mortality replaces (up to a point) another kind of mortality in animal populations.

9. The major mortality factors affecting deer populations include starvation, predation, hunting, disease, and accidents.

10. Emergency feeding of starving deer is not a sound management practice because it facilitates the spread of disease and is very costly.

11. Waterfowl populations are adversely affected by loss of habitat, oil and chemical pollution, hunting and lead poisoning, disease, and acid rain.

12. The most productive "duck factory" on the North American continent is located in the grassland biome of Manitoba, Saskatchewan, Alberta, the Dakotas, western Minnesota, and northwestern Iowa. More than 50% of the waterfowl in North America come from this region, primarily raised in tiny, 0.5- to 1-hectare (1.25- to 2.5-acre) potholes.

13. The Comprehensive Everglades Restoration Plan, part of the Water Resources Development Act of 2000, is a $7.8 billion federal–state project intended to restore, protect, and preserve the defining ecological features of the original Everglades and other parts of south Florida.

14. Waterfowl destruction by oil pollution may increase as more offshore oil wells are drilled and tanker traffic increases.

15. Lead shot, which can cause lead poisoning, was completely phased out in the United States in 1991.

16. Botulism, which can cause severe disease outbreaks in waterfowl, may kill 100,000 waterfowl in California and Utah per year.

17. Two major wildlife management techniques are habitat development and regulation of wildlife populations.

18. Wildlife biologists can regulate the abundance of wildlife species by manipulating ecological succession.

19. Current hunting and trapping regulations place restrictions on the species and numbers of individuals taken, the season and time of day, and the type of firearm or trap used.

20. The 1937 Federal Aid in Wildlife Restoration Act, popularly known as the Pittman–Robertson Act, levies an excise tax on the sales of sporting arms and ammunition. Revenues from the federal tax are distributed to the states for habitat purchase and wildlife management and research.

21. Several voluntary government programs, such as the Wetlands Reserve Program, the Wildlife Habitat Incentives Program, and the Partners for Fish and Wildlife Program, help private landowners protect wildlife and improve their habitat.

22. The National Wildlife Refuge System includes 548 individual refuges and 37 wetland management districts, covering more than 39 million hectares (97 million acres).

23. Waterfowl habitat can be improved by creating openings in dense marsh vegetation, constructing artificial ponds and islands, developing artificial nests and nest sites, and establishing waterfowl refuges.

24. The mallard is one of the most highly prized ducks sought by hunters on the North American continent.

25. Analysis of recovery data from leg-banded waterfowl gives waterfowl biologists information concerning growth rate, life span, and mortality, as well as the length, speed, and route of migration.

26. Waterfowl cause damage to croplands, with the most serious example involving grain consumption in Canada.

27. Many people engage in recreation activities based on nongame wildlife, such as whale and bird watching or wildlife photography.

Key Words and Phrases

Acid Deposition	Disease
Additive Mortality	Dispersal of Young
Altitudinal Migration	Duck Stamps
Artificial Islands	Early-Successional Species
Artificial Nest Sites	Ecotone
Artificial Ponds	Edge
Botulism	Euryphagous
Browse Line	Everglades National Park
Climax-Associated Species	Flyway
Compensatory Mortality	Food
Comprehensive Everglades Restoration Plan	Food Security Act
	Game Animals
Conservation Reserve Program (CRP)	Habitat
	Habitat Fragmentation
Consumptive Use	Harvestable Surplus
Corridor	Home Range
Cover	Hunting

Latitudinal Migration
Lead Poisoning
Mass Emigration
Mid-Successional Species
Migratory Bird Hunting
  Stamp Act
National Audubon Society
Nature Conservancy
Nonconsumptive Use
Nongame Animals
Partners for Fish and Wildlife
  Program
Pittman–Robertson Act
Pothole

Rabies
Resilient
Shelterbelt
Stenophagous
Succession
"Swampbuster" Provision
Territory
Water
Wetlands Reserve Program
Wildlife
Wildlife Habitat Incentives
  Program
Wildlife Management
Wildlife Refuge

Critical Thinking and Discussion Questions

1. Compare stenophagous and euryphagous species. Give an example of each type.

2. Discuss the functions of territories.

3. Would you say that your college campus has a lot of edge? How does this edge increase survival for wildlife such as robins, squirrels, and rabbits?

4. What are the advantages of altitudinal and latitudinal migrations?

5. How can botulism be prevented or controlled in waterfowl?

6. What causes lead poisoning in waterfowl? How can it be prevented?

7. What is the difference between additive mortality and compensatory mortality?

8. How can ecological succession be manipulated to improve wildlife habitat?

9. Discuss the importance of cooperation between forestry officials and wildlife managers in developing high-quality habitat for ruffed grouse.

10. Discuss the ecological and economic effects of the American hunter. How do nonhuman predators differ from hunters as predators in their effect on wildlife populations?

11. Give an example of restrictive hunting regulations for deer. Give an example of liberal hunting regulations for deer.

12. Give an example of destruction caused by waterfowl. How may such destruction be controlled?

13. Why is the National Wildlife Refuge System important?

Suggested Readings

Bellrose, F. 1980. *Ducks, Geese, and Swans of North America.* Harrisburg, PA: Stackpole Books. The "bible" on North American waterfowl. Excellent coverage of distribution, feeding habits, nesting behavior, migration, and mortality factors, as well as detailed migration maps.

Berger, C. 2003. When Gardeners Grow Wild. *National Wildlife* 41(4): 28–34. A wonderful article about what homeowners can do to create wildlife habitat in their own backyards.

Bolen, E. G., and W. L. Robinson. 2003. *Wildlife Ecology and Management,* 5th ed. Englewood Cliffs, NJ: Prentice Hall. Informative introductory text on the management of wildlife populations and habitats.

Bookhout, T. A., ed. 1994. *Research and Management Techniques for Wildlife and Habitats.* Bethesda, MD: Wildlife Society. Comprehensive manual of wildlife management that wildlife professionals use in their daily work.

Brandt, E. 1993. How Much Is a Gray Wolf Worth? *National Wildlife* 31(4): 4–12. Description of how the emerging science of resource economics proposes that wildlife preservation can be "good business" to entrepreneurs.

Brick, P., D. Snow, and S. Van de Wetering, eds. 2000. *Across the Great Divide: Explorations in Collaborative Conservation and the American West.* Washington, DC: Island Press. Exploration of a new movement in conservation: collaborative conservation. A must-read for anyone interested in a career in environmental conservation or natural resource conservation.

Chadwick, D. H. 1996. U.S. National Wildlife Refuges. *National Geographic* 190(4): 2–35. Popular account and beautiful pictorial of our national wildlife refuges.

Kerasote, T., ed. 2002. *Return of the Wild: The Future of Our Natural Lands.* Washington, DC: Island Press. A collection of writings about the value of wilderness, current threats, and what can be done to truly protect it to protect wildlife and plants.

Krausman, P. R. 2002. *Introduction to Wildlife Management: The Basics.* Englewood Cliffs, NJ: Prentice Hall. A detailed text covering all the basic components of wildlife management, including its history and evolution, population dynamics and models, decimating and welfare factors, and an array of important issues at play in the field today.

Leopold, A. 1949. *A Sand County Almanac.* New York: Oxford University Press. A classic written in almost poetic prose by the founder of game management. Intriguing insights into the relations between humans and wildlife.

Levy, S. 2003. A Top Dog Takes Over. *National Wildlife* 41(6): 22–29. A great account of the ways in which wolf reintroduction into Yellowstone National Park is boosting biodiversity.

Maehr, D. S., R. F. Noss, and J. L. Larkin, eds. 2001. *Large Mammal Restoration: Ecological and Sociological Challenges in the 21st Century.* Washington, DC: Island Press. Important case studies in wildlife management.

McComb, B. C. 2008. *Wildlife Habitat Management: Concepts and Applications in Forestry.* Boca Raton, FL: CRC Press. Integration of silvicultural and forest-planning principles with principles of wildlife habitat ecology and conservation biology.

Morrison, M. L., B. G. Marcot, and R. W. Mannan. 1998. *Wildlife–Habitat Relationships: Concepts and Applications,* 2nd ed. Madison: University of Wisconsin Press. Broad but advanced explanation of habitat relationships applicable to all terrestrial species.

Sinclair, A. R. E., J. M. Fryxell, and G. Caughley. 2006. *Wildlife Ecology, Conservation and Management,* 2nd ed. Oxford, UK: Blackwell Publishing. Clear presentation of the best lessons that wildlife ecology, conservation biology, and habitat management have taught us over the past 25 years.

U.S. Department of the Interior, Fish and Wildlife Service, and U.S. Department of Commerce, U.S. Census Bureau. 2007. *2006 National Survey of Fishing, Hunting, and Wildlife-Associated Recreation.* Washington, DC: U.S. Department of the Interior, U.S. Government Printing Office, available at www.census.gov/prod/www/abs/fishing.html. Latest statistics and information (as of 2006) on fishing, hunting, and wildlife-associated recreation in the United States.

Willis, D. W., C. Scalet, and L. D. Flake, C. 2008. *Introduction to Wildlife and Fisheries,* 2nd ed. New York: W. H. Freeman.

Detailed introductory text on the science and management of wildlife and fisheries.

Wondolleck, J. M., and S. L. Yaffee. 2000. *Making Collaboration Work: Lessons From Innovation in Natural Resource Management.* Washington, DC: Island Press. Examination of collaborative efforts in natural resource conservation. Worthwhile reading.

Web Explorations

Online resources for this chapter are on the World Wide Web at: **http://www.prenhall.com/chiras** *(click on the Table of Contents link and then select Chapter 16).*

SUSTAINABLE WASTE MANAGEMENT

Waste is a by-product of all living organisms. In the animal kingdom, however, we humans hold the prize for the total amount and variety of waste produced daily. Worldwide, the human community produces trillions of gallons of sewage, millions of tons of solid waste, and millions of tons of hazardous waste each day! This chapter examines two types of waste: municipal solid waste and hazardous waste. (Chapter 11 examined sewage.) Here you will learn where our waste comes from, how we deal with it, and how to deal with it more sustainably.

17.1 Municipal Waste: Tapping a Wasted Resource

American cities and towns produce a mountain of garbage, or **municipal solid waste,** from our homes and businesses (not industries). In 2005, the latest year for which data is available, U.S. solid waste amounted to nearly 221 million metric tons (246 million tons), up from 209 million metric tons (232 million tons in 2005). Add to that the over 70 million metric tons of solid waste that is produced by industries each year, and it is clear that Americans have a major problem on their hands.

Although Americans are the undisputed leader in total annual solid waste production, on a per capita basis the Canadians are not far behind. While Americans produce a little more than 2 kilograms (4.5 pounds) per day, on average, Canadians produce about 1.7 kilograms (3.8 pounds) per person per day (Figure 17.1). The Germans, French, Spaniards, and Italians produce about half as much on a per capita basis.

As shown in Figure 17.2, most of the waste from U.S. cities and towns consists of paper, yard wastes—grass clippings and leaves—and food wastes with lesser amounts of metals, glass, and plastics making up the balance. Despite a marked increase in recycling, only 32.5% of the municipal solid waste produced in the United States is recovered for reuse and recycling—up from 17% in 1990—according to the U.S. Environmental Protection Agency (EPA). Incineration, which many view as a less than desirable option, skyrocketed from 4% in 1988 to 15% in 2000. In 2005, however, the percentage of trash incinerated had declined to 12.5%, according to the EPA. The remainder, approximately 55% of the nation's waste, is dumped in landfills, where it is covered with dirt and forgotten.

Solid waste is a problem from several perspectives. First, it is expensive. Americans, for instance, currently spend over $300 million a year just to landfill disposable diapers. In many cities, waste disposal is the second largest expenditure, exceeded only by education.

Second, solid-waste disposal takes up valuable and often costly land that could be put to better use. In crowded metropolitan areas, landfill space is often in short supply, and waste haulers must ship garbage many miles at huge expense. Many states, especially

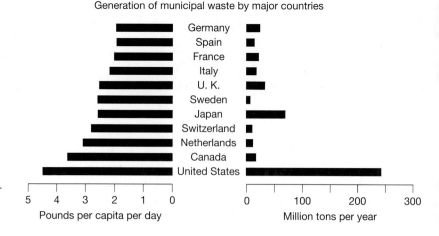

FIGURE 17.1   Comparison of municipal solid-waste production by country. Note that on a per capita basis, the United States and Canada are leading producers.

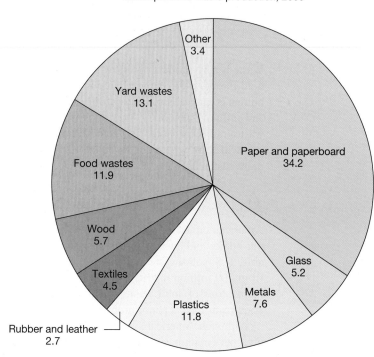

FIGURE 17.2   Composition of America's municipal sold-waste stream (before recycling and composting).

those in the northeastern United States, are facing a shortage of landfill space. Some states have taken to transporting their wastes to other states, often many hundreds of miles away, or even to other countries.

Third, garbage disposal and incineration waste millions of tons of valuable resources every year. (The term *waste,* which is commonly used to define this rather valuable and squandered resource, is wonderfully appropriate.) David Morris of the Washington-based Institute for Local Self-Reliance wrote, "A city the size of San Francisco disposes more aluminum [each year] than is produced by a small bauxite mine, more copper than a medium copper mine, and more paper than a good-sized timber stand." Our trash is a gold mine of wastes, offering a wealth of resources we can tap into—if only we're smart enough to take advantage of this opportunity.

Fourth, because solid-waste disposal and incineration squander valuable resources, more minerals must be mined,

more trees must be cut, and more oil must be extracted from the Earth (to make plastic or to power mining equipment, for example). Each of these activities produces enormous amounts of waste and has serious impacts on the environment.

Sustainable solid-waste management, which includes recycling and other measures, can eliminate these problems and help foster a sustainable society. The next section describes various aspects of waste management.

17.2   Managing Our Municipal Solid Wastes Sustainably

Most industrial societies have traditionally viewed municipal solid waste as something to get rid of—to dump in the ground or at sea as far away as possible—with little regard for the wealth of materials wasted in the process. When one dump site was full, city planners or private companies found another. Faced with shortages of suitable dump sites and growing transportation costs to haul the mountains of trash to outlying sites, however, many city planners have come to look skeptically at this approach. As conservationists and environmentalists have long said, the discard approach is unsustainable. This view was expressed by the World Commission on Environment and Development in a 1987 report entitled *Our Common Future.* "Many present efforts to guard and maintain human progress, to meet human needs, and to realize human ambitions," said the commissioners, "are simply unsustainable—in both the rich and poor nations. They draw too heavily, too quickly, on already overdrawn environmental resource accounts to be affordable far into the future without bankrupting those accounts." Looking at the impact of our actions on future generations, they wrote, "We borrow environmental capital from future generations with no intention or prospect of repaying. They may damn us for our spendthrift ways, but they can never collect on our debt to them." Some arguments for the present generation shouldering its debt, rather than passing it on to succeeding generations, are presented in Ethics in Resource Conservation 17.1.

Two approaches can be used to create a sustainable waste management system and a more sustainable society. The first strategy in a sustainable society is the **reduction approach,** which calls for lower levels of material consumption in society. The second is the **reuse and recycling approach,** which maximizes the life span of a material in the production–consumption cycle.

These approaches extend the planet's limited resource supplies, save energy, reduce pollution, and cut back on habitat destruction—in short, they reduce our destruction of environmental capital and create a more sustainable way of life. Some measures such as recycling also create jobs and can save money.

Reduction Approach

By reducing their per capita consumption of natural resources, modern societies can make tremendous inroads into the solid-waste problem—and greatly reduce the world's demand for natural resources. But this approach is often unattractive to individuals and businesses. "Why cut back on our pleasures and our profits to save a few birds?" some opponents ask.

In the discussion of sustainable ethics in Chapter 2, you learned that many people think other species have a right to exist, too. Protecting the environment by using less helps ensure a rich and diverse biological world. Cutting back today also benefits us by protecting the many free environmental services that can be lost when land is harvested or mined. Voluntary frugality, however, is also very likely inevitable as resources run low. Unfortunately, many powerful forces are at work in modern capitalistic economies that promote maximum material production and consumption.

Will frugality be forced upon those of us who live in the materialistic more-developed countries? As you shall see in Chapter 20, the supply of certain minerals is limited. More important, however, oil and natural-gas supplies are finite and fast on the decline. With the demand for oil and natural gas outstripping production today, we've witnessed a rapid increase in prices. The rapid increase in prices has forced many people to do with less—to drive less, to take fewer and shorter vacations, to buy fewer goods, and even to consume less food. To prevent further hardship and economic tumult that may result, why not begin by using less now? "Sustainable global development," according to the World Commission on Environment and Development, "requires that those who are more affluent adopt lifestyles within the planet's ecological means."

Individuals can reduce their consumption level in many ways, and there are many books to help citizens achieve this goal, including the *More Fun, Less Stuff Starter Guide* published by the Center for a New American Dream and the *Household Ecoteam Workbook* by David Gershon and Robert Gilman. There are even guides for parents, such as *Living Simply with Children* and *What Kids Really Want That Money Can't Buy.* The senior author, Dr. Chiras, has written a book that covers this and other related topics, titled *EcoKids: Raising Children Who Care for the Earth.* Here are some suggestions from these and other resources.

One strategy is to purchase items that are more durable. Better made clothing, tools, furniture, computers, and calculators outlast their cheap counterparts and greatly reduce demand for energy, minerals, water, land, and so on. Buying sturdy goods casts a vote for environmental protection that will not escape the attention of business leaders.

Individuals can also hold onto products longer. A car that is driven for seven to ten years rather than the typical three to five years, for instance, can greatly reduce your consumption of steel and other materials. If you live to be 80 and buy your first car at age 20, you would purchase 6 automobiles in your lifetime, if each one were held onto for 10 years. But if you purchase a new vehicle every 3 years, you would need 20 cars.

Faced with rising energy and mineral prices in the 1970s, many manufacturers found another route to reduce material consumption: miniaturization. Computer manufacturers, for instance, have greatly reduced the size of their computers over the past two decades, requiring less material per unit. (For more on this subject, see Chapter 20.) Buying or building smaller homes is one way people can help. A 1,500-square-foot home requires half as much building materials and much less energy to heat and cool than a 3,000-square-foot home.

Yet another approach to waste reduction is simply to consume less. We can lead simpler, less resource-intensive lifestyles. Like other critics of modern consumption, the authors of *Affluenza: The All-Consuming Epidemic* ask, Do we need all of the modern conveniences? Do they really enhance our lives? Or are they what the late actor-environmentalist Dennis Weaver described as simply something else to paint, repair, and insure?

By buying less, curbing our consumer passions, we can make tremendous inroads into the solid-waste dilemma and promote environmental sustainability. But swimming against the mainstream will not be easy in consumerist societies. U.S., Canadian, and Japanese societies have become cultures of mass

GO GREEN!

Avoid cheap products. When you buy a toaster or a microwave or even a blender, go for quality. It may cost more, but well-built products outlast their cheaper counterparts and reduce resource consumption and waste.

Every action you take has an impact on the environment and will affect the welfare of future generations. Although the impact of any one action may be insignificant, when combined with actions of others, individual acts can add up. But so what? All organisms have an impact on the environment.

Although it is true that all organisms affect their environment, humans have a much greater impact. Those of us who live in advanced technological societies have the greatest influence on the health and well-being of the environment. Given our disproportionately large impact, we face an important and difficult ethical question: Do we have an obligation to future generations to protect the planet? In other words, are we obliged by any ethical tenets to act in ways that respect the future of the planet, or is life a free-for-all?

Robert Mellert, who teaches philosophy at Brookdale Community College in New Jersey, argues that we do indeed have an obligation, which he bases on four ideas.

First, future generations will have many of the same needs that we have today. Although their priorities may differ, their needs for food, clothing, shelter, and recreation, among others, will remain. Mellert goes on to say that future generations, if they could be heard today, would claim a right to exist and to live well. To give them life without providing the means of meeting their needs would be cruel. To give them life without the means because of our carelessness and greed would be thoughtless and selfish.

Second, Mellert points out that an individual is born into a particular generation without his or her consent. None of us choose when to be born. He says, then, that because we have no claim to the time and place of our birth, justice requires that we have no more rights over the world and its resources than anyone else.

Third, Mellert contends that the survival of our species is more important than the survival of an individual member. Wildlife managers operate under this basic assumption. Their goal is to protect species and their habitat. That may require periodic culling—say, by sport hunters. No one is proposing this action, but just that individual rights are of lesser importance than the rights of many—now or in the future.

Fourth, he argues that even after we die, the effects of our lives continue. The present generation is the product of its ancestors—all that they did, died for, and believed in—but also the product of our own decisions. Future generations will be the result of what we are now and how they use what we leave them.

Mellert concludes by saying that if we accept these four truths, it is clear that we have an obligation to future generations. Our obligation, he says, is based on the fact that we are part of a much larger whole. We are inhabitants of a planet that will be used by many others who follow in our footsteps. We owe the future as much as we have received from the past.

Ponder these thoughts for awhile. Do you agree with them? Would you modify the case Mellert makes in any way? How? Why? Make a list of obligations we have to future generations.

---

consumption. Many other countries, including China, are following suit. Some critics argue that the United States and other countries have made a virtue out of consumption and materialism. In the United States, bumper stickers proclaim, "He who dies with the most toys wins" or proudly advise citizens to "Shop till you drop." Others gleefully announce that the owners of the vehicles were "born to shop." In the United States, people are no longer referred to as citizens, but as "consumers." And when the economy falters, we're badgered by the news media and government officials to spend, spend, spend to give the economy a boost. Some claim that it has become a disease, which they call *affluenza*. This disease, they say, dramatically alters our lives, forcing us to focus on consuming more and more in a fruitless battle to find satisfaction. Not only are our lives altered, but overconsumption has had a huge impact on the environment and the future of the planet.

Reuse and Recycling Approach

You have worn a pair of pants for a year, and even though they are in pretty good shape, you have grown tired of them. What do you do? If you are like many Americans, you throw them away. But why not drop your usable goods off at one of the many charitable groups like Goodwill or Disabled American Veterans or a local used-clothing store that will resell them?

Advocates of the reuse strategy point out that many products can be **reused**—put back into service in the same or a different application. Boxes, appliances, clothing, furniture, grocery bags, and so on can all be reused in many ways. Newspapers, for instance, can be ground up and used to make ceiling and wall insulation. They can be used to make animal bedding. Boxes can be used for moving or storing things or recycled to make new boxes. Glass can be crushed and mixed with asphalt and used for paving highways (Figure 17.3). Used office paper can be donated to schools for art projects. Even old eyeglasses can be donated to the less fortunate. Some cities and towns have public programs and private companies that reclaim construction waste. Habitat for Humanity, a nonprofit group that builds houses for the poor, has numerous stores that accept construction materials left over from projects and resells them to support their operations. Many commercial businesses

GO GREEN!

Hold on to your cell phone. Resist the temptation to upgrade every year or two!

B.

FIGURE 17.3   Glassphalt. (A) Ground-up glass becomes an ingredient of a road-paving material known as "glassphalt" (glass and asphalt). (B) Workers spread glassphalt along a busy street in Toledo, OH.

throughout the country sell old timbers and flooring removed from old barns or warehouses, which can be reused in new buildings. There's even a national organization that collects perfectly usable items from college students at the end of each academic year and sells them to incoming students the next year, donating the proceeds to charity.

By diverting trash-bound items to collection centers for reuse, we can greatly extend the useful life of many products and materials. In the process, we can reduce our demand for new materials. Such efforts, therefore, help reduce

GO GREEN!

At the end of the academic year, don't throw away perfectly usable items. Donate them through Freecycle or to local charities. You can also contact www.dumpandrun.org. This organization collects discards and sells them to incoming students the next year, donating the proceeds to charity.

resource consumption, habitat destruction, pollution, and waste disposal.

Reuse follows closely behind the reduction approach in its ecological appeal. For materials that cannot be reused, however, the next best approach is recycling. **Recycling** is another form of reuse but usually involves some kind of processing. For example, glass that is crushed, melted, and then used to make new glass is said to be recycled.

Recyclable materials can be extracted from municipal trash at central recycling stations, which is convenient for the citizen but often

FIGURE 17.4 Garbage, garbage everywhere. Workers at Ecocycle in Boulder, CO, sort recyclable materials that will be shipped to market and used to make new products at a fraction of the energy cost of their counterparts made from virgin materials, and with much less pollution.

costly. This process is called **end-point separation.** Much of the work is done by machines, although human labor also is needed. In Boulder, CO, low-risk prisoners from the county jail help separate the plastics on a large conveyor belt. Recyclables can also be separated at the source—at homes and factories—and picked up by recyclers or delivered to recycling centers by producers (Figure 17.4). This option is called **source separation.** It involves considerably more citizen participation than end-point separation but reduces the costs of separation facilities.

Once recyclable materials are collected, they are shipped off to recycling facilities. Paper, cardboard, aluminum, and plastics are often pressed into large bales. They are then shipped by truck or train to recycling centers, sometimes located hundreds or even thousands of miles from collection centers, sometimes nearby. (Much of America's paper waste is shipped to China.) At the recycling facilities, the materials are processed. Glass, for instance, is crushed and then used to make new glass. Metals are melted down and then used to make new metals. Paper products are shredded and then mixed with water. Inks are removed, and the recycled paper is turned into new paper products. Plastic is shredded, then melted, and used to make new products. The materials can be back on the shelves within a few weeks of being collected.

Recycling reduces our demand for virgin materials and thus cuts down on mining and other resource extraction activities. Doubling the rate of paper recycling worldwide would meet increased demand and dramatically reduce the need to cut timber. Recycling a 3-foot-high stack of newspapers saves a single 40-foot Douglas fir tree. Recycling a whole ton of paper saves 17 trees. As these statistics suggest, individual efforts added with those of others can have an enormous impact. If, for instance, the United States increased paper recycling 30%, an estimated 350 million trees could be saved annually. Enough electrical power would be saved to supply the homes of 10 million people.

Paper recycling saves other resources and reduces pollution. Each ton of paper recycled reduces our water demand by 230,000 liters (60,000 gallons) and saves 255 kilowatt-hours of

electricity—enough to run an energy-efficient refrigerator for a year. Recycling paper produces only 25% of the air pollution that comes from manufacturing paper from trees, and it cuts water pollution as well (see Table 17.1).

Similar, even larger, savings are available from recycling other materials. As shown in Table 17.1, producing an aluminum can by recycling uses 95% less energy than making it from aluminum ore (bauxite). Put another way, we can produce 20 aluminum cans from recycled scrap with the same energy it takes to make 1 can from raw ore. Quite a savings!

Another form of recycling is called **composting.** Composting occurs when organic matter such as kitchen waste, yard waste (leaves and branches), and even paper and cardboard are allowed to decompose to a stable, humus-like material. Aerobic (oxygen-requiring) bacteria and fungi in the waste decompose the organic matter. The humus-like material can be applied to flower beds, lawns, gardens, and even farm fields. This material, which contains inorganic nutrients as well, is essentially recycled back to

TABLE 17.1	Benefits of Recycling		
Material	Energy Savings (%)	Solid-Waste Reduction (%)	Air Pollution Reduction (%)
Paper	30–55	130*	25
Aluminum	90–95	100	95
Iron/Steel	60–70	95	30

*130% is possible because 1.3 pounds of waste paper are required to make 1 pound of recycled paper.

Source: William Chandler, *Materials Recycling: The Virtue of Necessity,* Worldwatch Paper 56 (Washington, DC: Worldwatch Institute, October 1989).

the Earth. Homeowners can compost their own organic matter to create soil conditioners for use in the yard or vegetable gardens. Cities and towns can compost waste. Because landfill space is limited, many states have banned the disposal of yard waste (grass clippings and branches) in landfills. There are currently around 9,350 municipal yard waste composting programs in the United States, according to the U.S. Composting Council: most of these are in states where yard waste disposal has been banned. Some zoos have even gotten into the act, most notably Seattle's Woodland Park Zoo, which produces compost from animal manure and sells it as Zoodoo (Figure 17.5).

Compost can be mixed with sewage sludge—a process called **cocomposting.** Bacterial decomposition releases heat, and in a few days, the compost pile may reach 66°C (150°F) inside, sufficient to destroy pathogenic (disease-causing) bacteria that might be present in the processed human waste.

In light of the progressive deterioration of soils here and abroad and the increasing demand for food, composting could be a popular recycling method. However, because the final product is relatively low in nitrogen and phosphorus, many farmers prefer to use commercial fertilizers. Another problem is the cost of transporting compost to farm fields.

GO GREEN!

Recycle used cell phones and rechargeable batteries. Some electronics and office supply stores, including Best Buy, Office Depot, and Staples, recycle old computers and other electronic equipment for free or a small fee. Cell phones can be recycled through certain charities.

Composting itself has problems. One of them is that, for composting to be successful, nonorganic material (for example, cans and glass) must be removed from the waste stream. Source separation could greatly reduce this problem and make composting more feasible. Still another problem is that large tracts of land are needed to compost organic wastes, and these areas must be situated away from people to avoid problems with insects and odor.

Composting is much more popular in Israel and in frugal European countries—in particular Italy, England, the Netherlands, and Belgium. In Holland, for example, a single company produces 180,000 metric tons of compost annually from the waste generated by 1 million people.

Model Recycling Programs

The United States recycles slightly more than 32% of its solid municipal waste, but the EPA believes we could nearly double that amount. A 60% to 80% recovery rate may eventually be achieved as energy prices rise and resource supplies fall. Many other nations have a tremendous lead on the United States. Japan, the Netherlands, Mexico, and South Korea lead the world in paper recycling. South Korea, Mexico, and China, in fact, now import wastepaper from other countries for recycling. (Paper and cardboard bound for recycling facilities in China is the number one export item leaving the port of New York City.)

Japan, a leading consumer of paper, intensified its paper-recycling program in the mid-1960s. Short on landfill sites, this tree-poor nation launched a major recycling program with prodding from environmentally concerned citizens. In Hiroshima, citizens separate their wastes and carry paper to local collection centers for recycling. Bottles, cans, and other items are also recycled. In Machida, Japan, an astounding 90% of the city's waste is recycled, thanks to citizens' efforts to separate trash, a highly computerized recycling system, and heavy fines.

The Japanese need to recycle paper and other goods because they import much of their energy, have little land for waste disposal, and lack the great forests of nations like the United States and Canada. But the success of their programs stems from more than necessity. The unity of the Japanese and their willingness to cooperate to solve a problem are crucial to their success. Their cultural ethic of frugality, their foresight, and strict laws also play a big role in their success. The government has helped out in many ways. In Fuchu City, a Tokyo suburb, the government purchased the costly recycling equipment and then turned it over to a private company, which operates it with profits earned from the sale of recycled materials.

The Netherlands is also a leader in paper recycling. Like Japan, it is short on land and forests. To reduce waste, the government established a **waste exchange,** a service that matches buyers and sellers of waste. The government also established a way to stabilize prices. Normally, the cost of recyclables vacillates wildly, according to supply and demand. In periods of low demand, recyclers may find few markets and prices too low to stay in business. To buffer against these destabilizing

FIGURE 17.5   Zoodoo. Manure and bedding of zoo animals at Seattle's Woodland Park Zoo are composted and turned into soil supplements, which are sold to area residents. This program cuts down on waste, reduces pollution, and makes the city a small profit.

cycles, the government of the Netherlands buys recyclables at a set price when demand or market values drop. When prices increase, the government sells off its supplies and replenishes the funds that keep this system and the country's recyclers in business.

The Netherlands also promotes source separation. But, unlike Japan, where source separation is merely promoted, in the Netherlands it is the law. The state of New Jersey and the town of Islip, New York, have similar mandatory recycling programs.

The United States is becoming a recycling nation. In 2006, the most recent year for which data are available, there were 8,660 curbside recycling programs, serving well over 130 million people, or nearly half of the population. Five states—California, Iowa, Maine, Oregon, and Minnesota—have achieved recycling rates over 40%. Seven states have recycling rates of 30% to 40%. Some materials with high recovery rates include newspaper (about 88%), corrugated cardboard (about 72%), major appliances (67%), steel cans (about 63%), and aluminum beverage cans (about 45%), according to the EPA. In addition, nearly 35% of all rubber tires were recycled, while others were retreaded or used as fuel.

The success of the United States' recycling efforts is largely the result of increased public awareness, the work of private companies, and can and bottle bills that require customers to pay a deposit on all beverage containers, which is returned to them when they return the container to a store. Eleven states now have container deposit bills in place, among them Oregon, Vermont, Maine, Michigan, Iowa, Connecticut, Massachusetts, Delaware, and New York. (Ten more have active campaigns to adopt bottle bills, according to BottleBill.org.) **Container deposit bills,** or more commonly **bottle bills,** require vendors to charge a small deposit for each bottle or can they sell—usually about 5 cents. The deposit is redeemed when the container is returned to a center or a machine in a grocery store, which scans the bar code and prints out a receipt, which can be redeemed for cash.

Bottle bills have played an important part in the U.S. recycling strategy. In Oregon, the state with the first bottle bill (enacted in 1972), 95% of the refillable bottles and 92% of the aluminum cans are returned. In Michigan, 96% of beverage bottles and cans are returned. Similar rates are found in other states as well. In states without aggressive recycling programs, the return rate is substantially below the rate in states with formal bottle bills.

The benefits of the container deposit programs are enormous. In states with such laws, the volume of roadside litter has been reduced by 35% to 40%. In New York State, where 400 million cases of beverages are sold each year, the Beer Wholesalers Association estimates that, in a two-year period, the deposit law saves the state $50 million on cleanup, $19 million in solid-waste disposal costs, and $50 million to $100 million on energy. The program created 3,800 net jobs—that is, 3,800 more jobs than were lost. In Michigan, 4,600 net jobs were created. A nationwide bill would create 100,000 net jobs, according to the U.S. General Accounting Office.

Commercial interests have responded well to the need for aluminum recycling. Coors, a Colorado-based brewery, has

recycling centers throughout the United States. Other private companies have installed automated aluminum-recycling machines, called **reverse vending machines** (Figure 17.6). Customers feed their aluminum cans into the machine, where they are weighed and crushed. The machine then pays the consumer for recycled aluminum.

In 1986, once again leading the nation in its efforts to promote recycling, Oregon passed the **Recycling Opportunity Act.** This law requires all cities with over 4,000 people to start curbside recycling programs that pick up materials at least once a month. For smaller communities, the law requires city officials to establish recycling centers at landfills.

Deepening Our Commitment to Sustainable Waste Management

Most conservationists applaud the remarkable gains in recycling and composting throughout the world. They argue, however, that much more must be done to create a sustainable system of waste management. Clues as to what needs to be done can be gained by looking at the commonly used recycling emblem. As you probably know, it consists of three arrows. The first stands for collection, the second for remanufacturing (the use of recycled materials in the production of new products), and the third for purchase (that is, the purchase of products made from recycled materials). All elements are essential to the success of recycling programs throughout the world.

Individuals, businesses, and governments can all play a significant role in promoting recycling by contributing to all three phases of this cyclical process. For the most part, however, efforts for many years have been focused primarily on collection. Thousands of collection programs have been started since the 1970s. Much less attention has been paid to

FIGURE 17.6 Reverse vending machine.

remanufacturing, that is, manufacturing new products with recyclable materials. Even less attention has focused on **procurement,** ensuring products made from recycled materials are purchased by individuals, businesses, and governments. We will consider each one separately.

Promoting Collection   Collection programs depend on all three key sectors of society: private business, government, and citizens. In some cities and towns, local government is responsible for solid-waste pickup and disposal. In others, private waste collection companies are responsible. Both can play a key role in promoting recycling. State and local governments as well as businesses, for example, can promote reuse and recycling campaigns through television advertisements, billboards, and pamphlets. Governments can also make recycling mandatory. Others may want to invest in the recycling equipment or in land for reuse and recycling facilities and either turn it over to private interests to run or enter into joint ventures with for-profit businesses. Governments and private industry can also help by setting up waste exchanges, that is, clearinghouses where consumers of reusable and recyclable materials find out what is available in their area. Clearinghouses link sources of reusable and recyclable items with folks interested in procuring them.

Governments can provide tax breaks to private trash haulers to help promote recycling. They can also remove subsidies from raw materials and energy, which will make the reuse and recycling option more attractive to commercial interests. Unfortunately, many current policies discriminate against recycled materials. For instance, the U.S. Forest Service sells timber each year at a substantial loss, which reduces the price of wood to consumers. The Wilderness Society estimates that the Forest Service loses about $265 million a year from below-cost timber sales. The Forest Service disputes this figure but admits that its accounting practices do not reflect all costs, such as road building into areas to be harvested.

This system sends false price signals to consumers—notably, that wood is much cheaper than it is. It discourages the efficient use of wood and wood pulp, which is used to make paper. Critics point out that this system shifts the true cost of wood products to the taxpayer (who is subsidizing cheap wood). Pricing more realistically would raise the price of wood, wood pulp, and other wood products and could be a shot in the arm to the recycling industry.

Cheap energy, a major factor in the production of goods from raw materials, also is heavily subsidized, which hinders recycling efforts. According to one estimate, subsidies to the U.S. energy industry—in the form of low-interest loans and tax breaks—cost U.S. taxpayers about $200 billion per year. Although energy subsidies, like those given to the oil industry, reduce the cost at the pump or in our homes, they artificially hold costs down, which can result in irresponsible energy use (waste). One sign of this relationship between artificially low price and waste was the emergence of gas-guzzling SUVs and trucks in the United States in the 1990s and early 2000s. Many Americans purchased low-mileage vehicles while gasoline prices were $1 to $2. Subsidies are used in other countries as well. In 2008, for instance, the Chinese government subsidized oil to keep the price at $3 per gallon to stimulate demand for more cars.

Without subsidies, which can be substantial, fuel prices would be much higher, and citizens would very likely act more responsibly. It's not just our choice in automobiles that would be affected, but choices in almost all aspects of our lives and in the business world as well. Recycling, for example, would be far more widespread if companies that produce materials and products paid the real cost of energy. "By underpricing energy and other natural resources," says the Worldwatch Institute's Cynthia Pollock, "governments subsidize the continuation of a throwaway society and the disruption of ecosystems."[1]

Recycling and reuse could also be stimulated by reducing the complexity of our waste stream. For example, manufacturers can eliminate containers such as ketchup bottles that are made of several different types of plastics, making them difficult to recycle. On the South Pacific island of Fiji, all soft drinks come in the same refillable glass container. Different brands have different paper labels, so glass bottles can be returned to any manufacturer to be washed and refilled. The governments of Denmark and Norway allow fewer than 20 different returnable containers for beer and soft drinks. Standardization facilitates reuse and eliminates the cost of transporting bottles to distant manufacturers.

Before we select one packaging over another, it is important to examine how much energy each requires. Not all containers are created equal. An aluminum can, for instance, requires an enormous 7,000 British thermal units (Btu) of energy to manufacture initially and 2,500 Btu to recycle. A glass bottle requires 3,700 Btu to manufacture and 2,500 Btu to recycle. But bear in mind that glass is a bit heavier, so it requires a bit more energy to transport.

Increasing Remanufacturing and Procurement   The strategies we have examined would surely help increase the supply of recyclable materials by increasing recovery rates. But increasing the recovery rate is only one-third of the battle, as noted earlier. Businesses, citizens, and governments must also find ways to incorporate recycled materials into their products.

To promote remanufacturing and stimulate demand, state and national governments sometimes require their agencies and other branches of government to buy recyclable materials. Or they may permit the purchase of such materials, if they are economically competitive with products made from virgin materials. In other instances, however, agencies are permitted to purchase recycled goods as long as they are within 5% to 10% of the cost of products made from virgin materials. Such actions greatly

GO GREEN!

Use the recycling facilities in your dorm and on your campus, and talk your friends into joining in. If your college doesn't have a recycling program, why not start one?

---

[1]The quotations from Cynthia Pollock in this chapter are from *Mining Urban Wastes: The Potential for Recycling,* Worldwatch Paper 76 (Washington, DC: Worldwatch Institute, 1987).

increase the demand for recycled products, with spillover effects on the general economy. That's because government purchases account for 20% of the gross national product of the United States. (The gross national product, as explained in Chapter 2, is the value of all goods and services produced by an economy.)

Recognizing the importance of recycling and the potential influence it could have on recycling, the U.S. Congress passed the **Resource Conservation and Recovery Act (RCRA)** in 1976. This law assigned to the EPA the task of drawing up guidelines for recycled materials. The guidelines, which were supposed to be completed within two years of the act's passage, were intended to establish recycled-material content that would be acceptable for use by various government agencies. However, much to the dismay of many conservationists, the EPA dragged its feet for over a decade and did not issue guidelines until it was brought to court by the Environmental Defense Fund, a national environmental group. Since then, the EPA has issued numerous guidelines for recycled content.

RCRA also permits governmental agencies to purchase products that are made from recycled materials, such as paper, but requires that they be of reasonable cost. Unfortunately, "reasonable cost" has been widely interpreted as meaning the cheapest. Nevertheless, many offices and departments are currently purchasing recycled products.

In response to the federal government's slow pace, Maryland and numerous other states have passed similar laws. In 1993, for instance, the state of Colorado passed a law requiring all attorneys filing documents with the state government to use recycled paper. Over the years, these laws have greatly increased the incorporation of recycled materials into new products.

Private business has also joined in by opening new plants that incorporate recycled materials into their products. For instance, in Canada and the United States, dozens of mills have opened in the past decade or so to make paper and paper products from secondary (recycled) materials.

Individuals can also make a difference by purchasing recycled products whenever possible and by writing manufacturers to request the use of recycled materials in their products.

Recycled products are becoming more and more common the world over. Numerous building products are now made from recycled material. Recycled milk jugs, for example, are used to make a safe, durable plastic "wood" used for decks, docks, and other purposes. One company is using recycled plastic from CDs and headlights to make pens and mechanical pencils. Rubber from automobile tires is used to make durable roof tiles. Carpeting is made from recycled plastic pop bottles. Numerous directories are now available to help builders and homeowners locate sources; one example is *GreenSpec* by the *Environmental Building News*.

Impressive as these efforts are, products made from recycled materials represent only a fraction of the total goods produced by the global economy. In a sustainable economy, just the opposite will be true.

Creating a Much More Environmentally Friendly Recycling System   Recycling is vital to creating a sustainable future, but it is still flawed. Many products, for instance, are **downcycled**

during the process, meaning they are recycled into lower-quality products. This is because many recyclable products, such as automobiles, contain a mixture of materials. When crushed and melted down, these materials form a lower-quality material that may not be as strong or as durable as the original components. They cannot be used to make the original product. Other products contain toxic materials that create problems for recyclers.

To circumvent these and other problems, William McDonough and Michael Braungart, authors of *Cradle to Cradle*, argue that we need to design products for complete recycling. Some products such as shoes, clothing, and packaging can be made from nontoxic organic materials—which McDonough and Braungart call biological nutrients. Such materials would contain no heavy metals or toxic dyes. As a result, these organic materials could easily be composted (converted into a soil amendment) without fear of toxic pollution. Other products are made of nonbiological materials, known as technical nutrients. They could be recycled back into the original product.

This concept, known as cradle-to-cradle manufacturing, is gaining popularity, and numerous products are now on the market, including fabrics such as those used to make the seats of the A380 jet liner. Products can even receive cradle-to-cradle certification. Certification provides a company with a means to credibly measure achievement in environmentally intelligent design and helps customers purchase products that meet these important goals: use of environmentally safe and healthy materials; design for material reutilization, such as recycling or composting; use of renewable energy and energy efficiency; efficient use of water and maximum water quality associated with production; and adherence to strategies for social responsibility.

Researchers are also examining a wide range of products from new adhesives to ceramics based on natural processes. Other products can be made from "technical nutrients"—materials that can be reused over and over again to make the same product. These efforts, in part inspired by the book *Biomimicry* by Janine Benyus, seek to find efficient, nontoxic, nonresource-intensive replacements for modern materials and could spark a revolution in modern industry. For example, production methods for ceramics based on methods used by shellfish could prove to have little, if any, impact on the environment—a far cry from modern ceramics manufacturing.

17.3   Waste Disposal: The Final Option

Ecologists envision an ideal world—some call it Ecotopia—in which there is no waste. In Ecotopia, virtually all plastics are either recycled or reused. Waste materials including paper, aluminum, steel, and glass are separated at their source and then shipped off to local recycling facilities, baled, and transported to regional manufacturers, who turn them into useful products once again. Yard wastes—grass clippings and leaves—either are composted by homeowners and used to enrich soils or are shipped to composting facilities, where they are piled in huge windrows and left to decay, turning into compost. Eventually, the rich organic materials are combined with wastes from municipal sewage treatment plants and sold as soil conditioners to gardeners and farmers. With all of this recycling and composting, there's no waste to speak of. Nothing needs to be landfilled or incinerated.

Ecotopia may be years, even decades, away. Until that time, modern industrial societies will undoubtedly continue to rely on landfills and incinerators.

Dumps and Sanitary Landfills

In the year 500 B.C., the Greeks and Romans hauled their trash outside the city walls and dumped it downwind, so as not to offend the residents. Flies and rats invaded the debris, and when the wind shifted, few people were spared the odorous onslaught.

Over 2,460 years later, in the United States, the most technologically advanced nation on Earth, many cities and towns still followed the same tradition. Open sores in the landscape, where rotting garbage swarmed with flies and rats, were common. Making matters worse, officials periodically burned the accumulating garbage to reduce its volume. Black smoke, filled with the toxic by-products of burning rubber and plastic, billowed out of dumps everywhere. Rain and snowmelt trickled through the garbage, carrying sometimes hazardous liquids into the underlying groundwater, threatening groundwater and drinking water supplies. Troubled by their own garbage, Americans called for changes.

In 1976, their demands were met when the U.S. Congress passed the Resource Conservation and Recovery Act (RCRA), which this chapter discussed earlier. Among its many key provisions, RCRA called for an end to **open dumps** by 1983. What replaced them was the **sanitary landfill,** an excavation or hollow in the ground in which garbage was dumped, compacted, and covered daily with a fresh layer of dirt (Figure 17.7).

Used today throughout the developed world, the sanitary landfill reduces odors caused by rotting garbage and problems with insects and rodents. Because a soil layer is placed over the trash, compacted, and generally sloped to reduce water percolation into the garbage, groundwater contamination can be greatly reduced or eliminated. The protective layer of soil also reduces insects and other pests that could carry disease.

Landfills offer many other advantages over the open dump. Besides being cleaner, they can be "reclaimed" after they've been filled to capacity—that is, they are covered with topsoil and returned to some previous use, or given over to some new use when they are filled. In Evanston, IL, city officials built a park with baseball fields, tennis courts, and toboggan runs on Mount Trashmore, a hill 30 meters (100 feet) high and underlain by garbage. In Maryland, wastes from 87 roadside dumps were hauled to abandoned coal strip mines. After the gullies were filled with waste, the site was covered with soil and replanted.

Despite their advantages over the open dump, landfills are still a primitive method of waste management. If the soil or rock below a landfill is permeable, pollutants may drain into underlying aquifers, polluting water needed by municipalities, farms, or industry. Rotting debris also produces methane, a potentially explosive gas. Ordinarily, methane production is greatest during the first two years of operation but is not high enough to be hazardous. However, if large concentrations build up, methane can pass through the soil into nearby buildings. If it reaches a high enough concentration, it can explode.

More important, though, landfills entomb valuable resources that could be reused or recycled. They also require large tracts of land. A town of 10,000 people, for instance, produces enough trash each year to cover a 0.4-hectare (1-acre) site 3 meters (10 feet) deep. Landfills are also expensive—and are growing more expensive by the day—especially in more-populated areas. Because of the growing shortage of suitable sites, city officials are often forced to haul their trash farther and farther from the site of production. As a result, the cost of landfilling garbage has skyrocketed. Philadelphia, for instance, has used up all of its landfill sites and must now transport its garbage to Ohio and Virginia. In eastern states, the cost of landfilling a ton of waste has climbed from $20 in 1980 to over $100 a ton today. With the high cost of gasoline and diesel needed by garbage trucks, this price is bound to skyrocket. In western states, where more open space is available, tipping fees (the cost of landfilling trash) are still on the low end of the scale, which is one reason why recycling efforts in many western states are way below the national average.

Worldwatch Institute's Cynthia Pollock notes, however, that in many areas, landfill fees are held artificially low by

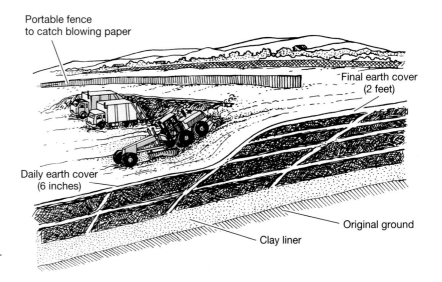

FIGURE 17.7 Sanitary landfill operation. The bulldozer spreads and compacts solid wastes. The scraper (foreground) is used to haul the cover material at the end of the day's operation. Note the fence that catches any blowing debris.

Portable fence to catch blowing paper

Final earth cover (2 feet)

Daily earth cover (6 inches)

Original ground

Clay liner

local governments. Because of this, trash removal companies and local governments have little incentive to make better use of these wasted resources.

Incineration

Another strategy now being used in the United States and abroad to reduce waste and capture part of its intrinsic value is **incineration**—burning municipal solid waste. Incinerators can burn unseparated trash—containing plastics, metals, paper, yard waste, and glass—or separated trash that is relatively free of noncombustibles such as glass and metal. The heat produced during combustion is often used to generate electricity or, less commonly, steam for industrial processes and heating buildings.

Incineration became popular during the oil crises in the 1970s, but proponents soon found that the technology had many problems—most notably, toxic air pollutants. Refinements in the technology have eliminated some of the problems, leading many U.S. cities to turn to this option. Currently, there are about 120 municipal incinerators operating in the United States, down from 170 in 1991. Incineration now handles 14.5% of the municipal solid waste output of the United States.

By comparison, there are over 350 incinerators in Western Europe, Japan, the former Soviet Union, and Brazil. Over half of the municipal solid waste in Japan, Sweden, Denmark, and Switzerland is burned.

Incineration is desirable for a number of reasons. First, it captures energy that would otherwise be lost. It also fits nicely with existing waste management practices and requires less land than landfills. No modifications of the pickup system are needed, as they are for recycling and composting. Incinerators can be designed to scale, providing flexibility to serve towns that produce as few as 100 metric tons of garbage a day, up to larger cities producing over 3,000 metric tons per day.

Despite their mushrooming popularity, incinerators are still viewed skeptically by many. In fact, at least 300 proposals for municipal solid-waste incinerators have been stopped in the United States because of public opposition. Why? Despite improvements in design and controls on air pollution, incinerators that burn plastics and other materials containing chlorine emit a dangerous class of compounds called **dioxins.** Dioxins have been linked to cancer, birth defects, and other problems. Recent evidence indicates that they may also weaken the immune system, making individuals more susceptible to cancer. Concerned about potential health problems, Sweden and Denmark have halted the construction of additional incinerators until their scientists can determine how substantial a risk they pose.

Toxic heavy metals such as mercury (contained in batteries) and acidic substances such as hydrochloric acid (produced when plastics are burned) also are emitted from waste incinerators. The EPA requires pollution-control devices on all municipal solid-waste incinerators to eliminate several toxic pollutants. The cost of retrofitting some facilities will probably put some incinerators out of business. At this writing, only a handful (six or so) of the United States' incinerators actually

meet emission standards. Requiring residents to separate nonburnables and plastics could help reduce toxic emissions.

Critics also note that pollution-control devices generate a hazardous residue called **fly ash**—materials that were removed from the smokestack gases. The ash in the bottom of the incinerator, called **bottom ash,** also is potentially toxic. It contains heavy metals and other pollutants. Consequently, some countries now classify ash from incinerators as a hazardous material that must be properly disposed of to protect human health. In the United States, however, the EPA has chosen not to classify municipal solid-waste incineration ash as a hazardous material, meaning that many tons of potentially hazardous waste are currently being landfilled alongside garbage each year.

17.4   Hazardous Wastes

During the 1960s and 1970s, growing discontent with solid-waste disposal techniques brought about sweeping changes in governmental policy on solid-waste disposal. But few people anticipated that another problem lay on the horizon, a problem brought to life by Love Canal, in Niagara Falls, NY. The canal had been the repository for over 20,000 metric tons of hazardous wastes, including dioxin, from 1947 to 1952 (Case Study 17.1). Troubles began in the late 1950s after the city pressured Hooker Chemical Corporation to turn over the land it had used as a hazardous-waste dump to build a school and residential community. As construction on the project began, workers broke through the clay cap that Hooker had placed over the dump site. Homes and a school were constructed nevertheless.

In the late 1950s, toxic chemicals began to ooze out of the ground from the rusted steel drums. Children playing in the toxic ooze became ill and suffered skin burns; some even died. Subsequent health studies showed a significantly higher incidence of birth defects, respiratory difficulties, and other illnesses in people living near the site. State and federal officials evacuated hundreds of families, and efforts began to clean up the site.

Love Canal was not an isolated case. The Netherlands, Austria, Hungary, and Sweden all witnessed similar problems. Many additional examples came to light after the downfall of the Soviet Union. Eastern European nations contained many horribly polluted industrial sites. One incident after another revealed the seriousness of the problem (Figure 17.8). As people began to realize that highly toxic materials had long been carelessly discarded on the land and in the water, they began to call for action. Governments and international agencies drafted rules and regulations to clean up the thousands of potentially harmful toxic-waste dumps already in existence and to prevent further tragedies. Despite many years of work on the problem, progress has been slow. New problems are also arising with E-waste, a popular, informal name for electronic products nearing the end of their "useful life." This waste includes discarded computers, televisions, VCRs, stereos, copiers, and fax machines. We'll discuss conventional hazardous waste and E-waste in this chapter.

THE CHEMICAL TIME BOMB AT LOVE CANAL

In the 1880s, industrialist William T. Love began constructing a canal that would provide water and electrical power to the growing city of Niagara Falls, NY. Connecting the Niagara River just above the falls to a point below the falls, he envisioned it as a focal point of industrial development. But because of economic troubles, the project was abandoned. Many parts were filled in as the city expanded, and by the 1900s, only a 900-meter (3,000-feet) section of the ill-fated canal remained—a mute testimony to one man's dream gone sour.

In 1942, Hooker Chemical Company entered into an agreement with the canal's owner, Niagara Power and Development Corporation. Their agreement would allow the company to use the remaining section as a hazardous-waste dump. There were no regulations for waste disposal at the time, and very little concern over the effects of burying steel drums containing hazardous wastes directly in the Earth.

In 1946, Hooker purchased the canal from Niagara Power, and for the next six years dumped thousands of steel drums containing 20,000 metric tons of hazardous materials into the canal. Then, in 1952, the city of Niagara Falls, which wanted land on which to build a school and residential community, began condemnation proceedings, which would allow the city to take the land from the company. Hooker caved in to the pressure and signed over ownership of the site for a nominal $1 payment. The company then sealed off the site with a clay liner and turned it over to the city, allegedly with a warning not to build on the dump site itself. In turn, the city signed an agreement releasing the company from any damages that might occur from use of the land.

In 1954, the city began building the 99th Street Elementary School right over the dump. Two hundred thirty-nine homes went up nearby, and another 700 soon skirted the chemical time bomb. During construction, though, workers broke through the protective clay cap and apparently did nothing to repair it.

A few years later, troubles began. Rusty and leaking barrels began to emerge in low spots (Figure 1). Chemical wastes pooled on the surface. Residents noticed harsh chemical smells, so powerful that they killed grass and vegetables and took the bark off trees. Children playing near the pools of toxic ooze suffered serious chemical burns; some became ill, and a few died.

Then, in 1977—20 years later—unusually heavy rain and snow converted the site into a sea of mud. Through the years, corrosion had turned the steel drums into leaking sieves. Toxic wastes came bubbling up to the surface, pooling in people's backyards and filling basements with black, smelly goo. Residents complained of strange odors. Pets began to die mysteriously. Many residents complained of severe headaches and rectal bleeding. Children playing in the area were seriously burned by chemicals. Concerned with the many complaints it was receiving, the New York State Health Department initiated a health study. To their surprise, they found a high incidence of liver, kidney, and respiratory disorders as well as epilepsy and cancer. In addition, the rate of miscarriages and birth defects among residents was found to be three times greater than the national average. In July 1978, the health department strongly recommended that pregnant women and children under two years of age move out of the area.

FIGURE 1 Toxic chemicals and rusted drums worked their way to the surface in Love Canal, forcing the evacuation of hundreds of families.

In an investigation of the site, more than 80 toxic chemicals were identified. At least one dozen were known carcinogens. As the evidence grew, city and state officials closed down the school, fenced it off, and evacuated several hundred families. In 1979, then-President Jimmy Carter declared Love Canal a disaster area. The federal government evacuated over 780 more and provided housing for them at a cost of $30 million.

In the fall of 1979, officials began a massive cleanup of the dump site. Some homes were removed, others bulldozed; still others were left intact, their doors and windows boarded up (Figure 2). As the cleanup proceeded, additional studies by the EPA revealed that the immediate vicinity was badly polluted but that the toxic wastes had not migrated much past the first two rows of houses on either side of the canal. The EPA concluded, therefore, that the 1980 evacuation was unwarranted. The study also showed that the chemical wastes had not migrated to deep aquifers and were unlikely to move much farther.

Today, Love Canal and the schoolyard where children once played is surrounded by a chain-link fence with bright yellow signs warning passersby of the hazardous wastes. The houses nearest the canal have been bulldozed. The dump site itself has undergone extensive work. Officials decided that the best strategy was to leave the hazardous waste in place and to attempt to contain it—that is, to restrict its movement. To this end, a huge ditch was dug

(continued)

FIGURE 2 One family took their home with them, leaving only the foundation as an eerie reminder of what had happened. Abandoned houses in the background are boarded up.

around the site. In the ditch, porous tile was installed. It's supposed to contain the waste—that is, keep it from migrating outward. The leachate collected by the drain system is pumped to an on-site treatment plant. Here, filters remove chemicals from the collected leachate. Water produced by the process is then released into storm water sewers. Critics point out that, while this may seem like a good strategy, chemicals such as mercury and other heavy metals that are not removed by the filters end up in the nearby Niagara River.

The site was also covered with a plastic liner and a clay cap to minimize the amount of water seeping into it from rainwater or snowmelt and to prevent chemicals from the site from evaporating into the air. The clay cap also prevents direct contact with contaminated soil.

In September 1998, the New York State Department of Health announced results of a five-year habitability study. Their conclusion: Portions of the Love Canal neighborhood are "as habitable as other areas of Niagara Falls." They did not declare these areas safe, however. The EPA weighed in on the issue, too, and deemed the homes outside the immediate site suitable for habitation. A public corporation, the Love Canal Revitalization Agency, soon took ownership of the abandoned properties. The agency renamed the area Black Creek Village, repaired the homes, and then sold them. Critics argue that this new development is foolhardy at best

because the "habitability" certification was improperly performed. In fact, the area was erroneously compared to two other badly contaminated sites. Lois Gibbs, a housewife turned activist who has been instrumental in forcing the government to take action at Love Canal, notes that the homes that were "reinhabited are still contaminated, still unsafe." She notes that there have been no cleanup measures taken around the homes, which were found to have several toxic chemicals in and around them. Only the creek and sewer systems were cleaned.

The canal itself and the region immediately surrounding it will probably remain fenced off forever, serving as a sort of national monument to our carelessness and as an impetus for a greater commitment to prevent future problems of this nature from ever happening again. In 1990, the first family moved into the outer region, but sales were slow at first because local banks were unwilling to lend money. In 1992, however, the Federal Housing Administration agreed to provide mortgage insurance to families who wish to purchase Love Canal homes. Today, all of the 239 homes have been sold, and while the EPA considers the area safe and new residents have no fear of toxic chemicals, only time will tell if the decision to resettle the area was a wise one. Costing over $227 million to clean up and for other expenses, Love Canal is a symbol of the cost of shortsightedness.

How Big Is the Problem?

Hazardous wastes are broadly defined as waste products from factories, businesses, and homes that affect a wide array of organisms in a variety of ways from nonthreatening symptoms to debilitating illnesses, even death. They include toxic organic chemicals and toxic heavy metals. Hazardous wastes also include corrosive materials, combustible materials, and highly reactive substances, for example, chemicals that can cause explosions when heated. Exposure and other dangers posed by hazardous wastes result from improper storage, disposal, and handling, as well as accidents during transportation. Estimating the global production of hazardous wastes is nearly impossible. Less-developed countries, for instance, have few laws governing hazardous materials, and government officials often have no idea how much is produced. Other countries may classify as hazardous a substance that another country

does not consider hazardous. Even states in the United States may classify materials as hazardous that the federal government does not consider hazardous.

Difficulties aside, experts estimate annual world production of hazardous materials to be about 600 million to 700 million metric tons. The United States, one of the few countries for which information is available, produces the lion's share of these potentially harmful materials. In 1991, hazardous-waste production peaked at about 275 million metric tons (306 million tons). Since then, production has declined dramatically, thanks to pollution prevention efforts. In 2005, the United States produced about 33 million metric tons (37 million tons), down from 37 million metric tons in 2001, according to the EPA.

For a century or more, hazardous wastes have been indiscriminately strewn about the landscape in the industrial nations of the world. Until quite recently in the history of these nations, wastes were frequently placed in large steel drums that were dumped

FIGURE 17.8 Workers clad in protective clothing and face masks pump toxic materials into a tank truck from barrels at an abandoned paint factory.

in landfills or on vacant lots and left to rust. Over time, leaks developed, releasing into the soil a steady trickle of harmful chemicals, which percolated into the underlying groundwater that drained into lakes and streams or was drawn to the surface in wells. In other instances, hazardous wastes were pumped into deep wells, into municipal sewage systems, or directly into lakes and streams. Some companies poured their wastes into sandy pits, creating evaporation ponds. From there, hazardous materials seeped into the ground. What is interesting is that all of these techniques were considered acceptable practice. Little concern was raised over the potential health effects or impact on the environment. As environmentally unsound as the acceptable practices were, some waste haulers practiced even worse disposal methods. Some unethical haulers, for example, pulled their trucks up to streams and discharged wastes under the cover of darkness. Others opened the spigots and drove along highways at night, spilling tons of toxic wastes.

Years of careless disposal have left a legacy of contaminated sites and polluted lakes, streams, and aquifers. In Europe and North America, hazardous-waste dumps have caused the uprooting of entire communities. Every country in Europe is plagued with toxic-waste sites—both old and new—needing urgent attention. In the United Kingdom alone, 5.5 million metric tons of hazardous wastes were discarded in 1980; three-quarters of this material was dumped into landfills without adequate liners to prevent it from leaking from the site. In Holland, an estimated 8 million metric tons of hazardous materials are buried in the soil, mostly in leaky steel drums. As noted earlier, with the fall of the Soviet Union and the liberation of Eastern Europe, reporters and scientists have discovered a toxic nightmare: thousands of abandoned dumps and contaminated sites.

Many countries have spent enormous amounts of money to locate and begin to clean up leaking landfills and contaminated industrial sites. In the United States, estimates vary as to the number of sites in need of cleanup and the eventual cost. The EPA currently includes over 1,300 sites on its National Priority List (NPL), a list of private and federal sites contaminated by hazardous wastes that pose or once posed a significant risk to human health or the environment and are eligible for cleanup under the Superfund program (discussed shortly). According to

the EPA, 1,010 Superfund sites are construction complete, meaning that all immediate threats have been eliminated and all long-term threats are under control. Even so, these sites may stay on the NPL for many years, as they require continuous monitoring and actions such as removal of groundwater, which is filtered to remove toxic chemicals. Unfortunately, these NPL sites may just be the tip of the hazardous-waste iceberg.

Soon after the Love Canal episode, the EPA estimated there were probably 10,000 additional sites in need of cleanup. In 1989, it increased its estimate to 39,000. The U.S. General Accounting Office estimates that the number could be much higher—as many as 100,000 to 300,000 sites. That estimate does not include the 17,000 toxic "hot spots" on military bases throughout the United States. The cleanup cost could come to hundreds of billions of dollars, a stern reminder of the economic sense behind prevention.

Alarmed at the number of hazardous-waste sites in need of cleanup, the U.S. Congress passed the **Comprehensive Environmental Response, Compensation, and Liability Act (CERCLA)** in 1980, commonly called the **Superfund Act.** This law called on the EPA to identify and clean up hazardous-waste sites with the assistance of state governments (which were required to chip in 10% of the cost). The Superfund Act initially created a $1.6 billion fund generated by taxes levied on polluters, notably petroleum-refining and chemical-manufacturing companies, between 1981 and 1985. During that period, however, only 13 sites were cleaned up—some of them, say critics, inadequately. Regulators soon found that cleaning toxic-waste dumps was much more difficult and costly than anticipated. Just analyzing the chemicals in a single site can cost upward of $800,000. Simple steps to stabilize a leaking site can cost $500,000.

The Superfund Act was renewed in 1986. The amendments expanded the fund, bringing the grand total to $16.3 billion, with money coming from federal and state coffers and an even broader business tax. Much of this money was earmarked for cleanup, but the 1986 law also provided money for research on new hazardous-waste treatment technologies.

The Superfund Act empowers the EPA to collect the cost of cleanup from the hazardous-waste dump or hazardous-waste site owners and operators, as well as from companies that paid to have their wastes dumped in them. Under the law, then, all participants are required to pay their portion. As one person put it, a company is liable for its waste forever.

While this provision may sound fair and reasonable, millions of dollars in Superfund money were spent to determine who was responsible for the hazardous waste at various sites. Because of the multiple-responsibility provision and other problems, cleanups occurred much more slowly than originally hoped. By 1992, only 71 contaminated sites had been cleaned up.

In response to widespread criticism over the slow rate of cleanup, the EPA found ways to cut the time for remediation by two years, and by 1999, 600 sites had been completely cleaned up and restored.

The Superfund Act is an important law that has helped address a critical problem, but it has been criticized for several reasons, besides the initial slow cleanup time. One criticism is that it provides money for cleanup and financial compensation for property damage but nothing to reimburse people for health effects. Critics also argue that toxic residues and contaminated soils excavated from hazardous-waste sites and transported to new landfills could eventually leak, even though the landfills are equipped with liners of clay and synthetic material, no matter how carefully the new landfills are planned and constructed. Leakage could create costly problems for future generations. Finally, critics point out that while CERCLA helps us clean up our polluted lands, many additional billions will have to be spent to purify the thousands of contaminated aquifers throughout the United States. (A considerable amount of money continues to be spent on cleaned-up sites to purify groundwater, which is why many sites remain on the National Priority List.)

The Superfund program remains in effect today, although the Superfund tax expired in 1995. As a result, the U.S. government no longer levies a tax on the petroleum and chemical-manufacturing industries to finance the cleanup of contaminated sites. Financial responsibility has shifted to taxpayers. Efforts are under way to overhaul the act, but progress has been slow. Some organizations are calling for a complete overhaul; others prefer a piecemeal approach, revising one section at a time. While some organizations are seeking to strengthen this law, industry groups have worked long and hard to weaken the law.

Brownfields: Converting Polluted Landscapes into Productive Ones

Cities and states throughout the world are littered with abandoned industrial facilities, some of which may be contaminated with toxic wastes. In the United States alone, the EPA estimates that there are around 450,000 contaminated sites. Rather than buy them and convert them back into productive use—that is, to rebuild factories or build new ones—many companies build on new sites in outlying areas, converting farm fields and forests into factories. Why?

Lenders, investors, and developers fear that if they become involved in the redevelopment of a site and discover contamination, they will become liable for cleaning up chemicals on the property. They might have to spend millions of dollars removing someone else's mistake. It's far cheaper to develop a pristine site, a **greenfield,** rather than a **brownfield**—a site that is contaminated or thought to be contaminated by hazardous materials.

The result of their fear is that many thousands of abandoned industrial sites are currently sitting idle, while companies bulldoze fields and cut down forests to build new facilities. Seeking to alleviate this problem, the EPA has launched a **Brownfields Economic Redevelopment Initiative,** commonly referred to as EPA's Brownfield Program. It is designed to empower cities, states, tribes, and others to redevelop abandoned sites, working together with community members and other stakeholders. The initiative seeks ways to clean up and sustainably reuse brownfields within the existing Superfund Act.

To this end, the EPA has published a list of 31,000 properties in the United States taken off the Superfund site inventory. Assessments of the properties have convinced the EPA that these properties are not subject to Superfund cleanup. The EPA will also enter into agreements with prospective purchasers of brownfields not to sue them for cleanup if a site is found to be contaminated. The EPA offers grants to assess sites for contamination and also provides substantial tax incentives to companies willing to redevelop brownfields that are hoped to bring thousands of abandoned or underutilized sites back into productive use. They also offer grants to pay for cleanup of sites should they prove to be contaminated. Further economic incentives come in the form of revolving loans offered by the states. A revolving loan is a low- or no-interest loan that companies can take to develop a brownfield. Their payments go back into a fund that is available for other companies.

Brownfield development helps reduce pressure on undeveloped land and puts a valuable resource (abandoned sites) back into service. Society and the environment, say proponents, are better served by developing these sites than by leaving them vacant. In most instances, contamination is slight to moderate. Interestingly, many of these properties are owned by cities, having been passed on to them by banks and businesses that simply stopped paying taxes on the property in a deliberate act to force their legal transfer. (Cities and counties can legally seize title to property if taxes are not paid within a certain period.) The upshot is that this land is often cheap. Even if some cleanup is required, it may be offset by the inexpensive price tag.

Brownfield development is occurring throughout the United States, with hundreds of projects under way, thanks to federal and state assistance. Since the beginning of the program, the EPA has awarded 1,255 assessment grants totaling more than $298 million, 230 revolving loan fund grants to develop brownfields totaling about $217 million, and 426 cleanup grants totaling $78.7 million as of 2008.

E-Waste

E-waste is the popular and informal name for electronic devices such as televisions, computers, printers, stereos, cell phones, and fax machines at the end of their useful life. Although many of these products could be reused, refurbished, or recycled, most are discarded. E-waste is one of the fastest growing segments of the United States' waste stream, according to the California Integrated Waste Management Board. Why worry about e-waste?

E-waste is not biodegradable and contains a lot of useful materials—substances that could be recycled. However, the greatest concern seems to be that certain components of electronic products contain potentially hazardous materials, for example, fire retardants and toxic heavy metals. Because of this, many European countries banned e-waste from landfills in the 1990s. The state of California currently classifies nonfunctioning CRTs (cathode ray tubes) from televisions and computer monitors as hazardous. Toxic materials they contain can be released when these products are deposited in landfills.

The European Union (EU) further advanced e-waste policy in Europe by implementing the Waste Electrical and Electronic Equipment Directive in 2002. This policy statement holds manufacturers physically or economically responsible for e-waste disposal or recycling at end of life under a policy known as **extended producer responsibility (EPR).** It is being implemented by various members of the EU. Similar legislation has been enacted in Asia.

The idea behind EPR came from Switzerland, which established the first e-waste-recycling system in 1991, beginning with collection of refrigerators. Since then, numerous other appliances and electronic devices have been added to the system. Since January 2005, it has been possible for the Swiss to return all electronic waste to the place of purchase and other collection points free of charge for recycling.

E-waste legislation in the United States has occurred at the state level as national efforts have stalled in the United States Congress. California, Maryland, Maine, Washington, Minnesota, Oregon, and Texas have all passed their own e-waste laws.

In more-developed countries, e-waste is usually dismantled into its parts—metal frames, power supplies, circuit boards, and plastics—and then recycled. E-waste is also mechanically shredded, and sophisticated equipment then separates the metal and plastic fractions, which are sold to various recycling companies. To cover the cost of e-waste recycling, the state of California instituted an Electronic Waste Recycling Fee on all new monitors and televisions in 2004. More and more electronic waste is reportedly being refurbished.

Increased regulation of e-waste in the United States has raised disposal costs. Unscrupulous recycling companies often export e-waste to developing countries. One nonprofit organization (Basel Action Network) estimates that about 80% of the e-waste that goes to recycling in the United States does not get recycled there but shipped to countries such as China, India, and Kenya, where it is recycled and/or dumped in landfills. Lower environmental standards and poorer working conditions—and lower salaries—make recycling e-waste much more profitable in these countries. E-wastes are sometimes burned to remove metals from electronic parts or physically smashed with hammers to remove recyclable plastic and metals. Studies show that uncontrolled burning, disassembly, and disposal are causing environmental and health problems, especially among workers. A recent study, for instance, showed that Chinese workers who recycle e-waste are often exposed to high levels of potentially harmful fire retardants, chemicals that are added to the plastic cases of computers and other electronic equipment to help prevent the electronic equipment from catching fire.

Individuals can help by holding on to electronic equipment, not replacing it just because there's a new model, and by fixing broken electronic devices, although the cost of replacement is often cheaper than repair. They can also help by donating or selling functional electronics to someone who can use it, as well as recycling electronic equipment.

Managing Hazardous Wastes: The Unmet Challenge

Cleaning up past mistakes is only half of the solution to our hazardous-waste problem. The other half is preventing this kind of thing from occurring again. "Unless the wastes currently produced are better managed," argues Sandra Postel of the Worldwatch Institute, "new threats will simply replace the old ones, committing society to a costly and perpetual mission of toxic chemical cleanups." To prevent the indiscriminate and illegal disposal of hazardous wastes, Congress added special hazardous-waste provisions to the RCRA of 1976. These provisions require all producers, transporters, and disposers of these materials to register them with the EPA. The materials can then be tracked from the site of production to the site of their disposal—from "cradle to grave," in the words of waste managers. It is more formally known as the **manifest system,** a manifest being a piece of paper used to denote the destination of cargo. Many other nations now have similar policies. RCRA also ordered the EPA to set standards for packaging, shipping, and disposal of wastes. To prevent further contamination, it required waste disposal companies to obtain licenses. Only licensed facilities may legally accept hazardous wastes.

Like municipal wastes, hazardous wastes can be dealt with in three basic ways, listed in order of desirability: the reduction approach, the reuse and recycling approach, and the discard approach. Hazardous wastes, however, are amenable to a fourth approach as well: detoxification. The discard approach has historically been the most widely used method until recently. Until the 1990s, recycling and reuse were rarely practiced. Source reduction was almost unheard of.

Reducing Hazardous Wastes Manufacturers have several options available to reduce hazardous wastes in factories, a process also called **source reduction.** The first line of attack in the source reduction mode, and often the cheapest, is **process manipulation** or **process redesign.** By modifying or redesigning the manufacturing processes that create hazardous wastes, companies can significantly reduce waste production. For instance, the Borden Chemical Company of California redesigned an equipment-cleaning procedure that once used toxic organic solvents and produced a dangerous sludge. The redesign reduced the company's discharge of toxic organic solvents by 93% and cut sludge wastes generated each year from 350 cubic meters (455 cubic yards) to 25 cubic meters (33 cubic yards). As an added benefit, the changes save the company about $50,000 a year.

3M Company (formerly Minnesota Mining and Manufacturing), a leader in waste reduction since 1975, has cut its waste production in half and, over a 30-year period, saved an estimated $1,000 million in the process (Figure 17.9). The company continues to save millions of dollars a year as a result of its pioneering program for pollution prevention. A large

FIGURE 17.9   3M Company has been a leader in pollution prevention since 1975.

chemical company in the Netherlands has installed a new manufacturing process that has cut its waste production by 95%.

Like 3M, other companies throughout the United States are learning that pollution prevention is not just good for the environment, it also makes extraordinary economic sense.

The EPA has also taken a lead in pollution prevention, starting with its 33/50 program. This voluntary program in the early 1990s called on manufacturers to cut their output of 17 hazardous wastes by 33% by 1992 and by 50% by 1994. Numerous companies participated and made significant progress toward reducing hazardous-waste production. In fact, the program was so successful that it reached its 50% reduction goals one year ahead of schedule and the next year cut the production of the targeted hazardous wastes another 10%.

To reduce their hazardous-waste output, companies can also substitute safer, nontoxic materials for more harmful ones. Numerous companies produce environmentally and people-friendly cleaning agents, paints, stains, and finishes. For some examples, see A Closer Look 17.1.

Conservatively, nationwide efforts to modify manufacturing processes and substitute nontoxic biodegradable compounds for toxic ones could reduce our toxic waste output by 15% to 30%. Many companies have pledged to reduce their output by 80% to 90% by process manipulation, substitution, and other methods.

Waste reduction represents a new way of thinking for businesses. For this idea to enter the mainstream, however, more companies must realize that waste is a sign of inefficiency that results in higher costs. Furthermore, top-level management must commit itself to a program of waste reduction, as 3M and other companies have. A surprisingly small effort can result in huge reductions and economic savings. USS Chemicals, for example, is a company whose management is committed to waste reduction. It has established a reward system for employees who suggest implementable waste-saving techniques.

Source reduction is extremely popular outside the United States, too. The governments of Canada, Japan, Sweden,

GO GREEN!

Use environmentally friendly cleaning agents like those discussed in A Closer Look to reduce the use of toxic chemicals, as well as your own and others' exposure to potentially harmful chemicals.

Germany, Denmark, and the Netherlands, for instance, actively promote nonwaste and low-waste technologies.

Reusing and Recycling Hazardous Wastes   Manufacturers can also make significant inroads into hazardous wastes by reusing or recycling toxic wastes. For example, companies can use relatively pure chemicals produced as a waste in one process as the raw material (input) for another process on site or at another facility. Or they can sell the "waste" to a willing buyer or even give it away, rather than paying exorbitant disposal fees. In some instances, hazardous wastes require some purification to make them reusable, but either way, the savings can be substantial.

To facilitate the exchange of hazardous wastes, the Netherlands put into operation a hazardous-waste clearinghouse or waste exchange. Established in 1969, it keeps track of as many as 150 different chemical substances produced by industry and links buyers and sellers. Numerous private and nonprofit clearinghouses now exist in many U.S. cities and throughout the rest of the developed world. The Northeast Industrial Waste Exchange in Syracuse, NY, operates a computerized network listing wastes from five different regions. Anyone with a computer and modem and the proper password can gain access to the files to find out what is available or to list wastes for sale.

As with other issues, the world may look to Japan for guidance. The Japanese produce about 200 million metric tons of industrial waste, including both nonhazardous and hazardous waste. They recycle over half of that material. Another 30% is incinerated. What is left—about 18% of the industrial waste stream—is disposed of, mostly in landfills.

GO GREEN!

One of the most important things you can do to reduce toxic pollution is to curb consumption, as the production of nearly all products, from cleaning agents to cars, results in the production of hazardous wastes.

A CLOSER LOOK 17.1   Green Cleaning Products

In January 2008, the Clorox Company announced the introduction of GREEN WORKS® cleaners, a line of natural cleaning products that are, according to the company, as effective as conventional cleaners but made from plant-based ingredients (Figure 1).

According to the company, GREEN WORKS® products are at least 99% natural and made from chemicals derived from coconuts and lemon oil. The cleaning agents smell great but are also biodegradable and non-allergenic. They're even packaged in bottles that can be recycled (number 1 plastic). The potential effects on humans were determined by tests not involving lab animals as is customary on many products humans come in contact with, such as cosmetics.

The GREEN WORKS® line includes natural cleaning products for use in the home, kitchen, and bathroom, including an all-purpose cleaner, a glass and surface cleaner, a toilet bowl cleaner, and a bathroom cleaner.

In lab and consumer in-home testing, GREEN WORKS® products performed as well as or better than leading conventional cleaners, according to the company. (I tried a couple before the product was released and was extremely impressed!)

GREEN WORKS® is the first line of natural cleaners developed by a major consumer products company and is an affordable alternative to natural cleaners. The company has even teamed up with the Sierra Club to promote the product. "One of the Sierra Club's primary goals is to foster vibrant, healthy communities with clean water and air that are free from pollution," said Sierra Club's executive director,

Carl Pope. "Products like GREEN WORKS® help to achieve this goal in the home. We're looking forward to working with Clorox and the GREEN WORKS® team to promote a line of natural cleaning products for consumers who are moving toward a greener lifestyle."

GREEN WORKS® natural cleaning products are available at grocery stores, drugstores, and mass retail outlets. To learn more, log on to www.greenworkscleaners.com.

Clorox is not the only company to go green. In February 2008, SC Johnson, a leader in corporate green, began to print a Greenlist logo on its eco-friendly products, starting with an old standby, Windex, its popular window cleaner. The Greenlist logo indicates that Windex is one of numerous cleaning agents that are now made with lower concentrations of volatile organic chemicals (VOCs), which can be hazardous to human health and also contribute to photochemical smog.

The company has also reformulated Pledge, a furniture polish, to increase its biodegradability while increasing cleaning power, and created Fantastik Orange Action, a cleaner with no VOCs.

Arm and Hammer also released a line of concentrated cleaning agents that are diluted with tap water. This reduces energy required to ship the products to market and reduces cost. Refills can be purchased without having to buy a new plastic spray bottle.

More and more North Americans are looking for ways to live greener lifestyles, and there are hundreds of things they can do to realize their goals. Green cleaning products like those from Clorox and SC Johnson are two simple, inexpensive steps we can take to create a healthier, more sustainable world.

FIGURE 1   Using environmentally friendly cleaning products like these from Clorox can help families reduce their environmental impact. GREEN WORKS® is a registered trademark of The Clorox Company. Used with permission. ©2010 The Clorox Company.)

Detoxification   Some hazardous wastes cannot be reused or recycled but can be detoxified, that is, chemically altered into less toxic or nontoxic chemicals. **Detoxification** may be affected by biological, chemical, and physical treatment. For example, organic wastes such as PCBs, DDT, and even dioxin can be burned in high-temperature incinerators. Incineration converts harmful organic substances into relatively harmless carbon dioxide (a global pollutant in its own right, but not responsible for adverse health effects) and water. Long criticized by many in the environmental community because they do not completely eliminate toxic emissions, incinerators are growing in popularity.

The EPA owns and operates a mobile incinerator that destroys 99.999% of the dioxin wastes in soils and liquids. The incinerator can be transported to waste sites, thus avoiding the transport of

hazardous materials to distant incineration facilities. At one time, six European nations burned some of their toxic waste at sea in huge oceangoing vessels equipped with high-temperature incinerators (Figure 17.10). This practice has now been banned.

Critics argue that land- and sea-based incineration would be acceptable only with much tighter controls on emissions. A newly developed plasma arc incinerator, for instance, burns toxic wastes at 25,000°C (45,000°F), destroying all traces of PCBs and other organic wastes. For comparison, the fire in a woodstove burns at 204°C (400°F) to 480°C (900°F). Plasma arc incineration may be the wave of the future.

Another promising development is the combustion of hazardous organic wastes in existing cement and lime kilns—huge furnaces in which cement and lime are heated. Typically

FIGURE 17.10 The Dutch incinerator ship *Vulcanus* burns hazardous waste at sea. Is this a suitable way to get rid of the many hazardous organic wastes industrial societies generate, or will it lead to widespread pollution of the air and water?

powered by oil, kilns may be an efficient and cost-effective alternative to traditional incinerators. In Sweden, for instance, kiln operators burn a mixture of oil and hazardous organic materials as fuel. Most kilns already have state-of-the-art pollution-control equipment. Basic materials in the kiln also neutralize acidic emissions. Cement companies are paid to incinerate wastes and are able to cut down on their fuel consumption in the process, saving money on operations.

On another front, geneticists have developed strains of bacteria that decompose chemical solvents such as benzene, toluene, and xylene, converting them into carbon dioxide and water. Scientists have also found naturally occurring bacteria that successfully degrade oil wastes in soil and water.

Numerous chemical methods are also available. For instance, ozone can be used to destroy organic compounds. Special ion-exchange columns can efficiently separate out toxic heavy metals, and various bases can be used to neutralize acids.

Proper Disposal of Hazardous Waste

In an ideal world, hazardous wastes could be reduced by 60% to 75% (perhaps more) by process manipulation, reuse, recycling, and detoxification. Some toxic substances would still remain, however. Heavy metals are an example. The remaining material must be disposed of by any one of a half dozen or so techniques such as in secured landfills, deep geological salt beds, surface impoundments, warehouses, and deep injection wells. Secured landfills and deep injection wells are the preferred methods and are discussed here.

Secured Landfills The most popular discard approach today is to deposit the waste in a **secured landfill**—a clay-lined pit designed to hold hazardous wastes (Figure 17.11). The thick, supposedly impermeable clay liners are often supplemented with synthetic liners. Special drain systems also pump liquid wastes from the bottom of the pit to treatment facilities, where they are detoxified, thus minimizing the migration of these substances out of the site. Monitoring wells are drilled as well. Groundwater samples taken from the wells help workers determine whether wastes are leaking out of the site. Careful siting (locating a site in a dry region where the water table is

not too close to the landfill) also minimizes the risk to groundwater and surface water. Grading and compaction of the soil over the site minimize the penetration of rain and snowmelt, thus reducing leaching.

Although extraordinary precautions are taken to prevent the escape of materials from landfills, critics are unconvinced that they can contain wastes over the long term—over periods of decades or centuries. Cracks in the liner and clay seal, they suggest, could emit wastes that drain into groundwater and contaminate aquifers. Earthquakes could tear asunder the careful controls. Lax monitoring could unleash a local environmental catastrophe.

Deep Injection Wells Perhaps the least talked about but most important problem in hazardous-waste management is that much of our hazardous waste is highly diluted in water. Removing hazardous materials from this mix is extremely expensive. Because of this, companies frequently dispose of their liquid wastes in deep wells or illegally in lakes and sewage systems.

For **deep injection wells,** deep wells are drilled into the Earth's crust to porous zones sandwiched between supposedly impermeable rock layers. In theory, the hazardous material remains in place forever. In practice, though, this is not always the case. Liquid wastes can migrate through unexpected fissures in the "impermeable" rock strata and contaminate aquifers. Cracks in the well casing can also result in leakage. Injecting large quantities of liquids into the Earth can destabilize rock layers, greatly increasing the frequency of minor earthquakes.

According to the latest statistics from the EPA, about 10% of all hazardous waste produced in the United States is injected into the ground. Because of the problems that can arise, many people would like to eliminate it entirely, preferring instead ways to remove hazardous materials for detoxification, incineration, or some other disposal method (see Case Study 17.2).

The NIMBY Syndrome: Taking Personal Responsibility Besides growing awareness of the widespread nature of the problem, action to reduce hazardous-waste production has been prompted by the unpopularity of hazardous-waste

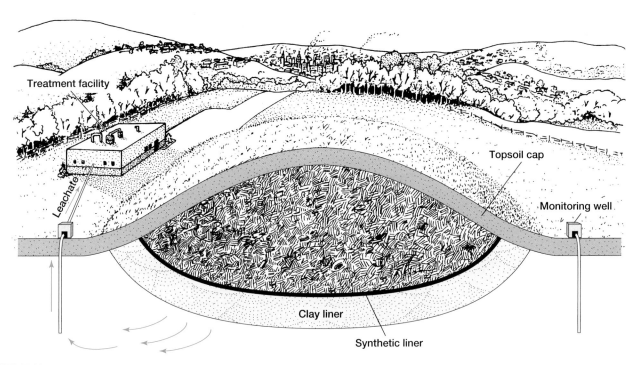

FIGURE 17.11 Secured landfill, showing monitoring wells, clay and synthetic liner, and treatment facilities for leachate.

CASE STUDY 17.2 EXPORTING TOXIC TROUBLES

In the 1970s and 1980s, U.S. laws and regulations for hazardous wastes were strengthened. Businesses found it increasingly expensive to dispose of their toxic wastes. Some unscrupulous companies began to dispose of their wastes illegally to avoid the high cost of waste disposal. Many others turned to overseas markets to dispose of their wastes. Wastes were often shipped to the cash-hungry less-developed nations of Africa or to Eastern Bloc nations, none of which had adequate laws requiring proper hazardous-waste disposal. European companies also joined in.

The problem with exporting waste was that many of the countries that received waste didn't know what was in the shipments or didn't know how toxic the materials really were. Moreover, most of them didn't have facilities to store it or dispose of the wastes properly. Hazardous wastes were more often than not disposed of recklessly. In trying to avoid the high costs of proper disposal, these companies, often knowingly, were contributing to serious environmental contamination of other countries.

As the problem became widely known, U.S lawmakers took action. In 1986, for example, the U.S. Congress amended RCRA, establishing procedures that require the United States to notify importing countries of shipments of hazardous waste and to obtain prior written consent. These regulations, however, were insufficient. EPA officials claimed that hundreds of tons of hazardous wastes were still being exported illegally.

Because exporting hazardous waste to a nation without its full consent goes against principles of international law, numerous African nations passed laws banning the import of hazardous wastes. In some countries, importing hazardous wastes is punished by stiff jail terms and multimillion-dollar fines. In Nigeria, an importer can be put to death. The Organization of Eastern Caribbean States and 22 Latin American countries joined forces to stop the dumping of hazardous wastes on their soils.

In 1990, the European Economic Community (EEC), a coalition of European nations, agreed to ban exports of toxic and radioactive waste to 68 former European colonies. Many of the less-developed nations that were part of the accord also agreed not to import hazardous wastes from non-EEC members. Today, 121 nations, including Canada, Mexico, and 13 European nations, have signed an agreement (the **Basel Convention**) that bans the transfer of hazardous wastes to less-developed nations. To date, the United States has refused to sign the agreement.

Although these are important steps forward, many less-developed nations are still open to exports, representing a potentially huge repository for hazardous wastes from the industrial nations. Signing an agreement will not stop the illegal flow.

In the United Nations, talks are under way for the development of global standards to regulate the shipment of hazardous wastes from one country to another. But what is and what is not a hazardous substance is a contentious issue. Global standards could be important, but some critics argue that the action may be a step in the wrong direction, for it could end up promoting export and discouraging waste reduction, a sustainable solution. Furthermore, regulating and enforcing such a program could prove to be extremely difficult. Thus, some proponents believe that a complete ban on the international movement of waste is the best answer. This would help protect the environment from inadequate disposal methods and would help nations develop long-term solutions that reduce hazardous-waste production—an essential element of a sustainable industrial design and a sustainable future.

facilities. Few people want a hazardous-waste landfill or a deep injection well near their homes, farms, or schools.

Public-policy makers have dubbed this the **NIMBY syndrome** (NIMBY stands for Not In My Back Yard). Ironically, most people want the amenities provided by manufacturers but want nothing to do with the waste that comes from them. Let some other community or some other country take it, they say.

This situation is not likely to change in the near future, but citizens can take actions to reduce the production of hazardous chemicals. One of the most obvious is to cut back on the use of such things as pesticides, herbicides, solvents, and cleaners. This simple step helps to reduce the volume of hazardous waste churned out by the many factories in the world. Nontoxic substitutes can be used. Numerous nurseries and retail outlets offer safer alternatives to the toxic chemicals they sell to control pests.

Because hazardous waste is produced in the manufacture of virtually all products, efforts to live with less or to buy more-durable goods, discussed earlier in this chapter as a means of reducing solid waste, also will reduce hazardous-waste output.

Individuals can also help by using nontoxic paints, stains, finishes, and cleaning agents and by properly disposing of **household hazardous wastes**—batteries, paint, paint thinner, pesticides, cleaning agents, and so on. Many cities have programs designed to pick up toxic household wastes. In some locations, these wastes are recycled. Used paint, for instance, may be given to companies that combine like colors, mix it with virgin paint, and then sell it. In other locations, you can acquire waste materials such as used paints directly through local waste exchanges. Whatever you do, don't dump hazardous materials down the drain; they'll poison a septic tank or contaminate sewage at sewage treatment plants. And don't dump them into storm sewers, either.

Beyond that, citizens can join environmental groups such as the Worldwatch Institute, the World Resources Institute, the Institute for Local Self-Reliance, and the National Coalition for Recyclable Waste, which are working on this and other issues. Citizens can write letters to government officials, asking for more efforts to reduce wastes and to reuse and recycle them. Without greater source reduction, reuse, and recycling, the hazardous-waste problem could worsen, overwhelming future generations as surely as it has plagued us.

Summary of Key Concepts

1. U.S. cities and towns generate 221 million metric tons of municipal solid waste each year. Only 32.5% of the solid waste is recovered for reuse and recycling, according to the Environmental Protection Agency. Approximately 12.5% is burned to generate energy.

2. Many modern industrial societies have traditionally viewed municipal waste as something to be rid of—to dump in the ground or at sea as far away as possible—with little regard for the wealth of materials it contains or the impacts of such actions on the environment. This is the discard approach.

3. Far more sustainable is the reduction approach to waste management—to reduce consumption. Reducing per capita consumption of natural resources may become an economic fact of life in materialistic societies as resource supplies, especially energy, decline.

4. Individuals can reduce consumption by avoiding unnecessary or frivolous purchases, purchasing more-durable products, and holding on to products for longer periods. Miniaturization has successfully helped to reduce resource consumption and could continue to be useful in the years to come.

5. Reuse and recycling are two additional sustainable approaches to waste management. They reduce solid-waste generation and conserve valuable resources. Many products can be reused and, if not, can be recycled.

6. Recycled materials can be extracted from municipal trash at central stations or separated out of the trash at the source.

7. Many nations now have successful recycling programs. Japan, the Netherlands, Mexico, and South Korea lead the world in recycling paper. The United States is on its way to becoming a leader in recycling, too, with can and bottle bills and numerous curbside recycling and composting programs.

8. A successful recycling effort, which is essential to creating a sustainable system of waste management, requires ways to promote remanufacturing and procurement.

9. Governments can play a key role in promoting all phases of recycling. Television ads, billboards, and pamphlets can encourage the public to take a more active role. New recycling laws, recycling centers, waste exchanges, and tax incentives can stimulate increased recycling as well.

10. Governments and businesses can also find ways to improve the market for recycled goods. Governments, for instance, can provide tax incentives for companies that use recycled materials in their manufacturing processes. They can also require their agencies to purchase recycled materials and can stipulate minimum recycled material content for products.

11. In an ideal world, wastes would first be reduced by the methods previously described. What is left would need to be discarded. For years, Americans discarded their wastes in open dumps that were periodically burned to reduce their volume. Then, in the 1960s, Americans began to revolt. In 1976, Congress passed the RCRA, which, among other things, called for a complete end to open dumps by 1983. The sanitary landfill replaced the dump.

12. Sanitary landfills are excavations or natural depressions in which garbage is dumped, compacted, and covered daily with a layer of soil to reduce pests.

13. Landfills offer many advantages over the open dump. Besides being cleaner, they can be reclaimed—returned to some previous use or given over to some new use. But they do have their problems. Pollutants may leak into underlying aquifers, contaminating groundwater. Rotting garbage produces methane, a potentially explosive gas. More importantly, landfilling squanders valuable resources that could be recycled or reused and also requires large tracts of

land, which are in short supply in and around growing cities and towns where huge quantities of waste are being produced.

14. Many local governments reduce their solid-waste disposal by composting, a form of recycling in which organic matter such as leaves and other yard wastes is piled, kept moist, and periodically turned. Aerobic bacteria decompose the organic matter to produce a stable humus-like material that can be used to condition and fertilize soil. Backyard composting is a popular strategy among some homeowners.

15. Another strategy now being used in the United States and elsewhere is incineration. Incinerators burn unseparated trash or glass- and metal-free trash to generate heat for electrical production, industrial processes, and heating.

16. Incineration is growing in popularity but is a source of air pollutants and ash, a hazardous waste. It also wastes perfectly recyclable materials.

17. Industrial nations also produce enormous amounts of hazardous waste. Hazardous wastes are waste products from factories, businesses, and homes that, if not properly stored, transported, handled, or disposed of, could adversely affect a wide array of organisms. Effects range from minor nonthreatening symptoms to serious debilitating illnesses, even death. Hazardous wastes include toxic organic chemicals, toxic heavy metals, corrosive materials, combustible chemicals, and highly reactive substances that could cause explosions or toxic fumes when heated.

18. Careless storage, handling, and disposal of such wastes have resulted in hundreds of thousands of hazardous-waste sites strewn about various nations.

19. Most nations face two problems in relation to hazardous wastes: (1) cleaning up the abandoned waste dumps and contaminated industrial sites, and (2) dealing with the millions of tons of hazardous waste produced each year by factories and other sources.

20. Identifying and cleaning up contaminated sites is one of the highest-priced items on the environmental agenda of many industrialized nations.

21. Although no one knows for sure, it is believed that about 600 million to over 700 million metric tons of hazardous waste are produced worldwide each year. The United States alone produces about 35 million metric tons per year.

22. In 1980, the U.S. Congress passed the Comprehensive Environmental Response, Compensation, and Liability Act, commonly called the Superfund Act. This new and important law calls on the EPA to identify and clean up hazardous-waste sites in the United States, using money largely from taxes levied on petroleum and chemical-manufacturing industries. The fund is reimbursed by culpable parties, that is, individuals responsible for the hazardous-waste site (producers and disposal companies).

23. Although Superfund cleanups initially occurred slowly, the pace has picked up, and 757 of the nearly 1,450 sites on the National Priorities List have been cleaned up to date.

24. The RCRA also includes measures to prevent further toxic contamination by establishing a system to monitor hazardous wastes from "cradle to grave." RCRA gave the EPA the power to set standards for packaging, shipping, and disposal of wastes and to license hazardous-waste disposal facilities.

25. Like municipal wastes, hazardous wastes can be dealt with in three basic ways, in order of desirability: reduction, reuse and recycling, and disposal. A fourth approach, detoxification, also is possible. Unfortunately, the most widely used method is the discard approach.

26. To reduce hazardous wastes, manufacturers can modify or redesign their processes. They can also substitute safe materials for more harmful ones. Both of these efforts reduce hazardous-waste output.

27. Manufacturers can also reduce the hazardous-waste output of their factories by reusing and recycling wastes. Hazardous-waste clearinghouses can help facilitate the exchange of wastes between businesses.

28. Some hazardous wastes that cannot be reused or recycled can be detoxified. For instance, organic wastes can be incinerated or decomposed by bacteria.

29. Some hazardous waste will inevitably be produced. This must be disposed of or stored safely for hundreds, perhaps thousands, of years.

30. Secured landfills, the most popular approach today, are seen by few as a permanent solution. Many critics believe that it is only a matter of time before they begin to leak, creating problems for future generations.

31. Deep injection wells are also popular for the disposal of liquid wastes, but their use is riddled with problems.

Key Words and Phrases

Basel Convention
Bottle Bill
Bottom Ash
Brownfield
Brownfield Development
Brownfields Economic
  Redevelopment Initiative
Cocomposting
Composting
Comprehensive
  Environmental Response,
  Compensation, and
  Liability Act (CERCLA)
Container Deposit Bill
Deep Injection Wells
Detoxification
Dioxins
Downcycle
End-Point Separation
E-waste
Extended Producer
  Responsibility (EPR)
Fly Ash
Greenfield

Hazardous Wastes
Household Hazardous Wastes
Incineration
Manifest System
Municipal Solid Waste
NIMBY Syndrome
Open Dumps
Process Manipulation
Process Redesign
Procurement
Recycling
Recycling Opportunity Act
Reduction Approach
Remanufacturing
Resource Conservation and
  Recovery Act (RCRA)
Reuse
Reverse Vending Machine
Sanitary Landfill
Secured Landfill
Source Reduction
Source Separation
Superfund Act
Waste Exchanges

Critical Thinking and Discussion Questions

1. Describe the three main techniques for managing municipal wastes. Discuss the pros and cons of each one. Give specific examples of each. What are the most sustainable measures, and why do they contribute to sustainability?

2. Outline a plan to reduce your or your family's trash. What obstacles stand in the way of reducing the volume by 50%? How can they be overcome?

3. Why is it theoretically possible to recycle only 60% to 80% of America's aluminum?

4. Debate the following statement: "In countries with large resource supplies, it is more economical to use raw ore than recyclable materials."

5. Outline a plan for your city or town to reduce its solid and hazardous wastes. Would you involve private citizens, and if so, how? Contact local officials, and ask them whether reuse and recycling programs exist and what the obstacles are to further waste reduction.

6. Describe ways to increase the demand for recycled goods.

7. Using your knowledge of this issue and of ecology in general, debate the following statement: "Governments have an obligation to help create markets for recyclable materials."

8. List the pros and cons of sanitary landfills, compost facilities, and incinerators.

9. What is hazardous waste? How can hazardous wastes best be reduced?

10. What problems does the Superfund Act address, and how does it address them?

11. What steps can individuals take to reduce hazardous-waste production at home and at factories?

12. What hazardous-waste problem(s) does the RCRA address, and how does it address the problem(s)?

13. Describe ways that manufacturers can reduce hazardous wastes.

14. List and describe several methods of hazardous-waste detoxification.

15. Debate this statement: "Secured landfills are the safest way to dispose of hazardous wastes."

16. Using your critical thinking skills and your knowledge of this issue, discuss the following statement: "It's up to manufacturers to control hazardous waste. They produce it, and they should be responsible for reducing it."

17. Review the arguments presented in the Ethics in Resource Conservation box in this chapter. Using your critical thinking skills, analyze and discuss each one.

Suggested Readings

Municipal Solid Waste

Abramovitz, J. N., and A. T. Mattoon. 2000. Recovering the Paper Landscape. In *State of the World 2000,* ed. L. Starke. New York: W. W. Norton. Examination of sustainable ways to produce paper, including recycling.

Benyus, J. M. 1997. *Biomimicry: Innovation Inspired by Nature.* New York: Perennial. A detailed exposé of the efforts now under way to create energy, food, and materials based on strategies derived from nature.

Carless, J. 1992. *Taking Out the Trash: A No-Nonsense Guide to Recycling.* Washington, DC: Island Press. A practical guide showing how individuals, businesses, and communities can help alleviate the solid-waste crisis.

Chiras, D. D. 1992. *Lessons from Nature: Learning to Live Sustainably on the Earth.* Washington, DC: Island Press. See Chapter 10 for a discussion of ways to build a sustainable waste management system.

Chiras, D. D. 2000. *The Natural House: A Complete Guide to Healthy, Energy-Efficient, Environmental Homes.* White River Junction, VT: Chelsea Green. Examination of many ways to build homes from natural and recycled materials.

Chiras, D. D. 2004. *The New Ecological House.* White River Junction, VT: Chelsea Green. Examination of ways to build homes that are economical, healthful, and safe to the environment.

Crampton, N. 2008. *Green House: Eco-Friendly Disposal and Recycling at Home.* New York: M. Evans and Co. A complete guide to disposing of and recycling all varieties of common household trash, including broken appliances, spent smoke detectors, and pet waste.

De Graaf, J., D. Wann, and T. H. Naylor. 2005. *Affluenza: The All-Consuming Epidemic.* San Francisco: Berrett-Koehler. A very popular treatment of overconsumption and its impact on our lives, based on a television documentary of the same name.

Environmental Protection Agency. 1995. *Municipal Solid Waste Factbook—Internet Version.* Washington, DC: Environmental Protection Agency. An extremely valuable resource, though slightly out of date. Available online at www.epa.gov/epaoswer/non-hw/muncpl/factbook/internet/.

Elgin, D. 1993. *Voluntary Simplicity: Toward a Life That Is Outwardly Simple, Inwardly Rich,* rev. ed. New York: William Morrow. A very popular book that describes how we can live simpler, less resource-intensive lives.

Gardner, G., and P. Sampat. 1999. Forging a Sustainable Materials Economy. In *State of the World 1999,* ed. L. Starke. New York: W. W. Norton. Great look at consumption and ways to decrease it.

Guelberth, C. R., and D. Chiras. 2003. *The Natural Plaster Book: Earth, Lime, and Gypsum Plasters for Natural Homes.* Gabriola Island, BC: New Society. Explanation of ways to make and apply environmentally friendly, natural plasters on new and existing homes.

Hamilton, C., and R. Denniss. 2006. *Affluenza: Why Too Much Is Never Enough.* New York: Allen and Unwin. An in-depth look at overconsumption and the many reasons behind it.

Kates, R. W. 2000. Population and Consumption: What We Know, What We Need to Know. *Environment* 42(3): 10–19. An important report on the subject.

McDonough, W., and M. Braungart. 2002. *Cradle to Cradle: Remaking the Way We Make Things.* New York: North Point Press. Exploration of a new way of making products to eliminate toxicity and enhance their recyclability.

Renner, M. 1992. Creating Sustainable Jobs in Industrial Countries. In *State of the World 1992,* ed. L. Starke. New York: W. W. Norton. Discussion of the jobs potential of a sustainable economy based in part on recycling.

Rosenblatt, R., ed. 1999. *Consuming Desires: Consumption, Culture, and the Pursuit of Happiness.* Washington, DC: Island Press. Collection of writings on one of the most challenging aspects of modern times, consumerism.

Taylor, B. 2001. *More Fun, Less Stuff: Starter Kit.* Takoma Park, MD: Center for a New American Dream. A simple guide for those interested in learning ways to reduce their impact on the planet.

Wann, D. 2007. *Simple Prosperity: Finding Real Wealth in a Sustainable Lifestyle.* New York: St. Martin's Griffin. Many solutions to living a more meaningful, less consumptive life.

Young, J., and A. Sachs. 1995. Creating a Sustainable Materials Economy. In *State of the World 1995,* ed. L. Starke. New York: W. W. Norton. Coverage of important issues regarding solid waste and recycling.

Hazardous Wastes

Carlin, A., P. F. Scodari, and D. H. Garner. 1992. Environmental Investments: The Cost of Cleaning Up. *Environment* 34(2): 12–20, 38–45. Summary of U.S. EPA report to Congress.

Fischhoff, B. 1991. Report from Poland: Science and Politics in the Midst of Environmental Disaster. *Environment* 33(2): 12–17, 37. Description of the dimensions of the hazardous-waste problem in Poland.

French, H. 1990. A Most Deadly Trade. *World-Watch* 3(4): 11–17. Documentation of the movement of hazardous materials to the developing countries and Eastern Europe.

Frosch, R. A. 1995. Industrial Ecology: Adapting Technology for a Sustainable World. *Environment* 37(10): 16–24, 34–37. A look at ways to reduce waste production by factories.

Frosch, R. A., and N. F. Gallopoulos. 1989. Strategies for Manufacturing. *Scientific American* 261(3): 144–152. Outline of the concept of the industrial ecosystem.

Gershon, D., and R. Gilman. 1991. *Household EcoTeam Workbook: A Six-Month Program to Bring Your Household into Environmental Balance.* New York: Global Action Plan. A handy guide for families who want to reduce their energy and water consumption and waste production.

Gibbs, L. M. 1998. *Love Canal: The Story Continues.* Gabriola Island, BC: New Society. Fascinating account of the containment of wastes, resettlement of part of the contaminated area, and more.

Krueger, J. 1999. What's to Become of Trade in Hazardous Wastes? The Basel Convention One Decade Later. *Environment* 41(9): 10–21. Examination of the effects of an important international agreement on hazardous-waste export to less-developed countries.

Mastny, L. *Purchasing Power: Harnessing Institutional Procurement for People and the Planet.* Worldwatch Paper 166. Washington, DC: Worldwatch Institute. Excellent overview of ways in which governments can help to promote environmentally responsible products through their massive purchasing power.

McGinn, A. P. 2000. Phasing Out Persistent Organic Pollutants. In *State of the World 2000,* ed. L. Starke. New York: W. W. Norton. A variety of strategies for reducing and eliminating the production and release of persistent organic chemicals, including those in hazardous waste.

McGinn, A. P. 2002. Reducing Our Toxic Burden. In *State of the World 2002,* ed. L. Starke. New York: W. W. Norton. A look at toxic chemicals such as persistent organic pollutants and heavy metals.

Probst, K. N., and T. C. Bierle. 1999. Hazardous Waste Management: Lessons from Eight Countries. *Environment* 41(9): 22–30. Valuable resource.

Rock, M. T., and D. P. Angel. 2007. Grow First, Clean Up Later? Industrial Transformation in East Asia. *Environment* 49(4): 8–19. Discussion of the impending waste and pollution problems created by growing east Asian economies.

Russell, M. E., W. Colglazier, and B. E. Tonn. 1992. The U.S. Hazardous Waste Legacy. *Environment* 34(6): 12–15, 34–39. Discussion of the cost of cleaning up America's hazardous wastes.

Sherlock, M. 2003. *Living Simply with Children.* New York: Three Rivers Press. A guide to voluntary simplicity for parents.

Selin, H., and S. D. VanDeveer. 2006. Raising Global Standards: Hazardous Substances and E-Waste Management in the European Union. *Environment* 48(10): 6–19. Discussion of the growing problem of waste from electronic and electrical devices.

Taylor, B. 2003. *What Kids Really Want that Money Can't Buy: Tips for Parenting in a Commercial World.* New York: Warner Books. An interesting look at ways to raise children without indulging in modern consumerism.

Ueta, K., and H. Koizumi. 2001. Reducing Household Waste: Japan Learns from Germany. *Environment* 43(9): 20–32. A very important reading on waste management in Japan and Germany.

 # Web Explorations

Online resources for this chapter are on the World Wide Web at: **http://www.prenhall.com/chiras** *(click on the Table of Contents link and then select Chapter 17).*

(18)

AIR POLLUTION

Human beings breathe in and out about once every four seconds, 16 times a minute, 960 times an hour—nearly 8.5 million times a year. Every year, we breathe nearly 4 million liters (1 million gallons) of oxygen-containing air from the Earth's atmosphere (Table 18.1).

In addition to being a vital source of oxygen, the Earth's atmosphere is of value to us in many other ways. The atmosphere, for instance, insulates the Earth. Without its atmosphere the Earth would be subjected to drastic day–night temperature changes very likely too extreme for life to exist. The atmosphere also helps to distribute heat, so the planet is more uniformly heated. (Warm tropical air moves toward the poles, warming continents along the way.) Without the atmosphere, sound vibrations could not be transmitted; the Earth would be silent. There would be no weather, no spring rains for crops and lawns, no snow, hail, or fog. Without its atmospheric shield, our planet would be more heavily bombarded with meteorites and would be exposed to much more potentially lethal ultraviolet radiation from the sun. Without an atmosphere, Earth would be as lifeless as the moon.

As valuable as the atmosphere is, though, we humans have treated it with almost total disregard until recently. This chapter examines our unwitting assault on the atmosphere, notably air pollution. You will see how we affect the atmosphere and how we can survive and prosper while protecting this vital component of the Earth's life support system.

18.1 Pollution of the Atmosphere

Pollutants in the atmosphere arise from two sources: natural and human. This section describes each type.

Natural Pollution

Long before humans evolved, the atmosphere was to some degree polluted, not from artificial sources but from natural causes. Smoke from lightning-triggered forest fires billowed across the land. Volcanoes spewed noxious gases into the atmosphere.

Natural pollution continues to be a problem. In May 1980, the massive eruption of Mount St. Helens in Washington State released thousands of tons of dust and ash into the air and briefly caused breathing problems for humans and wildlife downwind from the blast (Figures 18.1 and 18.2). Subsequent eruptions of other volcanoes including Mount Pinatubo in 1991 had a similar effect.

A given sample of today's atmosphere may contain a host of natural contaminants, including ragweed pollen, fungal spores, disease-causing bacteria, and minute particles of volcanic ash and salt, as well as a host of harmful gases from many different sources. The release of pollutants from natural sources may sometimes exceed emissions from human sources.

TABLE 18.1	Composition of Clean, Dry Air at Sea Level

Gas	Volume Percent
Nitrogen	78.08
Oxygen	20.94
Argon	0.9340
Carbon dioxide	0.0310
Neon	0.0018
Helium	0.0005
Methane	0.0002
Krypton	0.0001
Sulfur dioxide	0.0001

Note: Gases such as carbon dioxide, methane, and sulfur dioxide are normal constituents of clean air. However, they often reach much higher concentrations in polluted air, and may have adverse effects on the environment and/or human health.

Pollution Caused by Humans

Homo sapiens has been fouling the atmosphere ever since Stone Age people first roasted a deer over an open fire. The smoke smudged some of the magnificent cave-wall paintings in southern France—perhaps the first serious property damage caused by air pollution. In 1306, the English Parliament passed a law making it illegal to burn coal in a furnace in London; at least one violator was tortured for his offense. However, it was not until the Industrial Revolution that air pollution began to seriously affect the health of large segments of society. In 1909, more than 1,000 people died in Glasgow, Scotland, as a result of polluted air. In conjunction with this incident, the word **smog** was coined as a contraction of smoke and fog.

If the Earth is naturally polluted, why do we worry about pollutants coming from human sources? The reasons are several. First, human pollutants typically come from concentrated sources, usually cities or major industrial areas. With few exceptions, natural air pollutants typically come from widely dispersed sources. Therefore, even though natural sources may exceed human sources in terms of total annual output, the concentrated release of pollutants from human activities typically results in elevated levels, locally and regionally. High concentrations can have serious effects on people and the environment. Second, some human pollution sources can overshadow natural sources. Carbon dioxide releases from power plants, factories, homes, and automobiles, during the combustion of coal, oil, and natural gas, for example, have reached such high levels that most atmospheric scientists believe they are upsetting the Earth's energy balance, causing global warming. Many observers believe that global warming is having devastating effects on our environment, our economy, and our lives. This issue is discussed in Chapter 19. To understand the scope of the problem, let's begin by examining the major pollutants, that is, those produced in largest quantity by human activities.

FIGURE 18.1 Eruption of Mount St. Helens. Generally, pollution from natural sources is far less harmful than pollution from human sources. That's because natural sources tend to release small quantities over huge areas, which results in low ambient levels. The eruption of Mount St. Helens was an exception.

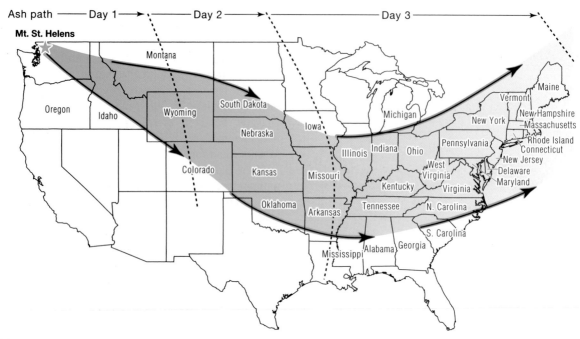

FIGURE 18.2 Approximate path (shaded area) of ashes emitted into the air by the eruption of Mount St. Helens.

18.2 Major Atmospheric Pollutants

The major air pollutants in the United States of direct concern to human health are carbon monoxide, oxides of sulfur, particulates, volatile organic compounds, and oxides of nitrogen (Figure 18.3, Table 18.2). These pollutants are all known as **primary pollutants** and are produced by a variety of sources. Some of these pollutants can be chemically converted into other, sometimes more harmful chemicals by natural processes. Nitrogen oxides, for instance, can combine with water, creating nitric acid. Pollutants produced from a primary pollutant are known as **secondary pollutants.**

Carbon Monoxide

Carbon monoxide (CO) is a colorless, odorless pollutant released during the incomplete combustion of organic material such as wood, coal, oil, natural gas, and gasoline. It is common in sometimes rather high levels above city streets and freeways.

Rather surprisingly, roughly 93% of the CO in the global atmosphere is derived from natural sources, such as the oxidation of methane (marsh gas), which is formed by the biological decay of organisms (mostly plants) in wetlands. However, this carbon monoxide does not build up to harmful concentrations because it is produced from widely dispersed sources and because it is quickly converted into carbon dioxide. (In this instance, carbon dioxide is considered a secondary pollutant.)

It might reasonably be asked, then, why we are so concerned with carbon monoxide as an atmospheric pollutant. The answer is that the seemingly insignificant 7% of the carbon monoxide generated by human activities, largely as the result of the incomplete combustion of fossil fuels, is concentrated in a relatively small volume of air in the world's major cities. In fact, the CO concentrations of urban areas are about 50 times greater than the worldwide average. The health effects, which are discussed shortly, can be substantial.

The amount of carbon monoxide released by U.S. sources has decreased substantially since the 1970s. According to the Environmental Protection Agency (EPA), carbon monoxide emissions have fallen from 183 million metric tons in 1970 to 96 million metric tons in 2005, the latest year for which data are available. Much of this decrease can be attributed to the improvements in car engines that were mandated by the Clean Air Act of 1970. The **Clean Air Act** and its amendments, which are discussed in A Closer Look 18.1, address a wide range of air pollution issues.

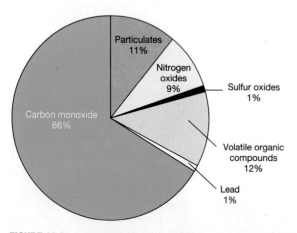

FIGURE 18.3 Major pollutants released into the air over the United States.

GO GREEN!

Keep your car well tuned, and keep the tires at the correct pressure. Both steps will increase your mileage, reduce air pollution, and save you money.

TABLE 18.2	Sources and Effects of Major Human-Caused Air Pollutants	
Pollutant	Description and Major Anthropogenic Sources	Human and Environmental Effects
Total suspended particulates	Solid or liquid particles produced by combustion and other processes at major industrial sources (e.g., steel mills, power plants, chemical plants, cement plants, incinerators).	Respiratory irritant; aggravates asthma and other lung and heart diseases (especially in combination with sulfur dioxide); many are known carcinogens. Toxic gases and heavy metals are adsorbed onto these particulates and are commonly carried deep into the lungs.
Sulfur dioxide	Colorless gas produced by combustion at power plants and certain industrial sources.	Respiratory irritant; aggravates asthma and other lung and heart diseases, reduces lung function. Sulfur dioxide damages plants and is a precursor to acid rain.
Carbon monoxide	Colorless gas produced by motor vehicles and some industrial processes.	Interferes with the blood's ability to absorb oxygen; can cause dizziness, drowsiness; impairs motor reflexes; may bring on angina.
Nitrogen dioxide	Brownish-orange gas produced by motor vehicles and combustion at major industrial sources.	Respiratory irritant; aggravates asthma and other lung and heart diseases.
Ozone	A colorless gas formed from a reaction between motor vehicle emissions and sunlight. It is the major component of smog.	Respiratory irritant; aggravates asthma and other lung and heart diseases; impairs lung functions. Ozone is toxic to plants and corrodes materials.
Volatile organic compounds (hydrocarbons)	Small quantities of hazardous pollutants emitted from industrial processes and diesel motor vehicle exhaust.	Linked to organ damage, serious chronic diseases, and various types of cancer.
Lead	Smelters.	Toxic to nervous and blood-forming systems; can cause brain and organ damage in high concentrations.

Oxides of Sulfur

Oxides of sulfur are gaseous pollutants that form whenever sulfur-containing fuels, such as coal and oil, are burned. During combustion, the sulfur combines with oxygen in the air to form **sulfur oxides,** denoted by the chemical formula SO_x. (The x can be a 2 or a 3.) In 2005, 16.6 million metric tons of sulfur oxides were released into the atmosphere from sources inside the United States. Colorless **sulfur dioxide** stings the eyes and burns the throat. About 1% of the U.S. population will develop chronic weariness, difficult breathing, sore throat, tonsillitis, coughing, and wheezing when exposed for lengthy periods to the concentrations of sulfur dioxide normally occurring in polluted urban air. Sulfur dioxide slows down or even halts the natural cleansing mechanism of the lungs. It also contributes importantly to such chronic diseases as bronchitis and emphysema.

As you shall see in the next chapter, sulfur oxides combine with water in air to form sulfuric acid, a potentially harmful secondary pollutant. It falls to the Earth in rain and snow and is deposited as sulfate particles. Chapter 20 discusses this problem in more detail.

A CLOSER LOOK 18.1   The Clean Air Act

The federal **Clean Air Act** is one of the most successful pieces of environmental legislation in U.S. history. But the Clean Air Act is not one law, but several. First passed in 1963, the law was fairly weak and ineffective. However, over the past four decades, the Clean Air Act has been amended three times, strengthening it considerably, in response to our growing air pollution problems and our improved understanding of the best ways to solve them.

The first amendments to the Clean Air Act occurred in 1970. These amendments resulted in the establishment of (1) emissions standards for automobiles, (2) emissions standards for new industries, and (3) ambient air quality standards for urban areas. The **national ambient air quality standards** established by the EPA covered six pollutants: carbon monoxide, sulfur oxides, nitrogen oxides, particulates, ozone, and hydrocarbons. These standards were designed to protect human health and the environment.

The 1970 amendments succeeded in reducing air pollution from automobiles, factories, and power plants. In addition, they stimulated many states to pass their own air pollution laws, some with regulations more stringent than federal ones. Despite these gains, the amendments created some problems. For instance, in regions whose air pollution exceeded national ambient air quality standards, the law prohibited the construction of new factories or the expansion of existing ones. As you might suspect, the business community objected. In addition, some of the wording of the 1970 amendments was vague and required clarification. Of special interest were provisions dealing with the deterioration of air quality in areas that had already met federal standards. Environmentalists worried that clean-air areas would deteriorate because of federal standards.

Because of these and other problems, the Clean Air Act was amended once again in 1977. To address the limits on industrial growth in areas that were violating the air quality standards, called **nonattainment areas,** lawmakers devised a strategy that allowed factories to expand and new ones to be built, but *only* if they met three provisions: (1) the new sources achieved the lowest possible emissions rates, (2) other sources of pollution under the same ownership in that state complied with emissions-control provisions, and (3) unavoidable emissions were offset by pollution reductions by the company in question or other companies in the same region.

The last provision, known as the **emissions offset policy,** forces companies to make reductions in their own facilities and requires newcomers to request existing companies in noncompliance areas to reduce their pollution emissions. In most cases, the newcomers pay the cost of air-pollution control devices.

The emissions offset policy is also used to produce an overall decrease in regional air pollution—because the air pollution emissions permitted from both the new and the existing facilities are set below preconstruction levels.

The 1977 amendments set forth rules for the **prevention of significant deterioration (PSD)** of air quality in **attainment regions,** regions where air quality meets federal standards. However, PSD requirements apply only to sulfur oxides and particulates. Many air pollution experts think the PSD requirements should be expanded to include other pollutants, such as ozone.

Another benefit of the 1977 amendments is that they strengthened the enforcement power of the EPA. In previous years, when the EPA wanted to stop a polluter, it had to initiate a criminal lawsuit. Violators would often engage in protracted legal battles, knowing that legal costs were often lower than the cost of installing pollution control devices. The 1977 amendments, however, allowed the EPA to levy **noncompliance penalties** without going to court. The logic behind this new power is that violators have an unfair business advantage over competitors that comply with the law. Penalties equal to the estimated cost of pollution control devices eliminate the cost incentive to pollute.

In 1990, the Clean Air Act was amended once more to address other important issues, among them acid rain. The 1990 amendments, for example, set deadlines for establishing emissions standards for 190 toxic chemicals from factories, a subject that had not been previously addressed. More important, it established a system of **pollution taxes** on toxic chemical emissions, charging manufacturers a tax on those chemicals they used that were potential toxic air pollutants. These provide a powerful incentive for companies to reduce their use of them.

The 1990 Clean Air Act amendments tightened emission standards for automobiles and raised the average mileage standards for new cars, a step that improved automobile efficiency and helped attack the pollution problem at its roots. In addition, the 1990 amendments established a market-based incentive program to reduce sulfur dioxide emissions, a primary contributor to acid deposition (Chapter 19). The law set up a system of **tradable** or **marketable permits** to companies throughout the United States. Each company is granted a certain number of allowances for sulfur dioxide release. (One allowance is equal to 1 ton of pollution.) These permits stipulated lower than present emission rates to improve air quality and could be bought and sold. Therefore, companies that found innovative and cost-effective ways to reduce pollutants below their permitted allowance could sell their unused credits at a profit. (These are bought and sold on the commodities market, starting in 1995.) This system encourages companies to develop cost-effective ways to prevent pollution. Thus, if a company can find inexpensive means of reducing its output of pollutants, it can benefit economically from the sale of its pollution credits. Most companies did. They switched to low-sulfur coal, which cost less than the coal they were using. So they not only saved money and cut sulfur dioxide emissions but also were able to sell their allowances to other companies. Finally, the 1990 amendments called for a phaseout of ozone-depleting chemicals, a topic discussed in the next chapter.

The Clean Air Act has worked admirably well and has grown more flexible as time has passed, giving companies greater leeway in how they control or eliminate pollution. Unfortunately, in recent years, many successful efforts have been mounted at the federal level to weaken the Clean Air Act in the name of economic prosperity. Coal-fired power plants (mostly in the Ohio River Valley) that were required to install pollutant control devices, for example, have been exempted, resulting in much higher emissions of pollutants that cause acid rain. With people driving larger cars and factories emitting more pollutants, the quality of our air can only worsen.

Particulate Matter

Particulate matter refers to small solid particles and liquid droplets suspended in the air. They range in size from very large to extremely small (Figure 18.4). These particles may remain suspended in the air for periods ranging from a few seconds to several months, depending on their size and weight. The finer the particulate, the longer it remains suspended in the atmosphere.

Most of the particulate matter in the atmosphere from human sources is emitted by facilities that burn coal, such as power plants, iron and steel mills, and foundries. Automobiles and other motorized vehicles also contribute to particulate pollution. Diesel trucks, trains, and jets also produce huge amounts of particulate matter. In many major metropolitan areas that sand roads in the winter, much of the atmospheric particulate pollution comes from sand on roadways that moving tires eject into the atmosphere.

Other sources of particulates include agricultural activities such as plowing, cultivating, and harvesting; slash-and-burn farming in Asia, Africa, and South America; debris burning by loggers; and strip-mine operations. In the United States, 14.6 million metric tons of particulate matter were released into the atmosphere in 2005, down from 22.4 million metric tons in 2000. The vast majority of the particulate matter (89%) was fugitive dust, larger particles from farming, mining, roadways, and wind erosion.

Studies show that fine particulates may cause a significant increase in mortality (death) from respiratory and heart malfunction. Extrapolating from studies conducted in Detroit and St. Louis, researchers estimate that high levels of particulates could be responsible for more than 60,000 deaths nationwide per year. Especially vulnerable are the elderly, asthma sufferers, and people recovering from recent heart attacks.

Particulates are dangerous in part because they may contain sulfates or nitrates (discussed shortly), which may enter the lungs and combine with water, forming sulfuric acid and nitric acids in the delicate tissues of the lung. Both can cause considerable damage.

Particulates also contain heavy metals such as **lead.** For years, lead was added to gasoline in many more-developed nations to enhance its octane rating, that is, the efficiency with which it burns. Lead added to gas was released with the exhaust gases. We humans inhaled this lead. A cumulative poison, lead also is ingested with food or water. It can damage the kidneys, blood, and liver. Moreover, it can damage the brains of children.

On the basis of lead concentrations in snows at high elevations in the Rockies, Clare C. Patterson, a California Institute of Technology geochemist, has suggested that lead concentrations in humans are 100 times the level of two centuries ago. In 1973, the EPA began restricting lead in gasoline, imposing a 90% reduction by 1985. The EPA eliminated lead from gasoline in 1995. In 2003, 1.5 million metric tons of lead were released into the atmosphere of the United States, down significantly since the 1970s, when 199 million were released annually into the atmosphere in the United States alone. Other industrial countries have eliminated lead additives from gasoline. Unfortunately, lead additives are still widely used in less-developed countries.

Another particulate of grave concern is mercury, which was discussed in Chapter 10.

Volatile Organic Compounds

A **volatile organic compound (VOC)** is, as the name suggests, an organic compound that evaporates readily. Organic compounds are composed of hydrogen and carbon. Examples are methane, benzene, and ethylene. In urban areas, humans may generate more than 200 kinds of VOCs. Many VOCs are chemically reactive. For example, some react with nitrogen oxides in the presence of sunlight to form **photochemical smog,** a potentially harmful mix of secondary pollutants (described below). Much of the VOC released results from the evaporation of gasoline from crankcases and gas tanks of cars and trucks. In addition, unburned hydrocarbons are released from the exhaust pipes of motor vehicles. In 2005, the United States released 16.9 million metric tons of VOCs, down from 24 million metric tons in 1980.

Oxides of Nitrogen

The atmosphere is 78% nitrogen. When air enters a combustion chamber—say, the cylinder of a car—the nitrogen reacts with oxygen to produce **nitrogen oxides (NO_x).** Major sources of the oxides of nitrogen are shown in Figure 18.5.

Nitric oxide is the first nitrogen oxide formed. Its chemical formula is NO. Nitric oxide is relatively harmless at ordinary concentrations. At unusually high concentrations, nitric oxide can be lethal. It causes death by asphyxiation because it combines with hemoglobin in the red blood cells, blocking their ability to transport oxygen. In fact, it binds 300,000 times more readily with red blood cells than does oxygen.

Nitric oxide also combines with atmospheric oxygen to form **nitrogen dioxide (NO_2),** a reddish brown gas with a pungent,

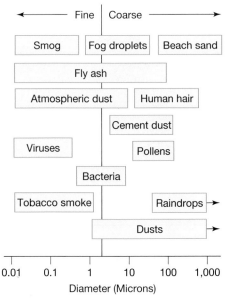

FIGURE 18.4 Size ranges of some airborne particulates in microns. Some particles in smog, fly ash, atmospheric dust, and tobacco smoke are only 0.01 micron in diameter, too small to be seen under an ordinary microscope.

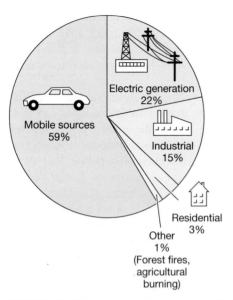

Electric generation
22%

Industrial
15%

Mobile sources
59%

Residential
3%

Other
1%
(Forest fires,
agricultural
burning)

FIGURE 18.5   Major sources of nitrogen oxides in the United States.

choking odor. Nitrogen dioxide is a major component of photochemical smog and gives it its distinctive orange-brown color. It causes a variety of human ailments, including gum inflammation, internal bleeding, emphysema, and increased susceptibility to pneumonia and lung cancer. Nitrogen dioxide is considered four times as toxic as nitric oxide. Nitrogen dioxide emissions have fallen since the 1980s, when they were about 28 million metric tons. In 2005, emissions were 13 million metric tons.

Ozone and Photochemical Smog   Ozone (O$_3$) is a gaseous pollutant and a major component of photochemical smog—the brownish haze that shrouds many urban areas during the hot,

GO GREEN!

Combine errands to reduce driving. Use your bike or walk to perform nearby errands.

sunny days of summer. This secondary pollutant is produced by chemical reactions between volatile organic compounds and nitrogen oxides. Because these reactions are powered by sunlight, the chemical soup produced in them is called photochemical smog. The greater the intensity of sunlight and the warmer the day, the larger the amount of ozone produced. Because VOCs and oxides of nitrogen are primarily generated by motor vehicles, the ozone levels in cities gradually rise in the morning (after the morning rush hour) and peak between noon and 4:00 P.M. (It takes that long for the morning pollution to react to form all of the chemicals in photochemical smog.) On cloudy days, ozone production is reduced; after sunset, it stops altogether.

Photochemical smog was first recognized in heavily populated Los Angeles, where it was a serious problem. However, during the 1970s and 1980s, it proved to be a daunting challenge to health officials in many other cities, among them Denver, Salt Lake City, Milwaukee, Chicago, New York, and Boston. During the extremely hot, dry summers of the 1990s, for example, many East Coast cities were plagued with record high ozone levels. In 2005, 345 counties failed to meet the EPA's ozone standard (Figure 18.6).

As many summer visitors to Los Angeles well know, the ozone in photochemical smog irritates the eyes, nose, and throat and makes breathing difficult (Figure 18.7). Recent scientific studies in laboratory animals, however, have shown that ozone pollution may have much more serious consequences to humans. For example, even low-level, short-term exposure to ozone can cause weight reduction and chromosome damage in rodents. Other animal studies have shown that chronic exposure to ozone causes permanent lung damage, including stiffening of the wall of the lung, which is normally associated with aging. Dr. Morton Lippmann, professor of environmental medicine at New York University, found that jogging in an ozone-polluted

FIGURE 18.6   Regions that at some time did not meet national ozone standards in 2005.

FIGURE 18.7 (Right) A view of Los Angeles on a clear day. (Left) Los Angeles under smog. The smog in this photograph is trapped by a temperature inversion. Inversions are present over the Los Angeles Basin about 320 days of the year!

urban environment may do the runner more harm than good. In fact, it can be as dangerous to one's health as smoking. The effect is cumulative. With each breath, the runner's lungs are damaged a little more. Said Lippman, "People won't fall over from exercising one or two days, but you may have breathing difficulties later on."

The health of farm crops is also adversely affected by ozone. For example, the U.S. Office of Technology has reported that an ozone-induced yield reduction of only four major crops—wheat, corn, soybeans, and peanuts—results in an estimated $3.2 billion annual loss to American agriculture (Table 18.3).

TABLE 18.3	The Effect of Ozone on Plants		
Plant	Ozone Concentration (ppm)	Duration of Exposure	Reduction in Weight or Height
Alfalfa	0.10	7 hr/day/70 days	51% total dry weight
Soybeans	0.10	6 hr/day/133 days	55% seed weight
Sweet corn	0.10	6 hr/day/64 days	45% seed weight
Wheat	0.20	4 hr/day/7 days	30% seed weight
Beets	0.20	2 hr/day/38 days	40% root weight
Ponderosa pine	0.10	6 hr/day/126 days	21% stem weight
Hybrid poplar	0.15	12 hr/day/102 days	58% height
Red maple	0.25	8 hr/day/6 wk	37% height

In the United States, the maximum concentration of ozone permitted under current EPA regulations is 0.075 parts per million (ppm) averaged over an eight-hour period. If the ozone in a city's air exceeds that level more than once in a given year, the city is subject to economic sanctions, such as the withdrawal of federal funds for highway construction.

18.3 Factors Affecting Air Pollution Concentrations

The concentration of air pollution in the air we breathe depends on the amount released into the atmosphere, but also on other factors. This section discusses two of them: thermal inversions and urban heat absorption.

Thermal Inversion

Have you ever noticed that on some days the air is cleaner than on others? One reason for the buildup of atmospheric contaminants to high levels on some days is a meteorological condition known as a **thermal inversion, or temperature inversion.** Let us explain how this forms before we define it; it's easier to understand this way. Under normal daytime conditions, the air temperature gradually decreases with altitude from ground level to a height of several miles above the Earth's surface (Figure 18.8, top). This pattern permits hot air from the surface to rise and disperses

pollutants at ground level as well. However, there are times when the temperature pattern deviates from this one. In such cases, temperature decreases somewhat with altitude and then begins to climb again. This is called a temperature inversion. During a thermal inversion, there's a layer of warm air over a region, which is often referred to as a warm air lid or thermal lid. In such instances, dispersion becomes impossible. Pollution becomes trapped near the surface, and levels increase. Before we show you why, let's learn a little bit more about thermal inversions.

Meteorologists have identified two basic types of inversions: radiation and subsidence.

Radiation Inversion At night, heat radiates from the Earth's surface into the atmosphere. The Earth is a better radiator than the atmosphere. As a result, both the ground and the air layer next to it cool off more quickly than the air above them. Consequently, a layer of cool air forms immediately above the Earth's surface. It may extend 300 meters (1,000 feet) above ground. Above this cooler layer is a warm layer, a thermal lid. The result is a **radiation inversion**—an inverted temperature profile caused by the radiative cooling of the Earth's surface (Figure 18.8, bottom). Because cool air does not rise, pollutants are trapped in this layer and remain relatively close to the ground. If winds are present, however, pollutants may disperse horizontally. When pollutants are not blown horizontally, they may build up to dangerous levels.

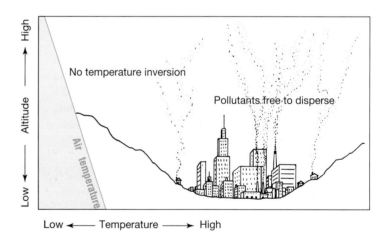

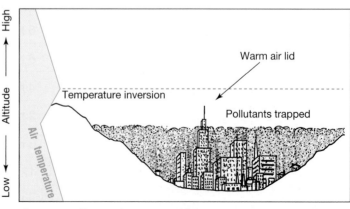

FIGURE 18.8 Effect of a radiation temperature inversion on the distribution of air pollution.

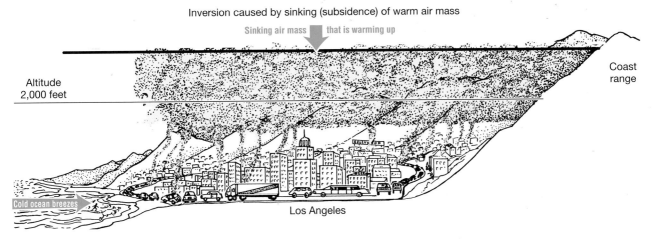

FIGURE 18.9   Nature of a subsidence inversion. In a subsidence inversion, warm air sinks down on an area, trapping cooler air and causing pollution levels to increase.

Radiation inversions are common in many areas of the United States, especially in mountainous regions. However, they are usually confined to small areas and usually dissipate by late morning, when the Earth's surface is warmed by the sun. The air immediately above the Earth then warms up as well. As the day advances, the inversion gradually disappears. Pollutants disperse vertically, and ground-level concentrations fall.

Subsidence Inversion   A **subsidence inversion** is formed when a warm, high-pressure air mass stalls over an area and sinks toward the ground, at times as low as 600 meters (2,000 feet) above the ground, forming a huge warm air lid. Although less common than radiation inversions, subsidence inversions usually last longer and may be much more extensive. Sometimes the subsidence inversion forms over several states. This type of inversion worsens the air pollution problems of Los Angeles in the summer (Figure 18.9). During the summer months, a warm, high-pressure air mass is constantly present above the Pacific Ocean off the California coast. This air mass occasionally moves inland over Los Angeles, Oakland, and other coastal cities and puts a lid on the pollutant-laden air near the ground that has been cooled by ocean currents moving along the coast. Coastal California experiences this type of inversion on nine out of every ten days in summer, which is one of the reasons why air pollution in San Diego and Los Angeles can be so bad in the summer.

Dust Domes and Heat Islands

Any motorist speeding toward the outskirts of Chicago, St. Louis, Des Moines, or any other large city has observed the haze of smoke and dust that frequently forms an "umbrella" over these cities. This shroud of pollutants is known as a **dust dome.** The dust dome is caused by a unique atmospheric circulation pattern that, in turn, depends on the marked temperature differences between the city proper and outlying regions. Let us explain. Cities tend to be warmer than the outlying areas, due to the massive amounts of concrete and pavement they contain. The average annual temperature of a city is typically around 0.98°C (1.7°F) higher than that of the surrounding rural areas. The city therefore forms a **heat island** (Figure 18.10).

FIGURE 18.10   Heat island effect of a city. The air is warmed up in the city because of large numbers of industrial furnaces, car motors, human bodies, heat-absorbing surfaces, and so on. This warm air moves upward and eventually cools. It may then sink, because of its greater density, move back into the city along with other cool air from the countryside, and be warmed again. As a result, a circular flow of air occurs. Any dust in the air will remain suspended above the city in a mushroom-like formation known as a dust dome.

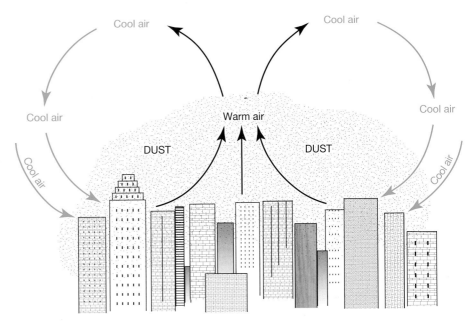
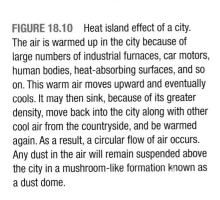

Contributing to the warmth of the city are such heat-generating sources as people, industrial furnaces, utility boilers, and motor vehicles, as well as the heat-absorbing and heat-radiating surfaces of streets, parking lots, and buildings. In rural areas, by contrast, heat-generating and heat-radiating structures are much less numerous. Moreover, there is more evaporative cooling in rural areas because of the vegetation. Evaporative cooling, you may recall, is the loss of water from transpiration. It cools the general vicinity.

Heat generated in a city causes air laden with pollutants to rise. Cool air from the countryside moves into the city to replace warm air rising from the urban center. As a result, smoke dust, nitrogen dioxide, and other aerial "garbage" tend to concentrate above the city, creating a dust dome. One thousand times as much dust may be present immediately over an urban industrial area as in the air of the nearby countryside.

When the air is calm, the dust dome persists, but when winds as slow as 12.8 kilometers (8 miles) per hour or more develop, the dome is pushed downward and horizontally into an elongated **dust plume.** Such plumes originating in cities such as Chicago are occasionally seen a distance of 240 kilometers (150 miles) away.

18.4 Effects of Air Pollution on Local Climate

Just as certain climatic factors such as inversions affect air pollution, growing evidence shows that air pollution can alter climate. Particulates, for instance, reduce the amount of sunlight penetrating the Earth's atmosphere, which decreases air temperature. Particulates also increase cloud formations, which can increase precipitation. Such dramatic climatic changes may disrupt terrestrial and aquatic ecosystems. Chapter 19 provides evidence suggesting that carbon dioxide is altering the climate on a global scale with many serious effects on the economy, the environment, and even human health and well-being. This chapter examines the well-documented local climatic effects.

Air Pollution and Precipitation

Local climate is influenced primarily by particulates. Particulates (soot and dust) in urban air from factories, power plants, and roadways help to seed clouds in the atmosphere downwind of cities. That is, they serve as **condensation nuclei,** small particles suspended in the air that absorb moisture in the atmosphere, forming tiny droplets. These droplets form clouds.

Because particulates promote cloud formation, they can result in an increase in rainfall downwind of cities and other sources of particulates. If human-generated air pollution induces rainfall, one would expect more rainy days in downwind areas from Monday to Friday, when factories are operating and pollutants are generated, than on weekends, when plants are closed. This is precisely the case in many instances. For example, in Paris, France, the average daily rainfall on weekdays is 31% higher than on weekends. Industrial contaminants, such as particulates, generated in Chicago are blown to the general region of La Porte, IN, 80 kilometers (50 miles) to the east, where they trigger a considerable increase in rain.

Air Pollution and Decreased Average Temperature

The total annual load of particulates generated directly or indirectly by human activities the world over amounts to 800 million metric tons. Because particulates can block sunlight, an increase in particulate matter may have a cooling effect on a large area. The cooling effect of particulate matter was impressively demonstrated in 1883, when a volcano on the island of Krakatoa in the Dutch East Indies injected many tons of fine dust particles high into the atmosphere. Over a period of years, these particles circled the globe several times. A short time after the eruption, the United States experienced a cooling trend; Bostonians, for example, had the rare privilege of throwing snowballs in June. More recent volcanic eruptions have had cooling effects, too, but much more modest.

Pollutants released into the atmosphere from natural and human activities continue to have a cooling effect, but overall, the Earth and its atmosphere are warming. A 2007 report by the International Panel on Climate Change, produced by the world's leading atmospheric scientists, showed that other forces, such as increased carbon dioxide pollution and deforestation, are offsetting the cooling effect of particulates.

18.5 Effects of Air Pollution on Human Health

Scientists estimate that 50,000 to 60,000 Americans die each year at least partly because of air pollutants. That is about the number of American soldiers who were killed in the Vietnam War. Tens of thousands probably also die in Europe and other countries, although estimates are difficult to come by. The European Environmental Agency, for instance, estimates that about 20,000 Europeans die each year as a result of exposure to low-level ozone in photochemical smog. Let's take a look at health effects, starting with some of the oldest known effects resulting from air pollution disasters.

Air Pollution Disasters

When many people die from air pollution in a short time, the episode is called an **air pollution disaster.** When most of the air pollution disasters are studied, a common pattern is revealed: (1) they occur in densely populated areas; (2) they occur in heavily industrialized centers where pollution sources are abundant; (3) they occur in valleys, which might serve as topographical receptacles for receiving and retaining pollutants; (4) they are accompanied by fog (it appears that the minute droplets of moisture are adsorbed on the surfaces of the pollutants); and (5) they are accompanied by a thermal inversion, which traps pollutants, causing levels to increase substantially.

The Donora Disaster   Forty-eight kilometers (30 miles) south of Pittsburgh, PA, in a horseshoe-shaped bend of the Monongahela River lies the industrial community of Donora, PA, with a population of 12,000. Almost encircled by hills rising to a height of 116 meters (350 feet), Donora is home to factories that manufacture steel, wire, sulfuric acid, and zinc, all crowded along the river margin for 5 kilometers (3 miles).

On October 26, 1948, a thermal inversion occurred. Soon afterward, a fog closed in on the valley. There was hardly a breath of air stirring. The black, red, and yellow smoke fumes that belched from the Donora smokestacks merged to form a multicolored blanket over the valley town. In a short time, the pungent odor of sulfur dioxide permeated the air. In addition to sulfur dioxide, the air over Donora contained high levels of nitrogen dioxide and hydrocarbons, which resulted from burning coal to provide heat and electricity for shops and homes. A sluggishly played football game between the Donora and Monongahela high schools was canceled in midplay when several of the players complained of chest pains and tortured breathing. Streets, sidewalks, and porches were covered with a film of soot. Motorists had to pull off to the side of the road because they couldn't see. People who had lived in Donora for over half a century got lost. The smog was so dense that it was extremely difficult to see from one side of the street to the other. The Donora fire department hauled oxygen tanks around the clock to people experiencing breathing difficulties.

Of the total population, roughly 43%—5,910 people—became ill, the most prevalent symptoms being nausea, vomiting, and severe headaches; nose, eye, and throat irritation; and labored breathing and constriction of the chest. Even pets and wildlife suffered. A veterinarian reported that the dense smoke caused the death of seven chickens, three canaries, two rats, two rabbits, and two dogs.

The complete death toll for Donora's "Black Saturday" was 17. Two more deaths occurred on Sunday. Then, on Sunday night, climatic conditions changed. A heavy rain washed some of the pollutants from the air. A breeze drove much of the smoke away. Visibility improved, and breathing became easier. The worst air pollution disaster in American history was over. Although the smog lasted for only five days, it left 20 people dead in its wake.

Incidents such as the one in Donora are dramatic examples of the effects of air pollution on humans. Similar examples have occurred in heavily industrialized cities, such as in London in 1952. In this episode, 4,000 people died as a result of pollution. Fortunately, such events are rare. Pollution controls have helped to reduce these events. More common, though, are chronic health effects resulting from long-term exposure to air pollutants. The next section discusses this problem.

Chronic Health Effects of Air Pollution

Many of us live in polluted areas, inhaling in pollution with every breath we take. However, over the years scientists have found that air pollution in modern cities and towns is not something about which we should be nonchalant. It can be lethal. But urban air pollution usually kills slowly and quietly, making the relationship between cause and effect difficult to detect—and equally difficult to respond to. Instead of stating that the deceased breathed in too much air pollution, the death certificates of many people simply state that death was caused by a heart attack, lung cancer, or emphysema.

Chronic Bronchitis   Many people, especially smokers and urban residents, suffer from a chronic irritation of the bronchial tubes, which carry air into the lungs. This persistent disease, caused by pollutants in tobacco smoke and urban air, is called **chronic bronchitis** (bron-kité-is). Symptoms include a persistent mucus buildup and cough. In some instances, individuals have difficulty breathing. Although cigarette smoking is the primary cause, urban air pollution is clearly a causative factor, even among children. Chronic bronchitis results from exposure to sulfur dioxide and nitrogen dioxide, which are found in both tobacco smoke and urban air. It is also caused by inhaling ozone, also present in urban air pollution. These chemicals irritate the respiratory passageways, causing mucus secretion.

Emphysema   Many people suffer from a debilitating disease known as **emphysema** (em-fi-zeé-mah), which leaves them short of breath. This disease results from a breakdown of the tiny air sacs, known as alveoli (al-vee-oh́-lie), in the lungs.

To understand this disease, you must first understand the structure and function of the lungs. When the air we breathe enters the lungs, it flows into millions of tiny air sacs in the lung, and oxygen passes through the ultrathin membranes of the tiny air sacs into capillaries, tiny vessels that carry blood (Figure 18.11, top). Oxygen is then distributed in the blood to the cells of the body. As oxygen flows into the blood, carbon dioxide flows out, passing from the blood in the capillaries surrounding the air sacs and into the alveoli. Carbon dioxide is then exhaled, which helps the body get rid of this waste product, a by-product of cellular energy production. The total respiratory membrane surface presented by each lung's 300 million air sacs is about the size of a tennis court. Each breath we take draws air into the lungs. Expelling that air, however, is usually a passive process that depends on elastic connective tissue in the walls of the alveoli. Like inflated balloons, the lungs recoil after filling with air. This forces air out of the lungs.

Three air pollutants—sulfur dioxide, nitrogen dioxide, and ozone in urban air—damage the alveoli. (Sulfur dioxide and nitrogen dioxide are also found in smoke from tobacco and marijuana.) These pollutants cause the walls of the alveoli to break down, creating larger and larger air sacs. As the alveoli become larger and larger, the surface area for exchange decreases, reducing oxygen uptake and carbon dioxide release. This condition is known as **emphysema.**[1]

Air pollution also causes the elastic tissue of the lungs to progressively deteriorate. In some cases, almost 50% of the lung's elastic tissue may be destroyed before the victim is aware of the problem. The decline in elasticity in emphysemic

---

[1]Emphysema and chronic bronchitis are two types of chronic obstructive pulmonary disease (COPD) that collectively kills over 120,000 Americans per year. COPD also includes asthma.

HEALTHY LUNG

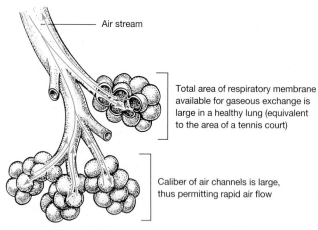

- Air stream

Total area of respiratory membrane available for gaseous exchange is large in a healthy lung (equivalent to the area of a tennis court)

Caliber of air channels is large, thus permitting rapid air flow

DISEASED LUNG (bronchitis, emphysema)

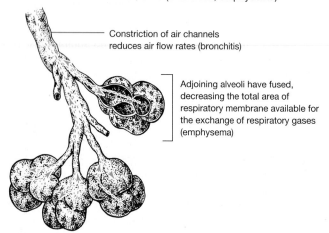

- Constriction of air channels reduces air flow rates (bronchitis)

Adjoining alveoli have fused, decreasing the total area of respiratory membrane available for the exchange of respiratory gases (emphysema)

FIGURE 18.11   Bronchioles and air sacs of a healthy lung compared with a lung diseased by air pollution. Note how air sacs break down in a diseased lung.

patients does not affect inhalation but makes exhalation very difficult indeed. (Elastic tissue causes the lungs to recoil (contract) after inhalation.) With each incoming breath, then, the air sacs become overinflated. This process is repeated many times until the sac "pops" like a burst balloon, resulting in further destruction of the respiratory membrane and blood capillaries. Eventually, the area available for the exchange of respiratory gases is greatly reduced (Figure 18.11, bottom).

In a severe case of emphysema, the cells throughout the body suffer from oxygen starvation. To counteract this, the victim's breathing accelerates in a vain attempt to aerate the blood properly. The heart speeds up to propel the blood more rapidly. Because the surface area for gaseous exchange is reduced, carbon dioxide levels in the blood remain higher than in healthy people. The skin of some sufferers turns slightly bluish (cyanosis) as a result.

Emphysema is most common in smokers. In fact, the rate of this disease is 13 times greater in smokers than in nonsmokers. However, studies show that air pollution also contributes to emphysema. Currently, 3 million Americans suffer from emphysema, according to the American Lung Association. An estimated 18,000 people die each year from emphysema in the United States alone, up from 1,500 in 1950—more than a tenfold increase. This increase in mortality coincides with increased levels of atmospheric pollution in areas where the fatalities occurred, although smoking is still the primary cause of emphysema.

Lung Cancer   Although smoking is the leading cause of lung cancer, several chemicals in air pollution could also contribute to this highly fatal disease. These **carcinogens** (cancer-causing agents) include benzopyrene, vinyl chloride, diesel exhaust, asbestos, nickel, and beryllium. The VOC benzopyrene can get into the lungs from the coal smoke released from industrial smokestacks. It is also present in cigarette smoke. The smoke generated when fat drippings from a charcoal-broiled steak spatter on the hot coals is also a source of benzopyrene. Just one smoked steak may contain as much benzopyrene as 600 cigarettes!

Researchers have shown in laboratory experiments on rats and mice that, although two air pollutants may not induce cancer independently, cancer results when an animal is exposed to both pollutants simultaneously. Such an effect is called **synergistic.** For example, lung cancers resembling those in humans have been induced in laboratory animals by first exposing the animals to influenza virus and then to artificial smog. Tumors have also been generated in laboratory animals by forcing them to inhale a combination of benzopyrene and sulfur dioxide.

Researchers have found that the lung cancer rate of men over 45 living on Staten Island, NY, is 155 per 100,000 in the most polluted areas, compared to only 40 per 100,000 in less polluted areas. To research biologists, such data indicate a cause-and-effect relationship between atmospheric pollution and lung cancer in humans. Studies of other cities are showing similar results. One study in the United Kingdom, for instance, showed that the rate of lung cancer was nearly a third higher among people living in areas with high levels of nitrogen oxides, mainly from cars and other vehicles, according to Cancer Research UK, a nonprofit dedicated to cancer research. Swedish researchers estimate that as many as one in ten lung cancers in the Swedish capital (Stockholm) may be caused by carcinogens in air pollution.

Workers in certain occupations are much more likely to develop lung cancer from carcinogens they inhale. Workers exposed to large amounts of diesel exhaust at work, for example, are more likely to develop lung cancer if exposed for long periods. Workers exposed to asbestos, metal dust, paints, polycyclic aromatic hydrocarbons, herbicides, some types of insecticides, and silica may also develop lung cancer. For most of these, however, the incidence is fairly low, and exposures must be fairly high.

Asbestosis   Asbestosis, a disease caused by the accumulation of asbestos fibers in the respiratory tract, is discussed in Case Study 18.1.

GO GREEN!

If you smoke, quit. You'll live longer and healthier, and you won't pollute the air of those around you.

CASE STUDY 18.1   ASBESTOS: THE DANGERS OF A USEFUL PRODUCT

Asbestos is a naturally occurring mineral that has found extensive industrial use, especially since World War II. Roughly 30 million metric tons were used in the United States during the period 1900–1990. Asbestos has been used in an estimated 3,000 to 5,000 products. It has been used as pipe and boiler insulation and ironing board pads. It has been used to manufacture brake linings, protective clothing for firefighters, and talcum powder. In the construction industry, it has been used to strengthen cement and plastics and to fireproof schools and skyscrapers (Figure 1). Its widespread use was made possible by the fiber's flexibility, great tensile strength, and resistance to heat, friction, and acid.

Dispersal Through the Environment

Wherever asbestos is mined or processed, or wherever asbestos products undergo wear, extremely minute fibers, asbestos dust, are released into the atmosphere. For example, a woman sets her iron down on her ironing board pad, inadvertently sending asbestos dust into the air. The driver of an older-model automobile puts his brakes on, sending asbestos dust into the air. What happens to the material worn from brake linings? It's still around, in pulverized form, some of it as asbestos dust, possibly floating in the air, possibly forming a thin film on roads and highways, or possibly adhering to the soft, delicate lining of human lungs. Even food, water, and beverages may contain some asbestos fibers. In Rockville, MD, large areas of the city became contaminated by asbestos-containing crushed rock applied to school playgrounds and city streets. One wonders how many millions of asbestos fibers were eventually inhaled by schoolchildren during recess or by motorists driving by with their windows open.

FIGURE 1   Asbestos fibers. They have been widely used in industry because of their flexibility, great tensile strength, and resistance to heat. However, when inhaled they may cause serious illness.

Some people claim that asbestos "time bombs" may be ticking away in thousands of schoolrooms throughout America. The reason? Starting in the 1940s, asbestos was mixed with paint and sprayed on ceilings and walls for fireproofing and sound insulation. Asbestos was also used to insulate pipes, boilers, and structural beams in schools. This practice continued into the early 1970s. (Such uses were banned by the EPA in 1973.) Dr. Lyman Condie, a toxicologist with the EPA, comments on the problem: "When sprayed surfaces are exposed to student activities—bouncing basketballs off gymnasium ceilings, or children running their hands along stairway ceilings—the asbestos can flake off into the air. Because the fibers are very small and light, they can move throughout the building, even though only a very small area was originally disturbed."

Of our nation's 87,000 school buildings, at least 30,000 contain asbestos in their walls and ceilings. In the worst cases, levels of airborne asbestos inside these buildings may be 100 times greater than ambient levels (levels you'd see in the air we breathe) but are three to four times lower than historic workplace levels associated with the well-documented asbestos-related diseases.

Human Illness Caused by Asbestos

Exposure to asbestos causes a disease known as asbestosis, lung cancer, and mesothelioma (another form of cancer). Currently an estimated 65,000 Americans suffer from **asbestosis,** the symptoms of which include breathlessness, coughing, chest pains, barrel-shaped chest, club-shaped fingers, and bluish discoloration of the skin. These symptoms may not appear until 20 to 30 years after the exposure to asbestos. More than 50% of the people suffering from asbestosis eventually die from lung cancer.

Health experts estimate that 3,000 to 12,000 asbestos-induced cancer deaths occur in the United States annually. Asbestos poses the greatest threat to cigarette smokers. In fact, a cigarette-smoking asbestos worker has 90 times the chance of developing lung cancer as a nonsmoker who has no contact with asbestos! As might be expected, surveys have revealed a relatively high incidence of asbestos-related lung cancers among the 120,000 people (miners, asbestos product processors, and so on) in the United States who worked directly with asbestos.

Mesothelioma (mez-oh-theel-ee-ome´-ah) is a cancer of the chest cavity lining. Although this disease was formerly quite rare, it has become much more common in the United States, especially among asbestos workers. (An estimated 3,000 new cases of mesothelioma are diagnosed in the United States every year.) Many of the victims or their families have sued the companies where their occupational exposure to the asbestos occurred. In Virginia alone, for example, asbestos product manufacturers were sued by 100 people for $300 million during a single year.

At this writing, it is still not clear whether children exposed to asbestos in school face any increase in health risk. Only time and further research will tell.

Control of Asbestos Emissions

In the 1970s, the EPA launched an aggressive program to remove asbestos from the walls and ceilings of schools, convention halls, theaters, and other public buildings. However, since then, discouragingly little progress has been made. Part of the problem was, and

(continued)

is, monetary. It would cost at least $2 billion for a comprehensive, nationwide cleanup in schools alone. Improper removal can pose a hazard to workers and can contaminate the building even more. Accordingly, many experts have argued in favor of stabilizing asbestos, rather than removing it. Exposed asbestos can be kept from flaking off by applying paint or some other type of sealant.

Sites where old buildings are being razed are also significant sources of asbestos emissions. Moreover, they far outnumber such sources as factories. The problem is complicated by the fact that demolition contractors frequently ignore federal regulations on asbestos removal. Federal law requires that all asbestos fibers be wetted down before removal. This material then must be placed in leakproof containers, such as plastic bags, and conspicuously marked.

In 1986, the U.S. Congress passed the **Asbestos Hazard Emergency Response Act,** which is part of the Toxic Substance Control Act. This new law required schools to inspect their buildings for asbestos-containing building materials and then required officials to prepare management plans that recommend the best way to reduce the asbestos hazard, including containment. Options include repairing damaged asbestos, spraying it with sealants, enclosing it, removing it, or keeping it in good condition so that it does not release fibers. Schools were also required to notify parents of any problems as well as actions they were taking. Many schools immediately began removing asbestos because of parents' concerns. In the hands of experts, asbestos is carefully removed, sealed, and then trucked off to an EPA-approved landfill. Unfortunately, the cost of removal and

dangers from improper removal caused many school districts to find alternatives to removal, especially stabilization. There are no laws requiring asbestos to be removed from schools. It is only when the asbestos cannot be maintained in good condition, when measurements show unacceptable levels in the air, or when demolition of a building is imminent that it must be removed.

The asbestos industry, under EPA regulations, has also taken steps to reduce occupational exposure to the life-threatening fibers. The asbestos concentration in the air inhaled by the workers must be less than one fiber per cubic meter of air. The problem has been further diminished by the use of vacuum devices, by the mandatory use of masks, and by automation of many manufacturing processes.

In addition, since the early 1970s, asbestos has been eliminated from many products, including drum brake linings, thanks to a ban imposed by the EPA.

According to the EPA, perhaps the most disturbing feature of the asbestos problem is that control measures have not prevented the gradual long-term development of cancer in persons who come into occasional, slight, or temporary contact with asbestos. Families of asbestos workers, for example, who inhaled asbestos fiber dust inadvertently brought into the home on the workers' clothing and shoes have an increased risk of disease. Such slightly exposed people also include those who live within 1.5 kilometers (0.93 mile) of an asbestos plant. Of the almost 2,000 autopsies on such people in New York City, one-half revealed asbestos fibers in their lungs. (If you're interested in learning more, see our Web site.)

Carbon Monoxide Poisoning Carbon monoxide is an air pollutant produced by the incomplete combustion of organic fuels, such as gasoline, natural gas, and wood. This pollutant can be dangerous because it combines 210 times more readily with the hemoglobin in red blood cells than does oxygen. As a result, it reduces the amount of oxygen carried in the bloodstream. At high doses, carbon monoxide is lethal. It is responsible for many fatalities—both intentional and unintentional—each year. Exposure to lower levels, say 80 ppm of carbon monoxide for eight hours, as might be encountered in a tunnel or a line at a toll booth, causes cellular oxygen starvation equivalent to losing one pint of blood. This can cause headaches and other symptoms.

Carbon monoxide is produced by the combustion of all fossil fuels, especially if the combustion device is inefficient. Cars are a major source of carbon monoxide, as are power plants (Figure 18.12). Furnaces, water heaters, woodstoves, fireplaces, and gas cooking stoves are common sources in our homes, so carbon monoxide is both an indoor and outdoor air pollutant.

CO is also produced by the combustion of tobacco in cigarettes, pipes, and cigars. Researchers have shown that cigarette smoke may contain 300 ppm of carbon monoxide. Carbon monoxide in tobacco smoke affects the smoker as well as nonsmokers in the vicinity. Carbon monoxide levels in a room of smokers can reach high levels, sufficient to deactivate roughly 10% of the hemoglobin in a nonsmoker in the same room.

The presence of carbon monoxide in the bloodstream of pregnant women has been suggested as a possible cause of stillbirths and deformed offspring. Certain conditions may render some people especially susceptible to **carbon monoxide poisoning.** They include heart disease, asthma, diseased lungs, high altitude, and high humidity.

Carbon monoxide may be the indirect cause of many fatal traffic accidents in the United States yearly. The effects of low-level carbon monoxide poisoning are similar to those of alcohol or fatigue; they impair the motorist's ability to control the vehicle. Because carbon monoxide is colorless and odorless, harmful levels may build up within the car without the driver being aware of them.

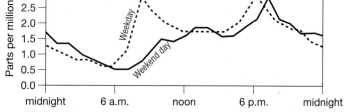

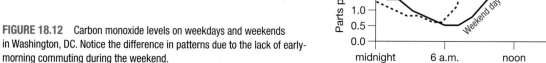

FIGURE 18.12 Carbon monoxide levels on weekdays and weekends in Washington, DC. Notice the difference in patterns due to the lack of early-morning commuting during the weekend.

Effects of Air Pollution on Other Organisms and Materials

Because it is so ubiquitous and damaging, air pollution affects many different species. Pollutants, for example, damage a wide assortment of plants—from trees along city streets to trees in forests downwind from urban centers. Air pollution even damages farm crops. Studies of farms in California and along the East Coast of the United States, for example, show that the pollutant ozone damages vegetable crops directly, reducing growth and thus affecting farm output. Like ozone, sulfur dioxide also damages plants directly. Sulfuric and nitric acid damage plants either directly or through changes in the soil.

Air pollutants can also damage materials such as metals, stone, concrete, clothing, rubber, and plastics. The four main culprits—once again—are sulfur dioxide, sulfuric acid, ozone, and nitric acid. Pollutants in cities often reach extraordinary levels and can cause considerable damage to buildings and valuable statuary. The Taj Mahal in India and the Statue of Liberty are two of the many victims of urban pollution. In fact, damage to the Statue of Liberty from sulfur dioxide and acids was so extensive that it required extensive repair in the 1990s, costing $35 million. Damage to crops and materials is estimated to be nearly $10 billion per year in the United States alone.

18.6 Air Pollution Abatement and Control

Because of its many effects on people, plants, animals, ecosystems, and materials, air pollution has long been the subject of concern. This section describes efforts to reduce and eliminate air pollution.

Pollution Control in Factories and Power Plants

Pollution can be reduced or eliminated either by pollution controls (devices that remove pollutants from smokestacks and other sources) or by preventive measures (energy efficiency measures that reduce energy consumption). This section will examine both strategies, beginning with the more traditional and often more costly—and less effective—approach: pollution control methods. Let's begin by examining the control of particulates.

Controlling Particulates   Three standard types of equipment for controlling particulates at stationary sources, such as power plants and factors, are available. The **fabric filter bag house** physically removes airborne particles from smokestacks via huge cloth bags suspended inside the device (Figure 18.13). A large filter bag house may consist of more than 1,000 elongated filter bags, each of which is several meters long. Up to 99.9% of the dust particles may be removed from the stack gases.

Another means of removing particulates is the **electrostatic precipitator.** This device is designed to remove small-diameter solid particles (dust, fly ash, asbestos fibers, and lead salts) less than 1 micron (one-millionth of a meter) in diameter. They're most commonly used in coal-fired power plants to remove particulates from gases escaping in the smokestack (Figure 18.14).

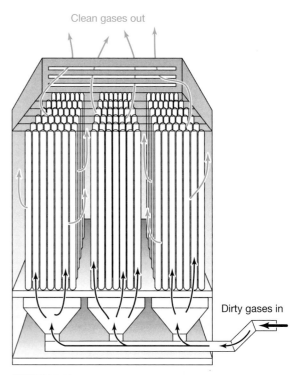

FIGURE 18.13   Filter bag house. Solid particles are removed from exhaust gases by long vacuum cleaner–type bags that pick up particulates as gases are passed through them.

In the electrostatic precipitator, pollutants pass between pairs of positively and negatively charged electrodes. The particles become negatively charged as they flow through the device and are then attracted to a positively charged collector electrode in the wall. From time to time, the device is switched off, so particulates that have accumulated on the wall can drop to the bottom and be removed. Although the initial cost of a large precipitator can be high, the power and maintenance costs are small.

The **cyclone filter** removes heavy dust particles with the aid of gravity and a downward-spiraling air stream (Figure 18.15). Particulate-laden air enters the device and travels through it. As it does, the particles strike the walls and fall to the bottom of the cylinder, where they can be removed.

Controlling Sulfur Oxide Emission   The release of sulfur oxides into the atmosphere from transportation, industrial, and residential sources adversely affects human health, wildlife, forests, and farm crops, as well as irreplaceable paintings, monuments, stonework, and statuary.

Sulfur oxide emissions can be reduced in several ways. Although no method will be satisfactory by itself, using the following approaches in combination can greatly reduce sulfur oxide emissions:

1. Shifting from high- to low-sulfur coal. Much of the industrial coal consumed before 1970 had a relatively high sulfur content (up to 3% or more). This coal came from mines in Pennsylvania, West Virginia, and Illinois. Once health officials become aware of the problems posed by oxides of sulfur, however, municipal, state, and federal regulations were passed to limit its use. Still, many companies continued to burn high- and medium-sulfur coal.

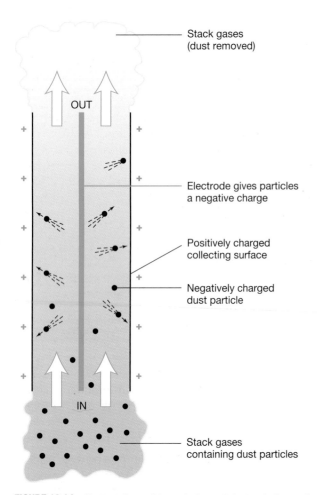

Stack gases
(dust removed)

OUT

Electrode gives particles
a negative charge

Positively charged
collecting surface

Negatively charged
dust particle

IN

Stack gases
containing dust particles

FIGURE 18.14 Electrostatic precipitator. As the soot, dust, and other particulates pass through the precipitator, they are given a negative charge. They are then attracted to the positively charged wall of the precipitator. After accumulating on this collecting surface, they are periodically released to a collecting chamber.

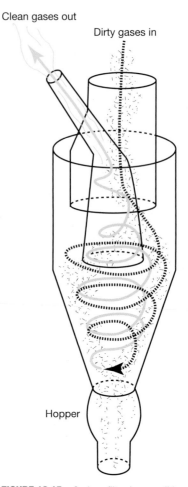

Clean gases out

Dirty gases in

Hopper

FIGURE 18.15 Cyclone filter. Large solid particles are removed by centrifugal force and are collected in a hopper.

However, when U.S. regulations were tightened further to control acid rain in the early 1990s, many companies switched to low-sulfur coal to avoid having to install pollution control devices. Fortunately, vast supplies of low-sulfur coal occur in many western states, such as Colorado, Montana, and Wyoming. These huge deposits can be removed by **strip mining,** a process in which the rock and soil overlying deposits of coal are removed in progressive strips. Strip mining can be very destructive to the environment. Costly reclamation efforts must be taken to restore the sites to their original condition (see Chapter 21 for a discussion).

2. Coal cleaning. Transporting low-sulfur coal a distance of more than 1,600 kilometers (1,000 miles) to eastern industrial plants adds greatly to the cost of coal. An alternative is to mine the high-sulfur coal available in the Midwest and East, where it is close to the industry that will use it, and then remove much of the sulfur before the coal is burned. Much of the sulfur in high-sulfur coal comes from an impurity known as iron pyrite (fool's gold). Because iron pyrite is heavier than the coal, it sinks to the bottom of a tank containing a mixture of pulverized coal and water and can easily be removed (Figure 18.16).

3. Removing sulfur from smokestack gases. The **flue gas desulfurization** process, more commonly known as **scrubbing,** is widely used in the United States and other more-developed countries to remove sulfur from smokestack gases of power plants and factories (Figure 18.17). A scrubber is a device through which smokestack gases pass. In the scrubber, a mist consisting of ground limestone and water is sprayed into the stack gases. The calcium in the limestone reacts with the sulfur to form calcium sulfate.

Scrubbers can remove up to 95% of the sulfur dioxide in stack gases. The calcium sulfate sludge may be used in road beds or drywall production, but is most commonly disposed of in landfills. A big advantage of scrubbers is that they can be added to existing plants as well as being used on new ones. The process is costly, however, and solves an air pollution problem by creating a solid-waste problem.

Although pollution control devices work, they typically result in other problems. Scrubbers, for example, produce toxic sludge. Electrostatic precipitators remove particulates from smokestack gases but contain an assortment of toxic heavy metals. More and more, policy makers and businesses themselves are seeking alternatives to these types of controls, notably pollution prevention measures. Why not, they

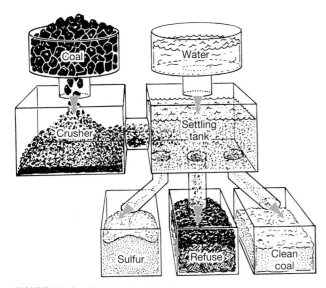

FIGURE 18.16 Coal-cleaning process.

say, avoid pollution altogether? If you can achieve the same goal without making pollution, you're much better off in the long run.

Control of Automotive Emissions

Automobiles, trucks, buses, and other types of motor vehicles are major sources of air pollution the world over. These **mobile sources** release

an assortment of air pollutants, among them carbon monoxide, carbon dioxide, sulfur oxides, nitrogen oxides, particulates, and VOCs. Emissions have been controlled in a variety of ways. This section surveys these approaches with an emphasis on the most sustainable ones.

Catalytic Converters To reduce the pollutant emissions from automobiles as required by the Clean Air Act, U.S. auto manufacturers have relied heavily on a device known as the catalytic converter (Figure 18.18). The **catalytic converter** is a muffler-shaped device that is incorporated into the exhaust system of the car. One type of converter has a honeycomb-like interior. The cells of the honeycomb are coated with a **catalyst,** a material that speeds up chemical reactions. In automobile catalytic converters, the catalyst is platinum, rhodium, or palladium. The products of incompletely burned gasoline such as carbon monoxide and VOCs pass through the cells of the catalytic converter. On the catalyst, these gases react with oxygen and are converted into carbon dioxide and water.

One of the drawbacks of the catalytic converter just described is that it doesn't remove nitrogen oxides. Fortunately, most cars are equipped with a three-way converter. This device reduces nitrogen oxides in automobile exhaust as well as carbon monoxide and hydrocarbons. In this converter, nitric oxide oxidizes carbon monoxide to carbon dioxide and converts hydrocarbons to carbon dioxide and water. During this process, the nitric oxide is broken down to nitrogen and oxygen. Three-way catalytic converters have been widely used in automobiles since 1981. Only more recently have they been required on large spark ignition (LSI) engines such as those found in construction equipment and forklifts.

Inspection and Maintenance Programs Poorly tuned automobiles burn gasoline less efficiently than their well-tuned counterparts, so they produce more pollution. But how does one monitor how well millions of individual automobile owners are keeping their vehicles in tune? The 1977 amendments to the U.S. Clean Air Act required regions of the United States in which air quality standards for carbon monoxide and ozone were not met by 1982 to set up **inspection and maintenance (IM) programs.** Under such programs, automobile emissions are checked each year by state-licensed inspectors. When levels of pollutants in the exhaust, such as carbon monoxide and hydrocarbons, are too high, the owner must have the engine adjusted or repaired to meet emissions requirements. In most cases, the vehicle needs only minor adjustments to the fuel injection system, which mixes fuel and air before it enters the cylinders where combustion takes place. The most common repairs, such as spark plug replacements and injector adjustments, may reduce pollutant emissions by 25%. In Portland, OR, an IM program resulted in a 40% reduction in emissions the very first year. The IM program in California, known as Smog Check, has cut smog-forming nitrogen oxides and VOCs by 18 and 90 metric tons per day, respectively, and emissions of carbon monoxide by 1,350 metric tons daily for the state as a whole.

GO GREEN!

Drive smart. Obey the speed limit, and accelerate gradually. If you see a stop sign or red light approaching, slow down gradually.

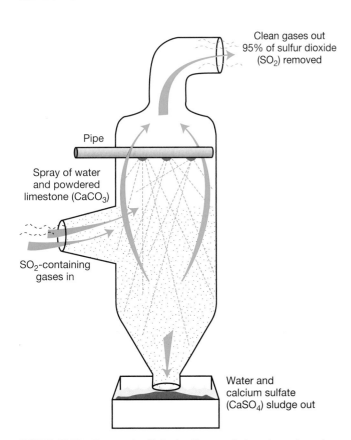

FIGURE 18.17 Flue gas desulfurization. Downward-streaming water and powdered limestone ($CaCO_3$) "scrub out" sulfur dioxide from stack gases. A sludge of calcium sulfate forms and must be removed.

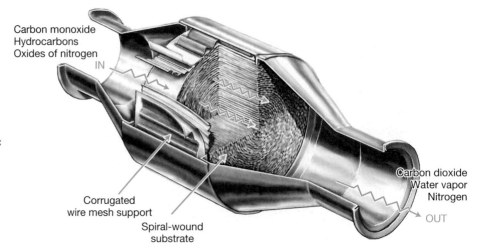

Catalytic Converter

Carbon monoxide
Hydrocarbons
Oxides of nitrogen

IN

Carbon dioxide
Water vapor
Nitrogen

OUT

Corrugated
wire mesh support

Spiral-wound
substrate

FIGURE 18.18 Cutaway illustration of a catalytic converter used by Ford Motor Company. Resembling a small muffler, the emission-control device converts hydrocarbons, carbon monoxide, and oxides of nitrogen into nitrogen, carbon dioxide, and water. The chemical reaction depends on the platinum and rhodium catalysts that coat the internal surfaces of the converter.

Increasing Fuel Efficiency Tuning automobiles makes them more efficient and cleaner-burning. But vehicles and their engines can also be redesigned to make them even more efficient and less polluting. In 1975, a U.S. federal law placed a federal excise tax on all cars of a particular model that did not attain a minimum standard of efficiency set by the EPA. It was commonly referred to as the gas-guzzler tax. The size of the tax was inversely related to the car's fuel efficiency. For example, a few years ago, the biggest "gas guzzler" on the American market was an Italian import, the Lamborghini Countach, which achieved only 2.5 kilometers per liter (6 miles per gallon) in the city—the lowest fuel efficiency ever recorded by the EPA. The tax on this car was a hefty $3,850. The gas-guzzler tax was designed to discourage typical consumers from buying energy-inefficient, pollution-producing vehicles. Unfortunately, President Clinton repealed the gas-guzzler tax in 1996. Soon after that, American auto manufacturers began producing very large sport utility vehicles (SUVs) with gas mileages only slightly better than the Lamborghini. These vehicles sold extremely well in many parts of the country—that is, until the cost of gasoline began to dramatically increase in 2007 and 2008. (In 2008 GM announced that it was closing four factories in which it built trucks, due to a sharp decline in demand.)

Another attempt to encourage the manufacturing and sale of energy-efficient vehicles in the United States is the **Corporate Average Fuel Efficiency (CAFE) standards.** These standards, established by the United States, set average efficiency goals for new autos sold in the United States, including imports and domestically manufactured vehicles. In 1994, the average efficiency required in new models was 27.5 miles per gallon (11.5 kilometers per liter) for city and highway driving. All attempts to raise the average fuel efficiency over the next 20-plus years were soundly defeated by auto manufacturers. (Several attempts were made to raise the fuel efficiency standard to 40 miles per gallon in the 1990s.) According to the *New York Times,* an EPA report in 2005 showed that loopholes in American fuel economy regulations allowed automakers to produce cars and trucks that are, on average, "significantly less

fuel-efficient...than they were in the late 1980s." As a result, the average fuel efficiency of new vehicles (cars, trucks, vans, and SUVs) began to decline. In 2000, for instance, new passenger cars averaged 22 miles per gallon, while SUVs, vans, and pickup trucks averaged 17.5 miles per gallon.

Improving fuel efficiency is not that difficult. In 2007, 113 cars sold worldwide averaged 40-plus miles per gallon. Only two, the Toyota Prius and Honda Civic Hybrid, were available in the United States, according to Nancy Taylor, author of *Go Green: How to Build an Earth-Friendly Community.* Interestingly, nearly two-thirds of these fuel-efficient cars sold overseas were produced by U.S. automakers, GM and Ford, or foreign manufacturers with large U.S. markets, such as Toyota and Volkswagen.

Some experts believe that in the not-too-distant future, a fuel efficiency of 33 kilometers per liter (80 miles per gallon) will be attained by American cars. The Japanese already have a model that seats five and gets 80 miles per gallon on the highway. Honda's VX series, now discontinued, achieved gas mileage of 56 miles per gallon. Chevy's Metro, formerly the Geo Metro, boasted similar mileage.

Managing and Reducing Traffic Volume City officials, collaborating with university researchers and departments of transportation, are working on ways to better manage traffic to make traffic flow more evenly. Such efforts prevent or reduce congestion and improve fuel efficiency of cars. Idling cars produce lots of pollution (and get 0 miles per gallon).

Computer systems that monitor traffic and flash messages to motorists are currently being used in many cities. These systems alert drivers about potential jams, allowing them to take alternative routes. On-ramp controls are also being used in many cities to produce a more seamless integration of oncoming traffic with existing traffic flows. Traffic lights on the on ramp stop cars and then let them proceed two at a time at spaced intervals, which allows them to enter traffic without interrupting its flow.

All of these measures, from much more efficient cars to traffic control, are effective, but a far more effective means to control urban air pollution is to reduce automobile traffic volume on highways (Figure 18.19). The EPA has proposed numerous was to reduce traffic volume in large urban areas, among them (1) terminating free parking facilities for employees by their employers; (2) placing an added tax on downtown parking fees; and (3) prohibiting commuters from driving to work one day per week, thus reducing commuter traffic by 20%. All of these could help promote alternative means of transportation such as buses, commuter trains, walking, carpooling, vanpooling, and bicycling. Mass transit can be boosted with free bus passes to employees of urban businesses and to students who attend urban colleges and universities.

New York City, which boasts the highest use of mass transit in the United States, does so in part because it (1) bans all street parking in the main business section of Manhattan; (2) restricts cruising by Manhattan taxis; (3) has placed several Manhattan bridges on the toll system; and (4) sets aside one lane on busy roads for the exclusive use of buses (Figure 18.20).

Replacing Gasoline With Ethanol Alternative fuels also can reduce air pollution. **Ethanol,** the same chemical found in alcoholic beverages, for example, is capable of powering motor vehicles. Ethanol can be produced from a renewable source, too—from fermented grains such as corn or from sugar cane. It can also be produced from switchgrasses and nongrain portions of plants—for instance, the stalks and leaves of corn plants. Ethanol can be added to gasoline, or it can be burned in pure form, although some engine modifications are necessary for this.

Ethanol not only comes from a renewable source, but also it pollutes much less than gasoline, producing less carbon monoxide, hydrocarbons, and sulfur oxides. Ethanol produces less carbon dioxide than fossil fuels because all the carbon dioxide released by ethanol-powered vehicles will be incorporated into plants to create more fuel for future generations. Ethanol is not a zero-carbon fuel, however, because it is grown on farms powered by diesel tractors and gasoline trucks supplied by oil. The consumption of these fuels produces carbon dioxide. Nonetheless, ethanol could help reduce the buildup of carbon dioxide in the atmosphere, a major cause of global warming (Chapter 19).

FIGURE 18.19 Traffic at a standstill in San Francisco. Like many cities, San Francisco faces a traffic crisis. Cars back up on the freeway during rush hour, making the commute a major ordeal and creating large amounts of air pollution.

FIGURE 18.20 In many urban centers, special lanes on busy thoroughfares are reserved for the exclusive use of commuter buses and other high-occupancy vehicles (car pools and van pools). This strategy has proved effective in reducing commuter traffic volume and air pollution.

Ethanol is used widely by motorists in Brazil today. Ethanol is currently less expensive in the United States than gasoline, although it contains about 30% less energy per gallon than gasoline. Researchers have developed more efficient means of producing ethanol—for instance, using switchgrass and plant material other than grains. As discussed in Chapter 23, this could help reduce one of the negative effects of ethanol as a fuel—the rise in food prices. (For more on ethanol, see Chapter 23.) As fossil fuel reserves decline over the next decade or two, ethanol could become even more popular.

In 1994, the EPA ordered that ethanol must be mixed with any gasoline used in the nation's nine smoggiest cities—Baltimore, Chicago, Houston, Los Angeles, Milwaukee, New York, Philadelphia, San Francisco, and Hartford (Connecticut). This fuel change, which started in 1995, was designed to reduce smog in these cities by 16%. California instituted a similar statewide program in 1996. Its program, which requires a greater percentage of ethanol in the fuel mix, was expected to cut California's smog by 40%.

Developing Alternatives to the Internal Combustion Engine   Since 1970, when the Clean Air Act was passed, emissions of hydrocarbons and carbon monoxide by the average car have been reduced by 90%, and emissions of oxides of nitrogen by 75%, resulting in cleaner air in many cities. Recently, however, federal analysts have been increasingly skeptical that further substantial overall reductions in urban air pollution can be made, even with the use of the emission-control strategies described earlier. The reason? The rapid increase in the number of motor vehicles on our highways. The number of motor vehicles has increased from 156 million in 1980 to 241 million in 2005, according to the U.S. Federal Highway Administration. Some people think that what is really needed is a shift from the traditional internal combustion engine to an engine powered by solar electricity or even steam.

Electric Cars   As gas prices increase throughout the world, more and more people are looking for less expensive alternatives. One of these is the **electric vehicle**—cars, trucks, and busses powered by electric motors instead of internal combustion engines. Electric cars are desirable because they do not burn gasoline and emit no pollutants. However, although the electric car itself is pollution free, we mustn't forget that the electricity generated to charge the batteries must come from power plants. Most of these power plants burn coal. Because electric motors are so efficient, however, electric cars get the equivalent of 100 to 150 miles per gallon and produce much less pollution overall than gas-powered vehicles.

Another big disadvantage of the electric car is that the lead acid batteries required to power the car are rather heavy and large. Electric cars using these batteries are usually good for distances of 60 to 100 miles before recharging, so they are good only for commuting. Fortunately, 90% of all car trips in the United States are shorter than 60 miles a day! New battery technologies, especially the lithium-ion battery, could boost the range considerably. The Tesla, an electric sports car made in the United States, operates on lithium-ion batteries and goes 220 miles between charges.

Electricity generated from clean, renewable sources such as wind generators and special solar electric cells (photovoltaics) could make the electric car an even more viable alternative to gas-powered vehicles.

A handful of companies produce electric cars, but many more are expected to join their ranks in the not-too-distant future, turning out affordable electric cars. In December 2007, for example, *Fortune* magazine reported that eleven new companies were planning to manufacture and sell electric cars within a few years, among them Mitsubishi, Dodge, and Nissan.

Hybrid Cars   One car that would make up for the limited range of the electric cars now on the road is the **hybrid car**—an automobile equipped with both an electric engine and a small internal combustion engine. For example, the Toyota Prius, which gets around 45 to 50 miles per gallon in combined city and highway driving, starts on its electric motor (Figure 18.21). Once it reaches speeds of around 10 to 17 miles per hour, the gas engine kicks in. On the highway, it operates on both gas and electricity. Electricity is supplied to the electric motor to boost power, for instance, to climb a hill or pass another vehicle.

Honda's hybrids start on gas—that is, start by turning on the gas engine. The electric motor kicks in to boost power. Until 2007, Honda produced a two-seater commuter vehicle, the Insight, which got nearly 70 miles per gallon. The company currently produces a manual transmission four-seater, the Hybrid Civic, rated at 46 miles per gallon in the city and 51 miles per gallon on the highway.

All automobile manufacturers produce hybrid vehicles, sometimes several models. Unlike electric cars, the hybrid car doesn't require a huge battery bank. Moreover, its batteries are recharged when the car is running on gasoline or when it is slowing down. When the car decelerates, the electric motor produces electricity that helps to keep the batteries full. Electric cars need to be plugged into an electric outlet to recharge.

Hybrid cars are an important step into a world of more efficient vehicles. Amory Lovins, an internationally renowned expert on energy who has helped design hybrids, believes that a fuel efficiency of 150 to 200 miles per gallon could be achieved

> 👣 **GO GREEN!**
> Whenever possible, walk, ride a bike, take the bus, or carpool with others to save gas and reduce pollution.

FIGURE 18.21   The Toyota Prius, one of many hybrid vehicles on the road today and currently the most fuel efficient. This car comfortably seats four to five passengers and averages around 45 miles per gallon, although the senior author routinely gets 48 to 52 miles per gallon in his because of his more conservative driving style.

by a hybrid car. But it will need to be made out of much lighter space-age materials that are durable and crash-resistant. Current hybrids are no more than conventionally made cars with steel bodies and frames with a hybrid power drive system. Such changes could dramatically boost their gas mileage.

Another, even more promising development is the plug-in hybrid. The **plug-in hybrid** has a larger electric motor than existing gas hybrids and also a larger battery bank. These features allow the vehicle to travel for up to 60 miles on electricity only—ideal for commuters. On trips longer than 60 miles, the car reverts to the gas-electric mode like that of a conventional hybrid. This would allow the vehicle to be used for commuting and long trips. Theoretically, an individual who travels fewer than 60 miles per day would never use any gasoline. Plug-in hybrids could average 100 to 125 mile per gallon.

Electric Cars Powered by Fuel Cells   Many analysts believe that the hybrid and plug-in hydrid cars are merely transitional vehicles. The cars of the future, they say, may be electric cars powered by fuel cells. **Fuel cells** look like batteries but function quite differently. They combine oxygen (from the air) and hydrogen (from the car's fuel tank) to produce electricity. The electricity is then used to run an electric motor that powers the car. Where will the hydrogen come from?

Hydrogen can be generated from water, for example, in a process called **electrolysis.** Electrolysis is the breakdown of water into oxygen and hydrogen. This occurs when an electrical current is passed through water. Electricity to make hydrogen can be generated from conventional power plants or from wind and solar sources, as discussed in Chapter 23. Hydrogen can then be stored and used to power the fuel cell to make electricity to run the car's electric motor.

Hydrogen can also be stripped from fossil fuels such as gasoline and natural gas by a special device known as a **reformer.** Unfortunately, such engines still produce lots of carbon dioxide and use finite resources.

While appealing on many levels, hydrogen vehicles do have some downsides. To replace gasoline, for instance, hydrogen must be available at convenient locations. Creating a hydrogen economy depends on creating a costly distribution system that rivals gasoline's. Hydrogen vehicles powered by fossil fuels, already being tested, appear to be more efficient than conventional cars, but they do produce a lot of carbon dioxide (caused by the production of electricity to split water). Another huge downside of hydrogen is that it's much more efficient (three to four times) to run a car directly from electricity than to use electricity to split water to make hydrogen to feed into a fuel cell to make electricity to power an electric motor of a fuel-cell car.

Mass Transit and Urban Growth Management   In the long-term future, sustainable cities will depend more on **mass transit,** modes of transportation that carry large numbers of people from their homes to work or other activities. One of the reasons mass transit will likely be more popular in the future is that buses and commuter trains or light rail (small trains that connect suburbs with the cities) carry passengers four to five times more efficiently than automobiles. As a result, they produce four to five times less pollution per passenger mile.

Mass transit may grow in popularity not just to reduce pollution and to reduce energy consumption, but also to quell growing traffic congestion.

To be acceptable, mass-transit systems must be designed to move people quickly to and from work. This will require special mass-transit lanes on major highways going in and out of cities, which are already common in many urban areas. They're known as high-occupancy vehicle lanes. Light-rail tracks could be laid in highway medians or alongside major rail lines. Some cities use existing rail lines. It is not inconceivable that over time, as pressure for mass transit grows, highway lanes, perhaps even entire highways, could be converted to rail lines.

Another component of improving transportation is growth management—efforts to restrict growth to certain areas, often to create denser human settlements as opposed to less dense sprawl, which is inefficient. Efforts to locate new businesses in certain areas such as existing central business districts and to concentrate people, rather than spread willy-nilly over the landscape, create high-density nodes that can be efficiently serviced by buses and light rail.

Growth management is now being practiced in several states, among them Oregon. For more information on this topic, see Chapter 5.

18.7   Indoor Air Pollution

When most people think of air pollution, they think of belching smokestacks and the exhaust fumes of cars and trucks. But how about your family living room or kitchen? Recently, scientists have been finding that the air we breathe in our own homes, as well as in schools, stores, and offices, may be more dangerous to our health than the smog-ridden air outside. **Indoor air pollution** comes from a variety of sources: gas ranges, gas furnaces, tobacco smoke, new carpeting, paint, and even furniture, hair sprays, and perfume.

Concern about indoor air quality is justifiable, first of all, because people spend much of their time indoors—for most people, 90% of their time is spent inside. Even construction workers and home builders, who spend a lot of time outdoors, are still inside about 60% of their life.

Indoor air pollution is a result of two factors. First, in an attempt to reduce their home heating bills, many homeowners and home builders have caulked and weatherstripped homes better to prevent warm air from leaking out of cracks in the building envelope—the walls, roof, window casings, and foundations. Although these measures are effective in reducing air leakage and energy losses, they can lock indoor air pollutants in. Before the energy crisis, the average residence time for a given molecule of gas inside the home was about one hour. However, in an airtight home, the residence time is about 400% greater. Because many homes are more tightly sealed, fresh air exchange occurs less frequently. Pollutants once purged from the house now can build up inside.

Scientists at the Lawrence Berkeley Laboratory in California measured concentrations of pollutants inside a tightly sealed, well-insulated house in which a gas stove and oven were used in ordinary meal preparation. What did they find? Concentrations of carbon monoxide and nitrogen dioxide in the kitchen, bedroom, and living room from the gas stove actually exceeded the levels of these gases outdoors! In fact, in some homes, the level of nitrogen dioxide may be two to seven times higher than is considered acceptable for outside air. Thus, for many pollutants such as particulates, oxides of nitrogen, carbon monoxide, and hydrocarbons (benzene, chloroform), indoor pollution is a much more serious health threat than outdoor pollution (outdoor pollution must be controlled to protect crops, forests, buildings, and other species).

The second major reason why indoor air pollution has come to the forefront in recent years—and the most important reason—is that new homes and buildings are made from products or furnished and decorated with products that release potentially toxic chemicals. Most brands of carpeting, for example, release volatile organic chemicals long after a family has move into its new home. (They come from the glue that holds the backing on.) Wood products such as cabinets and shelves and oriented strandboard, a sheathing material that is used for floors, exterior walls, and roof decking, also release a particularly harmful VOC, formaldehyde. Paints release VOCs as well. Latex paint releases mercury. Many paints also contain fungicides and other chemicals.

Chemicals released from building materials, paints, stains, finishes, and furnishings inside a house contaminate indoor air. This process is called outgassing. Unfortunately, many types of chemicals contaminate indoor air—too many for discussion here. (If you are looking for more information on the subject, see our Web site for references that may prove helpful.) This section will cover two of the major chemicals: formaldehyde and radon. Tobacco smoke is covered in Case Study 18.2.

Formaldehyde You probably remember **formaldehyde** as the fluid used to preserve frogs and fetal pigs in high school and college biology classes. You may be surprised to learn that it is commonly found in certain types of foam insulation, furniture, carpets, particle board, and plywood. In many products, including plywood, it is part of a resin used to hold the materials together. Unfortunately, formaldehyde is released from these products in our homes and can cause a variety of human health problems, including eye irritation, nausea, respiratory problems, and cancer.

The Occupational Safety and Health Administration (OSHA), the federal agency that establishes rules to protect workers and sets standards for new products, has set 3 ppm as the maximum formaldehyde concentration permitted inside industrial plants. Studies in Europe and the United States have shown that formaldehyde levels inside homes are often higher. In Mission Viejo, a large development in California, for instance, formaldehyde concentrations in a research house containing no furniture were relatively low. However, when furniture was added, formaldehyde levels increased almost threefold.

CASE STUDY 18.2 TOBACCO SMOKE: THE DEADLIEST AIR POLLUTANT

Although many of us are concerned about outdoor air pollution, it is a relatively minor problem (from a human health perspective) compared with tobacco smoke, which exposes both the participant and nonparticipants who are near smokers. Tobacco smoke is a witch's brew of toxic chemicals (at least 200 of them) that kills 440,000 Americans each year or 1,200 per day from lung and other forms of cancer, stroke, and heart attack. Worldwide, the death toll is about 5 million a year, according to the American Cancer Society (ACS).

"Cigarette smoke is a complex mixture of chemicals produced by the burning of tobacco and the additives," writes the ACS. "The smoke contains tar, which is made up of more than 4,000 chemicals, including over 60 known to cause cancer. Some of these substances cause heart and lung diseases, and all of them can be deadly." In additon, some of the chemicals in tobacco smoke, such as carbon monoxide, sulfur dioxide, oxides of nitrogen, and particulates, are serious pollutants of the air in our major cities. Smoking greatly increases the amounts of these substances that a smoker takes into his or her lungs, contributing to an increase in the incidence of emphysema and chronic bronchitis.

Medical research on the deadly effects of tobacco smoke is now firmly established. Today, more than 40,000 books and articles have been published on the effects of tobacco smoke. In 2005, the ACS estimated that nearly 164,000 Americans died of lung cancer.

Approximately 87% of all lung cancers are believed to be caused by smoking (Figure 1). Worldwide, the ACS estimates that 850,000 people die each year from lung cancer, and most of these cancers are caused by smoking.

Smoking is responsible for nearly one in five deaths in the United States, according to the ACS. Moreover, the society adds, "about half of all Americans who continue to smoke will die because of the habit." Many who smoke die in their fifties and early to mid-sixties of lung cancer. What's more, tobacco causes cancers of the mouth, throat, larynx, esophagus, pancreas, uterus, kidneys, and bladder. Smokers also greatly increase their chances of dying from strokes and heart attacks.

Smoking has other adverse health effects, too. A pregnant woman who smokes passes on the tobacco chemicals in her body to her developing fetus. Although much lower in prevalence, the effects are similar to those of cocaine. The effects on pregnant women and their offspring include spontaneous abortions, a slightly higher rate of sudden infant death syndrome (SIDS), reduced number of brain cells, behavioral problems, and learning difficulties.

Smokers are not the only ones who suffer ill effects from this action. Nonsmokers can be severely impaired if they inhale **secondhand smoke,** that is, smoke exhaled by smokers or smoke that issues from the tip of the cigarette itself. The EPA estimates that secondhand smoke causes the death of at least 3,000 Americans

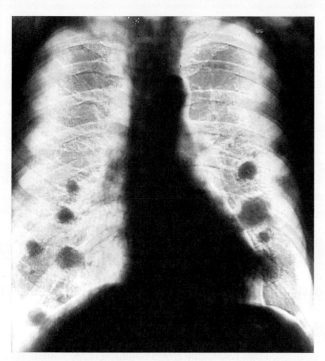

FIGURE 1 Frontal chest X-ray showing lung cancer. The dark areas in the lower half of each lung are cancerous masses.

every year. The agency also estimates that it causes 225,000 cases of bronchitis, pneumonia, and other respiratory infections in children up to 18 months old and increases the frequency of asthma attacks in 600,000 asthmatic children each year.

Rate of Smoking

Smoking has declined markedly in the United States since the 1950s as awareness of its ill effects have been publicized. Among adults in the United States, cigarette smoking has declined from about 50% of the population in 1950 to about 20.9% in 2005 (the latest year for which numbers are available), according to the American Cancer Society. Nonetheless, approximately 45 million adults currently smoke cigarettes (about 23.9% of men and 18.1% of women).

One of the biggest concerns in the past decade has been the increase in teenage smokers. Because smoking is addictive, many of these individuals will be hooked for life. "Anyone who starts smoking is at risk of becoming addicted to nicotine," notes the ACS. In fact, "studies show that cigarette smoking is most likely to become a habit during the teen years. The younger a person is when he or she begins to smoke, the more likely he or she is to develop nicotine addiction. Almost 90% of adult smokers started at or before age 19."

Although smoking is declining in more-developed countries, it is on the rise in less-developed countries. As markets in more-developed countries decline, tobacco companies have expanded efforts to market in less-developed countries.

Their efforts are paying off—at least for the tobacco companies. In China, for instance, more than 350 million people smoke, more than the entire U.S. population. "If these current patterns continue, there will be 2 billion smokers worldwide by the year 2030, half of whom will die of smoking-related diseases if they do not quit," notes Dr. Ahmedin Jemal of the American Cancer Society.

Control of Tobacco Use

Smoking is a private decision with public ramifications. Not only are nonsmokers exposed, but the cost of health care for many people who have smoked their whole life is paid through taxes—Medicaid. Smoking costs businesses millions of dollars a year in lost wages, too. Several strategies can be pursued to reduce smoking worldwide.

Smokers can be encouraged to use nicotine patches to help them kick the habit. When applied to the skin, these patches slowly release small amounts of nicotine into the bloodstream. The urge to smoke is therefore reduced. Twice as many smokers are able to end their nicotine addiction in this way as compared with those who do not use the patch technique.

The federal tax on tobacco products, some say, could be increased to help defray public health costs and cut down on smoking. An added benefit of such a tax would certainly be a reduction in the rate of tobacco use. After all, this tax would probably price many teenagers and children out of the cigarette market. (Note that tobacco companies have raised prices of cigarettes quite substantially in recent years to pay huge fines for class-action lawsuits, thus driving up the cost of cigarettes considerably.)

Restrictions on smoking could also help. Smoking has been restricted or banned in public places. Most states and hundreds of cities in the United States have done just that. The government of Thailand recently banned actors from smoking in movies, a measure designed to reduce smoking among children and teens.

Yet another way to reduce tobacco addiction would be to restrict or even ban all tobacco advertisements. For years, the industry has advertised its products with highly effective ads in newspapers, magazines, and on highway billboards at an annual cost of more than $4 billion. Drastic restrictions in advertising have gone into effect as a result of state and federal lawsuits against cigarette manufacturers. Campaigns directed to younger individuals have been eliminated.

Litigation: Will the Tobacco Industry Die?

Tobacco companies have faced extreme pressure from two types of lawsuits, class action suits filed on behalf of smokers or their surviving relatives and lawsuits by states. In 1999, the tobacco companies lost their first class-action lawsuit. Jurors in Miami found cigarette manufacturers guilty of defrauding smokers and forced them to pay billions of dollars in damages to smokers and their families. The Miami jury concluded that cigarette makers "engaged in extreme and outrageous conduct," concealing cigarettes' dangers, conspiring to hide their addictiveness, and making a product that caused more than a dozen deadly diseases from heart disease to lung cancer.

The states of Florida, Minnesota, Michigan, and Texas also successfully sued the tobacco companies to seek compensation for state funds that have gone into the medical treatment of smokers. To avoid an avalanche of similar suits, the federal government stepped in and came to an agreement with tobacco companies on behalf of the remaining 46 states, creating a huge settlement. The settlement will provide $206 billion to states, bans advertisement to youth and at sporting events, and has many other provisions. Full details can be obtained at our Web site.

Tobacco companies are also facing dozens of lawsuits in the United States and abroad. Lawyers have filed lawsuits against the tobacco industry on behalf of citizens in Argentina, Canada, Finland, France, Germany, Ireland, Israel, Italy, Japan, the Netherlands, Norway, Sri Lanka, Thailand, and Turkey. What will happen to the tobacco industry is uncertain. Only the future will tell whether this industry that causes so much death and disease will persist.

In 2007, health workers found that the FEMA trailers used to house many of the victims of Hurricane Katrina in 2005 had high levels of formaldehyde. Residents of those trailers are not alone however. Formaldehyde is outgassed (released) from building materials used to make many mobile homes. What's the risk to people?

The cancer risk from breathing formaldehyde fumes is especially high in more than 5 million mobile homes in the United States, due to a number of factors. This list includes (1) their relatively small air volume, (2) their poor air circulation, and (3) the relatively large amounts of formaldehyde-containing materials they contain, such as particleboard and plywood.

Fortunately, many manufacturers are trying to reduce formaldehyde. Fiberglass insulation manufacturers, for example, have eliminated it from their products. Builders can also purchase low- or no-formaldehyde oriented strandboard—used for subflooring and exterior sheathing in new-home construction. Some companies are even producing low- or no-formaldehyde glues for use during home building. (We'll discuss other ways of reducing indoor air pollution shortly.)

Radon Another potentially hazardous indoor air pollutant is **radon** gas. Radon gas is produced by uranium, a naturally occurring radioactive element that decays spontaneously to radon gas. Although uranium is widely distributed in small amounts in the rocks and soils of the United States, its abundance varies considerably from place to place. Radon gas diffuses from the soil directly into the outside air. However, since the radon gas quickly disperses, its concentration in the atmosphere is negligible, only about 0.5% of the EPA's recommended safe level of 4 picocuries per liter of air inside a building.

By contrast, radon levels inside homes and other buildings are much higher. Radon gas leaks into homes through cracks in the basement floor and foundation from the underlying soil and rocks. Residents of the home are generally unaware of the presence of radon, because it is tasteless, odorless, and colorless.

Inside our homes, radon gas eventually decays into radioactive lead and polonium. When radon in the lung decays to form lead, radioactive lead may be deposited in lung tissues, where it remains, emitting radiation that exposes nearby cells. This, in turn, may cause genetic mutations in these cells that can lead to cancer. In fact, the EPA estimates that 20,000 to 30,000 of the 167,030,000 lung cancer deaths in the United States each year are caused by radon.

Radon accounts for 55% of the annual radiation dose sustained by the average American from all sources, both natural (radon, cosmic rays) and artificial (television, X-rays for medical purposes, and so on). In fact, Anthony Nero, a nuclear scientist with the Lawrence Berkeley Laboratory, believes that hundreds of thousands of Americans living in homes with high radon levels are exposed to as much health-threatening radiation as the Russians who were living in the vicinity of the Chernobyl nuclear plant in 1986, the year of the disaster.

The radon levels in some American dwellings can reach alarmingly high levels. Take a home in Boyerstown, PA,

owned by the Watras family, as an example. The Watrases' home may have the dubious distinction of having the highest radon level of any dwelling in the United States—about 675 times higher than is considered acceptable even in a uranium mine. The lung cancer risk to the four occupants was equal to that of smoking 220 packs of cigarettes per day! Radon levels in some Wisconsin homes exceed 100 picocuries per liter—a higher level of radiation than one would get from 20,000 chest X-rays.

The EPA regulations for controlling levels of outdoor pollutants are usually set to keep the estimated risk of premature death below 1 in 100,000 (0.001%). In contrast, however, the estimated radon risk for most Americans in the United States is 4 in 1,000 (0.4%)—400 times higher.

For several years, it has been known that a region extending from Reading, PA, into New York and New Jersey has very high levels of radon. However, a ten-state survey recently conducted by the EPA revealed other "hot spots" as well. For example, 63% of the homes surveyed in North Dakota and 46% of those in Minnesota had radon levels above the EPA guideline.

In late 1988, Lee Thomas, then head of the EPA, announced that residential radon pollution in the United States was both sufficiently widespread and sufficiently serious that every home should be tested. Fortunately for the homeowner, this can be easily done with a radon detection kit that can be purchased for about $15 to $30 from home improvement centers or online.

Control of Indoor Air Pollution

How do you reduce or eliminate indoor air pollution? Indoor air pollution is addressed in many ways. Experts on the subject use a three-part rule: eliminate, isolate, and ventilate.

To *eliminate* means to get rid of potential sources in an existing structure and not to use such things in a new home or building. Leaky furnaces, for example, can be replaced with new, more-efficient models that bring in fresh air and exhaust all pollutants. Installing tile instead of carpet in a new home eliminates a potential source of indoor air pollution. In new homes, the use of nontoxic paints, stains, and finishes can greatly reduce indoor air pollution.

To *isolate* means to find ways to seal off polluters. Oriented strandboard, used for subflooring, for example, can be coated with a sealant before use. Asbestos pipes can be painted to stabilize the material so it doesn't flake off.

To *ventilate* means just that, to find ways to provide controlled ventilation to ensure a steady fresh air supply. Special systems can be installed to bring fresh air into homes. An **energy recovery ventilator (ERV),** for example, expels the polluted inside air to the outside and replaces it with fresh outside air. A large portion of the heat from the inside air is transferred to the incoming fresh air to save energy. ERVs are effective in controlling a wide number of indoor air pollutants and can be used in the winter and summer months. In new construction, builders can install subbasement vent systems, shown in Figure 18.22. It consists of a network of perforated pipe set in a gravel bed under the basement floor. The radon diffuses

A. Subslab Vent System

B. Drain Tile Vent System

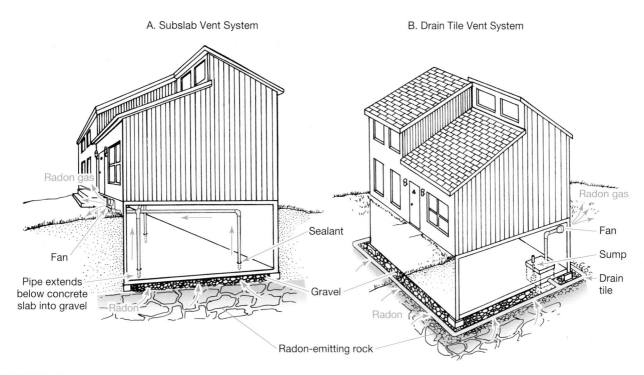

FIGURE 18.22 A subbasement system for venting cancer-causing radon gas from the home. Several options are available to the home builder. Both types require a porous gravel bed laid down before the basement is poured. (A) Radon is drawn out of the gravel through pipes that penetrate the floor slab. (B) Porous pipes laid in the gravel collect radon, which is pumped outside by a small pump in the basement.

from the soil and rock, passes through the perforations, and accumulates inside the pipe. The radon is then vented to the outside air. In homes overlying rock or soil that emit radon gas, this system is probably an absolute necessity. When buying a new home, laws require the seller of any lot to inform the prospective buyer of the level of radon gas in the house.

Indoor air pollution can be a significant problem, although no one knows for sure how much illness it causes. Even moist conditions in houses can cause molds to build up in wall cavities and other areas. Mold spores can contaminate indoor air, causing serious health problems. Expect to hear more about indoor air pollution in the years to come.

Summary of Key Concepts

1. Air pollutants come from natural and human sources.
2. Although air pollutants from natural sources are often produced far in excess of those from human sources, the latter are of greatest concern because they tend to be produced in limited areas, which results in local buildup of potentially harmful substances.
3. Pollutants are generally classified as either primary or secondary, with primary being those directly produced by some activity or natural event, and secondary being those produced from primary pollutants through chemical reactions occurring in the atmosphere.
4. Hundreds of air pollutants are released into the atmosphere, but the major pollutants are carbon monoxide,

carbon dioxide, sulfur oxides, nitrogen oxides, particulates, VOCs, and ozone.

5. Carbon dioxide is a greenhouse gas discussed in Chapter 19. Carbon monoxide is a colorless, odorless gas released from the combustion of fossil fuels and organic materials. It binds with hemoglobin in red blood cells and reduces the oxygen-carrying capacity of the blood.
6. Sulfur oxides are produced when naturally occurring sulfur in fuels such as coal and oil combine with oxygen. Much of the sulfur oxide generated by humans results from the burning of sulfur-containing fossil fuels in power plants and other industries.
7. Solid and liquid particles suspended in the air are known as particulates. Major sources are coal-burning facilities such as power plants, steel mills, fertilizer plants, jets, diesel trucks, construction sites, wind erosion, and roadways.
8. Lead is one of the particulates of concern. Lead was once added to gasoline in the United States to improve its octane rating. It was released from auto exhaust into the air we breathe. Lead can damage the kidneys, blood, and liver. Furthermore, it can impair proper brain development in children.
9. A volatile organic compound is an organic compound composed of hydrogen and carbon. VOCs are released from the incomplete combustion of fossil fuel and from evaporation. Good examples of hydrocarbon pollutants of the atmosphere are methane and benzene. VOCs react chemically to produce photochemical smog.

10. Photochemical smog is produced by the reaction between VOCs, nitrogen oxides, and atmospheric oxygen in the presence of sunlight. Ozone is one of the components of photochemical smog.

11. Concentrations of air pollution depend on the amount released by local sources, but also depend on certain climatic phenomena such as temperature inversions.

12. Radiation inversions occur primarily at night as heat radiates from the Earth's surface into the atmosphere, rapidly cooling both the ground and the layer of air next to it. This results in the formation of a warmer layer of air lying over a cooler layer containing pollutants that cannot escape by vertical mixing.

13. A subsidence inversion is formed when a high-pressure air mass sinks down and warms up. Here again, the warm air forms a lid on the cool air nearer the ground.

14. Cities also affect pollution patterns. Heat-producing processes and heat-absorbing structures usually result in slightly warmer temperatures inside cities, known as the heat island effect. The heat island effect causes pollutants to concentrate in a dust dome above a city.

15. Air pollution is affected by climate but also affects climate—locally, regionally, and even globally. Particulates, which act as condensation nuclei, can enhance local precipitation.

16. The major illnesses caused by air pollution include emphysema, chronic bronchitis, and lung cancer.

17. Pollution levels can reach very high levels during inversions, causing death and illness in large segments of the population. The first major air pollution disaster in the United States occurred in Donora, PA, in October 1948; it caused the deaths of 20 people. A more severe incident occurred in London several years later.

18. Asbestos was used in pipe and boiler insulation, brake linings, paint, plastics, and many other products and materials until its health effects became known. People who inhale asbestos fibers may develop lung cancer, mesothelioma, or asbestosis, characterized by coughing, chest pains, and bluish discoloration of the skin. Over 50% of the people who have asbestosis eventually die from lung cancer.

19. Two basic strategies exist for dealing with air pollution: pollution control and pollution prevention.

20. Many industries capture pollutants from smokestack gases. They are generally discarded in landfills. In some cases, though, air pollutants can be converted into useful products. For example, sulfur dioxide can be converted into sulfuric acid, and high-carbon fly ash can be used directly as fuel or can be converted into cinder blocks, paving materials, abrasives, and cement.

21. In some cases, pollution control devices convert pollutants into relatively harmless substances. Catalytic converters on automobiles, for instance, increase the rate at which carbon monoxide and hydrocarbons are oxidized to carbon dioxide and water, respectively.

22. Pollution control devices employed by utilities and industries include electrostatic precipitators, cyclone filters, fabric filter bag houses, and flue gas desulfurization systems, popularly known as scrubbers.

23. Flue gas desulfurization is the most widely used method for controlling sulfur dioxide emissions by industry and utilities.

24. Companies can also reduce air pollution by modifying processes or switching to nontoxic materials. These constitute a preventive approach and are desirable from the standpoint of sustainability.

25. Five general strategies for reducing auto emissions are: (1) ensuring a more steady flow of traffic, (2) improving the efficiency of the internal combustion engine, (3) replacing the internal combustion engine with electric engines powered by electricity directly or through fuel cells, (4) developing hybrid (combining an internal combustion engine and an electric engine) and plug-in hybrid cars, and (5) increasing the use of mass transit.

26. Another problem of growing concern is indoor air pollution. Recent attempts to make homes more airtight have worsened indoor air pollution problems from carbon monoxide, formaldehyde, radon gas, tobacco smoke, and other indoor air pollutants. Many new products and furnishings in our homes—and many products we use inside our homes, such as perfumes and hair spray—also contain toxic substances that are released into the air of our homes.

27. Addressing indoor air pollution requires a three-step process: eliminate, isolate, and ventilate.

Key Words and Phrases

Air Pollution Disaster	Flue Gas Desulfurization
Asbestos	Formaldehyde
Asbestos Hazard Emergency Response Act	Fuel Cell
Asbestosis	Heat Island
Attainment Regions	Hybrid Car
Carbon Monoxide	Indoor Air Pollution
Carbon Monoxide Poisoning	Inspection and Maintenance (IM) Programs
Carcinogens	Lead
Catalyst	Lung Cancer
Catalytic Converter	Marketable Permit
Chronic Bronchitis	Mass Transit
Clean Air Act	Mesothelioma
Condensation Nuclei	Mobile Pollution Source
Corporate Average Fuel Efficiency (CAFE) Standards	National Ambient Air Quality Standards
Cyclone Filter	Natural Pollution
Dust Dome	Nitric Oxide
Dust Plume	Nitrogen Dioxide
Electric Vehicle	Nitrogen Oxides
Electrolysis	Nonattainment Areas
Electrostatic Precipitator	Noncompliance Penalties
Emissions Offset Policy	Ozone
Emphysema	Particulate Matter
Energy Recovery Ventilator (ERV)	Photochemical Smog
Ethanol	Pollution Taxes
Fabric Filter Bag House	Prevention of Significant Deterioration (PSD)
	Primary Pollutants

Radiation Inversion
Radon
Reformer
Scrubbing
Secondhand Smoke
Secondary Pollutants
Smog
Stationary Pollution Source
Strip Mining

Subsidence Inversion
Sulfur Dioxide
Sulfur Oxides
Synergistic
Temperature Inversion
Thermal Inversion
Tradable Permits
Volatile Organic Compound
(VOC)

Critical Thinking and Discussion Questions

1. Briefly describe the biological significance of the major components of unpolluted air.

2. List three sources of natural pollution.

3. Using your knowledge of air pollution and your critical thinking skills, discuss the following statement: "Natural air pollutants are produced far in excess of those from human sources, so we really don't need to worry about pollution from cars and such."

4. Discuss the health effects of each of the following air pollutants: (a) carbon monoxide, (b) oxides of sulfur, (c) oxides of nitrogen, (d) hydrocarbons, (e) ozone, and (f) lead.

5. List the major human sources of carbon monoxide, oxides of sulfur, oxides of nitrogen, hydrocarbons, ozone, and lead.

6. List the pollutants responsible for the formation of photochemical smog.

7. Cite an example of how severe air pollution actually caused the "death" of a town.

8. List five conditions that are usually associated with an air pollution disaster.

9. Describe three conditions that contribute to the heat island phenomenon.

10. Which climatic conditions would be most conducive to the rapid dispersal of atmospheric pollutants? To their concentration in a localized area?

11. Discuss the advantages and disadvantages of the electric car.

12. Using your critical thinking skills, analyze the following statement: "Electric cars are pollution-free and therefore are a major boon to cities that are interested in cleaning up their air."

13. Discuss the advantages and disadvantages of replacing gasoline with ethanol.

14. What is a hybrid car? What's a plug-in hybrid? How are they similar? How are they different?

15. Discuss hydrogen cars. You may want to do some more research on this. How are they different from electric cars?

16. Now that you have studied this material on air pollution, what changes, if any, do you plan to make with respect to the use of your automobile?

17. Discuss the problem of indoor air pollution. Why is it a problem? Where do indoor air pollutants come from? What are the general ways of addressing them?

Suggested Readings

Amdur, M. O., J. Doull, and C. D. Klaassen, eds. 1993. Air Pollutants. In *Toxicology: The Basic Science of Poisons,* 4th ed. New York: Pergamon Press. Detailed account of air pollution and its effects on the body.

Beatly, T. 2000. *Green Urbanism: Learning from European Cities.* Washington, DC: Island Press. Description of, among other things, how to revamp our cities to reduce air pollution and other environmental problems.

Bower, J. 1997. *The Healthy House: How to Buy One, How to Build One, and How to Cure a Sick One.* Bloomington, IN: Healthy House Institute. Detailed reference.

Bower, J., and L. M. Bower. 1997. *The Healthy House Answer Book: Answers to the 133 Most Commonly Asked Questions.* Bloomington, IN: Healthy House Institute. Great introduction to the subject of healthy homes.

Buck, S. J. 1998. *The Global Commons.* Washington, DC: Island Press. Examination of management of global commons such as the atmosphere.

Chiras, D. D. 2004. *The New Ecological House.* White River Junction, VT: Chelsea Green. Chapters on building healthy homes to prevent indoor air pollution.

Chiras, D., and Wann, D. 2003. *Superbia! 31 Ways to Create Sustainable Neighborhoods.* Gabriola Island, BC: New Society Publishers. One of the senior author's most recent books, describing actions people can take to build a more sustainable world right in their own neighborhood, including measures that will help reduce their reliance on fossil fuels and their production of air pollution.

Clayton, T., G. Spinardi, and R. Williams. 1999. *Policies for Cleaner Technology: A New Agenda for Government and Industry.* London: Earthscan. A groundbreaking treatise on new ways to make industrial systems more sustainable by the businesses themselves.

Environmental Protection Agency. 2004. *National Air Pollutant Emission Trends, 1900–2001.* Washington, DC: U.S. Environmental Protection Agency. An annual publication that, in spite of being always three years behind the times, is a great resource.

Grubler, A. 2006. Doing More with Less: Improving the Environment Through Green Engineering. *Environment* 48(2): 22–37. A look at the importance of rethinking the way we solve problems in a growing world.

Harrington, W., and V. McConnel. A Lighter Tread? Policy and Technology Options for Motor Vehicles. *Environment* 45(9): 22–36. Detailed look at transportation and ways to improve the system.

Newman, P., and J. Kenworthy. 1999. *Sustainability and Cities: Overcoming Automobile Dependence.* Washington, DC: Island Press. Great ideas on ways to reduce the use of single-occupancy vehicles.

Roome, N. J. 1998. *Sustainability Strategies for Industry: The Future of Corporate Practice.* Washington, DC: Island Press. Great overview of the subject.

Sheehan, M. O. 2001. Making Better Transportation Choices. In *State of the World 2001,* ed. L. Starke.

New York: W. W. Norton. An excellent look at components of a more environmentally sustainable transportation system.

Sterner, T., et al. 2006. Quick Fixes for the Environment: Part of the Solution or Part of the Problem? *Environment* 48(10): 20–30. An insightful look into the difference between traditional approaches to pollution and sustainable solutions.

United Nations Environment Programme. 1999. *Global Environment Outlook 2000.* New York: Earthscan. Great reference book on global environmental trends and policy changes needed to address them.

Vig, N. 2006. Making the Grade. *Environment* 48(7): 40–42. A review of a report on environmental progress in the United States.

Wilson, A., and N. Malin, eds. 2003. *GreenSpec: The Environmental Building News Product Directory and Guideline Specifications.* Brattleboro, VT: E Build, Inc. One of five books on alternative building materials that are good for the environment and easy on our health.

Web Explorations

Online resources for this chapter are on the World Wide Web at: **http://www.prenhall.com/chiras** *(click on the Table of Contents link and then select Chapter 18).*

CHAPTER 19

GLOBAL WARMING AND CLIMATE CHANGE

In the summer of 1999, record heat and drought in the United States seriously damaged crops along the eastern seaboard, causing considerable hardship among farmers. Late that summer and throughout the fall, a record number of hurricanes battered the coast of the southeastern United States, causing flooding, killing dozens of people, and drowning countless livestock. The next year, in the spring of 2000, drought set in among the western states while the Northeast was inundated with rain. Farmers in western New York, for instance, couldn't get their crops in because the soils were so wet, or could only plant on high ground. In the early summer of the year 2000, long before the traditional dry season in the West, wildfires broke out in California, New Mexico, Colorado, and Montana, burning millions of acres of forest. That same year, researchers plying the waters of the Arctic Circle noticed a dramatic loss of ice. They were able to navigate the polar waters easily. Decades earlier, the area had been choked with ice, making ship travel nearly impossible.

In 2002, the drought in the nation's western half—then in its fifth year—triggered dramatic wildfires throughout the region. In Colorado, fires started in May, long before the typical fire season, which usually begins in August. And in January 2003, the state posted a wildfire alert in January—an occurrence unheard of in the state's long history! In the summer of 2003, wildfires ravaged southern California and many other western states. Even normally wet regions such as the Pacific Northwest suffered from drought, water shortages, and wildfires.

In the years following these incidents, droughts and violent storms plagued the United States and much of the rest of the world. In 2007 and 2008, for example, drought continued to make headlines in large sections of the United States. The southeastern United States, for instance, suffered crippling drought. Farmers' crops withered. Municipal reservoirs ran nearly dry. Extreme water-rationing measures were instituted in Georgia and other states. In drought-stricken California, wildlife burned out of control. Eastern and western Colorado faced massive drought as well.

During this period, biologists noted other peculiar occurrences. They discovered, for example, that plants are flowering earlier in the spring and that migrating species are showing up earlier in the spring than is customary. They have also found that the geographic ranges of birds, butterflies, and plants are also moving toward the poles. Scientists also recorded rapid glacial melting.

At first glance, it would appear that these events have little in common. It might seem as if they are attributable to bad weather. Nevertheless, many atmospheric scientists believe that these seemingly unrelated episodes may have a common thread: global warming. Most atmospheric scientists who have studied the subject believe that the planet has entered into a major period of global temperature increase, caused in large part by certain types of air pollution produced by human activities and deforestation.

Global warming, the increase in the temperature of the Earth's oceans and atmosphere, in turn, appears to have spawned another massive change, known as **global climate change.** Climate change is defined as a shift in the planet's **climate**—average weather conditions, including temperature, precipitation, and storms. We'll examine global warming in this chapter, answering two fundamental questions: (1) Are global warming and climate change occurring? and (2) Are they caused by human activities or the result of natural factors? We'll also look at the social, economic, and environmental effects of these phenomena. Before we begin, however, let's examine some important scientific concepts you'll need to know.

19.1 Global Energy Balance and the Greenhouse Effect

Each day the sun shines down on the Earth, bathing us in sunlight. Although only two-billionths of the sun's energy actually makes it to Earth, it still amounts to an amazing amount of energy. What happens to all that energy?

As shown in Figure 19.1, approximately one-third of the sunlight striking the Earth and its atmosphere is reflected back into space. It bounces off clouds, dust, and the Earth's surface, which reflect this energy back into space. The rest is absorbed by air, land, water, highways, parking lots, buildings, and plants.

As you may recall from Chapter 3, solar radiation consists of electromagnetic radiation ranging from high-energy, short-wave gamma rays to low-energy, long-wave radio waves (Figure 3.3). In between are X-rays, ultraviolet radiation, visible light, and heat (infrared radiation). All of this energy travels to Earth through space. However, the vast majority of the sun's energy consists of visible light (55%) and heat (about 40%).

Sunlight warms the Earth. Even visible light that's absorbed by surfaces is converted into heat or, more technically, infrared radiation. As shown in Figure 19.1A, this heat is eventually radiated back into outer space through the atmosphere. As a result, energy gained from the sun is balanced by energy radiated back into space.

Scientists have long known that certain chemical substances in the atmosphere can alter the Earth's energy balance. These include natural and human-produced chemicals. This list includes water vapor, carbon dioxide (CO_2), nitrous oxide (N_2O), methane (CH_4), and chlorofluorocarbons. To understand how these substances alter the Earth's energy balance, let's take a look at one chemical, carbon dioxide.

Carbon dioxide is a small, odorless molecule. It is released from the combustion of organic materials, including fossil fuels, wood, and vegetation (trees, for instance). It is also released by the natural decay of vegetation and animal waste. And lest we forget, it is released by all living animals during respiration.

Carbon dioxide molecules make up a tiny percent of the Earth's atmosphere—just under 0.04%. Nonetheless, these molecules have an amazing effect on the planet's temperature. Carbon dioxide, along with the others just listed, absorbs long-wave infrared radiation escaping from the Earth's surface and radiates it back to Earth. The carbon dioxide molecules act a bit like the glass in a greenhouse, which traps heat (Figure 19.1B).

This phenomenon, known as the **greenhouse effect,** helps maintain the Earth's surface temperature so essential to the continuation of life on Earth. Without carbon dioxide, scientists estimate, the Earth would be at least 30°C (55°F) cooler than it is today—making it inhospitable to nearly all life-forms, including humans. What exactly is the greenhouse effect?

You've very like observed the greenhouse effect in a car or house heated by the sun. The glass permits sunlight (consisting of heat and visible light) to enter a car's interior. There the visible light is absorbed by dark interior surfaces, such as the seat covers and dashboard. Visible light is then converted to heat, long-wave infrared radiation. Heat accumulates in the interior of the car because the glass—like the glass in a greenhouse—prevents it from escaping.[1]

As previously noted, carbon dioxide is just one of several natural and human-produced chemicals that increase the Earth's surface temperature. Each of these is known as a **greenhouse gas.**

On Earth, sunlight is absorbed by plants, soils, buildings, parking lots, and roadways and is converted into heat. The infrared radiation then rises, but some of it is trapped in the atmosphere by greenhouse gases.

Studies of the global energy balance show that the Earth's surface temperature is influenced by two main factors: those that alter the amount of sunlight striking the Earth's surface and those that change the rate at which heat is lost or retained. Both natural and anthropogenic factors influence these processes.

With this information in mind, let's take a closer look at other natural factors that increase global temperature, that is, factors other than greenhouse gases.

19.2 Natural Factors That Influence Global Temperature

Numerous studies show that the average temperature of the Earth and Earth's climate change naturally over time. Any student of geology will tell you that the Earth cycles between cold spells (ice ages) and interglacial warm periods. We're currently in an interglacial warm period.

Although not completely understood, these natural changes in climate are caused by at least five natural phenomena: (1) changes in the Earth's orbit around the sun, (2) changes in the Earth's tilt, (3) increases or decreases in solar activity, (4) increases or decreases in volcanic activity, and (5) chaotic interactions of the climate system. Let's start with the Earth's orbit around the sun.

GO GREEN!

Take time to learn more about global climate change. Be sure to read scientifically credible articles. Talk with friends about the problem and various solutions. Be sure to tune in to PlanetGreen for more ideas.

---

[1]The glass in a greenhouse and the glass in an auto prevent heat from escaping but also contain hot air, so the term *greenhouse effect* is not entirely accurate for the effect carbon dioxide has on the Earth's temperature.

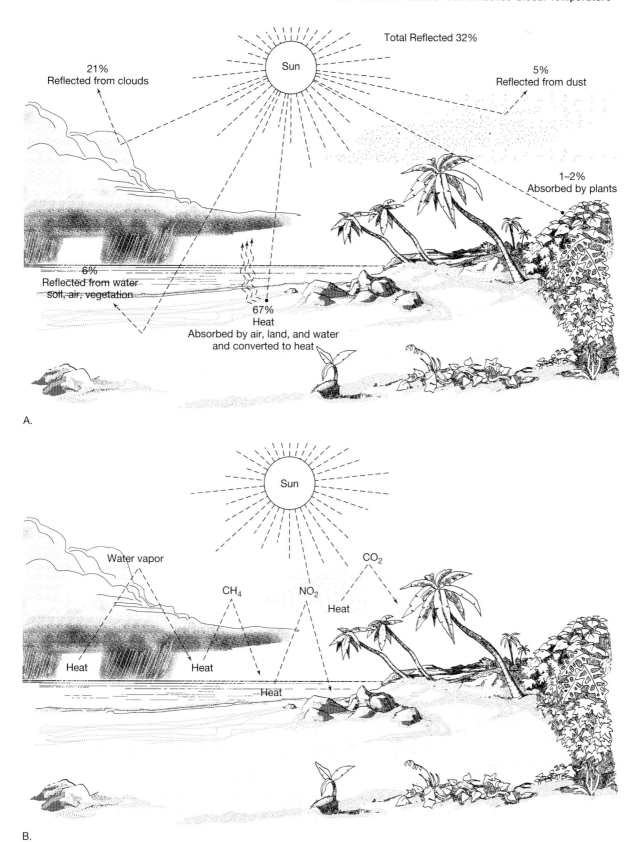

FIGURE 19.1   The Earth in balance. (A) This drawing shows the fate of incoming solar radiation. Note that all heat eventually dissipates into outer space. (B) Greenhouse gases absorb heat and radiate it back toward the Earth, raising the Earth's surface temperature.

As you probably know, the Earth orbits around the sun not in a circular path but in an elliptical one. That orbit, however, is not constant. It changes a bit over time, so that the Earth is sometimes closer to the sun and sometimes farther away. When the Earth's orbit brings it closer to the sun, it causes warming. When the sun is farther from the sun, it causes cooling.

The Earth is also tilted on its axis at an angle of approximately 23.5 degrees. The tilt angle, however, is not constant. It changes ever so slightly. This, in turn, alters the average temperature. If the Earth's tilt decreases, warming occurs. If the angle increases, cooling takes place.

Another wild card in the climate deck is solar activity. As you may have learned, the sun's output changes from time to time. In fact, scientists have found that the sun's output goes through an 11-year cycle. Increased output results in natural planetary warming. Decreased output results in cooling.

Volcanic activity also changes from time to time, resulting in warming and cooling. Increased volcanic activity, for example, can result in an increase in the amount of ash floating in the atmosphere. This, in turn, can reduce solar gain, cooling the Earth. Studies show that volcanic activity plays an important role in year-to-year climate variability.

Scientists recognize another factor that influences the Earth's temperature and climate: potentially chaotic interactions in the extremely complex climate system. Complex and inherently unpredictable interactions, they say, can take place between major parts of the Earth's climate system, such as the oceans and the atmosphere. Some scientists believe that these potentially unpredictable interactions could cause climate to shift, resulting in periods of either cooling or warming. These changes, they assert, could last for decades or even centuries.

19.3 Anthropogenic Factors That Alter Global Temperature

Although many natural factors can alter climate, various human activities also affect the Earth's average temperature, which can lead to significant changes in climate. The two activities that are most important are the release of greenhouse gases and deforestation.

As noted earlier, greenhouse gases are released from natural and anthropogenic sources. Interestingly, the release of greenhouse gases from natural sources has remained fairly constant over the past 100 years, while the emissions from human sources have increased substantially. The annual release of carbon dioxide from the combustion of fossil fuels, for example, has increased in the past 100 years from 534 million tons (of carbon) to about 7,600 million tons in 2006. Chapter 18 described many of the sources of carbon dioxide emissions.

While most greenhouse gases come from natural and human sources, some come solely from human activities. These include the chlorofluorocarbons (CFCs) and their replacements, the hydrochlorofluorocarbons (HCFCs), which are discussed in Chapter 20. These chemicals have no natural source. Like other greenhouse gases, their production and release have increased rather dramatically since 1950, although the release of CFCs has declined in recent years because of bans on their production and use.

Greenhouse gases are not all created equal, either. Some, like CFCs are more powerful greenhouse gases than others, such as carbon dioxide. That is, some trap heat more efficiently than others. As an example, one molecule of CFC is equivalent to 15,000 molecules of carbon dioxide when it comes to greenhouse warming. This is partly because CFCs last so long in the atmosphere. (They are not readily broken down.) Methane also is a more powerful greenhouse gas than carbon dioxide.

Even though carbon dioxide is not the most powerful greenhouse gas, it remains the single most important greenhouse gas we produce. That is, it has the greatest effect on global temperature. Why? Because it is produced in much greater quantities than other greenhouse gases. The sheer amount released makes carbon dioxide the most significant greenhouse gas.

Carbon dioxide is produced during the combustion of fossil fuels but also from the deforestation and torching of cleared forests. Why does deforestation increase in atmospheric concentrations of carbon dioxide?

As you may recall from your study of ecology (Chapter 3), all plants, trees included, absorb carbon dioxide from the atmosphere. They use it to produce plant matter and food molecules. As trees are lost through deforestation, the Earth's ability to absorb carbon dioxide declines, and atmospheric carbon dioxide levels rise.

Carbon dioxide levels have remained fairly constant for tens of thousands of years. Only in recent times has the concentration of this gas in the atmosphere begun to increase. Why have carbon dioxide levels remained constant until recently?

As noted in Chapter 3, carbon dioxide is produced by natural and human sources and continuously removed from the atmosphere by plants and algae (during photosynthesis). Plants use the carbon dioxide to make glucose and other molecules. Carbon dioxide is also absorbed by surface waters such as lakes and oceans, where it is taken up by aquatic plants and algae or is simply dissolved in water.

As you may also recall, carbon dioxide taken up by living organisms is released back into the atmosphere—for example, when plants and animals die and decompose. The breakdown of these organisms releases carbon dioxide into the atmosphere. While living, plants and animals also release carbon dioxide as they break down glucose to produce energy. Carbon dioxide is also released into the atmosphere when forests, grasslands, or any organic materials burn.

For thousands of years, before the beginning of the Industrial Revolution, carbon dioxide levels remained fairly constant, because production equaled absorption. That is, the amount of carbon dioxide entering the atmosphere equaled the amount removed from the atmosphere by various natural processes (Figure 19.2A). How do we know this?

Scientists have determined the concentration of carbon dioxide in the atmosphere of long ago by examining air bubbles trapped in glacial ice and by analyzing the wood of centuries-old trees. Such investigations have shown that carbon dioxide levels remained at about 270 parts per million (ppm) from the end of the last ice age (about 10,000 years ago) until the 19th century.

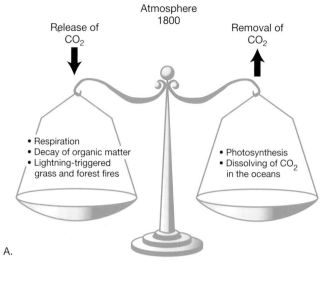

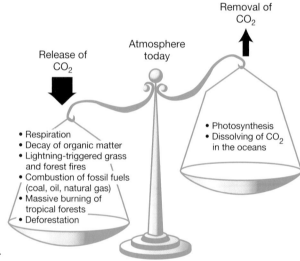

FIGURE 19.2 Upsetting the balance. (A) Before the Industrial Revolution, carbon dioxide emissions from natural sources were balanced by carbon dioxide absorption. (B) After the Industrial Revolution, this delicate balance has been dramatically upset. Far more carbon dioxide enters the atmosphere than the atmosphere can absorb.

As illustrated in Figure 19.2B, the increased combustion of fossil fuels, deforestation, and burning of forests have upset this balance. Studies have shown that, since 1860, the average concentration of carbon dioxide in the Earth's atmosphere has risen steadily from 290 ppm to over 380 ppm in 2007—a 31% increase (Figure 19.3A).

The greatest increase in atmospheric carbon dioxide levels since the last ice age has occurred since 1950. But further studies show that carbon dioxide levels do change naturally. Figure 19.4, for instance, shows carbon dioxide levels going back 450,000 years. As you can see, they rise and fall naturally.

If you study Figure 19.4 carefully, you'll see that while carbon dioxide levels do cycle, current concentrations of carbon dioxide in the atmosphere are 100 ppm higher than the highest levels seen in the last 450,000 years, most likely as a

result of human actions. Methane levels are 1,000 ppm higher than they've been in the same period. The increases in atmospheric carbon dioxide and methane are largely the result of human activities discussed earlier. As previously noted, carbon dioxide arises from the combustion of fossil fuels and the clearing and burning of forests the world over, especially tropical rainforests. Tropical rainforests continue to be cut down to make room for cattle ranches, farms, human settlements, and resorts. Rising methane levels come from an increase in humans and livestock, especially cattle, sheep, and pigs, used to feed the ever-growing human population.

According to the most recent estimates available, in 2005 human activities were responsible for the release of 7.6 billion tons of carbon dioxide into the atmosphere. Fossil fuel consumption was responsible for about 80% of these emissions, or about 6 billion metric tons—nearly 1 ton for every man, woman, and child on Earth, according to the Worldwatch Institute. The clearing and burning of tropical rain forests annually releases about 1.6 billion metric tons into the air, about 20% of the total amount released each year.

With this information in mind, we turn to two critical questions posed at the beginning of the chapter. First, are global warming and climate change really occurring? And second, are humans to blame?

19.4 Are Global Warming and Global Climate Change Occurring?

While many people believe that the Earth's oceans and our atmosphere are warming, others are not so sure. What is the truth? Are we warming or not?

As shown in Figure 19.3B, the Earth's temperature is indeed on the rise. How do we arrive at the Earth's temperature? The Earth's an awfully large body.

Scientists determine the Earth's temperature—actually the average surface temperature—by using data obtained from thousands of temperature sensors distributed over land and sea. These temperature readings are carefully averaged each year, giving an average temperature for the entire planet.

As Figure 19.3B shows, the Earth's surface temperature has risen from a little over 13.8°C in 1950 to 14.6°C in 2006, with only a small dip in temperature from 1940 to 1960. (Scientists believe that this transient decline may have been caused by particulate pollution from factories during the postwar surge in economic activity. Particulates reduce sunlight striking the Earth. Starting in the 1960s,

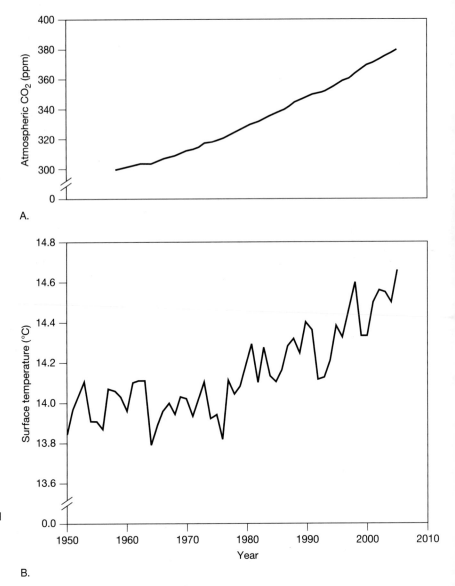

FIGURE 19.3 Carbon dioxide concentrations and global temperature. (A) Carbon dioxide concentrations in the atmosphere have grown steadily since 1950, which may be the main reason why (B) global average temperature has increased.

air pollution laws reduced particulate pollution, eliminating this cooling effect.)

The increase in global temperature actually started in the 1880s. Figure 19.5 shows a map of the world that indicates the magnitude of temperature change by comparing the average temperature for 2001–2005 with the average temperature for 1951–1980. As you can see, the average temperature has increased most profoundly over the poles, especially the North Pole. The increase in this region has been nearly 2°C.

Rising average global surface temperature indicates that the Earth is warming, but there's other evidence to support the conclusion that the Earth is warming and that the climate is changing. (Remember that climate is average weather conditions and includes temperature, rainfall, and storms.) Here is a summary of some of the evidence suggesting that the Earth's surface temperature is increasing and that climate is changing as a result:

- Rising ocean temperatures. Studies show that ocean temperatures are increasing at least to a depth of 3,000 meters (10,000 feet). The oceans currently absorb about 80% of the heat added to the climate system by global warming.

This is thought to be part of the reason for an increase in the intensity of storms, especially hurricanes.

- Rising sea level. Sea level has risen 15 to 20 centimeters (4 to 6 inches) over the past century. This is thought to be caused by two factors: (1) melting of glaciers in mountainous areas and in Antarctica: and (2) expansion of sea waters due to warming. According to the U.S. Environmental Protection Agency (EPA), approximately 2 to 5 centimeters (1 to 2 inches) of this rise in sea level has resulted from the melting of mountain glaciers. Another 2 to 7 centimeters (1 to 3 inches) has been caused by thermal expansion of ocean water. Whatever the cause, many island nations throughout the world, especially the low-lying islands in the South Pacific, are facing almost certain ruin. Their beaches are being flooded by seawater, and if the waters continue to rise, they could eventually be lost to the sea.

- Melting glaciers. Glaciers are melting at record rates throughout the world, yet another sign of global warming.

- Breakup of Antarctic ice sheet. Huge blocks of ice, often the size of small New England states, are breaking off the

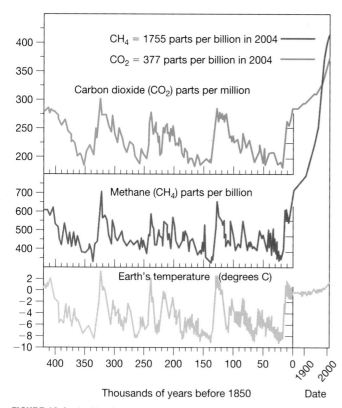

FIGURE 19.4   Looking back 450,000 years. This graph shows the change in global concentrations of carbon dioxide, methane, and global temperature. Cyclic changes in all three parameters occur, due to natural events, but methane and carbon dioxide levels are far higher today than they've been during this entire period.

Antarctic ice sheet, signaling a melting and breakup of this massive expanse of ice.
- A decrease in Arctic polar sea ice. The Arctic sea ice has been declining about 10% per decade during the past 25 years. Submarine data indicates that the ice mass is 40% thinner in the mid-1990s than 20 years earlier. A research ship (icebreaker) once took two years to traverse the North

Pole; in 2000, the ship completed its voyage in a few short months, due to the lack of polar sea ice.
- Heat waves, record-breaking temperatures, and drought are on the rise. Heat waves are becoming more commonplace in many parts of the world. Studies show that ten of the hottest years since 1880 (when data first began to be recorded), for instance, occurred between 1990 and 2003. The year 2003, in fact, was the second hottest year in this period. Although a couple of consecutive hot years are not unusual, the probability of having this many in a short period is extremely small. Record-breaking temperatures and drought are taking their toll on agriculture worldwide, causing severe economic problems in many areas. In the 2000s, severe drought plagued many parts of the United States. Heat waves are also responsible for numerous deaths. In 2003, an estimated 40,000 people died in Europe as a result of a record-breaking heat wave.
- Rise in tornados and violent storms. Heat energy stored in oceans and in the atmosphere spawns more violent storms. A recent study of the severity of storms in the United States showed that violent downpours are on the rise. Studies also show that the occurrence of tornadoes is on the rise in the United States. The average in the 1980s was 600 to 700 per year. In the 1990s, the average was 1,200 per year. In 2004, the United States experienced more than 1,400 tornadoes. Worldwide, the incidence of major weather events also has increased. In 1999, there were four times as many major storms and floods as there were in the 1960s, and damage was seven times greater.
- Changes in precipitation. Not only is the Earth getting warmer and are storms getting more violent, but precipitation is changing as well. Studies show that over the past century, precipitation has increased by about 1% over the world's continents. But the increase is not uniform. Rainfall appears to be on the decline in the tropics, while it is increasing in higher-latitude areas.
- Greater severity of hurricanes. Studies show that although the number of hurricanes has not increased, their severity

2001–2007 Mean surface temperature anomaly (°C)
Global mean = 0.54

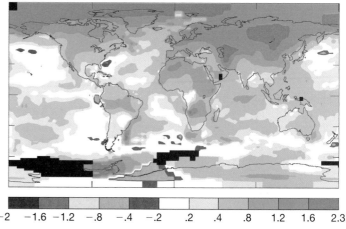

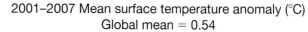

FIGURE 19.5   Rising temperature. This illustration shows the increase in temperature across the planet. Note that temperature has increased most dramatically at the poles.

has (Figure 19.6). Hurricane Katrina is a good example. This single event in 2005 killed 1,836 people and displaced 1 million residents in Louisiana and Mississippi. The hurricane destroyed 200,000 homes, costing the insurance industry $25.3 billion. It destroyed 526,000 hectares (1.3 million acres) of forest, costing $5 billion. The total economic impact of this event is estimated to be around $150 billion.

- Increase in forest fires. During the 16-year period from 1987 to 2003, forest fires have burned 6.5 times as much area per year as they did between 1970 and 1986. While the increase in forest and grass fires is often blamed on fire suppression and grazing policies, it is more likely caused by rising temperature.

- Spread of insects and disease. Warmer climates are also resulting in the spread of insects that carry diseases such as malaria and dengue fever to higher altitudes. In Africa, for example, malaria is spreading to higher-altitude regions in five countries, scientists believe, because of the warming climate. In Mexico and Costa Rica, dengue fever is spreading into the highlands as well. "The control issue looms largest in the developing world, where resources for prevention and treatment can be scarce," writes Dr. Paul Epstein, associate director of the Center for Health and the Global Environment at Harvard Medical School in an article in *Scientific American.* "But the technologically advanced nations, too, can fall victim to surprise attacks—as happened last year (1999) when the West Nile virus broke out for the first time in North America, killing seven New Yorkers. In these days of international commerce and travel, an infectious disorder that appears in one part of the world can quickly become a problem continents away if the disease-causing agent, or pathogen, finds itself in a hospitable environment." Since 1999, West Nile virus has

spread westward. It now exists in most states. In 2003, Colorado was hit hardest by the virus, as health officials reported nearly 2,600 cases. Fifty people died from the disease. Besides killing people, West Nile is also killing tens of thousands of infected birds.

19.5 Are Human Activities Causing Global Warming?

To most objective observers, the evidence is pretty convincing: the Earth's temperature—and its climate—are changing. Moreover, they are changing dramatically. But what's causing the change?

Scientists have gathered a lot of data to suggest that climate change is occurring as a result of human activities, especially the increase in global carbon dioxide levels. They point out that there's a close correlation between the increase in atmospheric concentrations of carbon dioxide, a known greenhouse gas, and the rise in the Earth's average surface temperature, shown in Figure 19.3B. That is to say, many atmospheric scientists view this correlation as evidence that humans are the principal cause of global warming and global climate change.

Critics of this view still argue that natural factors could be responsible for the changes we're seeing in climate. Is there any evidence to support their assertions?

In 2007, the UN-sponsored Intergovernmental Panel on Climate Change published a report that summarizes research on global warming and global climate change by the world's leading climatologists. Figure 19.7 summarizes the findings of the intense and costly research designed to determine how much of the current global warming is caused by humans and how much is caused by natural factors. Scientists expressed the rate of heating and cooling in watts per square meter. This is known as **radiative forcing.** A positive number indicates heating; a negative number indicates cooling.

As shown in Figure 19.7, natural radiative forcing is indeed occurring. Some factors are causing the Earth to cool, and others are causing the planet to warm up. As shown in the last two rows, however, both human activities and natural activities are warming the Earth's surface. Natural factors, however, pale in comparison with human factors.

Computer studies bear out this conclusion. Using complex computer models of global climate, scientists plugged in data on the natural and anthropogenic factors that affect global temperature going back as far as 1900. Based on this input, the computer programs then estimated the average temperature of various continents. The computer's estimates were compared with measured temperatures during this period and are shown in Figure 19.8. In this graph, the dark green shaded area indicates the average temperature that would be expected if only natural factors were involved in global temperature. The light green shaded areas indicate the predicted temperature if natural *and* human factors are factored in. The black lines represent the actual average temperature based on real measurements. As you can see, the actual temperature corresponds very well to predicted temperature only if both natural and human factors are taken into account.

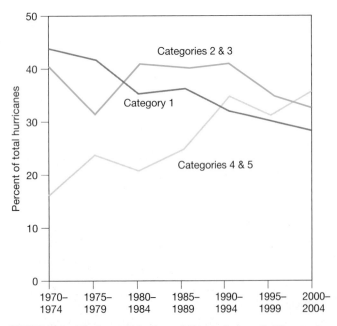

FIGURE 19.6 What's up with hurricanes? This graph shows that the severity of hurricanes is on the rise, very likely as a result of warmer sea temperatures, which provide energy to these storms.

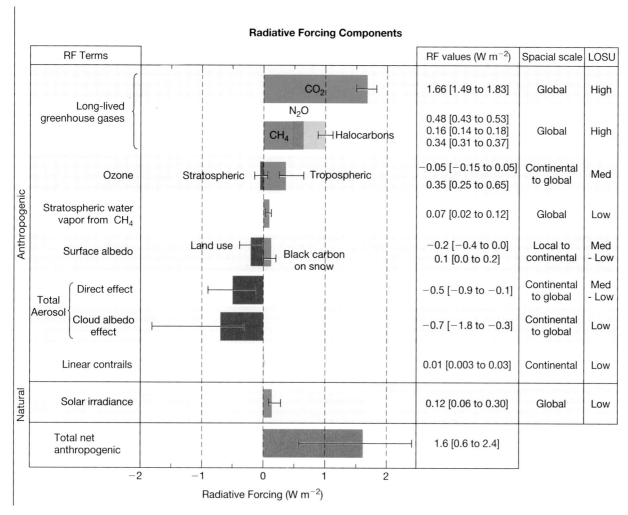

FIGURE 19.7 Responsibility for global warming. This graph shows the various factors that cause the Earth to heat up and cool down. Human activities far outweigh natural factors.

19.6 Projected Impacts of Global Warming

Growing evidence suggests that the planet is warming and that the climate is shifting. What will happen to us and our planet? No one knows for sure if the continued release of carbon dioxide and the loss of carbon dioxide-absorbing vegetation will result in further increases in global temperature and dramatic changes in our climate. However, most atmospheric scientists think that these changes are likely—and could be very costly. Estimates of how much temperature will rise vary and seem to change every few years as our understanding of the climate deepens and as computer models used to predict climate change improve.

In 2007, the Intergovernmental Panel on Climate Change (IPCC) conservatively predicted a 1.8°C to 4.0°C increase in average global temperature by 2100. Although an increase of 2°C to 4°C may seem insignificant, this change could drastically alter global climate and sea level. In 2007, IPCC scientists predicted that sea level could increase another 0.2 to 0.6 meter (8 to 24 inches) by 2100 as a result of the melting of glaciers and the land-based Antarctic ice pack and an expansion of the seas resulting from warmer temperatures.

Although carbon dioxide is the major greenhouse gas, as noted earlier, other pollutants contribute to greenhouse warming as well. Many of these are increasing at rates much higher than carbon dioxide. And many, like the CFCs, remain in the atmosphere for long periods—more than 100 years.

The following lists identify some of the social, economic, and environmental changes that might be expected. As you will notice, some effects may be positive or beneficial.

Beneficial Effects

1. In a warming climate, the cost of heating homes, offices, and commercial buildings during the winter in high-latitude regions might decline. Fossil fuel demand in these areas would be reduced, as would heating costs.

2. The far northern latitudes (tundra biome) might become more suitable for human settlement. Similarly, the ranges of many species of fish, birds, and mammals would be

Global and Continental Temperature Change

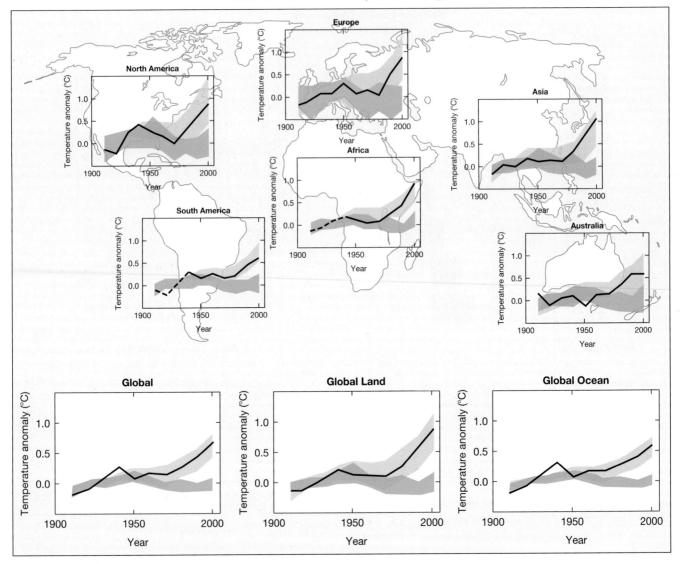

FIGURE 19.8 Measurements of actual temperature and expected temperature readings. Areas shown in dark green show temperature increases based on natural factors. Areas in light green represent predicted temperature increases based on natural and anthropogenic factors.

extended northward in North America, Europe, and Asia. Such has already been the case for the mockingbird, cardinal, and opossum as well as certain other birds and some butterflies and plants.

3. Global warming could increase rainfall in some areas. Increased rainfall and/or a longer growing season could increase food production in certain areas of the world, such as Canada, Mexico, Europe, northeastern Africa, and Southeast Asia. The political and economic clout of these nations could increase substantially.

4. For every 1% increase in atmospheric carbon dioxide, the rate of photosynthesis will increase 0.5%, increasing the yields of such crops as rice, corn, and wheat, provided that sufficient water and nutrients are available and temperatures don't get too high. (Photosynthesis stops when temperatures climb over 100°F.)

Harmful Effects

1. Although heating costs may decrease in northern latitudes, summer cooling demand is likely to increase. With more and more warm regions, summer cooling demand could far outweigh any gains from decreases in heating bills. Keeping buildings cool may require more fossil fuel consumption, increasing carbon dioxide emissions and further increasing global warming.

2. Many scientists believe that the increase in global temperature will further increase melting of the polar ice caps and glaciers. Scientists predict that further melting of the land-based ice—that is, glaciers and the massive Antarctic ice pack—will cause sea levels to continue rising. The warming of the Earth will also cause ocean waters to expand, further adding to the rise in sea levels. As noted earlier, IPCC scientists in 2007 predicted that

sea level could increase another 0.2 to 0.6 meter (8 to 24 inches) by 2100 as a result of global warming.

A rise in ocean levels would cause the seas to move inland along many portions of the world's coastlines. Coastal properties worth billions of dollars would be lost. The city of Charleston, SC, alone would suffer hundreds of millions of dollars in damage by 2035. Similar flood damage would probably be experienced by Boston, Philadelphia, New York City, Baltimore, Washington, DC, Norfolk, Miami, St. Petersburg, New Orleans, and many other coastal cities throughout the world.

3. Many of the world's most productive rice-growing areas (low-lying coastal areas), including the fertile, highly populated deltas of the Indus (Pakistan), Ganges (Bangladesh), and Yangtze (China) Rivers, would be destroyed by floodwaters.

4. A rising sea level could increase damage caused by hurricanes and tropical storms. When these violent storms come near shore, water levels rise. Seawaters are propelled inland, creating a storm, often the most damaging part of a hurricane. A rise in sea level would increase the distance that storm surges penetrate. As the seas rise, more and more inland property will be flooded during such events. Large tracts of agricultural land would be destroyed because of salts left behind by the receding waters. Many more homes would be destroyed, and many more lives would be lost.

5. Some scientists predict that global warming will also change weather patterns. Some areas of the world may become wetter, and others may become drier. Studies show that some parts of the United States are actually becoming cooler and experiencing more rainfall, while others are experiencing just the opposite. Crop production could fall drastically in both areas, as heavy rains can make planting difficult if they come too early, as is often the case.

6. Warmer weather in some areas could shift agriculture northward into northern Minnesota, Wisconsin, and Canada. But the shift in growing regions may not offset the loss in production farther south, because northern soils are much less fertile. Overall, corn yields would very likely decline. The U.S. wheat industry could suffer a $500 million yield decline annually because of increased heat and drought.

7. Extensive drought on the Great Plains of the United States and other areas of the world could trigger huge dust storms like the "black blizzards" of the Dust Bowl.

8. Many aquifers along the Gulf Coast and Atlantic Coast of the United States, which now serve as a source of drinking water for many American cities, could become contaminated by saltwater from the ocean.

9. The increased salinity of many American estuaries, such as the Chesapeake Bay and Delaware River estuaries, could destroy their value as breeding and nursery grounds for many valuable species of game and commercial fish.

10. The carbon dioxide buildup would inflict enormous damage on the global economy. Consider, for example, just one factor: the displacement of agricultural production to the north (in the Northern Hemisphere). Such a dramatic shift would require hundreds of billions of dollars for the construction of new flood control, irrigation, drainage, and grain storage systems. For example, if new irrigation systems were required for just 15% of the existing irrigated area in the world, the price tag would be $200 billion—$15 billion for the United States alone. Then there would be the cost of new dikes, dams, levees, and so on to prevent the flooding of coastal areas.

11. Many plants and animals could become extinct because of the rapid change in temperature. Although species can adapt to climate change, the problem with human-induced change is that it occurs much faster than the natural change that has occurred in the past. One species that could be wiped out as its habitat shrinks is the polar bear (Figure 19.9). The polar bear depends on winter ice to reach its prey, seals. As ice disappears, these bears are finding it more difficult to obtain the nutrition they require to tide them over during the warmer summer months. Numerous other species, including moose and penguins, are also threatened by changing climate due to global warming. The world's coral reefs have been especially hard hit as warmer temperatures kill coral, destroying what is one of the most diverse habitats on the planet. (Warm water kills the algae that form a symbiotic relationship with the corals. Algae provide food to the corals. Without them, corals die, and coral reefs turn white. This phenomenon is called bleaching.)

12. As the Northern Hemisphere warms, diseases that are restricted to the tropics and warm climates in general could spread northward into major population centers. As noted earlier, evidence of this is already starting to accumulate.

13. "Floods and droughts associated with global climate change could undermine health," writes Paul Epstein. "They could damage crops and make them vulnerable to infection and infestations by pests and choking weeds,

FIGURE 19.9 On thin ice? Arctic ice pack is shrinking due to global warming, stranding polar bears and making it difficult for them to find food at the end of their hunting season.

thereby reducing food supplies and potentially contributing to malnutrition. And they could permanently or semipermanently displace entire populations in developing countries, leading to overcrowding and the diseases connected with it, such as tuberculosis."

As you can see, global warming has its benefits, but the social, economic, and environmental costs of the adverse impacts clearly outweigh any gains we might make. In fact, most scientists believe that the consequences will be devastating. Already, several hundred Americans die each year during heat waves. In the summer of 2003, approximately 15,000 people—mostly elderly—died in France in a record-breaking summer heat wave. Another 25,000 died throughout Europe. Crop production in many parts of the world is suffering. Islands are being inundated as sea levels rise, and coastlines are being battered by violent hurricanes that kill dozens—sometimes tens of thousands—of people and create billions of dollars in property damage. Flooding is becoming more common in North America and Europe. Fires are raging in many parts of the world, burning millions of acres of forestland.

GO GREEN!

Turn your thermostat down when you leave home for an extended period of time in the winter, and turn it up during the summer when you leave for an extended period of time or during the night while you're sleeping.

19.7 Reducing or Eliminating Global Warming

Despite the repeated warnings of scientists that the greenhouse effect may eventually have highly adverse impacts on human society, the general public in the United States, as well as policy makers, have been largely unconcerned for many years. In 1988, however, Americans experienced unusual drought, searing heat, floods, and hurricanes—the precise events scientists predicted would occur with increasing frequency and severity as the levels of greenhouse gases rise. The climatic disasters of 1988 may or may not have been valid signals that the greenhouse effect was finally in evidence. Nevertheless, some segments of the public, at least, were aroused from their apathy.

In 1992, many nations of the world signed a voluntary agreement, known as the **United Nations Framework Convention on Climate Change,** to reduce greenhouse gas emissions in an effort to stabilize the release of greenhouse gases and prevent dangerous warming of the Earth's atmosphere. The Convention on Climate Change, however, was weak and largely ignored. Many saw it as a good beginning. Since then, the nations of the world have met numerous times to hammer out goals and strategies for reducing greenhouse gas emissions, In 1997, meeting in Kyoto, Japan, the world's nations developed the **Kyoto Protocol,** a treaty that legally obligates nations to cut emissions of greenhouse gases such as carbon dioxide to 5.2% below 1990 levels by dates between 2008 and 2012. As of January 2008, 105 nations had ratified the agreement. Only one more nation was required to put the agreement into effect. Unfortunately, the United States, Russia, and Australia declined to sign the agreement. Although many citizens and many scientists and

prominent businesses, including the insurance, and natural gas industries, and even large oil companies such as BP, support efforts to curb greenhouse emissions, powerful political forces, notably the coal, oil, and automobile industries, opposed such efforts in the United States. They and conservative government officials argued that reducing carbon dioxide emissions would have an adverse effect on the U.S. economy, despite considerable evidence that it won't and despite evidence that global climate change was costing us dearly in lost lives, decreased crop output, and property damage. In 2004, however, Russia agreed to ratify the agreement, thus putting it into effect. (The Ethics in Resource Conservation box discusses whether we have an obligation to reduce carbon emissions.)

Despite failure of the United States to ratify the agreement, some progress has been made. In Europe, Germany, the United Kingdom, and France posted declines in carbon dioxide emissions, although they were offset by increases in other European nations. Overall, in Europe, emissions rose 3.1% during this period, while carbon dioxide emissions in the United States increased nearly 12%. In Japan, carbon dioxide releases climbed nearly 6%. In less-developed nations, emissions rose nearly 40%. Clearly, much more work is needed if we're going to reach the goals of the Kyoto Protocol. For some encouraging examples in the United States, see A Closer Look 19.1.

This section considers some of the strategies to help slash carbon dioxide emissions.

Reducing Automotive Emissions

Carbon dioxide levels released into the atmosphere by motor vehicles could be substantially reduced by the emission-controlling strategies discussed in the previous chapter. As you recall, they include (1) reducing traffic volume, (2) instigating engine inspection and maintenance programs, (3) increasing fuel efficiency, (4) increasing the use of mass transit (buses and trains), (5) adopting alternative fuels such as biodiesel, hydrogen, and ethanol, and (6) replacing gasoline engines with those powered by electricity. Chapters 18 and 22 describe some of these ideas, including alternative renewable fuels such as hydrogen and biodiesel. Below are some specific strategies that expand on these ideas.

Convert from Gasoline to Ethanol   Motor vehicle fuel could be shifted from gasoline to ethanol (grain alcohol), a topic briefly mentioned in Chapter 18. Ethanol is a renewable resource produced from plants grown on "fuel farms." Corn, for instance, can be converted to ethanol. The use of ethanol has several advantages. Ethanol can be mixed with gas in small amounts (10 to 20% ethanol) and burned in cars. This requires no modification. Only slight modifications are required to burn 85% to 100% ethanol. Most important, ethanol burns fairly cleanly and could reduce carbon dioxide emissions. Why? The amount of carbon dioxide that is released when it is burned equals the amount removed from the air by photosynthesis in fuel crops. (If farm equipment and trucks used to transport ethanol were powered by ethanol, the system could be carbon neutral—that is, it would not increase carbon dioxide levels.)

ETHICS IN RESOURCE CONSERVATION 19.1   DEBATE OVER GLOBAL WARMING: DO WE HAVE AN OBLIGATION TO OTHER COUNTRIES?

International negotiations over limiting carbon dioxide emissions often focus on a disparity between more-developed countries and less-developed countries. The less-developed nations, the arguments go, use only a fraction of the world's fossil fuels (about 24%) but house 70% of the world's population. The United States, in contrast, has less than 5% of the world's population but uses over 25% of the world's fossil fuels. Representatives from less-developed nations also note that they have not enjoyed the benefits of the fossil-fuel-powered Industrial Revolution—and should be allowed to. Why should they be asked to find low-carbon alternatives to fuel economic growth they need to raise their people's standard of living? The more-developed nations are the major producers and should take the lead on such efforts. Less-developed countries should be allowed to develop using fossil fuels.

Many of the more-developed countries agree. In fact, they've agreed to an international effort (the Kyoto Protocol) that places much more stringent restrictions on carbon emissions from more-developed countries than those imposed on less-developed nations. Partly because of this, President George W. Bush and many senators and congressional representatives have opposed Kyoto. For many years, their position has been that the United States will not join international efforts to curb greenhouse gas emissions, not until the less-developed nations are forced to meet the same standards.

Do you think this stand is valid? Why or why not? What are the costs of such a stand? Do the more-developed nations such as the United States have an obligation to take steps and perhaps even help the less-developed nations pay the additional costs of developing alternative technologies? Why or why not? Is this an ethical debate?

Make a list of points in support of your viewpoint, and then take the opposing viewpoint and summarize what you think it might be. Does this exercise change your viewpoint at all?

Can you think of any alternative means of promoting the betterment of people's lives in the less-developed countries without financial assistance from the West?

A CLOSER LOOK 19.1   Going Green!

These days, green is in. Just about everyone is going green. From businesses to football teams to college campuses, green is in. Consider a few examples.

Like individuals, businesses, and governments, the Philadelphia Eagles football team has adopted a number of policies and practices to help reduce its impact on the environment. The team now prints its programs and tickets on recycled paper and requires vendors to use plastic cups made from corn, a renewable resource, for beverages sold on the premises. Plates used in the executive suites are made from plant products.

Even the lights in their stadium are powered by renewable energy. Overall, one-third of the energy the team consumes derives from renewable resources, mostly wind power. The team even subsidizes renewable-energy purchases by employees.

The team minimizes the use of potentially toxic chemicals and fertilizes its playing field with organic fertilizer. It also recycles much of the mountain of waste left over after each home game.

The team is also helping to educate its fans on the importance of little things we can all do to reduce our impact on the environment. The Eagles do this by sharing facts about their fans' contributions to the problem and ways they can help solve them.

Although the Eagles are leading the charge, they're not alone. The Rams are also taking steps to become a greener team. Their efforts have spurred interest in greening high school and college teams, not just in football, but in other sports as well. And while millions of football fans watched the undefeated New England Patriots battle the New York Giants in the Super Bowl in February 2008 in Glendale, AZ, few realized that this was the first-ever Super Bowl powered entirely by renewable energy generated from sunlight, wind, geothermal energy, landfill gas, and small hydropower facilities.

All across North America, colleges and universities are taking steps to green their campuses. Green is not a new color on college campuses. Many colleges and universities have been recycling waste and taking other measures to reduce their impact on the environment.

Today's green movement is much deeper and greener and is often aimed at creating a sustainable future. Over 300 colleges and universities in the United States, for instance, have joined the Campus Climate Challenge, aimed at reducing their contribution to global warming. They're buying renewable energy and implementing energy-efficiency measures that dramatically lower their carbon emissions.

Many colleges and universities are also building all new classrooms and other facilities to much higher, more energy-efficient standards using green building materials—often thanks to student insistence. The University of Vermont and the University of Denver, for instance, built green buildings to house their law schools. The Colorado College, where Dr. Chiras teaches, built a new green science building and a green building to house the art program. Dozens of others universities and colleges have followed suit (Figure 1).

(continued)

FIGURE 1 A model of green. The University of Denver Law School is one of many green-built college and university buildings. It is more energy efficient than other buildings, which saves on day-to-day operation costs, and is made from environmentally friendly products.

In January 2008, the Clorox Company announced the introduction of Green Works cleaners, a line of natural cleaning products that are as effective as conventional cleaners but made from plant-based ingredients. According to the company, Green Works is at least 99% natural and made from chemicals derived from coconuts and lemon oil. The cleaning agents are biodegradable, non-allergenic, packaged in bottles that can be recycled, and were tested on animals. The company also packages cleaning agents in recyclable (number 1) plastic. In the summer, Clorox introduced environmentally friendly dishwashing soap.

Green Works is the first line of natural cleaners developed by a major consumer products company and is an affordable alternative to natural cleaners. Clorox is not the only company to go green. In February 2008, SC Johnson, a leader in corporate green, began to print a Greenlist logo on its eco-friendly products, starting with an old standby, Windex, a window cleaner. Johnson's Greenlist logo indicates that its products are among the numerous cleaning agents that are now made with lower concentrations of volatile organic chemicals, which can be hazardous to human health and also contribute to photochemical smog. The company has also reformulated Pledge, a furniture polish, to increase its biodegradability while increasing cleaning power, and Fantastik Orange Action, a cleaner with no VOCs.

In 2008, computer chip giant Intel announced that the company would purchase more than 1.3 billion kilowatt-hours a year of renewable-energy certificates (RECs) from Sterling Planet to offset carbon dioxide emissions from its operations. The electricity will be generated by wind, solar, small hydroelectric facilities, and biomass. According to the EPA, this purchase makes Intel the single largest corporate buyer of green power—electricity generated from clean, renewable resources—in the United States.

Besides offsetting Intel's emissions, company officials hope their purchase will help stimulate other companies to follow suit and boost the market for green power. Their efforts, combined with those of other companies, could help lower the generating cost of electricity from renewable resources such as wind.

Companies are not the only entities that have made major commitments to purchase green power. The Commonwealth of Pennsylvania, the City of Houston, and the City of Dallas are all major purchasers, as are Sprint, Nextel, IBM, and the U.S. Department of Veterans Affairs.

To those of us who have been working on environmental issues for decades, these signs are exciting and extremely encouraging. They suggest that human society may have earnestly begun its shift to sustainability.

Most experts think ethanol (or some other environmentally safe fuels) will be needed sooner or later anyway to replace oil, the supplies of which are dwindling rapidly.

Unfortunately, ethanol production is not entirely without problems. For one, making ethanol from corn takes a lot of energy. However, newer processes under development now could dramatically improve the efficiency, making this fuel much more economically viable. Ethanol production also displaces food crop production and has been blamed for

rising food prices. For more on ethanol as a fuel, see Chapter 22.

Use More Methane Fuel Methane gas generates only about half as much carbon dioxide as coal when producing the same amount of energy. Also known as natural gas, methane is extracted from the Earth's crust. Methane can also be obtained from sanitary landfills by driving a pipe into the decomposing garbage. This is already being done at many sites in the United States, including the Fresh Kills Landfill on New York City's Staten Island. This fuel can then be used to power automobiles and various industrial processes. It can even be used to generate electricity.

Improve Fuel Efficiency: Taxing Fuels   Economists have long known that the price of fuel, whether it is home heating oil or gasoline, greatly affects the efficiency of its use. The cheaper the fuel, the less efficiently we use it. For this reason, some conservationists have proposed increasing the tax on gasoline in the United States and imposing a special carbon tax on all fuels that release carbon dioxide when burned. The greater the amount of carbon dioxide released from a given fuel, they suggest, the higher the fee. Such a tax would encourage industries and utilities to conserve fuel or switch from oil and coal to natural gas or solar energy. Motorists would very likely buy more efficient vehicles or choose other options for transport—for example, commuting by bus or light rail. Revenues from the tax could be used to develop and promote energy conservation and alternative fuels (Chapter 22).

Early in his first term, President Clinton proposed a carbon tax, but the idea was soundly rejected by senators from western states, those that produce coal and other fossil fuels. Other nations have imposed taxes. In Europe, it is not unusual to pay between \$5 and \$8 per gallon for gasoline at the pump. Much of the high cost is due to taxes. As the impacts of global warming become better known and the costs become evident, policy makers in the United States and Canada may become more open to the idea, but at this writing, it remains highly unpopular.

Some experts believe that high taxes are not necessary. High prices, caused by rising demand and shrinking supplies, and a host of other factors in 2007 and 2008 have begun to discourage wasteful practices. Many individuals and businesses are turning to more efficient transportation options. In 2007 and 2008, for example, the sales of large SUVs and trucks fell precipitously in the United States. In 2008, General Motors closed three of its large truck production facilities in the states. Ford converted one of its truck production facilities into a factory to manufacture small, energy-efficient cars. Sales of more fuel-efficient hybrids rose meteorically. The state of Utah put all of its state employees on a four-day, ten-hour-per-day workweek.

Halt the Deforestation in the Tropics

As described in Chapter 14, millions of square kilometers of tropical rainforests in Central and South America and Southeast Asia have been cut down and burned to clear areas for cattle ranches and farms. Much wood is also used as fuel. Some experts believe that 40% of the closed canopy (solid stands) of tropical rain forest has already been removed. At present rates of deforestation in Nepal (4% per year), the tropical rain forest will have been completely cleared away by 2012.

Tropical deforestation prevents the uptake of more than a billion tons of atmospheric carbon by photosynthesis each year. A halt to this activity, coupled with massive efforts to replant deforested areas, could have a major impact on global carbon dioxide levels. Individuals can help on a smaller scale by planting trees, cutting back on all forms of paper waste, and recycling.

GO GREEN!
Encourage your parents to plant trees on their property. Support non-profit organizations working on ways to protect forests (for example, the Wilderness Society and Sierra Club) and organizations that encourage reforestation (for example, the Arbor Day Foundation).

Reforestation

Many millions of hectares of tropical rainforest have been destroyed in the name of progress—an area equal to about half the land mass of the United States. At least two-thirds of this forestland has never been replanted. Replanting tropical rain forest and other devastated forests could go a long way toward helping reduce the threat of global warming. The American Forestry Association encourages the planting of 100 million trees yearly in the United States. Such projects will certainly help.

Although such efforts are important, they pale in comparison to the task at hand. Scientists, for instance, estimate that to remove the 7.6 billion tons of carbon dioxide released into the air each year as a result of the combustion of fossil fuels, the new forests would have to cover an area larger than that of the continental United States!

Sustainable Solutions

The previous strategies rely primarily on conservation (using energy more efficiently), renewable energy use, and restoration. Other strategies are also essential to reduce the threat of global warming. Population stabilization is an important step. Reducing the growth of the human population could have enormous benefits. Growth management strategies in urban areas also are important. They can greatly reduce vehicle miles traveled and fossil fuel combustion. Recycling also is vital to curbing greenhouse emissions. Chapter 20 discusses efforts to reduce ozone-depleting chemicals, which are greenhouse gases.

GO GREEN!
Keep the tires on your car properly inflated and your car's engine well tuned. Drive at the speed limit. Accelerate gradually. Remove roof racks when not in use. All these measures will save you money and reduce pollution.

Summary of Key Concepts

1. Global warming refers to the increase in the temperature of the Earth's oceans and atmosphere, which can be caused by natural and human factors.

2. Global warming can cause massive and devastating changes in the Earth's climate, known as global climate change. Climate is defined as the planet's climate, or average weather conditions, including temperature, precipitation, and storms.

3. Each day the sun shines down on the Earth, bathing us in sunlight. Approximately one-third of the sunlight striking the Earth and its atmosphere is reflected back into space by clouds, dust, and the Earth's surface. The rest is absorbed by air, land, water, highways, parking lots, buildings, and plants, warming the planet.

4. All heat is eventually radiated back into outer space through the atmosphere so that the energy gained from the sun is balanced by energy radiated back into space.

5. Certain chemical substances in the atmosphere can alter the Earth's energy balance. They absorb heat radiating toward outer space and radiate it back to the Earth's surface, warming the planet. This phenomenon is known as the greenhouse effect, and the gases that cause it are known as greenhouse gases.

6. Greenhouse gases include both natural and human-produced chemicals: water vapor, carbon dioxide (CO_2), nitrous oxide (N_2O), methane (CH_4), and chlorofluorocarbons. These gases are known as greenhouse gases.

7. Carbon dioxide is one of the most important greenhouse gases. It is released (1) from the combustion of organic materials, including fossil fuels, wood, and vegetation, (2) from the natural decay of vegetation and animal waste, and (3) by all living animals during respiration.

8. Without carbon dioxide, the Earth would be at least 30°C (55°F) cooler than it is today—making it inhospitable to nearly all life-forms, including humans.

9. Studies of the Earth's energy balance show that the surface temperature is influenced by two factors: those that alter the amount of sunlight striking the Earth's surface and those that change the rate at which heat is lost or retained. Both natural and anthropogenic factors influence these processes.

10. The average temperature of the Earth and the Earth's climate change naturally over time. These natural changes in climate are caused by: (1) changes in the Earth's orbit around the sun, (2) changes in the Earth's tilt, (3) increases or decreases in solar activity, (4) increases or decreases in volcanic activity, and (5) chaotic interactions of the climate system.

11. Although many natural factors can alter climate, human activities also can alter the Earth's temperature, which could lead to changes in climate. The two most important activities are the release of greenhouse gases and deforestation.

12. The release of greenhouse gases from natural sources has remained fairly constant over the past 100 years, while the emissions from human sources have increased substantially, rising from 534 million tons (of carbon) to about 7,600 million tons in the past 100 years.

13. Most greenhouse gases come from natural and human sources, but some come solely from human activities. These include the chlorofluorocarbons (CFCs) and their replacements, the hydrochlorofluorocarbons (HCFCs).

14. Not all greenhouse gases are created equal. Some, like CFCs and methane, are more powerful greenhouse gases than carbon dioxide. Because carbon dioxide is produced in much greater quantities than other greenhouse gases, it has the most significant impact on human-induced global warming.

15. Carbon dioxide is produced during the combustion of fossil fuels and also results from deforestation and the torching of cleared forests.

16. Studies show that the average concentration of carbon dioxide in the Earth's atmosphere has risen steadily from 290 ppm in 1860 to over 380 ppm in 2007—a 31% increase. The greatest increase in atmospheric carbon dioxide levels since the last ice age has occurred since 1950. Current concentrations of carbon dioxide in the atmosphere are 100 ppm higher than the highest levels seen in the past 450,000 years, most likely as a result of human actions. Methane levels are 1,000 parts per million higher than they've been in the same period.

17. In 2005, human activities were responsible for the release of 7.6 billion tons of carbon dioxide into the atmosphere. Fossil fuel consumption was responsible for about 80% of these emissions, or about 6 billion metric tons—nearly 1 ton for every man, woman, and child on Earth. The clearing and burning of tropical rain forests annually releases about 1.6 billion metric tons into the air.

18. While many people believe that the Earth's oceans and atmosphere are warming, others are not so sure. Measurements of global temperature show that Earth's average surface temperature has risen from a little over 13.8°C in 1950 to 14.6°C with only a small dip in temperature from 1940 to 1960.

19. Numerous additional evidence supports the assertion that the Earth's surface temperature is indeed rising: (1) rising ocean temperatures, (2) rising sea level, (3) melting of glaciers, (4) the breakup of the Antarctic ice sheet, (5) a decrease in Arctic polar sea ice, (6) more frequent heat waves and drought, (7) an increase in tornados and violent storms, (8) changes in precipitation, (9) an increase in the severity of hurricanes, (10) an increase in the incidence of forest fires, and (11) the spread of insects and diseases from tropical climates.

20. To most objective observers, the evidence is pretty convincing: the Earth's temperature—and its climate—are changing. Moreover, they are changing dramatically.

21. Numerous lines of evidence suggest that climate change is occurring as a result of human activities, especially the increase in global carbon dioxide levels and the correlation between increased concentrations of carbon dioxide in the atmosphere and global average temperature. Studies of natural and anthropogenic radiative forcing also suggest that human factors are primarily responsible for global warming.

22. Global warming could have some beneficial effects but poses many more costly harmful effects, including (1) the inundation of coastal areas, (2) billions of dollars of property damage in the United States and other countries, (3) saltwater contamination of aquifers, (4) drought-triggered dust storms on the Great Plains of North America and in other countries, (5) inundation of major rice-producing regions, and (6) termination of economic growth throughout the world.

23. Many efforts are under way to reduce greenhouse gas emissions, although progress has been slow. One of the most significant efforts is the Kyoto Protocol, an international agreement to curb emissions by industrial and former Eastern bloc nations. The United States refuses to sign it and has steadfastly opposed actions to curb domestic production of carbon dioxide.

24. Global climate change can be reduced by efforts to switch to more efficient combustion technologies and alternative non-fossil-fuel technologies such as wind and solar energy.

Key Words and Phrases

Carbon Dioxide
Chlorofluorocarbons
Climate
Coral Reefs
Global Climate Change
Global Warming
Greenhouse Effect
Greenhouse Gas
Hydrochlorofluorocarbons

Intergovernmental Panel on
  Climate Change
Kyoto Protocol
Methane
Nitrous Oxide
Radiative Forcing
United Nations Framework
  Convention on Climate
  Change

Critical Thinking and Discussion Questions

1. What is global warming? What is global climate change? How are they related? Is global warming really just a part of global climate change? Why or why not?

2. List and describe each of the natural factors that can alter the Earth's climate.

3. List and describe each of the human factors that can alter the Earth's climate. Which ones are the most important?

4. Debate the following statement: "There's no evidence showing the Earth is warming."

5. Using your critical thinking skills, discuss the following statement: "The global climate has gone through warming periods before, so we shouldn't be concerned about any climate change occurring now."

6. Is there a possible cause-and-effect relationship between air pollution and the northward extension of the ranges of the armadillo, opossum, cardinal, mockingbird, and sea otter in the United States?

7. Discuss both the benefits and the adverse effects of the carbon dioxide buildup in the global atmosphere.

Suggested Readings

Bird, E. 1993. *Submerging Coasts.* New York: Wiley. Examines the implications of rising sea level on coastlines and coastal communities.

Blair, T. 2003. Meeting the Sustainable Development Challenge. *Environment* 45(4): 20–26. What former British prime minister Tony Blair has to say about global climate change.

Doughman, P. M. 2007. California's Climate Change Policy: Raising the Bar. *Environment* 49(7): 34–42. A look at the very progressive policies of a state that often leads when it comes to environmental policies.

Dunn, S. 2002. *Reading the Weathervane: Climate Policy from Rio to Johannesburg.* Worldwatch Paper 160. Washington, DC: Worldwatch Institute. A great overview of efforts to create international agreements to control greenhouse gases.

Dunn, S., and C. Flavin. 2002. Moving the Climate Change Agenda Forward. In *State of the World 2002,* ed. L. Starke. New York: W. W. Norton. A must-read for students interested in understanding more about the interplay between politics and science as it pertains to climate change.

Dupree, J. 2007. Coral Crisis. *National Wildlife* 45(4): 22–30. Description of the plight of the world's coral reefs, which are being devastated by rising sea temperatures.

Flavin, C. 1994. Storm Warnings: Climate Change Hits the Insurance Industry. *World-Watch,* 7(6): 10–20. A telling tale of how seriously the insurance industry is taking warnings of global climate change.

Flavin, C. 1996. Facing Up to the Risks of Climate Change. In *State of the World 1996,* ed. L. Starke. New York: W. W. Norton.

Flavin, C., and O. Tunali. 1995. Getting Warmer: Looking for a Way out of the Climate Impasse. *World-Watch* 8(2): 10–19. Excellent look at what nations aren't doing to address global climate change.

Glick, D. 2007. On Thin Ice. *National Wildlife* 45(1): 22–30. Description of the plight of the polar bear.

Gardiner, D., and L. Jacobson. 2002. Will Voluntary Programs Be Sufficient to Reduce U.S. Greenhouse Gas Emissions? *Environment* 44(8): 24–33. Detailed examination of the Bush administration's proposal to reduce greenhouse gas emissions voluntarily.

Heiman, M. K., and B. D. Solomon. 2007. Fueling U.S. Transportation: The Hydrogen Economy and Its Alternatives. *Environment* 49(8): 10–27. A critical look at one possible non-carbon-based fuel that could help combat global warming.

Intergovernmental Panel on Climate Change. 2007. *Climate Change 2007: The Physical Science Basis.* Geneva: United Nations Environmental Programme. A must-read for anyone who wants to explore the science behind global warming and the impacts of human activities.

Kasemir, B., et. al. 2000. Involving the Public in Climate and Energy Decisions. *Environment* 42(3): 32–42. Exploration of an aspect of change we don't discuss often in this book.

Kates, R. W., and T. J. Wilbanks. Making the Global Local: Responding to Climate Change Concerns From the Ground Up. *Environment* 45(3): 12–23. Ideas on what can be done locally to reduce global warming.

Kutscher, C. 2006. What Is the Evidence and What Can We Do About It? *Solar Today* 20(4): 28–33. One of the best articles on the evidence behind global warming and ways we can solve this thorny issue.

Line, L. 2007. The Shrinking World of Penguins. *National Wildlife* 45(5): 30 H–30 O. Description of the plight of the world's penguin populations, which are now threatened by global warming.

Liverman, D. 2007. From Uncertain to Unequivocal. *Environment* 49(8): 28–32. An excellent review of a very important publication, *Climate Change 2007* by the IPCC.

Moore, C. A. 1997. Warming Up to the Hot New Evidence. *International Wildlife* 27(1): 21–25. A shocking article that looks at global-warming predictions and whether they are coming true.

Moran, J. M., and M. D. Morgan. 1994. *Meteorology: The Atmosphere and the Science of Weather,* 4th ed. New York:

Macmillan. Excellent nontechnical coverage of global pollution problems.

Newman, P., and J. Kenworthy. 2007. Greening Urban Transportation. In *State of the World 2005,* ed. L. Starke. New York: W. W. Norton. A detailed discussion of oil, its threat to national security, and ways renewable energy could help improve security.

Plotkin, S. 2007. Examining New U.S. Fuel Economy Standards. *Envrionment* 49(6): 8–19. An in-depth look at an important topic.

Prugh, T., C. Flavin, and J. L.Sawin. 2005. Changing the Oil Economy. Excellent ideas on how to reduce fossil fuel use in the transportation sector.

Sawin, J. L., and K. Hughes. 2007. Energizing Cities. In *State of the World 2007,* ed. L. Starke. New York: W. W. Norton. Excellent ideas on how to reduce fossil fuel use in our cities.

Skolnikoff, E. B. 1999. The Role of Science in Policy: The Climate Change Debate in the United States. *Environment* 41(5): 16–20, 42–45. Interesting reading.

Smith, D. 2007. The Mystery of the Disappearing Moose. *National Wildlife* 45(2): 46–50. Description of the plight of the polar bear.

Sovacool, B. K., and J. N. Barkenbus. 2007. Necessary but Insufficient: State Renewable Portfolio Standards and Climate Change Policies. *Environment* 49(6): 20–33. A look at the importance of federal policy to address global warming.

United Nations Environment Programme. 1994. *The Impact of Climate on Fisheries.* Nairobi, Kenya: United Nations Environment Programme. Examination of the impacts of climate change on fisheries, based on several case studies.

Wilbanks, T. J., et al. 2003. Possible Responses to Global Climate Change: Integrating Mitigation and Adaptation. *Environment* 45(5): 28–38. Examination of options for addressing global warming.

 Web Explorations

Online resources for this chapter are on the World Wide Web at: **http://www.prenhall.com/chiras** *(click on the Table of Contents link and then select Chapter 19).*

20

ACID DEPOSITION AND STRATOSPHERIC OZONE DEPLETION

"April showers bring May flowers"—or do they? Many scientists now believe that at least some April showers may bring death, not only to plants but also to fish, birds, mammals, and even humans. Why? Because April showers often contain acid. Rain containing acids is known as **acid rain.** But acid rain is actually only part of the problem. Acids are falling from the sky in snow and are deposited in dry form (described shortly). Together, these various forms of acid fallout are referred to as **acid deposition.**

Acid deposition is one of three serious global air pollution problems. The other two are global warming, discussed in Chapter 19, and ozone depletion. We'll look at acid deposition and ozone depletion in this chapter, focusing on their causes, their effects, and sustainable solutions to them. We begin with acid deposition.

20.1 Acid Deposition

Acid deposition originates from two major atmospheric pollutants: **sulfur dioxide** and **nitrogen dioxide.** These gases come from a variety of sources, both natural and human. In the atmosphere, they combine with water and oxygen to produce **sulfuric acid** and **nitric acid,** respectively. Globally, sulfur dioxide is responsible for about 60% to 70% of the acid deposition. Nitrogen dioxide is responsible for 30% to 40% of the rest. However, the contribution of each of these two gases varies from one region to the next. In the eastern United States, for example, sulfur dioxide has for years been responsible for most of the acidity. Why? Because it is released in copious amounts from coal-fired power plants in the upper Ohio River Valley. These plants often burn high-sulfur coal and have historically had inadequate pollution control devices. In the Rocky Mountains and along the Pacific Coast, however, oxides of nitrogen are the sources of acid deposition. These pollutants arise primarily from automobile exhaust.

As noted in the chapter introduction, acid deposition can be either wet or dry. Acids deposited in rain, snow, and fog represent **wet deposition. Dry deposition** occurs when dust particles containing sulfate or nitrate settle on Earth. These chemicals may then react with water to form sulfuric and nitric acids. Dry deposition also may occur when gases such as sulfur dioxide and nitric oxide come in contact with soil, trees, lakes, and streams, where they react with water to form acids.

Acidity is measured on the **pH scale.** As shown in Figure 20.1, pH ranges from 0 to 14. A substance with a pH of 7 is neutral—that is, it is neither acidic nor basic. Numerical values above 7 indicate nonacidic or basic substances. For example, lye, with a pH of 13, is extremely alkaline. Substances that are acidic have pH values below 7. Thus, battery acid,

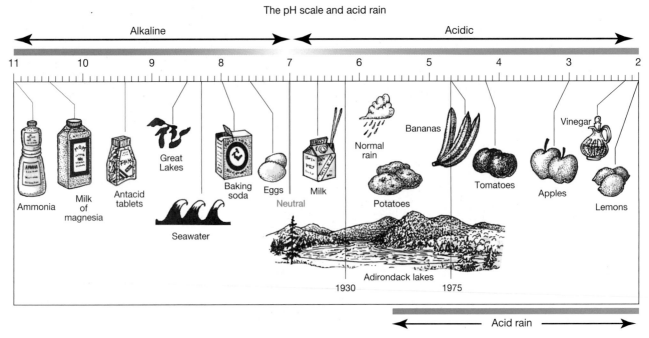

FIGURE 20.1   The pH of acid rain compared with that of fruits, vegetables, and common household substances. Note the sharp decrease in the pH of New York's Adirondack lakes between 1930 and 1975.

with a pH of 1, is extremely acidic. The pH scale is a special one. It is logarithmic. In logarithmic scales, a one-unit change results in a tenfold change in the item being measured. Thus, rain with a pH of 4 is 10 times as acidic as rain with a pH of 5, and 100 times as acidic as rain with a pH of 6. Rather surprisingly, normal unpolluted rain is weakly acidic, with a pH of 5.6. The reason is that carbon dioxide from the air reacts with the water to form a relatively weak acid known as carbonic acid.

Areas Affected by Acid Deposition

In 1980, the federal government established the **National Atmospheric Deposition Program.** It consists of acid-deposition-monitoring sites scattered throughout the United States. Through this program, a comprehensive assessment of acidity trends can be made, and maps can be drawn.

Figure 20.2 shows the pH of rainfall in the United States. As illustrated, the rain in the densely populated and highly industrialized eastern United States is quite acidic. Over the past five decades, two disturbing trends have become evident. First, the level of acidity has been increasing. That is, precipitation has become more and more acidic. Second, the area exposed to acid deposition has expanded considerably. These trends appear to be correlated with the increasing volume of sulfur and nitrogen oxides released into the air in the upper Ohio River Valley.

The Northeast has been the focus of many acid deposition studies. Recent work shows that many other states are now experiencing acid deposition. The California Air Resources Board, for example, has documented precipitation in the state that is among the most acidic found anywhere in the world. The pH of rain in California ranges from a low of 4.4 in San Jose and Pasadena to a high of 5.4 at Big Bear. Fog in California's

Central Valley had a pH of 2.6—1,000 times more acidic than nonpolluted rain. Of considerable interest is the fact that dry acid deposition is 15 times more prevalent than acid rain in parts of southern California where rainfall is scant and the nitrogen oxide emissions from motor vehicles are high.

The United States is not the only country suffering from acid deposition. In fact, virtually every area in the world downwind from a major industrial or population center experiences acid deposition. Scandinavia and the rest of Europe have been particularly hard hit. Thousands of lakes in Sweden and Norway have been made barren by acids deposited in rain and snow. The **acid precursors,** the substances that gave rise to the acids, are released by power plants and factories in England and the rest of Europe.

Canada also has been badly damaged by acid deposition, coming from sources within the nation and from the United States. At least 15,000 lakes in southeastern Canada have been made barren because of this problem, which was first identified in the early 1900s.

Sources of Acid Precursors

Like most pollutants, acid precursors come from natural sources and human or **anthropogenic sources.** Sulfur dioxide, for example, is released from volcanoes, swamps, and power plants. Human activity now accounts for about as much sulfur dioxide emission as does nature. However, most human-generated sulfur dioxide is released from only 5% of the Earth's surface—primarily the industrialized regions of the United States, Canada, Europe, and eastern Asia. Roughly 15 million metric tons (17 million tons) of sulfur dioxide are emitted into the atmosphere of the United States annually, 72% of it from coal-fired electric power plants. Industry and

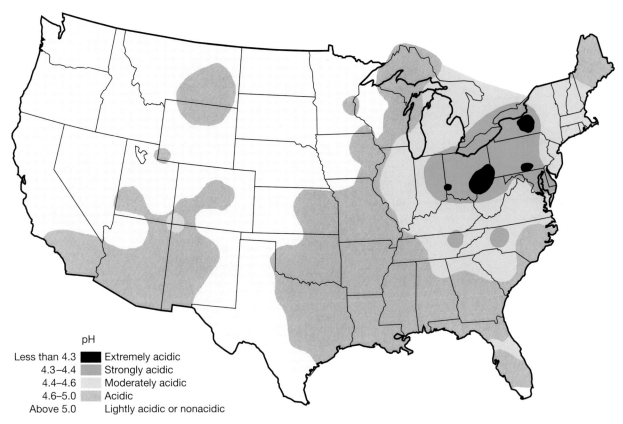

pH

Less than 4.3	■	Extremely acidic
4.3–4.4		Strongly acidic
4.4–4.6		Moderately acidic
4.6–5.0		Acidic
Above 5.0		Lightly acidic or nonacidic

FIGURE 20.2   Acidity (pH) of North American precipitation. Note the high acidity (low pH) in the northeastern United States and southeastern Canada.

motor vehicles contribute most of the remainder. Studies show that about 50% of the acid deposition in eastern Canada has its origin in the United States. In Ohio and nearby states, high-sulfur coal is king. As a result, Ohio smokestacks alone emit more sulfur dioxide into the air than those in New York, New Jersey, and the six New England states combined.

Nitrogen oxides also come from natural sources—for example, forest fires. But the major sources of concern are once again anthropogenic—primarily cars, trucks, buses, and power plants. In the United States, these sources release 18.4 million metric tons (20.4 million tons) a year.

Sensitive Areas and Long-Range Transport of Acids

Not all areas are equally vulnerable to acid deposition. Figure 20.3 maps the areas in the United States and Canada that are the most sensitive. As you can see, they include much of southeastern Canada, the eastern United States, northern Minnesota, Wisconsin, and Michigan. Sensitive areas also include certain areas in the Rocky Mountains, the Northwest, and the Pacific Coast. The vulnerability of a given region to acid deposition depends on the ability of the rocks and soils in the watershed (or the rocks on a lake bottom) to neutralize, or **buffer,** the acid. Soils derived from limestone, which are rich in calcium, are much more capable of neutralizing acids. Soils derived from granite, which are low in calcium, are highly vulnerable.

Researchers at Colorado State University have found that a given molecule of sulfur dioxide may remain in the atmosphere for up to 40 hours, and a sulfate particle may remain aloft for three weeks. Because of their relatively long residence in the atmosphere, these molecules may be carried hundreds of kilometers from their release point (Figure 20.4). Sulfur dioxide molecules that originate in an Arizona copper smelter may eventually fall in acid rain on a sensitive mountain lake in New York or Vermont. A molecule of sulfur dioxide that originated in Ohio may be transported to southern Ontario (Figure 20.4). Scientists estimate that 87% of the sulfate in New York and New Jersey and 92% of the amount in New England is delivered from outlying areas—most likely the upper Ohio River Valley (Figure 20.5). By contrast, California generates all of its own acid rain, receiving almost no inputs from outside its boundaries. This situation provides California legislators with a strong motive to control acid deposition, knowing that any reductions achieved in sulfur dioxide and nitrogen oxide emissions will result in a reduction in acid deposition in their state.

Superstacks, tall smokestacks that send pollutants high into the atmosphere, were once thought to be an effective solution to air pollution. Although they do reduce ground-level pollutants near major sources, they simply transfer the problem to ecosystems and people downwind from the sites. Many superstacks serve as mammoth sources of acid precursors. The superstack at the giant smelter at Sudbury, ON (Canada), towers to a height of 380 meters (1,140 feet) and releases 2,300 metric tons of sulfur dioxide into the air daily. It emits twice as much

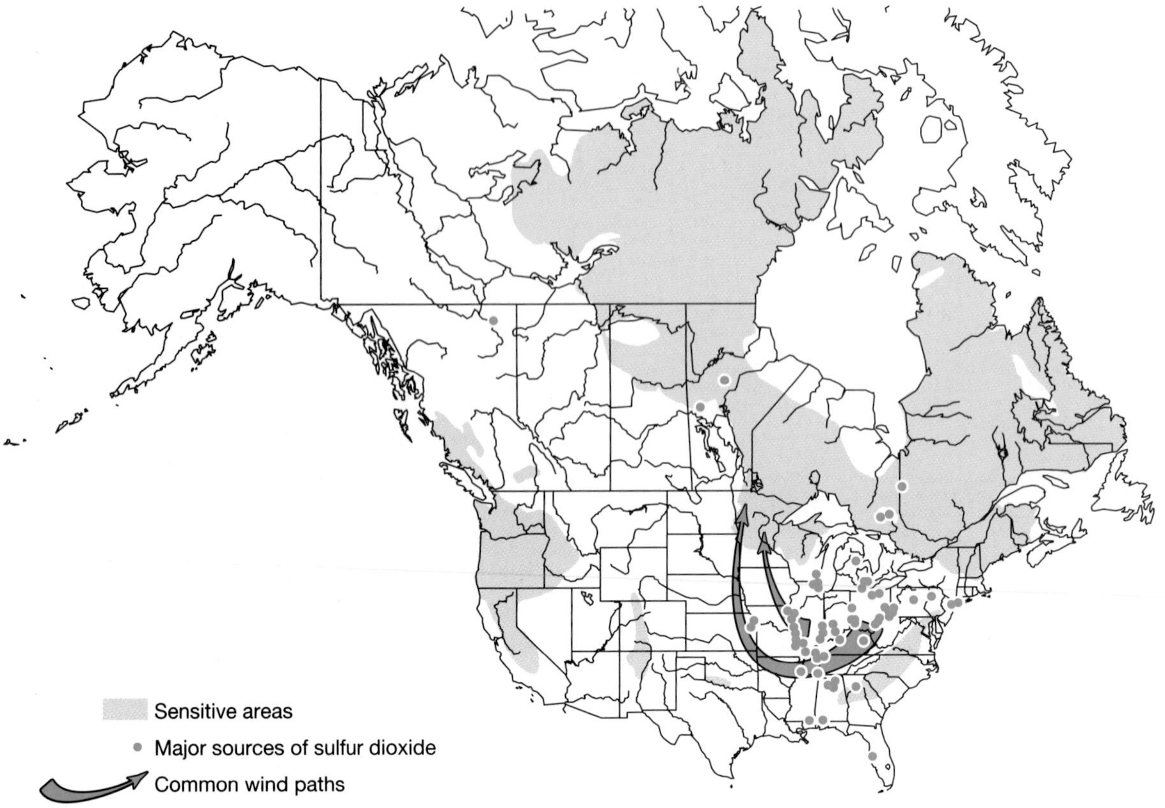

FIGURE 20.3 Areas of acid deposition in North America. Each dot represents an area that emits more than 100,000 tons of sulfur dioxide annually. Arrows indicate direction of major wind movements toward Canada. (Note that Figure 20.4 also shows wind currents that carry acid pollutants into the Northeast as well as into the Mid-Atlantic states.) Shaded areas are low in natural buffers such as limestone and therefore are particularly susceptible to acidification.

sulfur annually as did Mount St. Helens during its most active eruption year! It is the largest stack in the world and serves as a symbol of the acid rain problem in North America. Incredibly, this one stack gives off 1% of all the sulfur dioxide released worldwide!

Harmful Impacts of Acid Deposition

Acids have many adverse impacts on buildings and materials, soils and crops, ecosystems and the organisms within them, and human health. This section outlines the major impacts.

Impacts on Buildings and Materials   Acids erode the surfaces of structures made from sandstone, limestone, marble, and other materials (Figure 20.6). Priceless monuments of ancient Rome and Athens have been seriously damaged by acids, as have the Statue of Liberty, the Washington Monument, and the U.S. Capitol. The marble statues in front of the Field Museum in Chicago have undergone rapid deterioration due to acid rain. Numerous metallic structures such as bridges, rails, and industrial equipment have been corroded by acid rain. The U.S. Air Force spends millions of dollars annually to repair acid-rain-inflicted damage to planes. The damage to buildings and materials by acid rain exceeds $5 billion annually in 17 eastern states. No estimates of global destruction are available, but the figure is very likely quite substantial.

GO GREEN!

To reduce air pollution, including acid precursors, turn off and unplug laptop computers when not in use or charging. Also unplug chargers for electronic devices such as cell phones or iPods when not in use.

Impacts on Soils and Crops   Acid deposition adds hydrogen ions to the soil. Hydrogen ions displace nutrient elements including calcium, potassium, and magnesium from the soil particles to which they are bound. Such nutrients may then be leached from the soil and carried away by runoff water. Acid deposition also inhibits the activity of nitrogen-fixing bacteria. As a result of these two effects alone, acid deposition may substantially reduce fertility in certain soils within a decade. Heavy metals, such as aluminum, that normally are harmlessly bound to soil particles are also displaced by the hydrogen ions in acid deposition. These metals then become soluble in water and can be absorbed by the roots of farm crops or trees, often with harmful results (Table 20.1).

Changes in the soil chemistry affect plant growth. But plants can be directly affected by acids. Acids can leach minerals directly from leaves. They can also kill buds and stunt growth. Acids in soil can reduce germination of seeds and impair growth of seedlings. Crop damage in the United States is estimated to be at least $5 billion per year.

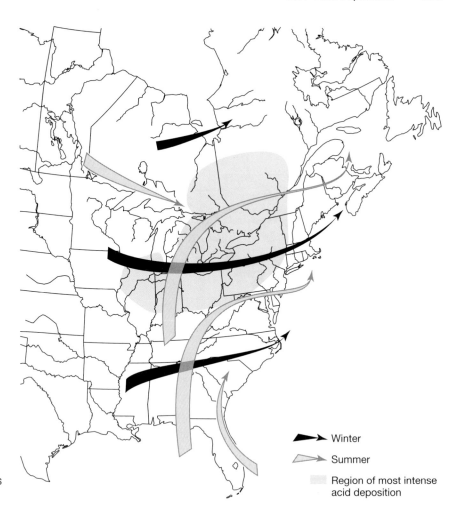

FIGURE 20.4 Wind transports acid to distant areas. Southeastern Canada and the northeastern United States are particularly vulnerable.

Winter

Summer

Region of most intense acid deposition

In forests, acids deposit on needles of conifer trees, killing them and reducing the photosynthetic capability of trees (Figure 20.7). Acids may also stimulate the growth of acid-tolerant mosses that grow on the forest floor. These impair tree growth by trapping moisture in the surface layers, drowning the feeder roots. Entire forests in the northeastern United States and Europe have perished because of acid deposition.

Impacts on Aquatic Ecosystems   Lakes and streams vary in their ability to cope with acid deposition. The most vulnerable are the soft-water lakes that contain relatively few dissolved minerals, such as calcium salts, that neutralize acid. The watersheds of these lakes have little or no soil, and what exists has little buffering capacity. Many lakes in the northeastern United States and in the Rocky Mountains and Sierra Nevada are of this type. By contrast, in the southern and central regions of the United States, most lakes are hard-water lakes and contain acid-buffering minerals. Their watersheds tend to have well-developed soils with good buffering capacity. In Canada, as in the United States, lakes and watersheds vary in their ability to neutralize acids. For example, many lakes in the Canadian Shield, a region in the southeastern portion of the nation, are underlain by granite. The lack of acid-neutralizing capacity of these lakes makes them vulnerable to acid deposition. Mountainous regions in the northern Rockies and central British Columbia are more resistant to acid deposition.

Acid deposition has poisoned thousands of lakes throughout the world. The hardest hit have been in Scandinavia, the province of Ontario, southeastern Canada, and the northeastern United States. In Canada, 300,000 lakes are vulnerable to acid deposition. At least 5,000 have become so acidic that they have suffered extreme fish loss. In Nova Scotia, the acidification of rivers has resulted in a dramatic decrease in Atlantic salmon, according to Environment Canada, Canada's equivalent of the U.S. Environmental Protection Agency. What's happening in the United States?

While acid rain can be found throughout the United States, the northeastern states have experienced the lowest-pH rains and the greatest impacts. In the Adirondack Mountains of New York, perhaps the hardest hit of all regions, 237 lakes have a pH of less than 5—an acidity level lethal to many species of fish.

Although some regions are highly vulnerable and have suffered substantial losses, a survey of lakes in the United States showed that nationwide the percentage is fairly small, according to the National Acid Deposition Assessment Program, a federal program involving representatives from more than a dozen federal agencies. That said, there are some regions that have witnessed alarming rates of acidification. The U.S. Office of Technology Assessment (OTA) conducted a survey of the acid rain problem in 27 eastern states. The OTA found that 80% of the lakes and streams in the

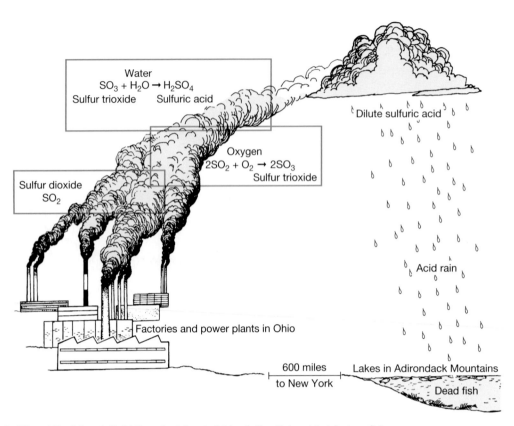

FIGURE 20.5 Much of the acidity of the rain that falls on the Adirondack lakes in New York and that destroys fish and other aquatic life has its origin in sulfur and nitrogen oxides (not shown) emitted from industrial smokestacks in the Midwest.

FIGURE 20.6 Acid rain is slowly causing the erosion of well-known statues in Europe and North America. Among them is this statue located in Versailles, France. The damage costs society billions of dollars a year.

Northeast and upper Midwest are vulnerable to acidification. More important, 17% of the lakes and 20% of the streams in 27 states have been damaged by acid deposition. Of the more than 17,000 lakes studied, nearly 3,000 have been acidified. Of the 187,876 kilometers (117,423 miles) of streams investigated, 39,500 kilometers (24,688 miles) have already been damaged.

Fortunately, reductions in both nitrogen and sulfur oxide emissions from power plants and other sources have resulted in noticeable declines in atmospheric concentrations. This, in turn, has resulted in a decrease in acid deposition and "the beginnings of recovery of acid-sensitive waters in some areas," according to the National Acid Deposition Assessment Program report to Congress in 2005. While this is good news, the authors of the report go on to say that research indicates that recovery will not be possible for many areas that continue to be exposed to acid deposition. Reductions in sulfur dioxide and nitrogen oxide emissions, they say, are insufficient to permit recovery or to prevent further acidification in some regions.

GO GREEN!

To reduce air pollution, especially acid precursors, when running errands, combine trips. Consider walking or riding a bike to run errands.

TABLE 20.1	Pollution-Related Forest Declines of the Past 50 Years

Widely assumed major role

Massive die-off of forests in Europe (*Waldsterben*)

Decline of ponderosa and Jeffrey pines in the San Bernardino Mountains of California

Regional decline of white pine in the eastern United States and Canada

Possible major role

Decline of red spruce and balsam and Fraser firs at high elevation in the Appalachian Mountains from Georgia to New England

Growth decline without other visible symptoms in loblolly, shortleaf, and slash pines in the Piedmont regions of Alabama, Georgia, and North and South Carolina

Growth decline without other visible symptoms in pitch and shortleaf pine in the Pine Barrens region of New Jersey

Widespread dieback of sugar maples in the northeastern United States and southeastern Canada

Declines related to biological or physical factors

Tannensterben (white fir decline) in Europe

Kiefersterben (Scotch pine decline) in East Germany and European Russia during the early 1970s

Oak decline in Germany and especially in France since the early 1900s

Decline of *Pinus pinaster* on the Atlantic coast of France since the early 1980s

Beech bark disease in the northeastern United States and southeastern Canada

Littleleaf disease of shortleaf pine in the southeastern United States

Birch dieback in the northeastern United States and southeastern Canada

Poleblight of western white pine in the Rocky Mountains

Maple decline in the northeastern United States and southeastern Canada

Oak decline in Pennsylvania, Virginia, and Texas

Ash dieback in the northeastern United States and southeastern Canada

Sweetgum blight in the southeastern United States

Note: Some of the declines attributed to biological or physical factors might involve toxic air pollutants, but studies are incomplete, and evidence is insufficient to make a strong connection.

Impacts on Lakes and Streams Acids have many impacts on aquatic ecosystems:

1. They alter reproduction in many species, including fish (Figure 20.8).
2. Acidified waters may disrupt the homing ability in sexually mature salmon. Research conducted at the University of New Hampshire has shown that when the pH of water is between 5.0 and 5.5, salmon apparently lose their ability to detect certain odors in the water. Since adult salmon depend on chemicals to guide them to their native streams so that they can spawn, the survival of the species in acidified streams is threatened.
3. Increased numbers of fish embryos and newly hatched young develop abnormally and eventually die in waters in which the pH is below 5.5.
4. Populations of important fish food organisms, such as crustaceans and insect larvae, rapidly decline.

FIGURE 20.7   Pines killed by acid rain in Czechoslovakia. Note defoliated and dead trees.

5. Game fish like bass and pike are replaced by more acid-tolerant but less desirable species such as bullheads and suckers.

6. Acids leach heavy metals such as aluminum, mercury, copper, zinc, and nickel from soils in watersheds. These toxic chemicals are then washed into streams and lakes. Many researchers believe that heavy metals, more so than acids themselves, are a major cause of fish deaths. Aluminum is toxic to fish at concentrations of less than 1 ppm. At this concentration, it causes mucus buildup on gills, and fish die by asphyxiation. Mercury builds up in the bodies of fish as a result of biological magnification in the food chain. In some acidic lakes in Wisconsin, predatory fish such as salmon, lake trout, and northern pike may have mercury concentrations above 2 ppm—more than twice the level considered safe for human consumption.

7. The skeletons of fish may become weakened by decalcification caused by acids. As a result, their bodies become distorted, and the fish lose their ability to swim. Death from starvation, disease, or predation soon follows.

8. Bacterial decomposition is inhibited. As a result, essential nutrients like nitrogen and phosphorus stay locked up in plant and animal remains and are not available to terrestrial and aquatic plants, the producer base of food webs.

Clearly, acid deposition can have a profound impact on lakes, ponds, streams, and rivers. What is more, the changes brought about by acid deposition tend to combine, making conditions difficult for living things and thus eliminating virtually all life from a lake. Some times of the year are worse than others, too. When snow containing acids begins to melt, for example, huge amounts of acid are released at the beginning of the snowmelt, creating an **acid shock.** Concentrations of acids in the initial meltwater can be five to ten times higher than during rainfall. Although adult fish are able to survive the sudden influx of acids, eggs and fry are more sensitive and often perish.

The impact of acid rain is not just biological. It is also economic. Acid deposition can literally destroy sport fishing in an area, eliminating millions of dollars in tourist revenues. Loss of revenues from the sales of fishing tackle and bait can also be substantial.

Impacts on Human Health   Fish are not the only organisms that are adversely affected by acid deposition. Humans are vulnerable as well. One of the greatest air pollution disasters the world has known occurred in London in 1952. The "killer fog" that shrouded the city for five days contributed to the death of more than 4,000 people from bronchitis, pneumonia, and heart disease. Although pH measurements of the fog were not recorded at the time, present-day scientists estimate that the fog was a highly diluted sulfuric acid mist with a pH ranging from 1.4 to 1.9! But that was more than six decades ago. Is acid deposition still a killer? According to the Office of Technology Assessment, more than 50,000 premature deaths are caused annually by sulfate-laden air. Acid precursors are the second largest cause of lung disease (after smoking), in the opinion of Dr. Phillip J. Landrigan of the Mount Sinai School of Medicine in New York City.

Toxic metals in drinking water can be harmful if ingested by humans. Today more than 40 million people in the United States are drinking water with lead levels above 20 parts per billion, the level considered safe by the Environmental Protection Agency (EPA). Where did the lead come from? It apparently was leached from water pipes and lead solder in copper pipes by the acidic drinking water. Such intake of lead by millions of Americans is cause for considerable concern because medical researchers have shown lead to be one of the factors responsible for high blood pressure and heart attacks in adults and brain damage in children.

Some health experts believe that acid deposition may also be indirectly responsible for many cases of **Alzheimer's disease.** This is a disease of the elderly characterized by degeneration of the brain and severe loss of memory. Chemical analyses of the brains of people who died from the disease have revealed relatively high levels of aluminum. Some researchers believe that aluminum may have caused the disease. If so, where did it come from? How did it get into the body? The involvement of acid deposition is strongly suggested.

In Canada, more than 80% of the population lives in regions in which precipitation is high in acid. According to Environment Canada, thousands of people in this area have respiratory problems that may be caused by acid precursors. Children are especially vulnerable. In communities whose air is polluted by acids, sulfates, and nitrates, children suffer from more chest colds, allergies, and coughs.

Controlling and Eliminating Acid Deposition

As with most issues, there are many ways to reduce the problem. This section looks at many of these solutions. It also examines ways to prevent the emission of acid precursors and eliminate the problem altogether. Many of these solutions will be further elaborated in Chapters 22 and 23, which discuss alternative energy strategies.

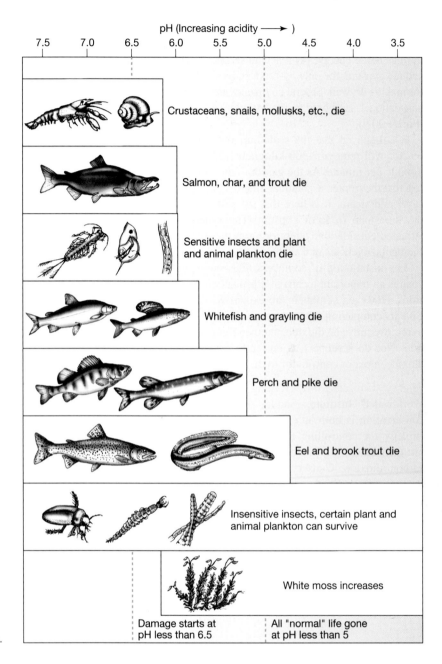

pH (Increasing acidity ⟶)

Crustaceans, snails, mollusks, etc., die

Salmon, char, and trout die

Sensitive insects and plant and animal plankton die

Whitefish and grayling die

Perch and pike die

Eel and brook trout die

Insensitive insects, certain plant and animal plankton can survive

White moss increases

Damage starts at pH less than 6.5

All "normal" life gone at pH less than 5

FIGURE 20.8 Effect of water acidity on aquatic organisms.

Switching From High-Sulfur to Low-Sulfur Coal The coal once burned in electric generating plants and industrial boilers varied greatly in sulfur content, from less than 1% to almost 6%. Switching to low-sulfur coal cuts emissions substantially, and many utilities in the United States and Canada have chosen this strategy. Rather than install expensive pollution control devices, American companies were able to cut sulfur emissions as required by the Clean Air Act of 1990 (Chapter 18). Many did so at a huge cost savings, too.

The question arises: Are our nation's supplies of low-sulfur coal sufficient for this purpose? Apparently, the answer is yes. A group of low-sulfur mining industries and environmentalists who have formed the Alliance for Clean Energy has determined that more than 14.1 billion metric tons of low-sulfur coal is available—adequate to meet the needs of our utilities and industries for many years.

As important as this strategy is, we must remember that switching to low-sulfur coal only minimizes sulfur dioxide. It does not affect nitrogen oxide emissions or carbon dioxide. At best, then, it is a temporary solution.

Using Smokestack Scrubbers As we discussed in Chapter 18 with regard to pollution control, **smokestack scrubbers** remove particulates and sulfur dioxide. Leonard Kreisle, professor of engineering at the University of Texas–Austin, recently announced the development of a new smokestack device for removing sulfur dioxide emissions. Known as a **synergistic reactor,** it has several advantages over

GO GREEN!

When staying in a hotel or motel in the winter, turn the heat down when you leave the room. In the summer, turn the thermostat setting up to save energy.

the smokestack scrubbers used to date: (1) it removes 100% of the sulfur dioxide; (2) it is much smaller; (3) it uses only one-third as much energy; (4) it acts in seconds rather than minutes; and (5) gypsum, the only by-product, has high commercial value for making drywall. Several companies are now using waste from conventional scrubbers to produce drywall for building homes (Figure 20.9).

It is believed that the scaled-up version of the synergistic reactor will remove 2,300 kilograms (5,000 pounds) of sulfur dioxide per minute. As the gaseous emissions pass through the reactor, they enter a chamber filled with finely ground limestone and steam. It is here that the sulfur dioxide reacts with the limestone to form gypsum. The synergistic reactor has attracted considerable attention from the EPA, as well as environmental agencies in Canada, Europe, Russia, and Japan.

The installation of scrubbers, like switching to low-sulfur coal, is an important means of drastically reducing sulfur emissions. However, scrubbers are expensive to build and operate. Like its companion strategy, scrubber installation does nothing to remove nitrogen dioxide, another component of acid deposition. Nor do scrubbers as currently designed remove carbon dioxide, a major contributor to the greenhouse effect.

Energy Conservation Writing for the highly respected Worldwatch Institute, Sandra Postel argues that energy conservation is both an effective and a relatively inexpensive strategy for controlling the emissions of sulfur dioxide and nitrogen oxide, the precursors of acid deposition, as well as carbon dioxide. Conservation is, in fact, one of the most important strategies for creating a sustainable society. Consider a simple example.

In 1987, the U.S. Congress passed the **National Appliance Energy Conservation Act.** It required manufacturers to greatly reduce energy consumption by common appliances such as air conditioners, refrigerators, and water heaters. According to Postel, this law reduces the nation's consumption of electricity by 70,000 megawatts per year, which is equal to the output of 140 large coal-fired power plants. Postel estimates that if the United States could

cut its electrical consumption by half, a goal attainable with current technology, the nation could reduce its coal combustion by 85 million metric tons per year. The net result: a 4-million-metric-ton annual reduction in sulfur dioxide emissions (in 2005, we released nearly 15 million metric tons). Nitrogen dioxide emissions also would plummet, as would carbon dioxide releases. Moreover, this would cost only 1% of the $5 billion to $10 billion expenditure required to install smokestack scrubbers on power plants, which would have the same effect. Additional energy conservation techniques are discussed in Chapter 23.

Renewable Energy Yet another preventive measure, and one that will become increasingly important in years to come as the human population and global economy expand, is renewable energy—solar energy, wind power, and the like. Although pollution is generated to make solar panels, wind turbines, and other devices required to capture renewable energy, once operating, these devices produce no carbon dioxide, nitrogen dioxide, or sulfur dioxide. Overall, they produce very little air pollution, so they are an excellent means of preventing pollution. Chapter 23 outlines many of the options and discusses their pros and cons.

Federal Legislation Environmental problems caused by acids in rain and snow have been understood since the 1880s. Lawsuits over acid rain caused by smelters were tried in the 1920s. It was not until the 1970s, however, that the true magnitude of the problem became well known. Still, it took nearly another 20 years for the federal government to enact legislation, the Clean Air Act Amendments of 1990, to control acid deposition. As noted in Chapter 18, the **Clean Air Act Amendments of 1990** called for reductions in sulfur dioxide emissions. This legislation set limits on emissions from major sources such as power plants and developed a marketable-permit system, allowing companies that were able to cut sulfur dioxide emissions below the levels required by the law to sell allowances to others.

Although this market-based system has been heralded as a success, most companies pursued other avenues to meet

FIGURE 20.9 Scrubber waste can be used to make drywall that is used to build and remodel homes and offices.

emission standards. Low-sulfur coal was the option of choice. Many utilities were able to buy it cheaper or at the same price they were paying for higher-sulfur coal. Because of the law, they were able to get out of contracts for high-sulfur coal that had built-in escalation clauses—wording that ensured a steadily increasing cost for the coal they were buying. In short, many businesses benefited from the law, as did our lakes, rivers, and people. Although industries that fought restrictions on acid precursors claimed that the Clean Air Act would cost $3 billion to $11 billion a year, the real cost was just slightly more than $800 million a year nationwide.

Several states passed acid rain legislation prior to the Clean Air Act Amendments of 1990. In 1991, after years of foot-dragging, the U.S. government signed an agreement with Canada, known as the **Air Quality Accord,** to control the flow of acid pollutants across the border. In this agreement, total emissions of acid pollutants are capped at 13.3 million metric tons of sulfur dioxide for the United States and 3.2 million metric tons of sulfur dioxide for Canada. The agreement also calls for a reduction in nitrogen oxide emissions. Canada, like the United States, limits the release of sulfur dioxide from industrial sources. Most companies have opted to use low-sulfur coal to meet emissions requirements. Such efforts have resulted in some promising results. Clearwater Lake near Sudbury, ON, was once heavily acidified, with a pH of 4.1. Today, the water measures 4.7. As noted earlier, efforts in the United States are also starting to show some modest returns: reduced concentrations of acid precursors, reduced acid deposition, and limited recovery of lakes and streams.

GO GREEN!

When doing laundry, use a cold-water detergent and the cold-water setting on the washing machine. Adjust the water volume appropriately on the washing machine, too, and consider hanging clothes on a line to dry rather than using the dryer.

20.2 Depletion of Stratospheric Ozone

Ozone is an atmospheric pollutant that baffles many people. At ground level, it is a component of photochemical smog (Chapter 18). Here, it irritates the eyes and causes damage to lungs. You'll witness its effects on a polluted summer day in virtually any major metropolitan area. It's often referred to as ground-level ozone. However, ozone is also found in the upper atmosphere. It's a natural component of the stratosphere that serves a valuable purpose.

As you may know from previous studies, the Earth's atmosphere is divided into several layers. The lowest region, which is rich in oxygen, is called the **troposphere** (tropo means to nourish). It extends from the Earth's surface up to 10 kilometers (about 6 miles) in altitude (Figure 20.10). Virtually all human activities take place in the troposphere, because it contains the oxygen we need to survive. Mount Everest, the tallest mountain on the planet, is only about 9 kilometers high. The next layer, the **stratosphere,** continues from 10 kilometers (6 miles) to about 50 kilometers (about 30 miles). Most commercial airline traffic occurs in the lower part of the stratosphere. Most atmospheric ozone is concentrated in a layer in the stratosphere 15 to 30 kilometers (about 9 to 18 miles) above the Earth's surface. This is called the **ozone layer.** It contains a slightly higher concentration of naturally occurring ozone. What is ozone?

Ozone is a molecule containing three oxygen atoms. Its chemical formula is O_3. Normal oxygen, the kind we breathe, has two atoms (O_2). Ozone gas is blue and has a strong odor. Gaseous oxygen is colorless and odorless. Of the two, ozone is much less common. Out of each 10 million air molecules, about 2 million are normal oxygen, but only 3 are ozone.

All life on Earth is dependent on the presence of this ozone layer. Why? Ozone molecules filter out a portion of the sun's longer-wave **ultraviolet (UV) radiation,** known as ultraviolet B (UVB) radiation (Figure 20.10). Were this absorption screen of ozone not present, or were it to be thinned out, much more UVB radiation would reach the Earth's surface. The results, say the experts, would be catastrophic, a topic we will discuss in more detail shortly. This chapter deals with ozone in the stratosphere. Ground-level ozone was discussed in Chapter 18.

Until a few years ago, the ozone layer was in a state of dynamic equilibrium. In other words, the amount of stratospheric ozone remained fairly constant, although concentrations might vary by 10% from year to year. Moreover, the amount of ozone in the stratosphere might be two or three times greater in one place than another.

In 1974, F. S. Rowland, of the University of California, startled the scientific community with a prediction that the ozone layer could be in danger. He speculated that chemicals known as **chlorofluorocarbons (CFCs)** could be depleting the ozone layer. These chemicals were used as propellants in spray cans containing deodorants, hair spray, shaving cream, and insecticides. They were also used as coolants in refrigerators and air conditioners. What is

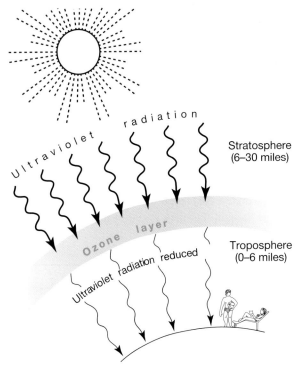

FIGURE 20.10 The ozone layer in the stratosphere shields the Earth from potentially lethal ultraviolet (UV) radiation.

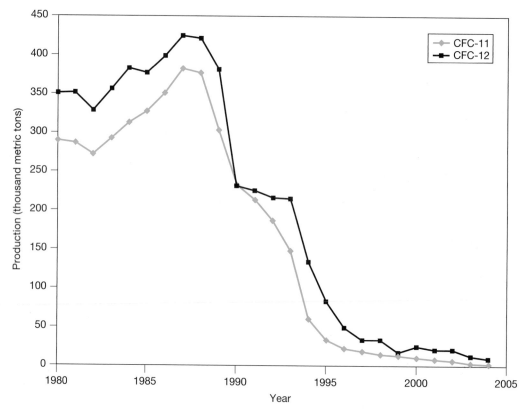

FIGURE 20.11 World production of CFCs, from 1980 to present.

more, it was previously thought that these chemicals were inert—that is, chemically unreactive.

Widespread concern soon developed, because the total amount of CFCs released into the global atmosphere was considerable. In 1988, for example, CFC production reached an all-time high of 1.1 million metric tons (1.2 million tons) (Figure 20.11).

Chlorofluorocarbons are rather simple chemicals. They have a central carbon atom to which are attached chlorine and fluorine atoms (Figure 20.12). Two CFC compounds were once commonly used in the United States and abroad, with DuPont being the major producer. One compound is CFC-11 ($CFCl_3$), a propellant used in spray cans, and the other is Freon-12 (CF_2Cl_2), a refrigerant. Both of these compounds are extremely stable and inert under ordinary environmental conditions close to Earth. As a result, they will not react chemically either with the contents of the spray can or with the can itself. It was thought that they wouldn't react with chemicals

in the environment either, a characteristic that made these substances desirable as propellants.

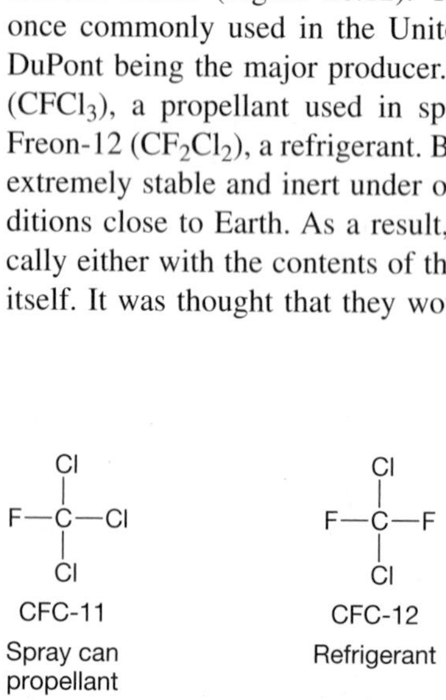

But the propellant molecules used by men to spray foamy shaving cream on their beards escaped through the bathroom window and then floated into the sky. Within five years or fewer, these CFC molecules made their way into the stratosphere.

Leaky refrigerator cooling systems or automobile air conditioners also emitted CFCs that escaped into the atmosphere. Once in the stratosphere, though, these ordinarily stable compounds were exposed to intense radiation from UV light. This caused the molecules to decompose and release chlorine free radicals, which are highly reactive atoms. We shall follow this process by using $CFCl_3$ as an example:

$$CFCl_3 + UV\ light \rightarrow Cl\cdot\ (chlorine\ free\ radical)$$

The chlorine free radicals then react with ozone (O_3), causing it to form into molecular oxygen (O_2). The reaction is as follows:

$$\underset{\text{(ozone)}}{O_3} + \underset{\substack{\text{(chlorine}\\\text{free radical)}}}{Cl\cdot} \rightarrow \underset{\substack{\text{(chlorine}\\\text{oxide)}}}{ClO} + \underset{\text{(oxygen)}}{O_2}$$

This reaction removes ozone from the ozone layer. Further reactions, too complex to include here, regenerate the chlorine free radical so that it can react with additional ozone molecules. As a result, one chlorine atom can cause the breakdown of 100,000 molecules of O_3.

CFC-11
Spray can
propellant

CFC-12
Refrigerant

FIGURE 20.12 Chemical formulas of two common CFCs, CFC-11 and CFC-12.

When the CFCs were first synthesized in the 1920s, they seemed too good to be true. They are stable, nonflammable, and nontoxic. Moreover, they vaporize at low temperatures. Furthermore, they could be produced very cheaply. As a result, they had many uses. Not only were they used in spray cans, refrigerators, and air conditioners, as noted earlier, they were used as solvents and as sterilants for medical equipment. They were even used as blowing agents (to produce bubbles) in plastic foams. For many years, the McDonald's fast-food chain used plastic foam containers. Most of the CFCs in the various products and applications have now found their way into the atmosphere. Over time, plastic foam burger containers released their CFCs, as did malfunctioning refrigerators. Junked refrigerators rusted and deteriorated, releasing CFCs into the air and eventually contributed to ozone depletion.

CFC Accumulation and Thinning of the Ozone Layer

The atmospheric concentrations of the ozone-depleting gases are measured each day by scientists in Oregon, Ireland, Tasmania (off the coast of Australia), Barbados (northeast of Venezuela), Samoa (in the South Pacific), New Zealand, and other countries. Figure 20.13 shows the results of these measurements. As you can see, the concentrations of ozone-depleting chemicals rose steadily from 1950 until the 1990s.

Because the average life span of a CFCl$_3$ molecule is about 75 years and that of CF$_2$Cl$_2$ is 110 years, CFCs can disperse throughout the stratosphere over the entire globe. As a result, they have become widely dispersed over both heavily industrialized and remote areas. In fact, concentrations in the heavily developed Northern Hemisphere are nearly identical to those in the much less developed and much less populated Southern Hemisphere.

Has the accumulation resulted in a decline in the ozone layer? Yes. Studies show that ozone concentrations have declined in both the Northern and Southern Hemispheres. As shown in Figure 20.14A, stratospheric ozone concentrations

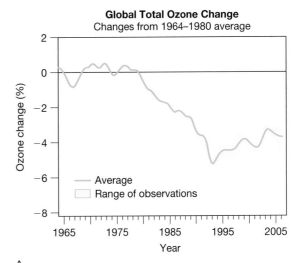
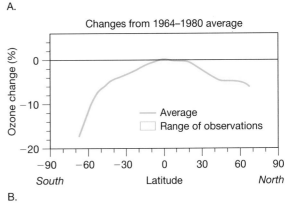

B.

FIGURE 20.14 (A) Ozone levels have declined throughout the world, primarily as a result of the release of CFCs. (B) Declines in the ozone concentration by latitude.

declined rapidly from 1980 to 1993. The greatest changes have occurred over the Southern Hemisphere and at the North and South Poles, as shown in Figure 20.14B.

The most noticeable declines have been in Antarctica. In 1979, atmospheric scientists were surprised and mystified by the appearance of a gigantic "hole" in the ozone layer during the fall and winter months over Antarctica (Figure 20.15). The level of ozone in this hole, which extended over the entire Antarctic continent, was 50% below normal. In 1988, the Ozone Trends Panel of the National Aeronautics and Space Administration (NASA) concluded, "The weight of evidence strongly indicates that man-made chlorine compounds are primarily responsible for the hole." Recent studies indicate that the ozone breakdown is accelerated in the presence of ice crystals and sunlight.

Although the greatest amount of ozone thinning has been caused by man-made CFCs, natural events, such as a volcanic eruption, sometimes contribute to the depletion of ozone. In 1992, for instance, Mount Pinatubo in the Philippines erupted and injected untold trillions of tiny sulfate droplets into the atmosphere. This caused a record decline in the ozone hole over Antarctica, according to David J. Hofmann of the National Oceanic and Atmospheric Administration. More than 70% of the ozone blanket was destroyed in a 23-million-square-kilometer area—almost the size of the entire North American continent. How? Scientists hypothesized that trillions of tiny

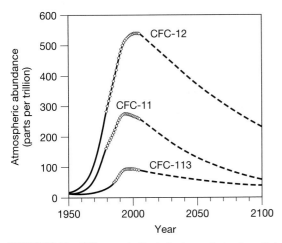

FIGURE 20.13 CFC concentration in the lower stratosphere. Note that CFC concentrations are now starting to decline.

Antarctic ozone hole

1979

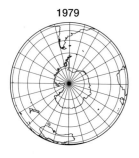

1986

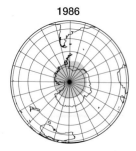

2000

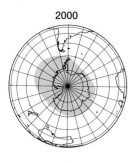

2006

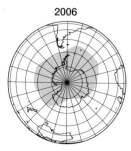

FIGURE 20.15   Growth of the ozone hole (shaded area) over Antarctica. The size of the ozone hole has remained more or less constant since 2000.

Changes in Surface Ultraviolet Radiation

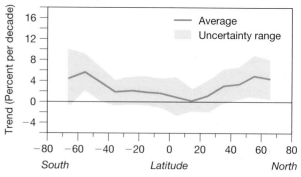

FIGURE 20.16   Increase in ultraviolet radiation striking the Earth.

ozone concentrations has resulted in a substantial increase in ultraviolet radiation at ground level. As shown, the increase varies. The greatest increases occur over the poles; the lowest increases occur near the equator. Check out the latitude of your home, and determine the average increase.

Harmful Effects of UVB Radiation

Because CFCs take many years to migrate into the ozone layer and because they last so long, it could take 50 to 100 years for the ozone layer to mend. Only time will tell for sure. In the meantime, what can we expect from the loss of ozone?

Like many natural components of our environment, UVB light is beneficial in small amounts. UV radiation tans light skin and stimulates vitamin D production in the skin. However, excess UVB exposure can cause serious problems. For example, it can cause serious skin burns, cataracts (clouding of the lens of the eye), and skin cancer in humans (Figure 20.17). Increased exposure to UV light may also suppress the human immune system, making us more susceptible to infectious diseases.

EPA researchers estimate that each 1% decrease in the concentration of ozone in the ozone layer will lead to a 0.7% to 2% increase in UVB light striking the Earth. This, in turn, would

sulfate droplets injected by the eruption provided the surface on which the ozone-depleting reactions could occur.

As the ozone layer declined, would the ground levels of dangerous ultraviolet radiation increase? Some scientists have maintained that even if the protective ozone blanket was partially destroyed, the increase of UVB radiation would be absorbed by clouds and atmospheric pollution. In some areas, this has been true. However, two Canadian scientists reported that the amount of UVB radiation over Toronto, ON, increased 5% each year from 1989 to 1993, coincident with the annual thinning of the ozone layer. More and more studies are showing similar results. Figure 20.16 shows that the decrease in

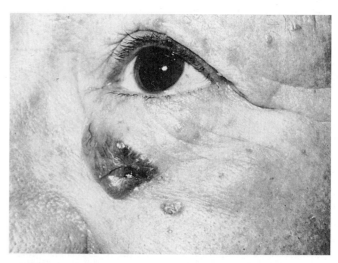

FIGURE 20.17   Skin cancer caused by excessive exposure to UV radiation in sunlight. Thinning of the ozone shield will cause a sharp increase in the incidence of such cancers.

lead to an increase in skin cancer rates of about 4%. The EPA estimates that ozone depletion will result in approximately 200,000 cases of skin cancer in the United States over the next five decades. Worldwide, the number is much higher. Especially hard hit will be countries in the Southern Hemisphere such as New Zealand and Australia. In northern Australia, one city has already reported a dramatic increase in skin cancer.

In November 1991, a panel of scientists convened by the United Nations released a report on the predicted effects of ozone depletion throughout the atmosphere. They estimated that a 10% decrease in stratospheric ozone concentrations would cause 300,000 additional cases of skin cancer per year worldwide. They predicted that it will also cause an additional 1.6 million cases of cataracts each year.

Studies of skin cancer show that light-skinned people are much more sensitive to UVB radiation than more heavily pigmented individuals. In addition, some chemicals commonly found in drugs, soaps, cosmetics, and detergents may sensitize the skin to UVB radiation. Thus, exposure to sunlight may increase the incidence of skin cancers among light-skinned people and users of many commercial products.

Most skin cancers have a low mortality, but because so many people will contract it, thousands of people will die from our use of CFCs. Many others will undergo surgery to remove tumors.

The study of the effects of UVB radiation on natural ecosystems such as forests, grasslands, lakes, streams, and estuaries has only recently begun. Studies of shallow-water ecosystems, however, indicate that UVB radiation can severely depress populations of phytoplankton, small crustaceans, and larval fish. Thus, say scientists, the depletion of the ozone layer will result in a diminished supply of animal protein for millions of hungry people in poor nations, where malnutrition is already a way of life.

Land- and water-dwelling plants could also suffer from increasing UVB radiation. Intense UVB radiation is usually lethal to plants; smaller, nonfatal doses damage leaves, inhibit photosynthesis, cause mutations, and stunt growth. Declining ozone and increasing UVB radiation could also cause dramatic declines in commercial crops such as corn, rice, and wheat, costing billions of dollars a year. It may also affect certain commercially valuable tree species such as the loblolly pine.

In a hearing before the U.S. Senate in November 1991, Susan Weiler, head of the American Society of Limnology and Oceanography in Walla Walla, WA, testified that studies in Antarctica by scientists have shown that populations of phytoplankton (algae and other free-floating photosynthetic organisms) decrease about 6% to 12% when stratospheric ozone concentrations over the region drop by 40%. Since phytoplankton form the base of the aquatic food chain, damage to them could cause widespread ecological problems. One scientist thinks that ozone depletion and subsequent effects on the food chain may be the reason why two species of penguin are declining in the Antarctic.

Finally, UVB light is harmful to many products. Paints, plastics, and other materials deteriorate when exposed to UV radiation. Losses from further decreases in the ozone layer could cost society enormous amounts of money.

The Ban on Ozone-Depleting Chemicals

In the 1970s, fears caused by early projections of ozone depletion moved several nations, including the United States, Sweden, Finland, Norway, and Canada, to cut back on CFC-11. In 1978, for example, the United States banned the CFC used in spray cans. CFC-12, a refrigerant, coolant, and blowing agent, was not affected by the ban.

But the continuing accumulation of scientific evidence on the decline of the ozone layer made it evident a decade later that worldwide cooperation was needed. In 1987, the United Nations sponsored negotiations aimed at reducing global CFC production. In September of that year, 24 nations signed a treaty, called the **Montreal Protocol,** which would cut production of five CFCs in half by 1999 and freeze production of halons (used in fire prevention systems) at 1986 levels. (In halons, bromine atoms replace some or all of the chlorine atoms.) Although halons are used in much smaller quantities worldwide, they are far more effective in destroying ozone than CFCs.

This agreement paved the way for a gradual decline in CFC production in the industrial nations. But critics argued that it had too many loopholes. Like so many other pollution control strategies, it would only slow the rate of destruction, not stop it. EPA computer projections showed that an 85% reduction in CFC emissions was needed to stabilize CFC levels in the atmosphere.

Before the Montreal Protocol went into effect, something unusual happened. In March 1988, an international panel announced that ozone levels had fallen throughout the world. Two weeks later, DuPont, a leading major producer of CFCs, called for a total worldwide ban on CFC production. Two weeks earlier, the company had said it would not support a ban.

Continuing bad news about ozone depletion brought negotiators to the table once again, this time in London, where in June 1990 they reached a new agreement. This treaty was signed by 93 nations and called for the complete elimination of CFCs and halons by the year 2000, if substitutes were available by then. The signatories also agreed to phase out other ozone-depleting chemicals, among them carbon tetrachloride, methyl chloride, and even the hydrofluorocarbons (HCFCs), a class of chemicals (described in the next section) once thought to be an excellent substitute for CFCs.

The news about the ozone layer continued to worsen. In 1992, a team of 40 scientists announced record-high concentrations of chlorine monoxide (ClO, which is produced from chlorine free radicals from the breakdown of CFCs) in the air above New England and Canada. Concentrations such as these had never been seen before, even in the Antarctic ozone hole. If levels of chlorine monoxide continue to climb, chances are good that a severe Arctic ozone hole will begin to appear with great regularity, exposing Canada and parts of the United States, Europe, and Asia to dangerous levels of UV radiation.

Aircraft measurements in 1992 also showed rather disturbing findings about global ozone outside the Arctic. In flights as far south as the Caribbean, scientists detected ClO concentrations of up to five times the amount they had anticipated.

In 1992, the nations of the world met in Copenhagen to sign another agreement calling for an acceleration of the phaseout of CFCs, carbon tetrachloride, and other ozone-depleting chemicals within four to nine years. The projected effects of global efforts to phase out CFCs are shown in Figure 20.18, a graph of projected concentrations of ozone-depleting compounds.

Nearly a decade later, some encouraging results are beginning to appear. The production of CFCs and other ozone-depleting compounds has been greatly curtailed, and CFC-11 concentrations in the lower atmosphere have begun to decline. CFC-12 concentrations appear to be leveling off.

Substitutes for Ozone-Destroying CFCs

The development of replacement chemicals played a big role in the dramatic change in public policy regarding CFC production. It gave industry options and, in some cases, opportunities to profit from the shift to less harmful chemicals. Manufacturers have pursued two basic options: the use of less stable CFC compounds, which break down before they reach the stratosphere, and the production of non-CFC chemicals to be used as substitutes.

Consider the first option. By adding a hydrogen atom to the stable CFC molecule, researchers can make a group of CFCs, technically known as HCFCs (hydrochlorofluorocarbons), that break up in the lower atmosphere. In theory, the chlorine atoms released during this process are less likely to reach the stratosphere. In practice, some do, but they cause less damage. That

is, they still deplete the ozone layer. Consequently, they are viewed as interim solutions.

Several less stable CFCs are already on the market. One of these, HCFC-22, is now used as a coolant in some home air conditioners. HCFC-22 is 20 times less destructive than the CFC-12 currently found in older refrigerators and automobile air conditioners. Because of their effect on the ozone layer, though, the HCFCs are slated to be phased out by 2030.

The second most widely used CFC was CFC-11. Until recently, it was primarily used as a blowing agent for foam and, outside of the United States and a few other countries, as a propellant in spray cans. HCFC-123 has been touted as a possible replacement. Manufacturers of some types of foam insulation have eliminated it entirely.

Perhaps one of the most difficult challenges is finding a replacement for CFC-113. This compound is an all-purpose cleaner for circuit boards produced for the computer industry. Because CFC-113 was not being considered for banning until a few years ago, the industry had not actively pursued replacements. At the signing of the Montreal Protocol, in fact, work to find a substitute had not even begun.

In January 1988, researchers announced the development of a compound called BIOCAT EC-7. It may be a partial replacement for CFC-113. This substance, isolated from orange peels, is very similar to kerosene and turpentine. EC-7 may replace a sizable share of the CFC-113 market, but it has its limitations. It is not versatile and is flammable. Industry representatives believe that no single compound will replace CFC-113 completely. Although options are available, intense research is under way to find ozone-friendly substitutes.

The Good News and Bad News About Ozone

The ozone story is one of humankind's greatest success stories. It not only illustrates how scientific knowledge can be used by society for the common good of humankind and all other species, it also shows how quickly changes can be made to bring about an end to actions responsible for the destruction of the global environment. Already, studies are showing that the rate of increase in the concentration of ozone-depleting chemicals is slowing. That's the good news. Unfortunately, however, many millions of tons of CFCs have already been released into the atmosphere. CFCs take 15 years or so to migrate into the stratosphere, and some CFCs last 75 to 100 years in the atmosphere. Because of this, scientists predict that the ozone layer will thin some more before it begins to improve. At least 100 years will be required to return the ozone layer to 1985 levels. Another 100 to 200 years may be needed for full recovery. A lot of people will contract skin cancer in the interim. Making matters worse, many ozone-depleting chemicals (and their replacements) are greenhouse gases and thus contribute to global warming and climate change.

Supporting the belief that we're not out of the woods yet, since 1979 ozone levels in the stratosphere have declined 4% to 6% over midlatitudes and 10% to 12% over higher southern latitudes. The ozone hole of 1997 and 1998 continues to

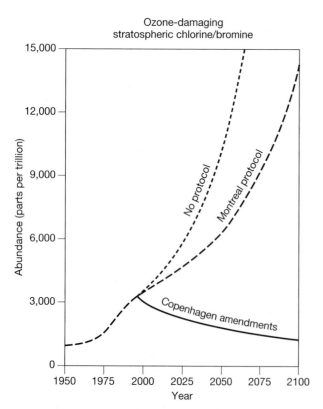

FIGURE 20.18 Projected concentration of CFCs as a result of no agreement, the Montreal Protocol, and the most stringent accord, the Copenhagen Amendments. This graph shows the value of a good treaty and concerted world effort to stop an environmental nightmare.

present itself, encompassing 27.3 million square kilometers (10.5 million square miles). The hole in 1998 was the largest ever. In 2001, it had shrunk slightly to 25 million square kilometers. Clearly, we've made a huge mess that will take a long time to correct itself.

Summary of Key Concepts

Acid Deposition

1. Acid deposition is the deposition of acidic compounds in either wet or dry form.

2. Wet deposition consists of acids deposited by rain, snow, fog, dew, or frost.

3. Dry deposition of acid-forming materials takes place when dust particles containing nitrates and sulfates settle on the soil, water, or vegetation. These chemicals may react with water to form sulfuric or nitric acid. Acid deposition also includes the absorption of nitrogen and sulfur dioxide gases onto solid surfaces. These then combine with water to form acids.

4. The pH scale ranges from 0 to 14. A substance with a pH of 7 is neutral, neither acidic nor basic. Substances with pH values above 7 are alkaline or basic. Substances with pH values below 7 are acidic.

5. The pH of normal, nonpolluted rain is about 5.6. This slight acidity is created when small amounts of carbon dioxide in air are dissolved in the atmospheric moisture, forming carbonic acid.

6. Rainfall in much of the northeastern United States, southeastern Canada, and Scandinavia has a pH of 4.5 or lower because of the release of oxides of sulfur and nitrogen from upwind power plants, cities, and industrial sites.

7. Acid precursors and acids may be transported hundreds of miles. Because of this, emissions generated in one country may be transported by winds to others. The control of the acid rain problem therefore requires international cooperation.

8. Among the adverse effects of acid deposition on aquatic ecosystems are the following: (1) a reduction or termination of reproduction in many aquatic organisms, (2) disruption of the homing ability of salmon, (3) abnormal development of fish embryos, (4) a decline in fish food organisms, (5) the loss of desirable species of fish such as bass and pike and an increase in less desirable species such as carp and bullheads, (6) increased levels of aluminum, which interferes with normal gill function in fish, (7) an increase in mercury levels in fish to the point that they cannot be safely eaten, and (8) an inhibition of bacterial decay, which locks up nutrient elements in the dead remains of aquatic organisms.

9. Acid deposition has been addressed by national and international agreements to limit sulfur dioxide emissions at power plants, factories, and other stationary sources. Most sources have opted to switch to low-sulfur coal, although some have installed smokestack scrubbers.

10. State, national, and international efforts to reduce acid deposition have resulted in a reduction in acid precursors and acid deposition. Some lakes and rivers have shown some improvement. However, research indicates that recovery will not be possible for many areas that continue to experience acid deposition. Reductions in sulfur dioxide and nitrogen oxide emissions are insufficient to permit full recovery and will not prevent further acidification in some regions.

Depletion of Stratospheric Ozone

11. Ozone is a pollutant at ground level, as explained in Chapter 18, but a desirable chemical in the stratosphere because it blocks dangerous ultraviolet radiation.

12. The UV radiation produced by the sun causes skin cancer in humans. However, under natural conditions, we are shielded from most of the sun's UV radiation by ozone in the ozone layer, part of the stratosphere.

13. In recent years, this ozone shield has begun to be depleted by gases such as chlorofluorocarbons (CFCs) from human sources.

14. CFCs have been used as spray-can propellants, refrigerants, cleaning agents, and blowing agents.

15. Widespread use of CFCs caused a dramatic increase in their concentration in the stratosphere, which has resulted in a substantial decline in ozone levels and an increase in ultraviolet radiation striking the Earth.

16. The thinning of the ozone layer has many adverse effects on people and the environment. It increases the incidence of skin cancer, increases the frequency of cataracts, and may reduce people's ability to fight off bacterial infections. It has a harmful impact on plant life. Corn, cotton, and wheat yields may drop. Aquatic ecosystems are harmed as well.

17. In 1987, many nations agreed to reduce CFC emissions, but further research showed that the goals set in this agreement were not sufficient to prevent widespread damage. The agreement was modified in 1990 and tightened again in 1992, at which time the United States and 23 other nations signed an agreement to halt CFC production.

18. While significant progress has been made in reducing the production of ozone-depleting compounds, it may take 50 to 100 years for the ozone layer to recover, because CFCs have an extremely long life span, lasting up to 100 years.

Key Words and Phrases

Acid Deposition
Acid Precursors
Acid Rain
Acid Shock
Air Quality Accord
Alzheimer's Disease
Anthropogenic Source of Air Pollution
Buffer
Chlorofluorocarbons (CFCs)
Clean Air Act Amendments of 1990
Dry Deposition
Montreal Protocol
National Appliance Energy Conservation Act
National Atmospheric Deposition Program
Nitric Acid
Nitrogen Dioxide
Ozone
Ozone Layer
pH Scale
Skin Cancer
Smokestack Scrubbers
Stratosphere
Sulfur Dioxide
Sulfuric Acid
Synergistic Reactor
Troposphere
Ultraviolet (UV) Radiation
Wet Deposition

Critical Thinking and Discussion Questions

1. What is acid deposition? How is it formed? What are the major sources? How do you contribute to this problem?

2. Discuss the harmful effects of acid deposition on aquatic ecosystems, soils, forests, materials, and human health.

3. Lakes A and B are located only 100 kilometers (60 miles) apart and receive the same amount of precipitation, the average annual pH of which is 4.5. Yet lake A is devoid of fish, whereas lake B abounds with them. Explain.

4. Discuss three strategies that can be used to bring the acid deposition problem under control. Which ones are the most sustainable? Why?

5. What can you do personally to reduce acid deposition? Will these efforts also help combat other forms of air pollution, such as greenhouse gas emissions and photochemical smog?

6. Using your critical thinking skills and knowledge you've gained in this course, analyze the following statement: "Solutions to regional and global environmental pollution are too costly. We can't afford them."

7. Discuss the following statement: "Ozone may be both beneficial and harmful to human health."

8. What is the ozone layer? How is it being altered?

9. What steps have been taken to protect the ozone layer?

10. Why is international cooperation needed to control the problems of global warming, acid deposition, and ozone thinning in the stratosphere?

11. Many steps have been taken to reduce the production and release of ozone-depleting compounds, but the ozone layer continues to thin. Why?

12. How long will it take for the ozone layer to heal? Why?

Suggested Readings

Anderson, S. O., and K. M. Sarma. 2005. *Protecting the Ozone Layer: The United Nations History.* New York: Earthscan. In-depth reading for anyone interested in learning about the history of this intriguing problem.

Driscoll, C., et al. 2003. Nitrogen Pollution: Sources and Consequences in the U.S. Northeast. *Environment* 45(7): 8–22. Detailed account of nitrogen pollution primarily from acid deposition.

Jenkins, J. C., K. Roy, C. Driscoll, and C. Buerkett. 2007. *Acid Rain in the Adirondacks: An Environmental History.* Ithaca: Cornell University Press. An excellent review of the history and science behind acid deposition in the northeastern United States.

Lisensky, G. C., R. Hulet, M. Beur, and S. Anthony. *Soil Equilibria: What Happens to Acid Rain.* New York: W. W. Norton. An advanced reading for those who want to study the effects of acid depositon on soils.

Martins, J. 2006. *Ultraviolet Danger: Holes in the Ozone Layer.* New York: Rosen Publishing Group. Worthwhile reading for those who want to learn more.

McDonal, A. 1999. Combatting Acid Deposition and Climate Change: Priorities for Asia. *Environment* 41(3): 4–11, 34–41. A great perspective on what other countries are doing and have to do to address these problems.

Moran, J. M., and M. D. Morgan. 1994. *Meteorology: The Atmosphere and the Science of Weather,* 4th ed. New York: Macmillan. Excellent nontechnical coverage of global pollution problems.

Morgan, S., and J. Vaughan. 2007. *Acid Rain.* London: Franklin Watts. A great book for teachers who want to teach this subject to youngsters.

Nierenberg, D. 2001. Nitrogen: The Other Cycle. *World-Watch* 14(2): 30–38. A must-read for those who want to learn more about the effects of nitrogen pollution, especially nitrogen oxides.

Parks, P. J. 2005. *Our Environment: Acid Rain.* New York: KidHaven Press. A great book for teachers whose students range from 9 to 12 years of age.

Parson, E. A. 2003. *Protecting the Ozone Layer: Science and Strategy.* New York: Oxford. A fascinating look at the social and scientific history of the effort to protect the ozone layer.

Roberts, L. 1991. Acid Rain Program: Mixed Review. *Science* 252(5004): 371. Penetrating analysis of the merits of the National Acid Precipitation Assessment program.

Web Explorations

Online resources for this chapter are on the World Wide Web at: **http://www.prenhall.com/chiras** (*click on the Table of Contents link and then select Chapter 20).*

21

MINERALS, MINING, AND A SUSTAINABLE SOCIETY

The Earth's mineral wealth has been tapped for thousands of years (Figure 21.1). However, where humans once mined the Earth's surface with primitive tools to extract valuable minerals, today huge mining machines extract mineral resources to support a society so dependent on minerals that a shortage of any one of a few dozen would bring it to its knees. The sheer magnitude of **mining** and its impact on the environment, combined with our dependence on minerals, make mining and mineral production an issue of extraordinary importance. Truly, the long-term future of modern civilization depends on the way we manage the Earth's mineral resources.

This chapter addresses two basic issues of concern: the supply of minerals and the impacts of mineral mining and processing. It answers four key questions crucial to modern society: (1) Are we running out of minerals? (2) Can we expand our mineral supplies? (3) What are the environmental impacts of our mineral-intensive lifestyle? (4) How can we create a more sustainable system of mineral extraction and production?

21.1 Supply and Demand

The automobile is, perhaps, the most visible sign of the industrialized world's dependence on minerals. In the United States, the automobile industry uses enormous amounts of metals, refined from minerals. Approximately 7% of all the copper, 10% of the aluminum, 13% of the nickel, 20% of the steel, 35% of the zinc, and 50% of the lead used by the United States each year goes into automobile manufacturing.

These minerals come from widely scattered parts of the world. Copper, for instance, comes from mines in Arizona, Chile, and Canada. Aluminum ore (bauxite) comes to the United States from Japan and Canada. Nickel is shipped from Australia, Norway, Botswana, and Canada. Iron ore comes primarily from U.S. mines, but also from mines in Canada, Liberia, and Brazil. Lead comes mostly from U.S. mines as well, with smaller contributions from other nations.

Some Features of Minerals

Unlike forests, wildlife, fisheries, and even soil, minerals are nonrenewable. That is, minerals are—like oil, natural gas, and coal—finite or limited resources. They're not being regenerated. Each aluminum can tossed in a garbage can and later buried in a landfill depletes the world's supply of aluminum. But unlike oil and coal, minerals and metal can

FIGURE 21.1 Open-pit mine in the western United States. This mine is a blatant reminder of the damage humans create in supplying their needs.

be recycled over and over, thus greatly extending their lifetimes. We say they're infinitely recyclable. Unfortunately, many countries recycle only a small percentage of their scrap metals.

The great majority of minerals now used by society are extracted from the Earth's crust, the outer layer of the planet, which extends 24 kilometers (15 miles) below the Earth's surface. Some valuable minerals in the Earth's crust, such as gold, occur in a pure or elemental form. Most minerals, however, exist as chemical compounds consisting of two or more elements. For example, copper exists most commonly in the form of copper sulfide (CuS). Aluminum exists as aluminum oxide (Al_2O_3), and lead as lead carbonate ($PbCO_3$). Most of these compounds are found in rock that contains other minerals as well. A rock containing important minerals is called an **ore**—for example, iron ore and aluminum ore (bauxite). Extraction of metals from ore requires crushing the rock and treating it with heat or chemicals.

Although there are over 2,000 minerals in the Earth's crust, only a small number of them are abundant enough to be economically worth extracting. A deposit rich enough to be mined is called an **ore deposit.** Ores may be classified as high or low grade, depending on their mineral concentration. For instance, copper ore containing 3% copper is considered a high-grade ore, whereas copper ore with only 0.3% copper is a low-grade ore.

U.S. Mineral Production and Consumption

The United States annually mines ores worth an estimated $64 billion. When refined into metals, these ores are worth $542 billion, or about 4% of the gross national product (the total value of all goods and services produced by the nation, including income of U.S. companies operating in other countries). The leading mining states are Texas, Louisiana, California, and West Virginia. The leading mining nations are Canada, Australia, Russia, and the United States.

Although the United States has less than 5% of the world's population, it consumes about 20% of the world's nonfuel minerals each year. Enormous amounts of steel, copper, aluminum, and other metals are used to produce a variety of products that give Americans one of the highest standards of living in the world. To raise the rest of the world to our level of material wealth would require staggering amounts of minerals. (Mathis Wackernagel from the California-based nonprofit organization Redefining Progress estimates that we'd need three more planets for everyone in the world to live by U.S. standards!)

The United States' prodigious use of minerals has made it a great and prosperous nation, as well as one of the most vulnerable, because many of the materials it depends on for commerce and defense are imported from politically volatile regions of Africa (Figure 21.2). Minerals of extraordinary importance to an economy and a nation's national defense are called **strategic minerals.** So important are they that they're stockpiled (held in strategic reserve) in case of a disruption in their supplies. The United States maintains a supply of numerous strategic metals, including lead, copper, cobalt, and bauxite. In most cases, the goal is to have a two-year supply.

The United States imports many minerals and metals in large quantity. Some experts warn that widespread dependence on these foreign sources—especially the politically volatile nations—could prove disastrous. Civil unrest or war between nations can temporarily halt the flow of valuable minerals.

Stockpiles may help in the short run but are not a long-term solution to vanishing mineral supplies. Nor is stockpiling an adequate solution to the possible development of mineral **cartels**—groups of mineral-exporting nations that band together to control supplies and prices of strategic minerals. China, for instance, controls most of the world's tungsten. What would happen if it and other tungsten-exporting nations joined forces to control tungsten exports to the United States, Japan, or England as the oil-producing nations have done? Or what would happen if Russia, which holds most of the world's palladium, decided to do the same?

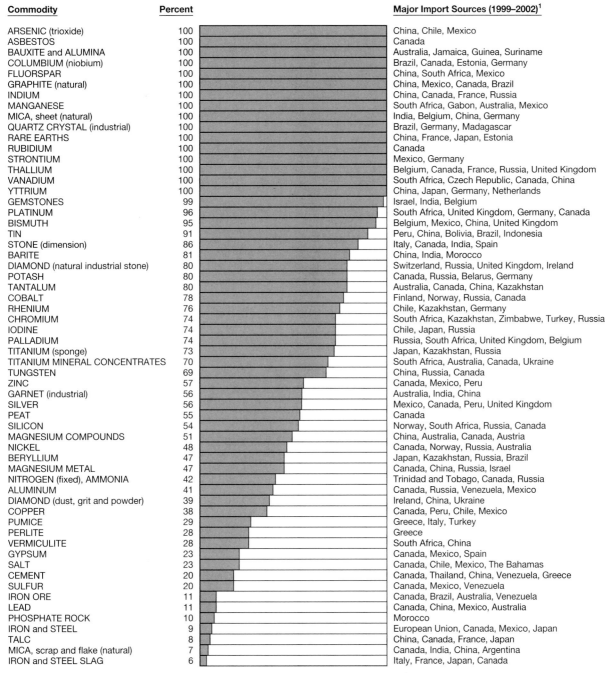

Commodity	Percent		Major Import Sources (1999–2002)[1]
ARSENIC (trioxide)	100		China, Chile, Mexico
ASBESTOS	100		Canada
BAUXITE and ALUMINA	100		Australia, Jamaica, Guinea, Suriname
COLUMBIUM (niobium)	100		Brazil, Canada, Estonia, Germany
FLUORSPAR	100		China, South Africa, Mexico
GRAPHITE (natural)	100		China, Mexico, Canada, Brazil
INDIUM	100		China, Canada, France, Russia
MANGANESE	100		South Africa, Gabon, Australia, Mexico
MICA, sheet (natural)	100		India, Belgium, China, Germany
QUARTZ CRYSTAL (industrial)	100		Brazil, Germany, Madagascar
RARE EARTHS	100		China, France, Japan, Estonia
RUBIDIUM	100		Canada
STRONTIUM	100		Mexico, Germany
THALLIUM	100		Belgium, Canada, France, Russia, United Kingdom
VANADIUM	100		South Africa, Czech Republic, Canada, China
YTTRIUM	100		China, Japan, Germany, Netherlands
GEMSTONES	99		Israel, India, Belgium
PLATINUM	96		South Africa, United Kingdom, Germany, Canada
BISMUTH	95		Belgium, Mexico, China, United Kingdom
TIN	91		Peru, China, Bolivia, Brazil, Indonesia
STONE (dimension)	86		Italy, Canada, India, Spain
BARITE	81		China, India, Morocco
DIAMOND (natural industrial stone)	80		Switzerland, Russia, United Kingdom, Ireland
POTASH	80		Canada, Russia, Belarus, Germany
TANTALUM	80		Australia, Canada, China, Kazakhstan
COBALT	78		Finland, Norway, Russia, Canada
RHENIUM	76		Chile, Kazakhstan, Germany
CHROMIUM	74		South Africa, Kazakhstan, Zimbabwe, Turkey, Russia
IODINE	74		Chile, Japan, Russia
PALLADIUM	74		Russia, South Africa, United Kingdom, Belgium
TITANIUM (sponge)	73		Japan, Kazakhstan, Russia
TITANIUM MINERAL CONCENTRATES	70		South Africa, Australia, Canada, Ukraine
TUNGSTEN	69		China, Russia, Canada
ZINC	57		Canada, Mexico, Peru
GARNET (industrial)	56		Australia, India, China
SILVER	56		Mexico, Canada, Peru, United Kingdom
PEAT	55		Canada
SILICON	54		Norway, South Africa, Russia, Canada
MAGNESIUM COMPOUNDS	51		China, Australia, Canada, Austria
NICKEL	48		Canada, Norway, Russia, Australia
BERYLLIUM	47		Japan, Kazakhstan, Russia, Brazil
MAGNESIUM METAL	47		Canada, China, Russia, Israel
NITROGEN (fixed), AMMONIA	42		Trinidad and Tobago, Canada, Russia
ALUMINUM	41		Canada, Russia, Venezuela, Mexico
DIAMOND (dust, grit and powder)	39		Ireland, China, Ukraine
COPPER	38		Canada, Peru, Chile, Mexico
PUMICE	29		Greece, Italy, Turkey
PERLITE	28		Greece
VERMICULITE	28		South Africa, China
GYPSUM	23		Canada, Mexico, Spain
SALT	23		Canada, Chile, Mexico, The Bahamas
CEMENT	20		Canada, Thailand, China, Venezuela, Greece
SULFUR	20		Canada, Mexico, Venezuela
IRON ORE	11		Canada, Brazil, Australia, Venezuela
LEAD	11		Canada, China, Mexico, Australia
PHOSPHATE ROCK	10		Morocco
IRON and STEEL	9		European Union, Canada, Mexico, Japan
TALC	8		China, Canada, France, Japan
MICA, scrap and flake (natural)	7		Canada, India, China, Argentina
IRON and STEEL SLAG	6		Italy, France, Japan, Canada

[1]In descending order of import share

FIGURE 21.2 U.S. net import reliance on selected minerals, metals, and other essential resources as a percentage of consumption. Shaded area represents imports.

Many experts argue that we need not fear mineral cartels, because most producer countries depend heavily on steady exports of minerals for foreign exchange. Zambia's mineral exports, for instance, provide over half of that country's national income. Many industrialized nations import minerals from less-developed countries, refine them, and sell the metals at ten times the price they paid for the minerals. Less-developed nations often feel robbed of potential economic gains and have long urged industrial nations to import more refined metals from them to offset this imbalance. Such actions on the part of the industrialized nations could reduce tensions that might inspire mineral-exporting nations to band together in the first place, and indeed such actions are taking place. The United States, for instance, has increased its imports of processed materials from $35 billion in 1994 to $132 billion in 2006.

Mineral Supplies: Are We Running Out?

With this background information in mind, let us look at an important question raised earlier in this chapter: Are we running out of minerals?

Determining the life span of existing mineral supplies is not easy. To do so, we must first determine the rate of consumption and estimate potential growth in consumption, bearing in mind that even a modest increase in growth rate can result in a dramatic increase in total annual consumption. For instance, a resource with a life span of 1 billion years will last only 580 years at a 3% annual increase in the rate of consumption.

Next, geologists must estimate the economically recoverable supply or **reserve** of each mineral in the Earth's crust. The reserve must not be mistaken for a similar figure, known as **total resources,** which is the total amount of mineral believed to be in the Earth's crust (Figure 21.3). The important difference is that the reserve includes only deposits that are feasible to mine, whereas total resources include all deposits and occurrences of a mineral, no matter how low the concentration may be.

The total resources of a mineral are often many times greater than its reserve. To understand why, consider copper. A copper deposit 10 miles below the surface of the Earth is part of the total resources but not part of the reserve, because it would be too costly to mine. The world's total resources for copper are currently estimated at 1,440 metric tons (1,600 million tons), whereas the **reserve base** is only 566 million tons.

Another factor to take into account is that the amount of a mineral reserve is not permanently fixed. That is, it can expand and contract, depending on a number of variables. For instance, mining depletes the reserve base. New discoveries of economically recoverable minerals can expand the reserve base—what is economically recoverable. Economic incentives, such as subsidies from the government, lower the cost of production to a company (but not to society). Such subsidies can make it more profitable for a mining company to extract marginal or subeconomic ores. This artificial adjustment of the cost will shift some of the total resource into the reserve category.

The price of energy, heavily used in mining and processing minerals, also affects the reserve base. When energy is cheap, marginal ore deposits may become economical, thus expanding the reserve base. In contrast, when energy prices rise, ores that were once economical to mine may become too costly. The reserve base thus shrinks.

Environmental and worker protection laws also can affect the reserve base. For instance, environmental laws that require companies to reduce pollution from mines and reclaim mined land add to the cost of mining. They can make marginally profitable reserves too costly to mine. Likewise, lax environmental laws—or the lack of them—in many less-developed nations stimulate mining and expand the reserve base, but typically at a substantial cost to the environment, safety, and worker health. This distortion of the market may also occur in more-developed countries, such as Canada, where government subsidies to the mining industry and lax laws lower the cost of production. In the United States, an antiquated mining law, the **Mining Act of 1872,** has a similar effect, according to many critics. This law, passed 140 years ago when the nation was young, offers huge tracts of federal lands to mining companies for extremely low prices. But it is not just U.S. companies that benefit from this near-giveaway; companies from Canada, Japan, and other countries can take advantage of the U.S. government's generous gift. What is more, companies are not required to pay any kind of royalty on minerals extracted from public lands.

Labor costs also can profoundly shape the economic picture of mining and influence the reserve base. New technologies may cause the reserve base to expand. Highly efficient processes to extract and refine minerals, for instance, can reduce the price of mining marginal or subeconomic ores. This, like a variety of other factors, can increase the reserve base.

Based on existing estimates of world reserves and projections of consumption, it appears that three-quarters of the 80 or

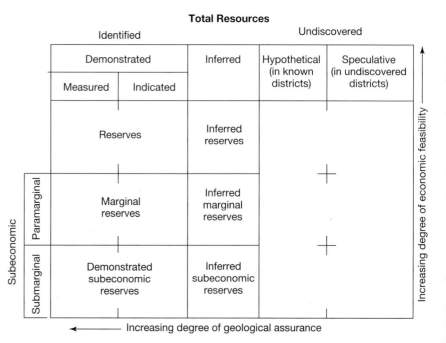

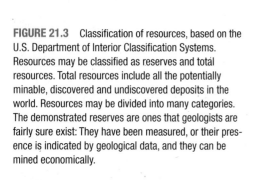

FIGURE 21.3 Classification of resources, based on the U.S. Department of Interior Classification Systems. Resources may be classified as reserves and total resources. Total resources include all the potentially minable, discovered and undiscovered deposits in the world. Resources may be divided into many categories. The demonstrated reserves are ones that geologists are fairly sure exist: They have been measured, or their presence is indicated by geological data, and they can be mined economically.

so economically important minerals are abundant enough to meet our needs for many years. However, at least 18 economically essential minerals are bound to fall into short supply even if nations greatly step up recovery and **recycling.** Gold, silver, mercury, lead, sulfur, tin, tungsten, and zinc belong to this group of endangered minerals. Don't be lulled into complacency by technological optimism; even if new discoveries and new technologies make it possible to mine five times the currently known reserves of these materials, this group will be 80% depleted by or before 2040, which is well within the lifetime of most of today's students.

21.2 Can We Expand Our Mineral Supplies?

Something must be done, and quickly, to forestall the depletion of key mineral resources. But what can we do?

Unfortunately, there is no consensus on ways to satisfy future demand and prevent the economic turmoil that could result from widespread shortages of essential minerals. Some people, in fact, flatly dismiss the possibility of future shortages. People with this viewpoint are called **technological optimists,** largely because they count on technological answers to this and a host of other environmental problems. The optimists are opposed by another group, often called the **pessimists,** who might be better labeled realists because they recognize the finite nature of the world's mineral resources and often propose innovative and cost-effective ways to avoid using up those resources.

This section looks at the views of the optimists and pessimists to answer the second question posed at the outset of this chapter: Can we expand our mineral supplies?

New Discoveries

A large portion of the Earth's crust has not been intensively explored for mineral deposits, say the optimists. By using current technologies, major finds are possible in Asia, Africa, South America, and Australia. The optimists point to substantial mineral discoveries in recent years as proof that current estimates of the world's mineral reserves fall short of the world's real reserves.

The pessimists, in contrast, argue that the extremely rich deposits needed to substantially expand our reserve base simply do not exist. Even if they did, a fivefold expansion of the world reserves of critical minerals would only slightly offset the rapid depletion, in large part because of the rapid growth in human population and economic development. Development of mineral resources could also lead to massive environmental damage in remote, pristine wilderness areas like the Arctic tundra or tropical rain forests, displacing wildlife and people who have lived on the land sustainably for years.

Extracting Minerals from Seawater

William Page, a researcher from Great Britain, summarizes the viewpoint of the optimists: "Seawater is estimated to contain 1,000 million years' supply of sodium chloride; more than one million years of molybdenum (used to harden steel), uranium, tin,

and cobalt; more than 1,000 years of nickel and copper. A cubic kilometer (0.25 cubic mile) of seawater contains approximately 11 metric tons each of aluminum, iron, and zinc." The oceans contain approximately 1,300 cubic kilometers (330 million cubic miles) of water, or about 14,000 million metric tons of each of these metals.

Not bad, say the pessimists, but there's a hitch. Even though the oceans contain vast quantities of dissolved minerals, most minerals are found in very low concentrations, except for bromine, magnesium, and table salt. The energy costs of extracting them would be prohibitive. In fact, to extract 0.003% of the zinc that the United States uses each year, a plant would have to process a water volume equal to the combined annual flows of the Hudson and Delaware Rivers! Clearly, in this instance, limiting factors such as energy and cost must be taken into account.

Minerals from the Sea Floor

Far-reaching optimists look to outer space to provide minerals for Earth, ignoring the exorbitant costs and energy requirements of these imaginative plans. Another group, looking closer to home, sees the ocean floor and the **continental shelf** as important sources of minerals for the future.

Minerals of the Continental Shelf   Because the continental shelf is an extension of the continent that happens to be under water, it is not surprising that it could yield many of the minerals that are now extracted from mines on dry land. By one estimate, the continental shelf contains about 15% of the world's minerals. Tapping the economically feasible deposits, say the optimists, could expand our reserve base and postpone the day of depletion. Mining it on a large scale could create an environmental disaster.

Minerals of the Sea Floor   In 1990, an American marine expedition discovered curious nodules, called **manganese nodules,** on the floor of the Pacific Ocean (Figure 21.4). Rich in manganese, these nodules also contain other minerals: nickel, iron, copper, cobalt, molybdenum, and aluminum. Manganese nodules cover about one-fourth of the ocean's floor, mostly in international waters. Most are the size of potatoes, although

FIGURE 21.4   Manganese nodules taken from the floor of the ocean. They contain manganese, iron, nickel, copper, and cobalt.

they vary in size from tiny granules to the size of cantaloupes. The aggregate weight of these nodules on the Pacific Ocean floor alone is estimated at about 1,500 billion metric tons. Thus, while the land-based supply of copper could last only a few more decades, tapping the copper of manganese nodules could extend our reserve for thousands of years.

Problems with Mining the Sea Floor   Mining sea floor mineral deposits and manganese nodules, although attractive, is fraught with problems. To remove manganese nodules, ships would have to scour the ocean with huge devices, not unlike large vacuum cleaners, that suck the nodules up from the bottom. Recovering the solid nodules would be more costly than extracting oil and natural gas from offshore wells. In addition, the environmental impacts of such activities could be far-reaching. Dredging or scooping up the minerals from the seabed would increase the turbidity of the water, possibly affecting a wide array of sea creatures. Clouding of shallow waters could increase water temperature and make it unfit for many organisms adapted to cooler waters. Ocean-mining equipment would require enormous amounts of energy and cooling water and add further to global climate change. Heated water released into the ocean could further raise the temperature of the water.

A final, and major, problem has more to do with politics than the environment. Most nations claim ownership of waters 330 kilometers (200 miles) off their coasts. They presumably own the minerals in the continental shelf. But what of the manganese nodules on the ocean floor outside of territorial waters? Who owns the minerals in international waters? Many people in the industrialized countries with the wealth and resources to mine manganese nodules believe they are legally entitled to these riches. However, the poorer nations wonder whether they shouldn't also share in the wealth, arguing that international waters are a common resource.

To settle this question of ownership and, more important, who will profit from the riches of international waters, the United Nations began extensive negotiations in 1958. But progress has been slow. In 1982, 100 nations signed the **Law of the Seas Treaty.** The treaty places deep-sea mining outside territorial waters under international regulation. It also calls for a tax on seabed minerals, the proceeds of which would go toward helping less-developed nations improve agriculture and develop economically. Believing that the seabed minerals belong to whomever can afford to mine them, many wealthy nations, including the United States, Germany, and Great Britain, have refused to sign. President Ronald Reagan refused to sign in 1981 because he felt the provisions of the treaty would jeopardize the ocean-mining interests of private U.S. firms. Unfortunately, the Reagan administration offered no alternatives. Today, the treaty has been signed by 109 countries, and efforts are under way to gain congressional approval in the United States. But at this writing (September 2008), the United States has still not signed the treaty.

Currently, little mining of manganese nodules or sea floor mineral deposits is going on. Economics don't support extraction at this time. But sea floor minerals may someday help us meet the mineral needs of the world. Experts warn, however, that the cost may be high, both environmentally and economically. Seabed mining will be energy-intensive and, in the light of declining fossil fuel supplies, the price of energy is bound to increase.

Improved Extraction Technologies

As new extraction technologies are developed, the optimists say, the mining industry will be able to extract more and more minerals from low-grade ores. The history of copper mining in the United States provides a good example. In the early 1900s, only very high-grade ore, containing 30 kilograms (66 pounds) of copper per metric ton, was mined. As high-grade ores were depleted, the efficiency of mining and processing improved so that progressively leaner ores could be used. Today, a very low-grade ore containing one-tenth as much copper (3 kilograms per metric ton) can be mined at a profit.

Pessimists admit that new technologies do permit the use of lower-grade ores. However, they point out that the lower the concentration of mineral in an ore, the greater is the amount of energy used for mining and processing. For example, twice as much energy is needed to produce a ton of copper from 10% bauxite as from 20% copper ore (Figure 21.5). In recent years, skyrocketing energy prices have caused the cost of all metals such as copper and steel to climb steeply. Rising costs, in turn, have affected numerous industries, including the wind energy sector. The increase in the cost of steel, for instance, has dramatically increased the cost of new towers for wind turbines. The dramatic increase in the cost of copper has increased the cost of wind generators, which use large amounts of copper. Even the cost of a penny is affected by rising energy costs and higher mineral prices. In 2008, a penny cost about 1.7 cents to manufacture. Greater energy use also increases environmental pollution and

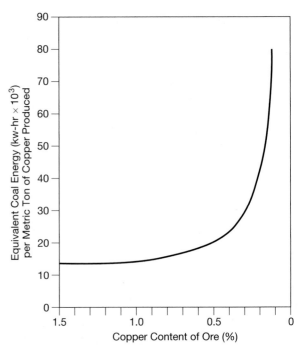

FIGURE 21.5   Amount of energy needed to extract a mineral as the concentration of the ore decreases. Note that once the concentration falls to a critical level, energy consumption increases drastically.

environmental damage from extraction and oil spills. This, in turn, increases economic costs (externalities) that are not included in the real cost of minerals.

Further reductions in mineral concentration will inevitably result in even larger increases in energy demand and greater environmental destruction. Making matters worse, the lower the concentration of the mineral, the greater is the amount of rock that must be mined and processed to produce a ton of mineral, and therefore the greater the amount of environmental damage, pollution from smelters, surface disruption at mines, and mine waste.

GO GREEN!
Start a compost pile if you can, or help your parents start one at home. You may need to read an online article or two to get started.

Pessimists also note that some substitutes are inferior to the materials they replace—as anyone who has used a plastic snow shovel can attest. Aluminum, iron, magnesium, and titanium are among the most abundant elements in the Earth's crust and have great potential for becoming substitutes. However, they do not stand up to the task in all situations.

Tapping Abundant Low-Grade Ores

As a general rule, optimists point out, for many minerals the total amount of ore increases as the grade decreases. Thus, the deeper miners dig into the Earth, the more ore they find. This principle was advanced by S. G. Lasky of the U.S. Geological Survey in 1950, and it appears to hold for a few vital minerals, among them iron and aluminum. Optimists then note that, as the technologies for deeper mining and for mineral extraction from lower-grade ores are developed, a literal bonanza of mineral wealth awaits the mining industry.

The pessimists respond that this principle does not apply to all ores, including nickel, copper, and manganese. The amount of copper ore, for instance, increases up to a point as the grade decreases. Thus, there is more 1% ore than 2% ore, but below 1%, the rule no longer holds. The 0.3% ore is one-fourth the amount of 1% ore.

Finding Substitutes

The substitution concept arose during World War II when, to save copper, government mints began turning out steel pennies. Within the past five decades, numerous substitutes have been devised. Plastics have been a key alternative and have replaced wood, steel, and paper in a wide variety of products from fishing rods and speedboats to bathtubs and automobiles. Plastics are now even being used to replace metals, sometimes to the chagrin of those who end up with the products. Many power tools, for instance, now come with plastic cases. Metal parts in vacuum cleaners, power tools, and numerous other products have been replaced with plastic parts that, regrettably, are often much less durable. Aluminum, another popular substitute, has nearly replaced steel in beverage cans and airplanes. Magnesium is being substituted for zinc in an increasing number of products, such as chemicals and pigments. Indeed, either naturally occurring or synthetic substitutes can be found for most metals used in today's society.

Pessimists argue that not all materials have adequate substitutes. For example, no other metal has the unique chemical and physical properties of mercury or the high melting point of tungsten, which makes it so valuable in high-speed tools and cutting edges. Furthermore, it may be impossible to find substitutes for the manganese used in desulfurized steel, the nickel and chromium used in stainless steel, and the tin used in solder. There's also no substitute for silver used in photographic film, but digital cameras may make film obsolete anyway!

21.3 Mineral Conservation Strategies

A final problem is that some substitutes are scarce as well. This is certainly true of the molybdenum that is now being used in place of tungsten. The cadmium and lead that have replaced mercury in some types of batteries are also in short supply.

The optimists would have us believe that our mineral future is secure, and that worries over mineral supplies are needless. They assure us that new discoveries, sea floor deposits, lower-grade ores, new mining technologies, and substitutions will come to the rescue. The pessimists, however, know that these avenues are limited at best and argue that we would be well advised to look for additional ways to meet our needs. One of the easiest and most cost-effective strategies is to reduce demand.

Reducing Demand

Reducing our demand for minerals can be achieved by reducing population growth, cutting back per capita consumption, decreasing the size of products, and increasing product durability.

Population stabilization was discussed in Chapter 4. It is, as we have pointed out before, the cornerstone of all sustainable strategies. Beyond that, individuals can reduce per capita consumption by reducing unnecessary purchases and avoiding disposable items, a topic discussed in Chapter 17.

Manufacturers can contribute by reducing wasteful packaging and eliminating throwaways such as disposable pens, razors, and lighters. Yet another effective means of cutting back on mineral demand is to increase **product durability**— that is, to make products that last longer. Today, however, many manufacturers crank out flimsy tools, toys, appliances, and automobiles for a price-conscious public that, in its blindness to quality, gobbles them up and then throws them away when they break or go out of fashion.

That said, there is a movement, albeit a small one, for companies to lease products and then make their money on service. One major elevator builder, for instance, is installing elevators in office buildings, but rather than installing an elevator for a flat fee, the company bears the construction costs itself and then rents the elevator to the business, servicing it as needed. Obviously, in this instance, it makes sense to install the most durable system possible, to minimize repair. A carpet manufacturer (Interface) introduced a similar idea. It leases carpet to businesses, and when a carpet becomes worn, the worn section is removed and replaced. (The carpet is laid down in squares that can be easily removed.) There's no need to throw away 100% of a carpet when only a few small sections in high-traffic areas have worn out.

In addition to increasing product durability, manufacturers could make products smaller, further stretching mineral supplies.

Manufacturers have produced a variety of household appliances such as clocks, toasters, telephones, and stoves that are smaller and lighter than those on the market two decades ago. Similarly, the first calculators were heavy and cumbersome. They would fit into a briefcase but might take up a third of the space. Today, solar-powered calculators slip into a breast pocket with room to spare. Computers have also undergone a dramatic reduction in size and weight (Figure 21.6).

GO GREEN!

Donate used clothing and other items in good working order to Goodwill and other charitable organizations. Consider shopping at resale stores for clothing and household necessities such as pots, pans, silverware, and plates.

Recycling

Ruben L. Parson, a geography professor at St. Cloud State University, Minnesota, wrote, "[During] World War II we [Americans] reclaimed anything metallic, from abandoned streetcar tracks and worn-out machinery to horseshoes and tin cans. Reclaimed metal gave us the machines that crushed Hitler's armored legions. We became scrap-conscious as never before. But we have too readily reverted to the reckless, wasteful discard of material that is typically American."

At least four factors are bound to spur greater interest in recycling in the United States and other countries: (1) a rise in energy prices (Chapter 22), (2) the depletion of strategic minerals, (3) the shortage of landfills (Chapter 17), and (4) environmental and economic concerns. Signs of this recycling revolution are already beginning to appear. In 1987, for instance, in response to a shortage of landfill sites, New Jersey passed a statewide law that required all communities to recycle at least three commodities. Other states quickly followed. (For more on recycling programs, see Chapter 17.) In 2006, the most recent year for which data are available, there were 8,660 curbside recycling programs in the United States, serving an estimated 130 million, according to the Environmental Protection Agency. Today, Americans recycle 32.5% of their trash, up dramatically from the past decade. Five states—California, Oregon, Iowa, Minnesota, and Maine—recycle more than 40% of their municipal solid waste, according to *Biocycle* magazine. Eight states have recycling rates of 30% to 40%.

GO GREEN!

Recycle everything you can, and be sure to buy recycled products. If your school doesn't have a recycling program, help get one started.

Although recycling has taken off remarkably well in the past decade, Americans have only begun to tap its potential. Several obstacles stand in the way of the full development of this strategy. In addition to the problems discussed in Chapter 17, notably a failure to develop adequate markets for recycled (secondary) materials, the federal government has proved to be a major impediment, for it still offers the mining industry economic incentives that put the recycling industry at a grave disadvantage. What are these incentives? First, the federal government provides billions of dollars in **depletion allowances,** tax breaks that go to mining companies as they deplete their reserves. The tax breaks were designed to help mining companies invest in the exploration needed to unearth additional mineral supplies. The net effect of such incentives is to make virgin minerals artificially cheap and give them a competitive advantage over recycled materials in some instances. Second, federally mandated freight costs for shipping raw materials are, by law, lower than rates for metals bound for recycling plants. Ending these two unfair practices would benefit the recycling industry enormously.

Before we move on, it is important to point out that many improvements are needed in recycling systems. As noted in Chapter 17, many products are mixtures of metals or plastics that are hard to recycle. Melting or breaking down such products may result in an inferior material—one that must be used for another, less essential product. In addition, some materials can be recycled only so many times. Paper fibers, for instance, become smaller and smaller after each recycling. After a half dozen or so recyclings, they are too short to be used. (In contrast, pure metals are infinitely recyclable. They can be melted down and reused ad infinitum.) Recycling may also produce potential toxic emissions—for example, the inks on aluminum cans when recycled can release toxic chemicals. In addition, toxic chemicals must be added to recycled materials to meet industrial standards for products. Solutions to these and other problems are discussed in *Cradle to Cradle* by William McDonough and Michael Braungart and in Chapter 17 of this book.

FIGURE 21.6 The MacBook Air, a computer so thin it can fit in a manila envelope. Computer manufacturers are designing smaller computers and monitors that require less material to make, thus reducing production costs and environmental impact.

Individual Efforts

Resource-conscious individuals can play a significant role in reducing mineral consumption by taking several simple measures. You can help reduce mineral consumption by cutting back on your consumption of unnecessary items. Most of us have closets and basements full of products that do little more than gather dust. Before you buy something, ask yourself if you really need it and if you will use it for more than a few weeks.

Individuals can contribute by purchasing durable clothes and goods. Avoid throwaways. Ditch your disposable pens (after you've used them up) and buy ballpoint pens, where all you have to throw away is the refill. You can buy products made from recycled materials—for example, recycled paper products.

When you have a choice between a recyclable and nonrecyclable good, choose the recyclable product. When you have a choice between a product made from a renewable resource (such as wood) and one made from nonrenewable materials (such as aluminum), choose the one made from a renewable resource.

And, of course, you can recycle. Take steps to recycle all waste from your household, including glass, aluminum, waste copper, and other metals (Figure 21.7). Donate used goods to Goodwill and similar groups that resell them.

The actions of individuals, when multiplied by many hundreds of thousands, can significantly reduce our reliance on minerals. If you are inclined toward action, you could organize a recycling program on your campus, in your community, or in your home. Armed with statistics on the benefits of recycling and with enthusiasm, you could become an instrument of social change in our transition to a sustainable society.

All of the techniques described in this section will help steer our society onto a sustainable course. But, as the next section points out, they are not enough to create a truly sustainable system of mineral supply.

21.4 Environmental Impacts of Mineral Production

Mining is one of the most environmentally destructive behaviors we humans engage in. A good example is Amax's huge molybdenum mine and waste dump in Climax, CO. Molybdenum (pronounced muh-lib′-de-num), or Molly-B for short, is a mineral that is used to harden steel in automobiles and other applications. At Climax, miners have torn down an entire mountain to extract this mineral (Figure 21.8). Found in a concentration of only about 0.2%, molybdenum is separated from the ore and trucked away for sale. The remaining material has for years been dumped into a once-beautiful valley near the mine. Over the years, waste has transformed a once-splendid mountain valley into a huge toxic moonscape, stark and ugly in a land otherwise breathtakingly beautiful. (In recent years, efforts have been taken to revegetate the valley where the mine wastes have been dumped, greatly reducing the visual impact.)

This is but one example of the impact of our dependency on minerals. This section takes a broad look at the many effects of mining and minerals processing and suggests ways to minimize them—that is, ways to create a more sustainable system of mineral supply.

Mining Impact

Ninety percent of our nonfuel minerals is extracted from surface mines, excavations in the Earth's surface that allow miners to access underlying deposits. Surface mines are particularly

FIGURE 21.7 Stackable bins, which allow the homeowner to separate recyclables for easy recycling.

FIGURE 21.8 A mountain that is being torn down by molybdenum miners in Climax, CO. The ore is then processed, and waste is dumped in a nearby valley. Molybdenum is used to harden steel.

destructive because the overlying soil and rock, called **overburden,** must first be removed and placed elsewhere. As a result, mining can quickly transform a scenic area into an ugly landscape. Reclamation is often difficult, because many open-pit mines extend deep into the Earth's crust. (Surface coal mining is discussed in Chapter 22.)

Underground mines provide a smaller percentage of our minerals. As with surface mines, wastes must be removed from the mine and dumped elsewhere. Waste piles from both types of mine cause additional environmental problems. If, for example, they are not vegetated and stabilized, heavy rains can wash the unstable soils into streams and lakes. Sediment in streams and lakes has many effects. It increases water temperatures, which lowers the concentration of dissolved oxygen. This, in turn, can kill fish and other aquatic organisms, disrupting aquatic food chains. Sediment destroys spawning beds and detracts from scenic beauty and recreational uses. Sedimentation increases flooding because it reduces a stream's water-carrying capacity. When heavy rains come, water flows over the banks more easily, flooding nearby towns, farms, and pastures, and creating social, economic, and environmental damage.

Each year, approximately 1.7 billion metric tons of mine waste are produced from U.S. surface mines, with 5% coming from mineral surface mines and the rest from coal surface mines. Canadian mines produce 650 million metric tons of waste. Mine wastes can contribute toxic metals to nearby waterways. For example, toxic metals such as zinc, arsenic, and lead may be leached from the spoils of iron ore mines by rainwater. In Canada and the United States, many mine wastes contain iron pyrite, a sulfur-containing compound. (Half of Canada's mine waste contains this chemical.) Rainwater combines with iron pyrite in gold and silver mine wastes in the West, creating sulfuric acid, which drains from the spoils as **acid mine drainage** into nearby streams, killing fish and other aquatic organisms.

The economic damage is staggering. Unstable spoil piles can also form dangerous landslides when rainfall is heavy.

Certain mineral deposits that contain water-soluble ores such as salt and potash may be removed by **solution mining.** In this technique, water is pumped into the deposit, where it dissolves much of the mineral, then is pumped back to the surface. Although safe, this technique can pollute groundwater supplies. It also results in the production of huge amounts of wastewater that must be properly treated before disposal. Some companies have used a more dangerous offshoot of the solution-mining technique. They spray a deadly cyanide solution over mine wastes to extract gold from rock. Many environmentalists are justifiably alarmed by the prospects for contamination. In fact, in Colorado, one such mine is now costing the state and citizens about $40,000 a day to prevent pollution from draining into nearby waterways.

Besides creating eyesores, enhancing erosion, and polluting nearby streams, mining activities compete with other uses of wildlands. Mining in U.S. national forests and wilderness areas, for instance, disturbs wildlife and outdoor recreationists. Mining can destroy valuable timberland and grassland as well.

Mining can also use tremendous quantities of water. The lower the grade of ore, the greater the volume of water required. In the **hydraulic mining** of gold and silver, powerful blasts of water are used to wash soils from hillsides, which are then treated to extract the precious metals. In some areas, then, mining can divert water from ranches, farms, businesses, and municipalities.

Processing Minerals

Many ores are heated to high temperatures in specially built ovens that separate metals from the ore. As a result of this process, called **smelting,** a variety of toxic materials may be released into the atmosphere (Figure 21.9). Smelters emit toxic

FIGURE 21.9 Aerial view of a mineral smelter in New Mexico. The tall stack emits toxic gases into the atmosphere.

substances such as arsenic, mercury, and zinc, which are dangerous to bees and other animals. Especially harmful is fluoride gas, which is released from phosphate smelters and settles on vegetation around smelters. Fluoride can be ingested by cattle and other livestock. In high enough levels, fluoride causes a disease called **fluorosis,** characterized by pain in an animal's joints and softened bones and teeth, which leads to inability to stand or move around.

Perhaps smelters' best-known effect, though, results from the release of sulfur dioxide, a corrosive gas that combines with atmospheric moisture and oxygen to produce sulfuric acid, described in Chapter 19. Sulfuric acid and sulfur dioxide are lethal to plants and aquatic life and are suspected of having adverse effects on human health as well.

The huge copper and nickel smelter in Sudbury, ON—long a major producer of sulfur dioxide—has turned the neighboring area into a barren wasteland. Similar devastation has occurred in Montana and Tennessee. Recognizing problems such as these in the 1950s and 1960s, governments pressed smelter owners to install tall smokestacks. Tall stacks, they thought, would inject pollutants higher into the atmosphere, so they would be more likely to disperse more widely and not be deposited in the immediate vicinity of the smelters. The idea was that tall stacks would help dilute pollutants. (The solution to pollution was dilution.) However, soon after they were installed, it became evident that taller stacks resulted in the pollution of distant landscapes and lakes, killing fish and other aquatic organisms hundreds of kilometers from the source. Fortunately, many of the world's smelters, including the one in Sudbury, have installed pollution-control devices and have cut sulfur dioxide emissions by 90%.

Creating a More Sustainable System of Mineral Production

Mining and mineral processing are two of the most environmentally damaging activities of humankind. They are also two of the world's most economically important industries. In many countries, mining and mineral processing are big business. Canada, for example, exports about 40 billion Canadian dollars' worth of minerals each year. The mining industry pumps over $200 million a year into Canada's economy, accounting for about 4% of the nation's gross domestic product (the total market value of all the goods and services produced within the borders of Canada during a specified period). Because mining is such big business, the industry carries a lot of weight in political decisions. Change is not easy.

By reducing their demand for minerals, citizens the world over can reduce their need for virgin materials and reduce the associated effects. But conservation is not sufficient. Minerals will continue to be extracted and fashioned into useful products, even with the best recycling and conservation efforts. Thus, society must find ways to reduce the impact of mining and mineral processing.

Mining companies in Canada and the United States are required to prepare **environmental impact statements** before they can mine on public lands. These reports list and describe all of the potential effects of mining and ways these effects are to be addressed. Unfortunately, critics point out, environmental impact statements often underestimate or ignore potential impacts. And despite this sometimes lengthy and costly procedure, companies sometimes ignore their plans to offset damage.

One important way to reduce the impact of mining is through **reclamation,** the rehabilitation of land altered by mining (Figure 21.10). Surface mines, for instance, can be filled, and the ground recontoured and planted to establish a vegetative cover that protects the soil. During mining, topsoil and wastes can be set aside and stabilized to minimize erosion, leaching, and landslides.

In 1977, the U.S. Congress passed the **Surface Mining Control and Reclamation Act,** which requires coal mining companies to reclaim all land used for surface mining. By law, companies must restore surface-mined land to its premining condition. Unfortunately, this law pertains only to coal mining; no specific federal legislation requiring reclamation on mineral lands exists, and state laws and regulations are often weak. According to the U.S. Bureau of Mines, between 1930 and 1980, only 8% of the land mined for metals and only 27% of the land mined for minerals was reclaimed. Showing how effective surface mining laws are, however, the bureau noted that 75% of the land on which coal was mined has been reclaimed.

FIGURE 21.10 Reclaimed land that had been surface-mined to remove coal.

By one estimate, recontouring and planting the vast unreclaimed lands and leveling the spoil piles from all surface and underground mines would cost the United States approximately $30 billion. In Canada, the price tag is estimated to be $6 billion in Canadian dollars. Given budget constraints, it is unlikely that the United States will make much progress toward refurbishing the unreclaimed lands.

In Canada, mineral mine reclamation falls under the purview of the provinces, each with its own set of rules. In the Northwest Territories and the Yukon, however, the Canadian federal government is in charge. Companies are required to develop and implement reclamation plans. As one source put it, "Mines are no longer abandoned; they are closed following legislated or provincial procedures." Because of this, they often have few downstream impacts.

Monetary shortages have also weakened the inspection and enforcement of the Surface Mining Control and Reclamation Act in the United States. Without oversight, mining companies could return to earlier practices that left the land in a shambles.

The United States needs strong enforcement and needs money to reclaim abandoned lands. Concerned citizens can write their representatives to support continued funding of reclamation programs. To raise money, governments could increase taxes on minerals; these monies could be earmarked for land rehabilitation.

Smelting operations can be made more efficient. Cleaner fuels, such as natural gas, can be used. Pollution-control devices can be installed to capture pollutants. Wherever possible, valuable minerals and other substances should be extracted from the sludge of pollution-control devices and sold to willing buyers. In Canada, Hudson Bay Mining installed a hydrometallurgical process to extract zinc from ore, rather than using a smelter, which cut sulfur dioxide emissions by 98%. Canada's federal government has started another program, known as the Accelerated Reduction/Elimination of Toxics program, to slash emissions at smelters. Nearly all of Canada's metal-producing companies have submitted plans that will cut emissions by 70%. These and other actions by the Canadian government and mining companies will help overcome a long history of environmental neglect by the industry.

In light of the limited supply of certain metals and minerals and the enormous environmental impact of mining, it is imperative that citizens the world over conserve what they have and find substitutes where they can. Reducing demand and recycling the metals can help us continue as we have. But eventually, many agree, unless adequate substitutes can be found, we will have to change our ways. These changes may involve reductions in the global population, developing less resource-intensive lifestyles, and shifting our dependency toward renewable resources, which if managed properly can ensure a sustainable society.

Summary of Key Concepts

1. The Earth's mineral wealth has been tapped for thousands of years. All around are the signs of our dependency on minerals and our often reckless exploitation of these resources. The sheer magnitude of human dependence on minerals and the effects that mineral mining and processing have on the environment make minerals an important environmental issue and crucial to building a sustainable society.

2. Unlike forests and wildlife, minerals are nonrenewable, but many can be recycled. Achieving higher rates of recycling is essential to creating a sustainable society.

3. Minerals come from the Earth's crust. Most minerals exist as chemical compounds consisting of two or more elements. A rock containing minerals is called an ore.

4. The United States annually mines ores worth an estimated $64 billion. When refined, they become more than ten times more valuable.

5. Although the United States has only 4.5% of the world's population, it consumes about 20% of the world's nonfuel minerals. To raise the world to U.S. levels of consumption would create an environmental disaster of epic proportions, and would be impossible because of a lack of reserves.

6. The United States' and other industrial countries' prodigious use of minerals has made them great and prosperous nations but also vulnerable ones, because many of the strategic minerals needed for commerce and defense come from unstable countries. To reduce this vulnerability, many governments have stockpiled strategic minerals.

7. Stockpiling may help in the short run but does nothing to protect nations from vanishing mineral supplies or the formation of cartels—groups of mineral-exporting nations that ban together to control the supplies and prices of minerals.

8. Many experts believe that cartels will not form around minerals because so many of the exporting nations need their mineral exports to produce foreign exchange.

9. Determining the life span of mineral reserves is not easy. First, scientists must determine the rate of consumption. Second, they must project future consumption levels, bearing in mind that increases in the consumption rate can greatly accelerate the depletion of a finite resource. Third, they must determine the economically recoverable supply, or reserve, of each mineral. The reserve must not be mistaken for the total resources, or the total amount of a mineral in the Earth's crust.

10. The reserve capacity of the world's mineral supply is not permanently fixed and can expand and contract, depending on a number of factors—for instance, new discoveries, economic incentives from governments, new technologies that allow miners to extract and process lower grades of ore more efficiently, the price of energy and labor, and the level of environmental controls.

11. Based on existing estimates, it appears that world reserves of three-quarters of the 80 or so economically important minerals are abundant enough to meet our needs for many years, or if they are not, adequate substitutes are available. However, at least 18 economically important minerals could fall into short supply, some within a decade or two.

12. Something must be done, and quickly, to provide much-needed minerals. Optimists believe that new discoveries, improved extraction technologies, the use of low-grade ores, and the use of substitutes will help to ensure sufficient supplies of important minerals.

13. Some observers believe that ocean floor minerals—deposits in the continental shelf and on the floor of the ocean—could help us stretch our mineral supplies. Manganese nodules, for instance, found on the ocean floor could expand our reserve by thousands of years. Manganese nodules cover about one-fourth of the ocean's floor and contain many important minerals, including copper and iron.

14. Although attractive, mining sea floor mineral deposits and manganese nodules would be energy-intensive and costly. It could increase the turbidity and temperature of ocean waters, and could upset the ecological balance of the oceans. Political problems also abound. The rich nations of the world want free access to the nodules in international waters and have refused to come to an agreement with the poorer nations, which want some of the proceeds of this international resource but have neither the money nor the resources to tap it.

15. Critics find serious fault in strategies aimed solely at increasing supply. They believe we must actively explore these strategies but must also tap the generous potential of conservation.

16. One of the chief ways to conserve minerals is to reduce mineral demand. This can be achieved by reducing population growth, per capita consumption, and the size of various products. Increasing recycling and product durability also can assist. Also, economic incentives for nonrecycled materials could be reduced or eliminated.

17. Considerable progress has been made in most of these areas in the past 25 years, but there is considerable room for improvement.

18. Individual efforts can also go a long way toward reducing mineral consumption.

19. The mining and processing of minerals have numerous impacts on the environment. Ninety percent of our nonfuel minerals is extracted from surface mines, which are particularly destructive because the overlying soil and rock, or overburden, must be removed to get to the mineral deposits. As a result, mining can destroy productive land or scenic beauty.

20. Underground mines and surface mines produce enormous amounts of waste that are trucked away from the mine and deposited in huge piles. These piles are an eyesore and, if not vegetated, may be eroded by wind and rain. Sediment from waste heaps can fill streams, increase water temperature, disrupt food chains, destroy spawning beds, kill aquatic life, and increase flooding. Mine wastes also leach toxic chemicals, such as sulfuric acid, arsenic, mercury, and zinc.

21. Many ores are heated to high temperatures in specially built furnaces to separate metals from the ore. A number of toxic substances may be released into the atmosphere during this process, called smelting. Perhaps the most significant is sulfur dioxide, which combines with oxygen and water in the atmosphere to produce sulfuric acid.

22. By reducing demand and increasing recycling, individuals can reduce their need for virgin materials and reduce the associated impacts, but conservation is not enough. Society must find ways to reduce the impacts created by mining and processing.

23. One important way to reduce the impact of mining is by reclamation, the restoration of all mined land.

24. Strong enforcement of state laws and funds to reclaim abandoned lands are both vital to this task.

Key Words and Phrases

Acid Mine Drainage	Pessimists
Cartels	Product Durability
Continental Shelf	Reclamation
Depletion Allowances	Recycling
Environmental Impact	Reserve
Statements	Reserve Base
Finite Resource	Smelters
Fluorosis	Smelting
Hydraulic Mining	Solution Mining
Law of the Seas Treaty	Stockpile
Manganese Nodules	Strategic Minerals
Metals	Substitution
Mineral	Surface Mining
Mining	Surface Mining Control and
Mining Act of 1872	Reclamation Act
Ore	Technological Optimists
Ore Deposit	Total Resources
Overburden	Waste Piles

Critical Thinking and Discussion Questions

1. Explain this paradoxical statement: "An American car is, in a sense, really a foreign car."

2. What is an ore? What is a mineral? What is a metal? How are ores, minerals, and metal related?

3. In what form are most minerals found?

4. What percentage of the world's population lives in the United States? What percentage of the world's minerals do Americans consume?

5. Do you agree with this statement? "The United States' heavy reliance on imported minerals also makes it highly vulnerable." Why or why not?

6. Define the terms *reserve* and *total resources*. How are they different? Why is it deceiving to calculate the life span of minerals based on total resources?

7. What factors cause the reserve base to shrink? What factors cause it to expand?

8. Debate the following statement: "Our mineral supplies are adequate for many years to come, so we need not worry about shortages."

9. List and describe the major ways in which we can expand the reserve base. Describe the pros and cons of each technique.

10. Describe how the high price of energy can affect the cost of metals, their consumption, and conservation efforts.

11. Describe the mineral conservation strategy. In what ways can you be a part of this strategy?

12. Why is recycling often at a competitive disadvantage when compared with the use of raw materials?

13. Describe the major impacts of surface and underground mining, and describe how they can be lessened.

14. Debate the following statement: "The wealthy countries can afford to mine manganese nodules and should be allowed to do so without giving developing nations any of the profits."

15. List and describe the impacts of manganese nodule mining.

16. Using your critical thinking skills and your knowledge of ecology and related issues, discuss the following statement: "Recycling creates pollution. It's no better than producing materials from virgin resources."

Suggested Readings

Draper, D. 2001. *Our Environment: A Canadian Perspective.* Toronto, ON: ITP Nelson. See the chapter on mining for an overview of Canadian issues and solutions.

Gardner, G., and P. Sampat. 1999. Forging a Sustainable Materials Economy. In *State of the World 1999,* ed. L. Starke. New York: W. W. Norton. Valuable reference.

Green, J. 2005. *Why Should I Recycle?* New York: Barron's Educational Series. A great resource for students interested in teaching environmental science at the high school and middle school level.

Kane, H. 1996. Shifting to Sustainable Industries. In *State of the World 1996,* ed. L. Starke. New York: W. W. Norton. Coverage of many important issues related to minerals and industry.

Lund, H. F. 2000. *McGraw-Hill Recycling Handbook,* 2nd ed. New York: McGraw-Hill. A great resource for students interested in a career in the recycling industry; provides practical advice on numerous topics, including waste stream management, separation and collection systems, processing facilities, and public awareness programs.

McDonough, W., and M. Braungart. 1999. *Cradle to Cradle: Remaking Ways We Make Things.* New York: North Point Press. Revolutionary new ideas on ways to improve the manufacturing and recycling of products in modern society.

Roca, N. 2008. *The Three R's: Reuse, Reduce, Recycle.* New York: Barron's Educational Series. A great resource for students interested in teaching environmental topics in middle school and high school.

Sampat, P. 2003. Scrapping Mining Dependence. In *State of the World 2003,* ed. L. Starke. New York: W. W. Norton. An insightful look into mining and ways to create a more sustainable system of mineral production.

U.S. Bureau of Mines. 2006. *Mineral Commodity Summaries.* Washington, DC: U.S. Government Printing Office. Important source of data for all minerals; published annually.

Young, J. E. 1992. Free-Loading off Uncle Sam. *World-Watch* 5(1): 34–35. Excellent discussion of the unfair advantages given to the mining industry.

Young, J. E. 1992. Mining the Earth. In *State of the World 1992,* ed. L. Starke. New York: W. W. Norton. Excellent source of information on the impacts of mining.

Young, J. E., and A. Sachs. 1995. Creating a Sustainable Materials Economy. In *State of the World 1995,* ed. L. Starke. New York: W. W. Norton. Excellent source of information on recycling.

 ## Web Explorations

Online resources for this chapter are on the World Wide Web at: **http://www.prenhall.com/chiras** (*click on the Table of Contents link and then select Chapter 21*).

CHAPTER 22

NONRENEWABLE ENERGY RESOURCES

ISSUES AND OPTIONS

Most college students today are only vaguely familiar with the twin oil crises of the 1970s. The first shock wave hit in 1973, when the **Oil Producing and Exporting Countries (OPEC)** imposed an embargo on oil, cutting back on exports and drastically raising the price. The brainchild of a Venezuelan multimillionaire, Juan Perez Alfonzo, who rode a bicycle to work and reportedly read by candlelight, the oil embargo was conceived in large part as a measure to reduce the waste of energy by the West and to slow down the rapid depletion of OPEC's vast but finite reserves. Iran's 1979 oil embargo, imposed largely for political reasons, dealt another blow to oil consumers.

In this perilous decade, crude-oil prices shot up from $3 per barrel to $35. Inflation set in, reaching more than 18% a year in the United States in 1978. High prices dampened consumer spending. One after another, manufacturers shut down operations and laid off workers, until 70% of U.S. industrial capacity lay idle.

The oil crises not only made apparent the extent to which our economy and our lives depend on oil, but also taught us that virtually everything we bought or did required energy—and lots of it. If oil was priced too high, inflation, unemployment, and recession would emerge with devastating power.

The oil crises not only raised awareness, they also spurred conservation efforts and exploration. These efforts brought the price of oil down—until recently.

In 2000, however, another oil crisis began to emerge, sending oil prices from $15 per barrel to $35 by September 2000. One of the causes of this crisis was overconsumption that led to demand that exceeded supplies. High demand for automobile fuel, especially in the United States, for example, sent gasoline prices skyrocketing from $1 per gallon to more than $2 per gallon in a year's time. Despite increases in production by certain OPEC countries, the price of oil remained high because of two essential shortages: a lack of refinery capacity and a lack of oil tankers to move oil from foreign sources to countries including the United States. Americans and others were simply using more oil and oil derivatives (such as gasoline) than could be shipped and processed.

As readers know, the current oil crisis has deepened in recent years, with oil prices reaching $145 per barrel in 2008. The most recent rise in oil prices can be attributed to several factors, one of the most important being a discrepancy between supply and demand. However, extremely high demand in the United States, China, and India were major factors in the rapid increase. Speculators who buy and sell oil through the commodities market also drove up the cost of oil.

In 2001 and 2002, Americans were stunned by another problem: dramatically increasing prices for natural gas. In 2001, natural gas shot up 100% due to shortages, although many months later, prices fell again. In 2003, prices roses again by nearly as much. The cause of these price shocks was a simple economic phenomenon: As was the case with oil, demand was outstripping supply. Despite intense efforts to produce more natural gas, supplies remained fairly constant while demand increased. Ironically, one of the causes of skyrocketing demand was America's effort to clean up its air. Many utilities had shifted to cleaner-burning natural gas to generate electricity. Rising demand for natural gas for power plants cut into limited natural gas supplies required to heat homes, cook food, and heat water.

This chapter is about energy—specifically, nonrenewable energy sources such as coal, oil, natural gas, and nuclear energy. It describes each resource and its projected supplies, and examines the pros and cons of each, especially the environmental pros and cons. It then looks at options for reducing our dependency on these fuel sources and steering onto what many observers believe is a more sustainable energy path.

22.1 Global Energy Sources: An Overview

In the industrialized world, **fossil fuels** are the major sources of energy. So named because they were formed from plant and animal remains buried in the Earth millions of years ago, fossil fuels provide 85% to 90% of the energy demand of the industrialized world (Figure 22.1). Even in the less-developed nations, fossil fuels are a main source of energy. China, for instance, gets 80% of its energy from coal. Three fossil fuels predominate: oil, natural gas, and coal. All of them are nonrenewable sources of energy. A small amount of the energy consumed in more-developed nations comes from nuclear fuel, which is also a nonrenewable source.

Of all the nonrenewable options, oil is the predominant energy source in the industrialized world, but oil supplies are on the decline. If current trends in population growth, economic growth, and energy use continue, most of us will live to see the end of oil.

In fact, many oil analysts claim that the beginning of the end is already at hand. They believe that oil production has peaked and that from this day forward, oil extraction rates will continue to decline, falling short of demand (Figures 22.2A and 22.2B).

After the oil crises of the 1970s, as briefly mentioned in the introduction, many countries turned to energy conservation, energy efficiency, and renewable energy resources to solve the energy crisis, but in the 1990s, many nations lost interest in these approaches. In this book, we use the term **energy conservation** to describe cutting back on consumption, for example, by shutting off lights and turning down thermostats. **Energy efficiency** means employing technologies that allow us to get the most from the energy we consume, for instance, driving efficient cars or using efficient refrigerators. Conservation and energy efficiency measures can save enormous amounts of energy, thus greatly extending the world's supplies of other fuels. John Herrington, former U.S. Secretary of Energy, calls energy conservation and efficiency "our single largest resource." Had conservation and

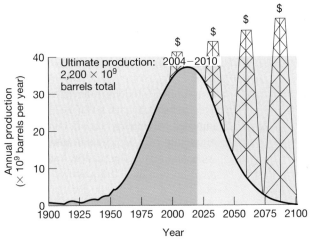

A. World

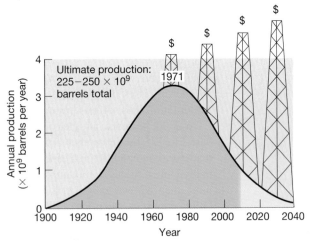

B. United States

FIGURE 22.2 Peak oil. (A) Theoretical world oil production curve. Many experts believe world oil production will follow a pattern similar to this, starting slowly, rising, then peaking and tapering off. Many analysts believe we are near the peak of production now. From this point forward, extraction rates could begin to fall, creating economic turmoil as supplies fail to keep up with demand. (B) Production of U.S. oil. Note that U.S. production peaked in 1971. We've depleted 75% to 80% of domestic supplies of oil.

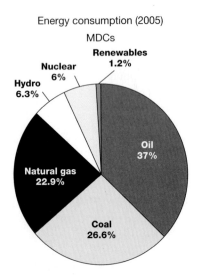

FIGURE 22.1 Energy consumption by the more-developed and less-developed countries.

efficiency been practiced since the 1970s, we could have freed up billions of gallons of gasoline. Had conservation, efficiency, and renewable energy been pursued in earnest, it's possible that we wouldn't have an energy crisis today.

Of the renewable energy sources in use today, only two have historically made a significant contribution to the energy needs of the more-developed countries. They are **hydropower,** or electricity generated from flowing rivers, and biomass, or crop residues, wood, and the like. Biomass resources can be burned or converted to gaseous or liquid fuels. In a few more-developed nations, such as Israel, rooftop solar hot-water panels provide a large percentage of the energy needed to heat water for domestic use. In recent years, though, our reliance on renewable energy resources, especially wind and solar, to generate electricity has grown considerably, as described in Chapter 23.

In sharp contrast, many of the less-developed nations rely heavily on renewable fuels—wood, charcoal, cattle dung, and crop residues. In fact, about half the world's population uses wood as its primary source of energy. Although wood is renewable, excessive exploitation—caused largely by overpopulation—has resulted in widespread deforestation and shortages. In many rural villages in India, Bangladesh, and Nepal, for example, trees around villages have been stripped bare or are missing altogether, having been cut down. Villagers must travel great distances to find a few sticks of fuel to cook the evening meal. Without wood, villagers have begun using dried cattle dung as a fuel, depriving farm fields of a rich source of fertilizer.

Struggling to industrialize, the less-developed nations have begun to use increasing amounts of coal and oil. China and India are two nations whose reliance on these fossil fuels has increased dramatically. However, as noted earlier, rising demand coupled with declining oil reserves are contributing to the dramatic increases in prices that could cause severe economic repercussions.

22.2   A Closer Look at Nonrenewable Energy Resources

This section examines our nonrenewable energy resources. It gives a brief history of their use, starting with coal; describes the benefits each provides; and looks at the impact of their use.

Coal

Coal in use today originated as plant matter that grew in hot, muggy regions of the Earth 225 million to 350 million years ago. These plants grew in lakes, streams, and coastal swamps and along their banks. Over the years, leaves and other plant matter fell into the water and accumulated on the bottom. Eventually, this rich organic material was covered by sediment that had eroded from the land. Over time, heat and pressure converted the organic material into peat, then coal, an organic material that, when burned, produces light and heat. The energy from coal, however, is ancient solar energy trapped by plants and stored in the bonds between the carbon atoms.

Today coal is burned by electric utilities and in some factories and homes, where it releases the solar energy captured in the leaves many millions of years ago by photosynthesis. Coal supplies about 23% of the energy consumed each year in the United States. Similar percentages apply to other industrial nations. In Canada, however, coal contributes only about 13% of the total energy demand. In China, a country with rich supplies of coal and badly polluted air, coal meets 80% of the total energy demand.

The value of coal as a fuel was first discovered in the 12th or 13th century by the inhabitants of the northeastern coast of England, who discovered that certain black rocks found along the shores, called "sea coales," burned. This monumental discovery led to coal mining for domestic heating and, much later, provided the impetus for the Industrial Revolution in the 1700s in England and the 1800s in the United States.

Coal mining started in the United States around 1860. Nevertheless, coal did not replace wood as a major fuel source until the early 1900s. Shortly after World War II, king coal began to be replaced by oil and natural gas because they were much easier to transport and were cheaper to extract (Figure 22.3).

Types of Coal   Geologists recognize three major types of coal: **lignite** (brown coal), **bituminous** (soft coal), and **anthracite** (hard coal). These types vary in several respects, the most important being their carbon content and their heat value, or the amount of heat they produce when burned per unit weight. Lignite has the lowest carbon content and heat value, anthracite the highest. As a general rule, the more heat and pressure that

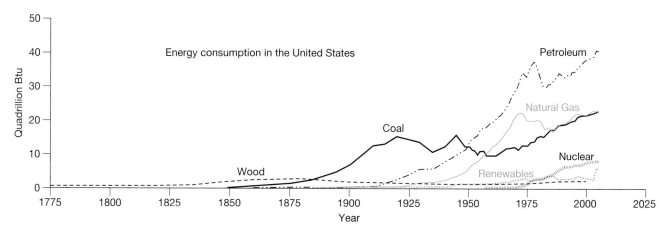

FIGURE 22.3   Pattern of U.S. energy consumption from 1850, including a question about our future patterns of energy use. As supplies of petroleum become exhausted, what alternative sources will be exploited?

were applied to a bed of coal in the prehistoric past, the higher the grade. Most of America's coal is bituminous. Coal also varies in sulfur content. Today, many more-developed countries such as Canada and the United States use low-sulfur coal to reduce sulfur dioxide emissions and acid deposition.

Coal Reserves   Because coal deposits lie just below the surface, scientists have been able to make fairly accurate estimates of worldwide coal reserves by drilling test holes in coal seams. A map of world coal deposits is shown in Figure 22.4. Surveys show that, of all the fossil fuels, coal is by far the most abundant. Worldwide, the **proven reserves** of coal (those we're certain exist) are nearly 790 billion metric tons. At the current rate of consumption, these proven reserves would last approximately 200 years. Undiscovered reserves—those thought to exist—are believed to be massive as well. Some experts believe that the world's recoverable coal supply may last close to 1,700 years at the current rate of consumption.

The United States has about 30% of the world's coal and is sometimes called the "Saudi Arabia of coal." It should be pointed out, however, that not all fossil fuels are of equal value. In fact, coal is of limited use. It cannot easily or efficiently be used to power automobiles (except maybe electric cars, indirectly) or jets, unlike liquid fuels derived from oil, the supplies of which are on the decline. Coal cannot easily or efficiently be used to heat homes or cook meals, unlike natural gas, whose supplies are limited. (Coal can be converted into synthetic gas and oil, as described later, but the processes are costly and dirty.)

Although coal is the world's most abundant fossil fuel, it is also the dirtiest. Mining it has an enormous impact on the environment. Burning coal to generate electricity produces incredible amounts of pollution and solid waste. Coal combustion, which occurs primarily in coal-fired power plants that generate electricity, is one of the main sources of the greenhouse gas carbon dioxide, which is causing global climate change

(Chapter 19). Global climate change, in turn, is having a devastating effect on people, the planet, and our economy. Unless we find much cleaner ways to burn coal, much of the world's impressive coal deposits may have to remain forever buried in the Earth's crust.

Environmental Impact of Surface Mining   America consumes over 1,047 million metric tons of coal each year, according to the Energy Information Office. About two-thirds of that coal comes from surface mines, of which there are two basic types: contour mines and area strip mines. In **contour mines,** which are found on hilly terrain in the eastern section of the country, coal seams are exposed by bulldozers or steam shovels that remove the overlying rock and dirt, called **overburden.** Before 1979, miners simply pushed the overburden over the side of the hill, destroying vegetation and creating serious erosion problems that often filled nearby streams with sediment (Figure 22.5A). Now, thanks to tougher mining regulations, miners must first remove the topsoil and set it aside to restore the site. Then they dig up the underlying layers and haul them away to a safe place, where they are held until they are trucked back to recontour the area when mining is completed.

Area strip mines are found on flat terrain in the Midwest and West in the United States and in the central provinces of Canada (Figure 22.5B). As in contour mines, topsoil is first removed by bulldozers and set aside for later use. Then huge shovels called **draglines** remove the overburden, which is placed in piles next to the excavation. The exposed coal seam is then dynamited, excavated, and loaded into trucks. After the coal has been removed from the first mined section, the process begins again, one strip at a time. The overburden from each new strip is placed in the previous excavation. Reclamation can begin on previously mined land after two or three strips have been cut. During reclamation, bulldozers first recontour the land and then spread the topsoil over the surface of the overburden. The land is then replanted, sometimes with native

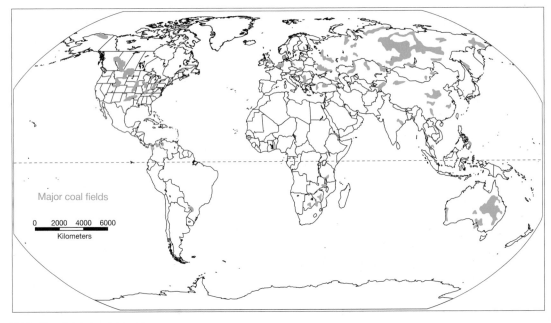

FIGURE 22.4   Distribution of global coal deposits.

FIGURE 22.5A A contour mine in the eastern United States. Overburden dumped over the hill creates a highly erodible surface. Soil erosion at this site during rainstorms and snowmelt leads to serious siltation in nearby streams.

FIGURE 22.5B Surface view of an area strip mine. Coal is extracted after a dragline removes overburden. Spoil piles will be recontoured and reseeded after mining.

species or fast-growing cover grasses that protect the soils from wind and rain.

Another kind of surface mining, called **mountain top removal,** is a kind of area strip mining that takes place in hilly or mountainous terrain. Miners remove the overburden from entire hilltops, often dumping it illegally in nearby valleys, causing enormous environmental damage. This environmentally devastating mining practice continues today, even though it violates federal mining laws.

Surface mining is a fast and efficient way of removing coal but creates serious visual impacts. In addition, mine waste (overburden) can erode if proper precautions are not taken. Eroded sediment spills into streams and lakes, destroying fish habitat, recreational sites, and reservoirs that supply water for human use. By one estimate, 0.6 million hectares (1.5 million acres) of unreclaimed coal-mined land in the United States are in need of reclamation, but only about a third of this land must be reclaimed by law. The rest was mined before the passage of the surface-mining legislation and is exempt from reclamation.

Sediment can also be eroded from access roads during rainstorms and by melting snow, further adding to the sediment load of nearby streams and lakes. Surface mining creates dust and noise and destroys wildlife habitat, often for many years. Surface mines can also cause groundwater levels to fall considerably, drying up municipal and agricultural wells in neighboring areas.

Interestingly, underground mines disturb about as much land as surface mines per ton of coal removed, because the rock removed to reach coal seams must be stored outside the mine.

In addition, underground mines can also collapse, killing workers and causing **subsidence,** a sinking of the surface above them. From 1900 through 2007, over 104,655 workers were killed and over a million were permanently disabled in underground mine accidents in the United States—a heavy price to pay for energy! Because safety standards vary, death rates vary considerably. China, which is very heavily dependent on coal for electrical generation and home heating, has the world's highest death rates. In 2004, the Mine Safety and Health Administration reported 6,027 deaths in China, compared with 28 in the United States. Subsidence caused by collapse of underground mines also damages buildings and roadways as mine shafts sink and cause huge cracks to develop in the surface. Imagine waking up one day to find your two-story house ripped in two because the ground underneath it had given way because an abandoned underground mine deep in the Earth below your home collapsed.

Cracks in the Earth's surface can divert stream water into coal seams. Water seeping naturally into mines or as a result of subsidence cracks combines with naturally occurring iron pyrite (a sulfur-bearing mineral, also known as fool's gold) and oxygen to produce **sulfuric acid.** If not captured and neutralized, the acidic outflow, called **acid mine drainage,** leaks out of mines and pollutes nearby streams. How serious is this problem? Abandoned underground mines in the United States produce an estimated 2.7 million metric tons of acid each year and pollute 11,000 kilometers (7,000 miles) of stream, most of which are in Appalachia. Stopping the flow from active mines is quite feasible, but acid mine drainage from abandoned mines has proved expensive and difficult, if not impossible, to control.

In 1977, the U.S. Congress passed the **Surface Mining Control and Reclamation Act,** which went into effect in 1979. This important law requires mining companies to reclaim their land—that is, restore the original contour of the land and replant it. It also calls on companies to control on-site erosion and acid contamination of nearby lakes and streams. In addition, the law established state and federal watchdog agencies to monitor reclamation. It also placed a federal tax on coal to finance the reclamation of land destroyed by previous strip-mining operations. Currently, most states monitor their own mine reclamation. The federal Office of Surface Mining Reclamation and Enforcement monitors these programs.

The Impact of Coal Combustion

A second major concern is the pollution produced by burning coal. Five pollutants are of particular interest: carbon dioxide, sulfur dioxide, nitrogen dioxide, particulates, and mercury.

As we pointed out in Chapter 19, the combustion of coal and other fossil fuels has resulted in the massive release of the greenhouse gas carbon dioxide. Much of it remains in the atmosphere. Although a little carbon dioxide is essential to maintaining the Earth's temperature, too much may cause a drastic increase in average global temperature that could disrupt our lives profoundly.

Sulfur dioxide and nitrogen dioxide, discussed in more detail in Chapter 20, are **acid precursors**—they form sulfuric and nitric acids, respectively, when they combine with oxygen and water in the atmosphere. These acids fall to the Earth in rain and snow and particulates, acidifying lakes and streams, killing trees and damaging crops, buildings, and statues. To control the emissions of sulfur dioxide, many utilities have switched to low-sulfur coal. This has helped the United States reduce sulfur dioxide emissions.

Unfortunately, controlling nitrogen dioxide is no easy matter. It comes from nitrogen in the air that combines with oxygen in a furnace under intense heat to form nitrogen dioxide. Unfortunately, this pollutant is not very soluble in water and therefore is not removed by typical smokestack scrubbers. The only measure that utilities use now is the control of combustion temperatures, but it has only a minor effect. Currently, U.S. motor vehicles, utilities, factories, and other sources produce about 13.3 metric tons (14.8 million tons) of nitrogen dioxide per year.

Coal combustion also produces enormous quantities of particulates. Like sulfur dioxide, they are relatively easy to control using various pollution control devices, discussed in Chapter 18. Even with the best pollution control, however, a single power plant supplying a million people produces 1,500 to 30,000 metric tons of particulates a year.

Coal-fired power plants also release mercury. Found naturally in small amounts in coal, mercury is released into the atmosphere when coal is burned in power plants and factories. It can be transported hundreds of miles and is deposited on land and in surface waters.

For years, it was assumed that the best means of reducing pollution from coal-fired power plants was by using pollution control devices. While they work well, pollution control devices merely capture gaseous pollutants and convert them into a solid waste. This essentially shifts pollution from one medium (air) to another (ground), trading one nuisance for another. Creating a sustainable future will ultimately require efforts to reduce emissions through conservation, efficiency measures, and greater reliance on renewable energy such as wind and solar energy.

GO GREEN!

To reduce energy use and pollution, be sure to close curtains, shades, and blinds in your dorm room or home at night on cold winter days. Uncovered windows are a major source of heat loss.

Fluidized Bed Combustion

Efficiency can occur at several levels. End-use efficiency refers to actions that users like you and I can take to use energy more efficiently. Combustion efficiency is another option. Both can be combined to create tremendous savings in energy demand and pollution.

Over the years, engineers and scientists have developed many ways to burn coal more efficiently. One of the most promising of the dozen or so alternative technologies is known as **fluidized bed combustion.** In this process, coal is crushed and mixed with bits of limestone and then propelled into a furnace in a strong current of air (Figure 22.6). The particles mix turbulently in the combustion chamber, ensuring very efficient combustion and therefore low carbon monoxide emissions. The furnace also operates at a much lower temperature than a conventional coal boiler, thus reducing nitrogen oxide emissions. The limestone reacts with sulfur oxide gases, producing calcium sulfite or calcium sulfate, reducing sulfur oxide emissions from the stack gases. These compounds, however, must be disposed of.

Several large demonstration projects are operating now in Colorado and Kentucky. Should they prove successful, they

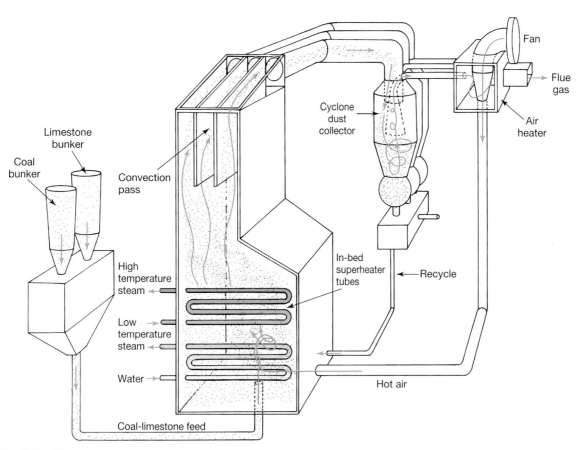

FIGURE 22.6   Fluidized bed combustion. This process burns crushed coal blown into a furnace mixed with tiny limestone particles. The air turbulence in the furnace ensures thorough combustion, thus increasing efficiency. The limestone reacts with sulfur oxide gas, removing most of it from the smokestack. Steam pipes in the furnace help maximize heat efficiency.

could pave the way for wider use of this technology. But don't expect an overnight transition: At least 30 to 50 years would be required for this technology to make a substantial contribution to our electrical generating capacity.

As signs of global warming become more evident, many nations have installed natural gas plants to generate electricity to meet peak loads—that is, energy requirements beyond those coal fired plants could provide. Peak loads often occur later in the afternoon, for example, when peoples' air conditioners are running, people are returning from work, and factories are still running. Natural gas is a much cleaner fuel than coal. It produces no particulates and much less carbon dioxide per unit of energy consumed. (Carbon dioxide emissions alone are 50% to 70% lower.) Natural gas is burned in turbines similar to jet engines, which are much more efficient than coal-fired power plants. Unfortunately, U.S. natural gas supplies have remained constant since the mid-1970s, and rising demand has caused a dramatic increase in prices. Many utilities rarely use their natural-gas-fired power plants anymore. Because of these problems and a desire to reduce carbon emissions, many nations are promoting end-user efficiency measures and are turning to environmentally friendly alternative fuels such as wind energy, solar energy, and clean-burning biofuels (for example, ethanol from corn). These measures do not contribute carbon dioxide to the atmosphere (Chapter 22).

Clean Coal   There has been a lot of talk lately about **clean coal.** Industry advocates use this term to indicate coal that is burned more efficiently, as in fluidized bed combustion, producing fewer pollutants. Cleaner coal combustion, they suggest, can also be achieved by capturing carbon dioxide and disposing of it, for example, in abandoned oil wells. This is known as **carbon sequestration.** This, say advocates, would allow countries like the United States to use our massive reserves of coal without contributing to costly global warming and climate change.

Critics point out that while carbon sequestration may sound good, it is impractical. Capturing carbon dioxide from smokestacks, for example, could be very difficult. It could also consume a lot of energy, as the gas would have to be condensed for ease of shipping. And how would all this carbon dioxide be shipped to oil wells? By truck? By pipeline? How much would this cost? How close are oil fields?

Critics argue that carbon sequestration is just a ploy used by the coal industry to promote further use of its product and that while advocates of coal are promoting clean coal, very little is done to implement more efficient coal-burning technologies on a broad scale.

Coal Gasification   Another possible technology is **coal gasification,** a process in which combustible gas is produced from coal by one of several methods. This technology is viewed as a means of putting to use the world's enormous supplies of coal. Proponents argue that it could provide a supply of gas to replace declining reserves of natural gas.

Coal gasification facilities can be designed to be cleaner than conventional coal boilers now in place throughout the country. In this process, a coal–water mixture, called a **slurry,** is injected with oxygen into a heated chamber, producing three combustible gases: carbon monoxide, hydrogen, and methane. The heated gas is then cooled a bit and purified. The resultant gas burns as cleanly as natural gas. Coal gasification, commonly used in the mid-1800s to produce gas from coal, could provide gas for home heating and other uses, a need that will become acute as domestic natural gas supplies become depleted. Like fluidized bed combustion, it produces less nitrogen oxide than conventional coal boilers and eliminates much of the sulfur dioxide as well. Both technologies, however, produce solid wastes that must be carefully disposed of to prevent groundwater contamination. In addition, further combustion of fossil fuels adds more carbon dioxide to the atmosphere.

Coal Liquefaction   Coal can also be treated to produce a thick, oily substance in a process called **coal liquefaction.** At least four major processes now exist, each of which adds hydrogen to coal to produce oil. The oil can then be refined, just as crude oil is, to produce a variety of products, among them jet fuel, gasoline, kerosene, and many chemicals used to manufacture drugs, plastics, and a host of other products.

Coal liquefaction plants currently in operation show that this technology, while feasible, will be costly. It also generates numerous potentially harmful pollutants, such as phenol. And it would do nothing to reduce carbon dioxide levels.

Oil

The year was 1859, and the place was Titusville, PA. When "Colonel" Edwin Drake's steel drill hit 20 meters (70 feet), a black, foul-smelling liquid came gushing from the well, signaling the dawn of a new energy era (Figure 22.7). After finds in several

other countries, including Russia, less than a century later, oil had become the world's most important source of energy.

Being liquid and relatively easy to transport long distances, either by ship or by pipeline, oil is often viewed as an ideal fuel, burning dirtier than natural gas but cleaner than coal. It provides about 40% of the United States' energy needs and a somewhat higher percentage of the demand in other industrial nations. Canada satisfies about 27% of its total energy demand with oil.

Oil Reserves   Unlike coal, oil is in short supply. According to analysts, the world's oil reserves could be gone by 2018.[1] Estimates of oil reserves are difficult and often inaccurate, however. Some oil geologists believe that the **ultimate production,** the amount of oil we will eventually be able to recover, is 2.5 times greater than once thought. Even so, rising demand (assuming a 5% per year increase) would wipe out these additional reserves, if indeed they exist, by 2038.

Although much of the debate has until recently been focused on when the world will economically deplete its oil resources, many point out that there's a more important factor to be concerned about: peak production or peak extraction (Figure 22.2). Many oil analysts believe that oil extraction will follow a bell-shaped curve, rising from a very low level, peaking, and then declining (Figure 22.2A). The end of oil could occur between 2018 and 2038. Peak extraction could be occurring right now. That is, global oil production may currently be at the top of the bell-shaped curve, where oil extraction tops out and demand exceeds supply. It's during this historic period that supply cannot meet up with demand, causing rising prices that could devastate the global economy.

The outlook for oil in the United States is even grimmer (Figure 22.2B). Oil extraction peaked in the early 1970s. Since

---

[1]Of course, this will never happen. As oil supplies fall, it will become prohibitively expensive. We'll never deplete the world's oil. We'll just run out of cheap oil and most likely switch to other energy sources.

FIGURE 22.7   America's first oil well, at Titusville, PA. This photograph was taken in 1864.

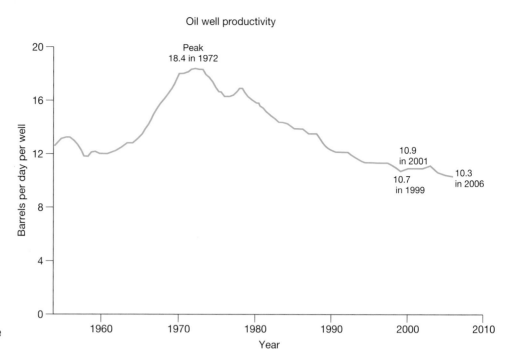

FIGURE 22.8   Oil well productivity. Note that oil production from U.S. wells has experienced a dramatic decline. Most of the good sites have been tapped out.

then, domestic production has fallen steadily, and geologists estimate that we've consumed about 75% to 80% of our ultimate production. Despite an increase in the number of wells drilled, daily output from domestic wells fell from 11 million barrels per day in 1973 to 5.8 million barrels in 2000, according to the Energy Information Administration. It continues to decline, falling to just a little over 5 million barrels in 2007. (Figure 22.8 shows this trend but in barrels per well per day.) Because of this, U.S. reliance on imported oil has increased dramatically (Figure 22.9). In 2005, the United States imported nearly 56% of its oil.

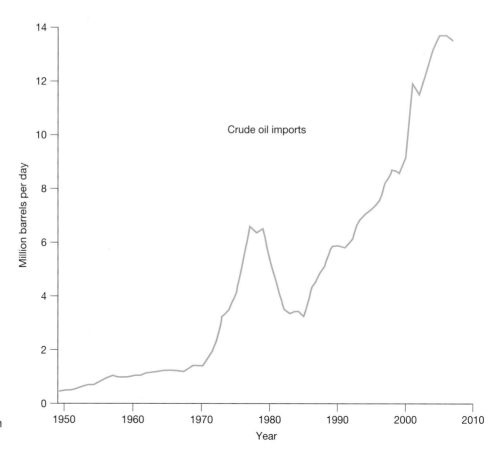

FIGURE 22.9   Crude oil imports. Because of declining domestic production, the United States has grown steadily more dependent on foreign oil, much of it from unstable sources in the Middle East.

Declining oil production in domestic wells means that the United States will become more and more dependent on imports. Unfortunately, much of the world's oil reserves are located in the politically unstable Middle East (Figure 22.10). Unless a clean, economical substitute is found, the United States could face frightening economic times, as could many other industrial nations that depend heavily on oil (Japan and Germany, for example).

The Impact of Oil Production and Consumption Oil comes from wells on land and sea (Figure 22.11). The impacts of oil on the sea are discussed in Chapter 10. Land-based drilling can have many impacts as well. Roads and well sites, for instance, destroy wildlife habitat and wilderness and disturb sensitive species. They also increase soil erosion. Leaks from wells or spills from pipelines can cover land with a thick, gooey residue that is difficult and costly to remove. Oil kills vegetation and seeps into the ground. For this reason, government efforts to permit oil and gas drilling in the 0.6-million-hectare (1.5-million-acre) coastal plain of the **Arctic National Wildlife Refuge (ANWR)** in Alaska continue to create a storm of controversy. By one estimate, development would force 20% to 40% of the caribou population to move out of this 8-million-hectare (19.5-million-acre) refuge or perish as a result of habitat loss. Huge flocks of snow geese, estimated at 300,000 birds, that use the coastal plain as a staging ground on their annual migration would also be forced to find new territory; many would likely die. Musk oxen, only just reintroduced in 1969 and 1970 and now numbering about 500, would feel the ill effects of oil development as well.

No one knows how much oil, if any, lies underneath the sensitive coastal plain. One estimate is that there are 4.2 billion barrels. Although that sounds like an impressive amount, it's only about a half year's consumption. (Americans consume about 7 billion barrels a year.) Full-scale development of the region would therefore supply only a fraction of America's projected oil demand. Although new plans for ANWR suggest that the oil can be removed with little environmental damage, experience in nearby Prudhoe Bay, lying 160 kilometers (100 miles) to the west, suggests that the damage could be significant. According to a report on Prudhoe Bay, between 400 and 600 oil spills are reported every year. In 1986, 240 million liters (64 million gallons) of toxic wastewater containing heavy metals, hydrocarbons, and chemical additives were released onto the tundra. Gas vapors and combustible liquids routinely burned in oil fields create plumes of black smoke that stretch 100 miles from the oil fields. Nitrogen oxide and sulfur dioxide emissions from plants, biologists fear, could acidify the fragile tundra ecosystem. Making matters worse, the landscape around Prudhoe Bay is littered with broken-down vehicles, junked airplanes, used batteries, foam insulation, tires, scrap metals, and other waste. The damage, some environmental scientists fear, may be irreparable.

All of this illustrates the extent to which our desire for energy affects the environment and disrupts the lives of the many species that share this planet with us. Because of public outcry, in 1991, the U.S. Congress defeated a bill that would have opened up the refuge for drilling, despite then-President George H. W. Bush's support. In ensuing years, oil companies and powerful representatives from oil-producing states have continued to lobby for passage of legislation that would allow them to drill in ANWR. During his administration, President George W. Bush supported oil development in this region, but proposals to open up a small area for development were continually defeated. At this writing, however, efforts are still under way to open the wildlife refuge for oil development.

GO GREEN!

If your home or apartment is equipped with ceiling fans, use them in the summer and the winter. Ceiling fans help keep a home warmer in the winter and cooler in the summer, and they reduce your heating and cooling bills. Ceiling fans can reduce cooling costs by up to 40% and heating costs by up to 10%.

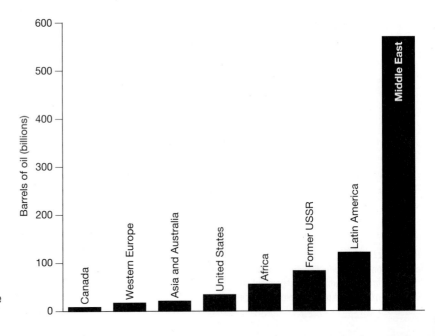

FIGURE 22.10 World oil reserves. This graph shows where the world's oil reserves are located.

FIGURE 22.11 Oil wells off the coast of Texas. Wells like these release oil into the ocean, which may soil nearby beaches and harm coastal wetlands.

Natural Gas

Natural gas is a combustible gas that consists primarily of methane (CH_4). Like coal and oil, it is a fossil fuel. Produced by decomposing plant and animal remains that were buried in the Earth for millions of years by sedimentary deposits, natural gas deposits often accompany coal and oil deposits.

Today natural gas supplies about 23% of the energy needs of the United States and 31% of Canada's energy supply. Easily transported within a country by pipeline, it is used primarily for heating buildings, cooking, heating water for domestic consumption, powering industrial processes, and generating electricity.

Natural Gas Reserves The picture for natural gas is a bit brighter than for oil globally, but not as good as that for coal. Estimates of the amount of natural gas that will be recovered from the Earth before supplies run out—the ultimate production—vary considerably. According to French researchers Peter Gerling and Hilmar Rempel, the ultimate recovery for conventional natural gas is about 4,870 trillion cubic feet, which is probably a conservative estimate. Gerling and Rempel predicted in 2003 that global natural gas production will last 64 years—that is, until around 2067—at the current rates of consumption and assuming no new discoveries are made. More important, they predict that natural gas production will peak around 2019 if current levels of consumption remain the same. From that point on, natural gas production will not be able to keep up with demand.

The outlook for natural gas in the United States is less bright. Some experts believe we have consumed over half of the nation's natural gas supplies. What remain are smaller pools that will run dry in a few years. Because of this, many new wells must be drilled just to keep production from falling. In west Texas, for instance, energy companies drill 5,500 new wells a year, just to keep production levels from dropping.

Because it burns so cleanly, natural gas could be an important transitional fuel—that is, a relatively clean fossil fuel that replaces oil and coal while the nation develops even cleaner, sustainable, renewable alternatives (Chapter 23). It could be used to power vehicles in regions. It could be used to generate electricity, heat homes, and power factories.

In the United States, however, the increasing use of natural gas has already resulted in excess demands on production capacity and dramatic increases in price. Importing more natural gas could help, but natural gas is a difficult fuel to import. The United States imports about 15% of its natural gas from Canada via pipelines (representing half of Canada's production). Natural gas can also be compressed and shipped in tankers. The United States currently imports about 2% of its natural gas this way. This process is costly and energy-intensive and is generally not considered a viable option. Important as it is in the short term, natural gas will very likely need replacement by clean, reliable, affordable renewable fuels.

Fortunately, many options exist; some are more sustainable than others. Among the least sustainable is coal gasification. More sustainable options are improved energy efficiency, solar, wind, and methane from organic wastes such as garbage, sewage, sludge, and livestock manure. These options are discussed in Chapter 23.

Oil Shale

Oil shale is a grayish brown sedimentary rock that was formed millions of years ago from the mud at the bottom of prehistoric lakes. Contained within the rock is a solid organic material known as **kerogen** (Figure 22.12). When heated to high temperatures, the rock gives off its oily residue, called **shale oil**. High-grade oil shale can produce up to 120 liters (30 gallons, or about three-fourths of a barrel) of oil per ton of rock. Like

petroleum, this oil can be refined to produce gasoline, jet fuel, kerosene, and a variety of feedstocks used by the chemical industry to produce fabrics, drugs, and plastics. The most valuable deposit in the United States is in the Green River Formation, a 43,000-square-kilometer (16,500-square-mile) region found in Colorado, Wyoming, and Utah (Figure 22.13). Eighty percent of this land is federally owned.

Scientists estimate that, with current technologies, 80 billion to 300 billion barrels of oil could be extracted from these deposits. Although that may seem like a lot, human society currently consumes about 23 billion barrels per year. Americans consume about 7 billion barrels of crude oil per year. This deposit, then, would provide enough oil to satisfy domestic demands (at the current rate of consumption) for 11.5 to 43 years. Doing so, however, would exact enormous economic and environmental costs.

Environmental Impact of Oil Shale Production Oil shale can be processed in **surface retorts,** large steel vessels in which crushed shale is heated to drive off the oil. Shale to supply surface retorts may come from either underground mines or surface mines, depending on the depth of the deposit. Surface-mining operations, however, tear up large tracts of land. Much of the land overlying the Green River Formation is prime habitat of mule deer. Hundreds of thousands of mule deer live in this area. In addition, the retorts produce enormous amounts of waste, or **spent shale** (shale that has been burned). A modest operation producing 50,000 barrels per day would also produce 53,000 metric tons of spent shale per day, which would have to be disposed of properly to avoid contaminating groundwater and surface waters. Adding to the problem, crushing the shale before it can be retorted increases the volume of the shale by about 12%, so not all of it can go back into the excavation.

Retorts are notorious polluters. Sulfur oxides, nitrogen oxides, heavy metals, and various organic pollutants, all toxic

FIGURE 22.12 Block of oil shale and a beaker of the oil extracted from the shale.

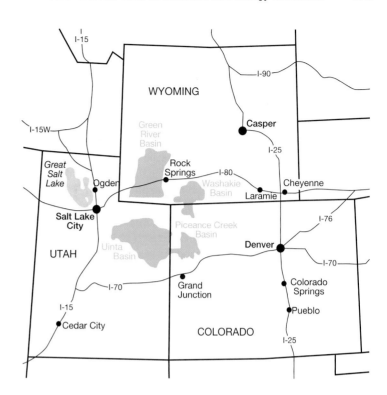

FIGURE 22.13   Location of the main U.S. oil shale deposits in Colorado, Utah, and Wyoming.

to humans and wildlife, could foul the western skies if oil shale becomes a viable business. Surface retorting would produce huge quantities of the greenhouse gas carbon dioxide. And surface retorts require tremendous volumes of cooling water—about 2.5 barrels per barrel of shale oil produced. Water, of course, is a hotly contested commodity in the rapidly growing and extremely arid West, which since the mid-1970s has been suffering from persistent drought, caused in part by global warming.

Some companies have experimented with another process called in situ retorting, to avoid the solid-waste problems. In an **in situ retort,** the oil shale deposit is fractured with dynamite and set afire underground. In theory, the fire burns the shale (consuming some of the useful fuel). Heat from combustion, however, drives off the rest of the kerogen, in a vapor state. It is pumped out of the deposit, condensed, and then refined. In situ retorting, tried in the late 1970s to avoid disturbing the land and generating solid wastes, has proved difficult. Fires go out easily because of incomplete fracturing. Complicating matters even more, groundwater often seeps into the oil shale bed, dousing the fire. Controlling pollution is also more difficult in situ. As a result, this technique has largely been abandoned.

Surface retorting and in situ retorting failed for a number of reasons, just cited, but also because of costs. High costs were due, in large part, to the fact that it requires huge amounts of energy to extract shale. Like tar sands, discussed next, oil shale has a low net energy efficiency. **Net energy efficiency** is a measure of the net energy balance—how much energy is used to extract a resource such as oil or oil shale versus the amount of energy available from the finished product. Some researchers believe that the net energy efficiency of oil shale is negative; that is, it takes more energy to extract the shale than we get from it.

To address these concerns, Shell Oil is currently experimenting with a new technique. In this method, workers drill holes into the oil shale and then insert electrical wires. Current flowing through the wires heats the shale, causing oil to be released. The shale oil is then pumped to the surface. The surrounding rock is cooled to prevent oil from migrating out of the heated zone.

Although still in the experimental phase, this technique is viewed as a promising alternative. Critics point out that it would require a huge amount of energy to heat and cool the shale and then extract the oil. New and large coal-fired power plants would be required to power the facilities. These power plants produce carbon dioxide and toxic solid wastes (fly ash, bottom ash, and sludge from the scrubbers). They also require enormous amounts of water in a region plagued by shortages.

Tar Sands

In some parts of the world, oil has migrated into neighboring layers of sandstone, creating **tar sands** or **oil sands.** Tar sands contain an organic material called **bitumen** that can be extracted from the sand. It is then refined like shale oil to produce a variety of fuels and chemicals. Significant deposits of bitumen are found in Alberta, Canada, and the United States. Although six states have economically attractive deposits, Utah is the leader, with over 90% of the commercially feasible tar sands.

Lying deep within the Earth's crust, U.S. deposits contain an estimated 27 billion barrels of bitumen but cannot be readily surface-mined. To recover the oil, hot steam is generally pumped down into the deposits to free the bitumen, which is then pumped to the surface in recovery wells. When all is said and done, only about 1 billion barrels are likely to be recovered from our deposits, equal to about one-sixth of the United States' annual oil consumption.

Canadian deposits are much greater than those in the United States. In addition, the richer Canadian deposits are readily surface-mined. The tar sands are processed in above-ground facilities. They contain an estimated 1.7 trillion barrels of oil. Some industry analysts believe that about 300 billion barrels of oil are recoverable, enough to supply world demand for 13 years at the current rate of consumption.

Unfortunately, tar sand production via surface mines is very energy-intensive. The energy needed to mine and process enough tar sand to produce a barrel of oil is equivalent to about 0.6 barrel of oil. The energy output of a fuel supply, or its net energy efficiency, takes into account what must be invested to get energy in the first place. Low net energy efficiency, say experts, will probably stifle tar sand production at least from surface mines. Surface mines will also require leveling of huge forests.

The Future of Fossil Fuels

The energy supply picture boils down to this: Oil, our most widely used fuel, is fast on the decline in the United States and abroad. Many oil analysts believe that demand is currently outstripping supply, causing a dramatic increase in price and severe economic problems.

The prospects for natural gas are better globally, although production could peak within a decade or so. The outlook for U.S. supplies is less rosy. Many analysts believe that domestic production has peaked and that we're quickly depleting remaining reserves.

The prospects for coal are the best. The world has abundant supplies of coal, and the United States has about one-third of the world's supplies. But coal combustion, no matter how efficient, increases atmospheric carbon dioxide levels, contributing to global warming. Moreover, simply reducing the emissions of acid-generating pollutants from coal-fired power plants only shifts the problem: Instead of going up the smokestack, the pollutants become a hazardous waste in need of careful disposal (Chapter 18).

It is time, say many experts, to find substitutes for oil that can be used to heat homes and produce liquid fuels for cars and other vehicles. Oil is also a source of numerous chemicals that are important in our society, including moth balls, resins used to make plastics, and chemicals used to make medicines and even chemical pesticides. Oil shale and tar sands are two alternatives, but their reserves are relatively small compared with global oil demand. They are also extremely costly to develop, both environmentally and economically. Low net energy efficiencies could hinder their production. Biofuels, such as ethanol from crops, may be an essential component of a new, sustainable energy strategy. Ethanol, a renewable fuel made from corn and other crops, could replace gasoline entirely. It reduces net carbon dioxide production. But corn ethanol has a low net energy efficiency, currently about 1.7 (meaning you get 1.7 units of energy for every unit of energy put into the process). The net energy efficiency of sugar cane ethanol is much higher, about 8. For every unit of energy invested in growing sugar cane and producing ethanol, you get 8 units of energy out. Efforts are under way to improve the net energy

GO GREEN!

Talk to your parents and other relatives about having an energy audit of their homes. Energy auditors identify numerous ways homeowners can reduce energy demand, saving them lots of money while making their home more comfortable. Contact local utilities or energy service companies in your area to locate an auditor.

efficiency of ethanol production. Researchers are working on a technique to produce ethanol from the cellulose from a variety of plants such as corn and switchgrass. Plants produce a lot more cellulose than starch.

Many energy advocates are also pinning their hopes on clean-burning hydrogen gas derived from the breakdown of water. Methane from trash, crop residues, and vast amounts of animal waste also could help, as could biodiesel. Electric cars and buses powered by solar and wind also could play a huge role in replacing oil. Energy efficiency, of course, will be essential to meeting our needs from renewable forms of energy.

Hydrogen gas can be generated from one of the Earth's most abundant substances, water. Passing electricity through water causes the water molecules to split; in so doing, they produce oxygen and hydrogen. Hydrogen, in turn, can be burned to produce heat, for instance, to power a vehicle, heat a home, or cook food. When hydrogen gas burns, it combines with oxygen, once again forming water. The only pollutant produced in the process is a small amount of nitrogen oxide. Hydrogen can also be fed into a fuel cell, discussed in Chapter 18, to produce electricity. Unlike burning of fossil fuels, hydrogen combustion produces no sulfur dioxide or carbon dioxide.

While hydrogen appears to be an ideal fuel, electricity is needed to make hydrogen. It would be more efficient to use the electricity directly—for instance, to power a car—than to use it to make hydrogen to feed into a fuel cell to make electricity to power a car. This energy option is described more fully in Chapter 23.

Another option that is sometimes proposed as an answer to the declining supplies of polluting fossil fuels is nuclear energy. Is it a good alternative?

22.3   The Nuclear Energy Option: Is It Sustainable?

The world's first atomic bomb was dropped on Hiroshima, Japan, in 1945. This momentous and catastrophic explosion, which helped conclude the long and costly Second World War, also ushered in the Atomic Age—an age of nuclear weapons and nuclear power. This section is primarily about nuclear energy and radiation. It examines the impacts of nuclear power as well as its benefits, starting with some background information.

Understanding Atomic Energy and Radiation

To understand nuclear energy, you must first understand a little bit about the atom.

Atomic Structure   All matter, whether solid, liquid, or gaseous, is composed of tiny particles called **atoms.** Put another way, atoms are the fundamental units of all living and nonliving matter. Each atom, in turn, is composed of a dense, centrally located **nucleus** containing positively charged particles called **protons** and electrically neutral particles called **neutrons.** The nucleus contains virtually all of the mass of an atom. Surrounding the nucleus is a region called the **electron cloud,** where tiny, almost massless, negatively charged particles, known as **electrons,** are found. They orbit around the nucleus, ever attracted to its positive charge and always in motion at nearly the speed of light.

Atoms combine with other atoms to form molecules, such as water (H_2O), which contains two atoms of the element hydrogen and one atom of the element oxygen.

Isotopes and the Origin of Radiation Early scientists thought the variety of matter they encountered resulted from the fact that pure substances, which they called **elements,** could combine to form a variety of different substances, a hypothesis that modern science has shown to be true. Today, scientists realize that elements are pure substances. Elements contain only one type of atom, and scientists have identified over a hundred different types of elements, 92 of which occur naturally in the Earth's crust. Scientists have found that each element differs from the next in the number of protons it contains in its nucleus. Thus, the element carbon always has 12 protons, and the element oxygen always has 16 protons. In fact, scientists classify the elements by the number of protons in their nucleus; this is called the **atomic number.** The atomic number of carbon is 12, and the atomic number of oxygen is 16.

In any given element, the number of electrons in the electron cloud of its atoms is always equal to the number of protons in the nucleus. Thus, atoms are electrically neutral. Although the number of electrons and the number of protons in the atoms of any given element are always the same, the number of neutrons may vary. For instance, all atoms of the element uranium have 92 protons in their nuclei. However, some uranium nuclei contain 143 neutrons, and others have 146 neutrons.

Scientists add the number of protons and neutrons to obtain the **atomic mass.** Each proton and neutron has a mass of one unit. Therefore, uranium atoms with 92 protons and 143 neutrons have an atomic mass of 235. They're written as U-235. Uranium atoms with 146 neutrons have an atomic mass of 238. These different forms of the same element are called **isotopes.**

Because atoms of a given element differ with regard to the number of neutrons, small but measurable differences in mass exist within a group of similar atoms. How is mass determined? For the most part, the mass of an atom is equal to the sum of the number of protons and neutrons. Although electrons are present, they are so light that they contribute little to the overall mass of an atom. To continue with the previous example, because some uranium atoms have more neutrons than others, their masses vary slightly.

Most elements are a mixture of isotopes. Scientists have found that many isotopes are stable but others are not. Unstable isotopes give off radiation and are referred to as **radioisotopes.**

The Nature of Radiation In 1896, the scientist Henri Becquerel accidentally discovered that the element radium, when placed over a photographic plate in the dark, exposed the film. Pierre and Marie Curie, two French scientists, later discovered that radium is radioactive—that is, it emits radiation, which was responsible for Becquerel's unexplained phenomenon. Three forms of radiation come from the nuclei of radium atoms: alpha particles, containing two protons and two neutrons; beta particles, negatively charged particles similar to electrons; and **gamma rays,** which are similar to X-rays.

Alpha particles are relatively heavy particles and can travel only 8 centimeters (3 inches) in air. They are easily stopped by material as thin as a sheet of paper or human skin (Figure 22.14). They are dangerous, however, if atoms that emit them are breathed into the lungs or ingested in contaminated food. In the lung or in the digestive tract, they can irradiate cells, causing mutations and cancer.

Beta particles are much lighter particles and can travel much greater distances. They can even penetrate skin to irradiate underlying tissues. Wood or a thin layer of lead will stop them dead in their tracks.

Gamma rays are potentially the most dangerous form of radiation, for they can pass through walls and can easily penetrate the skin and muscle, reaching internal organs. A thick concrete or lead shield is needed to block them.

Radioactive Decay Radioisotopes emit radiation—alpha particles, beta particles, gamma rays, and others of lesser importance—from their nuclei to become more stable. Thus, over time, a mass of radioactive material actually decreases. The measure of its decrease in mass is called its half-life. Technically, **half-life** is the time it takes a given mass of radioactive material to decay to one-half its present mass. A kilogram of a radioactive compound with a half-life of 10 years, for instance, will weigh

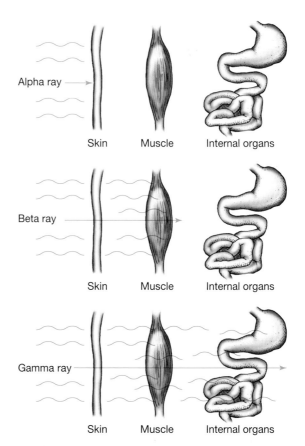

FIGURE 22.14 Relative ability of alpha, beta, and gamma rays to penetrate the body.

one-half of a kilogram if left to sit from 2000 to 2010. By 2020, it will weigh one-quarter of a kilogram.

The half-lives of isotopes vary from a fraction of a second to tens of thousands of years. For example, iodine-131 has a half-life of eight days; cesium-137, 27 years; strontium-90, 28 years; carbon-14, 5,600 years; and plutonium-239, 24,000 years.

Nuclear Power

With this background information in mind, we now turn our attention to the nuclear energy option.

Atomic Fission and Chain Reactions   **Nuclear reactors** and the first generation of nuclear bombs (the fission bombs) do not capture energy from **radioactive decay**, but instead draw their energy from **nuclear fission**— the splitting of certain atoms that occurs when they are struck by certain forms of radiation.

Nuclear power plants are fueled by a radioactive element known as **uranium.** Uranium-235 atoms are highly fissionable, or fissile. Uranium also gives off a form of radiation. When this radiation strikes a nearby uranium nucleus, it can cause the latter to split into fragments, called **daughter nuclei** (Figure 22.15).

GO GREEN!

Talk to your parents about installing a programmable thermostat in their home. When properly programmed, these simple devices can dramatically reduce energy bills by adjusting temperature to correspond to occupancy.

Enormous amounts of energy are given off during fission. So are additional neutrons. The neutrons create a cascade effect or **chain reaction** as they bombard additional uranium nuclei, causing still more fissions. In power plants, the rate of fission is carefully controlled by limiting the concentration of uranium in the fuel and by other means discussed later. Exceedingly complex in design, nuclear plants simply use the energy given off by this controlled chain reaction to boil water to make steam. Steam turns electrical turbines. In atomic bombs, uranium is more highly concentrated, and the chain reaction is uncontrolled or unmodulated, causing a huge explosion that releases enormous amounts of energy. Atomic explosions cannot occur in reactors because of the low uranium concentration. As we shall see in our discussion of the Chernobyl and Three Mile Island accidents, however, explosions at nuclear power plants can occur for other reasons and can rip apart a reactor, spewing radioactive material into the air.

Uranium contains lots of energy. In fact, 1 kilogram (2.2 pounds) of uranium produces the same explosive force as 9,000 metric tons of TNT. A kilogram of uranium releases as much energy as nearly 3,000 metric tons of coal. It is no wonder that modern civilization once set its sights on this technology.

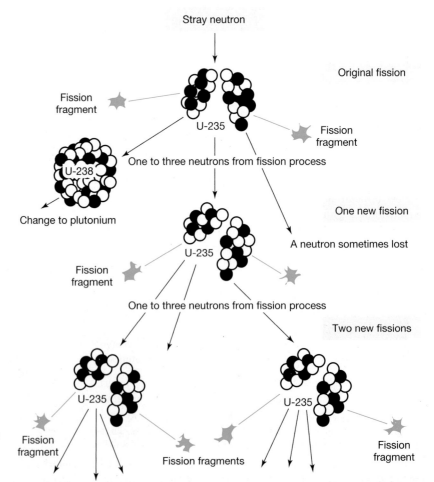

FIGURE 22.15   Nuclear fission. The chain reaction starts when a uranium-235 atom is struck by a neutron given off by other uranium-235 nuclei. Uranium is split into fission fragments, or daughter nuclei, and heat. Additional neutrons are given off, thus splitting other atoms.

The Anatomy of a Nuclear Reactor

The uranium fuel used in all large U.S. nuclear power plants is uranium-235. It comes from uranium ore. Uranium ore is a mixture of uranium-235, which is fissionable, and uranium-238, which is nonfissionable. In the United States, most uranium ore is found in the western states.

To create fuel for nuclear power plants like the one shown in Figure 22.16, the nuclear industry must first increase the concentration of uranium-235 to about 3% at a fuel enrichment plant. The enriched fuel, called **yellow cake,** is then formed into pellets. These are inserted into thin, stainless steel tubes that when filled are known as **fuel rods**. About 4 meters (12 feet) long and about 12 centimeters (0.5 inch) in diameter, fuel rods are then bundled together, 30 to 300 per bundle, and lowered into the core of the reactor (Figure 22.17).

The **reactor core** provides a site where nuclear fission can be carefully controlled. Bathed in water to draw off heat, the reactor core is contained by a **reactor vessel** with a steel casing that is 15 centimeters (6 inches) thick (Figures 22.18–22.20).

In most countries, the reactor vessel is usually housed in a **containment building,** a domelike structure with 1.2-meter (4-foot) walls of concrete (Figure 22.21). Its function is to contain the radiation in case the reactor vessel or pipes rupture during an accident. Unfortunately, historically not all reactors have had complete containment buildings. The nuclear reactor at Chernobyl, discussed shortly, for example, had only a partial containment building, which proved inadequate in the fateful accident in 1986. At the time, the same design was used in 20 of the former Soviet Union's 44 reactors and 1 U.S. reactor in the state of Washington (closed down shortly after the Chernobyl accident), used to make **plutonium** for nuclear weapons.

The intensity of the chain reaction is controlled primarily by neutron-absorbing **control rods.** Made of cadmium or boron steel, the control rods fit between the fuel rods. They can be raised or lowered to control the rate of fission. Raising them starts the chain reaction, and lowering them back into place can shut down the reaction almost entirely. Intermediate levels of fission occur between these two extremes.

The energy released from nuclear fission in the fuel rods heats the water bathing the reactor core. To avoid radioactive contamination, heat is then transferred to water circulating in huge pipes in the wall of the reactor vessel. The superheated water boils and produces steam (Figure 22.22). The steam is used to spin a turbine that turns an electrical generator.

Besides the reactor vessel and containment building, nuclear reactors are equipped with a host of safety devices, among them backup power units for the controls, redundant electrical wiring in case one circuit burns out, and redundant pipes in case one bursts. Perhaps the most important is the **emergency cooling system,** which provides cold water to the

FIGURE 22.16   The largest operating nuclear power plant in New England. This is the Millstone Nuclear Power Station, in Waterford, CT, located on the north shore of Long Island Sound.

FIGURE 22.17 A nuclear fuel assembly storage rack containing loaded fuel rods at the Yankee Nuclear Power Station.

FIGURE 22.18 An 800-ton reactor vessel, 72 feet long and 22 feet in diameter. Heat from the nuclear reactions contained in the vessel will produce steam sufficient to generate 800 megawatts of electrical power.

reactor core, should it become overheated. The result is a technology whose complexity exceeds most others known to modern civilization.

Unfortunately, this complex technology can yield some costly and hazardous surprises. In 1979, hydrogen gas built up in the Three Mile Island reactor, exposing the reactor core and threatening to blow a hole in the reactor vessel. A catastrophic explosion occurred in the Chernobyl reactor. Both were events that nuclear engineers had not foreseen. Interestingly, the explosion at Chernobyl was originally thought to have been caused by steam, but later accounts indicate the explosion may have been a nuclear one, albeit a weak one. In addition, studies suggest that more radiation was released into the environment than estimated by the government of the former Soviet Union.

Pros and Cons of Nuclear Fission Few technologies have been as hotly debated as nuclear power. The following discussion summarizes the major views for and against nuclear power.

View in Favor of Nuclear Power

1. Nuclear power can provide an abundant supply of electrical energy. Although uranium-235 is a limited resource, proponents of nuclear power believe that a family of reactors, called **breeder reactors,** which convert abundant uranium-238 into fissionable plutonium-239, could provide electrical energy for hundreds of years. In fact, some experts predict that the spent fuel rods and wastes from uranium-processing plants in the United States alone, which contain abundant amounts of uranium-238, could last 1,000 years.

   The breeder reactor fuel rods contain small amounts of plutonium-239 and larger amounts of uranium-238. Neutrons given off by plutonium-239 undergoing fission are absorbed by uranium-238 atoms, converting them to plutonium-239 (hence the name *breeder reactor*). Some of the new plutonium-239 undergoes fission, producing energy.

   In theory, breeder reactors should create more plutonium than they consume. In fact, for every 100 atoms of plutonium-239 that undergo fission, 130 atoms are produced. This fissionable fuel could be extracted and used to generate power in other reactors.

2. Nuclear power helps reduce our dependency on foreign oil. Proponents of nuclear power believe that our dependency on oil could be further reduced by switching to electricity generated by additional nuclear power plants.

3. Nuclear power plants cause low radiation exposure. Nuclear power plants, when operating correctly, release very little radiation. In fact, one study showed that coal-fired power plants emit more radiation from their smokestacks. (Coal may contain small amounts of radioactive materials.) On the average, Americans receive 0.003 millirem (a measure of exposure) a year from nuclear power plants but as much as 100 to 250 millirems from X-rays, television sets, and naturally occurring radioactivity. The added risk of cancer to an individual is equivalent to smoking one cigarette per year.

4. It is environmentally clean. A properly functioning nuclear power plant is a much cleaner source of energy

FIGURE 22.19 Nuclear fuel being lowered into its slot in the nuclear reactor. The reactor contains a grid guide structure (visible in the reactor vessel), control rods, and water.

FIGURE 22.20 Reactor vessel. This is the heart of the 540-megawatt nuclear-fueled electrical power generating plant near Monticello, MN. Cooling water keeps the heat generated by the vessel at about 540°F.

than a coal-fired power plant. In one year, a coal-fired plant producing 1,000 megawatts of electricity also produces 300,000 metric tons of ash, whereas a nuclear plant produces only about a ton of fission waste per year. Coal plants also release carbon dioxide, sulfur dioxide, nitrogen oxides, carbon monoxide, particulates, mercury, and other air pollutants. Solid wastes from coal plants contain selenium, mercury, benzopyrene, and some radioactive materials, such as uranium and thorium.

5. Nuclear power is a safe energy supply. The probability of an accident in the United States that would release large amounts of radioactivity, said supporters for many years, is not more than one in a million—about the probability of being struck by lightning. Other sources of electricity appear to pose far greater risks. For instance, each year, 30 to 110 people lose their lives in coal mine accidents, mishaps during coal transportation, and from air pollution resulting from coal combustion.

6. Nuclear power produces no carbon dioxide. Unlike a coal-fired power plant, proponents argue, nuclear power produces none of the greenhouse gas carbon dioxide.

FIGURE 22.21 Portland General Electric nuclear plant on the shores of the Columbia River at Rainier, OR. The dome-shaped containment building that shields the reactor is a characteristic feature of nuclear plants. Note the mammoth cooling tower to the left.

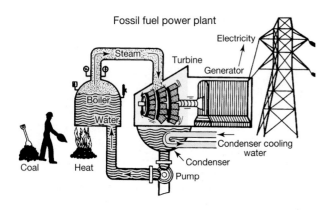

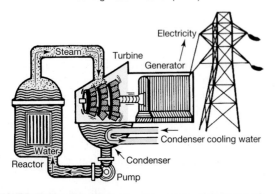

FIGURE 22.22 Comparison of cooling systems for fossil fuel power plant and nuclear power plant (boiling water reactor). Note that the water used to generate steam is not in direct contact with the coolant in the reactor core.

View Opposed to Nuclear Power

1. Despite what proponents say, nuclear energy will not provide an abundant amount of energy. Supporters once hoped that breeder reactors would replace conventional fission reactors by the year 2000, allowing the nuclear industry to tap the generous supply of nuclear wastes from uranium enrichment plants and spent fuel from conventional fission reactors. The U.S. breeder program was canceled in 1983 by the U.S. Congress because of high cost and other problems. France, which now has breeder reactors in operation, has found that they are costly and not as effective as once hoped. The average breeder reactor requires about 30 years to reach the breakeven point—that is, the point at which it produces as much plutonium fuel as it has consumed in the form of uranium-238. Because reactors have a 30-year life span at best, the breeder may never make good on its promise of generating more fuel than it consumes.

   In addition to all of the problems of fission reactors, discussed earlier, the breeder reactors now in use in France, supposedly the prototype for future development, use liquid sodium as a coolant instead of water. Liquid sodium reacts violently with water and burns in air. Therefore, even a small leak in the cooling system could result in a disastrous accident, possibly leading to a meltdown of the reactor's core.

   Despite the potential benefits of the breeder reactor, it seems to be a technology that may become obsolete before it has even gained a foothold. Without the breeder, the nuclear industry will be forced to rely on relatively scarce uranium-235. Some experts believe that uranium production has peaked. The current reserves may last only another 35 years or so.

2. Nuclear energy will not reduce the world's dependency on foreign oil. The argument that nuclear energy will reduce

our dependency on oil is a false one because very little oil is currently used to produce electricity. (In the United States, less than 3% of our electricity is produced from oil or oil by-products such as diesel fuel.) U.S. dependency on foreign oil has increased dramatically in recent years because of falling domestic oil supplies and despite an expansion in the nation's nuclear capacity. The plain truth is that oil and nuclear energy are two very different forms of energy that are used for very different purposes. Unless we convert to electric cars, nuclear power will not displace much oil.

3. Nuclear power plants release harmful radiation. Although the radiation released from the routine operation of a nuclear plant is very small, some isotopes released from nuclear plants accumulate in animal tissues. For instance, cesium-137 concentrates in muscles, and iodine-131 concentrates in the milk of cows and in the human thyroid gland. Thus, low background levels can result in significant exposures to internal organs. Significant amounts of radiation are also released from uranium mines and mills, uranium mill tailings piles, processing plants, waste dumps, and transportation accidents, resulting in significant exposure.

Serious accidents at nuclear power plants can release enormous amounts of radiation. For instance, the 1986 disaster at the Chernobyl reactor in the former Soviet Union released 7,000 kilograms (15,400 pounds) of radioactive material, which contaminated many European countries. That accident injured 1,000 people, killed 31, and caused at least a

$3 billion to $5 billion loss to the former Soviet Union. By various estimates, it may result in 5,000 to 1 million cases of cancer in the former Soviet Union and Europe. If the Chernobyl accident had taken place at one of the many nuclear power plants closer to major urban centers, the damage would have been substantially higher. As a point of reference, a reactor accident at the Salem 2 nuclear reactor in Salem, NJ, could kill 100,000 people from radiation poisoning within a few months of the accident. Approximately 40,000 cases of cancer would result, and the economic damage would come to $150 billion (Table 22.1).

4. It is not environmentally clean. Nuclear power plants are generally cleaner than coal-fired power plants but produce 40% more thermal pollution per kilowatt-hour of electricity than a coal-fired power plant, because fission releases more heat than coal, oil, and natural gas.

5. Nuclear power poses a safety risk. Several minor and two major nuclear reactor accidents have already occurred in the 50-plus-year history of the nuclear industry. A near meltdown of the reactor resulted from a fire in 1975 at the Brown's Ferry plant in Alabama. A study by a prominent nuclear engineer and nuclear advocate, Norman Rasmussen of the Massachusetts Institute of Technology, however, determined that the probability of an accident was not more than one in 10,000 reactor years. Although that may sound reassuring, consider what it really means. In 2003, 440 nuclear reactors were operating in 31 countries, with 30

TABLE 22.1   Potential Severity of Nuclear Reactor Accidents

Plant*	Location	Early Deaths (Thousands)	Cases of Cancer (Thousands)	Financial Losses (Billion 1980 Dollars)
Salem 2	Salem, NJ	100	40	150
Peach Bottom 2	Peach Bottom, PA	72	37	119
Limerick 1	Montgomery, PA	74	34	213
Waterford 3	St. Charles, LA	96	9	131
Susquehanna 1	Berwick, PA	67	28	143
Shoreham	Wading River, NY	40	35	157
Three Mile Island 1	Middletown, PA	42	26	102
Indian Point 3	Buchanan, NY	50	14	314
Millstone 3	Waterford, CT	23	38	174
Dresden 3	Morris, IL	42	13	90

*Plants listed are those in densely populated areas.

Source: Sandia National Laboratory, "Estimates of the Financial Consequences of Nuclear Reactor Accidents." Prepared for the Nuclear Regulatory Commission (Washington, DC, November 1982); Critical Mass Energy Project.

more under construction in 11 countries. If the Rasmussen report estimate is correct, we can expect a major meltdown every 20 years or so. The Three Mile Island and Chernobyl accidents, discussed shortly, occurred only 7 years apart. To be fair, there haven't been any major accidents since Chernobyl.

The Rasmussen report is no longer considered valid. But don't be lulled into complacency. It was discredited largely because it greatly *underestimated* the probability of a nuclear accident. Moreover, it is turning out that many nuclear power plants were built in areas in which evacuation plans are ill conceived or inadequate. Many contain faulty or overly optimistic assessments of how rapidly and efficiently a population can evacuate a region after an accident.

6. Nuclear power production does produce carbon dioxide. Although proponents rightly note that nuclear power plants do not release carbon dioxide, the mining, processing, and transportation of nuclear fuels do. So does the disposal of nuclear wastes. So does the construction of a power plant. That said, nuclear power plants' carbon dioxide emissions, when considered over their entire life cycle, produce considerably less carbon dioxide than coal and natural gas power plants, according to a study by a Swedish utility, Vattenfall, which uses all of these and other sources to generate electricity. The Vattenfall study considered carbon dioxide emissions, averaged over the entire life cycle of their nuclear plant, including uranium mining, milling, and enrichment; plant construction, operation, and decommissioning; and waste disposal. These emissions totaled 3.3 grams per kilowatt-hour of electricity. Emissions from Vattenfall's natural gas plant were 400 grams per kilowatt-hour, and from coal, they were 700 grams per kilowatt-hour. (A kilowatt-hour is 1,000 watts of power over a period of one hour.)

While converting to nuclear power might help lower global carbon dioxide emissions, the Nuclear Energy Information Service points out, "Most experts agree that major action must take place in the next 5–10 years" to combat global warming. They go on to say that building the number of nuclear power plants needed to combat global warming would take decades. Even if the world built the 8,000 nuclear power plants needed to replace coal, world carbon dioxide levels would still increase 65% over the next 30 years.

7. The cost is high. When nuclear power was first conceived, proponents argued it would be so cheap to produce that metering it would not be economical. Fifty years later, that dream is far from reality. Nuclear power is, in fact, one of the most expensive forms of electricity now commercially available, costing 10 to 15 cents per kilowatt-hour, compared with 5 to 7 cents per kilowatt-hour for coal. One of the chief reasons why nuclear power plants are so costly is that they are expensive to build. A large plant costs $3 billion to $6 billion. A comparable coal-fired power plant costs $800 million to $1 billion. Investments in energy efficiency in these ranges would result in far more energy saved than could be produced by either type of plant! Moreover, repairs of nuclear power plants are extraordinarily costly and time-consuming as well,

because of the danger of radiation exposure to workers. Thus, what would be a simple and inexpensive repair in another type of power plant may take months and may cost millions of dollars, further adding to the cost of electricity. Finally, at the conclusion of a plant's useful life span of 20 to 30 years, it must be disassembled and disposed of. The estimated costs for "decommissioning" nuclear plants are about $500,000 to $1 million per megawatt of power generation capacity. A 1,000-megawatt plant would cost an additional $500 million to $1 billion to decommission. Unfortunately, the costs of decommissioning have not been added to fuel bills and will invariably be included as the older plants now reaching obsolescence are disassembled, making nuclear power even more expensive.

8. Public and private support is lacking. In 1975, long before the Chernobyl accident, 64% of the American population polled supported nuclear power; in 1986, after this tragic and costly accident, public support nearly vanished, dropping to 19%. Similar trends were witnessed in other countries as well. In 2008, however, public support of nuclear power had climbed to 67%, according to a poll by Zogby International. High fuel prices and concern over global climate change were two key reasons for the change in public opinion.

Long before Chernobyl, however, U.S. banks and insurance companies had withdrawn their support for the nuclear power industry. It was, they said, too costly and too risky an endeavor to support. Because of this, the U.S. nuclear power industry collapsed in 1979. For over two decades, not one new nuclear plant was ordered in the United States. Many new nuclear power plants were canceled, either in the planning stages or during construction. Despite this trend, the United States still produces 20% of its electricity from nuclear power plants (and nearly 50% from coal-fired power plants). Canada currently derives about 15% of its total annual energy consumption from nuclear power plants, and France generates nearly 80% of its electricity from nuclear power. These and other countries remain strong supporters of this technology.

In recent years in the United States, nuclear power has begun to make a comeback, thanks to strong financial support from the Bush Administration. Several new plants have already been permitted. In 2006, for instance, General Electric and Hitachi announced a joint venture to build two nuclear power plants in Texas. The plants are scheduled to begin operation in 2014. Proponents argue that the new plants are safer and that nuclear power can help combat global warming. Tens of thousands of claims have been established for possible uranium mines in Colorado and Utah.

Health Effects of Radiation

In this section, we discuss two broad categories of effects: nongenetic and genetic effects. The effects of radiation, in general, vary with the age of the individual. Fetuses and newborn infants, for example, are much more sensitive to radiation than adults. Sensitivity also varies according to the type of tissue or organ. Fast-growing cells, such as those lining the intestines, those that produce hairs, and those in the bone marrow, are the most sensitive. Cells that do not divide, such as those in cartilage, muscle, and nervous tissue, are much more resistant.

The effects of radiation also vary with the intensity of the exposure. The higher the dose, the more serious the effects, and the more quickly they are manifested. Radiation doses are frequently measured in **rads,** a measure of energy absorption by the tissue (rad stands for radiation absorbed dose). A rad is equal to about 100 ergs of energy per gram of tissue. An erg is an extremely small amount of energy, approximately the amount of energy imparted by a mosquito alighting on your arm.

Nongenetic Effects   Radiation causes many effects, death being one of the most pronounced. As shown in Table 22.2, individuals exposed to 100 to 250 rads or less withstand such exposure levels, although they generally suffer a higher incidence of cancer later on in life. As the dose increases to 400 to 500 rads, however, only 50% of the people exposed will survive three weeks. At doses of 900 rads or greater, all people die within three weeks of exposure.

What are some of the symptoms displayed by people exposed to radiation? A person receiving less than 25 rads would not be aware of changes in his or her immediate health. (The likelihood of cancer, however, is increased considerably.) When the dose rises to 100 rads, the individual may suffer from weakness, fatigue, vomiting, and diarrhea. Eventually, however, these symptoms disappear, and the individual begins to function normally. At a radiation exposure of 400 to 500 rads, individuals experience extreme nausea, vomiting, hair loss, and fatigue. Making matters worse, this level of radiation impairs production of red blood cells, resulting in severe anemia. White blood cell production also is severely hampered. Because white blood cells help protect the body against infections, radiation victims are susceptible to infectious diseases caused by bacteria and viruses. This dose also destroys blood platelets, tiny structures in the blood that promote blood clotting. The result is massive hemorrhaging and excessive blood loss. Individuals exposed to 400 to 500 rads therefore usually die from one or a combination of the following problems: anemia, hemorrhage, and severe bacterial infections.

Much of the information on radiation's effects comes from studies of the survivors of the nuclear bomb blasts in

TABLE 22.2	Noncancer Effects of Radiation on Humans	
Radiation Dosage in Rads	Percent Who Will Die	Other Symptoms
10,000	100	Death in 2 days due to brain and heart damage
3,000	100	Permanent damage of brain, spinal cord, bone marrow, and intestines; bleeding; infections
2,000	100	Permanent destruction of bone marrow and intestines; severe dehydration; bleeding; infections
1,000	100	Permanent destruction of bone marrow; permanent damage to intestines
600	60	Permanent destruction of bone marrow
400–500	50	Fatigue; nausea; vomiting; loss of hair; reduced number of blood platelets and white blood cells; bleeding; infections
100–250*	0	Fatigue; nausea; vomiting; loss of hair; diarrhea; ulcers of gut; some may suffer permanent loss of fertility
25–100	0	No visible effects; lower number of white blood cells
0–25	0	No measurable effects

*Note: All individuals exposed to 100 rads or more will experience fatigue, nausea, vomiting, diarrhea, and hair loss.

Hiroshima and Nagasaki at the end of World War II. Thousands of people were exposed to radiation doses ranging from 100 to 150 rads. The immediate effects included burns, fever, loss of hair, fatigue, intestinal bleeding, vomiting, and diarrhea. If the radiation damaged their bone marrow, many delayed effects were observed, including anemia, leukemia (cancer of the blood), and infections due to suppression of white blood cells.

The children of pregnant mothers who were within 1,200 meters (3,960 feet) of the center of the explosion were born with severe birth defects, including mental retardation and deformations of the skull, heart, and skeleton. Many of the survivors and members of their families are shunned today by Japanese, fearful of the long-term genetic effects of radiation.

Genetic Effects: Cancer, Birth Defects, and Mutations

Experiments on laboratory animals have shown that radiation causes cancer in a variety of organs, including the lymph glands, breasts, lungs, ovaries, and skin. Years of research confirm the connection between radiation and various forms of cancer in humans. Consider the following examples:

1. Skin cancer was prevalent among radiologists in the early days of the profession before adequate safeguards were instituted.

2. Bone cancer appeared in many women who painted radium dials on watches in the 1940s. Apparently, the women ingested radium when drawing paint brushes to a point with their lips.

3. Uranium miners have a high incidence of lung cancer, largely from breathing radioactive radon gas found in the air in and around mines.

4. Children irradiated when still in their mother's uterus (during routine maternal pelvic X-ray examinations) have a much higher cancer rate—50% higher—than children who were not X-rayed during gestation.

5. The rate of leukemia was substantially higher among Japanese citizens who had survived the bomb blast than in unexposed Japanese citizens from other cities. Leukemia often showed up 30 years after the bombings.

Most tumors result from genetic changes in the cells of the body. In general, these changes, known as **mutations,** may be caused by biological agents such as viruses, chemical agents such as toxic chemicals, or physical agents such as radiation. These mutagenic agents result in genetic changes that may cause body cells to divide uncontrollably, forming tumors. The resultant tumors grow, debilitating individuals and often spreading through the lymphatic vessels to other parts of the body, where secondary tumors form. Exhausted and depleted by the rapidly growing tumor, the individual eventually dies—unless, of course, the tumor is excised or destroyed by drugs or radiation treatment.

Scientists have long known that radiation can also damage the chromosomes of germ cells—the sperm and ova. These mutations can be passed to an individual's offspring and manifest themselves as birth defects or as childhood cancer. Background radiation probably results in one mutation for every million sperm or ova produced. Additional radiation, such as from X-rays or from an accident at a nuclear power plant, can greatly increase the rate of mutation. James F. Crow, an eminent University of Wisconsin geneticist, suggests that any amount of radiation is potentially damaging to genetic material.

New and controversial research shows that exceedingly low levels of radiation have a more significant effect than previously thought. Studies also indicate that radiation is cumulative—that is, low levels add up over many years. As a result, a study by the National Academy of Sciences suggests that the current health standard—17 rads per year for the general public—may be too high. This level of exposure, the authors of the study assert, could increase cancer rates by 2% and increase the incidence of genetic defects by about 1 per 2,000 newborns. Against strong opposition, the academy and other scientists have suggested reducing the allowable level of radiation to 0.017 rad per year.

Reactor Safety: Two Case Studies

To understand why nuclear energy has faltered, one need only look at the two most horrifying experiences with it in recent history. Both accidents raise many questions and throw into question this approach to energy production.

Three Mile Island: The Beginning of the End?

On March 28, 1979, disaster struck at the Three Mile Island nuclear power plant 10 miles from Harrisburg, Pennsylvania (Figure 22.23). The three-month-old reactor ran amok when a valve failed to close. This failure in turn triggered a series of events that resulted in the most costly and potentially dangerous accident in the history of the U.S. nuclear industry.

The details of the accident are difficult for all but trained nuclear engineers to understand, but the results were clear: Radioactive steam was dumped into the containment building of the reactor and into the air outside the plant. A cloud of radioactive steam drifted through the valley, sending residents scurrying for safety. Bursting pipes spilled radioactive cooling water into two buildings. In one of them, a million liters of radioactive water accumulated ankle-deep.

Something unusual then followed: Mysteriously, hydrogen gas began to accumulate in the reactor vessel, threatening to expose the core and cause a meltdown. Some experts feared that the bubble might result in an explosion of the reactor vessel and possibly the containment building that would release even larger amounts of radioactive material. Fortunately, the utility was able to draw off the hydrogen to prevent an explosion. A year later, photographs taken in the core by a remote camera showed that the nuclear fuel had melted down, rendering the reactor useless and adding to the cleanup costs.

The human toll—besides the agony and fear of those close to the reactor—was minimal, unless you were one of the victims. Eight workers inside the plant were exposed to high levels of radiation and are likely candidates for cancer. Fifty thousand people within the vicinity of the plant were evacuated, and 130,000 in outlying regions were warned to stay indoors to avoid being exposed to radiation. So far, no studies

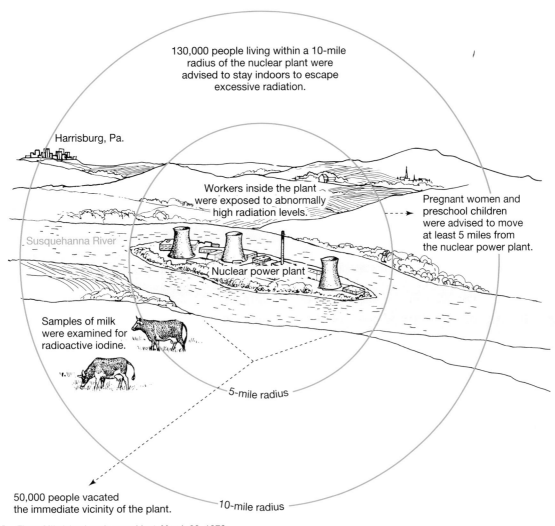

130,000 people living within a 10-mile radius of the nuclear plant were advised to stay indoors to escape excessive radiation.

Harrisburg, Pa.

Workers inside the plant were exposed to abnormally high radiation levels.

Pregnant women and preschool children were advised to move at least 5 miles from the nuclear power plant.

Susquehanna River

Nuclear power plant

Samples of milk were examined for radioactive iodine.

5-mile radius

50,000 people vacated the immediate vicinity of the plant.

10-mile radius

FIGURE 22.23 Three Mile Island nuclear accident, March 28, 1979.

have shown any measurable increase in cancer as a result of this accident.

After several days of feverish efforts, nuclear engineers from the Nuclear Regulatory Commission and Metropolitan Edison, the company that owns the reactor, were able to shut down the plant, thus ending the release of radiation. Soon thereafter, the company began the long and expensive cleanup, which cost an estimated $1 billion.

Studies of the accident showed that human error, mechanical failures, and design flaws were the causes. Inexperienced operators working late at night, for instance, turned off the emergency core-cooling system at the wrong time, closed valves on an emergency cooling system when they should have left them open, and disengaged water pumps that should have been left operable.

The Three Mile Island accident was a disaster of minor proportions, say utility officials and supporters of nuclear energy in general. No one died during the accident, and many health officials believe that the radiation released into the air and water will have little, if any, effect on local residents. Two noted (and controversial) radiation experts, however, believe that 300, possibly as many as 900, cases of cancer and leukemia will result from the incident. Many residents, they

say, were exposed for 100 hours or longer to low levels of radiation, enough to noticeably increase the incidence of cancer in the region. Unfortunately, only time will tell if they are correct.

The Chernobyl Accident Many thought that the accident at the Three Mile Island nuclear plant marked the beginning of the end of the nuclear power industry, then burdened by high construction costs and growing public skepticism. The accident at Chernobyl, they asserted, sealed the fate of this troubled industry forever.

At 1:00 A.M. on April 26, 1986, the nuclear operators at the Chernobyl nuclear power plant in the former Soviet Union had just completed a full day of testing on the fourth and newest reactor. In the course of their test, the operators shut down safety systems and violated a number of operating procedures.

By 1:23 A.M. the reactor's power had fallen to only 6% of its operating level. With the emergency cooling system and other safety systems shut off and the control rods only part of the way out of the reactor core, power began to increase. However, when the operators pushed the button to drop the control rods back into place, the rods wouldn't budge. The core, experts now believe, had already begun to melt down, preventing the rods from falling into place. A few seconds later, shock waves about as strong as a mild earthquake tremor shook the plant. Two large explosions then shook the complex (Figure 22.24).

Experts believe the explosions occurred when the fuel rods split open in the intensely hot reactor core. Enormous amounts of energy were released from the fuel rods, superheating the reactor cooling water and suddenly converting it into steam. The steam ripped open the partial reactor containment building. The 1,000-ton concrete slab above the reactor was thrown aside as steam, nuclear fuel, and graphite (one of the components of this nuclear reactor) were hurled skyward.

Helicopter pilots arriving soon after the explosion reported that they could see down into the glowing red core of the reactor through a gaping hole in the reactor building. To put out the fire and stop the spread of radiation, the pilots dropped 36 metric tons of boron carbide, 720 metric tons of limestone, 2,160 metric tons of lead, and thousands of metric tons of sand and clay on the damaged reactor over the next few days. The fire went out soon, but the reactor smoldered for several days, continuing to release millions of curies of radiation a day for nearly two weeks. On the 11th day, liquid nitrogen was pumped under the reactor in an emergency cooling system installed after the accident. The molten fuel cooled enough to end the release of radiation.

News of the Chernobyl accident reached the West on April 29, several days after the explosion, and only after Sweden and other countries had reported high levels of radiation wafting in from the former Soviet Union. The world press picked up the story on the 29th and hastily put together the few facts they could uncover while the world looked on in terror. The earliest reports from diplomats in the Soviet Union suggested that at least 2,000 people had been killed by the accident, the most serious accident in the history of nuclear energy.

When the dust settled and Western reporters were allowed access to the facts, it became clear that the initial reports of human fatalities were grossly exaggerated. Four months after the accident, the death toll had reached 31. Most of the victims were workers who had been exposed to high levels of radiation. All told, 237 people were hospitalized with acute radiation poisoning and burns. The high levels of radiation to which they were exposed will greatly increase their risk of cancer.

Some 135,000 people were evacuated from a 30-square kilometer (11.6-square-mile) region, mostly north of the plant. Most of them will never return home but live in a new city built by the government. A quarter of a million schoolchildren were sent away from Kiev, a city 80 kilometers south of the plant, on an early summer vacation a few weeks after the explosion. By various estimates, 150 square kilometers (60 square miles) of prime farmland around the plant were so badly contaminated with radiation that this land will lie fallow for many decades. The reactor itself has been sealed in concrete to prevent further radiation leakage, and it may be several hundred years before workers can safely remove what is left of the reactor core.

The cost of this accident is difficult to determine. Soviet officials estimated their damages to be about $3 billion to $5 billion. But outside experts believed that the total costs would exceed $10 billion, when all the indirect effects were taken into account. Outside the former Soviet Union, the costs were beginning to be known. In the United Kingdom, sheep farmers, whose livestock were contaminated by radiation and thus rendered unsuitable for food, estimated that they lost $15 million. Swedish officials estimated their country's losses amounted to $145 million, and the German government said it would pay farmers $240 million for lost crops and livestock.

In all, 20 countries were dusted with radiation, the long-term health effects of which no one really knows. Although the number is difficult to estimate, scientists believe that the

FIGURE 22.24   Aerial view of the damaged Soviet nuclear reactor building at Chernobyl after the fateful accident in 1986, which killed 31 people and may cause cancer in tens of thousands of others.

Chernobyl accident will result in 5,000 to 100,000 additional cases of cancer in the country. Outside of the former Soviet Union, there may be as few as 2,500 cancers up to as many as 300,000, according to some experts. Radiation specialist John Gofman believes that the total number of cancer cases could range from 600,000 to 1 million. Half of these, he predicts, will be fatal. Russian scientists have already reported a thyroid cancer epidemic with rates 100 times greater than normal in areas near the accident.

The accident at Chernobyl raised questions about the wisdom of nuclear power. Although many experts believe that an accident of this magnitude could not happen in the United States, others are not so sure. Meanwhile, antinuclear sentiment skyrocketed throughout the world. About half of the European population favors shutting down existing nuclear power plants. In the former West German state of Hesse, government officials have decided to push for a phaseout of nuclear power. In Poland, the nuclear program has been slowed. In Sweden, the government promised to phase out nuclear energy, while Swiss officials, long committed to nuclear power, have postponed further expansion of their nuclear power capability.

The Nuclear Waste Issue

After accidents, one of the biggest concerns of the environmental community, the public, and many government officials is radioactive-waste disposal, not only from power plants but also from weapons production facilities. The latter,

not discussed in this chapter, could cost billions of dollars in the United States alone.

In nuclear power plants, fission by-products begin to build up in fuel rods within a year or two of the time they are installed. These radioactive materials absorb free neutrons. As free neutrons build up, they slow the rate of fission, reducing a plant's operating efficiency. Therefore, each year about one-third of the old fuel rods (40 to 60 fuel rods) are replaced with new ones. But what happens to the spent fuel?

In many countries, the spent rods are stored underwater and then transported to reprocessing plants that remove the highly radioactive by-products (Figure 22.25). Reprocessing plants recover uranium-235 that has not been fissioned and also plutonium-239, which forms from uranium-238. At the reprocessing plant, the contents of the spent rods are dissolved in acid. The usable uranium-235 and plutonium-239 are separated from the waste. During this process, however, about 460 liters (120 gallons) of highly radioactive liquid waste is produced for each metric ton of spent fuel. Reprocessing plants also emit approximately 100 times as much radiation as a properly operating nuclear power plant.

One of the greatest problems of the U.S. nuclear industry is that there are no reprocessing plants in operation in the country. As a result, many tens of thousands of metric tons of highly radioactive waste are now stored in water tanks at nuclear plants all over the country (Figure 22.26). Delays in establishing a permanent waste repository for high-level radioactive

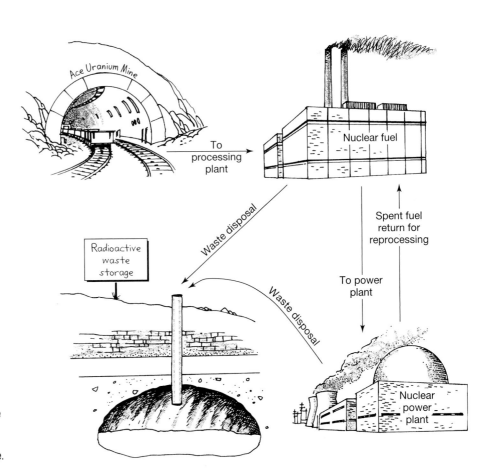

FIGURE 22.25 Nuclear fuel and waste cycle (simplified view). Radioactive contamination of the human environment is most likely to occur during the transport of spent fuel for reprocessing and during the transport and disposal of nuclear waste.

FIGURE 22.26   Spent nuclear fuel "cooling off" underwater. The spent fuel assemblies from the Savannah River Plant in Georgia illuminate a cooling basin. In the photograph, spent fuel of three different "ages" is shown under 20 feet of water. The brightest assemblies (top) were just discharged from the reactor. In front of them are assemblies that were discharged a month earlier. Barely visible on the extreme lower left are assemblies that have "cooled" by radioactive decay for three months.

materials have turned reactors into long-term disposal sites. The mounting high-level waste potentially increases the severity of accidents. Making matters worse, many reactors are now reaching the end of their useful lifetime. "To continue the temporary storage of fuel rods at these sites would be asking for trouble," writes then-Congressman George Brown Jr. of California. Since almost all of these soon-to-be-closed reactors are near rivers and lakes for easy access to cooling water, a catastrophic accident at any one could result in contamination of local waterways.

In an effort to solve America's high-level waste problem, the U.S. Congress passed the **Nuclear Waste Policy Act** in 1982. It called on the Department of Energy to choose two sites for the disposal of high-level radioactive waste by 1987—one in the East and one in the West. Most experts believe that high-level wastes should be disposed of in deep caverns carved out of ancient geologically stable salt beds, granite, volcanic tuff, or basalt. The experts hope that the waste will remain in these sites many thousands of years without leaking into ground water and threatening human health.

Despite much initial optimism, site selection proved extraordinarily difficult. In 1986, the Department of Energy proposed three western sites: the Hanford nuclear reservation in southeastern Washington State, Yucca Mountain in southwestern Nevada, and Deaf Smith County on the high plains of western Texas. Each of these sites, selected from a far larger field, was then to be studied in detail at a cost of $1 billion apiece to determine the best one. To save money, though, Congress decided to focus efforts on the most likely candidate, the Yucca Mountain site in Nevada. It appeared to be free from earthquakes and is dry. The repository could be constructed above the water table, which lies 700 meters (2,000 feet) below the surface. Unfortunately, Nevada's governor and three-quarters of the people of the state oppose the site. They are afraid that it is not as stable as government scientists believe, and they want Nevada to avoid the stigma of being the country's nuclear wasteland.

This site would store nuclear waste from utilities and defense installations—mostly factories where atomic weapons were once made. After extensive research, officials are finding many problems that make even this site questionable.

Although many people believe it was a mistake to build nuclear power plants in the first place, approximately 60,000 metric tons of high-level waste are currently stored at U.S. reactors. We must solve the waste problem that we have created. We have no other choice.

22.4 Fusion Reactors

The fission reactor is clearly a technology in decline in the United States. Although the nuclear power industry has proposed smaller, safer plants, high costs and skepticism on the part of the general public, insurance companies, and banks have not translated these new dreams into realities. Despite President George W. Bush's support for nuclear power, its fate within the United States is still questionable. For many years, the nuclear power industry has also banked some of its hopes on another type of reactor, known as a **fusion reactor.** What are fusion reactors, and what are their prospects?

The sunshine that helps brighten your day has its origin in nuclear fusion reactions occurring in the sun. In a sense, **nuclear fusion** is the opposite of fission. Fission involves a splitting apart of large atomic nuclei, whereas fusion unites two smaller atomic nuclei to form a new one. In nuclear fission, heavy nuclei like those of uranium are used because they tend to be unstable. In nuclear fusion, two lightweight nuclei are brought together by overcoming the mutual repulsion resulting from the positively charged protons in each nucleus. Light nuclei are used because they repel each other much less than larger nuclei. When two nuclei fuse, energy is released—lots of energy. It can be used to generate electricity.

Hydrogen atoms are used in experimental fusion reactors because they are the lightest elements. Hydrogen atoms have one proton and no neutrons in their nuclei. The repulsive force is therefore quite small. Another advantage of hydrogen is that it is extremely abundant.

Hydrogen has two less-common isotopes, deuterium and tritium. Deuterium has a proton and a neutron in its nucleus; tritium has a proton and two neutrons. A deuterium nucleus may be fused with another deuterium nucleus or with a tritium nucleus.

To overcome the repulsive force, the nuclei must be supplied with enormous amounts of energy. This can be done by bombarding a fuel pellet containing deuterium and tritium with a laser beam. Thus energized, the fuel turns into a hot mixture of nuclei and electrons known as **plasma.** In the plasma, nuclei can speed toward each other, collide, and fuse. The fusion of two nuclei releases enormous amounts of energy, potentially much more than is needed to initiate the reaction.

The 40 million °C heat required to start the reaction poses some design problems. No known metal can withstand temperatures anywhere near those required by fusion reactors. Therefore, scientists have proposed two major designs. The most popular is magnetic confinement. In this

technique, a magnetic field suspends the plasma long enough for fusion to occur (Figure 22.27). The heat given off is picked up by liquid lithium, transferred to liquid potassium, and then transferred to water, which boils to generate steam and electricity.

Nuclear fusion offers several key advantages over nuclear fission. First, the fuel is abundant and inexpensive. The deuterium in only 1 cubic kilometer of seawater could provide us with as much energy as 1,500 billion barrels of oil, or 1.5 times as much oil as the world has consumed in the history of human civilization. Nuclear fusion is believed to be much safer as well. If a fusion reactor malfunctioned, the reaction would simply come to a stop. There is, say experts, no chance of an explosion. An accident would not release the massive quantities of radiation released by a fission reactor. Finally, because a fusion reaction is much more efficient than a fission reaction, less waste heat is released. As a result, a fusion reactor's potential for thermal pollution is considerably less, and less cooling water would be needed.

Fusion has some drawbacks—some so significant that they could forever keep this form of energy out of our grasp. One of the most important hazards is tritium, the radioactive isotope of hydrogen used as a fuel. At high temperatures, tritium is extremely difficult to control and can pass right through metal. A second major problem is that, despite more than five decades of research, scientists have been unable to reach the breakeven point, that is, they have been unable to produce as much energy as the experimental reactors consume. To be economical, fusion reactors must generate electricity at a price we can afford. Even though the costs cannot be predicted with great accuracy, it is quite possible that commercial fusion reactors could cost $12 billion to $20 billion, far in excess of conventional nuclear reactors and even breeder reactors. That money invested in energy efficiency and renewable sources such as wind, say critics, could produce enormous amounts of clean energy.

Fusion reactors also produce highly energetic neutrons that will bombard the containment vessel, weakening the metal and requiring frequent replacement. The containment vessel would also become radioactive. Furthermore, if a vessel burst, it would release radioactive tritium and molten lithium, the coolant, which burns spontaneously and vigorously in air.

In 1980, Congress passed the **Magnetic Fusion Energy Emergency Act** to promote fusion energy. This law authorized the Department of Energy to spend roughly $1 billion per year from 1980 to 2000 on fusion research and development. The hard economic times of the 1980s, however, put an end to much of this money, further dimming the prospects for fusion energy. In 1996, the United States spent about $250 million on fusion research, a fraction of what is being spent in Europe and Japan. Plans to have a demonstration plant on line in the United States by 2025 have been abandoned because of these economic cutbacks.

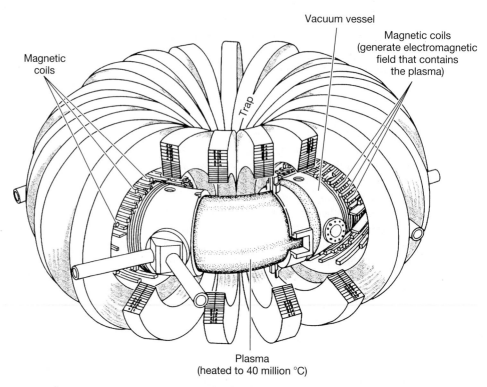

Vacuum vessel

Magnetic coils
(generate electromagnetic
field that contains
the plasma)

Magnetic
coils

Trap

Plasma
(heated to 40 million °C)

FIGURE 22.27 Nuclear fusion reactor. (Top) Fusion research device. This diagram shows one research device that could bring the United States closer to controlling fusion, the force that powers the sun and the stars. Because it is easier to work with, this device will use hydrogen as its fusion fuel. Later-generation devices will employ deuterium and tritium as fuels. This device is known as the Princeton Large Torus (PLT). (Bottom) Nuclear fusion research. Magnetic coils used to confine and compress fusion plasma are visible in this fusion device. Microwave power levels are increased for the early-stage testing of a technique to heat plasma to the very high temperatures needed for fusion reactions. This device, known as the Elmo Bumpy Torus, is located at the Department of Energy's Holifield National Laboratory.

22.5 America's Energy Future

The world is running out of oil, and natural gas is also on the decline. The costs of both fuels are rising rapidly, threatening the global economy, people's jobs, and our standard of living. Moreover, our heavy reliance on fossil fuels is wreaking havoc on the environment through global climate change. All the while, our energy needs are expanding. Fortunately, the world community has many options for replacing these costly and environmentally unfriendly fuels. Nuclear fission is one of

them, but given its high costs, potential shortages of fuel, and the risks it poses to our safety, fission may not play a huge role in our long-term future. Nuclear fusion offers the promise of abundant energy but because of costs may never become commercially feasible. What about coal? Enormous deposits of this fossil fuel lie beneath the Earth's surface, but serious environmental obstacles lie in the way of further development of this fuel—global warming and acid rain being the most significant.

That leaves energy conservation, energy efficiency, and the host of renewable energy resources like solar energy, wind energy, geothermal energy, hydropower, and biomass energy, all discussed in the next chapter. These sources of energy are currently underutilized and are likely to become increasingly more popular and less expensive as conventional fuel sources decline and improvements in technologies are brought on line. Even major oil companies such as BP (formerly British Petroleum) and Royal Dutch Shell have made significant investments in renewable energy. They both sell solar electric systems. Even General Electric, which builds wind turbines and nuclear power plants, sells solar electric systems. Many major corporations, including Wal-Mart, are installing solar electric systems on their facilities. In 2008, for instance, Wal-Mart announced that it will install solar electric systems on 22 of its stores in California and Hawaii, which will provide 30% of each store's power. Widespread government support of these technologies, akin to that now given to the fossil fuel and nuclear industries, say proponents, could help realize the promise of these environmentally clean and renewable energy resources.

Summary of Key Concepts

1. The twin oil crises of the 1970s awakened the world to the extraordinary cost of our dependency on oil. It showed us that everything we buy or do requires energy, and lots of it. High inflation created a near crippling of the world economy. In 2000, another energy crunch began to unfold, primarily as a result of levels of consumption that exceeded demand.

2. In more-developed countries, fossil fuels such as coal, oil, and natural gas are the predominant sources of energy. Nuclear energy, conservation, and renewable energy sources have grown in importance since the 1970s.

3. In the less-developed nations, renewable fuels, especially wood, form the mainstay of the energy diet, although industrialization in these countries is increasing their dependence on fossil fuels.

4. Coal supplies about 23% of the United States' and 15% of Canada's energy needs. It is burned today in electric utilities and in some factories and homes, where it releases the solar energy captured in plants many millions of years ago by photosynthesis.

5. Proven global coal reserves are estimated to be about 700 billion metric tons, enough to last about 200 years. Undiscovered global reserves could supply the world's needs for another 1,700 years. U.S. coal supplies could last 200 years at the current rate of consumption, making coal a likely energy source for many years to come.

6. Unfortunately, coal cannot easily or efficiently be substituted for oil and natural gas, and it is a dirty fuel. Mining has an enormous impact on the environment.

7. Because of environmental problems, the coal industry is promoting clean coal, burning coal more efficiently to reduce pollution and sequestering carbon dioxide in deep wells to prevent further global warming.

8. One more efficient coal-burning technology is fluidized bed combustion. Crushed coal is mixed with bits of limestone and propelled into a furnace in a strong current of air. The limestone reacts with the sulfur oxide gases, eliminating them from the smokestack gases.

9. Carbon sequestration is a multistep process that begins with the removal of carbon dioxide from combustion gases of a coal-fired power plant. The gas must then be compressed and shipped to sites where it can be injected into abandoned oil wells. The energy required to build pipelines and the cost of this proposed process could be quite high.

10. Another technology to put coal to use to meet our needs is coal gasification, in which a combustible gas is produced from coal. New developments in this technology have greatly improved its efficiency.

11. Coal can be treated to produce a thick, oily substance in a process called coal liquefaction. The oily product can be refined in the same way that crude oil is refined to produce a variety of useful products, such as jet fuel, gasoline, kerosene, and the chemicals needed to make drugs and plastics.

12. Unlike coal, oil is in short supply. Many experts believe that oil production may have reached an all-time high. They call this peak oil, which is short for peak oil extraction. Peak oil represents a historic moment after which demand cannot be met by supply. This could cause potentially crippling inflation and economic stagnation, even deep recession that could last many years.

13. Natural gas burns much more cleanly than other fossil fuels and may become an important transitional fuel as we expand our reliance on renewable energy resources. The picture for natural gas is brighter than for oil, but not as good as for coal. Some analysts predict that global natural gas production could peak around 2019.

14. Oil shale is a sedimentary rock that was formed millions of years ago from the mud at the bottoms of lakes. It contains a solid organic material known as kerogen that, when heated, is released from the rock, producing shale oil, which can be refined much as petroleum is refined.

15. The largest and most valuable deposits of oil shale in the United States occur in Colorado, Wyoming, and Utah. However, oil shale development is economically and environmentally costly.

16. Tar sands contain a thick, oily residue called bitumen that can be extracted and refined much as shale oil is refined. U.S. tar sand deposits are not extensive. Canadian tar sand deposits are much larger. Unfortunately, tar sand production is very energy-intensive. Surface mining of tar sands causes a considerable amount of environmental damage.

17. An immediate substitute for oil is needed. It is also time to start developing alternatives to natural gas and coal, and especially to find some energy sources that do not add to global carbon dioxide levels. One possibility is nuclear energy. Nuclear energy is produced when atoms are split apart.

18. To understand nuclear energy, one must understand the structure of atoms. All matter is composed of atoms. Each atom is composed of a dense, centrally located nucleus containing protons, positively charged particles, and neutrons, particles without a charge. Surrounding the nucleus is the electron cloud, where tiny, negatively charged particles, the electrons, are found.

19. Over 100 elements are known to science, each one differing from the next in the number of protons in its nucleus. An element is a pure form of matter. In any given element, the number of protons in the atoms is always the same. Because the number of electrons in the electron cloud always equals the number of protons, atoms are electrically neutral.

20. Although all atoms of a given element have the same number of protons, atoms of the same element may vary in the number of neutrons. Thus, an element may have several alternative forms of atoms, called isotopes. If these forms are unstable, they are called radioisotopes.

21. Radiation consists of particles or electromagnetic waves given off by the nuclei of radioisotopes in an attempt to reach stability. The three common forms of radiation are alpha particles, beta particles, and gamma rays. Over time, radioactive elements decrease in mass as a result of the release of radiation. The measure of this decrease is called the half-life.

22. Nuclear reactors (and the first generation of nuclear bombs) capture energy from controlled nuclear fission, the splitting of atoms. In reactors, fission occurs when uranium atoms in fuel rods are bombarded by neutrons given off by other uranium atoms in the same rod. Enormous amounts of energy are released during this process.

23. Additional neutrons are also released during fission. These neutrons continue to bombard other nuclei, causing still other fissions in a chain reaction.

24. Nuclear power plants use the energy given off by controlled chain reactions to boil water to make the steam that turns electrical generators.

25. The uranium fuel used in all large nuclear power plants is a mixture of uranium-235 and uranium-238 that is packed in pellets and inserted in fuel rods. Fuel rods are inserted into the core of the nuclear reactor. The rate of fission is moderated by neutron-absorbing materials in the control rods. The entire assemblage of control and fuel rods is housed in a reactor vessel within a containment building.

26. Supporters of nuclear power argue that this technology offers many advantages that make it a desirable energy source: (1) Should the breeder reactor become feasible, the fuel supply would be abundant. (2) Nuclear power could help us reduce our dependency on foreign oil. (3) Properly operating plants release little radiation; in fact, a coal-fired power plant releases more radiation than a nuclear plant. (4) Nuclear plants release fewer solid wastes and air pollutants, including the greenhouse gas carbon dioxide. (5) The probability of an accident, supporters once argued, was quite slim.

27. Opponents take exception. They note the following disadvantages: (1) Nuclear power plants are extremely expensive to build. (2) The breeder reactor is costly, fraught with safety problems, and plagued by a long payback period. (3) Nuclear power will replace very little imported oil. (4) Nuclear power will do little to reduce overall carbon dioxide emissions unless thousands of costly plants are built. (5) Accidents at nuclear power plants can release enormous amounts of radiation with potentially devastating effects. (6) Nuclear power plants operating normally release more thermal pollution than coal-fired power plants. (7) Accidents at nuclear power plants occur with a rather high frequency. (8) Nuclear power is an expensive form of energy and is bound to increase in cost as we grapple with waste disposal and ways to decommission reactors.

28. The effects of radiation vary with the age of the exposed individual. Fetuses and newborn infants are much more sensitive than adults. Fast-growing cells are much more sensitive than cells that do not divide.

29. Radiation exposure can be measured in rads (radiation absorbed dose). Individuals receiving 400 to 500 rads of radiation suffer extreme mortality; only half of them will be alive after three weeks. Individuals receiving doses less than 25 rads would not be aware of changes in their health. When the dose rises to 100 rads, however, they may experience weakness, fatigue, vomiting, and diarrhea. Eventually these symptoms disappear. These seemingly low levels of exposure nonetheless often result in an increase in cancer and birth defects in offspring.

30. Numerous studies show that radiation increases the incidence of a number of cancers, including leukemia, bone cancer, lung cancer, and skin cancer. Scientists believe that most tumors result from genetic changes in the cells of the body, causing them to divide uncontrollably. Radiation can also damage the chromosomes of germ cells; the resulting mutations can be passed to an individual's offspring and may show up as birth defects or cancer early in life.

31. One of the biggest concerns of many people is the release of radiation during normal, routine operations and, more important, during accidents.

32. Many people are concerned about radioactive-waste disposal. In the United States, highly radioactive reactor and military waste has been building up for over four decades. The United States has neither the reprocessing facilities to extract usable fuel from the waste nor any high-level radioactive-waste disposal sites. To solve this dilemma, the U.S. Congress passed the Nuclear Policy Waste Act in 1982. It called on the Department of Energy to choose two sites for disposal, but site selection was plagued with difficulties. The biggest problem was that few people want a high-level radioactive-waste dump in their state. Although many people believe that it was a mistake to build nuclear power plants in the first place, the simple truth is that in 2007, 60,000 metric tons of high-level waste were stored at commercial reactors. We must solve the waste problem that we have created.

33. One alternative to nuclear fission is the fusion reactor. Fusion reactors are fueled by deuterium and tritium, isotopes of hydrogen that unite to form larger nuclei. Making them fuse, however, requires enormous amounts of energy.

34. One of the main attractions of fusion power is that the fuel supply is exceptionally abundant. Also, fewer wastes would be produced than from a conventional fission reactor. However, fusion reactors are likely to be prohibitively expensive and could be dangerous, since the coolant, molten lithium, burns on contact with air. A tiny leak could destroy a reactor.

Key Words and Phrases

Acid Mine Drainage
Acid Precursors
Alpha Particle
Anthracite Coal
Arctic National Wildlife
  Refuge (ANWR)
Area Strip Mine
Atom
Atomic Mass
Atomic Number
Beta Particle
Bitumen
Bituminous Coal
Breeder Reactor
Carbon Sequestration
Chain Reaction
Clean Coal
Coal
Coal Gasification
Coal Liquefaction
Coal Reserves
Containment Building
Contour Mine
Control Rods
Daughter Nuclei
Dragline
Electron
Electron Cloud
Element
Emergency Cooling
  System
Energy Conservation
Energy Efficiency
Fluidized Bed Combustion
Fossil Fuels
Fuel Rods
Fusion Reactor
Gamma Ray
Half-Life
Hydropower
In Situ Retort
Isotope
Kerogen

Lignite Coal
Magnetic Fusion Energy
  Emergency Act
Mountain Top Removal
Mutation
Natural Gas
Net Energy Efficiency
Neutron
Nuclear Energy
Nuclear Fission
Nuclear Fusion
Nuclear Reactor
Nuclear Waste
  Policy Act
Nucleus
Oil
Oil Producing and Exporting
  Countries (OPEC)
Oil Reserves
Oil Sands
Oil Shale
Overburden
Plasma
Plutonium
Proton
Proven Reserves
Rad
Radioactive Decay
Radioisotope
Reactor Core
Reactor Vessel
Shale Oil
Slurry
Spent Shale
Subsidence
Sulfuric Acid
Surface Mining Control and
  Reclamation Act
Surface Retort
Tar Sands
Ultimate Production
Uranium
Yellow Cake

Critical Thinking and Discussion Questions

1. Describe the major sources of energy used by more-developed countries. Which ones are in short supply? Which ones do you expect to be plentiful 50 years from now?

2. Using your critical thinking skills, analyze the following statement: "The less-developed countries rely primarily on renewable energy. Since it is renewable, they have nothing to worry about."

3. Define the following terms: proven reserves, ultimate reserves, and undiscovered reserves.

4. How large is the global proven reserve of coal? How long will it last at the current rate of consumption?

5. Describe the environmental impacts of surface and underground mining of coal.

6. Describe fluidized bed combustion, and draw a schematic diagram of the process. Why does it produce more energy and less airborne pollution than a conventional coal-fired power plant? Is it a sustainable solution to our nation's energy needs? Why or why not?

7. Define coal gasification and coal liquefaction.

8. Coal is an abundant fossil fuel resource and could power society for many years. Is this a good idea? Why or why not?

9. Go online and research what proponents of clean coal technology are proposing. Write a paragraph summarizing their views and the view of the critics.

10. How does exponential growth affect global oil supplies?

11. What does the term *peak oil production* mean? What are the consequences of peak oil, both positive and negative?

12. Discuss the pros and cons of drilling for oil in ANWR. Use your critical thinking skills to analyze the various points. Do you think it can be carried out in an environmentally sensitive manner?

13. Describe oil shale and tar sands. What are they? Where did they come from? How big are the deposits? What are the effects of their development?

14. What does net energy efficiency mean? Why is this an important measure of the value of renewable and nonrenewable fuels?

15. Describe the structure of an atom, and define the terms *isotope, radioisotope, radiation,* and *half-life.*

16. Which cells of the body are most sensitive to radiation?

17. Describe how a nuclear power plant works, being sure to mention fission, neutrons, daughter nuclei, uranium-235, fuel rods, control rods, reactor core, chain reaction, reactor vessel, containment building, and emergency cooling system.

18. Do you agree with the following statement? "Nuclear fission is a clean, safe alternative to coal-fired power plants." Why or why not?

19. How does a breeder reactor differ from a conventional fission reactor? What are the pros and cons of the breeder reactor? What unique problems does it have?

20. How does a fusion reactor operate? What is its fuel? What are the pros and cons of nuclear fusion?

21. Suppose that a utility company proposed building a nuclear power plant near your campus. The construction of the plant, the utility asserts, would revitalize the local economy. Critics agree that it would create many jobs, helping the economy, but argue that most of the workers would be brought in from outside. What would be your response to the proposed project?

Suggested Readings

Chiras, D. D. 1992. *Lessons From Nature: Learning to Live Sustainably on the Earth.* Washington, DC: Island Press. Description of transitional steps to a renewable energy economy.

Chiras, D. D. 2005. *The Homeowner's Guide to Renewable Energy.* Gabriola Island, BC: New Society. Highly popular book that illustrates all of the major renewable energy technologies for homes.

DeCarolis, J. F., R. L. Goble, and C. Hohenemser. 2000. Searching for Energy Efficiency on Campus: Clark University's 30-Year Quest. *Environment* 42(4): 8–20. Interesting story about efforts that could be carried out on many college campuses.

Flavin, C. 1999. Rethinking the Energy System. In *State of the World 1999,* ed. L. Starke. New York: W. W. Norton. Part of an ongoing series on ways to restructure global energy systems that are good for people and the environment—a very important reference.

Gardener, G., and E. Assadourian. 2004. Rethinking the Good Life. In *State of the World 2004,* ed. L. Starke. New York: W. W. Norton. An interesting look at our current overly consumptive way of life and alternatives for a more sustainable future.

Heinberg, R. 2004. *Powerdown: Options and Actions for a Post-Carbon World.* Gabriola Island, BC: New Society. Description of the need to reduce global demand for fossil energy.

Heinberg, R. 2005. *The Party's Over.* Gabriola Island, BC: New Society. A popular book for those interested in learning more about oil supplies and problems that may arise as reserves are depleted.

Kammen, D. M. 1999. Bringing Power to the People: Promoting Appropriate Energy Technologies in the Developing World. *Environment* 41(5): 10–15, 34–41. A look at some of the key issues involved in creating more-sustainable energy systems in less-developed countries.

Lee, N., and D. A. Shalmon. 2007. Searching for Oil: China's Initiatives in the Middle East. *Environment* 49(5): 8–21.

A detailed account of the efforts China is making to find more oil to fuel its growing economy.

Prugh, T., C. Flavin, and J. L. Sawin. 2005. Changing the Oil Economy. In *State of the World 2005.* New York: W. W. Norton. A valuable read for anyone interested in ways we can shift from our dependence on oil.

Renner, M. 2004. Moving Toward a Less Consumptive Economy. In *State of the World 2004.* New York: W. W. Norton. A look at the many ways societies can reduce material consumption and reduce our consumption of raw materials and energy.

Roberts, P. 2005. The End of Oil: On the Edge of a Perilous New World. New York: Mariner Books. Description of the end of cheap oil and the impacts it could have on our society.

Roodman, D. M. 1997. Reforming Subsidies. In *State of the World 1997,* ed. L. Starke. New York: W. W. Norton. A look at the ways subsidies give fossil fuels and other environmentally unsound activities an advantage and how changes in policy can make other, more environmentally acceptable strategies affordable.

Sawin, J. L. 2004. Making Better Energy Choices. In *State of the World 2004.* New York: W. W. Norton. A look at the many ways individuals can save on energy.

Segerstahl, B., A. Akleyev, and V. Novikov. 1997. The Long Shadow of Soviet Plutonium Production. *Environment* 39(1): 12–20. A look at the impacts of another aspect of radiation pollution, notably weapons production.

Youngquist, W. 1997. *Geodestinies: The Inevitable Control of Earth Resources Over Nations and Individuals.* Portland, OR: National Book Company. Fun to read and a thought-provoking look at renewable and nonrenewable resources.

 ## Web Explorations

Online resources for this chapter are on the World Wide Web at: **http://www.prenhall.com/chiras** *(click on the Table of Contents link and then select Chapter 22).*

CREATING A SUSTAINABLE SYSTEM OF ENERGY

EFFICIENCY AND RENEWABLE ENERGY

Imagine a world powered by clean energy from the sun or wind. Imagine, too, a prosperous economy and a world free of oil spills, toxic air pollution, acid rain, and greenhouse gases. Idyllic, you say? Utopian and far from reality?

Not so, say proponents of energy efficiency and renewable energy sources. In fact, in a publication of the American Solar Energy Society, Dr. H. M. Hubbard asserted that by 2030, 50% of U.S. energy supplies could come from renewable sources. In subsequent decades, Americans could wean themselves nearly completely from fossil fuels. Why would this be desirable? Proponents of a renewable-energy future note that fossil fuel use, despite its many benefits, lies at the root of many serious environmental problems, among them global warming, acid deposition, urban air pollution, habitat destruction, and water pollution. Several countries, among them Germany and Great Britain, have already started down the path to a renewable-energy future.

Environmentalists and scientists interested in creating an Earth-friendly energy system envision a variety of renewable energy resources, many of which are being used today. In addition to energy conservation and energy efficiency, these alternatives include numerous forms of solar energy, wind energy, biomass, hydropower, geothermal energy, and others. This chapter discusses the main options for a sustainable-energy future, explaining each technology, its potential, and some of its main challenges. We begin with energy conservation and energy efficiency measures—ways to reduce energy use and ways to use energy more efficiently, respectively.

23.1 Energy Conservation and Energy Efficiency

In most of the more-developed nations, rising energy needs over the past 200 years have been met by producing more fuel, mostly fossil fuels in the past 100 years. Many people who are feeling the crunch of higher energy bills support a continuation of that strategy today by producing more coal, oil, and natural gas and developing oil shale, tar sands, coal gasification, and coal liquefaction. Over the past 200 years, little attention has been given to finding ways to reduce energy consumption by using less (conserving energy and being more frugal) or to use energy more efficiently. In the Harvard Business School report entitled *Energy Future,* conservation is described as "no less an energy alternative than oil, gas, coal, or nuclear. Indeed in the near term," say the authors, "conservation could do more than any of the conventional sources to help the country deal with the energy problem."

U.S. Conservation and Efficiency Efforts: A Brief History

Energy conservation is not foreign to Americans. Following the oil crises of the 1970s, briefly described in Chapter 22, the United States took drastic measures to slash energy waste—at home, at work, in the air, and on the highways. Consumers added insulation to their attics, replaced leaky single-pane windows, caulked cracks around doors and windows, bought smaller cars, and turned down their thermostats, saving enormous amounts of energy (Figure 23.1). The average gas mileage of a new car increased from a paltry 14 miles per gallon in 1974 to 26 miles per gallon in 1986, thanks to congressional action. Over the same period, factories cut their energy use, and consequently, the amount of energy needed to produce a dollar of gross national product dropped by 25%. Between 1949 and 1999, the amount of energy required to produce $1 of gross national product (GNP) decreased by 47%.

After a period of initial interest, however, the nation's fervor for energy conservation seemed to wane. Unsympathetic administrations under Presidents Ronald Reagan, George H. W. Bush, and George W. Bush greatly slowed progress in energy conservation. Under Reagan, for example, federal energy conservation programs were cut 91% between 1981 and 1987. Presidents Reagan and George H. W. Bush resisted measures that would have improved energy efficiency in automobiles. By the time President Bill Clinton took office in 1992, the average fleet mileage for new vehicles produced in the United States stood at 27.5 miles per gallon.[1] During Clinton's term of office, attempts to increase automobile energy efficiency were thwarted year after year through congressional action—riders on bills that forbade the Department of Energy even from con-

sidering further increases in gas mileage! Cars got bigger, and new-car gas mileage began to slip (Figure 23.2). Under George W. Bush, energy conservation continued in disfavor. The federal government even gave businesses hefty tax incentives to buy super large sport utility vehicles. Vice President Dick Cheney continually dismissed conservation efforts, saying the administration would not support any measures that would "change the American way of life."

The loss of interest in conservation is frustrating to those who recognize its many social, economic, and environmental benefits. One of the most pronounced, if least appreciated, benefits is the way conservation efforts stem the rise in oil prices and help national economies retain their vigor. In the early 1980s, conservation efforts in the United States and other countries, including Canada, lowered demand for gasoline, resulting in a temporary glut of oil. This, in turn, stabilized oil prices, as one might predict from the supply-and-demand theory. Stable oil prices helped stabilize the cost of consumer goods, quelling the economic turmoil that had strangled the United States and much of the rest of the industrial world in the 1970s and early 1980s.

Another factor that contributed to the fall in oil prices was the fact that Great Britain and many other nations outside of the Oil Producing and Exporting Countries (OPEC) increased exploration and oil production, causing OPEC's share of the world oil market to fall. In 1977, OPEC nations supplied two-thirds of the world's oil. By 1985, their share had dropped to one-third. To combat their loss in market share, in December 1985, the OPEC nations announced they would cut prices to regain their position in the market, and by April 1986, crude oil prices had been slashed in half. Today, OPEC's share of the world oil market is again on the rise, once again creating a potentially costly dependency, for the OPEC nations currently house a majority of the world's oil reserves.

---

[1]Nonetheless, energy consumption still increased dramatically, rising over 200% during that period while population rose only a little more than 80%.

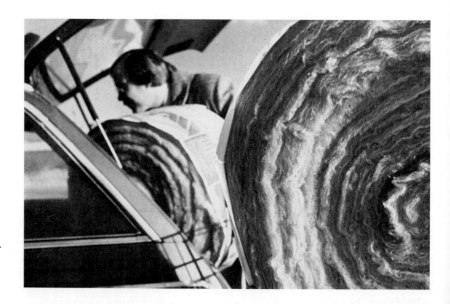

FIGURE 23.1   Getting ready for winter. Adding insulation yourself will save both energy and money. Many families in the northern states—for example, New York, Pennsylvania, and Michigan—invested in insulation, caulking, weather stripping, and storm windows in the 1970s and early 1980s. Their homes will be snug despite the icy blasts of winter. Don't forget, however, that insulation also works in hot climates to keep a house cool.

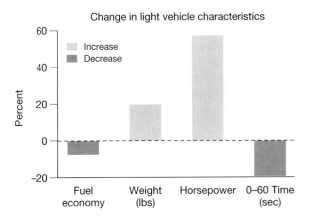

Change in light vehicle characteristics

FIGURE 23.2 American cars are getting heavier, more energy-intensive, and have lower 0–60 mph acceleration times.

FIGURE 23.3 The Ford Excursion represents the epitome of fuel consumption. This sport utility vehicle, which weighs nearly 4 tons, gets less than 10 miles per gallon. It was voted "Big Dumb Car of the Millennium" by NPR's *Car Talk* hosts, Tom and Ray Magliozzi, aka Click and Clack, the Tappet Brothers, in 2000.

Through the latter half of the 1980s, oil prices remained low, and while that was good for many sectors of the economy, it eroded much of our nation's resolve to conserve energy. As a sign of our nonchalance, in 1987 the U.S. Congress passed a popular law that allowed states to increase the speed limit on rural sections of interstate highways to 65 miles per hour, a measure that greatly increased fuel consumption. In 1996, the speed limit was raised again. In addition, in 1996, President Clinton lifted the extra tax on gas-guzzling cars. U.S. and foreign auto manufacturers began producing larger and larger vehicles with lower and lower gas mileage (Figure 23.3). Their sales skyrocketed until the early 2000s, when our profligate ways began to backfire.

As pointed out in Chapter 22, the inefficient use of energy, especially in the United States, combined with rising demand from China and India, began to wreak havoc on world energy prices. In 2008, gasoline reached $4 per gallon as the price of oil hit $140 per barrel, a record high. This crisis, also spurred on by commodity traders (speculators) who drove up the price of oil over fears of shortages, renewed the world's interest in energy conservation and energy efficiency.

Today, the United States and other countries are heading into stormy seas. With domestic oil nearly gone and production falling, U.S. oil imports constitute the lion's share of the nation's total annual consumption (Figure 23.4). High levels of consumption, say some experts, are unsustainable. Not only do they exceed the

U.S. Petroleum production, consumption, and imports

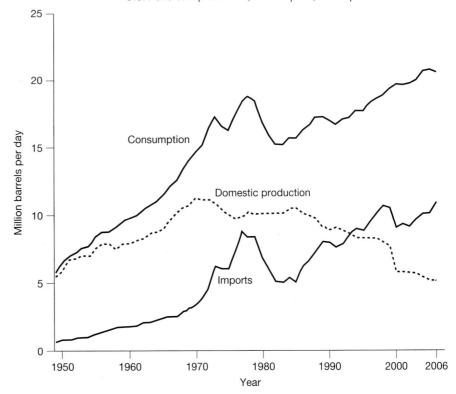

FIGURE 23.4 U.S. oil imports are climbing rapidly again, making us more and more vulnerable to OPEC.

oil production system's ability to transport and refine oil, but they also very likely exceed the world's ability to extract oil.

Oil is costly in environmental and economic terms. Oil and oil by-products are a major source of the greenhouse gas, carbon dioxide. America spends billions of dollars protecting oil as well—money that could be invested in creating much more sustainable sources of fuel. In 2008, the United States was spending $10 billion a month to fund military operations in Iraq and Afghanistan, an action many believe is largely due to our interest in stabilizing the Middle East to ensure a steady stream of oil to the United States. In the early 1990s, the world's nations spent $61 billion to wage war in the Persian Gulf, a war fought largely to protect Kuwait's oil supplies from invading Iraqis. Although few people realize it, the United States today spends $50 billion a year for the military protection of oil tankers moving in and out of the Persian Gulf, according to the Rocky Mountain Institute. If this subsidy were included in the price of Middle Eastern oil, it would cost about $500 per barrel, much higher than the approximately $140 per barrel price tag on the open market in mid-2008. If citizens of the United States paid the real cost of gasoline from Persian Gulf oil, it would cost over $25 per gallon. If the cost of the military actions in the Middle East were taken into account, the cost of oil would be even higher—much higher. Overall, according to some estimates, the U.S. government subsidizes the oil, coal, and natural gas industry to the tune of $200 billion a year, compared with $200 million a year for renewable energy.

Getting Back on Track

Faced with high fuel costs, many people assert that Americans must get back on the efficiency and conservation track. We must, they say, conserve energy and become more efficient in energy use. And we must develop and implement clean, non-polluting alternatives.

We should point out that the benefits of efficiency and conservation measures, the topic of this section, are more than simply a matter of conserving finite resources. Conservation and efficiency help families save money so they can have more to do the many things they need to do—such as pay for a college education. These efforts help businesses produce goods and services at a lower cost. They help American companies compete with more-efficient companies in Germany and Japan. And they can also help the world sustain lower oil prices, helping the global economy create economic stability. Energy conservation also creates jobs.

Energy efficiency and energy conservation can also drastically reduce many forms of pollution discussed in previous chapters, such as acid deposition, greenhouse gas emissions, urban air pollution, and even water pollution. By helping us stretch world oil supplies, efficiency and conservation give societies additional time to develop and more time to install clean, affordable, and reliable sustainable alternatives to oil and other fossil fuels.

The Untapped Potential

Energy conservation is to our economy what preventive medicine is to health care. Despite what some individuals claim, most nations have only begun to tap efficiency and conservation. Enormous opportunities still exist in buildings, industry, motorized vehicles, and appliances.

Consider some examples. Plug-in hybrid cars, mentioned in Chapter 18, could average 100 miles per gallon or more. The U.S average new-car mileage, 23.8 miles per gallon, is the lowest in the developed world.[2] By increasing the average fuel mileage to 40 or 50 miles per gallon, which some experts believe is easy to achieve, America could greatly reduce pollution, including greenhouse gas emissions, and stretch existing oil supplies. This, in turn, will create more time in which to develop even more efficient vehicles and efficient renewable liquid fuels.

On still another front are enormous savings from lighting. The United States has approximately 4.7 billion square meters (50 billion square feet) of commercial space, requiring roughly 100 power plants just to supply it with electricity for lighting. In Seattle, officials predict they could cut their demand for electric lighting by 80%. How? By switching to 18-watt compact fluorescent light bulbs that produce as much light as 75-watt bulbs, by changing to energy-efficient light switches, by using brighter interior surfaces, and by using special sensors that shut off interior lights when natural light is sufficient. Nationwide, the economic savings and environmental benefits of such a plan boggle the mind.

New energy-efficient compact fluorescent light bulbs (CFLs) can be used in the home as well. The bulbs screw into incandescent sockets, making conversion a snap. These bulbs cost a bit more but outlast nine to ten standard incandescent bulbs and save enormous amounts of energy over their seven- to ten-year life span (Figure 23.5). The 15-watt bulb, for instance, replaces a 60-watt incandescent bulb and yields as

FIGURE 23.5 Compact fluorescent light bulb. This bulb produces the same amount of light as a 100-watt incandescent bulb but with 75% less energy. Although it costs more, the bulb saves $30 to $50 over its lifetime, so it actually saves you money.

---

[2]According to the Environmental Protection Agency, the average mileage for domestic cars is currently 28.1 miles per gallon, and 20.3 miles per gallon for light-duty trucks.

much light. The $2 bulb will save $30 to $50 in electricity costs over its lifetime, depending on the cost of electricity in your locale. If you're paying 10 cents per kilowatt-hour, for example, the bulb will save $45. And, if every home in America replaced just four of its frequently used light bulbs with these bulbs, the nation would save more electricity than is produced by six large nuclear power plants.

A new generation of even more efficient light bulbs is now emerging on the market. Known as **LED lights** (LED stands for light-emitting diodes), they use much less energy—90% less electricity than compact fluorescents (Figure 23.6). These bulbs are widely used in taillights on cars and trucks but are just starting to be used in homes. They currently cost a lot more than CFLs, but costs are bound to tumble in upcoming years as demand and production increase.

Finally, consider appliances. One of the few major gains in government-sponsored conservation nationwide in the 1980s was the **National Appliance Energy Conservation Act of 1987.** The act called on manufacturers to produce appliances that use 20% less energy than 1987 models, starting in 1992. In 20 years' time, this law alone will eliminate the need for 22,000 megawatts of electricity—the amount of energy produced by 22 large nuclear power plants. It will save Americans well over $28 billion in electric bills! Even greater improvements in appliance efficiency have occurred since 1992.

Energy efficiency makes good economic sense for the consumer and producer. It's a lot cheaper to save energy than to buy new energy to run a light bulb or refrigerator or dishwasher. Homeowners can cut utility bills by half, saving hundreds of dollars a year, by investing in simple, inexpensive insulation, weather stripping, caulking, insulation for hot-water pipes, an insulated water heater jacket, and a half dozen compact fluorescent light bulbs.

Resetting Our Priorities

Building a sustainable society will inevitably require a massive reduction in the use of fossil fuels—a worldwide cutback—say proponents of renewable energy. Reductions will require actions on the part of all citizens, businesses, and governments. To stir action, many observers argue that we must first raise awareness of the key issues, notably the potential seriousness of global warming, caused in large part by carbon dioxide emitted from the combustion of fossil fuels. We must raise awareness about the limited supplies of oil as well. Through colleges and universities, through public schools, through government study and political leadership, say proponents, we must also raise the consciousness of the world's people regarding the dangers of our nation's dependence on oil.

The federal government and the states can also take leadership roles in making energy conservation and energy efficiency a national priority. Our long-term economic security depends on bringing energy conservation back into the American mainstream. Fortunately, there are some positive signs. In 1991, for example, the Environmental Protection Agency (EPA) instituted a nationwide energy conservation program aimed at voluntarily reducing electrical demand by the nation's largest corporations and by government buildings. Called the **Green Lights program,** this project has signed on hundreds of companies that, with the aid of the EPA, are working to cut electrical demand. They do so by installing compact fluorescent light bulbs, energy-efficient fluorescent tubes and ballasts, and other technologies.

The EPA started another highly successful voluntary project, the **Energy Star program.** Now run by the EPA and Department of Energy, this program calls on manufacturers of computers, printers, and monitors to reduce the energy demand of their machines, primarily by providing power-down options—that is, automatic switches that lower energy demand when the machines are not in use. Such machines, for example, go into sleep mode when operators are on the phone or when they leave the room for a while. Because many operators (about 30% nationwide) leave their machines on all night, this feature could save enormous amounts of energy. (It's much more efficient to shut a computer off entirely when not in use. This simple measure can reduce the amount of energy consumed by a computer by up to 40%.) Some companies have also found ways to reduce energy demand while computers are in use. The new MacBook Air, for instance, uses a solid-state hard drive, like those found in flash drives (aka memory sticks), rather than a rotating hard drive. This dramatically reduces the computer's energy demand and could become the wave of the future for both laptop and desktop computers.

FIGURE 23.6 LED lights. Although CFLs are becoming more and more popular, LED lights may someday replace them. They are extremely efficient, although still rather pricey.

The Energy Star program has been extended to many other appliances and electronic devices, such as televisions, stereos, washing machines, refrigerators, and even cordless telephones. (There are about 50 different classes of product for which Energy Star ratings are available.) An Energy Star label on a product indicates that it is among the most energy efficient in its category.

The Energy Star program is now being used in homes. For homes, builders seek ways of building new homes that use at least 30% less energy than homes based on the national Model Energy Code (already a more efficient home than those built in earlier years). After a home's energy performance has been verified by an independent expert, it is given the Energy Star label.

Personal Actions

There is almost no end to the things an individual can do to reduce energy use. This abundance of options is both a blessing and a problem. It's a blessing because you have a wide assortment of options to choose from; it's a curse because, faced with such choices, many people don't know which what to choose, and so do nothing.

Accordingly, we have limited the following discussion to individual actions that do the most good—that is, actions that cost very little yet make significant reductions in energy use. You may want to read through this material and then draw up a plan to reduce your own resource consumption.

Energy Conservation and Energy Efficiency Around the House   As individuals, we can make significant contributions to resource conservation with little effort and monetary investment. You can begin by turning down the thermostat, which is one of the easiest and most cost-effective strategies available. A 6°F drop in room temperature can reduce fuel consumption by 15% to 20%. Keeping your furnace thermostat at 68°F during the winter (when you are home) can save enormous amounts of energy. (You can turn the thermostat down even more when you're gone or when you're sleeping, saving even more energy and more money.)

Try turning your thermostat down gradually. Drop it, on average, 1°F every few weeks until you reach the desired setting. This will give your body a chance to adjust. To help counterbalance the drop in temperature, you may want to dress more warmly. Thermal underwear, wool sweaters, and heavy socks can help you stay comfortable. Putting on a sweater is equivalent to raising the room temperature nearly 4°F. Wrapping up in a blanket while you are watching television or reading has the same effect.

As noted, you can drop the thermostat even further during the winter when you leave during the day to save on energy. This saves energy, despite the bad advice to leave the thermostat setting high while you're gone. It only takes 20 minutes or so for a home or apartment to heat up. You can also turn the thermostat down at night during the winter and stay perfectly warm without an electric blanket. Better yet, your body produces a large amount of heat (about as much as a 70- to 100-watt light bulb), and you can keep comfortable even in the

coldest weather with blankets. Using a ceiling fan during the winter also helps push hot air that accumulates near ceilings down to floor level, saving energy.

You can install automatic thermostats to adjust daytime and nighttime temperatures to save heat. Automatic thermostats, in fact, pay for themselves in reduced energy bills in as few as two or three months to as long as three years, depending on who installs them and other factors.

To some people, turning down the thermostat means they will end up being cold and uncomfortable. There's a good reason for this: Many houses are too drafty, and lowering the thermostat setting reveals the frigid internal winds created by leaks. In the 1970s, many people who tried to conserve energy in their homes found them frightfully uncomfortable. To compensate, they reverted to old ways: cranking up their heat to fight drafts. Many complained that conservation was a bad idea. This strategy is shortsighted and wasteful. It's far cheaper and far wiser to seal up the cracks through which air escapes with caulk and weather stripping. And it's amazing how many leaky spots you can find with a little effort.

Caulking and weather stripping have an astonishing return on investment. In cold climates, caulking and weather stripping pay for themselves in reduced energy bills in a few months if you do the work yourself, a little longer if you hire someone else to do the job for you. Reducing drafts will make your home more comfortable.

If your home or apartment is air-conditioned, turn the thermostat to 78°F while you're home in the summer. Turn it up even higher when you're gone—say, to 82°F—to save energy and reduce your fuel bills. Using a ceiling fan will allow you to set the thermostat a bit higher than 78°F, saving even more energy and money. That's because ceiling fans use a lot less energy than air conditioners.

Turning down the setting on your water heater is another simple measure to reduce energy consumption. Most water heaters are set too high. By turning your water heater to 140°F if you have a dishwasher (without a water-heating function) and 120°F if you don't, you can save a ton on water heating every year. A study by Oak Ridge National Laboratory, for instance, showed that lowering the setting from 160°F to 140°F saves 400 kilowatt-hours a year, as much electricity as many households use in a month. That's a savings of about $4 to $5 per month or $48 to $60 per year. Lowering the heater setting to 120°F will save even more. Lowering your water heater thermostat is fairly easy and won't affect personal hygiene at all.

GO GREEN!

When purchasing any electronic device, always select the one with the Energy Star label. These products are the most efficient in their category.

GO GREEN!

Take shorter showers to save energy. Install a water-efficient showerhead. These measures reduce water demand and energy consumption. A water-efficient showerhead can pay for itself many times over in a single year and will give you a good, invigorating shower.

The measures just described achieve significant energy savings with virtually no expenditure. For individuals willing to invest a little money to cut energy demands, four additional approaches with relatively quick paybacks are recommended. The first is ceiling insulation. Because most heat escapes through the ceiling, upgrading the insulation in your attic to R-38 or higher (R is a measure of heat resistance) can help cut energy consumption drastically. Depending on where you live and how warm you like your house, insulation can pay for itself in three to seven years, according to the EPA. A three-year payback is equivalent to a 33% return on your investment.

The second measure is storm windows. Self-installed storm windows may take five to seven years to pay for themselves. If someone else does the work, count on a payback period that is doubled. However, the payback for insulation and storm windows is immediate in one respect: comfort. If you care about comfort, you'll be rewarded immediately. Ceiling insulation, by far the easiest and cheapest to install in many homes, is one of the most effective ways of reducing heat loss. Ceiling insulation creates a much cozier domicile almost instantly—a payback few people seem to consider when debating whether they can afford extra insulation. Storm windows have the same effect. They eliminate cold spots and drafts and greatly increase comfort levels. What is more, they help reduce energy consumption, with all its attendant environmental costs. You save money and help reduce environmental deterioration—not a bad investment!

The third inexpensive energy conservation measure, which also rapidly pays for itself, is insulation for water heaters and hot-water pipes. Now available for $20, insulating blankets help hold in heat and reduce energy demand. Easy-to-install insulation for hot-water pipes has a similar effect. A $20 investment and a few minutes of your time will save an average home about $20 per month, which represents a 100% monthly return on investment.

The fourth measure is energy-efficient light bulbs. Now available in most grocery stores or discount stores, the General Electric Miser series saves approximately 10%. As mentioned earlier, however, many manufacturers produce screw-in compact fluorescent bulbs that use 75% less energy than conventional incandescent bulbs. Many hardware and building supply stores and many discount stores—even the corporate giant Wal-Mart—carry these bulbs. CFLs are designed to produce a very appealing yellowish color of light, not the harsh blue of typical fluorescent bulbs.

Energy Conservation in Transportation   Nationwide, automobiles consume approximately 540 billion liters (140 billion gallons) of gasoline each year, according to the U.S. Department of Energy. Much of that fuel is wasted by people driving erratically and at excessive speeds. By driving reasonably and sticking to the speed limit (both conservation measures), you can cut personal gasoline consumption by 10%, saving money and time spent at the pump, reducing pollution, and reducing the environmental impact of oil production. A study by *Consumer Reports,* for instance, showed that gas mileage drops 12.5% as one speeds up from 55 to 65 miles per hour and plummets 25% when one travels at 75 miles per hour, rather than 55. Short trips that can be combined into one trip or replaced by walking or riding a bicycle could help further cut energy use.

Perhaps one of the biggest consumers of fuel is the daily commute—driving back and forth to work and school. Each year, commuters travel billions of kilometers. Often only one passenger rides in a car. This common practice burns more fuel than is necessary. It also creates unnecessary pollution and crowding on our highways. Individuals can help reduce these and other problems by joining car or van pools or taking buses to work or school. Van pooling and buses are, on average, nearly five times more energy efficient than the automobile. For example, it takes 400 kilojoules (a unit of energy) to move a passenger 1 kilometer (0.6 mile) by van pool, train, or bus. Passenger cars consume 1,800 kilojoules, which is about the same for the newer energy-efficient jet airplanes. In other words, it is 4.5 times more efficient to carpool than to drive your automobile. You're getting more or less the same service, only using much less energy and creating much less pollution. If you can join a van pool or ride the bus even occasionally, you should consider doing it. You may also want to consider walking or riding a bike, if that's practical, at least once in a while. Your contribution, combined with that of other like-minded individuals, can add up quickly.

One of the wisest steps is purchasing an energy-efficient vehicle. Certain models on the road, such as the Toyota Prius Gen II, get around 50 miles per gallon (combined city and highway driving) and provide a useful alternative for those who want to reduce their carbon footprint and save gas. If you must have a larger vehicle, at least find the model that gets the best mileage—as close as you can get to 40 miles per gallon.

This section provided many useful tips to those interested in saving energy and helping solve the energy crisis and address global warming. If you'd like to do more, there are plenty of resources to provide advice, including one of Dr. Chiras's newest books, *Green Home Improvement: 65 Projects That Will Cut Utility Bills, Protect Your Health, and Help the Environment.*

23.2   Renewable-Energy Strategies

Energy conservation and energy efficiency are vital to building a sustainable energy future. These efforts, most experts agree, must also be accompanied by a shift to clean, renewable energy technologies. This section discusses the many exciting options.

Solar Energy

Sunlight is the ultimate source of energy on planet Earth. It powers the entire global ecosystem, and it is destined to last 5 billion years, compared with oil's 50-year life span. Sunlight comes to us free of charge and is a clean source of fuel. All we need to do is find ways to capture this **solar energy** and put it to use.

The thought of harnessing solar energy has piqued the interest of humans for ages. The Greeks used it 2,000 years ago to heat their homes—and considered those who didn't orient their homes to the south to tap the sun's generous heat to be barbarians. One thousand years ago, the Anasazi Indians of the desert Southwest used solar energy to heat their villages by building their homes in south-facing rock walls protected from the summer sun but open to the low-angled winter sun.

The amount of solar energy striking the Earth's surface on a cloudless day is over 100,000 times greater than the world's currently installed electrical capacity. The sunlight striking an area the size of Connecticut each year could provide all of the energy needed by the entire United States, yet Americans have tapped only a tiny portion of the sun's potential (Figure 23.7).

The sun's awesome power is amply illustrated by the world's largest solar furnace, in the French Pyrenees (Figure 23.8). This power plant has a 45-meter (150-foot) mirror that focuses sunlight on a boiler. Concentrated sunlight raises the temperature to 3,500°C (6,300°F), sufficiently hot to melt a 0.3-meter (1-foot) wide hole in a steel plate 0.9 cm (3/8 inch) thick in 60 seconds. Several similar power plants have recently been built in Spain.

Solar energy has a huge potential. To understand how solar power can help us create a sustainable energy future, let's look at the most promising technologies meant to satisfy our needs. It's important that you understand each technology and its pros and cons.

Solar Heating   All buildings are solar heated to some extent. Without the sun, the Earth's average temperature would be a chilling −230°C (−450°F), so in many ways, we already use the sun for heating, even if unintentionally. The intentional use of sunlight for heating takes two forms: active solar heating and passive solar heating.

Active solar systems, like the one shown in Figure 23.9, are typically used to heat water for domestic uses. They can also be used to heat homes and other buildings. Active solar hot-water systems consist of solar collectors or solar panels. They are generally mounted on rooftops of homes or other buildings. A solar collector is an insulated box with a glass face. The interior is painted black to absorb light. Sunlight penetrating the glass is converted to heat. The heat is then removed from the interior of the collector by water or some other fluid that flows in pipes in the panels (Figure 23.10). It is then transferred directly to the room, as in the case of air collectors, or transported to a storage device (e.g., a water tank) for later use.

Solar domestic hot-water systems can provide 50% to 100% of a person's hot-water needs, depending on one's location. Good, reliable systems are now available. When you take into account the cost of the system and the cost of its operation, domestic solar hot-water systems are often cheaper than electric water heaters and water heaters powered by propane. They cost a little more than hot water heated by natural gas.

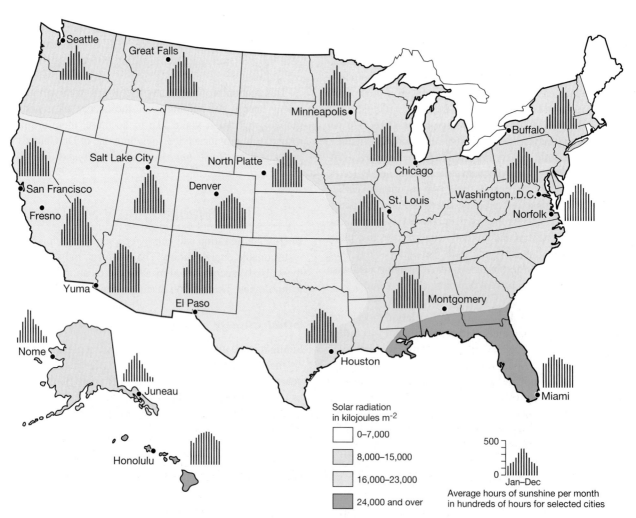

FIGURE 23.7   Solar radiation and hours of sunshine. Many parts of the United States and other countries have abundant sunlight for solar energy, both solar electricity and passive solar heating.

FIGURE 23.8 French solar furnace. This solar furnace, located near Odeillo in the Pyrenees Mountains of southern France, was built to test materials under extremely high temperatures. In the foreground, an array of 63 mirrors (heliostats), each measuring 6 meters by 7 meters, reflects sunlight onto the curved mirror surface of the office building in the background. This in turn focuses the sunlight on an aperture in the tower at the center, where temperatures of 6,300°F can be produced—enough heat to melt any known material.

Passive solar systems are designed for space heating. In such systems, buildings become the solar collectors. Passively heated homes are so named because they do not require collectors or pumps or fans to move heat around. The passive solar home shown in Figure 23.11, belonging to Dr. Chiras, is oriented to the south. South-facing windows let the low-angled winter sun enter the house. Sunlight strikes the walls and floors and is converted into heat. Special concrete or block walls in the inside store the

FIGURE 23.9 Mount Rushmore goes solar. Mount Rushmore, in South Dakota, site of the famed sculpted faces of former presidents, has a new solar energy system. Solar collectors on the roof of the Visitors Center transform the rays of the sun into energy for heating and air conditioning. The solar panels were developed by Honeywell. The system provides energy for 53% of the heating and 41% of the cooling for the 9,250-square-foot building.

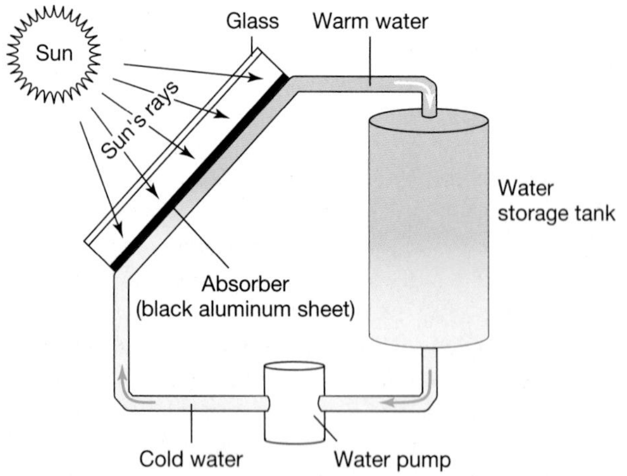

FIGURE 23.10   A type of active solar space-heating system. It relies on solar panels to absorb sunlight and produce heat.

FIGURE 23.11   Passive solar home at 8,000 feet in the foothills of the Rockies featuring south-facing windows and skylights that let the winter sun in. Superinsulation and passive solar keep winter heat bills low—about $200 a year.

heat during the day and radiate it out into the room at night when interior temperatures drop. An overhang on south-facing windows prevents the high-angled summer sun from entering the house, keeping it comfortable all year round.

Passive solar homes must be airtight and well insulated; windows generally have double or triple panes and thick curtains to prevent heat loss during the night. Surprisingly, well-insulated passive solar homes can function efficiently in fairly cold climates. One home built by the engineering department at the University of Saskatchewan had an annual heating bill of $40. By comparison, a conventional home of the same size would have a heating bill of $1,400. Excellent passive solar homes now operate in Maine, Vermont, and Wisconsin, among other places.

Passive solar energy can be used to heat commercial buildings as well. Ontario Hydro, a Canadian energy company, built a mammoth office building in Toronto that relies entirely on solar energy and a system that captures waste heat from people, lights, and equipment and reuses it (Figure 23.12). Despite frigid winter weather, the building stays warm and comfortable inside. In Soldier's Grove, WI, the entire business community moved out of the floodplain to avoid the periodic flooding that

FIGURE 23.12   Ontario Hydro's energy-efficient, solar office building in Toronto.

wreaked havoc on their town. The town's citizens decided to convert the town into a solar town. They now have a solar post office, fire station, library, gas station, woodworking shop, American Legion hall, grocery store, and other community buildings. During the first three years of the new town's existence, the grocery store's heating bill was zero despite harsh winters—thanks to passive solar energy, superb insulation, and a system that captures waste heat from compressors and pumps it back into the 7,000-square-foot store. Remarkably, the entire system paid for itself in those three years.

Interestingly, many of the features that are used to create a passive-solar-heated building also contribute to natural cooling, or **passive cooling.** Orienting the long axis of the house on an east–west plane, which helps the house gather sun in the winter, reduces summertime heat gain. Airtight construction prevents heat from entering. Thick insulation in walls and ceilings, which keeps houses warm in the winter, also keeps them cool in the summer.

Other passive cooling techniques also can help. For instance, in desert climates, where nighttime temperatures drop dramatically, windows can be opened to let in cool air to flush out the accumulated heat from the house, recharging it for the next day. **Earth sheltering** a home, that is, pushing dirt up against back walls, sometimes covering the back walls and roof entirely with a layer of earth, also helps solar homes operate more efficiently. Because the earth in most areas stays a constant 10° to 12.8°C (50°F to 55°F) below the frost line, the temperature of a house sheltered by earth remains fairly constant year-round. During the winter, it is necessary to add only a small amount of heat to bring it into the comfort zone. Solar energy can do that admirably well. In the summer, the home stays naturally cool. The earth keeps it cool.

Solar Electricity

Solar energy can also be used to generate electricity. The French solar tower, described earlier and shown in Figure 23.8, heats water with sunlight and creates steam that runs an electric generator. This technology is therefore called **solar thermal electricity.** Somewhat similar systems are now operating in Sandia, NM, and Barstow, CA.

As a rule, these prototype power towers, completed in 1977, are small plants, producing only 1 megawatt of electricity—much smaller than conventional coal-fired power plants, which typically produce around 500 megawatts. The power tower stands about 15 stories high and is located in a large field of movable mirrors. Controlled by a computer, the mirrors track the sun across the sky and focus their beam of energy on the top of the tower. The intense heat produced there boils water, which is converted to steam to run an electric generator (Figure 23.13).

In recent years, engineers have created much simpler designs, which are proving quite successful (Figure 23.14). One company in California, for instance, developed a technology that produces electricity for 8.5 cents per kilowatt-hour, which is more expensive than coal (not counting coal's external costs) but cheaper than nuclear energy.

Electricity is more commonly produced by photovoltaic cells. **Photovoltaic cells (PVs)** are thin sheets primarily made of silicon from silica, sand, or quartz rock. They convert sunlight energy striking them directly into electricity. Photovoltaic cells are mounted in small panels called modules. The modules are mounted on racks on rooftops, on poles, or on the ground (Figure 23.15).

Some are mounted on racks that track the sun across the sky from sunrise to sunset, which can increase their electrical output. In recent years, several companies have begun producing roofing shingles and other roofing materials that contain PVs (Figure 23.16). That way, you get a roof and solar electricity. Companies are also producing glass for windows and skylights that contains PV material. Building materials that incorporate solar electric generating capacity are referred to as **building integrated photovoltaics.**

Developed in 1954 by Bell Laboratories, photovoltaic cells were first used in 1958 to provide power for *Vanguard I,* America's second satellite. Since that time, photovoltaic cells have provided power for almost all other satellites and are now being used on Earth. One of the first terrestrial applications was in remote locations where electricity is not available or is too expensive to install. River flow monitors in remote country, mountaintop radio relays, remote irrigation pumps, highway signs in unpopulated areas, lighthouses, and buoys all use them to provide energy. Photovoltaic cells also have some uses much closer to home, such as in calculators, watches, and other electronic devices. In 1980, photovoltaic cells powered the first flight of the *Gossamer Penguin,* a one-person solar airplane.

Electricity generated by PVs is the second fastest growing source of energy in the world. Despite this, photovoltaics currently contribute very little to our overall energy consumption. As prices go down, however, experts predict that they will become more widely used. Dr. Chiras currently generates most of his home's electricity from a small array of solar panels and a small wind generator. Batteries, panels, and generator cost about $17,000.

Photovoltaic cells are an attractive energy source. The fuel is free and virtually inexhaustible. It is a clean technology as well, the only pollution being created during their manufacture. Widespread use would reduce acid deposition, greenhouse gas emissions, urban air pollution, strip mining, and all the other impacts created by the use of coal. They would also reduce the need for nuclear power, with all its effects. Solar electricity could be used to recharge batteries in electric cars and could be used to generate hydrogen fuel that could be used in cars of the future as well.

So why haven't photovoltaic cells made it to the rooftop of every house in America? The answer is cost. Although the silicon used to make solar cells comes from silicon dioxide, one of the most abundant chemical compounds on Earth, the making of the most widely used solar cells on the market requires a fair amount of energy and is therefore rather costly. Also, the solar cells commercially available are fairly inefficient. A 1-square-meter panel is only enough to power a single 120-watt light bulb, and that panel costs about $300 to $500. Providing energy for a whole house with inefficient appliances and lighting might require a large array of solar cells costing $40,000 to $50,000. Making a house efficient in how it uses electricity could cut the cost of a system to $10,000 to $20,000. Even as such, solar electricity costs two to three times more than commercially available electrical power.

What is needed, say proponents, are more-efficient and less-expensive solar cells, both of which are currently being developed, to help make this technology more competitive with other sources of electricity. The first solar cells were cut from a pure crystal of silicon, which took a huge amount of energy to produce. To cut costs, researchers developed a cheaper substitute known as a multicrystalline solar cell. Multicrystalline wafers are now widely

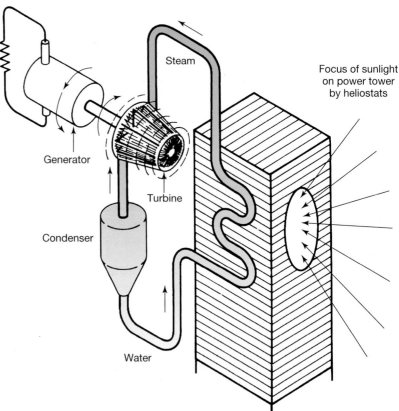

FIGURE 23.13   (Top) View of solar power tower located at the Department of Energy's test facility at Sandia Laboratories in Albuquerque, NM. (Bottom) The solar power tower. The concentrated rays of the sun convert water to steam, which in turn propels the turbine.

used. To reduce costs and the amount of material required to make a solar cell, researchers developed a thin-film or solar module. In this technology, the photosensitive material is deposited on a backing in a very thin layer. Additional work is now under way to produce solar cells from very inexpensive polymer (plastic) and other inexpensive materials, which could dramatically lower costs and make solar electricity very inexpensive.

Tracking devices, mentioned earlier, can also improve the efficiency of a solar electric system. In 1986, researchers at Stanford University announced a new design that uses a parabolic mirror to concentrate sunlight on solar cells. This could boost the efficiency from 10% to nearly 30%, a tremendous boon for the industry. Researchers are also working on other forms of concentrating solar electric modules that use lenses to concentrate light on PV cells, boosting their efficiency and electrical output.

Another means of lowering prices is mass production. For many years, photovoltaic modules were manufactured by small companies that could not capitalize on economies of scale.

FIGURE 23.14   Solar thermal electric system. Sunlight reflected from these mirrors heats oil to 735°F. This heat is used to generate steam that drives a turbine to produce electricity.

Photovoltaic expert Jack Stone at the National Renewable Energy Laboratory in Golden, CO, has pointed out that photovoltaic technology experienced a catch-22. Small companies could not justify building large plants without market demand, but market demand, in turn, was thwarted because prices were too high. Today, production of solar electric cells is skyrocketing. Many companies are expanding their operations by leaps and bounds, a trend that could create an economy of scale that may help lower PV prices.

Photovoltaics could provide Americans with much of their electric needs. Theoretically, all of the electric power required by the United States could come from 12,000 square kilometers of solar cells—roughly the area occupied by all rooftops in the lower 48 states. Brown University professor Joseph Loferski estimated that photovoltaics mounted on only 20% of Rhode Island's rooftops would provide for all of the electric power

needs of the state. The electric power would be generated only a few feet from where it would be used, thus removing the need for costly, inefficient, and unsightly transmission lines.

Photovoltaics do have some drawbacks. Sunlight is inconsistent, and some kind of storage system would be necessary. Batteries are the common means of storing electricity. While a PV array might last 50 years, batteries have to be replaced about every seven to ten years, depending on how well they're maintained. Backup systems such as gasoline-powered generators are also needed to provide energy for long, sunless periods; some people use wind generators to provide additional power in such instances, but wind is also inconsistent. Some photovoltaic production practices also generate hazardous wastes.

Solar electricity—and other renewables, for that matter—can be tied into the grid, that is, the commercial power lines that supply electricity from other sources. Grid-connected systems use the utility lines as a battery, an option that is cheaper, less resource intensive, and more efficient. For these reasons, grid-connected systems are the most popular ones in use today.

Despite these problems, there are areas where solar electricity is thriving. In the less-developed countries, for instance, solar electricity is proving to be a valuable source of electricity for thousands of villages located many miles away from power plants. Government officials in these countries have found that it is far cheaper to import this technology than to string power lines to remote villages. Today, the less-developed nations are the world's leading market for solar electricity. Solar electricity is also popular in remote areas of more-developed countries, and if homes are being built more than a few tenths of a mile away from a power line, installing a good solar electric system may be cheaper than stringing wires.

Solar electricity is also thriving in states and nations that offer incentives for businesses and homeowners to install systems. These incentives take many forms. Some pay a portion of the system's cost; others exempt systems from property tax and/or sales tax. The U.S. government also offers tax incentives for residential and business systems. Business PV systems receive much more generous incentives. Incentives can cut the cost of solar electricity in half, making it cost competitive with conventional electricity.

FIGURE 23.15   Photovoltaic cells. The solar modules on these homes consist of thin wafers of silicon absorb sunlight and generate an electrical current. They produce electricity with little impact on the environment and last 30 to 50 years.

FIGURE 23.16 Solar shingles. One of the newest developments, the solar shingle does double duty: It protects the home from rain and provides electricity.

Most states also offer net metering programs. **Net metering** programs require utilities to pay homeowners or businesses with grid-connected systems for surplus electricity they deliver to the electrical grid. Net metering requires the utility to pay at the same rate they charge for electricity. Surpluses, if any, can help lower the cost of a solar electric system.

Despite its difficulties, solar electricity is the world's second fastest growing source of energy, after wind power. In the 1990s, sales of PVs increased on average about 16% per year. Sales continue to be very strong in the 2000s, although the total capacity of the world's photovoltaic system is still quite small compared to coal or nuclear energy. Another interesting fact about solar electricity that foreshadows its success is that Solarex, once one of the world's leading suppliers of PVs, is now owned by BP, the world's third largest oil company. Royal Dutch Shell also has committed to spending $1 billion on renewable energy. A Closer Look 23.1 describes the efforts of many additional businesses to go solar.

Our Solar Future

"The transition to the solar future," says Professor Kurt Hohenemser of Washington University, represents "a gigantic technological enterprise that could require a century to complete." By increasing efficiency and conservation measures and supporting renewable energy sources, governments can become a positive force in the transition to a sustainable society. This section presents many of our options.

Geothermal Energy

The Earth stores enormous amounts of heat, or **geothermal energy.** This energy comes from two principal sources: (1) the radioactive decay of naturally occurring radioactive substances in the Earth's crust and (2) molten rock in its interior (Figure 23.17). In some places, geothermal heat boils groundwater in the Earth's interior. Heated water and steam may spew to the surface through cracks in remarkable displays called **geysers.** In other places, the water merely bubbles up, filling pools (hot springs) or trickling into nearby streams.

A CLOSER LOOK 23.1 **Solar Electricity Enters the Mainstream**

"Retailers are typically obsessed with what to put under their roofs, not on them," notes Stephanie Rosenbloom in the August 11 issue of the *New York Times.* "Yet the nation's biggest store chains are coming to see their immense, flat roofs as an untapped resource."

"In recent months," she goes on to say, "chains including Wal-Mart Stores, Kohl's, Safeway and Whole Foods Market have installed solar panels on roofs of their stores to generate electricity on a large scale."

"So far, most chains have outfitted fewer than 10 percent of their stores. Over the long run, assuming Congress renews a favorable tax provision (which it did) and more states offer incentives, the chains promise a solar construction program that would ultimately put panels atop almost every big store in the country," Rosenbloom adds (parentheses added by author).

"In the coming months," Rosenbloom notes, "85 Kohl's stores will get solar panels; 43 already have them." But they are not alone, she adds. Macy's has installed solar electric systems on 18 stores and will install them on another 40 stores by the end of 2008. Safeway, a grocery chain, is planning to mount solar modules on 23 of its stores. The list goes on. Whole Foods Market, REI, and others are taking the leap; Microsoft and Toyota also have installed large solar arrays on some of their facilities. Even corporate giant Wal-Mart is retrofitting many of its stores and distribution centers with solar electric systems. Some day hundreds of their stores may be powered, at least in part, by electricity generated by the sun.

Some companies are taking another route. They're buying green power—electricity generated by renewable energy sources such as wind. In 2008, for instance, computer chip giant Intel announced that the company will purchase more than 1.3 billion kilowatt-hours a year of renewable-energy certificates (RECs) from Sterling Planet to offset carbon dioxide emissions from their operations. The electricity will be generated by wind, solar, small hydroelectric facilities, and biomass. According to the EPA, this purchase makes Intel the single largest corporate buyer of green power—electricity generated from clean, renewable resources—in the United States.

Besides offsetting their emissions, Intel officials hoped their purchase would help stimulate other companies to follow suit and hence boost the market for green power. Their efforts, combined with those of other companies, could help lower the generating cost of electricity from renewable resources such as wind.

To learn of other companies, check out EPA's Green Power Partners list online. At this writing, Intel takes the number one spot, and PepsiCo is number two. Companies are not the only entities that have made major commitments to purchase green power. The Commonwealth of Pennsylvania, the City of Houston, and the City of Dallas are all major purchasers, as are Sprint Nextel, IBM, and the U.S. Department of Veterans Affairs.

FIGURE 23.17 Geothermal areas in the world. Most are located in regions of volcanic activity, past or present.

These regions are called **hydrothermal convection zones.** In other regions, heated groundwater may be trapped by impervious rock layers and cannot escape. These sites are called **geopressurized zones.** Heated by the molten rock underneath, the steam and superheated water can only be tapped by drilling holes into the pockets of steam. In still other areas, magma heats overlying rock, forming **hot rock zones.** Water can be pumped into these zones and then pumped out. Heat stripped from the hot rock zone can then be used for heating buildings or generating electricity.

Most geothermal energy today comes from hydrothermal convection zones. They are the easiest and cheapest to tap (Figure 23.18). Steam or hot water from these zones can be used to heat a wide assortment of buildings. For example,

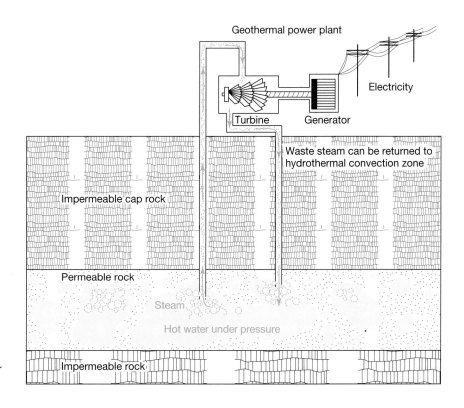

FIGURE 23.18 Schematic view of a geothermal power plant operation.

almost all of the houses in Reykjavik, the capital city of Iceland, are heated by geothermal steam. Icelanders grow a variety of vegetables in steam-heated greenhouses as well. In the United States, at least 300 communities use geothermal heat for one reason or another. One rancher in South Dakota uses it to heat his home and several buildings. He also uses it to dry several thousand bushels of grain each year. All told, geothermal energy saves him over $5,000 a year in fuel bills.

Geothermal steam or hot water can be used, as mentioned earlier, to generate electricity. First used successfully in 1904 in Larderello, Italy, geothermal electricity is now produced in New Zealand, Japan, Mexico, the United States, the Philippines, Italy, Iceland, and Russia. One of the world's largest projects is in northern California on the slope of an extinct volcano. Started in 1960, the project has a 900-megawatt electrical capacity, supplying nearly a million people.

All in all, however, geothermal energy produces very little electricity worldwide. The total global capacity is now only about 8,000 megawatts, equivalent to eight large nuclear power plants, but increases are predicted. According to one estimate, the United States alone could produce 27,000 megawatts of electricity from geothermal sources—enough electricity for 27 million people, or nearly one-tenth of the population.

Geothermal power is relatively inexpensive and much cleaner than coal-fired and nuclear power plants. Capital investment—the money needed to start a plant—is 40% less than for a coal-fired power plant and 70% less than for a nuclear power plant. However, there are some drawbacks. First, minerals dissolved in the steam can corrode pipes and turbine blades. Second, geothermal plants can be very noisy. Third, geothermal plants release gaseous pollutants such as hydrogen sulfide, carbon dioxide, ammonia, and methane that often occur in the steam and require pollution control devices. Fourth, steam cannot be transported long distances without losing its heat. Finally, most of the hydrothermal convection zones in the United States are located along the Pacific Coast. The highly populated East Coast, where the demand for energy is high, would have to rely on hot rock zones, which are found throughout the United States but are more costly and difficult to tap.

Hydropower

Humans have tapped the power of flowing rivers, or **hydropower,** for thousands of years. In the United States, hydropower was used extensively in the 1800s to grind wheat and corn, saw logs, and power textile mills. Today, hydropower is often used to generate electricity via turbines through which water flows, usually from the bottom of a dam. Hydroelectricity provides about 12% to 13% of America's total electricity, or about 3.6% of the nation's total energy consumption (Figure 23.19).

Hydropower offers several advantages: it is (1) relatively inexpensive, (2) pollution free, and (3) a potentially renewable source, provided reservoirs can be kept from filling with sediment. According to some estimates, enormous hydropower resources still lie untapped. Today, only about 17% of potential global hydropower is being used. You might assume, therefore, that when the supply of fossil fuels falls so low that it becomes too expensive, the world could turn to hydropower.

FIGURE 23.19   Aerial view of Hoover Dam and Lake Mead on the Arizona–Nevada border. This world-famous dam spans the Colorado River. Built in 1935, the dam provides multiple benefits, such as flood protection, water storage for irrigation purposes, and hydroelectric power.

Unfortunately, it is not that simple. Several problems cloud the prospects for hydropower: (1) The greatest hydropower potential lies in less-developed countries in Africa, South America, and Asia. Because many of them lack financial and other resources, their hydropower potential is not likely to be developed. (2) The increasing cost of dam construction, with price tags ranging from $300 million to $2 billion, makes it increasingly less cost-effective. In addition, dams often have a short life span because of heavy sedimentation caused by poor land management—deforestation, overgrazing, and poor farming practices. This problem threatens many projects. For example, Egypt's Aswan Dam, the world's largest, probably will have a functional life of less than 200 years. In the United States, over 2,000 reservoirs have been filled with sediment, and some succumbed in fewer than 20 years. Pakistan's $1.3 billion Tarbella Dam, which took nine years to build, could be filled by sediment in 20 years. (3) Dams destroy the scenic beauty of wild canyons, used by anglers, kayakers, rafters, canoeists, and a variety of others. (4) Reservoirs behind dams inundate forests, farmland, wildlife habitat, and habitat of indigenous human populations. (5) Dams reduce the natural flow of sediment-bearing nutrients to estuaries and disrupt aquatic food chains in these ecologically and economically important zones. (6) Dams can interfere with the migration of fish, such as salmon, and fluctuating water levels can kill fish eggs laid in shallow water near shore. (7) In arid regions, reservoirs accelerate evaporation. In some cases, evaporation reduces the amount of the available irrigation water.

Despite the disadvantages, hydropower is here to stay, and its use is likely to increase, although only modestly, in the near future. But the generous estimates of untapped potential should be viewed cautiously. For instance, the United States currently has 75,000 megawatts of hydroelectric-power-generating capacity and, according to estimates, has an additional 160,000 megawatts of untapped potential. However, much of our untapped energy lies in out-of-the-way places where dams may be economically and environmentally unfeasible. In fact, half of our untapped potential lies in Alaska, far from industrial and population centers. Canada, a world leader in hydropower, currently generates 60% of the electricity the nation produces from hydropower and has untapped potential, too. However, as with the United States, most of this resource lies in the northern reaches of this huge country, far from Canada's population, which is concentrated along the southern portion of the country near the U.S.–Canada border. Much of the untapped potential in Africa, Asia, and South America is in remote locations as well.

Most hydropower projections concern themselves with large dams, with little consideration of smaller projects, those that could produce 1 to 10 megawatts of energy. It is these small projects, some proponents argue, that could provide enormous amounts of energy in countries throughout the world. China, which produces more hydroelectricity than any other nation in the world, for instance, has 90,000 small dams on streams and rivers providing electricity to remote rural villages—equivalent to one-third of the country's total electric production. France and the United States have numerous small dams as well.

Although these structures are inexpensive to build and operate and provide energy to consumers where it is needed without huge transmission losses, they do significantly alter streams, affecting fish and other aquatic organisms. Some environmentalists believe that a far better strategy would be to retrofit the 50,000 nonhydroelectric dams in the United States with small turbines to generate electricity. The environmental damage, they argue, has already been done, so why not make a dam built for recreation, water supply, or flood control, for instance, do double duty—that is, generate electricity as well?

Wind Power

For many decades, Great Plains farmers harnessed the winds sweeping across their rolling grasslands and wheat fields with a relatively simple device, the **windmill** (Figure 23.20). Early windmills were used to pump water. Windmills were even used to fill water towers needed to replenish the early steam locomotives that traveled across the nation. Many farms also installed windmills that generated electricity to light barns and to power radios and kitchen appliances, such as toasters. However, the use of wind power declined in the 1930s as rural electric lines began to spring up (part of a nationwide rural electrification program). These lines connected remote farms with central power plants many miles away, forming an extensive electrical grid. Soon thereafter, the sight of windmills silhouetted against the blue prairie skies began to fade. In the 1970s, however, windmills began to pop up all across the nation, this time on special **wind farms** or in the backyards of rural and sometimes even suburban homes.

Wind energy contributed little to global energy needs 20 years ago. In fact, in 1980, the global capacity was a paltry 10 megawatts. In 1986, however, 13,000 wind turbines in California provided 1,100 megawatts of power—enough electricity for

FIGURE 23.20 The windmill of the past. Many-bladed windmills like these were once landmarks on American farms. They were used to pump water. Farmers stopped using them following the successful rural electrification program of the 1930s. Now farmers need alternative power sources to pump irrigation water because of high prices for gas and other fuels.

about 1 million people. Most of these and newer turbines are sited within specially built wind farms, located on mountain passes and connected to the existing electrical grid. In recent years, wind energy has grown by leaps and bounds in the United States and several other countries. Large wind farms have been built in numerous states, including Maine, Vermont, New York, Wisconsin, Minnesota, Iowa, Wyoming, Colorado, Texas, Montana, Oregon, and Washington.

Global capacity, including that in the United States, climbed from 7,600 megawatts in 1997 to nearly 32,000 megawatts in 2002, or enough to supply nearly 32 million people (Figure 23.21). By 2007, it had increased to 94,100 megawatts, making wind the fastest-growing source of energy in the world, bar none!

Wind farms are being built in many countries. Among the leaders are Germany, the United States, Spain, India, China, and Denmark. The Worldwatch Institute predicts that wind energy could eventually provide 20% to 30% of the electricity needed by many countries. And some experts believe that wind pumps—similar to the windmills of the past—could be used in remote rural villages of India and Africa to provide drinking and irrigation water.

Many versions of the windmill are currently being installed, ranging from huge units with massive blades to much smaller wind machines that provide electricity for individual homes, businesses, farms, and ranches. One large wind generator in Boone, NC, for example, sits atop a 40-meter (140-foot) tower. Its huge blades rotate 35 times a minute in a 50-kilometer (30-mile)-per-hour wind; the wind generator supplies electricity to 500 homes. But today, the wind generator is something of a midget. Newest machines stand even taller—100 meters (330 feet)—and have blades nearly 100 meters (300 feet) long. Many companies are now manufacturing smaller wind turbines, used for homes and farms.

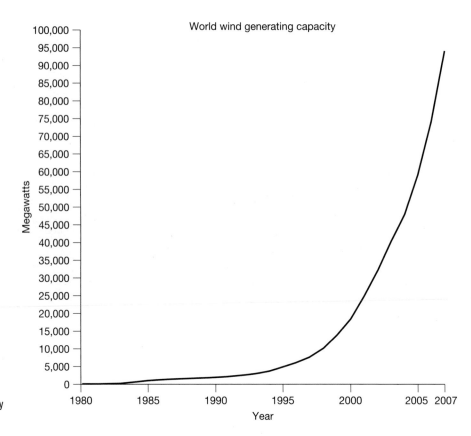

FIGURE 23.21   Growth in wind energy. Wind energy is the world's fastest-growing source of energy!

Wind energy comes free of charge, is clean, and is renewable. It requires little land and does not preclude many other uses, such as grazing. Electricity from wind energy (wind farms in particular) currently costs 5 to 8 cents per kilowatt-hour and is therefore competitive with coal (5 to 7 cents) and cheaper than electricity from nuclear power (10 to 15 cents per kilowatt-hour). Moreover, wind resources are fairly large in many parts of the world. In the state of Holstein, Germany, for example, wind satisfies 15% of the total electrical demand. In the United States, North Dakota, South Dakota, and Texas have sufficient wind capacity to provide electricity to the entire country. Today, many utility customers can purchase wind power through their local utilities. They typically pay a bit more for it, but utilities are finding customers ready, willing, and able to pay the small additional fee, and most programs have waiting lists of customers who want to be added to the legion of citizens who are getting their power from clean, renewable sources.

But wind power is not a panacea. It has its own unique set of problems. Large generators are viewed by some as unsightly. Wind machines also result in the death of many birds—about 25,000 per year, according to a study by Wally Erickson in 1991. Although this seems like a lot, it is far fewer than bird deaths from other sources (Table 23.1). Scientists estimate that about 60 million birds die each year in the United States after being struck by motorized

TABLE 23.1	Estimated Annual Bird Mortality in the United States by Source
Activity/Source of Bird Mortality	Estimated Annual Mortality
Killed by cats	270 million or more
Collisions with and electrocution by electrical transmission wires	130 million–170 million
Collisions with windows	100 million–900 million
Poisoning by pesticides	67 million
Collisions with motor vehicles	60 million
Collisions with communications towers	40 million–50 million

Source: American Wind Energy Association.

vehicles, according to the American Wind Energy Association's report "Facts About Wind Energy and Birds." Another 100 million to 900 million perish after flying into plate glass windows, according to another report. Yet another 10 million to 40 million birds perish after flying into communications towers and the guy wires that support them. Pesticides kill an estimated 67 million birds each year. Another 1.25 million die as a result of collisions with tall structures such as buildings, smokestacks, and towers.

Cats are a lethal force in the lives of birds. Scientists who have studied bird deaths from cats in Wichita, KS, found that a single cat kills, on average, 4.2 birds per year. Other studies suggest even higher numbers. According to one study, a feral cat kills as many birds in one week as a large commercial wind turbine does in 1 to 2 years. Declawing a cat doesn't seem to help much. According to one estimate, the majority of cats (83%) kill birds; even declawed and well-fed cats prey on wild birds. Neutering or spaying a cat does not seem to cut down on hunting, either. With more than 64 million cats in America alone, what's the total loss? Nationwide, the number is estimated to be around 270 million, very likely many many more.

Another downside of wind energy is that winds are not constant, although wind companies place their turbines in areas where the winds blow 65% to 85% of the time. In North America, wind is abundant much of the year in the Great Plains and in many northern states. It is also a year-round source of energy along the nation's coastlines, for example, along the Pacific and Atlantic Oceans and along the shores of North America's Great Lakes. Tapping into the windiest locations in United States—for example, in North Dakota, South Dakota, and Texas—could produce enough electricity to supply *all* of the nation's electrical needs.

Electricity generated from windy locations could be shipped to neighboring areas, with solar electric systems and existing natural gas, coal, and nuclear power plants making up for shortfalls. For more on the pros and cons of wind, you may want to take a look at the American Wind Energy Association's Web site and one of Dr. Chiras's newest books, *Power From the Wind.*

Biomass

Throughout the world, many people rely on **biomass,** that is, wood, manure, crop wastes, and other forms of organic material. Biomass can be burned directly or can be converted to gaseous or liquid fuels. For example, manure decomposes to form methane gas, and sugars from corn and sugar cane can be converted into ethanol. Today, biomass supplies about 14% to 19% of the world's energy. In many less-developed nations, where fossil fuel consumption is still low, biomass may provide up to 90% of the total energy demand.

What makes biomass so attractive is that it is renewable, helps cut down on wastes, is often inexpensive, and on balance, could lower global carbon dioxide emissions. Furthermore, biomass has a wide range of applications. As just noted, it can be burned directly or converted into gaseous fuels to heat our homes and even into liquid fuels to power vehicles. Biomass fuels are also labor-intensive—they require many workers to produce them. In the less-developed nations, where employment is a major problem, labor-intensive fuels could provide work and income for people currently out of work.

Wood   Wood is the most widely used form of biomass. In more-developed countries, such as the United States, Norway, and Sweden, wood supplies about 10% of the home heating fuel. In the United States, 3% to 4% of the total energy consumed each year comes from wood. In Canada, wood provides about 6%. Most of this wood is burned by the wood and wood products industry to power their facilities in place of costly fossil fuels.

Unfortunately, wood is a somewhat dirty fuel. Burned in wood stoves, it becomes a major source of particulates and other pollutants in many of America's urban areas. Now many areas that do not meet National Ambient Air Quality Standards restrict wood burning on high-pollution days. Some restrict the number of woodstoves and fireplaces, and some counties even require special EPA-rated woodstoves with catalytic converters or other design innovations that result in drastic decreases in pollution. Because of concerns about pollution, many woodstove manufacturers have improved designs of their stoves to boost efficiency and cut pollution.

In the less-developed nations, where as many as 2.5 billion people depend on wood for cooking and heating, shortages are becoming widespread. A report by the Food and Agriculture Organization of the United Nations, for instance, noted that 1.3 billion people are meeting their needs for wood by depleting existing supplies, cutting trees down faster than they can be replenished. Two-thirds of these people live in Asia, especially near the Himalayas. Many of the rest live in arid parts of Central Africa and the Andean plateau of South America. In Africa, women and children may travel 50 kilometers (30 miles) a day in search of fuel wood. Depletion of wood creates enormous human suffering and also leads to widespread ecological damage, including erosion, desertification, flooding, and habitat destruction.

What can be done? First and foremost, say some experts, population growth must be curbed to slow the demand for wood. Second, forests can be planted near villages and managed to produce a sustained yield. In the Philippines, for instance, the government launched a program to replant marginal rural land with trees to provide fuelwood for numerous small power plants aimed at providing outlying villages with electricity. Third, new, more efficient cooking stoves and heating stoves can be used by villagers to cut demand. Fourth, substitute energy sources such as wind energy, photovoltaic cells, and small hydroelectric power generators could be developed with the assistance of the United Nations. Solar cookstoves can be used in sunny, wood-short areas of the world.

Wood can also be burned more efficiently in more-developed countries. Pellet stoves, which burn wood pellets made from sawdust from sawmills, burn hot and efficiently. Better yet are **masonry heater** (Figure 23.22A). A masonry heater is a high-mass woodstove made from brick or stone. This type of stove burns wood in a highly insulated combustion chamber. It burns so hot that it results in a nearly perfect burn, which eliminates unburned gases and prevents creosote from building up in the chimney. Creosote is a black, gooey material formed from volatile chemicals released when wood is burned. These chemicals deposit in flue pipes and in chimneys and can catch fire. Because these gases are burned in a masonry heater, more heat can be wrung from the wood. Because creosote does not accumulate in the chimney, the chances of a chimney fire are eliminated.

A.

B.

FIGURE 23.22 Masonry heaters. These stoves are efficient and clean. You can heat a good-sized room with a single burn. (A) Photo of masonry stove. (B) Drawing showing the interior. Note that the flue gases must pass through an extensive labyrinth to escape. In the process, they lose much of their heat, which is absorbed by the masonry and then radiated slowly into the room.

The other secret of a masonry heater is that it has high mass. Unlike conventional woodstoves made of cast iron or steel, the thick masonry walls absorb heat given off by the fire and radiate slowly into the room. There's no baking in your easy chair while the woodstove is on! In addition, unlike conventional woodstoves, where the heat escapes right up a straight pipe, the flue of a masonry stove is a labyrinth—that is, an elaborate, elongated pathway for hot gases to escape. As the hot gases escape, the heat is transferred to the masonry. Because of this, much of the heat is captured and radiated into the room (Figure 23.22B).

Other Biofuels   Wood is only one of many forms of biomass that can be used to produce energy. Garbage can be incinerated to produce steam heat and electricity (Chapter 17). Manure, human wastes, and other organic refuse also can be used to produce methane gas, which is combustible. When mixed with water in a closed container, organic material is degraded by anaerobic bacteria. Methane given off in the process can then be burned to produce heat and electricity. In fact, many sewage treatment plants in the United States and abroad now capture methane that was once vented to the atmosphere; they burn it and use the heat for offices or to generate electricity needed by the plant, and often sell the excess to local utilities. Many companies now tap methane released from landfills and use it to generate electricity.

Plant material can also be used to generate ethanol, a liquid fuel that can be mixed with gasoline in a ratio of 1:9 and burned in automobiles. Called **gasohol,** it results in more complete combustion and fewer pollutants. It is used in many cities to combat wintertime air pollution.

Alcohol can also be burned without dilution in slightly modified cars and in factories. Most gas stations that sell ethanol, though, sell E85, an 85% mixture of alcohol and gasoline. It's typically cheaper than straight gasoline but contains about 20% to 30% less energy per gallon.

Most ethanol in the United States currently comes from corn, but virtually any crop material when properly fermented can produce alcohol. Brazil, for example, uses sugarcane to produce alcohol and cut its imports of foreign oil. Sugarcane is now a major source of fuel for Brazilian trucks, cars, and buses. Companies are even using the sugarcane wastes to produce notebook paper sold through Staples in the United States. Researchers are also working on ways to produce ethanol from the leaves and stems of corn. Just like starch, cellulose molecules in these nonfood parts of the plant are long-chain compounds made of many smaller glucose molecules. The glucose molecules are then fermented to produce ethanol.

Ethanol has been widely criticized for a number of reasons. For one, widespread ethanol production from corn has driven up the price of corn as well as food products made from corn, such as tortillas. While farmers like the higher prices, some consumers are suffering, especially in less-developed countries. (A good part of the increase in price has been the rising cost of energy to grow crops, but that's not often brought up in debates.) **Cellulosic ethanol** production, the production of ethanol from cellulose—from stems and leaves, and production from sugarcane could help ease this problem.

Some critics wrongly point out that it takes more energy to make ethanol than we get from it. This is based on an old data. Contemporary ethanol production has a net energy efficiency of 1.7, meaning you get 1.7 units of energy out of ethanol for every unit required to produce it. Although that's not high, it's probably not much different from gasoline production. Cellulosic ethanol production, if successful, is projected to have a much higher net energy efficiency—around 8 to 10. Ethanol produced from sugarcane has a net energy efficiency of around 8.

Fuel farms could provide much of America's liquid fuel in the coming years. Crop wastes—excess corn and wheat, for instance—could be used to produce substantial amounts of liquid fuel. As many developing nations become self-sufficient in food production, countries such as Canada, Australia, and the United States, which currently export large amounts of food, could become major producers of ethanol fuel.

Hydrogen and Fuel Cells

Many people believe that hydrogen will be the fuel of the future. Is their optimism justified?

Hydrogen is a combustible gas that can be used to heat our homes, cook food, and heat water for showers. It burns just like natural gas, without producing any carbon dioxide. In fact, when hydrogen burns, it produces heat, light, and water. You couldn't ask for a cleaner-burning fuel.

Hydrogen could also be used to power automobiles. It can be stored under pressure in specially built tanks (already available) and burned directly in specially designed engines (already available). Or it can be fed into a device called a fuel cell, discussed in Chapter 18. Fuel cells are small batterylike contraptions that combine hydrogen with oxygen (from the air) to create electricity. Electricity generated in a fuel cell can then be used to power electric motors that power our cars, buses, and trucks. The water produced in the process can be captured and recycled.

While all this sounds fine, bear in mind that there are no underground reservoirs of hydrogen. Hydrogen gas does not exist free in the atmosphere. So where will hydrogen come from? Hydrogen can be extracted from a variety of sources. For example, it can be stripped from the organic molecules in various conventional fossil fuels—natural gas, propane, gasoline—or any hydrocarbon molecule (ethanol or methanol, for example) that contains hydrogen. Unfortunately, stripping hydrogen from fossil fuels still produces carbon dioxide—and lots of it. And why would we want to utilize finite supplies of fossil fuels to make hydrogen? Wouldn't this put us right back where we started? Why not obtain the hydrogen we need to power our cars from renewable resources—water, for instance?

Not only is water abundant, it is renewable. Moreover, it is also not under the control of wealthy sheiks or tyrannical leaders in the politically volatile Middle East.

Hydrogen is made from water by running electricity through it. When electricity flows through water, the electrical current splits the water molecules into their components, hydrogen and oxygen. Both hydrogen and oxygen are gases.

Besides powering our cars, hydrogen gas can be burned directly in water heaters, stoves, and even furnaces, replacing declining supplies of natural gas. If it could be pumped through the pipelines and underground pipes that currently carry natural gas to our cities and homes, we'd be able to make use of an elaborate and costly supply system that's already in place.

Unfortunately, there's a problem with this scheme. Hydrogen is a small molecular weight gas, much smaller than natural gas. Because of its small molecular weight, hydrogen can't be contained in existing pipelines. Some proponents of a hydrogen economy suggest we would have to rebuild all of the pipelines that carry natural gas across North America. Unfortunately, we would also have to install new pipes in cities to every home to deliver hydrogen.

As we research the prospects of hydrogen gas as a fuel of the future, we will also need to ask how we're going to generate all of the electricity that's required to split water to make this fuel. And where are we going to get the water, especially in water-short areas of North America?

The most recent Bush Administration embraced the idea of the hydrogen economy but maintains that we should obtain hydrogen from fossil fuels, described previously. The administration also supports the use of nuclear energy and coal to generate electricity needed to split water to produce hydrogen.

Renewable-energy advocates disagree with this approach. Richard Engle, a hydrogen proponent at the Schatz Energy Research Center, argues, "The nuclear and coal industries are…latching onto hydrogen as a means to assure their continued dominance of energy markets." Dominic Crea, a physics teacher who supports renewable energy, argues, "A fossil-fueled hydrogen economy is a gamble at best and a nightmare at worst."

Renewable-energy advocates such as Engle argue that we should be using the nation's clean and abundant wind and solar energy resources to make electricity from water. It should then be used to generate hydrogen. But is this the best use of electricity generated from renewable resources?

As we pursue a path to make hydrogen, it is important to remember one thing about hydrogen: that hydrogen is *not* an energy source. That is, it is not a primary fuel like natural gas or oil. Companies won't be drilling into pockets of hydrogen gas in the Earth's crust. They don't exist. Rather, hydrogen is a fuel we manufacture. Hydrogen is a secondary fuel made by splitting water with electricity. In other words, making hydrogen takes energy. Dominic Crea, mentioned earlier, is a critic of hydrogen use. His calculations suggest that it makes much more sense to use the electricity generated from renewable-energy technologies (like wind) to run cars directly than to use it to split hydrogen to power fuel cells. According to Crea, a system that generates electricity to power electric cars is three times more efficient than a system that generates electricity from renewable resources to produce hydrogen to power fuel cell cars.

Unfortunately, existing electric cars require large, heavy battery banks and can't run very far without needing recharging. In other words, they have a short range—60 to 120 miles on a single charge. Charging takes a long time, too. You can't just stop at a gas station and recharge your car's batteries in five minutes. Until new, higher-capacity batteries are available, electric cars could primarily serve as commuter vehicles for urban residents. While that may seem insignificant, commuting is the predominant use of the automobile in modern society.

Ninety percent of all Americans drive fewer than 60 miles per day—to and from work. Researchers are also working on new batteries that could provide much greater range.

Clearly, hydrogen's future is not as rosy as some would have you believe. Some critics question whether we should be even pursuing hydrogen.

Crea believes that a renewably based hydrogen economy will require the installation of $40 trillion worth of photovoltaic panels. His advice: go with a renewably fueled grid system. "The decision to go with a renewably fueled utility grid system, as opposed to the hydrogen system, would save enough money in photovoltaic panels alone to provide every American family with an electric car and the photovoltaic panels to run it," says Crea. "Furthermore, enough would be left over to handle all of the electricity, heating, and cooling needs for an entire house—for free!"

Will hydrogen have a bright future or any at all? At this time, it is difficult to say whether hydrogen will become a dominant fuel source or whether it will play much of a role in our energy future. We don't know enough to make wise choices.

Tidal Power

President Franklin Roosevelt often watched the rise and fall of the tides near his summer home on the Bay of Fundy. He was impressed with the potential power of the surging tides of Fundy, the largest on Earth—up to 16 meters (50 feet). He wondered, no doubt, if the energy of those tides could somehow be captured, perhaps by finding a way to channel their energy through turbines similar to those used in power plants, where spinning blades produce electric energy (Figure 23.23).

The world's first tidal-electric installation was built in 1966 in the La Rance estuary in France. It has a capacity of about 240 megawatts, equivalent to a small coal-fired power plant. A similar plant has been constructed in Russia. The power plant channels the tides through openings in a massive dam built at the mouth of the river. The water spins an underwater turbine, generating electricity.

Tidal power is a type of hydropower that converts tidal flows into electricity. Tidal power consists of two main types: (1) tidal stream systems and (2) barrages.

Tidal stream systems capture energy of underwater power turbines, much the same way that windmills capture energy from moving air (winds) (Figure 23.23). It is cheaper and poses fewer environmental problems than the second type of system, barrages.

The **barrage system** involves building a structure (called a barrage) across a bay or river that experiences significant tides (Figure 23.24). Turbines installed in the barrage generate electricity as water flows in and out with the tides.

Tidal power has many advantages. It is renewable, being generated by the gravitational pull of the moon on the Earth's oceans. Tides are also much more predictable than wind and solar energy. Electricity can be generated when tides are moving in and out. Tidal power is also relatively inexpensive to tap, produces no toxic wastes or pollutants, and is renewable.

On the downside, tidal power may be limited. There are, for instance, only about 24 good sites worldwide, and their potential is fairly low. They would, for instance, meet only

A.

B.

FIGURE 23.23   These structures convert tidal energy into electricity.

about 5% of the United States' current electrical demand. Underwater turbines are exposed to a fairly hostile salty environment. Barrages used to direct water through the turbines could impair ship travel and may upset aquatic ecosystems. Like hydropower, this technology may be of limited value in the future.

Researchers at Florida Atlantic University are currently experimenting with underwater turbines that could be placed in the Gulf Stream and anchored to the sea floor, capturing the energy of this massive, powerful, and reliable current.

Ocean Thermal Energy Conversion

The ocean is a vast storehouse of heat, which could be tapped by specially designed **ocean thermal energy conversion (OTEC)** plants. These facilities exploit the temperature difference between warm surface waters and cold bottom waters in tropical areas. Each OTEC plant consists of a

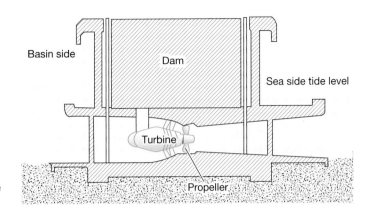

FIGURE 23.24 Tidal power plant. Water flows through small openings in the dam, spinning a propeller that generates electricity.

floating platform with enormous pipes that extend to a depth of 900 meters (3,000 feet). On the surface platform, ammonia, which boils at a relatively low temperature, is circulated through a series of tubes. The warm surface waters convert the ammonia into a gas that drives the blades of a turbine that drives a generator (Figure 23.25). The gas is then recondensed by cool water pumped up from the depths to start the cycle over again.

Electricity produced by OTEC plants can be transmitted to shore. Some plants could use it to desalinate salt water, thus providing drinking water for local populations.

OTEC plants could theoretically operate day and night in tropical waters, providing energy for nearby cities and towns. However, like many other technologically feasible ways to generate energy, this one has some problems. First is the question of efficiency. Because large amounts of water must be used, a considerable amount of energy must be spent to pump water to the surface. An experimental plant off the coast of Hawaii, for example, used 80% of the electricity it generated to pump water, giving it a low net energy yield. Second, suitable locations for OTEC plants are limited to a few places, such as the Gulf Coast and the waters off Hawaii, Guam, and Puerto Rico. These plants would also release

considerable amounts of carbon dioxide into the atmosphere as the cold waters in which this gas was once dissolved are heated at the surface. Cooling surface waters could affect fisheries and local climate, increasing rainfall. Released on the surface, deeper waters, which are rich in nutrients, could also stimulate the growth of phytoplankton, upsetting the ecological balance.

23.3 Summary

No greater challenge exists for a democratic nation besieged with pressing problems than to sort out among the frantic rush of minor crises the long-term problems that could potentially cripple it—and to act on them. Energy is one of those problems that, because of years of neglect, even foolhardy wasteful policies, has escalated to a major crisis. Supplies of several critical fossil fuels are limited, and rising prices are causing serious economic problems the world over. Fossil fuels are also at the root of some of our most threatening environmental problems—global climate change being the most important. It could be argued that our heavy dependency on fossil fuels is at the root of the current crisis of unsustainability.

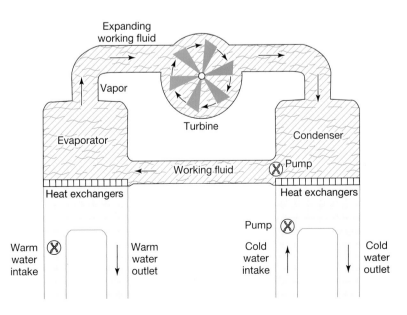

FIGURE 23.25 Schematic diagram of a closed-cycle ocean thermal power plant.

At no time in our history have present actions been so important. Faced with impending oil shortages and a massive buildup of greenhouse gases, many observers believe that we must revamp our energy system to build a sustainable energy system. By combining energy efficiency and renewable energy with other important measures, such as population control, recycling, and restoration of habitats, we can steer onto a sustainable path. The responsibility belongs to all of us, and to each of us. As Marshall McLuhan once wrote, "There are no passengers on planet Earth, we are all crew." And lest we forget, good planets are hard to find.

Summary of Key Concepts

1. Throughout history, satisfying the demand for future energy needs has primarily involved producing more fuel or developing alternatives. Very little attention has been given to finding ways to reduce energy consumption through conservation or efficiency measures.

2. Nevertheless, conservation and efficiency have created some enormous fuel savings in the recent past. The average gas mileage of a new car sold in the United States, for instance, increased from 14 miles per gallon in 1974 to 26 miles per gallon in 1986 to around 28 miles per gallon in 1995. In the first half of the 2000s, however, it fell to around 23 miles per gallon, largely because of the popularity of large SUVs, vans, and trucks.

3. Despite what many think, most of the world's people have only begun to tap the full potential of energy conservation. Enormous opportunities to improve efficiency exist in buildings, industry, motorized vehicles, lighting, appliances, and electronic devices.

4. Conservation and efficiency save money for families and businesses. For businesses, efficiency and conservation measures lower the production costs of goods and services. Efficiency and conservation help American companies compete with more-efficient companies in Germany and Japan. And they also help the world sustain lower oil prices, helping the global economy create economic stability. Energy conservation also creates jobs.

5. Energy conservation makes good economic sense for consumers and producers. It's much cheaper to save energy than to make energy.

6. Energy conservation will require actions by citizens, businesses, and governments.

7. Individuals can help by turning down their thermostats and wearing more clothes inside in the winter, turning up their thermostats in the summer, making their homes more airtight by caulking and weather-stripping, turning down the setting on water heaters, insulating water heaters, driving at or below the speed limit, car pooling, taking mass transit, and buying energy-efficient vehicles.

8. Another important step required for the development of a sustainable energy system is the installation of clean, affordable renewable alternatives to fossil fuels. The most likely renewable candidates include solar energy, wind energy, biomass, hydropower, and geothermal energy.

9. The sunlight that strikes an area the size of Connecticut each year could provide all of the energy needed by the United States, yet Americans have tapped only a tiny portion of the sun's potential.

10. Solar energy can be used to heat water for homes and industries and to heat building interiors. Two major heating systems are in use today: active and passive.

11. Active solar systems consist of solar collectors that absorb sunlight and convert it to heat. The heat is then carried away and used to heat building interiors or water.

12. Passive solar systems are designed strictly for space heating. The building itself becomes a collector and heat storage device. South-facing windows and skylights let the winter sun penetrate the interior of the house. Sunlight strikes the walls and floors and is converted into heat. Cement or block walls store the heat and radiate it out into the room at night.

13. Solar energy can also be used to generate electricity. Sunlight may be concentrated by mirrors to boil water to run an electrical turbine. This technology is called solar thermal electricity.

14. Electricity may also be generated by photovoltaic cells, thin wafers made primarily of silicon that, when struck by light, produce an electrical current. Solar energy trapped by photovoltaic cells could become a major source of energy to replace costly and polluting fossil fuels, but the cells are high-priced. Numerous efforts are under way to reduce the cost of solar electric systems to make them cost competitive with fossil fuels. Solar cells make good economic sense at remote sites, however. They may also make sense if state and federal incentives are available to offset the initial systems cost. More-efficient solar cells and mass production would make them competitive with other forms of electrical generation.

15. The federal government currently provides approximately $200 billion a year in subsidies to the nuclear, coal, oil, and gas industries but almost no support to developing alternative energy sources. Renewable-energy technologies receive a subsidy that is 1/1,000 as much, or about $200 million a year. To build a sustainable future, we must level the playing field—either provide subsidies for renewable energy resources or eliminate subsidies for nonrenewables.

16. The Earth stores enormous amounts of heat, or geothermal energy, which comes from the radioactive decay of naturally occurring radioactive substances and from molten rock in the Earth's interior.

17. Most commercial geothermal energy today comes from hydrothermal convection zones, areas where heated underground water is naturally ejected to the surface, emerging as geysers or hot springs. Steam or hot water from hydrothermal convection zones can be used to heat buildings and generate electricity. Despite the ease with which it can be used, however, global geothermal capacity is small.

18. Hydropower makes a significant contribution to global energy and is bound to increase in the future. Unfortunately, much of the untapped hydropower is in regions far from population centers and industry. Some environmentalists believe that retrofitting existing nonhydroelectric dams with small generators could provide additional electricity without the environmental effects created by building new dams.

19. Large-scale wind systems are contributing more and more to global electricity needs. In fact, wind is the fastest-growing source of energy in the world, with massive potential in many of the more-developed countries, the United States and Germany being the top two. Wind energy may also provide electricity to rural villages in less-developed nations where large electric generating plants are unfeasible.

20. Wood, manure, crop wastes, and other forms of biomass supply 14% to 19% of the world's energy demands. In less-developed nations, where fossil fuel consumption is low, biomass may provide up to 90% of the total energy needed.

21. Wood is the most widely used form of biomass. In developed countries, including Norway and the United States, wood supplies about 10% of the home heating fuel. But most of the wood in these countries is burned by the forest products industry.

22. In developing nations, 2.5 billion people depend on wood as their primary fuel. Over half of them are depleting local wood reserves faster than they can be replenished.

23. Other forms of biomass can also be used to produce energy. Manure, human wastes, and other organic wastes, for instance, can be used to produce methane gas. Plant material can be used to generate ethanol. Fuel farms that produce crops to make ethanol could provide much of America's liquid fuel in coming years if the net energy yield of the process can be improved.

24. Hydrogen gas, produced from water, is a renewable fuel that burns very cleanly in cars and in stoves. Unfortunately, making hydrogen takes a fair amount of energy in the form of electricity. Critics believe that it might be better to use electricity directly, rather than using it to make hydrogen. This is a much more efficient use of energy.

25. The tides can be tapped by building special dams or towers that channel water flowing in and out with the tides through openings equipped with turbines. Unfortunately, the global capacity of tidal energy is rather small.

26. The ocean is a vast storehouse of heat, which could be tapped by specially designed OTEC plants. They use the warm surface waters to vaporize ammonia gas, which then runs an electric turbine. The gas is cooled by cold bottom waters that are pumped to the surface. Unfortunately, the energy yield of OTEC plants is poor, making the electricity uncompetitive.

Key Words and Phrases

Active Solar Systems	Geothermal Energy
Barrage System (tidal power)	Geysers
Biomass	Green Lights Program
Building Integrated Photovoltaics	Hot Rock Zones
	Hydrogen
Cellulosic Ethanol	Hydropower
Earth Sheltering	Hydrothermal Convection Zones
Energy Star Program	LED Lights
Fuel Farms	Masonry Stove
Gasohol	National Appliance Energy Conservation Act of 1987
Geopressurized Zones	

Net Metering	Solar Energy
Ocean Thermal Energy Conversion (OTEC)	Solar Thermal Electricity
	Tidal Power
Passive Cooling	Tidal Stream System
Passive Solar Systems	Wind Farm
Photovoltaic Cell (PV)	Wind Power
Solar Electricity	Windmill

Critical Thinking and Discussion Questions

1. Make a list of the conventional energy sources, and describe each one. Write a list of criteria to judge the sustainability of all potential energy resources. Now study your list of energy sources, and rate each one for its sustainability.

2. Make a list of the different types of technologies that capture solar energy, and describe each of their purposes—for example, to generate electricity or provide space heat.

3. What is an active solar energy system? Give an example. What is a passive solar energy system? How are they similar? How are they different?

4. Describe two methods by which electricity can be generated from sunlight.

5. An engineering professor has discovered a new source of energy. Drawing on your critical thinking skills, what would you want to know about it before promoting it?

6. What is geothermal energy? What form of geothermal energy is the most readily accessible?

7. Using your critical thinking skills, debate the following statement: "The United States has vast untapped hydropower potential that could produce enormous amounts of electricity."

8. Describe the potential of wind energy. Is it competitive with conventional fuel sources? What are the real and perceived problems associated with wind energy? Which ones have been blown out of proportion? How can real problems be overcome? (You may want to go online to check the pros and cons of wind.)

9. What is biomass energy? How important is it to our energy future?

10. Describe the pros and cons of tidal power, hydrogen as an energy source, and ocean thermal energy conversion.

11. Do you agree with the following statement? "Conservation and efficiency are no less energy alternatives than oil, gas, coal, and nuclear energy and are often much cheaper."

12. Discuss this statement: "Energy efficiency can be considered a form of renewable energy."

13. Make a list of ways you can cut your energy consumption by 25%. What are the most important things you can do?

14. You are appointed head of your city's energy department and asked to devise a long-term energy strategy that will carry your city well into the middle of the next century. Prepare your plan and defend it.

15. Make a list of criteria to judge the sustainability of all potential energy resources.

Suggested Readings

Amann, J. T., A. Wilson, and K. Ackerly. 2007. *Consumer Guide to Home Energy Savings.* Gabriola Island, BC: New Society. An excellent overview of home energy savings.

Chiras, D. D. 1992. *Lessons From Nature: Learning to Live Sustainably on the Earth.* Washington, DC: Island Press. Description of steps to create sustainable energy systems.

Chiras, D. D. 2002. *The Solar House: Passive Heating and Cooling.* White River Junction, VT: Chelsea Green. Description of ways to heat and cool homes naturally.

Chiras, D. D. 2004. *The New Ecological House.* White River Junction, VT: Chelsea Green. See the chapters on green power and passive solar heating and cooling to learn more about sustainable home energy systems.

Chiras, D. D. 2005. *The Homeowner's Guide to Renewable Energy.* Gabriola Island, BC: New Society. Description of numerous ways to save energy around the home and all of the renewable energy systems that can be used to meet our needs.

Chiras, D. D. 2008. *Green Home Improvement: 65 Projects That Will Cut Utility Bills, Protect Your Health, and Help the Environment.* Kingston, MA: RS Means. Numerous projects that help homeowners and renters green their homes, especially by reducing energy and water use.

Chiras, D. D. 2009. *Power From the Wind: A Practical Guide to Small-Scale Energy Production.* Gabriola Island, BC: New Society. One of the senior author's latest books, which discusses wind-electric systems for homes and small businesses.

Chiras, D. D., and D. Wann. 2003. *Superbia! 31 Ways to Create Sustainable Neighborhoods.* Gabriola Island, BC: New Society. One of the senior author's most recent books, which describes actions people can take to build a more sustainable world right in their own neighborhood, including many steps to improve energy efficiency and reliance on renewable energy.

Dunn, S. 2000. The Hydrogen Experiment. *World-Watch* 13(6): 14–25. Discussion of how Iceland is planning to convert to hydrogen power by 2030.

Dunn, S. 2001. Decarbonizing the Energy Economy. In *State of the World 2001,* ed. L. Starke. New York: W. W. Norton. Important reading on renewable energy.

Flavin, C. 1995. Harnessing the Sun and the Wind. In *State of the World,* ed. L. Starke. New York: W. W. Norton. Important reading.

Flavin, C., and N. Lenssen. 1994. Reshaping the Power Industry. In *State of the World 1994,* ed. L. Starke. New York: W. W. Norton. Important insights into changes under way in the power industry, especially those related to energy conservation; ideas that are as relevant today as they were in the early 1990s.

Freudenberger, R. 2009. *Alcohol Fuel: Making and Using Ethanol as a Renewable Fuel.* Gabriola Island, BC: New Society. Lessons on how to make alcohol to power your own car.

Greer, J. M. 2008. *The Long Descent: A User's Guide to the End of the Industrial Age.* Gabriola Island, BC: New Society. Description of the end of the world as we know it and things we can do to survive the impending crises.

Kutscher, C. 2007. Tackling Climate Change in the U.S. *Solar Today* 21(2): 26–29. A very important paper on the potential for efficiency and renewable energy to reduce carbon dioxide emissions.

Perlin, J. 1999. *From Space to Earth: The Story of Solar Electricity.* Ann Arbor, MI: aatec publications. A wonderfully readable history of solar electricity.

Renner, M. 2000. Creating Jobs, Preserving the Environment. In *State of the World 2000,* ed. L. Starke. New York: W. W. Norton. Details of the economic and employment benefits of a renewable energy strategy.

Renner, M. 2001. Employment in Wind Power. *World-Watch* 14(1): 22–30. A fascinating look at the employment boom in the wind energy sector.

Rifkin, J. 2002. *The Hydrogen Economy.* New York: Tarcher/Putnam. An interesting but overly simplified study of the potential for building an energy future based on hydrogen.

Roodman, D. M. 2000. Reforming Subsidies. In *State of the World 2000,* ed. L. Starke. New York: W. W. Norton. Extremely important reading.

Savin, J. 2003. Charting a New Energy Future. In *State of the World 2003,* ed. L. Starke. New York: W. W. Norton. A good overview of renewable-energy potential and what other nations are doing to create a sustainable energy future.

Schaeffer, J. 2008. *Solar Living Sourcebook,* 13th ed. Ukiah, CA: Real Goods. Superb reference! Showcases renewable-energy products but also provides a great deal of background information on renewable energy.

Scheckel, P. 2005. *The Home Energy Diet: How to Save Money by Making Your House Energy Smart.* Gabriola Island, BC: New Society. A superb book that describes many of the things we can do to make our homes more energy efficient.

Short, W., and N. Blair. 2003. The Long-Term Potential of Wind Power in the U.S. *Solar Today* 17(6): 28–29. Important study of wind power's potential.

Swain, J. 2003. Charting a New Energy Future. In *State of the World 2003,* ed. L. Starke. New York: W. W. Norton. Examination of the potential of renewable energy resources and many other important subjects.

Totten, M. 2007. China's Bold Initiative. *Solar Today* 21(2): 38–41. Description of China's efforts to utilize renewable energy.

Werner, C. 2008. Revolving the Biofuels Dilemma. *Solar Today* 22(4): 38–42. Discussion of ways to resolve some of the problems with biofuels.

 ## Web Explorations

Online resources for this chapter are on the World Wide Web at: **http://www.prenhall.com/chiras** *(click on the Table of Contents link and then select Chapter 23).*

AFTERWORD

Over the course of the semester, you have learned a great deal about the environment and pressing environmental and resource issues. If you've read the entire book, you have learned about ecology, economics, and ethics. You've studied human population growth, agricultural problems, pest control, water management, and water pollution. You've examined fisheries conservation, rangeland management, forest management, plant and animal extinction, and wildlife management. You've studied solid waste, hazardous waste, urban air pollution, global warming, and mining. You have studied nonrenewable and renewable energy.

You may have been already familiar with some of these issues. You may have had no idea the other issues existed. No matter what your prior understandings were, we are confident that you have come to new realizations. We hope you've developed a deeper and broader understanding of the challenges that lie ahead, and discovered new ways to address the many pressing environmental problems.

After you've finished reading this book, we hope you recognize that environmental issues vary. Some are local; others are regional or statewide, and still others are global in scope. Some problems are minor; others are more severe, so severe that they threaten our well-being and our economy. Many problems have an impact on nonhuman species—in some cases, driving them to extinction.

In the course of your studies, you have learned about the many causes of environmental and resource issues, and the plethora of solutions available to individuals, businesses, and governments. You have learned that solving issues often requires a cooperative approach with the participation of many stakeholders working together to find sustainable solutions. In some cases, solutions require international cooperation.

Sustainable solutions are often preventive in nature. Renewable energy such as solar and wind power, for instance, prevents pollution from entering the atmosphere. Crop rotation reduces the need for insecticides, which in turn decreases the potential for pollution of ground and surface waters. It can also reduce soil erosion and the need for fertilizers, which protects cropland and nearby surface waters.

For solutions to be most effective, many experts in sustainable development emphasize that they must address the root causes of the problems—including our inefficient use of resources, linear thinking and linear systems design, heavy dependence on nonrenewable resources, failure to restore damaged ecosystems, failure to manage systems for the long term, and ever-expanding human population.

Sustainable resource management seeks ways for us to thrive within the Earth's very real limits—that is, limits in the Earth's capacity to supply us with resources and assimilate our wastes. In this book, we've talked about learning to live within the carrying capacity of the ecosystems upon which we all depend. A good example is regulations on commercial fishing operations. They seek to limit harvest so that fisheries can remain productive well into the future. Likewise, timber management seeks to establish limits on the commercial harvest of trees so we don't deplete our forests and destroy the soils upon which a healthy forest depends.

Sustainable solutions are designed to achieve the long-term health and stability of natural systems as well as many human-dominated systems. Sustainable grazing, for instance, is carried out in ways that ensure the long-term health of grassland soils as well as the productivity of forage upon which our livestock depends. Agricultural practices seek to maintain the health of the topsoil, too, so that farmers can produce affordable, nutritious food ad infinitum without poisoning nearby waterways or beneficial species such as birds. Air pollution prevention seeks to ensure healthy air in our cities.

Another key in creating sustainable solutions is a shift toward management of entire ecosystems and watersheds. In the past, resource management has suffered from a kind of myopia, dealing with problems in limited ways. As a result, we didn't always devise solutions that could address the complex ecosystems from which many of our resources were derived. Ecosystem and watershed management broaden the scope of our actions and serve to protect entire systems with far better results.

Policies and practices that prevent problems, promote ways to live within limits, and ensure long-term health of ecosystems fully recognize our dependence on the Earth and its resources. They recognize that the Earth and its many and varied ecosystems are the biological infrastructure of our society. They're not frivolous concerns, as some people would like you to believe. Such policies also promote intergenerational equity (fairness to future generations) and ecological justice (fairness to the millions of species that share this planet with us).

Ultimately, to be sustainable, solutions must make sense from three perspectives: social, economic, and environmental. The challenge before us is to give equal priority to all three, not to draft policy that favors one over another. Hopefully, you have found in this book that there are many sustainable solutions to many problems that make sense from these three perspectives. Unfortunately, a great many people have not realized the importance of natural systems to human welfare. They persist in old thinking, putting human interests, especially economic pursuit, above all else. You can help address the old thinking and can become a positive influence on those around you. We need you to help spread the word that the rules of environmental protection and resource management have changed. We can live well, with little impact on the environment, if we seek sustainable solutions.

At the heart of sustainable solutions are five guiding tenets: conservation, recycling, renewable-resource use, restoration, and population control. As the operating principles of nature, they helped ensure sustainability of undisturbed ecosystems for thousands of years; they can help us redesign the systems that provide us with the goods and services we require.

To create a sustainable future, we also need a new ethic—a sustainable ethic that recognizes humans are part of nature, the Earth and its resources are limited, and success comes from cooperation with natural forces rather than their domination and control.

We urge you to take an active part in building a sustainable future. Recycle. Buy recycled products. Use energy and other resources efficiently. Limit your consumption. Walk or ride a bus. When you drive, drive an energy-efficient vehicle. Contribute to organizations that seek to restore and protect the environment. Remember that planet care is the ultimate form of self-care.

GLOSSARY

Abortion. The premature expulsion of the fetus from the uterus.

Abstinence. Refraining from sexual intercourse either during key times of a woman's menstrual cycle (around ovulation) or until after marriage.

Abyssal Zone. The bottom zone of the ocean, characterized by darkness, close to freezing temperatures, and high water pressures.

Accelerated Erosion. Wearing away of the land surface primarily as a result of human activities or, in some cases, animal activities. Accelerated erosion operates at a much faster rate than geological erosion.

Accelerated Eutrophication. Accelerated increase in the concentration of plant nutrients in water bodies caused by human actions (for example, excess fertilizer application on farm fields).

Acid Mine Drainage. Sulfuric acid produced by underground coal mines in areas with high levels of iron pyrite in the soil.

Acid Precipitation. Deposition of acids in rain, snow, mist, and fog.

Acid Precursors. A chemical such as sulfur dioxide that gives rise to acids (in this case, sulfuric acid) after reacting with other chemicals in the atmosphere.

Acid Rain. Rain that has a lower pH than "normal" rain—in other words, lower than pH 5.7. It is caused by the release of oxides of sulfur and nitrogen into the atmosphere.

Activated Sludge. The solid organic waste that has been intensively aerated and "seeded" with bacteria (in a secondary or tertiary sewage treatment process) to promote rapid bacterial decomposition.

Active Fishing Gear. Devices and methods for fishing that have mobility and are able to seek out and capture the fish or shellfish. Examples include trawl nets, purse seine nets, and large drift nets.

Active Solar System. System that gathers energy from the sun and stores it for heating water or rooms.

Adaptive Management. Scientific management of natural resources in which management strategies may be modified as a result of scientific findings on the effectiveness of existing management strategies.

Additive Mortality. The concept that one kind of death of an animal population simply adds to the deaths caused by other factors.

Adsorption. The attraction of ions or compounds to the surface of a solid; for example, the attraction of ions to clay particles in soil.

Agent Orange. Defoliant used by American forces during the Vietnam War.

Age Structure Diagram (or Population Histogram). Graphical representation of a population according to age and sex. This type of diagram aids in forecasting population trends.

Aggregates. Soil particles grouped into a single mass or cluster.

Agroforestry. The practice of growing trees among crops or on pastureland.

Algal Bloom. Dramatic increase in algal growth in a lake or stream, resulting from high levels of nutrient pollution.

Alpha Particle. A positively charged particle (proton) that is emitted from the nucleus of a radioactive atom.

Alternative Agriculture. Nonconventional approaches to agriculture that include, but are not limited to, organic, ecological, biodynamic, integrated, low-input, and no-till farming.

Altitudinal Migration. Seasonal movement of birds (e.g., grosbeaks and finches) and mammals (e.g., elk and bighorn sheep) up and down mountain slopes.

Ammonification. The process by which the bacteria of decay convert complex nitrogenous compounds occurring in animal carcasses and the excretions of animals, as well as the dead bodies of plants, into relatively simple ammonia (NH_3) compounds.

Anadromous Fish. Fish that begin life in freshwater, travel to and mature in the sea, and return to their native stream to reproduce and die. An example is the Pacific salmon.

Animal Unit Month. The amount of forage needed to keep one head of cattle healthy for 1 month.

Annular Ring. A concentric ring, visible in the cross section of a tree trunk, that is useful in determining the age of the tree.

Anthropogenic Pollutants. Pollution from human sources; for example, the combustion of fossil fuels. Compare with natural pollution.

Antimycin. A toxic substance that has been extensively used by fisheries biologists to eradicate carp.

Aquaculture. Cultivation of fish and other aquatic organisms in freshwater or saltwater for food, sport, and other goods and services.

Aquifer. A subterranean layer of porous water-bearing rock, gravel, or sand.

Aquifer Recharge Zone. A region in which surface water seeps into groundwater, replenishing it.

Artificial Insemination.   The technique employed by cattle breeders in which sperm from a bull of one breed, such as a Hereford, might be refrigerated and used, over a period of time, to fertilize the eggs of the same breed or other breeds of cattle, possibly from widely separated localities.

Artificial Propagation.   The captive breeding of fish, such as in hatcheries.

Artificial Reef.   A reef constructed of housing debris, rubble, junked automobile bodies, tires, sunken ships, and so on, frequently placed in relatively shallow water near the coast. Artificial reefs increase the number of breeding sites and provide habitat for marine fish.

Asbestos.   A naturally occurring fibrous mineral once used in numerous applications, for example, for sound insulation and heat insulation in buildings. When inhaled, asbestos fibers can cause lung cancer and other diseases, which are most prevalent in smokers.

Aseptic Container.   A sterile chamber.

Backyard Wildlife Habitat.   Habitat for wild species in rural and urban environments designed to provide food, shelter, and nesting sites for wild species such as songbirds and butterflies.

Barrier Islands.   Accumulations of coastal sediments parallel to and near the shore created by coastal wave, wind, and current action. These low-elevation islands are found worldwide.

Bathyal Zone.   An ocean region of semidarkness in which photosynthesis cannot occur and green plants cannot survive. It is located between the euphotic and abyssal zones.

Beach Nourishment.   An expensive human management practice of transporting sand from other sources and depositing it on a beach, temporarily replenishing eroded beach sand.

Bioaccumulation.   Concentration of a chemical substance in an organism.

Biochemical Oxygen Demand (BOD).   *See* Biological Oxygen Demand.

Biogeochemical Cycle.   *See* Elemental Cycle. Also known as a nutrient cycle.

Biological Control.   Means of controlling pests using natural enemies or other potentially less environmentally harmful measures than chemical pesticides.

Biological Magnification (or Biomagnification).   The increase in concentration of a chemical substance in a food web as it passes from lower levels to highest levels.

Biological Oxygen Demand (BOD).   Measure of organic matter in water samples. Assesses oxygen used by decomposing bacteria.

Biome.   Region of the Earth with characteristic climate and characteristic community of living organisms.

Biotic Potential (BP).   The theoretical reproductive capacity of a species.

Birth Rate.   Number of births per 1,000 people in a population.

Black Lung Disease.   An occupational disease frequently contracted by coal miners.

Blackwater.   Wastewater from houses that contains feces and urine. Kitchen sink water, which carries large amounts of food debris, also is considered to be blackwater.

Blister Rust.   A fungus-caused disease of the white pine that is characterized by the appearance of orange "blisters" on the bark.

Bog.   A shallow depression filled with organic matter; for example, a glacial lake or pond basin filled with peat.

Botulism.   A waterfowl disease caused by a bacterium and characterized by eventual respiratory paralysis and death.

Breeder Reactor.   A nuclear reactor that uses a relatively small amount of uranium-235 as a "primer" to release energy from the much more abundant uranium-238. This type of reactor produces plutonium-239 that can be used as a fuel in other nuclear reactors.

Brownfield.   A site that has been previously contaminated by hazardous materials.

Brown Lung Disease.   An occupational disease frequently contracted by textile workers.

Browse Line.   A line delimiting the browsed from unbrowsed portions of shrubs and trees in an area where the deer population exceeds the carrying capacity of the range.

Buffer.   A chemical substance that protects soils and water from increases in pH.

Bycatch.   Captured marine organisms, including fish, shellfish, oceanic birds, and marine mammals that are not the target species of a fishery.

Carbon Absorption.   A process employed by a tertiary sewage treatment plant in which dissolved organic compounds are removed from the effluent as they pass through a tower packed with small particles of carbon.

Carbon Dioxide.   A chemical produced by the combustion of organic materials, including oil, coal, natural gas, and wood.

Carbon Dioxide Fixation.   The incorporation of carbon dioxide into glucose molecules during the process of photosynthesis.

Carbon Monoxide.   A clear, colorless, odorless gas produced by the incomplete combustion of organic substances such as coal or oil. In high concentrations, carbon monoxide exposure can be lethal.

Carcinogen.   A cancer-causing chemical.

Carrying Capacity.   The capacity of a given habitat to sustain a population of animals for an indefinite period of time.

Catadromous Fish.   A type of fish that grows to sexual maturity in freshwater but migrates to the ocean for spawning. An example is the American eel.

Catalytic Converter.   A device in the exhaust system of automobiles that oxidizes hydrocarbons to carbon dioxide and water and converts carbon monoxide to carbon dioxide.

Catch-and-Release-Only Restrictions.   Fisheries management strategy for reducing angler pressure on a particular fish population by requiring the fish to be released (essentially unharmed) back into the water body from which it was caught.

Cellular Respiration. The complete breakdown of glucose in the cells of the body to produce energy needed by various cellular processes.

Central Arizona Project. A multimillion-dollar project to alleviate water shortage problems in Arizona by transporting water from the Colorado River.

Certified Forest. A forest that has been certified by one of several nonprofit organizations as having met criteria that promote more sustainable timber production and harvest. The Forest Stewardship Council is believed to promote the most rigorous and potentially sustainable set of criteria for certifying forests.

Chain Reaction. The sequence of events in a nuclear reactor and nuclear bomb that occurs when neutrons emitted from a radioactive atom bombard another atom and cause it to emit neutrons, which in turn bombard yet other atoms, and so on. In nuclear reactors, the chain reaction is controlled to prevent it from resulting in an explosion.

Channelization. The process by which a natural stream is converted into a ditch for the ostensible purpose of flood control. Attendant environmental abuse is severe.

Chlorinated Hydrocarbon. A family of nondegradable pesticides such as DDT, dieldrin, and toxaphene. Hydrocarbons may have a harmful effect on nontarget organisms such as fish and birds. They persist for a long time in the environment and undergo biological magnification as they move through food chains.

Chlororganics. Potentially toxic organic compounds that form in water treated with chlorine. Good examples are chloroform and carbon tetrachloride.

Clear-Cutting. A method of harvesting timber in which all trees are removed from a given patch or block of forest. This is the method of choice when harvesting a stand composed of a single species in which all trees are of the same age.

Climax Community. The stable terminal stage of an ecological succession.

Closed-Cycle Cooling System. A method of cooling power plants in which the cooling water is continuously recirculated instead of being discharged into a stream and causing thermal pollution.

Closed Seasons. Fisheries management strategy that limits the take of a particular fish species at critical times (e.g., spawning, mating season) by prohibiting fishing during that particular season.

Coal. An organic mineral produced from plant matter 250 million to 300 million years ago. When burned, it produces energy that is harnessed to produce electricity.

Coal Gasification. Production of combustible gas from coal.

Coal Liquefaction. Production of oil from coal.

Cocomposting. Mixing of compost with sewage sludge.

Coliform Bacteria. Bacteria that occur in the human gut. The coliform count is used as an index of the degree to which stream or lake water has been contaminated with human sewage.

Command Economy. An economy in which the production of goods is subject to central control, as in Cuba or the former Soviet Union.

Commons. A public area over which no single nation has sovereignty, such as the open ocean.

Community. All species living in a given area. Examples are the community of an oak woods, an abandoned field, or a cattail marsh.

Compensation Depth. The depth in a lake at which photosynthesis balances respiration. This level delimits the upper limnetic zone from the lower profundal zone.

Compensatory Mortality. The concept that one kind of mortality replaces another kind of mortality in animal populations.

Compost. Partially decomposed organic matter that can be used as a soil conditioner and fertilizer.

Condensation Nuclei. Particulates in the air that absorb moisture and can facilitate cloud formation.

Conservation District. The administrative and operative unit of the U.S. Department of Agriculture's Natural Resource Conservation Service. Conservation districts are organized and run by farmers and ranchers.

Conservation Tillage. The practice of restricting plowing of the soil to reduce erosion and leave enough of the previous crop residues so that at least 30% of the soil surface is covered when the next crop is planted.

Consumer. Any animal link in a food chain.

Consumptive Use. The removal and alteration of natural resources.

Contour Farming. Plowing, seeding, cultivating, and harvesting at right angles to the direction of the slope, rather than down it.

Contour Mine. Mine used for coal and other minerals in hilly terrain. Cuts are made along the contour of the land.

Control Group. In a scientific experiment, the control group is the untreated group. It is identical to the experimental group in every way except the lack of treatment.

Conventional Farming. An agricultural system that relies heavily on agrochemicals, new varieties of crops, and labor-saving, energy-intensive farm machinery.

Cooling Tower. Device used to reduce thermal pollution before releasing cooling water from power plants and factories into lakes and streams.

Coral Reefs. Elaborate ocean structures, found primarily in tropical and subtropical waters, formed from the bodies of animals having calcareous skeletons (made from calcium materials) and certain species of algae that provide the sediment or "cement" that seals the coral framework.

Core Reserve. An area set aside for wildlife protection in which no human activity is allowed. The core reserve is surrounded by a buffer zone, a region permitting modest human activity.

Corridor, Terrestrial. A narrow strip of land that differs, usually in terms of dominant vegetation (such as forest or grassland), from the surrounding areas.

Creel (or Catch) Limits. Fisheries management regulation that controls fishing pressure on a certain fish species by restricting the size or the number of fish an angler may take home.

Critical Population Size. Population size below which recovery is impossible.

Critical Thinking. A process by which one analyzes facts, assertions, and conclusions, attempting to discern their validity.

Crop Rotation. A planned sequence of various crops growing in a regularly recurring succession on one field.

Cross-Media Contamination. The movement of a pollutant from one medium, such as air, to another, such as water.

Crown Fire. A fire that spreads from treetop to treetop.

Cultivar. Type of plants cultivated to produce food or fiber.

Cyanosis. A disease that is characterized by a bluish discoloration of the skin and is caused by the impaired effectiveness of hemoglobin to carry oxygen. An infant who drinks water carrying too high a level of nitrates may undergo chemical changes of his or her hemoglobin that in turn will result in cyanosis of the skin.

Cyclic Population. A population that peaks and troughs at regular intervals. Good examples are the 4-year cycle of the lemming and the 10-year cycle of the ruffed grouse.

Cyclone Filter. A type of air-pollution control device that removes particulate matter (dust) with the aid of gravity and a downward-spiraling airstream.

Death Rate. Number of deaths per 1,000 people in a population.

Decreasers. A category of highly nutritious and extremely palatable range plants that generally decrease even under moderate grazing pressure.

Deepwater Swamp. Freshwater wetlands that support woody vegetation, primarily various species of cypress, gum, and tupelo trees, and remain flooded all or most of the year.

Deferred-Rotation Grazing. A grazing management system in which livestock are rotated between two or among more range areas to increase the long-term efficiency of range conversion into livestock production.

Demographic Transition. A change in a population that is characterized by decreasing birth and death rates. It usually occurs when a nation becomes industrialized.

Denitrification. The decomposition of ammonia compounds, nitrites, and nitrates by bacteria, which can result in the release of nitrogen into the atmosphere.

Density-Dependent Factor. A population-regulating factor, such as predation or infectious disease, whose effect on a population depends on the population density.

Density-Independent Factor. A population-regulating factor, such as a storm, drought, flood, or volcanic eruption, whose effect is independent of population density.

Depletion Time. The time required until 80% of the available mineral supply is consumed.

Desalinization (or Desalination). The removal of salt from seawater in order to make it usable by humans, crops, and wildlife.

Desertification. The conversion of rangeland, rainfed cropland, or irrigated cropland to desert-like conditions caused by natural factors (climatic changes) and artificial factors (human-induced activities).

Desert Pavement. The stony surface of some deserts, caused by excessive erosion of the thin topsoil resulting from water and wind action.

Detritus. Organic matter derived from dead bodies of animals, insects, plants, and so on.

Detritus (or Decomposer) Food Chain. Sequence of organisms, each feeding on the one before it, starting with dead organic material (waste or animal and plant remains).

Deuterium. An isotope of hydrogen that contains one proton and one neutron. It can serve as fuel in nuclear fusion reactions.

Dioxin. An extremely toxic chemical occurring in the herbicide 2,4,5-T. In some areas, it is suspected of causing birth defects and miscarriages.

Discards. The fish and other aquatic bycatch species fish thrown back for various reasons, such as being nontarget species, juveniles, endangered species, wrong size, inferior quality, or surplus to quotas.

Discount Rate. A number used by economists and business-people to determine the present economic value of various business strategies.

Drainage Basin. *See* Watershed.

Drawdown. Lowering of the water level in a reservoir for the purpose of flood control or hydropower.

Drawdown Phase. The occurrence of periodic low water levels common to many inland wetlands in regions that experience very little precipitation during the summer months or alternate wet and dry years.

Dredge Spoil. The sediment that has been scooped from harbor and river bottoms to deepen channels for navigation.

Dust Dome. A shroud of dust particles characteristically found over urban areas. It is caused by the unique atmospheric circulation pattern that results from the marked temperature differences between the urban area and outlying farmlands.

Dynamic Equilibrium. Condition in a system in which change that occurs is corrected by natural mechanisms, so that the system remains more or less the same over long periods.

Ecological Island. Habitat cut off from the surrounding area by natural features such as water or by farms, cities, roads, and so on. This habitat is highly vulnerable to species loss.

Ecological Justice. The concept that species other than humans have a right to exist.

Ecology. Study of the interrelationships that occur between organisms and their environment.

Economic Externality. An economic cost that is not factored into the determination of the cost of goods and services; for example, the economic cost of health effects of air pollution from factories.

Economic Incentives. Inducements to business and individuals to encourage environmentally friendly actions, for example, energy conservation or installation of renewable energy systems.

Ecosphere. The total area in which living organisms occur.

Ecosystem. A contraction for *ecological system.*

Ecosystem Simplification. The intentional or unintentional elimination of species from an ecosystem. Such changes may destabilize an ecosystem.

Ecotone (or Edge). A transition zone within a landscape between two distinct ecosystems that usually has its own unique soil, vegetation, and hydrologic characteristics.

Electrolysis. Breakdown of water using electricity. Electrolysis results in the formation of hydrogen, a useful fuel.

Electromagnetic Spectrum. Range of energy given off by the sun. At the lower end are low-energy radio waves; at the higher end are high-energy gamma rays. Visible light falls in the middle of the spectrum.

Electron. A negatively charged particle occurring in the orbit of an atom.

Electron Cloud. Region that surrounds the atomic nucleus and contains electrons.

Electrostatic Precipitator. Device to remove particulates from smokestack gases.

Elemental Cycle. The cycle of movement of an element (e.g., nitrogen, carbon) from the nonliving environment (air, water, soil) into the bodies of living organisms and then back into the nonliving environment.

Emphysema. A potentially lethal disease characterized by a reduction in the number of alveoli in the lungs as well as a reduction in total respiratory membrane area.

Endangered Species. A species that is in immediate danger of extinction.

End-Point Separation. Process in which recyclables are removed from municipal trash at central stations.

Energy. Defined by physicists as the ability to do work. Two basic forms exist: potential energy and kinetic energy.

Energy Laws. Also known as the laws of thermodynamics. Principles that describe the origin and behavior of energy. *See* First Law of Energy; Second Law of Energy.

Energy Pyramid. Graphical representation of the energy in the various trophic layers in a food chain.

Entropy. Disorder or randomness in any system.

Environmental Disruptors. Chemicals from human sources that disrupt the endocrine system—for example, by mimicking natural hormones or blocking hormone release by the body, causing potentially serious problems with reproduction and other bodily functions.

Environmental Justice. Attempts to ensure that disadvantaged groups of individuals do not share a disproportionate amount of risk posed by human activities or technologies.

Environmental Limitations. Natural and human-induced occurrences that can cause mortality in fish populations. Natural environmental limitations include storms, soil mass movements, debris jams, winterkill, and predation by animals.

Human-induced limitations include dams, pollution, riparian vegetation removal, and competition by exotic fish species.

Environmental Resistance (ER). Any factor in the environment of an organism that tends to limit its numbers.

Ephemeral Stream. A stream or creek that flows only during the wet season.

Epilimnion. The upper stratum of a lake, characterized by a temperature gradient of less than $1°C$ per meter of depth.

Essential Element. A chemical element required for the normal growth of plants.

Estuary. Semi-enclosed inlet (bay) that forms a transitional zone between a coastal river and the sea.

Euphotic Zone. The open-water zone of the ocean, characterized by sufficient sunlight penetration to support photosynthesis. The euphotic zone is located just above the bathyal zone.

Euryphagous species. An organism having a highly varied diet, such as a pheasant or opossum, in contrast to an animal having a narrow (stenophagous) diet, such as an ivory-billed woodpecker.

Eutrophication. The enrichment of an aquatic ecosystem with nutrients (nitrates, phosphates) that promote biological productivity (growth of algae and weeds).

Everglades. Large, unique, freshwater marsh wetlands located near the southern tip of Florida and dominated by a sedge called saw grass.

E-waste. A name for discarded electronic products such as computer monitors and CPUs, televisions, VCRs, stereos, copiers, fax machines, and cell phones. These products contain toxic chemicals to which unprotected employees of recycling facilities, often found in less developed countries, can be exposed.

Exclusive Economic Zone (EEZ). The 330-kilometer (200-mile) belt of sea space extending from the coastline of each nation over which each respective nation has economic jurisdiction.

Exotic (Nonnative) Fish Species. Fish that have been introduced, either intentionally or accidentally, from a different part of the country or from a foreign country into a water body in which they are not native.

Experimental Group. In a scientific experiment, the group that receives the treatment or exposure under study.

Exponential Growth. Growth of any entity (such as population or resource demand) that occurs by a fixed annual percentage when the annual growth is added to the base amount.

Fabric Filter Bag House. An air-pollution control device that operates somewhat like a giant vacuum cleaner in removing solid particles from industrial smokestacks.

Fall Overturn. The thorough mixing of lake waters during autumn.

Faunal Collapse. Dramatic decrease in numbers and diversity of animal species.

Fibrous Root System. A complex root system, such as that of grass plants, in which there are several major roots and a great number of primary, secondary, and tertiary branches. This type of root system binds soil in place, preventing erosion.

Field Capacity.   The water remaining in a soil two to three days after it has been saturated and after free drainage has practically stopped.

First Law of Energy (or Thermodynamics).   The principle that energy can be neither created nor destroyed but can be converted from one form to another.

Fishery.   A place where fish or other aquatic species such as shrimp and sponges are caught. Also the species caught and the occupation or industry—and technologies—involved in catching fish and other aquatic species. The most common usages have to do with place and species.

Fish Ladders.   A series of step-like rungs designed and built into hydroelectric dams for the purpose of providing safe passage to fish as they migrate upstream to their spawning grounds.

Floodplain.   Flat areas of fine sediment deposition extending outward from the bank tops of lowland streams that receive floodwater and accompanying sediment when rivers and streams periodically overflow their banks.

Fluidized Bed Combustion.   Process in which coal is crushed, mixed with bits of limestone, and air-blown into a furnace, where it is burned.

Fluorocarbons.   A group of chemical compounds containing the elements carbon, chlorine, and fluorine. One type of these compounds, manufactured by DuPont under the trade name Freon, has been used in refrigerators, air conditioners, and aerosol spray bombs.

Fluorosis.   A disease in animals caused by fluoride poisoning. Symptoms in livestock include thickened bones and stiff joints.

Flyway.   One of the major migration pathways used by waterfowl. Examples include the Atlantic flyway and Pacific flyway.

Food Chain.   The flow of nutrients and energy from one organism to another by means of a series of eating processes.

Food Web.   An interconnected series of food chains.

Forage.   Grasses, grasslike plants (sedges and rushes), forbs, and shrubs that supply food and energy for domesticated animals such as cattle, sheep, and horses, and for wild animals such as deer, giraffes, and wildebeests.

Fossil Fuel.   Organic fuels derived from ancient plant or animal matter, including coal, oil, shale oil, and natural gas.

Freon.   A type of fluorocarbon compound manufactured by DuPont for use in refrigerators, air conditioners, and aerosol spray bombs. Freon compounds have contributed to the breakdown of the shield of ozone in the upper stratosphere that protects humans from ultraviolet radiation.

Freshwater Aquaculture.   The controlled culturing, or "farming," of fish and other aquatic food that is practiced in inland, freshwater ponds.

Freshwater Marsh.   An inland wetland characterized by nonwoody hydrophytes, shallow water depths, and shallow accumulations of peat material. The term describes a diverse group of inland wetlands, including prairie potholes, vernal pools, the Everglades, the Great Lakes marshes, and the Nebraska sandhill wetlands.

Fuel Cell.   A device that produces electricity from oxygen and hydrogen.

Fuel Crops.   Crops grown to generate ethanol or some other fuel that could replace traditional fossil fuels or other energy resources.

Fuel Rods.   Rods packed with enriched uranium used for fueling nuclear power plants.

Fusion.   *See* Nuclear Fusion.

Game Animals.   Species harvested for recreational purposes.

Gamma Ray or Radiation.   An intense type of radiation, similar to X-rays, that easily penetrates human tissues.

Gas-Liquid Chromatograph.   Instrument used to determine the source of an oil spill.

Gene.   A segment of DNA that regulates some characteristic of an organism, including behavior.

Generalist.   Species that can live anywhere and eat many different types of food.

Genetic Diversity.   A term used to indicate a great variety of organisms (many different species) occupying a given area.

Genetic Engineering.   A complex process that involves the isolation of genes, synthesis of genes, and their insertion into cells from the same or a different organism or species.

Genetic Erosion.   A harmful effect on the gene pool of a given species when a large number of artificially propagated fish are used to enhance a population. Research has shown that the offspring of a hatchery fish bred with a native parent have a lesser chance of survival than pure native offspring.

Geographic Information Systems.   A computer system consisting of hardware and software designed to assemble and store geographical reference information about the Earth, including vegetation, land disturbance, and human settlement patterns. This information can be displayed, manipulated, and analyzed.

Geological (Natural) Erosion.   Wearing away of the Earth's surface by water, ice, wind, or other geological agents under natural environmental conditions.

Geopressurized Zones.   Zones where heated groundwater is trapped by layers of impervious rock.

Geothermal Energy.   Heat produced by the Earth from naturally occurring radioactive decay and from magma. This heat energy can be tapped to heat buildings or to produce electricity.

Glasphalt.   A type of road-surfacing material that employs crushed glass in its manufacture rather than sand. It is more durable than ordinary asphalt.

Glassification.   A method of disposing of radioactive waste by concentrating it and enclosing it in solid ceramic bricks.

Global Climate Change.   Changes in the average weather conditions or climate of the planet brought about by pollutants such as carbon dioxide and methane produced by human activities.

Global Warming. The warming of the Earth's atmosphere and oceans caused by a buildup of greenhouse gases such as carbon dioxide and chlorofluorocarbons produced from human sources.

Grassland. Land in which grasses are the predominant form of vegetation. Grassland is flat or rolling terrain with few or widely distributed trees.

Graywater. Wastewater from showers, washing machines, and sinks.

Grazer Food Chain. Sequence of organisms, each feeding on the one before it, starting with plants or algae.

Great Barrier Reef. The largest chain of coral reefs in the world, located off the coast of Queensland, Australia, reaching a length of over 2,000 kilometers (1,240 miles), and containing over 2,500 individual reefs.

Greenhouse Effect. The warming influence caused by the increased concentration of carbon dioxide and several other pollutants in the Earth's atmosphere.

Greenhouse Gas. Any one of several gases that trap heat escaping from the Earth's surface, making the Earth's atmosphere warmer.

Green Manure. The practice of plowing under or surface-mulching plants (usually a grass or legume) while green, or soon after maturity, for improving the soil.

Green Revolution. The increased food production capability made possible in recent years because of selective breeding, the increased use of fertilizer, the development of seed banks, and the more intensive use of herbicides and insecticides.

Groins. Piers of stone spaced about 30 meters (100 feet) apart that extend into the sea at right angles to the shoreline for the purpose of retaining beach sediments.

Gross Domestic Product (GDP). The sum total of all goods and services produced by an economy.

Gross National Product (GNP). The sum total of expenditures by governments and individuals for goods, services, and investments.

Gross Primary Production. The sum of all biomass production in an ecosystem, not taking into account the losses due to cellular respiration.

Groundwater. Water that has infiltrated the ground, in contrast to runoff water, which flows over the ground surface.

Groundwater Overdraft. Removal of groundwater at rates that exceed its natural replenishment.

Gully Reclamation. The mending of a gully by either physical methods (check dams of boulders or cement) or vegetational means (planting of rapidly growing shrubs on the slopes).

Gyptol. The sex attractant produced by the female gypsy moth that serves to attract the male moth from considerable distances.

Habitat. The immediate environment in which an organism lives. A habitat includes such components as cover, food, shelter, water, and breeding sites.

Habitat Conservation Plans. Documents outlining steps to protect habitat to protect endangered or threatened species.

Habitat Fragmentation. The breaking up of contiguous areas of wildlife habitat, often by home building or farming.

Habitat Restoration. The combination of various strategies—such as adding spawning gravels, revegetating streambanks and lake edges, establishing brush shelters, restoring the historic meanders of rivers, creating pools and riffles, and placing in-stream woody debris—for the purpose of improving habitat for fish and other aquatic life.

Half-Life. The time required for one half of the radioactivity of a given radioactive isotope (e.g., uranium, strontium) to be dissipated.

Hardwood. A species of tree, such as oak, hickory, and maple, that has relatively hard wood, in contrast to the soft woods of the conifers such as spruce and pine.

Harvestable Surplus. The portion of a population that can be taken by humans (hunters) without adversely affecting subsequent populations of that animal.

Hazardous Wastes. Substances produced by homes and factories that pollute the air, water, and soils and can harm the environment and human health.

Headwater Stream. The beginning segment of a river system that is usually confined by a narrow valley, shallow bedrock, and coarse sediments and thus tends to have a narrow, relatively straight channel pattern instead of one that is wide and meandering.

Heartwood. The dark, central portion of a tree trunk characterized by the presence of dead xylem cells that have become filled with gums and resins.

Heat Island. The tendency for the atmosphere of a city to be warmer than the air in the surrounding farmlands. This is partly the result of the greater number of heat-generating sources (autos, factories, human bodies) in the city.

Home Range. The total area occupied by an animal during its life cycle—that is, the area required for feeding, breeding, loafing, and securing refuge from the weather and from predators.

Horizon. A layer of soil with specific properties and characteristics. The major horizons from the ground surface downward to bedrock are designated as horizons O (organic layer), A (topsoil), E (subsurface), B (subsoil), C (parent material), and R (bedrock).

Host. An organism that contains parasites.

Humus. The semistable, dark-colored material that represents the decomposition products of organic residues and materials synthesized by microorganisms.

Hybrid. The offspring that results from a cross of two different species or strains of animals or plants. For example, the Santa Gertrudis cattle resulted from a series of crosses involving two parental types, the Brahmin cattle and the shorthorn.

Hydraulic Mining. A mining technique used to extract gold and silver in which a powerful stream of water is directed against the face of the rock containing the minerals.

Hydrological Cycle. The circular movement of water from the ocean reservoir to the air (clouds), to the Earth in the form of rain and snow, and finally back to the ocean reservoir via streams and estuaries. Also known as the water cycle.

Hydrolysis.   A type of chemical reaction in which a compound is broken down into simpler components by the action of water.

Hydroperiods.   The various fluctuations in the duration and depth of flooding events throughout the year in an inland freshwater marsh.

Hydrophytes.   Water-loving plants or plants adapted to saturated or flooded soil conditions, including emergent, floating, and submergent species.

Hydroponics.   The technique of growing crops in an aqueous nutrient solution without soil.

Hydropower.   Power generated by the flow of water. Hydropower is usually tapped by dams.

Hydroseeder.   A machine employed to disperse grass seed, water, and fertilizer on steep banks.

Hypolimnion.   The bottom layer of a lake.

Hypothesis.   A tentative explanation of a particular phenomenon or observation.

Imprinting.   The psychological or behavioral process by which migratory fish assimilate environmental clues to aid their return to their stream of origin as adults.

Incidental Take.   The inadvertent harvest of a species—for example, the harvest of turtles during shrimp-fishing operations.

Increasers.   A category of somewhat palatable but still highly nutritious range plants that tend to increase (at least temporarily) when a range is heavily grazed.

Individual Transferable Quotas (ITQs).   A current marine fisheries management strategy that gives fishers a guaranteed portion of the catch. In theory, ITQs essentially transform a common property resource into a private one; fishers have an economic incentive to protect the resource in order to maintain the same quota in subsequent years.

Indoor Air Pollution.   Contamination of the air inside homes, office buildings, and factories. Caused by combustion by-products, paints, stains, finishes, and off-gassing from building materials, carpets, and other materials.

Industrial Fixation.   The "fixing" of nitrogen (in other words, combining it with hydrogen to form ammonia) by industrial means rather than by natural methods such as bacterial action.

Input Control.   Any strategy that seeks to reduce the production of pollutants, such as modifying processes and reducing demand for goods.

In Situ Retort.   An underground site in which an oil shale deposit is broken up with dynamite and set afire to drive off kerogen.

Integrated Pest Management.   A method of pest control that judiciously employs chemical, biological, and other methods (cultural), depending on the specific problem. The use of chemicals is minimized to avoid environmental damage.

Intercropping.   A cultural method of pest control in which the farmer mixes a number of different crops in a small area instead of devoting the entire area to a single crop.

Intergenerational Equity.   Fairness to future generations.

Introduction.   The bringing in of an exotic plant or animal to a new region—for example, the introduction of the European carp into the United States and of the coho salmon to Lake Michigan from the Pacific Coast.

Invaders.   A pioneer species of plant, usually a noxious weed, that has become established in an overgrazed pasture.

Irruption.   A sudden increase in the population of an organism, followed by a precipitous decline (crash). Frequently, the carrying capacity of the habitat is reduced for many years thereafter. An example is the Kaibab deer irruption.

Isotope.   A form of an element identical to the regular element except for a difference in atomic weight. Thus, deuterium is an isotope of hydrogen.

J-Shaped Curve.   An exponential growth curve shaped like the letter J.

Juvenile Dispersal.   The dispersal of young animals (such as bald eagles, ruffed grouse, and muskrats) from the general region of their hatching or birth site. The presumed function is to prevent overpopulation in the parental area.

Kepone.   A highly toxic insecticide manufactured by a company in Hopewell, Virginia. As a result of carelessness the chemical was allowed to contaminate the James River and destroy much aquatic life.

Kerogen.   The solid organic material that contains shale oil.

Krill.   The crustaceans and other small marine organisms that are used as food by baleen whales.

Kwashiorkor.   Deficiency of protein intake.

Land Capability Classification System.   A system that evaluates land according to its limitations for agricultural use. It classifies land into eight categories, with Class I having the fewest limitations and Class VIII having the greatest limitations.

Land Ethic.   A view put forth by Aldo Leopold calling on humans to respect the land and all living things.

Landfill.   A method of solid-waste disposal in which the waste is dumped, compacted, and then covered with a layer of soil.

Large Woody Debris.   Logs, branches, and root wads that have fallen into a stream (due to wind or erosional forces) that shape channel structure, provide fish habitat, and trap organic matter.

Latitudinal Migration.   The north–south migration characteristic of caribou, gray whales, and many birds.

Law of Supply and Demand.   Economic law describing the relationship of price, supply, and demand.

Law of Tolerance.   Principle that organisms can live within a range of conditions. When conditions exceed that range, the organisms are unable to survive.

Leach Field.   An underground system of porous pipes that serves as a means of disposing of liquid draining from household septic tanks.

Legume. A pod-bearing member of the Leguminosae family, including peas, beans, and clover, that can fix atmospheric nitrogen.

"Let It Burn" Policy (or Prescribed Natural Fire Policy). A management policy followed by U.S. government agencies that permits wildfires to burn under careful surveillance.

Levee. A dike composed of earth, stone, or concrete that is erected along the margin of a river for purposes of flood control.

Life Cycle Costs. All costs from the point of origin of a material to the point of disposal.

Limiting Factor. Any factor in the environment of an organism, such as radiation, excessive heat, floods, drought, disease, or lack of micronutrients, that tends to reduce the population of that organism.

Limnetic Zone. The region of open water in a lake, beyond the littoral zone, down to the depth at which there is just enough sunlight for photosynthesis.

Littoral Zone. The shallow, marginal region of a lake, characterized by rooted vegetation.

Lowland Stream. Downstream segments of a river system that are usually wider and have more freedom to meander across the finer sediments of floodplain materials.

Malnutrition. Dietary deficiency resulting from inadequate intake of nutrients.

Malthusian Overpopulation. Type of overpopulation, described by Robert Malthus, that results in an overtaxing of available food supplies and eventual mass starvation.

Manganese Nodules. Mineral-rich nodular accumulations found on much of the ocean floor. They contain manganese, nickel, iron, copper, cobalt, and other minerals.

Marasmus. Deficiency of proteins and calories.

Marine Aquaculture (or Mariculture). The controlled culturing, or "farming," of fish and other aquatic food that is practiced in shallow bays or estuaries.

Marine Sanctuary. A preserve set aside to protect valuable marine habitat and marine species.

Marketable Permit. *See* Tradable (or Marketable) Permit.

Market Economy. An economy that relies on monetary signals to control production and distribution of goods and services.

Mass Emigration. The mass movement of a given species from an area. An example is the mass movement of the snowy owl from the tundra to the United States during periods of lemming scarcity.

Mass Number. The number of protons and neutrons present in the nucleus of the atom of an element.

Maximization. The most efficient use of a resource that is possible with current technology. Waste is minimized.

Maximum Sustainable Yield (MSY). The highest sustained fish or shellfish catch (number or weight) that can be allowed each season without affecting the reproductive capacity of the species. MSY is a mathematical calculation based on factors such as spawning mass, annual recruitment, annual biomass growth, nonfishing mortality (natural death, predation), and fishing mortality.

Meandering Stream. The most common stream pattern, which results when a channel naturally migrates laterally, eroding one bank and depositing material on the opposite bank. The stream channel is straight only for relatively short distances between curves, resembling a continuous string of S shapes.

Metabolic Reserve. The lower half of a grass shoot, required by a grazed plant for survival. The metabolic reserve contains the minimum amount of photosynthetic equipment needed for food production.

Microcephaly. A birth defect characterized by an abnormally small brain. This condition is associated with mental retardation and may be induced by exposure to radiation.

Microhabitat. The immediate, localized environment of an organism.

Micronutrient. Any one of dozens of nutrients needed in minuscule amounts. Examples include zinc and vitamins.

Millirem. One-thousandth of a rem, which is a unit for measuring the effect of radiation on the body of a living organism.

Mineral. Useful substance obtained by mining or digging in the Earth.

Mineral Reserves. Mineral deposits that are practical to mine.

Monocropping. The practice of growing a crop on the same field year after year.

Monoculture. Greatly altered ecosystem consisting of one or a few species. Monocultures are generally more vulnerable to insects, disease, and adverse weather conditions than diverse ecosystems.

Monotype. An agricultural or forest planting composed of only one species.

Mulch. Dead plant material that accumulates on the ground surface. The amount of mulch found on rangeland is a reliable indicator of the condition of the range.

Multiple-Use Management. A management principle that requires meeting a number of different needs on the same area of land. These needs or uses can include timbering, grazing, agriculture, mining, oil and gas leasing, hunting and fishing, recreation, soil conservation, wildlife protection, and watershed management.

Mutation. Any one of several changes in the genetic material (chromosomes) of an organism. Mutations may be caused by radiation, chemicals in the environment, and other agents.

Mycelia. The branching "root system" of a fungus.

Mycorrhiza. An intimate relationship between the root systems of trees and soil fungi.

Myxoma Virus. The virus that was used to control the rabbit outbreak in Australia. Animals become infected when they consume contaminated forage.

National Park. An area of scenic beauty or historical importance that is to be maintained in unimpaired condition by a national government for the use of the people.

Native Fish Species. Fish that originated in the water body where they reside.

Natural Capital. A nation's ecological wealth.

Natural Resource. Any component of the natural environment—such as soil, water, rangeland, forest, wildlife, and minerals—that species depend on for their welfare.

Neritic Zone. The relatively warm, nutrient-rich, shallow water zone of the ocean that overlies the continental shelf, which is rich in fishlife and important for commercial fish production.

Net Energy Efficiency or Yield. Amount of energy produced by a system when taking into account the total amount invested in the first place.

Net Primary Production. The total energy incorporated into the body of a plant as a result of photosynthesis minus the energy required for respiration.

Neutron. The electrically neutral particle in the nucleus of an atom.

Niche. Habitat and total functional role of an organism in an ecosystem (the relationship to all biotic and abiotic factors).

NIMBY Syndrome. "Not in my backyard" syndrome; general and widespread resistance to siting waste dumps or other potentially dangerous or unsightly (e.g., wind farms) facilities near someone's place of residence or community.

Nitrate Bacteria. Bacteria that have the ability to convert nitrites into nitrates. These bacteria are essential in the cycling of nitrogen.

Nitrogen Dioxide. Air pollutant produced during combustion of any organic material. This compound combines with water to produce nitric acid.

Nitrogen Fixation. Process of converting atmospheric nitrogen into a form that can be used by plants.

Nitrogen Intoxication. A condition that can cause fish mortality when the fish come in contact with the turbulent, nitrogen-enriched waters immediately below hydroelectric dams.

Nonbiodegradable Material. Material not susceptible to decomposition by bacteria. For example, DDT and other chlorinated hydrocarbon pesticides are relatively nonbiodegradable.

Nonconsumptive Use. The use, without removal or alteration, of natural resources.

"No Net Loss" Wetland Policy. Federal land-use policy in the United States enacted in 1988 that protects wetlands from conversion to other land uses by requiring developers to create or restore wetland area and function in exchange for any wetland losses due to proposed development.

Nongame Animals. Species that are not harvested for sport purposes.

Nonrenewable Resources. Resources such as coal, oil, and minerals that cannot be replenished within a reasonable period by natural processes. Nonrenewable resources occur in a fixed amount.

North American Water and Power Alliance (NAWAPA). Scheme for transferring water from the water-rich, low-population areas of northwestern Canada to the water-deficient areas of the United States and Mexico.

No-Till Farming. A practice in which a crop is planted directly into a seedbed not tilled since harvest of the previous crop.

Nuclear Fusion. The generation of energy by causing the fusion of the nuclei of two atoms of a very light element such as hydrogen under temperatures of around 40 million degrees Celsius (72 million degrees Fahrenheit).

Nutrient Flush. The sudden increase in water fertility that results when soil is newly submerged in the creation of a reservoir.

Ocean Thermal Energy Conversion. A method of using the temperature differential between different levels of the ocean to alternately gasify and condense a working fluid such as ammonia and in this way propel a turbine to produce electricity.

Oil (or Petroleum). Thick liquid containing numerous organic molecules usually found in deep deposits in the Earth's crust. These molecules are refined to form jet fuel, gasoline, diesel, fuel oil, and many other useful products.

Opportunity Cost. Cost of lost economic opportunities.

Optimum Yield (OY). A management principle that considers the biological, economic, social, and political values of a given fishery in order to maximize benefits to society. The chosen optimum level of use established for any given fishery depends on the goals and resource objectives of the associated society. Ideally, an optimum yield would be one that allows the greatest extrinsic yield (economic, sociocultural, nutritional benefits) to society with minimum loss of intrinsic yield (health) to the ecosystem.

Ore. Rocks that contain important minerals.

Organic Compounds. Any of many thousands of chemicals made principally from carbon, hydrogen, and oxygen.

Organic Farming. A farming system that combines traditional conservation-minded farming methods with modern farming technologies but excludes such conventional inputs as synthetic pesticides and fertilizers, emphasizing instead the building up of soil with compost additions and animal and green manures, controlling pests naturally, rotating crops, and diversifying crops and livestock.

Organic Fertilizer. A soil supplement made from organic sources such as manure or compost.

Organic Phosphate Pesticides. A group of pesticides (malathion and parathion, for example) that is lethal to insects because they reduce the supply of cholinesterase at the junction (synapse) between two nerve cells in a nerve cell chain.

Organic Soil. Soil consisting of layers of plant material in various stages of decomposition. Organic soil contains at least 20% organic matter and is often referred to by the generic term *peat*.

Output Control. Any strategy that seeks to reduce pollution after it has been produced. Types of output control include pollution-control devices and disposal methods.

Overburden. The rock and dirt that overlies coal seams.

Overgrazing. Continued heavy grazing that exceeds the carrying capacity of the plant community and creates a deteriorated range.

Overnutrition. A condition in which one eats to excess. Overnutrition leads to obesity and numerous medical problems such as heart attacks and diabetes.

Oxidation. The chemical union of oxygen with metals (iron, aluminum) or organic compounds (sugars). The former process is an important factor in soil formation; the latter process permits the release of energy from cellular fuels (sugars, fats).

Oxygen-Demanding Waste. A type of water pollutant that is broken down by naturally occurring microorganisms that consume dissolved oxygen, lowering the level of oxygen available to other species.

Oyster Watch. A well-coordinated program employed along U.S. coasts that involves the use of oysters as biological monitors of marine pollutants.

Ozone. A gaseous component of the outer atmosphere, known as the stratosphere. Ozone in this region of the atmosphere is important to humans because it shields us from the ultraviolet radiation of the sun. Ozone is also produced in the lower atmosphere as when chemicals from car exhaust and power plants react in the presence of sunlight to produce photochemical smog. This is known as ground-level ozone.

Parasite. Any one of numerous organisms that live on or in other organisms, living off nutrients in blood or tissue fluids.

Parent Material. The unconsolidated (soft and loose), weathered mineral or organic material from which soils form.

Particulate Matter. Minute solid and liquid particles in the atmosphere; soot.

Passive Fishing Gear. Devices and methods for fishing that are stationary and, therefore, depend on the fish or shellfish to come into contact with them. Examples include pots, traps, cages, hooks and lines, set nets, pound nets, and anchored gill nets.

Passive Solar System. System or building designed to capture sunlight energy and produce heat for space heating.

Peat. Partially decomposed accumulations of organic matter under conditions of excessive moisture.

Peatlands. Land areas, such as bogs and fens, found in the colder northern, humid regions of the Northern Hemisphere. They are characterized by extremely slow plant decomposition and thick accumulations of peat.

Perennial Stream. Stream that flows year-round and is sustained by groundwater during dry periods.

Permafrost. A permanently frozen soil or soil horizon.

Permanent Wilting Point. The water content of a soil after plants have removed all the water they possibly can and have begun to wilt.

Pesticide Treadmill. The use of ever-increasing amounts of pesticides because of the pests' ever-increasing resistance to the pesticides.

Petroleum. *See* Oil.

Pheromone. Chemical substance released by insects that affect other insects. Some of the best-known examples are sex attractant pheromones released by females to attract males.

Phloem. The elongate food-conducting cells of the trunk and branches of a tree. These cells convey food from the leaves downward to the root system.

Photochemical Smog. The type of smog that has plagued Los Angeles and other cities worldwide. Photochemical smog forms as a result of the action of sunlight on the hydrocarbon and nitrogen oxide emissions from automobiles and other sources. At nightfall, the production of this type of smog ceases.

Photosynthesis. The process occurring in green plants by which solar energy is utilized in the conversion of carbon dioxide and water into sugar.

Photovoltaic Cell. *See* Solar Cell.

Phytoplankton. Minute plants, such as algae, that live in lakes, streams, and oceans and are passively transported by water currents or wave action.

Pioneer Community. First community to become established on barren land.

Pittman–Robertson Act. Federal law that levies an excise tax on the sales of sporting arms and ammunition; revenues from the tax are distributed to the states for habitat purchase and wildlife management and research. This law is also known as the Federal Aid in Wildlife Restoration Act.

Plankton. Tiny plants (such as algae) and animals (such as protozoa, small crustaceans, fish embryos, insect larvae) that live in aquatic ecosystems and are moved about by water currents and wave action.

Point Bar. A sediment deposition area that forms on an inside bend of a meandering stream.

Point Pollution. Pollution discharged from an extremely restricted area or "point," such as the discharge of sulfur dioxide from a smokestack or the discharge of carbon monoxide from the exhaust pipe of an auto. This type of pollution contrasts with nonpoint pollution, such as the runoff from a farm or urban area.

Pollution Control. Refers to measures like smokestack scrubbers that reduce pollution emitted by a pollution source. In many cases, pollution control devices merely capture pollutants, which then must be disposed of elsewhere.

Pollution Prevention. Any strategy designed to prevent the production of pollution. Examples include energy conservation and process changes in factories that eliminate pollutants.

Polychlorinated Biphenyls (PCBs). A class of chemicals used as electrical insulators. They persist in the environment and undergo biomagnification, causing problems in the food chain.

Pool. Deep area in a streambed that has been scoured out by storm flows. Pools are usually associated with the outside bends of a meandering stream.

Population.   The individuals of a species occurring in a given area—for example, the amount of deer in a cedar swamp or the amount of black bass in a lake.

Population Growth Rate.   Birth rate minus death rate plus net immigration multiplied by 100.

Pore Space.   The portion of a soil occupied by both air and water.

Precision Farming.   The integrated use of field observations, computers, geographic information systems (GIS), global positioning systems (GPS), remote sensing (aerial photography), and farm implements to practice site-specific management, so farmers can improve efficiency, reduce costs, increase yields, and reduce environmental hazards.

Prescribed (Controlled) Burning.   A type of surface burning (used by foresters, wildlife biologists, and ranchers) to improve the quality of forest, range, or wildlife habitat.

Primary Pollutant.   A chemical pollutant produced by any one of many sources. It may undergo chemical reactions that result in the production of a new pollutant, called a secondary pollutant.

Primary Production.   The total chemical energy produced by photosynthesis. On a global basis, it amounts to about 270 billion tons annually.

Primary Sewage Treatment.   A rudimentary sewage treatment that removes a substantial amount of the settleable solids and about 90% of the biological oxygen demand (BOD).

Primary Succession.   An ecological succession that develops in an area not previously occupied by a community—for example, a succession that develops on a granite outcrop or on lava.

Process Redesign.   A strategy in which businesses adjust manufacturing processes to reduce or eliminate hazardous waste and pollution.

Producer.   A plant that can carry on photosynthesis and thus produce food for itself and indirectly for other organisms in the food chain of which it is a part.

Profundal Zone.   The bottom zone of a lake, extending from the lake bottom upward to the limnetic zone and characterized by insufficient sunlight for photosynthesis.

Proton.   A positively charged particle in the nucleus of an atom.

Purse Seine.   A seine used by commercial fishers that closes to entrap fish somewhat as a drawstring purse closes to "trap" money.

Put-and-Take Stocking.   A management practice common in urban-area water bodies where fishing pressure is heavy but there is no natural production. Fish of catchable size are planted at the beginning of fishing season so that most of them are caught in that same season.

Pyramid of Biomass.   The graphic expression of the fact that there is a progressive reduction in total biomass (protoplasm) with each successive level in a food chain.

Pyramid of Energy.   The graphic expression of the second law of thermodynamics as applied to the energy transfer in food chains. A certain amount of energy is lost in the form of heat as it moves through the links of a food chain. The greatest amount of energy is present at the base of the food chain (producers), and the least is present in the top of the food chain (carnivores).

Pyramid of Numbers.   The graphic expression of the fact that the number of individuals in a given food chain is generally greatest at the producer level, with fewer at the herbivore level and fewest at the carnivore level.

Pyrolysis.   The destructive distillation of solid waste.

Quadrillion.   The number 1 followed by 15 zeros.

Quaking Bog.   Natural successional stage in the "filling in" of small, shallow lakes. Vegetation attaches to the lake edge and grows outward into the middle of the lake, resulting in the creation of a cushionlike, floating mat that bounces or "quakes" when walked upon.

Rabies.   A disease caused by a virus that lives in the saliva of the host (carrier) animal.

Rad.   A unit devised to measure the amount of radiation absorbed by living tissue. One rad is equal to 100 ergs (an erg is a unit of energy) absorbed by 1 gram of tissue, or about 1 roentgen, which is about the amount of radioactivity received from a single dental X-ray.

Radiation Inversion.   Temperature inversion that occurs because the ground cools faster than the air above it. This inversion results in the buildup of air pollution at ground level.

Radioisotope.   Radioactive form of an atom.

Radon.   Naturally occurring radioactive gas in certain types of rock. Radon can leak from the basement or slab into houses and other buildings, where it can cause lung cancer.

Range or Rangeland Condition.   The current state of vegetation of a range site in relation to the potential natural plant community. In a sense, rangeland condition is an estimate of how close a particular rangeland is to its productive potential.

Rangelands.   Areas of the world that are a source of forage (such as grasses and shrubs) for free-ranging native and domestic animals, as well as a source of wood products, water, energy, wildlife, minerals, recreational opportunities, natural beauty, open space, and wilderness.

Range Management.   Methods used to manage rangeland, usually to maximize livestock or wild herbivore productivity without degrading rangeland quality or to improve the quality of rangeland.

Range of Tolerance.   The range of physical and/or chemical conditions in which a species can survive. These include moisture, temperature, radiation, micronutrients, and oxygen.

Redd.   A "nest" formed in gravel by a spawning fish.

Rem.   A unit of absorbed radiation dose, taking into account the relative biological effect of various types of radiation; about 1 roentgen, which is roughly the amount of radiation received from a single dental X-ray.

Renewable Resources.   Resources that replenish naturally through some biological or geophysical process. Examples include solar energy, trees, grass, and fish.

Replacement-Level Fertility. Number of children a couple must have to replace themselves.

Reseeding. A practice where ranchers improve their rangelands by periodic seeding.

Reservoir. The artificial lake created behind a dam as river water is ponded and stored for use in hydropower generation and flood control.

Resident Fish. Fish that live their entire life in freshwater, sometimes confined to a lake, particular stream reach, or single tributary.

Resource Conservation and Recovery Act (RCRA). A federal law that gave the Environmental Protection Agency full authority to control pollution by solid waste, just as it has authority for controlling air and water pollution.

Respiration. The process by means of which cellular fuels are burned with the aid of oxygen to permit the release of the energy required to sustain life. During respiration, oxygen is used up, and carbon dioxide is given off.

Retort. Vessel used to heat crushed oil shale rock to produce shale oil.

Reverse Vending Machine. Automated recycling machine that pays the operator for materials put into the machine.

Rhizobium. One of a genus of nitrogen-fixing bacteria that lives in the root nodules of legumes (alfalfa, clover).

Rhizome. An underground stem that occurs in grasses. It permits vegetative reproduction because the tip of the rhizome may develop a bud that can develop into a new plant.

Rhizosphere. The soil in the immediate vicinity of a plant's root system.

Rhythm Method. A fairly unreliable form of contraception that relies on abstinence during a period just prior to and after ovulation.

Riparian Wetlands. Wetlands that form along rivers and streams due to periodic flooding of these aquatic systems.

Riparian Zone. The entire linear strip of land on either side of a river or stream that forms a transitional area between the adjacent aquatic and upland ecosystems. A riparian zone has unique soil, vegetation, and hydrologic characteristics. These zones include riparian wetlands as well as land that may not meet the criteria for a jurisdictional wetland.

Risk Assessment. Process of identifying potential risks from human activities and natural events and determining the potential impacts.

Root Nodule. Small bulbous attachment on the roots of certain plants, notably legumes (peas, beans, vetch, etc.). Root nodules contain bacteria that fix atmospheric nitrogen—that is, convert atmospheric nitrogen to a form plants can use.

Rotenone. A poisonous substance derived from the roots of an Asiatic legume. It has been extensively used to control rough fish populations.

Rough Fish. Undesirable trash fish such as garpike and carp.

Runoff Water. The water that flows over the land surface after rainfall or snowmelt and eventually forms streams, lakes, and marshes.

Salinization (or Salination). An adverse aftereffect of irrigating land that has poor drainage properties. As a result, especially in the arid western states, evaporation of the salty water leaves a salt accumulation on the land, which renders the soil unsuitable for crop production.

Saltwater Intrusion. The contamination of freshwater aquifers with saltwater as the result of excessive exploitation of those aquifers in coastal regions near the ocean.

Sanitary Landfill. A dump for municipal garbage. Refuse is covered daily with a layer of dirt to reduce flies, odors, and rodents.

Sapwood. The lighter, moister, more porous layer of xylem tissue immediately ensheathing the heartwood, composed of water and nutrient-transporting xylem cells.

Scrubber (or Smokestack Scrubber). Device that removes particulates and sulfur dioxide from exhaust air in smokestacks.

Seawall. An expensive structure, usually made of rock or concrete, placed along a beachfront to prevent erosion caused by the force of ocean waves.

Secondary Pollutant. A chemical pollutant derived from another pollutant after reacting with sunlight, moisture, or other pollutants.

Secondary Sewage Treatment. An advanced type of sewage treatment that involves both mechanical and biological (bacterial action) phases. Although it is superior to primary treatment, many of the phosphates and nitrates remain in the effluent.

Secondary Succession. An ecological succession that occurs in an area that at one time already supported living organisms—for example, a succession developing in a burned-over forest or in an abandoned field.

Second Law of Energy (or Thermodynamics). The principle stating that during energy conversions, entropy increases.

Secured Landfill. Clay-lined landfill designed to hold hazardous wastes.

Seed-Tree Method. A silvicultural system in which trees are left during cutting to seed the next tree generation.

Selective Cutting. Procedure in which only certain trees are cut from a forest. Contrast to clear-cutting.

Septic Tank. A part of a rudimentary type of sewage treatment system used commonly by families in rural areas. The sewage flows into the subterranean septic tank and is gradually decomposed by bacterial action.

Sewage. Human waste consisting of feces and urine.

Sewage Treatment Plant. A facility designed to remove most or all of the pollutants from sewage from homes.

Shade-Tolerant Plants. A species of plant that reproduces well under conditions of reduced light intensity. An example is the sugar maple.

Shale Oil.   Oil produced by heating (retorting) oil shale. Shale oil can be refined and made into a variety of combustible products and chemicals used to make products like plastic in much the same way that crude oil is refined.

Sheet Irrigation.   A type of irrigation in which water flows slowly over the land in the form of a sheet.

Shelterbelt (or Windbreak).   Rows of trees and shrubs arranged at right angles to the prevalent wind for the purpose of diminishing the desiccating and eroding effects of the wind on cropland and rangeland. This type of planting is commonly employed in the Great Plains.

Shelterwood Method.   A silvicultural system that uses rotational selective cutting designed to leave a seed source and protective cover for forest regeneration.

Short-Duration Grazing.   A grazing management system in which each pasture is grazed briefly and intensively, followed by a long period of nonuse.

Sigmoidal Growth Curve.   The S-shaped curve commonly followed by the population of an animal (deer, grouse, rabbit) when it has been newly introduced into a habitat with good carrying capacity.

Siltation.   The filling up of a stream or reservoir with waterborne sediment.

Slash-and-Burn Agriculture.   A type of agriculture maintained by natives of tropical rain forest regions in which a patch of forest is cut and burned, crops are grown in the clearing for a few years until the fertility of the soil is exhausted, and then the area is deserted when the farmers move to another part of the forest to repeat the process.

Smelter.   Device that melts ore to separate metals from the ore.

Smog.   Term coined to refer to a combination of smoke and fog. Now used more broadly to include urban pollution.

Softwood.   A species of tree that has softer wood (such as spruce, pine, and fir) than hardwood trees (such as oak and hickory).

Soil.   A natural system with four components: mineral matter, organic matter, water, and air.

Soil Classification.   The systematic arrangement of soils into groups on the basis of their characteristics.

Soil Erosion.   A three-step process by which soil particles are detached from their original site, transported, and eventually deposited at a new location.

Soil Fertility.   The ability of a soil to supply nutrients in amounts and forms required for maximum plant growth.

Soil-Forming Factors.   The five factors responsible for kind, rate, and extent of soil development: (1) climate, (2) parent material, (3) organisms, (4) topography, and (5) time.

Soil Organic Matter.   The living or dead plant and animal materials in the soil. It includes plant and animal residues at various stages of decomposition, small animals, microorganisms, and humus.

Soil Profile.   A vertical section of the soil, extending through all its horizons and into the parent material.

Soil Reaction.   The degree of acidity, neutrality, or alkalinity of a soil. Soil reaction depends on the relative amounts of hydrogen (H^+) and hydroxide (OH^-) ions and is expressed as a pH value.

Soil Structure.   The arrangement or grouping of soil particles into clusters or aggregates.

Soil Taxonomy.   The soil classification system used in the United States based on the physical and chemical properties of soils as they are found in the landscape.

Soil Texture.   The relative proportions of sand, silt, and clay in a particular soil.

Solar Cell.   A platelike device composed of two layers of silicon that converts solar energy directly into electricity.

Solar Energy.   The radiant energy generated by the sun that "powers" all energy-consuming processes on Earth, whether biological or nonbiological.

Solution Mining.   Type of mining in which water is pumped into the deposit and dissolves the mineral. The mineral is then pumped to the surface.

Source Reduction.   Efforts that decrease the amount of waste at the source through reduced consumption of resources, efficiency measures, process manipulation, recycling, or other measures.

Source Separation.   A process in which recyclables are segregated by category at a source (home, factory, etc.).

Spawning.   Major life cycle stage of a fish; egg deposition.

Specialist.   Organism that feeds on only one or a few types of food and can live only in a certain habitat.

Splash Dams.   Early logging practice of building dams to create a large volume of water in which to float logs downriver.

Spoil Bank.   Mounds of overburden that accumulate during a strip-mining operation. Unless properly limed and then vegetated, spoil banks may become an important source of acid mine drainage and erosion.

Spring Overturn or Turnover.   The complete top-to-bottom mixing of water in a lake during the spring of the year when all the water is of about the same temperature and density.

Stamen.   The club-shaped, pollen-producing part of a flower.

Stenophagous.   Having a very specialized diet. Examples are the ivory-billed woodpecker, which consumes beetle larvae secured only from recently dead trees, and the Everglade kite, which feeds almost exclusively on the snail *Pomacea caliginosa*.

Sterilization.   A method of human population control that involves cutting and tying the sperm ducts in the male and the oviducts in the female.

Stocking.   The planting of artificially propagated fish into ponds, lakes, reservoirs, or streams.

Strategic Minerals.   Minerals that are essential to the economic well-being and/or security of a nation.

Stream Order.   A classification system, developed by Horton, based on the position of a stream within the entire network of tributaries.

Strip Cropping. The practice of growing crops that require different types of tillage, such as row and sod, in alternate strips, usually along the contours or across the prevailing direction of the wind.

Strip Cutting. A method of harvesting timber in wide (about 80-meter) strips, allowing natural regrowth.

Strip Mining. The type of mining in which a mineral such as coal or iron is scooped from the ground by giant earth-moving machines.

Subsidence. Land collapse that results from mining underground water reservoirs (aquifers).

Subsidence Inversion. Temperature inversion caused when a mass of high-pressure air settles over an area. The inversion causes pollutants to build up at ground level.

Succession. The replacement of one community by another in an orderly and predictable manner. The succession begins with a pioneer community and terminates with a climax community.

Sulfur Oxide. One of two gases made from sulfur and oxygen and released during the combustion of sulfur-containing compounds such as those found in oil and gasoline.

Superfund Act. The federal law that provided a fund to the Environmental Protection Agency to clean up extremely hazardous waste sites.

Surface Fire. A type of forest fire that moves along the surface of a forest floor, consuming litter, herbs, shrubs, and seedlings.

Surface Mining Control and Reclamation Act. A federal law, effective in 1979, under which mining companies must restore a strip-mined area to its original condition, within the limits of available technology.

Surface Retort. A large steel vessel in which crushed oil shale is heated to drive off oil.

Surface Water. Any water body above the ground surface, including lakes, rivers, and ponds.

Suspended Sediment. Sediment carried in the water column.

Sustainable Agriculture. A system-level approach to understanding the complex interactions within agricultural ecologies. For a farm to be sustainable, it must produce adequate amounts of high-quality food, conserve resources, and be environmentally safe, profitable, and socially responsible.

Sustainable Society. Society based on conservation, recycling, the use of renewable resources, and population control. A sustainable society respects nature and seeks to find a sustainable relationship between humans and other species.

Sustainable Yield. The concept that a forest or wildlife resource can be managed in such a way that a modest crop can be harvested year after year without depleting the resource as long as annual decrements are counterbalanced by annual growth increments.

Synergistic Effect. A condition in which the toxic effect of two or more pollutants (copper, zinc, heat) is much greater than the sum of the effects of the pollutants when operating individually.

Taiga. The northern coniferous forest biome, typically composed of spruce, fir, and pine.

Taproot. The type of root system characterized by one large main root—for example, that of a beet or carrot.

Technological Overpopulation. Excessive population level that results in massive pollution and resource exhaustion because of the high technological level maintained by the population. People die because of toxic contamination of the environment rather than because of food shortages.

Telemetry. The electronic technique involving a transmitter–receiver system in which the movements and behavior of animals (deer, elk, grizzly bear, salmon) are monitored from a distance.

Temperature Inversion. *See* Thermal Inversion.

Terrace. Abandoned floodplain that no longer receives overflow discharge. Terraces lie adjacent to but at a higher elevation than a current floodplain.

Terracing. A soil conservation technique in which steep slopes are converted into a series of broad-based "steps." The velocity of runoff water is thus retarded, and soil erosion is arrested.

Territory. An area defended by a member of one species against other members of that same species.

Tertiary Sewage Treatment. The most advanced type of sewage treatment, which removes not only the BOD and the solids but also the phosphates and nitrates. The installation of such a plant at Lake Tahoe has arrested eutrophication of the lake.

Thermal Inversion. An abnormal temperature stratification of the lower atmosphere in which a layer of warm air overlies a layer of cooler air. Such an inversion frequently occurs at heights of 30 to 100 meters (100 to 3,000 feet), resulting in stagnation of the air mass below the inversion. It contributes to air pollution, especially in industrial areas.

Thermal Plume. The warm-water zone in a steam or lake caused by the release of heated water.

Thermal Pollution. An increase (or decrease) in water temperature that harms aquatic organisms.

Thermocline. The middle layer of water in a lake in summer, characterized by a temperature gradient of more than 1°C per meter of depth.

Threatened Species. Species that is likely to become endangered in the near future.

Throughput. In economics, the amount of materials being produced and consumed in a given society.

Tidal Freshwater Marsh. A marsh located far enough inland to escape saltwater intrusion, but near enough to the coast to experience tidal fluctuations.

Tidal Salt Marsh. A marsh formed on the coast near a river mouth, in a bay, on a coastal plain, around a lagoon, or behind barrier islands. Tidal salt marshes are characterized by high salinity, tidal fluctuations, tidal creeks, salt-tolerant nonwoody vegetation, and extreme changes in daily and seasonal temperatures.

Time Preference.   An economic term that describes a person or business's preference for timing of a return on investments for various actions such as soil conservation. A short time preference indicates a preference for receiving income in the near term. The term comes from the field of economics.

Tolerable Soil Loss (or Soil Loss Tolerance).   The maximum combined water and wind erosion that can take place on a given soil without degrading that soil's long-term productivity.

Topsoil.   The uppermost and usually the most fertile part of the soil.

Total Fertility Rate.   Projected number of children that females in a population will produce in their lifetimes, given current trends.

Toxic Substances Control Act (TOSCA).   A federal law that makes it mandatory for a company to notify the Environmental Protection Agency 90 days in advance of its intention to manufacture a new chemical. If the EPA concludes that the chemical may be harmful to human health or may be environmentally destructive, it can deny the company permission to produce the chemical.

Tradable (or Marketable) Permit.   A permit issued to a company that generates pollution, allowing the release of a certain amount of pollution. A company that can reduce its pollution below the allotted amount can sell the emission permit to another company.

Translocation.   A practice used to restore a native fish population to its original water body. Large numbers of fertilized eggs are obtained directly from spawning adults (in their natural environment) of the target fish species that are well established in an ecologically similar water body. The eggs are then immediately transferred to and placed in the original water body, thus creating a new population.

Transpiration.   The evaporation of water from the breathing pores of a plant leaf.

Tree Farm.   A private land area used to grow trees for profit under sound management principles.

Trickling Filter.   Device used in secondary sewage treatment to reduce organic matter as well as nitrogen and phosphorus levels in human waste. Sewage is dripped over a bed of stones or bark coated with decomposer organisms.

Tritium.   A radioactive isotope of hydrogen that contains one proton and two neutrons; a potential fuel for a nuclear fusion reactor.

Tropical Rain Forest.   A forest with 200 centimeters (79 inches) of annual rainfall spread evenly enough through the year to support broad-leaved evergreen trees, typically arrayed in several irregular canopy layers dense enough to capture more than 90% of the sunlight before it reaches the ground.

Tundra.   The type of biome occurring in northern Canada and Eurasia north of the timberline. It is characterized by fewer than 10 inches (4 cm) of annual rainfall; subzero weather in winter; low-lying vegetation composed of grasses, dwarf willows, and lichens; and fauna consisting of lemmings, Arctic foxes, Arctic wolves, caribou, and snowy owls, among other species. The growing season lasts only 6 to 7 weeks.

Turbidity.   A measure of suspended sediment in water.

2,4-D.   An herbicide that kills a weed because it mimics the plant's growth hormones, causing more rapid growth than can be sustained by its supply of oxygen and food materials.

2,4,5-T.   An herbicide that operates on the same principle as 2,4-D. It kills a weed because it mimics the plant's growth hormones, causing more rapid growth than can be sustained by its supply of food, moisture, and oxygen.

Two-Story Reservoir.   A reservoir that is deep enough to become thermally stratified during the summer, complete with epilimnion, thermocline, and hypolimnion.

Ultimate Production.   Amount of a natural resource that will probably have been removed after total supplies have been depleted.

Ultraviolet Radiation.   One type of radiation produced by the sun. It is essential in small amounts, for it increases vitamin D production in the skin of humans. Excess levels can cause sunburn, cataracts, and skin cancer.

Undercut Bank.   Bank shaped in the form of a C in cross section, maintained by rooted riparian vegetation, providing excellent overhead cover and shade for fish.

Undernutrition.   Dietary deficiency resulting from an inadequate intake of food.

Universal Soil Loss Equation (USLE).   An equation used to predict the average annual soil loss per acre per year. Now revised, it is called the revised universal soil loss equation.

Upwelling.   The movement of nutrient-rich cold water from the ocean bottom to higher levels by means of vertically moving currents.

User Fees.   Levies placed on raw materials that are paid by the producers.

Vessel Element.   The elongated xylem cell in a tree trunk or branch that has water transportation as a major function.

Vibert Box.   A device designed to enhance trout populations by allowing eggs to develop under natural conditions while being protected from predation. This plastic box, filled with about 100 trout eggs, is planted in the gravel bed of a stream. Slots on all sides permit the free flow of stream water.

Volatile Organic Compounds.   Any one of dozens of organic molecules produced during the incomplete combustion of fossil fuels and other organic materials such as wood.

Water Diversion.   Removal of water from surface waters, especially streams, to areas of need, often by pipes or canals.

Waterlogging.   Saturation of the root zone with water. Waterlogging chokes plants and kills them. It is usually caused by excess irrigation in poorly drained soils.

Watershed.   The total area drained by a particular stream or river. It may range from a few square miles in the case of a small stream to thousands of square miles in the case of the Mississippi River.

Water Table.   The upper level of water-saturated ground.

Weathering.   The physical and chemical changes produced in rocks and soils by temperature and precipitation.

West Nile Virus. A viral strain that was introduced into the eastern United States in the 1990s and is transmitted by certain species of mosquito. The virus, which has spread to many states, infects birds, humans, and other animals. It has been responsible for countless bird deaths and several human deaths in the United States.

Wetland Delineation. The practice of defining the specific boundaries of a wetland based on field indicators of hydric soils, hydrophytic vegetation, and wetland hydrology.

Wetlands. Areas that are inundated or saturated by surface or ground water at a frequency and duration sufficient to support, and that under normal circumstances do support, a prevalence of vegetation typically adapted for life in saturated-soil conditions. Wetlands generally include swamps, marshes, bogs, and similar areas.

Whirling Disease. A condition currently infecting wild trout populations in the intermountain West. Whirling disease is caused by the fish parasite *Myxobolus cerebralis,* believed to have been unintentionally introduced in 1955 from Europe. Visible symptoms of the disease are whirling behavior, black tail, and cranial and skeletal deformity.

Wilderness. A recognized, protected area where the Earth and its community of life are untrammeled by humans, where humans themselves are visitors who do not remain.

Wildlife. All plants and animals on Earth that are not domesticated. As it is generally used, the term is restricted to birds and mammals.

Wildlife Corridor. A protected area that connects two or more wildlife preserves, allowing species to migrate to new habitat to find food, mates, or nesting.

Wildlife Management. The planned use, protection, and control of wildlife by the application of ecological principles.

Winterkill. The fish kills that can occur during the long winters of the northern states when an icy barrier seals off lakes from atmospheric oxygen. Snow can form an opaque barrier that prevents sunlight from penetrating through the lake surface, so photosynthetic activity decreases. Then the concentration of dissolved oxygen may drop to a level below which fish can survive.

Xerophytes. Specialized plants that are well adapted to survive in arid regions because of such water-conserving features as reduced leaves, recessed stomata, thick cuticles, accelerated life cycles, periodic dormancy, and the presence of water-storing (succulent) tissues.

Xylem. A type of tissue occurring in the trunks and branches of trees (and other plants) that serves to transport water and nutrients from the roots to the leaves and also provides support.

Zero Population Growth. Condition in which the growth rate of a country equals zero.

Zone of Deposition. The B horizon, or subsoil, which receives and accumulates the soluble salts and organic matter carried downward from the A horizon by percolating water.

Zone of Leaching. The A horizon, or topsoil, from which many soluble salts are carried downward, or leached from this horizon, to the B horizon below it.

Zooplankton. Minute animals (such as protozoans, crustaceans, fish embryos, and insect larvae) that live in a lake, stream, or ocean and are moved by water currents and wave action.

ILLUSTRATION ACKNOWLEDGMENTS

1.2A Kelvin Aitken/Peter Arnold
1.2B Argus Fotoarchiv/Peter Arnold
1.3 Photodisc/Getty Images
1.4 Ulrike Welsch/Photo Researchers
1.5 istockphoto
1.8, 3.15, 5.9, 5.13, 7.24, 8.3, 8.7, 8.8, 8.15, 8.16, 13.24, 14.3, 14.16, 22.20 U.S. Department of Agriculture
1.9, 17.10 UPI/Corbis/Bettmann
1.10 Terry Ashe/Getty Images
1.11 Nysted Offshore Wind Farm
1.12 Courtesy of the Library of Congress
1.13 Chip Vanai/Getty Images, Inc.—Liaison
2.1 Newscom
2.2 Mike Wintroath/AP Wide World Photos
2.4, 5.8A, 7.2 Courtesy of www.istockphoto.com
2.7 Tri-Met (Tri-County Metropolitan Transportation District of Oregon)
2.8 Lee Prince/Shutterstock
2.9 Courtesy of the Aldo Leopold Foundation Archives
3.29 J. Malcolm Greany/U.S. Fish and Wildlife Service
3.30 USDA/NRCS/Natural Resources Conservation Service
Case Study 3.1 Figure 1, 21.5B U.S. Geological Survey, Denver
4.8 Reuters/UPI/Corbis/Bettmann
4.9 World Food Programme
4.10 Hank Morgan/Science Source/Photo Researchers
5.2, 5.3 P. Pitter/FAO Food and Agriculture Organization of the United Nations
5.4 UN/DPI Photo by John Isaac
5.5 Tim McCabe/State of Virginia
5.6, 5.14, 5.15 FAO Food and Agriculture Organization of the United Nations
5.8B Stephen Mcsweeny/Shutterstock
5.10 D. Mason/FAO Food and Agriculture Organization of the United Nations
5.11 Scott Bauer/ARS/Scott Bauer/USDA ARS
7.6, 7.8, 7.11, 7.16A, 7.16B, 7.20, 7.21 USDA/NRCS/Natural Resources Conservation Service
7.10, 7.23 John P. Reganold
7.13 Jeff Vanuga/USDA/NRCS/Natural Resources Conservation Service
7.14 Quentin P. Bennett/USDA/NRCS/NCGC/National Cartography and Geospatial Center
7.22 The Land Institute
7.25 John Lee, USDA-Natural Resources Conservation Service, Mississippi
8.1 USDA/ARS/Agriculture Research Service
8.9 Nigel Cattlin/Photo Researchers, Inc.
8.11, 13.25 Department of Interior, Bureau of Sport Fisheries and Wildlife
8.12, 8.13, 15.8, 16.9 Michigan Department of Natural Resources
8.17, 8.19 USDA/APHIS Animal and Plant Health Inspection Service
9.3 Guana Tolomato Matanzas National Estuarine Research Reserve, Florida Department of Environmental Protection
9.7, 12.7, 15.1, 16.2 U.S. Fish and Wildlife Service
9.19 Aerial photograph SW-H-69 71-2-14 used by permission of the Engineering Division Resource Mapptin Section, Washington Department of Natural Resources
9.21 ©James Marshall/Corbis

9.22 Marc Pesetsky/Teuters/Corbis/Bettmann
9.23 U.S. Coast Guard photography by Petty Officer 2nd Class Kyle Niemi/U.S. Army Photography
9.25 Courtesy of Dr. Erik Zobrist, Restoration Center, NOAA Fisheries Service
9.26 H. E. Rodenhizer III/North Carolina Division of Coastal Management
10.8, 10.17, 10.22, 10.25 E. E. Hertzog/Bureau of Reclamation
10.9 Sustain Dane
10.10 Jeff Greenberg/Visuals Unlimited
10.11, 11.23 Sun-Mar-Corp., www.sun-mar.com
10.13, 17.2, 17.8, 18.7A, 18.7B, 22.3 AP Wide World Photos
10.15 Ron Curbow/Photo Researchers
10.16 Liz Roll/FEMA News Photo/FEMA
10.18 Jim McConnell/USDA/NRCS/Natural Resources Conservation Service
10.19 Bureau of Reclamation photo by Glade Walker
10.26 Brian Slobe
11.1, 18.20 Getty Images
11.3 USDA Natural Resources Conservation Service, National Soil Survey Center
11.4 W. Banaszewski/Visuals Unlimited
11.6 H. L. Personius/Bureau of Reclamation
11.7 Erwin W. Cole/USDA/NRCS/Natural Resources Conservation Service
11.9 F. Scarpace and T. Green, Dynamic surface temperature structure of thermal plumes. Water Resource Research 9(1): 138–153, February 1973. ©1973 by the American Geophysical Union.
11.10A, 11.10B, 11.10C Phototake/Carolina Biological Supply Company
11.11 Kristin Smith/Shutterstock
11.19 Bureau of Reclamation
11.21 John Todd Research and Design, Inc.
11.24 Milwaukee Metropolitan Sewage District
11.25 The City of New York, Department of Sanitation
11.26, 12.16, 13.12, 13.16, 13.21, 13.23, 13.26, 19.17, 20.8, 22.10 Daniel D. Chiras
11.29 John Gaps III/AP Wide World Photos
11.30, 11.31 Nick Didlick/Reuters/Corbis/Bettmann
11.32 Kevin R. Morris/Corbis/Bettmann
12.4 Natalie Fobes/Getty Images
12.8 Jeri Gleiter/Getty Images
12.11 Department of Natural Resources Bureau of Information and Education State of Minnesota
12.29 Joe Raedle/Getty Images, Inc.—Liaison
Case Study 12.1 Figure 1, Case Study 12.1 Figure 2 University of Michigan Sea Grant Program
Case Study 12.1 Figure 3 Michigan Conservation Department
Case Study 12.2 Figure 1 Affordable Photo Stock
Case Study 12.2 Figure 2 Courtesy of Portland District Army Corps of Engineers
Case Study 12.2 Figure 3 Earl Roberge/Photo Researchers
13.2 Dr. Linda H. Hardesty
13.3 Ryan Morgan/Shutterstock
13.11 Cartoon –"Tragedy of the commons" (Redrawn by John N. Smith, based on Bernard J. Nebel, Environmental Science, 2nd Edition, ©1987, p. 485. Reprinted with permission of Prentice-Hall, Inc., Englewood Cliffs, NJ.)

INDEX

Common Conversions

Metric Measure	Multiply by	English Equivalent
Length		
Centimeters (cm)	0.3937	Inches (in.)
Meters (m)	3.2808	Feet (ft)
Meters (m)	1.0936	Yards (yd)
Kilometers (km)	0.6214	Miles (mi)
Nautical mile	1.15	Statute mile
Area		
Square Centimeters (cm^2)	0.155	Square inches (in.^{2})
Square meters (m^2)	10.7639	Square feet (ft^2)
Square meters (m^2)	1.1960	Square yards (yd^2)
Square kilometers (km^2)	0.3831	Square miles (mi^2)
Hectares (ha) (10,000 m^2)	2.4710	Acres (a)
Volume		
Cubic centimeters (cm^3)	0.06	Cubic inches (in.^{3})
Cubic meters (m^3)	35.30	Cubic feet (ft^3)
Cubic meters (m^3)	1.3079	Cubic yards (yd^3)
Cubic kilometers (km^3)	0.24	Cubic miles (mi^3)
Liters (L)	1.0567	Quarts (qt), U.S.
Liters (L)	0.88	Quarts (qt), Imperial
Liters (L)	0.26	Gallons (gal), U.S.
Liters (L)	0.22	Gallons (gal), Imperial
Mass		
Grams (g)	0.03527	Ounces (oz)
Kilograms (kg)	2.2046	Pounds (lb)
Metric ton (tonne) (t)	1.10	Short ton (tn), U.S.
Velocity		
Meters/second (mps)	2.24	Miles/hour (mph)
Kilometers/hour (kmph)	0.62	Miles/hour (mph)
Knots (kn) (nautical mph)	1.15	Miles/hour (mph)
Temperature		
Degrees Celsius (°C)	1.80 (then add 32)	Degrees Fahrenheit (°F)
Celsius degree (C°)	1.80	Fahrenheit degree (F°)

ADDITIONAL ENERGY AND POWER MEASUREMENTS

1 watt (W)	=	1 joule/sec	1 W/m^2	=	2.064 cal/cm^2/day
1 joule	=	0.239 calorie	1 W/m^2	=	61.91 cal/cm^2/month
1 calorie	=	4.186 joules	1 W/m^2	=	753.4 cal/cm^2/year
1 W/m^2	=	0.001433 cal/min	100 W/m^2	=	75 kcal/cm^2/year
697.8 W/m^2	=	1 cal/cm^2/minute			

Solar constant:
1,372 W/m^2
2 cal/cm^2/minute